Fundamentals of Water Pollution

Quantifying Pollutant Formation, Transport, Transformation, Fate, and Risks

Fundamentals of Water Pollution

Quantifying Pollutant Formation, Transport, Transformation, Fate, and Risks

Daniel A. Vallero
Department of Civil and Environmental Engineering, Pratt School of Engineering, Duke University, Durham, NC, United States

ELSEVIER

Elsevier
Radarweg 29, PO Box 211, 1000 AE Amsterdam, Netherlands
125 London Wall, London EC2Y 5AS, United Kingdom
50 Hampshire Street, 5th Floor, Cambridge, MA 02139, United States

ISBN: 978-0-443-28987-3

For information on all Elsevier publications
visit our website at https://www.elsevier.com/books-and-journals

Publisher: Candice Janco
Acquisitions Editor: Jennette McClain
Editorial Project Manager: Ashi Jain
Production Project Manager: Rashmi Manoharan
Cover Designer: Mark Rogers

Typeset by STRAIVE, India

Contents

Preface

The genesis of this book was that engineers and scientists needed a book like my *Air Pollution Calculations*. Much like air pollution, experts and others interested in water pollution also needed some helpful examples of the principles that drive water pollution. I started writing a helpful set of example calculations and solutions related to water quality. However, as my research progressed, I became increasingly convinced that there is an even greater need for a companion to a *Fundamentals of Air Pollution*. As a result, this book combines theory and practice. It explains the thermodynamics and motion principles that are the basis of pollutant formation, transport, transformation, fate, and risks *and* real-world examples of these principles. Hence, the title now includes "fundamentals."[a]

I have spent a half century working on ways to understand and improve the environment. I can think of no better way to spend a career. I have seen much progress. Many problems that seemed intractable in the 1970s have progressively been addressed. When I started my education, levels of detection of a few parts per million were considered quite precise. Now, pollutants are measured in parts per trillion, at resolutions six orders of magnitude better. Pollution control technologies continue to enhance the efficiency and effectiveness of removing pollutants from water, air, sediment, and soil. These and other first principles of science must be understood and applied to water quality. Systems range from the cell to habitats to human populations to the planet. Environmental scientists and engineers consider water pollution to be a collective outcome of changes in these systems. This book uncovers these fundamentals from both a description of the water pollution problems and the keys to solving these problems.

The book is designed to be a resource for engineers, hydrologists, hydrogeologists, biologists, meteorologists, chemists, environmental managers, public health experts, and faculty and students in water quality and water pollution courses. The level of interest will vary. Those directly involved in protecting water quality will, hopefully, find the entire book useful in day-to-day needs and interests. Researchers will find specific chapters useful in explaining the basics of water pollution directly related to their chosen topics. Students from high school to PhD candidates can enhance their understanding of the scientific fundamentals as a foundation and rationale for their work. For example, teachers and students of an AP environmental and earth science high school class will find the first three chapters particularly useful. A math class may use the ample examples in Chapter 2 to explain statistics and contingent and Bayesian probabilities.

Select chapters and sections will serve a wide range of interests and usefulness depending on the scientific discipline. Undergraduate, senior-level environmental and chemical engineering courses can assign select chapters, but especially Chapters 3–5, 8, 9, 14, 16, and 17. Ecologists and environmental biologists will find Chapters 10, 11, 13–15, 17, and 18 most relevant to their research and courses. Meteorology is covered intensively in Chapter 7. Courses in the history of science can draw from Chapter 1.

One of my aims is to cover even the most complicated material so that a range of readership will understand the salient points, but not to water down (pun intended) the content at the expense of rigorous science. To do so, I introduce the topic and progressively treating the complexities, with an increasing use of equations, reactions, coefficients, variables, and contingencies. After all, this is the purpose of a book with "Fundamentals" in its title. Certain subject matter, then, will require a few rereadings, as I had to do in researching many of the topics.

I have been blessed by the wisdom that I received and continue to receive from formal and informal mentors. Yes, even at 70 years old I continue to listen, or at least try, to others who know more than I about a particular subject. I remain grateful to my colleagues at Southern Illinois, Kansas, Rutgers, North Carolina Central, Cal-Berkley, KwaZulu-Natal, and Duke Universities; the U.S. Environmental Protection Agency; Elsevier Publishing; the International Society of Exposure Science; the National Institutes of Health; and the University of North Carolina Newman Catholic Center, to name a few. I would be remiss if I did not mention a few by name, especially Jeff Peirce, Aarne Vesilind, Henry Petroski,

a. My lovely spouse, Janis, and I laughed over coffee that my real goal was to make discussions of water quality both "fun" and "mental." I hope, to some extent, I achieved this since learning is indeed a joyful experience.

Henri Gavin, and Sy Mauskopf at Duke; Ross E. McKinney and Steve Randke at Kansas; Loran Marlow and Charles Hess at SIUE; Ed Vest, Tim Kubiak, Tony Wayne, Bernie Burnham, Gene Ramsey, Ralph Langemeier, and Peter Egeghy at EPA; Trevor Letcher at KwaZulu-Natal; Paul Lioy and Panos Georgopoulos at Rutgers; and many more. Some have passed, but not before leaving behind great legacies of a cleaner planet and shared knowledge to ensure the advances continue in the next generations.

Henry Petroski, my colleague at Duke, died a year before I completed this book. Most readers will recognize his name. He was arguably the greatest author of our time of engineering subject matter. No one could make seemingly mundane topics more interesting, whether it was the life cycle of a pencil, paper clip, or door, or why engineering is such a uniquely human enterprise [1–6]. I was honored recently to say a few words at his memorial. Henry was the department chair when I arrived as a PhD student at Duke, so was one of those who thought me worthy of joining this august institution. I am ever grateful to him and hope that my writings will do justice to Henry's suggestions that engineering must be inclusive of thinking often avoided by engineers, including history and culture.

I am gratified to that my career began at a time of the preliminary advances in environmental science, engineering, and technology, and that in 50 years these have progressed to increase the environmental knowledge-building that supports the exciting innovations that are happening now. I often share with my younger colleagues that many of the solutions to problems that were too computationally expensive can now be run on a smart phone. Sensors are becoming more accurate, less expensive, and provide real-time results that had taken weeks to collect, transport, and analyze in the lab. I am privileged to work every day with the new generation of scientists and engineering who fearlessly employ emerging models and measurement technologies that I could not imagine a half century ago. As my career evolved, I increasingly applied computational tools to estimate pollutant exposures and risks. These now include machine learning, artificial intelligence, and informatics [7].

I have become increasingly adherent to systems sciences, which I explored in depth in *Environmental Systems Science*. As such, I have incorporated many topics that are not usually covered in water pollution texts. Pollutants spend much of their time in other environmental compartments before and after they are water pollutants. This means I included discussions of pollutants and their precursors in the air, soil, sediment, and biota. As such, if the reader needs additional depth in the handshake between water and air in Chapters 7 and 16, I recommend that they refer to *Fundamentals of Air Pollution*.

Humility is the topic of one of my favorite talks that I share with students, faculty, professional groups, and anyone who will listen. That can be a challenge. After all, scientists are trained to be staunch adherents to the scientific method and often have received rigorous educations and experiences on very focused investigations, sometimes for decades. Scientific rigor is certainly commendable but can evolve into hubris and polarization of ideas. When that happens, scientists may confuse opinions with facts and trend toward closed-mindedness, holding tightly to a specific paradigm of thinking. We must remind ourselves to consider other perspectives. There is always so much to know that I cringe when I hear journalist say that any science is settled. Indeed, we hold strongly to the principles I share in this book, but the application of the principles is where hubris can arise. There is much knowledge and wisdom available beyond any scientist's sphere of knowledge. We so-called experts know only a small fraction of what needs to be known to confront the grand challenges of water pollution.

This book is designed to have a long shelf life, especially regarding pollution models. Many excellent titles on my bookshelf are now obsolete because the central theme relied on a model written with a version of software no longer used or in code that is difficult to update or translate into a current version. Indeed, hyperlinks to some are no longer active. Many texts included long appendices of the scripts that are no longer useful. That said, I have provided the scientific underpinning for any model, but not the code for any water pollution model. As evidence, my discussions of KABAM and EXAMS models are limited to their parameters since these are the variables and coefficients in equations for bioaccumulation and bioconcentration. All equations in this book can be coded into a model or to support the understanding of what is coded in an off-the-shelf model. Indeed, many models have user-defined space for which the equations can be the basis.

Those of us who live in areas with plentiful clean water may lose sight of the fact that most of the world suffers from poor water quality and the prevalence of waterborne diseases. My hope is to increase a few wrinkles in your and my gray matter [8] and that you, the reader, will apply the book's principles and examples to improve the quality of water in your community and in the world.

Daniel A. Vallero
Duke University
September 2024

References

[1] H. Petroski, Success Through Failure: The Paradox of Design, Princeton University Press, 2013.

[2] H. Petroski, The Essential Engineer: Why Science Alone Will Not Solve Our Global Problems, Vintage, 2011.

[3] H. Petroski, The Evolution of Useful Things: How Everyday Artifacts—From Forks and Pins to Paper Clips and Zippers—Came to Be as They Are, Vintage, 1994.

[4] H. Petroski, The Pencil: A History of Design and Circumstance, Knopf, 1992.

[5] H. Petroski, To Engineer Is Human, St. Martin's Press, New York, 1985.

[6] H. Petroski, To Engineer Is Human: The Role of Failure in Successful Design, St. Martin's Press, 1985.

[7] J.F. Wambaugh, J.C. Bare, C.C. Carignan, K.L. Dionisio, R.E. Dodson, O. Jolliet, et al., New approach methodologies for exposure science, Curr. Opin. Toxicol. 15 (2019) 76–92.

[8] T. Tallinen, J.Y. Chung, J.S. Biggins, L. Mahadevan, Gyrification from constrained cortical expansion, Proc. Natl. Acad. Sci. U. S. A. 111 (35) (2014) 12667–12672.

Chapter 1

Introduction

1 Water pollution defined

Water pollution can be understood and described from many perspectives. Researchers often choose a very specific aspect of water pollution and delve deeply into its meaning. They need to understand mathematics and the basic sciences to learn what is already known and to extend this knowledge to their needs.

Water pollution is commonly defined as the presence of contaminants or substances in the water that interfere with human health or welfare, or produce other harmful environmental effects [1]. Put simply, anything in the water that is or can be harmful is water pollution. However, harm is a value-laden concept [2–5]. As such, a threshold exists between desired and undesired states, i.e., the difference between clean and polluted water. In 1949 by G.R. Scott of the Tennessee Valley Authority (TVA) observed [6]:

> *Mention of the word "water" engenders a wide variety of mental pictures. Just now, inspired by a discussion of stream pollution, the picture might be of a foul, grossly polluted stream, teeming with pathogenic organisms. Later in the year, during vacation time, the picture might be a glass of cool, sparkling water, a swimming pool, a place in which to fish, or to be used for other recreational purposes. At other times, depending upon one's current interest, it might represent power, transportation, or any one of the other uses to which water can be put....*

These socially acceptable thresholds vary according to cultural norms. Thus, water pollution's definition requires context, i.e., something is a water pollutant if it causes conditions to deviate from a desired state. At a minimum, a water pollutant interferes with a benchmark of a desired condition that provides for water quality that supports human and other life. However, the definition goes beyond health and extends to welfare. Laws like the Clean Air Act in the U.S. specifically include in the conception of "welfare" acceptable ecological conditions as a key part of societal well-being, whereas the Clean Water Act is more subtle in its call for fishable and swimmable waters. From a systems perspective, the sum of these desired states is, in fact, a single desired state. Pope Francis put this succinctly in his encyclical letter, *Laudato Sí* [7]:

> *What kind of world do we want to leave to those who come after us, to children who are now growing up? This question not only concerns the environment in isolation; the issue cannot be approached piecemeal.... When we fail to acknowledge as part of reality the worth of a poor person, a human embryo, a person with disabilities—to offer just a few examples—it becomes difficult to hear the cry of nature itself; everything is connected.... Our world has a grave social debt towards the poor who lack access to drinking water, because they are denied the right to a life consistent with their inalienable dignity.*

When the general population accepts a desired condition, this is merely the first step in addressing water pollution. One of the challenges of a book on water pollution is blending the qualitative observations with quantitative metrics needed to assess and improve the quality of water. The determination of acceptability must be based in objective science. Indeed, only the most extreme conditions of pollution are detected without the aid of advanced equipment and expertise. Once conditions reach a point of obvious changes in the smell, taste, look, or feel of the water, the concentrations are relatively high. Most water pollutants are not detectable with unaided human senses until they reach very toxic levels. Detecting these substances at or below thresholds of toxic concentrations requires sophisticated equipment. Too often, people who have been exposed at harmful levels may not even know they have been exposed until disease symptoms arise and epidemiological studies link them to the presence of contaminants. Indeed, exposure to pollutants that cause cancer and other chronic diseases usually occur many years before the diseases themselves symptomatically. This is known as the latency period [8–13]. Thus, for most water pollution, perception is not reality.

Complicating things further is that pollution can be indirect. A pollutant may have to go through several steps before an adverse effect is observed in humans and other organisms. The outcome only occurs after a series of events. For example, a lake may receive loads of organic matter, which is not necessarily inherently hazardous, but after bacteria break it down

Fundamentals of Water Pollution. https://doi.org/10.1016/B978-0-443-28987-3.00013-8

and, in the process, dissolved oxygen concentrations in the water decrease. The lower O_2 concentrations then lead to fish kills or loss of specific aquatic species. As such, water pollution can be understood in terms of dynamics, both thermodynamics and fluid dynamics. The laws of thermodynamics apply both to the solvent or matrix and solute. For water pollution, the solvent is usually the water, the matrix is commonly the sediment, and the solute is the pollutant. This aqueous solubility is one a compound's inherent properties, as is solubility in other solvents. For example, handbooks and computational dashboards publish aqueous solubility and solubility in organic solvents like dimethyl sulfoxide (DMSO) for a specific chemical compounds (see Table 1.1), along with other inherent properties like vapor pressure, and partitioning coefficients that express the affinity of the pollutant to various compartments in an organism or in the environment [14–22].

Producing something of value often leads to a waste product that is released to the water. How a substances changes after its release into the environment is determined by its inherent properties and the conditions of the environment it enters. Thermodynamically, a water pollutant, or any pollutant for that matter, is the result of inefficiency. The pollutant is mass or energy that is exiting a control volume. The environment is subdivided into finite compartments, but with constant traversing of mass and energy among these compartments. The mass of the contaminant entering and the mass leaving a control volume must be balanced by what remains within the control volume. Each environmental compartment may gain or lose the mass, but the overall mass must balance. The generally inclusive term for these compartmental changes is fugacity, i.e., the chemical compound's "fleeing potential." A substance's fugacity is its propensity to escape from one compartment and to join another. The fate of a chemical may be modeled by combining the partitioning taking place within and between environmental compartments [23–25].

TABLE 1.1 Select physicochemical properties of dimethyl sulfoxide (DMSO).

Property	Experimental average	Predicted average	Predicted median	Experimental range	Predicted range	Unit
Polarizability	–	8.00	8	–	8	Å
Property	1.51e–9	1.51e–9	1.51e–9	1.51e–9	1.51e–9	atm-m^3 mol^{-1}
Ready biodegradation	–	1.00	1	–	1	Binary 0 or 1
Boiling point	189	169	180	189	129–189	°C
Flash point	–	86.9	86.9	–	85.0–88.9	°C
Melting point	18.4	5.48	19	18.0–18.5	−49.7 to 47.2	°C
Molar volume	–	71.1	71.1	–	71.1	cm^3
Viscosity	–	0.695	0.695	–	0.695	cP
Surface tension	–	43.7	43.7	–	43.7	dyn cm^{-1}
Density	–	1.12	1.12	–	1.10–1.14	Gcm^{-3}
Liquid chromatography retention time	–	0.00	0	–	0	min
Vapor pressure	0.610 (1)	5.10	0.805	0.61	0.603–13.9	mmHg
Water solubility	12.8	9.09	12.8	12.8	1.58–12.9	mol L^{-1}
Index of refraction	–	1.48	1.48	–	1.48	–
Log K_{oa}: octanol-air	4.96	4.83	4.83	4.96	4.83	–
Log K_{ow}: octanol-water	−1.35	−1.21	−1.29	−1.35	−1.35 to −0.913	–

Data from U.S. Environmental Protection Agency, Dimethyl sulfoxide—67-68-5 | DTXSID2021735. Available from: https://comptox.epa.gov/dashboard/chemical/properties/DTXSID2021735, 2024.

The linkages among physical, chemical, and biological processes affect the fate and transport of substances as they undergo transfer between and among environmental compartments, e.g., soil, sediment, air, or water. The inherent properties are influenced and changed by the properties of the water, sediment, and other environmental media. The fate of a water pollutant, for example, is determined by many abiotic and biotic processes in the water, air, water, soil and sediment, as well as within organisms residing in these media. For example, when the metal mercury (Hg) is released into water, it is changed by abiotic processes like reduction, oxidation, and hydrolysis in the water, but also by reduction and metabolic processes in the microbes, fish, and other organisms living in the water. The latter processes are suspected of being the major route to the formation of highly toxic methylmercury compounds [26–29].

Biotic media include all levels of biological organization, from subcellular and cellular matter in organisms to communities as a result of many processes discussed in later chapters, but especially the coefficients of sorption (distribution coefficient, K_D, or solid-water partition coefficient, K_p), dissolution or solubility, air-water partitioning (the Henry's law (K_H) constant), and organic carbon-water (K_{oc}) [30]. Based solely on a compound's inherent properties, a substance is suspected to have an affinity for particular environmental locations in air, water, soil, sediment, and biota. However, even within a single compartment, a compound may exist in various phases, e.g., dissolved in water and sorbed to a particle in the solid phase. The physical interactions of a contaminant at the interface between each compartment are a determining factor in the fate of the pollutant. Within a compartment, a contaminant may remain unchanged for a few seconds or for decades. In time, it will move physically and/or be transformed chemically into another substance. A fraction of the mass can remain unmoved and unchanged, while another fraction remains unchanged but is transported to a different compartment, and another fraction becomes chemically transformed. All remaining products staying in the compartment where they were generated. Finally, a fraction of the original contaminant is transformed and then moved to another compartment. Thus, upon release from a source, the contaminant moves as a result of thermodynamic processes [30].

Fugacity requires that at least two phases be in contact with the contaminant. For example, the chemical compound's octanol-water coefficient (K_{ow}) indicates a compound's likelihood to exist in the organic versus aqueous phase. If a substance is dissolved in water and the water comes into contact with another substance, for example, octanol, the substance is likely to move from the water to the octanol. Its octanol-water partitioning coefficient reflects just how much of the substance will move until the aqueous and organic solvents will reach equilibrium. So, the release of equal amounts of the polychlorinated biphenyl, decachlorobiphenyl (log K_{ow} of 8.23), and the pesticide chlordane (log K_{ow} of 2.78), the PCB has about five orders of magnitude greater affinity for the organic phases than does the chlordane. This does not necessarily dictate than a great amount of either of the compounds is likely to stay in the water column, since they are both hydrophobic, but it does tell us that they will vary in the time and mass of each contaminant moving between phases. The kinetics differ, so the time it takes for the PCB and chlordane to reach equilibrium will be different.

When phases contact one another, a contaminant will escape from one to another until the contaminant reaches equilibrium among the phases that are in contact with one another. The kinetics of the process determines the time it takes for equilibrium to be achieved. After it is released into water, a substance may attach to or infuse surfaces of solid phase particles, that is, become sorbed. Sorption is the process in which a chemical compound becomes associated, physically or chemically, with a solid substance. Sorption is an important transfer process that determines the bioavailability or toxicity of a compound in surface waters.

Not all scientists are researchers. Indeed, some of the best scientists are those who apply what researchers have discovered. Such scientists work in national, state, provincial and local government agencies with missions to protect the environment. They write regulations, rules and guidance and issue permits and licenses on discharges and water quality protections. Other scientists work in the private sector or in environmental and health advocacy groups. Even when they all adhere to the scientific method and avail themselves to the same data, there is always various levels of certainty about environmental harm. Thus, establishing water quality standards or deciding on the best water pollution control approach is an iterative and multifactorial process [31–36].

Often in combating water pollution the researchers, regulatory scientists, and engineers work closely together. For example, in the U.S. the Environmental Protection Agency (EPA) has three basic functions: (1) To conduct research; (2) To enforce environmental rules and regulations; and (3) To provide funding for other entities to help with the first two functions. EPA's water pollution enforcement and compliance work are conducted by the Office of Water (OW), the Office of Drinking Water (ODW), and its 10 regional offices, and its research is conducted by the Office of Research and Development (ORD). The Clean Water Act [37] gives OW the principal statutory authority to regulate discharges to the waters of the U.S. The Safe Drinking Water Act [38,39] gives ODW the authority to regulate public water supplies, especially setting maximum contaminant levels (MCLs) for water provided by drinking water supplies. Other laws aim to reduce water pollution, such as the Comprehensive Environmental, Response and Compensation Act (CERCLA), which is also known as Superfund, and the Resource Conservation and Recovery Act (RCRA). Both address hazardous waste that may contaminate surface water and ground water, but Superfund addresses abandoned waste sites and RCRA addresses active processes of generating, transporting, treating, storing, and disposal of hazardous wastes. As such, RCRA addresses contaminants from the so-called cradle to grave. Water, especially ground water, can be polluted during any of these stages.

There are other perspectives in addition to research and enforcement, including teaching, journalism, and policy making. Those who educate others, write about, and promulgate polices and laws must understand water pollution in different ways, but all with a modicum of science.

Understanding the principles of water pollution and the calculations involved is something we all share, although the extent and specificity of these calculations vary. For example, a research scientist may need to explain in detail all the reactions in a combustion facility that lead to the generation of a toxic compound, e.g., benzo(*a*)pyrene (BaP). The water emissions permit writer may not need to know all the reactions to the extent of the research, but enough to know which ones may be most difficult to change. The high school chemistry teacher may need to know only the most basic chemical reactions, but the water pollution engineering college professor may need to know just as much as the researcher knows to tailor their lectures properly to the target student. The journalist who can "do the math" and grasp the scientific principles has an advantage over her colleagues that only know the essence of the science. The local water pollution authority manager may need to know as much as the permit writer, since he supervises the permit writers. The politician who serves on a technical committee responsible for writing water pollution legislation will need to know more than the other members of the legislature, perhaps even grasp the math and science as well as the permit writer and the researcher, at least to the extent of knowing whether to increase or decrease funding.

Therefore, every water pollution perspective requires an aptitude and appreciation for characterizations and calculations. This book attempts to provide examples of most of the types of calculations and problem-solving one may encounter, whether to succeed in one's career or simply to trust what one reads in a report or article.

2 Growing appreciation for water quality

Not long ago, making environmental decisions based on and underpinned by sound physical and biological science was not the norm compared to those made for other reasons [40]. At first, the weight of environmental factors was based on theory and limited, isolated studies. Indeed, science-based water pollution decision making tracked directly with the experiences of contagious diseases. Even after germ theory took hold in the biological research community, there was substantial time before public health decisions became predominantly based on contagion data. A rather large body of theory testing occurred before John Snow was able to make the famous decision to intervene and insist on infection controls by banning the use of contaminated Thames River water to abate the outbreak of the "Broad Street" cholera epidemic in the mid-19th century [41,42].

Concerns about water pollution predated Snow's work. For centuries, people have known intuitively that something was amiss when their air was filled with smoke or when they smelled an unpleasant odor. But, for most pollutants, those that were not readily sensed without the aid of sensitive equipment, a baseline had to be set to begin to act. One way to look at the interferences mentioned in the water pollution definition is to place them within the context of "harm." Harm can be acceptable, so long as it is sufficiently unavoidable and arises only due to obtaining a more favored good or service. The objects of the harm have received varying levels of interests. In the 1960s, the perception of harm to ecosystems grew as larger number of people perceived that the very survival of certain biological species was threatened and that these threats extended to humans. Indeed, the direct harm to humans was perceived earlier than those to ecosystems, especially in terms of diseases directly associated with obvious episodes, such as respiratory diseases and even death associated with combinations of weather and pollutant releases [43].

Myriad pollutant emissions have accompanied the industrial and technological eras. Nuclear power plants are associated with the possibilities of meltdown and the release of waterborne radioactive materials. Burning fossil fuels releases products of incomplete combustion, e.g., halogenated dioxins and polycyclic aromatic hydrocarbons. Even complete combustion releases the greenhouse gas, carbon dioxide (CO_2). Leaks from chemical, pesticide and other manufacturing facilities may cause exposures to toxic substances, such as the infamous release of methyl isocyanate in Bhopal, India. In the last quarter of 20th century, these apprehensions led to the public's growing wariness about "toxic" chemicals added to the more familiar "conventional" pollutants like soot, carbon monoxide, and oxides of nitrogen and sulfur. Toxic chemical exposures were increasingly linked to cancer and threats to hormonal systems in humans and wildlife, neurotoxicity (notably from lead and mercury exposure in children), and immune system disorders.

Some substances thought to be almost exclusively with food and nonfood ingestion were found in the air and water. For example, the source of blood lead (blood-Pb) levels in children were initially thought to be pica, i.e., eating chips of lead paint. However, the Pb concentrations declined precipitously in children living near roadways following the decreased use of tetraethyl-lead and other Pb organometallic compounds in gasoline, which may have accounted for more than half of blood-Pb in the 1970s [44]. Blood-Pb levels were increasingly associated with drinking water with Pb leaching from household pipes.

Growing numbers of studies have provided evidence linking disease and adverse effects to extremely low levels of certain particularly toxic substances. For example, exposure to dioxin at almost any level above what science could detect could be associated with numerous adverse effects in humans.

At the threshold of the new millennium, the need to consider water pollution within the context of larger-scale environmental systems became obvious, including the loss of aquatic diversity in lakes due to deposition of acidic precipitation, so-called "acid rain." Acid deposition was also being associated with the corrosion of materials, including some of the most important human-made structures, such as the pyramids in Egypt and monuments throughout the world. Presently, global pollutants have become the source of public concern, such as those that seemed to be destroying the stratospheric ozone layer or those that may affect the global climate.

A major event that raised awareness of environmental pollution occurred on June 22, 1969, near Cleveland, Ohio [45]. Like many cities in North America and Europe, Cleveland suffered from water and land pollution resulting from almost unbridled industrial production but lacking sufficient regulations and authority to curb the release of pollutants into the Cuyahoga River, one of the most polluted rivers in the world. At about noon that day oil debris between two wooden trestles in the river in the southeastern area of Cleveland became ignited by a passing train's sparks. The fire burned for about a half-hour, reaching heights of nearly 20m. Seemingly defying logic, the river had caught fire at least 13 times since 1868. Indeed, previous fires led to loss of life and major property damage but were not covered as extensively by the news media as the 1969 fire. Other rivers had previously caught fire due to large discharges of chemicals and sewage, including the Chicago River in Illinois, Buffalo River in New York, and the River Rouge in Detroit [46].

The Cuyahoga fire actually was a major impetus behind passage of the National Environmental Policy Act (NEPA) signed into law on January 1, 1970. The U.S. law created the Council on Environmental Quality (CEQ) in the President's office [47] and required that environmental impact statements (EISs) be written by all U.S. federal agencies for major projects that could significantly affect the quality of the environment [48,49]. It was also a major impetus for establishing the Environmental Protection Agency (EPA), as well as other environmental agencies around the world and in states, provinces, and local jurisdictions. In subsequent years, many new and amended laws were passed to address pollution, including the Clean Water Act Amendments of 1970 and 1977 [50–52]. Water pollution concerns, like burning rivers, were among the major reasons for the passage of laws and regulations to address water pollution problems more aggressively in the 1970s. Certainly seeing the plumes rise from flowing water was an amazing site, but it was the awareness that waters were so polluted to be flammable that was the real call for action.

In the 20th century, sanitary engineer became a separate branch of civil engineering, mainly to design structures for treating water pollution. The sanitary engineering profession had been forward-thinking for some time. For example, in 1909, sanitary engineer William Paul Gerhard (1854–1927) observed that microbes were effective in degrading pollutants [53]:

> *The purifying action, which takes place in the two stages of treatment mentioned, is due to the different kinds of bacteria, known respectively as the aerobic, the anaerobic and the facultative bacteria. None of the methods of treatment are automatic, and all require intelligent management and constant supervision.*

This was a big step in applying biological process for pollution control and treatment, especially microbial oxidation, reduction, and hydrolysis [54–57]. Microbes like the soil bacterium *Pseudomonas* spp. can be coaxed to use pollutants as their exclusive food source by withholding other preferred food sources, making the microbes to provide their energy by using their metabolic processes [58]. Of course, the prominent applications of microbial treatment have traditionally been water, sediment, and soil treatment, but numerous water pollutant control technologies now deploy Gerhard's observations [59–63].

In the second half of the 20th century, sanitary engineering programs in many universities were transformed into environmental engineering programs. This was a move beyond sanitary engineering's primary focus on structures, e.g., design of sanitary landfills, sewage treatment facilities, drinking water systems, and pollution control equipment to additional systematic attention to water quality and ecosystems, as well as applying engineering expertise in support of environmental assessments and impact statements [47,48,64–69]. It was also a first move to toward systems engineering, especially what is now known as green engineering and sustainable design, including water quality models, like physical or dynamic models, computational fluid dynamics (CFD) models, exposure and risk models, and optimization and predictive models to compare and chose among different outcomes expected to result from alternative actions and pollution control options [70–89].

Water pollution control technologies grew rapidly from the 1950s through the 1970s, focusing predominantly on the removal and treatment of biological and chemical pollution from industries and municipalities. This coincided with advances in solid waste technologies, especially sanitary landfills, and air pollution controls to remove particulate matter (PM), to treat oxides of sulfur (SO_x) pollution by sulfur oxide removal from flue gases and fuel desulfurization, and to emissions of control oxides of nitrogen, especially nitrogen dioxide (NO_2) produced in combustion processes [43].

Environmental protection became almost universally accepted in the 1970s. Organizationally, this has taken the form of departments or ministries of the environment in governments at all levels throughout the world. In addition to national agencies, states, provinces, counties, and cities took on greater responsibility for water and water quality, sold waste sanitation, noise abatement, and control of the hazards associated with radiation and the use of pesticides. This was paralleled in industry, where formerly diffuse responsibility for these areas has become increasingly the responsibility of an environmental protection departments. Similar changes became evident in research and education institutions.

New professions were formed commensurate with water pollution awareness. Indeed, water pollution hydrology and micrometeorology were becoming distinct academic and professional disciplines [90]. By 1980, mathematical models of the pollution of the atmosphere were being energetically developed. A start had been made in elucidating the chemistry of water pollution. Water quality monitoring systems became operational throughout the world. A wide variety of measuring instruments became available, many of which are very similar to those being used today.

Much of the early water pollution research was devoted to localized effects, but later began to be increasingly directed to the larger scale problems, that were multimedia in nature, i.e., not specifically water or air pollution, but that affect more than one environmental medium. Notably, the buildup of greenhouse gases in the atmosphere changes the distribution and availability of water on the planet. The depletion of the stratospheric ozone layer by chlorofluorocarbons (CFCs) changed the amount and types of electromagnetic energy reaching the earth's surface, the long-range transport of persistent bioaccumulating toxics (PBTs) persistent organic pollutants (POPs) contaminated water, air, and food in distant, sensitive populations and ecosystems. Conversely, pollution control measures in other media helped to improve water quality, such as the prevention of significant deterioration (PSD) by particulate matter (PM) and, especially, sulfur dioxide (SO_2) actions against acidic deposition incrementally improved water quality [91].

More recently, disasters have highlighted the importance of air-water transport exchanges, especially the long-range transport of pollutants on the quality of distant waterbodies [30,92–100]. For example, the meltdown of the Fukushima nuclear power facility in Japan was feared to have potential for attendant migration of radioisotopes created across the ocean, even as far as the West Coast of North America [101–103]. Like the PBT pollution of Inuit peoples [104–109], Fukushima served as another reminder that pollutants can travel thousands of miles in winds aloft. Wildfires in North and biomass burning in South America [110,111] and dust transported from Saharan Africa have released large amounts of PM and other pollutants that can find their way to surface and ground water systems [112,113]. Even what would be considered exclusively water disasters, the huge oil spills in the Gulf of Mexico and around the world had releases of volatile organic compounds and other potential waterborne contaminants [114,115].

As mentioned, combustion of fossil fuels and other heat-content substances is a major source of pollution. One of the indicators of thermal treatment success is the production of carbon dioxide (CO_2). The presence of water and CO_2 means complete combustion is achieved and the production of many very toxic and persistent water pollutants is avoided, such as polycyclic aromatic compounds (PAHs), known as products of incomplete combustion (PICs). More recently, however, the desired product of complete combustion, CO_2, is a greenhouse gas.

There are many successes in water pollution prevention and control over recent decades, but industrial pollution continues to be a worldwide and growing problem. Hopefully, developing nations will not repeat the 20th century problems experienced by Western nations and will apply lessons without having to suffer the consequences of side effects from economic development. Sound water pollution practice and planning will be needed. Water quality programs will need [61]:

- to describe and document environmental conditions accurately and precisely;
- to consider the central factors leading associated with and causing water pollution events;
- to evaluate the data and modeling results and to articulate completely the decisions to be made based on sound science; and,
- to uphold objectivity to arrive at comprehensive, fair, and consistent decisions that lead to suitable water pollution control and prevention actions, outcomes and consequences.

Understanding water pollution begins with the physical sciences. The hydrosphere is one of the spheres of importance to environment. As such, all physical processes are at work in the hydrosphere, including gravity, pressure, friction, and exchanges of matter and energy. The hydrosphere provides drinking water, food, and other essential ingredients to survival of humans and other organisms in the biosphere. It interacts with the atmosphere, which is the major system by which substances are transported and transformed into either essential or detrimental compounds that we breath. Thus, descriptions and calculations related to the structure of the atmosphere are crucial to science that underpins water pollution decisions. The troposphere, i.e., the lowest atmospheric layer, is the major location of pollution exchanges to surface waters and soil [116–118].

Like most water pollution texts, chemical water pollutants dominate this book. However, pollutants can also be physical and biological. Water pollutants may also be biological, as when bacteria and viruses are released to the water from medical facilities. Other biological water pollutants include irritants and allergens, such as pollen and spores and other materials released by molds (i.e., bioaerosols).

This book approaches water pollution from a question-and-answer perspective. Questioning is the very essence of water quality. Some of the questions go back to the earliest times of human existence, like "why does the water taste so bad?" or "when I drink water from this well why do I get sick the next day"? Modern questions are often different only in details, such as "what chemicals make the water taste this way" or "what pathogen or chemical contaminant is being ingested"?

There are myriad questions about water pollution. Some can be readily and completely answered. Most can be answered only under stipulated scientific conditions. Some have more than one answer. Some have answers that will differ depending on who is asking and on whom one asks.

3 Recent trends

Engineers, scientists, professionals, regulators, and managers have contributed to astonishing advances in water quality in the recent decades. Public health has been the principal driver for assessing and controlling water contaminants, but water pollution abatement laws and programs throughout the world have also recognized that effects beyond health are also important, especially as previously mentioned, the need to protect and enhance the welfare of the public and indeed of all creation. Protecting ecosystems is one of the chief welfare considerations since they receive the pollution and manifest the effects of this contamination. In addition, welfare embodies the need to prevent contaminants that diminish the integrity of structures and other engineered systems by corrosion. For example, corrosive agents reduce the efficiency of water and wastewater treatment systems, but also more general systems like refrigeration units. A recent example of corrosion that affected public health was the change in water supplies in Flint, Michigan, in which the new water's higher acidity caused pipes to release greater amounts of the metal lead in connections between the public water system and home water supplies [119,120]. This meant the people's exposure to lead increased dramatically in some cases, especially in socioeconomically disadvantaged neighborhoods [121–124]. Thus, from a water pollution perspective, there is a cascade of hazards from biotic systems of human health and ecosystems to abiotic (i.e., nonliving) systems.

4 Spatial scale

Characterizing and controlling water pollution varies by the size of the problem. Water is polluted at all geographic scales, from the individual receptor, e.g., a person, to local communities, to the globe. A town engineer or city manager, for example, may rely on water pollution data and reports at the urban or regional scale. Pollution can be subdivided into five ascending scales: local; urban; regional; continental; and global. For situations at finer scales, e.g., truck leaking diesel fuel during road repairs, this information may need to be augmented with data on the concentration of certain water pollutants at microenvironmental scale (e.g., indoor water in a home, office, or garage) or personal scale (i.e., at the individual's hands and mouth). The spheres of influence of the water pollutants themselves range from molecular (e.g., in living cells) to planetary (e.g., diffusion of greenhouse gases throughout the troposphere). The local scale is up to about 5 km of the earth's surface. The urban scale extends to the order of 50 km. The regional scale is from 50 to 500 km. Continental scales range from several 1000 km. Of course, the global scale is planet-wide [125].

5 Stressor-receptor paradigm

The main aim in controlling water pollution is to prevent adverse responses to receptors exposed to stressors in the water. Stressor may be human, animal, plant, and material. These adverse outcomes occur over various periods of time, from short-term (i.e., seconds or minutes) to intermediate-term (i.e., hours or days) to long-term (i.e., months or years). To elicit no adverse responses, the pollutant concentration in the water must be lower than the concentration level at which these responses occur.

Plant damage is an example. Damage can be measured biologically or socioeconomically. Biological damage can be measured as stress on growth and survival. Socioeconomic damage can be measured as 0% loss when there is no loss of the sale value of the crops or ornamental plants but a 100% loss if the crop is damaged to the extent that it cannot be sold. Such responses are related to dose, i.e., concentration times duration of exposure. Many manifestations of damage, e.g., increased acidity of rainfall on an ecosystem or increased rates of corrosion of pipes at high acidity, require an exposure duration sufficient to elicit economically significant effects. That is, attack for minutes may not affect the function of an ecosystem or the utility of a material from its intended use, but in several days that threshold may be crossed.

The biological response line for an acute disease is the dose-response curve which for a constant concentration becomes a duration-response curve. The shape of this curve reflects the ability of the human body to cope with short-term exposures and the overwhelming of the body's defenses by continued exposure. Short-term effects include acute poisoning, e.g., pesticides at sufficiently high concentrations to induce anything from a headache to death. Long-term effects include chronic diseases like various types of cancer, damage to the kidney, liver, central nervous systems, and other organs, endocrine disruption, and birth defects.

Water pollutants, like most pollutants, under other circumstances would be "resources," such as nitrogen and sulfur compounds and organic compounds. The next chapters discuss the biogeochemical cycles of water, nutrients, carbon, and metals. In the soil, nitrogen compounds are essential nutrients. As water pollutants, however, certain chemical species of nutrients can cause developmental and other diseases. Thus, simply "removing" pollutants is often less feasible, if even possible, than managing systems to ensure that optimal conditions for health and environmental quality exist. For example, the oxidation state of elements often determines the difference between an essential and toxic compound. Even an essential compound in excess can become a pollutant. For example, water that contains too little and too much molecular oxygen, i.e., supersaturation, causes gas bubble trauma (GBT), which is deadly to fish [126]. Such conditions can occur downstream from the turbines of a hydroelectric generation plant [127]. Thus, the quantity and location of a substance are determinants of whether it will be is a pollutant.

6 Systems thinking and sustainability

Before venturing into the details of water pollution physics, chemistry, and biology, we need to consider how the approach has changed in addressing water pollution and environmental problems in general.

As mentioned, the 20th century saw an explosion of control technologies developed and deployed to address the water pollution that accompanied world-wide industrial expansion. The predominant technologies were specific to the extent of pollution and type of pollutant, ranging from no controls to increasingly sophisticated technologies designed to address specific pollutants.

Traditionally, water pollution has focused on controls at the source, e.g., a discharge to a stream, lake, or ocean or polluted water running off farmland. Engineering solutions were designed to fit the discharge. The factory is an example of a point source, and the farm is an example of a nonpoint source. The releases from these sources had to be connected to the water quality in a neighborhood, town, or larger area. It became increasingly obvious that the combinations of individual contributions of point and nonpoint sources led to increasingly unacceptable concentrations of pollutants in the water. However, at first, we did not understand the mechanisms of transport and transformation. Indeed, we still cannot explain many of these mechanisms, with so many poorly understood "black boxes" in water pollution engineering. Thus, thermodynamics indicated that if the discharged mass of a pollutant could be decreased, the water quality would be expected to improve correspondingly. In growing numbers over the past two centuries, water pollution experts became aware that thresholds were being crossed [128]. Streams were not ever-available neutral sinks into which pollutants could be sent without major ecological and human health repercussions. G.R. Scott of the Tennessee Valley Authority (TVA) observed this in the first half of the 20th century [6]:

> *(A)s the water leaves the homes and industries, it carries with it the waste products of the metabolism or activity of the community. As these wastes enter the stream, natural forces are brought into play which, if given sufficient time, will remove or stabilize these pollutants. When this is accomplished, the water is again ready for use by a downstream community, and it may undergo numerous other such cycles before it finally reaches the ocean. If these pollutants are not removed, the communities downstream must search elsewhere for a suitable water supply. If they fail, they will be unable to realize their full potentialities in the developing economy, and it appears that failure on the part of a community to keep in step with progress results in retrogression. Water, therefore, is a necessity, and it must be considered a valuable resource or even a community's life blood. It should be protected and utilized wisely for the optimum development of our civilization.*

And, in 1954, M.A. Churchill, also of the TVA, wrote [129]:

> *The ultimate purpose of all stream sanitation activities is to establish and maintain a reasonable and economic balance between man's many uses of streams which do not lower water quality, and his use of streams for waste disposal. One of the prerequisites for establishing this balance is the determination of the stream's capacity for absorbing pollution.*

The central goal of water pollution work has been to decrease harm. The metric for harm is risk. The health risk from a water pollutant is proportional to amount of the pollutant reaching a person, i.e., exposure. Water pollution managers and engineers are increasingly applying systems thinking and looking for greener, more sustainable approaches. We still rely heavily on proven water pollution control, measurement, and models, but are increasing adding systems tools, such as design for the environment (DfE), design for disassembly, human exposure modeling, and multicriteria decision analysis (MCDA). One of the means of achieving this is by expert elicitation, where professional judgment is used as a first step in screening for potential exposure and risk of large numbers of chemicals that are either water pollutants or which may be transformed into pollutants [130].

Systems thinking begins with an appreciation of how science and engineering is changing to address water quality needs. To date, water pollutants, especially the so-called water toxics, have been addressed using proven control technologies. However, personal exposure to most chemicals predominantly occurs within near-field exposure scenarios, e.g., exposure while using a product that contains the chemical or contact with an article treated with the chemical [87]. This is leading to new screening models and tools. As mentioned these include MCDA and expert elicitation to prioritize pollutants based on potential exposure [131].

Systems tools, like MCDA, allow for the consideration of numerous variables, from various information sources. For near-field water pollution, the two main categories of variables are: (1) those associated with product use; and (2) the physical and chemical properties of the pollutant.

Thus, one of the important gaps in data and tools needed for life cycle and risk-based decision making to address and to prevent water pollution is improved exposure predictions. For example, human health characterization factors (CFs) in life cycle assessments (LCAs) have benefited from improved hazard, especially toxicity, information [132]. The CFs are weighting factors used in an important component of the LCA process, i.e., the life cycle impact assessment (LCIA). A water pollution LCIA converts emissions into impact scores for various impact categories [133].

An impact score is a weighted sum of the damage due to all water pollutant emissions to one of these or other impact categories [134]. LCAs generally and LCIAs specifically are important are means of taking a systematic view of the costs and benefits of an entire process rather than a single stage of the life cycle. For water pollution, LCAs allow the engineer or process designer to compare among various alternatives. For example, a chemical compound may appear to be the best choice for manufacturing of product, but the LCIA may indicate that it will produce a toxic water pollutant in later stages, which will then have to be treated. The treatment will add costs and risks that can be prevented by changes in process design or selection of safer chemicals that, if substituted, may obviate, or greatly reduce the treatment costs. Likewise, the LCA may indicate that a choice of a substance may increase water and other pollutants in earlier stages, such as the extraction of ores, which may not be necessary if another substance with less extraction-related pollution is chosen.

Until recently, LCA's have underweighted exposure and relied more heavily on inherent toxicity estimates to indicate risks posed by substances in the human health CFs [135]. Knowing the toxicity and other hazards of a water pollutant is necessary, but not sufficient, for water pollution decision making. Reliable exposure predictions can be combined with sound exposure estimates to predict the risks of various scenarios and technologies. Unlike the inherent physical and chemical properties of a pollutant, exposure predictions rely on numerous factors that differ among water pollutants. These factors go beyond physical and biological sciences and must also incorporate the social sciences, e.g., the use of products or daily activities that increase exposures [130,136].

7 Conclusions

With this introduction to concepts commonly used in water pollution science, engineering, technology, and management, we are now ready to discuss how a water pollutant is formed. In subsequent chapters, we will address how substances change after being released, how they are exchanged within the hydrosphere and between the other environmental media, how they are physically transported, the various ways they are chemically transformed, and their environmental fate.

We are now ready to consider the various ways that pollutants form, change, and move into, within, and out of surface and ground water systems.

References

[1] U.S. Environmental Protection Agency, Terms of environment, 2018, Available from: https://ofmpub.epa.gov/sor_internet/registry/termreg/searchandretrieve/glossariesandkeywordlists/search.do?details=&vocabName=Terms%20of%20Env%20(2009)#formTop. (Accessed 18 June 2009).

[2] J.S. Mill, Utilitarianism, Parker, Son and Bourn, London, 1863.

[3] R. White, Environmental Harm: An Eco-Justice Perspective, Policy Press, 2013.

[4] S. Lenton, E. Single, The definition of harm reduction, Drug Alcohol Rev. 17 (2) (1998) 213–219.

[5] V.G. Zartarian, W.R. Ott, N. Duan, Basic concepts and definitions of exposure and dose, in: W.R. Ott, A.C. Steinemann, L.A. Wallace (Eds.), Exposure Analysis, CRC Press, Boca Raton, Florida, 2007.

[6] G.R. Scott, Stream sanitation, Sew. Work. J. 18 (2) (1946) 284–287.

[7] P. Francis, Laudato Sí: On Care for Our Common Home: Our Sunday Visitor, 2015.

[8] P.A. Oliveira, A. Colaço, R. Chaves, H. Guedes-Pinto, P.L.F. De-La-Cruz, C. Lopes, Chemical carcinogenesis, An. Acad. Bras. Ciênc. 79 (4) (2007) 593–616.

[9] U.S. Environmental Protection Agency, Proposed guidelines for carcinogen risk assessment, US EPA (Ed.), Office of Research and Development, National Center For Environmental Assessment, Research Triangle Park, NC, 2015.

[10] U.S. Environmental Protection Agency, EPA's approach for assessing the risks associated with chronic exposure to carcinogens background document 2 January 17, 1992, U.S. EPA, Washington, DC, 2018. Available from: https://www.epa.gov/iris/epas-approach-assessing-risks-associated-chronic-exposure-carcinogens. (Accessed 7 March 2018).

[11] M. Casas, Air pollution exposure and interstitial lung diseases: have we identified all the harmful environmental exposures? Thorax 74 (2019) 1013–1014.

[12] M.B. Rice, W. Li, J. Schwartz, Q. Di, I. Kloog, P. Koutrakis, et al., Ambient air pollution exposure and risk and progression of interstitial lung abnormalities: the Framingham Heart Study, Thorax 74 (11) (2019) 1063–1069.

[13] F.W. Lipfert, Air pollution and mortality: timing is everything, Atmosphere 11 (12) (2020) 1274.

[14] U.S. Environmental Protection Agency, Chemistry dashboard, 2020, Research Triangle Park, NC. Available from: https://www.epa.gov/chemical-research/chemistry-dashboard. (Accessed 2 April 2020).

[15] R. Krieger, Hayes' Handbook of Pesticide Toxicology, Academic Press, 2010.

[16] D.W. Green, R.H. Perry, Perry's Chemical Engineers' Handbook, eighth Illustrated ed., McGraw-Hill, New York, 2007.

[17] D. Mackay, W.-Y. Shiu, K.-C. Ma, S.C. Lee, Handbook of Physical-Chemical Properties and Environmental Fate for Organic Chemicals, CRC Press, 2006.

[18] U.S. Environmental Protection Agency, Chapter 3: What is our current situation? and Chapter 4: What is best for our community? in: Handbook of Managing Onsite and Clustered (Decentralized) Wastewater Treatment Systems: An Introduction to Management Tools and Information for Implementing EPA's Management Guidelines, Office of Wastewater Management, Cincinnati, OH/Office of Water, U.S. EPA, Washington, DC, 2005, pp. 7–25, In this issue.

[19] H. Perry, D. Green, Perry's Chemical Engineers Handbook, Mc-Graw Hill, 1997.

[20] D. Lide, H. Frederikse, Handbook of Chemistry and Physics, CRC Press Inc., Boca Raton, FL, 1996, p. 9.

[21] H.M. Freeman, Standard Handbook of Hazardous Waste Treatment and Disposal, 1989.

[22] U.S. Environmental Protection Agency, Dimethyl sulfoxide—67-68-5 | DTXSID2021735, 2024, Available from: https://comptox.epa.gov/dashboard/chemical/properties/DTXSID2021735.

[23] T. McKone, J. Daniels, M. Goldman, Uncertainties in the link between global climate change and predicted health risks from pollution: hexachlorobenzene (HCB) case study using a fugacity model, Risk Anal. 16 (3) (1996) 377–393.

[24] D. Mackay, S. Paterson, Fugacity revisited, Environ. Sci. Technol. 16 (12) (1982). 654A-60A.

[25] V. Misra, F.N. Jaffery, P. Viswanathan, Risk assessment of water pollutants, Environ. Monit. Assess. 29 (1994) 29–40.

[26] P. de Almeida Rodrigues, R.G. Ferrari, L.N. Dos Santos, C.A.C. Junior, Mercury in aquatic fauna contamination: a systematic review on its dynamics and potential health risks, J. Environ. Sci. 84 (2019) 205–218.

[27] United Nations Environment Programme, Global mercury assessment 2018, UN Environment Programme, Chemicals and Health Branch, Geneva, Switzerland, 2019.

[28] J. Munthe, K. Kindbom, O. Kruger, G. Petersen, J. Pacyna, Å. Iverfeldt, Emission, deposition, and atmospheric pathways of mercury in Sweden, Water Air Soil Pollut. (2001).

[29] G. Rice, R. Ambrose, O. Bullock, J. Swartout, Mercury study report to Congress, in: Volume 3. Fate and Transport of Mercury in the Environment, Environmental Protection Agency; Office of Air Quality Planning and Standards, Research Triangle Park, NC (United States), 1997.

[30] D.A. Vallero, T.M. Letcher, Unraveling Environmental Disasters, Newnes, 2012.

[31] D.-T. Avgerinopoulou, Science-Based Lawmaking: How to Effectively Integrate Science in International Environmental Law, Springer Nature, 2019.

[32] D.M. Byrd III, R.C. Cothem, Introduction to Risk Analysis: A Systematic Approach to Science-Based Decision Making, Government Institutes, 2000.

[33] S.M. Capalbo, J.M. Antle, C. Seavert, Next generation data systems and knowledge products to support agricultural producers and science-based policy decision making, Agric. Syst. (2016).

[34] J.A. Arnot, D. Mackay, Policies for chemical hazard and risk priority setting: can persistence, bioaccumulation, toxicity, and quantity information be combined? Environ. Sci. Technol. 42 (13) (2008) 4648–4654.

[35] A. Fairbrother, D. Muir, K.R. Solomon, G.T. Ankley, M.A. Rudd, A.B. Boxall, et al., Toward sustainable environmental quality: priority research questions for North America, Environ. Toxicol. Chem. 38 (8) (2019) 1606–1624.

[36] J. Shatkin, S. Qian, Classification Schemes for Priority Setting and Decision Making. Comparative Risk Assessment and Environmental Decision Making, Springer, 2004, pp. 213–243.

[37] US EPA, Clean Water Act of 1972, 92-500, 1972.

[38] Safe Drinking Water Act Amendments of 1996, Stat. 104-182, 1996.

[39] Safe Drinking Water ACt, Pub. L. No. 93–523, 1974.

[40] R.M. Alexander, Social aspects of environmental pollution, Air Resources Center, Water Resources Research Institute, Oregon State University, Corvallis, OR, 1971.

[41] S.J. Snow, Commentary: Sutherland, Snow and water: the transmission of cholera in the nineteenth century, Int. J. Epidemiol. 31 (5) (2002) 908–911.
[42] S. Newsom, Pioneers in infection control: John Snow, Henry Whitehead, the Broad Street pump, and the beginnings of geographical epidemiology, J. Hosp. Infect. 64 (3) (2006) 210–216.
[43] A.C. Stern, H.C. Wohlers, R.W. Boubel, W.P. Lowery, Fundamentals of Air Pollution, Academic Press, New York and London, 1973.
[44] J. Schwartz, H. Pitcher, The relationship between gasoline lead and blood lead in the United States, J. Off. Stat. 5 (4) (1989) 421.
[45] Ohio History Connection, Cuyahoga River Fire, 2023, Available from: https://ohiohistorycentral.org/w/Cuyahoga_River_Fire. (Accessed 31 January 2022).
[46] R. Adams, New scholarship, Environ. Hist. 16 (4) (2011) 742–766.
[47] Council on Environmental Quality, Regulations for Implementing the Procedural Provisions of the National Environmental Policy Act, 2005.
[48] US Congress, The National Environmental Policy Act of 1969 (and following amendments), US Council on Environmental Quality, 1969. Available from: https://ceq.doe.gov/laws_and_executive_orders/the_nepa_statute.html.
[49] L.G. Luther (Ed.), The National Environmental Policy Act: background and implementation, Congressional Research Service, Library of Congress, 2005.
[50] M. Greenstone, The impacts of environmental regulations on industrial activity: evidence from the 1970 and 1977 clean air act amendments and the census of manufactures, J. Polit. Econ. 110 (6) (2002) 1175–1219.
[51] P.G. Rogers, The Clean Air Act of 1970, EPA J. 16 (1990) 21.
[52] U.S. Environmental Protection Agency, Clean Air Act National Stack Testing Guidance, US EPA, Office of Enforcement and Compliance Assurance, Washington, DC, 2009.
[53] W.P. Gerhard, Sanitation and Sanitary Engineering, The Author, 1909.
[54] R.E. McKinney, Discussion of population dynamics in biological treatment plants, 1966.
[55] R.E. Mckinney, Microbiology for Sanitary Engineers, McGraw Hill Book Co., New York, 1962.
[56] R. McKinney, Microbial relationships in biological wastewater treatment systems. The structure and function of fresh-water microbial communities, Virginia Polytechnic Institute and State Union, Blacksburg, Virginia, 1973, pp. 165–179.
[57] W. Eckenfelder, Biological treatment of waste water, Adv. Hydrosci. 3 (1966) 153–190.
[58] D.A. Vallero, J.J. Peirce, Engineering the Risks of Hazardous Wastes, Butterworth-Heinemann, Burlington, MA, 2003. xxi, 306 pp.
[59] C. Kennes, E.R. Rene, M.C. Veiga, Bioprocesses for air pollution control, J. Chem. Technol. Biotechnol. 84 (10) (2009) 1419–1436.
[60] R. Muñoz, S. Villaverde, B. Guieysse, S. Revah, Two-phase partitioning bioreactors for treatment of volatile organic compounds, Biotechnol. Adv. 25 (4) (2007) 410–422.
[61] D. Vallero, Environmental Biotechnology: A Biosystems Approach, Elsevier Science, 2015.
[62] J.W. van Groenestijn, P.G. Hesselink, Biotechniques for air pollution control, Biodegradation 4 (4) (1993) 283–301.
[63] N.J.R. Kraakman, J. González-Martín, C. Pérez, R. Lebrero, R. Muñoz, Recent advances in biological systems for improving indoor air quality, Rev. Environ. Sci. Biotechnol. 20 (2) (2021) 363–387.
[64] L. Luther, The national environmental policy act: streamlining NEPA, Congressional Research Service, Library of Congress, 2006.
[65] U.S. Environmental Protection Agency, Environmental impact statement rating system criteria, 2017, Washington, DC. Available from: https://www.epa.gov/nepa/environmental-impact-statement-rating-system-criteria. (Accessed 7 April 2017).
[66] B.B. Marriott, Environmental Impact Assessment: A Practical Guide, McGraw-Hill Inc., 1997.
[67] S.-O. Ryding, ISO 14042. Environmental management • Life cycle assessment • Life cycle impact assessment, Int. J. Life Cycle Assess. 4 (6) (1999) 307.
[68] National Research Council, Risk assessment in the federal government: managing the process, Committee on the Institutional Means for Assessment of Risks to Public Health (Ed.), National Academy Press, Washington, DC, 1983.
[69] National Research Council, Human Exposure Assessment for Airborne Pollutants: Advances and Opportunities, National Academies Press, 1991.
[70] M.C. Barber, K.K. Isaacs, C. Tebes-Stevens, Developing and applying metamodels of high resolution process-based simulations for high throughput exposure assessment of organic chemicals in riverine ecosystems, Sci. Total Environ. 605 (2017) 471–481.
[71] M. Borysiewicz, M. Borysiewicz, Models and techniques for health and environmental hazard assessment and management, 20, Institute of Atomic Energy, Swierk, Poland, 2006, p. 2010 (Retrieved March 2006).
[72] N. Brandon, K. Dionisio, K. Isaacs, D. Kapraun, W. Setzer, R. Tornero-Velez (Eds.), A novel framework for characterizing exposure-related behaviors using agent-based models embedded with needs-based artificial intelligence (CSSSA2016), 17–20 November 2016, The Computational Social Science Society of the Americas, Santa Fe, NM, 2016.
[73] E.L. Cussler, Diffusion: Mass Transfer in Fluid Systems, third ed., Cambridge University Press, Cambridge, 2007.
[74] Federal Register, Revision to the guideline on air quality models: adoption of a preferred general purpose (flat and complex terrain) dispersion model and other revisions; final rule. November 9, 2005, U.S EPA (Ed.), 2005.
[75] P.J. Gleckler, K.E. Taylor, C. Doutriaux, Performance metrics for climate models, J. Geophys. Res. Atmos. (2008) 113(D6).
[76] T. Li, J. Ban, R.M. Horton, D.A. Bader, G. Huang, Q. Sun, et al., Heat-related mortality projections for cardiovascular and respiratory disease under the changing climate in Beijing, China, Sci. Rep. 5 (2015) 11441.
[77] M. Markiewicz, Modelling of the air pollution dispersion. MANHAZ, in: Models and Techniques for Health and Environmental Hazards Assessment and Management IEA, Otwock-Swierk, Poland, 2006, pp. 303–347.
[78] S. Mitragotri, Y.G. Anissimov, A.L. Bunge, H.F. Frasch, R.H. Guy, J. Hadgraft, et al., Mathematical models of skin permeability: an overview, Int. J. Pharm. 418 (1) (2011) 115–129.

[79] N. Moussiopoulos, Ambient air quality, pollutant dispersion and transport models, European Environment Agency, 1997.
[80] J.D. Neelin, A. Bracco, H. Luo, J.C. McWilliams, J.E. Meyerson, Considerations for parameter optimization and sensitivity in climate models, Proc. Natl. Acad. Sci. U. S. A. 107 (50) (2010) 21349–21354.
[81] A. Salski, Fuzzy knowledge-based models in ecological research, Ecol. Model. 63 (1–4) (1992) 103–112.
[82] D.M. Sexton, J.M. Murphy, Multivariate probabilistic projections using imperfect climate models. Part II: robustness of methodological choices and consequences for climate sensitivity, Clim. Dyn. 38 (11) (2012) 2543–2558.
[83] H.-M. Shin, A. Ernstoff, J.A. Arnot, B.A. Wetmore, S.A. Csiszar, P. Fantke, et al., Risk-based high-throughput chemical screening and prioritization using exposure models and in vitro bioactivity assays, Environ. Sci. Technol. 49 (11) (2015) 6760–6771.
[84] R. Tan, K. Aviso, D. Ng, Optimization models for financing innovations in green energy technologies, Renew. Sustain. Energy Rev. 113 (2019) 109258.
[85] U.S. Environmental Protection Agency, Air emissions models for waste and wastewater, US Environmental Protection Agency, Washington, DC, 1994.
[86] D.A. Vallero, Air Pollution. Kirk-Othmer Encyclopedia of Chemical Technology, John Wiley & Sons, Inc., 2000.
[87] J.F. Wambaugh, R.W. Setzer, D.M. Reif, S. Gangwal, J. Mitchell-Blackwood, J.A. Arnot, et al., High-throughput models for exposure-based chemical prioritization in the ExpoCast project, Environ. Sci. Technol. 47 (15) (2013) 8479–8488.
[88] P.R. Williams, B.J. Hubbell, E. Weber, C. Fehrenbacher, D. Hrdy, V. Zartarian, An overview of exposure assessment models used by the US Environmental Protection Agency, in: Modelling of Pollutants in Complex Environmental Systems, vol. 2, 2010, pp. 61–131.
[89] S. Yanilmaz, D. Baskak, M. Yucesan, M. Gul, Extension of FEMA and SMUG models with Bayesian best-worst method for disaster risk reduction, Int. J. Disaster Risk Reduct. 66 (2021) 102631.
[90] K. Heidorn, A chronology of important events in the history of air pollution meteorology to 1970, Bull. Am. Meteorol. Soc. 59 (12) (1978) 1589–1597.
[91] S. Hull, Nondeterioration and the Protection of high quality waters under Federal Water Pollution Control law, Utah L. Rev. (1977) 737.
[92] D. Alexander, K. Smith, Environmental hazards: assessing risk and reducing disaster, Prog. Phys. Geogr. 17 (1993) 504.
[93] S.C. Bagby, C.M. Reddy, C. Aeppli, G.B. Fisher, D.L. Valentine, Persistence and biodegradation of oil at the ocean floor following Deepwater Horizon, Proc. Natl. Acad. Sci. U. S. A. 114 (1) (2017) E9–E18.
[94] T. Chouhan, I. Jaising, C.A. Alvares, Bhopal: The Inside Story: Carbide Workers Speak Out on the World's Worst Industrial Disaster, Apex Press, 1994.
[95] National Research Council, Exposure Science in the 21st Century: A Vision and a Strategy, The National Academies Press, Washington, DC, 2012. 196 pp.
[96] K. Poljanšek, A. Casajus Valles, M. Marin Ferrer, A. De Jager, F. Dottori, L. Galbusera, et al., Recommendations for national risk assessment for disaster risk management in EU, Publications Office of the European Union, Luxembourg, 2019.
[97] K. Smith, Environmental Hazards: Assessing Risk and Reducing Disaster, Routledge, London, UK, 1992.
[98] V. Srinivasan, G. Halada, J. Quinn, Learning from engineering disasters, Department of Materials Science and Chemical Engineering, Stony Brook University, Stony Brook, New York, 2018. Available from: http://dol1.eng.sunysb.edu/disaster/. (20 December 2022).
[99] D.A. Vallero, T.M. Letcher, Engineering risks and failures: lessons learned from environmental disasters, Leadersh. Manag. Eng. 12 (4) (2012) 199–209.
[100] D.A. Vallero, P.J. Lioy, The 5 Rs: reliable postdisaster exposure assessment, Leadersh. Manag. Eng. 12 (4) (2012) 247–253.
[101] H. Fu, G. Han, W. Li, X. Zhang (Eds.), Investigation of transport paths and times of Fukushima nuclear pollutants in the North Pacific surface using surface drifting buoys, in: 2012 2nd International Conference on Remote Sensing, Environment and Transportation Engineering, IEEE, 2012.
[102] H. Fu, W. Li, X. Zhang, G. Han, X. Wang, X. Wu, et al., A study of transport and impact strength of Fukushima nuclear pollutants in the North Pacific surface, J. Ocean Univ. China 13 (2) (2014) 183–190.
[103] G. Han, W. Li, H. Fu, X. Zhang, X. Wang, X. Wu, et al., An ensemble estimation of impact times and strength of Fukushima nuclear pollution to the east coast of China and the west coast of America, Sci. China Earth Sci. 56 (8) (2013) 1447–1451.
[104] P.I. Bank-Nielsen, M. Long, E.C. Bonefeld-Jørgensen, Pregnant Inuit women's exposure to metals and association with fetal growth outcomes: ACCEPT 2010–2015, Int. J. Environ. Res. Public Health 16 (7) (2019) 1171.
[105] M. Long, M. Wielsøe, E.C. Bonefeld-Jørgensen, Time trend of persistent organic pollutants and metals in Greenlandic Inuit during 1994–2015, Int. J. Environ. Res. Public Health 18 (5) (2021) 2774.
[106] A.S. Gerber, D.P. Green, Field experiment and natural experiments in: R. Goodin (Ed.), The Oxford Handbook of Political Science, Oxford University Press, Cambridge, UK, 2013, pp. 1108–1132 (Chapter 50), https://doi.org/10.1093/oxfordhb/9780199604456.013.0050. ISBN: 9780199604456, In this issue.
[107] Arctic Monitoring Assessment Program, AMAP assessment 2002: persistent organic pollutants in the Arctic, 2004.
[108] A.A. Dudarev, J.O. Odland, Forty-year biomonitoring of environmental contaminants in Russian Arctic: progress, gaps and perspectives, Int. J. Environ. Res. Public Health 19 (19) (2022) 11951.
[109] R. Gioia, R. Lohmann, J. Dachs, C. Temme, S. Lakaschus, D. Schulz-Bull, et al., Polychlorinated biphenyls in air and water of the North Atlantic and Arctic Ocean, J. Geophys. Res. Atmos. 113 (D19) (2008).
[110] K. Ballesteros-González, A.P. Sullivan, R. Morales-Betancourt, Estimating the air quality and health impacts of biomass burning in northern South America using a chemical transport model, Sci. Total Environ. 739 (2020) 139755.
[111] J. Mendez-Espinosa, L. Belalcazar, R.M. Betancourt, Regional air quality impact of northern South America biomass burning emissions, Atmos. Environ. 203 (2019) 131–140.

[112] C. Ichoku, L.T. Ellison, K.E. Willmot, T. Matsui, A.K. Dezfuli, C.K. Gatebe, et al., Biomass burning, land-cover change, and the hydrological cycle in Northern sub-Saharan Africa, Environ. Res. Lett. 11 (9) (2016) 095005.
[113] M. Osborne, F.F. Malavelle, M. Adam, J. Buxmann, J. Sugier, F. Marenco, et al., Saharan dust and biomass burning aerosols during ex-hurricane Ophelia: observations from the new UK lidar and sun-photometer network, Atmos. Chem. Phys. 19 (6) (2019) 3557–3578.
[114] National Oceanic and Atmospheric Administration, Deepwater horizon, U.S. Department of Commerce, Silver Springs, Maryland, 2022. Available from: https://darrp.noaa.gov/oil-spills/deepwater-horizon.
[115] H. Rajabi, M.H. Mosleh, P. Mandal, A. Lea-Langton, M. Sedighi, Emissions of volatile organic compounds from crude oil processing—global emission inventory and environmental release, Sci. Total Environ. 727 (2020) 138654.
[116] D. Vallero, Air Pollution Calculations: Quantifying Pollutant Formation, Transport, Transformation, Fate and Risks, Elsevier, 2019.
[117] D.A. Vallero, Air Pollution Physics. Methods and Calculations in Environmental Physics, AIP Publishing LLC, Melville, New York, 2022, pp. 7-1–7-70.
[118] D.A. Vallero, Biotic Physics. Methods and Calculations in Environmental Physics, AIP Publishing LLC, Melville, New York, 2022, pp. 9-1–9-40.
[119] H. Zhong, Y. Tang, H. Yan, Y. Zhang, L. Dong, B. Wang, Corrosion of pipelines in urban water systems: current research status and future trends based on bibliometric analysis, J. Water Process Eng. 56 (2023) 104288.
[120] A. Ghosh, H. Adams, S. Ash, M. Southard, Operators need to know how to maintain corrosion control conditions, Opflow 49 (7) (2023) 18–22.
[121] J.A. Hamm, J.S. Carrera, J.A. Van Fossen, K.D. Key, S.J. Woolford, S.B. Bailey, et al., Conceptualizing trust and distrust as alternative stable states: lessons from the Flint Water Crisis, Ecol. Soc. 28 (3) (2023).
[122] B.J. Pauli, The Flint water crisis, Wiley Interdiscip. Rev. Water 7 (3) (2020) e1420.
[123] L.J. Butler, M.K. Scammell, E.B. Benson, The Flint, Michigan, water crisis: a case study in regulatory failure and environmental injustice, Environ. Justice 9 (4) (2016) 93–97.
[124] M. Hanna-Attisha, J. LaChance, R.C. Sadler, A.C. Schnepp, Elevated blood lead levels in children associated with the Flint drinking water crisis: a spatial analysis of risk and public health response, Am. J. Public Health 106 (2) (2016) 283–290.
[125] D.A. Vallero, Fundamentals of Air Pollution, fifth ed., Elsevier Academic Press, Waltham, MA, 2014. 999 pages cm p.
[126] A. Velázquez-Wallraf, M.J. Caballero, A. Fernández, M.B. Betancor, P. Saavedra, H.W. Hemingway, et al., Biomarkers related to gas embolism: gas score, pathology, and gene expression in a gas bubble disease model, PLoS One 18 (7) (2023) e0288659.
[127] B. Zhang, X. Fu, K. Li, R. Li, X. Guo, Generation and release mechanism and abatement measures for gas supersaturation downstream of hydropower dams: a review, Water Resour. Res. 59 (2023) e2022WR034007.
[128] A. Markham, A Brief History of Pollution, Martin's Press, New York, NY, 1994, Routledge, Boca Raton, Florida, 2021. ISBN: 9780367362539.
[129] M. Churchill, Analysis of a stream's capacity for assimilating pollution, Sewage Ind. Waste. (1954) 887–904.
[130] D.A. Vallero, Air pollution monitoring changes to accompany the transition from a control to a systems focus, Sustainability 8 (12) (2016) 1216.
[131] M.D. Wood, K. Plourde, S. Larkin, P.P. Egeghy, A.J. Williams, V. Zemba, I. Linkov, D.A. Vallero, Advances on a decision analytic approach to exposure-based chemical prioritization, Risk Anal. 40 (1) (2020) 83–96.
[132] P.A. Schulte, L.T. McKernan, D.S. Heidel, A.H. Okun, G.S. Dotson, T.J. Lentz, et al., Occupational safety and health, green chemistry, and sustainability: a review of areas of convergence, Environ. Health 12 (1) (2013) 1.
[133] J. Bare, TRACI 2.0: the tool for the reduction and assessment of chemical and other environmental impacts 2.0, Clean Technol. Environ. Policy 13 (5) (2011) 687–696.
[134] R. van Zelm, M.A. Huijbregts, H.A. den Hollander, H.A. Van Jaarsveld, F.J. Sauter, J. Struijs, et al., European characterization factors for human health damage of PM 10 and ozone in life cycle impact assessment, Atmos. Environ. 42 (3) (2008) 441–453.
[135] J.C. Bare, D.A. Vallero, Incorporating Exposure Science into Life-Cycle Assessment, AccessScience, McGraw-Hill Education; Yearbook of Science & Technology, 2014.
[136] D.A. Vallero, C. Brasier, Sustainable Design: The Science of Sustainability and Green Engineering, John Wiley & Sons, 2008.

Chapter 2

Water math

1 Introduction

Mathematics is the language of the sciences whether it is physical, chemical, biological, or engineering science. So, it is no surprise that water pollution engineering and scientific problems, concepts, and solutions are most precisely explained using equations, reactions, statistics, indices, and models. Even at the highest levels of science-based decision making, arithmetic is the source of many errors and misinterpretations of data and information.

During my decades of academic and professional work, I have made more than my share of math mistakes. Arguably, the most extreme example was my error in calculating the amount of carbon monoxide (CO) released using a simple model at the beginning of my book, *Environmental Contaminants* [1], in which I used the opposite sign for an exponent. The exponent was supposed to be +3 but was shown as −3. Thus, in this hypothetical example, an error of six orders of magnitude carried through the example, underestimating the amount of carbon monoxide in the air by a factor of a million! In other words, if you make the same mistake I did, you will have what is known as a "false negative," i.e., appearing not to have a problem with CO in your basement, but there is a million times more CO in the basement air than the model shows. I included an errata sheet with the book, *Environmental Contaminants,* since I discovered this error shortly after its printing. If you happen to have the book, please make sure that it includes this sheet in your copy or see the correction in the e-book of *Environmental Contaminants* and second edition of my book, *Air Pollution Calculations: Quantifying Pollutant Formation, Transport, Transformation, Fate, and Risk*, released on September 17, 2023 (Paperback ISBN: 9780443139871; e-Book ISBN: 9780443139888).

Confession is truly good for the soul, but that is not the major point here. I am reminded of a conversation with my son, Daniel Joseph Vallero, who was at that time an undergraduate engineering student at North Carolina State University and is now an accomplished professional engineer. He was taking a coding class and told me that in coding courses every question is a trick question. His example was the use of punctuation, such as comma versus a semicolon, could lead to calculation errors and a wrong answer on an exam. This is an example, albeit a painful one, of how a sound education like that provided by certain programs offered by N.C. State. Duke, the University of North Carolina, the University of Kansas, M.I.T., Caltech, and other colleges, is a first step in preparing engineers and scientists to be rigorous in their calculations. Students must first become aware of the underpinnings of computational and quantitative tools in a "safe" environment and of what will be expected of them before they must properly apply the sciences in the messy, real world. Indeed, this is analogous with steps needed to become a proficient driver of a motor vehicle. First, the nascent driver must learn the rules of the road, "book learning" if you will. Then, the student gets behind the wheel with a licensed driver, i.e., no actual real road experience, but permitted to be mentored. Finally, the new driver receives a driver's license and can now venture on her or his own. This does not mean the driver is truly proficient since it takes some time to become comfortable merging onto high-speed higher and other maneuvers. After some time, the driver becomes sufficiently independent, just as the engineer in training or future engineer must spend years with a licensed P.E. before becoming one. With that in mind let us delve into the basics, starting with foundational and straightforward examples.

Certainly, the calculator helps with the more common arithmetic mistakes, such as errors in carrying numbers in division or misreading handwritten values. However, the calculator can be misused and can introduce other type errors, especially regarding significant figures. For example, when dividing 3.2 by 1.5 on my smartphone calculator shows 2.13333333. However, each of the numbers has only two significant figures, so the correct answer usually carried into calculations should be simply 2.1. You will note that several calculation examples in this book do not use the equal sign, =, but rather the symbol $\cong$ or $\approx$, both meaning approximately equal to, because the calculation is limited to the representative significant figures on the left-hand side of the equation. Like my CO model mistake, the calculator or computer program will simply do what is commanded. The proper level of uncertainty and validity is a human decision. Computational and

Fundamentals of Water Pollution. https://doi.org/10.1016/B978-0-443-28987-3.00009-6

numerical tools will obediently string out the answer as far as its programming will allow. If that means 20 digits to the right of the decimal point, that is what it will show.

Thus, understanding and explaining water pollution requires not only applying mathematical tools, but also adherence to calculation limitations. As such, a modicum of mathematics knowledge is needed before we venture into the physical, chemical, and biological properties and mechanisms involved in water pollution. Thus, we begin our water pollution quest with some sample calculations of the chemical principles and the laws of motion and thermodynamics.

1.1 Reasoning and logic

Mathematics can be defined as the abstract science of numbers, quantity, and space. Pure mathematics, i.e., math studied for its own sake, is a noble endeavor, but is beyond the need of most environmental calculations. Indeed, water pollution calculations is an example of applied mathematics. I have had conversations with mathematicians, scientists, and engineers about the difference between engineering and applied math. In my opinion, the difference lies in logic. Mathematics tends to be somewhat limited to deductive reasoning. That is to say, math works from the general to specifics. Conversely, engineers tend to use three types of reasoning. Like mathematicians, they reason deductively. However, engineers also reason inductively, i.e., from the specific to the general. They also employ engineering intuition.

When scientists and engineers deduce, they use a general principle or fact to derive information about a specific situation. For example, the first principles of physics determine gravity. This provides the general conditions that can be applied to a specific settling rate in a town's wastewater treatment tank. Similarly, low dissolved oxygen levels in a waterbody present a limiting factor in supporting the survival of certain fish species. Thus, oxygen level can be deduced to be the main or contributing reason for a specific fish kill. This demonstrates that synthesis is a product of deductive reasoning. Properly applying general theories, laws, and experiential information provides accurate information about the problem and is a first step in solving that problem. Thus, the simplest example of deductive reasoning can be expressed as:

$$\begin{aligned} &\text{If A} = \text{B and,} \\ &\text{B} = \text{C,} \\ &\text{Then A} = \text{C.} \end{aligned} \tag{2.1}$$

This simple example indicates that deductive reasoning depends on correct fact. For example, if a stream's dissolved oxygen concentrations decrease downstream from a point source releasing an oxygen-depleting chemical, the connection between the oxygen sag and the point source release is only valid if the release indeed contains that oxygen-depleting chemical. For example, if the factory no longer uses that chemical the link is invalid. Perhaps, the factory is releasing another chemical that also depletes oxygen, or an upstream source is releasing a chemical, or a natural weather event is depressing oxygen levels. These are three examples of not getting the facts right in deductive reasoning. The first example is what is known as an "regrettable substitution," i.e., the chemical with an unforeseen hazard replaces a chemical with one with a known hazard [2]. The second example is known as a problem in source apportionment, i.e., the effect is well document, but the suspected source is wrong [3–5]. The third example is also a type of source apportionment error but is also an example of not properly accounting for background conditions in linking causes with effects [3–9]. Deductive reasoning builds from facts to conclusions, i.e., synthesis [10]. All water pollution deductions are limited by the knowledgebase.

Water quality data in themselves do not completely inform a water pollution decision. In our example, A never completely equals B, so it is not necessarily equal to C. Even something as straightforward as the connection between cloud cover and the temperature of the upper layers of lake water. It is likely that the temperature will be depressed compared to a sunny day, but other factors like inflows from tributaries and turbulence will affect that assertion. Thus, deductive reasoning begins with scientific principles that show that incoming solar radiation is decreased with cloud cover, conversion of shortwave to longer wave radiation induces heat in the lake water, and movement of water from high to low elevation. However, other factors induce uncertainty and must be accounted for in the knowledge synthesis [11–19].

When scientists and engineers engage in inductive reasoning they are moving in the opposite direction from synthesis. They are conducting an analysis. Since inductive reasoning moves from the specific to the general it is a way to link an effect to the cause. The fish kill can be linked to the low dissolved oxygen levels via deductive reasoning, but a complete analysis needs a more in-depth investigation. For example, results from the presence of certain substances supports the increase in the biomass of microbes that consume the O_2. Experiments in microcosms help explain specific, well-defined, and well-controlled aspects of a system. From these experimental observations, larger principles can be established beyond the specific study. Robert Boyle, for example, conducted experiments with the air pump that led to his gas laws. The peril of induction is that any conclusion must be limited. For example, an experiment may establish a direct relationship between an independent and dependent variable, but the relationship beyond the controlled environment of the laboratory is uncertain. Increasing the amount of variable Y results increased growth of microbe Z, but what happens in the presence of A, B, and C, or the removal of variables P, Q, and R? The scientific method requires that experimenters be arbiters of what meaningful and care must be given before extending knowledge, especially when applying specific results to derive new laws.

Inductive reasoning is also called abstraction because it begins with concrete facts to form an abstract ideal. Induction takes specific facts, findings or cases and generally applying them to construct new concepts and ideas. Abstraction omits specific details, coalescing them into a whole that adheres to a defined principle. For example, a brown-furred bear, a white-furred bear, and a spotted-furred bear can all be integrated because each is a bear, notwithstanding the differences. The fur color therefore can be eliminated under the principle or criterion of being a bear (i.e., "bearness"), i.e., color is not "relevant." A brown bear, brown duck, and brown grocery bag can be integrated according to the criteria of having brown color. The features other than "brownness" are irrelevant here, so they are ignored. The18th Century Scottish philosopher, David Hume, coined the "problem of induction" [20]. Hume postulated that there is no justification in using induction because the conclusion of any inductive argument has no basis in validity [21]. Hume likely would not accept the following inductive syllogism:

> Every time the metal nickel (Ni) is added to activated sludge, the rate of growth of a particular strain of bacteria increases significantly. Therefore, the next time Ni is added to the sludge, this strain of bacteria's growth rate will concomitantly increase.

Scientists can offer strong reasons why the Ni addition may not lead to increased growth, e.g., different strains even from the same species may not have adapted a physiological need for Ni, temperature changes may encourage changing behaviors that render the Ni ineffective, and incomplete mixing prevents access of the microbes to the Ni. However, scientists also know that under the laboratory studies and conditions in the treatment plant that the micronutrient Ni did indeed induce growth rates consistently, which is a strong indicator that under similar conditions the results with appear again. It is possible that the result will differ, but the greater the number of repeated results occur under the same conditions, i.e., the association becomes increasingly strong, the likelihood of the same result should be progressively expected. For example, pressing the brake pedal should be associated with the moving vehicle being stopped. Indeed, even subtle differences in condition will change the probability. For example, an undetected leak in the brake cylinder or brake line could lead to a hydraulics failure that could cause the vehicle not to stop when the pedal is depressed. This would be an example of incorrect deductive reasoning for this specific case. In water treatment engineering, for instance, optimal treatment falls within a range of values for numerous variables. Adding x kilograms of alum to every incoming water source may lead to problems since the water differs in its chemical constituencies, e.g., hardness and pH. Too much or too little alum under real conditions detracts from the treatment efficiency.

Over a career, a professional's expertise is enhanced by experience. Each individual builds and edits the knowledgebase that supports professional intuition. Engineers and scientists build expertise that begins with formal education but grows continuously over a career. General knowledge consists of subject matter and rules. Rules provide the instructions for extending knowledge in both directions, i.e., from general to specific and from specific to general. Thus, intuition may well be deductive and inductive reasoning long forgotten by the individual professional [10,22–24]. Perhaps, intuition can best be defined of subconscious reasoning based on an ever-growing experiential knowledgebase. The sage engineer may not remember why a particular equation works but is somewhat certain it will again if the conditions are right. The problem occurs when the conditions are not properly accounted for. The old saw that when you have a hammer everything looks like a nail is worth pondering, especially when conditions are even slightly different this time from all previous applications.

2 Numbers

All calculations are numerical endeavors. Numbers are well understood by scientists and engineers, especially of more kinds and usage than most of the public. From a very young age we have been introduced to number concepts. Number usage rules may be long forgotten or simply part of our calculation intuition, so let us start with the basics.

An integer is a whole number. The numbers 1, 2, and 3 are the first three positive integers and the numbers -1, -2, and -3 are the first negative integers. Zero is also considered to be a whole number, so it is also an integer. A factor of an integer is any integer that divides it exactly, i.e., no remainder. For example, 3 is a factor of 9; 5 and 3 are factors of 15. A prime number has no factors other than itself and 1. The numbers 1, 3, 5, 7, 11, 13, 17, and 19 are prime numbers. So is 97. Conversely, a composite number is a nonprime number, that is, a composite number has factors other than itself and 1.

A common factor of two or more numbers is a factor that will divide each of them exactly. The greatest common factor (GCF) is the largest real number shared between two integers. Thus, it is the whole, real number that two integers share so that when broken down to their lowest multiples, it is the greatest integer that is shared between the two numbers. For example, 10 and 15 are broken down as:

10: 1, 2, **5**
15: 1, 3, **5**, 15

Thus, for 10 and 15 the greatest prime integer shared is 5, so the GCF in this case is 5.

The lowest common multiple (LCM) of two numbers is the lowest shared multiple between the two numbers. The GCF must be a prime number, but the LCM must be composite number because the GCF is based on the integer that divides evenly into two numbers whereas the LCM is based upon the integer that two numbers share in a list of multiples. The numbers 6 and 9 can be broken down as:

6: 6, 12, **18**, 24, 30, 36, …
9: 9, **18**, 27, 36, 45, 54 …

Thus, 18 is divisible by 1, 6, 9 and itself. So, 18 is the LCM in this case.

The sequence of mathematical operations must follow rules:

Rule 1. In additions, the terms may be placed in any order and grouped in any way without changing the result, i.e., the sum. For example, $5+3+13.3=21.3$ and $3+13.3+5=21.3$.
Rule 2. For subtractions, the result, i.e., difference, can be changed by the sequence and grouping of terms. For example, $10-3=7$, but $3-10=-7$. Also, $[10-3]-1=6$; however, $10-[3-1]=8$.
Rule 3. If grouping is not shown in an equation, the subtractions follow the order written, left to right. For example, $10-5.2-4=0.8$; whereas $4-5.2-10=-11.2$.
Rule 4. In multiplication series, factors can be in any order and in any grouping. For example, $10\times5\times4.2\times8.1=1701$ and $8.1\times4.2\times5\times10=1701$ and $5\times(8.1\times10)\times4.2=1701$.
Rule 5. In a series of divisions, the order and groupings matter. For example, $7\div15\div(2\div8.3)\approx1.9$, but $(8.3\div2)\div7\div15\cong0.04$.
Rule 6. When operations are mixed and no grouping in shown, the order multiplications and divisions follows the written order, then followed by additions and subtractions in the order shown.

2.1 Notation

Scientific notation uses coefficients and exponents. A coefficient is a numerical or constant quantity that is multiplied times the variable in an algebraic expression. It is usually greater than 0 and less than 10. An exponent represents the power to which a given quantity is to be raised, shown as a raised symbol beside the number or expression of 10, such as 10^7 or 10^{-3}. Scientific notation uses powers of 10. For example, using scientific notation, the number 1,530,000 is 1.53×10^6. The number 0.00037 is 3.7×10^{-4}. Scientific notation can also be expressed as "e notation". Thus, number $0.00037=3.7\times10^{-4}=3.7\text{e-}4$.

Epidemiologists and others who need to express population statistics often select one multiple of 10 in their scientific notation to be consistent. The number of cases of cancer in a population are often expressed as cases per 100,000 or cases per million, or 10^{-4} and 10^{-6}, respectively. For example, in 2020 Mongolia had 85.6 cases per 100,000 population (85.6×10^{-4}) age-standardized rate (ASR) of liver cancer incidence, i.e., the number of new cases. This, incidentally, was the highest in the world. The world ASR for liver cancer was 9.5×10^{-4} [25].

Scientific notation example

To demonstrate the use of scientific notations, at a rate of 1 atom per second, let us calculate the number of years needed to count the number of atoms in 12 g of carbon.

Answer:

Time is equal to the total number of atoms (N), divided by the counting rate (R):

$$t=\frac{N}{R} \tag{2.2}$$

$$N = \frac{N_A}{12\text{ g}} \times 1\text{ g} = 5.02 \times 10^{22}\text{ atoms} \quad (2.3)$$

$$\therefore t = \frac{N}{R} = \frac{5.02 \times 1022\text{ atoms}}{1\text{ atom}^{-s}} = 5.02 \times 10^{22}\text{ s} = 1.59 \times 10^{15}\text{ years}$$

Thus, it takes 1,590,000,000,000,000 years, or about 100,000 times the estimated age of the universe.

One challenge for scientific calculations is the ability of the computational tools used. So, if a computer becomes more efficient and a million times faster, i.e., it can count 1 million atoms per second, the result would be:

$\frac{1.59\times10^{15}}{10^6} = 1.59 \times 10^9$ years, or *only* 1,590,000,000 years.

Whereas this example may seem of interest to subatomic physicists, such endeavors do show up in environmental analytical sciences, which are concerned with quantum mechanics and other nanoscale and finer resolutions. For example, not so long ago, reporting parts per million (ppm) was an analytical challenge for many water pollution investigations, but environmental calculations have steadily improved, with measurements commonly reported in the low parts per billion (ppb), and now for many water pollutants, in parts per trillion (ppt) or less. For example, regulators have recently proposed a maximum contaminant level of 4.0 ppt for perfluorooctanoic acid (PFOA) and perfluorooctane sulfonic acid (PFOS) in drinking water supplies. PFOA and PFOS are *per-* and polyfluorinated alkylated substances (PFAS). This group of chemicals has been commonly used in non-stick and stain-resistant consumer products, food packaging, fire-fighting foam, and industrial processes [26].

Volume examples

A wastewater treatment plant operator is assessing the efficiency of dewatering of sludge in a municipal facility. Recent test results show that the dewatering system can remove 1.2 kg solids per cubic meter of wastewater (kg m^{-3}). The plant is treating 1,900,000 L day^{-1}. Calculate the amount of solids that would be expected to be removed from the water each day.

First, we need to know that $1\text{ m}^3 = 1000\text{ L}$.

Thus, the plant is treating 1900 m^3 of wastewater per day. At the rate of 1.2 kg m^{-3}:

$1.2\text{ kg m}^{-3} \times 1900\text{ m}^3 = 2280\text{ kg}$. This is about 2.5 tons of sludge that must be managed each day.

A construction crew is digging a drainage ditch with a rise of 200 mm per 35 m length (run). Calculate the slope of the ditch (S).

The slope of a line is calculated as rise over run, and usually expressed as a percentage:

$$S = \frac{\text{rise}}{\text{run}} \times 100\%$$

$$S = \frac{200\text{ mm}}{35\text{ m}} \times 100\%$$

$$1\text{ m} = 1000\text{ mm, so } 1\text{ mm} = 0.001\text{ m}$$

$$S = \frac{200\text{ mm}}{35\text{ m}} \times \frac{1\text{ m}}{1000\text{ mm}} \times 100\% \cong 0.57\%$$

The city engineer estimated that this grade is insufficient to produce the head needed, so she advised the crew to change the ditch slope to be 200 mm per only 25 m. What will the new grade be?

$$S = \frac{200\text{ mm}}{25\text{ m}} \times \frac{1\text{m}}{1000\text{ mm}} \times 100\% = 0.8\%$$

Efficiency example

A water treatment plant has purchased 5 electric motors rated at 745 watts (W). However, the plant operators found that the motors averaged 0.4 horsepower (hp) each. What is the average electrical efficiency (E) of the motors?

$$1\text{ hp} = 745.7\text{ W}$$

$$E = \frac{\text{power output}}{\text{power input}} = \frac{0.4\text{ hp}}{745\text{ W}} \times \frac{745.7\text{ W}}{1\text{ hp}} \times 100\% \cong 40\%$$

The level of confidence and validity of a calculation must be commensurate with the quality needed for a specific scientific and engineering endeavor. As PFAS is an example of a need for an extremely low level of detection (LoD), but many water pollution examples do not require such resolution. In some cases, ppm or even parts per hundred, i.e., per cent, may suffice to answer a water pollution question or to make a properly informed decision.

Macroscopic analytical example

A farm family needs to know the amount of water that has been added to the soil on their land, so Abuela asks each of her three grandchildren to record the rainfall for a week. The results are shown below. What is the total rainfall?

Grandchild	Days	Rainfall amount (cm)
Samuel	1–7	2
Alexander	8–14	3.009
Chloe	15–21	4.10002

A calculator indicates the total to be 9.10902 cm. However, Abuela recognizes that value appears overly precise, and that since this information's usefulness depends on its validity, which is how close these values are to the actual amount of water reaching the soil. She knows that Samuel, the youngest, probably was not nearly as careful as his older relatives and that Chloe and Alexander may have been a bit too precise in their interpretation of the rain in the gauge. So, she decides to apply the rules of significant figures.

2.2 Significant figures

Before calculators, scientists used slide rules and long division to quantitate pollution. The precision that these produced made numbers quite a bit easier than the results now generated by calculators and computers. Although the results on the display indicate precision 12 or more digits to the right of the decimal point, the actual results being calculated are seldom that precise. This calls for rules for significant figures when reporting water pollution results.

Significant Figure Rule 1

When adding or subtracting, the answer's significance must be only as high as the lowest value in the set. In this case, given Samuel's rainfall result the answer is simply **9 cm**.

Abuela loves her grandchildren so she wants to use the data but being a good farmer, decided that the next time, the team would implement and follow a standard operating procedure (SOP). She trained her team on the proper use of the rainfall gauge and otherwise improved the way the results were recorded, including waterproof tablets, stopwatches, color-coding, and audits. The results for the next period are shown below. What is the total rainfall?

Team member	Days	Rainfall amount (cm)
Samuel	22–28	8.2
Alexander	29–35	5.05
Chloe	36–42	3.24

The calculator gives a total of 16.49 cm. Applying Rule 1, she knows that once again Samuel's results, albeit more precise this time, is the least significant figure. However, she knows that she cannot simply truncate the "9." Thus, she applies the next rule.

Significant Figure Rule 2

To round off, a significant figure stays the same if the next digit is less than 5, but is increased by 1, if the next digit is greater than 5. Thus, she should report that, for this 3-week period, the total rainfall was 16.5 cm.

Microscopic analytical example

Abuela would like to know the total rainfall on her property for the 6-week period. If she has 1.579 hectares (ha) of land, of which 12% is impervious surfaces (e.g., buildings and concrete drives), and assuming the rainfall results are the same over the entire 88% soil surface area, how much water reached the soil?

Each hectare is 10^4 m^2. The rainfall height can be multiplied by the area receiving the rainfall to ascertain the volume of water. Abuela must also convert her rainfall results from cm to m, i.e., divide by 100, to get an answer in cubic meters. Assuming that the water is diverted and will not reach the soil under the impervious surfaces, the problem can be stated as:

$$\frac{(1.579 \times 0.88)\text{m}^2 \times 10^4 \times (9 \times 16.5)\text{m}}{100} = 20{,}634.372\,\text{m}^3$$

Significant Figure Rule 3

However, we know that this is an overly precise number. The third rule of significant figures states that when multiplying or dividing, the significant figure of the answer is the same as the quantity in the calculation with the fewest significant figures. This means that since Samuel's first answer has but one significant figure, so should our answer. Using the three rules, we should report the answer as $\mathbf{2 \times 10^4\ m^3}$.

Rainfall soil enrichment example

Abuela now wants to estimate the amount of nitrate (NO_3) that is reaching the soil from rainfall. Using the map in Fig. 2.1, she decides that there must be about $0.5\,\text{mg}\,\text{L}^{-1}$ NO_3 in the average rainfall in her county in North Carolina. How much NO_3 reached her soil in during the 6 weeks?

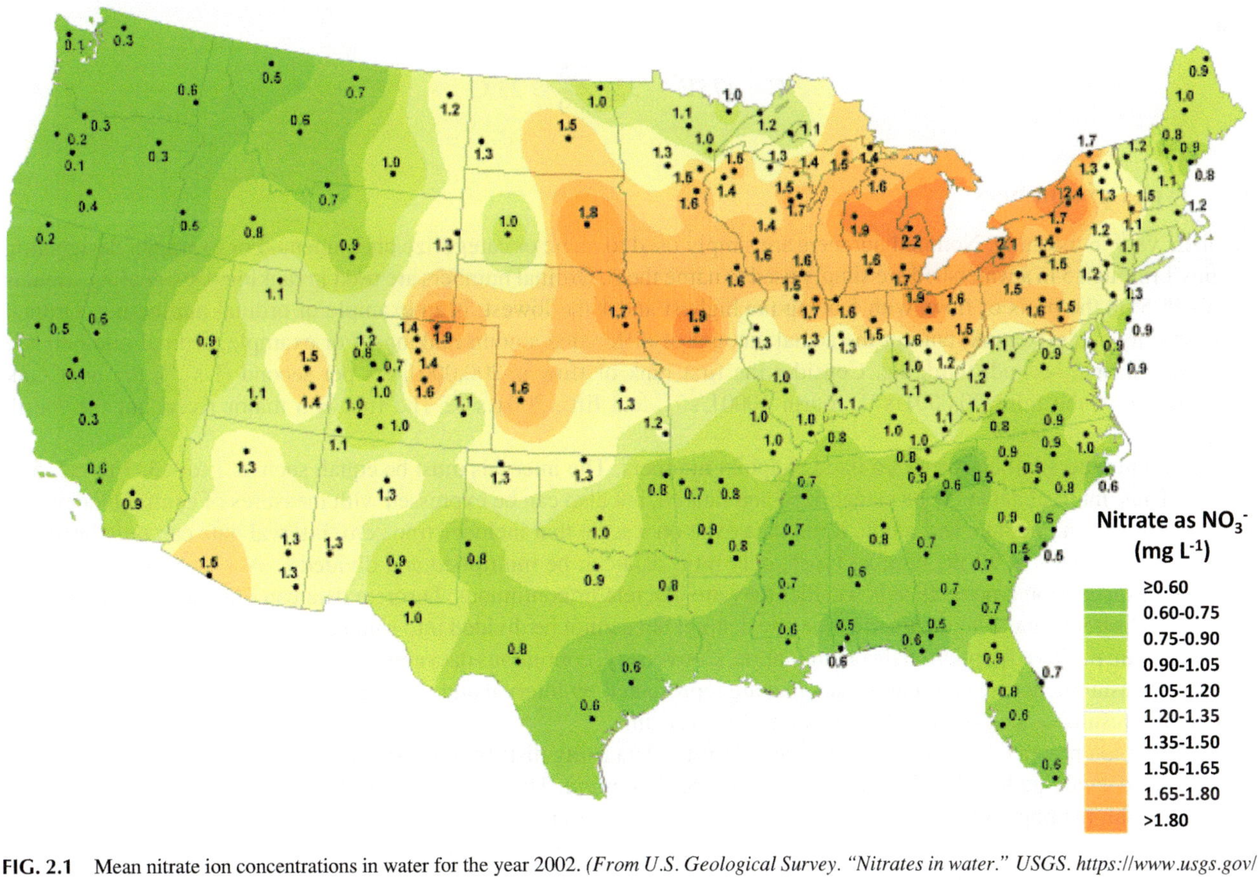

FIG. 2.1 Mean nitrate ion concentrations in water for the year 2002. *(From U.S. Geological Survey. "Nitrates in water." USGS. https://www.usgs.gov/special-topics/water-science-school/science/nitrogen-and-water (accessed July 12, 2024).)* Map lines delineate study areas and do not necessarily depict accepted national boundaries.

2.3 Concentration

Concentration of a water pollutant is commonly expressed in units of mass per volume. This is also common for air pollutants. Conversely, soil and sediment pollutants are often expressed as mass per mass, e.g., mass of nitrate per mass of soil. In the previous example, the source of the concentration data is the National Atmospheric Deposition Program. Nitrate (NO_3^-) is considered a water pollutant here. To convert this concentration to a total mass, we must multiply the concentration by the total volume of water that we found in the previous question:

$$\text{Concentration} \times \text{Volume} = \text{Total Mass of Pollutant} \tag{2.4}$$

Since the concentration is expressed in liters, we must also convert L to m^3.

$(0.5\,\text{mg}\,\text{L}^{-1}\ NO_3) \times (2 \times 10^4\,\text{m}^3) \times (10^3\,\text{L}\,\text{m}^{-3}) = 10^7\,\text{mg} = 10^4\,\text{g} = \mathbf{10\,kg\ NO_3}$.

It should be noted that almost every environmental measurement must consider temperature, as dictated by the gas laws that will be discussed in upcoming chapters.

2.4 Temperature

In our hypothetical example, Abuela noted that the average daily temperature during the first set of measurements was 68° Fahrenheit, i.e., 68 °F. The average daily temperature during the second set of measurements was 73 °. Fahrenheit values can be converted to Celsius:

$$^{\circ}\text{C} = \left(\text{F}^{\circ} - 32\right) \times \frac{5}{9} \tag{2.5}$$

So, in this case:

$$^{\circ}\text{C} = \left(68\,\text{F}^{\circ} - 32\right) \times \frac{5}{9} = 20^{\circ}\text{C}$$

And

$$^{\circ}\text{C} = \left(73\,\text{F}^{\circ} - 32\right) \times \frac{5}{9} \approx 23^{\circ}\text{C}$$

2.5 Types of numbers

Numbers vary in meaning. Nominal numbers are simply used to name or categorize phenomena. For example, designating streams 1 through 5 in a watershed are simply used to name them. Ordinal numbers are used in ranking. Designating streams 1 through 5 on the basis of flow, with 5 being the highest and 1 the lowest, is an example of ordinal numbering. Ordinal numbers only can give rank, but not the actual amount of difference among rankings. For example, ordinal numbers are expressed as first, second, third, etc. Consider the first rank of flow is 200,000 liters per second ($\text{L}\,\text{s}^{-1}$), the next rank 1,900,000 $\text{L}\,\text{s}^{-1}$, the third 50,000 $\text{L}\,\text{s}^{-1}$, fourth 1000 $\text{L}\,\text{s}^{-1}$, and fifth 300 $\text{L}\,\text{s}^{-1}$. The first is a major river, but the fourth and fifth are small creeks.

Interval numbers fall between two fixed endpoint numbers. The intervals must be equal, such as 1 to <3, 3 to <5, 5 to <7, etc. Thus, interval data may be categorized and ranked, but also can be expressed with measures of central tendencies, including the mean, median, and mode. Ratio data exist on a scale that includes true zero. Interval data can only be added and subtracted. Thus, of all the number types, only ratio data can be multiplied and divided.

Another distinction among data, is whether they are discrete or continuous. Discrete data can only be counted, but not measured. Discrete data's values are separate and distinct but cannot be divided into small parts. An example is the number of tests that are positive for microbial contamination. Conversely, continuous data consist of measurements. They cannot be counted. Measurements are continuous data that are represented by interval and ratio numbers. An example of continuous data are the results of size of algal blooms in a lake over time.

It is common practice to analyze and display scientific data using distributions, specifically probabilistic distributions. Sometimes, data are gathered without a specific hypothesis in mind. These are sometimes referred to a fishing expedition conducted with the hope others will evaluate the data to look for trends and to develop their own hypotheses. When well structured, these data gathering efforts are quite valuable. For example, surveys conducted by environmental agencies and institutions provide information on diseases, ecosystem condition, and other factors that lead to research and hypothesis generation by health and environmental scientists. Notably, the National Health and Nutrition Examination Survey

(NHANES) gathers data related to the health and nutritional status of adults and children in the United States by combining interviews and physical examinations. The survey continuously adds and updates the factors it collects regarding public health and nutrition measurements, such as the addition of emerging pollutant measurements, most recently collecting concentrations of some *per*- and polyfluoroalkyl substances (PFAS) in blood. The survey collects representative samples of about 5000 persons annually in 15 counties [27,28].

Examples of aquatic ecosystem water pollution data gathering efforts by United States Geological Survey (USGS) are the National Water Information System (NWIS), Water Quality Portal, Water Quality Watch, Tracking Water Quality of the Nation's Rivers and Streams, and the Health-Based Screening Levels for Evaluating Water-Quality Data [29].

3 Water pollution analysis

The word "analysis" appears often in water quality literature. It has several connotations. For one, it can be the opposite of synthesis, i.e., analysis means the breaking down of something whereas synthesis means building parts into a whole. However, in water pollution studies, analysis mainly means either chemical analysis or mathematical analysis. We will address chemical analysis in detail in subsequent chapters, but for now, it includes all the ways that chemical concentrations in water, soil, sediment, air, and biota are determined. This includes ways to analyze concentrations in the field using direct analytical tests and methods, as well as in the laboratory, e.g., separation science like liquid and gas chromatography and detection methods like mass spectroscopy. Conversely, mathematical analysis includes all the quantitative, semi-quantitative, and qualitative methods for identifying and vetting the data from measurements and models and evaluating these data to estimate the amount and extent of pollution and water quality.

When addressing water pollution, or any environmental problem, situations that are ostensibly identical can lead to diverse outcomes. When trying to replicate outcome data from a field study or lab experiment, or when analyzing an undesirable outcome to find ways to avoid it, complex mathematics, and statistics are used, such as nonlinear and chaotic approaches and nontraditional statistical methods. Combinations of analytical methods are deployed, depending on the type of data available and the scientific question being answered.

As mentioned, the major distinction among water pollution studies is whether the methods needed are qualitative, semi-quantitative, or quantitative. The analytical methods depend on the type and quality of decisions being supported. For example, if the question is whether a lake surface is lighter or darker this month, qualitative methods, e.g., photography, may be all that is required. If the decision is whether a pollutant is increasing or decreasing in a water body, semi-quantitative methods may be appropriate. If we need to know how much the pollutant concentration is increasing or decreasing, we will need quantitative methods.

3.1 Data interpretation methods

Data are generated from measurements and from models. These data are selected based on the question being answered. The answers come from various ways to understand and manipulate the data, usually starting with equations. Equations are at the heart of all water quality calculations. The equations can generate results in the form of numbers that can then be input into tables and graphs. Equations range from very simple arithmetic to highly complex mathematical equations. An example of a simple equation is finding a chemical dose. Drinking water dose is merely the product of a substance's concentration in the water and the intake rate. So, a person drinking 2 L of water per day that has 5 milligrams (mg) concentration of a contaminant would have dose of $2 \times 5 = 10$ mg of the contaminant per day, i.e., $10\,\text{mg}\,\text{L}^{-1}\,\text{day}^{-1}$. Note that throughout this book, it is preferable to use exponents rather than slashes, i.e., $10\,\text{mg}\,\text{L}^{-1}\,\text{day}^{-1}$ rather than 10 mg/day. This is because slashes also indicate division, and this can become unwieldy for calculations with several steps and components.

Many water quality features and phenomena can be analyzed and modeled mathematically using algebra, calculus, and differential equations. An equation that contains at least one variable that is squared is a quadratic equation, in the form:

$$ax^2 + bx + c = 0 \tag{2.6}$$

The standard form of a quadratic equation is set to have zero on the right-hand side. The terms *a*, *b*, and *c* are real numbers and *a* cannot equal zero. The example shows a constant term, *c*, but quadratic forms do not have to have a constant, such as $ax^2 + bx = 0$. The equation can also be expressed in factor form, such as $(x+2)(5x-9)=0$. This can be converted to standard form using the FOIL rule: first, outside, inside, last. The first terms are multiplied, then the outside terms, then the inside terms, then the last terms, and then like terms are combined to give the answer. In this case, this is $(x \times 5x)+(x \times 9)$ $+(2 \times 5x)+(2 \times 9)=0$, which is $5 \times 2+19x+18=0$.

Quadratic equations underly calculations involving time and length, so are highly useful in water pollution and other environmental fate models, e.g., predicting the rate of change of pollutant plume in an estuary after a spill. Indeed, there can be two solutions derived from a quadratic equation for an equation in the form:

$$x = \frac{-b \pm \sqrt{b^2 - 4ac}}{2a} \tag{2.7}$$

To a mathematician this means the problem is not well-posed since its solution is not unique. If one of the solutions is negative, it can be disregarded for length and time, i.e., there is no negative length or time in the real world. Many environmental models are based on quadratic relationships, generating graphs as outputs that display pollutant transport estimates and predictions.

Aquatic ecosystem functional models like survival rates, predator-prey relationships and primary productivity commonly apply calculus and differential equations. For example, the population of rainbow trout can be evaluated over time to estimate the total number of fish in a lake.

Calculus is the study of change, such as the count of rainbow trout, i.e., abundance, over time. We can first make observations of the population over a period of time, estimate the total number of organisms, and make assumptions to build a mathematical model for range of scenarios. Factors can include the chemical quality of the lake, threshold values, available food, and effectiveness of predators. Differential equations are key tools to help to predict how characteristics are expected to change over time. As the complexity of these characteristics increases so do the differential equations needed in the models.

4 Statistics

Water pollution data from measurements and models are usually limited in scope and size compared to the areas of concern. Indeed, information is needed to describe conditions of a whole stream, lake, estuary, or aquifer, but the actual measurement data may be taken from a few samples of water or sediment. Thus, these measurement data need to be extended. Water quality information must often be inferred from samples of whole water systems. Statistics are used to explain and predict outcomes in water pollution in numerous ways, especially to estimate values that describe a particular population of people, aquatic biota, hydrological phenomena, and chemical and biological agents. Statistics are also needed to test hypotheses. Statistics involves the collection and analysis of samples. The larger the population of concern, e.g., a large lake's dissolved oxygen concentrations, the greater number of samples that will be needed, i.e., the sampling density must increase with the size of the population from which conditions are to be inferred from samples.

4.1 Central tendencies

Often the first steps in interpreting data are to find and express the measures of central tendencies and expressions of the spread of the data around these tendencies. The first step in describing data is the sum. Consider the example of the O_2 concentrations in five streams:

Stream 1 = 7.0 mg L^{-1}
Stream 2 = 7.5 mg L^{-1}
Stream 3 = 5.5 mg L^{-1}
Stream 4 = 9.1 mg L^{-1}
Stream 5 = 7.8 mg L^{-1}

In this case the sum is expressed as:

$$\sum_{i=1}^{5} X_1 = 7.0 + 7.5 + 5.5 + 9.1 + 7.8 = 36.9 \text{ mg L}^{-1}$$

Thus, summation of values from 1 through n can be generalized as:

$$\sum_{i=1}^{n} X_i = X_1 + X_2 + X_3 + \ldots X_n \tag{2.8}$$

And:

$$\overline{X} = \frac{\sum_{i=1}^{n} X_i}{n} \tag{2.9}$$

The mean, median, and mode are the most commonly applied measures of central tendency in environmental science. The arithmetic mean is simply the sum of all the values of observations divided by the number of observations. In our example above, the sum is 36.9 mg L^{-1}, and the number of values is five, so the mean is calculated as:

$$\frac{36.9}{5} \cong 7.4$$

Thus, the mean of dissolved oxygen concentrations in these five streams is about 7.4 mg L^{-1}. Note that simply expressing the mean, like other measures of central tendencies, loses information about the data. In this case, the range is between 5.5 and 9.1 mg L^{-1}. Generally, for freshwater streams and lakes, DO values between 6.5 and 8.0 are considered acceptable. So, we have one streams that has diminished environmental quality, yet the mean for the streams is well within the acceptable range. The tradeoff between summing and averaging versus presenting the complete dataset primarily lies in the amount of data. For example, comparing the water quality of all the streams in the State of North Carolina or in Italy, would be aided by summation and averaging, since looking at all the data would be tedious for most readers. However, for scientists conducting research or evaluating the success of a water pollution control program, it would be advantageous to look at every value in the dataset.

The arithmetic mean of a sample from a population is expressed as:

$$\widehat{T} = N\overline{X} \tag{2.10}$$

Where, $\widehat{T}$ is the frequency that expresses the estimated value versus the true yet unknown population value (the ^ above the T is known as hat or circumflex); N is the number of units in the populations, and $\overline{X}$ is the sample mean.

Inferential statistics extends samples to a larger group, i.e., a population. Conversely, a parameter is a characteristic of an entire population. A census collects information from the entire population whereas a statistic is a sample or subset of the population from which conditions are inferred. For example, if every person in a town is interviewed about the quality of the town's drinking water, this is a parameter. If 100 people are interviewed from which information about drinking water quality of the town's entire 100,000 population, this is a statistic. The statistic is considered significant if it helps to explain the difference between a chance outcome and one that is linked to a causative or independent variable. Physical, chemical, and biological sciences, medicine, economics, and engineering apply a so-called P-value to express the reliability that a null hypothesis can be rejected and to show the strength of association between causes and effects. The P-value does not say anything about precision, i.e., we must also know the confidence of a finding. The confidence interval can be calculated around the point estimate to give a defined level of confidence that the true value is certain to exist. A tighter confidence interval means greater precision. A wide confidence interval means the certainty of a value being true is low. Accepting statistical significance at a P-value of 0.05 or 5%, confidence intervals can be calculated at a confidence level of 95%. Generally, if an observed result is statistically significant at a P-value of 5%, the null hypothesis fails within the 95% confidence interval.

Another important measure of central tendency is the median, which is simply the middle value of all values after they have been sorted by size. For example, our five streams could be reordered as:

Stream 3 = 5.5 mg L^{-1}
Stream 1 = 7.0 mg L^{-1}
Stream 2 = 7.5 mg L^{-1}
Stream 5 = 7.8 mg L^{-1}
Stream 4 = 9.1 mg L^{-1}

Thus, the median value in this example is **7.5 mg L^{-1}**. Thus, the mean and median values in this small dataset are similar, with the mean equal to 7.4 and the median just 0.1 mg L^{-1} higher. However, there are times when median is preferable to the mean, especially when a few values will substantially skew the distribution of the data. A famous example is the so-called "Michael Jordan Fallacy." The major with the highest mean starting salary of all graduates from the University of North Carolina at Chapel Hill in 1986 was geography, making about $250,000 ($680,000 in 2023), but in the rest of the

United States, the mean starting salary for geography majors was about \$22,000 (\$62,000 in 2023). However, Mr. Jordan's first year salary in 1986 from the National Basketball Association was \$700,000 (\$1.9 million in 2023), which exaggerated the mean, but which have had no effect on the median.

Mean-median example

Let us transform this interesting nugget into a water example. Assume there are six streams with flows measured at noon on August 15 as:

$$\begin{aligned} A &= 26.01\ \mathrm{L\,s^{-1}} \\ B &= \text{dry} - 0.00\ \mathrm{L\,s^{-1}} \\ C &= 17.55\ \mathrm{L\,s^{-1}} \\ D &= 19.33\ \mathrm{L\,s^{-1}} \\ E &= 21.12\ \mathrm{L\,s^{-1}} \\ F &= 23.01\ \mathrm{L\,s^{-1}} \end{aligned}$$

The mean would be 17.84 $\mathrm{L\,s^{-1}}$.

Sort to find the median:

0.00, 17.55, **19.33**, **21.12**, 23.01, 26.01

Note that with an even number of items, there is no middle number. The rule is to add the two middle numbers together and divide by 2. So, the median is $\frac{19.33 + 21.12}{2} = 20.23$.

In this example, the median is much larger than the mean because the zero-value outlier skewed the data in the negative direction. The warmer temperatures and drier conditions caused the stream B to dry. This is known as an intermittent stream or a seasonal stream. Streams may also cease to flow other times in the year, such as those that only flow after a rain event or snow melt. These are known as ephemeral streams and are more common in arid regions than more humid regions. The rest of the streams in the example are known as perennial streams, which flow year-round. The water available to streams is mainly provided by upstream waters and ground water, although snow melt and precipitation of various types supplement the flow [30].

In both cases, the single outlier salary or flow skewed the results significantly.

The third type of central tendency is the mode, which is the value that appears most frequently in the data. For datasets that include only integers, this is simply the integer that appears most frequently. However, for other types of data, the mode is usually a range of values, for example, for interval data, the mode would be interval of numbers that fall between two fixed endpoint numbers that appears most often. For example, if the interval that equals 5 and is less than 10 appears more than any other interval in the distribution, this is the mode. Modes are quite useful in describing total distributions. A distribution that is bimodal means that two intervals appear most often, i.e., two peaks appear. The peaks are seldom the same height, but both are obviously higher than the rest of the distribution. For example, a study of the size of particles that make up the sediment beneath an estuary may find that particle size peaks at about 0.9, 90, and 500 μm. In this case, the particle size distribution would be trimodal (See Fig. 2.2).

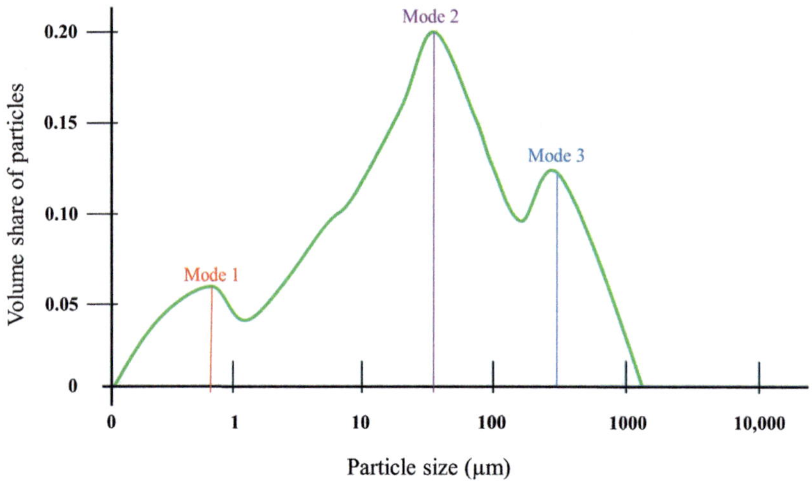

FIG. 2.2 Trimodal distribution of a hypothetical estuary's sediment grain size.

When the term mean is mentioned, it is usually taken to be the arithmetic mean. However, environmental data are sometimes better presented using geometric means. The geometric mean uses the product of the values of real numbers (whereas the arithmetic mean uses their sum). The geometric mean $\left(\prod_{i=1}^{n} x_i\right)$, is the *n*th root of the product of *n* numbers. Thus, for a set of numbers $x_1, x_2, \ldots, x_n$, the geometric mean is:

$$\left(\prod_{i=1}^{n} x_i\right) = \sqrt[n]{x_i + x_2 \ldots x_n} \tag{2.11}$$

Incidentally, the upper-case pi (Π) is the symbol denoting the need to conduct a series of multiplications. Obviously, zero values must be ignored when calculating a geometric mean.

Thus, ignoring the 0 value in our earlier example of the numbers 17.55, 19.33, 21.12, 23.01, 26.01, the geometric mean can be calculated as:

$$\left(\prod_{i=1}^{n} x_i\right) = \sqrt[5]{17.55 \times 19.33 \times 21.12 \times 23.01 \times 26.01} = (4{,}288{,}050.19)^{\frac{1}{5}} \cong 21.21$$

Note again that the 0 value was omitted from the values under the radical, since the product would have been 0, incorrectly implying that the central tendency would be 0. Thus, using a geometric mean is limited to exponential nonzero values that are intended to be multiplied together with indication of exponentiality. To avoid this limitation, another way to express the geometric mean is as the arithmetic mean in logscale:

$$\mathrm{GM} = \exp\left(\frac{1}{n}\sum_{i=1}^{n} \ln x_i\right) \tag{2.12}$$

Our simple examples indicate that in addition to central tendencies, the spread or variance of the data is also important. The first indication, as mentioned, is the range of the data, i.e., the difference between the lowest and highest values.

Distributions can show the expected outcomes of possible values generated by a computational tool. The distribution is a mathematical function that generates the probabilities of the occurrence of various possible outcomes from a study or experiment. Thus, most of the graphical presentations of data in environmental reports are of interval and ratio data, such as line graphs, histograms, and pie charts. The major exceptions include dots on a map, which are discrete data points. Incidentally, given that digital data are binary, i.e., either a one or a zero, continuous data are often converted to discrete data in a process known as discretization. This is done by dividing the continuous variable into ranges of values, followed by assigning the same discrete value to every value of the continuous variable that falls within a defined range. For example, a digital flow scale can assign all flow, which are continuous data, between 1206 and 1215 $\mathrm{L\,s^{-1}}$ to 1210.

4.2 Root mean square

Sometimes, data values have a probability of being both positive and negative, so neither the arithmetic mean nor the geometric mean, can accurately reflect the central tendency of the data set. For example, if the positive and negative values are equally likely, the arithmetic mean is close to zero.

For example, gas molecules in a container move in different directions, which means the molecule velocities can be either positive and negative values, so the arithmetic and geometric means would be zero. Obviously, this is not the case as each molecule is moving. One way to address this is to use exclusively the absolute values of each velocity. Another approach is to calculate the root mean square (rms) of the data to overcome the signs. The rms is the square root of the sum of all the squares divided by the total number of values, in this case, the number of molecules.

Root mean square example 1

Consider a gas molecule with its molar mass = *m*, which is confined to cubic box with perfectly elastic walls, i.e. there is no loss of kinetic energy when a gas molecule collides with the wall. The molecule is moving at speed *u*, which can be resolved to the three dimensions of the box edges, *x*, *y* and *z*, i.e., u_x, u_y and u_z. Assuming perfect elasticity purports that after the molecule hits the wall it will rebound, but not lose speed, remaining at *u*, and will move in the opposite direction, normal to the wall. Thus, the change in the molecule's momentum is [31]:

$$\Delta mu = (\Delta mu_x)_f - (\Delta mu_x)_i = -(mu_x) - (mu_x) = -2\,mu_x \tag{2.13}$$

Where, *f* and *i* subscripts represent final and initial states, respectively.

The total momentum in this system is conserved, so the molecule will transfer 2*mu* momentum onto a surface (A). If after rebounding the molecule avoids all other gas molecules and reaches another surface (A′), it would take $l\,u^{-1}$ time to cross the cube, and the round trip back to surface A, would take a total of $2l\,u^{-1}$, so that the gas molecule's rate of momentum transferred onto the surface A is:

$$\frac{\Delta mu}{\Delta t} = 2mu_x\frac{u_x}{2l} = \frac{mu_x^2}{l} \quad (2.14)$$

This rate of momentum transfer onto surface A is the force exerted on the surface according to Newton's law of momentum. The transfer rate for all gas molecules onto A can be found by summing the $\frac{mu_x^2}{l}$ values (n):

$$F_{total} = \frac{m}{l}(u_{x1} + u_{x2} + \ldots u_{xn}) \quad (2.15)$$

Because pressure is force divided by area, pressure (P) is:

$$P = \frac{F_{total}}{A} = \frac{m}{l^3}(u_{x1} + u_{x2} + \ldots u_{xn}) = \frac{mN}{V}\left(\frac{u_{x1} + u_{x2} + \ldots u_{xn}}{N}\right) \quad (2.16)$$

Where N is the total number of gas molecules in the box and V is box volume. mN is the total gas mass in the box, so $\frac{mN}{V}$ is the gas density (ρ), i.e. mass per unit volume. Also, $\frac{u_{x1} + u_{x2} + \ldots u_{xn}}{N}$ is the average u_x^2 for all gas molecules in the box, i.e. $\overline{u_x^2}$. Therefore, Eq. (2.16) can be rewritten as:

$$P = \rho\overline{u_x^2} \quad (2.17)$$

For any molecule, $u^2 = u_x^2 + u_y^2 + u_z^2$. The large number of molecules are moving randomly, so the average values of u_x^2, u_y^2, and u_z^2 are equal. Hence, each is one-third of u^2. The movement has no direction preference, so the pressure exerted in the wall is:

$$P = \rho\overline{u_x^2} = \frac{\rho\overline{u^2}}{3} \quad (2.18)$$

The square root of $\overline{u_x^2}$ is the gas molecules' root-mean-square speed, i.e., u_{rms}, which is the average molecular speed, expressed in m s^{-1}. Assuming that Eq. (2.18) accounts for collisions among the molecules and these collisions are elastic, the collisions occur between identical molecules, and the time of the collision is negligible compared to time between collisions, this equation would be true [31]. Thus, the relationship between macroscopic pressure and microscopic rms speed can be established and expressed as:

$$u_{rms} = \left(\overline{u^2}\right)^{1/2} = \left(\frac{3P}{\rho}\right)^{1/2} \quad (2.19)$$

If the gas pressure and density are known, the root-mean-square speed of the gas molecules can be calculated. Within the time and density constraints of most environmental scenarios, Eq. (2.19) reasonably approximates molecular speed. The equations can be further simplified to find molecular speed with the universal gas constant (R) and time (T).

These calculations show that u_{rms} is directly proportional to the square root of temperature in K°. Thus, increasing temperature by a factor of 100 means that the molecular speed (velocity) will increase by a factor of 10, increasing T by a factor 9 results in a gas velocity increase by a factor of 3, etc.

Increasing the molar mass of the gas decreases the gas velocity proportionately. Thus, increasing mass by a factor of 9 means that the molecular speed will decrease by a factor of 3, increasing m by a factor of 16 results in a gas velocity decrease by a factor of 4, and so on.

Root mean square example 2

Find the molecular speed of molecular oxygen at room temperature and use this to demonstrate how gas velocity is indeed expressed in meters per second.

The kinetic energy of a single gas molecule (E_g) is:

$$E_g = \frac{mu^2}{2} \quad (2.20)$$

Given the large number of molecules and collisions, the average kinetic energy (KE) of a gas $\left(\overline{E}_g\right)$ depends entirely on temperature:

$$\overline{E}_g = \frac{3RT}{2} \quad (2.21)$$

Since kinetic energy is expressed as joules, m is expressed in kg, and velocity in m s^{-1}, the two energy equations need to be converted to be expressed in moles by introducing the inverse of number of molecules $\left(\frac{1}{n}\right)$ to KE:

$$\overline{E}_g = E_g \cdot \frac{1}{n} \quad (2.22)$$

$$\frac{3RT}{2} = \frac{mu^2}{2} \cdot \frac{1}{n} \tag{2.23}$$

$$3RT = \frac{mu^2}{n} \tag{2.24}$$

Rearranging gives:

$$u = \sqrt{\frac{3RT}{m}} \tag{2.25}$$

Calculating gas velocity requires that temperature be in degrees Kelvin, so:
Assuming room temperature is 25 °C, this is 298.15 K.
Molar mass of O_2 is 32 g mol^{-1} or 0.032 kg mol^{-1}.

R contains joules (J), the energy unit equal to Newton (N) times meters (m). N is kg m s^{-2} and J is derived as kg m^2 s^{-2}. Thus, we must convert the molar mass to kg mol^{-1}.

Thus, replacing joules with its derivation units:

$$u_{rms} = \sqrt{\frac{3\left(8.3145\,\frac{\text{kg}\,\text{m}^2}{\text{mol}\,\text{K}\,\text{s}^2}\right)298.15\,\text{K}}{0.032\,\frac{\text{kg}}{\text{mol}}}} = \sqrt{\frac{3\left(8.3145\,\frac{\text{kg}\,\text{m}^2}{\text{mol}\,\text{K}\,\text{s}^2}\right)298.15\,\text{K}}{0.032\,\frac{\text{kg}}{\text{mol}}}} = \sqrt{2.24 \times 10^5\,\frac{\text{m}^2}{\text{s}^2}} \approx 473\,\text{m}\,\text{s}^{-1}$$

The square root of m^2 s^{-2} is m s^{-1}, this shows that gas velocity is indeed expressed as length per time, e.g., m s^{-1}.

Thus, at room temperature, the velocity of molecular oxygen gas is about faster by 130 m s^{-1} than the speed of sound, which is 343 m s^{-1}.

5 Probability

A random experiment produces a definite outcome that cannot be predicted with complete certainty. An event is an outcome or set of outcomes that can result from a study or otherwise drawn from a population. The sample space (Ω) is the set of all possible outcomes, so an event is a subset of the sample space:

$$\Omega = (s_1, s_2, \ldots, s_n) \tag{2.26}$$

Where, each *s* is an outcome. The outcomes must be mutually exclusive. They must also be collectively exhaustive, i.e., every sample space will always be an outcome in an experiment or trial.

In other words, an event *E* occurs in a particular situation or trial of an experiment or study if the outcome that is observed is an element of the set *E*. Outcomes can be labeled according to the number of possible outcomes.

All probabilities range from zero to one or, if stated as a percentage, 0–100%. The probabilities of all possible outcomes add up to 1. The concept of probability is commonly applied in water pollution and other environmental studies and reports. Since probability is unitless, a one-in-ten risk of developing a disease after consuming contaminated water is simply stated as a risk outcome that has a 0.10 or 10% probability. Low probabilities are a feature of many water pollution outcomes. Unacceptable cancer risk from exposure to a water pollutant, for example, is often expressed as not exceeding one in a million new cases (10^{-6}). For example, in a coin toss, there is usually thought to be only two possible outcomes, i.e., heads or tails. However, there are actually three possible outcomes when adding the very low probability of the coin standing on its edge, i.e., heads, tails, and edge. All outcomes together form a finite sample space:

$$E = [e_1, e_2, \ldots, e_n] \tag{2.27}$$

Furthermore, the probability of that event is expressed as:

$$P(E) = P(e_1) + P(e_2) + \ldots, P(e_n) \tag{2.28}$$

The edge outcome in a coin toss is a very low probability event that is often, by convention, ignored. For example, a coin is tossed at the beginning of overtime in a football game to decide which team will receive the ball, and thus have the first opportunity to score. When the referee tosses the coin, there is little concern about any outcome other than heads or tails. However, I can imagine that there is a statistician sitting in the stands who is hoping for an edge outcome.

In a water pollution scenario involving exposure to a carcinogen, *e* could represent a case of cancer. Here, we are looking for an outcome that is much like an adverse "edge" outcome in a coin toss. As mentioned, if a population of 1 million people is exposed to a carcinogenic pollutant over a specific time period, with one additional cancer attributed to that exposure: the probability of *e*, i.e., $p(e) = 10^{-6}$. This one-in-a-million event has the same probability as the hypothetical event of a coin landing on its edge. Thus, even a rare event will occur in a sufficiently large population. Indeed, environmental, and public health professionals and researchers commonly address probabilities of disease outcomes that are, thankfully, quite low when dealing with toxic substances, sometimes even lower than $p(e) = 10^{-6}$. For example, disease risks in a population can be expressed as the product of probability and an outcome, i.e., a frequency (F) at which the adverse outcome occurs:

$$F = \frac{X}{N} \tag{2.29}$$

Where, X = number of adverse events; N = number of individuals in the population. For example, if 10 people contract a disease that is randomly distributed in a population of 100,000, frequency of the disease is $\frac{10^1}{10^5} = 10^{-4}$, or one in a ten thousand. The F term is usually referred to as a probability (P). Environmental and health studies often compare populations, e.g., workers exposed to a particular pollutant compared to the general population's exposure to that same pollutant. This comparison is known as a relative risk, which is calculated as:

$$\frac{P_{exposed}}{P_{unexposed}} = \frac{X_{exposed}\left(N_{exposed}\right)^{-1}}{X_{unexposed}\left(N_{unexposed}\right)^{-1}} \tag{2.30}$$

Relative risk for pollutants is commonly stated as a ratio, particularly as a standard mortality ratio (SMR) or a standardized morbidity ratio (SMbR).

$$\text{SMR} = \frac{D_{observed}}{D_{expected}} = \frac{P_{exposed}}{P_{unexposed}} \tag{2.31}$$

Where, D = number of deaths for standardized mortality rates or number of diseases for standardized morbidity rates. Thus, SMbR is calculated the same way as SMR:

$$\text{SMbR} = \frac{D_{observed}}{D_{expected}} = \frac{P_{exposed}}{P_{unexposed}} \tag{2.32}$$

For example, if 15 out of 10,000 people in a part of town who are exposed to water polluted with a metallic compound have died of a type of liver cancer that occurs at a rate of 15 per million in the U.S. population, the SMR for this hypothetical cohort of exposed workers would be: $\frac{15\left(10^4\right)^{-1}}{15\left(10^6\right)^{-1}} = 10^2$. This means that the people living in that part of town have a relative risk of 100 times that of the general population of dying from this particular type of cancer. Such an observation would be considered to be a "cancer cluster."

The SMR is used as a specific measure of the strength of association between a population's exposure to a pollutant and deaths that may be attributed to that exposure. The mortality rate in a community without a specific exposure or intervention is called the baseline rate, which represents the expected level of the mortality. Note that the denominators in Eqs. (2.16)–(2.18) include "unexposed" subscripts. This is a relative term since most populations are seldom completely unexposed to the pollutant of interest.

In the metallic compound example above, the workers live in the general population, so those 10,000 persons would have to be subtracted from the million. Also, unless the compound is exclusively found in that part of town, others in the general population would be exposed, albeit often at much lower concentrations.

Workplace exposures can differ substantially from environmental exposures. The two types may not only differ in the size of the concentrations of certain pollutants, but in the ways that people are exposed. For example, the toxic metal cadmium (Cd) is a pollutant released during the production of nickel-cadmium (NiCd) batteries, paints, and coatings, as well as during metal work, e.g., welding, soldering, and plating. Concentrations of pollutants in the workplace can be much higher than in the general environment. As evidence, the Occupational Safety and Health Administration (OSHA) sets permissible exposure limits (PELs) for some substances that are orders of magnitude lower than average concentrations in an unpolluted environment [32–36].

A good way to think about probability is by constructing a Venn diagram of the sample space, i.e., all possible outcomes that could occur in an experiment, study, or situation (Fig. 2.3).

A is the occurrence of an event and $\overline{A}$ is the nonoccurrence of that event [37]. Obviously, $\overline{A}$ is complementary to A. Therefore, recalling that a maximum probability is 1, the standard format for this probability is:

$$P(A) = 1 - P(\overline{A}) \tag{2.33}$$

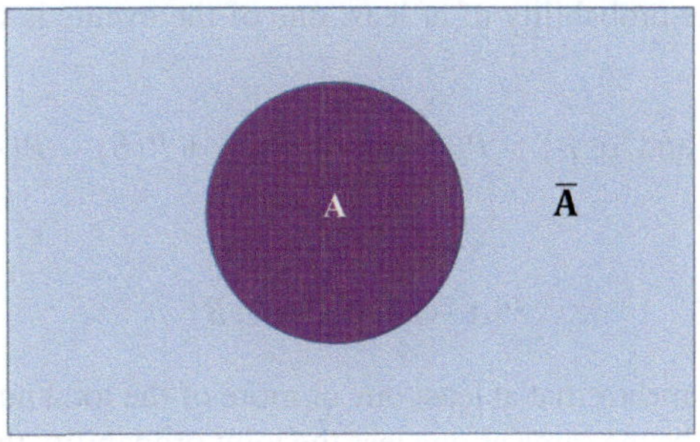

FIG. 2.3 Venn diagram of sample space and events A and Ā.

A joint probability occurs when two independent events occur in the sample space, which is the product of both events' probabilities:

$$P(AB) = P(A)P(B) \tag{2.34}$$

A joint probability is the intersection of the two events:

$$P(A \cap B) = P(AB) = P(A)P(B) \tag{2.35}$$

Furthermore, Fig. 2.4 shows the sample space of independent events *A* and *B*. In this case, the events are mutually exclusive, which means that only events *A* or *B* can occur in the sample space, but not both *A* and *B*. Thus, a mutually exclusive probability is the sum of the probabilities:

$$P(A \text{ or } B) = P(A) + P(B) \tag{2.36}$$

Conversely, if *A* and *B* can occur within the same sample space, they are not mutually exclusive events (Fig. 2.5). Non-mutually exclusive probabilities are expressed as:

$$P(AB) \neq 0 \tag{2.37}$$

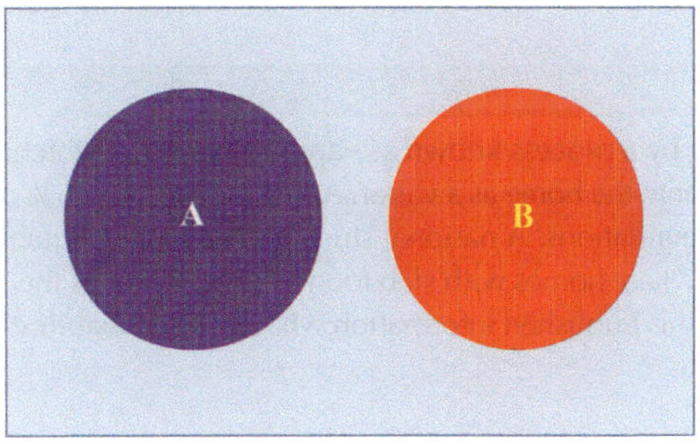

FIG. 2.4 Venn diagram of sample space for events *A* and *B* that are mutually exclusive.

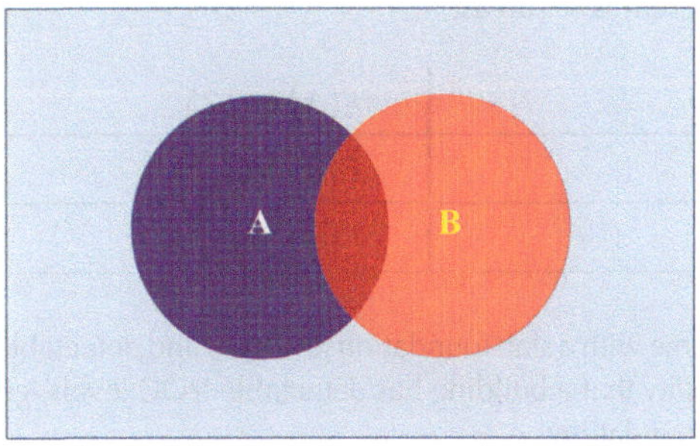

FIG. 2.5 Venn diagram of sample space for events *A* and *B* that that are not mutually exclusive. The darker *red* region is the intersection of the occurrence of the two events.

Thus, when not mutually exclusive, the probability of at least one of the events is the sum of the individual probabilities minus the joint probability:

$$P(A \text{ and/or } B) = P(A + B) = P(A) + P(B) - P(AB) \tag{2.38}$$

This is the union of two events:

$$P(A + B) = P(A \cup B) \tag{2.39}$$

Extending this to more than two events means that at least one or more of the total number of events will occur. Thus, if the events occur independently, this would be the same as one less the joint probability that none of the events would occur. For example, for three non-mutually exclusive events, the probabilities are expressed as:

$$\begin{aligned} P(A + B) &= 1 - P(\overline{A})P(\overline{B})P(\overline{C}) \\ &= 1 - [(1 - P(A))(1 - P(B))(1 - P(C))] \\ &= P(A) + P(B) + P(C) - P(AB) - P(BC) - P(AC) + P(ABC) \end{aligned} \tag{2.40}$$

In many water pollution scenarios, events are not independent, so that the occurrence of one event will influence the likelihood of other events. Thus, the probabilities are said to be conditional. For a situation where the conditional probability of event B given that event A has occurred is expressed as $P\{A|B\}$, which is derived as:

$$P(A|B) = \frac{P(AB)}{P(A)} \tag{2.41}$$

Over recent decades, environmental scientists and engineers have increasingly deployed probabilities to explain pollution and other phenomena. Fig. 2.5 will change after event A has happened. Thus, event B can only occur as AB. The elements of event B without AB must be excluded. Thus, solving the conditional probability expression for $P\{AB\}$, the product of the conditional probability of A given that we know event B has occurred must be:

$$P(AB) = P(B|A)P(A) \tag{2.42}$$

Hypothetical probability example

Volatile compounds can enter a building by a process known as vapor intrusion in which the ground water that has been contaminated reaches the building and migrates into the home as a vapor. A subdivision has 25% of its homes built on slab foundations and 75% of homes that have other types of foundations. A national study shows that detectable concentrations of a particular volatile organic compound (VOC) migrate into 77% of homes with slab foundations, whereas the VOC migrates in only 15% of the homes with foundations other than slabs. Using this published information what is the probability of finding a home with a slab foundation that has vapor intrusion of this VOC?

A=home with slab event.
$\overline{A}$=home without slab event.
B=home with detectable VOC.
$\overline{B}$=home with no detectable VOC.
Using the given percentages, the problem is set up as:

$P(A) = 0.25$	$P(\overline{A}) = 0.75$
$P(B\|A) = 0.77$	$P(\overline{B}\|A) = 0.23$
$P(B\|\overline{A}) = 0.15$	$P(\overline{B}\|\overline{A}) = 0.15$

This means that the probability of a home with a slab foundation (event A) and detectable levels of the VOC (event B) is expressed as the product of the conditional probability that a building has detectable VOC levels, given that its foundation is a slab, and the probability that the building has a slab foundation:

$P(B|A)P(A) = P(AB) = (0.77)(0.25) = 0.1925$. Given that these values have only two significant figures, this probability is about 19%.

The probability that a building chosen at random has detectable levels of the VOC and does not have a slab foundation is expressed by the product of the conditional probability of VOC detection given that the building has a foundation other than a slab, and the probability that the build does not have a slab foundation:

$P(B|\overline{A})P(\overline{A}) = P(\overline{A}B) = (0.15)(0.75) = 0.1125$. Given that these values have only two significant figures, this probability is about 11%.

Since the ground water contamination causing the occurrence of this VOC in a home that has a slab foundation and the occurrence of VOC detection in a home that does not have a slab are mutually exclusive events, the probability of detecting the VOC, regardless of the type of foundation, will be the union of these two events. Thus, it is expressed as the sum of the joint probabilities of the two possible scenarios:

$P(B) = P(B|A)P(A) + P(B|\overline{A})P(\overline{A}) = 0.1925 + 0.1125 = 30.5$. Given these values have only two significant figures, this probability is about 31%.

So, 31% of the houses would be expected to have detectable levels of this VOC in the indoor air.

5.1 Bayes theorem

As mentioned, quantitative analysis can be based on empirical evidence, in which case the meaning of the underlying probability of events is based entirely on data. In the case of the six-sided die, this approach might involve rolling the die repeatedly and estimating the probability of each outcome to be its observed relative frequency. This is seldom possible in water quality analysis since the variables and outcomes are so complex and the data about them so limited. This frequentist inference approach was almost exclusively taught in most introductory statistics courses for most of the 20th century (including three I took in undergraduate and graduate school). An advantage of frequentism is its objectivity. For example, concepts like P-value are consistent. An assigned $P\text{-value} = 0.05$ means that if that value is met, the null hypothesis is rejected.

Another method commonly employed analytical method builds from early hypotheses of Thomas Bayes, English preacher and mathematician, who held that the knowledge of prior events is essential in predicting future events. Bayes' theorem was published 2 years after his death in 1761 [38]:

$$P(A|B) = \frac{P(B|A)P(A)}{P(B)} \tag{2.43}$$

Where $P(B) \neq 0$, and where $P(A)$ and $P(B)$ represent the marginal probabilities of events A and B.

$P(A|B)$ and $P(B|A)$ represent the conditional probabilities of event A given that B has occurred, and event B given that A has occurred, respectively. Within the Bayesian framework, the event $A|B$, is the posterior probability of event A. Bayes' theorem states that the posterior probability of A. This means that the probability of A given that event B has occurred is equal to the likelihood [written $P(B|A)$] times the prior probability distribution of A, which is $P(B)$, divided by the marginal distribution of B. Thus, the prior probability distribution, the likelihood, and the posterior probability distribution together provides both the framework and the necessary elements of a Bayesian statistical problem. The theorem assigns a contingent probability of an outcome that was based on logic and history. This is very valuable in scientific investigation and modeling by applying previous knowledge, i.e., "priors," to calculate posterior distributions [39,40].

Conditional probability can be extended further as an important tool in analysis and prediction of water pollution problems. In the conditional probability $P(B|A)$, B is known as the posterior event and A is the prior event. Therefore, $P(A)$ is the prior probability and $P(B)$ is the posterior distribution. Bayes' theorem (also known as Bayes' Rule) indicates that the probability of event A, given that event B has occurred, equals the likelihood multiplied by the prior probability distribution of event A divided by the marginal distribution of event B.

Applying Bayes theorem requires that the initial probability distribution, the likelihood, and the posterior probability distribution be known. For example, researchers have used the theorem to understand many environmental problems, including describing ecosystem function and structure [41], evaluating pollutant removal infrastructure performance [42,43], predicting water quality condition [44], and guiding environmental management decisions [45–47].

In our introduction to probability, probabilities are assumed to express the specific random events that occur after enough repeated trials, which is the basis for classical or frequentist statistics. Conversely, Bayesian statistics includes updates based on prior evidence that changes the outcomes, i.e., posterior distributions are generated using the new information. As such, frequentists attempt to eliminate uncertainty by giving estimates, whereas Bayesians attempt to preserve and to refine uncertainty by correcting the distribution to account for the new evidence. Expressed in probability terms, Bayesian inference is:

$$P(\theta|D) = \frac{P(D|\theta)P(\theta)}{P(D)} \tag{2.44}$$

Where, $P(\theta)$ is the prior, $P(\theta|D)$ is the posterior, $P(D|\theta)$ is the likelihood, and $P(D)$ is the evidence.

As the name implies, frequentist statistics has data that are a frequency that is a repeatable random sample, whereas Bayesian statistics observes data from the real sample. Frequentist statistics has fixed parameters that remain constant during the repeatable process, but Bayesian statistics has data that are fixed and unknown parameters that are described probabilistically [48].

Bayesian inference example

The accuracy of a hypothetical device that detects lead (Pb) in a home's drinking water lead. The device changes color when the Pb concentrations are at or above $3\,\mu g\,L^{-1}$. Assume that studies have shown that 95% of the time the device has correctly identified water with concentrations above this level. A false positive outcome occurs 10% of the time, i.e., the device shows that the water contains greater than $3\,\mu g\,L^{-1}$ when in fact it was lower than that concentration. Also, assume that the 10% of homes in a town are expected to exceed $3\,\mu g\,L^{-1}$, based on previous data.

If the device is used in a study in the town and a positive ($\geq 3\,\mu g\,L^{-1}$) color indication is recorded in a home, what is the probability that there is indeed an exceedance of the $3\,\mu g\,L^{-1}$ concentration?

Let us assume that a home that has water above $3\,\mu g\,L^{-1}$ has a real Pb problem and is denoted as *A*. If the device shows a problem, this is denoted as *B*.

A = real problem event.

B = device reading $3\,\mu g\,L^{-1}$ event.

Thus:

$P(A) = 0.10$	$P(\overline{A}) = 0.90$
$P(B/A) = 0.95$	$P(\overline{B}/A) = 0.05$ (false negative)
$P(B/\overline{A}) = 0.10$ (false positive)	$P(\overline{B}/\overline{A}) = 0.90$

If the device shows a positive reading in a home, what is the probability that a real Pb problem is there, based on the known error of the device? This means that we are looking for the conditional probability that a real Pb problem exists, given that the device displays this problem, i.e., the prior conditional probability $P(A|B)$. Thus:

$$P(A|B) = \frac{1}{1 + \frac{P\left(B|\overline{A}\right)P\left(\overline{A}\right)}{P(B|A)P(A)}} = \frac{1}{1 + \frac{0.10 \times 0.90}{0.95 \times 0.10}} \cong 0.51$$

This means that the observed positive reading with Bayesian inference changed the probability of a real lead problem in the home from the original $P(A)$ estimate of 0.10 to 0.51 that there is truly an exceedance of the $3\,\mu g\,L^{-1}$ concentration.

As mentioned, common statistical approaches to environmental predictions can be classified within three categories, each involving a different approach to calculating the likelihood of an event relative to a set of all possible events. The first approach applies *a priori* beliefs, e.g., in the case of a single roll of a six-sided die, might reflect an expectation that the die is fair in generating an outcome, which means that the probability of each of the six possible outcomes (that is, 1, 2, ..., 6) is exactly 1/6. Incidentally, as the limits of detection decrease and the concomitant levels of precision decrease, this assumption becomes increasingly unacceptable. For example, the side of the die that contains only one dot would be slightly heavier in weight than the side with six dots if the dots are concave but would be slightly lighter if the dots are merely painted on the die surface (more paint on the six-side means more weight). Also, the depressions may change the way the dice roll. These changes may mean little at tolerances of percentages (10^{-2}) but must be accounted for if the tolerances were 1 in a billion (10^{-9}). Analogously, needing to know whether the dissolved oxygen (DO) of as stream falls below 4 parts per million (ppm) requires six orders of magnitude less precision than determining whether drinking water contains less than 4 parts per trillion (ppt). Both are key indicators of water quality but require quite different levels of precision.

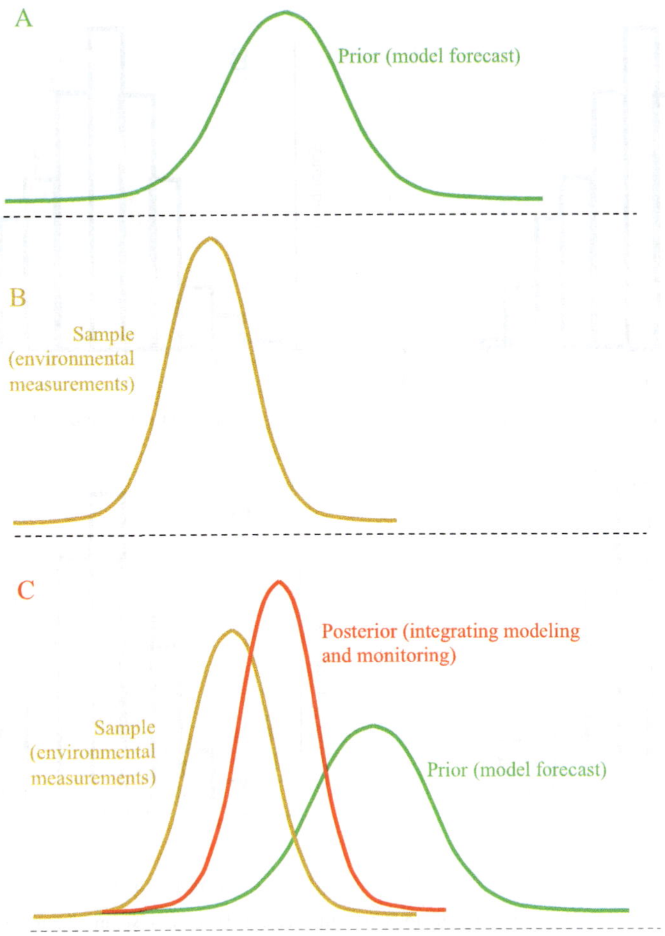

FIG. 2.6 Construction of a posterior distribution with environmental sampling data and priors according to Bayes Theorem. *(From DA.Vallero, Environmental Systems Science: Theory and Practical Applications, first ed., Elsevier, Amsterdam, NV, 2020 May 1, used with permission.)*

In practical terms, Bayes' theorem allows scientists to combine *a priori* beliefs expressed in the model (See Fig. 2.6A) that is based on scientific knowledge about the probability of an event or an environmental condition, or other metric with empirical evidence from real-world measurements (See Fig. 2.6B). The resulting output can provide a more robust view, i.e., the posterior distribution in Fig. 2.5C.

A hypothetical example of the steps in building a posterior probability distribution is shown in Fig. 2.7 [49]. Consider the need to design and operate a stormwater management infrastructure system to remove suspended solids (SS) that is presently entering a lake. In addition to being a pollutant in its own right, SS carries nutrients, metals, and other contaminants. The fraction of sediment removed by a stormwater management system is represented as the prior, i.e., θ. The progression of this understanding in a Bayesian framework begins with the development of a prior probability distribution for θ based on pollutant removal rate values in a dataset documenting hundreds of studies. The data arefirst expressed as a histogram of these historic values (See Fig. 2.7A), and then as a line approximating the pollutant removal rate prior probability distribution (Fig. 2.7B). Next, the hypothetical sediment removal rates from a recent study at the lake are entered through a likelihood function (Fig. 2.7C). Finally, the posterior probability distribution is calculated using Bayes' theorem and represented by the red line in Fig. 2.7D.

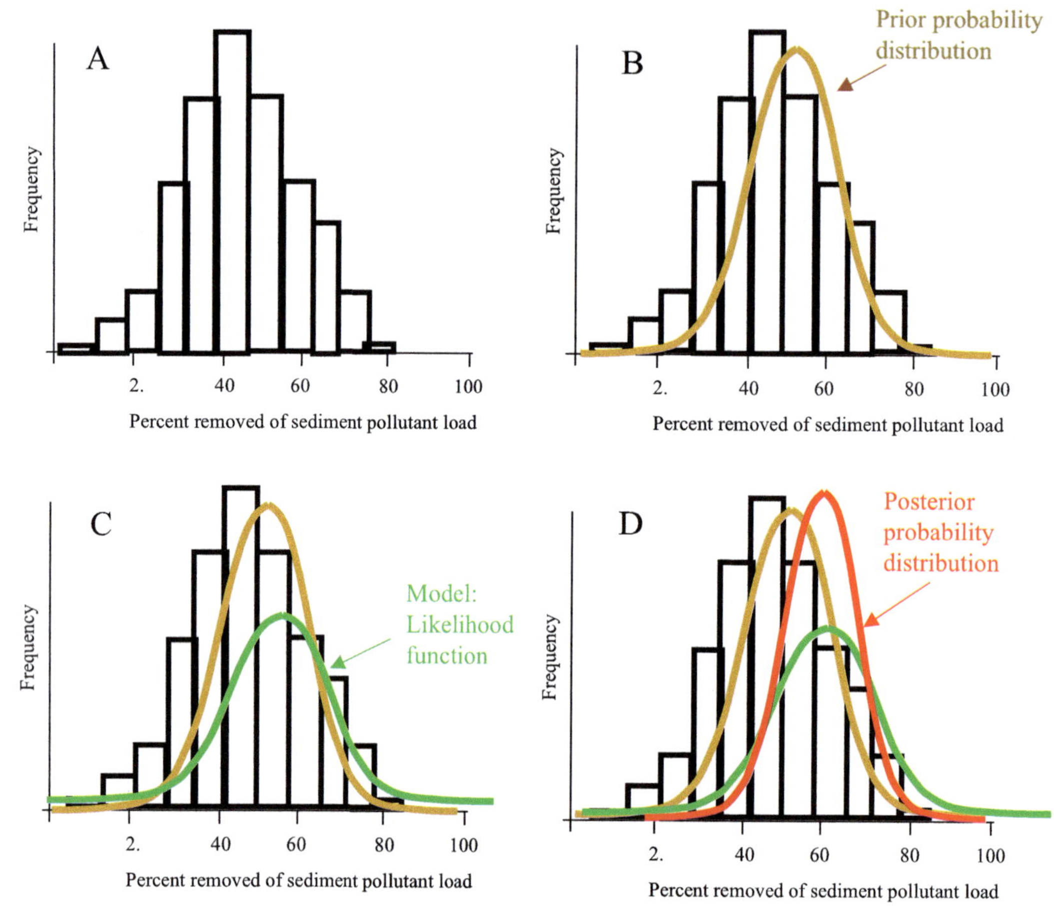

FIG. 2.7 Construction of a posterior distribution according to Bayes' theorem. A. Histogram summarizing historical and other measurement data; B. Approximation of associated prior pollutant load removal rate probability distribution; C. Application of a likelihood function that represents evidence from hypothetical site-specific measurements; and D. Posterior probability distribution of sediment removal rates. The frequency on the *y*-axis shows the number of times each value was observed in the historical dataset. *(From DA.Vallero, Environmental Systems Science: Theory and Practical Applications, first ed., Elsevier, Amsterdam, NV, 2020 May 1, used with permission.)*

5.2 Bayesian networks

By integrating dissimilar types of knowledge, logic, and rules in systemic entities, Bayesian networks (BNs) can aid in evaluating effects and potential contributors to the effect, which supports identifying and examining potential contributing causes to an adverse outcome like a fish kill, a cancer cluster, or an acute human disease outbreak. Such linkages between causes and effects in complex environmental and public health scenarios usually have many dependencies among variables. These dependencies are represented by BN models as conditional probability distributions that address the uncertainty in different parts of the analyzed system. BNs can be extended into influence diagrams (IDs) comprised of the different decision options and valuations of the various outcomes [50]. The IDs can be applied in decision analysis to identify the formally optimal management approaches under various water pollution scenarios (See Discussion Box 2.1).

Discussion Box 2.1 Bayesian networks and water quality

Given that water quality is expressed by the concentration of one or more pollutants, and the degree of change in an environmental resource, e.g., surface water body or ground water aquifer, the expression of quality reflects the water's suitability for desired purposes, For example, these include a water body's suitability as a water supply source, recreational area, or productive agricultural area. The suitability threshold can be set at a concentration that above which the measured pollutant concentrations exceed a quantitative water quality standard [49]. When all relevant water pollutants cannot be measured directly for an entire site, surrogate indicators are developed and applied. Depending on the pollutant, the strength of the relationship between an indicator concentration and the concentration of the pollutant it represents varies extensively. For example, the quality of waters used for recreation and shellfish-harvesting throughout the United States is based on

Discussion Box 2.1 Bayesian networks and water quality—cont'd

the concentration of fecal indicator bacteria (FIB), e.g., *Escherichia coli*. These bacteria provide a conservative indicator of fecal contamination and of the potential presence of harmful waterborne pathogens. Although they are directly linked to human and environmental health, measurements of FIB are difficult and costly to measure. No matter the specific pollutant and its associated indicator, both the pollutant-indicator relationship, and the spatiotemporal frequency of sampling and other factors can collectively contribute to uncertainty and variability in environmental condition forecasts. To illustrate a Bayesian approach to evaluate these uncertainties and variability, consider an example whereby water quality conditions of a shellfish harvesting area are assessed using fecal coliform concentration (organisms $100\,\text{mL}^{-1}$) measurements.

Like many pollutants, FIB concentrations commonly follow a lognormal probability distribution, with a log-concentration mean (μ) and a log-concentration standard deviation (σ). This probability model recognizes inherent spatial and temporal variability in FIB dispersion patterns, However, like other simple probability models, this model often does not explicitly acknowledge other subtle sources of variability. Some are due to intrinsic sources resulting from the FIB concentration measurements and the calculation methods. All sources of uncertainty and variability propagate into greater overall uncertainty in FIB concentration predictions, and to uncertainty in probability distribution parameters μ and σ. A Bayesian framework explicitly acknowledges these uncertainties by first placing a prior probability distribution on the population parameters μ and σ. This may account for *a priori* beliefs about their potential values and enables a likelihood function for μ and σ based on empirical evidence. In this example, the evidence is water quality samples. Ultimately, a joint posterior probability distribution is derived for both. Results of this procedure are a smoothed contour plot of the joint posterior probability density for the fecal coliform log-concentration mean (μ) and standard deviation (σ) for a sample site in eastern North Carolina (See Fig. 2.8).

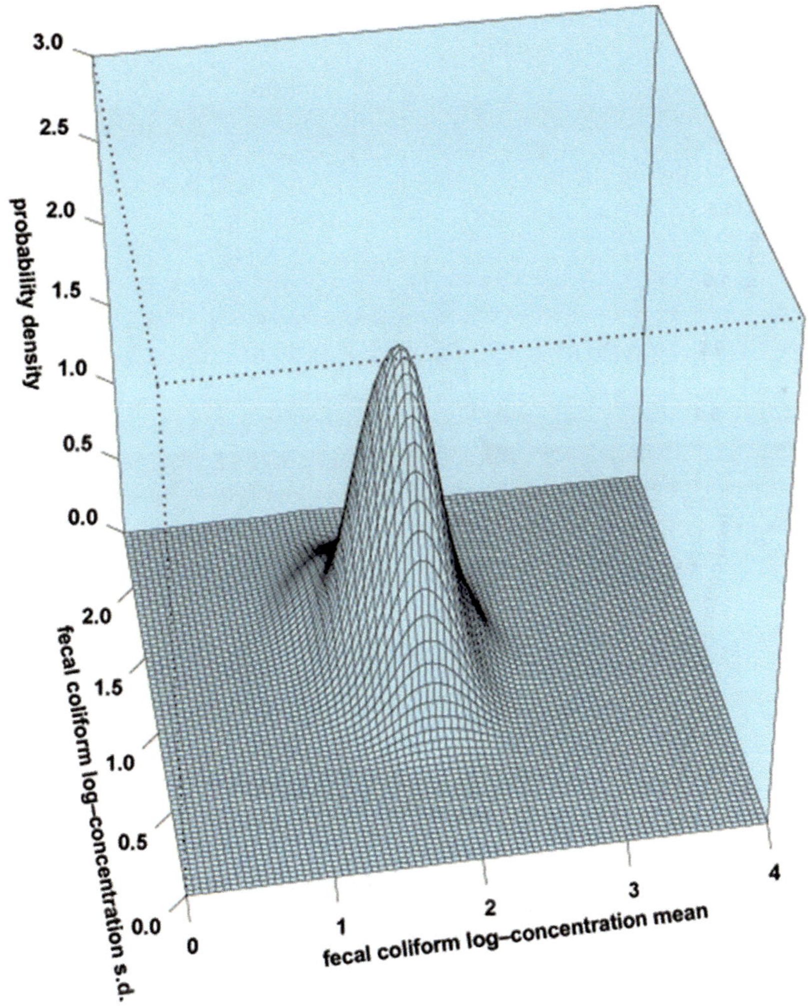

FIG. 2.8 Smoothed contour plot of the joint posterior probability density function for the fecal coliform log-concentration mean μ and log-concentration standard deviation σ at a shellfish-harvesting area water quality monitoring site in eastern North Carolina. Using Bayes' theorem to develop posterior probability density functions and presenting the results graphically (as shown here) provides regulatory and planning agencies with valuable information regarding the relationship between variability in experimental information and uncertainty in future water quality conditions. *(Data from A. Gronewold, D. Vallero, Application of Bayes' Theorem for Predicting Environment Damage, Access Science Yearbook of Science and Technology, 2010.)*

Continued

Discussion Box 2.1 Bayesian networks and water quality—cont'd

In addition to determining uncertainty in water quality predictions, it is important to understand how that uncertainty propagates into water quality-based management decisions. For example, the predicted conditions presented in Fig. 2.8 may guide decisions on ecological and a health standard, e.g., a failure to meet a water quality standard is a strong indication of potential threat to human and environmental health. In this example, water quality standards for shellfish-harvesting waters may indicate unsafe consumption of harvest shellfish when either the fecal coliform concentration median, geometric mean, or 90th percentile of a minimum of 30 water quality samples exceeds 14, 14, and 43 organisms $100\,m^{-1}$, respectively. When water quality sample concentrations exceed these quantitative limits, the corresponding management decision would be to close the area to shellfish-harvesting and to post signs warning of potential health risks [49].

To improve the understanding of the uncertainty in fecal coliform concentration predictions, these quantitative limits are translated into corresponding maximum allowable combinations of the fecal coliform log-concentration mean (μ) and log-concentration standard deviation (σ). These maximum allowable μ, $-\sigma$ pairs can be projected onto the three-dimensional joint $\mu-\sigma$ posterior probability space (red dotted line in Fig. 2.9), which indicate how likely the water quality conditions are to yield a water quality sample in violation of the given standards. The dotted line istruncating a portion of the three-dimensional joint probability space to the bottom left of the figure from the relative volume of this portion, i.e., the confidence of compliance, which can be thought of as the degree of confidence that the water body will comply with water quality standards. In this example, the confidence of compliance is about 0.03 (or 3%).

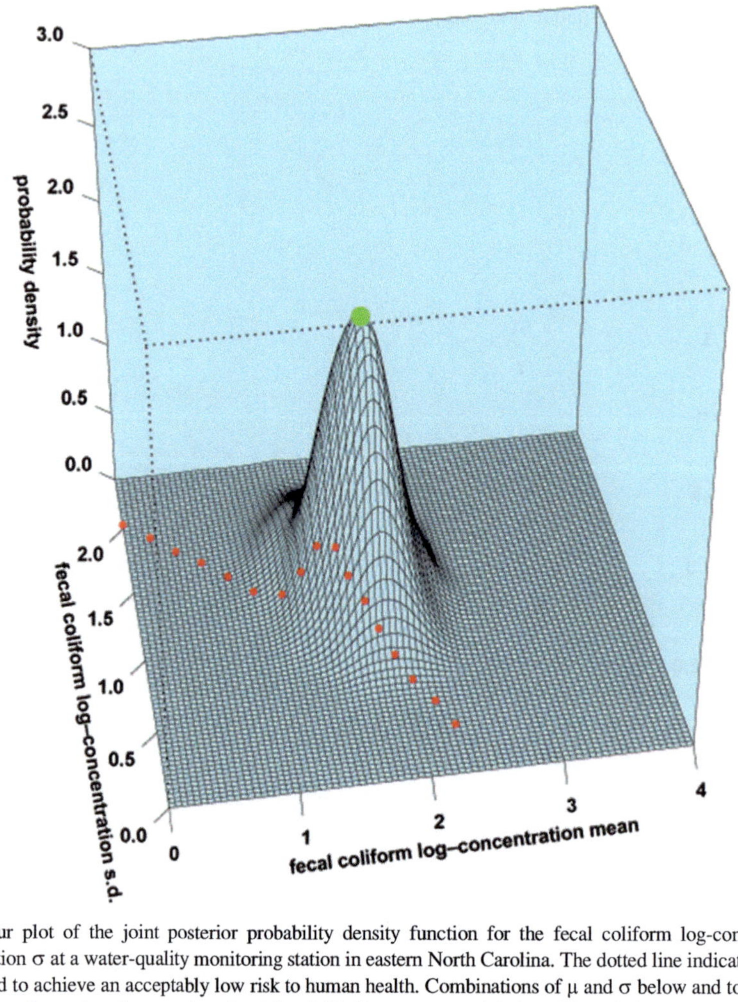

FIG. 2.9 Smoothed contour plot of the joint posterior probability density function for the fecal coliform log-concentration mean μ and log-concentration standard deviation σ at a water-quality monitoring station in eastern North Carolina. The dotted line indicates maximum allowable combinations of μ and σ expected to achieve an acceptably low risk to human health. Combinations of μ and σ below and to the left of this line represent compliance, and combinations of μ and σ above and to the right of this line represent violations. The relative volume of the three-dimensional joint probability space "sliced off" by the red dotted line represents the confidence of compliance with the proposed water-quality standard. In this example, the confidence of compliance is about 0.03 (or 3%). As with Fig. 8.5, this figure demonstrates graphically the relationship between experimental information and uncertainty in future water-quality conditions, but this figure also demonstrates the implicit (and often ignored) relationship between uncertainty in future water-quality conditions and water-quality-based management decisions (as indicated by the *dotted line*). *(Data from A. Gronewold, D. Vallero, Application of Bayes' Theorem for Predicting Environment Damage, Access Science Yearbook of Science and Technology, 2010.)*

Discussion Box 2.1 Bayesian networks and water quality—cont'd

To contrast the Bayesian-based confidence of compliance result with non-Bayesian strategies, the green dot plotted in Fig. 2.9 is a potential point estimate of the most likely combination of μ and σ. Deterministic predictions of water quality conditions would possibly be based solely on these point estimates, which ignores much of the potential variability in the future fecal coliform concentrations and could lead to a decision based on oversimplifying assumptions. The decision would have to have been based not on a confidence of compliance, but on a simple statement of whether the water body violates the standard. In the case of the assessment results presented in Fig. 2.9, the deterministic approach could lead to a prediction that future conditions will violate the given standard. A summary of monitoring assessment results for the station presented in Figs. 2.8 and 2.9, along with other neighboring water quality monitoring stations, is presented in Table 2.1. Thus, a Bayesian approach to predicting environmental conditions and to guiding management decisions provides a relatively robust approach to quantifying risk and protecting human and environmental health.

TABLE 2.1 Estimated confidence of compliance and deterministic assessment of future violations for monitoring stations in an Eastern North Carolina estuary.

Station	Bayesian assessment (confidence of compliance, %)	Deterministic assessment (will standard be violated?)
3	52	No
4	44	Yes
7	<1	Yes
8	14	Yes
9	93	No
25	3	Yes
28	96	No
29	<1	Yes
35	80	No
41	<1	Yes
84	13	Yes

By propagating experimental information variability into uncertainty in future water quality conditions, the Bayesian assessment allows regulatory and planning agencies to understand the relative risks associated with restricting (or allowing) public access to a water-resource area. In contrast, the deterministic assessment indicates only whether the water quality standard is violated without any indication of how severe the violation is and, consequently, the magnitude of potential risks to human health.
Data from A. Gronewold, D. Vallero, Application of Bayes' Theorem for Predicting Environment Damage, Access Science Yearbook of Science and Technology, 2010.

6 Cause and effect calculation

Cause and effect diagrams, also known as Ishikawa diagrams, identify and characterize the totality of causes or events that contribute to a specified outcome event. The "fishbone" diagram (See Fig. 2.10) assembles the categories of as many causative factors as reasonable according to their importance, i.e., their portion of the cause. The composition of this diagram begins with the failure event to the far right (i.e., the "head" of the fish), followed by the fish spine (flow of events leading to the failure). The "bones" are each of the contributing categories to the cause. This can be an effective tool in explaining failures to stakeholders, clients, and the public, e.g., explanation of the steps in a root cause analysis of a pollution event. More preferably, the decision-maker may construct the diagrams in "real time" at a design meeting. This will help to open the design for peer review and will help get insights, good and bad, early in the design process [51].

The premise behind cause-and-effect diagrams like the fishbones and other failure logic graphics is that all the causes must connect through a logic gate. This is not always the case, so another more qualitative tool may need to be used, such as the Bayesian belief network (BBN). Like the fishbone, the BBN starts with the failure (See Fig. 2.11). Next, the most

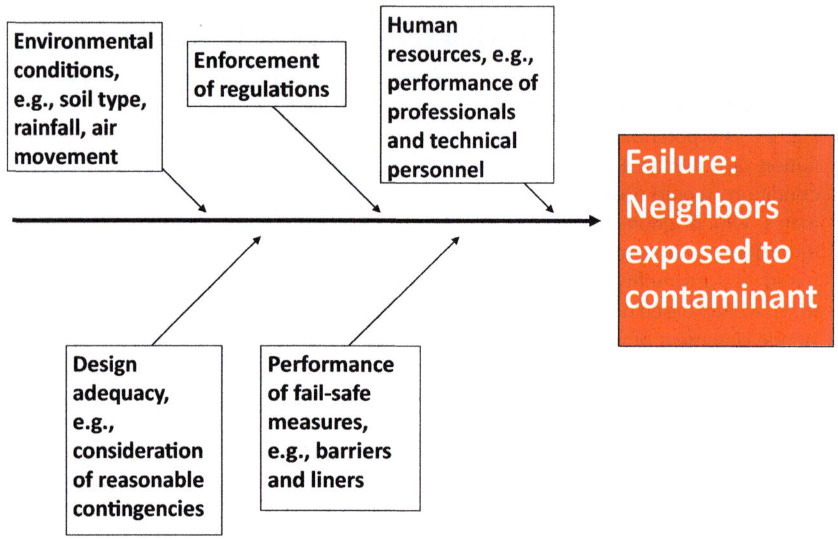

FIG. 2.10 Fishbone reliability diagram showing contributing causes to an adverse outcome (exposure to an environmental contaminant). *(Data from D.A. Vallero, P.A. Vesilind, Socially Responsible Engineering: Justice in Risk Management, John Wiley & Sons, Hoboken, NJ, 2007.)*

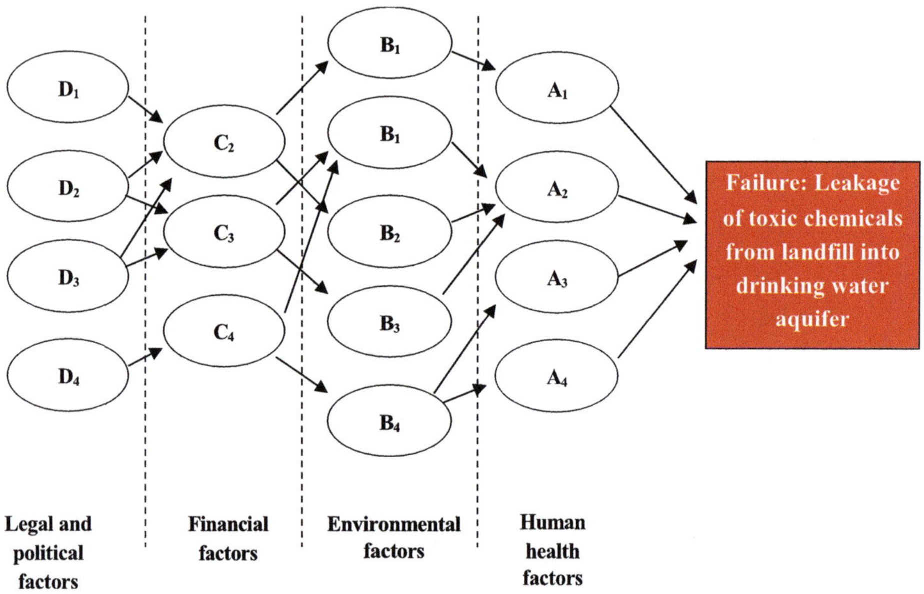

FIG. 2.11 Bayesian belief network, with three groups of contributing causes leading to a failure. *(Data from D.A. Vallero, P.A. Vesilind, Socially Responsible Engineering: Justice in Risk Management, John Wiley & Sons, Hoboken, NJ, 2007.)*

immediate contributing causes are linked to the failure event. The next group of factors that led to the immediate causes are then identified, followed by the remaining contributing groups. This diagram helps to identify and weight the contributing factors, allowing for investigations of how one group of factors impacts the others.

As mentioned, Figs. 2.10 and 2.11 are tools that can be applied during a root cause analysis of a water pollution outcome, e.g., a disease in a human population from a waterborne pathogen or a fish kill. Generally, a root cause analysis consists of a series of steps beginning with a direct statement of the analytical objectives, for example, to identify the factors and occurrences that led to the adverse event and to identify and implement corrective actions. The next steps are to define the failure that needs to be addressed, to summarize as many of the events as possible in an event tree, to conduct the analysis, and to identify the lessons learned. Finally, the realistic options for remediating the problem are described, the optimal option is chosen, and the schedule for completing the actions needed are approved by decision makers before the actions can be implemented [52–54]. For example, various industries experience combustible dust explosion hazards, especially agriculture, chemicals, food, grain, and wood and paper processing. The initiator of many seem to result from mechanical friction, smoldering products, and heated surfaces that lead to combustion followed by an explosion. However, the root cause is very often due to humans' failure to follow safety rules [55,56]. The cause-effect diagrams must in these cases give a sense of the relative weight of each contributing cause to the adverse outcome.

As discussed in Chapter 17, like risk, the engineering and scientific communities often define reliability in nuanced ways. Environmental engineering and other empirical sciences also use the term "reliability," to indicate quality, especially for data derived from measurements, including health data. In this use, reliability is defined as the degree to which measured results are dependable and consistent with respect to the study objectives, e.g., stream water quality. This specific connotation is sometimes called "test reliability" in that it indicates how consistent measured values are over time, how these values compare to other measured values, and how they differ when other tests are applied. Test reliability, like engineering reliability, is a matter of trust. As such, it is often paired with test validity, that is, how near the measured value is to the true value (as indicated by some type of known standard). The less reliable and valid the results, the less confidence scientists and engineers have in interpreting and using them. This is very important in engineering and scientific communications generally, and risk communications specifically.

It is important to keep in mind that, even if the water and wastewater actions are efficient, there is always the potential for attendant, undesired outcomes. Furthermore, even if a project is acceptable, there can still be an unacceptable cumulative outcome when viewing it in combination with all the other "acceptable" projects. In other words, water pollution decision making must define measures of success and failure. A multi-step process can be used to approach the various metrics for success in light of the complexities of ecosystems and human populations that [57]:

1. Properly accounting for the various forms of uncertainty [58];
2. Understanding the limitations of using secondary data sources, by applying rigorous systematic reviews and evidence mapping [10,46,59–70];
3. Comparing successes and failures that may result consistently among the alternative actions [71,72];
4. Employing a tiered, intelligent, adaptive, and flexible approach to success metrics; beginning with a higher-tier, scoping effort and adding needed details as information is continuously re-evaluated [73,74];
5. Evaluating the value of information on which decisions will be based, including assignment of quality and relevance thresholds below which the data and information should be avoided or, at a minimum, given appropriate caveats [75].
6. Validating and evaluating the models and other tools used to assess success [76]; and,
7. Considering the cumulative impacts, i.e., physical, chemical and biological [77,78], in water and wastewater decisions and assessments and the effect any proposed actions will have on the integrity of ecosystems in time and space. A practical means of estimating cumulative outcomes is to apply ecosystem indices and models [79].

7 Conclusions

The information in this chapter sets the stage for a wide range of calculations and expressions of the complexities involved in water pollution. The amount and extent of water pollution is mainly determined quantitatively, which means mathematics and statistics must be applied. The examples of mathematics deployed in estimating, calculating, and reporting water pollution are the basis for quantitating the formation, transport, transformation, fate, exposure, and risks posed by water pollution.

References

[1] D.A. Vallero, Environmental Contaminants: Assessment and Control, vol. xxxix, Elsevier Academic Press, Amsterdam; Boston, 2004, p. 801.

[2] A. Maertens, E. Golden, T. Hartung, Avoiding regrettable substitutions: green toxicology for sustainable chemistry, ACS Sustain. Chem. Eng. 9 (23) (2021) 7749–7758.

[3] K. Rahman, S. Barua, H. Imran, Assessment of water quality and apportionment of pollution sources of an urban lake using multivariate statistical analysis, Clean. Eng. Technol. 5 (2021) 100309.

[4] A. Azzellino, R. Salvetti, R. Vismara, L. Bonomo, Combined use of the EPA-QUAL2E simulation model and factor analysis to assess the source apportionment of point and non point loads of nutrients to surface waters, Sci. Total Environ. 371 (1–3) (2006) 214–222.

[5] P. Soonthornnonda, E.R. Christensen, Source apportionment of pollutants and flows of combined sewer wastewater, Water Res. 42 (8–9) (2008) 1989–1998.

[6] M. Baborowski, V. Simeonov, J.W. Einax, Assessment of water quality in the Elbe river at flood water conditions based on cluster analysis, principle components analysis, and source apportionment, Clean–Soil, Air, Water 40 (4) (2012) 373–380.

[7] X. Cao, W. Li, S. Song, C. Wang, K. Khan, Source apportionment and risk assessment of soil heavy metals around a key drinking water source area in northern China: multivariate statistical analysis approach, Environ. Geochem. Health 45 (2) (2023) 343–357.

[8] K. Chen, Q. Liu, W. Peng, X. Liu, Source apportionment and natural background levels of major ions in shallow groundwater using multivariate statistical method: a case study in Huaibei plain, China, J. Environ. Manag. 301 (2022) 113806.

[9] M.C. Thorne, Background radiation: natural and man-made, J. Radiol. Prot. 23 (1) (2003) 29–42.

[10] D.A. Vallero, Translating Diverse Environmental Data into Reliable Information: How to Coordinate Evidence from Different Sources, Academic Press, 2017.

[11] M.B. Van Asselt, Perspectives on uncertainty and risk, in: Perspectives on Uncertainty and Risk, Springer, 2000, pp. 407–417.

[12] J. Shepherd, Geoengineering the Climate: Science, Governance and Uncertainty, Royal Society, 2009.

[13] I. Scoones, What Is Uncertainty and Why Does it Matter?, 2019.

[14] R.L. Kodell, Managing uncertainty in health risk assessment, Int. J. Risk Assess. Manag. 5 (2) (2005) 193–205.

[15] P. Hamel, A.J. Guswa, Uncertainty analysis of a spatially explicit annual water-balance model: case study of the cape fear basin, North Carolina, Hydrol. Earth Syst. Sci. 19 (2) (2015) 839–853.

[16] J. Greskowiak, E. Hamann, V. Burke, G. Massmann, The uncertainty of biodegradation rate constants of emerging organic compounds in soil and groundwater—a compilation of literature values for 82 substances, Water Res. 126 (2017) 122–133.

[17] A.M. Finkel, Confronting Uncertainty in Risk Management; a Guide for Decision Makers a Report, Resources for the Future, Washington, DC (EUA), 1990.

[18] R.A. Camacho, J.L. Martin, T. Wool, V.P. Singh, A framework for uncertainty and risk analysis in total maximum daily load applications, Environ. Model Softw. 101 (2018) 218–235.

[19] P. Abbaszadeh, H. Moradkhani, D.N. Daescu, The quest for model uncertainty quantification: a hybrid ensemble and variational data assimilation framework, Water Resour. Res. 55 (3) (2019) 2407–2431.

[20] D. Hume, The problem of induction, in: Philosophy of Science: An Historical Anthology., vol. 15, 2009, p. 219.

[21] M. Lange, Hume and the problem of induction, in: Handbook of the History of Logic, vol. 10, Elsevier, 2011, pp. 43–91.

[22] M.R. Taleyarkhan, A.M. Lucietto, N.L. Hobson, T.M. Azevedo, Approach to problem solving and use of intuition by engineering technology students, J. Glob. Educ. Res. 7 (1) (2023) 81–98.

[23] E. Miskioğlu, K.M. Martin, A.R. Carberry (Eds.), Work in progress: experts' perceptions of engineering intuition, 2020 ASEE Virtual Annual Conference, 2020.

[24] R.M. Hogarth, Educating Intuition, University of Chicago Press, 2001.

[25] World Cancer Research Fund International, Liver Cancer Statistics, WCRF, London, UK, 2020. Available from: https://www.wcrf.org/cancer-trends/liver-cancer-statistics/.

[26] U.S EPA, Per- and Polyfluoroalkyl Substances (PFAS): Proposed PFAS National Primary Drinking Water Regulation, U.S. EPA, Washington, DC, 2023. updated September 22, 2023. Available from: https://www.epa.gov/sdwa/and-polyfluoroalkyl-substances-pfas.

[27] Centers for Disease Control and Prevention NCfHS, National Health and Nutrition Examination Survey (NHANES) Atlanta, Georgia, 2022, updated June 30, 2022. Available from: https://www.cdc.gov/nchs/nhanes/index.htm.

[28] National Center for Health Statistics, About the National Health and Nutrition Examination Survey, NCHS, Atlanta, GA, 2022. Available from: https://www.cdc.gov/nchs/nhanes/about_nhanes.htm.

[29] US Geological Survey, Data Resources Reston, VA, USA, 2019, Available from: https://www.usgs.gov/science/data-resources.

[30] US Environmental Protection Agency, Streams under CWA Section 404, U.S. EPA, Washington, DC, 2023. updated February 17, 2023. Available from: https://www.epa.gov/cwa-404/streams-under-cwa-section-404.

[31] Y. Zhang, Indoor Air Quality Engineering, CRC Press, Boca Raton, FL, 2004.

[32] K. Eisenreich, in: U.S EPA (Ed.), Existing Chemical Exposure Limit (ECEL) for Occupational Use of Carbon Tetrachloride, Office of Chemical Safety and Pollution Prevention, Washington, DC, 2021.

[33] D. Wang, T. Chen, Z. Fu, L. Yang, R. Li, S. Sui, et al., Occupational exposure to polybrominated diphenyl ethers or decabromodiphenyl ethane during chemical manufacturing: occurrence and health risk assessment, Chemosphere 231 (2019) 385–392.

[34] C.M. Steinmaus, Perchlorate in water supplies: sources, exposures, and health effects, Curr. Environ. Health Rep. 3 (2) (2016) 136–143.

[35] K. Ashley, NIOSH manual of analytical methods 5th edition and harmonization of occupational exposure monitoring, Gefahrst. Reinhalt. Luft. 2015 (1–2) (2015) 7.

[36] OSHA, Permissible Exposure Limits (PELs), U.S. Department of Labor OSaHA, 2013.

[37] W.R. Ott, Environmental Statistics and Data Analysis, Lewis Pubishers, CRC Press, Inc., Boca Raton, FL, 1995.

[38] T. Bayes, LII. An essay towards solving a problem in the doctrine of chances. By the late Rev. Mr. Bayes, FRS communicated by Mr. Price, in a letter to John Canton, AMFR S, Phil. Trans. R. Soc. Lond. 53 (1763) 370–418.

[39] A. Gronewold, K. Reckhow, D. Vallero, Improving human and ecological exposure assessments: a Bayesian network modeling approach, Epidemiology 19 (6) (2008) S228–S229.

[40] D.A. Vallero, K.H. Reckhow, A.D. Gronewold (Eds.), Application of multimedia models for human and ecological exposure analysis, in: International Conference on Environmental Epidemiology and Exposure, International Society of Exposure Analysis, Durham, NC, 2007.

[41] K. McDonald, D. Ryder, M. Tighe, Developing best-practice Bayesian belief networks in ecological risk assessments for freshwater and estuarine ecosystems: a quantitative review, J. Environ. Manag. 154 (2015) 190–200.

[42] V. Jurado-Davila, J.T. De Oliveira, D. Estumano, L.A. Féris, Fixed-bed column for phosphate adsorption combining experimental observation, mathematical simulation, and statistics: classical and Bayesian, Sep. Purif. Technol. 317 (2023) 123914.

[43] T. Knappenberger, A.D. Jayakaran, J.D. Stark, A Bayesian modeling framework to predict stormwater pollutant reduction in bioretention media, Ecol. Eng. 178 (2022) 106582.

[44] J. Zhao, J. Jin, Q. Guo, Y. Chen, M. Lu, L. Tinoco, Forewarning model for water pollution risk based on Bayes theory, Environ. Sci. Pollut. Res. 21 (2014) 3073–3081.
[45] J. Liu, R. Liu, Z. Zhang, Y. Cai, L. Zhang, A Bayesian network-based risk dynamic simulation model for accidental water pollution discharge of mine tailings ponds at watershed-scale, J. Environ. Manag. 246 (2019) 821–831.
[46] S.I. Watson, R.J. Lilford, Integrating Multiple Sources of Evidence: A Bayesian Perspective, 2016.
[47] J.K. Kruschke, H. Aguinis, H. Joo, The time has come Bayesian methods for data analysis in the organizational sciences, Organ. Res. Methods 15 (4) (2012) 722–752.
[48] G. Casella (Ed.), Refresher on Bayesian and frequentist concepts: models, assumptions, and inference, 37th Meeting of the American College of Clinical Pharmacology, American College of Clinical Pharmacology, Philadelphia, PA, 2008.
[49] A. Gronewold, D. Vallero, Application of Bayes' Theorem for Predicting Environment Damage, Access Science Yearbook of Science and Technology, 2010.
[50] L. Kaikkonen, T. Parviainen, M. Rahikainen, L. Uusitalo, A. Lehikoinen, Bayesian networks in environmental risk assessment: a review, Integr. Environ. Assess. Manag. (2020).
[51] D.A. Vallero, P.A. Vesilind, Excerpted and Adapted from: Socially Responsible Engineering: Justice in Risk Management, Wiley Online Library, 2007.
[52] D.A. Vallero, Chapter 8—Decision support tools, in: D.A. Vallero (Ed.), Environmental Systems Science, Elsevier, 2021, pp. 293–357.
[53] M.A. Latino, R.J. Latino, K.C. Latino, Root Cause Analysis: Improving Performance for Bottom-Line Results, CRC Press, 2019.
[54] B. Andersen, T. Fagerhaug, Root Cause Analysis, Quality Press, Birmingham, AL, 2006.
[55] Occupational Safety and Health Administration, Combustible Dust: An Explosion Hazard, U.S. Department of Labor, Washington, DC, 2023 (Report No.: SHIB 07-31-2005 Contract No.: February 14, 2023).
[56] D.A. Vallero, T.M. Letcher, Unraveling Environmental Disasters, second ed., Elsevier Academic Press, Amsterdam, The Netherlands, 2023.
[57] R.A. Hill, C. Sendashonga, General principles for risk assessment of living modified organisms: lessons from chemical risk assessment, Environ. Biosaf. Res. 2 (02) (2003) 81–88.
[58] Environmental Risk Management Authority NZ, Preparing Information on Risks, Costs and Benefits for Applications Under the Hazardous Substances and New Organisms Act 1996, 2000, See http://www.ermanz.govt.nz/resources/publications/pdf/ER-TG-03-1.pdf.
[59] A. Gatherer, Evidence, values, and 'right versus right' dilemmas in public health practice, J. Public Health Policy 32 (4) (2011) 504–507.
[60] D.L. Weed, Weight of evidence: a review of concept and methods, Risk Anal. 25 (6) (2005) 1545–1557.
[61] N.M. DeLuca, M. Angrish, A. Wilkins, K. Thayer, E.A.C. Hubal, Human exposure pathways to poly-and perfluoroalkyl substances (PFAS) from indoor media: a systematic review protocol, Environ. Int. 146 (2021) 106308.
[62] P. Whaley, B.J. Blaauboer, J. Brozek, E.A.C. Hubal, K. Hair, S. Kacew, et al., Improving the quality of toxicology and environmental health systematic reviews: what journal editors can do, ALTEX 38 (3) (2021) 513–522.
[63] U.S. Environmental Protection Agency, in: USEPA (Ed.), Draft Systematic Review Protocol Supporting TSCA Risk Evaluations for Chemical Substances Version 1.0: A Generic TSCA Systematic Review Protocol with Chemical-Specific Methodologies, Office of Chemical Safety and Pollution Preventrion, Washington, DC, 2021.
[64] Z. Noorimotlagh, S.A. Mirzaee, N. Jaafarzadeh, M. Maleki, G. Kalvandi, C. Karami, A systematic review of emerging human coronavirus (SARS-CoV-2) outbreak: focus on disinfection methods, environmental survival, and control and prevention strategies, Environ. Sci. Pollut. Res. 28 (1) (2021) 1–15.
[65] K. Dack, M. Fell, C.M. Taylor, A. Havdahl, S.J. Lewis, Mercury and prenatal growth: a systematic review, Int. J. Environ. Res. Public Health 18 (13) (2021) 7140.
[66] S. Ohlwein, R. Kappeler, M.K. Joss, N. Künzli, B. Hoffmann, Health effects of ultrafine particles: a systematic literature review update of epidemiological evidence, Int. J. Public Health 64 (4) (2019) 547–559.
[67] C. Ewbank, S. Gupta, B.T. Stewart, A.L. Kushner, A. Charles, A systematic review of oil tanker truck disasters: identifying prevention targets, Burns 45 (4) (2019) 905–913.
[68] R.P. de Almeida, R.G. Ferrari, L.N. Dos Santos, C.A.C. Junior, Mercury in aquatic fauna contamination: a systematic review on its dynamics and potential health risks, J. Environ. Sci. 84 (2019) 205–218.
[69] Z. Bi, Finite Element Analysis Applications: A Systematic and Practical Approach, Academic Press, 2017.
[70] F. Pianosi, K. Beven, J. Freer, J.W. Hall, J. Rougier, D.B. Stephenson, et al., Sensitivity analysis of environmental models: a systematic review with practical workflow, Environ. Model Softw. 79 (2016) 214–232.
[71] B. Hong, T.L. Fisher, T.S. Sult, C.A. Maxwell, J.A. Mickelson, H. Kishino, et al., Model-based tolerance intervals derived from cumulative historical composition data: application for substantial equivalence assessment of a genetically modified crop, J. Agric. Food Chem. 62 (40) (2014) 9916–9926.
[72] P. Macdonald, Developing Workable Regulatory Frameworks for the Environmental Release of Transgenic Plants, Collection of Biosafety Reviews, 2012.
[73] Environmental Risk Management Authority NZ, Assessment of Effects of Hazardous Substances and New Organisms on Human Health, 2000, See http://www.ermanz.govt.nz/resources/publications/pdf/ER-TG-02-1pdf.
[74] R.A. Hill, From Science to Decision-Making: The Applicability of Bayesian Methods to Risk Assessment, 1996.
[75] V.F. Reyna, F. Farley, Risk and rationality in adolescent decision making: implications for theory, practice, and public policy, Psychol. Sci. Public Interest 7 (1) (2006) 1–44.

[76] B. Breckling, H. Reuter, U. Middelhoff, M. Glemnitz, A. Wurbs, G. Schmidt, et al., Risk indication of genetically modified organisms (GMO): modelling environmental exposure and dispersal across different scales: oilseed rape in northern Germany as an integrated case study, Ecol. Indic. 11 (4) (2011) 936–941.

[77] J.G. Lundgren, J.J. Duan, RNAi-based insecticidal crops: potential effects on nontarget species, Bioscience 63 (8) (2013) 657–665.

[78] C.V. Rider, K. Boekelheide, N. Catlin, C.J. Gordon, T. Morata, M.K. Selgrade, et al., Cumulative risk: toxicity and interactions of physical and chemical stressors, Toxicol. Sci. 137 (1) (2014) 3–11.

[79] Y. Li, J. Qiu, B. Zhao, M. Pavao-Zuckerman, A. Bruns, S. Qureshi, et al., Quantifying urban ecological governance: a suite of indices characterizes the ecological planning implications of rapid coastal urbanization, Ecol. Indic. 72 (2017) 225–233.

Chapter 3

Scientific principles

1 Introduction

Water pollutants may be classified in many ways. The primary distinction of types of pollution scientists and engineers, however, lies in a pollutants' inherent properties, beginning with their physical attributes. These provide first clues as to how a substance will move and change in water and the damage it may ultimately cause. All physical and biological sciences differentiate matter based on thermodynamics.

Under environmental conditions, water exists in the three major physical phases in the hydrosphere. It moves into the atmosphere when it evaporates and out of the atmosphere when it precipitates. At the global scale, the gas concentrations are usually described by volume, rather than by weight. The presence and concentrations of water is always a consideration in environmental sciences and engineering. Water in solid, liquid, and gas phases is found at widely varying concentrations in most of the lower atmosphere. The vapor phase of water in the atmosphere is commonly referred to as humidity. Given the large ranges of humidity, meteorologists, and atmospheric scientists often express gas concentrations as dry volume, i.e., no humidity. By dry volume, 99.997% of the atmosphere consists of four gases, molecular nitrogen and oxygen (N_2 and O_2, respectively), argon (Ar) and carbon dioxide (CO_2). Chemically, Ar is inert (nonreactive) since it is a noble gas. The other three compounds are also very stable and nonreactive under most lower atmospheric conditions of temperature and pressure, so they remain very stable components of the troposphere (the lowest level). Approximately 99% of the mass of the atmosphere lies within 50 kilometers (km) of the Earth's surface, i.e., in the troposphere and stratosphere. This is where most water pollution occurs. The major exception is the stratospheric ozone layer, which is affected by human activities at the earth's surface, especially the use of halogenated chemical compounds like the chlorofluorocarbon (CFC) and hydrochlorofluorocarbon (HCFC) compounds that find their way to the stratosphere [1–4].

Matter exists in equilibrium between phases. The equilibrium is controlled by environmental conditions. In the laboratory, these conditions are usually kept at or adjusted to laboratory-defined specifications, especially standard temperature and pressure (STP), which are 0 °C and 1 bar (0.986923 atm). Outside of the confines of the laboratory, STP is not usually the best way to adjust or standardize conditions, for example, when taking measurements of water in a lake or wastewater being discharged at an outfall structure. On the earth's surface, a compound's phases fall within a relatively tight range of temperatures and atmospheric pressures. Thus, even slight changes in environmental conditions can substantially change the thermodynamics, kinetics, and equilibria of a compound in water, soil, sediment, or the atmosphere.

2 Basic scientific principles

Water quality protection, like any scientific endeavor, must abide by the laws of motion and the laws of thermodynamics. Newton's laws of motion are:

- An object at rest remains at rest, and an object in motion remains in motion at constant speed and in a straight line unless acted on by an unbalanced force.
- The acceleration of an object depends on the mass of the object and the amount of force applied.
- Whenever one object exerts a force on another object, the second object exerts an equal and opposite on the first.

These are the basic principles of motion, which underpin discussions of transport in subsequent chapters.

The laws of thermodynamics that apply to water pollution are:

- Matter and energy can neither be created nor destroyed, but only changed from one form into another.
- The processes described in the First Law are always inefficient, and there is an inescapable loss of heat.
- A process cannot reach a state of no activity—no exchange of heat—without reaching the environmentally unattainable absolute zero.

Fundamentals of Water Pollution. https://doi.org/10.1016/B978-0-443-28987-3.00018-7

3 Environmental and experimental conditions

For most environmental situations, calculations involving the real gases and vapors deviate very little from ideal-gas behavior. At STP (223.15 K and 1 bar, i.e., 0.986923 atm), 1 mol of gas occupies 22.4 L, that is, the molar volume is 22.4 mol L^{-1}. At normal temperature and pressure (NTP; 293.15 K and 1 atm), 1 mol of gas occupies a volume of 22.7 L. Under standard conditions (SC; 298.15 K and 1 atm), the molar volume is 24.5 mol L^{-1}. At SC, the higher temperature further increases the molar volume, that is, 1 mol of gas occupies a volume of 24.47 L (24.47 mol L^{-1}).[a]

The standard atmosphere is like STP. A standard atmosphere is a vertical depiction of temperature, pressure, and density that is representative of atmosphere and used to compare and calibrate observed conditions. A standard atmosphere has 1 atm pressure, temperature equal to 15 °C, and ρ equal to 1.225 g m^{-3}, at sea level.

Another standard condition concept is the virtual temperature (T_v), which is a way to compare and calibrate various moisture conditions in the air. Under common tropospheric conditions, air moisture ranges from almost completely dry, that is, 0% water vapor, to a few percent water vapor. Thus, the gas constant, R, is larger for a mass of moist air than R for the same mass of dry air. The difference increases with increasing moisture. Thus, to facilitate calculations, the R for dry can be used by assuming a fictitious temperature, that is, T_v, in the ideal-gas equation. The value of T_v is the temperature that the dry air must be to have the same density of moist air at the same pressure. Each molecule of lighter water vapor (atomic weight of 16) displaces a heavier molecule of air (atomic weight of 29), so moist air is less dense than dry air. Therefore, T_v is always larger than actual temperatures.

T_v in degrees Kelvin can be found by:

$$T_v = \frac{T}{1 - \left(\frac{e}{P}\right)(1 - \varepsilon)} \tag{3.1}$$

Where, T is the actual temperature (Kelvin), e is the partial pressure of water vapor, P is the total pressure (partial pressure of dry air + e), and ε is the ratio of gas constants for dry and moist air, that is, $\frac{R_{dry}}{R_{H2O}}$, which is 0.622.

Virtual temperature example

A moist air sample collected at 25 °C and 1.1 atm shows the water vapor partial pressure to be 0.01 atm. The next day, a sample was collected at 22 °C and 0.9 atm and water vapor pressure measured at 0.02 atm. On day 3, the temperature was 25 °C and 0.9 atm and water vapor pressure measured at 0.02 atm. Calculate the virtual temperatures under these conditions.

Solution:

$$\text{Day 1}: T_v = \frac{298.15\text{K}}{1 - \left(\frac{0.01}{1.1}\right)(1 - 0.622)} = 299.17\ \text{K} = 26.02^{\circ}\text{C}$$

$$\text{Day 2}: T_v = \frac{295.15\text{K}}{1 - \left(\frac{0.02}{0.9}\right)(1 - 0.622)} = 297.65.\text{K} = 24.5^{\circ}\text{C}$$

$$\text{Day 2}: T_v = \frac{298.15\text{K}}{1 - \left(\frac{0.02}{0.9}\right)(1 - 0.622)} = 300.68.\text{K} = 27.5^{\circ}\text{C}$$

Therefore, day 3 had the highest virtual temperature; even though the actual temperatures for days 1 and 3 were the same.

Often, several pollutants are measured simultaneously at single site. All measurements are adjusted to SC or are otherwise standardized condition when reported.

To this point, we have used vapor to describe water's gas phase but have referred to N_2 and O_2 as gases. A gas is matter that has a single thermodynamic state within these typical atmospheric ranges, for example, NTP, RTP, or SC. On the other hand, a vapor is one that exists in more than one phase in these ranges. For example, O_2 and argon (Ar) are gases because every O_2 and Ar molecule is in the gas phase at and near SC. The boiling point, the temperature at which O_2 changes from a

a. SC is also known as room temperature and pressure (RTP).

gas to a liquid at 1 atm is −183 °C. Conversely, gas-phase benzene is a vapor because at SC, a fraction of benzene molecules behaves as a gas and the remaining benzene molecules behave as a liquid.

Oxymoronically, liquefied gases, especially N_2, in tanks and cylinders in the laboratory undergo a phase change to be stored. This is important because water tests and studies use liquid N_2. Cryogenic liquids are gases that have an STP boiling point less than −90 °C. At STP, N_2 has a boiling point equal to −196 °C. Typically, a gas liquifying system has a cryogenic storage tank, one or more vaporizers, and a pressure-temperature control system. The system is designed to repel heat by enclosing the liquid in a separate compartment within an inner vessel. If the N_2 is needed in its liquid state, this is all that is needed. Otherwise, the vaporizers convert the liquid to its gaseous state. A pressure control manifold mediates the pressure during the feed process. This is dramatic evidence that gases can be compressed and can be expanded to fill their containers uniformly.

At STP, NTP, and SC, a gas will remain in the gaseous state notwithstanding concentration, that is, it will not condense. Molecular nitrogen (N_2) will not condense at STP, even at 100%. Vapors at STP will exist as both gases and aerosols, but at SC, many of the vapor-phase pollutants rarely reach saturation.

Of course, the distinctions between gases and vapors become less clear outside of the lower troposphere. For example, within reactors and combustors and even within stacks, temperatures and/or pressures are highly elevated, so substances that are pure liquids and solids under SC will volatilize. However, within the atmosphere, the two major phases apply.

The gas laws discussed in detail in Chapters 4 and 5 state that density is a function of pressure and temperature since density is indirectly proportional to gas volume. The density of water and is important to pollution, especially because water density and the density of the pollutant gas are factors in transport. Water density is the mass of water in a given volume. Density varies with temperature and pressure per the ideal-gas law:

$$p = \rho RT \tag{3.2}$$

Where, p is pressure in atmospheres (atm), R *is* the universal gas constant $\left(\frac{0.08212\ \mathrm{L \cdot atm}}{\mathrm{mol \cdot K}}\right)$, T is temperature in K, and ρ is density.

Typically, air density is about 1.23 kg m^{-3} at sea level.

4 Energy

Electromagnetic radiation (EMR) is energy that can travel through a vacuum and through space. Whenever an electric charge oscillates or is accelerated, an electric and magnetic field propagates outward from it. In a vacuum, EMR travels at the speed of light. EMR exhibits a wave-like behavior as it travels through space, and the wavelength and frequency are related through the speed of light:

$$\mathrm{c} = v\lambda \tag{3.3}$$

where c is the speed of light (3.00×10^8 m s^{-1}), v is the frequency, and λ is the wavelength.

Fig. 3.1 shows the EMR spectrum. The range of EMR is extremely large, ranging from gamma radiation (λ values of about 10^{-12} m) to X-rays (about 10^{-10} m), far ultra violet (about 10^{-8} m), near UV (about 10^{-7} m) to visible light, to near

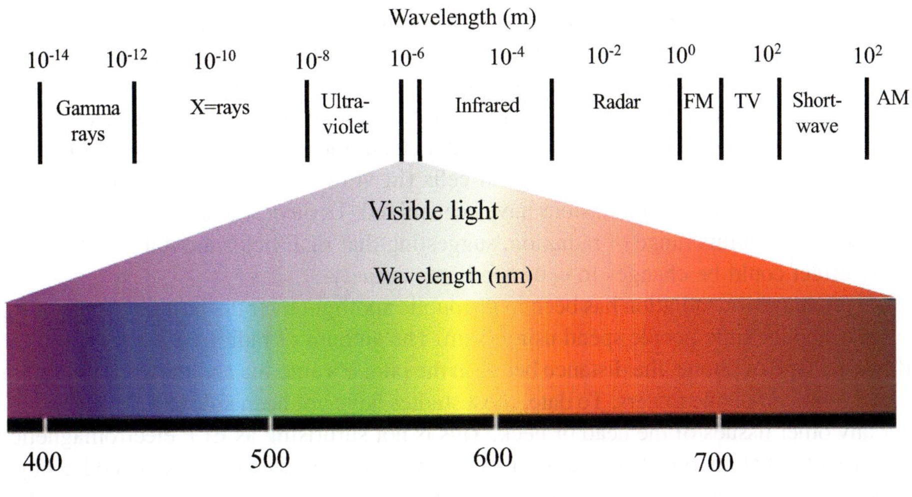

FIG. 3.1 Electromagnetic spectrum.

infra-red hot bodies (about 10^{-6} m), to far-infrared hot bodies (about 10^{-4} m), to microwaves (about 10^{-2} m), to radar (about 10^{-1} m), to TV, FM, and AM and long radio wavelengths (3–3000 m), and to even higher wavelengths for some electronic devices and rotating machinery. The visible region is a small range of EMR from violet (400 nm) through yellow (600 nm) to red (750 nm).

The energy (E) of the waves is related to the frequency by:

$$E = hv = \frac{hc}{\lambda} \tag{3.4}$$

Where, h is the Planck's constant, i.e., 6.63×10^{-34} J s.

EMR can also be considered to have a particle nature and the term used is "photon." A photon of gamma radiation has a greater energy than a photon in the visible region of the electromagnetic spectrum. This is a very important fact when discussing pollution from nuclear fission.

With energy inversely proportional to the wavelength, the UV range has more energy than the IR range and that γ radiation has even more energy. The effects of EMR upon living cells, including those in humans, depend upon on this energy. When the radiation is absorbed by the cell, the effect is very small for low-frequency radiation (radio waves to visible light) and most, if not all, effects involve simple heating. In this range, the frequency dictates the radiation penetration into the organism (for example, microwaves penetrate better than infrared).

At higher frequencies starting with ultraviolet radiation, the effects of individual photons of the radiation begin to become important, and these have enough energy to damage biological molecules. The UV frequencies do more damage to biological systems than do IR heat waves. The far ultraviolet, X-ray, and gamma radiation are referred to as ionizing radiation due to the ability of the photons of this radiation to produce ions in living tissue [5].

Ionizing radiation refers to the type of radiation that creates ions. It does this by knocking out an electron from an atom, thereby creating a positively charged atom or ion. Such radiation can damage human cells. Ionizing radiation includes high energetic electromagnetic radiation (X-rays, gamma rays, and cosmic rays), with wavelengths of less than about 120 nm: that is radiation with wavelengths of less than that of far ultraviolet radiation. The shorter the wavelength, the larger the energy. Thus, gamma radiation ($\lambda = 0.1$–1 picometer [pm]) is more energetic than X-rays ($\lambda = 1$ pm to 10 nm), which in turn are more energetic than radiation in the far ultraviolet ($\lambda \approx 120$ nm) spectrum.

Ionizing radiation also includes particulate radiation such as alpha particles (helium nuclei, He^{2+}), beta particles (high energy electrons, e^-), and neutrons.

By contrast, nonionizing radiation includes ultraviolet radiation ($\lambda = 10$–400 nm), visible light ($\lambda = 420$–700 nm), microwaves ($\lambda = 1$ mm to 10 cm), radio waves ($\lambda > 10$ cm). Nonionizing radiation does not have the energy to knock out an electron from atoms so poses little threat to creating mutations and cancer. It does, however, have the energy to cause molecular excitation, which is really a heating effect. This is confined to special chemicals where the incident energy is matched by the energy of the bonds that bind the chemical species. For example, the energy of microwave radiation, as found in microwave ovens, matches the energy of the bond between the oxygen atom and the hydrogen atom in water. Hence, this is the heating effect experienced when a cup of water is placed in a microwave oven. In general, metals reflect microwaves, whereas glass and ceramics transmit microwaves, and water absorbs microwave energies.

The extremely low frequency (ELF) electromagnetic emanations (all nonionizing radiations) from a cell phone, high-voltage current in power transmission lines, and from some household and industrial electrical equipment, are the least understood types of radiation, arguably making them the most controversial forms of radiation [6]. This is particularly true for the effects of ELF electromagnetic radiation on biological material (especially human cells). The issue of potential biological damage has arisen in recent decades because of the introduction of very high-voltage electric power transmission lines (>440 kV), cell phones, television, and other electrical and electronic equipment. Many biological studies have been carried out with ELF electromagnetic fields on biological cells (in vitro) and in living animals (in vivo), as well as in humans (epidemiological studies), but the conclusions have been mixed. To date, statistically significant chromosome aberrations have not been observed in this range of radiation, suggesting that ELF fields are unlikely to cause cancer or mutations. The most likely effect could be changes to cell membrane activity.

Cell phone ELF electromagnetic radiation has been in the public spotlight over the past few years, triggered by the huge popularity of the phones and the time people spend using them. The amount of radiofrequency energy a cell phone user is exposed to depends on the type of phone, the distance between the phone's antenna and the user, the extent and type of use, and the user's distance from cell phone towers. To date, most studies have not been able to link cell phone use to cancers of the brain, nerves, or any other tissues of the head or neck. This is not surprising as ELF electromagnetic radiation is very much a nonionizing form of radiation, but also because such chronic diseases often have extended latency periods, i.e., time needed between exposure and effects. With any new technology that introduces a whole new type of exposure, the risk clock only starts running when the technology reaches the marketplace. Indeed, cell phone technology has changed numerous times since its introduction and continues to do so. Each change may introduce a new threat and risk profile [7].

In many ways the approach scientists have adopted toward ELF electromagnetic radiation is the reverse of climate change (see Chapters 10 and 11). That is, the scientific community seems to want to adopt the precautionary approach to climate, i.e., the need to act even with uncertainties in projections, but the same scientific community's consensus toward ELF electromagnetic radiation is to abide by an evidence-based, risk approach and do nothing until and unless enough data prove that there is indeed a risk. It could be a disaster of global proportions should many people contract different types of effects associated with ELF exposure, perhaps neurological or cancer, as well as other adverse conditions that are presently presymptomatic.

Ecosystem impacts from ELF exposures is even more likely than effects on most human populations, with sufficient evidence to indicate adverse effects in nonhuman species at local and biosphere levels across a broad range of species. The rising background levels of anthropogenic EMF from 0 Hz to 300 GHz. These species appear to be extraordinarily sensitive to natural and anthropogenic EMF, with ELF and radiofrequency (RF) ranges of the nonionizing electromagnetic spectrum affecting fauna and flora. The few existing exposure standards apply exclusively to humans, leaving constantly exposed wildlife virtually unprotected [8].

4.1 Nuclear radiation

Radioactive decay takes place when an unstable atomic nucleus breaks up by emitting ionizing radiation. This emission is spontaneous and no heat or any interaction is required to make it happen. In many such unstable nuclei, the process is predictable from the known values of the decay constant or from a knowledge of the half-life of the nuclide. It must be remembered that the decay is concerned almost always with the nucleus alone.

Nuclei are only radioactive if there is some instability in the nucleus between the protons (unit positive charge and unit mass) and the neutrons (no charge and unit mass).

For example, there are many known nuclides of oxygen, including ^{15}O, ^{16}O, ^{17}O, ^{18}O, and ^{19}O (being oxygen, they all have eight protons). The number of neutrons here varies from 7 to 11. The ^{15}O and ^{19}O isotopes are unstable. The former has a half-life of 122 s and decays with a positron β^+, while the latter has a half-life of 26 s and decays with β^- emission. Small nuclei such as oxygen usually decay with beta emissions. Larger nuclei often decay with alpha particle emission. These are helium nuclei with a 2+ charges ($^4_2He^{2+}$). The alpha particle has two protons and two neutrons.

Radioactive decay takes place by:

1. beta emission (β^-) [this is an electron emission ($^{0}_{-1}e$)];
2. neutron emission (^0_1n);
3. positron emission (β^+) [this is a positron emission (^0_1e)];
4. alpha emission (α) [this is a helium nuclei emission ($^4_2He^{2+}$)]; and
5. gamma radiation (γ), EMR with a very short wavelength.

One radioactive isotope that was very much in the news after the Chernobyl disaster is strontium-90 (^{90}Sr).[b] It has a half-life of about 29 years, which makes is very dangerous. It undergoes B^- decay forming yttrium-90. In this process, a neutron in the strontium atom has decayed into a proton (which remained in the nucleus), and an electron is emitted with a certain amount of energy. As a result, the protons have increased from 38 to 39 (producing Y) and the mass number has remained the same at 90.

These decay products, i.e., alpha and beta particles, are all charged and can cause serious trouble when striking living material. These particles can damage DNA, for example, and cause a breakdown of cells that they encounter. Gamma radiation is very much more penetrating than beta or alpha radiation. It is like X-rays in that it forms part of the electromagnetic spectrum and has a very short wavelength. As such, gamma rays can be considered as being made up of photons. Like all EMR, gamma radiation travels at the speed of light and will pass through the human body very quickly, affecting all organs and tissue. It leaves an ionizing trace which means that exposed body parts can become ionized.

Radiation exposure to the body targets rapidly dividing cells of the gastrointestinal tract and reproductive regions. The effect on humans can be summarized as follows:

1. Hair falls out with dose of 2.00 sievert (Sv) or more.

---

b. Even before, I recall my mother, Berniece Vallero, in the 1950s, warning me not to drink rainwater and worrying about ^{90}Sr in milk because of the nuclear bomb testing.

2. With levels of around 1 Sv, the blood's lymphocyte cell count begins to drop, making the person susceptible to infection. This is mild radiation sickness. This can, over many years, lead to leukemia and lymphoma.
3. The thyroid is particularly sensitive to radioactive iodine and KI can be used as an antidote to reduce the effect of the exposure.
4. With an exposure of more than 2.00 Sv, the lining of the intestinal tract will cause vomiting and diarrhea.
5. With levels as low as 2.00 Sv, the cells of the reproductive organs can be damaged leading to sterility with long-term exposures.
6. The brain and heart do not involve rapidly dividing cells so are not affected until the levels reach 10–20 Sv. A sievert is a unit of dose equivalent and has units of J kg^{-1}.

Nuclear power plants are based on the heat released from nuclear reactions. When ^{235}U atoms are hit by neutrons, the atom splits into two unequal parts, releasing gamma radiation and two neutrons. Other pairs of isotopes are also released, and they include ^{128}I, ^{95}Y, and ^{140}Cs with ^{92}Rb.

The pair of neutrons released reacts with more U atoms and if there are enough U atoms present, the reactions continue as a chain reaction. In a nuclear reactor, the heat of this reaction (ΔH) is used to heat water and the resulting steam is used to turn turbines and hence produce electricity. The energy from this nuclear reactor, per unit mass of fuel, is more than a million times the energy from burning coal (mean energy from coal is 2.4×10^7 J kg^{-1} and for the nuclear reaction above, it is 7.3×10^{13} J kg^{-1}). The secret to harnessing nuclear power is to control the uranium so that just the right amount of heat is produced. In nuclear plants, the uranium is enriched (by increasing the ratio of the isotope ^{235}U to ^{238}U) but not to the extent that is required for an atomic bomb. In a nuclear reactor, the uranium in the form of pellets is embedded in rods of zirconium which can be lowered or raised into the core of the reactor with other rods. Zirconium metal is chosen because it allows neutrons to pass through. To control the reaction, control rods of metals such as cadmium are used. This metal absorbs neutrons and, like the uranium rods, can be inserted in or withdrawn from the reactor core. In this way, the rate of the nuclear reaction can be adjusted when required. The reactor core is cooled using a variety of methods, including cooling with water. The reactor core is shielded by thick steel and concrete walls to prevent the release of radioactive material [9].

Small modular reactors (SMRs) are smaller versions of the nuclear reactors just described. Their usual maximum output is about 300 Megawatt electric (MWe), generating 7.2 million kWh per day. SMR output ranges can be less than 10 MWe. Coolants include light water, liquid metal or molten salt. SMRs based on nonlight water technology are known as advanced modular reactors (AMRs). Arguably, SMRs are safer than their larger counterparts, given their simpler design that includes less components that may fail, lower core power, larger coolant fraction, and less damage area if a failure were to occur.

4.2 Heat

The addition of energy into an ecosystem can be problematic. Living systems operate within a narrow range of light and heat. Direct heating of water is a type of water pollution, such as harm to sensitive aquatic life like larva and young-of-the-year fish. It is particularly harmful to cold-water fish, e.g., trout (genus *Salvelinus*) and salmon (genera *Salmo* and *Oncorhynchus*). Conversely, increased temperatures favor rough fish like carp (genus *Cyprinus*) which will dominate species abundance in aquatic systems in direct proportion to water temperature. Microclimatological changes can occur due to the release of heated plumes released from combustion facilities. Soil adjacent to water systems that receives added heat can alter habitats (e.g., changes to freeze-thaw cycles, seasonal variations, and selectivity), especially of certain soil biota (e.g., bacteria, plant root systems and fauna like earthworms, burrowing and nesting animals). Such thermal pollution needs to be treated, such as by cooling towers before heated water reaches surface water and other sensitive habitat.

Heat can initiate cumulative impacts, such as the heat exchange and changing conditions of receiving water bodies [10]. In fact, the value of a fishable stream can be directly related to water temperature since it defines the dissolved oxygen (DO) content, which is a limiting factor of the type of fish communities that can be supported by a water ecosystem. A trout stream is a highly valued resource that is adversely impacted if mean temperatures increase. Rougher, less valued fish (e.g., carp and catfish) can live at much lower ambient water body temperatures than can salmon, trout, and other cold-water fish populations. Thus, net increase in heat may directly stress the game fish population. That is, fish species vary in their ability to tolerate higher temperatures, meaning that the less tolerant, higher value fish will be inordinately threatened [11]. Much can be explained by the concomitant decrease in the stream's DO concentrations, which increasingly renders the water body hostile to the fish survival and reproduction. Even if the adult fish can survive at the reduced DO levels, their reproductive capacities diminish; or if the reproduction is not adversely affected, survival rates of juvenile fish decline [10].

Chemical reactions are temperature-dependent. Thus, with increasing temperatures, chemical substances become more soluble, bioavailable, and toxic. Aquatic organisms' exposure to harmful substances increases with increasing temperature. For example, greater concentrations of mercury and other toxic metals will occur with elevated temperatures. The lower DO concentrations will lead to a reduced environment where the metals and compounds could form sulfides and other compounds that can be more toxic to the fish than elemental and oxidized forms. Therefore, the change in temperature, the resulting decrease in DO and increasing metal concentrations, and the synergistic impact of the combining the hypoxic water and reduced metal compounds is a cascade of harm to a stream's ecosystems [12].

Abiotic chemical mechanisms, e.g., hydrolysis and redox, combine with and influence biotic response, e.g., photosynthesis and respiration. Notably, the growth and metabolism of the bacteria results in even more rapidly decreasing DO levels. Algae both consume DO for metabolism and produce DO by photosynthesis. As mentioned the increase in temperature increases their aqueous solubility and the decrease in DO is accompanied by redox changes, e.g., formation of reduced metal species, such as metal sulfides. These reactions also being mediated by the bacteria. The decrease in DO changes the redox potential of the water and sediment, initiating or increasing the rate of the reduction of metals. However, the opposite is true in the more oxidized regions, i.e., where the metals are forming oxides. The increase in the metal compounds combined with the reduced DO, exacerbated by the increased temperatures, can act synergistically to make the conditions toxic for higher animals, e.g., a fish kill [13]. This is an example of how an outcome can become a new stressor.

To recap, the first-order abiotic effect, i.e., increased temperature, results in an increased microbial population. However, the growth and metabolism of the bacteria results in decreasing the DO levels but, as mentioned, the growth of the algae both consume DO for metabolism and produce DO by photosynthesis. Meanwhile a combined abiotic and biotic response occurs with the metals. The increase in temperature increases their aqueous solubility and the decrease in DO is accompanied by redox changes, e.g., formation of reduced metal species, such as metal sulfides. This is also being mediated by the bacteria, some of which will begin reducing the metals as the oxygen levels drop (reduced conditions in the water and sediment). However, the opposite is true in the more oxidized regions, i.e., the metals are forming oxides. The increase in the metal compounds combined with the reduced DO, combined with the increased temperatures can act synergistically to make the conditions toxic for higher animals. The movement of mass and energy downstream, both figuratively and literally, results in an adverse aquatic outcome, e.g., lower reproduction rates, harm to juvenile fish and, ultimately, a fish kill [13]. Predicting the likelihood of a fish kill, however, can be quite complicated, with many factors that either mitigate or exacerbate the outcome. This is a classic depiction of Bayesian thinking [13] discussed in Chapter 2, i.e., prior events must update predicted outcomes, e.g., posterior probabilities [14].

The first law of thermodynamics requires that releasing heated water in any amount to a water body, even at the permitted level, would increase the overall temperature of the receiving stream. Heating processes generate or aid in generating almost every class of pollutants. Indirect thermal pollution usually involves changing the physical, chemical, and biological integrity of a system. One important type of thermal pollution is indirect thermo-chemical pollution that results from changes in the heat budgets within these control volumes. For example, the secondary gases generated in thermal technologies can cause or exacerbate the generation of other chemical pollutants. This can be mediated by control technologies, e.g., wastewater treatment, secondary combustion chamber for more complete combustion of incinerator gases, or wet scrubbers for their removal, as well as electrostatic precipitators and fabric filters to remove particulate matter that may ultimately contaminate surface and ground water. Preferably, designing systems that require less heat or wherein heat is managed properly would obviate the need for control equipment.

During combustion, the total mass and the total concentrations of thermally generated compounds increase with temperature, but the types of compounds change [15,16]. This occurs until complete combustion is reached, with only CO_2 and water remaining. Of course, the presence of other elements in the fuel, will lead to other oxidized compounds up to complete combustion, e.g., if N is present, oxides of nitrogen will form and if chlorine is present, at times during temperature increase the Cl will react to form chlorinated compounds, which will themselves be oxidized until the Cl completely leaves the reaction [17,18]. This is particularly important to the waste incineration since the feedstock is so diverse. At some point, the temperature, pressure, sorbent and oxygen conditions will transform these compounds to form others that may be even more toxic, e.g., the halogenated dioxins and furans [16,19]. The optimal combination of factors in the thermal process must be identified to convert as much of the organic mass to nontoxic products [20].

The earth's atmospheric temperature varies with altitude, as does the density of the substances comprising the atmosphere. The earth is warmed when incoming solar radiation is absorbed at or near the earth's surface and reradiated at longer electromagnetic wavelengths (infrared). In general, the air grows progressively less dense with increasing altitude moving upward from the troposphere through the stratosphere and the chemosphere to the ionosphere. In the upper reaches of the ionosphere, the gaseous molecules are few and far between as compared with the troposphere.

The ionosphere and chemosphere are of interest to space scientists because they must be traversed by space vehicles *en route* to or from the moon or the planets, and they are also regions in which satellites travel in the earth's orbit. These regions are also of interest to communications scientists because of their influence on radio communications. However, these layers are of interest to water pollution scientists primarily because of their absorption and scattering of solar energy, which influence the amount and spectral distribution of solar energy and cosmic rays reaching the earth and water surfaces.

The mixing ratio is the amount of a single substance divided by the total amount of all substances in a mixture. This can be a mole ratio (r_i):

$$r_i = \frac{n_i}{n_{total} - n_i} \tag{3.5}$$

Where, n_i = the amount of the single substance and n_{total} = the total amount of the mixture. The amount is usually expressed as the International System of Units (SI) unit, which is the mole. A mole, which is abbreviated "mol" is expressed using Avogadro's number. That is, a mol contains $6.02214076 \times 10^{23}$ elementary entities. For water pollutants the elementary entity is usually a molecule or an atom, but can also include other entities, especially an ion. Thus, 3 mol of N_2 would contain $3 \times 6.02214076 \times 10^{23}$ molecules, which is about 1.81×10^{23} molecules. In fact, 3 mol of O_2 would also contain 1.81×10^{23} molecules. 3 mol of argon (Ar) would contain 1.81×10^{23} atoms, and so on.

The mixing ratio may also be expressed as a mass ratio (ζi):

$$\zeta = \frac{m_i}{m_{total} - m_i} \tag{3.6}$$

Where, m_i = the mass of the single substance and m_{total} = the total mass of the mixture.

The density of water varies horizontally, vertically, and within water systems. Water density is the volume of water divided by the mass of water [21].

4.3 Atmospheric pressure

Atmospheric pressure is the force exerted on an object by the weight of the atmosphere above it. It can be visualized as a column atop the object on the earth's surface, extending all the way to the top of the atmosphere. Atmospheric pressure results from the transfer of momentum by the random motion of water molecules.

Unequal (differential) heating of the earth's surface makes the water move because of temperature and pressure differences. The laws of potentialities state that movement of mass or energy is from higher potentials to lower potentials (e.g., height, energy, pressure).

4.4 Standard conditions

Standardization is crucial to good science, since sound and credible science depends on repeatability. Diversity is a wonderful thing in most human enterprises. "Variety is the spice of life." However, if each scientist were to use a unique method, generalizations would be impossible. Science depends on logical progressions. Most commonly, science is deductive, i.e., knowledge is inferred from the general to specifics. Scientific laws are the general rules that dictate the applications (the specifics).

Of course, even these laws are subject to change, given enough countervailing evidence, e.g., inductive reasoning by a sufficient body of specifics can override outdated scientific principles. Such paradigm shifts come only after much evidence and can run into resistance from the *status quo* [22,23]. But, to make science repeatable, we need standards. Thus, to compare the corpus of water pollution literature and knowledge and to apply meaning to water research and investigations, we can begin with how water pollution scientists understand pressure and temperature.

The standard atmosphere is unit of pressure. 1 Standard atmosphere (atm) = 101,325 pascals (Pa) = 101.325 kilopascals (kPa) = 1013.25 hectopascals (hPa) = 1013.25 millibars (mb). Water pollution publications use these and other pressure units (e.g., mm Hg). Engineers in the U.S. use a mix of pressure units, depending on the task, e.g., designing a water mover, installing a pump, or testing indoor water. Most other nations use common metric systems exclusively. Meteorologists use atm as the preferred pressure unit, but also use other units like hectopascal (hPa), inches mercury (in Hg), and mm Hg. Water pollution texts and manuals may prefer using the kilopascal (kPa) most often because this unit is commonly used in environmental and engineering literature to express vapor pressure and partial pressures. Some water pollution partitioning coefficients include pascal (Pa) units, e.g., Henry's law coefficients, but the Pa is usually too large to be used in most water pollution literature, given that frequently pollutants thankfully have low partial pressures and concentrations under most environmental conditions. Also, where water pollution precursors are at very high concentrations, e.g., in an wastewater outfall structure, a reactor, or an industrial stack air emission, the chemical concentrations may be expressed as in Pa units. The key is that secondary data sources useful to water pollution calculations vary in the units used for pressure and other factors, so conversions are often needed.

Standard temperature and pressure (STP) are 273.15 K and 1 atm. It is used to allow for consistent comparisons of results from scientific studies. For many engineering and scientific studies and experiments, normal temperature and pressure (NTP) is preferred, because NTP is 293.15 K (20 °C), which is closer to "room temperature." Studies conducted at different temperatures and pressures need to be adjusted to these standards. For example, if a study were done at 290 K and 1.5 atm, the results must be corrected to 273.15 K and 1 atm for STP. Other factors are particularly important for air emissions, especially moisture content and oxygen levels. In stack tests, corrections to STP are a means of ensuring that the amount of any pollutant that can leave a stack is consistently applied to all similar stacks in a jurisdiction. In testing water flow, corrections to NTP allow for consistent comparisons for other water studies conducted at NTP.

For water pollution studies specifically, NTP is often used. For example, many academic and private sector studies are conducted at 20 °C and 1 atm.

Substances vary in their affinity for various compartments within a single organism or in the environment. Water solubility is generally inversely related to K_{ow}, because as hydrophobicity increases the compound's affinity for the octanol compared to it affinity for the water. However, the inverse relationship is not true for all organic compounds given the wide range of aqueous solubilities. Certainly, organic compounds can have affinities for neither, either, or both the organic and the aqueous phases. Most compounds are not exclusively associated with either phase; i.e., they have at least a modicum amount of amphiphilicity. As mentioned, the fluid representing the organic component, octanol, is amphiphilic, since it, like all alcohol, contains the very polar hydroxyl group (OH^-).

Seemingly minor structural attributes of a molecule can change phase partitioning and density. Even in isomers, i.e., compounds having the same chemical composition with a different arrangement, vary in their K_{ow} values and densities. As evidence, the "1,1" versus "1,2" arrangements of chlorine atoms on 1,1-dichloroethane, and 1,2-dichloroethane cause the former to have a slightly decreased density but twice the K_{ow} value than the latter. The location of the chlorine atoms alone accounts for a substantial difference in water solubility in the two compounds.

5 Scientific advancement

Water pollution engineers, scientists and managers have done a remarkable job in the past half-century. Public health has been the principal driver for assessing and controlling water contaminants, but water pollution abatement laws and programs throughout the world have also recognized that effects beyond health are also important, especially welfare protection. One of the main welfare considerations is that ecosystems are important receptors of contamination. Another welfare concern is that contaminants ieffect the integrity of structures and other engineered systems by corrosion. Thus, from a water pollution perspective, there is a cascade of hazards from human health to ecosystems to abiotic (i.e., nonliving) systems.

Certainly, water pollution characterization and treatment require the understanding and application of sound science. We often think of researchers in the lab or in the field as "scientists," but most scientists are not researchers. Indeed, some of the best scientists are those who apply what researchers have discovered. Such scientists work in national, state, provincial and local government agencies with missions to protect the environment. They write regulations, rules and guidance and issue permits and licenses on discharges and water quality protections. Other scientists work in the private sector or in water quality and health advocacy groups. Even when they all adhere to the scientific method and avail themselves to the same data, there is always various levels of certainty about environmental harm. Thus, establishing water quality standards or deciding on the best water pollution control and prevention approach is an iterative and multifactorial process [24–29].

As mentioned, pollution is best and most precisely explained according to the scientific principles of motion and thermodynamics. The subdiscipline of motion, mechanics, describes the energy and forces that effect of the movement of bodies, i.e., the way things from one place to another. Given that the movement of pollutants is mainly via water and air, pollutant transport can be explained by fluid mechanics. Fluid mechanics addresses both statics, i.e., fluids at rest, and dynamics, i.e., fluids in motion. Fluids can be the pollutant itself or the medium in which the pollutant resides. An example of the former is the movement of benzene as it infiltrates the soil and enters ground water. An example of the latter is that after the benzene reaches the ground water, the water in the soil and other unconsolidated material carries the benzene as a plume that moves down gradient. Similarly, surface hydrology explains the movement of water on the earth's surfaces, especially streams, oceans, lakes, and estuaries. Hydraulics addresses fluid motions in conveyances like pipes and conduits, including those that carry wastewater to treatment plants and those that carry clean water in water mains.

One of the key considerations of fluid dynamics is flow. This is important to hydrology and its subdiscipline hydrogeology, which explains the flow of water below the earth's surface. For example, the channelization of a stream can greatly increase the velocity of water, as can the drainage of wetlands. These changes can have profound effects on watersheds. As evidence, prior to a hurricane, the loss of wetlands and replacement with greater areas of impervious surfaces changes the fluid dynamics and increases the effects of wind and waves due to less immediate infiltration, more runoff, and the scouring of sand and other barriers [30]. Also, flow rates affect recharge of aquifers, such as the decrease in infiltration rates of less permeable surfaces during and after precipitation events. Even in more permeable surfaces, rapid precipitation events allow less infiltration compared to slower and extended rain events.

Given millennia of scientific discovery, engineering applications and technology have incrementally enhanced scientific knowledge with ever increasing precision, accuracy, and representativeness:

> *Technology is the outcome of engineering; it is rare that science translates directly to technology, just as it is not true that engineering is just applied science [31].*

Researchers try to explain what has previously been unexplained or at least incompletely explained by using the scientific method. For example, fewer "data-rich" pollutants exist in the environmental and aquatic sciences, compared to the wealth of "data-poor" pollutants. Presently, computational and informatic tools have been developed to build upon information from a data-rich chemical compound to draw in knowledge about other data-poor chemicals that share a few attributes with the data-rich chemical. This is the basis for quantitative structure-activity relationship (QSAR) modeling [32–35].

Every water pollution event or episode is initiated and worsened by a combination of anthropogenic and natural factors. Indeed, wastewater treatment and prevention are relatively recent concepts. A hundred years ago, treating a waste before releasing it into a stream was not normal. In fact, these concentrations of wastes may not have even been particularly harmful if the natural microbial populations could degrade the wastes efficiently. Also, until the 20th century, most of the chemical compounds released were not synthetic, so that the microbes had already adapted an ability to biodegrade them. Once the petrochemical and industrial expansion took hold, these natural degradation processes were increasingly unable to withstand the onslaught of synthetic chemical loads.

Environmental and aquatic scientific advancement shares aspects of other advances. Today's water quality knowledge was the result of several paradigm shifts, a term coined in the late 20th century by Thomas S. Kuhn. Actually, the term existed long before Kuhn's idea, but he extended it to apply to an accepted specific set of scientific practices [22,23]. The scientific paradigm is made up of what is to be observed and analyzed, the questions that arise pertaining to this scientific subject matter, to whom such questions are to be asked, and how the results of the investigations into this subject matter will be interpreted. The paradigm is unacceptable if theories and information that are erroneous continue to be accepted by the scientific and engineering communities. Excessive comfort with the *status quo* may cause a scientific community to protect wrong thinking, i.e., so-called "groupthink" [36], and gate-keeping to prevent the acceptance of new, correct concepts over old, incorrect ones [37]. The bottom line is that scientists must be skeptical and ready to accept new knowledge that is supported by sufficient, objective data and information.

There are two basic forms of scientific research, basic and applied. Basic research contributes to general knowledge and seeks to explain truths, whereas applied sciences use these and other knowledgebases to solve problems and provide societal value [38]. For water pollution science, both forms are important, with a growing application of scientific principles by teams and cooperative investigation [39–42]. Understanding water pollution and addressing its harm require collaborative scientific research [43].

Basic research is guided by the quest to further knowledge that develops new technologies. For example, remote sensing was an outgrowth of a better understanding of particle-wave phenomena in basic physics research. Similarly, Louis Pasteur's work in advancing germ theory, i.e., basic research, led to his application of this knowledge to decrease the rate of spoilage of milk, vinegar, beer, and wine [38].[c]

The continuum from basic to applied research is captured in the linear knowledge flow model. Each step depends on outcomes from the previous stage (see Fig. 3.2).

This linear model is the embodiment of technology transfer. It describes the movement from basic science to technology. First is basic research that sets the groundwork for practical applications. For example, Niels Bohr's focus on the atom falls within the far-left box. Next, applied research builds on the basic knowledge, wherein scientists convert the basic knowledge into useful applications of Bohr's atom, such as quantum physics applications to develop detection technologies like mass spectroscopy [44]. In the final stage, scientists methodically adapt research findings into beneficial devices, systems, methods, processes, and materials [30,38].

---

c. My mentor at the University of Kansas, Ross E. McKinney, received his doctorate in microbiology but used the basics of microbiology to become one of the most prominent environmental engineers in the world in advancing biological wastewater treatment methods.

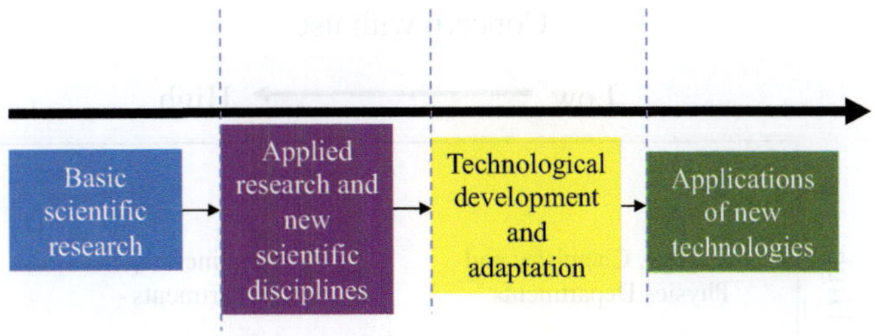

FIG. 3.2 The linear knowledge flow from basic research to applied research to development of technologies. *(From D.A. Vallero, T.M. Letcher, Unraveling Environmental Disasters, second ed., Elsevier Academic Press, Amsterdam, The Netherlands, 2023; based on information from D.E. Stokes, Pasteur's Quadrant: Basic Science and Technological Innovation, Brookings Institution Press, 2011.)*

The one-way flow in the linear knowledge model has weaknesses. In fact, there are not always clear delineations between basic and applied research:

> *The terms basic and applied are, in another sense, not opposites. Work directed toward applied goals can be highly fundamental in character in that it has an important impact on the conceptual structure or outlook of a field. Moreover, the fact that research of such a nature that it can be applied does not mean that it is not also basic [38].*

Science does not always lead to engineering which subsequently drives the need for technology. Engineering has often driven basic science, e.g., biology's so-called "black boxes" never become completely understood by bioscience researchers who must learned from practicing biomedical engineers and physicians. The application of basic biological principles to the recent COVID-19 pandemic, for example, was limited. Understanding the structure of DNA and RNA was not sufficient to arrive at safe and effective mRNA vaccines and viral therapies [45]. Shrinking these knowledge gaps is an ongoing, incremental process [46–48]. Often, technology drives advances in both science and engineering. A new device allows scientists to investigate new areas of research, e.g., the PCR in rDNA research, and engineers to conceive new designs, e.g., the DNA markers that provide information that leads to improved bioremediation of water pollutants.

Technology driving science is not a recent phenomenon. The telescope led to astronomical understanding and the microscope advanced microbiological discovery beyond what the naked eye could provide. More recently, fractal dimensions and rDNA technologies led to mathematical and medical breakthroughs respectively [38]. Biotechnology and nanotechnology have led to scientific and engineering advances in wastewater treatment [10,49–62].

Water pollution must be addressed with credible science, combining theory and practice, with the former seeking knowledge for its own sake and the latter seeking knowledge to meet some need. Interestingly, one of the scariest aspects in my pursuit of a Ph.D. at Duke University was that my research had to conduct original research and to add to the existing body of engineering knowledge. This was daunting in that I was supposed to add to a knowledgebase built over thousands of years. With the help of my mentor, Jeffrey Peirce, and the other brilliant faculty members on my dissertation committee, I was able to add a very small kernel of knowledge in how a pesticide moves from the soil to the air. I relied heavily on the basic principles discovered and augmented by the scientists who conducted pure research, but even more heavily on the scientists and engineers who applied these principles to real-world water, air, and soil pollution challenges. This construct is demonstrated at most universities (see Fig. 3.3). The science departments are concerned with knowledge-building, the engineering departments with applied knowledge to understand how to solve society's problems, and the university designer and facilities director in finding innovative ways to use this knowledge. For example, a U.S. Army Corps of Engineers civil engineer applies general scientific principles to upgrade levee designs but uses historical knowledge and personal experience to improve water retention and soil strength. Like Thomas Edison, who was most interested in utility, the engineer is less interested in knowledge for knowledge's sake compared to the need to solve a particular problem. The engineer must work closely with land use planners, administrators, and managers who have decided the need for the levees. Innovations certainly also come from the southwest box in Fig. 3.3, but people in this box have measures of success more strongly tied to investments in operation and maintenance. Indeed, every quadrant has feedback loops to the other three [38].

North Carolina's Research Triangle demonstrates this view. Duke University, the University of North Carolina—Chapel Hill, and North Carolina State University all have engineering programs, but with different emphases. Duke is recognized as a world leader in basic research, but its Pratt School of Engineering place particular emphasis on application of these sciences, e.g., clean water in developing nations and more efficacious drug delivery. This calls for collaboration among Duke's schools of engineering, environment, and medicine. The School of Public Health houses the engineering programs at the University of North Carolina. As expected, this engineering research tends to focus on addressing and

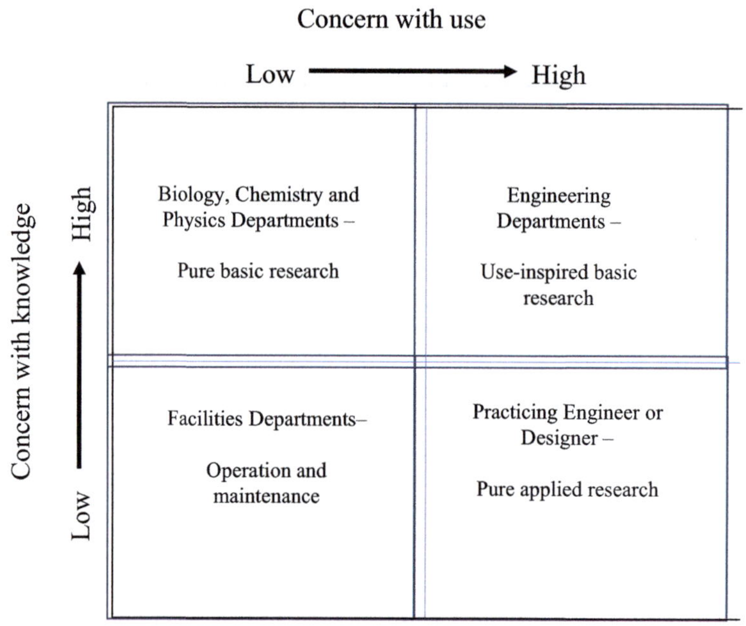

FIG. 3.3 University research categorized according to knowledge and utility objectives. *(Data from D.A. Vallero, T.M. Letcher, Unraveling Environmental Disasters, second ed., Elsevier Academic Press, Amsterdam, The Netherlands, 2023.)*

preventing environmental problems that would result from societal problems, such as damage to confined animal feeding operations and nutrient loading of rivers. As its land grant college, North Carolina State is the first place that the State of North Carolina looks for designers of public projects like wastewater treatment plants, drinking water plants, and roads. Engineers graduating from NC State are ready to design as soon as they receive their diplomas. However, NC State also has excellent engineering research programs that apply the basic sciences. All this occurs within the scientific community of the Research Triangle, represented by Research Triangle Park (RTP), which includes knowledge and research centers supported by private and public entities. The RTP researchers and other scientists look for improved products and better processes. Thus, the RTP can be envisaged to reside in the "Edison" quadrant, although research in the other two quadrants is also ongoing in the RTP laboratories [38].

Research that is committed to potential utility *and* to advancing fundamental understanding is represented as "use-inspired" research [38]. Water quality sciences benefit from both.

The last quarter of the 20th century saw substantial advances and new applications of science, engineering, and associated technologies to characterize and begin to address water pollution problems. The knowledge coalesced into a new environmental ethos, at least new to most of Western Civilization. Ancient cultures on all continents, including the Judeo-Christian belief systems, had innately cautioned that humans will waste the resources bestowed upon us lest the view as stewards and caretakers of the environment be taken seriously.

Most practitioners and researchers see their work as justified by the general societal need for better medicine, food, NS environmental quality, safer products, and more efficient manufacturing. The societal demands are compelling. Science explains the physical world, while engineering encompasses applications of science to achieve results. Trial and error learning certainly has advanced our understanding of water pollution, but often at a great cost in terms of the loss of lives and diseases associated with blunders, lack of knowledgeable decision making, and the insufficient appreciation of downstream effects.

The scientific community seldom has complete consensus on anything except the basic principles. Even these are occasionally suspect given advances in knowledge like the rethinking caused by quantum mechanics and subatomic particle research. Indeed, there is arguably consensus about many aspects of water pollution, but we should be open to improvements with each new research discovery.

The march toward credible environmental science can be meticulously slow. In water quality assessments, sometimes the only sure statement is that a factor either makes the water quality better or worse. For example, increased development along shorelines will likely worsen the conditions of coastal wetlands. The exact amount of worsening is impossible to predict with complete precision. However, there will undoubtably be a greater number of vulnerable habitats. Also, the decrease in buffering and changes in hydrology due to development will likely increase the ecological harm.

Fortunately, modern society has come to a greater appreciation of the systematic relationship among the sciences, engineering, and technologies. Most of society embraces the concept of "spaceship earth," i.e., our planet consists of a finite life support system and that our air, water, food, soil, and ecosystems were not infinitely elastic in their ability to absorb humanity's willful disregard [63]. Scientific curiosity and the need to provide stronger evidence about our hypotheses pushes researchers to revisit some previously neglected details that will continue to improve ways to address water pollution.

Combating water pollution requires that researchers, regulatory scientists, and engineers work closely together. For example, in the U.S., the Environmental Protection Agency (EPA) has three basic functions: (1) To conduct research; (2) To enforce environmental rules and regulations; and (3) To provide funding for other entities to help with the first two functions. EPA's water pollution enforcement and compliance work are conducted by the Office of Water (OW), the Office of Drinking Water (ODW), and its 10 regional offices, and its research is conducted by the Office of Research and Development (ORD). The Clean Water Act [64] gives OW principal statutory authority to regulate discharges to the waters of the U.S. The Safe Drinking Water Act [65,66] gives ODW the authority to regulate public water supplies, especially setting maximum contaminant levels (MCLs) for water provided by drinking water supplies. Other laws aim to improve water quality, such as the Comprehensive Environmental, Response and Compensation Act (CERCLA), which is also known as Superfund, and the Resource Conservation and Recovery Act (RCRA). Both address hazardous waste that may contaminate surface water and ground water, with Superfund addressing abandoned waste sites and RCRA addressing active processes of generating, transporting, treating, storing, and disposal of hazardous wastes. As such, RCRA addresses contaminants from the so-called cradle to grave. Water, especially ground water, can be polluted during any of these stages.

There are other perspectives in addition to research and enforcement, including teaching, journalism, and policy making. Those who educate others, write about, and promulgate polices and laws must understand water pollution in different ways, but with a modicum of scientific knowledge.

Understanding the scientific principles of water pollution and the methods needed to estimate and document the amount of pollution involved is something all environmental scientists and engineers share, although the focus, extent, and specificity of these characterizations will vary. For example, a research scientist may need to completely explain all the reactions in a waste facility that cause the generation of pollutants, e.g., the metal lead (Pb) and its ionic forms. The water discharge permit writer may not need to know all the reactions to the extent of the researcher, but enough to know which ones may be most difficult to change. For example, the permit writer may need to know the total mass of Pb released from the facility. Likewise, the high school chemistry teacher may need to know only the most basic chemical reactions, but the water pollution engineering college professor may need to know just as much as the researcher to tailor their lectures and projects properly to the target the engineering syllabus. The journalist who can "do the math" and grasp the scientific principles has an advantage over her colleagues that only know the essence of the science. The local water pollution authority manager may need to know as much as the permit writer, since he supervises the permit writers. The politician who serves on a technical committee responsible for writing water pollution legislation will need to know more chemistry and other scientific principles compared to the other members of the legislature, perhaps even grasping the math and science to the same extent as the permit writer and the researcher, at least when it comes to deciding whether to increase or decreasing funding for research and compliance.

Therefore, every water pollution perspective requires an aptitude and appreciation for the scientific principles underpinning water quality. This book attempts to provide examples of most of the types of problems and opportunities expected to be encountered in water pollution-related careers, whether to succeed in one's vocation or simply to trust and be confident in the quality of the data, information, and calculations in what one reads in a report or article written by others [67].

6 Conclusions

With this basic introduction to some of the key principles of water pollution science, engineering, technology, and management, we are now ready to explore how a water pollutant is formed and changed. In subsequent chapters, we will address how these and other principles based on the laws of motion, thermodynamics, chemistry, and biology change the concentrations and locations of a pollutant after being released. The principles also underpin how chemical compounds can be precursors of pollutants and how pollutants are exchanged between the surface and ground water and other environmental media, how they are physically transported, and their fate, and the magnitude and duration of exposure to human populations or in ecosystems, the risks that need to be addressed.

References

[1] United Nations Environmental Programme, Panel report pursuant to article 6 of the Montreal protocol on substances that deplete the ozone layer under the auspices of the UNEP, United Nations Environment Programme, Nairobi, Kenya, 1991.

[2] M. Molina, The probable role of stratospheric ice clouds: heterogeneous chemistry of the ozone hole, in: The Chemistry of the Atmosphere: Its Impact on Global Change, 1994, pp. 27–38.

[3] National Aeronautic and Space Administration, Ozone facts: history of the ozone hole, National Aeronautics and Space Administration, 2013.

[4] G.H. Bernhard, R.E. Neale, P.W. Barnes, P. Neale, R.G. Zepp, S.R. Wilson, et al., Environmental effects of stratospheric ozone depletion, UV radiation and interactions with climate change: UNEP Environmental Effects Assessment Panel, update 2019, Photochem. Photobiol. Sci. 19 (5) (2020) 542–584.

[5] D. Ebbing, General Chemistry, fifth ed., Houghton Mifflin Company, Boston, MA, 1996. ISBN: 0-395-76493-9, In this issue.

[6] P. Héroux, I. Belyaev, K. Chamberlin, S. Dasdag, A. De Salles, C. Rodriguez, L. Hardell, E. Kelley, K. Kesari, E. Mallery-Blythe, R. Melnick, Cell phone radiation exposure limits and engineering solutions, Int. J. Environ. Res. Public Health 20 (7) (2003) 5398, https://doi.org/10.3390/ijerph20075398. In this issue.

[7] D. Davis, L. Birnbaum, P. Ben-Ishai, H. Taylor, M. Sears, T. Butler, T. Scarato, Wireless technologies, non-ionizing electromagnetic fields and children: identifying and reducing health risks, Curr. Probl. Pediatr. Adolesc. Health Care 53 (2) (2023) 101374, https://doi.org/10.1016/j.cppeds.2023.101374. In this issue.

[8] B.B. Levitt, H.C. Lai, A.M. Manville, Low-level EMF effects on wildlife and plants: what research tells us about an ecosystem approach, Front. Public Health 10 (2022) 4654.

[9] Y. Wang, W. Chen, L. Zhang, X. Zhao, Y. Gao, V. Dinavahi, Small modular reactors: an overview of modeling, control, simulation, and applications, IEEE Access 12 (2024) 39628–39650, https://doi.org/10.1109/ACCESS.2024.3351220. In this issue.

[10] D. Vallero, Environmental Biotechnology: A Biosystems Approach, Elsevier Science, 2015.

[11] E. Dohner, A. Markowitz, M. Barbour, J. Simpson, J. Byrne, G. Dates, 5.2. Monitoring and assessing water quality, in: Volunteer Stream Monitoring: A Methods Manual, 1997.

[12] N.T. Shears, P.M. Ross, Toxic cascades: multiple anthropogenic stressors have complex and unanticipated interactive effects on temperate reefs, Ecol. Lett. 13 (9) (2010) 1149–1159.

[13] D.A. Vallero, K.H. Reckhow, A.D. Gronewold, International Society of Exposure Analysis, Application of multimedia models for human and ecological exposure analysis, in: ISEA, et al. (Eds.), International Conference on Environmental Epidemiology and Exposure, International Society of Exposure Analysis, Durham, NC, 2007.

[14] K. McDonald, D. Ryder, M. Tighe, Developing best-practice Bayesian Belief Networks in ecological risk assessments for freshwater and estuarine ecosystems: a quantitative review, J. Environ. Manag. 154 (2015) 190–200.

[15] D.A. Vallero, Fundamentals of Air Pollution, fifth ed., Elsevier Academic Press, Waltham, MA, 2014. 999 pages cm p.

[16] R. Hawley-Fedder, M. Parsons, F. Karasek, Products obtained during combustion of polymers under simulated incinerator conditions: I. Polyethylene, J. Chromatogr. A. 314 (1984) 263–273.

[17] U.S. Environmental Protection Agency, SW846 Method 8290A-1: polychlorinated dibenzodioxins (PCDDs) and polychlorinated dibenzofurans (PCDFs) by high resolution gas chromatograph/high resolution mass spectrometry (HRGC/HRMS). Revision 1 ed, 2007.

[18] US Environmental Protection Agency, Method 1613: tetra-through octa-chlorinated dioxins and furans by isotope dilution HRGC/HRMS (Rev. B), UEP Agency (Ed.), Washington, DC, 1994.

[19] B.C. Willis, M.M. Howie, R.C. Williams, Public health reviews of hazardous waste thermal treatment technologies: a guidance manual for public health assessors, US Department of Health and Human Services, Agency for Toxic Substances and Disease Registry, Division of Health Assessment and Consultation, 2002.

[20] U.S. Environmental Protection Agency, Guide for conducting treatability studies under CERCLA: thermal desorption, Washington, DC, 1992.

[21] R.M. Goody, Y.L. Yung, Atmospheric Radiation: Theoretical Basis, Oxford University Press, 1995.

[22] T. Kuhn, What Are Scientific Revolutions, MIT, Boston, 1981.

[23] T.S. Kuhn, D. Hawkins, The structure of scientific revolutions, Am. J. Phys. 31 (7) (1963) 554–555.

[24] D.-T. Avgerinopoulou, Science-Based Lawmaking: How to Effectively Integrate Science in International Environmental Law, Springer Nature, 2019.

[25] D.M. Byrd III, R.C. Cothern, Introduction to Risk Analysis: A Systematic Approach to Science-Based Decision Making, Government Institutes, 2000.

[26] S.M. Capalbo, J.M. Antle, C. Seavert, Next generation data systems and knowledge products to support agricultural producers and science-based policy decision making, Agric. Syst. (2016).

[27] J.A. Arnot, D. Mackay, Policies for chemical hazard and risk priority setting: can persistence, bioaccumulation, toxicity, and quantity information be combined? Environ. Sci. Technol. 42 (13) (2008) 4648–4654.

[28] A. Fairbrother, D. Muir, K.R. Solomon, G.T. Ankley, M.A. Rudd, A.B. Boxall, et al., Toward sustainable environmental quality: priority research questions for North America, Environ. Toxicol. Chem. 38 (8) (2019) 1606–1624.

[29] J. Shatkin, S. Qian, Classification schemes for priority setting and decision making, in: Comparative Risk Assessment and Environmental Decision Making, Springer, 2004, pp. 213–243.

[30] D.A. Vallero, T.M. Letcher, Unraveling Environmental Disasters, second ed., Elsevier Academic Press, Amsterdam, The Netherlands, 2023.

[31] National Academy of Engineering, The Engineer of 2020: Visions of Engineering in the New Century, National Academies Press, Washington, DC, 2004.

[32] I.A. Balabin, R.S. Judson, Exploring non-linear distance metrics in the structure–activity space: QSAR models for human estrogen receptor, J. Cheminform. 10 (1) (2018) 47.

[33] U.S. Environmental Protection Agency, Quantitative structure activity relationship, 2015, Available from: http://www.epa.gov/nrmrl/std/qsar/qsar.html.

[34] D. Wang, J. Yu, L. Chen, X. Li, H. Jiang, K. Chen, et al., A hybrid framework for improving uncertainty quantification in deep learning-based QSAR regression modeling, J. Cheminform. 13 (1) (2021) 69.

[35] H. Zhu, T.M. Martin, L. Ye, A. Sedykh, D.M. Young, A. Tropsha, Quantitative structure−activity relationship modeling of rat acute toxicity by oral exposure, Chem. Res. Toxicol. 22 (12) (2009) 1913–1921.

[36] I.L. Janis, Groupthink and group dynamics: a social psychological analysis of defective policy decisions, Policy Stud. J. 2 (1) (1973) 19.

[37] E.M. Smith, Reimagining the peer-review system for translational health science journals, Clin. Transl. Sci. 14 (4) (2021) 1210–1221.

[38] D.E. Stokes, Pasteur's Quadrant: Basic Science and Technological Innovation, Brookings Institution Press, 2011.

[39] D. Beaver, The many faces of collaboration and teamwork in scientific research: updated reflections on scientific collaboration, COLLNET J. Scientomet. Inf. Manag. 7 (1) (2013) 45–54.

[40] G.P. Bush, L.H. Hattery, Teamwork and creativity in research, Adm. Sci. Q. (1956) 361–372.

[41] W.O. Hagstrom, Traditional and modern forms of scientific teamwork, Adm. Sci. Q. (1964) 241–263.

[42] J.J. Winnink, R.J. Tijssen, A.F. van Raan, Theory-changing breakthroughs in science: the impact of research teamwork on scientific discoveries, J. Assoc. Inf. Sci. Technol. 67 (5) (2016) 1210–1223.

[43] J. Fernandez, Understanding group dynamics, Business Line (2002) 2.

[44] H. Kragh, Niels Bohr and the Quantum Atom: The Bohr Model of Atomic Structure 1913–1925, OUP, Oxford, 2012.

[45] J.K. Lee, H.S. Han, Coronavirus disease 2019 and mRNA vaccines: what's next—miRNA? Clin. Exp. Pediatr. 65 (6) (2022) 302–303.

[46] S. Olofsson, M. Mehrian, R. Calandra, L. Geris, M.P. Deisenroth, R. Misener, Bayesian multiobjective optimisation with mixed analytical and black-box functions: application to tissue engineering, IEEE Trans. Biomed. Eng. 66 (3) (2018) 727–739.

[47] R. Langer, N.A. Peppas, Advances in biomaterials, drug delivery, and bionanotechnology, AICHE J. 49 (12) (2003) 2990–3006.

[48] N. Ogata, S.W. Kim, J. Feijen, T. Okano, Advanced Biomaterials in Biomedical Engineering and Drug Delivery Systems, Springer Science & Business Media, 2012.

[49] K. Jain, A.S. Patel, V.P. Pardhi, S.J.S. Flora, Nanotechnology in wastewater management: a new paradigm towards wastewater treatment, Molecules 26 (6) (2021) 1797.

[50] U.S. Environmental Protection Agency, Nanotechnology: applications for environmental remediation, 2015, Available from: http://clu-in.org/techfocus/default.focus/sec/Nanotechnology%3A_Applications_for_Environmental_Remediation/cat/Overview/.

[51] D. Yang, M.R. Hartman, T.L. Derrien, S. Hamada, D. An, K.G. Yancey, et al., DNA materials: bridging nanotechnology and biotechnology, Acc. Chem. Res. 47 (6) (2014) 1902–1911.

[52] X. Qu, P.J. Alvarez, Q. Li, Applications of nanotechnology in water and wastewater treatment, Water Res. 47 (12) (2013) 3931–3946.

[53] A. Ahmad, F. Banat, H. Alsafar, S.W. Hasan, Algae biotechnology for industrial wastewater treatment, bioenergy production, and high-value bio-products, Sci. Total Environ. 806 (2022) 150585.

[54] Y. Liu, A. Su, J. Li, R. Ledesma-Amaro, P. Xu, G. Du, et al., Towards next-generation model microorganism chassis for biomanufacturing, Appl. Microbiol. Biotechnol. (2020) 1–14.

[55] D.A. Vallero, C.K. Gunsch, Applications and implications of emerging biotechnologies in environmental engineering, J. Environ. Eng. 146 (6) (2020) 03120005.

[56] H. Yi, M. Li, X. Huo, G. Zeng, C. Lai, D. Huang, et al., Recent development of advanced biotechnology for wastewater treatment, Crit. Rev. Biotechnol. 40 (1) (2020) 99–118.

[57] D. Jullesson, F. David, B. Pfleger, J. Nielsen, Impact of synthetic biology and metabolic engineering on industrial production of fine chemicals, Biotechnol. Adv. (2015).

[58] E.U. von Weizsäcker, The environmental dimension of biotechnology, in: Ernst Ulrich von Weizsäcker, Springer, 2014, pp. 57–65.

[59] B.E. Rittmann, P.L. McCarty, Environmental Biotechnology: Principles and Applications, Tata McGraw-Hill Education, 2012.

[60] P.K. Jain, V.K. Gupta, V. Bajpai, Recent Advances in Environmental Biotechnology, 2011.

[61] B.E. Rittmann, Microbial ecology to manage processes in environmental biotechnology, Trends Biotechnol. 24 (6) (2006) 261–266.

[62] D.H. Pieper, W. Reineke, Engineering bacteria for bioremediation, Curr. Opin. Biotechnol. 11 (3) (2000) 262–270.

[63] B.R. Fuller, Operating Manual for Spaceship Earth, Lars Muller Publishers, Baden, Switzerland, 1969.

[64] U.S. Congress, Federal Water Pollution Control Act Amendments of 1972 (33 U.S.C. 1251 et seq. - P.L. 92-500), Government Printing Office, Washington, DC, 1972.

[65] U.S. Congress, Safe Drinking Water Act Amendments of 1996. (P.L.104-182), Government Printing Office, Washington, DC, 1996.

[66] U.S. Congress, Safe Drinking Water Act. 88 Stat. 1660 (P.L. 93-523), Government Printing Office, Washington, DC, 1974.

[67] M.C. McKinnon, S.H. Cheng, R. Garside, Y.J. Masuda, D.C. Miller, Sustainability: map the evidence, Nature 528 (7581) (2015) 185–187.

Chapter 4

Biogeochemistry

1 Introduction

Although the focus of this book is water quality, pollution is never confined to a single medium. Water pollutants and their precursors move and change throughout the environment. Matter and energy cycle into and out of various compartments in the environment. Many physical, chemical, and biological processes are at work in the formation of a water pollutant. Indeed, whether a substance is a pollutant is determined by where it is. For example, the same soil particle that is needed for the growth of an essential plant becomes a pollutant if it enters the stream as a suspended solid or in the air as an aerosol.

The quantity, the chemical form, and the fate of a pollutant determine whether a chemical substance is essential or detrimental. Compounds of nitrogen (N), for example, are vital nutrients for vascular plants, but the plants need specific chemical forms that are sufficiently bioavailable. Depending on countless conditions, such as excessive N concentrations, abundance and diversity of microbes and other biota in the water and soil, and hydrological conditions, N can be transformed into water, soil, and air pollutants.

Both matter and energy are generated, transformed, and cycle throughout the environment, which means that an understanding of water quality and quantity depends on knowledge of the physical mechanisms by which matter and energy move and change. These mechanisms are related to the inherent properties of a substance to quantify and predict the extent to which that substance is relatively likely to be transported and its likely environmental fate, including in water above and below the earth's surface. The physical and chemical mechanisms and properties also are determinants of the mobility and toxicity of the substance and its transformation products.

Characterizing biological aspects of the biogeochemical cycle is a systematic method to address pollution by considering how matter and energy move through trophic states within water systems and how the non-living components of a system interact with the living components. That is, biogeochemistry's key consideration is the interrelationships between abiotic and biotic systems, respectively. Thus, biogeochemical interactions can be studied, documented, and analyzed to track the earth's mass and energy balances, both thermodynamically and fluid dynamically [1].

This chapter specifically addresses the cycles of water, carbon, and the environmental nutrients, especially N and sulfur (S), as well as the metal/metalloid cycles. Energy is discussed in this chapter as it pertains to biogeochemical cycling, especially the role of light in cycling mechanisms like photosynthesis, electrical energy's actions in ion exchange, and, as introduced in Chapter 3, heat's role in the solubility and volatilization of water pollutants.

2 The hydrologic cycle

Water is a part of every environmental phenomenon. The water molecule configuration (Fig. 4.1) is a key aspect of geomorphology, biology, chemistry, and society. Water directly and indirectly shapes the earth's surface, as it interacts with the lithosphere, i.e., the thin outermost layer of the earth. Water does so by facilitating countless physical, chemical, and biological processes (See Fig. 4.2).

During physical weathering, water's molecular configuration engenders substantial expansion by volume as it freezes or sublimates (i.e., change of state from gas directly to solid). This initiates frost heaving, one of the mechanisms needed to break down rock formations. The resulting unconsolidated materials eventually become primitive soil [2]. Chemically, water's molecular configuration makes for water's ability to dissolve so many chemical compounds, including acids that react with compounds in solid rock, e.g., within fissures and cracks, changing the geomorphology of the terrain. The alignment of a water molecule's oxygen and two hydrogen atoms gives a slightly negative charge at the oxygen end and a slightly positive charge at the hydrogen ends. In chemistry "like dissolves like," so polar substances have an affinity

Fundamentals of Water Pollution. https://doi.org/10.1016/B978-0-443-28987-3.00004-7

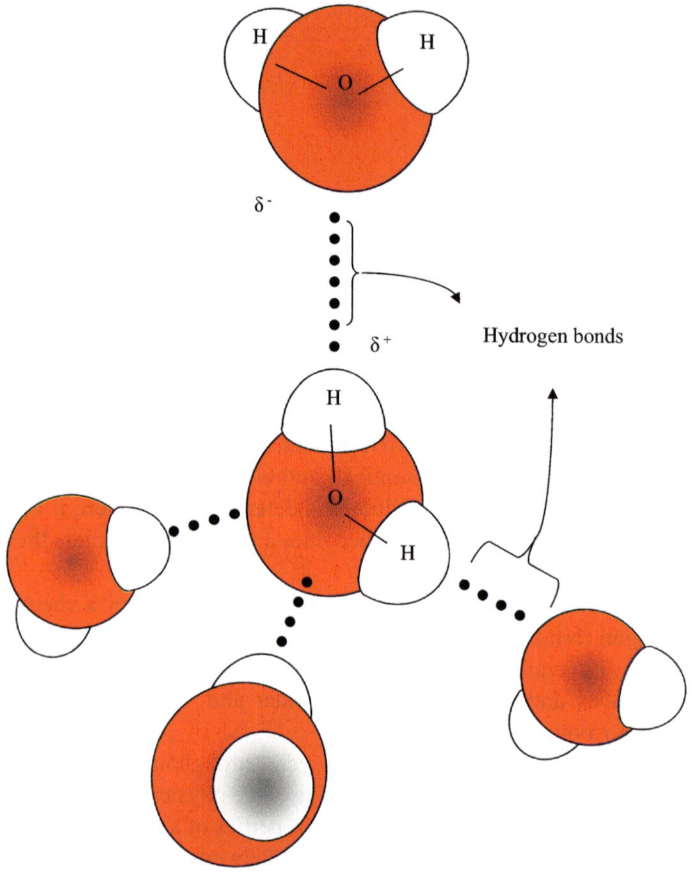

FIG. 4.1 Configuration of the water molecule, showing the electronegativity (δ) at each end. The hydrogen atoms form an angle of 105° with the oxygen atom. *(Data from D.A. Vallero, Fundamentals of Air Pollution, fifth ed., Elsevier Academic Press, Waltham, MA, 2014, p. 999 pages cm.)*

to become dissolved in water, and nonpolar substances resist being dissolved in water. The hydrogen atoms are at an angle of 105° with the oxygen atom. The asymmetry of the water molecule leads to a dipole moment in the symmetry plane toward the more positive hydrogen atoms, which induces the increase in volume during freezing.

The many physical and chemical processes make the original rock material suitable for microbial and root growth. Lichen, a symbiotic living complex between a fungus and an algae, and other simple plant systems further break down the rock and its unconsolidated materials, setting the stage for larger plant growth. As a result, the terrain hardly resembles the original rock formation.

The hydrosphere is the discontinuous stratum around the earth that holds its water in many forms, including ground water beneath the surface, surface water, water in soil and biota, and moisture in the atmosphere. The fluid properties of air and water combine in the hydrosphere, not only making weather, but to move and transform all types of chemical compounds, including pollutants. The atmosphere holds 0.0055% of the earth's total amount of water (0.04% of freshwater) compared to the 96.5% in the oceans (See Table 4.1). However, the surface waters, aquifers, soils, and the atmosphere account for much of water's activity on earth, from weather systems to nutrient cycling to pollutant transformation (e.g., hydrolysis), as well as pollutant transport, transformation, and deposition [3].

2.1 The hydrosphere

The life cycle of a pollutant mainly takes place in the hydrosphere, whether it is exchange of pollutants between ground and surface waters and the atmosphere, the storage of pollutants in soil water, or the release and the biota's uptake of pollutants after their release (See Fig. 4.3). For example, water allows microbial populations to grow and break down wastes in landfills and other facilities. Pollutants that are released from various sources can be transformed by hydrolysis and other chemical reactions in water, whether in receiving water bodies, soil, roots, or droplets suspended in the atmosphere.

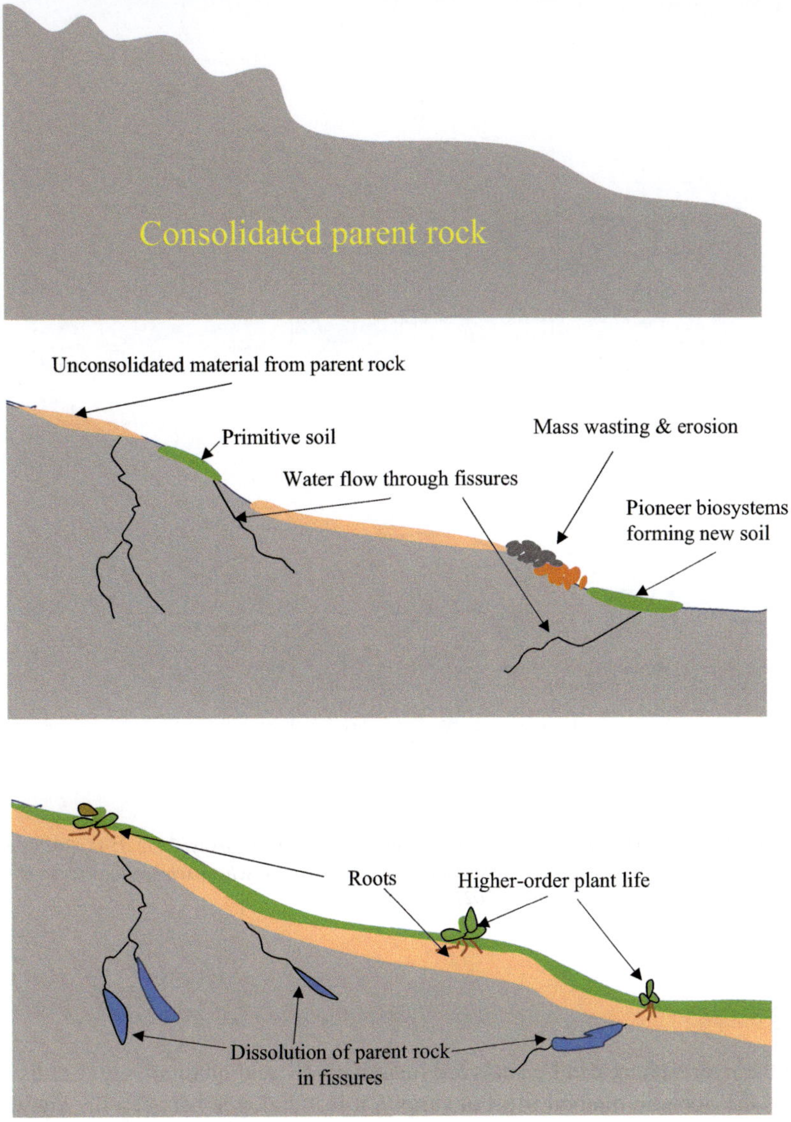

FIG. 4.2 Side view example of some of the many physical, chemical, and biological mechanisms that change terrain. (A) Initial rock formation. (B) Erosion and weather form pockets of unconsolidated material as matrices for early soils. Water flows into fissures and widens them by chemical reactions, e.g., carbonic acid in limestone. (C) Early microbial and plant root systems continue to deepen the soil horizons and to smooth the terrain.

TABLE 4.1 Water volume in the hydrosphere.

Water source	Water volume ($\times 1000\,km^3$)	Percent of freshwater	Percent of total water
Oceans, seas, & bays	1,338,000	–	96.5
Ice caps, glaciers, & permanent Snow	24,064	68.6	1.74
Ground water	23,400	–	1.7
Fresh	10,530	30.1	0.76
Saline	12,870	–	0.93
Ground Ice & permafrost	300	0.86	0.022
Lakes	176.4	–	0.013
Fresh	91	0.26	0.007
Saline	85.4	–	0.007
Soil moisture	16.5	0.05	0.001
Atmosphere	12.9	0.04	0.001
Swamp water	11.5	0.03	0.0008
Rivers	2.1	0.006	0.0002
Biological water	1.1	0.003	0.0001

Sources: P.H. Gleick, Water in Crisis: A Guide to the Worlds Fresh Water Resources, 1993 and U.S. Geological Survey, The World's Water, https://water.usgs.gov/edu/earthwherewater.html (Accessed 28 February 2018).

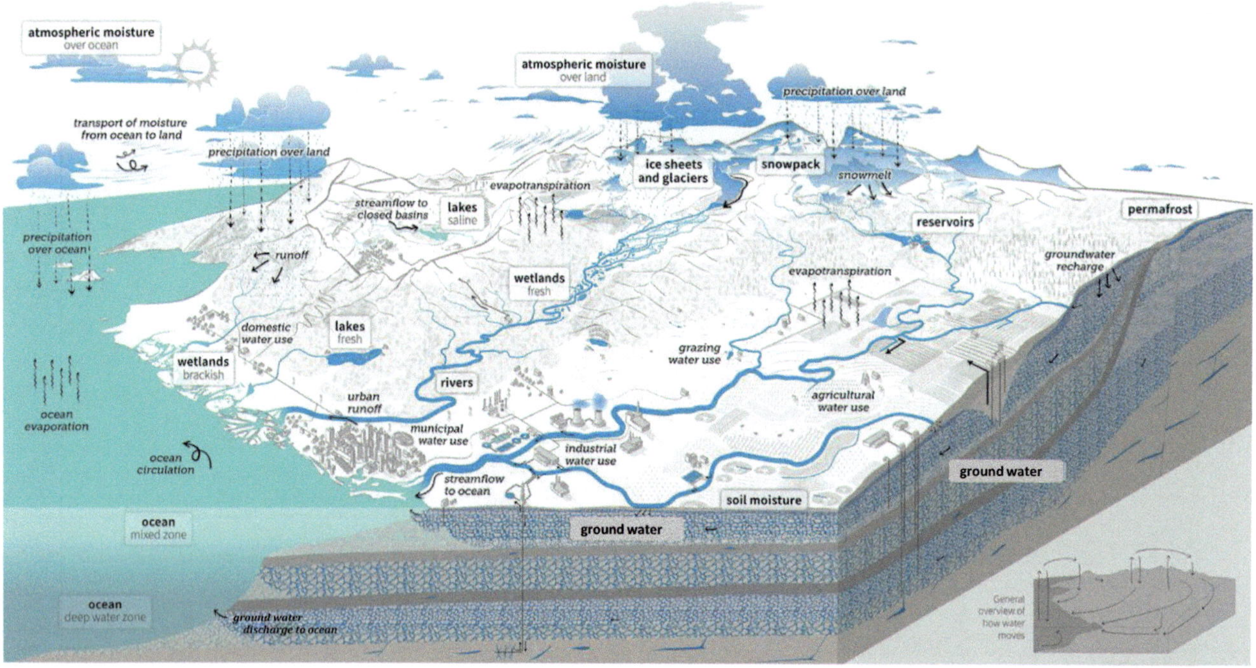

FIG. 4.3 The hydrological (water) cycle. *(Data from H.R. Corson-Dosch* et al., *The Water Cycle, US Geological Survey, 2332-354X, 2023 and U.S. Geological Survey. The Fundamentals of the Water Cycle, https://www.usgs.gov/special-topic/water-science-school/science/fundamentals-water-cycle?qt-science_center_objects=0#qt-science_center_objects (Accessed 4 November 2021).)*

3 Socrates, a biogeochemist?

The origin and fate of water pollutants in the hydrosphere can be analyzed quantitatively and qualitatively. Let us begin with the latter using an approach known as the Socratic method [4]. Put simply, it is a dialogue between the knowledge holder (KH) and the willing learner (WL) comprised of series of questions. The answer to each question leads to the subsequent question until, ideally, WL ultimately has as much knowledge about a topic as KH. The following hypothetical discussion between KH and WL is paraphrased from [4].

KH: Assume it has been raining steadily for 2 days in a nearly flat area covered mainly by wheat fields outside of Topeka, Kansas in July. The total 2-day rainfall event is 5 cm. Where does the water from the rainfall go?
WL: I assume most seeps into the soil.
KH: What assumption must you make to say that the 5 cm of rainfall soaks into the wheat fields?
WL: The ground's soil must have enough void spaces that are connected to allow the water to soak into it.
KH: Correct. This is known as permeability. The hydrogeological definition of permeability is the extent to which a material allows a fluid to pass through it. Incidentally, a fluid is any substance that flows, so it is either a liquid or gas.
WL: Okay, I say that the soil is sufficiently permeable to all the 5 cm of rainfall to pass through it.
KH: That may be, but, thinking about your answer, could the water have gone somewhere else in addition to the soil permeability? What about the atmosphere?
WL: Maybe.
KH: Right. Let us conduct a thought experiment. What if you put an uncovered bucket of water on a patio table in the sunny location on a July afternoon in Kansas for a few hours?
WL: Some of the water would evaporate.
KH: I agree. Do you think some of the water from the rainfall event could have evaporated before it reached the wheat field surface?
WL: That makes sense, so my answer is yes.

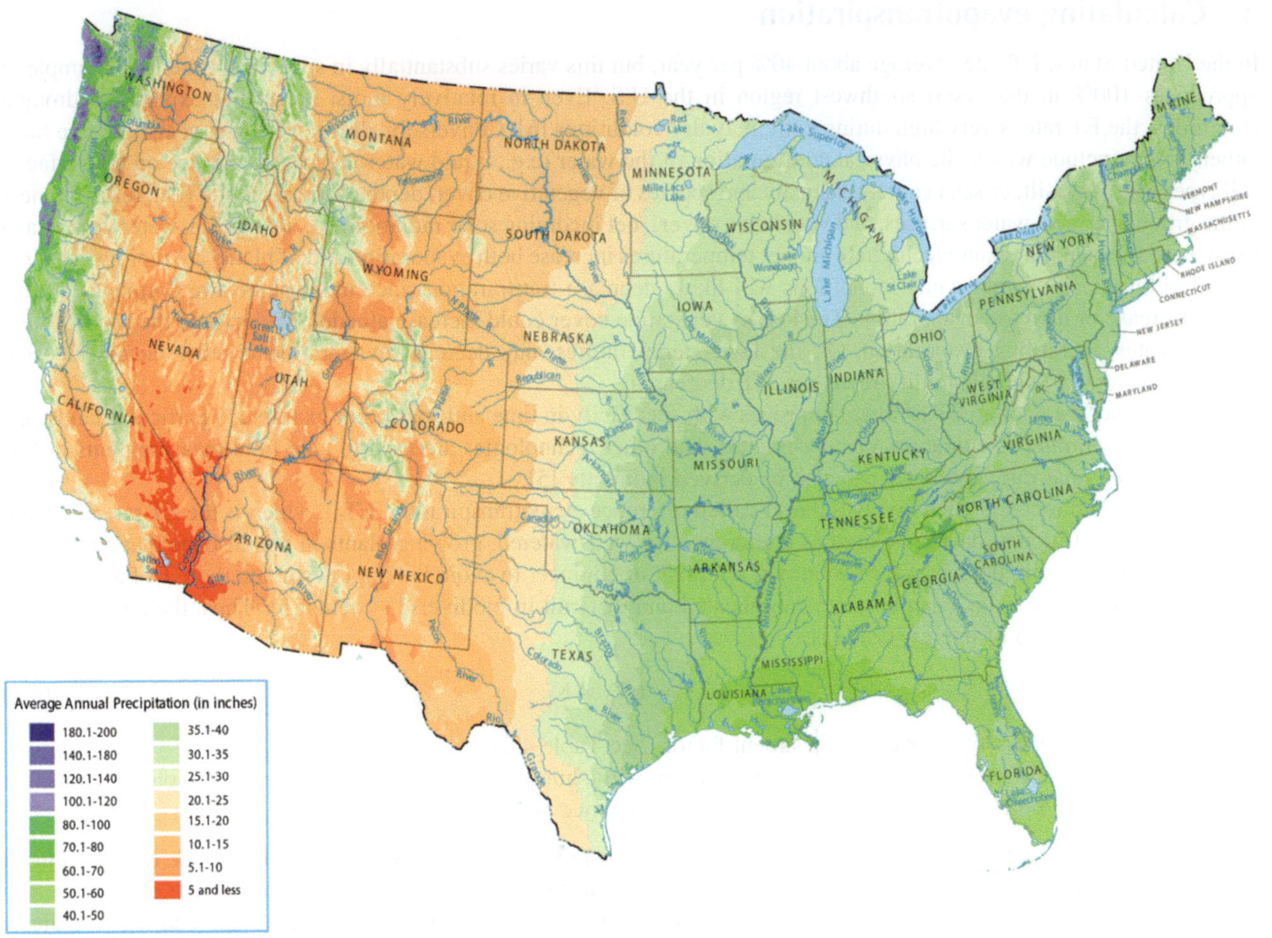

FIG. 4.4 Average annual precipitation in the United States. *(From GISGeography, US Precipitation Map, 2024, https://gisgeography.com/us-precipitation-map/ (Accessed 4 July 2024).)* Map lines delineate study areas and do not necessarily depict accepted national boundaries.

KH: Look at the annual rainfall map for the United States (Fig. 4.4). It shows that precipitation is much lower in the Great Plains than in the central or eastern U.S. Topeka is a light green region. What is average precipitation rate for Topeka?

WL: It seems to be about 40 in. per year.

KH: Yes, that is about right. If we take that number to represent the amount of water that will find its way to earth's surface, and we think about the bucket of water that evaporated, what can we say about the water cycle?

WL: That the amount of water that will soak into the ground depends on how much water falls from the atmosphere minus the amount that evaporates back to the atmosphere as water vapor.

KH: Correct. However, is evaporation the only way that water vapor returns to the atmosphere? Is there another mechanism that needs to be included in this equation for the available water to permeate the soil?

WL: I seem to remember from biology class that plants' leaves emit oxygen and water after photosynthesis and respiration.

KH: Excellent. This is known as transpiration, i.e., the loss of water from living plant surfaces, so the amount of ground covered by plant life and the type of plants also affect the amount of water returned to the atmosphere. In fact, scientists usually combine evaporation and transpiration as the total loss of water emitted to the atmosphere. This is called evapotranspiration (ET). To meet their physiological and metabolic needs, plants absorb water from the soil through their roots. The water eventually is released to the atmosphere as vapor via the minute, pore-like structures that open and close on the surfaces of leaves. These are known as stomata. The uptake of water at the roots and movement of water through plant tissues results in the transpiration [5].

WL: Thank you. I believe that I am ready to calculate ET.

4 Calculating evapotranspiration

In the United States, ET rates average about 40% per year, but this varies substantially in space and time. For example, it approaches 100% in the desert southwest region in the U.S. Even in relatively moist areas that experience drought conditions, the ET rate is very high during extremely dry conditions [6]. Conversely, during very wet years, the rate falls. Other factors include winds, the physical configuration of the water (e.g., liquid water versus snow or ice on the surface), soil type (e.g., clay, silt, or sand content), density and species of vegetative cover, depth of roots, albedo (i.e., reflectiveness of the ground surface, water surfaces, and vegetative cover), net incoming solar radiation, i.e., insolation, water versus land surface, and season. For example, higher summer temperatures increase both evaporation and, if plants are rapidly growing, transpiration rates, whereas winter conditions will likely decrease transpiration since more plants are dormant but may either decrease or increase evaporation. The loss of vegetative cover could increase albedo and decrease the roots' ability to hold water, increasing evaporation. Or, the lower temperatures may decrease evaporation if other factors are held constant. Generally, when humidity increases, ET decreases.

Thus, the accuracy of an ET estimate improves with specificity in time and space. For example, a particular landscape will have its own rate. Evaporation pans, lysimeters, and other technologies are used to take field measurements of ET. These yield the total rate. Estimates can also be derived indirectly [5].

A common indirect approach involves the so-called reference evapotranspiration (ET_0), which is the ET rate of a plant species that has sufficient data to be used as a reference for well-watered, growing plants in a landscape that completely shade the soil. By using the ET_0 allows a farmer, scientist, or engineer to avoid having to estimate the ET of all the plant species in a landscape, which can be quite complicated and difficult in biodiverse systems [5]. Thus, the equation for a landscape plant or crop of interest (ET_L) is:

$$ET_L = ET_0 \times K_L \tag{4.1}$$

Where, K_L is the specific plant or crop adjustment factor (See Table 4.2).

The K_{MC} value represents the site-specific conditions of wind, humidity, temperature, insolation, and other factors that would differ from reported conditions, e.g., on maps and reference tables [5]:

- Low K_{MC} values (0.5–0.9) occur if the area is shaded for most of the afternoon, e.g., the northeast side of a building in the Northern Hemisphere. Also, the K_{MC} would be low if substantially protected by wind.
- Midrange K_{MC} value (1.0) occurs if conditions are about the same as reported, e.g., typical wind and temperature.
- High value K_{MC} (1.1–1.4) occurs when winds are higher than expected, e.g., the so-called wind tunnel effect caused by hills or buildings, where there are low albedo and high heat absorption surfaces, e.g., paved parking lots on the south side of buildings in the Northern Hemisphere, or conversely, in high albedo areas.

The K_L is affected by the vegetative density and microclimate, so the factor can be expanded:

$$K_L = K_S \times K_D \times K_{MC} \tag{4.2}$$

Where, K_S is the species factor, K_D is the vegetative density factor, and K_{MC} is the specific microclimate adjustment factor.

The K_S value reflects the variation in the amount of water needed by different plant species. Examples of species factors are given in Table 4.3. Note that the values differ substantially from the two sources [7,8], which is not surprising since they are both based on field observations. It emphasizes how even seemingly small differences during the observation can lead to large differences in K_S. This is an example of the importance of variable sensitivity to an outcome. When running a model, for example, a water quality model, sensitivity analyses must be conducted to determine which of the variables has the greater influence on the modeled result [9,10].

K_D considers the collective areas of leaves that affect water loss, i.e., decreasing leaf cover lessens water loss [5]:

- Low K_D values (0.5–0.9) occur in sparsely planted areas, with only on plant category, e.g., only shrubs or turf. The tree canopy shades less than 70% of the ground, or for shrubs and ground covers less than 90% of the ground is covered.
- Midrange K_D value (1.0) occurs if characteristics indicating the need for adjustment downward are counterbalanced by characteristics indicating the need for adjustment upward. For example, if only trees, the canopy shades 70% to 100% of the ground. If only shrubs and ground covers, the canopy shades 90% to 100% of the ground. If the canopy is mixed, the plant types are dominated by trees shading 70% to 100% or shrubs and ground covers together shading 90% to 100% of the ground.
- High value K_D (1.1–1.3) occurs for densely planted cover, with two or more plant types. Tree canopy shades 70% to 100% of the ground. For shrubs and ground covers, 90% to 100% of the ground is shaded.

TABLE 4.2 Various plant adjustment factors (K_L) used in calculating evapotranspiration rates.

Plant species	K_L
Arbutus unedo "Compacta" (compact strawberry tree)	0.18–0.36
Arctostaphylos uva-ursi "Pacific Mist" (bearberry)	0.18–0.36
Artemisia "Powis Castle" (workwood)	0–0.36
Baccharis pilularis "Twin Peaks" (Twin Peaks coyote bush)	0.20
Calliandra haematocephala (pink powder puff)	0.18–0.36
Cassia artemisioides (feathery cassia)	0–0.36
Cerastium tomentosum	0.25
Cistus x purpureus (orchid spot rock rose)	0–0.36
Correa alba "Ivory Bells" (white Australian correa)	0.18–0.36
Drosanthemum hispidum (pink iceplant)	0.20
Echium fastuosum (pride of Madeira)	0–0.36
Escallonia x exoniensis "Fradessii" (Frades escallonia)	0.18–0.36
Ficus microcarpa nitida	0.20
Galvezia speciosa (bush snapdragon)	0–0.36
Gazania hybrida	0.25–0.50
Gazania rigens v. leucolaena "Yellow Cascade" (trailing gazania)	0.50–0.80
Grevillea *"Noelli"*	0–0.36
Hedera helix "Needlepoint" (Needlepoint English ivy)	0.20–0.30
Heteromeles arbutifolia (toyon)	0–0.36
Hibiscua rosa-sinensis (rose of China)	0.40–0.30
Lantana montevidensis (trailing lantana)	0.18–0.36
Leptospermum scoparium (New Zealand tea tree)	0.18–0.36
Leucophyllum frutescens "Green Cloud" (Texas ranger)	0–0.36
Ligustrum japonicum "Texanum" (Texas privet)	0.40–0.60
Liquidambar styraciflua	0.20
Myoporum "Pacificum" (prostrate myoporum)	0–0.36
Otatea acuminata (Mexican bamboo)	0.18–0.36
Phormium tenax (New Zealand flax)	0.18–0.36
Pittosporum tobira (mock orange)	0.18–0.36
Potentilla tabernaemontani (spring cinquefoil)	0.70–0.80
Prunus caroliniana (carolina laurel cherry)	0–0.36
Pyracantha koidzumii "Santa Cruz" (firethorn)	0–0.36
Quercus ilex	0.20
Rhaphiolepis indica (Indian hawthorn)	0.18–0.36
Sedum acre	0.25
Teucrium chamaedrys (germander)	0.18–0.36
Vinca major (periwinkle; myrtle)	0.30–0.40
Westringia rosamarinaformis (rosemary bush)	0.18–0.36
Xylosma congestum (shiny xylosma)	0.18–0.36

Data from WateResuse Foundation, Salinity Management Guide, https://watereuse.org/salinity-management/ew/ew_2.html (Accessed 11 December 2023).

TABLE 4.3 Observation-based species factor (K_S) for selected trees in Southern California, United States.

	Factor (K_S) for South coast		Factor (K_S) for South inland	
Plant species	**Source: [7]**	**Source: [8]**	**Source: [7]**	**Source: [8]**
Acacia farnesiana (sweet acacia)	0.25	0.10–0.30	0.25	0.10–0.30
Acacia pendula (weeping myall)	0.25	0.40–0.60	0.50	0.40–0.60
Acacia salicina (willow acacia)	0.25	0.10–0.30	0.50	0.40–0.60
Acacia smallii	0.25	0–0.10	0.25	0.10–0.30
Acacia stenophylla (shoestring acacia)	0.25	0.10–0.30	0.25	0.10–0.30
Aesculus californica (California buckeye)	0–0.25	0–0.10	0–0.25	0.10–0.30
Agonis flexuosa (peppermint tree)	0.50	0.10–0.30	0.50	0.40–0.60
Albizia julibrissin (silk tree)	0.25	0.40–0.60	0.50	0.40–0.60
Angophora cordifolia or *cordistata* (gum myrtle)	0.25	0.10–0.30	0.25–0.50	0.40–0.60
Arbutus unedo (strawberry tree)	0.25	0.10–0.30	0.50	0.10–0.30
Brachychiton populneus (Kurrajong bottle tree)	0.25	0.10–0.30	0.50	0.10–0.30
Callistemon citrinus (lemon bottlebrush)	0.25	0.10–0.30	0.50	0.10–0.30
Callistemon viminalis (weeping bottlebrush)	0.25	0.40–0.60	0.50	0.40–0.60
Calocedrus decurrens (incense cedar)	0.50	0.40–0.60	0.50	0.40–0.60
Casuarina cunninghamiana (river she-oak)	0.25	0.10–0.30	0.50	0.10–0.30
Casuarina stricta (drooping she-oak), a.k.a. *Allocasuarina verticillata* (coast beefwood)	0–0.25	0.10–0.30	0.25	0.10–0.30
Cedrus atlantica (atlas cedar)	0.50	0.10–0.30	0.50	0.40–0.60
Cedrus deodara (deodar cedar)	0.50	0.10–0.30	0.50	0.40–0.60
Ceratonia siliqua (carob)	0–0.25	0.10–0.30	0–0.25	0.10–0.30
Cercis canadensis (eastern redbud)	0.50	0.40–0.60	0.50–0.75	0.40–0.60
Cercis occidentalis (western redbud)	0.25	0.10–0.30	0.50	0.10–0.30
Chilopsis linearis (desert willow)	0.25	0–0.10	0.25	0.10–0.30
Chitalpa tashkentensis (chitalpa)	0.25	0.10–0.30	0.50	0.10–0.30
Cupressus arizonica (Arizona cypress)	0.25	0–0.10	0.25	0.10–0.30
Cupressus sempervirens (Italian cypress)	0.25	0.10–0.30	0.25	0.10–0.30
Dodonaea viscosa (hopseed bush)	0.25–0.50	0.10–0.30	0.25–0.50	0.40–0.60
Dracaena draco (dragon tree)	0.25	0–0.10	0.25	0.10–0.30
Eucalyptus cinerea (argyle apple, ash leaved gum, silver dollar tree)	0–0.25	0.10–0.30	0.25–0.50	0.40–0.60
Eucalyptus citriodora (lemon-scented gum)	0.25	0.10–0.30	0.50	0.40–0.60
Eucalyptus cladocalyx (sugar gum)	0–0.25	0.10–0.30	0.25	0.10–0.30
Eucalyptus erythrocorys (red cap gum)	0.25	0.40–0.60	0.50	0.40–0.60
Eucalyptus ficifolia (red flowering gum)	0–0.25	0.40–0.60	0.25–0.50	0.40–0.60
Eucalyptus lehmannii (Lehmann's Mallee, bushy yate)	0–0.25	0.10–0.30	0.25–0.50	0.10–0.30
Eucalyptus leucoxylon (white ironbark)	0–0.25	0.10–0.30	0.25–0.50	0.10–0.30

TABLE 4.3 Observation-based species factor (K_S) for selected trees in Southern California, United States—cont'd

Plant species	Factor (K_S) for South coast		Factor (K_S) for South inland	
	Source: [7]	Source: [8]	Source: [7]	Source: [8]
Eucalyptus maculata (spotted gum)	0–0.25	0.40–0.60	0.25–0.50	0.40–0.60
Eucalyptus microtheca (coolihan)	0–0.25	0.10–0.30	0.25	0.10–0.30
Eucalyptus nicholii (willow peppermint, Nichol's willow leaf peppermint)	0–0.25	0.40–0.60	0.25–0.50	0.40–0.60
Eucalyptus polyanthemos (silver dollar gum)	0–0.25	0.10–0.30	0.25–0.50	0.10–0.30
Eucalyptus rudis (flooded gum)	0–0.25	0.10–0.30	0.25–0.50	0.10–0.30
Eucalyptus sideroxylon (red ironbark)	0–0.25	0.10–0.30	0.25–0.50	0.10–0.30
Eucalyptus spathulata (narrow-leaved gimlet, swamp mallee)	0.25	0.10–0.30	0.25–0.50	0.10–0.30
Eucalyptus torquata (coral gum)	0.25	0.10–0.30	0.25–0.50	0.40–0.60
Eucalyptus viminalis (manna gum)	0.25	0.10–0.30	0.25–0.50	0.40–0.60
Feijoa sellowiana (pineapple guava), a.k.a. *Acca sellowiana*	0.25	0.10–0.30	0.25–0.50	0.40–0.60
Geijera parviflora (Australian willow)	0.25	0.10–0.30	0.25–0.50	0.40–0.60
Grevillea robusta (silky oak, silk oak)	0.25	0.10–0.30	0.25–0.50	0.40–0.60
Juglans californica (Southern California black walnut)	0–0.25	0.10–0.30	0–0.25	0.10–0.30
Lagerstroemia indica (crape myrtle)	0.25	0.40–0.60	0.25–0.50	0.40–0.60
Laurus nobilis (sweet bay)	0.25	0.10–0.30	0.25–0.50	0.10–0.30
Leptospermum laevigatum (Australian tea tree)	0–0.25	0.10–0.30	0.25	0.10–0.30
Leptospermum scoparium (New Zealand tea tree)	0.25–0.50	0.40–0.60	0.50	0.40–0.60
Lyonothamnus floribundus (Catalina ironwood)	0–0.25	0–0.10	0.25–0.50	0.10–0.30
Melaleuca armillaris (drooping melaleuca, bracelet honey myrtle)	0.25	0.10–0.30	0.25–0.50	0.40–0.60
Melaleuca linariifolia (flaxlead paperbark)	0.25	0.10–0.30	0.50	0.10–0.30
Melia azedarach (chinaberry)	0–0.25	0–0.10	0–0.25	0.10–0.30
Metrosideros excelsus (New Zealand Christmas tree)	0.25–0.50	0.40–0.60	0.50	0.40–0.60
Morus alba (white mulberry)	0–0.25	0.40–0.60	0–0.25	0.40–0.60
Myrica californica (Pacific wax myrtle)	0.25–0.50	0.10–0.30	0.50	0.40–0.60
Olea europaea (olive)	0–0.25	0.10–0.30	0.25	0.10–0.30
Parkinsonia aculeata (Mexican palo verde, Jerusalem thorn)	0.25	0.10–0.30	0.25–0.50	0.10–0.30
Pinus brutia (Calabrian pine)	0.25	0.10–0.30	0.25–0.50	0.10–0.30
Pinus canariensis (Canary Island pine)	0.25	0.10–0.30	0.25–0.50	0.40–0.60
Pinus coulteri (Coulter pine)	0.25	0.10–0.30	0.25	0.10–0.30
Pinus eldarica (Afghan pine), a.k.a. *Pinus brutia* species *eldarica* (eldarica pine)	0.25	0.10–0.30	0.25–0.50	0.10–0.30
Pinus halepensis (Aleppo pine)	0–0.25	0.10–0.30	0–0.25	0.10–0.30
Pinus sabiniana (Digger pine, foothill/Gray pine)	0.25	0–0.10	0.25	0.10–0.30
Pinus torreyana (Torrey pine)	0–0.25	0.10–0.30	0.25–0.50	0.40–0.60
Pittosporum phillyraeoides (willow pittosporum)	0.25	0.10–0.30	0.25–0.50	0.10–0.30

Continued

TABLE 4.3 Observation-based species factor (K_S) for selected trees in Southern California, United States—cont'd

	Factor (K_S) for South coast		Factor (K_S) for South inland	
Plant species	**Source: [7]**	**Source: [8]**	**Source: [7]**	**Source: [8]**
Planatus racemosa (Western sycamore, California sycamore)	0.50	0.40–0.60	0.50–0.75	0.40–0.60
Prunus caroliniana (Carolina laurel cherry)	0.25	0.40–0.60	0.25–0.50	0.40–0.60
Prunus lyonii (Catalina cherry)	0–0.25	0.10–0.30	0.25–0.50	0.10–0.30
Punica granatum (pomegranate)	0.25	0.40–0.60	0.25–0.50	0.40–0.60
Quercus agrifolia (coast live oak)	0–0.25	0.10–0.30	0–0.25	0.10–0.30
Quercus douglasii (blue oak)	0–0.25	0–0.10	0–0.25	0.10–0.30
Quercus engelmannii (mesa oak)	0–0.25	0.10–0.30	0–0.25	0.10–0.30
Quercus ilex (holly oak)	0.25	0.10–0.30	0.25–0.50	0.10–0.30
Quercus suber (cork oak)	0.25	0.10–0.30	0.25	0.10–0.30
Rhus lancea (African sumac)	0.25	0.10–0.30	0.25–0.50	0.10–0.30
Robinia ambigua (locust)	0.25	0.10–0.30	0.25–0.50	0.10–0.30
Sambucus (elderberry)	0–0.25	0.10–0.30	0–0.25	0.10–0.30
Schinus polygamus (Peruvian pepper)	0.25	0.10–0.30	0.25	0.10–0.30
Tristania conferta (Brisbane box), a.k.a. *Lophostemon confertus*	0.50	0.40–0.60	0.50–0.75	0.40–0.60
Vitex angus-castus (chaste tree)	0.25	0.10–0.30	0.25–0.50	0.40–0.60
Xylosma congestum (shiny xylosma)	0.25–0.50	0.40–0.60	0.25–0.50	0.40–0.60

Data from WateResuse Foundation, Salinity Management Guide, https://watereuse.org/salinity-management/ew/ew_2.html (Accessed 11 December 2023).

Evapotranspiration example

An ecosystem in the South Coast region of California has two dominant species, the chaste tree and Catalina cherry tree. It is also in an area that experiences very high winds, The tree canopy covers only 30% or the ground. Estimate the evapotranspiration expected for this landscape in July.

Solution: From Table 4.3, the K_L for the chaste tree ranges between 0.10 and 0.30 and the cherry tree between 0 and 0.3. Thus, the K_L range for the two trees would be 0 to 0.36. There is no need for a reference value if the K_L is 0:

$$ET_L = ET_0 \times K_L$$
$$ET_L = ET_0 \times 0 \tag{4.3}$$
$$ET_L = 0$$

In this case, there is no ET. The rainfall is completely taken up by the soil and roots.

At the high end, K_L is 0.36. Values for ET_0 are available for various climates. In this case we find the South Coast ET_0 in July to be 7.44 in. (Zone 9 in Table 4.4). Therefore:

$$ET_L = ET_0 \times 0.36$$
$$ET_L = 7.44 \times 0.36 \approx 2.7\text{ in.}$$

If we were to calculate K_L from Eq. (4.2), the microenvironment appears to be highly important given the winds, so we can assign K_{MC} to equal to 1.4. Conversely, the vegetative density factor is low, so we can assign a K_D equal to 0.5. There is no K_S value for the chaste tree in Table 4.2, and the cherry tree ranges from 0 to 0.36. Plugging these values into Eq. (4.2), the K_L:

$$K_L = K_S \times K_D \times K_{MC} \tag{4.3}$$
$$\text{If } K_S = 0, \text{then } K_L = 0 \times K_D \times K_{MC} = 0, \text{which means ET} = 0$$
$$\text{If } K_S = 0.36, \text{then } K_L = 0.36 \times 0.5 \times 1.4 \approx 0.25$$
$$\text{And, } ET_L = ET_0 \times K_L$$
$$ET_L = 7.44 \times 0.25 \approx 1.9\text{ in.}$$

Thus, from these calculations, we can estimate that the evapotranspiration will range from 0 to 2.7 in., depending on available data and the calculation method employed.

TABLE 4.4 California monthly average reference evapotranspiration by ET0 zone (inches/month).

Zone	Jan	Feb	Mar	Apr	May	Jun	Jul	Aug	Sep	Oct	Nov	Dec	Total
1	0.93	1.40	2.48	3.30	4.03	4.50	4.65	4.03	3.30	2.48	1.20	0.62	32.9
2	1.24	1.68	3.10	3.90	4.65	5.10	4.96	4.65	3.90	2.79	1.80	1.24	39.0
3	1.86	2.24	3.72	4.80	5.27	5.70	5.58	5.27	4.20	3.41	2.40	1.86	46.3
4	1.86	2.24	3.41	4.50	5.27	5.70	5.89	5.58	4.50	3.41	2.40	1.86	46.6
5	0.93	1.68	2.79	4.20	5.58	6.30	6.51	5.89	4.50	3.10	1.50	0.93	43.9
6	1.86	2.24	3.41	4.80	5.58	6.30	6.51	6.20	4.80	3.72	2.40	1.86	49.7
7	0.62	1.40	2.48	3.90	5.27	6.30	7.44	6.51	4.80	2.79	1.20	0.62	43.3
8	1.24	1.68	3.41	4.80	6.20	6.90	7.44	6.51	5.10	3.41	1.80	0.93	49.4
9	2.17	2.80	4.03	5.10	5.89	6.60	7.44	6.82	5.70	4.03	2.70	1.86	55.1
10	0.93	1.68	3.10	4.50	5.89	7.20	8.06	7.13	5.10	3.10	1.50	0.93	49.1
11	1.55	2.24	3.10	4.50	5.89	7.20	8.06	7.44	5.70	3.72	2.10	1.55	53.1
12	1.24	1.96	3.41	5.10	6.82	7.80	8.06	7.13	5.40	3.72	1.80	0.93	53.4
13	1.24	1.96	3.10	4.80	6.51	7.80	8.99	7.75	5.70	3.72	1.80	0.93	54.3
14	1.55	2.24	3.72	5.10	6.82	7.80	8.68	7.75	5.70	4.03	2.10	1.55	57.0
15	1.24	2.24	3.72	5.70	7.44	8.10	8.68	7.75	5.70	4.03	2.10	1.24	57.9
16	1.55	2.52	4.03	5.70	7.75	8.70	9.30	8.37	6.30	4.34	2.40	1.55	62.5
17	1.86	2.80	4.65	6.00	8.06	9.00	9.92	8.68	6.60	4.34	2.70	1.86	66.5
18	2.48	3.36	5.27	6.90	8.68	9.60	9.61	8.68	6.90	4.96	3.00	2.17	71.6

Data from California Irrigation Management Information System, Reference Evaporation Zones, https://wwwcimis.water.ca.gov/Content/pdf/CimisRefEvapZones.pdf (Accessed 12 December 2023).

5 Biogeochemical aspects of water

Hydrospheric processes help to degrade the pollutants but can also render some pollutants even more toxic. Aerosols can react and be removed from the atmosphere before being deposited by several physical and chemical processes. Once deposited, the particles and their chemical constituents can move into various water systems, during which chemical reactions occur. Both homogeneous and heterogeneous reactions occur in environmental systems. If the reaction only occurs in one phase, e.g., all substances are in the liquid phase, the reaction is homogeneous. If the reaction involves an interface with a surface of a sediment or soil particle in water, or an aerosol, or within a liquid droplet, it is heterogeneous. Certain chemical species of sulfur (S) and N have sufficient aqueous solubility to be dissolved by water. Thus, the liquid-phase compounds are acidic when S, N, and C compounds become dissolved in water, producing increased concentrations of H_2SO_4, H_2NO_3, H_2CO_3* and other acids. The movement of these acidic waters can affect pH levels throughout the environment. For example, an acidic water droplet that reaches the earth's surface can lead to acidic conditions in soils and surface waters (See Fig. 4.5)

The total amount of water in the hydrosphere is constant, but its location and quality vary in time and space, given variations in weather, climate, and other hydrological conditions. Mean global and local temperature increases due to increased atmospheric concentrations of global greenhouse gases should be factored into scientific studies, designs, and facility citing decisions [11,12]. Water systems are affected by changes in hydrological and hydraulic conditions, e.g., the salt water wedge encroaching inward, as discussed in Chapters 10 and 11. The density difference between fresh and saltwater can be a slowly unfolding disaster for the health of people living in coastal communities and for marine and estuarine ecosystems. Salt water contains a significantly greater mass of ions than does freshwater (See Table 4.5). The denser saline water can wedge beneath freshwaters and pollute surface waters and ground water. This phenomenon, known as saltwater intrusion, can significantly alter an ecosystem's structure and function, and threaten freshwater organisms. It can also pose a huge challenge to coastal communities who depend on aquifers for their water supply. The freshwater Everglades, for example, recharge the Biscayne aquifer, a natural underground layer that collects water. The Biscayne aquifer is the primary water supply to the Florida Keys. Thus, contamination can result in the loss of a crucial water supply.

The effect in sensitive coastal habitats like those in South Florida would be exacerbated by the rising sea level, which could submerge low-lying areas of the Everglades and other aquifers and ecosystems in southern Florida. Although a small part of the aquifer is beneath salty mangrove area, most of it is recharged by the freshwater Everglades, rendering the area

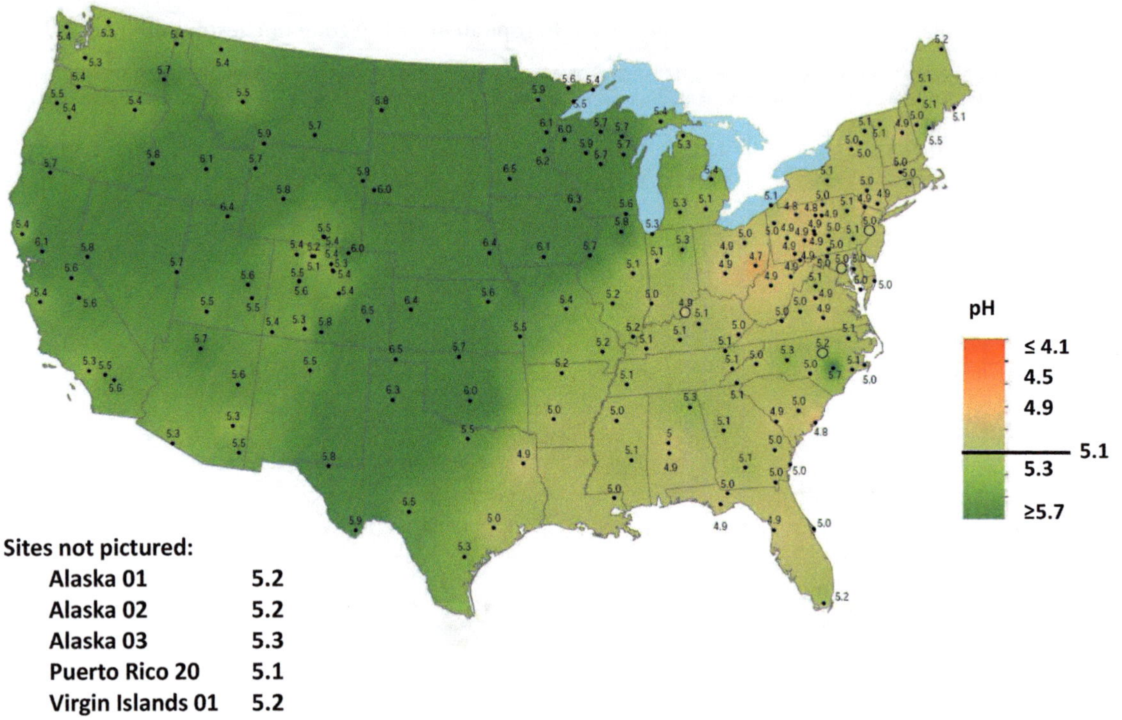

FIG. 4.5 Precipitation-weighted mean pH in wet deposition for the year 2012. *(Source: Illinois State Water Survey, University of Illinois at Urbana-Champaign, NADP Data Report 2013–01, National Atmospheric Deposition Program: 2012 Annual Summary, 2013.)* Map lines delineate study areas and do not necessarily depict accepted national boundaries.

TABLE 4.5 Ionic composition classifications of freshwaters and marine waters. M = molarity (moles per liter solution).

Composition	River water	Salt water
pH	6–8	8
Ca^{2+}	$4\times10^{-5}M$	$1\times10^{-2}M$
Cl^-	$2\times10^{-4}M$	$6\times10^{-1}M$
HCO_3^-	$1\times10^{-4}M$	$2\times10^{-3}M$
K^+	$6\times10^{-5}M$	$1\times10^{-2}M$
Mg^{2+}	$2\times10^{-4}M$	$5\times10^{-2}M$
Na^+	$4\times10^{-4}M$	$5\times10^{-1}M$
SO_4^{2-}	$1\times10^{-4}M$	$3\times10^{-2}M$

Data from D. Vallero, Environmental Biotechnology: A Biosystems Approach, Elsevier Science, 2015, P.M. Gschwend, Environmental Organic Chemistry, John Wiley & Sons, 2016, J.P. Kim, K.A. Hunter, M.R. Reid, Factors influencing the inorganic speciation of trace metal cations in fresh waters, Mar. Freshw. Res. 50(4) (1999) 367–372, and R.P. Schwarzenbach, P.M. Gschwend, D.M. Imboden, Organic acids and bases: acidity constant and partitioning behavior, in: Environmental Organic Chemistry, 1993, pp. 245–274.

vulnerable to saltwater intrusion and increased salinity of both surface and ground water sources of drinking water [13]. The rising sea water could force salty waters upstream into coastal areas, thus threatening surface water supplies. Furthermore, any waste disposal practices in these sensitive areas would be sources of additional contaminants. Similar problems could also occur in Northeastern U.S. aquifers that are recharged by fresh portions of streams that are vulnerable to increased salinity during severe droughts [11]. Indeed, similar ground water-surface water–sea water interactions could be affected across the globe. Unfortunately, planning of water and wastewater facilities often lacks the methods and tools to make

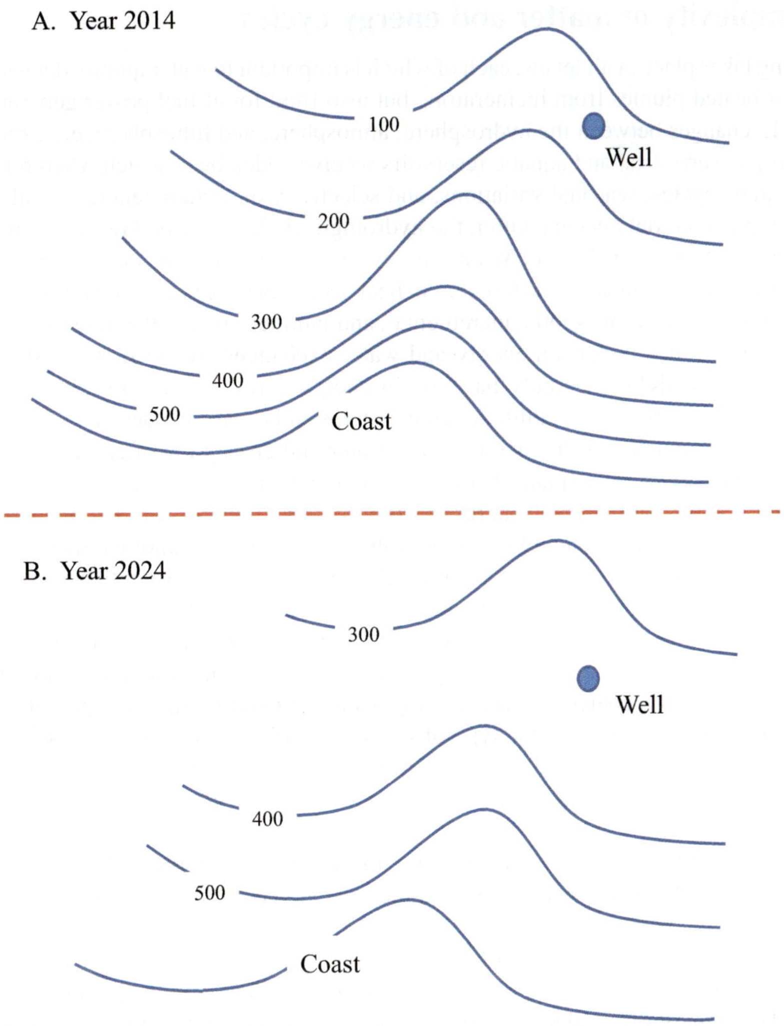

FIG. 4.6 Hypothetical isopleths of total dissolved solids ($mg\,L^{-1}$) concentrations in ground water (e.g., 30m depth). The increases over the decade indicate saltwater intrusion and potential contamination at the drinking water well site. *(From D. Vallero, Environmental Biotechnology: A Biosystems Approach. Elsevier Science, 2015.)*

decisions to optimize siting alternatives and to avoid these problems, although geographic information systems (GIS) and other tools are increasingly assisting in these efforts [14].

Salinity of water is a relative term. The values in Table 4.5 are at best averages and target concentrations. Salinity is the total concentration of dissolved solids (TDS). Indeed, in the United States, the drinking water standard for TDS is a "secondary" or nuisance standard, $500\,mg\,L^{-1}$. Note that the classifications using two dominant ions, Na^{+} and Cl^{-}, differ by three orders of magnitude between "fresh" and "saline" waters. Consider the hypothetical example in Fig. 4.6 of these ionic strengths, indicated by TDS, before and after saltwater intrusion. Although the water near the well is not "saltwater," nor does it currently violate the secondary drinking water standard, the increase in ionic concentrations is a cause for concern and a warning that the trend is likely to be toward even higher salinity in the well water.

The previous discussion illustrates how an indirect effect like atmospheric warming from emissions of greenhouse gases can lead to hydrospheric impacts like warming and sea level rise can worsen freshwater contamination. Adding another potential stressor like runoff and leaching from a municipal landfill further exacerbates and complicates this pollution. The interconnectedness of the atmosphere, hydrosphere and biosphere are complex and extensive [12]. A seemingly minor change in one small part of the spheres can lead to unforeseen and potentially disastrous outcomes. Thus, hydrologic cycling specifically and biogeochemical cycling generally are sensitive to initial conditions and, as such, must be treated as chaotic systems [15,16].

6 Scale and complexity of matter and energy cycles

Matter and energy cycling takes place at all levels, each of which is important to water quality decisions. Energy cycling can begin with the release of heated plumes from incinerators, but also from fossil fuel power generation plants, composting facilities, and landfills. Exchanges between the hydrosphere, atmosphere, and lithosphere, e.g., soil and sediment, creates sinks and sources of energy. Terrestrial and aquatic reservoirs receive added heat, which when released can alter habitats (e.g., changes to freeze-thaw cycles, seasonal variations, and selectivity of certain genera of soil bacteria).

Mass and energy cycling commonly occurs within the hydrologic cycle. The first law of thermodynamics dictates that these cycles be both mass and energy balanced. Water-energy relationships can be obvious and direct, like the thermal inversions that cause urban air pollution due to differences in heat energy in water vapor at various layers in the troposphere. Some water-energy relationships are more subtle, incremental, and indirect, such as the transfer of energy between trophic states in an ecosystem, which relies in part on energy and water exchanges among biotic tissue and abiotic substrates, e.g., aqueous-phase phosphorous [K] compounds that transfer energy during photosynthesis.

Water-energy cycling is also important in anthropogenic systems and is part of a design of pollution control equipment selection and application. For example, at the facility-scale, water and energy are often addressed together, such as the management of water and energy at a power plant. Managing water and managing energy occur together. Mismanagement of water and energy systems at the facility-level can result from improper tradeoffs or ignorance of what happens after the water and energy move from one compartment, like cooling water in the facility, to another, like a stream or the atmosphere. Again, the first law of thermodynamics requires, for example, that allowing heated water to be released in any amount, even the permitted level, would increase the overall temperature of the receiving waters.

The heat from boilers and other industrial operations will be exchanged intercompartmentally. Facility design determines in large part where the energy goes (See Fig. 4.10). In fact, a pollution control and heat control design can directly affect dissolved oxygen (DO) content of the receiving water temperature since temperature is directly proportional to DO content. The DO is a limiting factor of the type of fish communities that can be supported by a water body (See Tables 4.6 and 4.7). The resulting net increase in heat may directly stress the biotic integrity of a surface water ecosystem, e.g., fish species vary in their ability to tolerate higher temperatures, meaning that the less tolerant, higher value fish will be inordinately threatened.

Increasing surface water temperature can also increase the aqueous solubility of substances toxic to organisms. Concentrations of mercury and other toxic metals, for example, will increase with increasing temperatures. The depressed DO concentrations will affect redox conditions, i.e., a reduced environment in which metal sulfides and other compounds are formed. These reduced compounds are notoriously toxic to the aquatic life and can induce elevated toxic metal concentrations up the food chain, including humans [17–33]. Thus, the change in temperature, the resulting decrease in DO and increasing metal concentrations, and the synergistic impact of the combining the hypoxic water and reduced metal compounds is a cascade of harm to the stream's ecosystems.

Both abiotic and biotic responses to heat occur continuously in the environment. The growth and metabolism of the bacteria are particularly sensitive to changing DO levels. As the oxygen concentrations in water decline, anaerobic bacteria growth increases. Fungi are mainly restricted to aerobic conditions, but specific fungi have been found in freshwater, as well as in deep sediment, landfill leachate, and most commonly in rumen (cud) of cattle and other ruminating herbivores [34–37]. Algae both consume dissolved O_2 for metabolism and respiration while simultaneously producing O_2 by photosynthesis. Increasing temperature enhances the solubility of most substances. As mentioned, decreasing dissolved O_2 in water is accompanied by redox changes, e.g., formation of reduced metal species, such as metal sulfides. Conversely, increasing concentrations of O_2 in water increases the concentrations of metal oxides. This is also being mediated by the bacteria, some of which will begin reducing the metals as the oxygen levels drop (reduced conditions in the water and sediment). However, the opposite is true in the more oxidized regions, i.e., the metals are forming oxides. The increase in the metal compounds combined with the reduced DO, combined with the increased temperatures can act synergistically to worsen conditions for higher animals, e.g., a fish kill or diminished survival rates of sensitive species [1].

Physical, chemical, and biological conditions integrate to determine the formation and fate of a pollutant Both energy and matter are part of this fate. For example, heat can be a water pollutant [38]. The control of energy exiting a source like the cooling water from an electric generating plant must be optimized based on the transfer of energy. Excessive heat in an ecosystem causes a primary abiotic response, i.e., decreased dissolved oxygen (DO) concentrations in the water. The primary response leads to biotic processes that either increase or decrease DO and increase the toxic responses. These secondary responses include abiotic changes, such as the change in redox due to decreased molecular oxygen in the water. The tertiary responses can also be abiotic, e.g., change in the valence state of metals, and biotic, e.g., greater bioavailability and toxicity of metal compounds in the water [2].

TABLE 4.6 Relationship between water temperature and maximum dissolved oxygen (DO) concentration in water (at 1 atmosphere [atm]).

Temperature (°C)	DO (mg L^{-1})	Temperature (°C)	DO (mg L^{-1})
0	14.60	23	8.56
1	14.19	24	8.40
2	13.81	25	8.24
3	13.44	26	8.09
4	13.09	27	7.95
5	12.75	28	7.81
6	12.43	29	7.67
7	12.12	30	7.54
8	11.83	31	7.41
9	11.55	32	7.28
10	11.27	33	7.16
11	11.01	34	7.16
12	10.76	35	6.93
13	10.52	36	6.82
14	10.29	37	6.71
15	10.07	38	6.61
16	9.85	39	6.51
17	9.65	40	6.41
18	9.45	41	6.41
19	9.26	42	6.22
20	9.07	43	6.13
21	8.90	44	6.04
22	8.72	45	5.95

Data from D. Vallero, Environmental Biotechnology: A Biosystems Approach, Academic Press, 2010 and E. Dohner, A. Markowitz, M. Barbour, J. Simpson, J. Byrne, G. Dates, 5.2. Monitoring and assessing water quality, in: Volunteer Stream Monitoring: A Methods Manual, 1997, (Chapter 5).

The initiating abiotic effect (i.e., increased temperature) also results in a rising microbial population. The responses are not unidirectional. For example, the associated decline in DO, however, is a net decline between photosynthetic and non-photosynthetic microbes. The growth and metabolism of the bacteria result in decreasing the DO levels, but the growth of the algae both consume DO for metabolism and produce DO by photosynthesis.

TABLE 4.7 Typical temperature tolerances of aquatic organisms.

Organism	Taxonomy	Range in temperature tolerance (°C)	Minimum dissolved oxygen (mg L^{-1})
Trout	*Salma, Oncorhynchus and Salvlinus spp*	5 – 20	6.5
Smallmouth bass	*Micopterus dolomieu*	5 – 28	6.5
Caddisfly larvae	*Brachycentrus spp.*	10 – 25	4.0
Mayfly larvae	*Ephemerella invaria*	10 – 25	4.0
Stonefly larvae	*Pteronarcys spp.*	10 – 25	4.0
Catfish	Order Siluriformes	20 – 25	2.5
Carp	Cyprinus spp	10 – 25	2.0
Water boatmen	*Notonecta spp.*	10 – 25	2.0
Mosquito larvae	Family Culicidae	10 – 25	1.0

From D. Vallero, Environmental Biotechnology: A Biosystems Approach, Academic Press, 2010. Data from V. Corporation, Computer 19: Dissolved Oxygen in Water, http://www2.vernier.com/sample_labs/BWV-19-COMP-dissolved_oxygen.pdf (Accessed 5 March 2018).

Meanwhile a combined abiotic and biotic response occurs with the metals. The increase in temperature increases their aqueous solubility and the decrease in DO is accompanied by redox changes, e.g., formation of reduced metal species, such as metal sulfides. These changes can increase the exposure of human populations to toxic substances. The increase in temperature in Fig. 4.10, for example, leads to increased solubility of metals in water. With rising temperature, the suspended or settled solids will begin to release metals at a faster rate. The process is also being mediated by the bacteria, some of which will begin reducing the metals as the oxygen levels drop (reduced conditions in the water and sediment). However, the opposite is true in the more oxidized regions, where the metals are forming oxides. The increase in the metal compounds combined with the reduced DO, in the presence of increased temperatures can act synergistically to make the conditions toxic for higher animals, e.g., a fish kill [39]. Predicting the likelihood of ecosystem change and adverse events like fish kills can be quite complicated, with many factors that either mitigate or exacerbate the outcome (See Fig. 4.7). The increase in metal concentrations in water not only increases the potential exposure to and risks from all metals, but also the transformation of metals into toxic chemical species. If these new compounds have higher bioavailability, then exposure, dose and body burden of the metals will also likely increase (e.g., methyl mercury in fish ingested by humans).

Almost any industrial or municipal facility will release substances, which can become pollutants into the environment. After release of a substance, many processes and exchanges occur among and within the environmental compartments (See Fig. 4.8); as well as those at the interface between the organism and the environment. The equilibrium and speciation of elements in the atmosphere, hydrosphere and biosphere determines their importance to water quality.

7 Water in the lithosphere

The earth's atmosphere is often treated as a closed system in pollution models, but in fact has many inputs from and outputs to various environmental compartments. Many of the reactions and mechanisms that produce water pollutants begin and end in the air, soil, sediment, and biota above and below the earth's surface.

Soil formation, i.e., pedogenesis, involves numerous physical, chemical, and biological processes. Parent rock is broken down into particles. The particles continue to be degraded and decomposed by microbes and plant roots.

As mentioned, physical weathering owes much to water's molecular configuration (Fig. 4.1), which generates substantial expansion by volume as it freezes or sublimates, i.e., change of state from gas to solid phase. The H_2O configuration of oxygen and two hydrogen atoms allows for frost heaving, i.e., the expansion of water in rock cracks. Temperature, pressure, water content, porosity, and pore size influence the rate and intensity of frost heaving pressure [40].

Eventually, the unconsolidated materials resulting from physical weathering become soil [2]. Once dissolved or suspended in the pore water, chemical compounds undergo hydrolysis, reduction, oxidation, and other chemical reactions in the pore water and on the soil particle. The troposphere is the source of some compounds like carbonic acid (H_2CO_3),

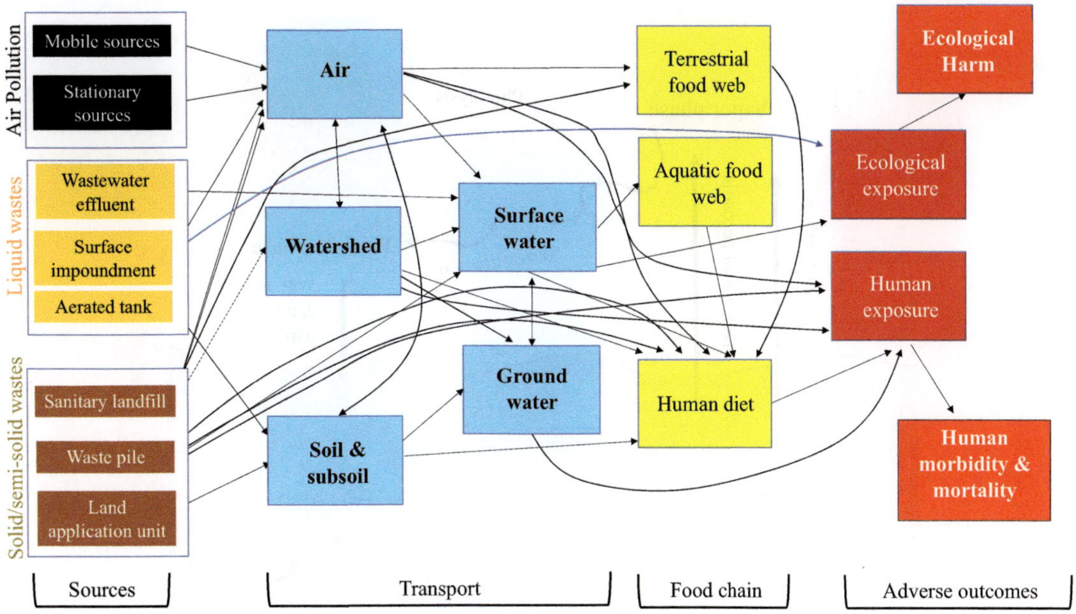

FIG. 4.7 Environmental transport pathways can be affected by net heat gain. Compounds (nutrients, contaminants), microbes and energy (e.g., heat) follow the path through the environment indicated by arrows. The residence time within in any of the boxes is affected by abiotic conditions, including temperature. *(Data from D. Vallero, Translating Diverse Environmental Data into Reliable Information: How to Coordinate Evidence from Different Sources. Academic Press, 2017. Adapted from D.A. Vallero, K.H. Reckhow, and A. D. Gronewold, Application of multimedia models for human and ecological exposure analysis, in: International Conference on Environmental Epidemiology and Exposure, International Society of Exposure Analysis, Durham, NC, 2007.)*

sulfuric acid (H_2SO_4), and nitric acid (HNO_3) that increase the hydrogen ion concentration, i.e., lower the pH, which increases the rates of dissolution of the carbonates in rocks, especially limestone and dolomite. For example, sulfuric acid reacts with solid phase calcium carbonate in these rocks to form the divalent calcium cation, divalent sulfate ion, water, and carbon dioxide:

$$CaCO_2(s) + H_2SO_4(aq) \rightarrow Ca^{2+}(aq) + SO_4{}^{2-}(aq) + H_2O + CO_2 \tag{4.5}$$

Many other physical and chemical processes set the stage biochemical processes, e.g., microbial degradation and root growth. As mentioned, lichen, for example, grow on rocks, which gives access to larger plant growth [41–47]. Wind and water erode and transport the rock and soil material. Erosion and mass wasting move the disintegrated material, as shown in Fig. 4.2. The flow of liquid water, ice, and wind move small particles eroded from the parent rock. Larger portions of rock are moved by mass wasting, caused by gravity. The disintegration of coarse rock particles continues by exfoliation, water erosion, wind erosion, freezing, and thawing [48]. These mechanisms continue and are enhance by transport and transformation processes to form a new landform [49–54]. These processes work together to create terrain, i.e., geomorphology. Of course, much more rapid geomorphological processes occur when induced by human activities like agriculture, road construction, urban development, and waterway modifications like the installation of locks, dams, and levees.

Sediments move with flowing water and air [55]. These natural soil and sediment formation and modification are eventually joined by anthropogenic actions [56,57]. These include, nutrient enrichment by fertilization, contouring, tilling, and highly impactful practices like bulldozing and mineral extraction, which with industrialization and economic development, have accelerated exponentially [58,59].

Soil and sediment consist of particles in a matrix. Typically, water and air pollution are expressed as the concentration of a pollutant in the fluid. Conversely, soil and sediment pollution are usually expressed as concentrations within the matrix. Soil and sediment pollution are also characterized by the pollutant concentration within the water and air in the pore spaces of the matrix. Also, there is an ongoing exchange between the fluid-filled pore spaces since the particles sorb or desorb pollutants depending on the environmental conditions.

Thus both soil and sediment are porous and permeable. The pore spaces must be sufficiently interconnected to allow fluids into and out of the matrices. Sandstone is highly porous, but the pore spaces are not completely interconnected throughout the rock, so it is generally not permeable.

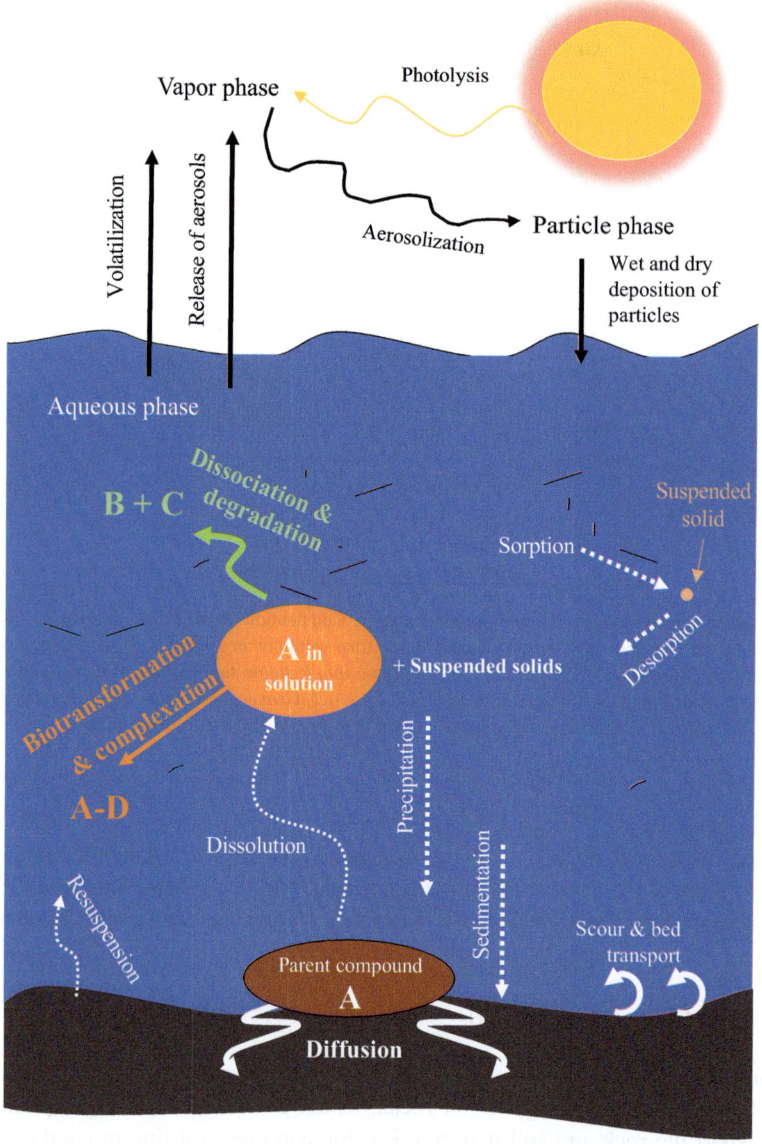

FIG. 4.8 Transport and transformation of chemicals in a water system. Contaminants from a pollutant source may reach a water system in various ways. Upon reaching the water, numerous transformation processes, including dissociation and degradation to form metabolites and degradation products (B, C and D), are brought about by both abiotic (e.g., hydrolysis and photolysis) and biotic (i.e., biodegradation) processes. *(From D. Vallero, Air Pollution Calculations: Quantifying Pollutant Formation, Transport, Transformation, Fate and Risks, Elsevier, 2019. Adapted from W. Lyman, Transport and transformation processes, in: G.M. Rand, Fundamentals of Aquatic Toxicology. Effects, Environmental Fate, and Risk Assessment, second ed., Taylor and Francis, Washington, DC, 1995, pp. 449–492, used with permission.)*

Sediment is a component of surface water and wetland ecosystems. Soil and sediment are key components of land systems. The land-water-air system is a three-dimensional, thermodynamic control volume into which substances enter, cycle, are transformed and are exchanged by chemical processes, especially, redox, hydrolysis, cation and anion exchanges and, when algae and vascular plants are present, photosynthesis. There are myriad relationships between soil water and the troposphere. For example, the troposphere is mostly molecular nitrogen by volume, so it is the source of nutrients for nitrogen-fixing plants like legumes. Sediment's connection to the troposphere is less direct than that of soil since sediment resides in both water and land-based systems, e.g., wetlands, but is often simply considered a layer beneath the water column of a surface water. Thus, pollutants and other substances must move from sediment through a water column before reaching the troposphere.

The composition of land varies considerably spatially. Climate is one determinant of land types. This determines the type and abundance of vegetation that is and can be supported by land. Desert lands differ from temperate lands, which

TABLE 4.8 Select soil properties and associated activities important to chemical exchanges between soils and the hydrosphere and the troposphere.

Substance or property	Description	Optimal range
Phosphorus	An essential nutrient for plants and animals	30 to 50 mg-P/kg-soil (ppm)
Nitrogen	An essential nutrient for plants and animals	25–50 mg-N/kg-soil
Major cations	Calcium (Ca), magnesium (Mg), potassium (K) are essential nutrients for plants and animals; these cations may affect soil properties (e.g., crusting with excess sodium) or soil processes (e.g., buffering)	Ca: > 300 mg/kg-soil; Mg: > 35 mg/kg-soil; K: 40–80 mg-kg-soil; sodium: less than 15% of soil CEC
Sulfur	An essential nutrient for plants and animals	Optimum concentration is 30–40 mg-S/kg-soil
Trace metals	Some trace metals and metalloids are essential for plants and animals (molybdenum (Mo), copper (Cu), boron (B), manganese (Mn), iron (Fe), zinc (Zn)); most trace metals are toxic above certain concentrations	Mo: 2 mg/kg-soil; b: 0.5–4 mg/kg-soil; Cu: 2–50 mg/kg-soil; Zn: 1–200 mg/kg-soil; Mn: 2–25 mg/kg-soil; Fe: 20–30 mg/kg-soil. Non-essential trace metals vary in their toxic effects on vegetation.
Cation exchange capacity (CEC)	Affects the availability of nutrients and pollutants	12–40 meq/100 g-soil
Conductivity	An indicator of salt concentration in soil. Salts affect availability and transport of nutrients and pollutants	Less than 8 dS/cm is desirable
Organic matter	Effects include storage of nutrients, improving soil. Structure, porosity, infiltration, retarding pollutant transport, and stabilizing soil pH, and added buffering capacity	2–8% of soil, by weight
Base saturation	An indicator of the extent to which cation exchange capacity is attributable to calcium, magnesium, and potassium. Low base saturation may reflect soils high in exchangeable aluminum, which can be detrimental to plants	40–80% for calcium, 10–40% for magnesium (Mg), and 1–5% for potassium (K)
Enzymes	Enzymes mediate many chemical processes in soil	Varies with specific enzyme and soil process being targeted
Salinity	High salinity reflects soils with high sodium concentrations, which negatively impacts soil structure	Take electrical conductivity readings, with values greater than 16 dS/cm indicating sodic soils
Sodium adsorption ratio	An indicator of sodium concentrations in soil. Elevated sodium concentrations negatively impact soil structure	Values great than 13 indicate sodic soils
Acidity	Affects soil processes like microbial activity, and availability and mobility of nutrients and pollutants	pH between 6 and 8

Optimal ranges vary by soil type, geographic location, vegetation, and other physical factors. Note: ppm = parts per million; dS/cm = deciSiemens per centimeter; meq = milliequivalent, the amount of a substance that reacts with one mole of another substance in a chemical reaction.
Data from Minnesota Pollution Control Agency, Minnesota Stormwater Manual, https://stormwater.pca.state.mn.us/index.php?title=Soil_chemical_properties_and_processes (Accessed 1 March 2023).

differ from tropical lands in terms of thickness and organic matter content of soil horizons, ability to support various types of root systems, microbial populations, and moisture profiles. Of course, within these climate-determined soils, there is great diversity of these and other physical, chemical, and biological properties.

Food supply and agricultural efficiency depend on optimal concentrations of macronutrients, like nitrogen and phosphorous, and micronutrients, like iron (See Table 4.8). These and other compounds in excess supply become pollutants.

Soils typically consist of a profile of layers or horizons that vary in thickness (See Fig. 4.9). The O horizon at the surface has the highest concentration of soil organic matter (SOM) of all soil layers because plant and animal cells and wastes are deposited, then are decomposed by bacteria and fungi. Decomposition produces molecules of organic compounds in humic material, giving the O horizon a dark brown or black color [60]. Most chemical exchanges between the troposphere and soil are between the O and A horizons. For example, the largest source of mercury (Hg) transport to atmosphere from terrestrial ecosystems results from Hg evasion from soil [33]. However, during and following mineral extraction, surface mining, road

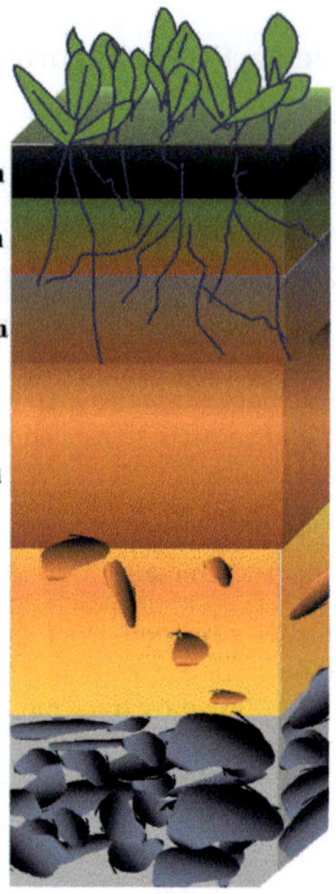

FIG. 4.9 Soil profile consisting of O, A, E, B, C, and R horizons Movement of chemical compounds occurs in the top layers, with most interactions with the troposphere in the A and E layers. The C and R horizons usually only interface with the atmosphere when the soil layers are disrupted, e.g., after surface mining. *(Data from NRCS, From the Ground Down, [Online] Available: https://www.nrcs.usda.gov/Internet/FSE_DOCUMENTS/nrcseprd366047.pdf, 2015.)*

construction, and dredging activities, chemical exchanges occur between water bodies and the atmosphere with rock, soil, and sediment layers [61–63].

Below the relatively thin O horizon lies the A horizon, which consists of topsoil. It is called the humus layer because microbial decomposition forms dark-colored humic materials, especially humic acid (HA). Within the microbial cell, HA is produced enzymatically by chemical oxidative condensation following cellular autolysis, producing various aromatic and aliphatic compounds containing carbon, oxygen, nitrogen, hydrogen, phosphorus, and other elements [64]. Higher organisms like earthworms living in the A horizon also contribute to SOM. Thus, HA is not a specific chemical compound, but an accumulation of recalcitrant end products from various abiotic and biotic reactions [64–68]. HA provides energy to many biochemical processes in soil [69,70] and regulates the rates at which nutrients are taken up, retained, transferred, and cycled, i.e., the nutrient dynamics of the soil. The percentage of humic materials in a soil or sediment affects the carbon-to-nitrogen ratio (C:N). The topsoil in the A horizon has high water retention given its porosity.

Some soil horizons have an E horizon that consists of silty grains that are lighter in color and lower clay content than those of the A horizon. The E horizon is a leaching zone, e.g., water moving downward collects aqueous soluble substances, leading to eluvial loss of silicate clay, iron, aluminum, and nutrients from the O and A horizons [60].

The B horizon is the subsurface layer with greater clay content than the upper layers. It is denser than the overlying layers since it contains less soluble minerals, organic matter, and humus. Compounds of less aqueous solubility deposit here, e.g., oxides of iron and other metal salts. It is common for the pH of B horizon's pore water to differ from the water in the O and A horizons [60].

The C-Horizon lacks organic matter and is made up of broken bedrock. This layer is also known as saprolite. The parent material in the C horizon may have been deposited by wind, e.g., loess, by water flows, i.e., alluvium, or by glaciers, i.e., glacial till. The R horizon is the compact, cemented layer of bedrock that has not yet been chemically altered [60].

In describing the soil horizons, terms like silt and clay were used. These describe the size of soil particles, known as soil texture (See Fig. 4.10). Texture is based on the size of the individual particles in the soil (See Table 4.9). Soil texture is classified by separates, i.e., clay separates, silt separates, and sand separates. Note that gravel is excluded from Fig. 4.11, given that soils are generally limited to particles with diameters less than 2 mm [71].

A particle's relative surface area, i.e., a particle's specific surface area (a_s), increases with decreasing particle size. Surface area has units in area per unit mass (e.g., $m^2\ g^{-1}$) [72,73]. This means that as measurements depend the experimental method, e.g., the probe size [74]. Thus, surface area can differ substantially for materials with the same total volume and total mass. Soils with high relative surface area values like clayey soils have high water retention rates, increased ability to sorb pollutants, and large swelling potential.

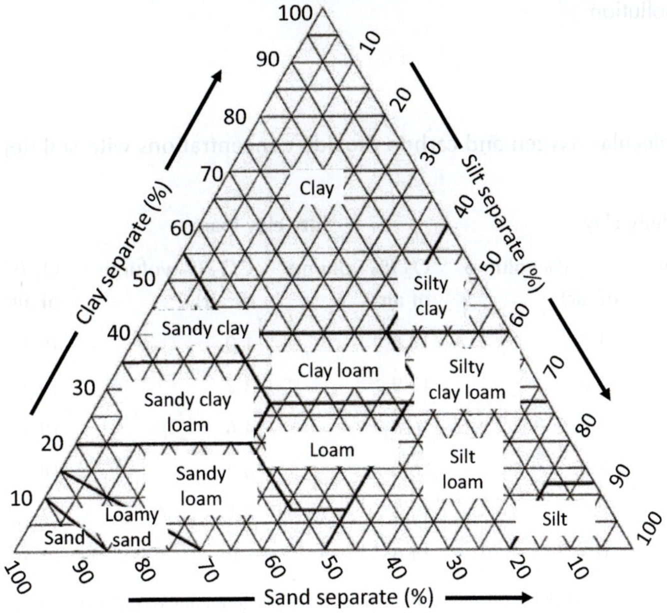

FIG. 4.10 Soil texture triangle. *(Data from Natural Resources Conservation Service Soils, Soil Texture Calculator, NRCS Soils, https://www.nrcs.usda.gov/wps/portal/nrcs/detail/soils/survey/?cid=nrcs142p2_054167 (Accessed).)*

TABLE 4.9 Commonly used soil texture classifications.

Soil separate	Size range (mm)
Gravel	>2.0
Very coarse sand	1.0–1.999
Coarse sand	0.500–0.999
Medium sand	0.250–0.499
Fine sand	0.100–0.249
Very fine sand	0.050–0.099
Silt	0.002–0.049
Clay	<0.002

Data from T.E. Loynachan, K.W. Brown, T.H. Cooper, M.H. Milford, Sustaining our Soils and Society, American Geological Institute, 1999.

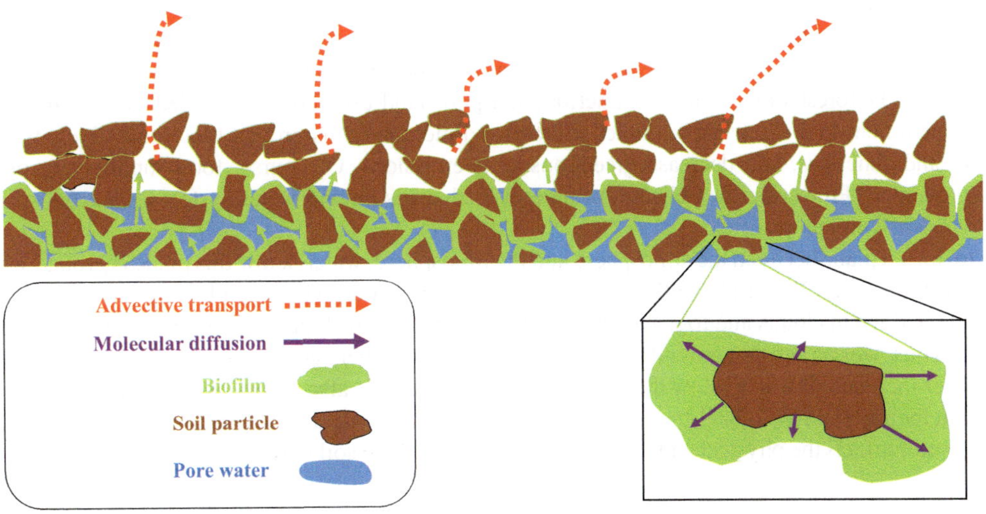

FIG. 4.11 Transport mechanisms in soil and sediment. *(Sources of information: D.A. Vallero, Soil and sediment physics, in: Methods and Calculations in Environmental Physics, AIP Publishing LLC, Melville, New York, 2022, pp. 8-1–8-34 and D.A. Vallero, Translating Diverse Environmental Data into Reliable Information: How to Coordinate Evidence from Different Sources, Academic Press, 2017.)*

TABLE 4.10 Example of molecular oxygen and carbon dioxide concentrations with soil depth by grain size from one field study.

	Silty clay		Silty clay loam		Sandy loam	
Depth from surface (cm)	O_2 (% volume of air)	CO_2 (% volume of air)	O_2 (% volume of air)	CO_2 (% volume of air)	O_2 (% volume of air)	CO_2 (% volume of air)
30	18.2	1.7	19.8	1.0	19.9	0.8
61	16.7	2.8	17.9	3.2	19.4	1.3
91	15.6	3.7	16.8	4.6	19.1	1.5
122	12.3	7.9	16.0	6.2	18.3	2.1
152	8.8	10.6	15.3	7.1	17.9	2.7
183	4.6	10.3	14.8	7.0	17.5	3.0

Data from V. Evangelou, Environmental Soil and Water Chemistry: Principles and Applications (no. 631.455 E6), 1998.

At the molecular scale, gaseous substances move through soil pores either advectively or by molecular diffusion (See Fig. 4.11). Microbes mainly reside on the particles, so soil texture and moisture are major determines of the types and size of microbial populations. For example, lower soil moisture lessons the thickness of the biofilm and lowers the activity of microbes in large part owing to diminished diffusion and advective transport of nutrients and water [75]. This then can impair the microbial communities' ability to produce biofilms [76].

Water, sediment, soil, and air contain solutions and suspensions of inorganic and organic compounds. The soil matrix is predominantly solid by volume, having various volumes of soil water and other liquids and gases including molecular nitrogen, molecular oxygen, methane, and carbon dioxide. The sediment matrix is similar. In many sediments the matrix when nearly dry resembles that of most soils, and when wet the solid volume contains a substantial amount of saturated organic matter.

Soil gas concentrations differ from those in the atmosphere with depth from the surface (See Table 4.10) [71]. Notwithstanding grain size, CO_2 concentrations increase with depth whereas O_2 concentrations decrease. Much of this owes to distance from the reaeration layers near the surface. The re-aeration is a function of the amount of oxidizable matter. The intensity of soil reaeration is proportional to the square root of the specific surface area of the inert particles and is directly proportional to dispersion [77].

Sediment is an assemblage of particles that have settled on the bottom of water bodies. Aerobic microbial species prefer oxygenated zones near the water surface, with anaerobic microbial populations increasing with depth owing to decreasing dissolved oxygen levels and chemical reduction.

Soils and wetlands provide ecosystem goods and services to society, including water storage essential to drought moderation, flood control, and carbon and nutrient sequestration that affects agriculture, water supplies, and climate change moderation. Indeed, pollutants that are filtered by wetlands and marshes, for example, would otherwise be transported to water bodies and the atmosphere. This retention potential provides a sink for tropospheric carbon, especially carbon dioxide and methane. Deforestation, mining, extraction, and poor land use choices have led to dramatic and often irreversible soil and sediment loss. Conversely, intact soil and sediment systems protect biological diversity, improve ecosystem structure and functions, and decrease greenhouse gas emissions. Of course, food supplies are protected when topsoil loss is prevented. Protecting the delicate balance of exchanges among water systems, the troposphere, and land systems is a "win-win-win" proposition.

If we were to draw a plane between the atmosphere and the lithosphere, water below this plane is ground water and water on the plane is surface water. Water above this plane is atmospheric water, which can be any physical phase, i.e., water vapor, condensate and ice in clouds and fog, and precipitation in the form of liquid rain, sleet, snow, and mixtures of solid and liquid water. Thus, we can say that ground water resides in the lithosphere and surface waters reside on the lithosphere. Since ground water includes all water below the surface, water in the soil is also ground water, but scientists from different disciplines vary in their definitions of soil-bound water. Engineers commonly differentiate water in soil from ground water because the soil water affects the physical and mechanical properties of the soil horizons commonly manipulated by human activities, especially construction. Farmers tend to focus almost entirely on the water for crops from the ground water that they draw from wells. Soil moisture is crucial to agriculture because, for example, it affects the farmer's ability to till the soil, who differs from the road engineer who must be able to calculate gel strength. By contrast, environmental science and engineering are principally concerned with the soil's role in storing and transporting pollutants and to support microbial populations. Thus, environmental publications frequently describe soil water according to the amount of void space filled, or the water-filled pore space (WFPS),[a] the percentage of void space containing water. The WFPS is another way to express saturation [78].

Almost all environmental and hydrological science professions classify the water below the soil layer based upon whether the unconsolidated material (e.g., gravel and sand) is completely saturated or unsaturated. The saturated zone lies under the unsaturated zone. Depending on the jurisdiction, the movement of contaminants in each section of Fig. 4.3 is regulated. For example, a landfill's leachate collection system may have to meet local, state, provincial or national standards to limit the mass and volume of contaminants that reach the ground water. The landfill may also have to meet air quality standards for volatile organic compounds (VOCs) by the installation and operation of a landfill gas management system [4,79]. There are often numerous specifications required by governments, including restrictions to protect surface waters, biota in ecosystems, and soils after the landfill closure.

Hydrogeologists refer to the unsaturated zone as the "vadose zone." It is also referred to as the zone of aeration. Another type of ground water, albeit rare, is the Karst ground water system, which is made up of underground lakes and streams that flow through fractured limestone and dolomite rock strata. Small cracks in the rock erode over time to allow rapidly flowing water to move at rates like those of water on the earth's surface (See Chapter 10). Much of the water flow in Karst aquifers does not adhere to Darcy's Law (See Chapter 11). Darcian flow assumes that the flow is laminar, but non-Darcian flow in the conduits created by caves and caverns cause that the water to flow rapidly enough to be turbulent [80,81]. However, in most aquifers, a contaminant moves through and is stored in unconsolidated material, e.g., sand. As such, the pollutant may remain in the interstices for decades. Changes in meteorological conditions, soil disturbance, or continued leakage from a source are among the factors that change the rates of ground water flow [82].

Soil scientists have struggled with uniformity in the classification and taxonomy of soil. Much of the rich history and foundation of soil scientists has been associated with agricultural productivity. The very essence of a soil's "value" has been its capacity to support plant life, especially crops. Even forest soil knowledge owes much to the agricultural perspective, since much of the reason for investing in forests has been monetary. To foresters, a stand of trees is indeed a "standing crop." In the United States, for example, the National Forest Service is an agency of the U.S. Department of Agriculture. The agricultural and engineering perspectives have provided valuable information about soil that environmental professionals can use. The information is certainly necessary, but not completely sufficient, to understand how pollutants move through soils, how the soils themselves are affected by the pollutants (e.g., loss of productivity and diversity of soil microbes), and the ways that soils and contaminants interact chemically (e.g., changes in soil pH will change the chemical and biochemical transformation of organic compounds). Water pollution calls for modicum of understanding of how soils are classified according to their texture or grain size (See Table 4.11), ion exchange capacities, ionic strength, pH, microbial populations, and soil organic matter content.

Air and water are fluids that reside and move through the interstices of matrices in soil and sediment. It can be difficult to separate the fluid from the matrix. For example, sediment resembles soil in that it is a matrix made up of various components, including organic matter and unconsolidated material, but some of these matrix substances also move as fluids, such as fulvic and humic acids. The liquids in the interstices are chemically known as "substrates," which hold solutes (e.g., contaminants and nutrients).

The predominant volume of substrate in this matrix is usually water that contains varying amounts of solutes. At least under most environmental conditions, air and water are solutions of very dilute amounts of compounds. For example, air's solutes represent small percentages of the solution at the highest (e.g., water vapor, usually ranging between 0.3% and 0.4%), whereas most other solutes represent parts per million (CO_2 is about 425ppm). Most "contaminants" in air and water, thankfully, are found in the parts per billion range, if found at all. On the other hand, soil and sediment themselves are conglomerations of all states of matter. Soil is predominantly solid, but frequently has large fractions of liquid (soil water) and gas (soil air, methane [CH_4], and CO_2) that make up the matrix. The composition of each fraction is highly variable. For example, soil gas concentrations are different from those in the atmosphere and change profoundly with depth from the surface, as demonstrated in Table 4.10.

a. During my defense of my Ph.D. dissertation, Fred Boadu of Duke University's Civil and Environmental Engineering Department, disagreed with my use of the term water-filled pore space, since the water I was describing was polluted and may have contained other fluids. Instead, he recommended I use the term, Fluid-filled pore space (FFPS). I did, so if you read my dissertation, you will see FFPS rather than WFPS.

TABLE 4.11 Commonly used soil texture classifications.

Name	Size Range/mm
Gravel	>2.0
Very coarse sand	1.0–1.999
Coarse sand	0.500–0.999
Medium sand	0.250–0.499
Fine sand	0.100–0.249
Very fine sand	0.050–0.099
Silt	0.002–0.049
Clay	<0.002

Data from T.E. Loynachan, K.W. Brown, T.H. Cooper, M.H. Milford, Sustaining our Soils and Society, American Geological Institute, 1999.

Ecosystems are amalgamations of all the previously mentioned media. For example, a wetland system consists of microbes and vascular plants that grow in soil, sediment, and water. The water flows through living and non-living materials. Microbial populations live in the surface water, with aerobic species assembling near the water surface and anaerobic microbes increasing with depth due to the decrease in oxygen levels, due to the reduced conditions. Air is not only important at the water and soil interfaces, but it is a vehicle for nutrients and contaminants delivered to the wetland. The ground water is fed by the surface water during high water conditions and feeds the wetland during low water conditions.

Thus, one of the profound challenges in regulating water quality is that they involve many aspects of the environment beyond the water itself. Environmental laws most often address a single environmental medium, e.g., the Clean Water Act, the Safe Drinking Water Act, the Clean Air Act, the solid waste and hazardous waste statutes, and consumer and worker protection laws. Pollutants are formed and move through all the media, which may be envisioned as thermodynamic compartments, each with its own boundary conditions, kinetics, and partitioning relationships. Energy and matter are shared and converted among the compartments. Chemicals, whether nutrients or contaminants, change due to the time spent in each compartment. The engineer's, scientist's, planner's, and manager's challenge is to describe, characterize, and predict the behaviors of pollutants and other substances as they move through the media. These are based on the physics, chemistry, and biology of the system.

8 Carbon equilibrium and cycling

Many carbon-based molecules exist in equilibrium with one another. For example, Fig. 4.12 demonstrates the equilibrium among carbonates, bicarbonates, organic compounds, carbonic acid, and carbon dioxide. On a global scale, uncontaminated rain's mean pH is about 5.6, owing to its dissolution of carbon dioxide, CO_2. As the water droplets fall through the air, the CO_2 in the atmosphere becomes dissolved in the water, setting up an equilibrium condition:

$$CO_2 \text{ (gas in air)} \leftrightarrow CO_2 \text{ (dissolved in the water)} \tag{4.6}$$

The CO_2 in the water reacts to produce hydrogen ions, as

$$CO_2 + H_2O \leftrightarrow H_2CO_3 \leftrightarrow H^+ + HCO_3^- \tag{4.7}$$

$$HCO_3^- \leftrightarrow 2H^+ + CO_3^{2-} \tag{4.8}$$

Assuming the mean partial pressure of CO_2 in the air is 4.2×10^{-4} atm, the pH of water in equilibrium can be calculated. Such chemistry is always temperature-dependent. Henry's law states that the concentration of a dissolved gas is directly proportional to the partial pressure of that gas above the solution:

$$p_a = K_H[c] \tag{4.9}$$

Where, K_H = Henry's law constant; p_a = partial pressure of the gas; and $[c]$ = molar concentration of the gas.

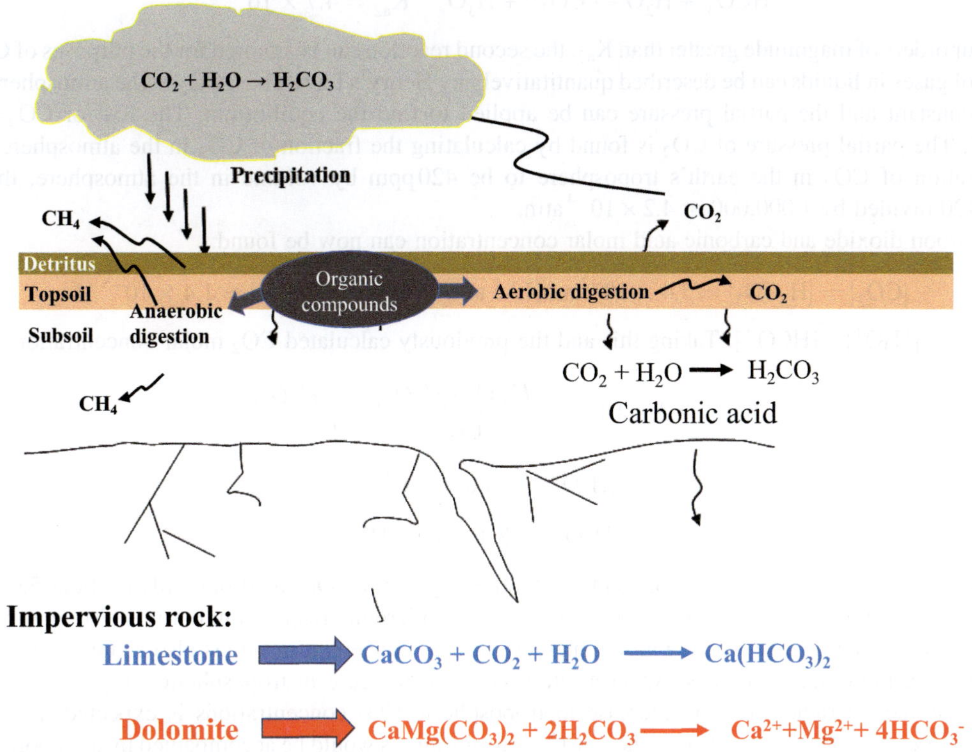

FIG. 4.12 Biogeochemistry of carbon equilibrium. The processes that release carbonates are responsible for much of the buffering capacity of natural soils against the effects of acid rain. *(From D.A. Vallero, Fundamentals of Air Pollution, fifth ed., Elsevier Academic Press, Waltham, MA, 2014, p. 999 pages cm.)*

Or,

$$p_a = K_H\, C_W \tag{4.10}$$

Where, C_W is the concentration of gas in water.

Henry's Law is a function of a substance's solubility in water and its vapor pressure and expresses the proportionality between the concentration of a dissolved contaminant and its partial pressure in the open atmosphere at equilibrium. That is, the Henry's Law constant is an example of an equilibrium constant, the ratio of concentrations when chemical equilibrium is reached in a reversible reaction, the time when the rate of the forward reaction is the same as the rate of the reverse reaction.

The CO_2 concentration of the water droplet at equilibrium with air is obtained from the partial pressure of Henry's law constant:

$$p_{co_2} = K_H[co_2]_{aq} \tag{4.11}$$

The change from carbon dioxide in the atmosphere to carbonate ions in water droplets follows a sequence of equilibrium reactions:

$$co_{2(g)} \overset{K_H}{\longleftrightarrow} co_{2(aq)} \overset{K_r}{\longleftrightarrow} H_2co_{3(aq)} \overset{Kal}{\longleftrightarrow} HCO^{-}_{3(aq)} \overset{Ka2}{\longleftrightarrow} co^{2-}_{3(aq)} \tag{4.12}$$

The processes that release carbonates increase the buffering capacity of natural soils against the effects of acidic water ($pH < 5$). The ionic strength of the receiving soil or surface waters determines the actual change in pH. For example, the carbonate-rich soils like those central North America can withstand even elevated acid deposition compared to the thin soil areas, such as those in the Canadian Shield, the New York Finger Lakes region, and much of Scandinavia.

The concentration of CO_2 is constant, since the CO_2 in solution is in equilibrium with the air that has a constant partial pressure of CO_2. The two reactions and ionization constants for carbonic acid are:

$$H_2CO_3 + H_2O \leftrightarrow HCO_3^- + H_3O^+ \quad K_{a1} = 4.3 \times 10^{-7} \tag{4.13}$$

$$HCO_3^- + H_2O \leftrightarrow CO_3^{-2} + H_3O^+ \quad K_{a2} = 4.7 \times 10^{-11} \tag{4.14}$$

Since K_{a1} is four orders of magnitude greater than K_{a2}, the second reaction can be ignored for the purposes of C equilibrium. The solubility of gases in liquids can be described quantitatively by Henry's Law. Thus, CO_2 in the atmosphere at 25 °C, the Henry's Law constant and the partial pressure can be applied to find the equilibrium. The K_H for $CO_2 = 3.4 \times 10^{-2}$-$mol\,L^{-1}\,atm^{-1}$. The partial pressure of CO_2 is found by calculating the fraction of CO_2 in the atmosphere. Assuming a mean concentration of CO_2 in the earth's troposphere to be 420 ppm by volume in the atmosphere, the fraction of CO_2 must be 420 divided by 1,000,000 or 4.2×10^{-4} atm.

Thus, the carbon dioxide and carbonic acid molar concentration can now be found:

$$[CO_2] = [H_2CO_3] = 3.4 \times 10^{-2}\,mol\,L^{-1}\,atm^{-1} \times 0.000420\,atm \approx 1.4 \times 10^{-5}\,M$$

The equilibrium is $[H_3O^+] = [HCO^-]$. Taking this and the previously calculated CO_2 molar concentration gives:

$$K_{a1} = 4.3 \times 10^{-7} = \frac{[HCO_3^-][H_3O^+]}{CO_2} = \frac{[H_3O^+]^2}{1.4 \times 10^{-5}}$$

$$[H_3O^+]^2 = 6.0 \times 10^{-12}$$

$$[H_3O^+] = 2.4 \times 10^{-6}\,M$$

The log of 2.4×10^{-6} is approximately −5.63, so the negative log is 5.63. Or, the droplet pH is about 5.6.

Carbon dioxide, with water, is the ultimate product of aerobic microbial respiration, but it is also an important greenhouse gas. From the preceding discussion, a global increase in CO_2 concentrations must also change the mean acidity of precipitation. For example, many models expect a rather constant increase in tropospheric CO_2 concentrations. If the current trend continues, a rather constant increase in tropospheric CO_2 concentrations is expected. For example, the increase from the present 420–500 ppm tropospheric CO_2 concentrations would be accompanied by a proportional decrease in precipitation pH. The molar concentration can be adjusted using the previous equations:

$3.4 \times 10^{-2}\,mol\,L^{-1}\,atm^{-1} \times 0.000500\,atm = 1.7 \times 10^{-5}$ M, so $4.3 \times 10^{-7} = \frac{[H_3O^+]^2}{1.7 \times 10^{-5}}$ and $[H_3O^+]^2 = 7.3 \times 10^{-12}$ and $[H_3O^+] = 2.7 \times 10^{-6}\,M$

Thus, average water droplet pH would be decrease from the present 5.63 to about 5.56. This means that the incremental increase in atmospheric carbon dioxide can be expected contribute to greater acidity of natural rainfall. The precipitation rates themselves would also be affected if greenhouse gas concentrations continue to increase, so any changes in atmospheric precipitation rates would also, on average, be expected to be more acidic. This is an interesting example of how the earth is actually a very large bioreactor. Changing one variable can profoundly change the entire system; in this instance the release of one gas changes numerous physical (e.g., temperature) and chemical (e.g., precipitation pH) factors, which in turn evoke a biological response (biome and ecosystem diversity).

Carbon dioxide is a prominent greenhouse gas, so it is logical to expect the increases in its atmospheric concentrations to be associated with changes in climate. Modeling climate change is difficult given the many drivers and constraints involved (See Fig. 4.13). Even slight changes can affect biological systems. Ecological structure and function can be particularly sensitive. The effects can worsen or improve depending on the interactions of these drivers and constraints. For example, wetland and stream sediment structures may change if anaerobic microbial decomposition increases, which would result in increasing releases of CH_4, which would mean increasing global temperatures, with all other factors being held constant. However, if greater biological activity and increased photosynthesis is triggered by the increase in CO_2, and wetland depth is decreased, CH_4 global concentrations would fall, leading to less global temperature rise.[b] Conversely, if this increased biological activity and photosynthesis leads to a decrease forest floor detritus mass, then less anaerobic activity may lead to lower releases of CH_4. Given the diversity of most ecosystems, there will be increases and decreases and changes in biological activity within the same ecosystem at various scales.

Another important carbon-based greenhouse gas is methane (CH_4), the product of anaerobic decomposition from myriad natural and human activities and facilities, such as from landfills. Methane also is emitted during the combustion of fossil fuels and cutting and clearing of forests. The concentration of CH_4 in the atmosphere had been steady at about 0.75 for over a thousand years, and then increased to 0.85 ppm in 1900. Since then, in the span of only a hundred years,

b. I am reminded by some of my ecologist colleagues that many ecosystems are not carbon-limited, but are often nitrogen and phosphorous-limited, so an increase in CO_2 concentrations would have little effect on these systems.

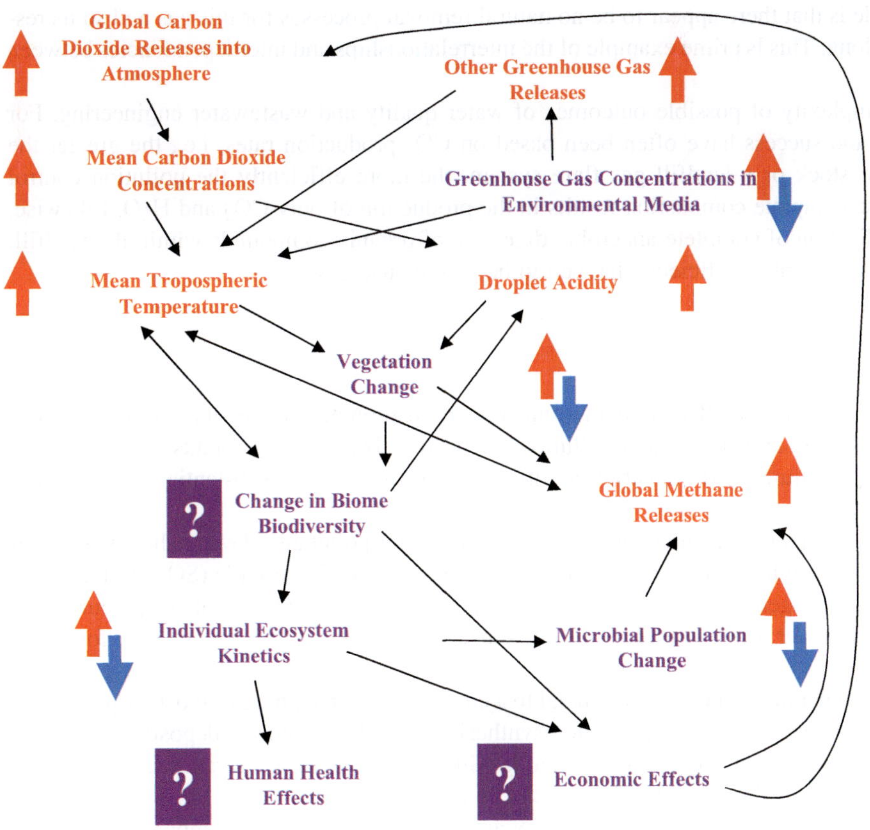

FIG. 4.13 Systematic view of changes in tropospheric carbon dioxide. Thick *arrows* indicate whether this factor will increase *(up arrow)*, decrease *(down arrow)*, or will vary depending on the specific scenario. For example, certain greenhouse gas releases have decreased, e.g., the chlorofluorocarbons, and certain substances have cooling effects, e.g., sulfate-containing aerosol. Question mark indicates that the type and/or direction of change are unknown or mixed. *(Data from D. Vallero, Environmental Biotechnology: A Biosystems Approach, Elsevier Science, 2015.)*

CH_4 concentrations have skyrocketed to 1.7 ppm. Methane is removed from the atmosphere by reaction with the hydroxyl radical (•OH) as

$$CH_4 + \cdot OH + 9O_2 \rightarrow CO_2 + 0.5H_2 + 2H_2O + 5O_3 \tag{4.15}$$

This indicates that the reaction creates carbon dioxide, water vapor, and ozone, all of which are greenhouse gases, so the effect of one molecule of methane profoundly alters the the greenhouse effect.

The difference in gas concentrations and the exchange coefficients between the atmosphere and surface waters determines how quickly a molecule of gas can move across the ocean-atmosphere boundary. It takes about 1 year to equilibrate CO_2 in the ocean surface with atmospheric CO_2, thus large atmosphere-ocean differences in CO_2 concentrations are common. Biota and ocean circulation account for most of the difference. The oceans contain vast C reservoirs to and from which the atmosphere exchanges occur since CO_2 reacts with water to form carbonic acid and its dissociation products. With the increased atmospheric CO_2 concentrations, the interaction with the ocean surface alters the chemistry of the seawater resulting in ocean acidification [83,84].

Ocean uptake of anthropogenic CO_2 is primarily a physical response to increasing atmospheric CO_2 concentrations. Increasing the partial pressure of a gas in the atmosphere directly above the body of water causes the gas to diffuse into that water until the partial pressures across the air-water interface are equilibrated. The effects are complex, e.g., increasing CO_2 also modifies the climate which in turn may change ocean circulation, which changes the rate of ocean CO_2 uptake. Marine ecosystem changes also alter the uptake [84]. The vulnerability of ocean systems to slight changes should encourage precaution when considering large-scale ocean projects, i.e., geoengineering, as possible solutions to climate change [85].

Halocarbons, the classes of compounds involved in the destruction of the stratospheric ozone layer, are also promote global warming. The most effective global warming gases are CFC-11 and CFC-12, both of which are no longer manufactured, and the banning of these substances has shown a leveling off and decrease in the size of the so-called stratospheric "ozone hole."

Many, but not all, greenhouse gases are C-based. Nitrous oxide is emitted to the atmosphere predominantly due to human activities, especially the cutting and clearing of tropical forests, which are vital sources of atmospheric oxygen.

The greatest problem with nitrous oxide is that there appear to be no natural removal processes for this gas and so its residence time in the stratosphere is quite long. This is prime example of the interrelationships and interdependencies between the C and N cycles.

The cycling of C indicates the complexity of possible outcomes of water quality and wastewater engineering. For example, pollution control efficiency and success have often been based on CO_2 production rates, i.e., the greater the amount of CO_2 leaving an incinerator stock or a landfill gas flare system, the more efficiently the pollution control equipment is operating. This is because complete combustion results in the production of only CO_2 and H_2O. Likewise, the production of CH_4 has been an indication of complete anaerobic digestion of organic compounds within the landfill. Unfortunately, both indicators of pollution control efficiency happen to be greenhouse gases.

9 Nutrient cycling

Nitrogen (N) and sulfur (S), along with phosphorous (P) and potassium (K), are macronutrients given plant life's dependence on them for survival and growth. However, from a water pollution perspective, N and S compounds deserve special attention, given that many municipal, agricultural, and industrial facilities generate and release substantial amounts of N and S compounds into the environment.

Nitrate and nitrite compounds, as well as ammonia and amines are important water pollutants. Two of the six National Ambient Air Quality Standards (NAAQS) pollutants are oxides of these elements, i.e., sulfur dioxide (SO_4) and nitrogen dioxide (NO_2). In addition, the oxides of nitrogen (NO_x), made up of NO_2 and nitric oxide (NO), are major precursors of another NAAQS pollutant, i.e., ozone. However, some of these same and other nutrient compounds are essential for crops, woodlands, and other ecosystems.

In addition to N and S, the cycling of other nonmetals is also crucial to a complete understanding of most air pollutants. In fact, P molecules are essential parts of energy transfer during photosynthesis. When P compounds deposit onto surface waters, the increased P concentrations accelerate eutrophication and, along with increasing N concentrations, contribute to the so-called dead zones in large water bodies, such as the Chesapeake Bay. In fact, P and K are intricately woven into the N and S cycles. P is a component of atmospheric deposition, along with N and S compounds. The P component is usually higher when wet deposition is greater than dry deposition since P is usually bound to aerosols. Thus, in areas where large amounts of P are applied as a fertilizer to land or in arid regions with strong winds to transport soils, airborne P can be substantial, albeit still less than airborne N [86]. Most of the total P in the atmosphere is in mineral form as aerosols, but there are also large global sources of P from biomass burning, and even less from energy and industrial combustion sources [87].

Phosphorous is also indirectly responsible for pollution since the production of P fertilizer releases fluoride compounds. The movement of fluoride through the atmosphere and into a food chain illustrates an air–water interaction at the local scale ($<100\,km$) [88]. Other industrial sources of fluoride include aluminum processing and glass manufacturing plants. Domestic livestock in the vicinity of substantial fluoride sources are exposed to fluoride by ingestion of forage crops. Fluoride released into the air by industry is deposited and accumulated in vegetation. Its concentration is sufficient to cause damage to the teeth and bone structure of the animals that consume the crops. Thus, if a waste stream has fluoride, the amount should be known by the engineers and managers.

Potassium is also part of the biogeochemistry of N and S. For example, when base cations like K deposited on soil and other surfaces, they increase surface's pH. This alkalinity buffers the acidity generated by N and S acidic compounds (e.g., H_2NO_3 and H_2SO_4, respectively).

9.1 The nitrogen cycle

Most of the Earth's nitrogen resides in the atmosphere, but is also stored in reservoirs that include living organisms, soils, sediment, and surface waters [89]. About 79% of the molecules in Earth's atmosphere are molecular nitrogen (N_2). From these sinks, N cycles between the biosphere and the atmosphere (See Fig. 4.14). All organisms must metabolize N into larger molecules; notably amino acids, proteins, and deoxyribonucleic acid (DNA). Amino acids contain the functional group amine (NH_2) and are the building blocks of proteins. DNA consists of adenine, cytosine, guanine, and thymine (T), which are comprised of rings of carbon and nitrogen atoms, with various side chains.

The N_2 is not directly available to most organisms (a major exception being those that have N-fixing bacteria found in legume roots). The atmospheric N_2 is converted to bioavailable N species (e.g., nitrates; NO_3^-) abiotically (e.g., lightning strikes or fires) and biotically (e.g., by the aforementioned N-fixing bacteria known as diazotrophs).

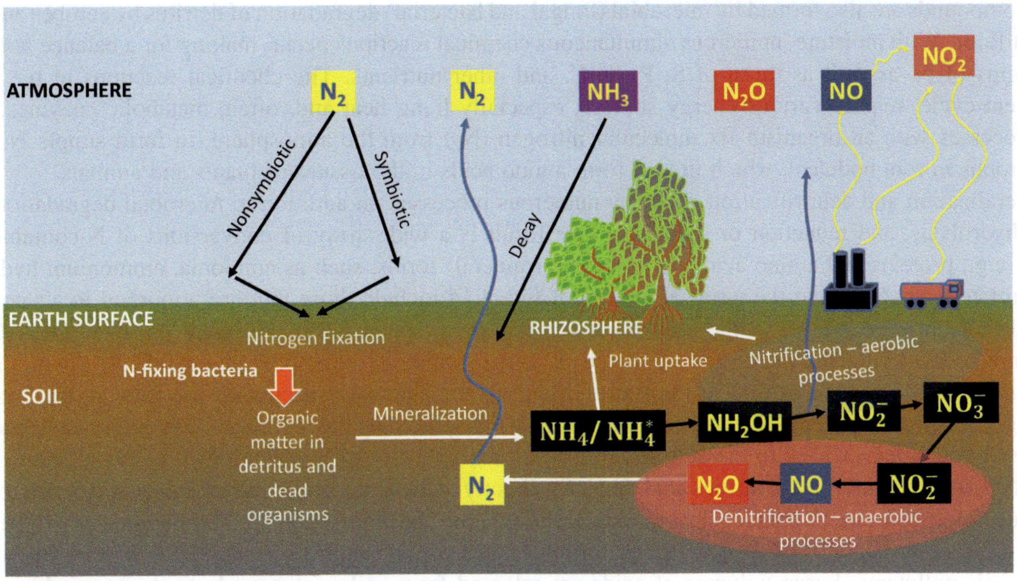

FIG. 4.14 Nitrogen cycling in the hydrosphere, biosphere, atmosphere, and lithosphere.

Many ecosystems are N-limited due to N demands of the flora and microorganisms living within these systems. Indeed, along with P, the amount of N is the major limiting factor on the rate of net primary production of biomass in most ecosystems. That is, an increase in bioavailable N is associated with an increase in production of biomass of these organisms in an ecosystem. For example, increasing the amount of carbon in a lake system will have little effect on the production of biomass, since most systems have ample mass of bioavailable C. Conversely, increasing bioavailable N in most lake systems increases total biomass. Thus, N is a limiting factor for plant growth.

The N-cycle mirrors the trophic state and energy cycles in ecosystems. Producers convert mineral forms of N to N-containing molecules, e.g., amino acids, and DNA. Consumers then receive N by consuming producers and forming even larger N-containing molecules, e.g., proteins (built from amino acids). Higher trophic state consumers receive N from consuming lower-level organisms, i.e., both consumers and producers. Decomposers convert N in the opposite direction, i.e., mineralization, converting the N-containing, organic molecules of dead organisms to inorganic compounds, e.g., ammonium (NH_4^+) salts. The NH_4^+ salts have high aqueous solubility and readily absorb to soil particles. At these sites, bacteria and other microbes can oxidize the NH_4^+ into nitrite (NO_2^-) compounds and in turn transform to nitrate (NO_3^-) compounds. This oxidation process is known as nitrification. The NO_3^- compounds are quite bioavailable to most plant life (i.e., macrophytes). These dissolved oxidized forms of N are reduced (i.e., denitrified) by bacteria known as denitrifiers. This is a major process by which N is returned to the atmosphere.

Ionization is an important part of N and S cycling, as it is for all nutrient cycling. This is due to the configuration of electrons in an atom. The arrangement of the electrons in the atom's outermost shell, i.e., valence, determines the ultimate chemical behavior of the atom. The outer electrons become involved in transfer to and sharing with shells in other atoms, i.e., forming new compounds and ions. An atom will gain or lose valence electrons to form a stable molecule that have the same number of electrons as the noble gas nearest the atom's atomic number. As mentioned, the N cycle includes three principal forms that are soluble in water under environmental conditions: the cation (positively charged ion) ammonium (NH_4^+) and the anions (negatively charged ions) nitrate (NO_3^-) and nitrite (NO_2^-). Nitrates and nitrites combine with various organic and inorganic compounds. Once taken into the body, NO_3^- is converted to NO_2^-. Since NO_3^- is soluble and readily available as a nitrogen source for plants (e.g., to form plant tissue such as amino acids and proteins), farmers are the biggest users of NO_3^- compounds in commercial fertilizers (although even manure can contain high levels of NO_3^-).

Nitrogen in several forms finds its way from the atmosphere to the soil through abiotic and biotic processes. As mentioned, NO_3^- compounds are formed from lightning and by nitrogen-fixing bacteria in legumes' root nodules. The Rhizobia (genera *Rhizobium*, *Sinorhizobium* and *Bradyrhizobium*) are gram-negative soil bacteria that are motile, rod-shaped, and aerobic. These bacteria that infect legume roots in a symbiotic relationship with legumes, e.g., *Medicago, Melilotus* and *Trigonella*. The symbiotic relationship results from the bacteria fixing atmospheric nitrogen, providing ammonium for protein production in the plant. In exchange, the bacteria obtain energy from the plant [90].

Nitrate compounds are also formed by microbial (fungal and bacterial) degradation of detritus by aerobes and anaerobes on the forest floor. With moisture, numerous simultaneous chemical reactions occur, making for a balance among various chemical forms of N, as well as those of S, P, K, C, and other nutrients. The chemical reactions in the N-cycle, as in any nutrient cycle, require various energy sources, especially light, heat and, often, metabolic enzymes. Some biochemical processes with an organism fix molecular nitrogen (N_2) from the atmosphere, to form simple N-compounds (e.g., diazotrophs in root nodules), which in turn form amino acids in the tissues of plants and animals.

The mineralization and denitrification occur by numerous processes, in addition to microbial degradation, including photolysis, hydrolysis, and reduction or oxidation. The result is a wide array of conversions of N-containing organic compounds (e.g., proteins and amino acids) to inorganic (mineral) forms, such as ammonia, ammonium hydroxide, and nitrite and nitrate salts. Note that the gases at the top of Fig. 4.14 include those that are important in air pollution, and the ions and molecules at and below the earth's surface are important to water pollution. For example, NO is one of the compounds involved in the photochemistry that leads to the formation of the pollutant ozone (O_3) in the troposphere. Note also that macrophytes are central in the figure. Much of the chemistry occurs on the floor in the detritus where microbes degrade complex molecules. Nutrients in the soil are transported by the root's capillary action to plant cells. Gases are transpired through leaves back to the atmosphere.

In soils, NO_2^- compounds release nitrous acid (HONO) directly to the atmosphere. When soil contains elevated amounts of nitrates due to fertilization, HONO is released to the atmosphere. The more acidic the soil the greater will be the release of HONO. In the troposphere, HONO leads to the formation of hydroxyl radicals, which both degrade and increase the deposition of air pollutants. Large volumes of acids are released from soil continuously to the atmosphere. Soils with high N concentrations form the acids from NO_2^- and NO_3^- ions. These anions are first released into the soil by microbes that have transformed ammonium and nitrate ions into nitrite ions. Increasing soil acidity produces high nitrite concentrations, leading to greater concentrations of HONO emitted to the troposphere [91]. About 30% of the primary OH radical production is attributed to the photolysis of HONO.

Thus, what may at first appear as a soil-water interaction, e.g., fertilizer incorporated into soil increase crop yield, is indeed a soil-water-biota process, followed by an air-water interaction that involves chemical transformation of air pollutants, followed by precipitation, which is a link between the atmosphere and the lithosphere and hydrosphere at the earth's surface. In this instance, water and soil pollution increase while air pollution decreases, at least in the case of ammonium, nitrates, nitrites, and oxides of nitrogen.

When organic compounds are degraded biotically, especially by microbes like nitrifying bacteria, oxidized forms of nitrogen (NO_x) are released to the atmosphere. Thus, the flux of nitric oxide (NO) and nitrogen dioxide (NO_2) from the soil to the lower troposphere is inversely related to the rate of degradation of organic compounds in the soil.

Ion exchange is an example of sorption; that is, the movement of a chemical species from the liquid or gas phase to the solid phase. Vascular plants and microbes grow as a function of available nutrients and other cycles within the forest ecosystem [92]. With this growth, compounds of N and other nutrients find their way to the atmosphere.

Nitrogen has three common oxidation states, +3, +4 and +5. Like carbon, N can form many compounds, even with a single, other N element. For example, two N atoms form the diatomic molecule, N_2. The N atom can also combine with oxygen to form N_2O, NO, NO_2, or N_2O_5, all important air pollutants or precursors to air pollutants. The N-atom can combine with hydrogen to form ammonia and ammonium compounds. Indeed, the N atom can react to form nitrogenous organic compounds, such at the N-polycyclic aromatic hydrocarbons (N-PAHs).

As mentioned, nitrite is an intermediate compound in both nitrification and denitrification. The N in the nitrate anion (NO_3^-) has an oxidation state of +5 (i.e., three O atoms each with -2 charge plus 5 yields the nitrate charge of -1). The nitrite anion has a trivalent N, i.e., N^{3+} (i.e., two O atoms each with -2 charge plus 3 yields the nitrite charge of -1). The major mechanism for producing N^{3+} in soil is by biological nitrification and denitrification processes. Nitrifying microbes produce nitrite ions from ammonium. Denitrifying microbes produce nitrite from nitrate [91].

The N compounds leave water, soil, wastes, and fuels to enter the troposphere by several mechanisms. In addition to the ordinary concentrations of molecular N, various nitrogen compounds are formed from reactions ranging from very fast (especially combustion) to quite slow, multistage (microbial) reaction rates. The two principal air pollutants regulated throughout the world are nitric oxide (NO) and nitrogen dioxide (NO_2). Nitric oxide (NO) is a colorless, odorless gas and is essentially insoluble in water. NO_2 has a pungent acid odor and is somewhat soluble in water.

Air pollution experts now refer to NO and NO_2 collectively as NO_x. This is in part because the NO lifetime after emission is quite short (> minute < hour) [93]. The simple oxidation of molecular N at high temperature is:

$$N_2 + O_2 \xrightarrow{\Delta} 2NO \tag{4.16}$$

The emitted NO rapidly undergoes photochemical transformation to NO_2:

$$2NO + O_2 \rightarrow 2NO_2 \tag{4.17}$$

Indeed, NO an NO_2 can interconvert, depending on redox and photochemical conditions in the troposphere and stratosphere, For example, in sunlight, ultraviolet radiation splits NO_2 into NO and O. So, it is often preferred to consider the dynamic relationship of the two molecules rather than to ascribe a single speciation, especially for characterizing the atmosphere. However, the two compounds must be distinguished in risk assessments given their very different health effects.

Mobile sources (i.e., motor vehicles, trains, boats, and aircraft) comprise the largest contributor to NO_x emissions in the United States (59%; 12.7 million tons in 2002). However, there is a likely shift toward stationary sources as more electric vehicles enter the marketplace and electricity generating plants providing the energy are sources of NO_x. Indeed, the U.S.'s largest contributing stationary source category is electric generation (22%; 4.6 million tons in 2002), followed by industrial factories (11%; 2.4 million tons in 2002). The eastern and midwestern U.S. regions had the highest emissions. Europe has similar emission profiles for NO_x, with the largest percentage coming from mobile sources (road and non-road transport), and the largest stationary source being energy production and distribution. Although both NO and NO_2 can be harmful to humans and ecosystems, NO is also actually essential to metabolism, so is harmful at much higher concentrations than that of NO_2. On the other hand, NO_2 is inherently toxic, causing respiratory problems. Both compounds are precursors to tropospheric ozone formation, so contribute indirectly to health problems when the smog and O_3 are inhaled. The NO_x reaction from NO to NO_2 likely involves water and an intermediate, i.e., nitrous acid (HONO):

$$4NO + O_2 + 2H_2O \rightarrow 4HONO \tag{4.18}$$

The larger suite of N air pollution compounds is known as total reactive nitrogen, denoted as NO_y. This suite includes NO and NO_2, plus their oxidation products [94]. The major vapor-phase and particulate constituents of NO_y include: NO, NO_2, nitric acid (HNO_3), peroxyacetyl nitrate (PAN), HONO, organic nitrates and particulate nitrates:

$$NO_y = NO_2 + NO + HNO_3 + PAN + 2N_2O_5 + HONO + NO_3 + NO_3^- \text{ compounds} + NO_3^- \text{ aerosols} \tag{4.19}$$

Every NO_y compound is derived directly from NO_x emissions or through transformations in the troposphere [95]. Most of the compounds originate from combustion processes. As noted, the atmosphere itself is the source of much of the nitrogen leading to the formation of nitrogen compounds. Because molecular nitrogen (N_2) is relatively non-reactive under most environmental conditions, it seldom enters into chemical reactions. However, under pressure and at very high temperatures, such as in an internal combustion engine or industrial boiler, the molecular N will react with O_2, i.e., reaction (4.16). This N_2 precursor is known as "thermal NO_x" since the oxides form at high temperatures, such as near burner flames in combustion chambers. Approximately 90% to 95% of the nitrogen oxides generated in combustion processes are in the form of NO. Other nitrogen oxides can also form at high heat and pressure, especially NO_2. Most motor vehicles around the world still employ temperature/high pressure internal combustion engines. Thus, as mentioned, such mobile sources contribute largely to the tropospheric concentrations of NO_x, making it a major mobile source air pollutant in terms of human health directly (e.g., respiratory toxicity of NO_2) and indirectly (i.e., NO_x as the key components in tropospheric ozone production). These conditions of high temperature and pressure can also exist in boilers such as those in power plants so NO_x is also commonly found in high concentrations leaving fossil fuel power generating stations.

Nitrogen oxides that form from the fuel or feedstock are called "fuel NO_x." Unlike the sulfur compounds, which mainly exit stationary source stacks as vapor-phase compounds, e.g., sulfur dioxide (SO_2) and other oxides of sulfur, a significant fraction of the fuel nitrogen burned in power plants and other stationary sources remains in the bottom ash or in unburned aerosols in the gases leaving the combustion chamber, i.e., the fly ash. Nitrogen oxides can also be released from nitric acid plants and other types of industrial processes involving the generation and/or use of nitric acid (HNO_3).

At temperatures well below those needed for combustion, such as those often present in the ambient atmosphere, NO_2 can form the molecule NO_2-O_2N or simply N_2O_4 that consists of two identical simpler NO_2 molecules. This molecular configuration is known as a dimer. The dimer N_2O_4 is distinctly reddish-brown and contributes to the brown haze that is often associated with photochemical smog incidents.

In addition to the health effects associated with NO_2 exposure, much of the concern for regulating emissions of nitrogen compounds is to suppress the reactions in the atmosphere that generate the highly reactive molecule ozone (O_3). Nitrogen oxides play key roles in O_3 formation. Ozone forms photochemically (i.e., the reaction is caused or accelerated by light energy) in the troposphere, where people live. Nitrogen dioxide is the principal gas responsible for absorbing sunlight needed for these photochemical reactions. So, in the presence of sunlight, the NO_2 that forms from the NO incrementally stimulates the photochemical smog-forming reactions because nitrogen dioxide is very efficient at absorbing sunlight in the ultraviolet portion of its spectrum. This is why ozone episodes are more common the summer and in areas with ample sunlight. Other chemical ingredients, i.e., ozone precursors, in O_3 formation include volatile organic compounds (VOCs), and carbon monoxide (CO). Governments regulate the emissions of precursor compounds to diminish the rate at which O_3 forms.

Cyanide (CN^-) is another N anion that is a water and air pollutant. Cyanide in the bloodstream impairs oxidative phosphorylation, a process by which oxygen is taken up to produce essential cellular energy in the form of adenosine triphosphate (ATP). This process transfers electrons from nicotinamide adenine dinucleotide (NADH) to form water from H^+ and O_2, through a series of reactions catalyzed by enzymes. With less O_2 available for the cytochrome C with which to react and hence complete the electron transport process, then ATP production is diminished. The high binding affinity of CN^- to the ferric ion in hemoglobin is responsible for the decrease in O_2 carried in the blood. Thus, CN^- binds preferentially to hemoglobin and prohibits oxygen binding.

Methemoglobinemia is a particularly troublesome outcome resulting from endogenous production of the CN^- anion results from infants drinking nitrate-containing water. Ingesting high concentrations of nitrates, e.g., in drinking water, can cause serious short-term illness and even death in infants 6 months or younger. Contributing factors infant methemoglobinemia susceptibility include higher pH in the stomach of the newborns compared to adults, which may encourage the growth of nitrate-reducing bacteria and the concomitant increase in the chemical reduction of NO_3^- to NO_2^- and higher plasma methemoglobin levels. Also, NO_3^- is more easily converted to NO_2^- than in adult hemoglobin and circulatory system is too immature in small children to return to normal hemoglobin. Small children are also susceptible given their greater fluid intake to body weight ratio compared to adults, as well as lower enzyme levels needed to convert methemoglobin to hemoglobin [96]. As a result, the increased NO_2^- concentrations interfere with the oxygen-carrying capacity of the blood. Especially in small children, nitrates compete successfully against molecular oxygen, causing the blood to carry methemoglobin as opposed to healthy hemoglobin, giving rise to clinical symptoms. This acute condition can deteriorate a child's health rapidly over a period of days, especially if the water source continues to be used.

Long-term, elevated exposures nitrates and nitrites can cause an increase in the kidneys' production of urine (diuresis), increased starchy deposits and hemorrhaging of the spleen [97]. Some studies suggest that elevated NO_3^- and NO_2^- in exposures may also elicit other effects, e.g., increased miscarriage rates and anencephaly. Nitrate exposure may also cause hypothyroidism by mimicking and blocking iodide from reaching the thyroid [98].

9.2 Interactions between sulfur and nitrogen

From the standpoint of water pollution, the biogeochemical cycles must not be limited to single elements. After a compound is released, it reacts with numerous other compounds and changes physically and chemically. It is transported advectively, is dispersed, and continues to undergo physical, chemical, and biological changes before reaching a water body or other environmental fate. This is demonstrated by the interactions of N and S in acid deposition, as well as by the integration of myriad atmospheric chemical species that influence and are influenced by N and S [95]. This includes both the formation of water pollutants like sulfuric acid and sodium hydroxide (NaOH), and air pollutants, likelike ammonia, hydrogen sulfide (H_2S), ozone and particulate matter. Nitrogen and sulfur reactions also influence the transformation and changes in bioavailability of toxic pollutants, e.g., changing the pH of water that increases the rates of formation of reduced forms of mercury (Hg) and the bioavailability of semivolatile organic compounds (SVOCs) like dioxins and organochlorine pesticides.

The oxidized chemical species of sulfur and nitrogen, such as SO_2 and nitrogen dioxide (NO_2) respectively, form acids when they react with water. This can occur in any media, e.g., in mining waste and ash runoff to surface and ground water and in the atmosphere. The lowered pH is responsible for numerous environmental problems, popularly known as acid rain. Many compounds contain both N and S along with the typical organic elements (C, H and O). The reaction for the combustion of such compounds, in general form, is:

$$C_aH_bO_cN_dS_e + 4a + b - 2c) \rightarrow aCO_2 + \left(\frac{b}{2}\right)H_2O + \left(\frac{d}{2}\right)N_2 + eS \tag{4.20}$$

Reaction (4.20) demonstrates the incremental complexity as additional elements enter the reaction. In the real world, pure reactions are rare. The environment is filled with mixtures and heterogeneous reactions. Reactions can occur in sequence, parallel or both. For example, a feedstock to a municipal incinerator contains myriad types of wastes, from garbage to household chemicals to commercial wastes, and even small (and sometimes) large industrial wastes that may be illegally dumped. The N-content of typical cow manure is about 5 kg per metric ton (about 0.5%). If the fuel used to burn the waste also contains S along with the organic matter, then the five elements will react according to the stoichiometry of Reaction (4.20).

Unquestionably, combustion specifically and oxidation generally are very important processes that lead to N and S pollutants. But they are certainly not the only ones. As mentioned, oxidation and reduction of N and S occur in ecosystems affect trophic state and energy levels. The formation of sulfur dioxide (SO_2) and nitric oxide (NO) by acidifying molecular sulfur is a redox reaction:

$$S(s) + NO_3^-(aq) \rightarrow SO_2(g) + NO(g) \tag{4.21}$$

The designations in parentheses give the physical phase of each reactant and product: "s" for solid; "aq" for aqueous; and "g" for gas.

The oxidation half-reactions for this reaction are:

$$S \rightarrow SO_2 \tag{4.22}$$

$$S + 2H_2O \rightarrow SO_2 + 4H^+ + 4e^- \tag{4.23}$$

The reduction half-reactions are:

$$NO_3^- \rightarrow NO \tag{4.24}$$

$$NO_3^- + 4H^+ + 3e^- \rightarrow NO + 2H_2O \tag{4.25}$$

Consequently, the balanced oxidation-reduction reactions are:

$$4NO_3^- + 3S + 16H^+ + 6H_2O \rightarrow 3SO_2 + 16H^+ + 4NO + 8H_2O \tag{4.26}$$

$$4NO_3^- + 3S + 4H^+ \rightarrow 3SO_2 + 4NO + 2H_2 \tag{4.27}$$

9.3 The sulfur cycle

Compounds of sulfur (S), as those of N, exist at atmospheric concentrations well in excess of what would be expected from equilibrium geochemistry in an atmosphere with 21% O_2 [99]. Sulfur is released to the atmosphere as either reduced forms, e.g., hydrogen sulfide, or oxidized forms, e.g., sulfur dioxide. Both forms include air pollutants. Hydrogen sulfide is oxidized to sulfur dioxide in a three-step process. Note that the hydroxyl radical initiates the transformation from hydrogen sulfide to sulfur dioxide:

$$H_2S + HO^\bullet \rightarrow HS^\bullet + H_2O \tag{4.28}$$

$$HS^\bullet + O_2 \rightarrow HO^\bullet + SO^\bullet \tag{4.29}$$

$$SO^\bullet + O_2 \rightarrow SO_2 + O^\bullet \tag{4.30}$$

The atmospheric reactions of SO_2 are very complex and proceed through three different pathways to the sulfate ion (SO_4^{2-}). Sulfur dioxide can react with the hydroxyl radical to form the HSO_3 radical, which then can react with another hydroxyl radical to form water and SO_3 or H_2SO_4. Sulfur dioxide has sufficient aqueous solubility to dissolve in water droplets where it can react with oxygen gas to form SO_4^{2-}. The third pathway to sulfate occurs when sulfur dioxide reacts with hydrogen peroxide to form sulfuric acid:

$$HO^\bullet + SO_2 \rightarrow HOSO_2^\bullet \tag{4.31}$$

$$HOSO_2^\bullet + HO^\bullet \rightarrow H_2SO_4 \tag{4.32}$$

$$SO_{2(aqueous)} + [O]_{(aqueous)} \rightarrow H_2SO_4 \tag{4.33}$$

$$SO_2 + H_2O_2 \rightarrow H_2SO_4 \tag{4.34}$$

With sufficient residence time in the atmosphere, S will be oxidized to the sulfate ion, usually as sulfuric acid (H_2SO_4). Ammonia (NH_3), the most common base in the atmosphere, reacts with $H_2SO_4$to form ammonium bisulfate (NH_4HSO_4) and ammonium sulfate ($(NH_4)_2SO_4$). Sulfuric acid, NH_4HSO_4 and $(NH_4)_2SO_4$ all are hydroscopic substances, i.e., readily dissolved in water. Thus, they wash out of the atmosphere during precipitation events [18].

TABLE 4.12 Theoretical amounts of various agents required to oxidize $1\,mg\,L^{-1}$ of sulfide ion.

Oxidizing agent	Amount ($mg\,L^{-1}$) needed to oxidize $1\,mg\,L^{-1}$ of S^{2-} based on practical observations	Theoretical stoichiometry ($mg\,L^{-1}$)
Chlorine (Cl_2)	2.0–3.0	2.2
Chlorine dioxide (ClO_2)	7.2–10.8	4.2
Hydrogen peroxide (H_2O_2)	1.0–1.5	1.1
Potassium permanganate ($KMnO_4$)	4.0–6.0	3.3
Oxygen (O_2)	2.8–3.6	0.5
Ozone (O_3)	2.2–3.6	1.5

Data from O.T.F. Water Quality Association, Ozone for POU, POE & Small Water System Applications, Lisle, Illinois, 1999.

A reduced form of sulfur that is highly toxic and an important pollutant is H_2S. Certain microbes, especially bacteria, reduce nitrogen and sulfur, using the N or S as energy sources through the acceptance of electrons. For example, sulfur-reducing bacteria can produce H_2S by chemically changing oxidized forms of sulfur, especially sulfates (SO_4). To do so, the bacteria must have access to the sulfur, i.e., it must be in the water, which can be in surface or ground water, or the water in soil and sediment. These sulfur-reducers are often anaerobes, i.e., bacteria that live in water where concentrations of molecular oxygen (O_2) are deficient. The bacteria remove the O_2 molecule from the sulfate leaving only the S, which in turn combines with hydrogen (H) to form gaseous H_2S. In ground water, sediment, and soil water, H_2S is formed from the anaerobic or nearly anaerobic decomposition of deposits of organic matter, e.g., plant residues. Thus, redox principles can be used to treat H_2S contamination, i.e., the compound can be oxidized using several different oxidants (See Table 4.12). Strong oxidizers, like molecular oxygen and hydrogen peroxide, most effectively oxidize the reduced forms of S, N, or any reduced compound.

Sulfur is also an important component of aerosols (See Table 4.13). For example, diesel particulate matter is formed by several simultaneous physical processes during cooling and dilution of vehicle and other internal combustion exhaust, i.e., nucleation, coagulation, condensation, and adsorption (See Fig. 4.15). The core of the particles is formed by nucleation and coagulation from primary spherical particles consisting of solid carbonaceous matter (known as elemental carbon; EC) and ash (metals and other elements). By coagulation, adsorption, and condensation, various organic and S compounds (e.g., sulfates) are added and combined with other condensed material [100].

Combustion of fuels and the emission of particulate matter is an example of the formation of an air pollutant that ultimately reaches water bodies. For example, the small diameter of diesel aerosols (i.e., aerodynamic diameter $<0.5\,\mu m$)[c] makes for very large surface areas. These particles have a very large surface area per gram of mass, which allows them

TABLE 4.13 Typical chemical composition (percent) of fine particulate matter ($PM_{2.5}$).

	Eastern U.S.	Western U.S.	Diesel $PM_{2.5}$
Elemental carbon	4	15	75
Organic carbon	21	39	19
Sulfate, nitrate, ammonium	48	35	1
Minerals	4	15	2
Unknown	23	–	3

Data from N.C.f.E. Assessment, Air Quality Criteria for Particulate Matter, US Environmental Protection Agency, Office of Research and Development..., 1996.

c. The aerodynamic diameter of a particle is defined as a sphere with the density of water, which is $1\,g\,cm^{-3}$ and which settles in stationary air at the same velocity as the particle velocity.

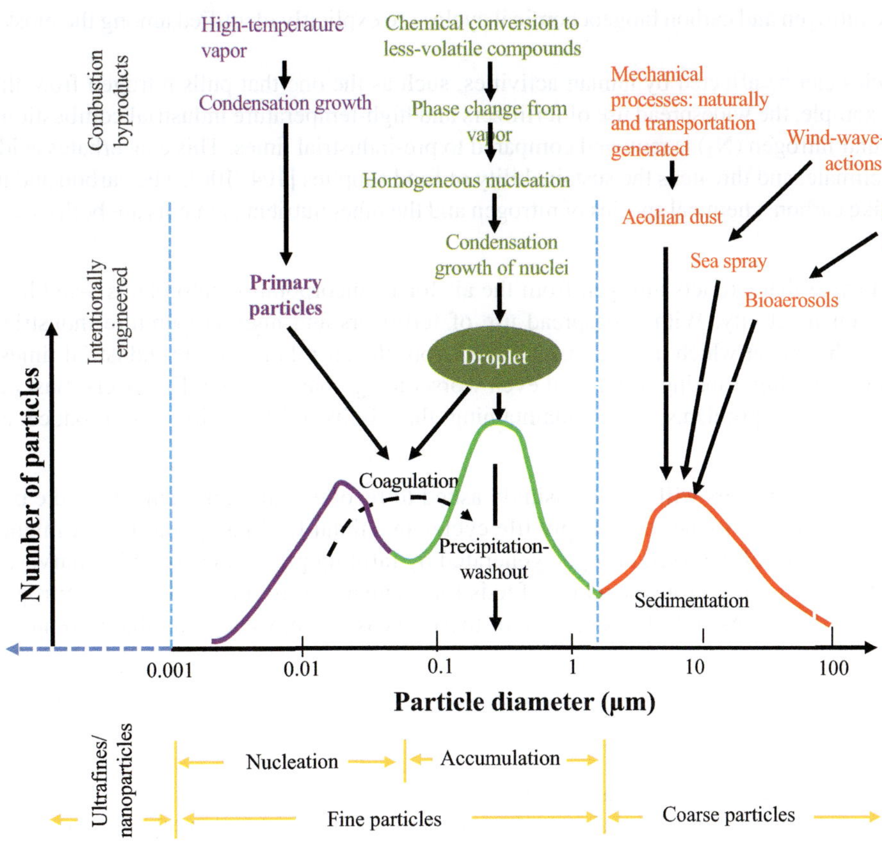

FIG. 4.15 Trimodal size distribution of aerosols in the atmosphere. *(Data from D.A. Vallero, Fundamentals of Air Pollution, fifth ed., Elsevier Academic Press, Waltham, MA, 2014, p. 999 pages cm, D.A. Vallero, Air Pollution Calculations: Quantifying Pollutant Formation, Transport, Transformation, Fate and Risks, Elsevier, 2019, D.A. Vallero, Air pollution physics, in: Methods and Calculations in Environmental Physics, AIP Publishing LLC, Melville, NY, 2022, pp. 7-1–7-70, X. Fang, et al., Study on particle size distribution characteristics in urban roadside microenvironment using an ELPI+, Water Air Soil Pollut. 233(11) (2022) 467, K. Schifftner, Air Pollution Control Equipment Selection Guide, CRC Press, 2021, D. Vallero, Air pollution, in: Kirk-Othmer Encyclopedia of Chemical Technology, John Wiley & Sons, Inc., Hoboken, NJ, 2015, K.B. Schnelle Jr., R.F. Dunn, M.E. Ternes, Air Pollution Control Technology Handbook, CRC Press, Boca Raton, FL, 2015, U.S. Environmental Protection Agency, Basic Air Pollution Meteorology Student Guidebook (SI 409), U.S. EPA, Research Triangle Park, NC, 2005, R.W. Boubel, D.L. Fox, B.Tuner, A.C. Stern, Fundamentals of Air Pollution, third ed., Academic Press, Amsterdam; Boston, 1995, and W. Boubel, D. Fox, D. Turner, A. Stern, Sources of air pollution, Fundamentals of Air Pollution, 1994, pp. 72–96.)*

to adsorb large quantities of ash, organic compounds, and sulfate. The organic constituents originate from unburned fuel, engine lubrication oil, and small quantities of partial combustion and pyrolysis products. During combustion, S compounds in the fuel are oxidized to sulfur dioxide (SO_2). From 1% to 4% of fuel S is oxidized to form H_2SO_4. Upon cooling, this sulfuric acid and water condense into an aerosol that is nonvolatile under ambient conditions [100]. The aerosols eventually directly deposit onto land and waterbody surfaces or react in the atmosphere and become other sulfur compounds. Both are processes that change water pH and deliver pollutants to water [101,102].

Thus, N and S compounds are ubiquitous in the hydrosphere. They are constituents of both essential and detrimental compounds. As such, these and other complete biogeochemical cycles must be understood to explain the formation and fate of water pollutants.

10 Biogeochemical cycles and decision making

Water quality decisions are often made among competing interests and perspectives, with many involving tradeoffs. For example, remediating a contaminated site, may call for excavation of soil, which is then incinerated. The complete incineration will generate and emit greenhouse gases. Determining whether this approach is successful and appropriate depends on the extent and quality of options from which these emissions occur. For example, if dioxin-laden soil is incinerated, this may have been the only viable approach to detoxify a very toxic and persistent compound. Releasing CO_2 in this case is truly a measure of success. The National Academy of Engineering has identified the most important challenges to the future of engineering. Both the nitrogen and carbon biogeochemical cycles are explicitly identified among the most pressing engineering needs [103].

Biogeochemical cycles can be affected by human activities, such as the one that pulls nitrogen from the air to incorporate into plants. For example, the widespread use of fertilizers and high-temperature industrial combustion have doubled the rate at which molecular nitrogen (N_2) is removed compared to pre-industrial times. This exacerbates acid rain, pollutes drinking water, affects climate, and threatens the sustainability of food supplies [104–106]. The carbon and nutrient cycles are intimately linked. Like carbon, chemical species of nitrogen and the other nutrient elements are both essential and toxic, depending on its form:

> The biogeochemical cycle that extracts nitrogen from the air for its incorporation into plants—and hence food—has become altered by human activity. With widespread use of fertilizers and high-temperature industrial combustion, humans have doubled the rate at which nitrogen is removed from the air relative to pre-industrial times, contributing to smog and acid rain, polluting drinking water, and even worsening global warming. Engineers must design countermeasures for nitrogen cycle problems, while maintaining the ability of agriculture to produce adequate food supplies [107].

Environmental scientists and engineers will be increasingly asked to recommend improvements to the biogeochemical cycling of C, such as enhancements to the food supply life cycles in animal feeding operations, farmlands, and rangelands. As evidence, less than half of the fixed nitrogen generated by farming practices is stored in harvested crops, with less than half of the nitrogen in those crops becoming foods that humans consume. Thus, fixed nitrogen escapes the system at various stages in the process [108]. Prevention of these releases is always preferable to removing them after the fact.

References

[1] D. Vallero, Environmental Biotechnology: A Biosystems Approach, Elsevier Science, 2015.

[2] D.A. Vallero, Fundamentals of Air Pollution, fifth ed., Elsevier Academic Press, Waltham, MA, 2014. p. 999 pages cm.

[3] T. Chouhan, I. Jaising, C.A. Alvares, Bhopal, the Inside Story: Carbide Workers Speak Out on the World's Worst Industrial Disaster, Apex Pr, 1994.

[4] D. Merritts, Hydrosphere: Questions and Answers. https://serc.carleton.edu/introgeo/socratic/examples/hydrosphere/questions.html. (Accessed 8 December 2023).

[5] WateResuse Foundation, Salinity Management Guide. https://watereuse.org/salinity-management/ew/ew_2.html. (Accessed 11 December 2023).

[6] R.L. Hanson, Evapotranspiration and droughts, US Geological Survey Water-Supply Paper, vol. 2375, 1991, pp. 99–104.

[7] R.C. Perry, Landscape Plants for Western Regions: An Illustrated Guide to Plants for Water Conservation (no Title), Land Design Publications, Pomona, CA, 1992.

[8] L.R. Costello, WUCOLS, Water Use Classification of Landscape Species: A Guide to the Water Needs of Landscape Plants, California Department of Water Resources, 1994.

[9] R.K. Rosenbaum, et al., USEtox human exposure and toxicity factors for comparative assessment of toxic emissions in life cycle analysis: sensitivity to key chemical properties, Int. J. Life Cycle Assess. 16 (8) (2011) 710–727.

[10] J.D. Neelin, A. Bracco, H. Luo, J.C. McWilliams, J.E. Meyerson, Considerations for parameter optimization and sensitivity in climate models, Proc. Natl. Acad. Sci. 107 (50) (2010) 21349–21354.

[11] U.S. Environmental Protection Agency, Climate Impacts on Water Resources. https://19january2017snapshot.epa.gov/climate-impacts/climate-impacts-water-resources_.html. (Accessed 1 March 2018).

[12] T.R. Karl, J. Melillo, T. Peterson, Global climate change impacts in the United States, in: Global Climate Change Impacts In The United States, 2009.

[13] U.S. Environmental Protection Agency, Saving Florida's vanishing shores. http://www.epa.gov/climatechange/Downloads/impacts-adaptation/saving_FL.pdf. (Accessed 1 November 2013).

[14] H. Sarptas, N. Alpaslan, D. Dolgen, GIS supported solid waste management in coastal areas, Water Sci. Technol. 51 (11) (2005) 213–220.

[15] J. Liu, J.-X. Chang, W.-G. Zhang, Groundwater level dynamic prediction based on chaos optimization and support vector machine, in: *2009 Third International Conference on Genetic and Evolutionary Computing*, IEEE, 2009, pp. 39–43.

[16] G.B. Pasternack, Does the river run wild? Assessing chaos in hydrological systems, Adv. Water Resour. 23 (3) (1999) 253–260.

[17] J.P. Kim, K.A. Hunter, M.R. Reid, Factors influencing the inorganic speciation of trace metal cations in fresh waters, Mar. Freshw. Res. 50 (4) (1999) 367–372.

[18] R. Foust, Atmospheric Reactions of Sulfur and Nitrogen, Northern Arizona University. http://jan.ucc.nau.edu/~doetqp-p/courses/env440/env440_2/lectures/lec37/lec37.htm (Accessed 2024).

[19] Y. Li, et al., Coupled carbon, sulfur, and nitrogen cycles mediated by microorganisms in the water column of a shallow-water hydrothermal ecosystem, Front. Microbiol. 9 (2018) 2718.

[20] K. Dack, M. Fell, C.M. Taylor, A. Havdahl, S.J. Lewis, Mercury and prenatal growth: a systematic review, Int. J. Environ. Res. Public Health 18 (13) (2021) 7140.

[21] P. de Almeida Rodrigues, R.G. Ferrari, L.N. Dos Santos, C.A.C. Junior, Mercury in aquatic fauna contamination: a systematic review on its dynamics and potential health risks, J. Environ. Sci. 84 (2019) 205–218.

[22] J.A. Dittman, et al., Mercury dynamics in relation to dissolved organic carbon concentration and quality during high flow events in three northeastern US streams, Water Resour. Res. 46 (7) (2010).

[23] W. Dong, L. Liang, S. Brooks, G. Southworth, B. Gu, Roles of dissolved organic matter in the speciation of mercury and methylmercury in a contaminated ecosystem in Oak Ridge, Tennessee, Environ. Chem. 7 (1) (2010) 94–102.

[24] J. Kuss, C. Zülicke, C. Pohl, B. Schneider, Atlantic mercury emission determined from continuous analysis of the elemental mercury sea-air concentration difference within transects between 50° N and 50° S, Glob. Biogeochem. Cycles 25 (3) (2011).

[25] T.M. Letcher, Mercury pollution-effect of covering mercury with water and other liquids, S. Afr. J. Sci. 75 (2) (1979) 80–81.

[26] C.-J. Lin, S.O. Pehkonen, The chemistry of atmospheric mercury: a review, Atmos. Environ. 33 (13) (1999) 2067–2079.

[27] O. Lindqvist, et al., Mercury in the Swedish environment—recent research on causes, consequences and corrective methods, Water Air Soil Pollut. 55 (1–2) (1991) xi–261.

[28] J. Munthe, K. Kindbom, O. Kruger, G. Petersen, J. Pacyna, Å. Iverfeldt, Emission, deposition, and atmospheric pathways of mercury in Sweden, in: Water, Air and Soil Pollution, 2001.

[29] M. Pascual, J.A. Dunne, Ecological Networks: Linking Structure to Dynamics in Food Webs, Oxford University Press, 2006.

[30] G. Rice, R. Ambrose, O. Bullock, J. Swartout, Mercury study report to Congress. Volume 3. Fate and transport of mercury in the environment, Environmental Protection Agency, Research Triangle Park, NC (United States), 1997. Office of Air Quality Planning and Standards.

[31] F. Slemr, W. Seiler, C. Eberling, P. Roggendorf, The determination of total gaseous mercury in air at background levels, Anal. Chim. Acta 110 (1) (1979) 35–47.

[32] S.X. Wang, et al., Mercury emission and speciation of coal-fired power plants in China, Atmos. Chem. Phys. 10 (3) (2010) 1183–1192, https://doi.org/10.5194/acp-10-1183-2010.

[33] J. Zhou, Z. Wang, X. Zhang, C.T. Driscoll, C.-J. Lin, Soil–atmosphere exchange flux of total gaseous mercury (TGM) at subtropical and temperate forest catchments, Atmos. Chem. Phys. 20 (24) (2020) 16117–16133.

[34] M. Ivarsson, A. Schnürer, S. Bengtson, A. Neubeck, Anaerobic fungi: a potential source of biological H_2 in the oceanic crust, Front. Microbiol. 7 (2016) 674.

[35] P. Khejornsart, M. Wanapat, Diversity of rumen anaerobic fungi and methanogenic archaea in swamp buffalo influenced by various diets, J. Anim. Vet. Adv. 9 (24) (2010) 3062–3069.

[36] A.S. Liggenstoffer, N.H. Youssef, M. Couger, M.S. Elshahed, Phylogenetic diversity and community structure of anaerobic gut fungi (phylum Neocallimastigomycota) in ruminant and non-ruminant herbivores, ISME J. 4 (10) (2010) 1225–1235.

[37] Y. Nagano, T. Nagahama, Fungal diversity in deep-sea extreme environments, Fungal Ecol. 5 (4) (2012) 463–471.

[38] D.A. Vallero, Thermal pollution, in: D.A. Vallero, T.M. Letcher (Eds.), Waste, Elsevier, Amsterdam, NV, 2019, pp. 381–404.

[39] D.A. Vallero, K.H. Reckhow, A.D. Gronewold, Application of multimedia models for human and ecological exposure analysis, in: International Conference on Environmental Epidemiology and Exposure, International Society of Exposure Analysis, Durham, NC, 2007.

[40] P. Wang, G. Zhou, Frost-heaving pressure in geotechnical engineering materials during freezing process, Int. J. Min. Sci. Technol. 28 (2) (2018) 287–296, https://doi.org/10.1016/j.ijmst.2017.06.003.

[41] L. Taylor, J. Leake, J. Quirk, K. Hardy, S. Banwart, D. Beerling, Biological weathering and the long-term carbon cycle: integrating mycorrhizal evolution and function into the current paradigm, Geobiology 7 (2) (2009) 171–191.

[42] J. Quirk, J.R. Leake, D.A. Johnson, L.L. Taylor, L. Saccone, D.J. Beerling, Constraining the role of early land plants in Palaeozoic weathering and global cooling, Proc. R. Soc. B Biol. Sci. 282 (1813) (2015) 20151115.

[43] J.A. Matthews, G. Owen, Endolithic lichens, rapid biological weathering and Schmidt hammer R-values on recently exposed rock surfaces: Storbreen glacier foreland, Jotunheimen, Norway, Geogr. Ann. Ser. A Phys. Geogr. 90 (4) (2008) 287–297.

[44] J. Leake, et al., Biological weathering in soil: the role of symbiotic root-associated fungi biosensing minerals and directing photosynthate-energy into grain-scale mineral weathering, Mineral. Mag. 72 (1) (2008) 85–89.

[45] R.D. Finlay, et al., Reviews and syntheses: biological weathering and its consequences at different spatial levels–from nanoscale to global scale, Biogeosciences 17 (6) (2020) 1507–1533.

[46] R. Finlay, et al., The role of fungi in biogenic weathering in boreal forest soils, Fungal Biol. Rev. 23 (4) (2009) 101–106.

[47] M. Andrews, J. Leake, B. Palmer, S. Banwart, D. Beerling, Plant and mycorrhizal driven silicate weathering: quantifying carbon flux and mineral weathering processes at the laboratory mesocosm scale, Appl. Geochem. 26 (2011) S314–S316.

[48] T.M. Letcher, D.A. Vallero, Waste: A Handbook for Management, Academic Press, 2019.

[49] D. Vallero, Environmental Biotechnology: A Biosystems Approach, Academic Press, 2010.

[50] W.J. Weber Jr., F.A. DiGiano, Process Dynamics in Environmental Systems, John Wiley & Sons, Inc., New York, NY, 1996.

[51] D.A. Vallero, Environmental Systems Science: Theory and Practical Applications, first ed., Elsevier, Amsterdam, NV, 2020.

[52] J.V. Leeds Jr., The Systems Approach to Environmental Pollution Control, vol. 53, Rice Institute Pamphlet-Rice University Studies, 1967. Special Issue.

[53] Environmental Resources Limited, Effects on aquatic ecosystems and fish, in: Acid Rain: A Review of the Phenomenon in the EEC and Europe, Unipub, New York, NY, 1983, pp. 89–111.

[54] R.C. Ahlert, Process dynamics in environmental systems, Environ. Prog. Sustain. Energy 16 (1) (1997) S6.

[55] W.M. Lewis, Ecological zonation in lakes, in: G.E. Likens (Ed.), Encyclopedia of Inland Waters, Academic Press, Oxford, 2009, pp. 416–422.

[56] J. Howard, Anthropogenic Soils, Springer, 2017.

[57] C. Dazzi, G.L. Papa, Anthropogenic soils: general aspects and features, Ecocycles 1 (1) (2015) 3–8.

[58] A. Lehmann, K. Stahr, Nature and significance of anthropogenic urban soils, J. Soils Sediments 7 (4) (2007) 247–260.
[59] W. Cannon, J.D. Horton, Soil geochemical signature of urbanization and industrialization–Chicago, Illinois, USA, Appl. Geochem. 24 (8) (2009) 1590–1601.
[60] NRCS, From the Ground Down, 2015, [Online] Available: https://www.nrcs.usda.gov/Internet/FSE_DOCUMENTS/nrcseprd366047.pdf.
[61] M. Schindler, M. Santosh, G. Dotto, L.F. Silva, M.F. Hochella Jr., A review on Pb-bearing nanoparticles, particulate matter and colloids released from mining and smelting activities, Gondwana Res. 110 (2022) 330–346.
[62] R. Herrington, Mining our green future, Nat. Rev. Mater. 6 (6) (2021) 456–458.
[63] R.B. Kaunda, Potential environmental impacts of lithium mining, J. Energy Nat. Resour. Law 38 (3) (2020) 237–244.
[64] L.W. Szajdak, I. Życzyńska-Bałoniak, K. Wegner, Free, bounded, and included in humic acids amino acids: thermal properties of humic acids from cropping systems, in: Bioactive Compounds in Agricultural Soils, Springer, 2016, pp. 25–52.
[65] T. Senn, A.R. Kingman, A review of humus and humic acids, Res. Ser. 145 (1973) 1–5.
[66] K. Kumada, Studies on the colour of humic acids: part 1 on the concepts of humic substances and humification, Soil Sci. Plant Nutr. 11 (4) (1965) 11–16.
[67] M. Kononova, I. Alexandrova, Formation of humic acids during plant residue humification and their nature, Geoderma 9 (3) (1973) 157–164.
[68] N. DiDonato, H. Chen, D. Waggoner, P.G. Hatcher, Potential origin and formation for molecular components of humic acids in soils, Geochim. Cosmochim. Acta 178 (2016) 210–222.
[69] D. Gołębiowska, et al., The influence of illumination, oxidation and pH on optical absorption spectra of humic acids taken from four types of soil in Poland, in: N. Senesi, T.M. Miano (Eds.), Humic Substances in the Global Environment and Implications on Human Health, Elsevier Science BV, 1994, pp. 481–486.
[70] H. Dziadowiec, Szybkość rozkładu różnych ściółek leśnych w świetle badań stacjonarnych (Speed of decomposition of various forest litter in the light of stationary studies), in: Material Conference on Humus of Forest Soils, vol. 201, Polish Geographical Society, Warsaw-Toruń, Poland, 1979, p. 210.
[71] D.A. Vallero, Soil and sediment physics, in: Methods and Calculations in Environmental Physics, AIP Publishing LLC, Melville, New York, 2022. pp. 8-1–8-34.
[72] A.D. McNaught, A. Wilkinson, Compendium of Chemical Terminology. IUPAC Recommendations, 1997.
[73] A.D. McNaught, A. Wilkinson, Compendium of Chemical Terminology, Blackwell Science Oxford, 1997.
[74] P. Klobes, R.G. Munro, Porosity and Specific Surface Area Measurements for Solid Materials, 2006.
[75] L.V. Rinaudi, W. Giordano, An integrated view of biofilm formation in rhizobia, FEMS Microbiol. Lett. 304 (1) (2010) 1–11.
[76] P.C. Bogino, M.d.l.M. Oliva, F.G. Sorroche, W. Giordano, The role of bacterial biofilms and surface components in plant-bacterial associations, Int. J. Mol. Sci. 14 (8) (2013) 15838–15859, https://doi.org/10.3390/ijms140815838 (in Eng).
[77] A. Znamensky, M. Filippova, Reaeration of soils, Pedology (2) (1940) 43–51.
[78] D.A. Vallero, Dicarboximide Fungicide Flux to the Lower Troposphere from an Aquic Hapludult Soil (Dissertation), Duke University, Durham, North Carolina, May, 2000.
[79] J.N. Meegoda, H. Hettiarachchi, P. Hettiaratchi, Landfill design and operation, in: Sustainable Solid Waste Management, American Society of Civil Engineers, Reston, VA, 2016, pp. 577–604.
[80] Laboratory of Ecohydrology (ECHO), Chapter 5: Darcy's Law. https://echo2.epfl.ch/VICAIRE/mod_3/chapt_5/main.htm. (Accessed 13 January 2024).
[81] Y.-F. Chen, et al., Non-Darcian flow effect on discharge into a tunnel in karst aquifers, Int. J. Rock Mech. Min. Sci. 130 (2020) 104319, https://doi.org/10.1016/j.ijrmms.2020.104319.
[82] D.A. Vallero, Air Pollution Calculations: Quantifying Pollutant Formation, Transport, Transformation, Fate and Risks, Elsevier, 2019.
[83] D.A. Vallero, Engineering aspects of climate change, in: Climate Change, second ed., Elsevier, 2015, pp. 547–568.
[84] National Oceanic and Atmospheric Administration, Carbon education tools, 2018. https://www.pmel.noaa.gov/co2/file/Carbon+Cycle+Graphics. (Accessed 2 March 2018).
[85] K. DePryck, M. Boettcher, The rise, fall and rebirth of ocean carbon sequestration as a climate solution, Glob. Environ. Change 85 (March) (2024) 102820, https://doi.org/10.1016/j.gloenvcha.2024.102820. In this issue.
[86] R.A. Duce, The impact of atmospheric nitrogen, phosphorus, and iron species on marine biological productivity, in: The Role of Air-Sea Exchange in Geochemical Cycling, Springer, 1986, pp. 497–529.
[87] N. Mahowald, et al., Global distribution of atmospheric phosphorus sources, concentrations and deposition rates, and anthropogenic impacts, Glob. Biogeochem. Cycles 22 (4) (2008).
[88] J.L. Shupe, Fluorides: Effects on Vegetation, Animals and Humans, Paragon Press, Inc, 1983.
[89] National Center for Atmospheric Research, Biogeochemical Cycles, 2015. http://scied.ucar.edu/longcontent/biogeochemical-cycles. (Accessed 4 March 2015).
[90] EPA/630/R-94/003, A Review of Ecological Assessment Case Studies From a Risk Assessment Perspective: Volume II, 1994.
[91] H. Su, et al., Soil nitrite as a source of atmospheric HONO and OH radicals, Science 333 (6049) (2011) 1616–1618.
[92] T. Green, G. Brown, L. Bingham, Environmental Impacts of Conversion of Cropland to Biomass Production, Tennessee Valley Authority Southeastern Regional Biomass Energy Program, Muscle Shoals, AL, 1996.
[93] UK Air Pollution Information System, Nitrogen Oxides (NOx). http://www.apis.ac.uk/overview/pollutants/overview_NOx.htm. (Accessed 2 March 2018).

[94] B. Finalyson-Pitts, J. Pitts, Chemistry of the Upper and Lower Atmosphere: Theory, Experiments, and Application, Academic Press, San Diego, CA, 1999.
[95] EPA-452/R-11-005a, Policy Assessment for the Review of the Secondary National Ambient Air Quality Standards for Oxides of Nitrogen and Oxides of Sulfur, 2011.
[96] L. Fewtrell, Drinking-water nitrate, methemoglobinemia, and global burden of disease: a discussion, Environ. Health Perspect. 112 (14) (2004) 1371.
[97] U.S. Environmental Protection Agency, National Primary Drinking Water Regulations. https://www.epa.gov/ground-water-and-drinking-water/national-primary-drinking-water-regulations. (Accessed 27 February 2018).
[98] S.M.A. Adelana, Nitrate health effects, in: Water Encyclopedia, 2005.
[99] W.H. Schlesinger, Biogeochemistry, Elsevier, 2005.
[100] US Environmental Protection Agency, Health Assessment Document for Diesel Engine Exhaust, National Centre for Environmental Assessment, Washington, DC, 2002.
[101] G. Greeves, I.M. Khan, G. Onion, Effects of water introduction on diesel engine combustion and emissions, in: Symposium (International) on Combustion, vol. 16, Elsevier, 1977, pp. 321–336. no. 1.
[102] B. Comer, E. Georgeff, L. Osipova, Air emissions and water pollution discharges from ships with scrubbers, Consult. Rep. Int. Counc. Clean Transp, 2020, pp. 12–13.
[103] National Academy of Engineering, 14 Grand Challenges for Engineering in the 21st Century, NAE, Washington, DC, 2018.
[104] C. Mote Jr., D.A. Dowling, J. Zhou, The power of an idea: the international impacts of the grand challenges for engineering, Engineering 2 (1) (2016) 4–7.
[105] W. El Maraghy, Future trends in engineering education and research, in: Advances in Sustainable Manufacturing, Springer, 2011, pp. 11–16.
[106] J.-Y. Yoon, M.R. Riley, Grand challenges for biological engineering, J. Biol. Eng. 3 (1) (2009) 16.
[107] National Academy of Engineering, Introduction to the Grand Challenges for Engineering, National Academy of Sciences. http://www.engineeringchallenges.org/challenges/16091.aspx (Accessed 4 July 2024).
[108] J. Galloway, et al., Human alteration of the nitrogen cycle–threats, benefits, and opportunities, UNESCO-SCOPE Policy Briefs, 2007.

Chapter 5

Scientific basis for water quality criteria and standards

1 Introduction

A direct water pollutant is intrinsically harmful to humans and other organisms. An indirect pollutant's harm is a secondary result from another contingent event. For example, the release into an estuary of a chemical compound toxic to several fish species is a direct pollution. Indirect effects occurs if an estuary receives loads of organic matter, which is not necessarily inherently hazardous, but after bacteria break down the organic matter down dissolved oxygen levels will decline. The lower O_2 concentrations then lead to problems in the estuary's aquatic ecosystem. Both pollutant types may have the same outcome, e.g., a fish kill, but the first is direct and the second is indirect.

Water quality is affected by both thermodynamics and fluid dynamics. The laws of thermodynamics apply to the open water and to water in the interstices of a soil or sediment matrix. The inherent properties of a pollutant include its solubility in water and in other solvents, e.g., trichloroethane, benzene, and toluene, its vapor pressure, and its partitioning coefficients (see Chapter 14). These and other factors determine the affinity of the pollutant to various tissues in an organism or to different compartments in the environment.

Humans and other organisms live in a very small optimal range of conditions. Small changes can threaten the delicate balances within these optimal ranges of the water regimes. Achieving acceptable water quality requires optimization. Acceptable water quality falls within a range of biological, chemical, and physical parameters. A chemical compound that is essential for certain purposes at one concentration becomes toxic or otherwise harmful at a higher concentration. For example, certain metals are toxic to humans at high concentrations, but deficiency occurs at excessively low concentrations (see Fig. 5.1).

2 Effluent limits

To protect the quality of waters, there must be limits on the amounts and types of pollutants that can enter a water body. This usually is accomplished by a process in which potential pollution sources request approval to discharges into waterbodies. In the U.S., for example, the National Pollutant Discharge Elimination System (NPDES) was established under the Federal Water Pollution Control Amendments and the Clean Water Act to prohibit the release of pollutants from point sources into the waters of the U.S. unless that source operator obtains a permit. The permit requires that the pollutant source limit the amount and type of substances being discharged and verify that the release meet the permit requirements by reporting the results of ongoing, regular monitoring [2,3].

There are numerous details on when a permit is required. As evidence, in the United Kingdom, an effluent permit is required for discharges into surface waters into or on the ground or discharging treated sewage effluent to ground through an infiltration system, i.e., ground water activities [4]. Wastewater includes poisonous, noxious, or polluting matter, polluting substances, waste matter, and trade or sewage effluent. A permit is not required when to discharge clean rainwater from roofs or other uncontaminated water or small areas of hardstanding, e.g., a parking lot, through a properly maintained oil separator, to surface water. A permit is also not required for enclosed lakes or ponds that hold water throughout the year, other than in extreme weather conditions, nor for an outfall that connects to a watercourse or an outfall that only discharges in extreme weather conditions, nor if the waterbody is sealed or lined to prevent water draining into the ground or soaking into the surrounding soil. The permit decision assumes that appropriate pollution prevention measures will be taken [4].

Fundamentals of Water Pollution. https://doi.org/10.1016/B978-0-443-28987-3.00015-1

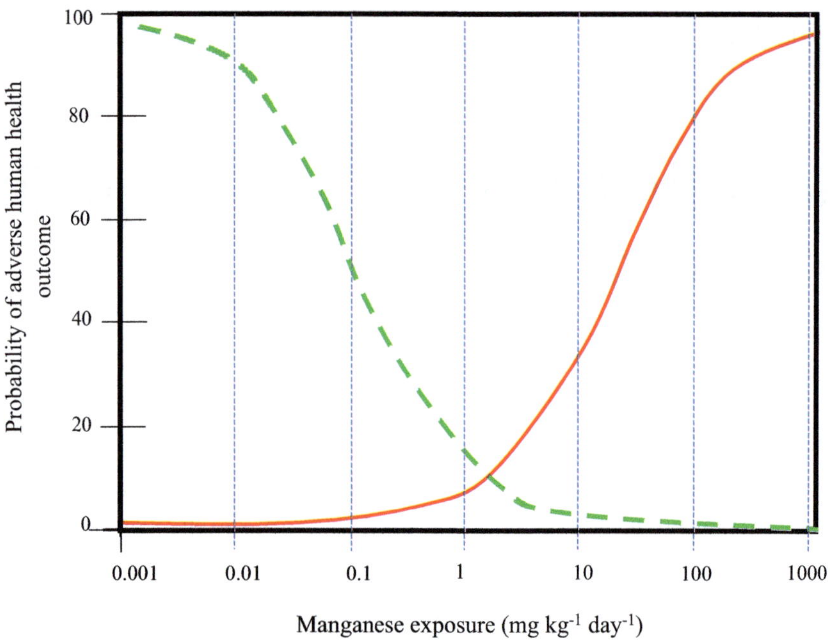

FIG. 5.1 U-shaped dose response curve for the metal manganese (Mn). These fitted curves indicate the probability of an adverse event based on excess and deficiency data derived from [1] compared to exposure rates of milligrams Mn per kilogram body weight per day ($mg\,kg^{-1}\,day^{-1}$). The dashed green curve shows the predicted probability of an adverse deficiency event based on the joint model for excess and deficiency (JMED). The solid red curve shows the predicted probability of an adverse excess intake event based on the JMED mode. *(Data from B. Milton, et al., Modeling U-shaped dose-response curves for manganese using categorical regression, Neurotoxicology 58 (2017) 217–225, https://doi.org/10.1016/j.neuro.2016.10.001.)*

3 Physicochemical characteristics of water pollutants

Water pollutant concentrations result for the mass of a substance entering the waterbody and the inherent physicochemical properties of the substance. Some pollutants maintain their properties after entering the water, such as neutral organic compounds with low aqueous solubility. Others change almost immediately after entering the water, such as those that are readily hydrolyzed, or which undergo enzymatic catalysis as does triphosphate ($H_5P_3O_{10}$) [5,6]. Often, the changes require more than one chemical reaction that occur in steps. For example, after triphosphate enters water, it undergoes a first reaction to form diphosphate ($H_4P_2O_7$), and the second dissociates the diphosphate into orthophosphate (H_3PO_4) [7]. The two hydrolytic reactions for triphosphate are:

$$H_5P_3O_{10} + H_2O \rightarrow H_3PO_4 + H_4P_2O_7 \tag{5.1}$$

which is followed by:

$$H_4P_2O_7 + H_2O \rightarrow 2H_3PO_4 \tag{5.2}$$

where the first is a much faster reaction than the second [7].

Other reactions are quite slow, resisting degradation, such as the persistent organic pollutants (POPs), the stable, toxic compounds that resist photolytic, biological, and chemical degradation. They are highly lipophilic, so they accumulate in the lipid fragments of soil, sediment, and biota, and magnify in concentrations in food chains [8–13].

Although water quality standards are reasonably similar around the world, each jurisdiction sets them based on factors related to the resources being protected, such as the difference between warm-water and cold-water fish communities, with the latter often being more protective, especially to keep dissolved oxygen (DO) concentrations comparatively high. In the U.S., for example, each state, territory, and tribe has unique methods of adopting water quality standards. Often, standards are developed with deliberation and public involvement. The standards are not promulgated until the public has had a legally required opportunity to review and comment on the proposed standards. Several of the factors to be considered when deciding on the desired integrity of a waterbody are given in Table 5.1.

Standards are designed to ensure that the physicochemical and biological characteristics of water can meet so-called designated uses [14]. These are based on the desired conditions of the waterbody and the function that the water supports, such as the ability to swim safely in the water or to consume aquatic organisms taken from the water, i.e., the water is swimmable and fishable in the language of the U.S. Clean Water Act [15–18].

For example, the habitat suitability index (HIS) listed in Table 5.1 represents the habitat's capacity to support a selected species (see Fig. 5.2). The models hypothesize species-habitat relationships and are not based on proven cause and effect

TABLE 5.1 Typical factors to consider when analyzing a waterbody's need for water quality limits.

Physical	Chemical	Biological
Stream size (mean width/depth)	Dissolved oxygen	Fish abundance
Flow velocity	Toxicants	Macroinvertebrate abundance
Annual hydrology	Suspended solids	Microinvertebrate abundance
Total volume	Suspended solids	Phytoplankton abundance
Reaeration rates	Nutrient concentrations	Periphyton abundance
Gradient/pools/riffles	Sediment oxygen concentrations	Macrophyte abundance
Temperature	Salinity	Biodiversity indices
Sedimentation	Hardness	Habitat suitability index
Channel modifications	Alkalinity	Aquatic tissue condition
Channel stability	pH	Recovery and resilience indices
Substrate composition and characteristics	Dissolved solids	Species intolerance
Channel debris	Tissue pollutant concentrations	Omnivore-carnivore comparison
Sludge deposits		Other biological comparisons
Riparian characteristics		
Downstream characteristics		

Data from U.S. Environmental Protection Agency, Designation of uses, in: Water Quality Standards Handbook, U.S. EPA, Office of Water Regulations and Standards, Washington, DC, 1983 (Chapter 2).

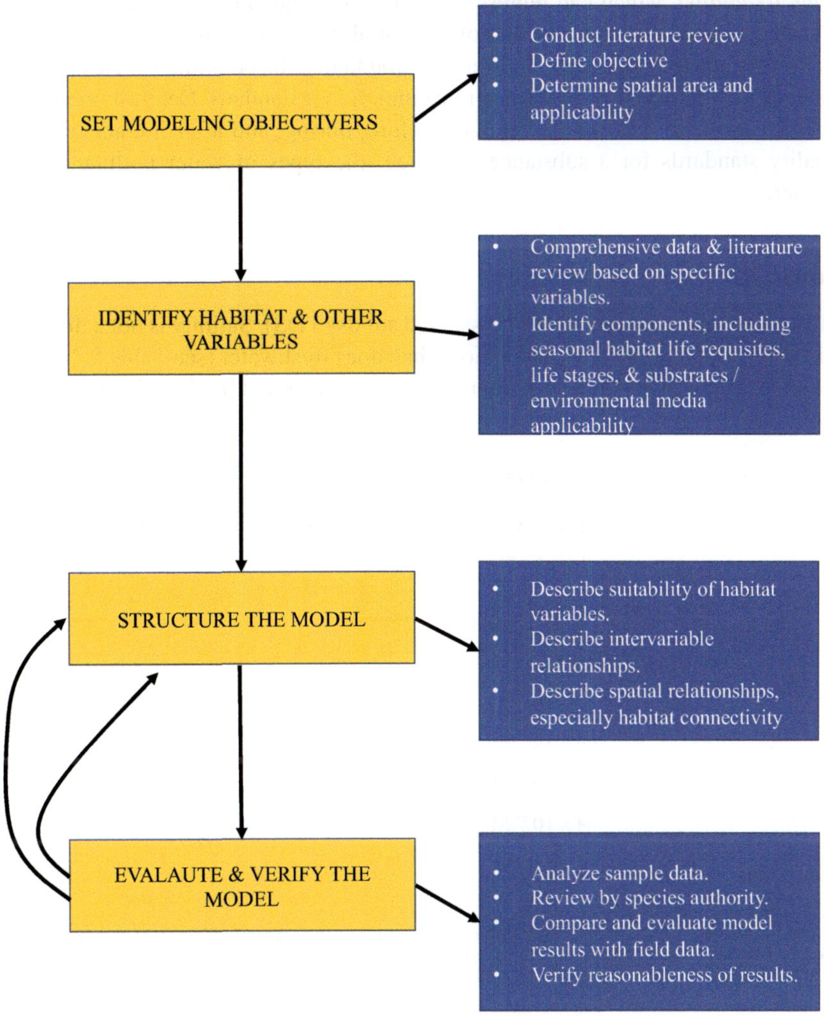

FIG. 5.2 Steps to develop a habitat suitability model. *(Data from U.S. Environmental Protection Agency, Habitat Suitability Index (HSI). https://archive.epa.gov/aed/html/research/scallop/web/html/hsi.html, 2016 (Accessed 26 January 2024).)*

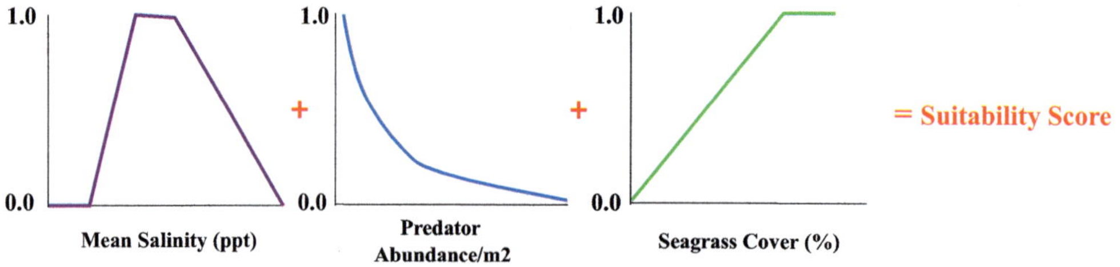

FIG. 5.3 Salinity, predator dominance, and vegetative cover graphs for a hypothetical estuary that display the types of output from a habitat suitability model. Along the *x*-axis of each graph is a habitat variable, usually a stressor. The *y*-axis shows a suitability score from 0 to 1, with 1 representing the optimal habitat and 0 the least suitable habitat. *(Data from U.S. Environmental Protection Agency, Habitat Suitability Index (HSI). https://archive.epa.gov/aed/html/research/scallop/web/html/hsi.html, 2016 (Accessed 26 January 2024).)*

relationships. The HSI results represent the interactions of the habitat characteristics and the ways that each habitat relates to a given species. These can help to elucidate several species-habitat relationships [19].

Thus, the HSI results identify the extent to which a habitat is suitable for target species by combining the interactions of key environmental variables on a species' vital rates and survival (see Fig. 5.3).

Scoring of HIS results are from 0 to 1, representing the combined interactions of all species-habitat relationships. A result of 1 represents the optimal habitat, with 0 representing the least suitable habitat. These are not absolute numerical results. For example, a species may be well supported by a habitat with a 0.1 value but might have better survival rates in a habitat with a value of 0.7. The HSI combines individual scores into one index to predict the ability of a habitat to support a species. Indeed, the HSI is measuring the carrying capacity or the population size that a habitat can be support with its available resources. Habitat maps can be drawn from HSI values of habitat patches for a specific species, such as the bay scallop (*Argopecten irradians*), which can represent the health of the entire aquatic habitat, i.e., it can be a sentinel species [20]. Remote sensing, aerial photography, and other spatial methods are used to determine the habitat type so that the suitability indices may infer the distribution and relative abundance of potential species [19].

No matter the administrative requirements, science must underpin a standard. Determining the physicochemical properties of a substance under consideration and its behavior before, during, and after it enters the water is the first step in establishing water quality standards for a substance. As such, the types of water pollutants can be characterized by how they behave in water.

3.1 Ionic pollutants

Salinity is an expression of the total concentration of dissolved solids (TDS). Salinity is not a direct expression of pollution since salt water contains a significantly greater mass of ions than does freshwater (see Table 5.2). Pollution occurs when the ionic constituents are in the wrong place in the environment, as demonstrated by the saltwater intrusion in coastal waters,

TABLE 5.2 Important general ionic composition classifications of freshwaters and marine waters.

Composition	River water	Salt water
pH	6–8	8
Ca^{2+}	4×10^{-5}M	1×10^{-2}M
Cl^-	2×10^{-4}M	6×10^{-1}M
HCO_3^-	1×10^{-4}M	2×10^{-3}M
K^+	6×10^{-5}M	1×10^{-2}M
Mg^{2+}	2×10^{-4}M	5×10^{-2}M
Na^+	4×10^{-4}M	5×10^{-1}M
SO_4^{2-}	1×10^{-4}M	3×10^{-2}M

M=molarity (moles per liter solution).
Data from J.P. Kim, K.A. Hunter, M.R. Reid, Factors influencing the inorganic speciation of trace metal cations in fresh waters, Mar. Freshw. Res. 50 (4) (1999) 367–372. R.P. Schwarzenbach, P.M. Gschwend, D.M. Imboden, Environmental Organic Chemistry, John Wiley & Sons, 2016.

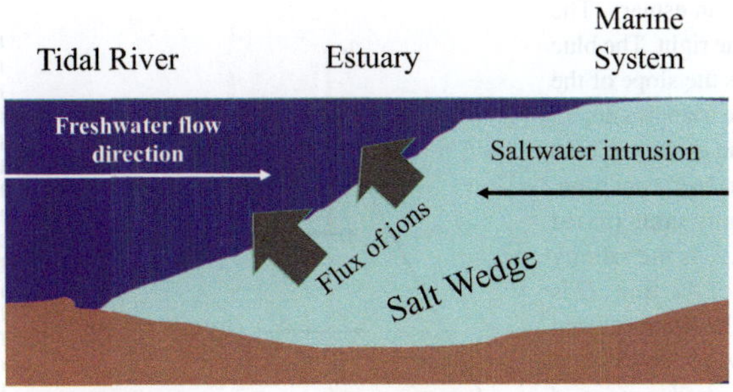

FIG. 5.4 Saltwater intrusion into freshwater systems. Denser saltwater submerges under the lighter freshwater system. Since the concentration of ions is higher in the marine system than in the tidal river, the ion flux is in the direction of the freshwater, which means the ocean is the pollutant source in this scenario. The same phenomenon can occur in coastal aquifers when saline water intrudes into freshwater stored underground into the zone of saturation. *(Data from V. Arnold, A note on the Ghyben-Herzberg formula, Bull. Int. Assoc. Sci. Hydrol. (1968) 43–46. M.P. Anderson, W.W. Woessner, R.J. Hunt, Applied Groundwater Modeling: Simulation of Flow and Advective Transport, Academic Press, 2015.)*

i.e., denser saline water can wedge beneath freshwaters and pollute surface waters and ground water (see Fig. 5.4). Saltwater intrusion can significantly alter an ecosystem's structure and function and threaten freshwater organisms. It can also pose a huge challenge to coastal communities who depend on aquifers for their water supply. The freshwater Everglades, for example, recharge the Biscayne aquifer, a natural underground layer that collects water. The Biscayne aquifer is the primary water supply to the Florida Keys.

Saltwater intrusion [21] into an aquifer can be expressed as:

$$\rho_s g z = \rho_f g (z + h) \tag{5.3}$$

Where, ρ_f is the density of freshwater, ρ_s is the density of the saline water, h is the thickness of the freshwater zone above sea level, z is the thickness of the underground zone of saline water below sea level, and a g is the gravitational constant ($6.67408 \times 10^{-11}\ \mathrm{m^3\,kg^{-1}\,s^{-2}}$).

Rearranging the variables and solving for z removes g:

$$z = \frac{\rho_f}{\left(\rho_{s-}\rho_f\right)} h \tag{5.4}$$

At 20°C, ρ_f is the density of freshwater and ρ_s is the density of the saline water. ρ_f is about $1.000\ \mathrm{g\,cm^{-3}}$, while ρ_s is about $1.025\ \mathrm{g\,cm^{-3}}$. The higher density of saline water creates the flux of ions that intrude into the freshwater aquifer. Given these density values, z at 20°C and 1 atm equals 40, i.e., $z = 40\ h$. Saltwater intrusion is usually irreversible given the higher ionic strength of ocean water that is a continuously available ionic concentration gradient in the direction of the freshwater, whether above or below land surfaces [22,23]. The interfaces in the surface water and in the aquifer migrate with tides, drawdown of ground water by pumping, recharge, and other factors, so that the interface is a transition or dispersion zone [22,24].

The intrusion gradient is a function of energy dissipation. The entering freshwater contains potential energy that is greater than that of potential energy of the more quiescent estuary. A gradual transition from the ocean of saline water to freshwater along with the attendant rise in the estuary's water level all join at the salt wedge. The kinetic energy of the tide and the potential energy of the river flow and to a lesser extent wind and wave kinetic energy are mixed. In so doing, the mixing dissipates the potential energy of the freshwater. The various mixing mechanisms of the estuary maximize the power represented by a dispersion coefficient of all combined mixing mechanisms. The resulting differential equation can be solved by combining the advective and dispersive forces [25]:

$$A\frac{\partial S}{\partial t} + Q\frac{\partial S}{\partial x} - \frac{\partial}{\partial x}\left(AD\frac{\partial S}{\partial x}\right) = 0 \tag{5.5}$$

Where, S is the salinity expressed in practical salinity units (based on the electrical conductivity of the water), A is cross-sectional area of the flow, D is the dispersion coefficient, t is the time, x is the distance from the estuary mouth, and Q is the water flow in the estuary. The first term is the salinity time gradient caused by the balance between the advective flow, i.e., Q, and the mixing of water having different salinity by the dispersive exchange flow, i.e., $\frac{\partial}{\partial x}\left(AD\frac{\partial S}{\partial x}\right)$. Without other salinity sources, the sum of the terms equals zero [25]. Averaging the equation over a complete tidal period would reflect a more

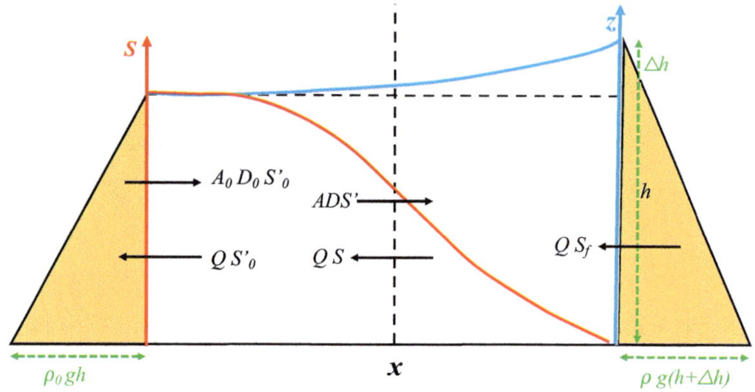

FIG. 5.5 Saline and freshwater mixing in an estuary. The seaside is on the left and the riverside is on the right. The blue line depicts the water surface; the red line is the slope of the resultant salinity distribution; and the orange triangles depict the hydrostatic pressure on both sides of the estuary. The 0 subscript represents the downstream boundary condition. S is the salinity expressed in practical salinity units (based on the electrical conductivity of the water), S_f is the salinity of the freshwater, A is cross-sectional area of the flow, D is the dispersion coefficient, t is the time, x is the distance from the estuary mouth, and Q is the water flow in the estuary. *(Data from Z. Zhang, H.H. Savenije, Thermodynamics of saline and fresh water mixing in estuaries, Earth Syst. Dyn. 9 (1) (2018) 241–247.)*

extended, long-term change in salinity due to the freshwater advection and the flows of the average tidal exchanges, which would eliminate the first term of Eq. (5.5) [25]. Thus, to reach steady state, the equation is integrated with respect to x:

$$Q\left(S - S_f\right) - AD\frac{dS}{dx} = 0 \tag{5.6}$$

Where $\frac{dS}{dx}$ is the upstream boundary and S_f is the salinity of the freshwater; and with the conditions that $\frac{dS}{dx} = 0$ and $S = S_f$. Under this steady-state condition, the water flow, Q, equals the freshwater upstream flow, so that the saltwater-freshwater wedge can be visualized as the mixing and the slopes of the salinity distribution and the hydrostatic pressure (see Fig. 5.5). In this steady-state condition, the salt that is flushed out by the fresh river flow is balanced by the exchange of saline and freshwater from various mixing processes, causing an upstream salt flux. The blue line has an upstream gradient due to decreasing salinity. Given the density difference, hydrostatic pressures on both sides are not equal. The water level at the toe of the salt intrusion curve is higher by Δh, which results in a seaward pressure difference toward the surface and an inland pressure difference toward the estuary floor. The integrals of the hydrostatic pressure distributions are equal and opposed in steady state, but they have different working lines $\frac{\Delta h}{z}$ which triggers angular momentum.

Saltwater intrusion is an example of the laws of thermodynamics and motion at work in the environment. Dissipative systems are a means of conserving mass and energy, while exporting entropy [26].

In addition to the type of saltwater intrusion shown in Fig. 5.5, rising sea level hypothesized to occur due to changes in climate could submerge lower-lying areas, like those of the Everglades in Florida, exacerbating saltwater contamination of the underlying aquifer. In addition, the rising sea water could force salty waters upstream into coastal areas, thus threatening surface water supplies. Similar problems could also occur in Northeastern U.S. aquifers that are recharged by fresh portions of streams that are vulnerable to increased salinity during severe droughts [27]. Similar interactions among ground water and surface waters with estuaries and ocean water could occur across the globe [28].

Table 5.2 indicates that the classifications using two dominant ions, Na^+ and Cl^- differ by three orders of magnitude between "fresh" and "saline" waters. In the U.S., the drinking water standard for TDS is a "secondary" or nuisance standard, which is $500\,mg\,L^{-1}$. Although the water near the water supply is likely not "saltwater," its ionic concentrations may show trend of increasing salinity. The degree of harm depends on the species. For example, some aquatic species can tolerate only small amounts of ionic strength, i.e., most freshwater species. Others can tolerate larger ionic strength, living in brackish waters. Others require minimum amounts of ionic strength, i.e., marine species (see Fig. 5.6).

Ions have an electrical charge that can attract or repel each other. Ionic strength is the concentration of ionic charge in solution. The interactions that occur between the ions of solutions and the ions in water may lead to shifts in ecological community composition and function based on preferences for certain species.

Measurements of electrical conductivity, salinity, and TDS often represent the ionic strength of water. Since ionic compounds are natural constituents of both inland and marine systems, they are not inherently harmful until their concentrations fall outside the tolerance ranges of aquatic organisms. Usually, a range of concentrations of certain ionic compounds are essential to the survival of aquatic organisms. Increases in ionic strength that occur in the presence of other stresses will act synergistically by increasing the organismal uptake of toxic substances by organisms [29,30].

Increases in ionic strength can also alter soil structure and decrease permeability [31]. The structural stability, dispersion, swelling, crusting, electrical conductivity, and osmotic potential of soils are affected by increases in ionic strength, which in turn increase pollutant runoff and soil erosion [32].

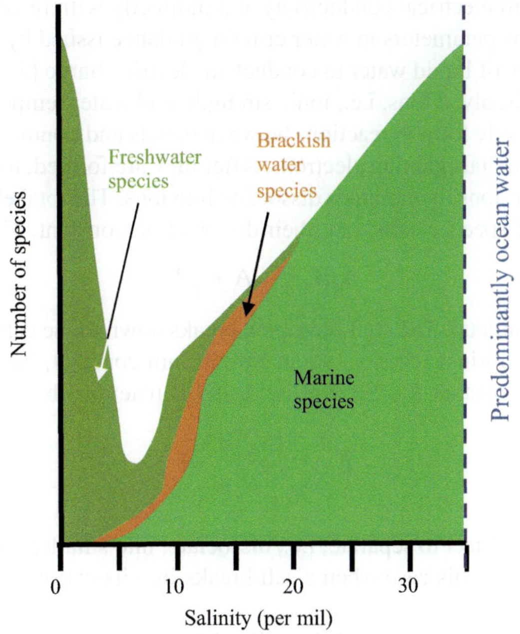

FIG. 5.6 Effect of salinity on species richness. Freshwater systems generally have salinity values below 0.5 parts per thousand (per mil), ocean water about 35 per mil. *(Data from A. Remane, Ecology of brackish water (physiological features of life in brackish water). Part I, in: Biology of Brackish Water. Die Binnengewässer, vol. 25, 1971, p. 2. R. Wetzel, Limnology: Lake and River Ecosystems, Academic Press, New York, USA, 2001. U.S. Environmental Protection Agency, Ionic strength. https://www.epa.gov/caddis-vol2/ionic-strength#listing, 2021 (Accessed 19 December 2021).)*

This discussion is an example of how pollution differs among different types of water, i.e., fresh, brackish, and saline. It suggests a way that water pollutants can be classified, i.e., determining the difference between polluted water and unpolluted water by visualizing a gradient from higher to lower concentrations of a substance. In the case of saltwater intrusion, this was represented by the gradient from saline-to-freshwater in the estuary. The ionic strength or TDS in the ocean's water does not represent pollution. It is what one would expect in the ocean water. However, the migration of ions and dissolved compounds from ocean water to freshwaters, which should have much lower TDS, represents water pollution. The larger the TDS, the more saline the freshwater has become.

TDS is an example of so-called general or conventional water pollution, as are pH, total suspended solids (TSS), and biochemical oxygen demand (BOD). For example, permits specify effluent limits on TSS, BOD, other conventional pollutants, and toxic substances, as well as pathogenic microbes, before industries, commercial operations, and wastewater treatment plants are allowed to release their effluents to surface waters or applied to land surfaces. Drinking water limits and standards also limit these and other substances, e.g., radionuclides, that leave treatment facilities prior to the treated water being distributed in the water supply system. Often, these substances are again tested at the point-of-use, e.g., a tap in a residence, to ensure that even though the water had been treated at the plant that it did not subsequently become contaminated by infiltration of pollutants into the water lines. General trends in tap water quality are often measured by surveys ranging in scope from municipalities to nationwide.

Svante Arrhenius is credited with establishing the relationship between electrical and chemical properties of molecules. He observed that some chemical compounds, subsequently named electrolytes, could conduct electricity when dissolved in water, while other chemical compounds do not. He also noted difference in chemical reaction times, from instantaneous to very extended times. Finally, he observed that particular chemical compounds showed extremely strange colligative properties[a] while others were consistent with Raoult's Law, which states that the solvent's vapor pressure in an ideal solution is equal to the product of the mole fraction of the solvent and the vapor pressure of the pure solvent. The four colligative properties of solutions are the elevation of boiling point, the depression of freezing point, the decreasing of vapor pressure, and osmotic pressure. Arrhenius' observations provide a clue to the behavior of ions in solution.

---

a. Properties of solutions that depend exclusively on the total number of molecules or ions, and not on inherent properties of solute molecules or ions in solution.

Since TDS correlates directly with electrical conductivity and indirectly with resistance, water quality standards often include both conductivity and TDS as parameters in water criteria guidance issued by governments throughout the world. Electrical conductivity is the capacity of liquid water to conduct an electric charge [33–35]. The amount of conductivity is dependent on the concentration of dissolved ions, i.e., ionic strength, and water temperature, since both add energy to the water. Ionic compounds most commonly form in reactions between metals and nonmetals, so electrons are gained and lost, with metals losing electrons and nonmetals gaining electrons. After they are formed, ionic compounds are held together by ionic bonds. When dissolved in water, ionic compounds dissociate into ions. The total electrolyte concentration in this ionic solution determines the solubility of specific salts, i.e., their dissociation constant. The water dissociation constant is:

$$A_xB_y \leftrightarrow {}_xA + {}_yB \tag{5.7}$$

Where A is the metal and B is the nonmetal. The A_xB_y complex breaks down to the cation (${}_xA$) and the anion (${}_yB$). Note by the double arrow that this is an equilibrium reaction, so an equilibrium constant, i.e., the dissociation constant for ionic reactions is known as an ionization constant K_D. So, for the A and B reaction above K_D is determined by:

$$K_D = \frac{[A]^x[BA]^y}{[A_xB_y]} \tag{5.8}$$

K_D indicates the tendency of a larger object to separate, i.e., dissociate, into smaller components, such as when a complex breaks into component molecules, or in this case when a salt breaks into its cation and anion. This ionization reaction is irreversible.

The dissolved ions concentration is usually measured as TDS. EC can be measured easily and inexpensively *in situ* by a portable water quality checker. On the other hand, the analysis of TDS is more difficult and expensive as it needs more equipment and time [9]. However, TDS analysis is very important and principal because it can illustrate ground water quality, particularly in understanding the previously mentioned effect of seawater intrusion better than the EC analysis [10]. Hence, researchers have done various investigations to find the precise mathematical correlation between these two parameters, so TDS concentration can be directly calculated from the EC value [36]:

$$\text{TDS}\left(\text{mg L}^{-1}\right) = k \times \text{EC}\left(\text{Siemens per centimeter, } \mu\text{S cm}^{-1}\right) \tag{5.9}$$

The value of k will increase directly with the increase of ions in the water. However, the relationship between conductivity and TDS is not linear; but depends on the activity of specific dissolved ions average activity of all ions in the liquid, and ionic strength.

Ions form throughout the environment. The hydrosphere, atmosphere, and biosphere interconnect in complex and extensive ways. A small disruption in one part of the spheres can lead to unanticipated outcomes in and among the spheres. The hydrosphere has many features that are sensitive to initial conditions, so considerations of thermodynamic chaos must be considered. Understanding the behavior of these complex systems requires not only an understanding of each sphere and its parts, but also the interplay of every part [37–42].

3.2 Dissolved pollutants

Chemical compounds in liquid or gas phases can become dissolved pollutants. Since a solution is a physical process, not a chemical reaction, it is a mixture in which the dissolved substance remains intact and held in a solvent. Nonfluids dissolve in fluids. Fluids dissolve in other fluids. Thus, solubility is the measure of the amount of chemical compound that can dissolve in water or any other solvent.

Gases dissolve in other gases, which are expressed as partial pressures, e.g., methane in air. Gases dissolve in liquids, e.g., CO_2 in water. Liquids dissolve in liquids, octanol in water. Solubility is an inherent, saturation property of that compound that is commonly expressed in units of mass of solute per volume of solvent, e.g., mg L^{-1}, or as the mass of a solute in mass of solvent, e.g., g kg^{-1}.

Often, the term "solubility" means aqueous solubility, i.e., solubility in water. When the solvent is not water, the solvent is specified, e.g., the pollutant tetrachlorodibenzo-*para*-dioxin (TCDD) dissolved in dimethyl sulfoxide (DMSO), benzene, methanol, or hexane at a specified temperature. Solubility as volume per volume can be represented as parts per million (ppm), parts per billion (ppb), or parts per trillion (ppt). Occasionally, solubility is expressed as parts per thousand (per mil)[b] or parts per hundred (percent). Thankfully solubility is almost never expressed in percent for water pollutants given that

b. Parts per thousand is better expressed as "per mil" since "ppt" is now mainly used in environmental literature to be parts per trillion.

pollutant concentrations are usually too low. Percent concentrations are usually reserved for nutrients and essential gases, e.g., percent carbon dioxide in water or percent water vapor in the air.

Solubility is important to environmental exposure and risk assessments. For example, mercury in its zero-valence metal form is relatively insoluble in water and is much less toxic when ingested than ingesting methylmercury chloride, which is much more polar and soluble in the water in the tissues of fauna, including humans. Thus, environmental scientists are quite concerned when mercury is present in a sample, but even more concerned when it is reduced to methylated species, which happens within aquatic organisms and within reduced, low oxygen parts of the environment like sediment.

Because solubility is the saturation concentration of a dissolved solute in a solvent in equilibrium with undissolved solute at a fixed temperature and pressure [43], the solubility of a compound can be understood by defining an ideal solubility, i.e., a solute-solvent system that obeys Raoult's law. As mentioned, Raoult's law states that for an ideal solution, the equilibrium partial pressure (p) of a pure substance at constant temperature is the product of the substance's vapor pressure and its mole fraction (x_A) in the liquid. Chemical solutions that obey Raoult's law at every concentration and at every temperature are ideal solutions. For example, mixing two fluids, A and B, will result in intermolecular forces:

- A-A intermolecular forces
- B-B intermolecular forces
- A-B intermolecular forces

When all three types of intermolecular forces are nearly equal, the solution is said to be an ideal solution. The ideal solubility for a solute A at temperature T, and at a partial pressure p of 1 atm is given by [44]:

$$x_A = \frac{p}{p_A^*} \tag{5.10}$$

Where p_A* is the vapor pressure of the pure solute at temperature T. Note that x_A and p are linearly related.

The calculated ideal solubilities for some solutes in water are given in Table 5.3.

The solubilities of the gases in Table 5.3 vary by more than five orders of magnitude when expressed as mole fractions. One molecule of hexafluoroethane (C_2F_6) is saturated per million water molecules, whereas it takes 350,000 molecules of HCl to reach saturation in 1 million molecules of water. Water molecules are strongly associated, which relates to the values that show strong B-B interactions. For nonpolar substances like propane (C_3H_8) and sulfur hexafluoride

TABLE 5.3 Solubility of gas (A) in water (B) at 25 °C and 101.325 kPa (1 atm).

Gas	Mole fraction, $x_A \times 10^5$	Ideal solubility, $x_A \times 10^5$
Helium (He)	0.71	–
Argon (Ar)	2.53	–
Hydrogen (H_2)	1.41	–
Nitrogen (N_2)	1.18	–
Oxygen (O_2)	2.30	–
Methane (CH_4)	2.55	–
Propane (C_3H_8)	6.99	11,600
Sulfur hexafluoride (SF_6)	0.44	4000
Hexafluoroethane (C_2F_6)	0.10	–
Carbon dioxide (CO_2)	61	1780
Fluoromethane (CH_3F)	106.2	3000
Sulfur dioxide (SO_2)	2460	2500
Hydrochloric acid (HCl)	25,900	2000

Data from R. Battino, T.M. Letcher, An introduction to the understanding of solubility, J. Chem. Educ. 78 (1) (2001), 103. J. Lorimer, P. Gujral, Solubility data series, in: Alkaline Earth Hydroxides in Water and Aqueous Solutions, Citeseer, 1979.

(SF_6), the solute-solvent association is so weak that the experimental solubility is vastly lower than the ideal solubility for these compounds. For a highly polar compound like hydrochloric acid (HCl) there are strong interactions between the polar HCl and the water leading to a solubility that is larger than either B-B or A-A interactions, resulting a very high aqueous solubility for HCl [43].

The water molecule is unique, given its small size and its strong self-association through hydrogen bonding [45]. Thus, discrepancies between experimental and ideal solubilities for water systems are more common than for any other solvent.

The variability of solubility of gases in liquids is aptly demonstrated by the carbon dioxide (CO_2) (see Table 5.4). Other than the value for water, solubility in the liquid does not vary a great deal in organic solvents, especially when compared to the values in Table 5.3. Also, the ideal solubility of CO_2 is reasonably close to experimental values. This is likely due to the similar strengths of the associations, i.e., A-A=A-B=B-B, which is common for nonpolar compounds like CO_2 and the organic solvents. The very strong H-bonds between the water molecules and relatively weak CO_2-H_2O and CO_2-CO_2 associations make the solubility of CO_2, $x_{CO_2}=6.1$, which is substantially less than the ideal solubility, $x^*_{CO_2}=178$. Generally nonpolar compounds, which include most hydrocarbons, have low solubility in water, whereas ionic compounds like HCl have a very high aqueous solubility.

For solids dissolved in liquids, the ideal solubility can be found by [46]:

$$\ln x_A = -\frac{\Delta H_{\text{fusion,A}}(T_m)}{RT}\left[1-\frac{T}{T_m}\right] \tag{5.11}$$

Where, x_A is the solubility mole fraction of the solute at temperature T, T_m is the melting temperature of the solute, and $\Delta H_{\text{fusion,A}}$ is the enthalpy of fusion, i.e., melting, of the solute at T_m. For example, Table 5.5 shows the solubility of lightest polycyclic aromatic hydrocarbon, naphthalene, calculated from Eq. (5.11) in various solvents. $\Delta H_{\text{fusion, A}}$ and the T_m for naphthalene are $18.58\,\text{kJ}\,\text{mol}^{-1}$ and 353.20 K, respectively [47].

Since "like dissolves like", nonpolar compounds readily dissolve in organic nonpolar solvents. Polarity is extremely important in environmental systems. For example, the disaster at Bhopal involved solubilities as the water that poured into the tank filled with the highly nonpolar compound methylisocyanate (MIC) dissolved very little of the MIC mass, so that the reaction was largely at the interface of the two liquids. The small amount of MIC that did dissolve reacted exothermically and heated the solution to form CO_2, building up pressure as the heat was expediting the reaction. When the tank ultimately ruptured, it filled the air with largely unreacted and highly toxic MIC [43].

A dipole exists when two atoms form a molecule in which atoms share electrons unevenly in a covalent bond, i.e., part of the molecule becomes slightly positively charged and part becomes slightly negatively charged, as indicated by the symbol δ in Fig. 4.1 in Chapter 4. Such molecules are known as polar molecules. Water is a very polar molecule and, since, as mentioned, like dissolves like, it is a very efficient solvent for other polar compounds. For example, water and ethanol

TABLE 5.4 Solubility of carbon dioxide in liquids at 20 °C and 1 atm.

Solvent	Mole fraction, x_{CO_2}
Amyl acetate	270
Pyridine	129
Ethylene chloride	125
Chloroform	123
Toluene	107
Carbon tetrachloride	100
Ethyl alcohol	70
Carbon disulfide	22
Water	6.1
Ideal solubility ($x^*_{CO_2}$)	**1.78**

Data from K.G. Denbigh, K.G. Denbigh, Equilibria of reactions involving gases, in: The Principles of Chemical Equilibrium: With Applications in Chemistry and Chemical Engineering, fourth ed., Cambridge University Press, Cambridge, United Kingdom, 1981 (Chapter 4).

TABLE 5.5 Solubility of naphthalene in various organic solvents at 20 °C.

Solvent	Solubility mole fraction, x_A
Chlorobenzene (C_6H4C_{l2})	0.256
Benzene (C_6H_6)	0.241
Toluene ($C_6H_5CH_3$)	0.224
Carbon tetrachloride (CCl_4)	0.205
Hexane (C_6H_{14})	0.090
Ideal solubility[a]	**0.273**

[a]*The ideal solubility was determined from the $\Delta H_{fusion,NaCl}$ and T_m for NaCl, which was 28.2 kJ mol^{-1} and 1074 K, respectively, where T_m is the melting temperature of the solute and $\Delta H_{fusion,NaCl}$ is the enthalpy of fusion (melting) of sodium chloride at T_m.*
Data from K.G. Denbigh, K.G. Denbigh, Equilibria of reactions involving gases, in: The Principles of Chemical Equilibrium: With Applications in Chemistry and Chemical Engineering, fourth ed., Cambridge University Press, Cambridge, United Kingdom, 1981 (Chapter 4). D.A. Vallero, T.M. Letcher, Engineering risks and failures: lessons learned from environmental disasters, Leadersh. Manag. Eng. 12 (4) (2012) 199–209.

(C_2H_5OH) are completely miscible, i.e., they dissolve each other to form a homogeneous mixture. The hydroxyl group (—OH) on the ethanol renders it polar, i.e., it is the same as the —OH group on water (HOH).

Conversely, nonpolar compounds covalently bond to share their electrons equally, so no dipole exists, which means a nonpolar compound has no net electrical charge. Methane (CH_4) is formed from covalent bonds, so it is highly nonpolar, i.e., solubility of only about 0.023 g of methane per kilogram of water at 25 °C and 1 atm. Conversely, when a hydroxyl group replaces one of methane's hydrogen atoms to form methanol (CH_3OH), a dipole forms and the molecule becomes completely dissolved in water in any amount.

The diversity of solubilities in various solvents is a first indication of the affinity of a compound for certain environmental compartments. For example, the various solubilities of the most toxic form of dioxin, tetrachlorodibenzo-*para*-dioxin (TCDD) are provided in Table 5.6. Dioxins have a much greater affinity for sediment, organic particles, and the organic fraction of soils than to water. The low water solubility indicates that dissolved TCDD in the water column should be present at only extremely low concentrations. However, other processes, such as co-solvation (wherein the dioxin

TABLE 5.6 Solubility of tetrachlorodibenzo-*para*-dioxin in water and organic solvents.

Solvent	Solubility (mg L^{-1})
Water	1.93×10^{-5}
Water	6.90×10^{-4} (25 °C)
Methanol	10
Lard oil	40
n-Octanol	50
Acetone	110
Chloroform	370
Benzene	570
Chlorobenzene	720
Orthochlorobenzene	1400

Data from IARC, Monographs on the Evaluation of the Carcinogenic Risk of Chemicals to Man, 1972–Present, World Health Organization, Geneva, Switzerland, 1995. R.T. Podoll, H.M. Jaber, T. Mill, Tetrachlorodibenzodioxin: rates of volatilization and photolysis in the environment, Environ. Sci. Technol. 20 (5) (1986) 490–492. H. Fiedler, K.-W. Schramm, QSAR generated octanol-water partition coefficients of selected mixed halogenated dibenzodioxins and dibenzofurans, Chemosphere 20 (10–12) (1990) 1597–1602.

TABLE 5.7 Solubility, octanol-water partitioning coefficient, and density values for some environmental pollutants at 25 °C and 1 atm.

Chemical	Water solubility ($mg\,L^{-1}$)	K_{ow}	Density ($kg\,m^{-3}$)
Atrazine	33	724	
Benzene	1780	135	879
Chlorobenzene	472	832	1110
Cyclohexane	60	2754	780
1,1-Dichloroethane	4960	62	1180
1,2-Dichloroethane	8426	30	1240
Ethanol	Completely miscible	0.49	790
Toluene	515	490	870
Vinyl chloride	2790	4	910
Tetrachlorodibenzo-*para*-dioxin (TCDD)	2×10^{-4}	6.3×10^{6}	

Data from H.F. Hemond, E.J. Fechner, Chemical Fate and Transport in the Environment, Elsevier, 2014. National Toxicology Program, Report on carcinogens: 2,3,7,8-tetrachlorodibenzo-p-dioxin, Research Triangle Park, North Carolina. Available: https://ntp.niehs.nih.gov/ntp/roc/content/profiles/tetrachlorodibenzodioxin.pdf, 2016 (Accessed 20 December 2021).

dissolves in something else like benzene that itself has much higher aqueous solubility) will change the behavior of a contaminant (see Chapter 14). This is a good example of the need to consider contamination from a systematic perspective.

If a compound is readily dissolved in water versus most organic solvents under normal environmental conditions of temperature and pressure, the compound is classified as hydrophilic. From the previous discussion, since hydrophilic compound is polar, it will also not dissolve well in lipids and fats, i.e., it is lipophobic. A substance not easily dissolved in water under these conditions, is hydrophobic and lipophilic. This affinity for water versus affinity for lipids is an important indicator of environmental partitioning, i.e., the octanol-water partition coefficient (K_{ow}). The K_{ow} is the ratio of a nonionized substance's concentration in octanol ($C_7H_{13}CH_2OH$) to that substance's concentration in water at equilibrium (see Chapter 14). Octanol has affinity for both polar and nonpolar compounds, that is, octanol is amphiphilic. So, the mixture of water and octanol allows for measuring concentrations of a compound in a two-phase system. Since the ratio forming the K_{ow} is $[C_7H_{13}CH_2OH]:[H_2O]$, then the larger the K_{ow} value, the more lipophilic the substance. Values for solubility in water and K_{ow} values of some important environmental compounds, along with their densities, are shown in Table 5.7.

Table 5.7 illustrates that water solubility is generally inversely related to K_{ow}, because greater hydrophobicity means greater affinity for the octanol phase compared to the aqueous phase. Minor structural differences between otherwise similar molecules can lead to substantial differences in phase partitioning. Indeed, isomers, i.e., compounds with the same chemical composition but with a different arrangement, vary in their K_{ow} values. For example, moving a chlorine atom one position away on the 1,1-dichloroethane and 1,2-dichloroethane molecules causes the former to have only a slight decrease in density but twice the K_{ow} value than the latter. In this case, it appears that the lower aqueous solubility of 1,1-dichloroethane accounts for its higher K_{ow}. In other cases, substitutions account for greater or lesser K_{ow} values. For example, a log-normal relationship between the number of chlorine atom substitutions and an organic compound's K_{ow} (see Fig. 5.7). This type of information is not only important in determining solubility but is also valuable in decisions on treatment of water and wastes that are contaminated with organic pollutants. For example, dehalogenation can be an effective treatment of wastes contaminated with halogenated solvents and persistent organic pollutants [48–50].

3.3 Multiphase pollutants

The inherent properties of a substance combine with the properties of the environment to drive pollutant transport, as shown in Fig. 5.8. The transport of the nonaqueous phase liquids (NAPLs) through the vadose zone is affected by their extremely high K_{ow} values and extremely low water solubility. NAPLs, therefore, have greater affinity for lipids than for water. Substances with higher hydrophilicity are more likely to become completely mixed with the water, so its flow will increasingly

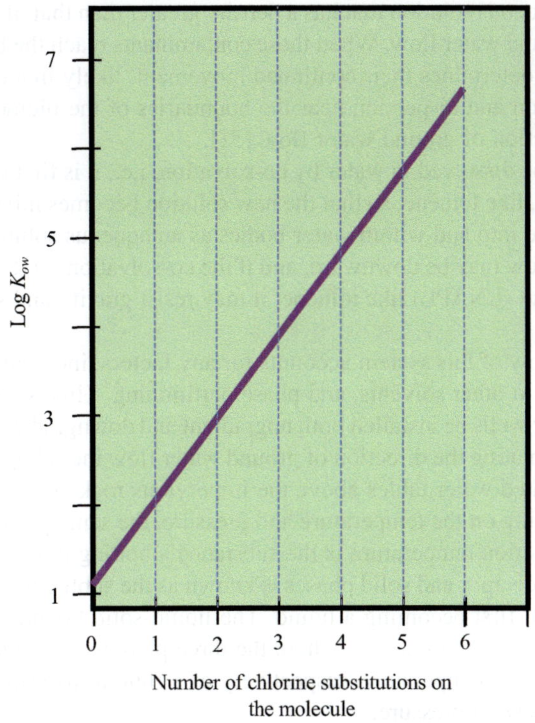

FIG. 5.7 Hydrophobicity as represented by the octanol-water coefficient (K_{ow}) resulting from the number of chlorine atom substitutions on a benzene molecule. *(Data from Molecular Design Research Network, Aqueous and lipid solubility, Yale University. https://modrn.yale.edu/education/undergraduate-curriculum/modrn-u-modules/aqueous-and-lipid-solubility, 2021 (Accessed 20 December 2021).)*

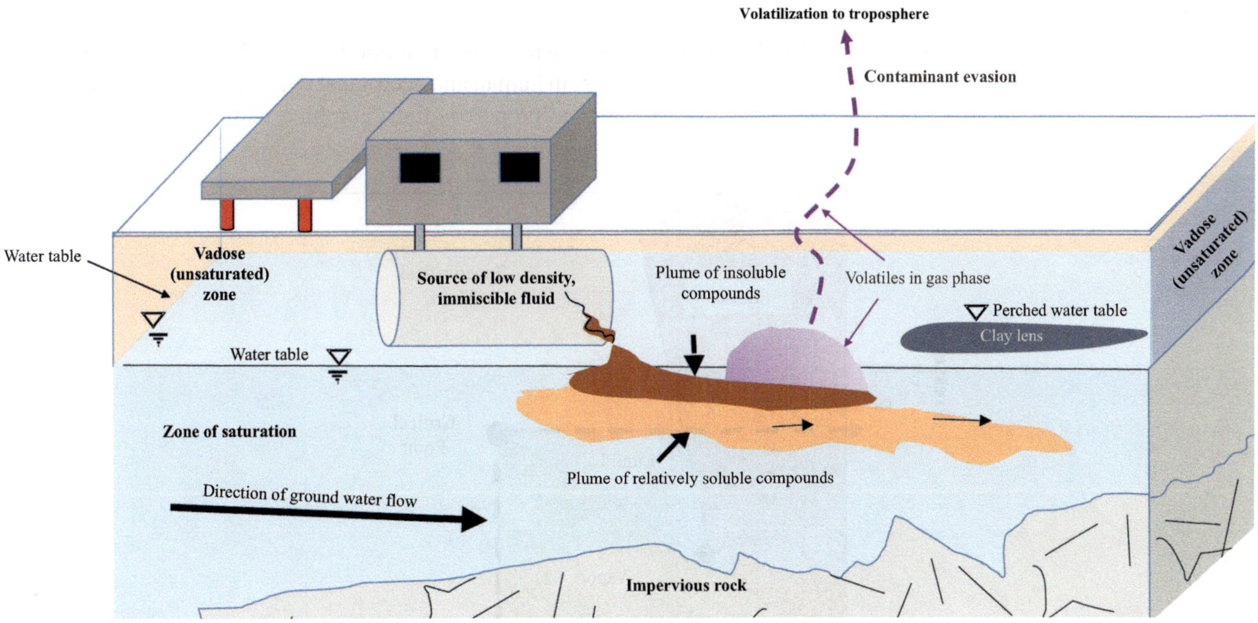

FIG. 5.8 Hypothetical plume of fluids varying in lipophilicity, aqueous solubility, density, and vapor pressure.

follow the water flow lines. Conversely, NAPLs and other more lipophilic compounds, i.e., those with higher K_{ow} values, will not necessarily move directly in the direction of the water flow.

Dense fluids that seep into the zone of saturation will move downward if their densities are much greater than that of the water. If the fluid is immiscible, e.g., a dense nonaqueous phase liquid (DNAPL), the fluid will independently move downward in the zone of saturation. According to Raoult's law, the molecules that join the water solution may also move downwardly but will move as an aqueous solution that has a density greater than that of pure water. If completely miscible, the solution will move with the ground water flow. When these contaminants reach the bottom of the aquifer, the shape and slopes of the impervious boundary determines their continued movement, likely in a direction other than the flow of the ground water in the aquifer. Solution and dispersion near the boundaries of the plume may form secondary plumes that commonly follow the general direction of ground water flow [51].

As hydrophobic substance can be dissolved in water by co-solvation, i.e., it is first dissolved in an amphiphilic, like an alcohol, or slightly lipophilic liquid, like toluene, so that the new solution becomes mixed with water. By co-solvation this hydrophobic compound can migrate into and within water bodies as an aqueous solution in the alcohol or toluene. If the mixture is heavier than water, the flow may be downward, and if the co-solvation mixture is lighter than water, e.g., mixed with a light nonaqueous phase liquid (LNAPL) like toluene, it may resist gravity and stay near the water table and in the vadose zone [52].

The physics of predicting the flow of this system accounts for any factors, including the densities of all fluids that are present, the solubility in water and in other solvents, and phase partitioning. This is one reason that environmental regulations often require that monitoring wells be installed both upgradient and downgradient from the pollutant source [53,54]. Other reasons include errors in estimating the direction of ground water flow including miscalculating the depth to impervious rock, or the presence of perched water tables above the impervious rock zone [55].

As indicated in Fig. 5.9, depending on the temperature and pressure, the same pollutant can exist in vapor, liquid, and solid phases. The vapor-liquid transition temperature is the substance's boiling point, and the pressure is the substance's vapor pressure. The line between the vapor and solid phases is known as the sublimation point, i.e., the phase change from solid to gas, or gas to solid, without first becoming a liquid. The liquid-solid boundary line temperature is freezing and melting point. T_{tp} and P_{tp} define the triple point, i.e., where the three phases of pure substance can exist in equilibrium. $T_{critical}$ and $P_{critical}$ are the pressure and temperature, respectively of a fluid at its critical point, i.e., the point at which a gas cannot be liquefied by an increase of pressure.

The movements of a pollutant within and between environmental compartments and to, within, and from organisms are strongly related to the pollutant's physical and chemical properties. The potential exposure to the pollutant is driven by this movement (see Fig. 5.10). A substance's K_{ow} is a key determinant in the likelihood that a water pollutant will have affinity for aqueous and lipid substrates in an organism's tissue, including that of humans, other animals, plants, and microbes (see Figs. 14.9 and 14.10 in Chapter 14).

As mentioned, the vapor pressure and molecular weight are among the inherent properties of a chemical compound that give clues to the behavior of a pollutant and are common parameters in human and ecological exposure and risk models (see Chapter 17). However, these are merely clues. The physicochemical properties must combine with substrate and matrix

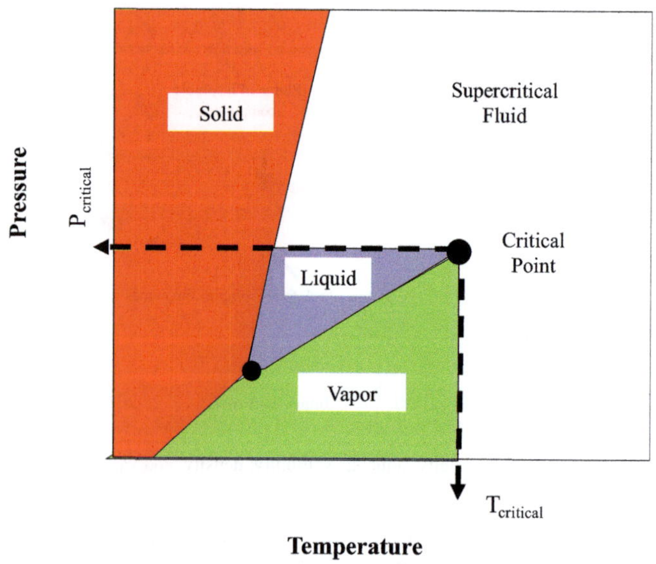

FIG. 5.9 Phase diagram for a hypothetical substance. All phase regions shown in the diagram will exist for any substance, but their shapes and slopes of the polygon sides and points of phase changes will differ depending on the inherent properties of the substance and external conditions in the environment, especially temperature and pressure.

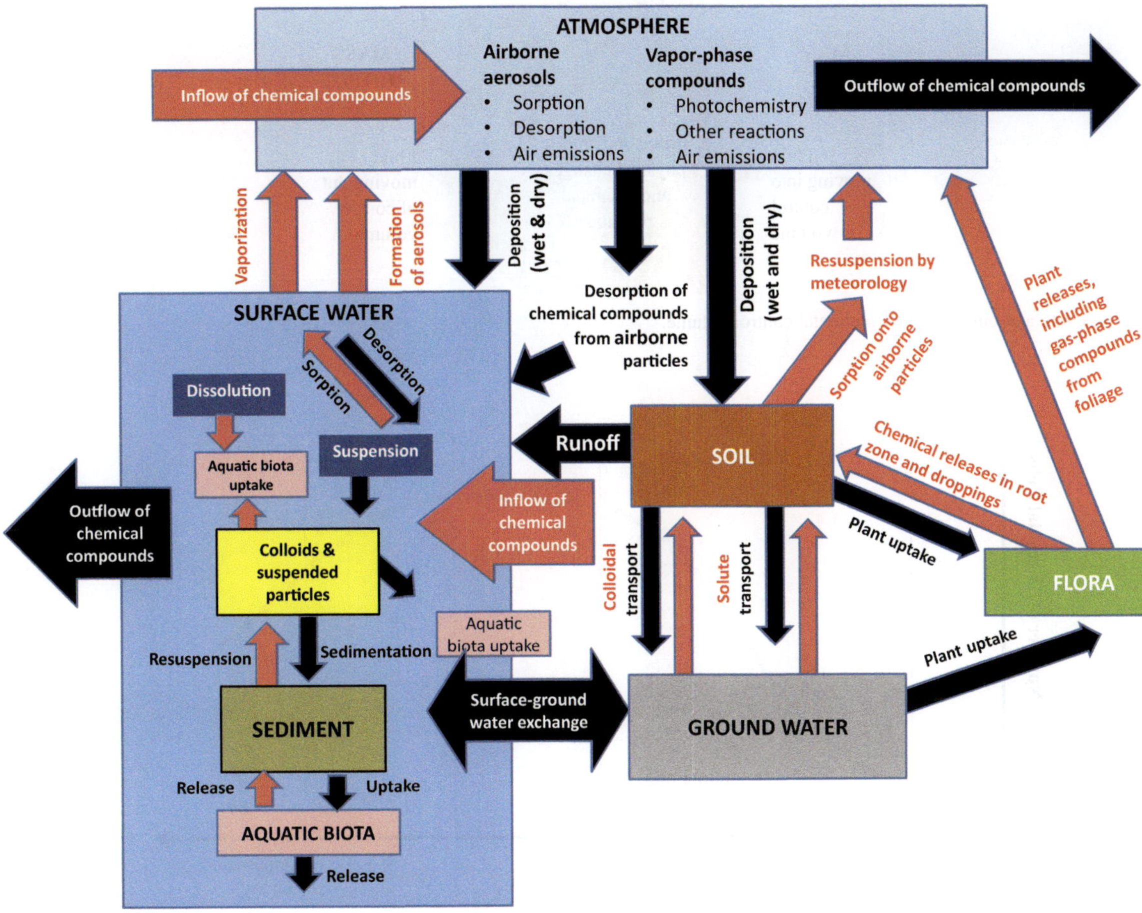

FIG. 5.10 Physical, chemical, and biological factors at work after pollutant release into the environment, uptake by three classes of fauna, ultimately leading to exposure. *(Data from D. Vallero, Environmental Biotechnology: A Biosystems Approach, Elsevier Science, 2015. T. McKone, W. Riley, R. Maddalena, R. Rosenbaum, D. Vallero, Common issues in human and ecosystem exposure assessment: the significance of partitioning, kinetics, and uptake at biological exchange surfaces, Epidemiology 17 (6) (2006) S134. R.K. Rosenbaum, et al., USEtox human exposure and toxicity factors for comparative assessment of toxic emissions in life cycle analysis: sensitivity to key chemical properties, Int. J. Life Cycle Assess. 16 (8) (2011) 710–727. C.M. Eichler, J.C. Little, A framework to model exposure to per-and polyfluoroalkyl substances in indoor environments, Environ. Sci. Process. Impacts 22 (3) (2020) 500–511. J.C. Little, C.J. Weschler, W.W. Nazaroff, Z. Liu, E.A. Cohen Hubal, Rapid methods to estimate potential exposure to semivolatile organic compounds in the indoor environment, Environ. Sci. Technol. 46 (20) (2012) 11171–11178. D. Vallero, Air Pollution Calculations: Quantifying Pollutant Formation, Transport, Transformation, Fate and Risks, Elsevier, 2019.)*

characteristics, along with other environmental conditions to determine the fate of the pollutant. As shown in Fig. 5.10, after entering the water, air, soil, and biota, a chemical compound or microbe will either remain in a compartment or move to another. Both chemical and biological agents have relative preferences for different compartments, but no compartment is an exclusive residence. Similarly, for human exposure models, exposure occurs in various places, so-called microenvironments, in space and time [56–59]. Whether in various compartments of an ecosystem or as exposures during a human's daily activities, a fraction of the mass will remain there, with the remainder moving to one or more other compartments, according to the laws of motion. The propensity for this distribution is known as the partitioning described in Chapter 14. While residing in a compartment, abiotic chemical reactions will occur, e.g., photolytic and hydrolytic, as well as biochemical reactions by microbes, plants, and animals.

Microbial populations can either degrade and worsen pollution. Soil and other bacteria, for example, can detoxify pollutants when they break them down to gain energy. Conversely, conditions in waterbodies and soil pore spaces can lead to the growth of pathogens. Thus, each environmental compartment is a thermodynamic control volume, with fluids in gas and liquid phases carrying abiotic and biotic mass that enters the volume. Abiotic and biochemical reactions occur within the volume, and converted and unconverted mass exits the volume.

For chemical pollutants, the exiting mass will be a combination of the parent chemical compound and its degradation products (see Fig. 5.11). For microbes, the exiting mass will consist of net cellular biomass, i.e., the difference between the

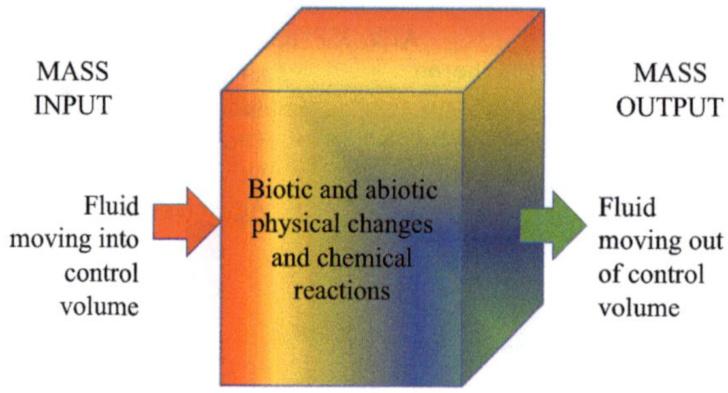

FIG. 5.11 Flow into and out of an environmental control volume.

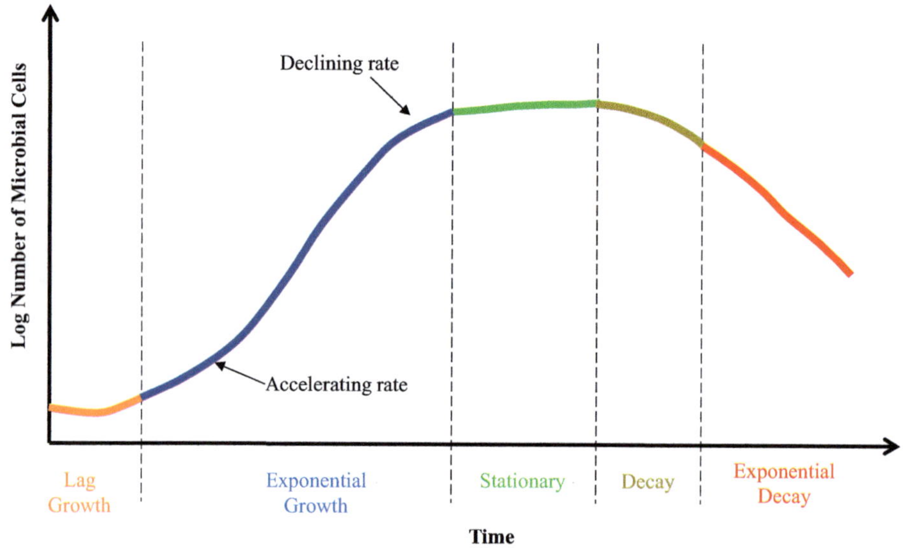

FIG. 5.12 Prototypical growth and decay curve for bacteria. *(Data from S.S. Suthersan, J. Horst, M. Schnobrich, N. Welty, J. McDonough, Remediation Engineering: Design Concepts, CRC Press, 2016.)*

initial microbial population and net cell growth. Time is a key parameter in microbial degradation. Microbial populations undergo a lag in growth after food is available, followed by exponential increases in number of organisms before leveling off and beginning to decay [60] (see Fig. 5.12). Each stage will vary in duration for a given species and under different environmental conditions.

The microbial biomass exiting a control volume may be beneficial or harmful, depending on the acceptability and hostility of the control volume. In a completely hostile control volume, the microbial population could die off, i.e., the exiting living biomass would be zero. The Monod equation can be used to calculate and express the rate of microbial biomass:

$$\mu = \frac{\mu_{max} S}{K_S + S} \tag{5.12}$$

where μ = the specific growth rate of the microbe; $\mu_{\max}$ = maximum specific growth rate; and K_s = the Monod growth rate coefficient representing the substrate concentration at which the growth rate is half the maximum rate. The $\mu_{\max}$ is reached at the higher ranges of substrate concentrations. K_s is an expression of the affinity of the microbe for a substance, i.e., as K_s decreases the greater the affinity that microbe has for that substance (as expressed by the concomitantly increasing μ). Thus, wastewater treatment must bring the wastewater into contact with the microbes and enhance the degradation environment, such as by activated sludge (see Chapter 15). Engineered systems may also add activated microbes, if indigenous microbes are not already breaking down the chemical compound. As these conditions change, the microbes undergo a series of stages [11]:

1. Lag phase: Upon initial exposure of the microbes to the chemical contaminant, sufficient time is needed for the organisms to become acclimated.

2. Accelerated growth rate phase: Following acclimation, the microbes propagate at an increasing rate. There are two major processes that allow for the degradation of chemical compounds by microbes.
3. Exponential phase: The cell mass and the number of cells are growing exponentially by binary fission.
4. Declining growth phase: Cell mass and numbers of cells continue to grow, but at a decreasing rate. This is usually due to depletion of the food and electron source (i.e., the contaminant, hopefully). Other limiting factors could also come into play, such as some of the newly generated chemicals ("degradates") inhibiting the growth of the microbes due to their toxicity.
5. Stationary phase: Cell decay is about equal to cell propagation during this time.
6. Decay: Cell decay now exceeds cell propagation.
7. Exponential death phase: Cells are dying exponentially as cells no longer grow or propagate. Under optimal conditions, this indicates successful bioremediation because the food source (toxic organics) is entirely metabolized by the microbes.

The cycle will repeat if the microbes are again meet and degrade organic compounds.

3.4 Flow

One of the key aspects of effluent and water quality limits is flow. As discussed in Chapter 10, the pollutant concentration in a stream (C_r) can be used to allocate waste loads discharged to surface waters and to calculate the concentration in the stream reach after complete mixing occurs [2]. Government regulations use equations to calculate water quality-based limits, such as:

$$C_r = \frac{Q_d C_d + Q_s C_s}{Q_r} \tag{5.13}$$

Where: Q_d = waste discharge flow in units commonly used by U.S. environmental engineers, i.e., million gallons per day (MGD) or cubic feet per second (cfs); C_d = pollutant concentration in waste discharge in milligrams per liter (mg L^{-1}) [61]; Q_s = background stream flow in MGD or cfs above point of discharge; C_s = background in-stream pollutant concentration in mg L^{-1}; and, Q_r = resultant in-stream flow after discharge in MGD or cfs.

If a value of zero is used for the stream concentration for a pollutant (C_s), the agency can establish discharge concentration limit, however, this would not account for upstream background concentrations of the pollutant and other sources of the pollutant like nonpoint runoff, so these would need to be added to determine whether a stream truly meets water quality criteria [2]. Surface water flow is discussed in detail in Chapter 10.

The ways that water penetrates the earth's surface into the zones of unconsolidated zones and ultimately into aquifers in the covered in detail in Chapter 11.

4 Drinking water standards

Drinking water standards are generally classified as either primary or secondary. In the U.S., the National Primary Drinking Water Regulations set the primary standards, i.e., those that legally require public water systems to limit levels of contamination of drinking water. These include pathogenic microorganisms, chemical disinfectants, disinfection byproducts, inorganic chemical compounds, organic compounds, and radionuclides [62].

The microorganisms of concern include *Cryptosporidium* species, *Giardia lamblia*, *Legionella* species, select viruses, and coliform bacteria including *Escherichia coli* and other fecal coliforms.

The chemical disinfectants are those used by water systems to kill microbes, which include chlorine (Cl_2), chloramines (measured as Cl_2), and chlorine dioxide (Cl_2). Unfortunately, addition of these disinfectants to water can result in the production of toxic byproducts. Disinfectant byproducts include bromate and chlorite compounds, haloacetic acids, and total trihalomethanes.

Regulated inorganics include heavy metals, lead (Pb), cadmium (Cd), chromium (Cr), and mercury (Hg), selenium (Se), thallium (Tl), and barium (Ba), the metalloids arsenic (As) and antimony (Sb), fluoride compounds, cyanide, and asbestos fibers. There primary standards include are more than 70 regulated organic compounds, including chlorinated solvents like tetrachloromethane (carbon tetrachloride), trichloroethane, and trichlorobenzene, and halogenated organic pesticides like 2,4-D and dibromochloropropane (DBCP).

Radionuclides include alpha particle, beta particles and photon emitters, combined radium-226 (^{226}Ra) and radium-228 (^{228}Ra), and total uranium (U).

In the U.S., drinking water standard metrics are:

- Maximum Contaminant Level Goal (MCLG)—The contaminant concentration in drinking water below which there is no known or expected risk to health. MCLGs allow for a margin of safety and are not enforceable public health goals [62].
- Maximum Contaminant Level (MCL)—The highest concentration of a contaminant allowed in drinking water. MCLs are set as close to MCLGs as feasible based on the best available treatment technology. MCLs are enforceable.
- Maximum Residual Disinfectant Level Goal (MRDLG)—The drinking water disinfectant concentration below which there is no known or expected risk to health. MRDLGs do not account for the benefits of the use of disinfectants to control microbial contaminants.
- Treatment Technique (TT)—The process needed to decrease the concentration of a contaminant.
- Maximum Residual Disinfectant Level (MRDL)—The highest concentration of a disinfectant allowed in drinking water.

In the U.S., secondary standards are not enforceable but are set to caution about possible adverse effects like tooth and skin discoloration, taste, odor, or color of tap water, and other aesthetic effects. However, individual states and other jurisdictions may indeed require water supplies to meet these standards [63]. Table 5.8 shows the list of secondary drinking water standards. Fluoride falls between a primary and secondary standard since the federal government requires special notice be given to the water users if the secondary standard of $2.0\,mg\,L^{-1}$ is exceeded. If the fluoride concentration is greater than $2.0\,mg\,L^{-1}$ but less than the primary standard of $4.0\,mg\,L^{-1}$, the users must be notified no later than 12 months after the finding [63].

The World Health Organization (WHO) has established similar guidelines to be considered globally as an authoritative basis for nations and regions to set drinking-water policies and actions. The guidelines are also intended to help nations and regions establish comprehensive preventive health risk management frameworks to set risk-based standards. The WHO drinking water guidelines summarize the health implications and the ways that risk assessment and risk management can help prevent and control water-related diseases and offer effective options for decision makers to achieve clean water goals [64]. The listed contaminants are similar to those of the U.S. but are more inclusive. For example, the lists of

TABLE 5.8 U.S. National Secondary Drinking Water Standards.

Contaminant	Secondary standard
Aluminum	$0.05–0.2\,mg\,L^{-1}$
Chloride	$250\,mg\,L^{-1}$
Color	15 (Color units)
Copper	$1.0\,mg\,L^{-1}$
Corrosivity	Noncorrosive
Fluoride	$2.0\,mg\,L^{-1}$
Foaming agents	$0.5\,mg\,L^{-1}$
Iron	$0.3\,mg\,L^{-1}$
Manganese	$0.05\,mg\,L^{-1}$
Odor	3 Threshold odor number
pH	6.5–8.5
Silver	$0.10\,mg\,L^{-1}$
Sulfate	$250\,mg\,L^{-1}$
Total dissolved solids	$500\,mg\,L^{-1}$
Zinc	$5\,mg\,L^{-1}$

Data from U.S. Environmental Protection Agency, National Secondary Drinking Water Regulations, U.S. EPA. https://www.epa.gov/sites/production/files/2015-09/documents/rags_a.pdf, 2015 (Accessed 26 January 2024).

pathogenic bacteria and viruses are more exhaustive than those in the U.S. As evidence, the WHO list includes nine or more bacteria genera, six virus genera, six protozoa, and helminths known to be transmitted through drinking water and many more where there is less evidence but transmission is possible [64]. The chemical lists are like that of the U.S., but additional radionuclides are listed.

5 Conclusions

Water quality standards are written to protect public health and ecosystems. Developing a standard properly calls for sound scientific evidence to establish the thresholds between acceptable and polluted water. This means that standards are needed for what enters the water, i.e., effluent limits, and for the waterbody itself, i.e., water quality standards. Standards also are designed to protect designated uses of waterbodies, such as their recreational and drinking water values. This means that there are also point-of-use standards, especially the quality of water delivered by public water supplies. This chapter explored the numerous scientific principles that underpin water quality criteria and standards.

References

[1] U.S. Environmental Protection Agency, CatReg software for categorical regression analysis, Research Triangle Park, North Carolina, 2006.

[2] U.S. Environmental Protection Agency, Attachment B: water quality-based effluent limits—coal combustion waste impoundments, U.S. Environmental Protection Agency, 2017. #978. (Accessed 7 March 2018).

[3] U.S. Environmental Protection Agency, NPDES permit basics, U.S. EPA, 2023. https://www.epa.gov/npdes/npdes-permit-basics. (Accessed 26 January 2024).

[4] Department for Environment Agency, Food, and Rural Affairs, Discharges to surface water and groundwater: environmental permits, Gov.UK, 2016. https://www.gov.uk/guidance/discharges-to-surface-water-and-groundwater-environmental-permits#:~:text=If%20your%20domestic%20sewage%20discharge,right%20of%20appeal%20against%20them. (Accessed 26 January 2024).

[5] D.J. Halliwell, I.D. McKelvie, B.T. Hart, R.H. Dunhill, Hydrolysis of triphosphate from detergents in a rural waste water system, Water Res. 35 (2) (2001) 448–454, https://doi.org/10.1016/S0043-1354(00)00289-X.

[6] B. Zinder, J. Hertz, H. Oswald, Kinetic studies on the hydrolysis of sodium tripolyphosphate in sterile solution, Water Res. 18 (5) (1984) 509–512.

[7] S.L. Friess, The energetics of acid-catalyzed hydrolysis of triphosphoric and pyrophosphoric acids, J. Am. Chem. Soc. 74 (16) (1952) 4027–4029.

[8] A. Schecter, L. Birnbaum, J.J. Ryan, J.D. Constable, Dioxins: an overview, Environ. Res. 101 (3) (2006) 419–428.

[9] C. Li, et al., Long-term persistence of polychlorinated dibenzo-p-dioxins and dibenzofurans in air, soil and sediment around an abandoned pentachlorophenol factory in China, Environ. Pollut. 162 (2012) 138–143.

[10] D. Mackay, A. Fraser, Bioaccumulation of persistent organic chemicals: mechanisms and models, Environ. Pollut. 110 (3) (2000) 375–391.

[11] T.J. Harner, Measurement and partitioning of persistent organic pollutants in the environment: Application of octanol-air partition coefficients, Ph. D. Thesis, 1998, p. 276.

[12] F.A. Gobas, Assessing bioaccumulation factors of persistent organic pollutants in aquatic food-chains, in: Persistent Organic Pollutants: Environmental Behaviour and Pathways of Human Exposure, 2001, pp. 145–165.

[13] Arctic Monitoring Assessment Program, AMAP assessment 2002: persistent organic pollutants in the Arctic, 2004.

[14] J. Diamond, B. Tonning, K. Parry, C. Boschen, Integrating aquatic designated use protection and restoration strategies using an ecosystem goods and services framework, Integr. Environ. Assess. Manag. 15 (5) (2019) 808–818.

[15] U.S. Environmental Protection Agency, National recommended water quality criteria—aquatic life criteria table, U.S. EPA, 2016. https://www.epa.gov/wqc/national-recommended-water-quality-criteria-aquatic-life-criteria-table. (Accessed 21 January 2024).

[16] E. Dohner, A. Markowitz, M. Barbour, J. Simpson, J. Byrne, G. Dates, 5.2. Monitoring and assessing water quality, in: Volunteer Stream Monitoring: A Methods Manual, 1997 (Chapter 5).

[17] V. Novotny, Integrated water quality management, Water Sci. Technol. 33 (4–5) (1996) 1–7.

[18] F.E. Gostomski, Water Quality Criteria: Protection of Use Perspective, ASTM International, 1985.

[19] U.S. Environmental Protection Agency, Habitat Suitability Index (HSI), 2016. https://archive.epa.gov/aed/html/research/scallop/web/html/hsi.html. (Accessed 15 July 2024).

[20] W.A. Rourke, A. Jutason, J.L. Martin, C.J. Murphy, Shellfish toxin uptake and depuration in multiple Atlantic Canadian molluscan species: application to selection of sentinel species in monitoring programs, Toxins 13 (2) (2021) 168, https://doi.org/10.3390/toxins13020168. In this issue.

[21] V. Arnold, A note on the Ghyben-Herzberg formula, Bull. Int. Assoc. Sci. Hydrol. (1968) 43–46.

[22] M.P. Anderson, W.W. Woessner, R.J. Hunt, Applied Groundwater Modeling: Simulation of Flow and Advective Transport, Academic Press, 2015.

[23] P.M. Barlow, Ground water in freshwater-saltwater environments of the Atlantic coast, US Department of the Interior, US Geological Survey, Washington, DC, USA, 2003.

[24] M. Anderson, W. Woessner, R. Hunt, Model dimensionality and setting boundaries, in: Applied Groundwater Modeling: Simulation of Flow and Advective Transport, 2015, pp. 117–180.

[25] Z. Zhang, H.H. Savenije, Thermodynamics of saline and fresh water mixing in estuaries, Earth Syst. Dyn. 9 (1) (2018) 241–247.

[26] A. Kleidon, Thermodynamic Foundations of the Earth System, Cambridge University Press, 2016.

[27] U.S. Environmental Protection Agency, Water resources, 2013. http://www.epa.gov/climatechange/impacts-adaptation/water.html. (Accessed 1 November 2013).
[28] U.S. Environmental Protection Agency, Saving Florida's vanishing shores, 2002. http://www.epa.gov/climatechange/effects/coastal/saving_FL.pdf. (Accessed 1 November 2013).
[29] J.R. Bidwell, J.R. Gorrie, The influence of salinity on metal uptake and effects in the midge Chironomus maddeni, Environ. Pollut. 139 (2) (2006) 206–213.
[30] L. Zalizniak, B.J. Kefford, D. Nugegoda, Is all salinity the same? I. The effect of ionic compositions on the salinity tolerance of five species of freshwater invertebrates, Mar. Freshw. Res. 57 (1) (2006) 75–82.
[31] P. Rengasamy, Transient salinity and subsoil constraints to dryland farming in Australian sodic soils: an overview, Aust. J. Exp. Agric. 42 (3) (2002) 351–361.
[32] Environment Canada, Priority substances list assessment report for road salts, 2021. https://www.canada.ca/en/health-canada/services/environmental-workplace-health/reports-publications/environmental-contaminants/canadian-environmental-protection-act-1999-priority-substances-list-assessment-report-road-salts.html. (Accessed 15 July 2024).
[33] A. Marandi, M. Polikarpus, A. Jõeleht, A new approach for describing the relationship between electrical conductivity and major anion concentration in natural waters, Appl. Geochem. 38 (2013) 103–109.
[34] A. Logeshkumaran, N. Magesh, P.S. Godson, N. Chandrasekar, Hydro-geochemistry and application of water quality index (WQI) for groundwater quality assessment, Anna Nagar, part of Chennai City, Tamil Nadu, India, Appl. Water Sci. 5 (4) (2015) 335–343.
[35] A.F. Rusydi, Correlation between conductivity and total dissolved solid in various type of water: a review, IOP Conf. Ser. Earth Environ. Sci. 118 (1) (2018) 012019. IOP Publishing.
[36] M. Taylor, H.A. Elliott, L.O. Navitsky, Relationship between total dissolved solids and electrical conductivity in Marcellus hydraulic fracturing fluids, Water Sci. Technol. 77 (8) (2018) 1998–2004.
[37] B. Sivakumar, Chaos in Hydrology: Bridging Determinism and Stochasticity, Springer, 2016.
[38] A. Glushkov, Analysis and forecast of the hydroecological system pollution dynamics based on methods of chaos theory: new general scheme, in: Вісник Одеського державного екологічного університету, vol. 19, 2015, p. 12.
[39] P. Holmes, Poincaré, celestial mechanics, dynamical-systems theory and "chaos", Phys. Rep. 193 (3) (1990) 137–163.
[40] D.A. Vallero, Waste and biogeochemical cycling, in: Waste: A Handbook for Managers, Elsevier, Amsterdam, NV, 2019, pp. 91–125.
[41] National Center for Atmospheric Research, Biogeochemical cycles, 2015. http://scied.ucar.edu/longcontent/biogeochemical-cycles. (Accessed 4 March 2015).
[42] U.S. Environmental Protection Agency, Water cycle and water demand, US EPA, 2016. https://19january2017snapshot.epa.gov/climate-impacts/climate-impacts-water-resources_.html. (Accessed 15 July 2024. Last update on December 21, 2016.
[43] D.A. Vallero, T.M. Letcher, Unraveling Environmental Disasters, Newnes, 2012.
[44] P. Atkins, J. De Paula, Elements of Physical Chemistry, Oxford University Press, USA, 2013.
[45] R.P. Schwarzenbach, P.M. Gschwend, D.M. Imboden, Environmental Organic Chemistry, John Wiley & Sons, 2016.
[46] D.A. Vallero, T.M. Letcher, Engineering risks and failures: lessons learned from environmental disasters, Leadersh. Manag. Eng. 12 (4) (2012) 199–209.
[47] K.G. Denbigh, K.G. Denbigh, Equilibria of reactions involving gases, in: The Principles of Chemical Equilibrium: With Applications in Chemistry and Chemical Engineering, fourth ed., Cambridge University Press, Cambridge, United Kingdom, 1981 (Chapter 4).
[48] X. Zhu, X. Wang, N. Li, Q. Wang, C. Liao, Bioelectrochemical system for dehalogenation: A review, Environ. Pollut. 293 (2022) 118519.
[49] B.M. McDowell, Biotic and abiotic dehalogenation of halogenated methanes: trichlorofluoromethane, dichlorodifluoromethane, and tetrachloromethane, 2021.
[50] B. Heckel, et al., Reductive dehalogenation of trichloromethane by two different Dehalobacter restrictus strains reveal opposing dual element isotope effects, Environ. Sci. Technol. 53 (5) (2019) 2332–2343.
[51] D. Vallero, Environmental Contaminants: Assessment and Control, Academic Press, 2010.
[52] D. Vallero, Environmental Biotechnology: A Biosystems Approach, Elsevier Science, 2015.
[53] C.T. Miller, W.J. Weber Jr., Modeling organic contaminant partitioning in ground-water systems, Groundwater 22 (5) (1984) 584–592.
[54] H.A. Loaiciga, R.J. Charbeneau, L.G. Everett, G.E. Fogg, B.F. Hobbs, S. Rouhani, Review of ground-water quality monitoring network design, J. Hydraul. Eng. 118 (1) (1992) 11–37.
[55] U.S. Environmental Protection Agency, CRA ground-water monitoring technical enforcement guidance document, 1986, EC-G-2002-108, Washington, DC. Available: https://www.epa.gov/sites/default/files/documents/rcragwguiddoc-rpt_0.pdf.
[56] I.E. Medina-Ramirez, A. Jimenez-Chavez, A. De Vizcaya-Ruiz, Toxicity of nanoparticles, in: Antimicrobial Activity of Nanoparticles, Elsevier, 2023, pp. 249–284.
[57] S. Marinello, F. Lolli, A.M. Coruzzolo, R. Gamberini, Exposure to air pollution in transport microenvironments, Sustainability 15 (15) (2023) 11958.
[58] X. Fang, et al., Study on particle size distribution characteristics in urban roadside microenvironment using an ELPI+, Water Air Soil Pollut. 233 (11) (2022) 467.
[59] C. Schweizer, et al., Indoor time–microenvironment–activity patterns in seven regions of Europe, J. Expo. Sci. Environ. Epidemiol. 17 (2) (2007) 170–181.
[60] S.S. Suthersan, J. Horst, M. Schnobrich, N. Welty, J. McDonough, Remediation Engineering: Design Concepts, CRC Press, 2016.

[61] U.S. Environmental Protection Agency, Coal ash (coal combustion residuals, or CCR), 2010. https://www.epa.gov/coalash. (Accessed 8 March 2018).

[62] U.S. Environmental Protection Agency, National Primary Drinking Water Regulations, U.S. EPA, Washington, DC, 2024. https://www.epa.gov/ground-water-and-drinking-water/national-primary-drinking-water-regulations. (Accessed 15 July 2024. Last update on June 13, 2024.

[63] U.S. Environmental Protection Agency, National Secondary Drinking Water Regulations, U.S. EPA, Washington, DC, 2024. https://www.epa.gov/sdwa/secondary-drinking-water-standards-guidance-nuisance-chemicals. (Accessed 15 July 2024. Last update on June 14, 2024.

[64] World Health Organization, Guidelines for drinking-water quality: incorporating the first and second addenda, World Health Organization, Geneva, Switzerland, 2022.

Chapter 6

Climate and weather

1 Introduction

Climate can be thought of as "average weather". Pollutants are generally classified according to their physical and chemical structure, the harm they may cause, and their location in time and space [1–3]. Water in and from the atmosphere change these classifications. Thus, drawing a fine line between what may be a water pollutant and what defines an air pollutant is folly. The life cycle of many water pollutants some residence in in other compartments, including the atmosphere. For example, some of the mass of a mixture of chemical compounds emitted from a stack will be transformed physically and chemically before being deposited onto a land surface. Thereafter, certain substances move overland and, in the process, continue to change chemically, and become nonpoint sources of water pollution as they enter surface and ground water (see Chapters 10 and 11). Perhaps the substance is already a pollutant when it exits the stack or becomes a water pollutant when it volatilizes or when it is carried by an aerosol into the atmosphere. Indeed, many air pollutant in the atmosphere move to water bodies and vice versa. So, the residence time in the atmosphere is influenced by climate and weather [4].

Indeed, water quality is affected by meteorology. For example, increasing temperatures of water bodies and increasing flood and drought frequencies are forecasted to increase the rates of pollution in sediments in stream beds, increase the populations of pathogenic microbes, and exacerbate exposures to pesticides [5].The rise in sea level is projected to spread rates of salinization of ground water in coastal regions [5–8]. Increasing temperatures will also lead to thermal pollution [5], as water running off land surfaces will be warmer and more intensive solar radiation reaching water bodies could change trophic structures and threaten sensitive species, especially cold-water fish species (see Chapter 5).

In addition to the linkages between the hydrosphere and the atmosphere, discussions of pollutant fate and transport enhances discussion of water pollutant fate and transport. For example, both air and water are fluids, so equations and graphs in this chapter support many water pollution concepts, such as dispersion, mixing, molecular diffusion, and environmental fate. The transport of pollutants is affected by meteorology, so the primary property of a pollutant regarding weather is its vapor pressure. All the molecules of a substance are in the gas phase under normal meteorological ranges, the molecule is a gas. If most of the molecules are in the gas phase, but some are in liquid or solid phases under common meteorological conditions, the molecule is a vapor. For example, molecular nitrogen (N_2) is a gas, Thus, N_2 with a boiling point is $-196\,°C$, is a gas except under extreme temperatures. Under all normal meteorological conditions, which are much warmer than $-196\,°C$, N_2 is an atmospheric gas [1]. Conversely, the water molecule can exist in all phases under normal ambient conditions, i.e., as solid ice, liquid water, and water vapor. Although the pressure in the closed container in Fig. 6.1 is constant and the temperature is below the boiling point, molecules of the vapor continue to condense into the liquid phase and molecules of the liquid will continue to evaporate into the vapor phase. The system is at dynamic equilibrium and the conditions are at equilibrium vapor pressure. The rate of these two processes is equal, so no net change in the amount of vapor or liquid.

2 Meteorologically important pollutant properties

The vapor pressure (p^0) of a contaminant in the liquid or solid phase is the pressure that is exerted by its vapor when the liquid and vapor are in dynamic equilibrium. Pollutants with vapor pressures greater than 10^{-2} kilopascals (kPa) at 20 °C can be classified volatile [9], which means a substantial number of molecules readily enter the gas phase. Volatile organic compounds (VOCs) like benzene ($p^0 = 10\,kPa$) and toluene ($p^0 = 3\,kPa$) are water and soil pollutants with a substantial number of molecules also in the vapor phase, e.g., in sediment and soil pore spaces. Semivolatile compounds have p^0 values $< 10^{-2}$ kPa, $>10^{-5}$ kPa. These compounds have less attraction for the gas phase than volatile compounds, so are commonly sorbed to soils and sediments, attached to suspended solids in water, and as both aerosol and vapors in

Fundamentals of Water Pollution. https://doi.org/10.1016/B978-0-443-28987-3.00001-1

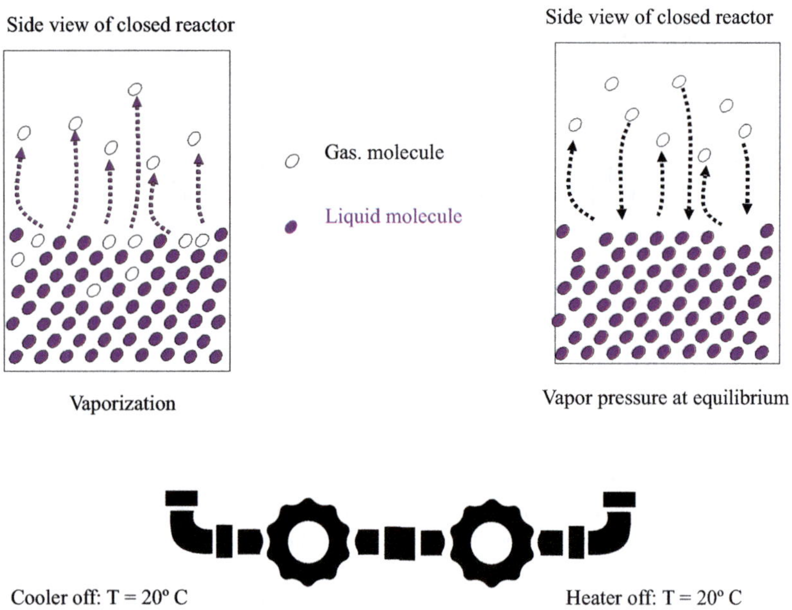

FIG. 6.1 Vapor pressure of a containerized fluid during vaporization (left) and ultimately reaching equilibrium (right). A portion of a substance in an evacuated, closed container with restricted headspace will vaporize. The pressure in the headspace above the liquid increases from zero and eventually stabilizes after reaching a constant value. This value is the vapor pressure of that substance. Substances not in a closed container, i.e., infinitely available headspace, will also vaporize, but will continue to vaporize until all the substance has partitioned to the gas phase. *(Data from D. A. Vallero, Air pollution physics, in: Methods and Calculations in Environmental Physics, AIP Publishing LLC, Melville, New York, 2022, pp. 7-1–7-70.)*

the atmosphere [10–15]. For example, semivolatile organic compounds (SVOCs) arguably include the most problematic water pollutants, including organochloride and organophosphate pesticides, polychlorinated biphenyls (PCBs), and polycyclic aromatic hydrocarbons (PAHs).

Nonvolatile compounds have an extremely small affinity for the vapor phase, i.e., x p^0 less than 10^{-5} kPa at 20°C. For example, the p^0 of the highly toxic compound tetrachlorodibenzo-*para*-dioxin (TCDD) is less than 4.8×10^{-9} kPa. This does not mean chemical compounds like TCDD cannot be air pollutants. To the contrary, TCDD is a product of incomplete combustion (PIC) that has been measured in the air downwind from fires, such as the one that burned for months at Ground Zero following the World Trade Center attacks on September 11, 2001 [16]. Semivolatile and nonvolatile chemical compounds can be sorbed to particles and transported advectively [17,18], as well as transported as suspensions along water droplets [19]. Thus, low volatility does not prevent a compound from entering the atmosphere, since they can evade by other mechanisms like aerosolization and colloidal air suspensions.

In many ways, p^0 is the most applied inherent property in characterizing gas-phase and vapor-phase pollutants. Pollutants are in the environment in various physical states, depending on meteorological conditions. Since the p^0 value expresses the partial pressure of a chemical substance in a gas phase that is in equilibrium with the nongaseous phases of the same substances and, according to Raoult's law, with other vapors. A chemical compound's reported p^0 value changes according to the concentrations of other substances in the atmosphere. Under meteorological conditions, a fraction of the molecules of many chemical compounds will be in liquid phase and a fraction in the gaseous phase. Under equal conditions, if molecule A has a p^0 higher than the p^0 of molecule B, then molecule A's liquid:gas ratio will be less than molecule B's liquid:gas ratio [1].

Gas-phase and vapor-phase pollutants are often assumed to act according to the ideal gas law. The thermodynamic equation of state expresses the relationships of pressure (p), volume (V), and thermodynamic temperature (T) in a defined quantity (n) of a substance. The ideal gas law expresses this relationship as:

$$pV = nRT \tag{6.1}$$

Where, R is the universal gas constant or molar gas constant, i.e., 8.31434 J mol^{-1} K^{-1}.

The ideal gas law can be used to convert p^0 into moles of vapor per unit volume:

$$\frac{n}{V} = \frac{p^0}{RT} \tag{6.2}$$

Where, V = volume of the container, n is the number of moles of chemical, R is the molar gas constant, and $\frac{n}{V}$ is the gas phase concentration (mol L^{-1}) of the chemical.

TABLE 6.1 Vapor pressures for select air pollutants.

Chemical	Vapor pressure (kPa) at 0 °C	Vapor pressure (kPa) at 20 °C	Vapor pressure (kPa) at 25 °C	Vapor pressure (kPa) at 50 °C
Atrazine		4.0×10^{-8}		
Benzene	3.3		12.7	36.2
Chlorobenzene			1.6	
Cyclohexane			13.0	36.3
1,1-Dichloroethane	9.6		30.5	79.2
1,2-Dichloroethane	2.8	9.2	10.6	31.4
Ethanol	1.5		7.9	29.5
Toluene			3.8	
Vinyl chloride	170	344	355	
Tetrachlorodibenzo-*para*-dioxin (TCDD)			2.1×10^{-10}	

Data from D. Lide, H. Frederikse, Handbook of Chemistry and Physics, vol. 9, CRC Press Inc, Boca Raton, FL, 1996. U.S. Environmental Protection Agency, 2,3,7,8-Tetrachlorodibenzo-p-dioxin—1746-01-6 | DTXSID2021315. https://www.epa.gov/chemical-research/chemistry-dashboardhttps://comptox.epa.gov/dashboard/chemical/details/DTXSID2021315, 2021 (Accessed 22 December 2021). H.F. Hemond, E.J. Fechner, Chemical Fate and Transport in the Environment, Elsevier, 2014.

The reported p^0 value is usually an expression of a substance's pure chemical form; that is, p^0 is expressed as the force per unit area exerted by a vapor in an equilibrium state with its pure solid, liquid, or solution at a given temperature (see Table 6.1). This situation is seldom encountered in meteorological scenarios, so that adjustments must be made to the published p^0 values for pollution calculations and estimates.

A common adjustment is for real gases and vapors. Recall that the ideal gas characterization applies exclusively to ideal gases, i.e., those with negligible spaces between the gas molecules. Real gas equilibrium relationships are adjusted to be expressed as:

$$(p + k)(V - nb) = nRT \tag{6.3}$$

Where, k is the factor for the decreased pressure on the walls of the container due to gas particle attractions, and nb is the volume occupied by gas particles at infinitely high pressure.

Furthermore, the van der Waals equation corrects for two properties of real gases, i.e., the excluded volume of gas particles and attractive forces between gas molecules. This van der Waals equation of state is expressed as:

$$k = \frac{n^2 a}{V^2} \tag{6.4}$$

Where a is a constant.

During atmospheric transport, chemical compounds may also undergo chemical reactions. The gas equilibrium K_p is quotient of the partial pressures of the products and reactants, expressed as:

$$K_p = \frac{p_C^z p_D^w}{p_A^x p_B^y} \tag{6.5}$$

Thus, K_p can also be expressed as:

$$K_p = K_{eq}(RT)^{\Delta v} \tag{6.6}$$

where Δv is defined as the difference in stoichiometric coefficients.

Therefore, p^0 is a measure of a substance's propensity to evaporate (Fig. 6.1), increasing exponentially with an increase in temperature (Fig. 6.2), so any p^0 is associated with a temperature. For example, the p^0 of trichloroethene at 21.0 °C is about 7.5 kPa, but at 25.5 °C rises to about 9.5 kPa. Thus, p^0 is a component of partitioning coefficients and volatilization rate constants [20].

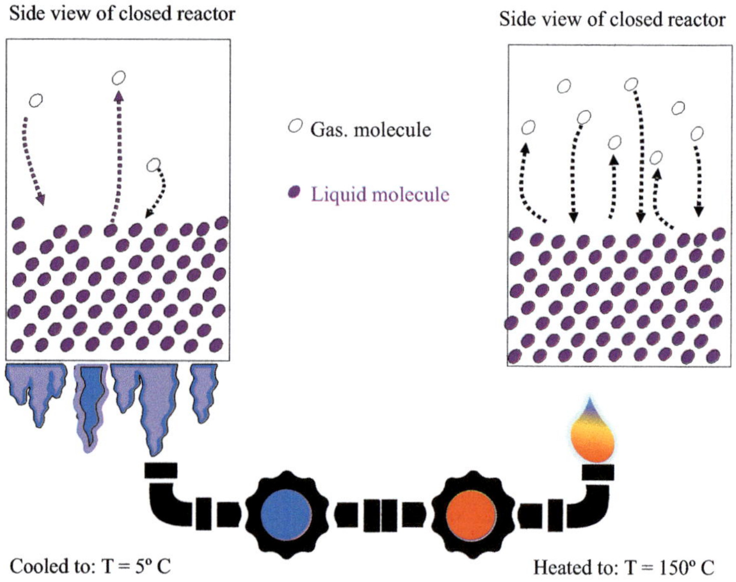

FIG. 6.2 Reactor vapor pressure increases in direct proportion with increasing temperature. Both reactors are at equilibrium, but at lower temperature, fewer molecules will be in the headspace above the liquid phase. Thus, readings of vapor pressure values must always be accompanied by the temperature at which each measured vapor pressure is occurring.

Any substance, depending upon the temperature, can exist in any phase. However, in most meteorological contexts, many chemical compounds can be excluded from certain phases. For example, molecular nitrogen (N_2) certainly exists in a liquid phase; however, it must be cooled below its boiling point (−196 °C). Nowhere on earth do ambient conditions meet this temperature requirement, so N_2 will be considered an atmospheric gas. Conversely, a vapor is a substance that is in its gas phase but under typical meteorological conditions exists as a liquid or solid under a given set of realistic environmental conditions. Although the pressure in the closed container in Fig. 6.1 is constant and below the boiling point, molecules of the vapor will continue to condense into the liquid phase and molecules of the liquid will continue to evaporate into the vapor phase. However, at equilibrium, the rate of these two processes is equal, meaning no net change in the amount of vapor or liquid. This is an example of dynamic equilibrium, or equilibrium vapor pressure.

At the boiling point temperature, a liquid's vapor pressure is equal to the external pressure. As mentioned, chemical compounds with high vapor pressures are classified as "volatile." Volatility is a measure of the propensity of a substance to form concentrations of vapor above the liquid, i.e., to boil. Therefore, boiling point is inversely proportional to vapor pressure. With decreasing boiling point there is an increasing likelihood that a chemical compound will be emitted into the air. The lower the boiling point, the more likely the compound will be emitted from a product or surface into the air. The boiling point also affects the way air pollutants can be measured, e.g., very volatile organic compounds (VVOCs) are so volatile that they must be treated as gases in the air, so their concentrations within materials or on surfaces are often unknown [21,22].[a] Conversely, the least volatile chemical compounds constitute a much smaller fraction in the air, with most found sorbed in or on material surfaces or dissolved or suspended in water and other liquids [13–15] (see Table 6.2).

Analytical laboratories take advantage of this relationship in separation sciences, especially gas chromatography (GC). Chemical compounds with the lowest boiling points, e.g., VVOCs exit the column early as the GC oven increases the temperature of the column (see Fig. 6.3). Thus, the VVOCs are the first to reach the mass spectrometer, electron capture, flame ionization, or other detector [23–27]. Next come the VOCs, followed by the SVOCs. Stated analytically, VVOCs have very short retention times (RTs) compared to SVOCs. Other chemical factors such as halogenation and sorption affect this relationship, so it is not a completely linear relationship, but as a rule, a chemical compound's boiling point is a good indicator of RT on a column. Other chemical factors such as halogenation and sorption affect this relationship, so this is not always the case, but generally, a chemical compound's boiling point is a good first indicator of residence time on a column.

Volatile chemical compounds are potential air pollutants when stored in vessels. As such their presence is always a threat to first responders. Even if not particularly toxic, any volatile compound that is flammable or explosive is dangerous. This is one reason why lower explosive limits (LELs) are so important for first responders, since released volatile compounds can concentrate in the air inside a confined space, such as a room, tunnel, drain, or vehicle tank, until they reach concentrations approaching LELs.

a. This can present a very real challenge to laboratory measurements. For example, if a chemical compound is so volatile that it loses mass quickly to the air, testing for microbial cultures is very difficult since the duration of time that microbes are exposed to the compound is quite short.

TABLE 6.2 Classification of organic air pollutants by boiling points.

Classification	Abbreviation	Boiling point range under meteorological conditions (°C)	Example chemical compounds
Very volatile (gaseous) organic compounds	VVOC	<0 to 50–100	Propane, butane, methyl chloride
Volatile organic compounds	VOC	50–100 to 240–260	Formaldehyde, D-limonene, toluene, acetone, ethanol, 2-propanol, hexanal
Semi volatile organic compounds	SVOC	240–260 to 380–400	Organochlorine pesticides, e.g., dichloro-diphenyl-trichloroethane (DDT) and chlordane, polychlorinated biphenyls (PCBs), plasticizers, e.g., phthalate compounds, fire retardants, e.g., phthalate esters and polybrominated diphenyl ethers (PBDEs), and polybrominated biphenyls (PBB)

Data from U.S. Environmental Protection Agency, Technical overview of volatile organic compounds. https://www.epa.gov/indoor-air-quality-iaq/technical-overview-volatile-organic-compounds#8, 2021 (Accessed 23 December 2021). World Health Organization, Indoor air quality: organic pollutants: report on a WHO meeting (no. 111), World Health Organization, Regional Office for Europe, 1989.

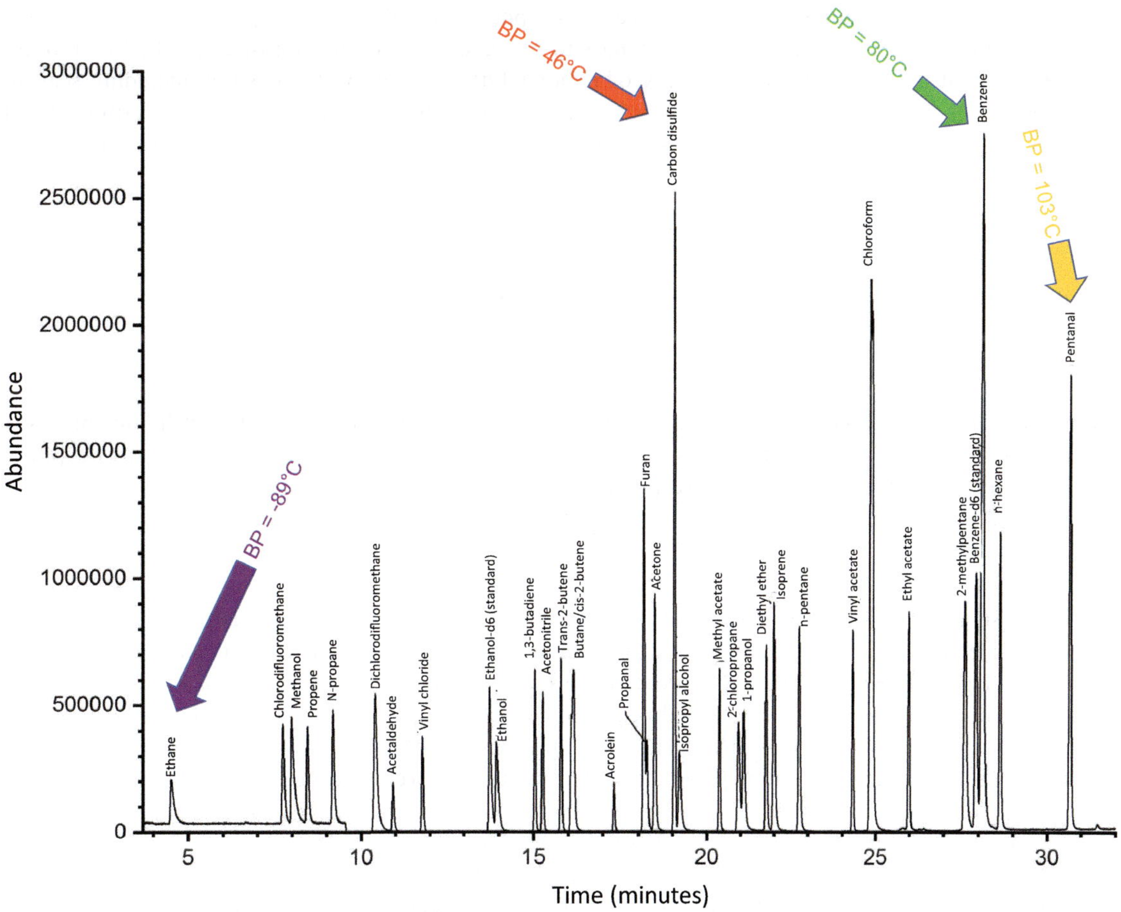

FIG. 6.3 Example of retention times for peaks of abundance for very volatile chemical compounds and volatile compounds from gas chromatographic/mass spectrometry analysis of an air sample, with boiling points for ethane, carbon disulfide, benzene, and pentanal at 1 atm shown. Note: gas chromatograph equipped with a split/splitless injector, an automated thermal desorption system using liquefied nitrogen cooling, and a mass selective detector. The GC column included a polystyrene-divinylbenzene phase suitable for the separation of low boiling chemical compounds that was installed and flushed with helium as the carrier gas. A particle trap was installed between column and MSD. The mass-to-charge (m/z) scan range was between 25 and 131. Abundance is a measure of how well the air samples ionized under the specific method the chromatographic and detection parameters that yielded the peaks. *(Modified from M. Richter, E. Juritsch, O. Jann, Determination of recovery rates of adsorbents for sampling very volatile organic compounds (C1C6) in dry and humid air in the sub-ppb range by use of thermal desorption gas chromatography-mass spectrometry, J. Chromatogr. A 1626 (2020) 461389, https://doi.org/10.1016/j.chroma.2020.461389; used with permission.)*

Volatility can combine with other inherent properties of a substance, which is important to know in pollution source characterization. As discussed in Chapter 11, ground water can be contaminated with nonaqueous phase liquids (NAPLs) of varying densities. Light NAPLs (LNAPLs) are more likely to remain in the upper reaches of aquifers than dense NAPLs (DNAPLs). The contaminated soil and ground water can become an air pollution source if an LNAPL includes a relatively insoluble substance that distributes between liquid and gas phases. The higher vapor pressures of portions of the plume will lead to upward movement of volatile chemical compounds in the gas phase. Thus, a water contamination scenario becomes soil and air pollution scenario because of the interplay of inherent properties, i.e., vapor pressure, solubility, and density of the fluid components. As always, the extent and severity of the pollution also must account for meteorological conditions, including temperature, atmospheric pressure, relative humidity, soil type (e.g., more permeable soil allows increased contaminant transfer), concentration of pollutant and presence of other substances. For example, if surfactants are present, the organic contaminant may have more affinity of the water and be less likely to escape to the atmosphere.

The Henry's Law Constant (K_H) is the ratio at equilibrium of the gas phase concentration to the liquid phase concentration of a gas. K_H combines two inherent properties, i.e., vapor pressure and solubility to estimate the likelihood that a substance will be exchanged between air and a liquid, especially water. Under precise conditions of pressure and temperature, the vapor pressure and solubility in water of a chemical compound are known. For example, the K_H measurements representing the partitioning of from in water at 298.15 °K vary from 0.12 to $0.22 \frac{\text{mol}}{\text{kg} \cdot \text{bar}}$, with even more variability using a temperature dependent method, i.e., with K_H ranging from 2200 to 5300 (unitless) [28].

Atmospheric residence time (τ) is the average time that a molecule spends in the atmosphere after it exits a source and before it reaches a sink. An indication of τ is a substance's atmospheric half-life is often expressed as its atmospheric half-life ($t_{1/2}$), i.e., the time it takes half of the molecular mass to be removed from the atmosphere. The loss processes that decrease τ are chemical reactions, radioactive decay of radionuclides, dry and wet deposition including scavenging of soluble vapor-phase chemical compounds and aerosols by clouds and precipitation, and gravitational settling of suspended coarse particles [29]. These steady state loss processes can be approximated using first-order rate laws for the mass concentration of a chemical $[C]$:

$$\frac{d[C]}{dt} = k[C] \tag{6.7}$$

Where, t is time and k is a first-order decay rate, which means that:

$$\tau = \frac{1}{k} \tag{6.8}$$

Given an ensemble of the loss processes mentioned above, the kinetic rates will differ for each contributing process (i). The global characteristic retention timescale, i.e., the net atmospheric residence time, can be calculated as the reciprocal of the sum of all first order rate constants for all the loss processes [29]:

$$\frac{1}{\tau} = \sum_i \frac{1}{\tau_i} \tag{6.9}$$

The units of k are seconds^{-1} and the units of τ and $t_{1/2}$ are seconds. Zero-valent mercury vapor (Hg^0), for example, has $\tau \approx 1$ year mainly because it does not readily react with other atmospheric constituents and it has low aqueous solubility [30,31]. However, Hg^0 can be transformed to more reactive species with much higher aqueous solubility, which means greater amounts of mercury will dissolve in water droplets and be removed by wet deposition. Mercury is released during combustion where it is readily vaporized. Thus, Hg concentrations in a flue gas of an incinerator or power plant will directly correlate with the Hg concentrations in in the fuel or waste. A power plant in China [32] with a coal Hg content of about 10 μg kg^{-1} was found to correlate with Hg emissions of about 2 μg m^{-3}; that is, combusting 150 μg kg^{-1} Hg coal correlated with Hg emissions of about 10 μg m^{-3}; 250 μg kg^{-1} coal correlated with Hg emissions of about 20 μg m^{-3}; and, 400 μg kg^{-1} coal correlated with Hg emissions of about 27 μg m^{-3}. The actual emissions will vary depending on type of coal. When normalized, combusting anthracite, for example, emits greater Hg concentrations than combusting bituminous coal. Other conditions that affect Hg emissions include the amount and types of catalysts present and fly ash concentrations.

Temperature and pressure conditions, and the presence of other chemical compounds determine the speciation during combustion. Even before leaving the stack, Hg, and other metals, for example, change forms when oxidized in the hot flue gases containing acidic compounds like hydrochloric acid (HCl) and oxides like sulfur dioxide (SO_2). Also, surface areas provide a platform for chemical reactions, e.g., substances can react heterogeneously on particulate matter, e.g., fly ash, thermo-chemically [32]. In the case of Hg, the zero-valence is converted to 2+, but a fraction of Hg^0 and Hg aerosols may also be present in varying proportions. After release into the atmosphere and deposition back to the earth's surface and waterbodies, organisms contact the metal and often alkylate them into organic forms, e.g., dimethyl mercury. Bacteria, for example, will alkylate metals anaerobically, but studies have shown that aerobic processes can also be involved [33,34]. Of course, metabolism can only progress if the microbes can resist metal toxicity [34]. These organic forms are often much more toxic than the inorganic species. The organic forms often also have greater bioaccumulation potentials. For example, anaerobic or facultative bacteria may reduce the oxidized Hg to methylated forms, which accumulates in the food chain [35,36]. Organic compounds can also become more toxic after being deposited from the atmosphere [37,38]. Indeed, heavy metal and persistent organic contamination often begins as air pollution before they become water, soil, or food pollutant.

Assuming a first-order loss rate, $t_{1/2}$ is the time when the mass concentration $[C]_0$ at t_0 is halved to become [C], this can be expressed as:

$$\ln \frac{[C]_0}{[C]} = \ln 2 = kt_{1/2} \tag{6.10}$$

Therefore, using the first-order k:

$$t_{1/2} = \frac{0.693}{k} \tag{6.11}$$

Thus, assuming a first-order loss rate:

$$t_{1/2} = 0.693\tau \tag{6.12}$$

Thus, if $t_{1/2}$ is 2 years it is the same as having a first-order rate constant k equal to 0.35 year^{-1}, and a $t_{1/2}$ of 10 years is the same as having a first-order rate constant of 0.0693. Increasing k correlates with a shortened amount of time to degrade a chemical compound. The percentage of mass concentration remaining after the number of half-lives are shown in Table 6.3. Theoretically, after $4t_{1/2}$, over 93% of the mass concentration has left the atmosphere.

Note that a substance's $t_{1/2}$ is merely an indication of τ, given that τ reflects the ratio of the total mass or volume of the air parcel to fluxes out of the parcel. Dilution and flushing of the substance dictate τ, along with numerous other physical and chemical processes. In addition, biological processes can affect τ if microbes are present in the plume and if they are suspended sufficiently long to degrade the substance and accelerate the removal.

Since atmospheric $t_{1/2}$ of a chemical compound is affected its inherent properties of vapor pressure and aqueous solubility, this is driven by its partitioning to water, indicated by its octanol-water coefficient (K_{ow}) and its partitioning to air, indicated by its Henry's law constant (K_H). Table 6.4 and Fig. 6.4 shows these relationships for some organic compounds.

TABLE 6.3 Percentage of mass concentration remaining after number of half-lives ($t_{1/2}$).

Number of half-lives	Percent remaining	
$1\ t_{1/2}$	$\frac{100\%}{2} = 50\%$	$\frac{1}{2} \times 100\% = 50\%$
$2\ t_{1/2}$	$\frac{50\%}{2} = 25\%$	$\frac{1}{2} \times \frac{1}{2} \times 100\% = 25\%$
$3\ t_{1/2}$	$\frac{25\%}{2} = 12.5\%$	$\frac{1}{2} \times \frac{1}{2} \times \frac{1}{2} \times 100\% = 12.5\%$
$4\ t_{1/2}$	$\frac{12.5\%}{2} = 6.25\%$	$\frac{1}{2} \times \frac{1}{2} \times \frac{1}{2} \times \frac{1}{2} \times 100\% = 6.25\%$
$n\ t_{1/2}$	$\frac{100\%}{2^n}$	$\left[\frac{1}{2}\right]^n \times 100\% = \left[\frac{1}{2}\right]^n \%$

TABLE 6.4 Atmospheric persistence associated with environmental partitioning as indicated by octanol-water coefficients (K_{ow}) and Henry's law coefficients (K_H).

Chemical compound	Half-life (days)	Log K_{ow}	Log K_H
Benzene	7.7	2.1	−0.6
Chloroform	360	1.97	−0.7
DDT	50	6.5	−2.8
Ethyl benzene	1.4	3.14	0.37
Formaldehyde	1.6	0.35	−5.0
Hexachlorobenzene	708	5.5	−3.5
Methyl chloride	470	0.94	−0.44
Methylene chloride	150	1.26	−0.9
PCBs	40	6.4	−1.8
1,1,1-Trichloroethane	718	2.47	0.77

Data from D. Di Toro, F. Hellweger, Long-range transport and deposition: the role of Henry's law constant, Final Report, International Council of Chemical Associations, Arlington, Virginia, 1999.

Therefore, both the chemical compound's inherent properties and meteorological conditions determine half-life. For instance, the lower $t_{1/2}$ unsubstituted aliphatic compounds also often have lower Bioconcentration factors (BCFs: see Chapter 14). For example, the *BCF* of the reactive compound acrolein is half of the chlorinated aliphatic compound chloroform; and is 4 orders of magnitude lower than the *BCF* of the chlorinated aromatic compound, 2,3,7,8-tetrachlorodibenzo-*para*-dioxin. Persistence potential also depends upon other partitioning coefficients, especially the *BCF* and bioaccumulation factor (*BAF*), which reflect a chemical compound's relationship to water media after the substance is deposited from the atmosphere. Chemicals with large *BCF* and *BAF* values are generally less persistent in water because they are readily taken up by organisms.

Half-life is an expression of a substance's persistence, bioaccumulation rate, and toxicity (PBT). The "P" and "B" in PBT interrelate. Values of $t_{1/2}$ indicate persistence directly and bioaccumulation indirectly. Persistence is both an intrinsic and extrinsic property of a substance. It is dependent upon the molecular structure of the chemical compound, such as the presence of aromatic rings, certain functional groups, isomeric structures, and especially the number and types of substitutions of hydrogen atoms with halogens, specifically chlorines and bromines. However, $t_{1/2}$ values are commonly reported for each compartment, which means that a chemical compound may be highly persistent in the air, still relatively reactive in water or soil [39–42]. This is because persistence and bioaccumulation fluctuate within each compartment due to myriad factors, especially temperature, pH and ionic strength, intensity and amount of sunlight, presence and diversity of microbial populations, and reaction rates of the chemical compound in the presence of oxidizing agents, free radicals and catalysts [42]. A persistent compound is also usually able to bioaccumulate because it is in the atmosphere or another meteorological compartment longer than a more reactive, less persistent compound would be.

As discussed in Chapter 5, a larger $t_{1/2}$ of a substance in a compartment, the larger the *BCF* will be. Recall that a chemical substance's *BCF* is the ratio of the tissue concentration of the chemical in an organism and its concentration in the environment, usually assumed to be water. As such, *BCF* indicates the chemical's potential to build up in tissue through direct uptake but excludes uptake from the organism's food. Conversely, the bioaccumulation factor (BAF) includes food; that is *BAF* indicates a compound's potential to accumulate in tissue through exposure to both food and water. A difference between *BCF* and *BAF* is that if food species abundance lessens due to exposure to a chemical with a high *BCF*. Even if higher trophic state organisms are not directly affected by the chemical, the biological balance would harm them indirectly due to the die-off of food species [43].

Physical, chemical, and biological transformations may render a chemical compound that was previously of low toxicity into a very toxic substance. Conversely, what may have been a precursor to a pollutant may have been changed into a less toxic chemical compound or compounds, as the pollutant undergoes chemical changes.

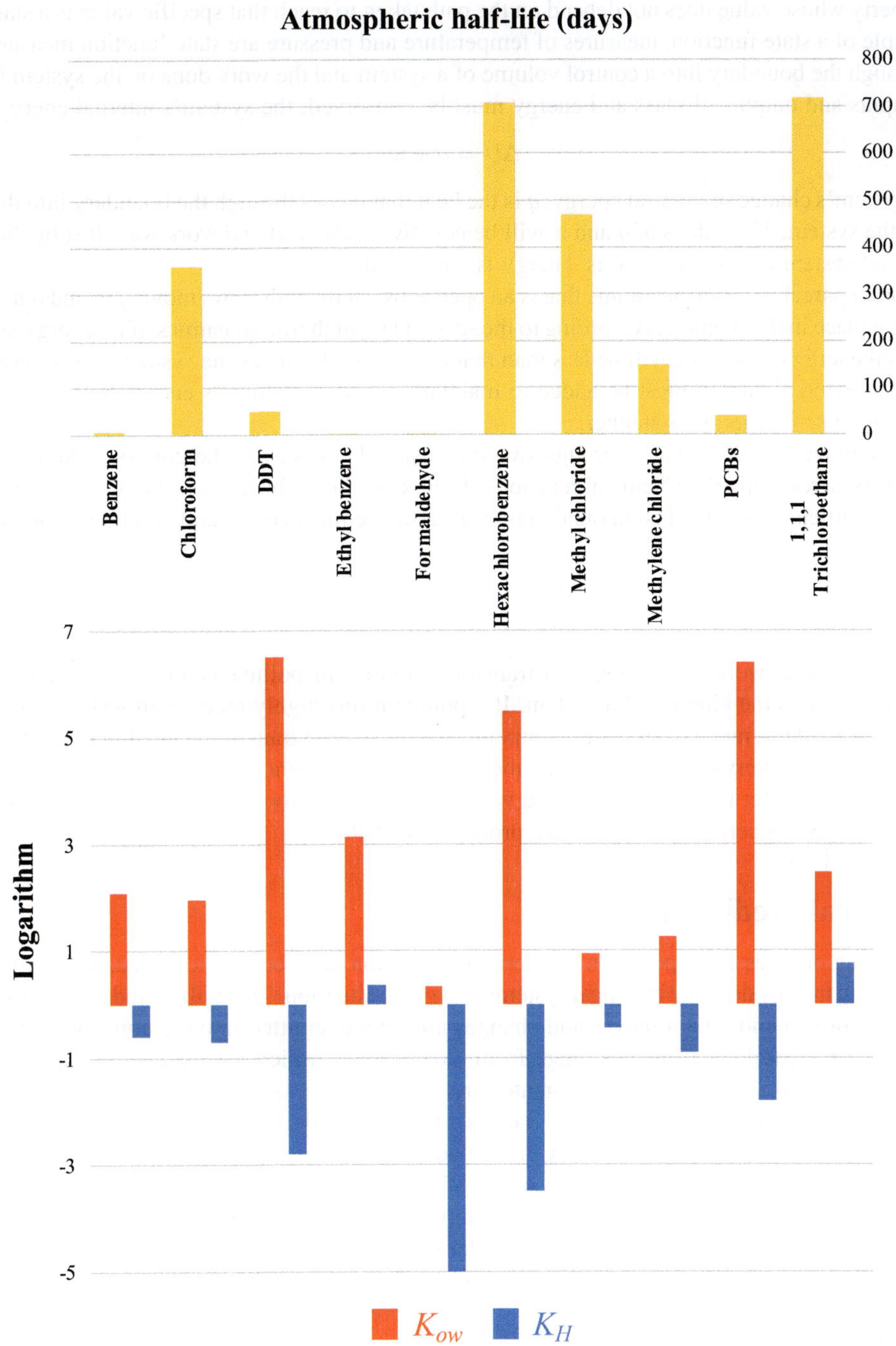

FIG. 6.4 Atmospheric persistence compared to octanol-water coefficient (K_{ow}) and Henry's law coefficient (K_H). *(Data from D. Di Toro, F. Hellweger, Long-range transport and deposition: the role of Henry's law constant, Final Report, International Council of Chemical Associations, Arlington, Virginia, 1999.)*

3 Entropy and enthalpy

For any thermodynamic system, the transfer of energy or mass is associated with mass and energy crossing the control boundary within the control volume. Matter and energy move into and out of a system's control volume, with changes taking place within the control volume. Work (w) is a product of a force **F** and a path element $d\mathbf{s}$ and that the first law of thermodynamics describes the total energy of a system (dE) as the sum of its change of heat (dq) and the change in work (dw). A property whose value does not depend on the path taken to reach that specific value is a state function. Since energy is an example of a state function, measures of temperature and pressure are state function measurements. The heat (q) that passes through the boundary into a control volume of a system and the work done on the system (w) are path functions. Since the inputs and outputs of mass and energy must be conserved, the system's internal energy is calculated as:

$$\Delta U = q + w \tag{6.13}$$

Where ΔU is the system's change of internal energy, q is the heat that passes through the boundary into the system and w is the work done on the system. The values of q and w will be negative if the heat and work were lost by the system. Internal energy of an isolated system is constant; that is, energy is conserved.

Thus, a water or air parcel is a control volume that is an open subsystem, with flows moving in and out of the volume and where reactions take place in that volume. According to the second law of thermodynamics, if no energy enters or leaves the system, the potential energy of the state will be less than that of the initial state as the system trends toward disorder. This entropy requires that external energy must be added to maintain a control volume's energy balance. All losses occur in conversions from one type of energy to another.

Mass and energy move perpendicularly through two-dimensional surfaces in the control volume, so these are mass fluxes and energy fluxes, respectively. Chemical reactions (R) of some or all the mass within the control volume are added or subtracted from a particular chemical compound mass (M), as a chemical compound is formed ($+R$) or degraded ($-R$), respectively:

$$\frac{dM}{dt} = M_{in} - M_{out} \pm R \tag{6.14}$$

Thus, thermodynamics are at work in the chemical transformation of air pollutants and their precursors. For example, raising temperature increases the kinetics of a system. If a pollutant this highly reactive in water vapor that is dispersed and remains in a plume for days, much or all of this compound will have been transformed within the water vapor. However, if a chemical compound has a very slow reaction rate, the volume of the parent compound would have changed very little during transport unless the water remains suspended a very long time. Environmental scientists are especially concerned about substances that raise concerns about their PBT properties [40,44].

4 Meteorological prediction

Often, before predicting or assessing a pollution problem, the situation must be characterized. This calls for a conceptual model to help to explain the physical and chemical systems of a given situation. This will include the factors and parameters of how a pollutant is formed and how it moves and changes after being emitted into the atmosphere [45–50].

Environmental and meteorological models range from "low-tiered," which are used at a simple screening-level, to "high-tiered," which are used for more precise estimates and predictions. Occasionally, screening level models are also high-throughput, i.e., they can run a large amount of data as a first step in deciding the importance of a pollution problem [51–55]. For example, a high-throughput model may predict possible exposures to an entire suite or class of chemical contaminants [56–58]. Screening models are often also conservative, i.e., they over-predict potential problems, attempting to detect a worst-case scenario of potential exposure of human populations and ecosystems to a pollutant. This is done by using protective default values and assumptions in the model's algorithms, e.g., predicting a low probability-high consequences outcome with the maximum adverse consequences. Predicting the emissions and the success of applications of pollution control equipment, for instance, is much less certain to predict the success of removing a data-poor compound than predictions of proven applications of technologies to remove conventional pollutants, like PM2.5 or NO_x. So, a worst-case scenario, low-tier model may be appropriate when uncertainties are high [59–63].

High-throughput screens provide a first approximation that disregards concentration levels of pollutants that are not likely to be of concern [24,51,55,64–66]. Conversely, higher-tiered models typically include algorithms that provide one or just a few specific chemical compounds, based on well-defined environmental conditions and time-activity patterns, and are based on relatively realistic values and assumptions. Such models require data of higher resolution and better quality data than the screening models and, as a result, provide more refined meteorological, exposure, and risk estimates [64].

The spatial domain of interest governs the selection of a model, e.g., within 1 km perimeter around a stationary source like an electric generating facility, a governmental jurisdiction not meeting an environmental quality standard, or a vulnerable ecosystem [67–71]. The boundary conditions must represent the influencing environment surrounding the domain. So-called "physical" or "dynamic" models are used to estimate the location where a chemical would be expected to move under controlled conditions, but on a much smaller spatial scale. For example, the U.S. Environmental Protection Agency's (EPA) employs chambers or full-size test houses to model the fate of pollutants indoors. The degree to which the actual conditions can be simulated and the quality of the information determines a dynamic model's accuracy [72].

A model of contaminant transport within the domain of a pollution source may begin with a qualitative assessment of the known and potential sources of contamination, the likely pathways of exposure, and the potentially affected human subpopulations and ecosystem receptors. This is often also where a risk assessment begins, i.e., exposure pathways are being identified in the conceptual model for further investigation and quantification [73]. This preliminary scoping is initially developed with secondary data from peer-reviewed literature and from other available and reliable information, with data refinement as new data are found and when actual samples are collected and analyzed. The scoping is designed to clarify the decisions, e.g., removal, remediation, treatment controls, baseline study, and assessment after actions are taken. It will inform the amount and types of data needed. Candidate site-specific modeling activities should be discussed during initial scoping stages, especially with decision-makers and stakeholder to help to ensure a high-quality and relevant risk assessment and that modeling results will indeed be able to be input to models, as well as to supplement the sampling efforts. These characterizations can be summarized and incorporated into the exposure and toxicity assessments to characterize baseline risk, both in quantitative expressions and qualitative statements. That is, the chemical-specific toxicity information can be compared to measured contaminant exposure levels and those levels predicted through fate and transport modeling to show and predict potential contaminants of concern, i.e., where potential exposures may exceed a harm threshold, as discussed in Chapters 9 and 10.

Table 6.5 includes selected parameters needed for environmental models [73]. It should be noted that these parameters can support more than one type of model, so care must be taken when using default values found in handbooks and model user guides [74]. For example, certain air emission models use a default value for the average wind speed at the site because representative site-specific measurements of wind speed can involve significant time, e.g., samples having to be collected over an entire year when a decision is needed in 1 month. However, winds vary considerably, such as the differences among wind speed and direction at the three major airports around Ground Zero of the World Trade Center site in 2001 and 2002. Due to the effects of buildings and other landscape features in and around Manhattan Island, the winds around all these airports varied considerably from those at Ground Zero [16,75,76]. The same is true for range and wildfires, where wind shifts can occur rapidly and create dangerous conditions. As such, response teams often require micrometeorological equipment at the response site [77].

TABLE 6.5 Example of parameters needing input data for models used in an environmental site investigation.

Type of modeling	Modeling parameters
Source characteristics	Geometry, physical/chemical conditions, emission rate, emission strength, geography
Soil	Particle size, dry weight, pH, redox potential, mineral class, organic carbon and clay content, bulk density, soil porosity
Ground-water	Head measurements, hydraulic conductivity (pump and slug test results), saturated thickness of aquifer, hydraulic gradient, pH, redox potential, soil-water partitioning
Air	Prevailing wind direction, wind speeds, stability class, topography, depth of waste, contaminant concentration in soil and soil gas, fraction organic content of soils, silt content of soils, percent vegetation, bulk density of soil, soil porosity
Surface water	Hardness, pH, redox potential, dissolved oxygen, salinity, temperature, conductivity, total suspended solids, flow rates, and depths for rivers/streams, estuary and embayment parameters such as tidal cycle, saltwater incursion extent, depth and area, lake parameters such as area, volume, depth, depth to thermocline
Sediment	Particle size distribution, organic content, pH, benthic oxygen conditions, water content
Biota	Dry weight, whole body, specific organ, and/or edible portion chemical concentrations, percent moisture, lipid content, size/age, life history stage

Data from U.S. Environmental Protection Agency, Risk assessment guidance for Superfund, in: Human Health Evaluation Manual (Part A): Interim Final, No. EPA/540/1-89/002. vol. I. https://www.epa.gov/sites/production/files/2015-09/documents/rags_a.pdf, 2020 (Accessed 13 October 2020).

Table 6.5 indicates some of the reasons that micrometeorological modeling is complex and unique to each pollution scenario. Winds always vary in speed and direction, air masses vary in stability, geomorphology of landforms affect the vertical and horizontal movement of air masses, and the air-soil-water interfaces affect the emissions and movement of pollutants. In addition, the location and number of monitoring sites and the choice of model must be based on the specific study objective, e.g., whether the results must comply with regulations, to screen for future modeling or measurement, to advance the state-of-the-science, or to be entered into other models, tools, or dashboards. Regulatory models may be defined, e.g., a state may be required to use a specific model, e.g., a state in the U.S. may select the American Meteorological Society/Environmental Protection Agency Regulatory Model (AERMOD) to characterize and predict concentrations of air pollutants as part of a State Implementation Plan (SIP) or to meet hazardous air pollutant rules required under clean air legislation [78–80].

Numerical models apply mathematical expressions to approximate three types of thermodynamic systems:

1. Isolated systems, in which no matter or energy crosses the boundaries of the system (i.e., no work can be done on the system).
2. Closed systems, in which energy can exchange with surroundings, but no matter crosses the boundary.
3. Open systems, in which both matter and energy freely exchange across system boundaries.

Isolated systems are usually only encountered in highly controlled reactors, so are important in modeling the formulation and manufacturing of chemicals like pesticides, but not directly pertinent to environmental modeling. In fact, most meteorological systems are open, but with simplifying assumptions, some subsystems can be treated as closed.

The need to address the constant chemical exchanges between the atmosphere and watersheds has called for linked models. For example, these models are being applied in the U.S. to estimate the total maximum daily loads of pollutants that impair water quality, discussed in Chapter 10. They are also used to estimate nonpoint-source impacts on watersheds. For example, the Community Multiscale Air Quality (CMAQ) model and its Watershed Deposition Tool (WDT) calculates pollutant loads for an average unit area and total deposition onto selected watersheds and sub-watersheds [4]. CMAQ generates estimates of the wet and dry deposition for pollutants in the vapor-phase and in aerosols, including ozone, sulfur compounds nitrogen compounds, secondary organic aerosols, and hazardous air pollutants at grids ranging from 4 to 36 km scales. The WDT has a graphical user interface that aids users in visualizing CMAQ gridded data and performing finer analyses on selected watersheds.

Pollutant transport and transformation models connect pollutant exposure to ecosystem and human health effects across levels of biological organization, in food chain and food webs, and within an organism, e.g., toxicokinetics within a human or other organisms [81]. The exposure pathway has five components:

1. The source of contamination (e.g., a leaking tank or fugitive dust blowing off an ash pile or waste site)
2. The transport mechanism (e.g., atmospheric and water partitioning)
3. A point of exposure (e.g., breathing zone, skin, or mouth)
4. A route of exposure (inhalation, nasal, dermal, and ingestion)
5. An individual receptor or population (humans or ecosystems having a potential for exposure).

Pollutant measurements and models are complementary. Models can predict, for example, the size and shape of a pollutant plume and deposition rates onto water and land surfaces. For example, coverage with models ranges from highly localized, e.g., one or two buildings [82], to urban scale, to continental scale and global scale [83].

Box models provide a straightforward way to model pollution, in which the volume of interest is treated as a rectangular box in which the mass of pollutant is conserved (see Fig. 6.5). The box is oriented so that the wind velocity (u) is incident to and normal to one box face. For further simplicity, box models assume that the pollutant is completely mixed within the box volume (V), i.e., the concentration (C) is homogeneous. This allows for a mass balance [84]:

$$\frac{dyCV}{dt} = QA + (\bar{u} \bullet C_{in} \bullet \mathrm{W} \bullet H_{mix}) - (\bar{u} \bullet C_{in} \bullet \mathrm{W} \bullet H_{mix}) \tag{6.15}$$

Where C_{in} is the concentration of the pollutant entering the box, Q is the pollutant emission rate per unit area of sources within the box, $\bar{u}$ is the average wind speed normal to the box, W is the width of the box, A is the horizontal area (length $\times$ W), and H_{mix} is the mixing depth.

The integration of this equation gives estimates of the pollutant concentration at steady state, assuming the dynamics of mixing depth are nearly stationary, and that Q is constant. The box model results are always average concentrations of large airsheds, given that they assume homogeneity and instantaneous mixing [84]. If H_{mix} is a function of downwind distance, x, these models may be applied to smaller airsheds. The atmospheric domain of interest can be improved with multiple boxes,

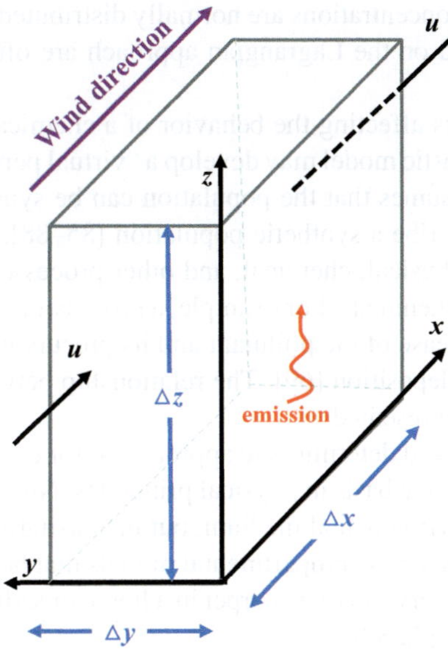

FIG. 6.5 Simple box model with dimensions $\Delta x \times \Delta y \times \Delta z$. Note: u is wind speed. *(Data from D.A. Vallero, Air Pollution Calculations: Quantifying Pollutant Formation, Transport, Transformation, Fate and Risks, Elsevier, 2019. Based on information from M. Markiewicz, Modelling of the air pollution dispersion, in: Models and Techniques for Health and Environmental Hazards Assessment and Management, IEA, Otwock-Swierk, Poland, 2006, pp. 303–347. M. Borysiewicz, M. Borysiewicz, Models and techniques for health and environmental hazard assessment and management, vol. 20, Institute of Atomic Energy, Swierk, Poland, 2006, p. 2010 (Retrieved March).)*

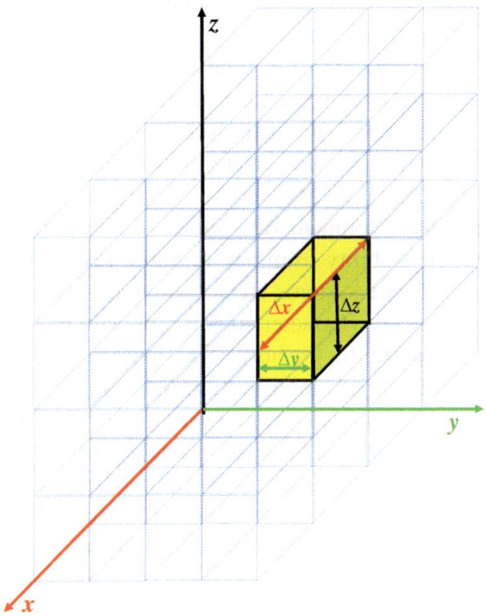

FIG. 6.6 Three-dimensional grid of interconnected boxes in multiple box model. *(Data from D.A. Vallero, Air Pollution Calculations: Quantifying Pollutant Formation, Transport, Transformation, Fate and Risks, Elsevier, 2019. Based on information from: M. Markiewicz, Modelling of the air pollution dispersion, in: Models and Techniques for Health and Environmental Hazards Assessment and Management, IEA, Otwock-Swierk, Poland, 2006, pp. 303–347.)*

i.e., an assemblage of boxes wherein the mass of the pollutant is exchanged (see Fig. 6.6). The mass balance of each box is linked using mass-in and mass-out terms [70]. Box models can screen for average pollutant concentrations, followed by more complex models if greater resolution is required.

Pollutant dispersion models are either statistical or deterministic. Statistical models track the movement of a control volume starting from the source to the receptor locations and often assume an idealized Gaussian distribution of pollutants from the point of release, i.e., the pollutant concentrations are normally distributed in both the vertical and horizontal directions from the source. Recent models based on the Lagrangian approach are often hybrid models, i.e., they incorporate descriptions of complex turbulence.

Stochastic models assume that the events affecting the behavior of a chemical compound are random, so they can be based on probabilities. For example, a stochastic model may develop a "virtual person" and randomize that person across an entire population or subpopulation. This assumes that the population can be synthesized from that person's information. Obviously, the randomization can only describe a synthetic population [85–88].

Deterministic models require that key physical, chemical, and other processes be sufficiently understood to be incorporated to reflect the movement and fate of chemicals. For example, a process-oriented model is based on the description of the physicochemical processes, from the release of the pollutant and its precursors, to the advection and dispersion within the environment, to transformation, and to deposition [69]. The relationship between and among the systems, such as the kinetics and mass balances, must also be represented.

Often, hybrid models use both statistical and deterministic approaches, for example, when one part of a system tends to be more random, while another has a very strong basis in physical principles. Numerous models are available to address the movement of chemicals through a single environmental medium, but increasingly, environmental scientists and engineers have begun to develop multimedia models, such as compartmental models that help to predict the behavior and changes to chemicals as they move within and among reservoirs (e.g., carpet in a home or sediment beneath a stream), in the air as dust and vapors, and in exchanges with surfaces [72,89].

Pollutants reach receptors by transport and transformation. As such, the outputs from pollutant models depend on meteorology. The variables and parameters vary by scale, e.g., large-scale models may use climatological data, whereas local and indoor models employ site specific and temporally explicit data and algorithms, e.g., to calculate penetration of outdoor pollutants into a microenvironment.

If air movement past a continuous pollutant source is gentle, pollutant concentrations in the plume moving downwind will be much higher than they would be if the air were moving rapidly past the source. If a plume continues to have pollution added to it, the concentration will increase. Generally, a source emits into different volumes of air over time. However, there can be a buildup of concentration over time even with significant air motion if there are many sources.

Low- and high-pressure systems differ in ventilation, i.e., total air volume moving past a location. Air commonly travels toward the center of a low-pressure system, i.e., cyclone, in the lower atmosphere, due in part to the frictional turning of the wind toward low pressure. This convergence causes vertical motion near the center of the low. Usually, winds near the center of the low are light and winds away from the center are moderate, resulting in increased ventilation rates. Low-pressure systems generally cover relatively small areas. They seldom remain the same at a given area for a significant period. Low-pressure systems are frequently accompanied by cloudy skies, which may cause precipitation. The cloudy skies minimize the variation in atmospheric stability from day to night. Primarily because of moderate horizontal wind speeds and upward vertical motion, low pressure systems are accompanied by an increase in ventilation.

High-pressure systems, i.e., anticyclones, are characteristically opposite of lows. The winds flow outward from the high-pressure center, so subsiding air compensates for the horizontal transport of mass. This sinking air causes a subsidence inversion. Skies are usually clear due in part to the subsiding vertical motion. This increases incoming solar radiation during the day and decreases outgoing radiation at night, inducing atmospheric instability during the day and stability at night, with frequent radiation inversions. Highs generally occupy large areas and are usually slow-moving.

Stagnation occurs when ventilation is low. The lack of motion often occurs in the weak pressure gradient near the center of the high. If the high-pressure system has a warm core, air movement is stifled near the center. This produces very light winds, few clouds, and the surface-based inversions at night. During the day, the mixing height does not reach high altitudes because of the subsidence inversion above the high-pressure system, even though the clear skies induce instability below the mixing height.

5 Meteorological phenomena

Pollutant concentrations and gradients are directly affected by meteorology. Substances move along with airflow, i.e., by advection. Concentrations in pollutant plumes can increase or decrease because of meteorological conditions. For example, particulate concentrations may accumulate in lower air strata during stagnation and inversions. Incoming solar radiation and visibility are decreased and concentrations of CO, tropospheric O_3, and other pollutants increase.

Wind speed affects rates of dilution after a pollutant is emitted. Whether in water, porous media, or the atmosphere, dilution occurs in the direction of plume transport with different vertical profiles. These profiles determine the elongation and spread of the plume. For example, consider the effect of wind speed for an elevated source with an emission of 6 mass

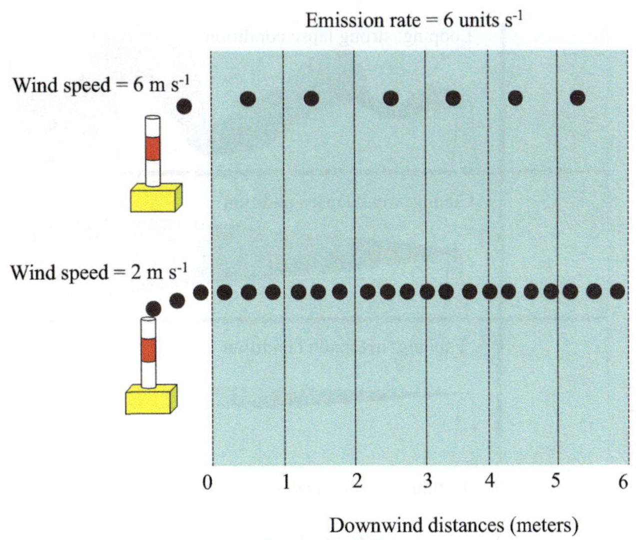

FIG. 6.7 Dilution by wind speed. *(Data from D.A. Vallero, Fundamentals of Air Pollution, fifth ed., Elsevier Academic Press, Waltham, MA, 2014, p. 999 pages cm.)*

units per second if the wind speed falls from 6 to $2\,\mathrm{m\,s^{-1}}$. Fig. 6.7 shows that there is 1 unit between the vertical parallel planes 1 m apart for a wind speed of $6\,\mathrm{m\,s^{-1}}$. When the wind is slowed to $2\,\mathrm{m\,s^{-1}}$, there are 3 units between those same vertical parallel planes 1 m apart.

Temperature and pressure determine wind speed and atmospheric stability. Eddy diffusion is the most important mixing process that causes pollutant dispersion [1,90]. The eddies in the atmosphere form in atmospheric parcels. The parcels with higher pollutant concentrations mix with less polluted air. The efficiency of eddy diffusion is highest when the eddy's scale approaches that of the pollutant puff or plume (Fig. 6.8). The shape of the plume is caused by both adiabatic and nonadiabatic conditions. The adiabatic lapse rate is a temperature change not caused by the gain or loss of heat entering or exiting the system.

The mixing of the pollutant within the atmosphere, i.e., dispersion, is predicted using either the Lagrangian or Eulerian method. A Lagrangian model applies statistical theory to turbulence, i.e., turbulent dispersion is a random process described by a distribution function. The Lagrangian model follows the individual random movements of molecules released into the plume, using statistical properties of random motions that are characterized mathematically. The Eulerian model is based on a differential volume. The velocity field is very important because the velocity field characterizes the motion of fluid particles and provides the means for computing these motions.

Both Lagrangian and Eulerian model frameworks track a substance in space and time. The difference is that the Eulerian approach uses a three-dimensional Cartesian grid as the frame of reference and the Lagrangian approach uses a probability theory as a moving frame of reference (see Chapter 2). Eulerian models place the observer in one location watching the air pass by. In Lagrangian models, the observer travels along with the parcel of air as it takes a random walk through the atmosphere. Within the Lagrangian frame, the path where each air parcel moves during a specified time is an ensemble mean field that relates to the displacement probabilities of the substance of interest.

Both types are used by regulatory agencies. Lagrangian models are usually easier to code, run and evaluate, can handle and readily modify explicit transport mechanisms, and are suitable for source apportionment since the random walk from the source to the receptor is defined. However, Lagrangian models are not well-suited for interactions among parcels, i.e., puffs, and when puffs become numerous the computation becomes increasing expensive. Examples of Lagrangian models include CALPUFF and the Reactive Plume Model (RPM) [90].

Eulerian models provide meteorological and transport information in three-dimensional space and time, so are more realistic than Lagrangian models. Concentrations can be defined over large spatial and temporal domains. Eulerian models currently in use include the Urban Airshed Model (UAM), CALGRID, Regional Oxidant Model (ROM) and the Community Multiscale Air Quality (CMAQ) model [90].

Like Gaussian ground water plume models, atmospheric Gaussian models define the change in plume size by standard deviations in cross-sectional x and y dimensions and the altitudinal z dimension, with the origin assumed to be at the earth's

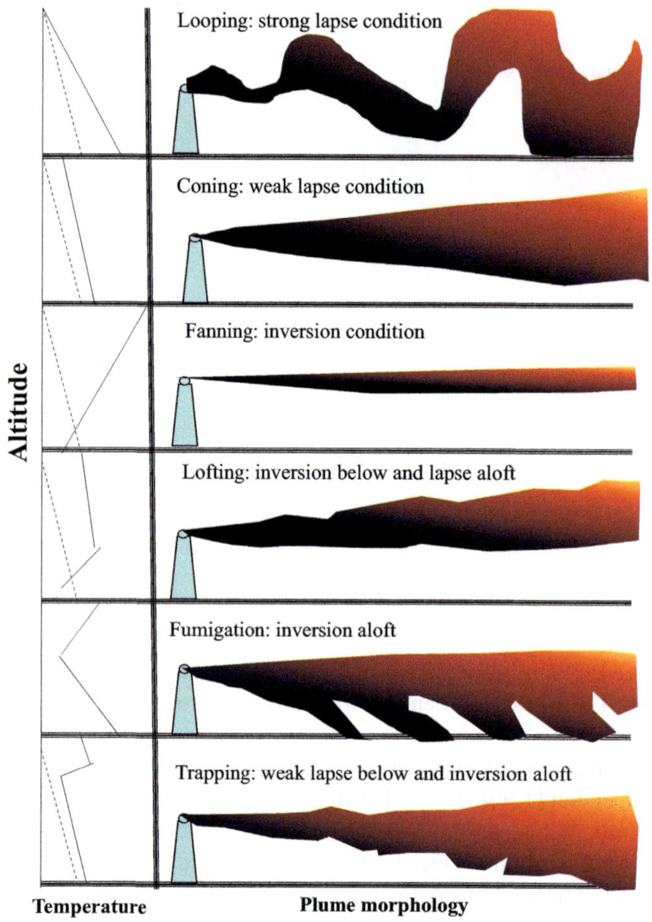

FIG. 6.8 Effect of vertical temperature profile on vertical expansion and shape change of continuous plumes after emission. Note: Dashed lines correspond to dry adiabatic lapse rate. *(Data from D.A. Vallero, Fundamentals of Air Pollution, fifth ed., Elsevier Academic Press, Waltham, MA, 2014, p. 999 pages cm.)*

surface (see Fig. 6.9). Like Eulerian models, the observer watches the pollutant move within a Cartesian grid, but the pollutant's concentrations are randomly distributed and the pollutant concentrations from a continuously emitting source are proportional to the emission rate. Winds are assumed to dilute pollutant concentrations at the point of emission at a rate inversely proportional to the wind speed. The standard deviations of the plume concentration in the x and y directions depend on turbulence in the atmosphere and the distance from the source. Often, Gaussian models assume the chemical compounds emitted do not undergo transformation and other removal processes during transport [91].

Air pollution models generate values for concentration (C) based on the emission rate (Q) expressed as mass per time. Other variables include:

- Wind speed (u, length per time);
- The cross-sectional dispersion coefficient (σ_y length), i.e., standard deviation of horizontal distribution of plume concentration, which is evaluated at the downwind distance x and for the appropriate stability;
- Dispersion coefficient in the vertical direction (σ_z length), i.e., standard deviation of vertical distribution of plume concentration, evaluated at the downwind distance x and for the appropriate stability;
- Mixing height (L, length);
- Physical stack height (h, length);
- Effective height of emission (H, length), which is the sum of h+plume rise (Δh);
- Downwind distance (x, length);
- Crosswind distance (y, length);
- Receptor height above ground (z, length).

A.

B.

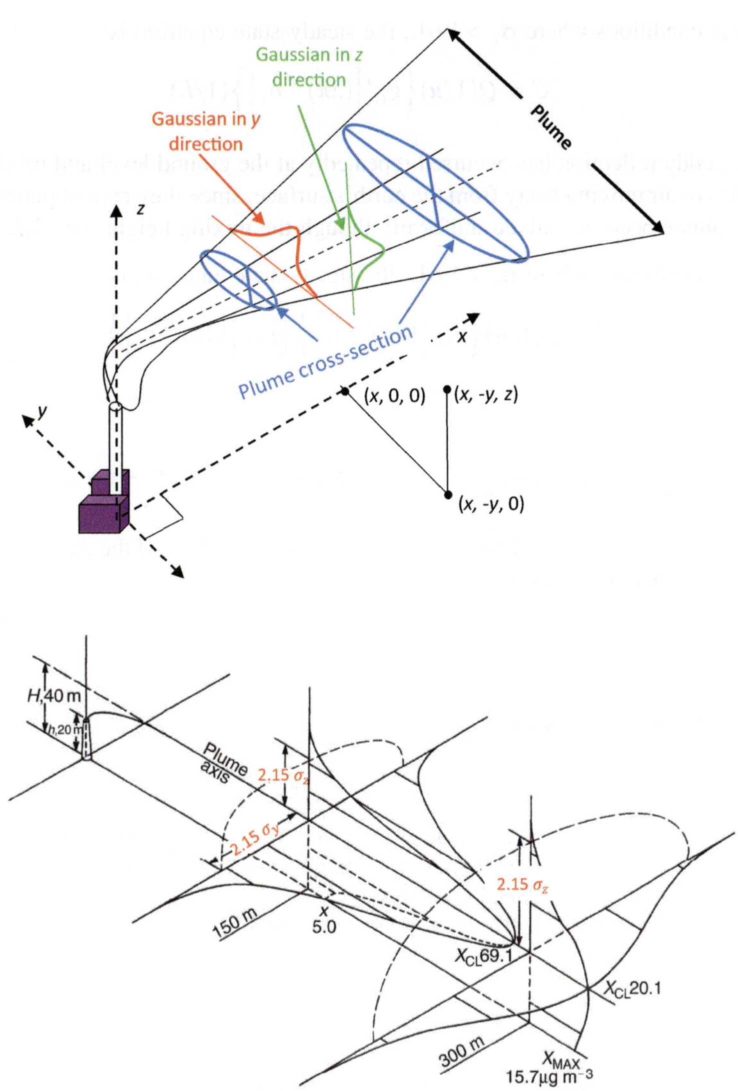

FIG. 6.9 (A) Gaussian plume aloft, i.e., not yet reaching the ground in the two cross-sections. (B) Two cross-sectional areas of a Gaussian plume showing the predicted total mass of a pollutant released at an effective height of 40 m. The cross-section under the dashed-line curves, with the maximum concentrations along the centerline. The time-averaged is about 1 h. *(Data from D.A. Vallero, Fundamentals of Air Pollution, fifth ed., Elsevier Academic Press, Waltham, MA, 2014, p. 999 pages cm.)*

The concentration at location coordinates x, y, z from a point source located at $(0, 0, H)$ is given by one of the three following equations [90].

1. The equation for stable conditions or unlimited vertical mixing (a very high mixing height) is:

$$C = Q(1/u)\left\{g_1/\left[(2\pi)^{0.5}\sigma_y\right]\right\}\left\{g_2/\left[(2\pi)^{0.5}\sigma_z\right]\right\} \tag{6.16}$$

Where,

$$g_1 = \exp\left(-0.5y^2/\sigma_y^2\right)$$

$$g_2 = \exp\left[-0.5(H-z)^2/\sigma_z^2\right] + \exp\left[-0.5(H+z)^2/\sigma_z^2\right]$$

If $y = 0$, or $z = 0$, or both z and H are 0, this equation is simplified. For locations in the vertical plane containing the plume centerline, $y = 0$ and $g_1 = 1$.

2. For unstable or neutral conditions where $\sigma_z > 1.6L$, the steady-state equation is:

$$C = Q(1/u)\left\{g_1/\left[(2\pi)^{0.5}\sigma_y\right]\right\}(1/L) \quad (6.17)$$

For these large σ_z values, eddy reflection has occurred repeatedly at the ground level and mixing height. Eddy reflection refers to the circular eddies of air moving away from the earth's surface, since they cannot penetrate that surface. Thus, the vertical extension of the plume becomes mixed uniformly though the mixing height, i.e., $1/L$.

3. For unstable or neutral conditions where $\sigma_z < 1.6L$, the steady-state equation is:

$$C = Q(1/u)\left\{g_1/\left[(2\pi)^{0.5}\sigma_y\right]\right\}\left\{g_3/\left[(2\pi)^{0.5}\sigma_z\right]\right\} \quad (6.18)$$

Where,

$$g_3 = \sum_{N=-\infty}^{\infty}\left\{\exp\left[-0.5(H - z + 2NL)^2/\sigma_z^2\right] + \exp\left[-0.5(H + z + 2NL)^2/\sigma_z^2\right]\right\}$$

These equations have been converted to calculators. Fig. 6.10 shows screenshots of the Atmospheric Dispersion Calculator [47], which summarizes the Gaussian equations [92]:

$$C(x,y,z) = \frac{Q}{2\pi u\sigma_y\sigma_z}e^{-\frac{y^2}{2\sigma_y^2}}\left(e^{-\frac{(z+H)^2}{2\sigma_z^2}} + e^{-\frac{(z-H)^2}{2\sigma_z^2}}\right) \quad (6.19)$$

The values are input into the model calculator.

Gaussian dispersion example 1

Find the pollutant concentration of a pollutant at 1.5 m above the ground, at a location 100 m downwind from a stack (effective stack height = 25 m) with an emission rate of $1\,\text{kg}\,\text{s}^{-1}$ and average wind speed of $2\,\text{m}\,\text{s}^{-1}$, if the standard deviation in cross-sectional standard deviation 100 m and in the vertical direction is 10 m.

Inserting these values into the calculator [92] gives a concentration of $0.045\,\text{g}\,\text{m}^{-3} = 45\,\text{mg}\,\text{m}^{-3}$ for this pollutant 100 m downwind.

Gaussian dispersion example 2

Calculate the pollutant concentration of a pollutant at 1.5 m above the ground, at a location 500 m downwind from the same stack and the same conditions as Example 1.

Inserting these values into the calculator [92] gives a concentration of $2.8 \times 10^{-7}\,\text{g}\,\text{m}^{-3} = 0.28\,\mu\text{g}\,\text{m}^{-3}$ for this pollutant 100 m downwind. Thus, at 400 m further downwind, the concentration decreased by over six orders of magnitude.

Gaussian dispersion example 3

Describe how the concentration at 500 m downwind would differ if the stack had been at 50 m effective height. Is it good practice to increase stack height?

Plugging this *H* into the calculator [92] the concentration is found to be $2.8 \times 10^{-11}\,\text{g}\,\text{m}^{-3}$. Doubling the stack height increased the area and volume downwind over which the pollutant is distributed, resulting in a decrease in concentration of four orders of magnitude. Of course, raising the stack height has local value by diluting the concentration, but it also increases the amount of the pollutant that reaches winds aloft, which can cause regional and continental scale problems,[b] e.g., acid rain [93].

b. This is but another example of what my mentor and colleague at Duke University, Jeff Peirce, says is the universally correct answer to every engineering question: "It depends".

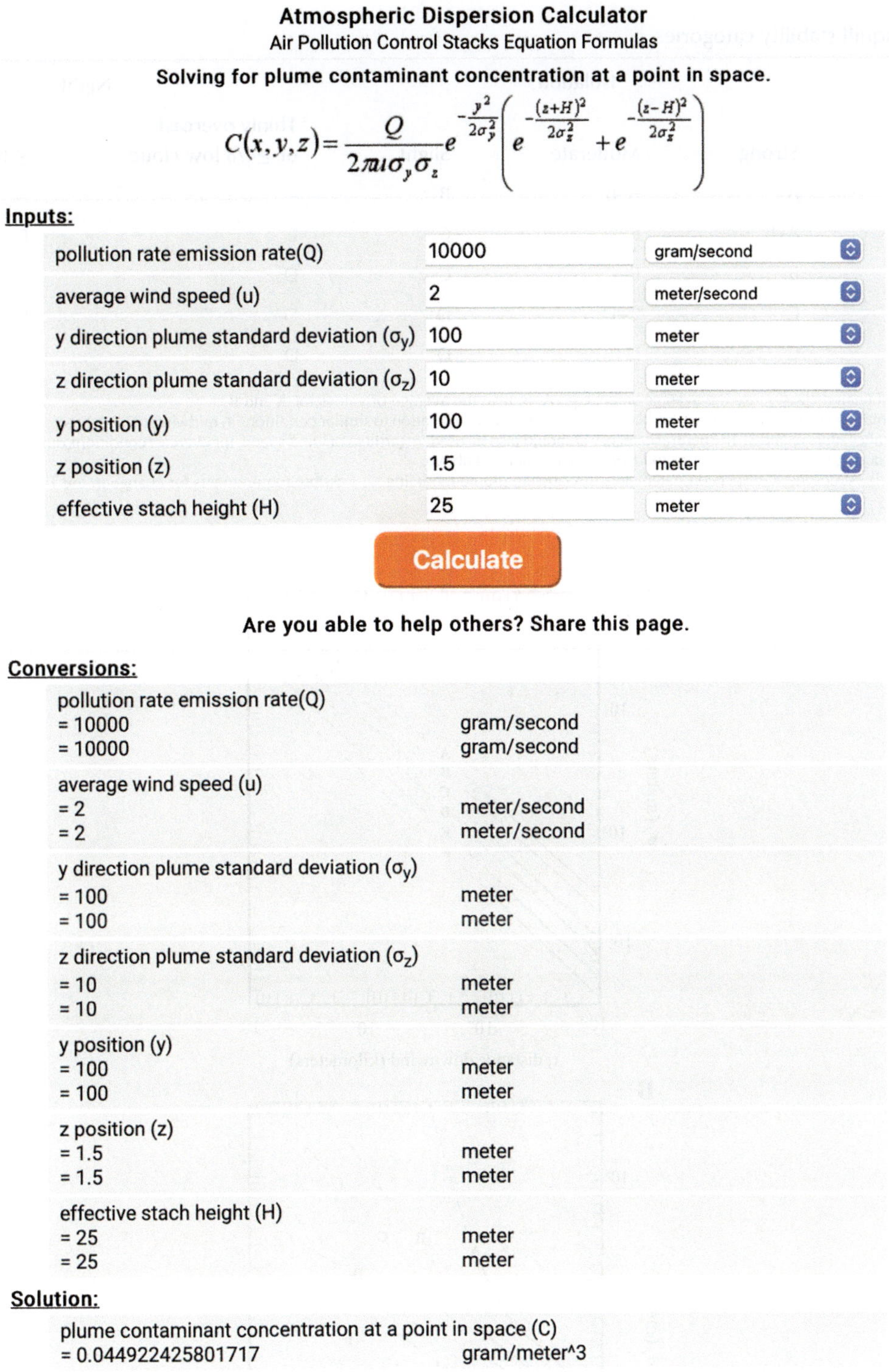

FIG. 6.10 Screenshots of Gaussian dispersion model. *(From J. Raymond, https://www.ajdesigner.com/phpdispersion/point_space_equation.php#ajscroll; accessed on September 11, 2024; based on J.J. Peirce, R. Weiner, R. Matthews, P.A. Vesilind, Environmental Engineering, Butterworth-Heinemann, 2003.)*

6 Atmospheric stability

Predicting pollutant dispersion must account for thermal stratification [94]. The necessary parameters for the scheme consist of wind speed, insolation, and cloudiness, which are basically obtainable from routine observations (see Table 6.6).

Pasquill's dispersion parameters can be stated in terms of σ_y and σ_z to allow their use in the Gaussian plume equations [71]. Parameters σ_y and σ_z are found by estimation from the graphs (see Fig. 6.11) as a function of the distance between

TABLE 6.6 Pasquill stability categories.

Surface wind speed (m s^{-1})	Isolation			Night	
	Strong	Moderate	Slight	Thinly overcast or ≥4/8 low cloud	≤3/8 Cloud
<2	A	A–B	B	–	–
2–3	A–B	B	C	E	F
3–5	B	B–C	C	D	E
5–6	C	C–D	D	D	D
>6	C	D	D	D	D

Table note: For hyphenated classes, use the average value, e.g., A–B, take the average of values for A and B.
Notes: 1. Strong insolation corresponds to sunny midday in midsummer; slight insolation to similar conditions in midwinter; 2. Night refers to the period from 1 h before sunset to 1 h after sunrise; and 3. The neutral category D should also be used, regardless of wind speed, for overcast conditions during day or night and for any sky conditions during the hour preceding or following night as defined above.
Data from F. Pasquill, Atmospheric dispersion parameters in Gaussian plume modeling. 2. Possible requirements for change in the Turner workbook values, 1976.

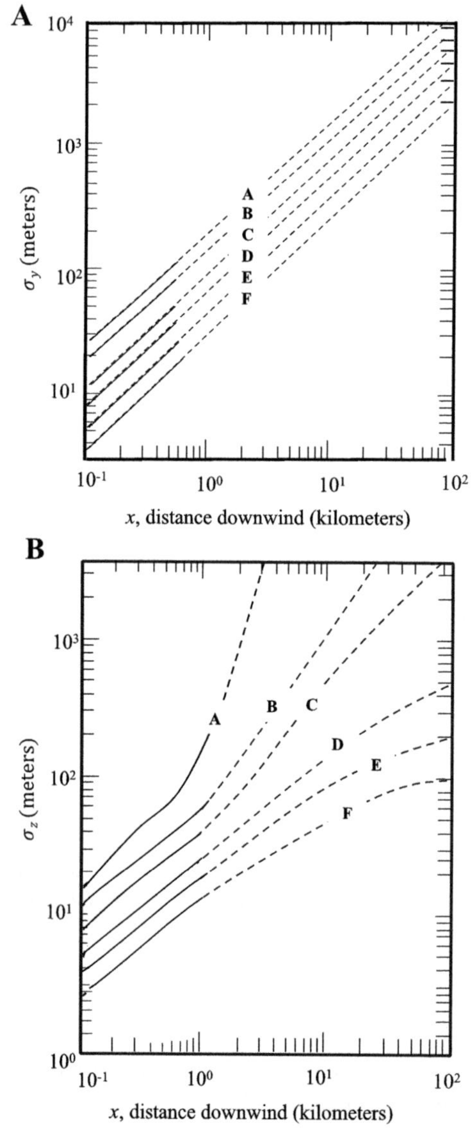

FIG. 6.11 Stability classes: A: σ_y and B: σ_z. *(Data from D.A. Vallero, Fundamentals of Air Pollution, fifth ed., Elsevier Academic Press, Waltham, MA, 2014, p. 999 pages cm; from F.A. Gifford Jr., Use of routine meteorological observations for estimating atmospheric dispersion, Nucl. Saf. 2 (1961) 47–51.)*

source and receptor, from the applicable curve, one for each stability class. Parameters σ_y and σ_z can also be calculated using the equations given in Tables 6.7 and 6.8. These are most applicable for releases within about 50 m height.

Values of σ_y and σ_z for each Pasquill stability class differ as a function of distance between source and receptor. For example, Tables 6.9 and 6.10 give these for urban and rural conditions, respectively.

TABLE 6.7 Horizontal dispersion parameters.

Stability	Parameter		
A	$T=24.167$	−2.5334	ln x
B	$T=18.333$	−1.8096	ln x
C	$T=12.5$	−1.0857	ln x
D	$T=8.3333$	−0.72382	ln x
E	$T=6.25$	−0.54287	ln x
F	$T=4.1667$	−0.36191	ln x

Table notes: σ_y (m) = 465.116x tan T; x is downwind distance in kilometers; T is one-half Pasquill's θ (degrees).
Data from D.A. Vallero, Fundamentals of Air Pollution, fifth ed., Elsevier Academic Press, Waltham, MA, 2014, p. 999 pages cm; from F.A. Gifford Jr., Use of routine meteorological observations for estimating atmospheric dispersion, Nucl. Saf. 2 (1961) 47–51.

TABLE 6.8 Pasquill-Gifford vertical dispersion parameter.

Stability	Distance (km)	a	b	σ_z[a]
A	>3.11			5000 m
	0.5–3.11	453.85	2.1166	
	0.4–0.5	346.75	1.7283	104.7
	0.3–0.4	258.89	1.4094	71.2
	0.25–0.3	217.41	1.2644	47.4
	0.2–0.25	179.52	1.1262	37.7
	0.15–0.2	170.22	1.0932	29.3
	0.1–0.15	158.08	1.0542	21.4
	<0.1	122.8	0.9447	14.0
B	≥35			5000 m
	0.4–35	109.30	1.0971	
	0.2–0.4	98.483	0.98332	40.0
	<0.2	90.673	0.93198	20.2
C	All x	61.141	0.91465	
D	>30	44.053	0.51179	
	10–30	36.650	0.56589	251.2
	3–10	33.504	0.60486	134.9
	1–3	32.093	0.64403	65.1
	0.3–1	32.093	0.81066	32.1
	<0.3	34.459	0.86974	12.1

Continued

TABLE 6.8 Pasquill-Gifford vertical dispersion parameter—cont'd

Stability	Distance (km)	a	b	σ_z
E	>40	47.618	0.29592	
	20–40	35.420	0.37615	141.9
	10–20	26.970	0.46713	109.3
	4–10	24.703	0.50527	79.1
	2–4	22.534	0.57154	49.8
	1–2	21.628	0.63077	33.5
	0.3–1	21.628	0.75660	21.6
	0.1–0.3	23.331	0.81956	8.7
	<0.1	24.260	0.83660	3.5
F	>60	34.219	0.21716	
	30–60	27.074	0.27436	83.3
	15–30	22.651	0.32681	68.8
	7–15	17.836	0.4150	54.9
	3–7	16.187	0.4649	40.0
	2–3	14.823	0.54503	27.0
	1–2	13.953	0.63227	21.6
	0.7–1.0	13.953	0.68465	14.0
	0.2–0.7	14.457	0.78407	7.9
	<0.2	15.209	0.81558	4.1

Note: σ_z (m) $= ax^b$; x is downwind distance in kilometers.
[a]*σ_z at boundary of distance range for all values except 5000 m.*
Data from D.A. Vallero, Fundamentals of Air Pollution, fifth ed., Elsevier Academic Press, Waltham, MA, 2014, p. 999 pages cm. F.A. Gifford Jr., Use of routine meteorological observations for estimating atmospheric dispersion, Nucl. Saf. 2 (1961) 47–51.

TABLE 6.9 Urban dispersion parameters by Briggs for distances between 100 and 10,000 m.

Pasquill type	σ_y (m)	σ_z (m)
A–B	$0.32x(1+0.0004x)^{-0.5}$	$0.24x(1+0.001x)^{-0.5}$
C	$0.22x(1+0.0004x)^{-0.5}$	$0.20x$
D	$0.16x(1+0.0004x)^{-0.5}$	$0.14x(1+0.0003x)^{-0.5}$
E–F	$0.11x(1+0.0004x)^{-0.5}$	$0.08x(1+0.0015x)^{-0.5}$

Data from F.A. Gifford Jr., Use of routine meteorological observations for estimating atmospheric dispersion, Nucl. Saf. 2 (1961) 47–51. D.A. Vallero, Fundamentals of Air Pollution, fifth ed., Elsevier Academic Press, Waltham, MA, 2014, p. 999 pages cm.

Unless the source is at ground level, the maximum concentration (C_{max}) of a pollutant along the x-axis occurs at some point downwind, i.e., after emission the chemical species require distance and time to reach ground level (see Fig. 6.12). For example, assuming the ratio of $\frac{\sigma_z}{\sigma_y}$ remains constant with distance, in the Pasquill stability class B, a stationary source emitting an organic chemical species at a rate of $0.37\,g\,s^{-1}$ from a stack with an effective height of 40 m and the wind speed

TABLE 6.10 Rural dispersion parameters by Briggs (for distances between 100 and 10,000 m).

Pasquill type	σ_y (m)	σ_z (m)
A	$0.22x(1+0.0001x)^{-0.5}$	$0.20x$
B	$0.16x(1+0.0001x)^{-0.5}$	$0.12x$
C	$0.11x(1+0.0001x)^{-0.5}$	$0.08x(1+0.0002x)^{-0.5}$
D	$0.08x(1+0.0001x)^{-0.5}$	$0.06x(1+0.0015x)^{-0.5}$
E	$0.06x(1+0.0001x)^{-0.5}$	$0.03x(1+0.0003x)^{-1}$
F	$0.04x(1+0.0001x)^{-0.5}$	$0.016x(1+0.0003x)^{-1}$

Data from D.A. Vallero, Fundamentals of Air Pollution, fifth ed., Elsevier Academic Press, Waltham, MA, 2014, p. 999 pages cm. F.A. Gifford Jr., Use of routine meteorological observations for estimating atmospheric dispersion, Nucl. Saf. 2 (1961) 47–51.

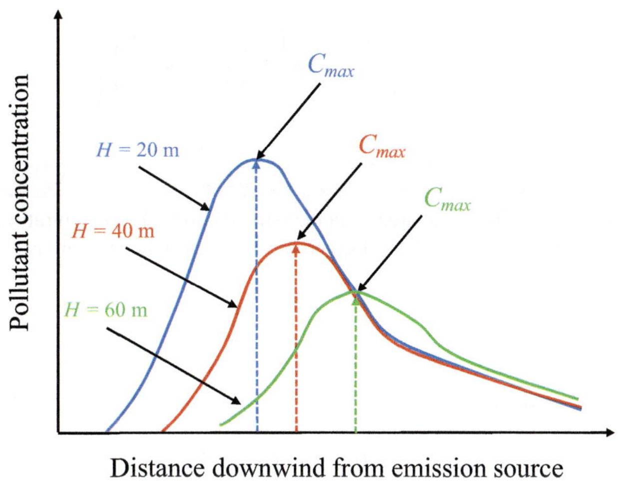

FIG. 6.12 Distance of maximum concentration (C_{max}) increases with increasing effective height of stack for hypothetical vapor-phase pollutant.

is averaging $2\,m\,s^{-1}$, the downwind location of the maximum concentration of this chemical can be found. The maximum concentration of an emitted substance is found at the distance where:

$$\sigma_{z^*} = \frac{H}{\sqrt{2}} \tag{6.20}$$

Thus, $\sigma_{z^*} = 28.3\,m$. At stability class B, this occurs at $x = 0.28\,km$, which means at this downwind position, using the values in Table 6.9, $\sigma_y = 49.0\,m$.

C_{max} can be estimated from:

$$C_{max} = \frac{2Q}{\pi \bullet u \bullet e \bullet H^2} \bullet \frac{\sigma_z}{\sigma_y} \tag{6.21}$$

Euler's number (e) is the base of natural logarithms and equals approximately 2.718, so:

$$C_{max} = \frac{2 \times 0.37}{\pi \bullet 2 \bullet e \bullet (40)^2} \bullet \frac{28.3}{49.0} = 1.56 \times 10^{-5}$$

To check whether the distance to C_{max} is reasonable, we can use the following equation at three reasonable distances for the standard deviations:

$$C = \left[Q/\left(\pi u \sigma_y \sigma_z\right)\right] \exp\left[-0.5(H/\sigma_z)^2\right] \tag{6.22}$$

This yields the following values:

x (km)	σ_z (m)	σ_y (m)	C (gm^{-3})
0.26	26.2	45.9	1.53×10^{-5}
0.28	28.2	49.0	1.56×10^{-5}
0.30	30.1	52.2	1.56×10^{-5}

Table notes: σ_z values obtained from Table 6.10; σ_y values obtained from Table 6.11.

This simple case shows that C_{max} occurs at 0.28 km downwind, ±20 m.

7 Meteorological models

Puff models assume that the source is episode, i.e., a puff of the pollutant is emitted at a specific time. Plume models assume that the source is continuous. For plumes, the previous equations can be stated as a mean concentration ($\overline{C}$) based on a mean emission rate ($\overline{Q}$):

$$\overline{C}(x,y,z) = \frac{\overline{Q}}{\pi\sigma_y\sigma_z u}\exp\left[-\frac{1}{2}\left(\frac{y^2}{\sigma_y^2}+\frac{z^2}{\sigma_z^2}\right)\right] \tag{6.23}$$

Puff model example

A tire fire is emitting NO_x at a rate of $8\,gs^{-1}$ in a rural area. Calculate the mean NO_x concentration expected downwind along the centerline 3 km from the fire on an overcast night and wind speed of $7\,ms^{-1}$ and the stability is class D. The background concentration of NO_x upwind from the fire is $1\,\mu gm^{-3}$.

Using Table 6.7 for stability class D,

$$\sigma_y = 0.08(3000)[1+(0.0001\times 3000)]^{-0.5} = 236.48$$

$$\sigma_z = 0.06(3000)[1+(0.0015\times 3000)]^{-0.5} = 990.00$$

Since the calculation is based on the centerline, i.e., y = 0, assuming that the fire is at ground level, i.e., z = 0, the calculation is:

$$\overline{C}(x,0,0) = \frac{\overline{Q}}{\pi\sigma_y\sigma_z u} = \frac{8}{\pi\bullet 236.48\bullet 990\bullet 7} = 1.5\times 10^{-6}\,gm^{-3} = 1.5\,\mu gm^{-3}\ NO_x$$

Thus, the average concentration at this location would have to also add the background, so the modeled concentration would be $1+1.5\,\mu gm^{-3} = 2.5\,\mu gm^{-3}$ NO_x.

There are many other ways to estimate and predict air pollution concentrations and movement and dilution within plumes besides the Gaussian methods, arguably the simplest and most used air pollution model type. The outputs from Gaussian models depend on the accuracy in calculating the dispersion coefficients, σ_y and σ_z.

Eulerian models include box, analytical, numerical and eddy simulation types [70]. As mentioned, the Eulerian model builds the mass balance around a differential volume. The velocity field is used to calculate the motion of the air. From the Eulerian perspective, the velocity field is a function of space and time:

$$\vec{v} = \vec{v}(x,y,z,t) \tag{6.24}$$

Where, $\vec{v}$ is the velocity field vector.

Thus, if the fluid properties and the flow characteristics at each position do not vary with time, t can be removed from the equation. In this case, the air flow is simply a function of space, yielding a steady or time-independent vector field:

$$\vec{v} = \vec{v}(x,y,z) \text{ and } \frac{\partial\vec{v}}{\partial x} = 0 \tag{6.25}$$

A time-dependent flow is an unsteady flow. Any flow with unchanging magnitude and direction of the velocity vector $\vec{v}$ is a uniform flow.

Recall that laminar flow has layers and turbulent flow has random movements of fluid particles in all directions. In incompressible flow, the variations in density are assumed to be constant, while the compressible flow has density variations that must be accounted for in flow calculations. Viscous flows must account for viscosity whereas inviscid flows assume viscosity is zero. Given that the atmosphere is seldom laminar, Eulerian conditions are assumed to be turbulent and dependent variables consist of both average and fluctuating components [50,84].

8 Electromagnetic energy[c]

The motion and change of air parcels depend on energy supplied by the sun. The atmosphere is approximately 76% nitrogen, 20% oxygen, 3% water, 0.9% argon, and 0.03% CO_2; the remaining mass consists of relatively inert gases such as neon, helium (He), methane (CH_4), krypton (Kr) and xenon (Xe). Other gases include nitrous oxide (N_2O) and hydrogen (H_2). Compared with the average radius of the earth, 6370 km, the atmosphere is merely a thin layer of gases, 90% of which is below 12 km altitude and 99% below 30 km altitude, with a total mass of the atmosphere is about 5×10^{18} kg.

The earth is a blackbody, i.e., an emitter and absorber of radiation at all wavelengths. Radiant flux over all wavelengths (F) from a unit area of a blackbody depends on the temperature of that body and is given by the Stefan-Boltzmann law:

$$F = \sigma T^4 \tag{6.26}$$

Where, σ is the Stefan-Boltzmann constant and equals 8.17×10^{-11} cal cm^{-2} min^{-1} deg^{-4} and T is the temperature in degrees K. Radiation from a blackbody discontinues at a temperature of absolute zero, 0 °K.

Absorptivity for a perfect blackbody is 1, since absorptivity is the radiant energy absorbed as a fraction of the total amount that reaches the object, which depends on both wave frequency and temperature. Emissivity is the ratio of the energy emitted by an object to that of a blackbody at the same temperature. Emissivity depends on both the properties of the substance and the frequency of electromagnetic radiation. Kirchhoff's law states that the emissivity of any substance at a given wavelength and temperature equals its absorptivity. The absorptivity and emissivity of a given substance can vary considerably for different frequencies.

From Eq. (6.29), the total radiation from a blackbody depends on the fourth power of its absolute temperature. The frequency of the maximum intensity of this radiation is also related to temperature through Wien's displacement law, which is derived from Planck's law:

$$v_{\max} = 1.04 \times 10^{11} T \tag{6.27}$$

where frequency v is in s^{-1} and the constant is in s^{-1} K^{-1}. The radiant flux is a function of frequency from Planck's distribution law for emission:

$$E_v \mathrm{d}v = c_1 v^3 \left[\exp\left(c_{2.} v \cdot T^{-1}\right) - 1\right]^{-1} \mathrm{d}v \tag{6.28}$$

Where, $c_1 = 2\pi \cdot h \cdot c_,^{-2}$ $h = 6.62 \times 10^{-27}$ erg s (Planck constant), $c = 3 \times 10^8$ m s^{-1} (speed of light, $c_2 = h \cdot k^{-1}$, and $k = 1.37 \times 10^{-16}$ erg K^{-1} (Boltzmann constant).

Blackbody radiation is continuous over the electromagnetic spectrum. The sun approximates a blackbody, with an effective temperature of about 6000 °K. The total solar flux is 3.9×10^{26} W. Per Wien's law, the frequency of maximum solar radiation intensity is 6.3×10^{14} s^{-1} [wavelength (λ) = 0.48 μm], which is within the visible portion of the spectrum; 99% of solar radiation occurs between the frequencies of 7.5×10^{13} s^{-1} (λ = 4 μm) and 2×10^{15} s^{-1} (λ = 0.15 μm) and about 50% in the visible region between 4.3×10^{14} s^{-1} (λ = 0.7 μm) and 7.5×10^{14} s^{-1} (λ = 4 μm). The intensity of this energy flux at the distance of the earth is about 1400 W m^{-2} on an area normal to a beam of solar radiation. This value is known as the solar constant.

The eccentricity of the earth's orbit as it revolves around the sun once a year causes the earth to be nearer to the sun in January (perihelion) than in July (aphelion). This causes a 7% difference in radiant flux at the outer limits of the atmosphere between perihelion and aphelion. The area of the solar beam intercepted by the earth is πE^2, where E is the radius of the earth, and the energy falling within this circle is spread over the area of the earth's sphere, $4\pi E^2$, in 24 h, so the average energy reaching the top of the atmosphere is 338 W m^{-2}. This average radiant energy reaching the outer limits of the atmosphere is depleted in its passage to the earth's surface [95]. Ultraviolet radiation with a wavelength less than 0.18 μm is strongly absorbed by molecular oxygen in the ionosphere 100 km above the earth; shorter X-rays are absorbed at even higher altitudes above the earth's surface. At 60–80 km above the earth, the absorption of 0.2–0.24 μm wavelength radiation leads to the formation of ozone; below 60 km there is so much ozone that much of the 0.2–0.3 μm wavelength radiation is absorbed. This ozone layer in the lower part of the mesosphere and the upper regions of the stratosphere shields life from

c. This section draws data, text, and figures from [90]. Used with permission.

much of the harmful ultraviolet radiation. The various layers warmed by the absorbed radiation reradiate in wavelengths dependent on their temperature and spectral emissivity. Approximately 5% of the total incoming solar radiation is absorbed above 40km. Under clear sky conditions, another 10–15% is absorbed by the lower atmosphere or scattered back to space by the atmospheric aerosols and molecules; as a result, only 80–85% of the incoming radiation reaches the earth's surface. With average cloudiness, only about 50% of the incoming radiation reaches the earth's surface, because of the additional interference of the clouds [91].

Sunspots and solar flares alter the amount of radiation, but this is predominantly in the X-ray and ultraviolet regions of electromagnetic energy, so does not change the amount in the wavelengths reaching the earth's surface. Therefore, the amount of radiation from the sun that can penetrate to the earth's surface are almost constant.

Albedo is the incoming radiation that is reflected and scattered.[d] Albedo is unitless and ranges between 0 and 1. The albedo of clouds, snow, and ice-covered surfaces is around 0.5–0.8, i.e., between 50% and 80% of the incoming solar radiation is reflected. The albedo of fields and forests is 0.03–0.3, and that of water is 0.02–0.05 except when the angle of incidence becomes nearly parallel to the water surface [96]. Albedo can also be expressed as percentages in which the above values are multiplied by 100. Table 6.11 shows the albedo values as percentages of incoming shortwave radiation for various materials on the earth's surface [97].

TABLE 6.11 Albedo values as percentage of incoming shortwave radiation for various surfaces and locations.

Surface	Albedo
Black soil, dry	14
Black soil, moist	8
Plowed field, moist	14
Sand, bright, fine	37
Dense, dry, and clean snow	86–95
Sea ice, slightly porous milky bluish	36
Ice sheet, covered by a water layer of 15–20cm	26
Woody farm covered with snow	33–40
Deciduous forest	17
Tops of oak	18
Pine forest	14
Desert shrub land	20–29
Swamp	10–14
Prairie	12–13
Winter wheat	16–23
Heather	10
Yuma, Arizona	20
Washington, DC (September)	12–13
Winnipeg, Manitoba (July)	13–16
Great Salt Lake, Utah	3

Data from U.S. Environmental Protection Agency, Basic air pollution meteorology student guidebook (SI 409), U.S. EPA, Research Triangle Park, North Carolina, 2005.

d. This section is an excerpt from [90]. Used with permission.

Albedo is affected by vegetative cover, materials used, and the paved surface areas. Albedo is a very important factor in microclimatology, e.g., it has a major impact on the urban heat island effect, where downtown areas have higher average summer temperatures than surrounding suburban and rural areas. Materials that reflect more and absorb less shortwave radiation can be used to ameliorate the heat island effect. For example, asphalt parking lots have an albedo between 5% and 8%; whereas green grass has between 18% and 22% albedo. Some roofing materials have albedos greater than 70% [90,98].

The angle of incidence of the sun's radiation determines the amount of energy received by the earth's surface. This angle is compared to the perpendicular to the surface affects the amount of radiation flux on an area [99]. The flux on a horizontal surface S_h is as follows:

$$S_h = S \cos Z \tag{6.29}$$

where S is the flux through an area normal to the solar beam and Z is the zenith angle (between the local vertical, the zenith, and the solar beam). The earth is tilted on its axis by 23.5° with respect to the plane of the earth's revolution around the sun, so the north pole is tilted toward the sun on June 21 and away from the sun on December 21. This tilt causes the solar beam to have perpendicular incidence at different latitudes depending on the date. The zenith angle Z is determined from:

$$\cos Z = \sin \varphi \, \sin \delta + \cos \varphi \, \cos \delta \, \cos \eta \tag{6.30}$$

where φ is latitude (positive for Northern Hemisphere, negative for Southern Hemisphere), δ is solar declination (see Table 6.12), and η is hour angle, 15° × the number of hours before or after local noon. The solar azimuth ω is the angle between south and the direction toward the sun in a horizontal plane:

$$\sin \omega 5(\cos \delta \, \sin \eta)/\sin Z \tag{6.31}$$

For surfaces receiving sunlight that are not horizontal, a slope at an angle i from the horizontal facing an azimuth ω' degrees from south experiences an intensity of sunlight (neglecting the effects of the atmosphere) of

$$S_s = S[\cos Z \, \cos i + \sin Z \, \sin i \, \cos(\omega - \omega')] \tag{6.32}$$

Where, ω and ω' are negative to the east of south and positive to the west. At angles away from the zenith, solar radiation must penetrate a greater thickness of the atmosphere. Consequently, it more scattering will occur due to interference from particles and greater light absorption due to this greater thickness. Most ultraviolet radiation is absorbed from the solar spectrum and does not reach the earth's surface, which means the peak of the solar radiation that reaches the earth's surface is in the visible part of the spectrum. The earth then reradiates like a blackbody at a mean temperature of 290 °K. The re-radiated infrared radiation is at wavelengths of 3–80 μm, with a peak at around 11 μm. The atmosphere absorbs and re-emits this longwave radiation primarily because of water vapor and CO_2 in the atmosphere. This absorption makes the atmosphere almost entirely opaque to wavelengths less than 7 μm and wavelengths greater than 14 μm and partly opaque between 7 and 8.5 μm and between 11 and 14 μm. The atmosphere loses heat to space directly through the nearly transparent window between 8.5 and 11 μm and through the absorption and successive re-radiation by layers of the atmosphere that contain substantial amounts of these absorbing gases [100,101].

TABLE 6.12 Solar declination.

Date	Declination degree	Date	Declination degree
January 21	−20.90	July 21	20.50
February 21	−10.83	August 21	12.38
March 21	0.0	September 21	1.02
April 21	11.58	October 21	−10.42
May 21	20.03	November 21	−19.75
June 21	23.45	December 21	−23.43

(Data from H.R. Byers, General Meteorology, fourth ed., McGraw-Hill, New York, NY, 1974.)

The earth operates as a heat engine, with hot-cold contrasts that generate power. Heat is transported by conduction, convection, and advection. During conduction, heat is transferred through matter without transferring the matter itself. Due to heat-induced density differentials, heat is transferred with matter moving vertically, i.e., conduction, or horizontally, i.e., advection [102].

Assorted landform surfaces reflect and absorb solar radiation in various ways. Sandy surfaces reach fairly high temperatures on a sunny day. The heat capacity and conductivity of sand are sufficiently low as to limit the heat penetration to 0.2–0.3 m, so that only a small amount of heat is stored compared to a waterbody in which the sun's rays penetrate several meters and slowly heat a rather deep layer. Fluidity also comes into play. Water movement and convection extends the heat more deeply than more rigid land that depends on conduction. The heat capacity of water is also considerably greater than that of sand and other soils. Sand's porosity and permeability mean that the pore spaces do not hold water as well as silt and clay, so a sandy landform has less heat capacity than these other soil systems. The synergistic factors allow considerably greater heat storage in waterbodies than in land systems.

Much more radiation penetrates and is stored by water near the equator than water in higher latitudes since the solar beam is more direct angle of incidence in equatorial regions. Recall that the first law of thermodynamics requires that the energy entering earth's atmosphere be balanced with that exiting the system. Heat is transferred poleward from the equatorial regions to reconcile a net outward transfer of heat near the poles (see Fig. 6.13). This heat is transferred by air and ocean currents as warm currents move poleward and cool currents move equatorward [99]. Considerable heat transfer occurs by the evaporation of water in the tropics and its condensation into droplets farther poleward, with the release of the heat of condensation. Enough heat is transferred to result in no net heating of the equatorial regions or cooling of the poles [103]. The poleward flux of heat across various latitudes is shown in Table 6.13.

For every 100 units of energy entering the atmosphere, 51 units are absorbed by the earth's surface, 19 units are absorbed by the atmosphere and 30 units are reflected and returned to space. Thus, 70 units will be eventually re-radiated as longwave radiation. The earth's heat exchange is mainly between and within the two most abundant fluids, i.e., air, and water, with the main exchanges within the ocean-atmosphere interaction. About 60% of the heat exchange is attributed to air movements, and approximately 40% is the result of ocean currents. In the oceans, warm water moves toward the poles along the western side of an ocean basin and the cold water flows toward the equator on the eastern side. At higher latitudes, warm water moves poleward in the eastern side of the ocean basin and cold water flows toward the equator on the western side [102].

Vertical air motions affect both weather and the mixing processes of importance to pollution. Upward vertical motions can be caused by lifting over terrain, lifting over weather fronts, and convergence toward low-pressure centers. Downward vertical motions can be caused by sinking to make up for divergence near high-pressure centers. One must know whether the atmosphere enhances or suppresses these vertical motions to assess their effects. When the atmosphere resists vertical motions, it is called stable; when the atmosphere enhances vertical motions, it is called unstable or in a state of instability.

In incompressible fluids the vertical structure of temperature very simply reveals the stability of the fluid. This applies to water, which in most engineering and environmental contexts is assumed to be incompressible. When the lower layer is warmer and thus less dense than the upper layer, the fluid is unstable and convective currents will cause it to overturn.

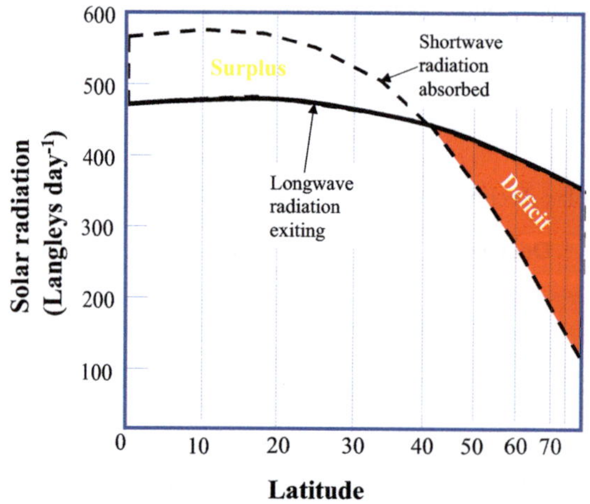

FIG. 6.13 Annual mean radiation by latitude. The latitude scale simulates the amount of the earth's surface area between the latitude bands. Incoming radiation is that absorbed by earth and atmosphere. Outgoing radiation is that leaving the atmosphere. A Langley is a unit of solar radiation equivalent to 1 thermal calorie per cubic centimeter or 41,840 joules per meter, or one gram calorie per square centimeter of irradiated surface.

TABLE 6.13 Poleward flux of heat across latitudes (10^{19} kcal year^{-1}).

Latitude	Flux
10	1.21
20	2.54
30	3.56
40	3.91
50	3.40
60	2.40
70	1.25
80	0.35

Data from W.D. Sellers, Physical Climatology, University of Chicago Press, 1965.

When the lower layer is cooler than the upper layer, the fluid is stable and vertical exchange is at a minimum. However, because air is compressible, the determination of stability is somewhat more complicated than that of water. The temperature and density of the atmosphere normally decrease with elevation; density is also affected by moisture in the air.

The bulk modulus (B) expresses a substance's resistance to compression [104,105]. It is the ratio of an infinitesimal pressure (p) increase to the resulting relative decrease in the substance's volume (V):

$$B = \frac{\Delta p}{\Delta V/V} = -V\frac{dp}{dV} \tag{6.33}$$

Given that volume is inversely proportional to density ($\boldsymbol{\rho}$), bulk modulus can also be expressed as:

$$B = \rho_1 \frac{dp}{d\rho} \tag{6.34}$$

Where, ρ_1 is the initial density and $\frac{dp}{d\rho}$ is the derivative of pressure with respect to density.

The reciprocal of the bulk modulus at fixed temperature is isothermal compressibility.

As mentioned, for atmospheric purposes, many liquids, including water, are defined as "incompressible." In fact, any liquid can be compressed [104]. Water's compressibility, expressed as change of volume per volume ($\Delta \frac{V}{V}$), is 4.6×10^{-5} per atm under atmospheric conditions. At standard temperature and pressure, air is about 20,000 times more compressible than water. Thus, compared to the compressibility of air, water can be considered incompressible.

The relationship between pressure p, volume V, mass m, and temperature T is given by the equation of state:

$$pV = RmT \tag{6.35}$$

Recall that R is a specific gas constant equal to the universal gas constant divided by the gram molecular weight of the gas. Since the density ρ is m/V, the equation can be rewritten as:

$$p = R\rho T \tag{6.36}$$

or considering specific volume $\alpha = \rho^{-1}$ as:

$$\alpha p = RT \tag{6.37}$$

These equations combine Boyle's law, which states that when temperature is held constant the volume varies inversely with the pressure, with the law of Guy-Lussac, which states that when pressure is held constant the volume varies in proportion to the absolute temperature.

If a volume of air is held constant and a small amount of heat Δh is added, the temperature of the air will increase by a small ΔT. This can be expressed as:

$$\Delta h = c_v \, \Delta T \tag{6.38}$$

where c_v is the specific heat at constant volume. In this case, all the heat added is used to increase the internal energy of the volume affected by the temperature.

If, instead of being restricted, the volume of air considered is allowed to remain at an equilibrium constant pressure and expand in volume, as well as changing temperature in response to the addition of heat, this can be expressed as:

$$\Delta h = c_v \Delta T + p \Delta v \tag{6.39}$$

By using the equation of state, the volume change can be replaced by a corresponding pressure change:

$$\Delta h = c_p \Delta T + v\, \Delta p \tag{6.40}$$

Where, c_p is the specific heat at constant pressure. For air, c_p is the average of specific heats for the air's constituents, weighted according to relative abundance, i.e., $m_T c_p = m_d \times (c_{pd} + m_{vap}) \times c_{pv}$; where m_T is the total mass of air, m_d is the mass of dry air, m_{vap} is the mass of water vapor, c_{pd} is the specicfic heat of air, and c_{pv} is the specific heat of water vapor.

8.1 Adiabatic processes

An adiabatic process is one with no loss or gain of heat to a volume of air. Conversely, if heat is supplied or withdrawn, the process is diabatic or nonadiabatic. Near the earth's surface, where heat is exchanged between the earth and the air, the processes are diabatic. However, at increasing altitude from the surface, processes frequently are adiabatic. For example, if a parcel of air is forced upward over a ridge, the upward-moving air will encounter decreased atmospheric pressure and will expand and cool. If the air is not saturated with water vapor, the process is called dry adiabatic.[e] Because no heat is added or subtracted, Δh in Eq. (6.41) can be set equal to zero, giving the hydrostatic equation:

$$-\Delta p = \rho g\, \Delta z \tag{6.41}$$

And, combining equations results in:

$$-\Delta T\, \Delta z^{-1} = g\, c_p^{-1} \tag{6.42}$$

Rising air cools and descending air warms. Since this assumes that the process is adiabatic then the value of $-\Delta T\, \Delta z^{-1}$ defines γ_d, which is the dry adiabatic process lapse rate, a constant equal to 0.0098 °K m^{-1} or nearly 1 °K 100 m^{-1} or 5.4 °F 1000 ft^{-1}. As an ascending air parcel reaches saturation, the addition of latent heat from condensing moisture will partially overcome the cooling due to expansion. Therefore, the saturated adiabatic lapse rate (γ_w) of cooling is smaller than γ_d.[f]

Latent heat is present within a substance without any change in the substance's temperature, e.g., molecular reconfigurations in the substance due to change of phase. Latent heat of vaporization (ΔH_{vap}) is the heat that must be added to convert a unit mass of water from liquid to vapor phase. Thus, ΔH_{vap} is the amount of energy (enthalpy) needed to transform a quantity of that substance into a gas. The enthalpy of vaporization is a function of the pressure at which this phase change occurs. Similarly, the latent heat of condensation (ΔH_{con}) is the energy released when a vapor condenses to form liquid droplets. To adhere to the laws of thermodynamics, ΔH_{con} must equal ΔH_{vap}, but with the opposite sign. This is because enthalpy changes of vaporization must always be positive. Heat is being absorbed by the substance during vaporization, but enthalpy changes of condensation are always negative because heat is being released by the substance. At standard, temperature, and pressure (STP), i.e., (273.15 K (0 °C)) and 100 kPa (1 bar), water's ΔH_{vap} and ΔH_{con} are both 2500 J g^{-1} [106]. Fig. 6.14 compares the molar enthalpy of water to those for three organic solvents, showing that water's latent heat is higher. A related concept in atmospheric physics is the temperature to which water must be cooled to become saturated, i.e., the dew point.

The increase or suppression of the vertical motion can be determined by comparing the density changes in a rising or descending parcel of air with the density of the surrounding environment. Since atmospheric pressure decreases with height, there is an upward-directed pressure gradient force. The force of gravity is downward. The difference between these two forces is the buoyancy force. Using Newton's second law of motion, which indicates that a net force equals an acceleration, the acceleration (a) of an air parcel at a particular position is given by:

$$a = g\left(T_p - T_e\right) T_p^{-1} \tag{6.43}$$

e. In atmospheric sciences, the term "dry" implies that no condensation or evaporation is occurring within the air parcel.

f. Saturated adiabatic lapse rate occurs within a cloud when the lapse rate is smaller than the dry adiabatic lapse rate (DALR) due the release of heat by condensation. In contrast to the DALR, the saturated adiabatic lapse rate depends on pressure and temperature [109].

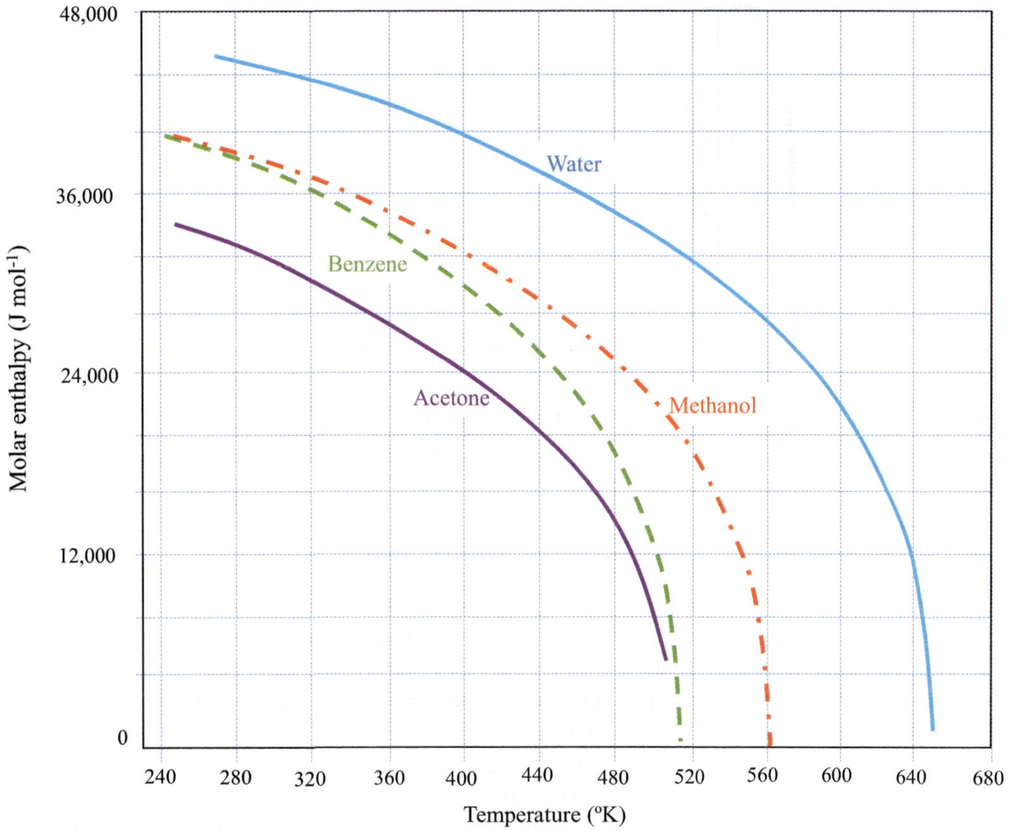

FIG. 6.14 Temperature dependence of molar enthalpy, i.e., latent heat of vaporization and condensation, for acetone, benzene, methanol, and water. *(Data from Dortman Data Bank, Enthalpy. http://www.ddbst.com/en/EED/PCP/HHH_C174.php, 2022 (Accessed 2 January 2022).)*

where g is the acceleration due to gravity ($9.8\,m\,s^{-2}$), T_p is the temperature of an air parcel that has undergone a temperature change according to the process lapse rate, and T_e is the temperature of the surrounding air at the same height, expressed in °K.

Fig. 6.15 shows the temperature change undergone by a parcel of air forced to rise 200 m in ascending a ridge. Assuming that the air is dry, and therefore that no condensation occurred, this Fig. 6.15 also represents the warming of the air parcel if the flow is reversed, so that the parcel moves downslope from B to A.

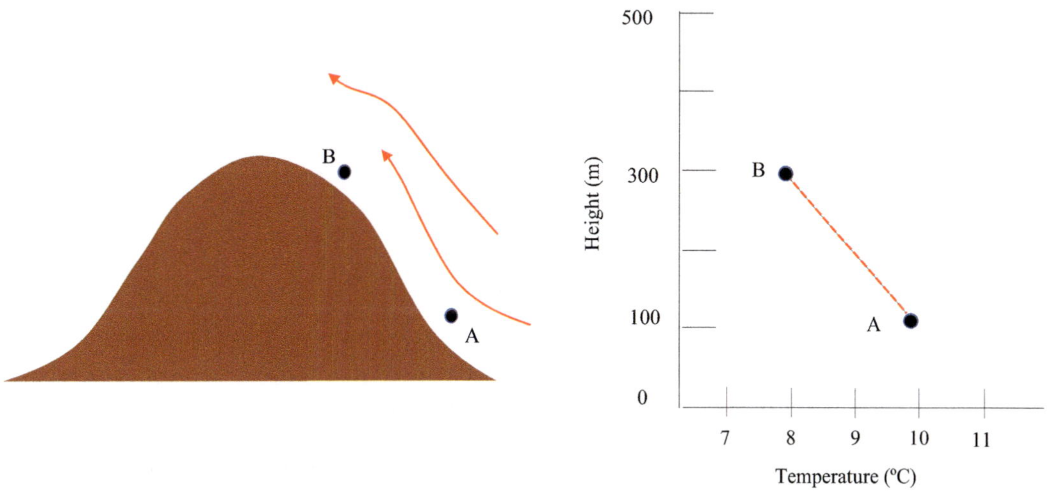

FIG. 6.15 Cooling of air ascending over a hill. Dry air forced to rise 200 m over a ridge cools adiabatically by 2 °C. *(Data from D.A. Vallero, Fundamentals of Air Pollution, fifth ed., Elsevier Academic Press, Waltham, MA, 2014, p. 999 pages cm.)*

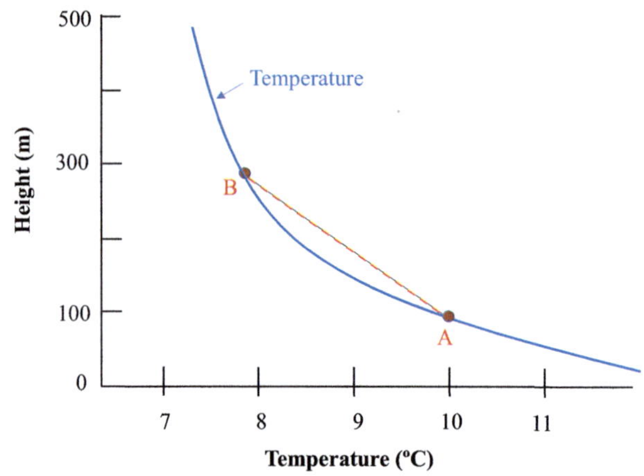

FIG. 6.16 Temperature of an air parcel forced to rise 200m compared to the superadiabatic lapse rate. Since the parcel is still warmer than the surrounding air, it will continue to rise. *(Data from D.A. Vallero, Fundamentals of Air Pollution, fifth ed., Elsevier Academic Press, Waltham, MA, 2014, p. 999 pages cm. U.S. Environmental Protection Agency, Basic air pollution meteorology student guidebook (SI 409), U.S. EPA, Research Triangle Park, North Carolina, 2005.)*

By rising from 100 to 300m, the parcel or air in Fig. 6.16 undergoes the temperature change of the dry adiabatic process lapse rate. The dashed line in Fig. 6.15 is a dry adiabatic line or dry adiabat. The ambient temperature structure is shown by the solid curve. This is a superadiabatic condition since the lapse rate of the surrounding environment in the lowest 150–200m is steeper than the adiabatic lapse rate. Given that the temperature drops more rapidly with height, this part of the atmosphere is thermally unstable. At 300m, the parcel is 0.2°C warmer than the surrounding air, so the resulting acceleration is upward, increasing the vertical motion and making the air unstable. The parcel of air continues to rise until it reaches 350m, where its temperature is the same as that of the surrounding air and its acceleration drops to zero. Above 350m the lapse rate of the surrounding air is not as steep as the adiabatic lapse rate, so this condition is subadiabatic, making the air thermally stable since it resists upward or downward motion [100].

Consider what happens if the temperature structure differs from that in Fig. 6.15 in the lower atmospheric layers so that a temperature inversion occurs, i.e., the air temperature increases with increasing height rather than decreasing, as shown in Fig. 6.17. During the forced ascent of the air parcel up the mountain slope, dry adiabatic cooling produces parcel temperatures that are everywhere cooler than the surrounding air, making downward acceleration and resisting displacement, so that the atmosphere becomes stable.

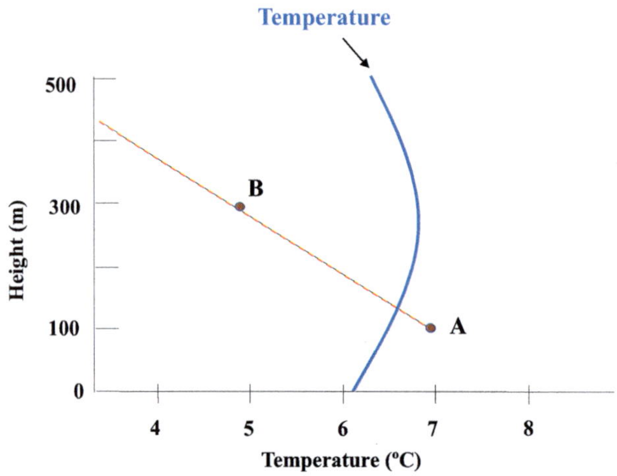

FIG. 6.17 Temperature of a parcel of air forced to rise 200m compared to an inversion lapse rate. Since the parcel is cooler than the environment, it will sink back to its original height. *(Data from D.A. Vallero, Fundamentals of Air Pollution, fifth ed., Elsevier Academic Press, Waltham, MA, 2014, p. 999 pages cm. U.S. Environmental Protection Agency, Basic air pollution meteorology student guidebook (SI 409), U.S. EPA, Research Triangle Park, North Carolina, 2005.)*

Thermodynamic diagrams which show the relationships between atmospheric pressure (rather than altitude), temperature, dry adiabatic lapse rates, and moist adiabatic lapse rates are useful for numerous atmospheric thermodynamic estimations. Standard meteorology provides additional details. IThe thermodynamic diagram may be used to find the current mixing height, i.e., the top of the neutral or unstable layer. The mixing height at a given time is estimated by applying a radiosonde ascent plotted on a thermodynamic chart. The surface temperature at the given time is plotted. Drawing a dry adiabat through this temperature, then the height aboveground at the point where this dry adiabat intersects the sounding, is the mixing height at that time. The maximum mixing height is the mixing height for the time of maximum temperature. This procedure assumes no advection since the time of the sounding [101].

Dynamic stability accounts for wind speed, whereas static stability does not. Increasing wind speed increases vertical mixing of air parcels so static stability assumes that buoyancy forces cause density differences. Dynamic stability is based on shear forces. Stability of an air parcel must include these and other factors. If the wind speed is $<2\,\mathrm{m\,s^{-1}}$ and incoming solar radiation is strong, for example, the atmosphere would be very unstable. Conversely, on a clear night, the same calm winds would make for a stable air mass. These conditions create very different air pollution plumes (see Fig. 6.18). To address these complexities, meteorologists have devised atmospheric the stability classifications.

The atmosphere is very dynamic in terms of temperature and pressure, which calls for a benchmark to compare existing conditions to other conditions. One layer of the atmosphere is always more or less stable than another, given various temperatures and pressures. Potential temperature (θ) relates the dry adiabat to a particular atmospheric combination of temperature and pressure. Potential temperature is the temperature of the air if the measured temperature were at sea level and 1 atm pressure [91,107]. It is the temperature that an unsaturated parcel of air would have if brought adiabatically and reversibly from its initial standard pressure (p_O):

$$\theta = T\left[\frac{p_0}{p}\right]^k \tag{6.44}$$

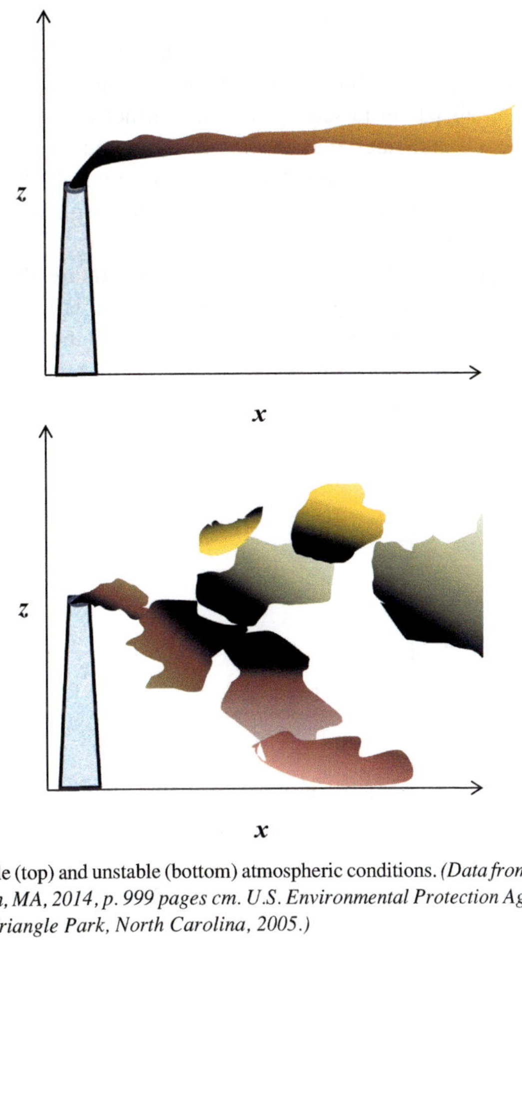

FIG. 6.18 Plumes (side views) under stable (top) and unstable (bottom) atmospheric conditions. *(Data from D.A. Vallero, Fundamentals of Air Pollution, fifth ed., Elsevier Academic Press, Waltham, MA, 2014, p. 999 pages cm. U.S. Environmental Protection Agency, Basic air pollution meteorology student guidebook (SI 409), U.S. EPA, Research Triangle Park, North Carolina, 2005.)*

Where, T= temperature in °K and k is the Poisson constant, which is usually assumed to be 2/7 ≈ 2.86, i.e., the ratio of the gas constant (R= 8,316,963 J mol^{-1} K^{-1}) to the specific heat capacity at constant pressure for an ideal diatomic gas like O_2 or N_2.

The potential temperature θ can also be calculated from the temperature (T) at a given pressure p:

$$\theta = T\left[\frac{1\ \text{atm}}{p}\right]^{0.286} \tag{6.45}$$

If p is in millibars (mb), Eq. (6.48) is:

$$\theta = T\left[\frac{1000}{p}\right]^{0.286} \tag{6.46}$$

This is the same as the temperature that a parcel of dry air would have if brought dry adiabatically to a pressure of 1000 mb.

As mentioned, if the potential temperature decreases with height, the atmosphere is unstable. If the potential temperature increases with height, the atmosphere is stable. The average lapse rate of the atmosphere is approximately 6.5 °C km^{-1}. So, on average, the potential temperature increases with height and the average state of the atmosphere is stable.

In a dry adiabatic, hydrostatic condition, the air's θ does not vary vertically:

$$\frac{\partial \theta}{\partial z} = 0 \tag{6.47}$$

Thus, in a dry adiabatic atmosphere, T must decrease with height. Near the earth's surface the atmosphere is continuously undergoing heating and cooling. Thus, the adiabatic assumptions are seldom met; nor are the hydrostatic, dry conditions. The lower portions of the troposphere are seldom neither adiabatic nor dry, which means that near the surface the atmosphere's slope of $\frac{dT}{dz}$ does not equal the dry adiabatic lapse rate. Air parcels are usually assumed to be under local adiabatic conditions, since they are commonly transported locally by a turbulent atmosphere that does not allow substantial energy exchange so long as the time of transport is short.

Potential temperature indicates stability. If θ increases with altitude, an air parcel raised adiabatically from its initial height will always be cooler, and therefore, denser than the surrounding air and will subside to its original pressure. Conversely, if θ decreases with altitude, an air parcel raised up will be warmer than its surrounding air and will accelerate upwardly by buoyancy; therefore, $\frac{\partial \theta}{\partial z} > 0$ indicates that the air is table and $\frac{\partial \theta}{\partial z} < 0$ indicates that the air is unstable [108]. If θ remains constant with altitude, the temperature at any height is precisely that which leads to an adiabatic ascent or adiabatic descent to the value θ at a pressure of 100 kPa. In this case, the atmosphere is in a state of adiabatic or convective equilibrium known as neutral stability [109].

The mixing of air in a vertical layer creates constant potential temperature throughout the layer. This mixing is mechanical over a rough surface. The solid line in Fig. 6.19 shows the initial temperature structure that is subadiabatic. Mixing leads to a mean potential temperature throughout the layer, shown by the dashed line. The lower part is dry adiabatic. The bottom part of the layer is warmed, and the top is cooled. Above the vertical extent of the mixing, an inversion is formed connecting the new cooled portion with the old temperature structure above the zone of mixing. If the initial layer

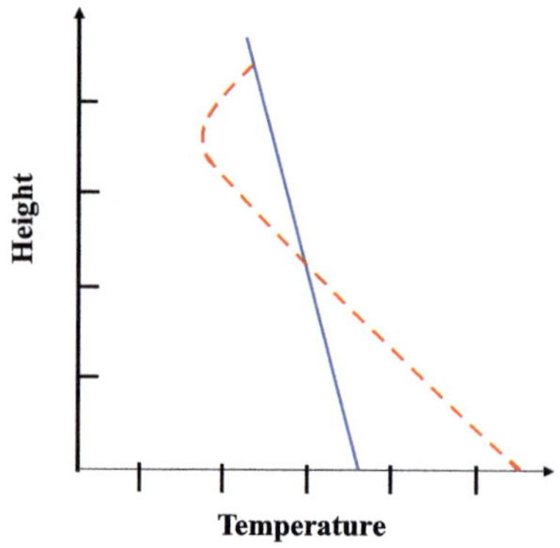

FIG. 6.19 Forced mixing (dashed line) on a subadiabatic lapse rate (solid line). An inversion is forming at the top of the mixed layer. *(Data from D.A. Vallero, Fundamentals of Air Pollution, fifth ed., Elsevier Academic Press, Waltham, MA, 2014, p. 999 pages cm. U.S. Environmental Protection Agency, Basic air pollution meteorology student guidebook (SI 409), U.S. EPA, Research Triangle Park, North Carolina, 2005.)*

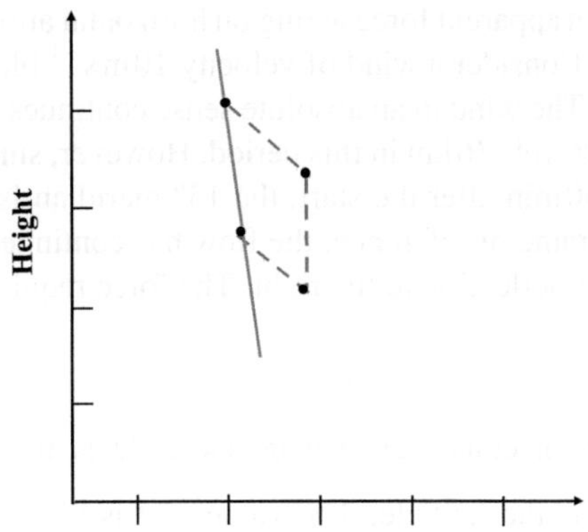

FIG. 6.20 Development of a subsidence inversion in subsiding air. Note the vertical compression of the sinking layer which is usually accompanied by horizontal divergence. *(Data from D.A. Vallero, Fundamentals of Air Pollution, fifth ed., Elsevier Academic Press, Waltham, MA, 2014, p. 999 pages cm. U.S. Environmental Protection Agency, Basic air pollution meteorology student guidebook (SI 409), U.S. EPA, Research Triangle Park, North Carolina, 2005.)*

has considerable moisture, although not saturated, cooling in the top portion of the layer may decrease the temperature to the point where some of the moisture condenses, forming clouds at the top. An example of this is the formation of an inversion and a layer of stratus clouds along the southwestern coast of a continent [91,107]. At night, inversions at the surface are often the result of radiational cooling of the ground, which subsequently cools the air just above the ground.

There is usually some descent (subsidence) of air above surface high-pressure systems. This air warms dry adiabatically as it descends, decreasing the relative humidity and dissipating any clouds in the layer. A subsidence inversion forms because of this sinking. Since the descending air compresses with the increased pressures lower in the atmosphere, the top portion of the descending layer will be further warmed due to its greater descent than will the bottom portion of the layer (see Fig. 6.20). Occasionally a subsidence inversion descends all the way to the surface, but usually its base is well above the ground.

Inversions are important to air pollution because of their stabilizing influence on the atmosphere, which suppresses the vertical motion that causes the vertical spreading of pollutants. In fact, some of the most notorious air pollution episodes, such as those in London, Donora, Pennsylvania, and the Meuse Valley in Belgium, have been exacerbated by inversions. In these cases, the inversions limited vertical mixing, which allowed the accumulation of air pollutants. Thus, the stable layer above the inversion trapped the polluted air in valley towns and cities for days [91].

8.2 Motion[g]

Although vertical motions certainly occur in the atmosphere and are important to both weather processes and the movement of pollutants, for convenience, wind velocity is often considered only as horizontal. Given that $F = ma$; if a force F is exerted on a mass m, the resulting acceleration a equals the force divided by the mass. This is also the rate of change of momentum of a body, which is equal to the sum of the forces that act on the body. Although Newton's second law applies to absolute motion, it is most convenient to consider wind relative to the earth's surface.

Atmospheric pressure equals mass times the acceleration of gravity. Considering a unit volume, $p = \rho g$; the gravitational force on the unit volume is directed downward. Due to horizontal temperature gradients, there are horizontal density gradients and, consequently, horizontal pressure gradients. The horizontal pressure gradient force is:

$$p_h = \frac{\Delta p}{\rho \Delta x} \tag{6.48}$$

Where, Δp is the horizontal pressure difference over the distance Δx. The direction of this force and of the pressure difference measurement is locally perpendicular to the lines of equal pressure, i.e., isobars, and is directed from high to low pressure.

g. This subsection is an excerpt from [90].

Because the earth is rotating, there is an apparent force acting on horizontal atmospheric motions when examined from a point of reference on the earth's surface. Consider a wind of velocity $10\,m\,s^{-1}$ blowing at time 1 in the direction of the 0° longitude meridian across the north pole. The wind in an absolute sense continues to blow in this direction for 1 hour, and a parcel of air starting at the pole at time 1 travels 36 km in this period. However, since the earth turns 360° every 24 h, or 15° h^{-1}, it has rotated 15° in the hour, and 60 min after the start, the 15° meridian is now beneath the wind vector.

As viewed from space, the absolute frame of reference, the flow has continued in a straight line, but viewed from the earth, the flow has undergone an apparent deflection to the right. The force required to produce this apparent deflection is the Coriolis force, which is:

$$D = vf \tag{6.49}$$

Where, f is the Coriolis parameter, which equals $2\Omega \sin \varphi$. Here, Ω is the angular speed of the earth's rotation, $\frac{2\pi}{24 \times 60 \times 60} = 7.27 \times 10^{-5} s^{-1}$, and φ is the latitude. The value of f is highest at the poles and zero at the equator. The deflecting force is to the right of the wind vector in the Northern Hemisphere and to the left in the Southern Hemisphere. For the present example, the deflecting force is $1.45 \times 10^{-3}\,m\,s^{-2}$, and the amount of deflection after the 36-km movement in 1 hour is 9.43 km.

Friction between the atmosphere and the earth's surface is often neglected at altitudes of about 700 m and higher. As such, large-scale air currents closely represent a balance between the pressure gradient force and the Coriolis force. Since the Coriolis force is at a right angle to the wind vector, when the Coriolis force is equal in magnitude and opposite in direction to the pressure gradient force, a wind vector perpendicular to both forces occurs. Its direction is along the lines of constant pressure. According to Buys Ballot's law, in the Northern Hemisphere, the low pressure is to the left of the wind vector and in the Southern Hemisphere, low pressure is to the right. A geostrophic wind is the theoretical wind that would result from the balance between the Coriolis force and the pressure gradient force. The geostrophic velocity is:

$$v_g = \frac{-\Delta p}{\rho f \Delta d} \tag{6.50}$$

When the isobars are essentially straight, the balance between the pressure gradient force and the Coriolis force results in a geostrophic wind parallel to the isobars. When the isobars are curved, an additional force, i.e., a centrifugal force outward from the center of curvature, enters the balance of forces. In the case of curvature around low pressure, a balance of forces occurs when the pressure gradient force equals the sum of the Coriolis and centrifugal forces, and the wind continues parallel to the isobars. In the case of curvature around high pressure, a balance of forces occurs when the sum of the pressure gradient and centrifugal forces equals the Coriolis force. To maintain a given gradient wind speed, a greater pressure gradient force, indicated by tighter spacing of the isobars, is required in the flow around low-pressure systems than in the flow around high-pressure systems.

The frictional effect of the earth's surface on the atmosphere increases as the earth's surface is approached from aloft. Assuming geostrophic balance aloft, the effect of friction with distance approaching the earth's surface is to slow the wind velocity, which in turn decreases the Coriolis force. The wind then turns toward low pressure until the resultant vector of the frictional force and the Coriolis force balances the pressure gradient force. The greater the friction, the slower the wind and the greater the amount of turning toward low pressure. The turning of the wind from the surface through the friction layer is called the Ekman spiral. This frictional effect will cause pollutants released at two different heights to tend to move in different directions.

In the friction layer where the isobars are curved, the effect of frictional drag is added to the forces discussed under gradient wind. The balance of the pressure gradient force, the Coriolis deviating force, the centrifugal force, and the frictional drag in the vicinity of the curved isobars results in wind flow around low pressure and high pressure in the Northern Hemisphere.

Although of much smaller magnitude than horizontal motions, vertical motions are important both to daily weather formation and to the transport and dispersion of pollutants.

8.3 Local winds

The sea breeze is a result of the differential heating of land and water surfaces by incoming solar radiation. Since solar radiation penetrates several meters of a body of water, it warms very slowly. In contrast, only the upper few centimeters of land are heated, and warming occurs rapidly in response to solar heating. Therefore, especially on clear summer days, the land surface heats rapidly, warming the air near the surface and decreasing its density. This causes the air to rise over the land, decreasing the atmospheric pressure near the surface relative to the pressure at the same altitude over the water surface. The rising air increases the pressure over the land relative to that above the water at altitudes of approximately 100–200 m. The air that rises over the land surface is replaced by cooler air from above the water surface. This air, in turn, is replaced by subsiding air from somewhat higher layers of the atmosphere over the water. Air from the higher-pressure zone several hundred meters above the surface then flows from over the land surface out over the water, completing a circular or cellular flow. Any general flow due to large-scale pressure systems will be superimposed on the sea breeze and may either reinforce or inhibit it. Ignoring the larger-scale influences, the strength of the sea breeze will generally be a function of the temperature excess of the air above the land surface compared to that above the water surface.

Heating in the daytime occurs more quickly over land than over water. Similarly, radiational cooling at night occurs more quickly over land. The pressure pattern tends to be the reverse of that in the daytime. The warmer air tends to rise over the water, which is replaced by the land breeze from land to water, with the reverse flow (water to land) completing the circular flow at altitudes somewhat aloft. Frequently at night, the temperature differences between land and water are smaller than those during the daytime, and therefore the land breeze has a lower speed.

Solar heating and radiational cooling influence local flows in terrain situations. Consider midday heating of a south-facing mountainside. As the slope heats, the air adjacent to the slope warms, its density is decreased, and the air attempts to ascend. Near the top of the slope, the air tends to rise vertically. Along each portion of the slope farther down the mountain, it is easier for each rising parcel of air to move upslope, replacing the parcel ahead of it rather than rising vertically. This upslope flow is the valley wind (see Fig. 6.21) [91].

At night when radiational cooling occurs on slopes, the cool dense air near the surface descends along the slope (see Fig. 6.22). This is the downslope wind. To compensate for this descending air, the air farther from the slope that is cooled very little is warmer relative to the descending air and rises, frequently resulting in a closed circular path. Where the down-slope winds occur on opposite slopes of a valley, the cold air can accumulate on the valley floor. If there is any slope to the valley floor, this pool of cold air can move down the valley, resulting in a drainage or canyon wind.

Variations of topography, especially those having slopes nearly perpendicular to the incoming radiation and others in deep shadow, lead to many variations of wind patterns. Any local flow can be modified by a larger-scale regional wind that will change pressure patterns. Table 6.14 gives characteristics of eight different situations depending on the orientation of

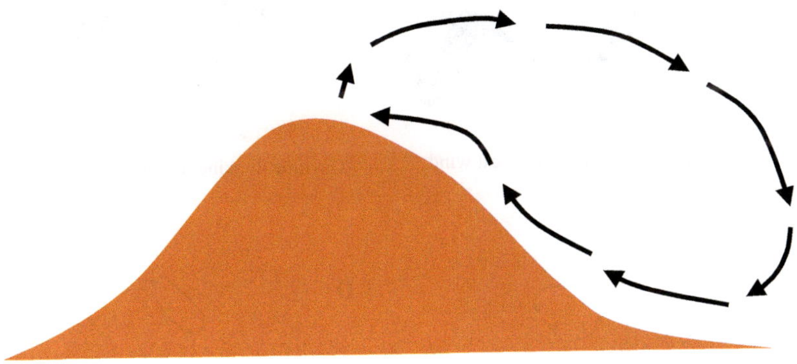

FIG. 6.21 Daytime upslope wind due to greater solar heating on the valley's side than in its center. *(Data from D.A. Vallero, Fundamentals of Air Pollution, fifth ed., Elsevier Academic Press, Waltham, MA, 2014, p. 999 pages cm. U.S. Environmental Protection Agency, Basic air pollution meteorology student guidebook (SI 409), U.S. EPA, Research Triangle Park, North Carolina, 2005.)*

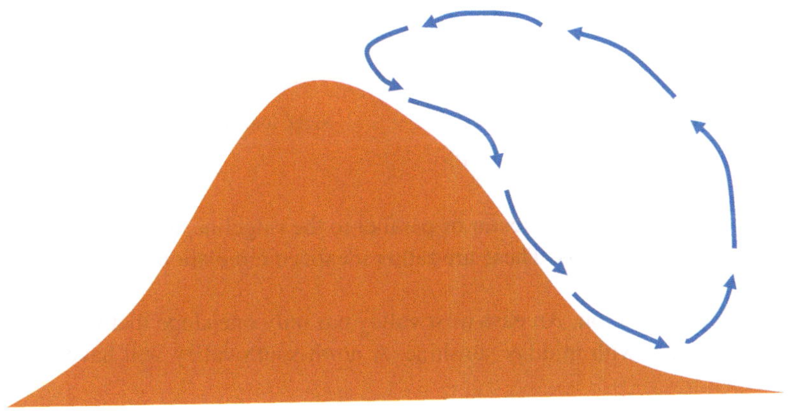

FIG. 6.22 Nighttime downslope wind due to more rapid radiational cooling on the valley's slope compared to its center. *(Data from D.A. Vallero, Fundamentals of Air Pollution, fifth ed., Elsevier Academic Press, Waltham, MA, 2014, p. 999 pages cm. U.S. Environmental Protection Agency, Basic air pollution meteorology student guidebook (SI 409), U.S. EPA, Research Triangle Park, North Carolina, 2005.)*

TABLE 6.14 Generalized mesoscale windflow patterns associated with different combinations of wind direction and ridgeline orientation.

Wind direction relative to ridgeline	Time of day	Ridgeline orientation: East–West	Ridgeline orientation: North–South
Parallel	Day	1. South-facing slope is heated—single helix	2. Upslope flow on both heated slopes—double helix
	Night	3. Downslope flow on both slopes—double helix	4. Downslope flow on both slopes—double helix
Perpendicular	Day	South-facing slope is heated	6. Upslope flow on both heated slopes—stationary eddy on one side of valley
		5a. North wind—stationary eddy fills valley	
		5b. South wind—eddy suppressed, flow without separation	
	Night	7. Indefinite flow—extreme stagnation in valley bottom	8. Indefinite flow—extreme stagnation in valley bottom

Examples of these ridgeline orientations at different times of day are shown in Fig. 6.22.
(Data from D.A. Vallero, Fundamentals of Air Pollution, fifth ed., Elsevier Academic Press, Waltham, MA, 2014, p. 999 pages cm and U.S. Environmental Protection Agency, Basic Air Pollution Meteorology Student Guidebook (SI 409), U.S. EPA, Research Triangle Park, North Carolina, 2005.)

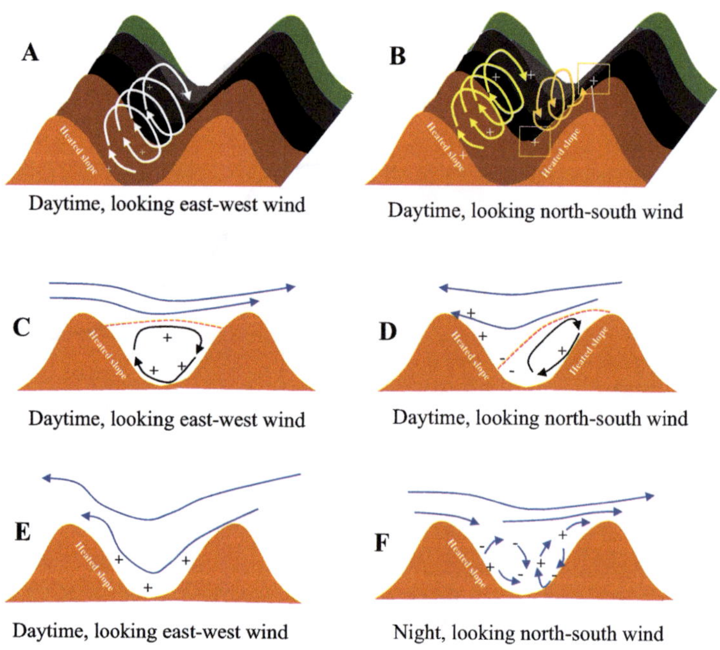

FIG. 6.23 Local valley-ridge flow patterns. *(Data from D. A. Vallero, Fundamentals of Air Pollution, fifth ed., Elsevier Academic Press, Waltham, MA, 2014, p. 999 pages cm. U.S. Environmental Protection Agency, Basic air pollution meteorology student guidebook (SI 409), U.S. EPA, Research Triangle Park, North Carolina, 2005.)*

the ridgeline and valley with respect to the sun, wind direction perpendicular or parallel to the ridgeline, and time of day. Fig. 6.23 shows examples of some of the mountain and valley winds. These are rather idealized circulations compared to observed flows at any one time.

The effect of solar radiation is different with valley orientation. An east-west valley has only one slope that is significantly heated—the south-facing slope may be near normal with midday sunshine. A north-south valley will have both slopes heated at midday. The effect of flow in relation to valley orientation is such that flows perpendicular to valleys tend to form circular eddies and encourage local flows; flows parallel to valleys tend to discourage local flows and to sweep clean the valley, especially with stronger wind speeds. When general flows lighten, the opportunity to generate local flows increases.

Complicated terrain such as a major canyon with numerous side canyons or a so-called urban canyon will produce complicated and unique flows, especially when side canyon drainage flows reinforce the drainage flow in the main valley. Urban areas have roughness and thermal characteristics that differ from those of their rural surroundings. Although the increased roughness affects both the vertical wind profile and the vertical temperature profile, the effects due to the thermal features are dominant.

The asphalt, concrete, and steel of urban areas heat quickly and have a high heat-storing capability compared to the soil and vegetation of rural areas. Also, some surfaces of buildings are normal to the sun's rays just after sunrise and before sunset, allowing warming throughout the day. The result is that the urban area becomes warmer than its surroundings during the day and stores sufficient heat that re-radiation of the stored heat during the night keeps the urban atmosphere considerably warmer than its rural surroundings throughout most nights with light winds (see Fig. 6.24). Under the lightest winds, the air rises over the warmest part of the urban core, drawing cooler air from all directions from the surroundings. Subsidence replaces this air in rural areas, and a closed torus, i.e., doughnut-shaped circulation, occurs with an outflow above the urban area. This circulation is referred to as the urban heat island, mentioned during the discussion of albedo. The strength of the resulting flow is dependent on the difference in temperature between the urban center and its surroundings (see Fig. 6.24).

When the regional wind allows the outflow to take place in primarily one direction and the rising warm urban air moves off with this regional flow, the circulation is termed the urban plume (see Fig. 6.25). Under this circumstance, the inflow to the urban center near the surface may also be asymmetric, although it is more likely to be symmetric than the outflow at higher altitudes.

Urban areas also emit heat through the release of gases and vapors from combustion, especially from vehicle emissions, power generation, and industrial processes. Compared to the heat received through solar radiation and subsequently released, the combustion and process heat is usually relatively small, but can be greater than 10% in major urban areas near the sources of emissions and during calm and light wind conditions [91,107]. Cities and urban areas also have complex terrain given tall and wide buildings and other structures. The flow induced by these structures is known as a street canyon effect. The different shapes and materials of urban structures modifies the radiation balance within the street canyon,

FIG. 6.24 Heat island effect under calm or light regional wind conditions. *(Data from D.A. Vallero, Fundamentals of Air Pollution, fifth ed., Elsevier Academic Press, Waltham, MA, 2014, p. 999 pages cm. U.S. Environmental Protection Agency, Basic air pollution meteorology student guidebook (SI 409), U.S. EPA, Research Triangle Park, North Carolina, 2005.)*

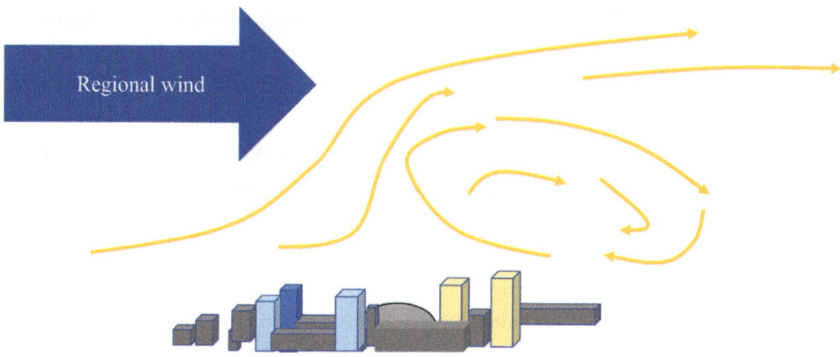

FIG. 6.25 Urban plume under moderate regional wind conditions. *(Data from D.A. Vallero, Fundamentals of Air Pollution, fifth ed., Elsevier Academic Press, Waltham, MA, 2014, p. 999 pages cm. U.S. Environmental Protection Agency, Basic air pollution meteorology student guidebook (SI 409), U. S. EPA, Research Triangle Park, North Carolina, 2005.)*

e.g., sunlit versus shaded walls. Emissions from walls dominate long-wave incoming radiation, whereas short-wave incoming radiation strongly depends on the orientation of the type of surface [110].

9 Climate change

Water pollution results from many earth processes. It may not be directly evident that changes in climate play many roles in water quality but indeed they do. The mechanisms of thermodynamics and motion discussed in previous sections of this chapter are influenced by climatic conditions, so a book on water pollution should include attention to climate change.

The earth is warmed by the greenhouse effect, wherein incoming shortwave radiation is converted to longwave radiation. Global greenhouse gases (GHGs) absorb this longwave energy and decrease the rate at which the energy escapes the atmosphere. These gases differ in their ability to absorb energy, i.e., they vary in radiative efficiencies. They also differ in atmospheric residence times. Thus, each GHG has a specific global warming potential (GWP), which allows comparisons of the amount of energy the emissions of 1 ton of a gas will absorb over a given time-period, usually a 100-year averaging time, compared to the emissions of 1 ton of CO_2.

Because CO_2 has a very long residence time in the atmosphere, its emissions cause increases in atmospheric concentrations of CO_2 that will last thousands of years [111]. Methane's (CH_4) average atmospheric residence time is about a decade but its capacity to absorb substantially more energy than CO_2 gives it a GWP ranging from 28 to 36. The GWP also accounts for some indirect effects, e.g., CH_4 is a precursor to another greenhouse gas, ozone.

The GWP of N_2O is 265–298 times that of CO_2, with an average residence time of 100 years. The CFCs, hydrofluorocarbons (HFCs), hydrochlorofluorocarbons (HCFCs), perfluorocarbons (PFCs), and sulfur hexafluoride (SF_6) are referred to as high-GWP gases, i.e., GWP can be up to tens of thousands. For a given amount of mass, they hold substantially more heat than does CO_2.

Global warming potential example 1

Explain what happens to the methane GWP if a 20-year averaging time is used instead of a 100-year averaging time.

A 20-year GWP is sometimes used as an alternative to the 100-year GWP. The 20-year GWP is based on the energy absorbed over 20 years, which prioritizes gases with shorter lifetimes, since it ignores any impacts that occur after 20 years from the emission. The GWPs are calculated relative to CO_2, so the GWPs that are based on an 80% shorter timeframe will be larger for gases with atmospheric residence times shorter than that of CO_2, and smaller for gases with residence times greater than CO_2.

Since CH_4 has a shorter atmospheric residence time than CO_2, the 100-year GWP is much less than the 20-year GWP. The CH_4 20-year GWP has been estimated [111] to be 84–87, compared to the 100-year GWP of 28–36.

Global warming potential example 2

What happens to the GWP for tetrafluoromethane (CF_4) if a 20-year averaging time is used?

The estimated atmospheric residence time for CF_4 is 50,000 years [111]. The 100-year GWP is estimated to range from 6630 to 7350, which is larger than the 20-year GWP of 4880–4950. Therefore, the shorter averaging time gives a 100-year GWP that is much larger than the 20-year GWP.

An alternate metric is the Global Temperature Potential (GTP). Unlike the GWP, which is a measure of the heat absorbed over a given time-period due to emissions of a gas, the GTP is a measure of the temperature change at the end of that time-period relative to CO_2. The calculation of the GTP is more complicated than that for the GWP, based on models of response of the climate system to increased greenhouse gas concentrations, i.e., climate sensitivity, as well as the amount of time it takes the system to respond, especially the oceans' capacity to absorb the heat [111].

Carbon dioxide equivalency (CDE) is an estimate of how much CO_2 would be needed for a mixture of emissions to have the same GWP, if measured over a defined period of time, often 100 years. Thus, CDE is the time-integrated radiative forcing of a quantity or rate of gas emissions to the troposphere. It is not an instantaneous value of the radiative forcing by a concentration of greenhouse gases in the atmosphere. The CDE is the product of the greenhouse gas mass (in million metric tons) and GWP.

Carbon dioxide equivalency example

A company is proposing to produce 500 kg per year of a refrigerant gas mixture containing 10% sulfur hexafluoride (SF_6), 40% perfluorohexane (C_6F_{14}), and 50% difluoromethane (CH_2F_2; also known as hydrofluorocarbon-32). Calculate the CDE for this mixture.

Based on data from the Intergovernmental Panel on Climate Change (IPCC) [112], the 100-year GWP values for these gases are:

SF_6: 22,800
C_6F_{14}: 9300
CH_2F_2: 679

Calculate the gas mixture GWP for the mixture ratio:

$$(0.1 \times 22{,}800) + (0.4 \times 9300) + (0.5 \times 675) = 2280 + 3720 + 337.5 = 6337.5$$

Next, multiply the mixture GWP times the mass to obtain the CDE. For industrial emissions, CO_2 equivalents are often expressed as "million metric tonnes of carbon dioxide equivalents" (MMTCDE):

$$500\,\text{kg} = 5 \times 10^{-7}\ \text{million tonnes}$$

$$\text{MMTCDE} = 6337.5 \times \left(5 \times 10^{-7}\right) = 3.17 \times 10^{-3}$$

This is equilibrated to CO_2, i.e., the 100-year GWP = 1, so the annual addition of this newly formulated gas mixture to the atmosphere is equivalent to 0.0317 million metric tons of CO_2.

10 Conclusions

Climate is essentially the average weather for a region, so water is influenced by climatic and meteorological conditions. Water pollutant transport and transformation result in part from meteorological phenomena, including winds, pressure systems, temperature gradients, and micrometeorological features. The laws of motion and thermodynamics explain the transport, transformation, and fate of pollutants. This chapter has delved into the first principles of physics from the perspective of inherent properties of the pollutants and their parent substances, as well as the properties and conditions of the air that carries these pollutants, including the major meteorological and climatic factors that affect water pollution.

References

[1] D.A. Vallero, Air pollution physics, in: Methods and Calculations in Environmental Physics, AIP Publishing LLC, Melville, New York, 2022, pp. 7-1–7-70.

[2] D.A. Vallero, Introduction: Importance of systems sciences and first principles, in: Methods and Calculations in Environmental Physics, AIP Publishing LLC, Melville, New York, 2022, pp. 1-1–1-20.

[3] D.A. Vallero, Applications and Calculations in Environmental Physics, AIP Publishing LLC, 2022.

[4] D.B. Schwede, R.L. Dennis, M.A. Bitz, The watershed deposition tool: a tool for incorporating atmospheric deposition in water-quality analyses 1, J. Am. Water Resour. Assoc. 45 (4) (2009) 973–985.

[5] Intergovernmental Panel On Climate Change, Climate change and water, in: IPCC Technical Paper V, vol. 30, IPCC Secretariat, Geneva, Switzerland, 2008. Available: https://archive.ipcc.ch/pdf/technical-papers/climate-change-water-en.pdf. (Acessed 5 February 2024).

[6] I. Abd-Elaty, S. Straface, A. Kuriqi, Sustainable saltwater intrusion management in coastal aquifers under climatic changes for humid and hyper-arid regions, Ecol. Eng. 171 (2021) 106382.

[7] S. Sithara, S. Pramada, S.G. Thampi, Impact of projected climate change on seawater intrusion on a regional coastal aquifer, J. Earth Syst. Sci. 129 (2020) 1–15.

[8] H. Xiao, et al., Saltwater intrusion into groundwater systems in the Mekong Delta and links to global change, Adv. Clim. Chang. Res. 12 (3) (2021) 342–352.

[9] R. Lewis, S. Gordon, Sampling for organic chemicals in air, Environmental Protection Agency, Research Triangle Park, NC, United States, 1994.

[10] J. Kvasnicka, E.A. Cohen Hubal, T.F. Rodgers, M.L. Diamond, Textile washing conveys SVOCs from indoors to outdoors: application and evaluation of a residential multimedia model, Environ. Sci. Technol. 55 (18) (2021) 12517–12527.

[11] C.M. Eichler, et al., Assessing human exposure to SVOCs in materials, products, and articles: a modular mechanistic framework, Environ. Sci. Technol. 55 (1) (2020) 25–43.

[12] J. Kvasnicka, E.A. Cohen Hubal, J. Ladan, X. Zhang, M.L. Diamond, Transient multimedia model for investigating the influence of indoor human activities on exposure to SVOCs, Environ. Sci. Technol. 54 (17) (2020) 10772–10782.

[13] C. Liu, S. Shi, C. Weschler, B. Zhao, Y. Zhang, Analysis of the dynamic interaction between SVOCs and airborne particles, Aerosol Sci. Technol. 47 (2) (2013) 125–136.

[14] C.J. Weschler, W. Nazaroff, SVOC exposure indoors: fresh look at dermal pathways, Indoor Air 22 (5) (2012) 356–377.

[15] Z. Liu, J. Little, Semivolatile organic compounds (SVOCs): phthalates and flame retardants, in: Toxicity of Building Materials, Elsevier, 2012, pp. 122–137.

[16] P.J. Lioy, et al., Characterization of the dust/smoke aerosol that settled east of the World Trade Center (WTC) in lower Manhattan after the collapse of the WTC 11 September 2001, Environ. Health Perspect. 110 (7) (2002) 703–714.

[17] G. Neumann, S. Fonselius, L. Wahlman, Measurements on the content of non-volatile organic material in atmospheric precipitation, Int. J. Air Pollut. 2 (1959).

[18] U.-U.T. Võ, M.P. Morris, Nonvolatile, semivolatile, or volatile: redefining volatile for volatile organic compounds, J. Air Waste Manage. Assoc. 64 (6) (2014) 661–669.

[19] K. Kawamura, I.R. Kaplan, Organic compounds in the rainwater of Los Angeles, Environ. Sci. Technol. 17 (8) (1983) 497–501.

[20] T. Boublík, V. Fried, E. Hála, The Vapour Pressures of Pure Substances, 1984.

[21] U.S. Environmental Protection Agency, Technical overview of volatile organic compounds, 2024. https://www.epa.gov/indoor-air-quality-iaq/technical-overview-volatile-organic-compounds#8. (Accessed 17 July 2024. Last update on March 5, 2024.

[22] M. Even, E. Juritsch, M. Richter, State of the art in the measurement of very volatile organic compounds (VVOCs) in indoor air by sorbent-based active sampling: identifying the gaps towards standardisation, Trends Anal. Chem. (2021) 116265.

[23] J.O. Okeme, T.F. Rodgers, J.M. Parnis, M.L. Diamond, T.F. Bidleman, L.M. Jantunen, Gas chromatographic estimation of vapor pressures and octanol–air partition coefficients of semivolatile organic compounds of emerging concern, J. Chem. Eng. Data 65 (5) (2020) 2467–2475.

[24] D.C. Hilton, R.S. Jones, A. Sjödin, A method for rapid, non-targeted screening for environmental contaminants in household dust, J. Chromatogr. A 1217 (44) (2010) 6851–6856.

[25] R. Singh, et al., Simultaneous determination of 8-oxo-2′-deoxyguanosine and 8-oxo-2′-deoxyadenosine in DNA using online column-switching liquid chromatography/tandem mass spectrometry, Rapid Commun. Mass Spectrom. 23 (1) (2009) 151–160.

[26] U.S. EPA, SW846 method 8290A-1: polychlorinated dibenzodioxins (PCDDs) and polychlorinated dibenzofurans (PCDFs) by high resolution gas chromatograph/high resolution mass spectrometry (HRGC/HRMS), 2007, Available: http://www3.epa.gov/epawaste/hazard/testmethods/sw846/pdfs/8290a.pdf.

[27] A. Zlatkis, C.F. Poole, Electron Capture: Theory and Practice in Chromatography, Elsevier, 1981.

[28] National Institute of Standards and Testing, NIST webbook: Benzene, 2023. https://webbook.nist.gov/cgi/cbook.cgi?ID=C71432&Mask=10#Solubility. (Accessed 17 July 2024).

[29] B. Sportisse, Fundamentals in Air Pollution: From Processes to Modelling, Springer Science & Business Media, 2009.

[30] F. Slemr, W. Seiler, C. Eberling, P. Roggendorf, The determination of total gaseous mercury in air at background levels, Anal. Chim. Acta 110 (1) (1979) 35–47.

[31] O. Lindqvist, et al., Mercury in the Swedish environment—recent research on causes, consequences and corrective methods, Water Air Soil Pollut. 55 (1–2) (1991). xi–261.

[32] S.X. Wang, et al., Mercury emission and speciation of coal-fired power plants in China, Atmos. Chem. Phys. 10 (3) (2010) 1183–1192, https://doi.org/10.5194/acp-10-1183-2010.

[33] K. Beijer, A. Jernelöv, Microbial methylation of lead, in: Biological Effects of Organolead Compounds, CRC Press, 2020, pp. 13–20.

[34] A. Jernelöv, A.-L. Martin, Ecological implications of metal metabolism by microorganisms, Ann. Rev. Microbiol. 29 (1) (1975) 61–77.

[35] W. Dong, L. Liang, S. Brooks, G. Southworth, B. Gu, Roles of dissolved organic matter in the speciation of mercury and methylmercury in a contaminated ecosystem in Oak Ridge, Tennessee, Environ. Chem. 7 (1) (2010) 94–102.

[36] B. Aberg, L. Ekman, R. Falk, U. Greitz, G. Persson, J.O. Snihs, Metabolism of methyl mercury (203Hg) compounds in man: excretion and distribution, Arch. Environ. Health 19 (4) (1969) 478–484.

[37] D.A. Abramowicz, Aerobic and anaerobic PCB biodegradation in the environment, Environ. Health Perspect. 103 (Suppl. 5) (1995) 97.

[38] S. Jensen, A. Johnels, M. Olsson, G. Otterlind, DDT and PCB in marine animals from Swedish waters, Nature 224 (5216) (1969) 247–250.

[39] S.E. Hale, H.P.H. Arp, I. Schliebner, M. Neumann, Persistent, mobile and toxic (PMT) and very persistent and very mobile (vPvM) substances pose an equivalent level of concern to persistent, bioaccumulative and toxic (PBT) and very persistent and very bioaccumulative (vPvB) substances under REACH, Environ. Sci. Eur. 32 (1) (2020) 1–15.

[40] Environmental Health Analysis Center, PBT profiler methodology, U.S. Environmental Protection Agency, 2006. http://www.pbtprofiler.net/Methodology.asp. (Accessed 1 January 2012).

[41] D.A. Vallero, Waste and biogeochemical cycling, in: Waste: A Handbook for Managers, Elsevier, Amsterdam, NV, 2019, pp. 91–125.

[42] D.A. Vallero, Engineering aspects of climate change, in: Climate Change, Elsevier, 2021, pp. 771–797.

[43] D. Vallero, Environmental Biotechnology: A Biosystems Approach, Elsevier Science, 2015.

[44] U.S. Environmental Protection Agency. PBT Profiler. Available: http://www.pbtprofiler.net/Results.asp, 2012.

[45] R. van Zelm, et al., European characterization factors for human health damage of PM 10 and ozone in life cycle impact assessment, Atmos. Environ. 42 (3) (2008) 441–453.

[46] M. Schultz, O. Lyapina, S. Schröder, O. Stein, D. Mallmann, Linking atmospheric composition data across data types and national boundaries, in: EGU General Assembly Conference Abstracts, vol. 18, 2016, p. 8558.

[47] J. Raymond, Atmospheric dispersion calculator: Air pollution control stacks equation formulas, 2015. https://www.ajdesigner.com/phpdispersion/point_space_equation.php. (Accessed 17 July 2024).

[48] National Center for Atmospheric Research, Biogeochemical cycles, 2015. http://scied.ucar.edu/longcontent/biogeochemical-cycles. (Accessed 4 March 2015).

[49] J. Munthe, K. Kindbom, O. Kruger, G. Petersen, J. Pacyna, Å. Iverfeldt, Examining source-receptor relationships for mercury in Scandinavia modelled and empirical evidence, Water Air Soil Pollut.: Focus 1 (2001) 299–310, https://doi.org/10.1023/A:1017580827549.

[50] R.A. Brown, Fluid Mechanics of the Atmosphere, Academic Press, 1991.

[51] X. Zhang, J.A. Arnot, F. Wania, Model for screening-level assessment of near-field human exposure to neutral organic chemicals released indoors, Environ. Sci. Technol. 48 (20) (2014) 12312–12319.

[52] A. Quintin, L. Fraiser, Comparison of international risk-based screening levels, in: Proceedings of the Annual International Conference on Soils, Sediments, Water and Energy, 15 (1), 2010, p. 24.

[53] Environment and Climate Change Canada, Screening assessment report: ethylbenzene, 2016. http://www.ec.gc.ca/ese-ees/default.asp?lang=En&n=FE722725-1. (Accessed 17 July 2024. Last update on April 29, 2016.

[54] P. Egeghy, N. Tulve, O. Adetona, L. Naeher, Using pesticide screening questions to identify the more highly exposed participants in a larger cohort, Epidemiology 19 (6) (2008) S200.

[55] Chemical Computing Group, Molecular operating environment: chemoinformatics and structure based tools for high throughput screening, Montreal Canada, 2013.

[56] J.F. Wambaugh, et al., High-throughput models for exposure-based chemical prioritization in the ExpoCast project, Environ. Sci. Technol. 47 (15) (2013) 8479–8488.

[57] J. Mitchell, et al., A decision analytic approach to exposure-based chemical prioritization, PLoS One 8 (8) (2013) e70911, https://doi.org/10.1371/journal.pone.0070911.

[58] J. Mitchell, et al., Comparison of modeling approaches to prioritize chemicals based on estimates of exposure and exposure potential, Sci. Total Environ. 458 (2013) 555–567.

[59] D.A. Vallero, Estimating and predicting exposure to products from emerging technologies, in: Synthetic Biology 2020: Frontiers in Risk Analysis and Governance, Springer, 2020, pp. 107–142.

[60] B. Trump, C. Cummings, J. Kuzma, I. Linkov, A decision analytic model to guide early-stage government regulatory action: applications for synthetic biology, Regul. Gov. 12 (2017) 88.

[61] C.L. Cummings, J. Kuzma, Societal risk evaluation scheme (SRES): scenario-based multi-criteria evaluation of synthetic biology applications, PLoS One 12 (1) (2017) e0168564.

[62] H. König, D. Frank, R. Heil, C. Coenen, Synthetic biology's multiple dimensions of benefits and risks: implications for governance and policies, in: Synthetic Biology, Springer, 2016, pp. 217–232.

[63] B.D. Trump, A comparative analysis of variations in synthetic biology regulation, Arizona State University, 2016.

[64] U.S. EPA, Guidelines for human exposure assessment: peer review draft, 2017, Available: https://www.epa.gov/sites/production/files/2016-02/documents/guidelines_for_human_exposure_assessment_peer_review_draftv2.pdf.

[65] A. Guy, C. Gauthier, G. Griffin, Adopting alternative methods for regulatory testing in Canada, in: Proceedings of the 6th World Congress on Alternatives & Animal Use in the Life Sciences. AATEX, vol. 14, 2008, pp. 322–327.

[66] R.S. Judson, et al., In vitro screening of environmental chemicals for targeted testing prioritization: the ToxCast project, Environ. Health Perspect. 118 (4) (2010) 485.

[67] D.A. Vallero, Air pollution, in: Kirk-Othmer Encyclopedia of Chemical Technology, John Wiley & Sons, Inc., 2000.

[68] J.P. Rishel, E.G. Chapman, F.C. Rutz, K.J. Allwine, Using DUSTRAN to simulate fog-oil dispersion and its impacts on local insect populations at Ft. Hood, Pacific Northwest National Lab (PNNL), Richland, WA, United States, 2006.

[69] N. Moussiopoulos, Ambient air quality, pollutant dispersion and transport models, European Environment Agency, 1997.

[70] M. Markiewicz, Modelling of the air pollution dispersion, in: Models and Techniques for Health and Environmental Hazards Assessment and Management, IEA, Otwock-Swierk, Poland, 2006, pp. 303–347.

[71] F.A. Gifford Jr., Use of routine meteorological observations for estimating atmospheric dispersion, Nucl. Saf. 2 (1961) 47–51.

[72] D.A. Vallero, S.S. Isukapalli, V.G. Zartarian, T.R. Mccurdy, T.E. Mckone, P. Georgopoulos, C.C. Dary, Modeling and predicting pesticide exposures, in: R. Krieger (Ed.), Hayes Handbook of Pesticide Toxicology, Elsevier Science, New York, NY, 2010, pp. 995–1020 (Chapter 44).

[73] U.S. Environmental Protection Agency, Risk assessment guidance for Superfund, in: Human Health Evaluation Manual (Part A): Interim Final, No. EPA/540/1-89/002, vol. I, USEPA, Washington, DC, 2020, EPA/540/1-89/002. https://www.epa.gov/sites/production/files/2015-09/documents/rags_a.pdf. (Accessed 17 July 2024).

[74] D.A. Vallero, Translating Diverse Environmental Data Into Reliable Information: How to Coordinate Evidence From Different Sources, Academic Press, 2017.

[75] D.A. Olson, G.A. Norris, Chemical characterization of ambient particulate matter near the World Trade Center: source apportionment using organic and inorganic source markers, Atmos. Environ. 42 (31) (2008) 7310–7315.

[76] L.-M. Yiin, et al., Comparisons of the dust/smoke particulate that settled inside the surrounding buildings and outside on the streets of southern New York City after the collapse of the World Trade Center, September 11, 2001, J. Air Waste Manage. Assoc. 54 (5) (2004) 515–528.

[77] J. Williams, Weather forecasters help keep wildfire fighters safe, Weatherwise 71 (4) (2018) 20–27, https://doi.org/10.1080/00431672.2018.1470883.

[78] L.G. Luther, The National Environmental Policy Act: Background and Implementation, Congressional Research Service, Library of Congress, 2005.

[79] New York Department of Environmental Conservation, State Implementation Plans, Albany, New York, 2009.

[80] U.S. Environmental Protection Agency, Final Rule: Implementation of the 2015 National Ambient Air Quality Standards for Ozone: Nonattainment Area Classifications Approach, 2018.

[81] E.A.C. Hubal, et al., Exposure science and the US EPA National Center for Computational Toxicology, J. Expo. Sci. Environ. Epidemiol. 20 (3) (2010) 231–236.

[82] D.A. Vallero, T.M. Letcher, Unraveling Environmental Disasters, Newnes, 2012.

[83] M.D. Cohen, et al., Modeling the global atmospheric transport and deposition of mercury to the Great Lakes, Elementa: Sci. Anthrop. 4 (2016) 000118, https://doi.org/10.12952/journal.elementa.000118.

[84] R.S. Collett, K. Oduyemi, Air quality modelling: a technical review of mathematical approaches, Meteorol. Appl. 4 (3) (1997) 235–246.

[85] M. Albareda-Sambola, E. Fernández, G. Laporte, Heuristic and lower bound for a stochastic location-routing problem, Eur. J. Oper. Res. 179 (3) (2007) 940–955.

[86] S.A. Csiszar, A.S. Ernstoff, P. Fantke, O. Jolliet, Stochastic modeling of near-field exposure to parabens in personal care products, J. Expo. Sci. Environ. Epidemiol. (2016).

[87] U. Ritzinger, J. Puchinger, R.F. Hartl, A survey on dynamic and stochastic vehicle routing problems, Int. J. Prod. Res. 54 (1) (2016) 215–231.

[88] R. Vandoorne, P.J. Gräbe, Stochastic modelling for the maintenance of life cycle cost of rails using Monte Carlo simulation, Proc. Inst. Mech. Eng. F J. Rail Rapid Transit 232 (4) (2018) 1240–1251.

[89] D. Vallero, Environmental Contaminants: Assessment and Control, Academic Press, 2010.

[90] D.A. Vallero, Air Pollution Calculations: Quantifying Pollutant Formation, Transport, Transformation, Fate and Risks, Elsevier, 2019.

[91] D.A. Vallero, Fundamentals of Air Pollution, fifth ed., Elsevier Academic Press, Waltham, MA, 2014. p. 999 pages cm.

[92] J.J. Peirce, R. Weiner, R. Matthews, P.A. Vesilind, Environmental Engineering, Butterworth-Heinemann, 2003.

[93] B. Fisher, The effect of tall stacks on the long range transport of air pollutants, J. Air Pollut. Control Assoc. 36 (4) (1986) 399–401.

[94] F. Pasquill, P. Michael, Atmospheric diffusion, Phys. Today 30 (1977) 55.

[95] A.C. Stern, H.C. Wohlers, R.W. Boubel, W.P. Lowery, Fundamentals of Air Pollution, Academic Press, New York and London, 1973.

[96] L.J. Battan, Fundamentals of Meteorology, Prentice-Hall, Englewood Cliffs, 1984.

[97] D.R. Bullard, et al., APTI course SI:409: basic air pollution meteorology: student guidebook, U.S. Environmental Protection Agency, Air Pollution Training Institute, Research Triangle Park, NC, 1982 (in English).

[98] National Aeronautics and Space Administration, Heat island, 2021. http://weather.msfc.nasa.gov/urban/urban_heat_island.html. (Accessed 30 December 2021).

[99] H.R. Byers, General Meteorology, fourth ed., McGraw-Hill College, New York, NY, 1974.

[100] R.W. Boubel, D.L. Fox, B. Tuner, A.C. Stern, Fundamentals of Air Pollution, third ed., Academic Press, Amsterdam; Boston, 1995.

[101] R.W. Boubel, D.L. Fox, D.B. Turner, A.C. Stern, T. Tromp, Fundamentals of air pollution, Pure Appl. Geophys. 145 (2) (1995) 393–394.

[102] U.S. Environmental Protection Agency, Heat balance in the atmosphere, in: SI 409: Basic Air Pollution Meteorology Student Guidebook, A. P. T. Institute, U.S. EPA, Research Triangle Park, North Carolina, 2005.

[103] W.D. Sellers, Physical Climatology, University of Chicago Press, 1965.

[104] R.A. Fine, F.J. Millero, Compressibility of water as a function of temperature and pressure, J. Chem. Phys. 59 (10) (1973) 5529–5536.

[105] R. Nave, Bulk modulus properties, Georgia State University, 2022. http://hyperphysics.phy-astr.gsu.edu/hbase/permot3.html. (Accessed 1 January 2022).

[106] The Engineering Toolbox, Triple points for some common substances, 2021. https://www.engineeringtoolbox.com/triple-point-d_1926.html. (Accessed 17 July 2024).

[107] U.S. Environmental Protection Agency, Basic air pollution meteorology student guidebook (SI 409), U.S. EPA, Research Triangle Park, North Carolina, 2005.

[108] D.L. Hartmann, Introduction to the climate system, in: D.L. Hartmann (Ed.), Global Physical Climatology, second ed., Elsevier, Boston, 2016, pp. 1–23 (Chapter 1).

[109] J. Monteith, M. Unsworth, Principles of Environmental Physics: Plants, Animals, and the Atmosphere, Academic Press, 2013.

[110] M.W. Rotach, P. Calanca, Microclimate, in: J.R. Holton (Ed.), Encyclopedia of Atmospheric Sciences, Academic Press, Oxford, 2003, pp. 1301–1307.

[111] U.S. Environmental Protection Agency, Understanding global warming potentials, 2018. https://www.epa.gov/ghgemissions/understanding-global-warming-potentials. (Accessed 17 July 2024. Last update on March 27, 2024.

[112] R.K. Pachauri, L. Meyer, Intergovernmental Panel on Climate Change, Contribution of Working Groups I, II and III to the Fifth Assessment Report of the Intergovernmental Panel on Climate Change, in: Climate Change 2014: Synthesis Report, 151, IPCC, Geneva, Switzerland, 2015, pp. 1–151. https://www.ipcc.ch/site/assets/uploads/2018/05/SYR_AR5_FINAL_full_wcover.pdf. (Accessed 17 July 2024).

Chapter 7

Water pollution physics

1 Introduction

The nature and properties of matter and energy are the province of physics. As such, chemistry and biology are subdisciplines of physics. Myriad physical mechanisms and processes contribute to water pollution. Physical principles are integral to most of the chapters in this book, but they are worthy of specific discussion here.

The most apparent water pollutants are substances in waterbodies above and below the earth surface, addressed in Chapters 10 and 11, but other environmental media besides surface and ground water are involved in the formation, transformation, transport, fate, and risks of water pollutants. Substances in the soil and sediment, for example, are indeed water pollutants because they are dissolved or suspended in soil and sediment pores and on particles that exchange waterborne pollutants with the pore water [1–3]. Pollutants and their precursor compounds reside in water droplets, on surfaces of aerosols, and in exchanges with water vapor in the air. Within biota, pollutants are dissolved or suspended in endogenous fluids, including blood, plant fluids, endoplasm, and water. Metal and organic contaminants are stored and carried in plant tissue [4–10] or dissolved, suspended, and transported in water of a human's blood serum [11,12].

Water pollutants are commonly thought of as chemical compounds or biological agents, such as pathogenic bacteria. Indeed, chemical substances represent most water pollutants, and pathogens are noticeable public health threats [13–16]. However, water pollutants can also be physical agents, such as electromagnetic radiation. As evidence, ultraviolet light exposure can threaten the health and survival of aquatic species, such as amphibians [17–19]. Heated water released into an ecosystem can alter biodiversity and the abundance of heat-sensitive organisms.

The physical causes and effects of water pollution may be direct or indirect. A direct pollutant is inherently harmful to the health and well-being of humans and other organisms. Conversely an indirect pollutant's harm occurs a secondary effect that follow the initial exposure [20,21]. For example, consider an estuary that receives of organic matter that has run off farmland. The organic matter is not inherently hazardous and beneficial in soil. However, after bacteria in the estuary break it down, the estuary's dissolved oxygen concentrations decrease due to the microbial respiration. The diminished O_2 concentrations in turn lead to fish kills or loss of aquatic species progressively according to their need for O_2. Thus, thermodynamics and, specifically, bioenergetics explains certain types of direct and indirect water pollution. In addition, fluid dynamics is involved in every instance of water pollution.

2 Physical phases

As indicated in Fig. 7.1, chemical compounds in the environment can exist in three phases, depending on the temperature and pressure. The vapor-liquid transition temperature is the compound's boiling point, and the pressure is the substance's vapor pressure. The line between the vapor and solid phases is the sublimation point (i.e., a phase change from solid directly to gas, or gas directly to solid, without first becoming a liquid). The liquid-solid boundary line temperature represents the freezing point and melting point. T_{tp} and P_{tp} define the triple point where the three phases of pure substance can exist in equilibrium. $T_{critical}$ and $P_{critical}$ are the pressure and temperature of a compound at its critical point, i.e., the point at which a gas cannot be liquefied by an increase of pressure [15].

3 Laws of motion

The first law of motion states that every object in a state of uniform motion tends to remain in that state of motion unless an external force is applied to it. This is a statement of inertia, which must be understood when considering forces. In water pollution science and engineering, the most common external force that changes the state of uniform motion is friction.

Fundamentals of Water Pollution. https://doi.org/10.1016/B978-0-443-28987-3.00005-9

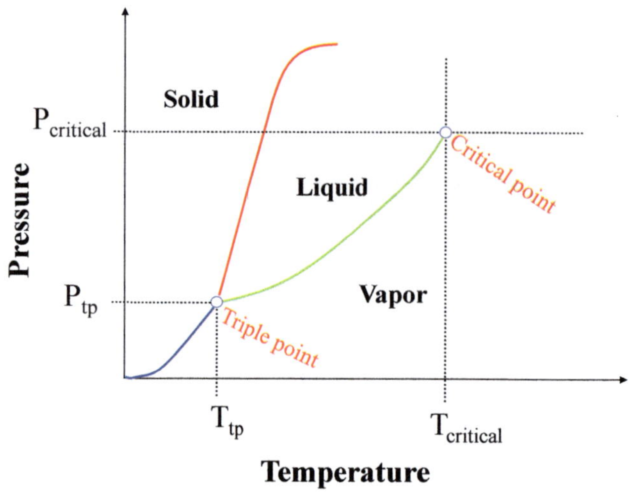

FIG. 7.1 Phase diagram. The phase regions shown in the diagram differ in shapes and slopes of the polygon sides and points for different compounds.

Mechanically speaking, friction must be addressed with lubricants and smooth surfaces to keep a body in motion. Conversely, it must be deployed to change directions of a body.

The second law of motion describes the relationship between an object's mass m, its acceleration a, and the applied force (F):

$$F = ma \tag{7.1}$$

Acceleration is the rate of change of velocity per unit of time. It and force are vectors; wherein the direction of the force vector is the same as the direction of the acceleration vector. With the second law, we can calculate unknowns from knowns. That is, if we know the mass of the structures and the applied force generated by wind we can calculate the acceleration of dislodged structures during a storm event.

The third law of motion says that for every action there is an equal and opposite reaction. Like the first law, this tells us that we can expect things to happen in response to what we do. If we apply a force, there will be an equal force in the opposite direction.

Analogously, the laws of chemistry remind us that what we put into a reaction always leads to a specific product. Thus, science tells us that so long as we understand the variables and parameters of any system, the outcome is predictable. This is obviously not completely possible for all water pollution episodes, but the more we know, the better our predictions and the more informed remedies will be. There are numerous ways to identify, characterize and analyze conditions that have led to pollution episodes to determine why they led to health and environmental impacts. The key to any approach is that it accurately describes the events in an objective and comprehensive way to evaluate events, outcomes, and consequences.

3.1 Fluids

Pollutants move in and by fluids, which are either liquid or gaseous. The most common environmental fluids are water and air, but also blood, lipids, plant extracts, and other fluids are at work in the environment. For example, a pollutant's fluid properties will determine how soluble it is in water and blood. A substance that is insoluble in water is often quite soluble in fats, so an organism's body burden of a lipophilic substance will increase with constant exposure. The fluid properties of an agent, whether chemical or biological, determines it likely fate.

A liquid is matter that is composed of molecules that move freely among themselves without separating from each other. A gas is matter composed of molecules that move freely and are infinitely able to occupy the space in which they are contained at a constant temperature. Engineers define a fluid as a substance that will deform continuously upon the application of a shear stress; i.e., a stress in which the material on one side of a surface pushes on the material on the other side of the surface with a force parallel to the surface.

Fluids can be classified according to observable physical characteristics of flow fields. Laminar flow is in layers, whereas turbulent flow has random movements of fluid particles in all directions. In incompressible flow, the variations in density are assumed to be constant, while compressible flow has density variations, which must be included in flow calculations. Viscous flows must factor in viscosity while inviscid flows assume viscosity is zero [22].

The time rate of change of a fluid particle's position in space is the fluid velocity (V). This is a vector field quantity. Speed (V) is the magnitude of the vector velocity V at some given point in the fluid, and average speed ($\overline{V}$) is the mean fluid speed through a control volume's surface. Velocity is a vector quantity of magnitude and direction, whereas speed is a scalar quantity of magnitude only. The standard units of velocity and speed are meters per second ($m\ s^{-1}$).

Velocity is involved in mixing rates after an effluent is discharged to a stream, the time it takes for an aquifer to become contaminated, and the ability of liners to slow the movement of leachate from a landfill toward the ground water, to name a few of the myriad calculations based on velocity. At the macroscopic scale, no distinction between velocity and speed is usually made. As discussed in detail in Chapter 10, surface water flow is known as stream discharge, Q, with units of volume per time. Discharge is derived by measuring a stream's velocity at numerous points across the stream.

Pressure (p) is another environmentally important fluid property, which is a force per unit area:

$$p = \frac{F}{A} \tag{7.2}$$

Pressure is a stress that is exerted uniformly in all directions. The standard unit of p is the Pascal (P), which is equal to $1\ N\ m^{-2}$. The preferred pressure unit in this and many other environmental texts is the kilopascal (kP), because P is very small in most environmental applications.

Potential and kinetic energy discussions consider the fluid acceleration due to gravity. In many ways, it seems that acceleration was a major reason for Isaac Newton's need to develop the calculus.[a] Known as the mathematics of change, calculus is the mathematical means of describing acceleration so it addressed Newton's need to express mathematically his new law of motion. The time rate of change in the velocity of a fluid particle is known as acceleration. In terms of calculus, it is a second derivative, i.e., the derivative of the velocity function. And a derivative of a function is itself a function, giving its rate of change. This explains why the second derivative must be a function showing the rate of change of the rate of change, which is readily apparent from the units of acceleration: length per time per time ($m\ s^{-2}$).

The relationship between mass and volume is important in both environmental physics and chemistry and is a fundamental property of fluids. The fluid's density (ρ) is defined as its mass per unit volume. Its metric units are $kg\ m^{-3}$. The density of an ideal gas is found using the specific gas constant and applying the ideal gas law:

$$\rho = p(RT)^{-1} \tag{7.3}$$

Where,

p = gas pressure,
R = specific gas constant,
T = absolute temperature.

The specific gas constant must be known to calculate gas density. For example, the R for air is $287\ J\ kg^{-1}\ K^{-1}$. The specific gas constant for methane (R_{CH4}) is $518\ J\ kg^{-1}\ K^{-1}$.

Density is an important fluid property for root cause analyses and risk assessments. For example, a responder must know the density of substances in an emergency. If a substance is burning, whether it is of greater or lesser density than water will be one of the factors on how to extinguish the fire. If the substance is less dense than water, the water will likely settle below the layer of water, making water a poor choice for fighting the fire. So, any flammable substance with a density less than water, like benzene or acetone, needs fire-extinguishing substances other than water. For substances heavier than water, like carbon disulfide (CS_2), water may be preferred [23]. Density is also important when designing facilities or protecting coastal water systems, since saltwater and freshwater have different densities [24,25]. As discussed in Chapter 11, denser saltwater submerges under less dense freshwater system [14].

The reciprocal of a substance's density is known as its specific volume (υ), i.e., the volume occupied by a unit mass of a fluid. The units of υ are reciprocal density units ($m^3\ kg^{-1}$). Stated mathematically, this is:

$$\upsilon = \rho^{-1} \tag{7.4}$$

a. It appears that Newton coinvented the calculus with Wilhelm Leibniz in the 17th Century. Both are credited with devising the symbolism and the system of rules for computing derivatives and integrals, but their notation and emphases differed. A debate rages on who did what first, but both giants had good reason to revise the language of science; i.e., mathematics, to explain motion.

The weight of a fluid per its volume is known as specific weight (γ). Scientists and engineers occasionally use the term interchangeably with density. Geologists and soil scientists more frequently refer to a substance's specific weight. A substance's γ depends upon the fluid itself and the local gravitational force:

$$\gamma = gp \tag{7.5}$$

Specific weight units are the same as those for density; e.g., kg m^{-3}.

The fractional change in a fluid's volume per unit change in pressure at constant temperature is the fluid's coefficient of compressibility. Any fluid can be compressed in response to the application of pressure (p). For example, water's compressibility at 1 atm is 4.9×10^{-5} atm^{-1}. This compares to the lesser compressibility of mercury (3.9×10^{-6} atm^{-1}) and the greater compressibility of hydrogen (1.6×10^{-3} atm^{-1}). A fluid's bulk modulus, E, is a function of stress and strain on the fluid, and is a description of its compressibility, and is defined according to the fluid volume (V):

$$E = \frac{\text{stress}}{\text{strain}} = -\frac{dp}{dV/V_1} \tag{7.6}$$

E is expressed in units of pressure (e.g., kP). Water's $E = 2.2 \times 10^6$ kP at 20 °C.

Surface tension effects occur at liquid surfaces (interfaces of liquid-liquid, liquid-gas, liquid-solid). Surface tension, σ, is the force in the liquid surface normal to a line of unit length drawn in the surface. Surface tension decreases with temperature and depends on the contact fluid. Surface tension is involved in capillary rise and drop. Water has a very high σ value, i.e., approximately 0.07 N m^{-2} at 20 °C. Of the environmental fluids, only elemental mercury has a higher σ (0.519 N m^{-2} at 20 °C) than water. The high surface tension creates a type of skin on a free surface, which is how an object that is denser than water (e.g., a steel needle) can "float" on a still water surface. It is the reason insects can sit comfortably on water surfaces. Surface tension is somewhat dependent upon the gas that is contacting the free surface. If not indicated, it is usually safe to assume that the gas is the air in the troposphere.

Capillarity is particularly important in ground water flow and the movement of contaminants between the unsaturated and saturation zones. The zone immediately above the water table is the capillary fringe. The spaces between unconsolidated material (e.g., gravel, sand, or clay) are interconnected, acting as conduits of moving water. Depending on the grain size and density of packing, the conduits will vary in diameter, ranging from large pore, i.e., macropores, to medium pore sizes, i.e., mesopores, to extremely small pores i.e., micropores. Capillarity is strongest in the micropores and fibrous root strands (see Chapters 5 and 11). Fluid pressures above the water table are negative with respect to atmospheric pressure, creating tension. A fluid rises in pores spaces for two reasons, its adhesion to a surface and the cohesion of fluid molecules to one another. Higher relative surface tension causes a fluid rise in a tube inversely to the diameter of the tube. Thus, capillarity increases with decreasing diameter of a tube. Fine grained clay, for example, has a much more elongate capillary fringe than that of sand. The rise is limited by the weight of the fluid in the tube. The rise ($h_{capillary}$) of the fluid in a capillary is expressed as:

$$h_{capillary} = \frac{2\sigma \cos \lambda}{\rho_w g R} \tag{7.7}$$

Where,

σ = fluid surface tension (g s^{-2})
λ = angle of meniscus (concavity of fluid) in capillary (degrees)
ρ_w = fluid density (g cm^{-3})
g = gravitational acceleration (cm s^{-1})
R = radius of capillary (cm)

Another property of polluted fluids is the mole fraction. If a composition of a fluid made up of two or more substances (A, B, C, ...), the mole fraction ($1_A, x_B, x_C, \ldots$) is number of moles of each substance divided by the total number of moles for the whole fluid:

$$x_A = \frac{n_A}{n_A + n_B + n_C + \ldots.} \tag{7.8}$$

The mole fraction value is always between 0 and 1. The mole fraction may be converted to mole percent as:

$$x_{A\%} = x_A \times 100 \tag{7.9}$$

For gases, the mole fraction is the same as the volumetric fraction of each gas in a mixture of more than one gas.

The amount of resistance to flow when it is acted on by an external force, especially a pressure differential or gravity, is the fluid's viscosity. This a crucial fluid property is used in numerous conventional and green engineering applications, but particularly in plume characterization, sludge management, wastewater and drinking water treatment and distribution systems.

3.2 Bernoulli's principle

Three forms of energy are at play at any point in the flow:

1. Kinetic energy from the motion of the fluid at that point.
2. Potential energy due to the positional elevation (h) above the point.
3. Pressure energy from the absolute pressure of the fluid at that point.

Adhering to the first law of thermodynamics, the total energy must remain constant, but it can be transformed within these three forms. Between two states, the net work on a fluid is equal to the gain in kinetic energy plus the gain in potential energy. Thus, from the perspective of potential energy, that is, head, flow work can be described as $\frac{p}{\rho g}$ and velocity head (i.e., kinetic energy) as $\frac{V^2}{2g}$, where g is the gravitational acceleration. This leads to the relationship of flow at two points, $x = 1$ and $x = 2$, at z height above each point. Thus, the Bernoulli equation states:

$$\left(\frac{p}{\rho g} + \frac{V^2}{2g} + z\right)_1 = \left(\frac{p}{\rho g} + \frac{V^2}{2g} + z\right)_2 \tag{7.10}$$

Bernoulli example

A wet scrubber is used to remove SO_2 from a power plant's stack. At 1 atm, the scrubber has a vertical nozzle that produces an upward 10 m s^{-1} jet of water that is 20 mm in diameter. If the jet retains a circular cross section, will its diameter be sufficiently large to traverse a 40 mm-diameter opening 3 m above the nozzle, assuming the density of water is 1000 kg m^{-3}?

Solution:

First apply the Bernoulli equation: $\left(\frac{p}{\rho g} + \frac{V^2}{2g} + z\right)_1 = \left(\frac{p}{\rho g} + \frac{V^2}{2g} + z\right)_2.$
Where, $z_1 = 0$, $z_2 = 4.5$ m
p_1 and p_2 are both 1 atm, so no change in pressure.
Apply the energy equation:

$$\frac{V_2^2}{2g} = \frac{V_1^2}{2g} + z_1 - z_2 \tag{7.11}$$

$$g = 9.8 \text{ m s}^{-1}$$

Therefore, $V_2\sqrt{V_1^2 - 2gh} = \sqrt{10^2 - 2(9.8)(4.5)} = 3.44 \text{ m s}^{-1}$
The diameter (D) is found by using the continuity equation:

$$r = \frac{D}{2} = \frac{0.02}{2} = 0.01$$

$$A_2 = \frac{A_1 \times V_1}{V_2} = \frac{\pi(0.01^2)(10)}{3.44} = 9.13 \times 10^{-4} \text{ m}^2$$

$$A_2 = \frac{\pi \times D_2^2}{4} \tag{7.11}$$

$$D_2 = \sqrt{\frac{4 \times 9.13 \times 10^{-4}}{\pi}} = 0.034 \text{ m}$$

Thus, the fountain has expanded from 0.02 to 0.034 m. Assuming the fountain keeps its circular cross section, the 40 mm (0.04 m) opening should be of sufficient diameter.

In the environment, flow will be reduced by friction and obstructions between the two points, expressed as the loss of elevation (head loss = Δh), that is, h_i:

$$\left(\frac{p}{\rho g} + \frac{V^2}{2g} + z\right)_1 = \left(\frac{p}{\rho g} + \frac{V^2}{2g} + z\right)_2 + h_i \tag{7.13}$$

Frictional losses increase with the length of the conveyance, for example, pipe or conduit. Frictional losses also increase due to the roughness of walls; the velocity of flow; the turbulence of flow; and changes in shape and sectioning, for example,

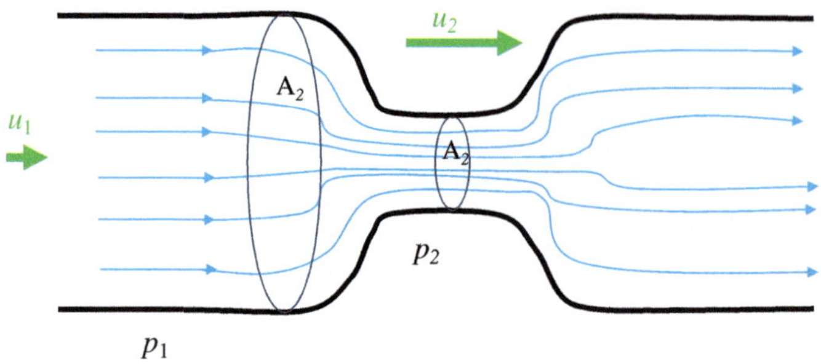

FIG. 7.2 The Bernoulli's principle and the effect of relationship between pressure (p), area (A), and velocity (u). As the cross-sectional area of flow decreases, the velocity increases, and the pressure decreases. At the center of the diagram, the increase in pressure results in decreasing ideal-gas velocity. *Note*: $p_2 < p_1$, $u_2 > u_1$, and $A_2 < A_1$.

valves, fittings, and bends, of the conveyance. In addition, frictional losses decrease with the diameter of the conveyance. For convenience, the head losses are included as fraction of the total energy at $x = 2$, even though the losses occur incrementally between points 1 and 2.

Engineers often restate the Bernoulli's equation when considering the flow from a pump, which increases the elevation coordinate (z), to a downflow point as:

$$\left(\frac{p}{\rho g} + \frac{V^2}{2g} + z\right)_1 + h_p = \left(\frac{p}{\rho g} + \frac{V^2}{2g} + z\right)_2 + h_i \tag{7.14}$$

where h_p is the head increase (gain) provided by the pump.

Adhering to the conservation of energy principle, a flowing fluid will maintain the energy, but velocity and pressure can change. In fact, velocity and pressure will compensate for each other to adhere to the conservation principle, so the Bernoulli's equation can be states as

$$p_1 + ½ p V_1^2 + pgh_1 = p_2 + ½ p V_2^2 + pgh_2 \tag{7.15}$$

This expression describes the "Bernoulli's effect" that occurs when increased fluid speed leads to decreased internal pressure (Fig. 7.2). Note that it includes the loss of head (Δh) from friction and obstructions within the flow between $x = 1$ (i.e., h_1) and $x = 2$ (i.e., h_2).

4 Fluid dynamics in pollutant transport[b]

Fluid dynamics is concerned with how liquids and gases flow. As mentioned, a fluid is a substance that will deform continually under shear stress, i.e., applied parallel to the exerted pressure. For a liquid at rest, the medium is a continuous distribution of matter. However, when considering p for a gas, it is an average of the forces against the vessel walls, i.e., gas pressure. Fluid pressure is a measure of energy per unit volume per the Bernoulli principle, which can be expressed as:

$$p + ½ p V^2 + pgh = \text{constant} \tag{7.16}$$

Where, p is pressure, V is the fluid velocity, h is the elevation or head, and g is the gravitational acceleration.

Fluids can be either Newtonian or non-Newtonian. A Newtonian fluid has a constant viscosity at all shear rates at constant temperature and pressure. Water, most organic solvents, and most environmentally important fluids are Newtonian fluids. Many contaminants can also be carried by non-Newtonian fluids or themselves be non-Newtonian fluids, i.e., those with viscosities not constant at all shear rates. For example, water bodies and surface and subsurface layers of soil and rocks can be contaminated with nonaqueous phase liquids (NAPLs). Indeed, NAPLs are the most common pollutants in contaminated soils and ground water [26], such as oil and drilling fluids. Another consideration is whether the fluid is compressible. So, when using software programs, it is important to define these and other factors unique to the fluid or to ensure that these and other factors appear as defaults that accurately represent the fluid being investigated.

Fundamental fluid dynamics theory [27] states that the motion of fluids is governed by the conservation of mass equation and conservation of momentum. For incompressible flow, the continuity equation is expressed as:

$$\nabla \bullet \boldsymbol{u} = 0 \tag{7.17}$$

b. This section draws data from chapter 4 of Ref. [15] and chapter 10 of Ref. [23].

Where, ∇ is a gradient operator, which is the vector differential operator "del," and $\boldsymbol{u}$ is the velocity vector field.[c] ∇f *is the* gradient operator of the scalar function $f(x_1, x_2, \ldots x_n)$. Although, as mentioned, all fluids are to some extent compressible, water is often assumed to be incompressible for calculations that assume environmental conditions.

In addition to assuming an incompressible fluid, the continuity also assumes a velocity of the fluid at the position $\mathbf{r}_0(x, y, z)$ at time t is $\boldsymbol{u}(x, y, z, t)$ and at the same time the velocity at a neighboring position $\mathbf{r}_0 + \mathbf{r}$ is $\boldsymbol{u} + \delta\boldsymbol{u}$. In matrix form, this is:

$$\begin{pmatrix} \delta u \\ \delta v \\ \delta w \end{pmatrix} = \begin{pmatrix} \frac{\partial u}{\partial x} & \frac{\partial u}{\partial y} & \frac{\partial u}{\partial z} \\ \frac{\partial v}{\partial x} & \frac{\partial v}{\partial y} & \frac{\partial v}{\partial z} \\ \frac{\partial w}{\partial x} & \frac{\partial w}{\partial y} & \frac{\partial w}{\partial z} \end{pmatrix} \begin{pmatrix} x \\ y \\ z \end{pmatrix} \tag{7.18}$$

The vector form of this matrix is:

$$\Delta \boldsymbol{u} = \mathbf{D} \bullet \mathbf{r} \tag{7.19}$$

Where, $\mathbf{D}$ is the deformation tensor, which defines how a nonuniform motion field is deforming the fluid element. This can be stated in three-dimensional coordinates:

$$\nabla f = \frac{\partial f}{\partial x}\mathbf{i} + \frac{\partial f}{\partial y}\mathbf{j} + \frac{\partial f}{\partial z}\mathbf{k} \tag{7.20}$$

Where, **i, j, k** are standard unit vectors.

For Newtonian fluids, this motion is expressed in the Navier-Stokes equation:

$$\rho\left[\frac{\partial \boldsymbol{u}}{\partial t} + (u \bullet \nabla)u\right] = -\nabla p + \eta \nabla^2 u + \mathrm{f} \tag{7.21}$$

Where, ρ is the fluid density, p is the fluid pressure, η is the dynamic viscosity of the fluid, ∇^2 is the Laplacian operator $[\nabla^2 u = \partial^2_{xx}u + \partial^2_{yy}u + \partial^2_{zz}u]$, and $\mathbf{f}$ denotes the body forces per unit volume. The solution to the Navier-Stokes equation is a flow field or a velocity field, i.e., the description of the velocity of the fluid at a specific point in space and time.

For cylindrical profiles, the gradient operator ∇ can be expressed in cylindrical coordinates:

$$\frac{\partial}{\partial r} + \frac{1}{r}\frac{\partial}{\partial \theta_\theta} + \frac{\partial}{\partial z} \tag{7.22}$$

Thus, the continuity equation can also be expressed cylindrically:

$$\frac{1}{r}\frac{\partial}{\partial r}(ru_r) + \frac{1}{r}\frac{\partial}{\partial \theta}u_\theta + \frac{1}{r}\frac{\partial}{\partial z}u_z = 0 \tag{7.23}$$

Where, ∇_r, ∇_θ and ∇_z are the components of the fluid velocity depicted in Fig. 7.3, i.e., radial (r), circumferential (θ) and axial (z) directions, respectively.[d]

These equations can be fitted to real-world conditions or idealized velocity profiles. Environmental and organism conditions include parabolic flow (laminar fully developed), plug flow (laminar undeveloped) and turbulent flow [27]. For example, in a lung, the upper tracheobronchial airways can be assumed to be turbulent, whereas in the pulmonary region, plug and parabolic profiles may be assumed [29].

Dynamics describes both the properties of the fluid and how it moves. Viscosity, compressibility, and turbulence affect continuum fluid mechanics. For example, the water flow is very slow in a most underground unconsolidated media, like sand and gravel, e.g., often less than a few meters per year. The flow in these aquifers is laminar and the

c. **Bold type** indicates a vector quantity.

d. For the derivation of the momentum equations mentioned here, see Ref. [28].

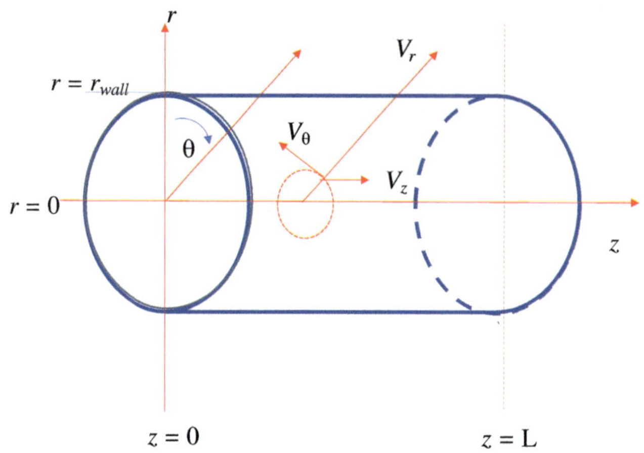

FIG. 7.3 Coordinate system for an ideal cylindrical airway, depicting velocity component at an arbitrary point. *(Data from K. Isaacs, J. Rosati, T. Martonen, L. Ruzer, N. Harley, Modeling deposition of inhaled particles, in: Aerosols Handbook, CRC Press, Boca Raton, FL, 2012, pp. 83–128.)*

water viscosity is low, about 10^{-3} kg m^{-1} s^{-1} at 15 °C. Conversely, the flow in an air mass is typically much faster, expressed in units of meters per second (m s^{-1}), turbulent, and the air viscosity is extremely low, i.e., 1.781×10^{-5} kg m^{-1} s^{-1} at 15 °C.

As velocity increases, a laminar flow field will reach a point when the flow becomes turbulent. The transition from laminar to turbulence is represented by the Reynolds number (Re or N_R), i.e., the ratio of inertial forces to viscous forces in a fluid:

$$\text{Re} = \frac{\text{Inertial Forces}}{\text{Viscous Forces}} \tag{7.24}$$

The inertial forces in the numerator are the tendency to avoid change and to continue to be in its existing state of rest or motion. The viscous forces in the denominator are the forces that resist deformation. During flow, the friction between fluid layers works against the free movement of the fluid. The friction is quantified according to the fluid's viscosity. Thus, the Reynolds number is the mathematical representation of opposition between the inertial forces to keep the fluid moving and the viscous forces to make motion cease. The inertial forces are proportional to the field's velocity, density, and a relevant length, e.g., the diameter of a pipe or any enclosed conduit. An increase in any of these factors will increase the momentum of the flow proportionally. The coefficient of viscosity or absolute viscosity (μ) represents the fluid's total viscous force, which means that the Reynolds number can be calculated as:

$$\text{Re} = \frac{d_e v \rho}{\mu} \tag{7.25}$$

Where, d_e is the conduit's equivalent diameter. Since $\mu\,\rho^{-1}$ is the kinematic viscosity υ, the Reynolds number has a relationship between the size of the conduit, the average fluid velocity v, and υ, which is expressed as:

$$\text{Re} = \frac{d_e v}{\upsilon} \tag{7.26}$$

For Newtonian fluids, the Re values of 2100 up to 2300 are commonly seen as a critical Reynolds number (see Fig. 7.4). This is merely a rule of thumb since the Reynolds number always depends on both the behavior and properties of a fluid. Petroleum drilling fluids, for example, can have Re critical values as high as 4000 [30]. When the flow has a low velocity, the mass will travel in discrete layers parallel to one another. The only movement across these fluid layers is molecular motion, which is the source of viscosity. This flow is laminar, i.e., an orderly flow.

Up to this point the Re values have indicated the fluid flow itself, which is very important to environmental physics. The water moving from a treatment plant into a stream, or the flow through pipes or ducts within an industrial facility are examples of flows that are described by Reynolds numbers. However, another very important aspect of fluid mechanics is explaining how an object moves within a fluid, e.g., the movement of suspended solids in a pond or river and the motion of particulate matter in the atmosphere. A particle Reynolds number describes the particles being carried by the flow of water, air, blood, or a carrier gas. Adding gas viscosity (μ_g) in units of g cm^{-1} s^{-1} to Eq. (7.26) gives a dimensionless particle Reynolds number (Re_p or N_{Rp}) that can be expressed in terms like those of the forces:

$$\text{Re}_p = \frac{d_p v_p \rho_g}{\mu_g} \tag{7.27}$$

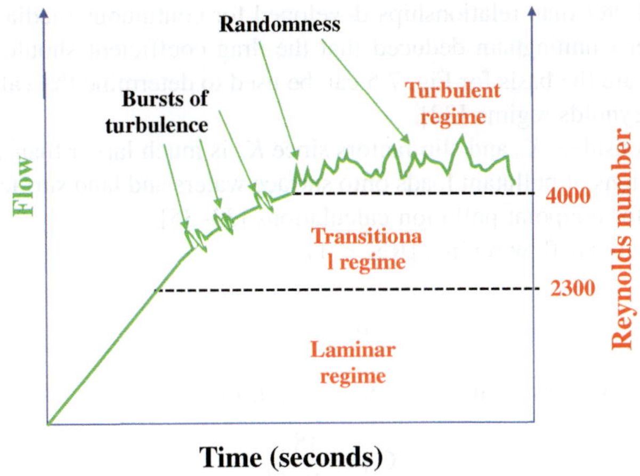

FIG. 7.4 Reynolds numbers and duct flow regimes. Flow is at rest at time zero the flow in a pipe will be at rest, so the Reynolds number is also zero. As flow begins, at low velocity the flow is laminar, such as low flows in and out of wastewater settling ponds. As the flow rate increases and the Reynolds number approaches 2300, bursts of turbulence appear more frequency in the transitional regime until the Reynolds number approaches 4000 when the flow becomes turbulent. *(Data from D.A. Vallero, Methods and Calculations in Environmental Physics, AIP Publishing, Melville, New York, 2022.)*

For low values of the particle Reynolds number ($Re_p < 1$), the flow is in the laminar or Stokes flow regime (see Fig. 7.5). Laminar flow moves in layers smoothly over an adjacent particle surface. For much greater values of the particle Reynolds number (e.g., $Re_p > 1000$), the flow is in the turbulent flow regime. Turbulent flow has erratic motion and an intense interchange of momentum throughout the fluid near the particle surface, i.e., a chaotic flow. Particle Reynolds numbers between 1 and 1000 indicate that the flow is in the transition regime, i.e., the flow can be either laminar or turbulent, depending on local conditions. Laminar flow in the transitional regime is interrupted intermittently by turbulence bursts. In most air pollution control applications, for example, particles less than 100 μm are in the laminar flow regime. The Reynolds number is a key design feature of pollution control equipment. For example, transition and turbulent flow conditions are relevant to the gravity settling in air pollution control systems.

To make the math manageable, a fluid is envisioned as a continuum, but in fact matter consists of particles. This assumption is generally acceptable, but as the scale gets smaller it becomes less correct, particularly related to drag forces. A fluid element is an infinitesimally small parcel of a fluid. So, fluid models include the Knudsen number (K_n) to address this. K_n is the ratio of the molecular mean free path length (λ) to a physical length (L) of the vessel holding the fluid:

$$K_n = \frac{\text{mean free path length}}{\text{physical length}} = \frac{\lambda}{\text{L}} \tag{7.28}$$

The previous equations apply to particles with diameters >3 μm, i.e., those where the fluid is continuously flowing around the particle. For smaller particles, the fluid appears as distinct molecules. Such small-diameter particles can slip between

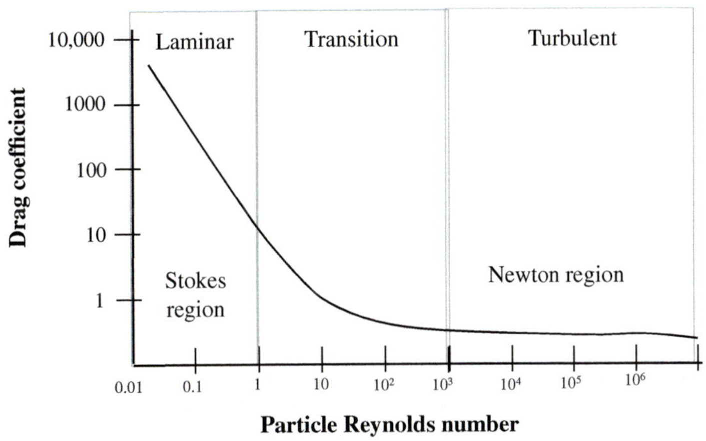

FIG. 7.5 Relationship between drag coefficient and particle Reynolds number. *(Data from W.J. Franek, L. DeRose, Control of particulate matter emissions. Chapter 8: Wet scrubbers, in: U.S. Environmental Protection Agency, Office of Air Quality Planning and Standards, Office of Air and Radiation (Eds.), Student Manual: APTI 413, U.S. EPA, Research Triangle Park, North Carolina, sec. Washington, DC, 2021.)*

the gas molecules, so they fall faster than relationships developed for continuous media predict. To adjust for this slip, English mathematician Ebenezer Cunningham deduced that the drag coefficient should be reduced for small particles [31]. The experimental data that are the basis for Fig. 7.5 can be used to determine the values of C_D and $\mathrm{Re_p}$ as expressed in equations for each particle Reynolds regime [32].

Surface water flow seldom considers K_n and slip factors since K_n is much larger than 1. However, since air and water interfaces are part of the calculations of pollutant loads onto surface waters and land surfaces, atmospheric pollutant deposition rates must be considered for nonpoint pollution calculations [33–35].

The drag coefficient in the laminar flow regime ($\mathrm{Re}_p < 1$) is:

$$C_D = \frac{24}{\mathrm{Re}_p} \tag{7.29}$$

The drag coefficient in the transition flow regime ($1 < \mathrm{Re}_p < 1000$) is:

$$C_D = \frac{18.5}{\mathrm{Re}_p^{0.6}} \tag{7.30}$$

The drag coefficient in the turbulent flow regime ($\mathrm{Re}_p > 1000$) is:

$$C_D = 0.44 \tag{7.31}$$

Thus, the drag coefficient equation for the laminar regime must be modified to include a term called the Cunningham slip correction factor (C_c):

$$C_D = \frac{24}{\mathrm{Re}_p \cdot C_C} \tag{7.32}$$

The slip correction (see Fig. 7.6) can be approximated based on the temperature and particle diameter:

$$C_C = 1 + \frac{6.21 \times 10^{-4} T}{d_p} \tag{7.33}$$

Where, T = absolute temperature (K) and d_p = particle diameter (μm).

Applying the slip corrections and substituting terms allow drag forces to be calculated for each flow regime. Accordingly, the drag force for the laminar regime becomes:

$$F_D = \frac{3\pi \mu_g v_p d_p}{C_C} \tag{7.34}$$

The drag force for the transition flow regime is:

$$F_D = 2.30\left(d_p \cdot v_p\right)^{1.4} \cdot \mu_g^{0.6} \cdot \rho_g^{0.4} \tag{7.35}$$

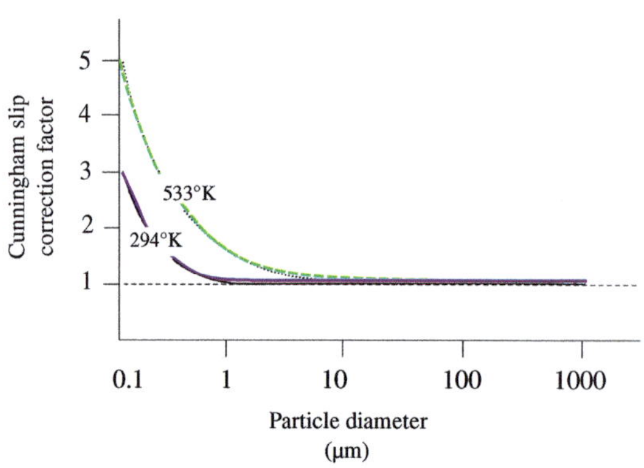

FIG. 7.6 Cunningham slip correction factor as a function of temperature and particle diameter. *(Data from D.A. Vallero, Fundamentals of Air Pollution, fifth ed., Elsevier Academic Press, Waltham, MA, 2014, p. 999 pages cm. U.S. Environmental Protection Agency, Particle collection mechanisms, in: J.R. Richards (Ed.), APTI Course 413: Control of Particulate Matter Emissions, third ed., ICES Ltd., Research Triangle Park, North Carolina, 2000. Available: http://www.4cleanair.org/APTI/413Combined.pdf (Chapter 4).)*

The drag force for the turbulent flow regime is:

$$F_D = 0.05\pi\left(d_p \cdot v_p\right)^2 \cdot \rho_g \tag{7.36}$$

Where, μ_g = gas viscosity (g cm^{-1} s^{-1}); d_p = particle diameter (cm); ρ_g = gas density (g cm^{-3}); v_p = particle velocity relative to the gas (cm s^{-1}); and C_C = Cunningham slip factor (dimensionless).

Ignoring buoyancy, the two forces dominating particle deposition for the atmosphere to water and land surfaces, as well as in pollution control equipment, are gravity and drag. As the particle accelerates, its velocity increases. Gravity and drag forces work against one another, either keeping the particle suspended and moving it toward settling. When a balance is achieved, it is expressed as:

$$\sum F = F_G - F_D = m_p \cdot a_p = m_p \frac{dv_p}{dt} = 0 \tag{7.37}$$

The drag force on the particle also increases with increasing fluid velocity. At a certain point, velocity will become sufficiently high so that the drag force will equal the gravitational force. At this point, the gravitational and drag forces will balance, i.e., the net force is zero, and the particle will no longer accelerate, reaching a constant velocity. This constant velocity is known as the terminal settling velocity (v_t):

Terminal velocities differ for the three regimes and can be derived by substituting the terms:

$$\text{Laminar flow}: v_t = \frac{g \cdot C_C \cdot \rho_p \cdot d_p^2}{18\mu_g} \tag{7.38}$$

$$\text{Transition flow}: v_t = \frac{0.153 g^{0.71} \cdot \rho_p^{0.71} \cdot d_p^{1.14}}{\mu_g^{0.43} \cdot \rho_g^{0.29}} \tag{7.39}$$

$$\text{Turbulent flow}: v_t = 1.74\left(\frac{g \cdot \rho_p \cdot d_p}{\cdot \rho_g}\right)^{0.5} \tag{7.40}$$

Where, v_t = terminal particle settling velocity (cm s^{-1}); g = particle acceleration due to gravity (980 cm s^{-1}); C_C = Cunningham slip correction factor (dimensionless); μ_g = gas viscosity (g cm^{-1} s^{-1}); d_p = particle diameter (cm); ρ_p = particle density (g cm^{-3}); ρ_g = gas density (g cm^{-3}); and v_p = particle velocity relative to the gas (cm s^{-1}).

The slip correction and Reynolds numbers can be used to determine the flow regime in which the particles are settling, such as in pollution control equipment. This is found by using a characteristic value, i.e., the K value, for each type of flow:

$$\text{K} = d_p\left(\frac{g \cdot \rho_p \cdot \rho_g}{\mu_g^2}\right)^{0.33} \tag{7.41}$$

The units for the variables are the same as the previous equations, with the gravitational acceleration (g) = 980 cm s^{-2}.

Each flow regime type has a K value range:

Laminar regime: K < 2.62
Transitional regime: 2.62 < K < 69.12
Turbulent regime: K > 69.12

To demonstrate, if the temperature is 20 °C, and the K value of a 10 μm particle in air (density = 1.2 × 10^{-3} g cm^{-3}; viscosity = 1.8 × 10) with a particle density of 1.0 g cm^{-3} the flow regime can be found by using Eq. (7.41):

$$\text{K} = 1.0 \times 10^{-3}\text{cm}\left(\frac{(980\text{cm sec}^{-2}) \times (1.0\text{g cm}^{-3}) \times \left(1.2\text{g} \times 10^{-3}\ \text{cm}^{-3}\right)}{\left(1.80 \times 10^{-4}\ \text{g cm}^{-1}\ \text{sec}^{-1}\right)^2}\right)^{0.33} \approx 0.3$$

This is in the laminar flow regime.

For more viscous gases, the K value is much smaller, because the term for gas viscosity in the denominator is squared, i.e., squaring numbers that are less than 1 decreases the product. More viscous gases have larger laminar flow regimes than less viscous gases, i.e., turbulence is less likely.

Since the flow is known to be laminar in the example, and if $C_C = 1$. the terminal velocity for the particle can also be calculated:

$$v_t = \frac{g \cdot C_C \cdot \rho_p \cdot d_p^2}{18\mu_g} = \left(\frac{(980\text{cm sec}^{-2}) \times (1.0\text{g cm}^{-3}) \times (10 \times 10^{-4}\text{ cm})^2}{18 \times (1.80 \times 10^{-4}\text{ g cm}^{-1}\text{ sec}^{-1})}\right) \approx 0.3\text{ cm sec}^{-1}.$$

This particle with a diameter of 10 μm is a coarse particle in the ambient air but is relatively small for gravity-based settling equipment.

To demonstrate the K value, flow regime and terminal velocity for a larger and denser particle, consider a particle with an aerodynamic diameter of 100 μm and $\rho_p = 2$ of 1.2 g cm^{-3}:

$$K = 1.0 \times 10^{-2}\text{cm}\left(\frac{(980\text{cm sec}^{-2}) \times (2.0\text{g cm}^{-3}) \times (1.2\text{g} \times 10^{-3}\text{ cm}^{-3})}{(1.80 \times 10^{-4}\text{ g cm}^{-1}\text{ sec}^{-1})^2}\right)^{0.33} \approx 38$$

This falls well within the transitional flow regime.

Therefore, the terminal velocity is calculated using Eq. (7.41):

$$v_t = \frac{0.153 g^{0.71} \cdot \rho_p^{0.71} \cdot d_p^{1.14}}{\mu_g^{0.43} \cdot \rho_g^{0.29}}$$

$$= \frac{[0.153 \times (980\text{ cm sec}^{-2}]^{0.71} \times (2.0\text{g cm}^{-3})^{0.71} \times (1.0 \times 10^{-2}\text{cm})^{1.14}}{(1.80 \times 10^{-4}\text{ g cm}^{-1}\text{ sec}^{-1})^{0.43} \times (1.2 \times 10^{-3}\text{ g cm}^{-3})^{0.29}} \approx 86\text{ cm sec}^{-1}.$$

With the greater terminal velocity, the denser and larger particle is likely to be removed from the flowing water or airstream by gravity. These equations have been used to calculate the terminal settling velocities of a wide range of particles with physical diameters from 10^{-2} to 10^5 μm. Terminal settling velocities are negligible for the very fine particles, i.e., diameters ≤2.5 μm and even for much of the coarse particle fraction, i.e., diameters >2.5 μm and <10 μm.

The less viscous the fluid and the higher the velocity of flow, the more likely the system will be turbulent. However, there are many factors around the Reynolds number and particle Reynolds number that will influence and change the transition from laminar to turbulent flow in open systems like channels [36,37], closed systems like ducts, pipes, and diffusers [36,38–40], as well as bodily fluid flows within humans and other organisms [41–43]. These factors need to be consulted prior to actual calculations.

4.1 Sorption

Sorption is the movement from a fluid phase to a solid phase. Desorption is the movement from the solid phase to the fluid. A pollutant or any substance can become sorbed to a particle that reaches water. Conversely, a substance may become desorbed from the sediment or suspended solid and released to the water column. Similarly, a soil particle may sorb microbes that degrade pollutants or on the media of a trickling filter wastewater treatment plant. The mass of sorbed solute is:

$$M_{sorb} = V \times (1 - \eta) \times \rho_{sorb} \times [C_{sorb}] \tag{7.42}$$

Where, M_{Sorb} is the mass of solute sorbed per dry unit weight of solid (kg kg^{-1}), V is the volume of solid (m^{-3}), η is the porosity (per cent, unitless), ρ_{Soil} is the density of soil (kg m^{-3}), and $[C_{sorb}]$ is the concentration of sorbed chemical (kg m^{-3}).

Sorption is classified as four general types:

1. *Adsorption* is the process wherein the chemical in solution attaches to a solid surface, which is a common sorption process in clay and organic constituents in soils.
2. *Absorption* occurs in porous materials so that the solute can diffuse (we will discuss diffusion later in this section) into the particle and be sorbed onto the inside surfaces of the particle.
3. *Chemisorption* integrates a chemical compound into porous materials surface via chemical reaction.
4. *Ion exchange* involves positively charged ions (cations) attracted to negatively charged particle surfaces or negatively charged ions (anions) attracted to positively charged particle surfaces. This process displaces ions on the particle

surface. Particles undergoing ion exchange can include soils, sediment, airborne particulate matter, or even biota, such as pollen particles.

Once the pollutant is sorbed or desorbed, its transport within the environmental compartments will result from the general mechanisms introduced here.

4.2 Diffusive flux

Flux is the movement of the mass or energy across a two-dimensional area. Flux, then, is either the number of atoms diffusing through unit area per unit time (atoms $m^{-2}\ s^{-1}$) or the mass per area per time, e.g., mg $m^{-2}\ s^{-1}$. Similarly, an energy flux is an amount of energy crossing a cross-sectional area per time, e.g., J $m^{-2}\ s^{-1}$, but this can be simplified to kg s^{-3}, since a joule is kg $m^{-2}\ s^{-2}$.

The flux of a contaminant is equal to the mass flux plus the dispersion flux and diffusion flux, as well as source and sink terms. Sources can be the result of a one-time or continuous release of a chemical compound from a reservoir or result from desorption of the chemical along the way. Sinks can be the result of sorption and surface processes. This means that even though the source contribution is known, there will be sorption occurring in soil, sediment and biota that will either remove the chemical compound from the fluid, or under other environmental conditions, the chemical will be desorbed from the soil, sediment, or biota. Thus, these interim sources and sinks must be considered in addition to the initial source and final sinks. For example, the pollutant may be released from a wastewater outfall (source) and move to the soil (interim sink), where it is metabolized by microbes (interim source), and then is taken up by a plant (interim sink and source), consumed by a human, after which a fraction of the mass is metabolized (interim sink), released into the bloodstream (interim source), and stored in fat cells (fate).

Once a substance is desorbed and fluxes across a surface, it can move as result of a concentration gradient. The concentration gradient (i_c) is the change in concentration (in units of mass per volume) with length (in meters), so the units of i_c are kg m^{-4}.

Diffusion is therefore analogous to the physical potential field theories, in which the flow direction is from high potential to low potential, just as higher-pressure fluids move to low pressure fields. The concentration gradient is observed in all phases of matter, whether solid, liquid or gas. In water quality assessments, diffusion is primarily limited to transport in soil or sediment, and is often ignored in surface water given that other processes, such as advection, generate a flow greater than 2×10^{-5} m s^{-1} [44]. Flows this low are common in aquifers, in lower layers of sediment, or at or through clay liners in landfills.

Diffusion can be an important process for source characterization, since it may be the principal means by which a contaminant becomes mixed in a quiescent container, such as a drum, in buried sediment, or in buried soils ("Turbulent Diffusion—an overview | ScienceDirect Topics"). Thus, diffusion is proportional to the concentration gradient as expressed by Fick's first law:

$$J_{Diffusion} = -d_o i_c \tag{7.43}$$

And,

$$i_c = \frac{\partial [C]}{\partial x} \tag{7.44}$$

Where, $[C]$ is the chemical concentration and x is the distance between the points of $[C]$ measurements.

Concentration is expressed in units of mass per mass, volume per volume, or mass per volume. For water pollutants, mass per volume is usually preferred, e.g., micrograms of a pollutant per liter of water (μg L^{-1}). Volume per volume is sometimes seen, e.g., parts per trillion (ppt) for certain chemical compounds in drinking water.

As mentioned, the flux of a chemical agent is the amount that passes through a surface perpendicular to the flow. Mass flux (J) is most easily calculated in one direction:

$$\mathrm{J} = \frac{m}{A\Delta t} \tag{7.45}$$

Where, m is the mass of the chemical species; A is the cross-sectional area of the flux surface; and t is time duration of the flux. Thus, flux can be expressed as μg $m^{-2}\ s^{-1}$. Other nomenclature and symbols are used for flux, e.g., J or N_x are commonly used to indicate mass flux and J is used to indicate diffusive flux.

The mass of a chemical compound moving through a cross-sectional surface varies by transport mechanism. For example, if the mechanism is advection, i.e., the passive entrainment of a chemical species carried by the fluid, the flux is readily related to concentration and fluid velocity:

$$J = \frac{m}{V} \times \frac{V}{A \bullet t} = [C] \bullet u \tag{7.46}$$

Where $[C]$ is the concentration of the chemical species and u is fluid velocity.

The other principal transport mechanism is diffusion, i.e., molecular motions lead to random movement of the chemical species. Commonly, advection and diffusion are the two main flux mechanisms, so the total flux would be the sum of the advective flux (Cu) and the diffusive flux (j). Therefore, flow (Q) is:

$$Q = [C] \bullet u + j \tag{7.47}$$

Molecular diffusion can be demonstrated with a two-compartment system (see Fig. 7.7). In compartment A, the chemical species concentration is lower than that in compartment B. C_A is the low concentration compartment and C_B is the high concentration compartment. Assuming there is no net flow of the fluid between the compartments, i.e., no advection, and that the fluctuating flow (u') is in one direction, then x, is compensated by flow of the same magnitude in opposite direction, also x. Then Δx is the distance between the compartment centers. The net diffusive flux (j) from A to B equals the flux from A to B minus the flux from B to A:

$$j = C_A \bullet u' - C_B \bullet u' = u'\Delta C \tag{7.48}$$

Where, ΔC is the concentration difference. Rearranging using Δx, the distance between the compartment centers and assuming the limit approaching infinitely small distances, the diffusive flux is expressed as:

$$\text{j} = (u'\Delta x)\frac{\Delta C}{\Delta x} \tag{7.49}$$

A diffusion coefficient, D, can now be derived as the limit of the product $u'\Delta x$ as Δx becomes small [45]. This gives a discrete form of Fick's law of diffusion:

$$\text{j} = -D\frac{\Delta C}{\Delta x} \tag{7.50}$$

The units of j are area per time, e.g., $m^2\ s^{-1}$. Eq. (7.50) is known as Fick's law of diffusion, which shows that diffusive flux is directly proportional to the concentration gradient. The difference in concentration drives the diffusion, i.e., flux increases according to the difference in concentration of the two compartments.

Fick's law can also be depicted in a continuous form:

$$j = -D_0 \,.\, i_c \tag{7.51}$$

Where D_0 is the proportionality constant for molar concentration, $[c]$ and i_c is the concentration gradient.

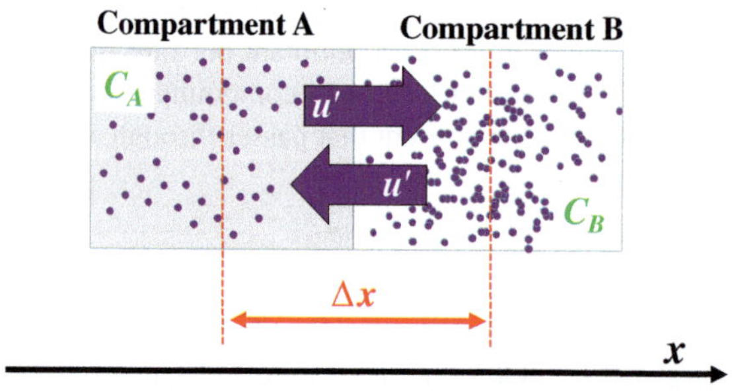

FIG. 7.7 Exchange of molecules in a system consisting of two distinct compartments. Note: In compartment A, the chemical compound concentration is less than that in compartment B. C_A is the concentration of chemical C in compartment A; C_B is the concentration of chemical C in compartment B and u' is the flow that is fluctuating in dimension x.

The negative sign of the gradient indicates that the transport is in the direction of greater concentration to lesser concentration. Fick's second law is relevant when the concentrations are changing with time. The change of concentrations with respect to time is proportional to the second derivative of the concentration gradient:

$$\frac{\partial[c]}{\partial t} = \frac{\partial^2[c]}{\partial x^2} \tag{7.52}$$

The diffusive flux is calculated by inserting the cross-sectional surface area (A):

$$\mathrm{j} = -\mathrm{AD}\frac{\Delta C}{\Delta x} \tag{7.53}$$

Flux example 1

Small, internal combustions are repaired in a building. The mean concentration of carbon monoxide (CO) in the building is 25 mg m^{-3}. The repair building is connected to another building that is used as an office and as a tailor shop. The connection is an enclosed hallway that is 3 m long, 1 m wide and 2 m high. The office/shop is ventilated so that the mean CO concentration is 2.5 mg m^{-3}. If diffusion accounts for the entire flux between the repair building and the office/shop, and the diffusion coefficient is 8×10^{-3} m^2 s^{-1}, calculate the CO diffusive flux.

Solution:

$$j = -AD\frac{\Delta C}{\Delta x} = 2\ \mathrm{m}^2 \times \left(8 \times 10^{-3}\ \mathrm{m}^2\ \mathrm{sec}^{-1}\right) \times \frac{25 - 2.5\ \mathrm{mg\ m}^{-3}}{3\ \mathrm{m}} \approx 0.26\ \mathrm{mg\ sec}^{-1} = 260\ \mu\mathrm{g\ sec}^{-1}$$

Flux example 2

Assume that the owners of the repair building have improved the heating, ventilation, and air conditioning (HVAC) system so that [CO] in the repair shop decreased from 25 mg m^{-3} to 3 mg m^{-3}. The decrease in concentration in the office/dress shop, [CO_{office}] can be calculated as:

$$0.26\ \mathrm{mg\ sec}^{-1} = -AD\frac{\Delta C}{\Delta x} = 2\ \mathrm{m}^2 \times \left(8 \times 10^{-3}\ \mathrm{m}^2\ \mathrm{sec}^{-1}\right) \times \frac{3 - [\mathrm{CO_{office}}]\ \mathrm{mg\ m}^{-3}}{3\ \mathrm{m}} = \frac{0.016 \times (3 - \mathrm{CO_{office}})}{3}$$

$$0.26 = 0.005 - (1 - \mathrm{CO_{office}})$$

$$1 - \mathrm{CO_{office}} = 0.255$$

$$\mathrm{CO_{office}} \approx 0.75\ \mathrm{mg\ m}^{-3}$$

Thus, the change in a nearby air volume has improved the air quality in the office/shop from [CO] = 2.5 mg m^{-3} to [CO] = about 0.75 mg m^{-3}.

Again, this assumes that the entire mass of CO is transported by diffusion and that the diffusion coefficient remains unchanged. This is highly unlikely given that diffusion in the air is secondary to other transport mechanisms.

Molecular diffusion is based on the random movement of the molecules. Fig. 7.8 shows two adjacent stacks of molecules with their horizontal center at $x = 0$. Each molecule moves in a random manner when heat energy is added to the system (Fig. 7.8B). This is known as Brownian movement. After a certain time, on average, these random movements make up half of the molecules moving to the left and half to the right. This leads to a decrease in the concentration of molecules because the molecular space has increased in size (Fig. 7.7A area is smaller than Fig. 7.7C area).

The concentration gradient unit has an exponent of −4, since volume's unit is length cubed, e.g., m^3 is the unit of volume. This is an expression of 3-dimensional space. This is where the substance resides as a concentration, e.g., 1 mg of chemical per cubic meter of fluid. Therefore, when we compare one location's concentration of a chemical species to another location's concentration of the same substance, expressing the change of number of molecules or mass per volume from the first location to the second:

$$i_c = \frac{(\mathrm{mass} \times \mathrm{length}^{-3}) - (\mathrm{mass} \times \mathrm{length}^{-3})}{\mathrm{length}} \tag{7.54}$$

Thus, the units represent mass per volume per distance, e.g., kg m^{-4}. These units indicate that diffusion is analogous to the physical potential field theories. The flow of matter or energy is from the direction of high potential to low potential, such as from high pressure to low pressure. This gradient is observed in all phases of matter, solid, liquid or gas. So, molecular diffusion is a major factor of transport in porous media, such as soil or sediment, but is often ignored if other bulk transport mechanisms are occurring, e.g., advection. Even in porous media, if the flow is 2×10^{-5} m s^{-1} or higher, molecular

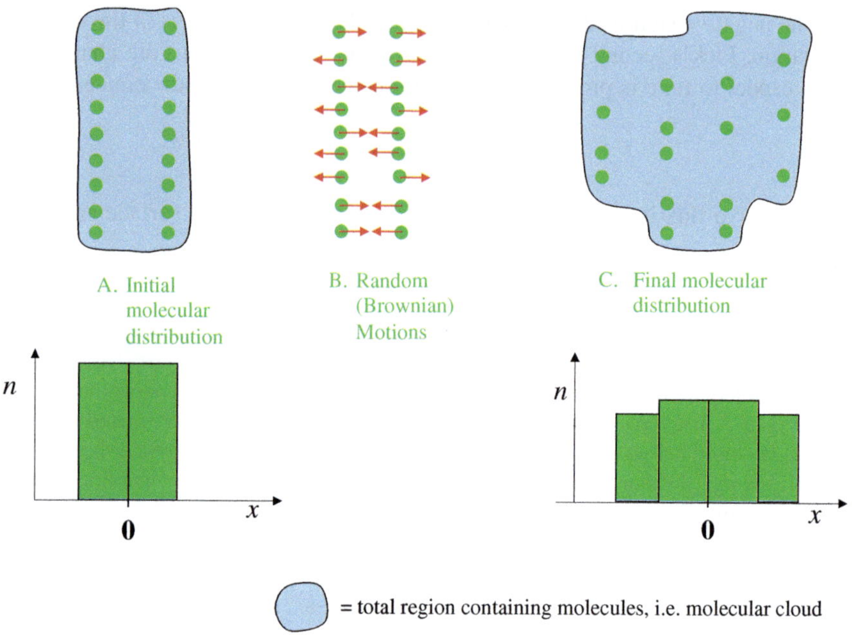

FIG. 7.8 Random motion of molecules depicting Fickian diffusion in one-dimension (x). Bottom graphs are histograms of the number of molecules (n) by location on the x-axis before and after the motion. Note that the size of the molecular cloud increases. The concentration of molecules decreases because diffusion has expanded the size of the cloud. This depiction can be extended to two dimensions by adding the y dimension and to three dimensions by adding the y and z dimensions. Pollution transport models commonly employ the z-axis. The concentration gradient (i_c) is the change in concentration (units of mass per volume, e.g., kg m^{-3}) with length (m), so the units of i_c are kg m^{-4}. *(Data from D. Vallero, Air Pollution Calculations: Quantifying Pollutant Formation, Transport, Transformation, Fate and Risks, Elsevier, 2019.)*

diffusion is often ignored as a transport mechanism [46]. Even in these systems, however, molecular diffusion can be an important process to describe diffusion within porous media. Diffusion may be the principal means by which a contaminant becomes mixed in a quiescent container, such as a drum, a buried sediment, or a covered pile, or at the boundaries near clay or artificial liners in landfill systems [28].

Since molecular diffusion dominates transport only when advection and dispersion are near zero, ambient air can most often be assumed to be turbulent. If molecular diffusion does dominate, it is usually in a small pocket for only a very short time-period. As mentioned, molecular diffusion should not be ignored in quiescent systems, such as within pollution control equipment or in wet soils and sediment, where concentration gradients serve as a major transport process. In these instances, diffusion is an important mechanism within pollution sources, such as contaminants like methane that are generated in the quiescent, chemically-reduced conditions in anoxic or hypoxic sediments and wetland soils before being released into and held with the interstices around sediment and soil particles.

Molecular diffusive transport can abiotic, i.e., occurring directly between a soil particle and water or air in soil or sediment pore spaces. Bacteria and other microbes form biofilm around particles, so that molecular diffusion occurs between chemical compounds in the biofilm surrounding soil and sediment particles and the void spaces in the matrix, i.e., interstices between particles. This is a key aspect of many wastewater treatment systems, especially those like trickling filters that require biofilms around the media so that microbes can access the wastewater.

Nondiffusive transport becomes dominant in the lesser quiescent regions, e.g., most streams. Fluid velocities increase above the thin layer so that the flow immediately becomes turbulent.

4.3 Dispersion

Dispersion is the mixing of the pollutant within the fluid body (e.g., aquifer, surface water or atmosphere). Generally, if reliable information is available, it is better to calculate the dispersion from physical principles using a deterministic approach, than to estimate the dispersion using probabilities. Mass balance is based around a differential volume in a Eulerian model. A Lagrangian model applies statistical theory of turbulence, assuming that turbulent dispersion is a random process described by a distribution function. The Lagrangian model follows the individual random movements of molecules released into the plume, using statistical properties of random motions that are characterized mathematically. Thus, this mathematical approach estimates the movement of a theoretical volume of chemical substance (particle[e]) from one

e. The term "particle" is used in many ways in environmental physics. In dispersion modeling, the term particle usually means a theoretical point that is followed in a fluid. The point represents the path that the pollutant is expected to take. Particle is also used to mean aerosol in atmospheric sciences. Particle is also commonly used to describe unconsolidated materials, especially in soil and sediment matrices. Particulate matter (PM) is commonly referred to as "particles." The physicist's particle-wave dichotomy is key to environmental analysis since the behavior of light is important in environmental chromatography.

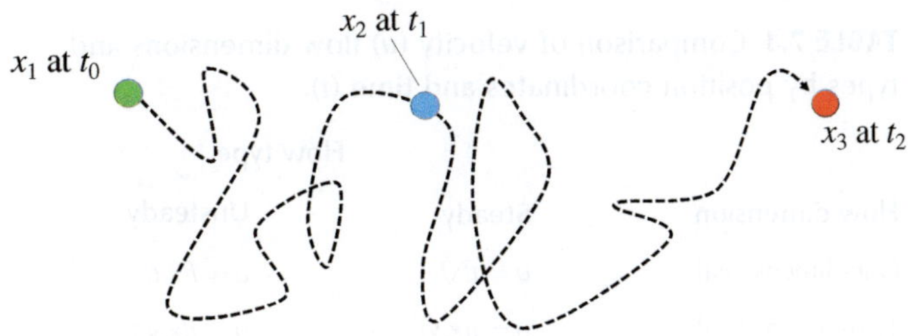

FIG. 7.9 Random walk representing the movement of a hypothetical particle, i.e., a point that moves in random paths from position x_1 to position x_2 during time interval $t_0 - t_1$, and from x_2 to x_3 during time interval $t_1 - t_2$. Lagrangian gas flow is based on this principle, which is used in plume models. *(Adapted from D.A. Vallero, Fundamentals of Air Pollution, fifth ed., Elsevier Academic Press, Waltham, MA, 2014, p. 999 pages cm.)*

point in the plume to another distinct point during a unit time. Physicists may refer to this theoretical volume as a "particle," i.e., a theoretical point that is followed in a fluid. The point represents the path that the pollutant is expected to take. Unfortunately, in air pollution engineering and atmospheric sciences, the term particle is often used to mean an aerosol. Particle is also commonly used to describe unconsolidated materials, such as soils and sediment. The path that each particle takes during this time is an ensemble mean field that relates to the particle displacement probabilities:

$$\overline{[c]}(x, y, z, t) = M_{Total} P(Dx, t) \tag{7.55}$$

Where, $D\,x = x_2\text{-}x_1$ = particle displacement (see Fig. 7.9), $P(Dx_2, t)$ = probability that the point x_2 will be immersed in the dispersing media at time t, M_{Total} = total mass of particles released at x_1, and $\overline{[c]}$ = mean concentration of all released particles = mass of particles the plume $dx \cdot dy \cdot dz$ around x_2.

The velocity field can characterize the motion of fluid particles and allows calculations of these motions. Using the Euler approach, fluid properties are written as functions of space and time, which means that the flow is determined by analysis of the behavior of functions. This differs from the Lagrange approach, which characterizes the fluid by tagging a piece of the fluid, e.g., a water or air parcel, and determining the flow properties by tracking the movement of these properties in time. This random walk provides the Lagrangian viewpoint, which is expressed mathematically as:

$$\vec{v} = [x(t), y(t), z(t)] \tag{7.56}$$

A Lagrangian plume model characterizes the plume by calculating the dispersion from statistics of the trajectories of many parcels, i.e., sufficient to represent the whole plume. Thus, statistics, i.e., standard deviations, are employed in to analyze the expected movement of a gas flow or a plume in the atmosphere with time in all three dimensions, i.e., the Gaussian model (see Chapter 6).

In pollutant transport, spatial and temporal scales interrelate. The concentration of a pollutant decreases over the distance (x) that the pollutant travels and also decreases with the time (t) that the pollutant remains in transport. Indeed, during transport in a fluid, the concentration (C) gradients of x and t are directly proportional:

$$\frac{dC}{dx} \approx \frac{dC}{dt} \tag{7.57}$$

Recall that concentration is simply the mass (m) of a substance per volume (V) of fluid, so:

$$\frac{d\left(\frac{C}{V}\right)}{dx} \propto \frac{d\left(\frac{C}{V}\right)}{dt} \tag{7.58}$$

By extension, the pollutant mass gradient is:

$$\frac{dm}{dx} \propto \frac{dm}{dt} \tag{7.59}$$

Thus, the distance (x) and time (t) in which the mass (m) movements are interchangeable. The mass gradient within a conveyance can be expressed as:

$$\frac{dm}{dt} = \rho \times \mathrm{A} \times u \tag{7.60}$$

Where, A is the cross-sectional area perpendicular to the flowlines, e.g., in a pipe, and u is the fluid velocity which has three-dimensional spatial coordinates, x, y, and z [23]:

TABLE 7.1 Comparison of velocity (*u*) flow dimensions and types by position coordinates and time (*t*).

Flow dimension	Flow type	
	Steady	**Unsteady**
One-dimensional	$u = f(x)$	$u = f(x,t)$
Two-dimensional	$u = f(x,y)$	$u = f(x,y,t)$
Three-dimensional	$u = f(x,y,z)$	$u = f(x,y,z,t)$

Data from D. Vallero, Air Pollution Calculations: Quantifying Pollutant Formation, Transport, Transformation, Fate and Risks, Elsevier, 2019.

$$u = f(x, y, z) \tag{7.61}$$

Transport models often apply a simplifying assumption of one-dimensional flow in the x direction only. This means that the velocity is constant over entire section, i.e., air, water, blood, and other fluids travel along parallel flowlines [23].

Flow may also be assumed to be steady or unsteady Steady flow is expressed solely by position, but unsteady flow is a function of both position and time (see Table 7.1).

4.4 Advective transport

Advection involves the transport of dissolved or suspended substances along with the fluid flow. Thus, if the fluid is moving at 1 m s^{-1}, so is the substance that the fluid is carrying. When a substance traverses a cross-sectional area, it is an advective flux ($J_{Advection}$). The mass is carried along with the flow through an interface, e.g., soil-air, water-air, skin-air, sediment-water, or leaf-air.

Many advective models are represented by the one-dimensional mass flux equation for advection to avoid the complexities of a three-dimensional transport. The one-dimensional mass flux equation for advection in soil or other unconsolidated, porous media can be expressed as:

$$J_{advection} = \overline{v}\eta_e[C] \tag{7.62}$$

Where, $\overline{v}$= average linear velocity (m s^{-1}), η_e =effective porosity (per cent, dimensionless), and $[C]$ = chemical concentration of the solute (kg m^{-3}).

Generally, the pollutants follow a four-step trek from the source to the receptor: desorption; diffusion; dispersion; and dilution [47]. From a physical motion perspective, advective transport is the most straightforward pollutant transport process, i.e., the transport of matter within the streamlines of the flow. The rate and direction of transport is completely determined by the rate and direction of the flow in the media.

Homogeneous advection, where only one fluid is carrying the contaminant, is the simplest bulk transport within one environmental medium. Thus, in terms of time and space, most of the transport of a water pollutant in a stream is by homogeneous advection., The three-dimensional rate of homogeneous, advective transport (N) is simply the product of the fluid medium's flow rate and the concentration of the contaminant in the medium:

$$N = QC \tag{7.63}$$

Where Q is the flow rate of the fluid medium (e.g., m^3 s^{-1}) and C is the concentration of the chemical contaminant being transported in the medium (e.g., μg m^{-3}). The units for three-dimensional advection are mass per time (e.g., μg s^{-1}).

When a secondary phase exists within the main advective medium, it is heterogeneous advection, such as the presence of suspended solid in a flowing stream. Heterogeneous advection involves more than one transport system within the compartment.

Another example of advective transport is deposition from the atmosphere. Chemical compounds dissolved in water droplets or sorbed onto aerosols suspended in the droplet is known as wet deposition. Sorption to solid particles is known as dry deposition.

The porosity term is crucial to unconsolidated material, like soil and underground aquifers, but is not needed for the open atmosphere air or water flow. Thus, for atmospheric and surface water fluxes, the equation can be stated in two dimensions:

$$J_{advection} = \overline{v}[C] \tag{7.64}$$

Two-dimensional fluxes define the transport of a contaminant across a unit area. This rate of transport is the mass flux density, which is the mass of the contaminant traversing a unit area per time. Fluid velocities vary considerably in time and space, e.g., upstream versus near the deltas of rivers, calm versus gusty wind conditions, or ground water near the surface compared to deeper portions of an aquifer. Estimating mass flux density for advection in a turbulent fluid usually calls for time-integration to derive average concentrations. Monitoring equipment may collect samples every minute, but if the model or calculation calls for an hourly value, the 60 values will be averaged to give a single integrated concentration of the pollutant.

Advection is the major means by which suspended particles move in water. Engineers and scientists often assume a mean velocity for an entire control volume, e.g., a well-mixed chamber, to simplify calculations of a substance's concentration compared to an unmixed chamber in which the substances and carrier gases move at multiple velocities. This assumption results in calculations of average concentrations within the entire chamber or plume. This is an example of the difficulty in scaling up laboratory or mesoscale findings to the real world. Plumes are likely to have higher concentrations near the source but will also have pockets of higher and lower concentrations given variability of stream profiles, discharges, temperature, and other features.

4.5 Eddy diffusion

Eddy diffusion is a transition between diffusion and bulk transport. It can be classified as a type of advection, given that eddy diffusion transport of substances moves with the flow. Fickian processes vary by scale. Turbulent motion can form eddies of various sizes. In the atmosphere, air flow is almost always turbulent. Laminar air flow is generally found only in engineered, controlled systems, like laminar flow hoods, or in buried sediment, e.g., quiescent regions beneath a lake.

Although they are more complex than the random, molecular-scale diffusion, the eddies produced by turbulence can also be modeled from the concentration gradients of Fick's first law. For this reason, turbulent dispersion is often called eddy diffusion, which can be modeled in one, two, or three dimensions, just as is done for molecular diffusion. The diffusion coefficient (*D*) remains the same with respect to direction in a one-dimensional model. The *D* terms is adjusted it is expected to vary substantially with spatial location and time. If so, *D* is anisotropic, which means it may vary in different directions or vertically in the air or water. In the environment, the concentration gradient is three-dimensional.

Eddy diffusion example 1

A 200 m^2 of soil has been contaminated by a leaking abandoned underground gasoline tank. At 3 m beneath the contaminated soil, the underlying soil has a concentration of 2 μg cm^{-3} total hydrocarbons (THC). If the diffusion coefficient is 0.1 $cm^2\ s^{-1}$ in this soil, calculate the mass flux density of the vapor and the rate of vapor release by molecular diffusion, assuming complete mixing and a one-dimensional flux.

Solution:

The vertical concentration gradient (upward on z axis). is:

$$\frac{dC}{dz} = (2 \times 10^{-6}\ \text{g cm}^{-3})/(300\ \text{cm}) = 6.7 \times 10^{-9}\ \text{g cm}^{-4}$$

The mass flux density is:

$$j = -D\frac{\Delta C}{\Delta z} = (10^{-1}\ \text{cm}^2\ \text{sec}^{-1}) \times (6.7 \times 10^{-9}\ \text{g cm}^{-4}) = 6.7 \times 10^{-10}\ \text{g cm}^{-2}\ \text{sec}^{-1}$$

This indicates the mass of the THC breaks through the soil surface. Applying the flux density to the 200 m^2 ($2 \times 10^6\ cm^2$), the penetration of the vapor into the air:

$$(6.7 \times 10^{-10}\ \text{g cm}^{-2}\ \text{sec}^{-1}) \times (2 \times 10^6\ \text{cm}^2) \times (3600\ \text{sec hr.}^{-1}) \times \left(24\ \text{hr day}^{-1}\right) = 116\ \text{g day}^{-1}$$

Eddy diffusion example 2

If instead of THC, the flux density of the single hydrocarbon must be determined. Assume that the compounds leaking from the gasoline tank are mainly benzene, toluene, ethylbenzene, and xylene (BTEX), a common indicator of gasoline leaks from underground tanks. If the same amount of each is leaking, take the average of the four diffusion coefficients, about 0.08 $cm^2\ s^{-1}$ in this soil (see Table 7.2). If the underlying soil has a concentration of 0.5 μg cm^{-3} of this compound, applying the same assumptions as above, calculate the flux density of the vapor and the rate of vapor release by molecular diffusion.

TABLE 7.2 Diffusion coefficients of selected hydrocarbons into the air.

Substance name	Molecular weight ($g\ mol^{-1}$)	Vapor pressure (Pa)	Water solubility ($mg\ L^{-1}$)	Diffusion coefficient in air ($cm^2\ s^{-1}$)
Benzene	78.1	12,700	1760	0.093
Toluene	92.1	3800	550	0.085
o-Xylene	106.2	880	180	0.073
m-Xylene	106.2	1110	160	0.069
p-Xylene	106.2	1170	200	0.067
1,2,3-Tri-methylbenzene	120.2	202	66	0.071[a]
1,3,5-Tri-methylbenzene	120.2	328	50–173	0.071[a]
1,2,4-Tri-methylbenzene	120.2	271	66	0.071[a]
Ethylbenzene	106.2	1270	170	0.076
1-Ethyl-2-methylbenzene	120.2	330	40–93	0.071[a]
1-Ethyl-4-methylbenzene	120.2	493	95	0.071[a]

[a] *Estimated based on the diffusion coefficient of the ethylbenzene.*
Data from V. Miljøstyrelsen, Guidelines on remediation of contaminated sites, Danish Environmental Protection Agency. https://www2.mst.dk/udgiv/publications/2002/87-7972-280-6/pdf/87-7972-281-4.pdf, 2002 (Accessed 14 December 2021).

Find the vertical concentration gradient (upward on *z*-axis):

$$\frac{dC}{dz} = (5 \times 10^{-7}\ \mathrm{g\ cm^{-3}})/(300\ \mathrm{cm}) = 1.7 \times 10^{-9}\ \mathrm{g\ cm^{-4}}$$

The mass flux density is:

$$j = -D\frac{\Delta C}{\Delta z} = \left(0.08\ \mathrm{cm^2\ sec^{-1}}\right) \times \left(1.7 \times 10^{-9}\ \mathrm{g\ cm^{-4}}\right) \cong 1.4 \times 10^{-11}\ \mathrm{g\ cm^{-2}\ sec^{-1}}$$

Applying the mass flux density equation to the 200 m^2 (2 × 10^6 cm^2), the penetration of the vapor into the air:

$$\left(1.4 \times 10^{-11}\ \mathrm{g\ cm^{-2}\ sec^{-1}}\right) \times \left(2 \times 10^6\ \mathrm{cm^2}\right) \times \left(3600\ \mathrm{sec\ hr^{-1}}\right) \times \left(24\ \mathrm{hr\ day^{-1}}\right) = 2.35\ \mathrm{g\ day^{-1}}.$$

Thus, the seemingly small amount of difference between in the diffusion coefficients of 0.1 and 0.08 translates into a BTEX vaporization rate that is about 50 times smaller.

Hmmm, these calculations raise questions about the hydrocarbons in THC besides BTEX. Finding the actual compounds and their diffusion coefficients, as well as collecting and analyzing additional samples is recommended. Also, keep in mind that *D* is but one of the coefficients in this scenario. Others, such as sorption and K_{ow}, may be as or more important than *D* within the soil. Oversimplifying assumptions can be dangerous in real-world scenarios.

4.6 Dispersion

Pollutant dispersion can be either dynamic dispersion or mechanical dispersion [48]. In water, hydrodynamic dispersion encompasses the combined results of molecular diffusion and mechanical dispersion. Hydrodynamic dispersion is estimated using a modified Fick's law of diffusion through a single-phase porous matrix [49]. These are not mutually exclusive terms. Indeed, mechanical dispersion is part of dynamic dispersion. The spread of a contaminant into multiple directions longitudinally in air is known as aerodynamic dispersion, and if in water it is hydrodynamic dispersion. The scattering of molecules results from physical processes that affect the velocity of different molecules in an environmental medium. For example, in aquifers, the process is at work when the contaminant traverses the flow path of the moving ground water, which involves both dynamic diffusion and mechanical dispersion.

Since dispersion is the mixing of the pollutant within the fluid, e.g., the atmosphere or waterbody, the question arises as to whether it is better to calculate the dispersion from physical principles, using a deterministic approach, or to estimate the dispersion using statistics and probabilities. The Eulerian model bases the mass balance around a differential volume. A Lagrangian model assumes that turbulent dispersion is a random process described by a distribution function. Individual random movements of molecules released into the plume are followed using statistical properties of random motions. The mathematical estimates of transport are based on a theoretical volume of a substance that is moving from one point in the plume to another definite point during a unit time.

The present discussion accounts for the effects of the theoretical particles (e.g., frictional) as the fluid moves through unconsolidated material. Unfortunately, the term "particle" means something different in many pollution discussions, i.e., a suspended solid in water or an aerosol in air. Thus, rather than use the term particle here, we will use the term molecule. Recall from the Lagrangian discussion that the path each molecule takes during this time is an ensemble mean field relates to the molecule's displacement probabilities:

$$\overline{C}(x, y, z, t) = M_{total} P(Dx, t) \tag{7.65}$$

where:

$Dx = x_2 - x_1$ = molecule displacement
$P(Dx_2, t)$ = probability that the point x_2 will be immersed in the dispersing media at time t
M_{Total} = total mass of molecules released at x_1
$\overline{C}$ = mean concentration of all released molecules = mass of molecules in the plume $dx \cdot dy \cdot dz$ around x_2.

Gaussian dispersion models assume a normal distribution of the plume (see Chapter 6). This is at best a first approximation of the actual dispersion within a biochemodynamic system [28].

Deterministic models assume that the dispersion includes mixing at all scales. At the larger scales, characteristics of strata and variability in the permeability of the layers must be described. So, a deterministic dispersion flux would be:

$$J_{Dispersion} = \underline{D} \cdot grad[c] \tag{7.66}$$

Where: $J_{Dispersion}$ = mass flux of solute due to dispersion (kg m^{-2} s^{-1}), $\underline{D}$ = dispersion tensor (m s^{-1}), and C = concentration of chemical contaminant (kg m^{-3}).

The $\underline{D}$ includes coefficients for each direction of dispersion [28], i.e., longitudinally, horizontally and vertically (D_{xx}, D_{xy}, D_{xz}, D_{yy}, D_{yz}, D_{zz}).

There is a lack of consistency in the usage of some terms in pollutant transport and even within the environmental and engineering community. For example, the net transport of substances by diffusion and advection in fluids is sometimes referred to as convection. However, convection is also defined as vertical motion because of buoyancy, and is contrasted with advection, which is the horizontal motion in the atmosphere driven by winds. Convection is often considered a local phenomenon, whereas advection is applied to long-range transport. Diffusion and dispersion are sometimes used synonymously. Given this lack of consistency, when addressing pollutant transport mechanisms, it is best to discuss each type of mass transport precisely.

4.7 Total mass transfer

Our depictions of chemical species transport represent the macro-scale between discrete compartments. However, the fluid flow and the coefficients for mass transfer and diffusion (D) must be connected.

The rate of mass transfer, i.e., the flux at the interface (N_i), is proportional to the product of interfacial area and the difference in concentration at the interface and in the bulk solution. So, the mass transfer coefficient is a proportionality constant (k). In a liquid this is:

$$N_i = k(C_l - C_{l-i}) \tag{7.67}$$

The flux, N_i, is a total flux that includes diffusion and advection. Note that the mass transfer coefficient allows for changing the size of the flux surface, which changes the amount of mass being transferred, while flux per area remains constant [50].

Tripling the surface area between two fluid compartments will triple the mass of a chemical compound that is being transferred to the adjacent compartment, but the chemical compound flux per area remains unchanged.

Mass transfer example

A 19.2-L closed tank is isothermal at 25 °C, so that the vapor pressure of water in the tank is 0.031 atm. The tank has 0.8 L water with 150 cm^2 surface area. In 3 min, the air is 5% saturated. From this information, calculate the mass transfer coefficient and the time it takes to reach 90% saturation:

Comparing the partial pressure of water to total pressure and adjusting temperature:

$$N_i = \frac{\left(C_{vapor} \times V_{air}\right)}{A_{liquid} \times t} = 0.05 \frac{\left(\frac{0.031\,\text{atm}}{1\,\text{atm}} \times \frac{1\,\text{mol}}{22.4\,\text{L}} \times \frac{273\,\text{K}}{298\,\text{K}} \times (19.2 - 0.8\,\text{L})\right)}{150\,\text{cm}^2 \times 180\,\text{sec}} = 4.3 \times 10^{-8}\,\text{mol}\,\text{cm}^{-2}\,\text{sec}^{-1}.$$

The air concentration immediately at the water surface is saturated. During short time-periods the concentration in the bulk solution (i.e., liquid water) is zero. So, calculate the difference in concentration at the water surface from the concentration at the bulk solution.

Solution: Applying the mass transfer equation and solving for k:

$$4.3 \times 10^{-8}\,\text{mol}\,\text{cm}^{-2}\,\text{sec}^{-1} = k\left(\frac{0.031\,\text{atm}}{1\,\text{atm}} \times \frac{1\,\text{mol}}{22.4 \times 10^3\,\text{cm}^3} \times \frac{273\,\text{K}}{298\,\text{K}}\right) - 0$$

$$k = \frac{4.3 \times 10^{-8}\,\text{mol}\,\text{cm}^{-2}\,\text{sec}^{-1}}{\left(\frac{0.031\,\text{atm}}{1\,\text{atm}} \times \frac{1\,\text{mol}}{22.4 \times 10^3\,\text{cm}^3} \times \frac{273\,\text{K}}{298\,\text{K}}\right)} = 0.034\,\text{cm}\,\text{sec}^{-1}.$$

At this transfer rate, the saturation percentage can be determined from the mass balance between accumulation in the vapor phase and the evaporation rate:

$$\frac{d}{dt} V_{C-vapor} = AN_i = Ak[C_{water}\,(\text{sat}) - C_{water}] \quad (7.68)$$

For dry air: $t = 0$ and $C_{water} = 0$

Integrating the mass balance with these values gives:

$$\frac{C_{water}}{C_{water}(\text{sat})} = 1 - e^{-\left(\frac{kA}{V}\right)t}$$

Thus:

$$t = \frac{V}{kA}\ln\left(1 - \frac{C_{water}}{C_{water}(\text{sat})}\right) = \frac{18.4 \times 10^3\,\text{cm}^3}{(0.034\,\text{cm}\,\text{sec}^{-1}) \times 150\,\text{cm}^2}\ln(1 - 0.9) = 8.3 \times 10^3\,\text{sec}.$$

This means that it will take over 2 h for the headspace of the tank to be 90% saturated [51].

If the flow is between fluids that differ substantially in density, e.g., gas versus liquid or aqueous-phase versus nonaqueous liquid, the interface between the high and low-concentration compartments is a boundary or interface with films on each side (see Fig. 7.10). As mentioned, molecular diffusion occurs in quiescent and stagnant conditions, whereas mass transfer by advection occurs in turbulence. The connections can be constructed from transfer theories. The two-film theory [52] represents transport of a chemical species across the films on either side of the compartment boundary, which is governed by molecular diffusion, and the bulk of the fluid outside of the film, which is governed by bulk, nondiffusive transport. The concentration profile in the two-film figure assume steady-state conditions, i.e., the diffusive flux is constant. Other assumptions are that there is linear concentration profile through each stagnant film, equilibrium is instantaneous, and solutions are dilute so that Henry's law applies, i.e., concentration of a dissolved gas is directly proportional to the partial pressure of that gas above the solution (See Chapter 4). Two-film theory also assumes that transport by bulk diffusion is not limiting. Concentration gradients disappear in the bulk phases due to turbulence.

The two-film theory is the simplest example for interfacial mass transfer but does not represent many systems because stagnant interface layers do not exist [53]. Another theory of mass transport, penetration theory [54], is more realistic in that it assumes diffusion to be unsteady due to eddies moving from the bulk phase to the interface and remaining there for a constant time interval of exposure, t_e. Unlike the two-film theory it assumes the film is sufficiently thick, i.e., large δ values, to allow diffusion to dominate in the z-direction, but convection is dominant in the x-direction. The chemical compound penetrates an eddy during t_e by unsteady molecular diffusion. At the same time, the chemical compound diffuses into the interface fluid. Penetration theory predicts that the mass transfer coefficient is directly proportional to the square root of the molecular diffusivity, i.e., the diffusion coefficient, D_{AB}.

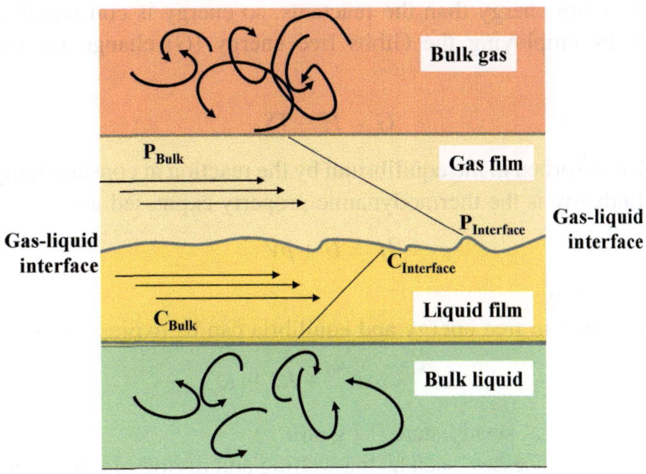

FIG. 7.10 Gas-liquid interface according to the two-film theory. *(Data from D.A. Vallero, Environmental Systems Science: Theory and Practical Applications, first ed., Elsevier, Amsterdam, NV, 2020.)*

These and other film theories clarify the understanding of the concentration gradient, but most are beyond the scope of this discussion. Empirical evidence indicates that the mass transfer coefficient, *k*, should vary with square root of the diffusion coefficient, *D*; and *k* should also vary with the two-third power of the fluid velocity, *v*. However, both of these film theories predict smaller variances [55]. This may in part be due to the nonuniform flow common in many environmental settings, especially in the troposphere. Some may also be due to the ongoing chemical kinetics that change the physical attributes of chemical compounds of interest.

5 Thermodynamics

The laws of motion and those of thermodynamics can combined in the previous discussion of the Bernoulli principle. The laws of thermodynamics apply to the carriers of pollution and to the pollutant's individual characteristics. The carrier is either a solvent or a matrix. The pollutant is either a solute if it is dissolved in the solvent, or a suspended particle. Water is the primary solvent for a water pollutant [56,57]. Sediment, soil and unconsolidated rock layers are common water-laden matrices, [58]. The inherent properties of a chemical pollutant include its solubility in water, solubility in other solvents like benzene, vapor pressure, and partitioning coefficients. These coefficients are expressions of the affinity of the pollutant to various compartments in an organism or in the environment.

As shown in Fig. 4.1 in Chapter 4, hydrogen atoms form an angle of 105 degrees with the oxygen atom. The asymmetry of the water molecule leads to a dipole moment in the symmetry plane pointed toward the more positive hydrogen atoms. Thus, the highly polar H_2O readily dissolves polar compounds. A dipole exists when two atoms form a molecule in which atoms share electrons unevenly in a covalent bond, i.e., part of the molecule becomes slightly positively charged and part becomes slightly negatively charged, as indicated by δ in in the water molecule. Such molecules are polar molecules. Water is a highly polar molecule and, since like dissolves like, is a very efficient solvent for other polar compounds. For example, water and ethanol (C_2H_5OH) dissolve each other to form a homogeneous mixture. This means that the two compounds are completely miscible. Ethanol's hydroxyl group (—OH) renders it polar. Ethanol and water each have an —OH group. Water scientists may show the water molecule as HOH in part to show that it can dissociate into H^+ and OH^- ions.

Conversely, nonpolar compounds covalently bond to share their electrons equally, so no dipole exists, which means a nonpolar compound has no net electrical charge. Methane (CH_4) is formed from covalent bonds, so it is highly nonpolar, i.e., only about 0.023 g of methane dissolves per kilogram of water at standard temperature and pressure (STP), whereas when a hydroxyl group replaces one of the hydrogen atoms to form methanol (CH_3OH), a dipole forms and the molecule becomes completely dissolved in water in any amount.

5.1 Enthalpy and entropy

Free energy is the measure of a system's ability to do work. In an exergonic reaction the reactants in a chemical reaction have greater free energy than the products, so energy is released from the reaction. An endergonic reaction occurs if the products from the reaction have more energy than the reactants, so energy is consumed. Equilibrium constants can be determined thermodynamically by employing the Gibbs free energy (G) change for the complete reaction. This is expressed as:

$$G = H - TS \tag{7.69}$$

where G is the energy liberated or absorbed in the equilibrium by the reaction at constant temperature (T). H is the system's enthalpy and S is its entropy. Enthalpy is the thermodynamic property expressed as:

$$H = U + pV \tag{7.70}$$

where U is the system's internal energy.

The relationship between a change in free energy and equilibria can be expressed by:

$$\Delta G^{*} = \Delta G_f^{*0} + RT \ln K_{eq} \tag{7.71}$$

where ΔG_f^{*0}= free energy of formation at steady state (kJ g mol^{-1}).

Free energy is involved in abiotic processes, such as the wetting and drying of porous materials like soil and sediment and in the process storing and releasing water pollutants. Free energy can be envisioned as the amount of work required to free a specific mass of water from the material from a pool of pure water at atmospheric pressure and a constant temperature. Free energy, i.e., water potential, also explains the relationship between relative humidity and water potential. Visualize a salt solution in an isothermal container at equilibrium as a closed thermodynamic system with a layer above the solution, i.e., head space, that has a partial pressure (e) is much lower than the saturation vapor pressure at the same temperature [$e_s(T)$]. The energy required to convert the liquid solution to vapor phase is the latent heat of vaporization (λ), i.e., from e_0 to $e_s(T)$ [5]:

$$\lambda = \frac{RT}{M_w} \ln\left[\frac{e_0}{e_s(T)}\right] \tag{7.72}$$

Where, M_w is water's molecular weight of water (18 g mol^{-1}).

Free energy is also at work, literally, in biological systems. For example, all cells need G for cell growth, cell division, and cell metabolism and health. The unusable energy is entropy (S), i.e., the disorder in the system, which tends to increase during the many conversion steps outside and inside of a system.

Heat exchange, partitioning, and other thermodynamic processes determine the amount of pollution generated, how a pollutant moves after being emitted, and the dynamics of a pollutant in water, air, soil, and sediment, as well as within the organism.

Entropy is key to how pollutants disrupt cellular metabolism, and to finding biological treatment technologies. All cells have two basic tasks, to survive and to grow. They must undergo biosynthesis, i.e., they must synthesize new biomolecules to construct cellular components. They must also harvest energy. Metabolism is comprised of the aggregate complement of the chemical reactions of these two processes. Thus, metabolism is the way that a cell derives energy from its surroundings and applies this energy to function and to construct more cellular material.

Catabolism consists of reactions that react with molecules in the energy source, e.g., carbohydrates. By degrading these larger molecules, cells gain energy. Anabolism is the synthesis of cell mass, which also requires energy. Anabolic reactions use the energy generated from the catabolic reactions [59–61].

The reactant and product concentrations and environmental conditions, like pH of the substrate, affect the observed free energy change (ΔG^{*}) values. If a reaction's ΔG^{*} is a negative value, the free energy is released and the reaction will occur spontaneously, and the reaction is exergonic. If a reaction's ΔG^{*} is positive, the reaction will not occur spontaneously. When the reverse reaction occurs, the reaction is endergonic.

As discussed in Chapter 5, each environmental compartment has a potential for a chemical compound to remain or leave, i.e., fugacity. In a compartment with a high fugacity and thus, a low fugacity capacity (Z), for a compound when there is no new source of that compound, the amount of mass of that compound will decrease precipitously with time. On the contrary, if a compartment has very low fugacity, that same chemical compound will have a strong affinity for that compartment, and it would require enough added energy for it to leave that compartment. Thus, a persistent substance will remain in sediment unless sufficient energy drives it into the water column [28,62]. This means that pollutant transport can be expressed by the amount of energy needed to overcome or reach equilibrium in each compartment. The processes that drive pollutant transport can be summarized using transfer coefficients, i.e., D values [63]. Fast processes have large D values and slow processes have small D values. Transport rate (N) is the product of fugacity (f) and the D value:

$$N = D \cdot f \tag{7.73}$$

The chemical compound concentration is $Z{\cdot}f$, so the transport rate can be expressed by this substitution and by adding a first-order rate constant k to give a first-order rate D value (D_R):

$$N = V[c]k = (V \cdot Z \cdot k) \cdot f = D_{R \cdot} f \tag{7.74}$$

Although the concentrations are shown as molar concentrations. They may also be represented as mass per volume concentrations. Diffusive and nondiffusive transport processes follow Fick's laws and can also be expressed with unique D values (D_D) that are related to the mass transfer coefficient (K) applied to area A:

$$N = K\mathrm{A}[\mathrm{c}] = (K \cdot \mathrm{A} \cdot Z.) \cdot f = D_{D \cdot} f \tag{7.75}$$

Nondiffusive transport, also known as advection or bulk flow, within a compartment with a flow rate (Q) has a different D value (D_A) that is expressed as:

$$N = Q[\mathrm{c}] = (Q \cdot Z \cdot)f = D_A \cdot f \tag{7.76}$$

Each chemical compound has a residence time in each compartment that is affected by many physical transport processes. In addition, its fate is determined by rates of abiotic and biotic transformation processes, discussed in Chapters 5 and 8. When a compound is degraded, the progeny compounds form with their own unique transport characteristic and D values (D_1, D_2, etc.), so the total rate of transport and transformation (N_{total}) is expressed as the product of fugacity times the sum of the D values:

$$N_{total} = f \cdot (D_1 + D_2 + \ldots\ D_n) \tag{7.77}$$

Water pollutant transport models improve the understanding the mechanisms for the movement of pollutants into and within surface and ground water [62,64–68]. Every molecule of mass moving into and out of the control volume must be accounted for, as well as any chemical changes to the pollutant that take place within a thermodynamic control volume. The first law of thermodynamics requires that any change in storage of a substance's mass in a control volume must equal the difference between the mass of the substance that moves into the system less the mass of the chemical transported out of the system. The reactions can generate greater mass of the compound (i.e., sources), or destroy chemical compound mass (i.e., sinks). The entering mass equals the inflow to the system that includes pollutant discharges, transfer from other control volumes and other media. The outflow is mass that is transported out of the control volume, which includes uptake by biota, volatilization to the atmosphere, deposition to surface waters and soil, and abiotic and biological chemical degradation [15]. Therefore, the rate of change of a chemical compound's mass in a control volume equals the rate of mass transported into the control volume minus the rate of mass transported out, plus the rate of production from sources, and minus the rate of elimination by sinks.

Chemical compounds exit and enter water by precipitation from the atmosphere, advective transport from other surface and ground water systems, diffusion from sediment, soil, and biota, and various other intercompartmental transfers [69]. As discussed in Chapter 5, partitioning can be applied to estimate and to model where a substance will go within a designed system like a trickling filter or sedimentation basin and after it is released to the environment, as well as to model natural processes, such as transport between surface waters and the atmosphere. Physical mechanisms involved in these transfers include sorption, solubility, volatilization, dissolution, and gravity. Interactions of these mechanisms occur within an environmental compartment. A mass fraction will remain unchanged in the compartment. Another fraction remains unchanged but is transported to a different compartment while another fraction is chemically transformed with all products remaining products staying in the compartment where the transformation products were generated. Finally, a fraction of the original pollutant mass is transformed and then moved to another compartment.

Physical processes affect water pollution. Increases in ambient temperatures increase evaporation rates into the atmosphere. Warmer air holds more water than cooler air, which means that increased evaporation rates can increase desiccation of soils and biota in the water loss areas. These and other changes threaten the optimal conditions of living creatures. Although the total amount of water in the hydrosphere will remain relatively constant, the specific locations of the water can change substantially with changes in environmental conditions. For example, water may move from impounded surface waters to the atmosphere with increased ambient temperatures, which in turn alter precipitation patterns, like the frequency and intensity of storm systems [70,71].

Density is a physical property. Salinity, expressed as the total concentration of dissolved solids (TDS), is a chemical property. Denser saline water can wedge beneath freshwaters and pollute surface waters and ground water (see Chapter 11). Saltwater intrusion can drastically alter an ecosystem's structure and function and threaten the habitat of freshwater organisms. It harms coastal communities dependent on aquifers for their water supply, whether the water source is a public treatment system or a private well [72,73]. During saltwater intrusion, denser saltwater submerges under the less dense freshwater. The concentration of ions is higher in the marine system than in the tidal river, creating an ion flux in the direction of the fresh surface waters. The same process occurs in coastal aquifers as saline water intrudes into freshwater stored underground into the saturated (vadose) zone [15,74–76].

The physics of saltwater intrusion [74] is expressed as:

$$\rho_s g z = \rho_f g (z + h) \tag{7.78}$$

Where, ρ_f is the density of freshwater, ρ_s is the density of the saline water, h is the thickness of the freshwater zone above sea level, z is the thickness of the underground zone of saline water below sea level, and a g is the gravitational constant (6.67408×10^{-11} m^3 kg^{-1} s^{-2}).

The gravitational term is removed by solving for z:

$$z = \frac{\rho_f}{\left(\rho_{s-}\rho_f\right)} h \tag{7.79}$$

ρ_f is the density of freshwater and ρ_s is the density of the saline water. ρ_f, which at 20 °C is about 1.000 g cm^{-3}, while ρ_s is about 1.025 g cm^{-3}. At these densities, z at 20 °C and 1 atm equals 40, thus $z = 40h$. Saltwater intrusion is usually irreversible due to the ionic strength of ocean water which means the ionic concentration gradient in the direction of the freshwater is continuous [75,77]. The interface is a transition or dispersion zone as additional physical properties of the surface water and the aquifer cause the ions to migrate due to drawdown of ground water by pumping and recharge, as well as tidal behavior [75,78].

Energy mixing occurs as the kinetic energy of the tide mixes with the potential energy of the freshwater stream flow and to a lesser extent wind and wave kinetic energy. Mixing dissipates the potential energy of the fresh water. The mixing mechanisms of the estuary maximize the power represented by a dispersion coefficient of all mixing mechanisms. The resulting differential equation is solved by combining the advective and dispersive forces [79]:

$$A\frac{\partial S}{\partial t} + Q\frac{\partial S}{\partial x} - \frac{\partial}{\partial x}\left(AD\frac{\partial S}{\partial x}\right) = 0 \tag{7.80}$$

Where, S is the salinity expressed in practical salinity units based on the electrical conductivity (EC) of the water, A is cross-sectional area of the flow, D is the dispersion coefficient, t is the time, x is the distance from the estuary mouth, and Q is the water flow in the estuary. The first term is the salinity time gradient caused by the balance between the advective flow, i.e., Q, and the mixing of water having different salinity by the dispersive exchange flow, i.e., $\frac{\partial}{\partial x}\left(AD\frac{\partial S}{\partial x}\right)$. Without other salinity sources, the sum of the terms equals zero [79]. Averaging the equation over a complete tidal period would reflect more extended, long-term change in salinity due to the fresh water advection and the flows of the average tidal exchanges, which would eliminate the first term of Eq. (7.80) [79]. Thus, to reach steady state, the equation is integrated with respect to x:

$$Q\left(S - S_f\right) - AD\frac{dS}{dx} = 0 \tag{7.81}$$

Where $\frac{dS}{dx}$ is the upstream boundary and S_f is the salinity of the fresh water; and with the conditions that $\frac{dS}{dx} = 0$ and $S = S_f$. Under this steady-state condition, the water flow, Q, equals the fresh water upstream flow, so that the saltwater-freshwater wedge can be visualized as the mixing and the slopes of the salinity distribution and the hydrostatic pressure (see Chapter 10). In this steady-state condition, the salt that is flushed out by the fresh river flow is balanced by the exchange of saline and fresh water from various mixing processes, causing an upstream salt flux. An upstream gradient can result from decreasing salinity. Given the density difference, hydrostatic pressures on both sides are not equal. The water level at the toe of the salt intrusion curve is higher by Δh, which results in a seaward pressure difference toward the surface and an inland pressure difference toward the estuary floor. The integrals of the hydrostatic pressure distributions are equal and opposed in steady state, but they have different working lines $\frac{\Delta h}{e}$ apart, which triggers angular momentum [79].

The laws of thermodynamics and motion are involved in saltwater intrusion, resulting in a dissipation that conserves mass and energy while exporting entropy [80]. A rise in sea level can submerge lower-lying coastal areas and exacerbate saltwater contamination of ground water, threatening surface water supplies. For example, aquifers that are recharged by freshwater in streams will have increased salinity during severe droughts [81]. Both surface elevation and aquifer settings can render a near-coastal area susceptible to increased salinity in both surface and ground water [82].

6 Heat energy

Exchanges between water and other media, e.g., air, soil, and sediment, create sinks and sources of energy. Heat reservoirs in terrestrial and aquatic systems receive added heat that can alter habitats, such as new freeze-thaw cycles, seasonal variations and selectivity of certain microbial species, and loss of heat-sensitive fish species [83].

Certain water-heat relationships are direct, e.g., the thermal inversions in lakes. Some are subtle and indirect, such as the transfer of energy between trophic states in an ecosystem, which relies in part on energy and water exchanges between biotic tissue and abiotic substrates. For example, photosynthesis requires aqueous-phase phosphorous compounds to transfer energy and also the optimal amount of heat to allow for metabolism and growth, which requires a balance of water, sunlight, and heat [84].

Latent heat relates to the rate of change of vapor pressure with temperature. Consider a chemical compound in its vapor phase at constant temperature in equilibrium with its liquid phase held in a cylinder that contains a piston. If the piston is pulled outward, at constant temperature the liquid will evaporate, and the pressure remain constant. If the piston is pushed inward, the vapor will condense at constant temperature and pressure. The cylinder's pressure and temperature remain constant, which means that Gibbs free energy of the system is constant. G_1 is the Gibbs free energy for the liquid and G_2 is the specific Gibbs free energy for the vapor [85]. If a liquid mass (*dm*) vaporizes, the Gibbs free energy for the liquid decreases by G_1dm while the Gibbs free energy for the vapor increases by G_2dm. However, the Gibbs free energy for the whole closed system is constant. This indicates that for a liquid in equilibrium with its vapor phase, i.e., its boiling point, the specific Gibbs free energies of liquid and vapor are equal [85]:

$$H_1 - TS_1 = H_2 - TS_2 \tag{7.82}$$

And,

$$T(S_2 - S_1) = H_2 - H_1 \tag{7.83}$$

Where, the enthalpy and entropy are specific. The left-hand side is the specific latent heat of vaporization, which is the difference in the specific enthalpies of liquid and vapor.

The above relationships can also be expressed as [85]:

$$U_1 - TS_1 + pV_1 = U_2 - TS_2 + pV_2 \tag{7.84}$$

And,

$$T(S_2 - S_1) = (U_2 - U_1) + p(V_2 - V_1) \tag{7.85}$$

The latent heat increases the internal energy upon vaporization and the works against the external pressure with increasing volume. The latent heat, then, can be divided into internal latent heat and external latent heat.

If a liquid vaporizes due to added heat at constant volume, the specific Gibbs free energies of both the liquid and vapor increase, but by the same amount. This is because when a liquid and its vapor are in equilibrium at the boiling point, their specific Gibbs free energies are the same [85]:

$$-S_1dT + V_1dp = -S_2dT + V_2dp \tag{7.86}$$

And,

$$\frac{dp}{dT} = \frac{S_2 - S_1}{V_2 - V_1} \tag{7.87}$$

The left-hand side shows the rate of increase of vapor pressure with temperature, while $S_2 - S_1$ is equal to $\frac{\lambda}{T}$ where λ is the latent heat of vaporization. This is an expression of the Clausius-Clapeyron equation [85]:

$$\frac{dp}{dT} = \frac{\lambda}{T(V_2 - V_1)} \tag{7.88}$$

Clausius-Clapeyron equation example (from Ref. [85])

Find the specific latent heat of vaporization if at 100 °C, the rate of increase in vapor pressure of vaporized water, i.e., steam, is 27.1 mmHg per °C, and 1 g of steam occupies a volume of 1674 cm³.

Solution:

$$\frac{dp}{dT} = \frac{\lambda}{T(V_2 - V_1)}; \text{so } \lambda - T(V_2 - V_1)\frac{dp}{dT}$$

Convert *°C to Kelvin:*

$$T = 373.15$$

$$V_2 - V_1 = 1.673 \text{ m}^3 \text{ kg}^{-1}$$

$$\frac{dp}{dT} = 1.36 \times 10^4 \times 9.81 \times 2.71 \times 10^{-2} = 3.616 \text{ pascals per degree Kelvin (Pa K}^{-1}\text{), so } \lambda = 2.26 \times 10^6 \text{ J kg}^{-1}.$$

Water-energy cycling is also important in anthropogenic systems and is part of a design of pollution control equipment selection and application. For example, at the facility-scale, water and energy are often addressed together, such as the management of water and energy at a factory or power plant. Managing water is a simultaneous process with managing energy. Indeed, mismanagement of water systems at the facility-level can involve tradeoffs between types of pollution. The first law of thermodynamics requires, for example, that allowing heated water to be released in any amount, even the permitted level, would increase the overall temperature of the receiving stream.

The water-energy exchanges will also scale upwardly. That is, heat can initiate cumulative environmental impacts, such as the heat exchange and balances that change the conditions of receiving water bodies [61]. Until the 1970s, most power plants near the major rivers of the U.S. released heated water into the river (see Fig. 7.11). The incremental effect of the

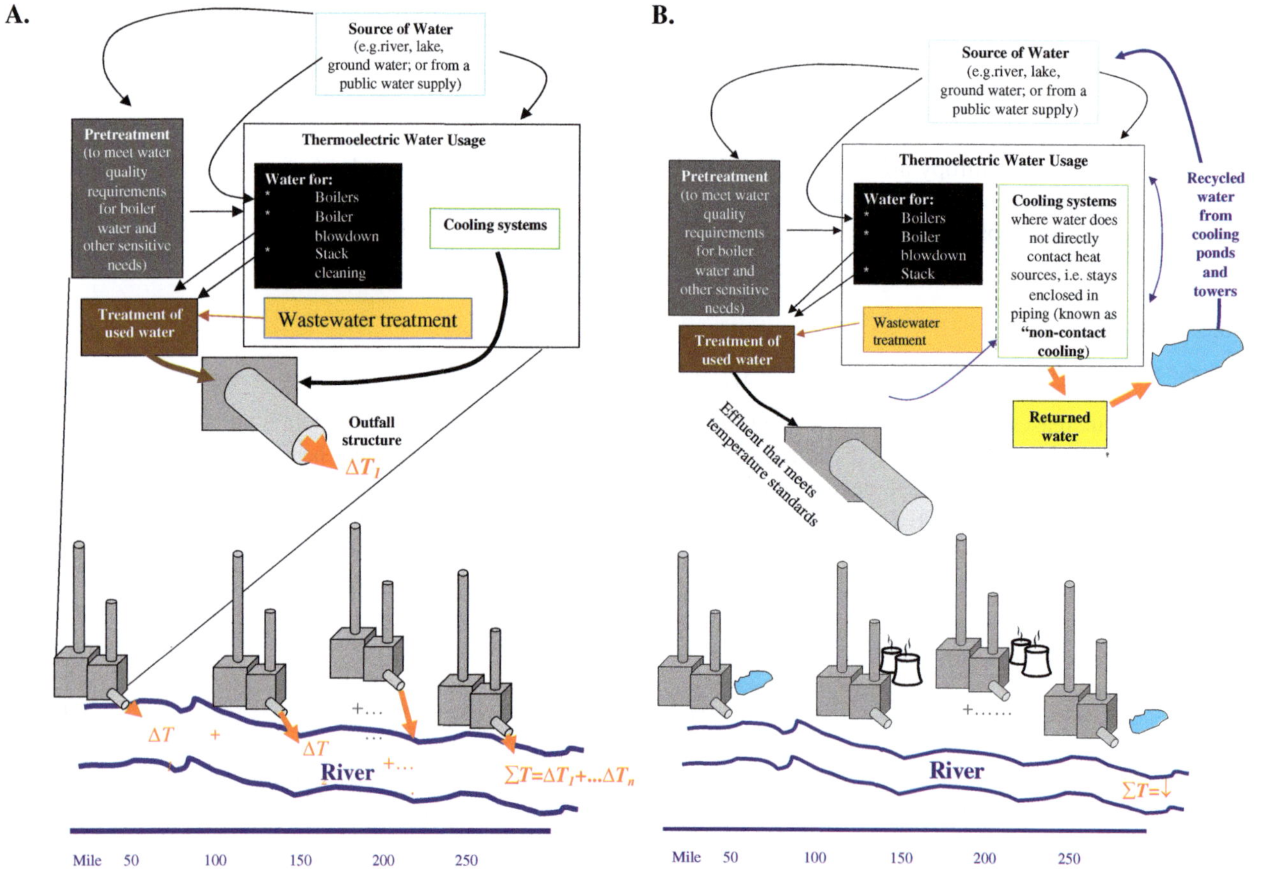

FIG. 7.11 Difference in cumulative heat contribution to a river from electric generating plants using a once-through cooling system (A) versus the same plants using a cooling water return system (B). The cumulatively added heat is greatly reduced with the closed water return systems compared to the once through cooling systems (hypothetical scenario). *(Data from D. Vallero, Environmental Biotechnology: A Biosystems Approach, Elsevier Science, 2015.)*

releases led to a cumulative increase in the temperature [86]. In the late 1970s, once-though cooling, i.e., cooling water passing through turbines and then discharged to adjacent streams, was being phased out in U.S. waters (Fig. 7.11A). With time, other cooling systems, e.g., cooling towers and cooling lakes were increasingly being installed and operated, so that power plant water systems became more closed, both from a fluid dynamics and thermodynamic perspective (Fig. 7.11B).

Heat has both direct and indirect effects on aquatic ecosystems. The increased temperature can become a hostile environment for sensitive biota, a direct impact. In addition, pollution control and heat control design can directly affect dissolved oxygen (DO) content of the receiving water temperature since temperature is directly proportional to DO content. The DO is a limiting factor of the type of fish communities that can be supported by a water body. The resulting net increase in heat may directly stress the biotic integrity of a surface water ecosystem, e.g., fish species vary in their ability to tolerate higher temperatures, meaning that the less tolerant, higher value fish will be inordinately threatened.. The increased temperature can also increase the aqueous solubility of substances toxic to organisms. For example, greater concentrations of mercury and other toxic metals will occur with elevated temperatures. The lower DO concentrations will lead to a chemically reduced environment where the metals and compounds will form sulfides and other compounds that can be toxic to the aquatic life. Thus, the change in temperature, the resulting decrease in DO and increasing metal concentrations, and the synergistic impact of the combining the hypoxic water and reduced metal compounds is a cascade of harm to the stream's ecosystems (see Fig. 7.12).

Biota also play a role in the heat-initiated effect. Combined abiotic and biotic responses occur. Notably, the growth and metabolism of the bacteria results in even more rapidly decreasing DO levels. Algae both consume DO for metabolism and produce DO by photosynthesis. The increase in temperature increases their aqueous solubility and the decrease in DO is accompanied by redox changes, e.g., formation of reduced metal species, such as metal sulfides. These reactions are also being mediated by the bacteria, some of which will begin chemically reducing the metals as the oxygen levels drop (reduced conditions in the water and sediment). However, the opposite is true in the more oxidized regions, i.e., the metals are forming oxides. The increase in the metal compounds combined with the reduced DO, combined with the increased temperatures can act synergistically to make the conditions toxic for higher animals, e.g., a fish kill.

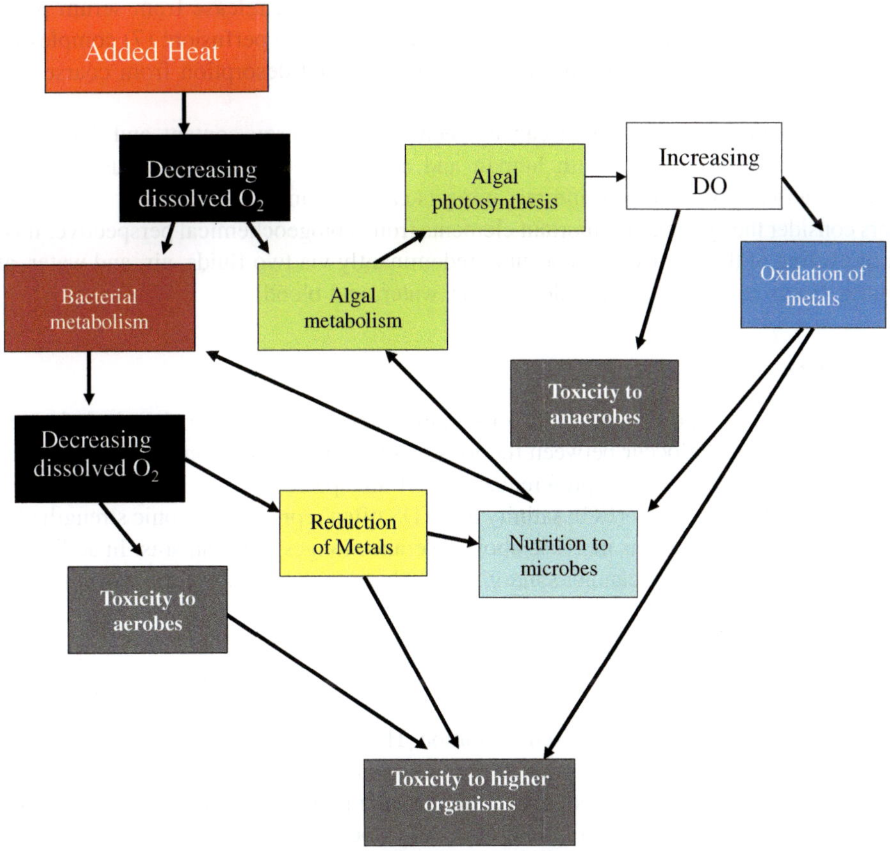

FIG. 7.12 Adverse effects in the real world usually result from a combination of conditions. In this example, both energy and matter must be considered in the control of air pollution. The control of matter leaving the combustion unit (upper left-hand) must be optimized with the transfer of energy (middle). The added heat results in an abiotic response (i.e., decreased dissolved oxygen (DO) concentrations in the water), which leads to biotic processes that either increase or decrease DO and increase the toxic responses (e.g., low DO and increase in metal bioavailability). *(Data from D. Vallero, Environmental Biotechnology: A Biosystems Approach, Elsevier Science, 2015.)*

The initiating abiotic effect (i.e., increased temperature) results in an increased microbial population. The associated decline in DO, however, is a net decline between photosynthetic and nonphotosynthetic microbes. The growth and metabolism of the bacteria result in decreasing the DO levels, but the growth of the algae both consume DO for metabolism and produce DO by photosynthesis.

As mentioned, another indirect heat effect is the combined abiotic and biotic exposure to metals. The increase in temperature increases their aqueous solubility and the decrease in DO is accompanied by redox changes, e.g., formation of reduced metal species, such as metal sulfides. These changes can increase the exposure of human populations to toxic substances. The increase in temperature in Fig. 7.11, for example, leads to increased solubility of metals. This is also being mediated by the bacteria, some of which will begin reducing the metals as the oxygen levels drop (reduced conditions in the water and sediment). However, the opposite is true in the more oxidized regions, i.e., the metals are forming oxides. The increase in the concentrations of metal compounds, combined with the reduced DO, combined with the increased temperatures can act synergistically to make the conditions toxic for higher animals, e.g., loss of sensitive aquatic species and fish kills [87].

Predicting the likelihood of ecosystem adverse events like fish kills is complicated, given the many factors that either mitigate or exacerbate the outcome. The increase in metal concentrations in water not only increases the potential exposure to total metal mass but may also transform the metals into chemical species with much higher bioavailability, which increases the dose and body burden of the metals in organisms up the food chain (e.g., methyl mercury in fish ingested by humans). Environmental transport pathways can be affected by net heat gain. Chemical compounds, microbes, and energy follow various pathways as they enter water, air, soil, sediment, and biota. The residence time within each of these compartments is determined substantially by various forms of energy, including heat, light, and mechanical energy. At the cellular scale, numerous processes and exchanges occur among and within the environmental compartments, as well as those at the interface between the organism and the environment. The gill-water exchanges by fish, for example are analogous to the lung-air exchanges in air-breathing animals.

Transfer of matter is a key part of the bioaccumulation process in a multiphase system (water, sediment, particles, and biota), as represented by the gill [88] and the lung. The mechanisms include: (1) water flow across membrane; (2) blood flow within organism; (3) chemical flux across membrane; (4) binding and release from serum proteins; (5) sorption/desorption to blood cells; (6) chemical mass transfer from blood to tissues by perfusion; (7) complexation to and decomplexation from organic carbon in a particulate phase; (8) sorption to and desorption from coarse particulate solids, in addition to internal diffusion within the particles.

Water is involved in biomass growth, which can be seen as heat energy content and flow (e.g., kcal fish^{-1} and kcal day^{-1}) [89–91]. Factors that lead to health human and ecological communities must account for atmospheric exchanges of mass and energy. The direct and indirect impacts can be dramatic.

The next chapters consider the cycling of important elements from a biogeochemical perspective. It is important to keep in mind that much of cycling of these substances occurs predominantly via two fluids, air, and water, at most scales, and predominately by three fluids at the organism scale, i.e., air, water, and blood.

7 Electrical energy

Chemical and physical properties of pollutants are at work in ions since an ion has an electrical charge that can attract or repel other ions. The interactions that occur between the ions of solutions and the ions in water may induce shifts in the composition of aquatic and other biota due to preferences for certain species.

Measurements of electrical conductivity (EC), salinity and TDS often represent the ionic strength of water. Ions are not inherently harmful until their concentrations increase above tolerance ranges of organisms. In addition, increases in ionic strength in the presence of other stressors can act synergistically by increasing an organism's uptake of toxic substances [92,93].

Increases in ionic strength can alter soil structure and decrease permeability [94]. The structural stability, dispersion, swelling, crusting, EC, and osmotic potential of soils are affected by increases in ionic strength, which in turn increase pollutant runoff and soil erosion [95].

As mentioned, TDS does not inherently represent pollution. Higher TDS is expected in the saltwater. However, the migration if ions and dissolved compounds from saltwater to freshwaters does indeed pollute the freshwater, which should have much lower TDS. Thus, TDS is an example of conventional water pollution that have acceptable ranges. Conventional water pollutants also include pH, total suspended solids (TSS), and biochemical oxygen demand (BOD), discussed in detail in Chapter 10. For example, effluent limits are set by environmental agencies for TSS, BOD, other conventional pollutants, and toxic substances, as well as pathogenic microbes. Before industries, commercial operations, and wastewater treatment plants are allowed to release the effluent to surface waters or apply it to land surfaces, they must ensure that they will meet standards. Water quality standards limit these and other substances, e.g., radionuclides, leaving water treatment facilities prior to the treated water being distributed to households and other users. To ensure the water meets these standards, they can be tested as they exit the facility, within the distribution system, and at the point-of-use, e.g., a tap in a residence. All too frequently, water quality is diminished as it travels from the plant to the user [96–99].

Svante Arrhenius observed that certain chemical compounds, subsequently named electrolytes, could conduct electricity when dissolved in water, while other chemical compounds do not. He also observed differences in chemical reaction times, from instantaneous to very long times. Finally, he observed that certain chemical compounds showed extremely strange colligative properties, but others adhered to Raoult's Law, which states that the solvent's vapor pressure in an ideal solution is equal to the product of the mole fraction of the solvent and the vapor pressure of the pure solvent. The colligative properties of solutions are the elevation of boiling point, the depression of freezing point, the decreasing of vapor pressure, and osmotic pressure.

Total dissolved solids values correlate directly with the water's EC and indirectly with resistance, so water quality standards often include both conductivity and TDS as criteria. Liquid water's EC is its capacity to conduct an electric charge [100–102]. The degree of conductivity depends on the concentration of dissolved solids and temperature, with both adding energy to the water. Ionic compounds most commonly form in reactions between metals and nonmetals in which electrons are gained and lost. Metals lose electrons and nonmetals gain electrons in these reactions. After they form, the salts are held together by ionic bonds. When dissolved in water, ionic compounds dissociate into ions. The total electrolyte concentration in the ionic solution governs the solubility of specific salts, i.e., their dissociation constant:

$$A_xB_y \leftrightarrow {}_xA + {}_yB \tag{7.89}$$

Where A is the metal and B is the nonmetal. In A_xB_y complex breaks down to the cation (${}_xA$) and anion (${}_yB$). The double arrow indicates that this is an equilibrium reaction. As such, an equilibrium constant is needed, i.e., the dissociation constant for ionic reactions is K_D. For the A and B reaction above K_D is determined by:

$$K_D = \frac{[A]^x[BA]^y}{[A_xB_y]} \tag{7.90}$$

K_D indicates the tendency of a bigger object to separate, i.e., dissociate, into smaller components, such as when a complex breaks into component molecules, or in this case, when the salt breaks into its cation and anion. Incidentally, this ionization reaction is irreversible.

Electrical conductivity can be measured *in situ* with portable meters. Conversely, the analysis of TDS is more difficult and expensive [10]. However, under specific scenarios, direct TDS analysis is preferred, such as to demonstrate ground water quality in a regulatory investigation [11]. Researchers have conducted many investigations to correlate between these two parameters to allow TDS concentration to be simply extrapolated from the EC value. The correlation of these parameters can be estimated by the following equation [102,103]:

$$\text{TDS} = k_e \times \text{EC} \tag{7.91}$$

Where, k_e is a correction coefficient, TDS is in units of mg L^{-1} and EC is in units of microSiemens per centimeter ($\mu S\ cm^{-1}$).

The value of k_e will increase directly with the ion concentration in water, but not linearly. The relationship between conductivity and TDS depends on the average activity of all ions in the liquid [102].

8 Solution physics

A water pollutant is ether suspended or dissolved in the water. Suspended pollutants retain their solid physical phase. Dissolution is a physical process, not a chemical reaction. A solution is a mixture in which the dissolved substance remains intact and held in a solvent. Salts dissolve in solutions. Fluids dissolve in other fluids. Gases dissolve in other gases, which are expressed as partial pressures, e.g., methane in air. Gases dissolve in liquids, e.g., CO_2 in water. Liquids dissolve in liquids, e.g., octanol in water. Solubility is an inherent, saturation property of that compound that is commonly expressed in units of mass of solute per volume of solvent, e.g., mg L^{-1}, or as the mass of a solute in mass of solvent, such as mg kg^{-1}.

In handbooks and texts, if the solvent of interest is not water, the solvent is specified, e.g., the solution of an organic compound in solvents like dimethyl sulfoxide (DMSO), benzene, methanol, or hexane. Solubility is temperature-dependent, so the reported values are at a specified temperature. Solubility may also be expressed as volume per volume, represented as parts per million (ppm), parts per billion (ppb), or parts per trillion (ppt).

Solubility is the saturation concentration of a dissolved solute in a solvent in equilibrium with undissolved solute at a fixed temperature and pressure [104], which means that the solubility of a compound can be explained by comparing it to the ideal solubility, i.e., a solute-solvent system that obeys Raoult's law which, as mentioned, states that for an ideal solution, the equilibrium partial pressure (p) of a pure substance at constant temperature is the product of the of the substance's vapor pressure and its mole fraction (x_A) in the liquid. Chemical solutions that obey Raoult's law at every concentration and at every temperature are ideal solutions. Mixing fluid A and fluid B will result in intermolecular forces:

- A-A intermolecular forces
- B-B intermolecular forces
- A-B intermolecular forces

When all three types of intermolecular forces are about equal, the solution is considered to be an ideal solution. The ideal solubility for a solute A at temperature T, and at a partial pressure p of 1 atm is given by [105]:

$$x_A = \frac{p}{p_A^*} \tag{7.92}$$

Where p_A* is the vapor pressure of the pure solute at temperature T. Thus, p and are x_A linearly related.

The solubilities of some of the gases in Table 7.3 differ by more than 5 orders of magnitude from each other. A molecule of hexafluoroethane (C_2F_6) is saturated at one per million water molecules, whereas it takes 259,000 molecules of HCl to reach saturation in 1 million molecules of water. Water molecules strongly associate, which relates to the values that show strong B-B interactions. For nonpolar substances like C_3H_8 and SF_6, the solute-solvent association is so weak that the experimental solubility is vastly lower than the ideal solubility for these compounds. For a highly polar compound like HCl there are strong interactions between the polar HCl and the water leading to a solubility that is larger than either B-B or A-A interactions, resulting a very high aqueous solubility for HCl [104].

Water is unique, given its strong self-association through hydrogen bonding [25]. Thus, inconsistencies between experimental and ideal solubilities for water systems are more common than for any other solvent.

TABLE 7.3 Solubility of gas (A) in water (B) at 25 °C and 101.325 kPa (1 atm).

Gas	Mole fraction, $x_A \times 10^5$	Ideal solubility, $x_A \times 10^5$
Helium (He)	0.71	–
Argon (Ar)	2.53	–
Hydrogen (H_2)	1.41	–
Nitrogen (N_2)	1.18	–
Oxygen (O_2)	2.30	–
Methane (CH_4)	2.55	–
Propane (C_3H_8)	6.99	11,600
Sulfur hexafluoride (SF_6)	0.44	4000
Hexafluoroethane (C_2F_6)	0.10	–
Carbon dioxide (CO_2)	61	1780
Fluoromethane (CH_3F)	106.2	3000
Sulfur dioxide (SO_2)	2460	2500
Hydrochloric acid (HCl)	25,900	2000

Data from R. Battino, T.M. Letcher, An introduction to the understanding of solubility, J. Chem. Educ. 78 (1) (2001) 103, J. Lorimer, P. Gujral, Solubility data series, in: Alkaline Earth Hydroxides in Water and Aqueous Solutions, Citeseer, 1979.

TABLE 7.4 Solubility of naphthalene in various organic solvents at 20 °C.

Solvent	Solubility mole fraction, x_A
Chlorobenzene	0.256
Benzene	0.241
Toluene	0.224
Carbon tetrachloride	0.205
Hexane	0.090
Ideal solubility[a]	**0.273**

[a]*The agreement between the experimental and ideal values is good considering the simple nature of the ideal model which again is based on the idea that A-A, B-B and A-B interaction are all the same.*
Data from D.A. Vallero, T.M. Letcher, Engineering risks and failures: lessons learned from environmental disasters, Leadersh. Manag. Eng. 12 (4) (2012) 199–209, K.G. Denbigh, K.G. Denbigh, Equilibria of reactions involving gases, in: The Principles of Chemical Equilibrium: With Applications in Chemistry and Chemical Engineering, fourth ed., Cambridge University Press, Cambridge, United Kingdom, 1981 (Chapter 4).

For solids dissolved in liquids, the ideal solubility is expressed as [106]:

$$\ln x_A = -\frac{\Delta H_{\text{fusion,A}}(T_m)}{RT}\left[1 - \frac{T}{T_m}\right] \tag{7.93}$$

Where, x_A is the solubility mole fraction of the solute at temperature T, T_m is the melting temperature of the solute, and $\Delta H_{\text{fusion, A}}$ is the enthalpy of fusion, i.e., melting, of the solute at T_m. Table 7.4 shows the solubility of naphthalene calculated from Eq. (7.93) in various solvents. $\Delta H_{\text{fusion, A}}$ and the T_m for naphthalene are 18.58 kJ mol^{-1} and 353.20 K, respectively [107].

A hydrophilic compound is highly water soluble, i.e., easily dissolved in water under normal environmental conditions of temperature and pressure. A hydrophobic substance is not easily dissolved in water under these conditions. Since many contaminants are organic chemical compounds that consist of molecules containing carbon-to-carbon bonds and/or carbon-to-hydrogen bonds, solubility can be differentiated by their lipid solubility. A lipophilic substance is readily dissolved in lipids. A lipophobic compound resists being dissolved in organic solvents under these conditions. Usually, a hydrophilic compound is lipophobic, and a hydrophobic compound is lipophilic.

This affinity for either water or lipids is the key to environmental partitioning. For example, the octanol-water partition coefficient (K_{ow}) is the ratio of a substance's concentration in octanol ($C_7H_{13}CH_2OH$) to the substance's concentration in water at equilibrium. In this case, octanol represents lipophilic solvents in general because it has degrees of affinity for both water and organic compounds, that is, octanol is amphiphilic.[f] Because the ratio forming the K_{ow} is [$C_7H_{13}CH_2OH$]: [H_2O], the larger the K_{ow} value, the larger the substance's lipophilicity will be.

Water solubility is inversely related to K_{ow}, but the inverse relationship is not universal as the solubility of organic compounds is widely ranging. Organic compounds can have affinities for neither, either, or both the organic and the aqueous phases. Most compounds are not completely associated with either phase.

The inherent properties of pollutants do not completely predict a substance's affinity for an environmental compartment. A substance's vapor pressure and molecular weight, for example, will give clues to the behavior of a pollutant, but are not definitive. These and other physicochemical properties combine to determine environmental fate.

9 Conclusions

Physical principles drive many aspects of water pollution. The laws of motion and thermodynamics determine the formation, transport, transformation, and fate of chemical and biological agents in the environment. The discussions in this chapter are preparation for chemical principles in the next chapters.

f. If strictly lipophilic solvent were used, it would be very difficult to measure K_{ow} since only lipophilic compounds would dissolve.

References

[1] Y. Zhang, X. Huang, Z. Zhang, J. Blewett, B.D.A. Naafs, Spatiotemporal dynamics of dissolved organic carbon in a subtropical wetland and their implications for methane emissions, Geoderma 419 (2022) 115876, https://doi.org/10.1016/j.geoderma.2022.115876.

[2] M. Sadat-Noori, W. Glamore, Porewater exchange drives trace metal, dissolved organic carbon and total dissolved nitrogen export from a temperate mangrove wetland, J. Environ. Manag. 248 (2019) 109264, https://doi.org/10.1016/j.jenvman.2019.109264.

[3] I.R. Santos, B.D. Eyre, M. Huettel, The driving forces of porewater and groundwater flow in permeable coastal sediments: a review, Estuar. Coast. Shelf Sci. 98 (2012) 1–15.

[4] A.J. McElrone, B. Choat, G.A. Gambetta, C.R. Brodersen, Water uptake and transport in vascular plants, Nat. Educ. Knowl. 4 (6) (2013).

[5] J. Monteith, M. Unsworth, Principles of Environmental Physics: Plants, Animals, and the Atmosphere, Academic Press, 2013.

[6] E. Wild, J. Dent, G.O. Thomas, K.C. Jones, Direct observation of organic contaminant uptake, storage, and metabolism within plant roots, Environ. Sci. Technol. 39 (10) (2005) 3695–3702.

[7] E. Steudle, J. Frensch, Water transport in plants: role of the apoplast, Plant Soil 187 (1) (1996) 67–79.

[8] E. Steudle, J. Frensch, Osmotic responses of maize roots, Planta 177 (3) (1989) 281–295.

[9] J.R. Philip, Plant water relations: some physical aspects, Annu. Rev. Plant Physiol. 17 (1) (1966) 245–268.

[10] W.R. Gardner, Dynamic aspects of water availability to plants, Soil Sci. 89 (2) (1960) 63–73.

[11] G.W. Yu, J. Laseter, C. Mylander, Persistent organic pollutants in serum and several different fat compartments in humans, J. Environ. Public Health 2011 (2011).

[12] G.W. Olsen, et al., Half-life of serum elimination of perfluorooctanesulfonate, perfluorohexanesulfonate, and perfluorooctanoate in retired fluorochemical production workers, Environ. Health Perspect. 115 (9) (2007) 1298–1305.

[13] D.A. Vallero, case studies, in: Applications and Calculations in Environmental Physics, AIP Publishing, Melville, New York, 2022, pp. 6-1–6-52. Chapter 6.

[14] D.A. Vallero, Physics of ecosystems, in: Methods and Calculations in Environmental Physics, AIP Publishing LLC, Melville, New York, 2022, pp. 5-1–5-38.

[15] D.A. Vallero, Methods and Calculations in Environmental Physics, AIP Publishing, Melville, New York, 2022.

[16] D.A. Vallero, Applications and Calculations in Environmental Physics, AIP Publishing LLC, 2022.

[17] C. Hird, E. Flanagan, C.E. Franklin, R. Cramp, Cold-induced skin darkening does not protect amphibian larvae from UV-associated DNA damage, J. Exp. Zool. A Ecol. Integr. Physiol. 341 (3) (2024) 272–281.

[18] J. Awkerman, et al., Guidance for developing amphibian population models for ecological risk assessment, Integr. Environ. Assess. Manag. 16 (2) (2020) 223–233.

[19] L.E. Licht, K.P. Grant, The effects of ultraviolet radiation on the biology of amphibians, Am. Zool. 37 (2) (1997) 137–145.

[20] P.S. Spencer, P.J. Lein, Neurotoxicity, in: P. Wexler (Ed.), Encyclopedia of Toxicology, third ed., Academic Press, Oxford, 2014, pp. 489–500.

[21] E.L. Johnston, M.J. Keough, Direct and indirect effects of repeated pollution events on marine hard-substrate assemblages, Ecol. Appl. 12 (4) (2002) 1212–1228.

[22] Research and Education Association, The Essentials of Fluid Mechanics and Dynamics, Research and Education Association, Piscataway, NJ, 1987.

[23] D. Vallero, Air Pollution Calculations: Quantifying Pollutant Formation, Transport, Transformation, Fate and Risks, Elsevier, 2019.

[24] J.P. Kim, K.A. Hunter, M.R. Reid, Factors influencing the inorganic speciation of trace metal cations in fresh waters, Mar. Freshw. Res. 50 (4) (1999) 367–372.

[25] R. Schwarzenbach, P. Gschwend, D. Imboden, Environmental Organic Chemistry, John Wiley & Sons, Hoboken, New Jersey, 2016.

[26] G.A. Pope, K. Sepehrnoori, M.M. Sharma, D.C. McKinney, G.E. Speitel, R.E. Jackson, Three-dimensional NAPL fate and transport model, Rep. EPA/600/R-99, vol. 11, 1999.

[27] K. Isaacs, J. Rosati, T. Martonen, L. Ruzer, N. Harley, Modeling deposition of inhaled particles, in: Aerosols Handbook, CRC Press, Boca Raton, FL, 2012, pp. 83–128.

[28] D.A. Vallero, Fundamentals of Air Pollution, fifth ed., Elsevier Academic Press, Waltham, MA, 2014. p. 999 pages cm.

[29] K. Isaacs, J.A. Rosati, T.B. Martonen, Modeling deposition of inhaled particles, in: Aerosols Handbook: Measurement, Dosimetry, and Health Effects, CRC Press, 2012 (Chapter 5).

[30] B. Rehm, Flow drilling: underbalance drilling with liquid single-phase systems, in: B. Rehm, A. Haghshenas, A. Paknejad, A. Al-Yami, J. Hughes, J. Schubert (Eds.), Underbalanced Drilling: Limits and Extremes, Gulf Publishing Company, 2012, pp. 39–108 (Chapter 2).

[31] E. Cunningham, On the velocity of steady fall of spherical particles through fluid medium, Proc. R. Soc. Lond. A 83 (563) (1910) 357–365.

[32] U.S. Environmental Protection Agency, Particle collection mechanisms, in: J.R. Richards (Ed.), APTI Course 413: Control of Particulate Matter Emissions, third ed., ICES Ltd., Research Triangle Park, North Carolina, 2000. Available: http://www.4cleanair.org/APTI/413Combined.pdf. (Chapter 4).

[33] X. Zhang, P. Chen, S. Dai, Y. Han, Analysis of non-point source nitrogen pollution in watersheds based on SWAT model, Ecol. Indic. 138 (2022) 108881.

[34] S.S. Kaushal, P.M. Groffman, L.E. Band, E.M. Elliott, C.A. Shields, C. Kendall, Tracking nonpoint source nitrogen pollution in human-impacted watersheds, Environ. Sci. Technol. 45 (19) (2011) 8225–8232.

[35] S. Chen, et al., Unexpected nitrogen flow and water quality change due to varying atmospheric deposition, J. Hydrol. 609 (2022) 127679.

[36] M. Escudier, Introduction to Engineering Fluid Mechanics, Oxford University Press, 2017.

[37] H. Rubin, Environmental Fluid Mechanics, CRC Press, 2001.

[38] D.F. Elger, B.A. LeBret, C.T. Crowe, J.A. Roberson, Engineering Fluid Mechanics, John Wiley & Sons, 2020.
[39] H. Song, Engineering Fluid Mechanics, Springer, 2018.
[40] W. Graebel, Engineering Fluid Mechanics, CRC Press, 2018.
[41] J.B. Grotberg, Respiratory fluid mechanics, Phys. Fluids 23 (2) (2011) 021301.
[42] S. Berger, L.-D. Jou, Flows in stenotic vessels, Annu. Rev. Fluid Mech. 32 (1) (2000) 347–382.
[43] T. Pedley, The Fluid Mechanics of Large Blood Vessels, Cambridge University Press, Cambridge, 1980.
[44] W. Tucker, L. Nelken, W. Reehl, D.H. Rosenblatt, Diffusion coefficients in air and water, in: W.J. Lyman (Ed.), Handbook of Chemical Property Estimation Methods, American Chemical Society, Washington, DC, 1990.
[45] B. Cushman-Roisin, Environmental Transport and Fate, Thayer School of Engineering Dartmouth College, University Lecture, 2012.
[46] W.A. Tucker, L.H. Nelken, Diffusion coefficients in air and water, in: Handbook of Chemical Property Estimation Methods: Environmental Behavior of Organic Compounds, American Chemical Society, Washington, DC, 1990, pp. 17.1–17.25. 7 tab, 28 ref.
[47] W. Schneider, Groundwater Modelling: Theory of Solute Transport, Technische Unversität Hambur, Harburg, Germany, 2014.
[48] D.A. Vallero, Translating Diverse Environmental Data Into Reliable Information: How to Coordinate Evidence From Different Sources, Academic Press, 2017.
[49] Y.-S. Wu, Multiphase Fluid Flow in Porous and Fractured Reservoirs, Gulf Professional Publishing, 2015.
[50] E.L. Cussler, Fundamentals of mass transfer, in: E.L. Cussler (Ed.), Diffusion: Mass Transfer in Fluid Systems, third ed., Cambridge Series in Chemical Engineering, Cambridge University Press, Cambridge, 2009, pp. 237–273.
[51] E.L. Cussler, Modified from "Fundamentals of mass transfer", in: E.L. Cussler (Ed.), Diffusion: Mass Transfer in Fluid Systems, third ed., Cambridge University Press, Cambridge, 2009, pp. 237–273.
[52] W. Lewis, W. Whitman, Principles of gas absorption, Ind. Eng. Chem. 16 (12) (1924) 1215–1220.
[53] G. McKay, M.J. Bino, A. Altememi, External mass transfer during the adsorption of various pollutants onto activated carbon, Water Res. 20 (4) (1986) 435–442, https://doi.org/10.1016/0043-1354(86)90190-9.
[54] R. Higbie, Penetration theory leads to use of the contact time in the calculation of the mass transfer coefficients in the two film theory, Trans. Am. Inst. Chem. Eng. 31 (1935) 365.
[55] E.L. Cussler, Diffusion: Mass Transfer in Fluid Systems, third ed., Cambridge University Press, Cambridge, 2007, pp. 237–273.
[56] F. Zhou, Z. Hearne, C.-J. Li, Water—the greenest solvent overall, Curr. Opin. Green Sustain. Chem. 18 (2019) 118–123.
[57] I.T. Horváth, Solvents from nature, Green Chem. 10 (10) (2008) 1024–1028.
[58] D.A. Vallero, Soil and sediment physics, in: Methods and Calculations in Environmental Physics, AIP Publishing LLC, Melville, New York, 2022, pp. 8-1–8-34.
[59] D.A. Vallero, Environmental Systems Science: Theory and Practical Applications, first ed., Elsevier Academic Press, 2021.
[60] D.A. Vallero, Estimating and predicting exposure to products from emerging technologies, in: Synthetic Biology 2020: Frontiers in Risk Analysis and Governance, Springer, 2020, pp. 107–142.
[61] D. Vallero, Environmental Biotechnology: A Biosystems Approach, Elsevier Science, 2015.
[62] X. Kong, F. Xu, W. He, W. Liu, B. Yang, Fugacity models, in: Ecological Model Types, 2016, pp. 181–210.
[63] D. Mackay, L. Burns, G. Rand, Fate modeling, in: Fundamentals of Aquatic Toxicology: Effects, Environmental Fate, and Risk Assessment, Taylor & Francis, Philadelphia, Pennsylvania, 1995, pp. 563–585.
[64] Y.-F. Li, et al., Treatment of particle/gas partitioning using level III fugacity models in a six-compartment system, Chemosphere 271 (2021) 129580.
[65] D. Mackay, J.A. Arnot, E. Webster, L. Reid, The evolution and future of environmental fugacity models, in: Ecotoxicology Modeling, 2009, pp. 355–375.
[66] D. Mackay, S. Paterson, Evaluating the multimedia fate of organic chemicals: a level III fugacity model, Environ. Sci. Technol. 25 (3) (1991) 427–436.
[67] D. Mackay, S. Patterson, Fugacity models, in: Practical Applications of Quantitative Structure-Activity Relations in Environmental Chemistry and Toxicology, 1990, pp. 433–460.
[68] S. Wang, Q. Wang, Z. Yuan, X. Wu, Application of the multimedia fugacity model in predicting the environmental behaviors of PCBs: based on field measurements and level III fugacity model simulation, Ecotoxicol. Environ. Saf. 263 (2023) 115286.
[69] L. Li, A. Sangion, F. Wania, J.M. Armitage, L. Toose, J.A. Arnot, Development and evaluation of a holistic and mechanistic modeling framework for chemical emissions, fate, exposure, and risk, Environ. Health Perspect. 129 (12) (2021) 127006, https://doi.org/10.1289/EHP9372. In this issue.
[70] J.B.d.F. Mesquita, I.E.L. Neto, Coupling hydrological and hydrodynamic models for assessing the impact of water pollution on lake evaporation, Sustainability 14 (20) (2022) 13465.
[71] K.E. Kunkel, et al., Monitoring and understanding trends in extreme storms: state of knowledge, Bull. Am. Meteorol. Soc. 94 (4) (2013) 499–514.
[72] J. Klassen, D. Allen, Assessing the risk of saltwater intrusion in coastal aquifers, J. Hydrol. 551 (2017) 730–745.
[73] I. Alameddine, G. Jawhari, M. El-Fadel, Social perception of public water supply network and groundwater quality in an urban setting facing saltwater intrusion and water shortages, Environ. Manag. 59 (2017) 571–583.
[74] V. Arnold, A note on the Ghyben-Herzberg formula, Bull. Int. Assoc. Sci. Hydrol. (1968) 43–46.
[75] M.P. Anderson, W.W. Woessner, R.J. Hunt, Applied Groundwater Modeling: Simulation of Flow and Advective Transport, Academic Press, 2015.
[76] D.A. Vallero, Air pollution physics, in: Methods and Calculations in Environmental Physics, AIP Publishing LLC, Melville, New York, 2022, pp. 7-1–7-70.

[77] P.M. Barlow, Ground Water in Freshwater-Saltwater Environments of the Atlantic Coast, US Department of the Interior, US Geological Survey, Washington, DC, USA, 2003.
[78] M. Anderson, W. Woessner, R. Hunt, Model dimensionality and setting boundaries, in: Applied Groundwater Modeling: Simulation of Flow and Advective Transport, 2015, pp. 117–180.
[79] Z. Zhang, H.H. Savenije, Thermodynamics of saline and fresh water mixing in estuaries, Earth Syst. Dyn. 9 (1) (2018) 241–247.
[80] A. Kleidon, Thermodynamic Foundations of the Earth System, Cambridge University Press, 2016.
[81] U.S. Environmental Protection Agency, Water Resources, 2013. http://www.epa.gov/climatechange/impacts-adaptation/water.html. (Accessed 1 November 2013).
[82] U.S. Environmental Protection Agency, Saving Florida's Vanishing Shores, 2002. http://www.epa.gov/climatechange/effects/coastal/saving_FL.pdf. (Accessed 1 November 2013).
[83] J. Elliott, J. Elliott, Temperature requirements of Atlantic salmon Salmo salar, brown trout Salmo trutta and Arctic charr Salvelinus alpinus: predicting the effects of climate change, J. Fish Biol. 77 (8) (2010) 1793–1817.
[84] E.P. Jeffree, Temperature—ranges and distributions of some vascular plants, Nature 176 (4484) (1955) 703–704, https://doi.org/10.1038/176703b0.
[85] J. Tatum, The Clausius-Clapeyron Equation. University of California, Davis Library, 2024, https://phys.libretexts.org/Bookshelves/Thermodynamics_and_Statistical_Mechanics/Heat_and_Thermodynamics_(Tatum)/14%3A_The_Clausius-Clapeyron_Equation#:~:text=dPdT%3DS,V2%E2%88%92VL) (Accessed 5 July 2024).
[86] M.D. Turvey, Incremental effect of heated water discharges to the Missouri River, in: D.A. Vallero (Ed.), Discussion with Author, Dates Are Uncertain, But I Learned Much From Mike About These Impacts, Since He Was an NPDES Inspector for the U.S. Environmental Protection Agency, Region 7, Kansas City, Missouri, Unpublished Work, Kansas City, Missouri, 1980.
[87] D. Vallero, K. Reckhow, A. Gronewold, Application of multimedia models for human and ecological exposure analysis, in: International Conference on Environmental Epidemiology and Exposure, International Society of Exposure Analysis, Durham, NC, 2007.
[88] A. Spacie, L. McCarty, G. Rand, Bioaccumulation and bioavailability in multiphase systems, in: Fundamentals of Aquatic Toxicology, CRC Press, 2020, pp. 493–521.
[89] J.A. Kushlan, S.A. Voorhees, W.F. Loftus, P. Frohring, Length, mass, and calorific relationships of Everglades animals, Fla. Sci. (1986).
[90] K.J. Hartman, S.B. Brandt, Estimating energy density of fish, Trans. Am. Fish. Soc. 124 (3) (1995) 347–355.
[91] K. Schreckenbach, R. Knosche, K. Ebert, Nutrient and energy content of freshwater fishes, J. Appl. Ichthyol. 17 (3) (2001) 142–144.
[92] J.R. Bidwell, J.R. Gorrie, The influence of salinity on metal uptake and effects in the midge Chironomus maddeni, Environ. Pollut. 139 (2) (2006) 206–213.
[93] L. Zalizniak, B.J. Kefford, D. Nugegoda, Is all salinity the same? I. The effect of ionic compositions on the salinity tolerance of five species of freshwater invertebrates, Mar. Freshw. Res. 57 (1) (2006) 75–82.
[94] P. Rengasamy, Transient salinity and subsoil constraints to dryland farming in Australian sodic soils: an overview, Aust. J. Exp. Agric. 42 (3) (2002) 351–361.
[95] Environment Canada, Priority Substances List Assessment Report for Road Salts, 2007. https://www.canada.ca/en/health-canada/services/environmental-workplace-health/reports-publications/environmental-contaminants/canadian-environmental-protection-act-1999-priority-substances-list-assessment-report-road-salts.html. (Accessed 5 July 2024).
[96] Z. Li, H. Liu, C. Zhang, G. Fu, Generative adversarial networks for detecting contamination events in water distribution systems using multi-parameter, multi-site water quality monitoring, Environ. Sci. Ecotechnol. 14 (2023) 100231.
[97] S.G. Vrachimis, R. Lifshitz, D.G. Eliades, M.M. Polycarpou, A. Ostfeld, Active contamination detection in water-distribution systems, J. Water Resour. Plan. Manag. 146 (4) (2020) 04020014.
[98] D.A. Lytle, T. Sorg, L. Wang, A. Chen, The accumulation of radioactive contaminants in drinking water distribution systems, Water Res. 50 (2014) 396–407, https://doi.org/10.1016/j.watres.2013.10.050.
[99] M.J. Davis, R. Janke, Patterns in potential impacts associated with contamination events in water distribution systems, J. Water Resour. Plan. Manag. 137 (1) (2011) 1–9.
[100] A. Marandi, M. Polikarpus, A. Jõeleht, A new approach for describing the relationship between electrical conductivity and major anion concentration in natural waters, Appl. Geochem. 38 (2013) 103–109.
[101] A. Logeshkumaran, N. Magesh, P.S. Godson, N. Chandrasekar, Hydro-geochemistry and application of water quality index (WQI) for groundwater quality assessment, Anna Nagar, part of Chennai City, Tamil Nadu, India, Appl. Water Sci. 5 (4) (2015) 335–343.
[102] A.F. Rusydi, Correlation between conductivity and total dissolved solid in various type of water: a review, IOP Conf. Ser. Earth Environ. Sci. 118 (1) (2018) 012019. IOP Publishing.
[103] M. Taylor, H.A. Elliott, L.O. Navitsky, Relationship between total dissolved solids and electrical conductivity in Marcellus hydraulic fracturing fluids, Water Sci. Technol. 77 (8) (2018) 1998–2004.
[104] D.A. Vallero, T.M. Letcher, Unraveling Environmental Disasters, second ed., Elsevier Academic Press, Amsterdam, The Netherlands, 2023.
[105] P. Atkins, J. De Paula, Elements of Physical Chemistry, Oxford University Press, USA, 2013.
[106] D.A. Vallero, T.M. Letcher, Engineering risks and failures: lessons learned from environmental disasters, Leadersh. Manag. Eng. 12 (4) (2012) 199–209.
[107] K.G. Denbigh, K.G. Denbigh, Equilibria of reactions involving gases, in: The Principles of Chemical Equilibrium: With Applications in Chemistry and Chemical Engineering, fourth ed., Cambridge University Press, Cambridge, United Kingdom, 1981 (Chapter 4).

Chapter 8

Water pollution chemistry

1 Introduction

Water pollutants form and change under combinations of chemical reactions. Each set of formation and transformation reactions has unique rates and orders of reactions. Chemical kinetics describes the rate of a chemical reaction, i.e., the rate at which the reactants are transformed into products. Since a rate is a change in quantity that occurs with time, this describes pollutant formation:

$$\text{Reaction rate} = \frac{\text{change in product concentration}}{\text{corresponding change in time}} \tag{8.1}$$

And,

$$\text{Reaction rate} = \frac{\text{change in reactant concentration}}{\text{corresponding change in time}} \tag{8.2}$$

These simple equations represent the basis for water pollution chemistry.

2 Reactions

The inherent properties of a pollutant are the first determinant of the amount that will be found in the water, how rapidly it will move into and out of the water, how it will transform in the water, and the amount and type of harm it will cause. Myriad environmental factors complement the inherent properties of a substance. Various physical, chemical, and biological processes within the water influence the residence time of a pollutant in the water. Thus, chemical inherency and extrinsic properties of the water and the organism that uptakes the pollutant determine the pollutant's fate [1–10]. When a chemical compound reacts with other substances, the change in product concentration decreases proportionately with the reactant concentration. Thus, for substance A the kinetics is:

$$\text{Rate} = -\frac{\Delta(A)}{\Delta t} \tag{8.3}$$

The negative sign indicates that the parent pollutant reactant concentration is breaking down and the concentration is decreasing. Thus, the mass of degradation product C resulting from the reaction also increases in proportion to the decreasing concentration of the contaminant A, and the reaction rate is:

$$\text{Rate} = \frac{\Delta(C)}{\Delta t} \tag{8.4}$$

The rate of reaction is the negative of the slope of the tangent to the concentration curve at that specific time (See Fig. 8.1). This rate is concentration-dependent; it is a first order reaction [11,12].

Fundamentals of Water Pollution. https://doi.org/10.1016/B978-0-443-28987-3.00012-6

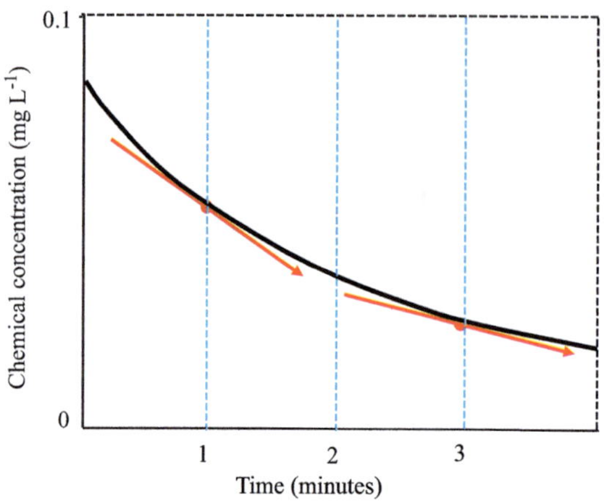

FIG. 8.1 The kinetics of the formation and/or transformation of a chemical compound. The rate of reaction at any time is the negative of the slope of the line tangent to the concentration curve at that time. The rate is higher after 1 min than after 3 min. This rate is concentration-dependent (first order). *(Data from J.N. Spencer, G.M. Bodner, L.H. Rickard, Chemistry: Structure and Dynamics, John Wiley & Sons, 2010. D.A. Vallero, Fundamentals of Air Pollution, fifth ed., Elsevier Academic Press, Waltham, MA, 2014, 999 pages cm.)*

Rate order example 1

A good way to consider these mixed chemical systems is pseudo-orders in the rate laws. Consider carbon dioxide (CO_2) as a product of complete combustion. Assume the average weight of a motor vehicle is 1800 kg, the average gas mileage is $10\,km\,L^{-1}$, and that gasoline weighs an average of $0.7\,kg\,L^{-1}$, with 85% of this weight as carbon. Calculate the CO_2 that the motor vehicle will produce from combusting the gasoline, which is a mixture of several organic compounds.

The question does not give the number of kilometers traveled per year. In 2021, the average U.S. driver logged 13,500 miles (21,726 km) [13], so the carbon in the total amount of gasoline combusted annually for the vehicle would be:

$$\frac{21726\text{ km}}{\text{year}} \times \frac{1\text{L}}{10\text{ km}} \times \frac{0.7\text{ kg}}{\text{l}} \times 0.85 \cong \frac{1292\text{ kg C}}{\text{year}}$$

The carbon dioxide emitted is:

$$\frac{1292\text{ kg C}}{\text{year}} \times \frac{44\text{ kg CO}_2}{12\text{ kg C}} \cong \frac{4737\text{ kg CO}_2}{\text{year}}$$

Thus, in 2021, the average vehicle emitted about 3.7 times more CO_2 than its weight.

Rate order example 2

From the previous calculation, estimate the annual global carbon emissions from vehicles assuming there are 1.5 billion vehicles in the world. Give your answer in tons, tonnes and teragrams. Use two significant figs.

$$4.74 \times 10^3\text{ kg CO}_2 \times \left(1.5 \times 10^9\right) = 7.11 \times 10^{12}\text{ kg} = 7.11 \times 10^{15}\text{ g}$$

1 tonne (metric ton; t) = 1000 kg; thus, vehicles would emit about *7.11×10^9 t CO_2 per year.*
1 teragram (Tg) = 10^9 g; thus, vehicles would emit about *7×10^6 Tg per year.*

Water pollution involves several types of chemical reactions:

2.1 Synthesis or combination

$$A + B \rightarrow AB \tag{8.5}$$

In combination reactions, two or more substances react to form a single product. Formation and hydration are combination reactions are particularly important to water quality. Formation reactions are those where elements combine to form a compound. Examples include the formation of ferric oxide and the formation of octane:

$$4Fe\ (s) + 3O_2\ (g) \rightarrow 2Fe_2O_3\ (s) \tag{8.6}$$

$$8C\ (s) + 9H_2\ (g) \rightarrow C_8H_{18}\ (l) \tag{8.7}$$

The g, l and s in parentheses indicate gas, liquid and solid, respectively.

Hydration reactions involve the addition of water to synthesize a new compound, for example, when calcium oxide is hydrated to form calcium hydroxide, and when phosphate is hydrated to form phosphoric acid:

$$CaO_2\ (s) + H_2O\ (l) \rightarrow Ca(OH)_2\ (s) \tag{8.8}$$

$$P_2O_5\ (s) + 3H_2O\ (l) \rightarrow 2H_3PO_4\ (aq) \tag{8.9}$$

The aq in parentheses indicates that is product is aqueous, which implies it is dissolved in water as the solvent.

2.2 Degradation

$$AB \rightarrow A + B \tag{8.10}$$

The general term for this type of reaction is decomposition, but when applied to the breakdown of organic compounds it is known as degradation. In this case, one substance breaks down into two or more new substances, such as in the decomposition of carbonates. For example, calcium carbonate breaks down into calcium oxide and carbon dioxide:

$$CaCO_3\ (s) \rightarrow CaO\ (s) + CO_2\ (g) \tag{8.11}$$

The mass of the calcium carbonate decreases in the environment while the two products increase. The same thing happens when a pesticide molecule degrades in the environment. Some pesticides degrade into progeny chemical compounds in a few days whereas other persistent pesticides may require decades to break down completely.

2.3 Single replacement

$$A + BC \rightarrow AC + B \tag{8.12}$$

Single replacement (or single displacement) commonly occurs when one metal ion in a compound is replaced with another metal ion, such as when trivalent chromium replaces monovalent silver:

$$3Ag\ NO_3\ (aq) + Cr\ (s) \rightarrow Cr(NO_3)_3\ (aq) + 3Ag\ (s) \tag{8.13}$$

2.4 Double replacement

$$AB + CD \rightarrow AD + CB \tag{8.14}$$

In double replacement reactions, metals are exchanged between the salts. These newly formed salts have different chemical and physical characteristics from those of the reagents and are commonly encountered in metal precipitation reactions, such as when lead is precipitated (indicated by the "(s)" following $PbCl_2$), as in the reaction of a lead salt with an acid like potassium chloride:

$$Pb(ClO_3)_2\ (aq) + 2KCl\ (aq) \rightarrow PbCl_2\ (s) + 2KCLO_3\ (aq) \tag{8.15}$$

Double replacement is also called metathesis and double displacement in some texts.

2.5 Precipitation reactions

Dissolved ions may react with one another to form a solid compound under environmental conditions of temperature and pressure. Salts are compounds that form when metals react with non-metals, such as sodium chloride (NaCl), and may also form from cation and anion combinations, such as ammonium nitrate (NH_4NO_3). Precipitation is a physicochemical process, where both physical and chemical processes are occurring together. In a precipitation reaction, soluble metals and inorganic compounds become insoluble metallic and inorganic salts. The dissolved forms are no longer dissolved and become solids.

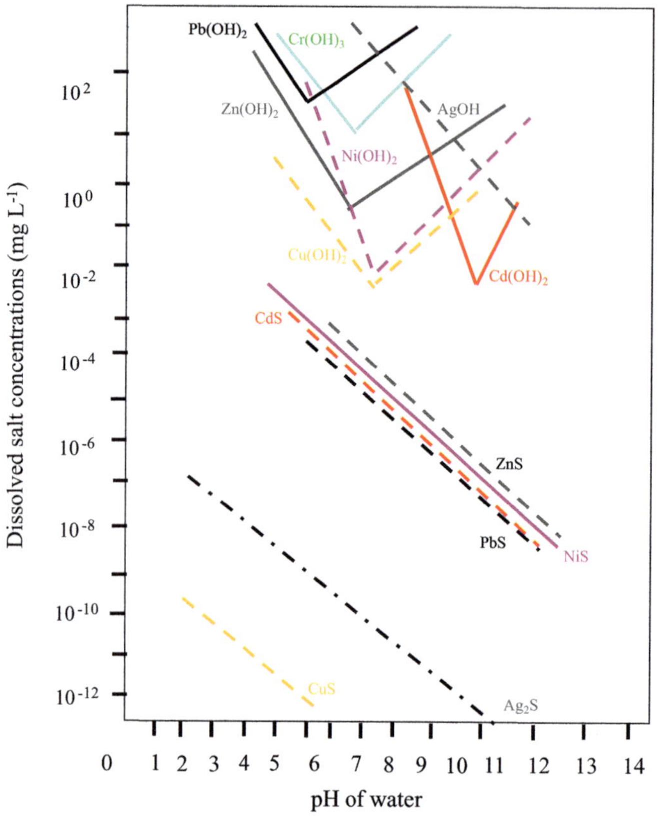

FIG. 8.2 Effect of pH on the solubility of metal hydroxides and sulfides. *(Data from U.S. Environmental Protection Agency, Summary report: control and treatment technology for the metal finishing industry; sulfide precipitation, in: U.S. EPA, EPA 625/8–80-003, vol. 625, 1980.)*

Chemical precipitation occurs within a defined pH and temperature range unique for each metallic salt. Usually in such reactions, an alkaline reagent is added to the solution, thereby raising the solution pH. Often, the higher pH decreases the solubility of the metallic compound, bringing about the precipitation (Fig. 8.2). An example occurs when caustic soda (NaOH) serves as the precipitating agent to lower the amount of soluble nickel, resulting in the product that is the much less water-soluble compound, nickel hydroxide precipitate (s):

$$Ni^{2+} + NaOH \Leftrightarrow Na^{+} + Ni\,(OH)_2\,(s) \tag{8.16}$$

2.6 Complete combustion

Complete or efficient combustion, also known as thermal oxidation, occurs when an organic compound is oxidized in the presence of heat (indicated by Δ). For example, when methane is thermally oxidized, the reaction is:

$$a(CH)_x + bO_2 \xrightarrow{\Delta} cCO_2 + dH_2O \tag{8.17}$$

Combustion is the combination of O_2 in the presence of heat (superscript Δ), as in burning a fossil fuel. Complete combustion of organic compounds produces CO_2 and H_2O, such as the combustion of octane:

$$C_8H_{18}\,(l) + 17O_2\,(g) \xrightarrow{\Delta} 8CO_2\,(g) + 9H_2O\,(g) \tag{8.18}$$

Complete combustion may also result in the production of molecular nitrogen (N_2) when nitrogen-containing organics are burned, such as in the combustion of methylamine:

$$4CH_3NH_2\,(l) + 9O_2\,(g) \xrightarrow{\Delta} 4CO_2\,(g) + 10H_2O\,(g) + 2N_2\,(g) \tag{8.19}$$

Incomplete combustion is particularly important in pollutant formation. Among the most notable products of incomplete combustion (PICs) are the polycyclic aromatic hydrocarbons (PAHs), chlorinated dioxins, furans, and CO. At least two observations can be made about these categories. First, all are kinetic, as denoted by the one-directional arrow (→). Second, in the environment, many reactions that are incomplete generate new compounds that are pollutants.

2.7 Respiration

During respiration an organism's cell gains energy by combining molecular oxygen (O_2) and a carbohydrate, e.g., glucose ($C_6H_{12}O_6$), thereby releasing CO_2, water, and adenosine triphosphate (ATP), the energy carrying compound in cells. Aerobic respiration occurs in the mitochondria, requiring oxygen and glucose, and producing CO_2, water, and energy:

$$C_6H_{12}O_6 + 6O_2 \rightarrow 6CO_2 + 6H_2O \tag{8.20}$$

Note this closely resembles a complete combustion reaction, but at substantially lower temperatures.

Both combustion and respiration produce CO_2 and water, but combustion requires a heat source to oxidize organic compounds (exothermic), whereas respiration is a biochemical process that converts O_2 to CO_2 and H_2O endogenously (i.e., cellular metabolism, often catalyzed by enzymes). Both processes follow the general oxidation reaction, e.g., for methane, this is:

$$CH_4(g) + 2O_2(g) \rightarrow CO_2(g) + 2H_2O(g) \tag{8.21}$$

For all hydrocarbons, the general stoichiometry is:

$$C_xH_y + zO_2 \rightarrow xCO_2 + \frac{y}{2}H_2O. \tag{8.22}$$

If the reactants in the chemical reaction are heated, the Greek symbol delta (Δ) is often shown above the arrow. Thus, the combustion reaction would be:

$$C_xH_y + zO_2 \xrightarrow{\Delta} xCO_2 + \frac{y}{2}H_2O \tag{8.23}$$

As mentioned, the respiration process is often catalyzed, so the reaction could denote the enzyme, e.g., cytochrome P-450 (cyp450):

$$C_xH_y + zO_2 \xrightarrow{cyp450} xCO_2 + \frac{y}{2}H_2O \tag{8.24}$$

Reaction example 1

Besides CH_4, what are some other pollutants potentially released by confined animal feeding operations and what can be done to address them?

The major pollutants from confined feeding operations are various forms of nitrogen. Organic compounds will break down ultimately to CO_2 in oxidized parts of the system, e.g., lagoons, whereas CH_4 will be released from reduced environments. The nitrogen can be released in the gas-phase as nitric oxide (NO), where oxidation is occurring (known as nitrification) or as ammonia (NH_3), in reduced (denitrification) environments, even when waste treatment is effective, e.g., nitrates in water can be aerosolized by aeration of farm lagoons. If lagoons become anaerobic, the sulfur-reducing bacteria (e.g., *Desulfovibrio* spp.) will release hydrogen sulfide (H_2S). Other pollutants can be released, e.g., pesticides and other farm chemicals. In addition, biological pollutants can be released as aerosols into the air from the animal waste (e.g., spores, cysts, bacteria, fungi, and viruses). In addition, these aerosols may sorb chemical compounds that are transported from the feeding site [14].

Reaction example 2

An engineer wants to increase the intensity of heat added to a diesel fueled boiler and needs to know the best option to increase the boiler's heat most rapidly. The options include increasing the concentration of molecular oxygen, adding more diesel fuel, or changing the fuel mix. What reactant is the best option?

The engineer is in fact asking which reactant can be added to give the highest reaction rate. Since the problem states that O_2 is present, this is a combustion reaction, not pyrolysis which occurs in the absence of O_2. Thus, the engineer must identify the combination of reactants, in concentrations, to produce the flame. Thus, this is a rate law question.

The rate of reaction is affected by the concentration of the reactants. Rate is the change over a time-period, i.e., $A+B \rightarrow \frac{\Delta A}{t}$. The rate equals $k[A]^m$, where k is the rate constant, and m is the reaction order. We also know:

If $m=0$, the reaction is zero order, meaning that the rate is independent of the concentration of A, i.e., [A].
If $m=1$, the reaction is first order, so the rate is directly proportional to [A].
If $m=2$, the reaction is second order, so the rate is proportional to the square of [A], i.e., $[A]^2$.
The order of the reaction can only be determined by experimentation, so the engineer doubles the amount of diesel fuel per minute and finds that the oxidation does not change. However, if the diesel is mixed with an additive, the reaction appears to be first order. If O_2 is increased, the rate is also not affected, i.e., the flame is receiving sufficient oxygen and is not O_2-limited. Thus, of these options, changing the mix with the additive is the one that gives the highest rate of combustion.

2.8 Photolysis

The sun's electromagnetic radiation at ultraviolet and visible wavelengths can induce chemical reactions directly and indirectly. Direct photolysis is the process where sunlight adds activation energy need to transform a compound. Indirect photolysis is the process by which an intermediate compound is energized, which in turn transfers energy to another compound.

Pollutants and other compounds are photochemically degraded in terrestrial, aquatic, and atmospheric environments. Photolysis can combine or interchange with other processes, such as in the degradation pathways for chlorinated organic contaminants. For example, the degradation pathway for 1,4-dichlorobenzene in air is a reaction with photochemically generated OH^- radicals and oxides of nitrogen. Upon reaching the soil and water, the degradation compounds are mainly degraded microbially. The combination of photolysis and other chemical reactions results in various products that in turn can be further degraded (See Fig. 8.3).

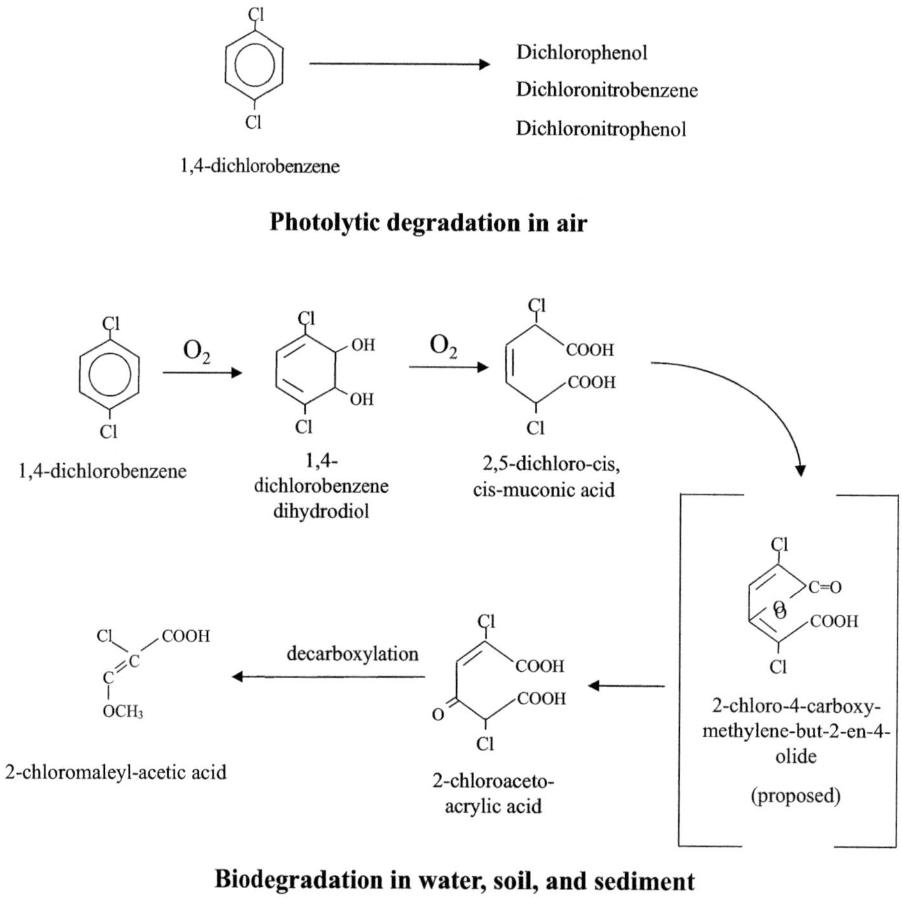

FIG. 8.3 Degradation of 1,4-dichlorobenzene within air, soil, and water. *(Data from Toxicological Profile for Dichlorobenzenes, 2006 [Online] Available: https://www.atsdr.cdc.gov/toxprofiles/tp10.pdf.)*

3 Rate law

For a reaction to occur, the molecules of the reactants must collide. High concentrations of a substance, given their greater number of molecules per volume are more likely to collide than low concentrations. Thus, the reaction rate must be a function of the concentrations of the reacting substances. The rate law is the mathematical expression of this function, which can be established based on experiments for any chemical compound. Varying the concentration of each reactant independently and then measuring the result will give a concentration curve. Each reactant has a unique rate law. The rate law for a chemical substance is an inherent property, i.e., one of a chemical compound's physicochemical properties [1].

In a reaction in which reactant A and reactant B yield product C (i.e., $A+B \rightarrow C$), the reaction rate increases in accordance with the increasing concentration of either A or B. If the amount of A is tripled, then the rate of this reaction triples. Thus, the rate law for such a reaction is:

$$\text{Rate} = k[\text{A}][\text{B}] \tag{8.25}$$

For a different reaction $X+Y \rightarrow Z$, in which the rate is only increased if the concentration of X is increased, thus changing the Y concentration has no effect on the rate law, the rate must be:

$$\text{Rate} = k[\text{X}] \tag{8.26}$$

Eqs. (8.25) and (8.26) indicate that the concentrations in the rate law are the concentrations of reacting chemical species at any specific point in time during the reaction. The rate is the velocity of the reaction at that specific time. The constant k is the rate constant, which is unique for every chemical reaction and is a fundamental physical constant for a reaction, as defined by the specific environmental conditions, including pH, temperature, pressure, and type of solvent [1].

The rate constant is the rate of the reaction when all reactants are present in a 1 molar (M) concentration. Consequently, the rate constant k is the rate of reaction under conditions standardized by a unit concentration.

Drawing a concentration curve for a contaminant that consists of an infinite number of points at each instant of time depicts an instantaneous rate that can be calculated along the concentration curve. At each point on the curve the rate of reaction is directly proportional to the concentration of the compound at that moment in time. This curve represents the kinetic order. The total kinetic order is the sum of the exponents of all the concentrations in the rate law. For the rate $k[\text{A}][\text{B}]$, the overall kinetic order is 2. Such a rate describes a second-order reaction because the rate depends on the concentration of the reactant raised to the second power. Other decomposition rates are represented as $k[\text{X}]$, and are first-order reactions because the rate depends on the concentration of the reactant raised to the first power. Thus, $k[\text{A}][\text{B}]$ is first-order for each reactant and $k[\text{X}]$ is first-order X and zero-order for Y for the reaction $X + Y \rightarrow Z$. In a zero-order reaction, compounds degrade at a constant rate independent of reactant concentration [1].

A parent compound is transformed into chemical progeny. For example, pesticide kinetics often concerns itself with the transformation of the active ingredient in the pesticide into its degradation products [15–17].

Chemical and physical processes, and often biological processes influence pollution formation and transformation. The physical process of solubility, for example, is the result of the chemical phenomenon polarity. Polarity is based upon electronegativity. Similarly, sorption and air-water exchanges involve numerous simultaneous chemical and physical processes that produce degradation products. In the environment, all physical, chemical, and biological processes are interrelated. Pollutant formation processes are a function of both the chemical characteristics of the environmental compartment, e.g., surface water and air, and the characteristics of the pollutant. The inherent properties of the pollutant are influenced and altered by the extrinsic properties of the water and other environmental media where the pollutant resides.

As mentioned, chemical compounds can be transformed by abiotic and biotic processes. The latter is known as biotransformation but involves the same rate laws as abiotic chemistry. For example, microbial degradation follows curves like the those in Fig. 8.4, in which the initial and ultimate pollution concentrations are equal, but each with different microbial biotransformation kinetics. Incidentally, the x-axis in Fig. 8.4 is time since degradation is the change in mass or concentration with time. However, if it is characterizing concentration in a plume, the x-axis can be distance.

The rate orders have been derived empirically, i.e., idealized curves have been adapted to the heterogeneous and highly variable conditions of the field [18,19]. Specific microbial growth, metabolism, and predictions of pollutant degradation rates vary in reports and handbooks, but are strong indications about the reaction rates of a particular microbial species [19]. The calculations are often adjusted with laboratory studies that can be scaled to represent actual conditions of the organism and environment of interest [20]. Computational models that combine fluid dynamics and microbial kinetics are increasingly improving in their ability to simulate the conditions and biochemical reactions with better accuracy and representativeness [21,22].

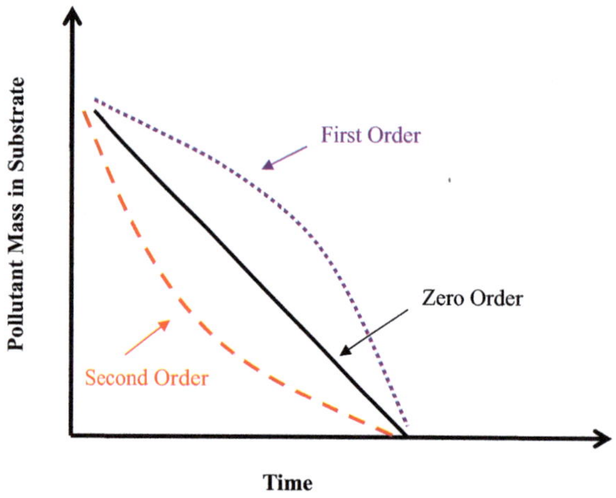

FIG. 8.4 Prototypical decay curve for microbial degradation of organic contaminants. *(Data from D.A. Vallero, Hazardous wastes, in: Waste: A Handbook for Management, 2011, p. 393.)*

The rate law is the basis for biodegradation:

$$aA + bB \rightarrow gG + hH \tag{8.27}$$

$$\text{Rate} = k[A]^x\,[B]^y \tag{8.28}$$

Where, like abiotic rates, k represents the rate constant for molar concentrations of the reactants and x and y represent the reaction rate order for the reactants. Like abiotic transformations, the overall order of the reaction is determined by summing the reactant orders. For example, experimental data about the biodegradation of nitrous oxide (N_2O_5) have shown that rate law for this reaction is first order:

$$\text{Rate} = k\,[N_2O_5]^1 \text{ or simply } k[N_2O_5] \tag{8.29}$$

Therefore, any first order reaction rate for any chemical compound is:

$$\text{Rate} = \frac{d[A]}{dt} = \frac{\Delta[A]}{\Delta t} + k[A] \tag{8.30}$$

As in abiotic transformation rate law, doubling [A] in a first order reaction doubles the reaction rate. A hundred-fold increase in concentration of A multiples the reaction rate by 100. Integrating the first order rate equation yields:

$$\ln\,[A] = \ln\,[A]_0 - k\Delta t \tag{8.31}$$

Correspondingly, integrating the rate equation for a second order reaction yields:

$$\frac{1}{[A]} = \frac{1}{[A]_0} + k\Delta t \tag{8.32}$$

Given specific environmental conditions, even data-rich variables like temperature or redox condition do not completely correlate, nor do they accurately predict the real-world rate order observed in the field. Scientific reviews recommend that laboratory and field-related variables cannot completely account for all the variables in the environment, so the degradation rates that are experimentally derived are merely estimates of what will occur [23–25]. When applying rate constants derived from an experiment or field site investigation to another scenario, be reminded that living systems are complicated and involve simultaneous reactions of various rate orders. A general rule, however, is that if the available energy source (food) in the substrate is not limiting, i.e., a sufficient mass of the pollutant exists in the water, media, and the microbial biofilm, the pollutant will be degraded as a function of the logarithmic cellular density of the microbes. In this case, the reactions follow zero-order kinetics, i.e., constant logarithmic growth and a logarithmic decay rate. When the rate of degradation of the chemical compound becomes directly proportional to the food mass, the decay will follow first-order kinetics [26].

Biotransformation scenarios can also flocculate between, zero, first and second orders, depending on environmental conditions, available food, competition among species, and other factors [25,27–29]. Fig. 8.5. indicates that over the life

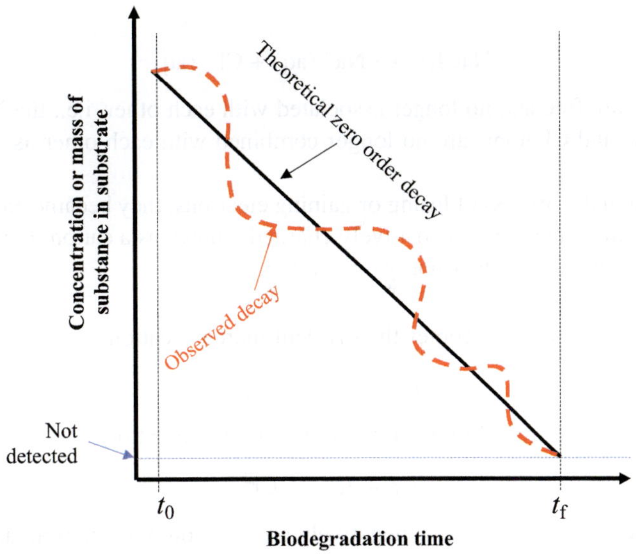

FIG. 8.5 Rate changes within an overall zero order decay rate. At various times during biodegradation, the orders will fluctuate, depending on abiotic and biotic conditions. Notes; t_0 is the time of the biodegradation measurement begins, e.g., a hazardous waste site cleanup project; it is the time when the concentration of the pollutant is no longer detected. The beginning and final concentration for both the solid and dashed lines are the same, but the rate orders needed to achieve the concentration fluctuate, i.e., the mean rate orders are the same. *(Data from D.A. Vallero, Hazardous wastes, in: Waste: A Handbook for Management, 2011, p. 393, I.S. Barkow, S.E. Oswald, H.J. Lensing, M. Munz, Seasonal dynamics modifies fate of oxygen, nitrate, and organic micropollutants during bank filtration-temperature-dependent reactive transport modeling of field data (in Eng), Environ. Sci. Pollut. Res. Int. 28(8) (2021) 9682–9700, https://doi.org/10.1007/s11356-020-11002-9, J. Greskowiak, E. Hamann, V. Burke, G. Massmann, The uncertainty of biodegradation rate constants of emerging organic compounds in soil and groundwater—a compilation of literature values for 82 substances (in Eng), Water Res. 126 (2017) 122–133, https://doi.org/10.1016/j.watres.2017.09.017, J. Regnery, A.D. Wing, M. Alidina, J.E. Drewes, Biotransformation of trace organic chemicals during groundwater recharge: how useful are first-order rate constants?, (in Eng), J. Contam. Hydrol. 179 (2015) 65–75, https://doi.org/10.1016/j.jconhyd.2015.05.008, T. Weidemeier, M. Swanson, J. Wilson, D. Kampbell, R. Miller, J. Hansen, Approximation of biodegradation rate constants for monoaromatic hydrocarbons (BTEX) in ground water, US Environmental Protection Agency Papers, 1996, p. 26, D.A. Vallero, Chapter 31—Hazardous wastes, in: T.M. Letcher, D.A. Vallero (Eds.), Waste, second ed., Academic Press, 2019, pp. 585–630, M. Alexander, Biodegradation and Bioremediation, Gulf Professional Publishing, 1999, D.A. Abramowicz, Aerobic and anaerobic PCB biodegradation in the environment, Environ. Health Perspect. 103(Suppl 5) (1995) 97, A. Mitra, Biology, genetic aspects, and oxidative stress response of Streptomyces and strategies for bioremediation of toxic metals, in: Microbial Biodegradation and Bioremediation, 2014, p. 287, Y. Ikada, H. Tsuji, Biodegradable polyesters for medical and ecological applications, Macromol. Rapid Commun. 21(3) (2000) 117–132, S. Arriaga, R. Muñoz, S. Hernández, B. Guieysse, S. Revah, Gaseous hexane biodegradation by Fusarium solani in two liquid phase packed-bed and stirred-tank bioreactors, Environ. Sci. Technol. 40(7) (2006) 2390–2395.)*

of the remediation project the realistic biodegradation rate may cycle through various orders [30]. Physical factors that affect the kinetics include dispersion rates of the water, dispersion and molecular diffusion in sediment and soil, dilution from ground water recharge, volatilization to the atmosphere, and sorption rates of soil, sediment, and suspended particles [24,25,31–34]. As such, microbial degradation models often assume nonlinear reactions so that biodegradation is fitted to experimental and field data that are inserted into the model and expressed as:

$$-d[\mathrm{A}] = k[\mathrm{A}]^n \frac{-d[\mathrm{A}]}{dt} = k[\mathrm{A}]_n \tag{8.33}$$

Where n is the rate curve fitting parameter.

4 Dissociation and association

Ionization determines the acidity and alkalinity of water and the production of acids and bases. When a salt is dissolved in water, it dissociates into ionic species. Ions that are dissolved in a solution can react with one another and can form solid complexes and compounds. For example, if the salts calcium sulfate and sodium chloride are added to water, $CaSO_4$ and NaCl are assumed to be dissolved and/or suspended in the water. In fact, the metals and nonmetals become associated with one another in equilibrium:

$$CaSO_4(s) \leftrightarrow Ca^{2+}(aq) + SO_4^{2-}(aq) \tag{8.34}$$

And,

$$NaCl(s) \leftrightarrow Na^+ (aq) + Cl^- (aq) \tag{8.35}$$

Thus, the four dissociated ions are free and no longer associated with each other, i.e., the Na^+, Ca^{2+}, Cl^-, and SO_4^- ions dissociate in the water. The Na and Cl atoms are no longer combined with each other as they were before they entered the water.

The atoms were neutral, but in the process of losing or gaining electrons, they became electrically charged species, i.e., ions. An atom that loses one or more electrons is positively charged, known as a cation. For example, the potassium atom loses one electron and becomes the monovalent potassium cation:

$$K - e^- \rightarrow K^+ \tag{8.36}$$

The mercury atom that loses two electrons becomes the divalent mercury cation:

$$Hg - 2e^- \rightarrow Hg^{2+} \tag{8.37}$$

A chromium atom that loses three electrons forms the trivalent chromium cation:

$$Cr - 3e^- \rightarrow Cr^{3+}$$

Conversely, an atom that gains electrons becomes a negatively charged ion, known as an anion. When the chlorine atom gains an electron it becomes the chlorine anion:

$$Cl + e^- \rightarrow Cl^- \tag{8.38}$$

When the sulfur atom gains two electrons, it becomes the divalent sulfide anion:

$$S + 2e^- \rightarrow S^{2-} \tag{8.39}$$

Note that the Greek prefix (mono-, di-, tri-, …) denoting the valence is the number of electrons that are in excess in the ion, the integer that expresses the difference from neutrality.

Ionic reactions occur when the ions of water-soluble salts react in aqueous solution to form salts that are often nearly insoluble in water, causing them to separate into insoluble precipitates:

$$\text{(Ions in Solution 1)} + \text{(Ions in Solution 2)} \rightarrow \text{Precipitate} + \text{Unreacted ions} \tag{8.40}$$

A salt in the solid phase exists in a lattice wherein ions are surrounded by one another. When the salt is dissolved in water, the ions become surrounded by the water, rather than by the other ions. Each ion now has its own hydration sphere, i.e., an envelope of water molecules surrounding it (See Fig. 8.6). An ion-association reaction is an ion-ion interaction between ions in an ion-containing solution, i.e., known as an electrolyte. An electrolyte conducts electricity when dissolved in water, yet other chemicals do not [35]. When the salt lattice enters the water, the ions assemble into couplets of separate oppositely charged ions, i.e., cations and anions, the so-called ion pairs. The pairs are held together by electrostatic attraction. Ion-

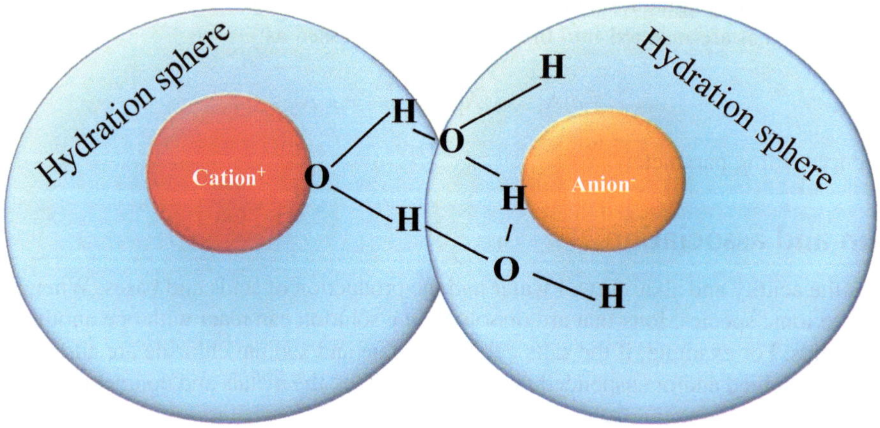

FIG. 8.6 Ion pair. *(Data from V. Evangelou, Environmental Soil and Water Chemistry: Principles and Applications, John Wiley Sons, Inc., Canada, 1998.)*

association is the reverse of dissociation where the ions separate from a compound into free ions. Ions or molecules surrounded by water exist in the aqueous phase.

Solvation is the interaction of solvent, like water, with its solutes, i.e., the molecules and ions dissolved in the solvent. Ionic compounds dissociate in water, breaking into their ionic components within the solvent interface of a chemical species that constitutes the solute, known as the solvation shell. In water, the solvation shell is a hydration shell or hydration sphere. The hydration number is the number of solvent molecules surrounding each unit of solute. For example, when water molecules configure around a metal ion, the electronegative oxygen atom of the water molecule is electrostatically attracted to the positive charge on the metal ion. The result is a solvation shell of water molecules that surround the ion (See Fig. 8.6).

Solutions can be a mix of strong electrolytes, weak electrolytes, and nonelectrolytes. Nonelectrolytes are nonionic molecular compounds that are neither acids nor bases. As such nonelectrolytes do not ionize. Examples of nonelectrolytes include sugars, alcohols, and most other organic compounds, as well as some inorganic compounds.

Weak electrolytes partially dissociate in water, usually dissociating less than 10% with greater than 90% remaining undissociated. Organic acids like acetic acid are generally weak electrolytes.

An example of dissociation is strontium carbonate dissolved in water:

$$SrCO_3\ (s) \rightarrow Sr^{2+}\ (aq) + CO_3^{\ 2-}\ (aq) \tag{8.41}$$

Conversely, the reverse reaction is known as a precipitation reaction that forms a solid that returns from the solution to re-form the lattice of ions surrounding ions. In the reaction above, the carbonate species is precipitated:

$$Sr^{2+}\ (aq) + CO_3^{\ 2-}\ (aq) \rightarrow SrCO_3\ (s) \tag{8.42}$$

The ionic product (Q) is the count of ions present in the solvent. The solubility product constant (K_{sp}) is the ionic product when the system is in equilibrium. As such, solubility is often expressed as the specific K_{sp}, the equilibrium constant for dissolution of the substance in water. Because K_{sp} is a chemical equilibrium constant, it is a state of balance of opposing reversible chemical reactions which proceed at rates that are constant and equal, with no net change in the system. Like sorption, Henry's Law, K_{ow}, and other equilibrium constants, solubility follows Le Chatelier's Principle, i.e., in a balanced equilibrium, if one or more factors change, the system will readjust to reach equilibrium. K_{sp} values and resulting solubility calculations for four important reactions are shown in Table 8.1. K_{sp} constants for many reactions are published in engineering, hydrogeology, and chemistry handbooks.

In the previous reaction, since $SrCO_3$ is a highly insoluble salt (aqueous solubility $= 6 \times 10^{-3}$ mg L^{-1}), its equilibrium constant for the reaction is quite small:

$$K_{sp} = \left[Sr^{+2}\right]\left[CO_3^{\ 2-}\right] = 1.6 \times 10^{-9} \tag{8.43}$$

This K_{sp} value reflects the low concentration of dissolved ions. As the number of dissolved ions approaches zero, the compound is increasingly insoluble in water. This does not mean that that an insoluble product will not be dissolved, merely that other conditions are needed for it to dissolve. For example, the addition of an acid will solubilize the Sr^{+2} ion.

The product of the concentration of the dissolved ions in the equilibrium expression must exceed the value of the K_{sp}. for precipitation to occur. The concentration of each of these ions need not be the same. For example, if $[Sr^{+2}]$ is 1×10^{-5} M, the carbonate ion concentration must exceed 0.0016 M for precipitation to occur because $(1 \times 10^{-5}) \times (1.6 \times 10^{-4}) = 1.6 \times 10^{-9}$, i.e., the K_{sp} for strontium carbonate.

TABLE 8.1 Solubility product constant versus solubility for four types of salts.

Salt	Example	Solubility product, K_{sp}	Solubility, *S*
AB	$CaCO_3$	$[Ca_{2+}][CO_3^{2-}] = 4.7 \times 10^{-9}$	$(K_{sp})^{1/2} = 6.85 \times 10^{-5}$ M
AB_2	$Zn(OH)_2$	$[Zn^{2+}][OH\text{-}]^2 = 4.5 \times 10^{-17}$	$(K_{sp}/4)^{1/3} = 2.24 \times 10^{-6}$ M
AB_3	$Cr(OH)_3$	$[Zn^{3+}][OH\text{-}]^3 = 6.7 \times 10^{-31}$	$(K_{sp}/27)^{1/4} = 1.25 \times 10^{-8}$ M
A_3B_2	$Ca_3(PO_4)_2$	$[Ca^{2+}]^3[PO_4^{3-}]^2 = 1.3 \times 10^{-32}$	$(K_{sp}/108)^{1/5} = 1.64 \times 10^{-7}$ M

Data from EM 1110-1-4012, Engineering and design precipitation/coagulation/flocculation, 2001.

Precipitation example

100 mL of 0.050 M NaCl is added to 200 mL of 0.020 M $Pb(NO_3)_2$. Will the lead chloride precipitate?

Calculate the ion product (Q) and compare it to the K_{sp} for the reaction:

$$PbCl_2(s) \rightarrow Pb^{2+}(aq) + 2Cl^-(aq)$$

When the two solutions are mixed, the unassociated ions are formed as:

$$[Pb^{2+}] = \frac{0.2\ L \times 2.0 \times 10^{-2}\ M}{0.3\ L} = 1.3 \times 10^{-2}\ M$$

and,

$$[Cl^-] = \frac{0.1\ L \times 5.0 \times 10^{-2}\ M}{0.3\ L} = 1.7 \times 10^{-2}\ M$$

The value for the ion product (Q) is calculated:

$$Q = [Pb^{2+}][Cl^-]^2 = [1.3 \times 10^{-2}][1.7 \times 10^{-2}]^2 = 3.8 \times 10^{-7}$$

The K_{sp} for this reaction is 1.6×10^{-5}.

$Q < K_{sp}$, which means that no precipitate will be formed. If the ion product were greater than the K_{sp} a precipitate would have formed.

The common ion effect states that compared to a solution in pure water, an ion's solubility is decreased in an aqueous solution that contains one of the ions that make up the compound. This allows a precipitate to form if the K_{sp} is exceeded. For example, soluble sodium carbonate (Na_2CO_3) in solution with strontium ions can cause the precipitation of strontium carbonate, given that the carbonate ions from the sodium salt contribute to their overall concentration in solution, thus reversing the solubility equilibrium of the "insoluble" compound, strontium carbonate:

$$Na_2CO_3\ (s) \rightarrow 2\ Na^+(aq) + CO_3^{2-}\ (aq) \tag{8.44}$$

$$SrCO_3\ (s) \rightarrow Sr^+(aq)^{2+}\ CO_3^{2-}\ (aq) \tag{8.45}$$

Also, a complexing agent, known as a ligand, may react with the cation of a precipitate, which enhances the solubility of the compound. In addition, several metal ions are weakly acidic and readily hydrolyze in solution. An example is when ferric ion (Fe^{3+}) reacts with water, i.e., hydrolyzes:

$$Fe^{3+} + H_2O \rightarrow Fe(OH)^{2+} + H^+ \tag{8.46}$$

When these metal ions hydrolyze, they produce a less soluble complex. The solubility of the salt is inversely related to the pH of the solution, with solubility increasing as the pH decreases. The minimum solubility is found under acidic conditions when the concentrations of the hydrolyzed species approach zero.

The ionization of water is crucial to pollution chemistry. The acidity and alkalinity of water is expressed as a solution's hydrogen ion $[H^+]$ concentration and hydroxide ion $[OH^-]$ concentration respectively. Unless deionized, water is the combination of the neutral molecular water and these ions. Water can act as either an acid or a base. In fact, water chemists sometimes show molecular water as HOH, perhaps to demonstrate ionization. The dissociation in water is a reversible reaction:

$$HOH(l) \leftrightarrow H^+(aq) + OH^-\ (aq) \tag{8.47}$$

However, instead of H^+, this equilibrium really involves the hydronium ion (H_3O^+), which is the chemical form of H^+ in an aqueous solution. When hydrogen loses its only electron, all that is left is a proton, which is highly reactive, so it immediately reacts with a water molecule, generating H_3O^+. An acid donates a proton to a solution, i.e., H^+, but this proton is extremely small, so it is more reasonable that it exists as hydronium. Therefore, the dissociation of water is a reversible reaction better described as:

$$2HOH(l) \leftrightarrow H_3O^+(aq) + OH^-(aq) \tag{8.48}$$

This dissociation is continuous and the number of ions dissociating at any time is known as the water dissociation constant, K_w, with the formula in same format as any equilibrium constant, i.e., the concentrations of products divided by the concentrations of reactants, all raised to the power of the coefficients of the balanced reaction:

$$K_w = \frac{[H_3O^+][OH^-]}{[HOH]^2} \tag{8.49}$$

However, the ions represent so small a fraction of the total mass in this reaction that the molecular water is assumed to be essentially pure water. Since pure substances are solvents without solutes, i.e., concentration is meaningless, they are not included in equilibrium calculations. At 25°C, this the dissociation reaction is:

$$K_w = [H_3O^+][OH^-] = 1.0 \times 10^{-14}\,\text{mol}\,\text{L}^{-1} \tag{8.50}$$

Recall that the brackets denote molar concentration. Because this equilibrium reaction is balanced, concentration proportions are 1:1. So, if $[H_3O^+]$ equals x, $[OH^-]$ must also be x. Thus, the equilibrium simplifies further to:

$$x^2 = 1.0 \times 10^{-14}$$

Therefore, $[H_3O^+] = [OH^-] = \sqrt{1.0 \times 10^{-14}}$ mol L^{-1}, and:

$$[H_3O^+] = [OH^-] = 1.0 \times 10^{-7}\,\text{mol}\,\text{L}^{-1}$$

Water chemistry refers to "p" as a negative log:

$$pX = -\log X \tag{8.51}$$

The pH of a substance is expression of the power of hydrogen in a solution, i.e., pH is a measure of the strength an acid or base in that solution. Much of the literature describes pH as negative of the base-logarithm ($\log_{10}$, or simply log) of $[H^+]$:

$$\text{pH} = -\log[H^+] \tag{8.52}$$

This is usually acceptable given the many other uncertainties in acidity calculations. However, as discussed above, pH is more accurately described as the negative of log $[H_3O^+]$:

$$\text{pH} = -\log[H_3O^+] \tag{8.53}$$

At equilibrium, $[H_3O^+] = 1.0 \times 10^{-7}$ mol L^{-1}. The negative log of 1.0×10^{-7} is 7, so water is neutral at pH = 7.

Since, at equilibrium, $[OH^-]$ is also 1.0×10^{-7} mol L^{-1}, then pOH is also equal to 7. Thus, the center of the pH scale is 7. Thus, the pH and pOH scales are reciprocal to one another, with both ranging from 0 to 14. The number of hydronium ions equals the number hydroxyl ions at pH = 7. The acidity increases logarithmically with decreasing pH (i.e., increasing number of hydronium ions, or protonation). The alkalinity increases logarithmically with increasing pH (i.e., increasing number of hydroxyl ions). This means that each pH integer represents a change in acidity by a factor of 10. A pH 5 solution is 100 times more acidity than neutral. Aa pH 2 solution has 100,000 times more hydronium ions than neutral water (pH 7), or $[H_3O^+] = 10^{12}$ versus $[H_3O^+] = 10^7$, respectively.

The Brönsted-Lowry model defines acids as proton donors and bases as proton acceptors. The Lewis model states that acids are electron-pair acceptors and bases are electron-pair donors. When acids react with bases, a double-replacement reaction takes place, resulting in neutralization. The products are water and a salt. One mole of acid neutralizes precisely one mole of base. Being electrolytes, a strong acid dissociates and ionizes 100% into H_3O^+ and anions. These anions are the acid's specific conjugate base. A strong base also dissociates and ionizes completely. As mentioned, this ionization results in hydroxyl ions and cations, known as the base's conjugate acid. Weak acids and weak bases dissociate less than 100% into the respective ions.

Numerous acids are involved in water pollution, but four strong acids are noteworthy, i.e., hydrochloric acid (HCl), nitric acid (HNO_3), sulfuric acid (H_2SO_4), and perchloric acid ($HClO_4$). Many weak acids are important, including carbonic acid (H_2CO_3), acetic acid (CH_3COOH), and phosphoric acid (H_3PO_4). Strong bases include sodium hydroxide (NaOH) and potassium hydroxide (KOH). Weak bases include ammonia (NH_3), which dissolves in the surface waters to become ammonium hydroxide (NH_4OH) and organic amines, i.e., compounds with the functional group, –NH.

Nonmetal oxides, such as carbonate (–CO_2) and sulfate (–SO_2), are generally acidic, forming carbonic acid and sulfuric acid, respectively in water. The metal oxides like those of calcium (e.g., CaO) and magnesium (e.g., MgO), are generally basic. These two metal oxides, for example, form calcium hydroxide (CaOH) and magnesium hydroxide (MgOH) in water, respectively.

Acid example

A laboratory has an aqueous solution of sulfuric acid, which is 0.05 M. What is the hydronium ion and chlorine ion molar concentration and the pH of the solution?

Since this is a strong acid, it should ionize and dissociate completely. Thus:

$[H_3O^+]=0.05$ and $[Cl^-]=0.05$, so none of the associated acid remains.

What is the pH of the solution above?

Since $[H_3O^+]=0.05$, and $pH=-\log[H^+]=-\log 0.05\,M=1.3$.

Hydroxyl example

What is the $[OH^-]$ of this solution?

$$K_w\ [H_3O^+][OH^-]=1.0\times10^{-14}$$

$$[OH^-]=\frac{K_w}{[H_3O^+]}=\frac{10^{-14}}{0.05}=2.0\times10^{-13}$$

Thus, even at an extremely low pH, with a very high relative concentration of hydronium ions in the acidic solution, there is still a small amount of hydroxyl ion concentration.

The strength of an acid or base is not affected by concentration, i.e., the molarity of the solution. The strength is the extent to which the acid or base dissociates when it enters water. Whether at a concentration of 6.0 M or at 0.00001 M, sulfuric acid will completely ionize in the water, whereas most organic acids will not completely ionize at any concentration. As evidence, Table 8.2 shows the pH for a few selected acids and bases with the same molar concentration, 0.1 M. The strongest acids rank at the highest to lowest from top to bottom. The weak acids and bases are in the middle of the table. Increasing acidic and basic strength moves from distilled water (blue) upward to the table's strongest acid, hydrochloric acid (red), and downward to the table strongest base, sodium hydroxide (green).

Most environmental acid-base reactions involve weak acids and bases. The amount of ionization in most environmental reactions, especially those in the ambient environment, compared to reactions deployed using chemical manufacturing and laboratory reactors, is usually well below 10% dissociation. Water is a weak acid autoionizing into 10^{-14} M concentration of hydroxyl and hydronium ions. At 25 °C, there are 55.35 mol water per liter. Half of the ions are hydronium ions, thus:

$$\frac{1.0\times10^{-7}M\ H_3O^+}{55.35M\ H_2O}=1.8\times10^{-9}\ \text{hydronium ions per water molecule} \quad (8.54)$$

This minuscule ratio provides enough H_3O^+ for many aquatic, terrestrial, and atmospheric reactions because the vast amount of water available and the highly reactive nature of the hydronium ion and the proton (H^+).

Generally, for every 1000 molecules of a weak acid, only about 50 molecules of the acid will dissociate into hydronium ions in the water, as evidenced by the acetic acid acid-base equilibrium reaction:

$$CH_3COOH\ (aq)+H_2O\ (l)\leftrightarrow H_3O^+\ (aq)+CH_3COO^-\ (aq) \quad (8.55)$$

The acetate ion (CH_3COO^-) is the reaction's conjugate base and the hydronium ion is the active acid chemical species. The reaction is in equilibrium, so all species exist together. This introduces another equilibrium constant for acid reactions, i.e., the acid constant (K_a):

$$K_a=\frac{[H_3O^+][CH_3COO^-]}{[CH_3COOH]} \quad (8.56)$$

At 25 °C, acetic acid's K_a is 1.8×10^{-5}. If the percent dissociation in an acid reaction is known, the product of this percentage and the initial acid concentration will give the molar concentration of protons, $[H^+]$. Thus, if a 0.1 M solution of cyanic acid (HOCN) is 2.8% ionized, the $[H^+]$ can be determined. Given that HOCN is a weak acid with a percent ionization is less than 100, well below 10% threshold between weak and strong acids. Thus, at 0.1 M, the proton molar concentration is:

$$[H^+]=2.8\%\times0.1\,M=0.0028\ \text{or}\ 2.8\times10^{-3} \quad (8.57)$$

Published K_a constants show the HOCN constant at 25 °C to be 3.5×10^{-3}; so environmental conditions, especially temperature, are slightly affecting the pH of the solution. This is another reminder that all equilibrium constants are temperature dependent.

TABLE 8.2 Experimentally derived pH values for 0.1 M solutions of acids and bases at 25 °C.

Compound	pH
HCl	1.1
H_2SO_4	1.2
H_3PO_4	1.5
CH_3COOH	2.9
H_2CO_3 (in saturated solution)	3.8
HCN	5.1
NaCl	6.4
H_2O (distilled)	7.0
$NaCH_3CO_2$	8.4
$NaSO_3$	9.8
NaCN	11.0
NH_3 (aqueous)	11.1
$NaPO_4$	12.0
NaOH	13.0

Data from J.N. Spencer, G.M. Bodner, L.H. Rickard, Chemistry: Structure and Dynamics, John Wiley & Sons, 2010.

Weak bases follow the same protocol as weak acids, but with a base equilibrium constant, K_b. A few acid and base equilibrium constants involved in water pollution are provided in Tables 8.3 and 8.4. A monoprotic acid can donate a single proton per acid molecule, but polyprotic acid can donate two or more protons per molecule. For example, the monoprotic nitrous acid, HNO_2, has but one H atom available to donate, but the polyprotic phosphoric acid, H_3PO_4, has three H atoms available to donate to the reaction.

TABLE 8.3 Equilibrium constants for selected weak monoprotic acids and bases at 25 °C.

Monoprotic acid	Dissociation reaction	K_a
Hydrofluoric acid	$HF + H_2O \leftrightarrow H_3O^+ + F^-$	7.2×10^{-4}
Nitrous acid	$HNO_2 + H_2O \leftrightarrow NO_2^- + H_3O^+$	4.0×10^{-4}
Lactic acid	$CH_3CH\ (OH)\ CO_2H + H_2O \leftrightarrow CH_3CH\ (OH)\ CO2^- + H_3O^+$	1.38×10^{-4}
Benzoic acid	$C_6H_5CO_2H + H_2O \leftrightarrow C_6H_5CO_2^- + H_3O^+$	6.4×10^{-5}
Acetic acid	$HC_2H_3O_2 + H_2O \leftrightarrow C_2H_3O_2^- + H_3O^+$	1.8×10^{-5}
Propionic acid	$CH_3CH_2CO_2H + H_2O \leftrightarrow CH_3CH_2CO_2^- + H_3O^+$	1.3×10^{-5}
Hypochlorous acid	$HOCl + H_2O \leftrightarrow OCl^- + H_3O^+$	3.5×10^{-8}
Hypobromous acid	$HOBr + H_2O \leftrightarrow OBr^- + H_3O^+$	2×10^{-9}
Hydrocyanic acid	$HCN + H_2O \leftrightarrow CN^- + H_3O^+$	6.2×10^{-10}
Phenol	$HOC_6H_5 + H_2O \leftrightarrow OC_6H_5^- + H_3O^+$	1.6×10^{-10}
Base	**Dissociation reaction**	K_b
Dimethylamine	$(CH_3)_2NH + H_2O \leftrightarrow (CH_3)_2NH_2{}^+ + OH^-$	5.9×10^{-5}
Methylamine	$CH_3NH_2 + H_2O \leftrightarrow CH_3NH_3{}^+ + OH^-$	7.2×10^{-4}
Ammonia	$NH_3 + H_2O \leftrightarrow NH_3{}^+ + OH^-$	1.8×10^{-5}
Hydrazine	$H_2NNH_2 + H_2O \leftrightarrow H_2NNH_3{}^+ + OH^-$	1.2×10^{-6}
Aniline	$C_6H_5NH_2 + H_2O \leftrightarrow C_6H_5NH_3{}^+ + OH^-$	4.0×10^{-10}
Urea	$H_2NCONH_2 + H_2O \leftrightarrow H_2NCONH_3{}^+ + OH^-$	1.5×10^{-14}

Data from M. Olia, A.S. Casparian, How to Prepare for the Fundamentals of Engineering, FE/EIT Exam. Barron's Educational Series, 2000.

TABLE 8.4 Equilibrium constants for selected environmentally important polyprotic acids at 25 °C.

Acid	Dissociation reactions	K_{a1}	K_{a2}	K_{a3}
Sulfuric acid	$H_2SO_4 + H_2O \leftrightarrow HSO_4{}^- + H_3O^+$	1.0×10^3		
	$HSO_4{}^- + H_2O \leftrightarrow SO_4{}^{2-} + H_3O^+$		1.2×10^{-2}	
Hydrogen sulfide	$H_2S + H_2O \leftrightarrow HS^- + H_3O^+$	1.0×10^{-7}		
	$HS^- + H_2O \leftrightarrow S^{2-} + H_3O^+$		1.3×10^{-13}	
Phosphoric acid	$H_3PO_4 + H_2O \leftrightarrow HPO_4{}^- + H_3O^+$	7.1×10^{-3}		
	$HPO_4{}^- + H_2O \leftrightarrow HPO_4{}^{2-} + H_3O^+$		6.3×10^{-8}	
	$HPO_4{}^{2-} + H_2O \leftrightarrow PO_4{}^{3-} + H_3O^+$			4.2×10^{-13}
Carbonic acid	$H_2CO_3 + H_2O \leftrightarrow HCO_3{}^- + H_3O^+$	4.5×10^{-7}		
	$HCO_3{}^- + H_2O \leftrightarrow CO_3{}^{2-} + H_3O^+$		4.7×10^{-11}	

Data from M. Olia, A.S. Casparian, How to Prepare for the Fundamentals of Engineering, FE/EIT Exam. Barron's Educational Series, 2000.

The atmosphere contains large amounts of CO_2, an average of about 427 ppm CO_2 in April 2024. Carbon dioxide becomes dissolved in suspended droplets, in surface water and in soil water. Indeed, CO_2 plays a large part in the acidity of natural rainwater, i.e., pH between 5.0 and 5.5. Therefore, the dissociation of CO_2 is one of the most important environmental acid-base reactions:

$$CO_2 + H_2O \leftrightarrow H_2CO_3{}^* \tag{8.58}$$

The asterisk (*) denotes that this compound is sum of two compounds, i.e., the dissolved CO_2 and the reaction product, carbonic acid H_2CO_3.

The carbonic acid that is formed has two hydrogen atoms it is a diprotic acid, so an additional equilibrium step reaction occurs in water. The first reaction forms bicarbonate and hydrogen ions:

$$H_2CO_3^* \leftrightarrow HCO_3^- + H^+$$

This is followed by a reaction that forms carbonate and hydrogen ions:

$$HCO_3^- \leftrightarrow CO_3^{2-} + H^+ \tag{8.59}$$

Each of the reactions in the two steps has its own acid equilibrium constant, K_{a1} and K_{a2}, respectively, as shown in Table 8.4. A triprotic acid has three unique constants. The constants decrease substantially with each step. Thus, most of the hydrogen ion production occurs in the first step.

Numerous water pollution reactions can be predicted from the relative strength of acids and bases, given that their strength results from the ability of the protons via the hydronium ion to be transferred from the acid and the electron to be transferred via the hydroxyl ion from the base. A weak acid's conjugate base must be strong. Conversely, if an acid is strong, its conjugate base must be weak. Similarly, if a base is weak, its conjugate acid must be strong, and if the base is strong, its conjugate acid must be weak. The tables show actual K_a and K_b constants. The values for pK_a and pK_b are often published. Because the "p" denotes negative logarithm, the larger the pK_a, the weaker the acid, and the larger the pK_b, the weaker the base will be.

Conjugate base and acid example

Hydrocyanic acid demonstrates K_a an indicator of relative strength of reactants and products:

Stronger base (OH⁻) **Weaker base** (CN⁻)

$$HCN + OH^- \Leftrightarrow CN^- + H_2O \tag{8.60}$$

Stronger acid (HCN) **(K_a = 6.2 × 10^{-10})** **Weaker acid** (H_2O) **(K_a = 1.8 × 10^{-16})**

The ratio of the K_a constant values of the two acids is a direct way to quantify the equilibrium. In the hydrocyanic acid instance above, the ratio is $\frac{6.2\times10^{-10}}{1.8\times10^{-16}} \cong 4 \times 10^6$.

This large quotient indicates that the equilibrium is far to the right. This means that if HCN is dissolved in a hydroxide solution (e.g., NaOH), the resulting reaction will produce much greater amounts of the cyanide ion (CN^-) than the amount of both the hydroxyl ion (OH^-) and the molecular HCN. Conversely, in an aqueous solution of sodium cyanide (NaCN), the water will only react with a very small amount of CN^-.

4.1 Henderson-Hasselbalch equation

Water is an amphoteric compound, i.e., it can act as either an acid or a base. As mentioned, this is why water is sometimes shown as HOH. When water acts as a base its $pK_b = -1.7$. When water acts as an acid its $pK_a = 15.7$.

The relationship between conjugate acid-base equilibrium and pH is important to water pollution chemistry, as well as other environmental, toxicological, and pharmacological fields. The Henderson-Hasselbach equation states this relationship:

$$pK_a = pH + \log\frac{[HA]}{[A^-]} \tag{8.61}$$

Thus, Henderson-Hasselbach states that when the pH of an aqueous solution equals the pK_a of an acidic component, the concentrations of the conjugate acids and bases must be equal (since the log of 1 = 0). If pH is 2 or more units lower than

pK_a, the acid concentration will be greater than 99% [36]. Conversely, when pH is greater than pK_a by 2 or more units, the conjugate base concentration will account for more than 99% of the solution [37].

This has strong implications for water pollution and other environmental treatment technologies since it shows that mixtures of acidic and non-acidic compounds can be separated with a pH adjustment. The application of this principle is also important in the transformation of environmental contaminants in the form of weak organic acids or bases because these compounds in their non-ionized form are much more lipophilic, meaning they will be absorbed more easily through the skin than when they exist in ionized forms. As a rule, the smaller the pK_a for an acid and the larger the pK_b for a base, the more extensive will be the dissociation in aqueous environments at normal pH values, and the greater the compound's electrolytic nature.

As mentioned, the composition of water can be molecular water, i.e., H_2O or HOH, hydronium ions (H_3O^+). and hydroxyl ions (OH^-). Two water molecules react to form hydronium and hydroxyl ions by self-ionization or auto-ionization:

$$2H_2O(l) \leftrightarrow H_3O^+(aq) + OH^-(aq)$$

Where, (l) denotes liquid and (aq) denotes aqueous phase.

Many contaminant reactions occur in water in the environment and within organisms, which underlies the importance of the relationship between conjugate acid-base equilibrium and the hydrogen ion. Thus, the Henderson-Hasselbach equation can be simplified as:

$$HA \leftrightarrow A^- + H^+ \tag{8.62}$$

Where HA is the acid and A^- is the base.

Acid rain example 1

As mentioned, natural rainwater is acidic due to the dissolved CO_2:

$$CO_2(aq) + H_2O(l) \Leftrightarrow H^+(aq) + HCO_3^-(aq)\ K_a = 4.3 \times 10^{-7} \tag{8.63}$$

Acidic precipitation contains the sulfuric acid (H_2SO_4) and nitric acid (HNO_3), largely from the combustion of fossil fuels that contain sulfur and the fact that the air contains molecular nitrogen that is oxidized during internal combustion in engines These release oxides of sulfur and nitrogen that react with water in the atmosphere. The formation of H_2SO_4 is stepped process. The emission contains sulfur dioxide when the elemental sulfur oxidized:

$$S + O_2 \rightarrow SO_2\ (g) \qquad (8.63)$$

The sulfur dioxide is then oxidized to sulfur trioxide:

$$SO_2(g) + ½\, O_2(g) \rightarrow SO_3\ (g) \qquad (8.63)$$

The sulfur trioxide reacts with water vapor, condensed water in clouds, water on particles to form sulfuric acid:

$$SO_3(g) + H_2O \rightarrow H_2SO_4\ (aq) \qquad (8.63)$$

The acidity of the rain and other precipitation increase as a net result of these and other reactions with the oxides of sulfur and nitrogen. Thus, the water pollution can be the overall pH change, or it can be measured as the concentration of the individual acidic compounds. The acidic conditions threaten aquatic life. Like other water pollutants, acid rain displays a concentration gradient. As evidence rainwater pH near a power plant or other stationary source has been measured at 3 which is 10,000 times more acidic than neutral water:

$$pH\,7 - pH\,4 = pH\,3$$

$$pH = -\log\ [H_3O^+], \text{so}\ [H_3O^+] = 10^{-pH}$$

Therefore, for pH 3, $[H_3O^+] = 10^{-3}$ and $[H_3O^+] = 1 \times 10^{-3}\,mol\,L^{-1}$.
For pH 7, $[H_3O^+] = 10^{-7}$ and $[H_3O^+] = 1 \times 10^{-7}\,mol\,L^{-1}$.
The difference of pH = pH 4, which $1 \times 10^4\,mol\,L^{-1}$, or 10,000 times the molar strength as neutral.

Acid rain example 2

When SO_2 dissolves in water, sulfurous acid (H_2SO_3, $K_{a1} = 1.7 \times 10^{-2}$, $K_{a2} = 6.4 \times 10^{-8}$) is formed. What is the reaction when sulfurous acid donates a proton to a water molecule? Identify the Brönsted-Lowry acid and base in this reaction. Surface waters have a natural buffering capacity, especially in regions where there is limestone which gives rise to dissolved calcium. For example, in the U.S., central Kansas is less at risk of acid rain's effects than is the Finger Lakes area of New York. What is the reaction of a minute amount of acid rain containing sulfuric acid reaching a lake containing carbonate (CO_3^{2-}) ions?

Since, $pH = -\log[H^+]$, then $[H^+] = 10^{-pH} = 10^{-3.7}$ 2.0×10^{-4} M in aqueous solution at 25 °C.

The sulfurous acid proton donation reaction is:

$$H_2SO_3\ (aq) + H_2O\ (l) \leftrightarrow HSO_3^-\ (aq) + H_3O^+(aq) \tag{8.67}$$

H_3O^+ is the stronger acid (i.e., $K_{a1} < 1$) and HSO_3^- is the stronger base.

The anion CO_3^{2-} is the conjugate base of the weak acid HCO_3^-, which means that the CO_3^{2-} is can react with the strong acid H_3O^+ in the solution:

$$CO_3^{2-}\ (aq) + H_3O^+\ (aq) \Leftrightarrow HCO_3^-\ (aq) + H_2O\ (l) \tag{8.67}$$

The $K = 1/K_{a2}$ of H_2CO_3, meaning a large K.

Similarly, HSO_4^- also reacts with CO_3^{2-}:

$$CO_3^{2-}\ (aq) + HSO_4^-\ (aq) \Leftrightarrow SO_4^{2-}\ (aq) + HCO_3^-\ (aq) \tag{8.67}$$

The $K = \frac{K_{a2}\ \text{of}\ H_2SO_3}{K_{a2}\ \text{of}\ H_2CO_3}$, which will be large.

Therefore, a HCO_3^-/CO_3^{2-} buffer is produced.

At some concentration and volume of acid rain that exceeds will overwhelm the buffering capacity of the soil, consuming all the CO_3^{2-} and HCO_3^-, and converting all the buffer to H_2CO_3. Thus, surface water and soil water at some point will approach the pH of the rainfall.

4.2 Hydrolysis

All acid-base reactions involve chemical species reacting with water. It is worth noting that even some so-called neutral compounds have acidic or basic properties. For example, metal acetates ($MeC_2H_3O_2)_x$) can dissolve in water and react with the water to form weak acids and hydrogen ions.[a] The resulting solutions are generally slightly basic (pH >7) since the acetate ion is the conjugate base of the weak acid. This process is known as hydrolysis (from Greek: *lysis*, which means breaking apart, in this case water).

Likewise, some compounds produce weak bases when dissolved in water. For example, when ammonium chloride (NH_4Cl) is dissolved in water, the solution becomes more acidic. The NH_4^+ cation serves as the conjugate base of ammonium hydroxide (NH_4OH), or molecular ammonia (NH_3) is the conjugate base of the NH_4^+ cation. Oxide gases like CO_2, nitric oxide (NO), nitrogen dioxide (NO_2), and sulfur dioxide (SO_2), can hydrolyze to form hydronium ions and anions:

$$CO_2\ (g) + H_2O\ (l) \Leftrightarrow H_3O^+\ (aq) + HCO_3^-\ (aq) \tag{8.70}$$

$$SO_2\ (g) + H_2O\ (l) \Leftrightarrow H_3O^+\ (aq) + HSO_3^-\ (aq) \tag{8.71}$$

Oxide reactions also occur within organic compounds, where the organic molecule, represented commonly as RX, reacts with water to form a covalent bond with OH^- and cleaves the covalent bond of the leaving group (X) in RX. This displaces X with the hydroxyl ion and an ion formed from the leaving group:

$$RX + H_2O \rightarrow ROH + H^+ + X^- \tag{8.72}$$

Amides, epoxides, carbonates, esters, organic halides, nitriles, urea compounds, and esters of organophosphate compounds are functional groups that are susceptible to hydrolysis. The process involves an electron-rich nucleus seeker, known as a nucleophile, attacking an electron-poor electron seeker, known as an electrophile, to displace the leaving group (such as a halogen). This is why hydrolysis is a method of dechlorination, a detoxification process to remove chlorine atoms on a chlorinated organic compound like a polychlorinated biphenyl (PCB) or polyvinyl chloride in contaminated water, sludge, soil, and sediment [38,39].

Hydrolysis may be abiotic or biotic and can eliminate or decrease the toxicity of hazardous contaminants by abiotic transformation and biotransformation, especially bacterial. Chemical catalysts and microbial enzymes are often part of hydrolytic reactions. This occurs both in the ambient environment, such as the hydrolysis of inorganic and organic

a. Me is often used in environmental literature to denote an unspecified metal.

compounds by soil bacteria, as well as in engineered systems, such as acclimating those same bacteria to break down chlorinated organic compounds in solid and liquid wastes.

Hydrolysis is also a common means of natural degradation, especially pH. Hydrolysis can be affected by acid and base catalysis. In acid catalysis, this H^+ ion catalyzes the reaction; and in base catalysis the OH^- ion serves as the catalyst. The effect of temperature on hydrolysis can be profound. Each 10°C incremental temperature increase results in a hydrolysis rate constant change by a factor of 2.5 [40]. Also, the hydrolysis rate is highest for reactive pesticides under basic conditions, i.e., commonly in the pH range of 8 to 9. For each pH point increase, the hydrolysis rate increases about 10 times. Thus, the increase in the molar concentration of hydroxyl ions in water can lead to concomitant increase in degradation rate of these pesticides.

Cations in water act like the Lewis acids given their propensity to accept electrons, while water behaves like a Lewis base because it makes its oxygen's two unshared electrons available to the cations. Thus, metal (Me) hydrolysis involving a strong water-metal reaction behaves as an acid-base interaction wherein the H^+ dissociates and hydronium ions form in a prototypical reaction:

$$Me^{n+} + H_2O \leftrightarrow MeOH^{(n+1)} + H^+ \tag{8.73}$$

These reactions account for much pollutant transformation in surface and ground water, sediment, soil water, solid wastes, and in any medium where water is present. Water in the atmosphere means that hydrolysis occurs in clouds and fog, and in droplet that condense around hygroscopic nuclei. In living cells, hydrolysis is a common process in metabolism, especially in the first phase where substances are broken down to provide energy to the cell.

5 Organic chemistry

A carbon atom can bond to itself and to other elements in a myriad of ways. It can form single, double, and triple bonds with itself. This makes for millions of possible organic compounds. An organic compound is a compound that includes at least one carbon-to-carbon or carbon-to-hydrogen covalent bond. Most water pollutants in number are organic. However, inorganic pollutants are larger in terms of total mass of reactants and products in the biosphere. For example, most hazardous waste sites are contaminated with organic contaminants, but mining, extraction, transportation, and agricultural activities have larger volumes and masses of metals and inorganic substances.

Organic compounds can be classified two basic groups: aliphatics and aromatics. Hydrocarbons are the most common type of organic compound. They are comprised exclusively of the elements carbon and hydrogen. In an aliphatic compound, each carbon atom normally forms four covalent bonds Aliphatic compounds are classified into a few chemical groups. The alkanes are hydrocarbons that form chains with each link comprised of the carbon. A single link is CH_4, methane. The carbon chain length increases with the addition of carbon atoms. For example, ethane's structure is:

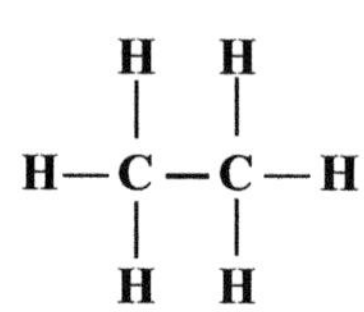

(8.74)

The prototypical alkane structure is:

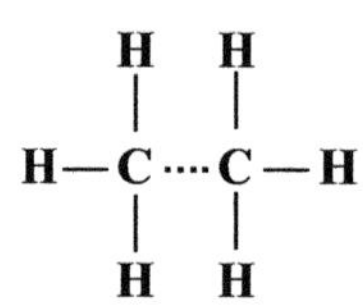

(8.75)

The alkanes contain a single bond between each carbon atoms, and include the simplest organic compound, methane (CH_4) and its derivative chains such as ethane (C_2H_6) and butane (C_4H_{10}). Alkenes contain at least one double bond between carbon atoms. For example, the gasoline-related pollutant 1,3-butadiene structure is $CH_2{=}CH\text{-}CH{=}CH_2$. The numbers "1" and "3" indicate the position of the double bonds. The alkynes contain triple bonds between carbon atoms, with the simplest being ethyne, $CH{\equiv}CH$, commonly known as acetylene, the gas used by welders.

The other major group of organic compounds are the aromatics, which are based on the six-carbon ring configuration of benzene (C_6H_6). The carbon-carbon bond in an aromatic compound shares more than one electron, providing the benzene structure with resonance among the double and single bonds, i.e., the bonds flip back and forth. Benzene is the average of two equally contributing resonance structures. The term "aromatic" comes from the observation that many compounds derived from benzene were highly fragrant, such as wintergreen oil, vanilla, and sassafras. Aromatic compounds contain at least one benzene ring. The carbon atoms in the ring are planar, that is, they remain in the same geometric plane as a unit. However, in compounds with more than one ring, such as the highly toxic PCBs, each ring is planar, but the rings may be bound together, but may or may not be planar. This is a very important property for toxic compounds where the hydrogen atom has been substituted by chlorine, bromine, or other halogen atoms. Often, planar aromatic compounds are more toxic than their non-planar counterparts, likely in part because living cells may be more likely to allow planar compounds to bind to them and to produce nucleopeptides that lead to biochemical reactions associated with cellular dysfunctions, such as cancer or endocrine disruption.

Both the aliphatic and aromatic compounds can undergo substitutions of the hydrogen atoms. These substitutions render new properties to the compounds, including changes in solubility, vapor pressure, partitioning, and toxicity. As mentioned, halogenation, the substitution of a hydrogen atom with a halogen, often makes an organic compound much more toxic than an unsubstituted compound. For example, trichloroethane is a carcinogenic liquid that has been found in drinking water supplies, whereas non-substituted ethane is a gas with relatively low toxicity. This is also why one of the means for treating the large number of contaminated sediment and soil that contains chlorinated hydrocarbons and chlorinated aromatic compounds involves dehalogenation techniques, including by hydrolysis, microbial degradation by methylotrophic bacteria, enzymatic intramolecular substitution, and reductive dehalogenation under aerobic conditions [38,39,41].

Some important functional groups that are part of many organic compounds are shown in Table 8.5.

Structures of organic compounds can induce unique physical and chemical characteristics, as well as the potential rates of bioaccumulation and toxicity of organic compounds. For example, the differences in structure between the estradiol and

TABLE 8.5 Structures of organic compounds.

Chemical Class	Functional group
Alkanes	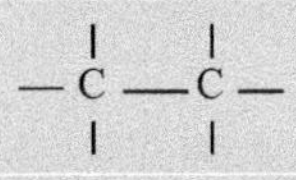
Alkenes	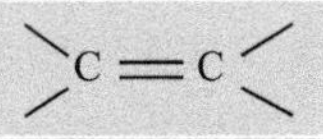
Alkynes	
Aromatics	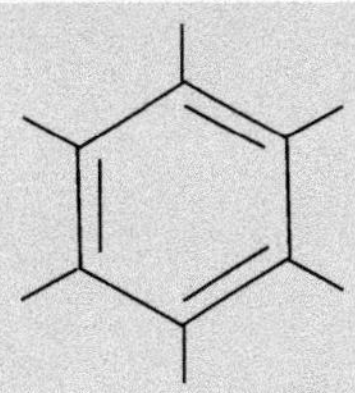
Alcohols	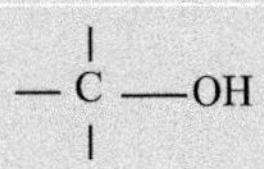
Amines	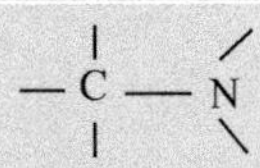
Aldehydes	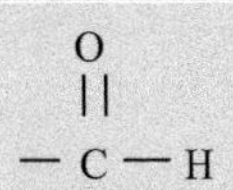
Ether	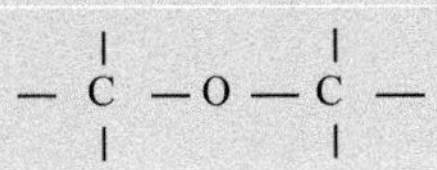

Continued

TABLE 8.5 Structures of organic compounds—cont'd

Chemical Class	Functional group
Ketones	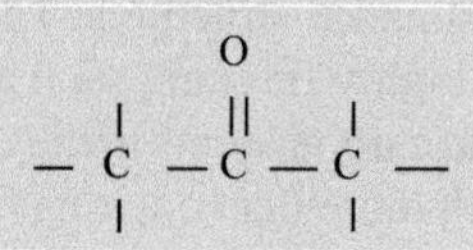
Carboxylic acids	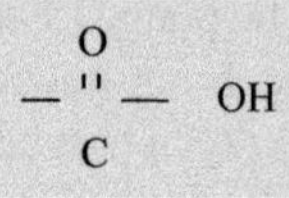
Alkyl halides	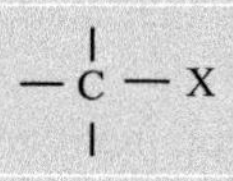
Phenols (aromatic alcohols)	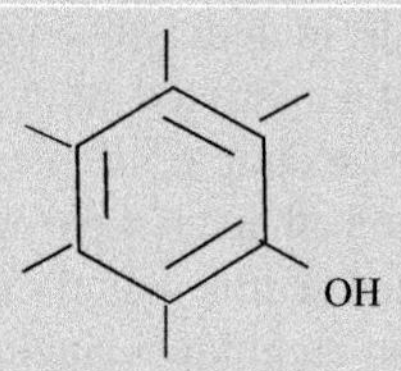
Substituted aromatics (substituted benzene derivatives):	
Nitrobenzene	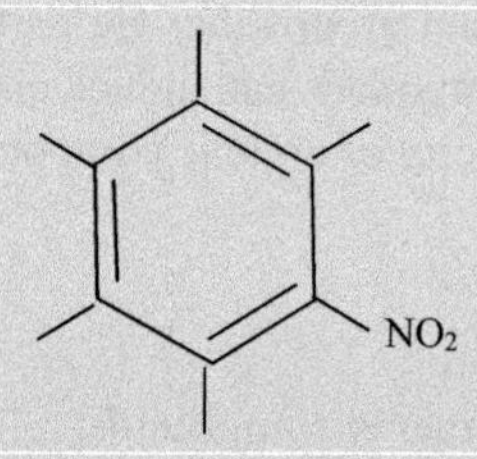
Monosubstituted alkylbenzenes	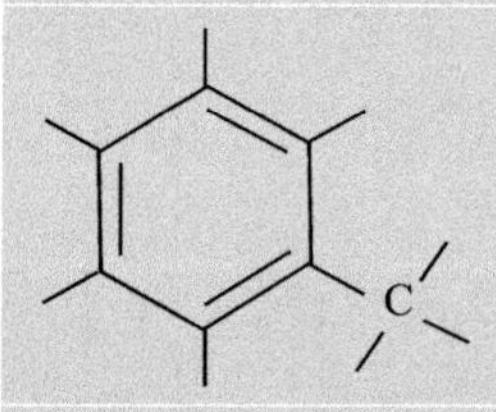
Toluene (Simplest monosubstituted alky benzene)	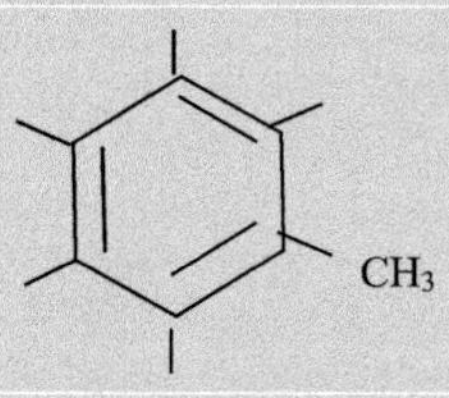
Polysubstituted alkylbenzenes	
1,2-Alkyl benzene (also known as ortho or *o*-...)	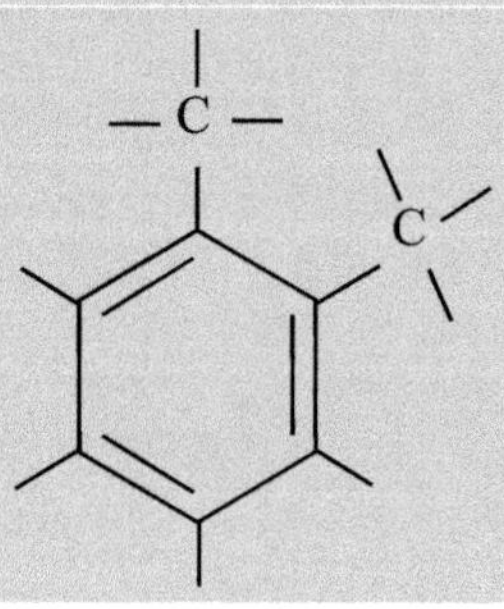

TABLE 8.5 Structures of organic compounds—cont'd

Chemical Class	Functional group
1,2-xylene or *ortho*-xylene (*o*-xylene)	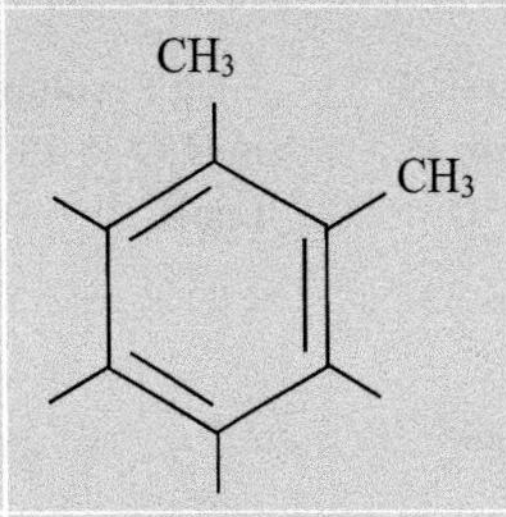
1,3-xylene or *meta*-xylene (*m*-xylene)	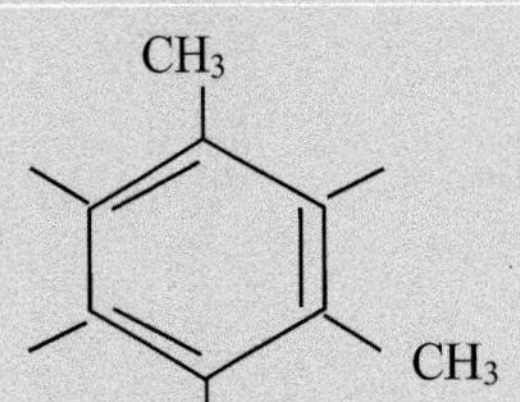
1,4-xylene or *para*-xylene (*p*-xylene)	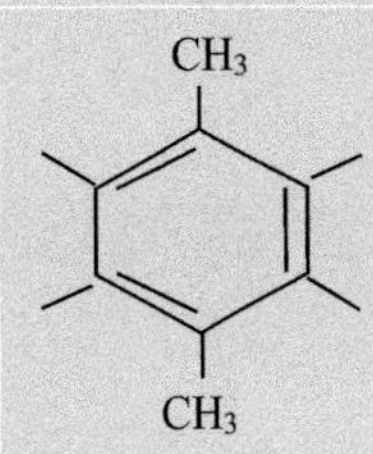
Hydroxyphenols do not follow general nomenclature rules for substituted benzenes:	
Catechol (1,2-hydroxiphenol)	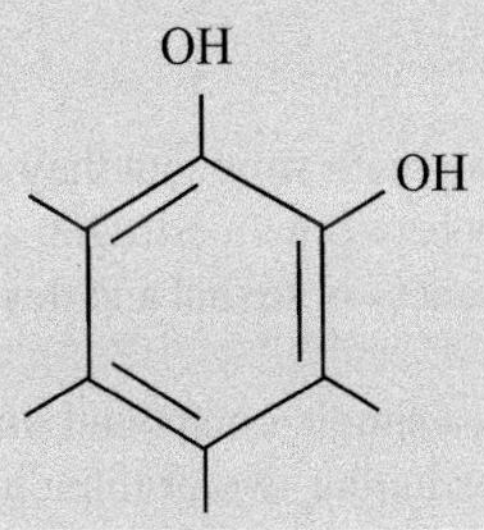
Resorcinol (1,3-hydroxiphenol)	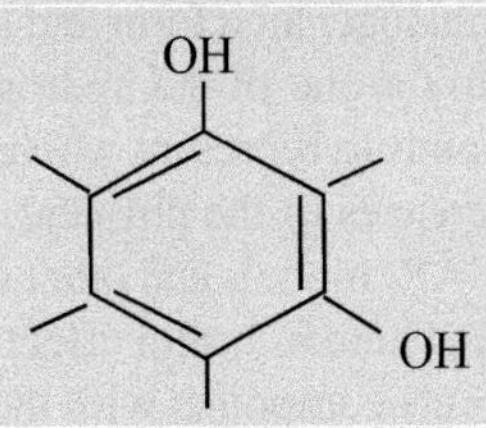
Hydroquinone (1,4-hydroxiphenol)	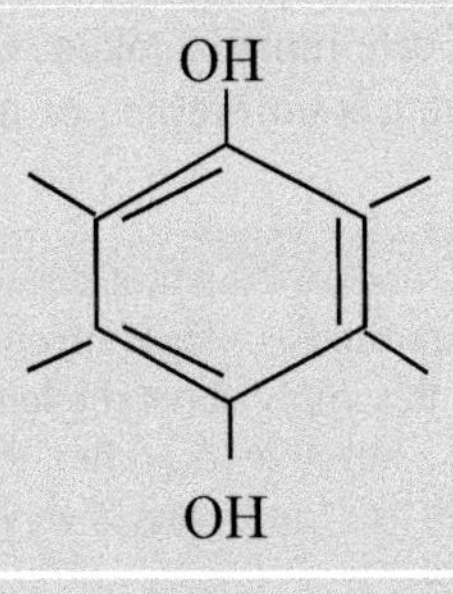

Note: Note that when consulting manuals and organic chemistry texts regarding the physicochemical properties of organic compounds, such as those for pesticides and synthetic chemistry, the "X" usually denotes a halogen, e.g., fluorine, chlorine, or bromine. At times, this text deviates from this convention because it is intended to be useful for many scientific and engineering disciplines involved in water pollution that often refer to X unknown variable or substance.

TABLE 8.6 Incremental differences in molecular structure leading to changes in physicochemical properties and hazards.

Compound	Physical state @ 25 °C	$-\log P^o$ solubility in H_2O @ 25 °C ($mol\,L^{-1}$)	−log vapor pressure @ 25 °C (atm)	Worker exposure limits (parts per million)	Regulating agency
Methane, CH_4	Gas	2.8	−2.4	25	Canadian Safety Association
Tetrachloromethane (carbon tetrachloride), CCl_4	Liquid	2.2	0.8	2 short-term exposure limit (STEL) = 60 min	National Institute of Occupation Health Sciences (NIOSH)
Ethane, C_2H_6	Gas	2.7	−1.6	None (simple asphyxiant)	Occupational Safety and Health Administration (OSHA)
Trichloroethane, $C_2H\,Cl_3$	Liquid	2.0	1.0	450 STEL (15 min)	OSHA
Benzene, C_6H_6	Liquid	1.6	0.9	5 STEL	OSHA
Phenol, C_6H_6O	Liquid	0.2	3.6	10 ppm	OSHA
Toluene C_7H_8	Liquid	2.3	1.4	150 STEL	UK Occupational & Environmental Safety Services

a testosterone molecule may seem small but they cause significant differences in the growth and reproduction of animals. The very subtle differences between an estrogen and an androgen, female and male hormones respectively, can be seen in these structures. The differences in sexual and developmental changes that result from this seemingly minor structural difference are dramatic.

Incremental changes to a simple compound like ethane can make for large differences in physical and chemical properties (See Table 8.6). Substituting two or three hydrogens with chlorine atoms makes for differences in the toxicity and other harm between the nonhalogenated form and the chlorinated form. The same holds for the simplest aromatic, benzene. Substituting a methyl group for one of the hydrogen atoms forms toluene.

Using surrogate compounds to represent whole groups of chemicals has many uncertainties since even a slight change can render substantial differences in the physical, chemical, and biological inherency of the molecule. However, quantitative structure activity (QSAR) models and other computational chemistry tools continue to grow in use and confidence in predicting environmental and human health effects. Also, the subtle differences in molecular structures in one compound and a potential, safer substitute compound with the same function, e.g., a solvent or plasticizer, is an important part of life cycle analysis [42–44]. Yet, care must be taken when selecting a substitute to avoid what is known as a "regrettable substitution" when the substitute has unforeseen or unknown hazards [45].

Organic chemistry example 1

If the aqueous solubility as expressed as $-\log P^o$ (in $mol\,L^{-1}$) of 1-butanol, 1-hexanol, 1-octanol, and 1-nonanol are 0.1, 0.9, 2.4, and 3.1, respectively, what does this imply about the length of carbon chains and the solubility of alcohols?

Solubility is expressed as a negative logarithm, thus lengthening the carbon chain decreases polarity and, therefore, aqueous solubility since "like dissolves like" and water is very polar. Thus, as shown in the Fig. 8.7, 1-butanol is orders of magnitude more hydrophilic than is 1-nonanol.

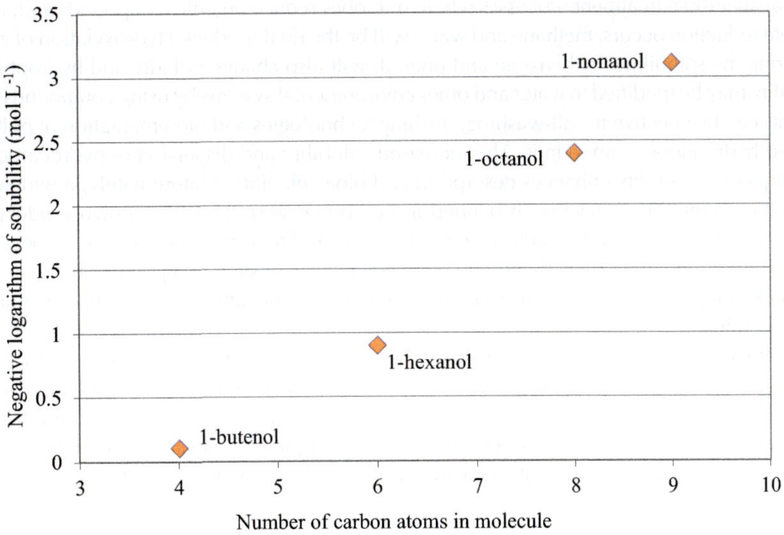

FIG. 8.7 Aqueous solubility of four alcohols. *(Data from D.A. Vallero, Fundamentals of Air Pollution, fifth ed., Elsevier Academic Press, Waltham, MA, 2014, 999 pages cm.)*

Organic chemistry example 2

If n-butane, n-hexane, n-octane, and n-nonane's aqueous solubilities as expressed as –log P° are, respectively: 3.0, 3.8, 5.2, and 5.9 mol L^{-1}, what effect does the substitution of a hydroxyl functional group for the hydrogen atom have on an alkane's polarity? Can this effect also be observed in aromatics?

First, compare P° for benzene, toluene to phenol in Table 8.6. Note that phenol is a hydroxylated benzene. This comparison shows a direct relationship between the increase in polarity and hydrophilicity when alkanes are hydroxylated into alcohols (See Fig. 8.8). This is why alcohols are miscible in water.

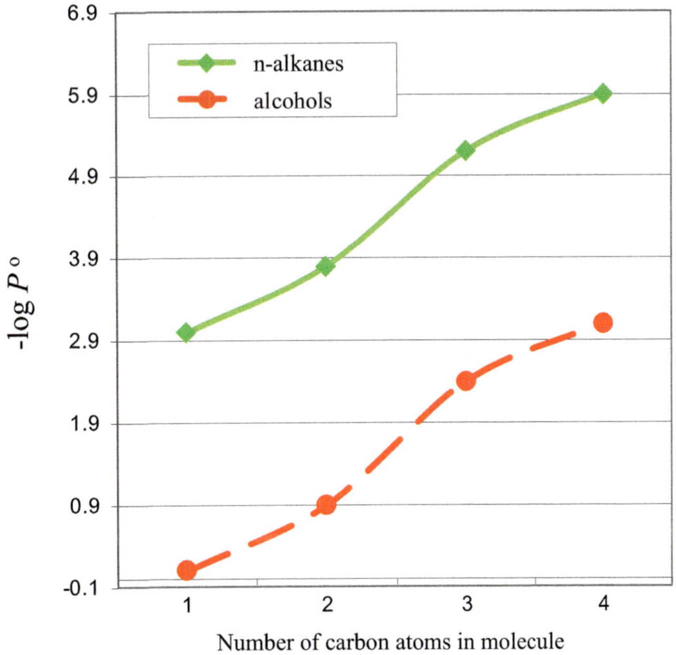

FIG. 8.8 Aqueous solubility of selected aliphatic compounds. *(Data from D.A. Vallero, Fundamentals of Air Pollution, fifth ed., Elsevier Academic Press, Waltham, MA, 2014, 999 pages cm.)*

This is a key aspect of anaerobic treatment processes where microbes reduce organic compounds and, in the process, generate alcohols and, if complete reduction occurs, methane and water will be the final product. Hydroxylation of an aromatic compound, as indicated by comparing the solubilities of benzene and phenol, will also change polarity and hydrophilicity.

A substance's solubility may be modified in water and other environmental systems by using compounds known as "surfactants" (See Fig. 8.9). Surfactants can be effective in soil-washing, flushing technologies and bioremediation of polluted soil, for example, by solubilizing adsorbed hydrophobic compounds. The increased solubility and dispersion of hydrocarbons and aromatic compounds with very low aqueous solubility enhances desorption and bioavailability. Unfortunately, a widely used group of surfactants, i.e., alkylphenolethoxylates (APEs) have been banned in Europe because their breakdown products can be very toxic to aquatic organisms. The APEs have been particularly of concern because of their hormonal effects, especially their estrogenicity. So, one must keep in mind not only the physical and chemical advantages of such compounds, but also any ancillary risks that they may introduce. This is a classic case of competition among values (i.e., easier and more effective cleanup versus additive health risks from the cleanup chemical).

Isomers are compounds with identical chemical formulae, but different structures. They are very important in environmental organic chemistry, because even slightly different structures can evoke dramatic differences in chemical and physical properties. So, isomers may exhibit different chemodynamic behavior and different toxicities. For example, the three isomers of pentane (C_5H_{12}) are shown in Fig. 8.10. The difference in structure accounts for significant physical differences. For example, the boiling points for *n*-pentane, isopentane, and neopentane at 1 atm are 36.1 °C, 27.8 °C, and 9.5 °C, respectively. Thus, neopentane's lower

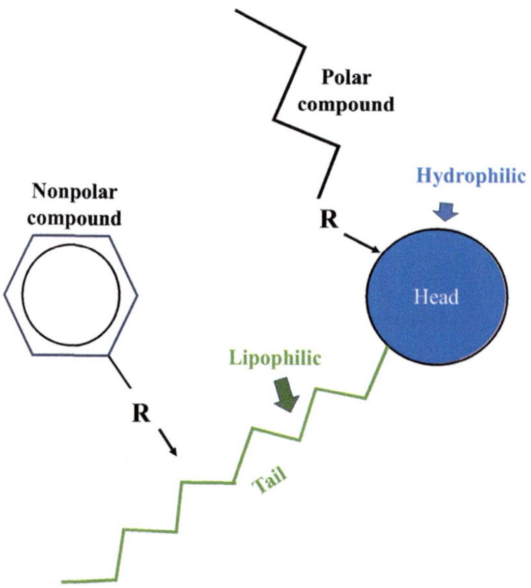

FIG. 8.9 Structure of a hypothetical surfactant. The head of the molecule has high relative polarity and the tail has low relative polarity, so that in the aqueous phase the hydrophilic compounds will react with the surfactant molecule at the head position, while the lipophilic compounds will react at the tail position. The result is that a greater amount of the organic compounds will be in solution when surfactants are present.

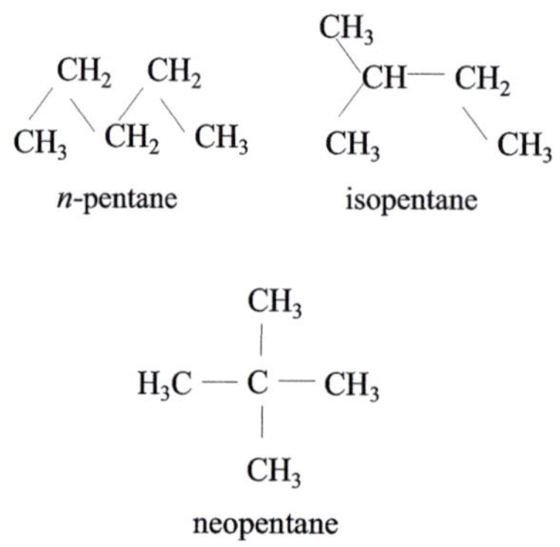

FIG. 8.10 Isomers of pentane. *(Data from D.A. Vallero, Fundamentals of Air Pollution, fifth ed., Elsevier Academic Press, Waltham, MA, 2014, 999 pages cm.)*

boiling point means that this isomer has a lower vapor pressure is more likely to enter the atmosphere than the other two isomers under the same environmental conditions [12].

Optical isomers, known as chiral molecules are important. A chiral's mirror images are not identical, i.e., they are not superposable. Enantiomers are a pair of molecules that are not identical but are mirror images of each other. Often, the left-hand chiral is much less toxic than the right-hand chiral of the same molecule. For example, the left-handed thalidomide molecule was relatively effective as a pharmaceutical, but the right-handed molecule is highly toxic. Indeed, it caused major birth defects in about 80% of the children born to mothers who took the drug during pregnancy. Most, if not all, of these may have been caused by the right-handed molecule. In the environment, chirality affects molecular interactions with enzymes. Thus, chirality can strongly effect the toxicity, mobility, and persistence of a chemical compound [46].

6 Conclusions

This chapter introduced several chemical principles related to water pollution. Calculating the fate, transport, exposure, and effects of water pollutants relies on sound chemistry. This applies to both abiotic and biotic factors in pollutants. Key aspects of chemistry were discussed as they relate to chemical compounds and biota, especially the biochemistry of microbes. These topics will be applied to specific problems and treatment scenarios in the next chapters.

References

[1] D.A. Vallero, Air Pollution Calculations: Quantifying Pollutant Formation, Transport, Transformation, Fate and Risks, Elsevier, 2019.

[2] H.F. Hemond, E.J. Fechner, Chemical Fate and Transport in the Environment, Elsevier, 2014.

[3] National Research Council, Exposure Science in the 21st Century: A Vision and a Strategy, The National Academies Press (in English), Washington, DC, 2012, p. 196.

[4] K. Gerber, User's guide and technical documentation KABAM version 1.0 (K OW (based) aquatic bioaccumulation model), Report of Environmental Fate and Effects Division Office of Pesticide Programs, US Environmental Protection Agency, Washington, DC, 2009.

[5] D. Mackay, W.-Y. Shiu, K.-C. Ma, S.C. Lee, Handbook of Physical-Chemical Properties and Environmental Fate for Organic Chemicals, CRC Press, 2006.

[6] R.H. Hesslein, Chemical Fate and Transport in the Environment, Wiley Online Library, 2000.

[7] D. Mackay, W.Y. Shiu, K.-C. Ma, Illustrated Handbook of Physical-Chemical Properties of Environmental Fate for Organic Chemicals, CRC Press, 1997.

[8] G. Rice, R. Ambrose, O. Bullock, J. Swartout, Mercury study report to Congress. Volume 3. Fate and transport of mercury in the environment, Environmental Protection Agency, Research Triangle Park, NC (United States). Office of Air Quality Planning and Standards, 1997.

[9] W. Lyman, Transport and transformation processes, in: Rand GM Fundamentals of Aquatic Toxicology. Effects, Environmental Fate, and Risk Assessment, second ed., Taylor and Francis, Washington, DC, 1995, pp. 449–492.

[10] D. Mackay, S. Paterson, Evaluating the multimedia fate of organic chemicals: a level III fugacity model, Environ. Sci. Technol. 25 (3) (1991) 427–436.

[11] J.N. Spencer, G.M. Bodner, L.H. Rickard, Chemistry: Structure and Dynamics, John Wiley & Sons, 2010.

[12] D.A. Vallero, Fundamentals of Air Pollution, fifth ed., Elsevier Academic Press, Waltham, MA, 2014. 999 pages cm.

[13] H. Autos, What is Average Mileage Per Year, 2024. https://www.caranddriver.com/auto-loans/a32880477/average-mileage-per-year/. (Accessed 6 January 2024).

[14] J.Y. Son, M.L. Miranda, M.L. Bell, Exposure to concentrated animal feeding operations (CAFOs) and risk of mortality in North Carolina, USA, Sci. Total Environ. 799 (10) (2021) https://doi.org/10.1016/j.scitotenv.2021.149407. In this issue.

[15] J. Wang, W. Yue, Y. Teng, Y. Zhai, H. Zhu, Degradation kinetics and transformation pathway of methyl parathion by δ-MnO_2/oxalic acid reaction system, Chemosphere 320 (2023) 138054.

[16] C. Meenambigai, K. Bhuvaneswari, Dissipation and degradation kinetics of commonly used pesticides and their metabolites in/on okra, Abelmoschus esculentus (L.) Moench, Curr. Sci. 124 (4) (2023) (00113891).

[17] S.B. Lartiges, P.P. Garrigues, Degradation kinetics of organophosphorus and organonitrogen pesticides in different waters under various environmental conditions, Environ. Sci. Technol. 29 (5) (1995) 1246–1254.

[18] S. Liu, Chapter 11—How cells grow, in: S. Liu (Ed.), Bioprocess Engineering, second ed., Elsevier, 2017, pp. 629–697.

[19] W. Gujer, M. Boller, Design of a nitrifying tertiary trickling filter based on theoretical concepts, Water Res. 20 (11) (1986) 1353–1362, https://doi.org/10.1016/0043-1354(86)90133-8.

[20] K. van't Riet, R.G.J.M. van der Lans, 2.07—Mixing in bioreactor vessels, in: M. Moo-Young (Ed.), Comprehensive Biotechnology, second ed., Academic Press, Burlington, 2011, pp. 63–80.

[21] X. Zou, J.-Y. Xia, J. Chu, Y.-P. Zhuang, S.-L. Zhang, Real-time fluid dynamics investigation and physiological response for erythromycin fermentation scale-up from 50 L to 132 m^3 fermenter, Bioprocess Biosyst. Eng. 35 (5) (2012) 789–800.

[22] J. Bao, Q. Ye, J.-J. Zhong, Bioreactor Engineering Research and Industrial Applications II, Springer, 2016.

[23] I.S. Barkow, S.E. Oswald, H.J. Lensing, M. Munz, Seasonal dynamics modifies fate of oxygen, nitrate, and organic micropollutants during bank filtration-temperature-dependent reactive transport modeling of field data (in Eng), Environ. Sci. Pollut. Res. Int. 28 (8) (2021) 9682–9700, https://doi.org/10.1007/s11356-020-11002-9.
[24] J. Greskowiak, E. Hamann, V. Burke, G. Massmann, The uncertainty of biodegradation rate constants of emerging organic compounds in soil and groundwater—a compilation of literature values for 82 substances (in Eng), Water Res. 126 (2017) 122–133, https://doi.org/10.1016/j.watres.2017.09.017.
[25] J. Regnery, A.D. Wing, M. Alidina, J.E. Drewes, Biotransformation of trace organic chemicals during groundwater recharge: how useful are first-order rate constants? (in Eng), J. Contam. Hydrol. 179 (2015) 65–75, https://doi.org/10.1016/j.jconhyd.2015.05.008.
[26] D.A. Vallero, Hazardous wastes, in: Waste: A Handbook for Management, 2011, p. 393.
[27] L. Wang, Z. Lei, S. Yun, X. Yang, R. Chen, Quantitative structure-biotransformation relationships of organic micropollutants in aerobic and anaerobic wastewater treatments, Sci. Total Environ. (2023) 169170.
[28] A. Berhanu, I. Mutanda, J. Taolin, M.A. Qaria, B. Yang, D. Zhu, A review of microbial degradation of per-and polyfluoroalkyl substances (PFAS): biotransformation routes and enzymes, Sci. Total Environ. (2022) 160010.
[29] B.A. Bekins, E. Warren, E.M. Godsy, A comparison of zero-order, first-order, and monod biotransformation models, Groundwater 36 (2) (1998) 261–268.
[30] S.S. Suthersan, J. Horst, M. Schnobrich, N. Welty, J. McDonough, Remediation Engineering: Design Concepts, CRC Press, 2016.
[31] T. Weidemeier, M. Swanson, J. Wilson, D. Kampbell, R. Miller, J. Hansen, Approximation of biodegradation rate constants for monoaromatic hydrocarbons (BTEX) in ground water, US Environmental Protection Agency Papers, 1996, p. 26.
[32] D.A. Vallero, Chapter 31—Hazardous wastes, in: T.M. Letcher, D.A. Vallero (Eds.), Waste, second ed., Academic Press, 2019, pp. 585–630.
[33] M. Alexander, Biodegradation and Bioremediation, Gulf Professional Publishing, 1999.
[34] D.A. Abramowicz, Aerobic and anaerobic PCB biodegradation in the environment, Environ. Health Perspect. 103 (Suppl 5) (1995) 97.
[35] S. Arrhenius, Electrolytic dissociation, J. Am. Chem. Soc. 34 (4) (1912) 353–364.
[36] R. Bronaugh, K.-C.T. Hoang, Dermal Exposure Assessment: Principles and Applications, US Environmental Protection Agency, Office of Health and Environmental…, 1992.
[37] USEPA, Dermal Exposure Assessment: Principles and Applications (EPA/600/8–91/011B), US Environmental Protection Agency, 1992.
[38] T. Kulikova, et al., Optimization of nucleophilic dechlorination of polychlorinated biphenyls: calculation and experiment, Int. J. Environ. Sci. Technol. 16 (2019) 3265–3274.
[39] F.-R. Xiu, Y. Lu, Y. Qi, DEHP degradation and dechlorination of polyvinyl chloride waste in subcritical water with alkali and ethanol: a comparative study, Chemosphere 249 (2020) 126138, https://doi.org/10.1016/j.chemosphere.2020.126138.
[40] R.C. Knox, D.A. Sabatini, L.W. Canter, Subsurface Transport and Fate Processes, CRC Press, 2018.
[41] K.-H. van Pée, S. Unversucht, Biological dehalogenation and halogenation reactions, Chemosphere 52 (2) (2003) 299–312, https://doi.org/10.1016/S0045-6535(03)00204-2.
[42] S.-O. Ryding, ISO 14042 Environmental management. Life cycle assessment, life cycle impact assessment, Int. J. Life Cycle Assess. 4 (6) (1999) 307.
[43] P.T. Anastas, R.L. Lankey, Life cycle assessment and green chemistry: the yin and yang of industrial ecology, Green Chem. 2 (6) (2000) 289–295.
[44] D. Dawson, et al., Assessment of non-occupational 1, 4-dioxane exposure pathways from drinking water and product use, Environ. Sci. Technol. 56 (8) (2022) 5266–5275.
[45] A. Maertens, E. Golden, T. Hartung, Avoiding regrettable substitutions: green toxicology for sustainable chemistry, ACS Sustain. Chem. Eng. 9 (23) (2021) 7749–7758, https://doi.org/10.1021/acssuschemeng.0c09435.
[46] M. Arenas, J. Martín, J.L. Santos, I. Aparicio, E. Alonso, Enantioselective behavior of environmental chiral pollutants: a comprehensive review, Crit. Rev. Environ. Sci. Technol. 52 (17) (2022) 2995–3034, https://doi.org/10.1080/10643389.2021.1900764.

Chapter 9

Water pollution biology

1 Introduction

As discussed in Chapter 4, matter and energy move through trophic states within water systems and affect how the non-living components of a system interact with the living components. There are countless interrelationships between abiotic and biotic systems, respectively. Thus, biogeochemical interactions can be studied, documented, and analyzed to track the earth's mass and energy balances, both thermodynamically and fluid dynamically [1]. Even the terrain of the ecosystem results in part due to biology. Physical and chemical processes make the original rock material suitable for microbial and root growth. Lichen and other simple plant systems further degrade solid rock and its unconsolidated materials, allowing roots of plants to take hold and expand.

Biota consist of fauna, flora, or microbial individual organisms and populations that can be receptors of water pollutant's effects via exposures. Pollutants are causative agents of environmental harm and human, animal, or plant diseases. Pollutants are stressors that can be transferred to, from, and within environmental media, e.g., a plant's roots extract a pollutant from the soil to plant tissues followed by the accumulation and transfer up to other trophic states from producers on to consumers in the food chain. Biota of various genera transform a pollutant biochemically within the organism, i.e., biotransformation, which can make a pollutant less toxic by depuration or more toxic by bioactivation. Biota can be seen as their own environmental compartment or part of the composition of other compartments, such as bacteria in river sediment and fish in the water column of an estuary.

As such, an organism can be envisioned as a control volume in which energy and matter enters and exists and where complex physical and chemical processes and mechanisms happen. Within the control volume, both abiotic chemical reactions and biochemical reactions take place continuously. Metabolic reactions are regularly catalyzed enzymatically [2,3] and mediated by microorganisms living within the organism [4–9]. Even the simplest one-cell organism is a chemical reactor in which molecules enter, are metabolized to produce energy and cellular molecules, and release new molecules as wastes from these processes [10–14].

The principal environmental hazard is toxicity, whether ecological toxicity[a] or human toxicity. The collective term, environmental toxicology, is inclusive of toxic substances in water, air, soil, sediment, food, or materials. The toxicity is differentiated by onset of an adverse event, e.g., cancer in humans or loss of species in a river from a fish kill. If the toxic response occurs soon after the exposure, it is an acute effect. If the adverse event occurs after a substantial amount of time, such as a cancer latency period, it is a chronic effect. Between the two, especially in health risk assessments, is a subchronic event.

Characterizing a hazard includes the sources, environmental conditions, and the time and dose needed to elicit an adverse event. These and other elements of risk are discussed in Chapter 17.

This chapter specifically addresses how organisms and biological systems affect water quality. Biological processes and mechanisms occur within organisms and in the organism's environment. From an engineering standpoint, these same biochemical can be used to remediate pollution [1,15–22].

a. Often shortened to "ecotoxicity."

Fundamentals of Water Pollution. https://doi.org/10.1016/B978-0-443-28987-3.00011-4

2 Biological organization

The water molecule configuration depicted in Fig. 4.1 in Chapter 4 is the basis for water pollution biology [23]. The cycle of biological processes includes the decomposition microbes that are part of the soil to plant life to the animals that consume plants to omnivores and carnivores, until they die and are decomposed by microbes. The cycle continues.

Thermodynamically, to overcome entropy, external energy must be added to keep biological processes proceeding. Indeed, life is constantly in need of new energy as it fights chaos [24]. Whereas physicists use thermodynamics to explain how mass and energy are distributed, ecologists consider the distributions and exchanges between mass and energy to explain the complex interrelationships between and within the compartments of the food webs and chains [24–26]. Humans are among the consumers [27]. Food chains represent the dynamics, complexity, and vulnerability of aquatic and other environmental systems (see Fig. 9.1). Species at a higher tropic level are predators of lower-trophic state species, which means that biotic matter and energy flow downward. The rates and quantity of the transfer of biomass and energy upwardly and downwardly to and from adjoining tropic levels vary in time and space. Stress from a pollutant entering an ecosystem alters these trophic interrelationships. For example, diminishing the overabundance of a primary producer like algae will modify predator-prey relationships, since the algal blooms may extract dissolved oxygen that in turn reduces the abundance of consumers and ultimately the top aquatic predators like seals and sharks, as well as top terrestrial predators like piscivorous birds, e.g., eagles and carnivorous mammals, including humans.

All organisms are made up of various molecular arrangements of the elements carbon (C), oxygen (O), hydrogen (H), and often nitrogen (N), in addition to various micronutrients. The C, O, H, and N are the biophile elements with an affinity for each other, which leads to their formation of complex organic compounds. Viruses, bacteria, and fungi are very efficient in using available organic material as sources of energy and carbon. Although a minority of microbial genera, pathogens have received most of the interest, at least from the general public, as agents of disease and, notably, epidemics and pandemics. However, the microbe is usually only one agent in the causal chains of diseases. Often, the cause of the epidemics involves numerous interrelationships among the trophic states. For example, when habitats are changed to allow advantages

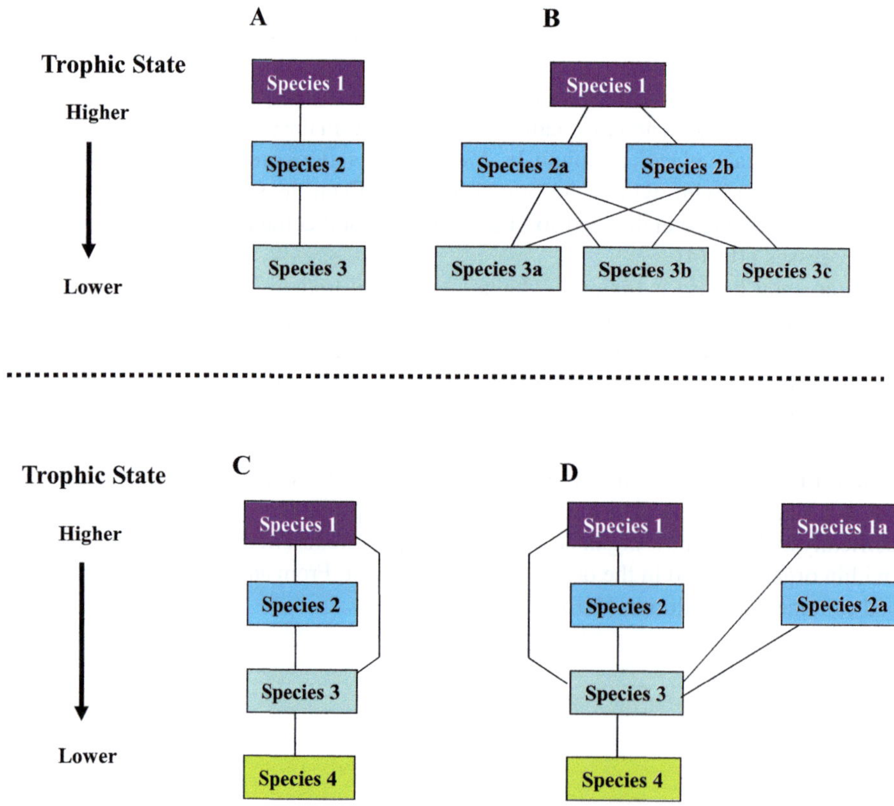

FIG. 9.1 Energy and matter flow in environmental systems from higher trophic levels to lower trophic levels. Lines represent interrelationships among species. (A) Linear biosystem; (B) Multilevel trophic biosystem; (C) Omnivorous biosystem; and (D) Multilevel biosystem with predation and omnivorous behaviors. An interference at any of these levels or changes in mass and energy flow can lead to disasters. *(From D. Vallero, Air Pollution Calculations: Quantifying Pollutant Formation, Transport, Transformation, Fate and Risks, Elsevier, 2019; used with permission.)*

to certain species, such as rats carrying the plague bacterium, *Yersinia pestis*, the epidemic cannot simply be attributed exclusively to the pathogen, but to the overall changes to the environment to allow the pathogen population to grow and to persist. Climatic shifts, poor water and air quality, and societal stressors like poverty, crime, and poor housing conditions can lead to outbreaks and worsen the outcomes of diseases in vulnerable human subpopulations [28–30].

Ecosystems and individual organisms and their components are thermodynamic systems. Beyond thermodynamics, a "system" is a method of organization, from smaller to larger aggregations. The "ecosystem" and the "organism" are examples of both connotations. They consist of physical phases and order (e.g., producer-consumer-decomposer; predator-prey; individual-association-community; or cell-tissue-organ-system). They also help to explain how matter and energy move and change within a parcel of matter. Within the context of thermodynamics, a system is a sector or region in space or some parcel of a sector that has at least one substance that is ordered into phases. Reactors, ground water plumes, the entire hydrosphere, communities, organisms, and cells have qualities of both closed and open systems. A closed system allows no material to enter nor to leave the system. The open system allows material to enter and to leave the systems, known as a control volume or, for engineers, a control mass [31].

Pollution results from complex interactions of substances in various states of matter in the various environmental compartments where substances are formed, transformed, and transported. The transformations may render a compound that was previously not inherently harmful to become toxic to one or more species in the ecosystem. Conversely, what may have been a precursor to a pollutant may have been altered to become a less harmful compound or compounds. During transport, a pollutant may also undergo chemical change in various parts of an ecosystem and have a selective effect on a habitat. These changes can be incremental and may not become obvious for years or decades.

3 Cells

Living things contain water, compounds of carbon, and deoxyribonucleic acid (DNA). These are all involved in cellular reproduction. Binary fission is the most common cellular reproductive mechanism (see Fig. 9.2), which begins with cell growth that causes the cell to enlarge and elongate to hold an increase in cellular components. The middle of the enlarged cell continues to enlarge and forms two progeny cells, each with a complete copy of the parent cell's genome and half of the cytoplasm. Cytoplasm is a gelatinous cellular liquid composed of water, salts, and organic molecules. Intracellular organelles, especially the nucleus and mitochondria, are bounded from the cytoplasm by membranes. Cytokinesis is the cytoplasmic division at the end of the cellular separation process. The replication of DNA begins where the chromosome is attached to the inner cell membrane. Ultimately, the cells separate completely [32].

Single-cell organisms are prokaryotes and taxonomically fall under the Bacteria and Archaea domains. The more complex eukaryotes fall under the Eukarya domain. All three domains contain unicellular organisms. Eukarya also contains multicellular organisms. Eukarya's single-cell organisms include fungi and protists. Protists include algae and protozoa like amoebas [32].

As mentioned, all cells contain DNA, and have cellular or plasma membranes, cytoplasm, and ribosomes, the organelles that manufacture proteins. A eukaryotic cell has a nucleus, whereas a prokaryotic cell does not. The prokaryotic cell's genetic material is suspended in the cell's cytoplasm.

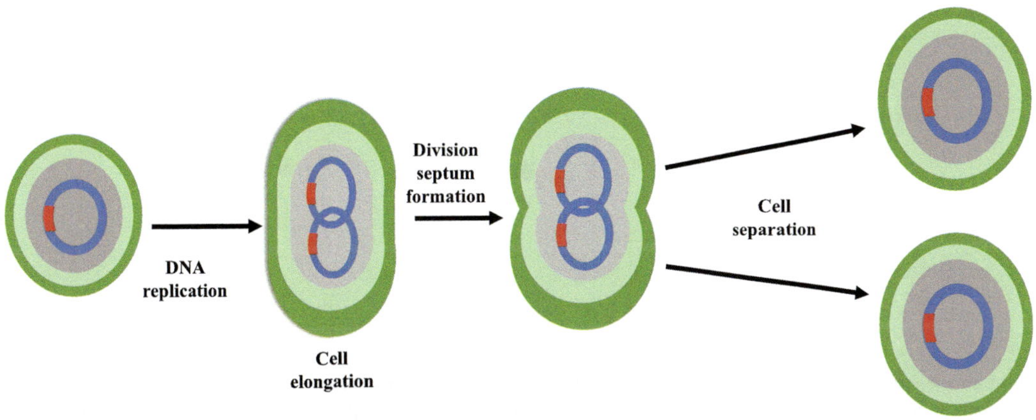

FIG. 9.2 Steps of cellular reproduction by binary fission. *(Data from D.A. Vallero, Biotic physics, in: Methods and Calculations in Environmental Physics, AIP Publishing LLC, Melville, New York, 2022, pp. 9-1–9-40.)*

Mitosis and meiosis are the two types of binary fission. Mitosis produces cells is the process shown in Fig. 9.2, in which a cell simply divides into two cells. Meiosis produces gametes that are needed to produce a cell or nucleus that contains two complete sets of chromosomes, i.e., a diploid. Chromosomes are threadlike structures comprised of nucleic acids and protein that carry genetic information to form genes made of DNA. Gametes have half of the genetic material needed to initiate a new diploid by fusion with a gamete from the opposite sex (see Fig. 9.3). The female gamete is an egg cell. The male gamete is a sperm. Most animals are diploid organisms, so that their somatic cells are diploid, and their gamete cells are produced through meiosis. Certain genera of insects are exceptions. In bees, wasps, and ants, the male is haploid because it develops from unfertilized eggs. Sexual reproduction distinguishes animals from other organisms like fungi, protists, and bacteria, that commonly or exclusively reproduce asexually. Some animals can also reproduce asexually, such as cnidarians, flatworm, and roundworms.

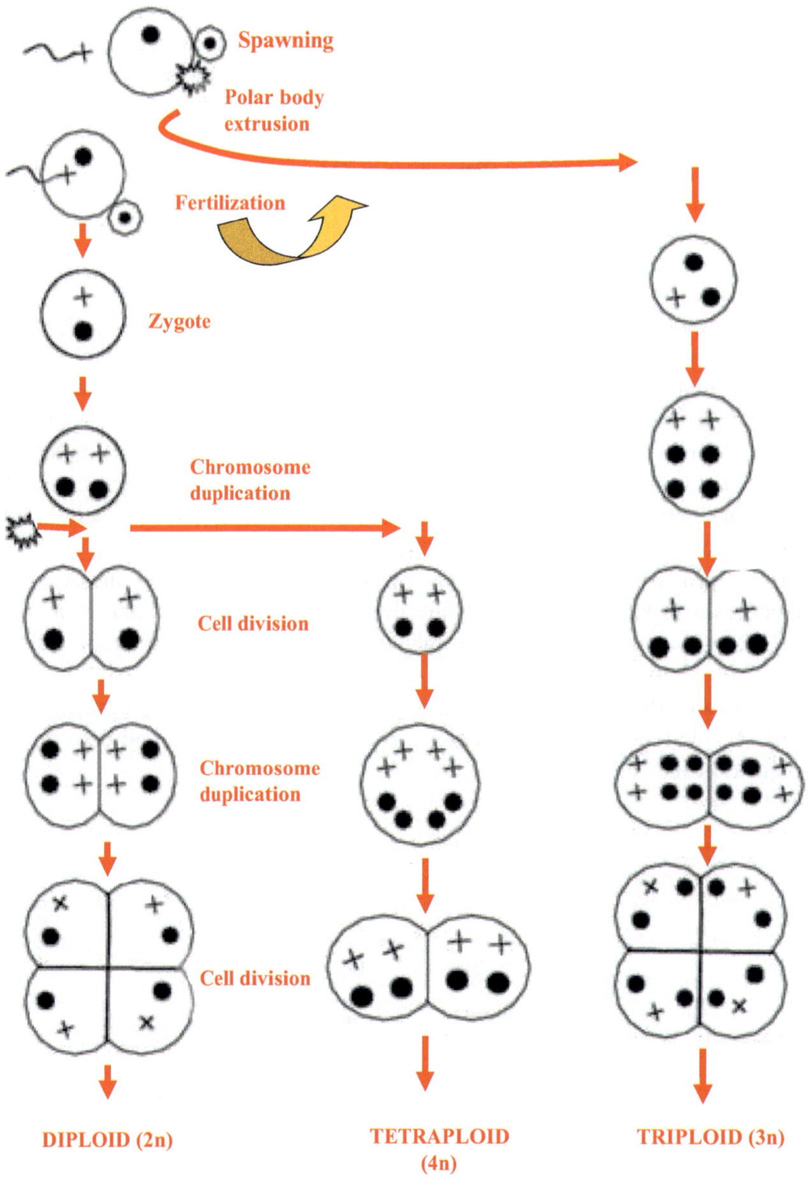

FIG. 9.3 Steps in gamete fertilization and cell division that lead to the normal diploid fish embryo. Induction of triploidy or tetraploidy occurs by for shocks (thermal, chemical or pressure) at a specific time after fertilization. Notes: ✸ denotes the point at which the shock is applied; ● denotes one haploid chromosome set derived from the female parent; + ● denotes one haploid chromosome set derived from the male. *(Data from National Research Council, Biological Confinement of Genetically Engineered Organisms, National Academies Press, 2004.)*

4 Invasive species

Invasion is special type of water pollution when opportunistic organisms can colonize a habitat. During the invasion, ecological structure and function are changed and become unbalanced, leading to the loss of diversity and other ecological damage. Invasive organisms can be any type of biota, that is, microbes, plants, and animals. Usually, their presence is the result of human activity.

An invasive organism has few or no predators during their entry into an ecosystem. As such, they can consume the food sources much faster than their native competitors [33]. For example, the Great Lakes basin has experienced major problems with invasive species. Many invasive species have been identified in the Laurentian Great Lakes drainage basin. Lake Erie watershed alone has 132 nonnative species, including 20 species of algae, 8 submerged plants species, 39 marsh plants species, 5 tree and shrub species, 3 pathogenic bacteria species, 12 mollusk species, 9 oligochaete species, 9 crustacean species, and 23 fish species. The increase is due in the most part to switching from solid to water ballast in cargo ships and the opening of the St. Lawrence Seaway in 1959. Among the nonnative invasive fish species recently invading Lake Erie is the Chinese bighead carp, *Hypophthalmichthys nobilis* [34].

One of the means of addressing invasive species is by using biocontrols, i.e., using one living organism to limit and eliminate another organism's effects. Introducing sterility has been used to control fish populations by interrupting the normal fish development process. The traditional sterilization approach change's the invasive fish's ploidy (i.e., affecting the complete set of chromosomes), by modifying the chromosome development using pressure, thermal, or chemical shock at the point of fertilization to disrupt the egg's normal extrusion of a polar body containing a haploid set of maternal chromosomes.

Inducing triploidy has been used effectively to produce sterile offspring by disrupting sexual reproduction (see Fig. 9.3). The retained polar body produces an embryo with two haploid chromosome sets from the female (rather than the normal one haploid set) as well as a third set from the male [35]. The odd sets of chromosomes appear to interfere with the mechanics of pairing of homologous chromosomes during each cell division, which disrupts normal gamete development. The resulting triploid varies from the normal diploid number of chromosomes. However, the major weakness of this method of sterilization is it can be incomplete, that is some of the offspring are not triploid. This can be ameliorated somewhat by producing tetraploids in the first generation.

An important environmental risk of this type of sterilization is the escape of these genetically modified organisms into the wild populations, since triploids still have sufficient levels of sex hormones to elicit normal courtship and spawning. Thus, the entire wild relative species' reproductive success would be in jeopardy, which is exacerbated if the wild relatives are endangered or threatened. In addition, given the large numbers of fish that escape fish farms, if many triploid transgenic fish enter the environment recurrently, some could survive to adversely affect fish diversity [35].

New genetic engineering methods have been developed to induce fish sterility, particularly the application of interference ribonucleic acid (RNAi), to produce what are known as sterile ferals [36]. The RNAi method involves the insertion of a transgene that blocks the expression of an endogenous gene needed to develop gametes and embryos. The blocker expression is controlled by an inducible promoter. This indicates that the transgenic approach has all the ecological concerns of traditional techniques, but with additional uncertainties, such as the presence of repressor chemicals in the water [36].

A biotechnological concern is the possible spread of deleterious genetic material. Controlling nonnative fish species invasions with transgenic fish of the same species as the targeted species is designed to be disruptive to the fish life cycle (see Fig. 9.3). These disruptions can occur from before the embryonic stage up to juvenile development. Several endogenous genes' expression can be disrupted in fish, including the aromatase, estrogen receptor, gonadotropin releasing hormone, protamine, vitellogenin and growth regulating genes.

Species can hybridize with closely related species, which increases the potential of transgene spread to near relatives of nontarget species. The amount of hazard and risk is a function of the proximity and availability of these near relatives, their proficiency to hybridize, their fertility, and viability of their offspring.

With transgenic species there is also the potential for unexpected expressions such as toxicity or allergenicity differences between the native and transgenic species. This could lead to human health effects. The bottom line is that the decision to use these technologies for even a worthwhile environmental effort (dealing with invaders), must be based on a thorough identification of these hazards and a reliable assessment of the risks that these technologies may elicit.

5 Plant physiology and pollution

Plants are very important to water quality, as they take in and release water and gases by their roots and leaves. In the water and air, various nutrients are dissolved and suspended, but so are other substances. In the leaves, pollutants can enter the plant via the stomata, which will interrupt processes in the chloroplasts. Chloroplasts are plant cell organelles that convert light energy into chemical energy via the photosynthesis. As such, they are needed to combine water and CO_2 in the presence of sunlight to form sugars and to release O_2:

$$6CO_2 + 6H_2O \xrightarrow[light]{} C_6H_{12}O_6 + 6O_2 \tag{9.1}$$

The sugar in this case is glucose ($C_6H_{12}O_6$). Transpiration moves water from roots to the leaves resulting in the emission of water vapor to the atmosphere. Nutrients move advectively and diffusively throughout the plant. Plants also respire to oxidize carbohydrates by O_2 to form CO_2 and H_2O and the release of energy, represented by adenosine triphosphate (ATP):

$$C_6H_{12}O_6 + 6O_2 \rightarrow 6CO_2 + 6H_2O + ATP \tag{9.2}$$

ATP is then hydrolyzed to adenosine diphosphate (ADP):

$$ATP + H_2O \rightarrow ADP + P_i + \text{free energy} \tag{9.3}$$

ADP then reacts with the phosphate molecule (P_i) to form ATP

$$ADP + P_i + \text{free energy} \rightarrow ATP + H_2O \tag{9.4}$$

Transpiration, photosynthesis, and respiration occur simultaneously as they transport O_2, CO_2, and H_2O throughout the plant. These mechanisms also move these and other chemical compounds into the troposphere, i.e., phytopartitioning. The presence of certain chemical compounds in plant tissues can cause direct or indirect harm to the plant, as well as adversely affect soil microbial populations, such as replacing beneficial bacteria with those that are toxic to the plant. Given that microbes are key decomposers that mineralize organic compounds needed by plants to survive and that first order consumers depend on specific plants, changes in microbial populations upset the balance of the entire ecosystem. Microorganisms already stressed from direct pollutant pollution, e.g., sulfur dioxide (SO_2), can be further stressed by ambient changes, such as droughts and fires. Foliar and root damage diminishes plant vitality, increasing the plant community's vulnerability to insect infestation, and weakens competition with weed species [23].

Approximately one-third of the earth's land surface has 20% or more crown tree cover [37]. Crown cover is measured as the canopy cover plus the amount of skylight obstructed by tree canopies. Human activities are particularly stressful to tropical rainforests. These include clearcutting, harvesting whole trees, and building roads that traverse habitats. Deforestation has global implications, e.g., concomitant increases the emissions of CO_2 and other greenhouse gases to the atmosphere. Soil and plants in tropical forests sequester 460–575 billion metric tons (tonnes) of carbon worldwide with each square kilometer of tropical forest holding 45,000 tonnes of carbon. From 1850 to 1990, deforestation released 122 billion tonnes of carbon into the atmosphere. Currently the deforestation carbon emission rate is 1.6 billion tonnes per year [38]. A secondary air pollution impact of deforestation is the increased emissions of fugitive dust [39].

Water uptake and storage is the key feature of the physics and biochemistry of vascular plants. As a solvent and medium for reactants and suspended matter, water provides the site for plant metabolism to builds cells. Water is the principal vehicle for heat dissipation within the plant and provides temperature homeostasis. Excess water vapor decreases the density of air, promoting circulation and assists in the availability of CO_2 for photosynthesis, in which water also acts as a reactant and reagent. Starches, for example, react directly with water, i.e., under hydrolysis, during metabolism resulting in sugars needed for healthy plant physiology. Water also serves physical needs, e.g., sufficient water content is needed to maintain turgidity, i.e., cell wall expansion and the osmoregulatory function of the vacuole to prevent wilting [40]. The plant must optimize available moisture as root uptake of water increases a vascular plant's water content, as transpiration simultaneously decreases water content [41].

Upon entry to the root, dissolved and suspended nutrients and other substances flow into the plant and enter several layers of differentiated cells. The water crosses these layers successively, i.e., composite transport [42], via pathways composed of cell walls and intercellular space (apoplastic pathway), plasmota (symplastic pathway), or water channels (osmotic, transcellular pathway) [43].

In addition to molecules of nutrients and pollutants, plants also take up ions. Ionic transport occurs between soil water and roots by active pumping across the root cell cytoplasm's plasma membrane. Conversely, during transpiration, roots allow water to pass through them, rather than being taken up like ions. The water potential (ϕ_T) gradient makes this movement possible [43–45]. The total water potential (ϕ_T) combines several potentials [46]:

$$\phi_T = \phi_p + \phi_m + \phi_z + \phi_o + \phi_{ob} + \ldots\phi_n \tag{9.5}$$

Where, ϕ_p is the pressure potential, ϕ_m is the matric potential, ϕ_z is the gravitational potential, ϕ_o is the osmotic potential, ϕ_{ob} is the overburden potential, and ϕ_n represents the potentials induced by other factors.

Pressure potential is the height of water above a designated point, i.e., the water pressure exerted by the saturated column of water above that point. Matric potential owes to capillary and sorptive forces. Gravitational potential is caused by the force of gravity measured from a reference point. Osmotic pressure is the solute pressure caused by water's attraction to salts dissolved in the water film around particles. Overburden potential is caused by the weight of soil on a given point in the soil column holding the roots [46,47].

Composite transport is affected by different types of movement through individual cells and tissues, with preferable routes for flow of low resistance controlled by the plant's geometry, properties of cell walls, and permeability of the various layers [43–45].

When entering the xylem, the transport becomes more extensive than in the root, with vessels carrying the water via the longest reaches of the plant. In a corn plant, for example, xylem hydraulics cause how water to be transported several meters compared to only a few millimeters distance from the root surface to the xylem. Likewise, the distance from the xylem to mesophyll cells is but a few millimeters. Thus, xylem transport is about three orders of magnitude longer [43]. The major limitations of xylem flow are friction and cavitation, according to Poiseuille's law and probability theory, respectively. Poiseuille's law states that the velocity of a fluid's steady flow via a narrow tube like the xylem varies directly as the pressure and the fourth power of the radius of the tube (r), and inversely as the length of the tube (L) and the coefficient of viscosity. Thus, the pressure drop is directly proportional to the length of the xylem vessel and volumetric flow rate, and inversely proportional to the xylem vessel radius and cross-sectional area [48,49]:

$$\Delta p = \frac{8\mu LQ}{\pi r^4} = \frac{8\pi\mu LQ}{A^2} \tag{9.6}$$

Where, Δp is the pressure difference between the two ends of the tube, μ is the dynamic viscosity, Q is the volumetric flow rate, and A is the cross-sectional area of the tube.

The formation of vapor-filled cavities, bubbles, and voids in a liquid, known as cavitation, is caused by static pressure that reduces the liquid's vapor pressure (see Discussion Box 9.1). Xylem cavitation results when the water tension inside the xylem increase enough that dissolved air within water expands to fill either an entire vessel or the xylem's tracheids, i.e., water-conducting cells that have no perforations in their cell walls. Embolism is the blocking of a xylem vessel or tracheid by an air bubble or cavity [60]. Subsequently being sorbed by the root, water reaches and crosses the epidermis. This flow is in the direction of the center of the root, traversing the cortex and endodermis, until it reaches the xylem [61].

5.1 Diffusive flux

The time that substance stays within plant tissue is the plant's mean residence time (MRT or τ):

$$\tau = \frac{M}{J_{out}} \tag{9.7}$$

Where, τ = MRT at steady state, M is the mass of the chemical species in a plant, and J_{out} is the mass of the chemical species leaving the plant. The concentration gradient (i_c) against the resistance of the chemical compound to exit the plant is described by Fick's law of diffusion:

$$J = \frac{i_c}{R} \tag{9.8}$$

Where, R is the resistance of the plant tissue. The exchange of gases between a plant's leaf and the atmosphere is chiefly determined by the size of the opening of stomata (Fig. 9.4), i.e., small pores in the plant's epidermis that are the sites of vapor and gas flux, mainly water vapor and carbon dioxide (CO_2). The stomata are "gatekeepers" for gas diffusions between the ambient air and the plant that balance gas exchange between the leaf interior and the bulk atmosphere [62]. The stomata limit transpiration by maximizing CO_2 flux into the leaf and minimizing H_2O flux out of the leaf. The gas flux is governed by a pathway consisting of a series of R values, beginning with the interface between the leaf and the air, i.e., the leaf boundary layer. The stomatal pore openings make available the dominant R between the air and the interior of the leaf. Simultaneously, CO_2 provides additional R as it encounters the aqueous and lipid boundaries within the leaf tissue, i.e., mesophyll cell and chloroplasts. Collectively, these processes comprise the mesophyll resistance. The outward flux of H_2O to the air results from all the forces in the R series except for mesophyll resistance. Guard cells regulate stomatal pores that expand or contract in response to external stimuli, like changes in temperature and humidity.

Discussion Box 9.1 Vapor pressure and pollution

From a water pollution perspective, arguably the most useful expression of vapor pressure is that it is the pressure at which a liquid and its vapor are in equilibrium at a given temperature. The vapor is pushing against the atmosphere, so it is one way a compound exits a waterbody. The greater the vapor pressure the faster liquids evaporate. A liquid is at its boiling point when vapor pressure reaches atmospheric pressure. Vapor pressure is measured in pressure units, including atmospheres (atm), millimeters of mercury (mmHg) or kilopascals (kPa). As a point of reference, normal atmospheric pressure is 1 atm (760 mmHg or 101.325 kPa).

A substance like molecular nitrogen (N_2) and oxygen (O_2) that is always in the gas phase at or near the earth's surface under typical conditions, i.e., permanent gases, has a vapor pressure of 1 atm (atm), which is 760 Torr or 101.325 kPa. Other compounds behave like permanent gases under environmental conditions, such as methane (CH_4), which becomes a liquid at about minus 82 °C, which is only reached occasionally in Antarctica. However, many compounds can be found in all three physical phases in the environment, so they are often classified according to their likelihood to reach the gas phase.

Vapor pressure is commonly used to categorize pollutants that are likely to move to the gas phase under environmental conditions. For example, volatile organic compounds (VOCs) readily move to the headspace of a container or to the atmosphere. Generally, these compounds have vapor pressures greater than 10^{-2} kPa. The semivolatile organic compounds (SVOCs) have vapor pressures between 10^{-5} and 10^{-2} kPa [50]. Others have proposed alternate ways to classify VOCs, such as evaporation rates over time, e.g., what remains after 6 months [51]. These values correspond to classifications that are based upon observations of the compounds' behaviors during air sampling [52].

Consider a scenario where 10 L of dichloromethane (CH_2Cl_2) is spilled to form 2 m^2 puddle of CH_2Cl_2. A slight 4 km h^{-1} wind is blowing. The temperature is 20 °C. Dichloromethane is a toxic organic solvent liquid with a vapor pressure of 0.46 atm at 25 °C.

Surface area is important an important factor because the amount of surface determines the rate of evasion due to vapor pressure. The greater amount of surface area, the greater the mass of the chemical compound that will exit the water. Temperature is also an important factor in vapor pressure since the greater the temperature, the more of the chemical compound that will leave the water.

Wind is also a factor because vapor pressure results from a kinetic process at the air-liquid interface that is driven by molecular interactions that allow energetic gas molecules in the air to transfer energy to volatile liquid molecules. This provides the liquid molecules of the chemical compound sufficient energy to change phase to a gas state. Increasing wind speed increases the probability of more energetic molecular interactions because a greater number of molecules traversing the surface of the standing water per time and therefore more chances for interaction.

Based on the scenario data, dichloromethane would vaporize relatively quickly, much faster than water will evaporate. Indeed, it would take about a half-hour for 10 L to completely evaporate under these conditions. Conversely, a 10 L-puddle of water would take hours to evaporate completely. The more volatile ethyl mercaptan (C_2H_6S; vapor pressure=0.59 atm at 25 °C) would evaporate in about 17 min. The low vapor pressure ethylene glycol ($C_2H_6O_2$; 6 $\times 10^{-5}$ atm at 25 °C) takes quite long to vaporize. Thus, most of the spill will persist until it is physically removed. However, other processes, including microbial [52] and photochemical [53] degradation, could break down some of the compound while it is on the surface. Unfortunately, if not remediated, some will infiltrate the soil and can contaminate ground water. Any of these chemical compounds' evasion rate would be slower due to other factors, especially if the compounds sorb to soil particles and other surfaces, if temperatures decrease, and if they are dissolved by dissolved humic substances (DHS) and organic solvents that have lower vapor pressures [54].

Pollution control and handling systems for volatile pollutants must include design and operation criteria that are based on their propensity to transfer to and from the soil, water, and air. Compounds can be deposited onto terrestrial and water surfaces if they are dissolved in raindrops and other precipitation or sorbed to particulate matter. Conversely, volatile compounds can be released into surface and ground water and can subsequently become air pollutants as they migrate toward the atmosphere. The SVOCs especially comprise some of the most toxic, bioaccumulating, and persistent pollutants. A particularly important aspect of SVOCs is that they may be transported from soil in the gas phase or as aerosols [55]. Thus, if SVOCs are present in the waste stream such as in a municipal landfill or hazardous waste storage facility, they must monitored as both in the gas phase and in and on soil and particulate matter.

One substantial source of VOCs and SVOCs is their emission to the atmosphere during thermal treatment of solid and hazardous wastes. For example, halogenated dioxins are generated when precursor compounds like pentachlorophenol and polyvinyl chloride are in wastes that is incinerated at certain temperatures [56–58]. In addition, ash and residues contain products of incomplete combustion, such as polycyclic aromatic hydrocarbons, dioxins, furans, and hexachlorobenzene, as well as many pesticides, phenolics, solvents, and pharmaceuticals, which fall under the SVOC chemical classification. Proper management of the life cycle of these chemical compounds requires an understanding of the factors that lead to the factors involved in their release, movement, and degradation [31,41,59].

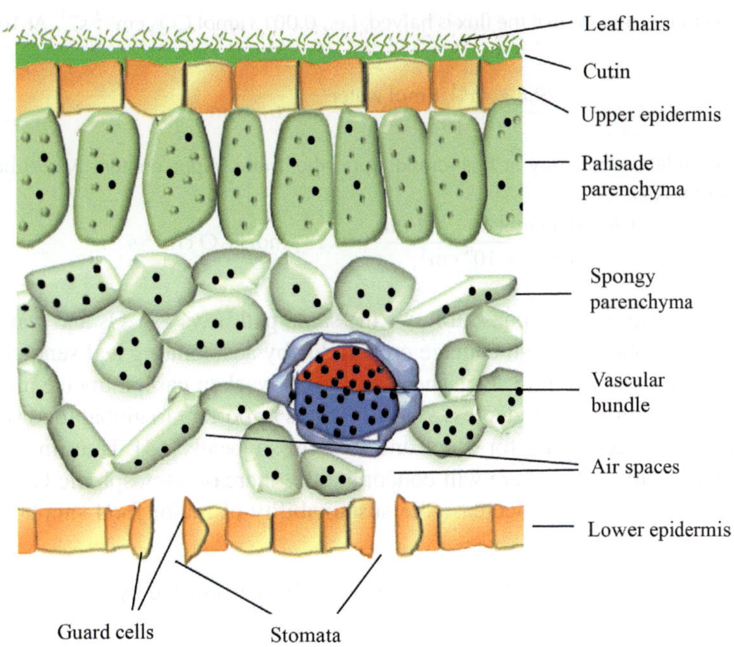

FIG. 9.4 Leaf cross-section. *(Data from D.A. Vallero, Fundamentals of Air Pollution, fifth ed., Elsevier Academic Press, Waltham, MA, 2014, p. 999 pages cm.)*

Stomatal pores make up only about 3% of the leaf surface, while being responsible for about 98% of CO_2 inward flux and H_2O outward flux [62].

Diffusive flux affects and is part of many biological processes, such as delivering carbon dioxide (CO_2) to a plant to begin photosynthesis.

Leaf flux example 1

Envision a plant leaf that contains CO_2 concentrations ranging from about 100 μL L^{-1} inside the leaf to 400 μL L^{-1} in the atmosphere [63]. The water vapor gradient of concentration ranges from saturation inside the leaf to the vapor pressure of the atmosphere. In this case, the vapor pressure is the point of equilibrium between the water in the plant and the atmosphere, i.e., when the water molecules leaving the leaf as a vapor equal the water molecules entering the plant as a liquid. The atmospheric vapor pressure is a function of the relative humidity and temperature of the air around the leaf. If the stomatal resistance is 5 s cm^{-1}, determine the flux of CO_2 into the leaf and water vapor exiting the leaf at 25 °C and 30% relative humidity.

First, calculate the amount CO_2 entering the leaf:

$$J = \frac{400 - 100\ \mu\text{L L}^{-1}}{5\text{scm}^{-1.}}$$

Convert the liters to cubic centimeters:

$$J = \frac{300\ \mu\text{L} \times 1000\ \text{cm}^{-3}}{5\text{scm}^{-1}} = 0.06\ \mu\text{L cm}^{-2}\ \text{s}^{-1}$$

Thus, by applying the gas laws, the amount of CO_2 entering the leaf is:

$$0.06\ \mu\text{L cm}^{-2}\ \text{s}^{-1} \times \frac{1\ \mu\text{mol}}{22.4\ \mu\text{mol}} = 0.0027\ \mu\text{mol CO}_2\ \text{cm}^{-2}\ \text{s}^{-1}.$$

Leaf flux example 2

Leaf flux is affected by many environmental conditions, including those of the water, soil, and air. For example, plants can be damaged in the summer from exposure to ozone (O_3), which forms when hydrocarbons and oxides of nitrogen react in the presence of sunlight. Thus, O_3 episodes usually occur during growing seasons. If the atmospheric concentration of O_3 increases and causes the stomatal aperture to shrink, which increases the resistance to 10 s cm^{-1}, what is the flux of CO_2 into the leaf and H_2O transpiration at 100% relative humidity and 25 °C?

Answer: The increased resistance means that the flux is halved, i.e., 0.0013 μmol CO_2 cm^{-2} s^{-1}. At 100% relative humidity and 25 °C, the air contains:

$$\frac{29.8\ \text{g m}^{-3}\ 1\ \text{mol water}}{18\ \text{g}} = 1.6\,\text{mol}\,\text{m}^{-3}.$$

Thus, if the air has 100% relative humidity at this temperature, the water vapor content would be 33% of 1.6, i.e., about 0.5 mol m^{-3}, so the leaf-to-air flux is:

$$\frac{1.6 - 0.50\ \text{mol m}^{-3} \times 1\ \text{m}^3}{5\text{scm}^{-1} \times 10^6\ \text{cm}^3} = 0.23\,\mu\text{mol H}_2\text{O cm}^{-2}\ \text{s}^{-1}.$$

The gas flux of any pollutant in vapor or gas phase follows the same pathways as CO_2 and water vapor. As shown in the example, elevated levels of tropospheric O_3 can damage crop yields by accelerating leaf senescence and detachment, i.e., abscission and by decreasing stomatal aperture size. These processes weaken carbon uptake and decrease photosynthetic carbon fixation. Elevated concentrations of O_3 depress yields of crops important to global food supply, e.g., average global yield reductions as high as 5.5% for corn, 14% for wheat and 15% for soybeans [64]. Thus, increasing emissions of hydrocarbons and oxides of nitrogen in the atmosphere will concomitantly increase atmospheric O_3 concentrations that induce stress on vascular plants, injure foliage, and decrease the sustainability of major food supplies. The stress can result in several problems, including [31]:

- Moderation of biomass growth by changing the amount of available carbon and by disrupting the translocation of fixed carbon to grains, fruits, pods, roots;
- Decrease in nutrient availability by redirection to synthesis of chemical protectants, or reduced transport capabilities by the phloem;
- Diminished carbon transport to roots that reduces nutrient and water uptake and affects anchorage;
- Harm to flowering;
- Pollen sterility;
- Ovule and/or grain abortion; and;
- Diminished ability of certain genotypes to withstand other environmental stresses like drought, high vapor pressure deficit, and increased photon ($h\nu$) flux density (i.e., larger numbers of photons passing through the stomata per second, $h\nu\ s^{-1}$) due to effects on stomatal control.

This is but another example of how pollution is a multicompartmental challenge. Water and soil pollutants can harm a plant at the root zone and within the plant. Malfunctioning stomata can result in substantial plant tissue loss. Atmospheric exchanges of sulfur dioxide (SO_2), oxides of nitrogen (NO_x) O_3, peroxyacetyl nitrate (PAN), and hydrocarbons with plant leaves are notable examples. Indeed, exposure to O_3 causes up to 90% of the air pollution injury to vegetation in the United States [65].

Generally, the pollutant gases follow the same diffusive pathways as CO_2, but are often followed by additional mechanisms, e.g., NO_x dissolves in cells and generates phytotoxic nitrite ions (NO_2) and nitrate ions (NO_3^-), which affect metabolism. Exposure to SO_2 closes stomata, which protects against further leaf injury but inhibits photosynthesis by diminishing CO_2 flux. Upon entering the plant, SO_2 dissolves in water and generates phytotoxic sulfite ions (SO_3^{2-}). Conversely, low concentrations of these ions can be beneficial by detoxifying the plant and providing sulfur, which is a plant nutrient. This is an example of an optimal range, below which is nutrient deficiency, and above which is toxicity. Of course, pollution-related crop loss varies in time and space, so assigning the fraction due to pollution is unclear. For example, when average daily ozone concentrations reach >50 ppb, yields of vegetables can be reduced by up to 15% [65]. Drought conditions that can worsen already threatened crops [23].

Root water uptake can be modeled. The Gardner model [66] assumes diffusivity is responsible for the uptake:

$$\frac{\partial\theta}{\partial t} = \frac{1}{r} \times \frac{\partial}{\partial r}\left[rD\frac{\partial\theta}{\partial t}\right] \tag{9.9}$$

Where, θ is the soil's volumetric water content, D is the soil diffusivity, t is time, and r is the radial distance from the root axis. The model has several simplifying assumptions, especially that every root is a cylinder of infinite length and of uniform radius, that all of the roots are separated by twice the radius of the cylinder, and that the water in the root moves only in the radial direction, not affected by gravity. The model also assumes that the initial soil water content (θ_0) is a uniform value that corresponds to the matric potential (Φ_0) in the soil. The initial conditions are [46]:

$$\theta = \theta_0; \Phi = \Phi_0; t = 0 \tag{9.10}$$

And, the boundary conditions are [46]:

$$2\pi\left(\frac{d\Phi}{dr}\right) = 2\pi D\left(\frac{d\theta}{dr}\right) = q, \quad r = a \tag{9.11}$$

Where, $\pi \approx 3.14$, a is the root radius, K is the capillary conductivity of the soil, and q is the water uptake rate by the root as the volume of water per unit length of root per unit time (units in $L^3/L/T$).

Other root water models account for some of the numerous relationships between roots and soils. For example, the Nimah and Hanks model accounts for the effect of salts on the soil and water potentials:

$$S(z,t) = \frac{[\Phi_{\text{root}} + (R_{\text{root}} \times z) - \Phi(z,t) - \Phi_n(z,t)]R(z) \times K(\theta)}{\Delta x \times \Delta x} \tag{9.12}$$

Where, Φ_{root} is the effective water potential in the root at the surface of the soil, soil depth (z) is zero, R_c is the flow coefficient in the root, which the model assumes to equal 0.05, and R_{root} is root resistance, which is equal to $1+R_c$.

Soil-plant-atmosphere-continuum (SPAC) systems assume downward water flows in a transpiration stream in a potential energy gradient form soil to root, from root to stem, and from stem to leaf, from which the water diffuses to the air [46,67]:

$$S_w = \frac{\Phi_{soil} + \Phi_{plant}}{R_{soil} + R_{roots}} \tag{9.13}$$

Where, S_w is the water extraction rate (flow) from a unit volume of soil, Φ_{plant} is the potential in the plant, R_{soil} and R_{roots} are resistances presented by the soil and roots, respectively, and Φ_{soil} is the total soil water potential, i.e., sum of matric (Φ_m), gravitational (Φ_g), and osmotic (Φ_π) potentials.

This movement is analogous to Ohm's law:

$$i = \frac{V}{R} \tag{9.14}$$

Where, R is the resistance, i is the flow of electrons, and V is the voltage. SPAC systems consider V to analogous to the water potential difference driving flow (S_w) and i is analogous to the flow of water across and within a plant segment [68]. Thus, the Ohm's law analogy allows the hydraulic conductance to be quantified as the inverse of hydraulic R of roots, stems, or leaves. It can also be used to quantify the hydraulic conductance from the soil to atmosphere through the whole plant.

These calculations provide an opportunity to introduce the concept of water-use efficiency (WUE), which is the ratio of water being used for plant metabolism to water lost to transpiration:

$$\text{WUE} = \frac{[C_{\text{CO2-fixed}}]}{[C_{\text{H2O-lost}}]} \tag{9.15}$$

Where $[C_{\text{CO2-fixed}}]$ is the mmoles of CO_2 fixed in the plant and $[C_{\text{H2O-lost}}]$ is moles of H_2O lost to evapotranspiration.

Elevated rates of evaporation lead to moles of water lost, which means they are not available for plant processes, especially photosynthesis. Thus, a declining WUE value indicates that stress is occurring or portends plant stress. For most vegetation, WUE ranges from 0.86 to 1.50, and varies by environmental conditions, with WUE increasing with decreasing stomatal conductance [69,70]. Stomatal conductance (g_s) is the rate of passage of CO_2 entering or water vapor exiting via the stomata of a leaf, usually measured in mmol m$^-$ s$^-$, a unit of diffusive flux [71].

Leaf flux example 3

In the previous example of an ozone air pollution episode, the leaf resistance increased to 10 s cm^{-1}. Calculate the flux of CO_2 into the leaf and H_2O transpiration at 100% relative humidity and 25 °C.

Recall that the stomatal resistance (r) was 5 s cm$^-$. Thus, the calculations show the water-use-efficiency (WUE) of this leaf when r=5 s cm$^-$, is about $\frac{0.0100\ \mu\text{mol}}{\mu\text{mo}} = 10$ mmol mol^{-1}. The gradient of water vapor between the leaf and the atmosphere is much steeper than the gradient for CO_2, meaning that the WUE is very low, and plants are less efficient at holding fixed CO_2, than the atmosphere is at extracting the water from the plant by diffusion. Thus, as tropospheric CO_2 rises, the rate of photosynthesis declines.

6 Applied biotics

Biotic cells convert nutrients and energy into biomass through the processes of photosynthesis, metabolism, and ion exchanges in microbes and plants [1,73,73]. Consequently, each organism metabolizes chemical compounds for energy and to build cells, but for pollution degradation the microbial decomposers are the main organisms that break down larger molecules into simpler ones. Bacteria, protozoa, fungi, algae, and even viruses can degrade organic molecules. Organisms at the lower trophic levels transform chemical compounds in the direction of mineralization, from large molecules like amino acids, proteins, and lipids, to become simpler and ultimately inorganic compounds that lack carbon-to-carbon and carbon-to-hydrogen covalent bonds (see Fig. 4.14 in Chapter 4). In the opposite direction, at higher levels of biological organization, consumers transform smaller molecules into larger ones, with producers doing some of each (see Fig. 9.1) [74].

Microbes decompose the accumulation of biomass in an ecosystem. Wastewater treatment engineers mimic these processes in wastewater treatment and hazardous waste treatment facilities and other bioreactors (see Chapters 14 and 16) [1,10–12,19,20,77–77]. These mimicked processes include physical and mechanical attributes to enhance biotransformation, such as the passage of water and air through pores unconsolidated media in an engineered trickling filter system. The rates of microbial growth is determined by physical factors like Fick's law of diffusion, coefficients for sorption, solubility and precipitation of the solutes that combine with the surface tension and capillarity of the media to determine the ability of the microbes to adhere to the surfaces of the media [79,79]. Flow of the wastewater through the pores in the trickling filter is affected by temperature, vapor pressure of the water's constituents, atmospheric pressure, and thickness of the biofilm. Thus, applied biotics combines biological factors with physicochemical factors. Important physicochemical factors include the polarity of the chemical compounds that form the solutes in the wastewater mixtures, hydrogen ion concentrations in the wastewater, i.e., pH, temperature, and catalysts determine the microbial degradation rate laws, e.g., zero, first, second reactions [81,81].

An extracellular polymeric substance (EPS) mediate microbial attachment [83,83]. The EPS adheres as a slime layer to the cell in a capsule or onto surrounding so that the bacteria have an anchor from which to degrade the organic content of the wastewater, i.e., to feed [79,83]. Biofilms become denser with depth due to decreasing pore volume [79].

Engineers and scientists have observed that microbes' ability to break down organic chemicals as a source of food and metabolic processes can be applied at the facility-scale by acclimating the microbes and limiting their food to other compounds, namely pollutants. As evidence, the sanitary engineering pioneer, William Paul Gerhard (1854–1927), took note of the importance of microbial degradation for pollution treatment as early as the year 1909 [84]:

> *The purifying action, which takes place in the two stages of treatment mentioned, is due to the different kinds of bacteria, known respectively as the aerobic, the anaerobic and the facultative bacteria. None of the methods of treatment are automatic, and all require intelligent management and constant supervision.*

Microbial metabolic processes of oxidation, reduction, hydrolysis, and other mechanisms could be tailored to degrade increasingly complex chemical compounds [10–12,85]. Microbial population can adapt to conditions of food scarcity and stress. Soil bacteria like *Pseudomonas* spp. acclimate to exposure to pollutants by withholding their preferred food sources, so that the microbes must adapt metabolic processes to use the pollutants instead for energy [86].[b]

Pollutants can be transformed biologically by three specific mechanisms: use of the compound as an electron acceptor; use of the compound as an electron donor; or co-metabolism. The processes can occur together and simultaneously in the same system, depending mainly on the availability of molecular oxygen in various regions of the system [1].

Electron acceptance occurs as redox conditions approach reduction. Reductive biotransformation requires a source of carbon as the electron donor for the growth and metabolism of an anaerobic bacterium. For example, if the compound contains chlorine (Cl), reductive dehalogenation causes the Cl to bind to carbon while other elements are removed. This is a common way to decrease the toxicity of a chlorinated compound like polychlorinated biphenyls (PCBs). When O_2 is depleted as the terminal electron acceptor (TEA), i.e., a compound that receives an electron during oxidation of a carbon source. Numerous anaerobic bacteria can use another TEA to generate intracellular energy [87]. Various types of organic matter can become sources of the electron donor carbon during cellular respiration or photosynthesis.

---

b. I recall in a wastewater treatment course at the University of Kansas, Professor Ross McKinney told us that the solution to most water pollution problems is "right under our feet." He meant that literally. Soil bacteria have the potential to destroy most organic pollutants; they just need a nudge. It has been a blessing to the environmental engineering community that Dr. McKinney decided to use his doctorate in microbiology from the prestigious Massachusetts Institute of Technology (MIT) to become a pioneer in applied microbiology, which today is better known as environmental biotechnology.

Electron donation occurs when a compound is used as the primary substrate, i.e., the electron donor, to allow the microbe to gain energy and organic carbon from the compound. The electron donor is a reducing agent, i.e., when it donates electrons it is oxidizing itself. Donation occurs under either aerobic or under certain anaerobic conditions. Lesser oxidized compounds like vinyl chloride or dichloroethylene are commonly biotransformed by electron donation.

When a compound is degraded co-metabolically, an enzyme already available in the microbe for other processes fortuitously catalyzes the reaction. The microbe has no benefit from the degradation of the compound, which indeed could harm or inhibit growth and metabolism a microbe. Co-metabolism has been most often observed in aerobic environments, but has potential under anaerobic conditions, as well [2].

The three degradation processes all involve reduction-oxidation reactions, i.e., redox, which is a key factor in the type of bacteria present in natural and engineered systems. As molecular oxygen (O_2^-) decreases in the consolidated media profile, the increasingly hypoxic conditions become less favorable for aerobes and the population of anaerobic bacteria increase. Wastewater, like much of the earth's biomass contains sulfur (S), so the low O_2 redox conditions induce bacteria-mediated reduction of oxidized forms to hydrogen sulfide (H_2S), with sulfate (SO_2^-) as the electron acceptor:

$$6SO_2^- + 37.5H^+ + 30e^- \rightarrow 3.75H_2S + 15H_2O \tag{9.16}$$

As discussed in Chapter 6, microbial populations degrade substrates and grow according to the empirically derived Monod equation:

$$\mu = \frac{\mu_{\max} S}{K_s + S} \tag{9.17}$$

Where μ = the specific growth rate of the microbe; $\mu_{\max}$ equals the maximum specific growth rate; S = concentration of limiting substrate; and K_s is the Monod growth rate coefficient representing the substrate concentration at which the growth rate is half the maximum rate. Recall that the $\mu_{\max}$ is reached at the higher ranges of substrate concentrations. K_s is an expression of the microbe's affinity for a nutrient. As K_s decreases, the greater the affinity microbe will have for that nutrient, as expressed by the correspondingly increasing μ.

The Monod-constant K typically ranges between about 1 and 10 g m^{-3} for organic substrates, but is much lower for O_2, i.e., $K \approx 1$ g m^{-3}. Given identical environmental conditions, the K will follow a zero-order growth curve (see Fig. 9.5). K will change according to changes in environmental factors like nutrient concentrations, pH, temperature, and concentrations of substances toxic to the species of bacteria [17].

For a microbe to gain energy from its food, e.g., wastewater, the microbes need to have contact with it. Engineers enhance this degradation environment, such as by injecting activated sludge and adding activated microbes. As these conditions change, the microbes undergo the series of stages discussed in Chapter 6 [12]:

1. Lag phase.
2. Accelerated growth rate phase.

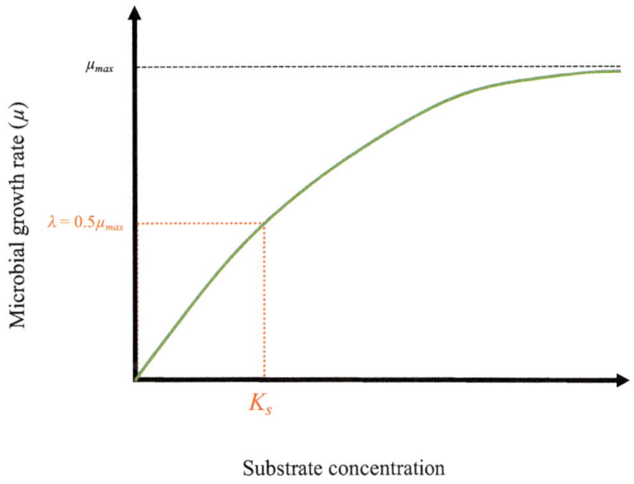

FIG. 9.5 Monod growth rate coefficient derived from many empirical studies showing the hyperbolic relationship between the microbial growth rate and the carbon source (organic compound) supporting the microbial growth. Curve represents no inhibitions. Notes: μ is the specific growth rate coefficient; λ is the maximum growth rate coefficient occurring at 0.5 $\mu_{\max}$; K_s is the Monod coefficient; and subscript S is the concentration of limiting nutrient, which can be represented by oxygen demand or total organic carbon [TOC]. *(Data from D.L. Russell, Practical Wastewater Treatment, John Wiley & Sons, 2019.)*

3. Exponential phase.
4. Declining growth phase.
5. Stationary phase.
6. Decay.
7. Exponential death phase.

The cycle returns when the microbes again contact the organic compounds in the food.

Sulfate compounds are reduced to H_2S by microbially mediated reactions:

$$6SO_{2^-} + 37.5H^+ + 30e^- \rightarrow 3.75H_2S + 15H_2O \tag{9.18}$$

In such redox reactions, electrons are transferred from an organic compound in the microbe's food or from the environment to an electron acceptor. In nature, the sediment at the bottom of a waterbody is deep enough to have a layer of very low O_2 concentrations. In an engineered system, it could involve designed O_2 removal. In either case, for an aerobic reaction, O_2 is the usual electron acceptor. Aerobic conditions mean that sufficient molecular oxygen is present. It does not mean, however, that O_2 is the exclusive electron acceptor in all aerobic microbial processes. For example, trivalent iron (Fe^{3+}) as iron sulfate, and carbon dioxide can also be the electron acceptors (see Table 9.1).

The first order biodegradation rate (R) can be calculated empirically from two representative points within the profile of the unconsolidated media:

$$R = \ln \frac{C_d}{C_u}(Dv) \tag{9.19}$$

TABLE 9.1 Redox reactions for benzene, xylene, and ethylbenzene.

Type	Reaction	Electron acceptor
Benzene redox reactions		
Oxidation	$C_6H_6+12H_2O \rightarrow 6CO_2+30H^++30e^-$	Oxygen
Reduction	$7.5O_2+30H^++30e^- \rightarrow 15H_2O$	
Reduction	$6NO_3^-+36H^++30e^- \rightarrow 3N_2+18H_2O$	Nitrate
Reduction	$15Mn^{4+}+30e^- \rightarrow 15Mn^{2+}$	Manganese
Reduction	$30Fe^{3+}+30e^- \rightarrow 30Fe^{2+}$	Iron
Reduction	$6SO_2^-+37.5H^++30e^- \rightarrow 3.75H_2S+15H_2O$	Sulfate
Reduction	$3.75CO_2+30H^++30e^- \rightarrow 3.75CH_4+7.5H_2O$	Methanogenic bacteria
Overall	$C_6H_6+7.5O_2 \rightarrow 6CO_2+3H_2O$	Oxygen
Overall	$C_6H_6+6H^++6NO_3^- \rightarrow 6CO_2+3N_2+6H_2O$	Nitrate
Overall	$C_6H_6+15Mn^{4+}+12H_2O^- \rightarrow 6CO_2+30H^++15Mn^{2+}$	Manganese
Overall	$C_6H_6+30Fe^{3+}+12H_2O^- \rightarrow 6CO_2+30H^++0Fe^{2+}$	Iron
Overall	$C_6H_6+3.75SO_4^{2-}+7.5H^+ \rightarrow 6CO_2+3.75H_2S+3H_2O$	Sulfate
Overall	$C_6H_6+4.5H_2O \rightarrow 2.25CO_2+3.75CH_4$	Methanogenic bacteria
Xylene and ethylbenzene redox reactions		
Oxidation	$C_8H_{10}+16H_2O \rightarrow 8CO_2+42H^++42e^-$	Oxygen
Reduction	$10.5O_2+42H^++42e^- \rightarrow 21H_2O$	
Reduction	$8.4NO_3^-+50.4H^++42e^- \rightarrow 4.2N_2+25.2H_2O$	Nitrate
Reduction	$21Mn^{4+}+42e^- \rightarrow 21Mn^{2+}$	Manganese
Reduction	$42Fe^{3+}+42e^- \rightarrow 42Fe^{2+}$	Iron
Reduction	$5.25SO_2^-+52.5H^++42e^- \rightarrow 5.25H_2S+21H_2O$	Sulfate

TABLE 9.1 Redox reactions for benzene, xylene, and ethylbenzene—cont'd

Type	Reaction	Electron acceptor
Reduction	$5.25CO_2+42H^++42e^-\rightarrow 5.25CH_4+10.5H_2O$	Methanogenic bacteria
Overall	$C_8H_{10}+10.5O_2\rightarrow 8CO_2+5H_2O$	Oxygen
Overall	$C_8H_{10}+8.4H^++8.4NO_3^-\rightarrow 8CO_2+4.2N_2+9.2H_2O$	Nitrate
Overall	$C_8H_{10}+21Mn^{4+}+16H_2O^-\rightarrow 8CO_2+42H^++21Mn^{2+}$	Manganese
Overall	$C_8H_{10}+42Fe^{3+}+16H_2O^-\rightarrow 8CO_2+42H^++42Fe^{2+}$	Iron
Overall	$C_8H_{10}+5.25SO_4^{2-}+10.5H^+\rightarrow 8CO_2+5.25H_2S+5H_2O$	Sulfate
Overall	$C_8H_{10}+5.5H_2O\rightarrow 2.75CO_2+5.25CH_4$	Methanogenic bacteria

The three isomers of xylene and ethylbenzene have the same formula, i.e., C_8H_{10}, but the xylenes' methyl group (—CH_3) is located on at three different carbon sites on the benzene ring and ethylbenzene's methyl group is located on a carbon bonded with the ring (—C—CH_3). Thus, the reactions are shown as the same, but the reaction rates differ.
Data from H. Rifai, C. Newell, J. Gonzales, S. Dendrou, L. Kennedy, J. Wilson, BIOPLUME III Natural Attenuation Decision Support System, Version 1.0, User's Manual, Air Force Center for Environmental Excellence (AFCEE), San Antonio, 1997.

Where C_d is the highest downgradient concentration of compound, C_u is the highest upgradient concentration of compound, D is the distance traveled, v is the velocity of the compound mass in the soil, sediment, or other unconsolidated media column.

Prototypical zero-order, first-order, and second-order microbial decay curves are similar Fig. 4.8 in Chapter 4. The initial and final pollution concentrations are equal, but the rates or kinetics of the three systems differ. The rate orders have been derived from experimental data, i.e., idealized curves that must be adapted to the heterogeneous and highly variable conditions of the field [81,88]. Zero-order decay is linear and first and second-order days are curved. However, estimates of expected microbial growth and metabolism and predictions of pollutant degradation rates under real-world conditions will differ from these theoretical decays and those in handbooks based solely on Monod equation, but provide strong clues about the reaction rates [81]. The Monod calculations are often adjusted with laboratory studies that can be scaled up increasingly to represent actual conditions [89]. In addition, computational models that combine fluid dynamics and microbial kinetics are increasingly becoming available to simulate the physics and biochemical reactions with increasing accuracy and representativeness [91,91].

7 Metrics of biotic condition

Healthy ecosystems have diverse and abundant living creatures. Biology and hydrology are key components of an ecosystem's condition. A biodiversity index is calculated as the number of species in area divided by the number of individuals in that area. It can also be described as species richness, which is the number of unique life forms, evenness, which expresses the equitability among the genera, and heterogeneity, which is the dissimilarity among the genera [92].

The diversity and abundance of biological species indicate ecosystem health. Another eco-indicator is the efficiency of the survival and thriving of biota [94,94]. The diversity and abundance of species and ecosystem efficiency must be maintained. Year-over-year generations of populations of organisms must have conditions that support reproduction and growth, i.e., ecosystem sustainability. For example, loss of canopy species in a forest and the impaction of soil may irreversibly affect the root systems of plants and decrease the soil interstices' support of bacterial and fungal species in soil organic matter (SOM). This, in turn, adversely affects the biotic conditions needed by higher-tier trophic levels of organisms.

Biodiversity is defined as the:

> *"... composition, structure, and function (that) determine, and in fact constitute, the biodiversity of an area. Composition has to do with the identity and variety of elements in a collection, and includes species lists and measures of species diversity and genetic diversity. Structure is the physical organization or pattern of a system, from habitat complexity as measured within communities to the pattern of patches and other elements at a landscape scale. Function involves ecological and evolutionary processes, including gene flow, disturbances, and nutrient cycling" [95].*

Expressions of biological community's diversity are relative and merely indicators, i.e., an index, that incorporate many uncertainties to describe the condition of habitat. One common metric of ecological condition is the Shannon-Weiner index of community diversity [96]:

$$H' = \sum_{i=1}^{m} P_i \ln P_i \tag{9.20}$$

This can also be expressed as:

$$H' = -1.44 \sum_{i=1}^{m} \left(\frac{n_i}{N}\right) \ln\left(\frac{n_i}{N}\right) \tag{9.21}$$

Where, H' = index of community diversity; $P_i = n_i/N$; n_i = number (i.e., density) of the ith genera or species; N = total number (i.e., density) of all organisms in the sample; $i = 1, 2\ldots m$; and m = number of genera or species.

The index is based on entropy, incorporating the number of species living in a habitat, i.e., species richness and their relative abundance, i.e., evenness. As the value of H' increases, the ecosystem's overall sustainability or stability also increases. The calculation of community diversity using the Shannon-Weiner index can demonstrate its value in comparing different habitats or changes in the same habitat over time.

Biodiversity example 1

A biology professor has conducted an ecological impact study that measured the abundance of microbes in a stream that receives effluent from an industrial plant. She found a total of 1510 organisms, represented by seven different species. The actual number count of each microbial species in stream community was 10, 50, 75, 125, 200, 350, 750 m L^{-1}. Using Eq. (9.21), the scientist constructed Table 9.2.

Generally, H' values range from about 1.5 to 4.5, but the diversity index in this ecosystem is approximately 2.1. The index is most useful when comparing systems. So, if the stream is 2.1 and surrounding streams are all around 4, there may be a problem. Further investigation is needed if these are biological indicators of the health of the stream and the facility is releasing pollutants that are known to stress ecosystems.

Now consider what happens if the facility installs new pollution control equipment, with every species quadrupling in abundance (highly unlikely, but let us assume it to keep the math simple). The measurements taken a year later are shown in Table 9.3.

TABLE 9.2 Hypothetical species abundance measurements needed to calculate community diversity H', using: $H' = -1.44 \sum_{i=1}^{m} \left(\frac{n_i}{N}\right) \ln\left(\frac{n_i}{N}\right)$.

i	n_i	n_i/N	$-1.44 \ln(n_i/N)$	$-1.44(n_i/N) \ln(n_i/N)$
1	10	0.006623	7.224883	0.047847
2	50	0.033113	4.907292	0.162493
3	75	0.049669	4.323423	0.21474
4	125	0.082781	3.587834	0.297006
5	200	0.13245	2.911028	0.385567
6	350	0.231788	2.105182	0.487956
7	700	0.463576	1.10705	0.513202
Σ	**1510**	**1**		**2.10881**

Data from D.A. Vallero, Biotic physics, in: Methods and Calculations in Environmental Physics, AIP Publishing LLC, Melville, New York, 2022, pp. 9-1–9-40.

TABLE 9.3 Hypothetical species abundance measurements 1 year after new pollution control equipment has been installed.

i	n_i	n_i/N	$-1.44\ \ln(n_i/N)$	$-1.44(n_i/N)\ \ln(n_i/N)$
1	40	0.006623	7.224883	0.047847
2	200	0.033113	4.907292	0.162493
3	300	0.049669	4.323423	0.21474
4	500	0.082781	3.587834	0.297006
5	800	0.13245	2.911028	0.385567
6	1400	0.231788	2.105182	0.487956
7	2800	0.463576	1.10705	0.513202
Σ	**6040**	**1**		**2.10881**

Data from D.A. Vallero, Biotic physics, in: Methods and Calculations in Environmental Physics, AIP Publishing LLC, Melville, New York, 2022, pp. 9-1–9-40.

Biodiversity example 2[a]

Even though the species abundance quadrupled, the index remains 2.1. The index indicates, correctly, that the total species abundance does not affect diversity.

[a]From reference [32].

Biodiversity example 3[b]

Consider now what would happen if the county encouraged more facilities to locate along the stream and several new sources discharged waste into the air and water nearby. A follow-up study is conducted 5 years later indicating that the species density has changed to 3000, 50, 35, 40, 30, 70, and 15 L^{-1} of the same species of microbes as the early study. Table 9.4 shows how the numbers and diversity changed in 5 years.

After 5 years, the actual number of microbes is more than doubling, but the diversity has been drastically reduced ($H'=0.6$ vs 2.1). This likely indicates that conditions are differentially favorable to one species, such as when concentration of a toxic chemical or a change in habitat as when a barrier like a road is built that interrupts migration. The chemical exposure or change in habitat would differentially affect the seven species, with one species appearing to have a much higher tolerance for the insult than the other six.

TABLE 9.4 Hypothetical species abundance measurements 5 years after new pollution sources locate near the stream.

i	n_i	n_i/N	$-1.44\ \ln(n_i/N)$	$-1.44(n_i/N)\ \ln(n_i/N)$
1	3000	0.92593	0.11082	0.10261
2	50	0.01543	6.00668	0.0927
3	35	0.0108	6.52029	0.07044
4	40	0.01235	6.32801	0.07812
5	30	0.00926	6.74227	0.06243
6	70	0.0216	5.52216	0.11931
7	15	0.00463	7.7404	0.03584
Σ	**3240**	**1**		**0.56144**

[b]From reference [32].

In many ecosystem condition assessments, a small number of rare species with only a few representative organisms should not affect a community's biodiversity. Thus, a community may appear healthy when in fact a sensitive, threatened species in being inordinately harmed. This calls for a dominance index like the Simpson index [97]. Again, P_i is the proportion (n/N) of individuals of one species counted in the community (n), which is divided by the total number of individuals in the community (N):

$$D = \frac{1}{\sum_{i=1}^{m} P_i^2} \tag{9.22}$$

Biodiversity example 4

Based on the professor's first community measurement, comparative Simpson indices can be developed (see Table 9.5).

Note that both indices stay the same albeit the species abundance being quadrupled, which indicates that the total species abundance does not affect diversity calculations in either index. However, doubling of microbes drastically reduced H' and increased D. The dominance of Species 1 appears to support the hypothesis that at 5 years ecological conditions began to favor Species 1 differentially.

TABLE 9.5 Calculation of Simpson indices for the previous H′ calculations for three microbial ecology scenarios.

i	n_i	n_i/N	$(n_i/N)^2$	H′
1	10	0.006623	0.000044	
2	50	0.033113	0.001096	
3	75	0.049669	0.002467	
4	125	0.082781	0.006853	
5	200	0.13245	0.017543	
6	350	0.231788	0.053726	
7	700	0.463576	0.214903	
Σ	**1510**	**1**	**0.296631**	**2.10881**
1	40	0.006623	0.000044	
2	200	0.033113	0.001096	
3	300	0.049669	0.002467	
4	500	0.082781	0.006853	
5	800	0.13245	0.017543	
6	1400	0.231788	0.053726	
7	2800	0.463576	0.214903	
Σ	**6040**	**1**	**0.296631**	**2.10881**
1	3000	0.92593	0.857346	
2	50	0.01543	0.000238	
3	35	0.0108	0.000117	
4	40	0.01235	0.000153	
5	30	0.00926	0.000086	
6	70	0.0216	0.000467	
7	15	0.00463	0.000021	
Σ	**3240**	**1**	**0.858427**	**0.56144**

Data from D.A. Vallero, Biotic physics, in: Methods and Calculations in Environmental Physics, AIP Publishing LLC, Melville, New York, 2022, pp. 9-1–9-40.

These and other diversity indices apply to microbial communities, macrophytic plants, fauna, and combinations of species in communities. They can be a first step in evaluating food chains and webs of various genera, functional types, and polymorphs, as well as biomass of a single species. They can be complemented with data regarding ecological conditions such species richness, i.e., the number of different species in a community. Often, it is much easier to observe the number of species than the abundance of species, so richness is often shorthand for biodiversity. As the tables above indicate, however, having but a few individuals suggests ecological vulnerability. For example, a few remaining top carnivores in a community is an indication of a threatened or endangered species, which calls for special protections.

Of course, consistency in measurements and modeling is fundamental to comparing ecosystem conditions among ecological communities. For example, there can be considerable intra-annual and seasonal variability in microbial abundance. Most microbes prefer warmer conditions, but some species compete more effectively in cooler waters. If the studies are comparable, further investigation is needed, but these findings are certainly an indication of a problem given that the expected H' value range is 1.5–4.5. Changes in diversity represent a first eco-risk screen and a forewarning of ecological harm. Generally, situations where diversity as measured by the index show a decline, this could indicate than an ecosystem is experiencing increasing stress.

Another ecosystem condition metric is productivity, i.e., how economical a system uses available energy. Productivity is the amount of biomass that is produced from abiotic resources like nutrients and minerals, and the efficiency of transfer of biotic resources from microbial populations to canopy plant species to top predators. Net primary productivity (NPP or P_I) is the difference between two energy rates:

$$P_I = k_p - k_e \tag{9.23}$$

Where k_p = rate of chemical energy storage by primary producers; and k_e = rate at which the producers use energy via respiration.

Sustainability describes the likelihood that the diversity and productivity can continue at present or improved rates in the future. Thus, predictions of ecological sustainability are difficult, relying on statistical and probabilistic methods. Often the extent of sustainability characterizations and predictions are semi-quantitative, extrapolating current trends of conditions to estimate whether things are improving or degrading, and whether the future conditions are sufficiently uncertain that precautions call for additional protective factors of safety to avoid irreversible problems.

Increasingly, evidence-based risk assessments are being augmented or even supplanted by precaution, especially if the decision involves a reasonable likelihood that an adverse effect is severe and irreversible [98]. The precautionary approach also calls for systems thinking and sustainable solutions. Recently developed screening models and tools, including multicriteria decision analysis (MCDA), give consideration of numerous variables and apply data from various sources. Tools like expert elicitation gather insights from members of specific disciplines within the scientific community on emerging environmental problems [99]. Emerging pollutants are often data-poor, lacking reliable historical data on the characteristics and potential risks of pollutants.

Wetlands provide are instructive of the need for ecosystem condition characterization. Large numbers of species exist and undergo change at the interfaces between surface and ground water, air, soil, sediment, and biota. For example, coastal wetlands like saltwater marshes undergo seasonal and interannual conditions as they are affected by tidal flooding and draining when the high salinity tidal water encroaches on fresher waters (see Fig. 9.6). Wetland habitats support fisheries, providing sustenance, refuge, and nursery habitat for various fisheries species. For coastal marshes in the northeastern regions of the U.S., for example, these include shrimp (order Decapoda), blue crab (*Callinectes sapidu*), and various finfish.

FIG. 9.6 Rhode Island saltwater marsh in Narragansett Bay National Estuarine Research Reserve. *(Photo courtesy of the U.S. National Oceanic and Atmospheric Administration.)*

The soils in wetlands consist of mixes of soil types and moisture. For saltwater marshes these are predominantly mud and peat horizons. Peat forms when decomposed plant biomass accumulates in place. Peat exists in many climate types, including arctic, taiga and boreal peatlands, as well as swamps in temperate and tropical climates [100].

Salt marshes provide protections against dune and other coastline erosion from wave actions, capturing sediment, reducing flooding by acting as a sponge that holds water that would otherwise back up. They play an important role as carbon sinks, continuously absorbing atmospheric carbon and holding it at an average carbon burial rate of 210 g C $m^{-2} year^{-1}$ [101].

Wetland plant biomass continuously undergoes decomposition by microbes. These reactions make for a reduced, low-oxygen environment, i.e., hypoxic, with depth. Tides can submerge marshes for long periods of time, allowing biomass to accumulate as organic matter is decomposed continuously by microbes, particularly bacteria. The decomposition depresses dissolved O_2 concentrations in the interstitial water, so that anaerobic microbial populations increase in the mud and peat that during hypoxic conditions [102]. This induces bacteria-mediated reactions that release CH_4, while more oxidized conditions are releasing CO_2. The anoxic environment also causes oxidized compounds in the plant debris to become reduced. Notably, sulfur in the plant biomass is reduced to H_2S.

In this case, sulfate (SO_4^{2-}) becomes the most abundant electron acceptor. This reaction's release of H_2S is responsible for the rotten-egg odor common to some marshlands. Both oxidation and reduction occur in saltwater marshes, but if normally aerobic or anaerobic microbial processes are interrupted by the introduction of pollutants onto the marsh soil or sediment, the balance can change. The change in redox of wetlands can have climatic impacts. In fact, wetlands hold large amounts carbon so are sinks for CH_4, CO_2, and other gas-phase carbon compounds as their plants and other biota sequester large amounts of CO_2 in underground and sediment biomass. Conversely, loss of wetlands creates a source of CH_4 and CO_2, which is a global greenhouse gas which, when concentrations increase sufficiently, will affect global temperatures and other climatic conditions [103].

Anaerobes in an inundated hypoxic wetland soil decompose carbon compounds more slowly than in higher oxygen soils because aerobic respiration is more thermodynamically efficient than anaerobic respiration. Physical stressors like the disruptions due to dredging and bulldozing of wetlands, causes some of the sequestered carbon to move to the atmosphere [104].

As mentioned, water and air pollutant exposure at sufficiently high concentrations can elicit physiological or biochemical effects in vascular plants., including changes in net photosynthesis rates, stomata response, and metabolic activity. Certain species become unable to survive in an ecosystem even with very subtle changes. For example, plant species exposed to O_3 even at concentrations below current regulatory standards experience decreased growth, and lower biomass production [23,105].

Other sensitive habitats share the biodiversity challenges of wetlands. Tropical rainforest acreage is lost due to clear-cutting and harvesting whole trees [106], as well as the construction of roads. These human activities diminish biodiversity by truncating migration routes and altering feeding zones, as well as increasing soil erosion and decreasing flood control capacity [108,108]. Human activities displace habitats and alter predator-prey, foraging, and breeding behaviors along with other ecosystem functions. About one-fourth of the total greenhouse gas production results from deforestation. Thus, one could argue that the bulldozer can be envisioned as a water and air pollutant, or at least an agent of environmental degradation.

Rainforest species are particularly vulnerable because they adapted to microhabitats that only exist in the unique conditions under ambient moisture and temperature regimes. Other species are lost to habitat destruction and deforestation. Ecosystems edges are particularly important to species survival. When they change due to climatic and anthropogenic stresses, canopy species can no longer survive. This leads to further destruction within the ecosystem, with rapid decreases in species abundance and loss of biodiversity. About half of the earth's species exist in rainforests while they cover merely 7% of the land surface [110,110]. About 25% of the world's total greenhouse gas production comes from deforestation [110].

Soil and plants in tropical forests sequester 460–575 billion metric tons of the earth's surface's carbon. Each square kilometer of tropical forest holds 45,000 tonnes of carbon. From 1850 to 1990, deforestation released 122 billion tonnes of carbon into the atmosphere. Currently the deforestation carbon emission rate is 1.6 billion tonnes per year [111,111].

In addition, deforestation increases emissions of fugitive dust, another example of the hydrosphere-atmosphere-lithosphere-biosphere interactions discussed in Chapter 4. The damage to human health from dust exposure is well documented, so other air-breathing wildlife populations would be expected to be weakened, resulting in decreased species abundance, diminished diversity, and changes to ecosystem function.

Biodiversity also has a social context. For example, individual farmers can suffer at the expense of large-scale corporate farms when the parameters for successful food production only include monetizable values. Indeed, smaller farms with greater biodiversity almost always provide more sustainable agriculture compared to crop monocultures [113,113]. Family farms also have nonmonetizable values, such as livability and psychological benefits to farm families and surrounding communities [114].

Net primary productivity can be a deceiving parameter. Plantation type agriculture consists of one or a few crops year after year, which usually requires large financial investments in fossil fuels to cultivate fields and harvest crops and to apply fossil-fuel based commercial fertilizers, genetically modified seeds, and pesticides. These add up to a lack of diversity and vulnerabilities to infestations and plant diseases [115,116]. Conversely, most small farms are often highly productive with fewer nonsustainable side effects than large-scale operations. For example, corn yields in traditional Mexican *chinampas*, i.e., raised beds in shallow ponds and wetlands, range from 3.5 to 6.3 tons per hectare, which means that up to 20 people can maintain a nutritionally acceptable diet from each hectare farmed [117]. Small farms are also sanctuaries for native species, which help to maintain biodiversity.

8 Conclusions

There are many biological aspects to water quality. This chapter has discussed the biological processes and mechanisms that lead to the formation, transport, transformation, and fate of substances that are or can become water pollutants. This and previous chapters set the stage for the discussions of surface and ground water in the next two chapters.

References

[1] D. Vallero, Environmental Biotechnology: A Biosystems Approach, Elsevier Science, 2015.

[2] D.P. German, M.N. Weintraub, A.S. Grandy, C.L. Lauber, Z.L. Rinkes, S.D. Allison, Optimization of hydrolytic and oxidative enzyme methods for ecosystem studies, Soil Biol. Biochem. 43 (7) (2011) 1387–1397.

[3] J.H. Makoi, P.A. Ndakidemi, Selected soil enzymes: examples of their potential roles in the ecosystem, Afr. J. Biotechnol. 7 (3) (2008).

[4] S.R. Tirkey, S. Ram, S. Mishra, Bacteria-mediated remediation: restoring polycyclic aromatic hydrocarbon (PAHs) contaminated marine ecosystems, in: Advanced Oxidation Processes for Effluent Treatment, Elsevier, Plants, 2021, pp. 119–136.

[5] J. Li, et al., Unraveling bacteria-mediated degradation of lignin-derived aromatic compounds in a freshwater environment, Sci. Total Environ. 749 (2020) 141236.

[6] N.-A.T. Tran, B. Tamburic, C.R. Evenhuis, J.R. Seymour, Bacteria-mediated aggregation of the marine phytoplankton Thalassiosira weissflogii and Nannochloropsis oceanica, J. Appl. Phycol. 32 (6) (2020) 3735–3748.

[7] Y. Li, et al., Coupled carbon, sulfur, and nitrogen cycles mediated by microorganisms in the water column of a shallow-water hydrothermal ecosystem, Front. Microbiol. 9 (2018) 2718.

[8] A. Hüsken, A. Dietz-Pfeilstetter, Pollen-mediated intraspecific gene flow from herbicide resistant oilseed rape (Brassica napus L.), Transgenic Res. 16 (5) (2007) 557–569.

[9] M.A. Rieger, M. Lamond, C. Preston, S.B. Powles, R.T. Roush, Pollen-mediated movement of herbicide resistance between commercial canola fields, Science 296 (5577) (2002) 2386–2388.

[10] R. McKinney, Microbial relationships in biological wastewater treatment systems, in: The structure and function of fresh-water microbial communities, Virginia Polytechnic Institute and State Union, Blacksburg, Virginia, 1973, pp. 165–179.

[11] R.E. McKinney, Discussion of Population Dynamics in Biological Treatment Plants, 1966.

[12] R.E. Mckinney, Microbiology for Sanitary Engineers, McGraw Hill Book Co, New York, 1962.

[13] R.E. McKinney, Civil engineering 876: wastewater treatment plant design, in: Course Lecture, Spring Semester, 1984, University of Kansas, 1984.

[14] D.S. Esser, J.H.J. Leveau, K.M. Meyer, Modeling microbial growth and dynamics, Appl. Microbiol. Biotechnol. 99 (21) (2015) 8831–8846, https://doi.org/10.1007/s00253-015-6877-6.

[15] A. Ahmad, F. Banat, H. Alsafar, S.W. Hasan, Algae biotechnology for industrial wastewater treatment, bioenergy production, and high-value bioproducts, Sci. Total Environ. 806 (2022) 150585.

[16] N.P. Cheremisinoff, Biotechnology for Waste and Wastewater Treatment, Elsevier, 1997.

[17] A. Dragt, J. van Ham, Biotechniques for Air Pollution Abatement and Odour Control Policies, Elsevier, 1992.

[18] P.K. Jain, V.K. Gupta, V. Bajpai, Recent Advances in Environmental Biotechnology, 2011.

[19] National Academies of Sciences, Engineering, and Medicine, Preparing for Future Products of Biotechnology, National Academies Press, 2017.

[20] J. Nielsen, et al., Engineering synergy in biotechnology, Nat. Chem. Biol. 10 (5) (2014) 319–322.

[21] B.E. Rittmann, P.L. McCarty, Environmental Biotechnology: Principles and Applications, Tata McGraw-Hill Education, 2012.

[22] H. Yi, et al., Recent development of advanced biotechnology for wastewater treatment, Crit. Rev. Biotechnol. 40 (1) (2020) 99–118.

[23] D.A. Vallero, Fundamentals of Air Pollution, fifth ed., Elsevier Academic Press, Waltham, MA, 2014. p. 999 pages cm.

[24] S. Mandal, S. Ray, S. Roy, S.E. Jørgensen, Order to chaos and vice versa in an aquatic ecosystem, Ecol. Model. 197 (3) (2006) 498–504, https://doi.org/10.1016/j.ecolmodel.2006.03.020.

[25] R.M. May, Chaos and the dynamics of biological populations, Proc. R. Soc. Lond. A Math. Phys. Sci. 413 (1844) (1987) 27–44.

[26] A. Hastings, T. Powell, Chaos in a three-species food chain, Ecology 72 (3) (1991) 896–903.

[27] T.E. Graedel, On the concept of industrial ecology, Annu. Rev. Energy Environ. 21 (1) (1996) 69–98.
[28] V. Mishra, et al., Health inequalities during COVID-19 and their effects on morbidity and mortality, J. Healthc. Leadersh. (2021) 19–26.
[29] J.B. Awotunde, E.A. Adeniyi, P.O. Kolawole, R.O. Ogundokun, Application of big data in COVID-19 epidemic, in: U. Kose, D. Gupta, V.H.C. de Albuquerque, A. Khanna (Eds.), Data Science for COVID-19, Academic Press, 2022, pp. 141–165 (Chapter 8).
[30] A. Shelke, S. Shelke, S. Acharya, S. Shukla, Synergistic epidemic or syndemic: an emerging pattern of human diseases, Cureus 15 (11) (2023).
[31] D. Vallero, Air Pollution Calculations: Quantifying Pollutant Formation, Transport, Transformation, Fate and Risks, Elsevier, 2019.
[32] D.A. Vallero, Biotic physics, in: Methods and Calculations in Environmental Physics, AIP Publishing LLC, Melville, New York, 2022, pp. 9-1–9-40.
[33] S. Pagad, P. Genovesi, L. Carnevali, R. Scalera, M. Clout, IUCN SSC Invasive Species Specialist Group: invasive alien species information management supporting practitioners, policy makers and decision takers, Manag. Biol. Invasions 6 (2) (2015) 127.
[34] L.E. LaMP, Lake Erie lakewide management plan (LaMP) 2004, in: Report prepared by the Lake Erie LaMP Work Group under the direction of the Lake Erie LaMP Management Committee, Environment Canada, Ontario Region, and the US Environmental Protection Agency, 2004.
[35] National Research Council, Biological Confinement of Genetically Engineered Organisms, National Academies Press, 2004.
[36] A.R. Kapuscinski, Genetic methods for biological control of non-native fish in the Gila River Basin: development and testing of methods, potential environmental risks, regulatory requirements, multi-stakeholder deliberation and cost estimates: final report to the US Fish and Wildlife Service, Institute for Social, Economic and Ecological Sustainability of the University of Minnesota (2005).
[37] M. Hansen, R. DeFries, J. Townshend, M. Carroll, C. Dimiceli, R. Sohlberg, Global percent tree cover at a spatial resolution of 500 meters: first results of the MODIS vegetation continuous fields algorithm, Earth Interact. 7 (10) (2003) 1–15.
[38] R. Lindsey, R. Simmon, Tropical Deforestation, NASA Earth Observatory (Ed.), NASA, Washington, DC, 2007.
[39] M.A. Bruns, K.M. Scow, DNA fingerprinting as a means to identify sources of soil-derived dust: problems and potential, in: Integrated Assessment of Ecosystem Health, CRC Press, 2019, pp. 193–206.
[40] H.D. Scott, Soil Physics: Agricultural and Environmental Applications, Iowa State University Press, Ames, Iowa, 2000.
[41] D.A. Vallero, Methods and Calculations in Environmental Physics, AIP Publishing, Melville, New York, 2022.
[42] Y.X. Kim, K. Ranathunge, S. Lee, Y. Lee, D. Lee, J. Sung, Composite transport model and water and solute transport across plant roots: an update, Front. Plant Sci. 9 (2018) 193.
[43] E. Steudle, C.A. Peterson, How does water get through roots? J. Exp. Bot. 49 (322) (1998) 775–788.
[44] E. Steudle, J. Frensch, Osmotic responses of maize roots, Planta 177 (3) (1989) 281–295.
[45] E. Steudle, J. Frensch, Water transport in plants: role of the apoplast, Plant Soil 187 (1) (1996) 67–79.
[46] M.K. Shukla, Soil Physics: An Introduction, CRC Press, 2013.
[47] Laboratory of Ecohydrology (ECHO), n.d., Chapter 4: Water potential in porous media, ECHO, https://echo2.epfl.ch/VICAIRE/mod_3/chapt_4/main.htm (Accessed July 19, 2024).
[48] H. Bruus, Theoretical Microfluidics, Oxford University Press, 2007.
[49] J. Pfitzner, Poiseuille and his law, Anaesthesia 31 (2) (1976) 273–275.
[50] R. Lewis, S. Gordon, Sampling for organic chemicals in air, Environmental Protection Agency, Research Triangle Park, NC, USA, 1994.
[51] U.-U.T. Võ, M.P. Morris, Nonvolatile, semivolatile, or volatile: Redefining volatile for volatile organic compounds, J. Air Waste Manage. Assoc. 64 (6) (2014) 661–669.
[52] C. McGahey, E. Bouwer, Biodegradation of ethylene glycol in simulated subsurface environments, Water Sci. Technol. 26 (1–2) (1992) 41–49.
[53] A. Ardila-Arias, E. Berrfo-Mesa, E. Arriola-Villaseñor, W. Alverez-Gomez, J. Hernandez-Maldanado, T. Zepeda-Partida, L. Ortiz-Frade, R. -Barrera-Zapata, Degradation of ethylene glycol through photo-Fenton heterogeneous system, Rev. Ing. Univ. Medellín 18 (35) (2019) 91–109, https://doi.org/10.22395/rium.v18n35a6.
[54] K. Ramus, F.-D. Kopinke, A. Georgi, Influence of dissolved humic substances on the mass transfer of organic compounds across the air–water interface, Chemosphere 86 (2) (2012) 138–143, https://doi.org/10.1016/j.chemosphere.2011.09.055.
[55] R.W. Williams, R.R. Watts, R.K. Stevens, C.L. Stone, J. Lewtas, Evaluation of a personal air sampler for twenty-four hour collection of fine particles and semivolatile organics, J. Expo. Sci. Environ. Epidemiol. 9 (2) (1999) 158.
[56] N. Ágnes, K. Rajmund, The environmental impact of plastic waste incineration, Acad. Appl. Res. Mil. Public Manag. Sci. 15 (3) (2016) 231–237.
[57] J. Thornton, M. McCally, P. Orris, J. Weinberg, Hospitals and plastics. Dioxin prevention and medical waste incinerators, Public Health Rep. 111 (4) (1996) 298–313 (in English).
[58] California Office of Environmental Health Hazard Assessment, 2,3,7,8-Tetrachlorodibenzo-p-dioxin and related compounds, 2024. https://oehha.ca.gov/chemicals/2378-tetrachlorodibenzo-p-dioxin-and-related-compounds. (Accessed 19 July 2024).
[59] D.A. Vallero, Air pollution physics, in: Methods and Calculations in Environmental Physics, AIP Publishing LLC, Melville, New York, 2022, pp. 7-1–7-70.
[60] E.J. Kleynhans, A.L. Atchley, S.T. Michaletz, Modeling fire effects on plants: from organs to ecosystems, in: E.A. Johnson, K. Miyanishi (Eds.), Plant Disturbance Ecology, second ed., Academic Press, San Diego, 2021, pp. 383–421 (Chapter 11).
[61] E. Steudle, The cohesion-tension mechanism and the acquisition of water by plant roots, Annu. Rev. Plant Biol. 52 (1) (2001) 847–875.
[62] T. Lawson, M.R. Blatt, Stomatal size, speed, and responsiveness impact on photosynthesis and water use efficiency, Plant Physiol. 164 (4) (2014) 1556–1570.
[63] E.S. Bernhardt, W. Schlesinger, Biogeochemistry: An Analysis of Global Change, Elsevier Inc, Waltham, MA, USA and Oxford, UK, 2013.
[64] S. Wilkinson, G. Mills, R. Illidge, W.J. Davies, How is ozone pollution reducing our food supply? J. Exp. Bot. 63 (2) (2012) 527–536.
[65] G.E. Brurst, Air pollution effects on vegetables, 2018. https://extension.umd.edu/learn/air-pollution-effects-vegetables. (Accessed 25 June 2018).

[66] W.R. Gardner, Dynamic aspects of water availability to plants, Soil Sci. 89 (2) (1960) 63–73.
[67] J.R. Philip, Plant water relations: some physical aspects, Annu. Rev. Plant Physiol. 17 (1) (1966) 245–268.
[68] A.J. McElrone, B. Choat, G.A. Gambetta, C.R. Brodersen, Water uptake and transport in vascular plants, Nat. Educ. Knowl. 4 (6) (2013).
[69] W.H. Schlesinger, E.S. Bernhardt, Biogeochemistry: An Analysis of Global Change, fourth ed., Academic Press, Amsterdam, NV, 2020, ISBN: 978-0128146088.
[70] C.B. Osmond, K. Winter, H. Ziegler, Functional significance of different pathways of CO2 fixation in photosynthesis, in: Physiological Plant Ecology II, Springer, 1982, pp. 479–547.
[71] L. Jeanguenin, A.P. Mir, F. Chaumont, Uptake, loss and control, in: L. Taiz, I.M. Møller, A. Murphy, E. Zeiger (Eds.), Plant Physiology and Development, Elsevier Academic Press, Amsterdam, NV, 2016, pp. 135–140, https://doi.org/10.1016/B978-0-12-394807-6.00087-3. In this issue.
[72] C. Yuan, C. Tebes-Stevens, E.J. Weber, A reaction library to predict direct photochemical transformation products of environmental organic contaminants in sunlit aquatic systems, Environ. Sci. Technol. (2020).
[73] D.A. Vallero, Environmental Systems Science: Theory and Practical Applications, first ed., Elsevier, Amsterdam, NV, 2020.
[74] D.A. Vallero, Environmental Contaminants: Assessment and Control, Elsevier Academic Press, Amsterdam; Boston, 2004. p. xxxix, 801 pp.
[75] D.A. Vallero, C.K. Gunsch, Applications and implications of emerging biotechnologies in environmental engineering, J. Environ. Eng. 146 (6) (2020) 03120005.
[76] M.M. Epstein, T. Vermeire, Scientific opinion on risk assessment of synthetic biology, Trends Biotechnol. 34 (8) (2016) 601–603.
[77] E.U. von Weizsäcker, The environmental dimension of biotechnology, in: Ernst Ulrich von Weizsäcker, Springer, 2014, pp. 57–65.
[78] B.E. Logan, J.R. Hunt, Bioflocculation as a microbial response to substrate limitations, Biotechnol. Bioeng. 31 (2) (1988) 91–101.
[79] J.P. Boltz, E.J. La Motta, J.A. Madrigal, The role of bioflocculation on suspended solids and particulate COD removal in the trickling filter process, J. Environ. Eng. 132 (5) (2006) 506–513.
[80] G. Vaughan, G. Holder, Substrate removal in the trickling filter process, J. Water Pollut. Control Fed. (1984) 417–423.
[81] W. Gujer, M. Boller, Design of a nitrifying tertiary trickling filter based on theoretical concepts, Water Res. 20 (11) (1986) 1353–1362, https://doi.org/10.1016/0043-1354(86)90133-8.
[82] M.J. Higgins, J.T. Novak, Characterization of exocellular protein and its role in bioflocculation, J. Environ. Eng. 123 (5) (1997) 479–485.
[83] X. Zhang, P.L. Bishop, B.K. Kinkle, Comparison of extraction methods for quantifying extracellular polymers in biofilms, Water Sci. Technol. 39 (7) (1999) 211–218.
[84] W.P. Gerhard, Sanitation and Sanitary Engineering, The Author, 1909.
[85] W. Eckenfelder, Biological treatment of waste water, Adv. Hydrosci. 3 (1966) 153–190.
[86] D.A. Vallero, J.J. Peirce, Engineering the Risks of Hazardous Wastes, Butterworth-Heinemann, Burlington, MA, 2003. p. xxi, 306 pp.
[87] M.P. Lau, M. Sander, J. Gelbrecht, M. Hupfer, Natural organic matter as electron acceptor: experimental evidence for its important role in anaerobic respiration, in: EGU General Assembly Conference Abstracts, 2014, p. 6136.
[88] S. Liu, How cells grow, in: S. Liu (Ed.), Bioprocess Engineering, Elsevier, 2017, pp. 629–697. second ed. (Chapter 11).
[89] K. van't Riet, R.G.J.M. van der Lans, 2.07—Mixing in bioreactor vessels, in: M. Moo-Young (Ed.), Comprehensive Biotechnology, second ed., Academic Press, Burlington, 2011, pp. 63–80.
[90] X. Zou, J.-y. Xia, J. Chu, Y.-p. Zhuang, S.-l. Zhang, Real-time fluid dynamics investigation and physiological response for erythromycin fermentation scale-up from 50 L to 132 m3 fermenter, Bioprocess Biosyst. Eng. 35 (5) (2012) 789–800.
[91] J. Bao, Q. Ye, J.-J. Zhong, Bioreactor Engineering Research and Industrial Applications II, Springer, 2016.
[92] Q. Yang, G. Liu, M. Casazza, F. Gonella, Z. Yang, Three dimensions of biodiversity: new perspectives and methods, Ecol. Indic. 130 (2021) 108099, https://doi.org/10.1016/j.ecolind.2021.108099.
[93] S.T. Allen, et al., Ecological resilience indicators for salt marsh ecosystems, in: Ecological Resilience Indicators for Five Northern Gulf of Mexico Ecosystems, NatureServe, Arlington, 2018, p. 53.
[94] B. Hermann, C. Kroeze, W. Jawjit, Assessing environmental performance by combining life cycle assessment, multi-criteria analysis and environmental performance indicators, J. Clean. Prod. 15 (18) (2007) 1787–1796.
[95] R.F. Noss, Indicators for monitoring biodiversity: a hierarchical approach, Conserv. Biol. 4 (4) (1990) 355–364.
[96] W. Strong, Biased richness and evenness relationships within Shannon–Wiener index values, Ecol. Indic. 67 (2016) 703–713.
[97] P.J. Somerfield, K.R. Clarke, R.M. Warwick, Simpson index, in: S.E. Jørgensen, B.D. Fath (Eds.), Encyclopedia of Ecology, Academic Press, Oxford, 2008, pp. 3252–3255.
[98] United Nations Environment Programme, Rio declaration on environment and development, Nairobi, Kenya, 1992.
[99] M.D. Wood, K. Plourde, S. Larkin, P.P. Egeghy, A.J. Williams, V. Zemba, I. Linkov, D.A. Vallero, Advances on a decision analytic approach to exposure-based chemical prioritization, Risk Anal. 40 (1) (2020) 83–96.
[100] Canadian Agricultural Services Coordinating Committee, Soil Classification Working Group, National Research Council Canada, Canada. Agriculture, Agri-Food Canada. Research Branch, The Canadian system of soil classification (no. 1646), NRC Research Press, Ottawa, Canada, 1998.
[101] J. Bai, et al., Depth-distribution patterns and control of soil organic carbon in coastal salt marshes with different plant covers, Sci. Rep. 6 (1) (2016) 1–12.
[102] R. Conrad, Methane production in soil environments—anaerobic biogeochemistry and microbial life between flooding and desiccation, Microorganisms 8 (6) (2020) 881.
[103] R.R. Lane, S.K. Mack, J.W. Day, R.D. DeLaune, M.J. Madison, P.R. Precht, Fate of soil organic carbon during wetland loss, Wetlands 36 (2016) 1167–1181.

[104] R.H. Day, S.T. Allen, J. Brenner, K. Goodin, D. Faber-Langendoen, K.W. Ames, Ecological resilience indicators for mangrove ecosystems, in: Ecological Resilience Indicators for Five Northern Gulf of Mexico Ecosystems, NatureServe, Arlington, Virginia, USA, 2018.

[105] C.S.B.W.W. Heck, Effects on vegetation, in: A.C. Stern (Ed.), Air Pollution, third ed., Academic Press, New York, NY, 1977, pp. 157–229.

[106] L. Bennett, Deforestation and Climate Change, vol. 1400, A Publication of Climate Institute, 2017.

[107] G. Ofomata, Impact of road building, urbanisation and general infrastructural development on the Nigerian rainforest ecosystem, Landsc. Plan. 8 (1) (1981) 21–29.

[108] S.C. Trombulak, C.A. Frissell, Review of ecological effects of roads on terrestrial and aquatic communities, Conserv. Biol. 14 (1) (2000) 18–30.

[109] D.A. Vallero, T.M. Letcher, Unraveling Environmental Disasters, Newnes, 2012.

[110] R. Lindsey, Tropical Deforestation. March 30, 2007; Adapted From R. Simmon. (2005). Millennium Ecosystem Assessment Biodiversity Synthesis, Goddard Space Flight Center, Earth Observatory, U.S. National Aeronautics and Space Administration, 2007. https://earthobservatory.nasa.gov/features/Deforestation. (Accessed 27 October 2021).

[111] NASA, Measurements of Pollution in the Troposphere (MOPITT) Data Langley Data Center Collection Guide, 2000, Available: http://eosweb.larc.nasa.gov.

[112] C.A. Monteiro, G. Cannon, The impact of transnational "big food" companies on the south: a view from Brazil, PLoS Med. 9 (7) (2012) e1001252.

[113] R. Patel, How society subsidizes big food and poor health, JAMA Intern. Med. 176 (8) (2016) 1132–1133, https://doi.org/10.1001/jamainternmed.2016.3068.

[114] K.M. Siegal, The tentative promise of social return on investment, Am. J. Eval. 43 (3) (2022) 438–457, https://doi.org/10.1177/10982140211072420. In this issue.

[115] D.A. Vallero, T.M. Letcher, Engineering risks and failures: lessons learned from environmental disasters, Leadersh. Manag. Eng. 12 (4) (2012) 199–209.

[116] D.A. Vallero, T.M. Letcher, Unraveling Environmental Disasters, second ed., Elsevier Academic Press, Amsterdam, The Netherlands, 2023.

[117] M.A. Altieri, Agroecology, small farms, and food sovereignty, Mon. Rev. 61 (3) (2009) 102–113.

Chapter 10

Surface waters

1 Introduction

In recent centuries, surface water pollution was one of the first to be recognized by urban dwellers. As mentioned in Chapter 1, among the first duties of civilian engineers was to address the worsening problem of river pollution. When I began my scientific career in the 1970s, the water quality of many of Western Civilization's streams was in that state of demise. I recall animal carcasses and oil sheen floating atop the Missouri River. Later, carp (*Cyprinus* genus), catfish (*Corydoras* genus), and other bottom dwelling fish caught in the Mississippi River were found to be laden with high levels of chlorinated pesticides and polychlorinated biphenyls (PCBs) [1]. Indeed, such surface water contamination had occurred well before scientifically credible analyses were available to verify it [2].

Water pollution science and engineering differentiate water by whether it is at or below the earth's surface. This chapter addresses water bodies at the surface, aptly known as surface waters. The focus of Chapter 11 is water that is below the earth's surface. The second classification distinction is whether a water body has freshwater or saltwater.

Freshwater and saltwater have important differences that affect the behavior of pollutants. The density difference between these two water types can be a slowly unfolding disaster for the health of people living in coastal communities and for marine and estuarine ecosystems. Salt water contains significantly greater mass of ions than does freshwater (see Table 10.1). The denser saline water can wedge beneath freshwaters and pollute surface waters and ground water (see Fig. 10.1). This phenomenon, known as "saltwater intrusion," can significantly alter an ecosystem's structure and function, and threaten freshwater organisms [3]. It can also pose a huge challenge to coastal communities who depend on aquifers for their water supply. Part of the problem and the solution to the problem can be found in dealing with the density differentials between fresh and saline waters.

A few reminders of fluid dynamics discussed in Chapter 7 are in order when discussing surface waters. A fluid is a collective term that includes all liquids and gases. In the liquid phase, molecules move freely among themselves without separating from each other. In the gas phase, move freely and can infinitely occupy the space of their container at a constant temperature. Engineers define a fluid as a substance that will deform continuously upon the application of a shear stress, i.e., a stress in which the material on one side of a surface pushes on the material on the other side of the surface with a force parallel to the surface.

For example, the reciprocal of a substance's density is its specific volume (υ), which is the volume occupied by a unit mass of water or any fluid. The units of υ are reciprocal density units ($m^3\ kg^{-1}$). Stated mathematically, this is:

$$\upsilon = \rho^{-1} \tag{10.1}$$

The weight of a fluid per its volume is known as specific weight (γ). Scientists and engineers often use the term interchangeably with density. Geoscientists commonly discuss a substance's specific weight. Note that γ is not an absolute fluid property but depends upon the local gravitational force on the fluid:

$$\gamma = gp \tag{10.2}$$

Specific weight units are the same as those for density, e.g., $kg\ m^{-3}$.

The fractional change in a fluid's volume per unit change in pressure at constant temperature is the fluid's coefficient of compressibility. Any fluid can be compressed in response to the application of pressure (p). For example, water's compressibility at 1 atm is $4.9 \times 10^{-5}\ atm^{-1}$. This compares to the lesser compressibility of mercury ($3.9 \times 10^{-6}\ atm^{-1}$) and the greater compressibility of hydrogen ($1.6 \times 10^{-3}\ atm^{-1}$). However, for most environmental engineering applications, water's compressibility is ignored.

Fundamentals of Water Pollution. https://doi.org/10.1016/B978-0-443-28987-3.00006-0

TABLE 10.1 Composition of freshwaters (river) and marine waters for some important ions.

Composition	River water	Saltwater
pH	6–8	8
Ca^{2+}	4×10^{-5} M	1×10^{-2} M
Cl^-	2×10^{-4} M	6×10^{-1} M
HCO_3^-	1×10^{-4} M	2×10^{-3} M
K^+	6×10^{-5} M	1×10^{-2} M
Mg^{2+}	2×10^{-4} M	5×10^{-2} M
Na^+	4×10^{-4} M	5×10^{-1} M
SO_4^{2-}	1×10^{-4} M	3×10^{-2} M

Data from J.P. Kim, K.A. Hunter, M.R. Reid, Factors influencing the inorganic speciation of trace metal cations in fresh waters, Mar. Freshw. Res. 50 (4) (1999) 367–372. R. Schwarzenbach, P. Gschwend, D. Imboden, Environmental Organic Chemistry, John Wiley & Sons, Hoboken, New Jersey, 2016.

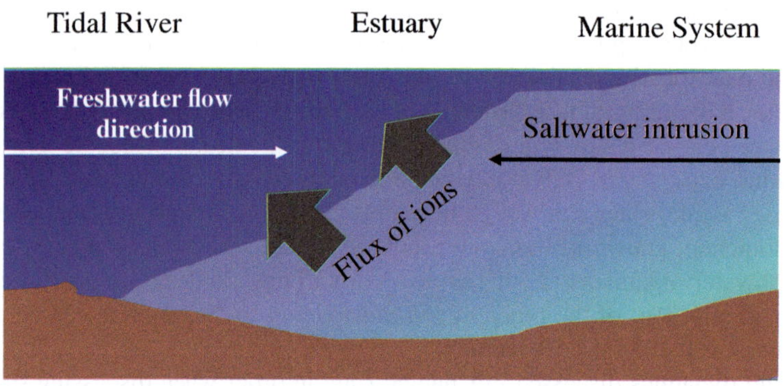

FIG. 10.1 Mixing of saltwater and freshwater in an estuary. Denser saltwater submerges beneath lighter freshwater.

A fluid's bulk modulus, E, is a function of stress and strain on the fluid, and is a description of its compressibility, and is defined according to the fluid volume (V):

$$E = \frac{\text{stress}}{\text{strain}} = -\frac{dp}{dV/V_1} \tag{10.3}$$

E is expressed in units of pressure (e.g., kP). Water's $E = 2.2 \times 10^6$ kP at 20 °C.

Surface water pollution begins with a release, which can be directly from a point source or indirectly from a nonpoint source (see Fig. 10.2). By gravity, the water carries the pollutant downstream. When a pollutant moves with the fluid, this is known as advective transport.

In addition to physical and abiotic chemical processes, the movement of organisms, especially microbes, also begins with point and/or nonpoint releases (see Fig. 10.3). During release, the microbes behave like particles, but with special considerations, e.g., surface properties like sorption will be affected by surrounding biofilm. After release, the properties change substantially, e.g., algal blooms that take on two- and three-dimensional changes, which affect light and other exchanges in ecosystems (see Fig. 10.3).

Fluids are generally divided into two types: ideal and real. The former has zero viscosity and, thus, no resistance to a force acting in a direction that's parallel to (over the top of) a surface or cross section of a body, i.e., shear. An ideal fluid is incompressible and flows with uniform velocity distributions. It also has no friction between moving layers and no turbulence (i.e., no eddy currents). On the contrary, a real fluid has finite viscosity, has nonuniform velocity distributions, is compressible, and experiences friction and turbulence. Real fluids are further subdivided according to their

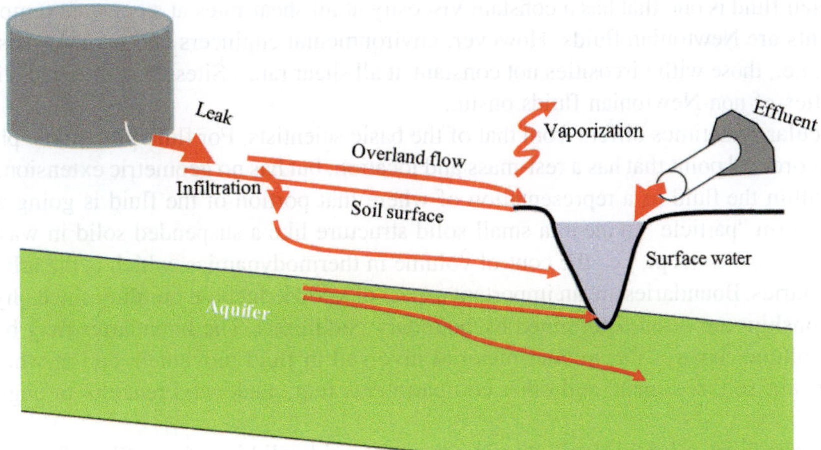

FIG. 10.2 Movement of a water pollutant after release. *(Data from D.A. Vallero, Fundamentals of Air Pollution, fifth ed., Elsevier Academic Press, Waltham, MA, 2014, 999 pages cm p.)*

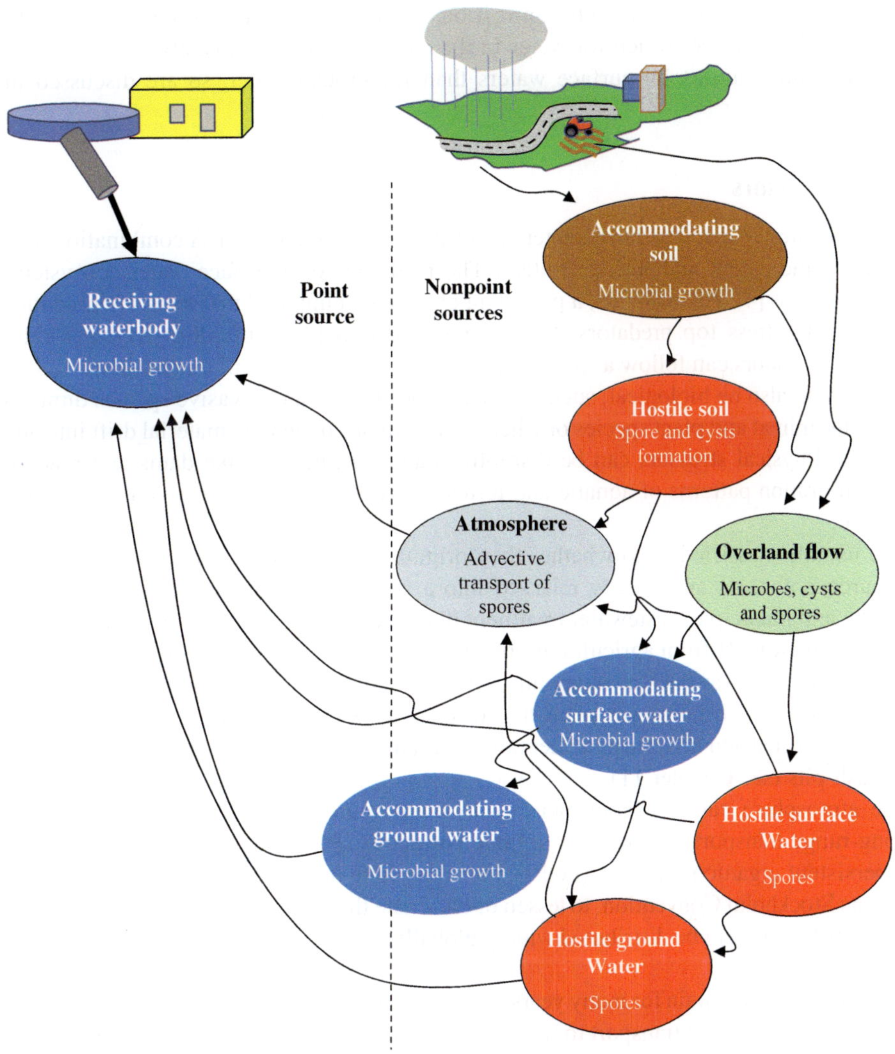

FIG. 10.3 Release of microorganisms to waterbodies by point and nonpoint sources. After release, the microbial population growth depends on the conditions in the water cycle and other biogeochemical cycles. *(Data from D. Vallero, Environmental Biotechnology: A Biosystems Approach, Elsevier Science, 2015.)*

viscosities. A Newtonian fluid is one that has a constant viscosity at all shear rates at a constant temperature and pressure. Water and most solvents are Newtonian fluids. However, environmental engineers and hydrologists are confronted with non-Newtonian fluids, i.e., those with viscosities not constant at all shear rates. Sites contaminated with drilling fluids and oils have large quantities of non-Newtonian fluids onsite.

Engineering vernacular sometimes differs from that of the basic scientists. For fluid properties, physicists use the term "particle" to mean a theoretical point that has a rest-mass and location, but has no geometric extension. We can observe this particle as it moves within the fluid as a representation of where that portion of the fluid is going and at what velocity. Engineers also use the term "particle" to mean a small solid structure like a suspended solid in wastewater.

This introduces the second concept, i.e., the control volume in thermodynamics, which is the arbitrary region in space that is defined by boundaries. Boundaries are an important aspect of knowledgebase building since physical, mathematical, and engineering relationships are often determined by boundary conditions. The boundaries may be either stationary or moving. The control volume conveys the amount material involved in fluid movement and at what rate the material is moving through water, air, soil, sediment, and other compartments (e.g., tanks and reactors in engineered systems, like treatment plants).

Surface tension effects occur at liquid-liquid, liquid-gas, and liquid-solid interfaces (See Chapter 7). Surface tension, σ, is the force in the liquid surface normal to a line of unit length drawn in the surface. Surface tension decreases with temperature and depends on the contact fluid. Surface tension is involved in capillary rise and drop. Water has a very high σ value (approximately 0.07 N m^{-2} at 20 °C). Of the environmental fluids, mercury (Hg) is the only one with a larger σ than water. Increasing surface tension creates a skin on a free surface, allowing an object that is denser than water to float on a still water surface. This is why insects can rest on water surfaces. Surface tension depends slightly upon the gas that is contacting the free surface, which for water in the environment is almost always air [3]. Surface tension and capillarity are less important features for surface waters than for ground water, so are discussed in greater detail in Chapter 11.

1.1 Surface water stressors

Stress is both a term of physics, as discussed in Chapter 7, and an environmental term. A combination of natural and human-induced stressors affects the extent and surface waters. The resilience of a surface water ecosystem is as affected by stressor-receptor relationships [4]. As evidence, a persistent chemical compound that resists degradation is likely to build up in the food web and to stress top predators. For example, dicofol [2,2,2-trichloro-1,1-bis(4-chlorophenyl)ethanol] exposure and toxicity in alligators can follow a spill and decades of accumulation of this persistent substance through food chains [5–9]. A stressor can also be biological, such as when opportunistic and invasive species diminish the survival and abundance of indigenous animal and plant species or when the chemicals or genetic material drift into adjacent habitats and supplant native species. Physical stressors can be disruptions and obstructions like dams and roads that interfere with feeding, mating, and migration patterns of aquatic and terrestrial species. Stress irreversibility is indirectly proportional to a system's resilience.

The stressors are usually classified as to whether they originate from a point source or from a nonpoint source. As mentioned, point source pollutants are directly released into a surface water body. Examples include discharges from factories, power plants, and municipal wastewater treatment plants. Conversely, nonpoint source pollutants indirectly find their way to surface waters. Runoff from agricultural operations, construction, roadways, mining, and electric generating stations' coal and ash piles are examples of nonpoint pollution. Indeed, one site may have combinations of point and nonpoint sources, such as a coal-fired power plant that discharges wastewater directly to a stream or lake while also allowing acidic runoff from coal and ash piles. Incidentally, the same site may also allow ground water to be contaminated from their piles and ash pits (see Chapter 11).

Pollutants in the atmosphere can also contribute to nonpoint source pollution, including acidic deposition, metal compounds, and long-range transport of persistent organic pollutants (POPs). The so-called Dirty Dozen is a list of the 12 most notorious persistent organic pollutants (POPs), including nine organochlorine pesticides. Several nations signed onto the United Nations Stockholm Convention, to lessen or eliminate the production, use, and/or release of these POPs (see Table 10.2). Persistent pollutants can be distributed widely globally, hundreds or thousands of miles from where they were used [10–13].

Persistent pollutants enter and remain for many years in food chains and can lead to exposures in humans, e.g., in infants via breast milk. Ecologically, pollutant transport to sensitive habitats obviously behaves according to thermodynamics and fluid dynamics. As evidence, pollutants can reach ecosystems by leapfrogging from one medium to another, e.g., air-to-water, soil-to-air, water-to-air, water-to-biota, etc. For example, streams deposit their loads into reef and other coastal

TABLE 10.2 Persistent organic pollutants listed in the Stockholm Convention.

Pollutant	Global historical use and source
Aldrin and dieldrin	Insecticides used on crops such as corn and cotton; also used for termite control
Chlordane	Insecticide used on crops, including vegetables, small grains, potatoes, sugarcane, sugar beets, fruits, nuts, citrus, and cotton. Used on home lawn and garden pests. Also used extensively to control termites
DDT	Insecticide used on agricultural crops, primarily cotton, and insects that carry diseases such as malaria and typhus
Endrin	Insecticide used on crops such as cotton and grains; also used to control rodents
Mirex	Insecticide used to combat fire ants, termites, and mealybugs. Also used as a fire retardant in plastics, rubber, and electrical products
Heptachlor	Insecticide used primarily against soil insects and termites. Also used against some crop pests and to combat malaria
Hexachlorobenzene	Fungicide used for seed treatment and an industrial chemical used to make fireworks, ammunition, synthetic rubber, and other substances. Also unintentionally produced during combustion and the manufacture of certain chemicals. It is also an impurity in certain pesticides
Polychlorinated biphenyls (PCBs)	Used for a variety of industrial processes and purposes, including in electrical transformers and capacitors, as heat exchange fluids, as paint additives, in carbonless copy paper, and in plastics. Also unintentionally produced during combustion
Toxaphene	Insecticide used to control pests on crops and livestock, and to kill unwanted fauna in lakes
Dioxins and furans	Unintentionally produced during most forms of combustion, including burning of municipal and medical wastes, backyard burning of trash, and industrial processes. Also, can be found as trace contaminants in certain herbicides, wood preservatives, and in PCB mixtures

Data from U.S. Environmental Protection Agency, Stockholm convention on organic pollutants, Washington, DC. Available from: https://comptox.epa.gov/dashboard/chemical-lists/STOCKHOLM/, 2023.

ecosystems. For example, aerosols in the atmosphere fall due to gravity onto land surfaces, with the pollutants sorbed onto the deposited particles becoming desorbed. The desorbed pollutants then reach a surface water body from nonpoint mechanisms, including mechanics and diffusion.

Pollutants may be deposited directly in water or may later flow over land and reach surface waters, e.g., emissions of mercury (Hg) to the air from combustion at coal-fired power plants has been shown to be transported and deposited in lakes long distances from industrial and power generating plants [14–21].

As mentioned, surface waters can be polluted by biological agents, such as pathogenic microbes or by invasive and opportunistic species of fauna and flora that alter habitats and crowd out native species and interrupt natural selection, mutations, and human interferences like breeding and other nonrandom mating [22,23]. In the latter case, new alleles are introduced into an aquatic fauna population because of natural processes like temperature and humidity changes, and by anthropogenic processes, e.g., insects' alleles having to adapt to pesticide exposure. The Hardy-Weinberg law of ecology states that the genetic variation in a population will persist from one generation to the next in the absence of disturbing factors [24]. Hardy-Weinberg equilibrium is altered in an aquatic habitat when breeding between two populations transfers new alleles into a population, which is known as gene flow [22,23]. Such genetic variations in nature indicate aberrations from a healthy aquatic ecosystem [25].

Hydrologic and atmospheric cycling occurs across interfaces between land and water bodies and the atmosphere. Within a surface water control volume, physical changes, abiotic chemical reactions, and biochemical reactions occur constantly. Various reactions occur in aquatic macrophytic plants and microbes, many of which are catalyzed enzymatically [26,27]. The thermodynamic processes of aquatic ecosystems are very complex, rendering most estimates to be overgeneralizations and simplification of the complex processes that drive what is indeed occurring in an ecosystem.

The rates of rainfall and other precipitation are major factors in the timing and amount of runoff and erosion, while other aspects of weather and climate influence heating, cooling, and mixing in lakes and estuaries, which affect the movement of contaminants and the cycling of nutrients. The mineral composition of bedrock and sediment helps determine whether a water body may be susceptible to acidification. For example, well-buffered soil and rock systems in central North America are less susceptible to acidic runoff into surface waters than the comparatively thin, less alkaline soils and rocks of the Adirondacks and other regions in the northeastern United States [28–31]. However, localized conditions can affect the acidification rates, such the comparatively greater acid-neutralization capacity of the Finger Lakes region in New York State compared to that of the Adirondacks, also in New York [32].

2 Saltwater systems

Most of the water on earth is in the oceans. The amount is so vast, it is tempting to think of the oceans as almost endless sinks that can absorb any quantity of pollutants that human activities can generate. This, however, is flawed logic because pollution is never evenly distributed in a water body. Rather, within a vast water body there are regions of elevated concentrations of pollutants much higher than the surrounding, larger volume of water. Also, like in any ecosystem, ocean organisms are part of complex set of webs and chains where even slight changes in one trophic level can lead to major problems in the entire system.

Some pollution is so large that it affects much of the ocean volume. For example, gas concentrations and exchange coefficients between the atmosphere and surface waters govern the rate of a gas molecule that crosses the ocean-atmosphere boundary. The carbon dioxide (CO_2) concentrations at the ocean surface takes about a year to equilibrate with atmospheric CO_2 concentrations. This causes large atmosphere-ocean differences in CO_2 concentrations between the air and ocean water. CO_2 reacts with large volumes of ocean water to form carbonic acid (H_2CO_3) and its dissociation products. In recent decades atmospheric CO_2 concentrations have increased dramatically, so the interaction with the ocean surface is progressively altering the chemistry that leads to ocean acidification [33,34].

The physicochemical mechanism for ocean acidification is the increase in the partial pressure of a gas, in this case CO_2, in the atmosphere directly above the body of water that causes the gas to diffuse into the water. This diffusion continues until the partial pressures across the air-water interface become equilibrated. Other factors are also at work, such as the increasing CO_2 atmospheric concentrations that affect large-scale climatic features. For example, the changing currents and winds may in turn alter ocean circulation, changing the rate of ocean CO_2 uptake. Migrations of marine organisms and other ecosystem changes are also affected [34].

Other ocean-scale problems result from global increases in the greenhouse gas emissions and releases of toxic substances, such as microplastics. Annual global plastic production in mid-20th century was 1.5×10^6 tonnes (t).[a] In 2002 this grew to 2×10^9 t. In 2021, it was 3.9×10^8 t, with a rate of increase of over 2% per year. Since 1950, 9×10^9 t of plastics has been produced [35].

Only about 9% of plastic is recycled each year, with 40% ending in landfill and 25% incinerated [36]. The remainder is mainly transported by surface waters, with much making it to the oceans [37]. An ocean has five large subtropical gyres, large systems of circular ocean currents formed by global wind patterns and the Coriolis effect. This effect is from the earth's rotation on its axis which deflects the atmosphere toward the right in north of the equator and to the left south of the equator, resulting in curved wind and water currents. The atmospheric deflection leads to complex global wind patterns that drive ocean currents that carry plastic matter along the atmosphere-ocean surface. The North Pacific Ocean Gyre has been monitored but little is known of the other gyres [38]. The plastic material breaks down physically into increasingly small particles until they become microplastic particles, which are 5 mm across at their widest dimension. Although the existence of microplastic material had been studied for several decades, the term was coined was 2004.

Humans ingest microplastics every day, usually without apparent symptoms. Despite their ubiquity, the impacts of microplastics on either organisms and their habitats have yet to be fully understood [39,40]. However, it is reasonable to be concerned since plastic material can contain harmful chemical additives which could have serious consequences, such as cancer, reproductive issues, and effects on the central nervous system to humans, animals and possibly plants.[b] Plastic additives include plasticizers like bisphenol or dibutyl phthalate, flame retardants, e.g., antioxidants like amine compounds, pigments, insecticides, and lubricants [41].

The atmosphere and ocean exchange matter and energy at the ocean's surface boundary layer. This interface is rich in biota, nutrients, and water pollutants. Models, including Lagrangian mixing models, account for the different scales of turbulence and physical principles to estimate travel time of the microplastic particles. Recently, marine turbulence has been combined with vertical turbulent diffusivity to estimate the subgrid turbulence occurring in the mixed layer [42]. This is based on characteristic scales calculated from meteorological and oceanographical models. The models support the analysis of the vertical distribution of microplastic particles along the water column, beaches, and seafloor zones where

a. Tonnes (t) are metric tons, which equal 1000 kg or 1.10231 U.S. tons.

b. This concern is analogous to exposure to aerosols in the air. The particle itself may cause lung and other diseases, but toxic chemical compounds can sorb to the particle. Thus, the aerosol and the microplastic particle can deliver toxic substances to the organism.

these particles tend to accumulate [43–45]. Thus, vertical turbulent diffusivity is an essential factor for the vertical dispersion of microplastics within the mixed layer.

Microplastics and ocean acidification are examples of how the oceans are too often the ultimate receptors of pollutants.

2.1 Coastal water

Coastal waters are the interface between the oceans and land. They support unique habitats like estuaries, coastal wetlands, seagrass meadows, coral reefs, and mangroves. As such, they provide breeding areas and habitat for myriad species of sea life and migratory birds. Parts of the habitat are surface and submerged vegetation that provides food and hiding places for many species [46].

The spatial extent of coastal waters can be diminished by both natural influences like dune erosion and storms and by human activities like dredging or filling wetlands for development [46]. Coastal area topographic levels can sink as the streams that feed them are channelized for navigation. The sediment loads that previously replenished deltas and other coastal zones decrease because the channelized stream has less floodplain area to load the stream sediment. Thus, this is an example of solving the human problems of flooding and improving navigation on a river that causes another problem downstream.

The condition of coastal waters reflects combinations of interrelated physical, chemical, and biological features. Coastal water condition can be affected by a variety of stressors including excessive concentrations of nitrogen and other nutrients and pathogenic bacteria and other microbes that are delivered to coastal and near-coastal waters by stormwater, runoff from farms, for example, and by point sources like outfalls from municipal wastewater treatment plants. The nutrient concentrations enhance the excessive growth of algae that decrease dissolved oxygen concentrations during algal blooms used by these and other microbes for respiration [47–51]. Ultimately, fish kills and other harm to aquatic life result. Note that these processes are the same as those of freshwater lakes discussed below.

Both biological and chemical agents in coastal waters can lead to human diseases caused by pathogenic bacteria like enteric disease-causing fecal coliforms, e.g., *Escherichia coli* and zoonotic parasite species like *Giardia*, *Toxoplasma*, and *Cryptosporidium* [52,53]. Compared to most other surface water bodies, coastal zones are especially vulnerable to hurricane and other tropical weather events which, in addition to physical changes to the coastal hydrology, may carry pathogenic organisms and cause standing water that can contribute to infestations of mosquitos and other disease vectors.

Chemical contaminants can enter coastal water bodies by both point and nonpoint sources. As mentioned, nonpoint sources are of many types that include agricultural runoff, industrial activities, and aerosols and vapors deposited from the atmosphere deposited directly on the water and on the land area of the coastal watershed. Food chain accumulation can affect every tropic level in the coastal ecosystem, including exposures to humans.

Weather events, such as floods and droughts change levels of salinity and water temperature and salinity. These can alter habitat quality and threaten native microbial, plant, and animal abundance and diversity and exacerbate conditions by synergistic processes when nutrient levels have already increased, leading to algal blooms and aquatic damage [46].

Shipping and other activities can carry nonindigenous species that can threaten the status of native communities. Invasive species can crowd out and displace native populations or otherwise alter coastal watersheds [46].

Overharvesting of sea life alters the biodiversity of coastal water ecosystems, as populations of fish, shellfish, marine mammals, and other species are unable to adapt to replenish these organisms by breeding and survival. Also, hydrological changes due to beach erosion and wetland destruction can affect chemical water pollutant concentrations and sediment levels. As a result, backwater and wetlands are less available to provide for breeding and protective cover [46].

3 Freshwater systems

Freshwater that is held in lakes, ponds, rivers, and streams comprise less than one thousandth of a percent of the earth's water but has been sufficient to sustain ecological systems and habitat for myriad species, from microbes to human populations. The various freshwater systems can be classified according to their extent and condition. The extent of fresh surface waters roughly equates to the quantity of water and reflects a variety of human and environmental stressors. It can be affected by direct withdrawal for drinking, irrigation, industrial processes, and other human uses, as well as by the withdrawal of ground water, which replenishes many surface waters [54].

Locks, dams, levees, and other hydromodifications fundamentally alter the extent of stream flow. Impervious and low permeability land covers, like road pavement, exacerbate the potential for flooding and runoff. Changes in climate alter rates and timing of precipitation, and alter evapotranspiration rates, all which influence surface water flow and retention.

The condition of a fresh surface water system determines its water quality, i.e., the physical, chemical, and biological characteristics that determine the potential as sources for drinking water, fishing, recreation, ecosystem sustainability, and other human and ecological utility and values. Temperature, clarity, and flow rates are examples of physical condition characteristic. Ion concentrations, especially salinity, nutrient concentrations like nitrogen and phosphorous compounds, concentrations of synthetic organic compounds, and chemical exchanges with sediments are examples of chemical condition characteristics. Abundance of fecal coliform and other microbial pathogens, cyanobacteria, algae, vascular plants, aquatic vertebrate abundance and diversity, beneficial microbial diversity, and overall habitat well-being are examples of biological condition characteristics [54].

Various faunal, floral, and microbial communities are affected by surface water hydrology, e.g., water level can strongly influence the health of riparian ecosystems. Other communities may have problems adapting to seasonal fluctuations in flow. Stressors that affect the extent of a freshwater body may eventually affect the condition of aquatic habitat, such as hydromodifications that restrict the migration of certain aquatic species [54].

Fresh surface water condition can be described with robust indicators. For example changes in stream flow patterns, an aspect of surface water extent, and freshwater withdrawal rates can reveal that a water body is vulnerable to pollution. Freshwater withdrawals alter the quantity of water. Other indicators included the physical condition of sediments, the condition of benthos (community of organisms on, in, or near the sediment at a water body's bottom), and the biological and chemical condition of the water, e.g., acidity, nutrient content, microbial populations, and pesticide concentrations [54].

3.1 Lakes

Lakes vary in area, depth, configuration, chemical composition, clarity, and ecosystems that they support.[c] The types of plant and animal life present. Some lakes, especially those near acidic wetlands like bogs, are stained with tannic acid that leaches from surrounding vegetation [55]. Natural lakes are frequently classified according to the quality needed as a potential source of water supply.

Drainage lakes have both an inlet and outlet where the main water source is stream drainage. A drainage lake having at least one-half of their maximum depth to an artificial dam is known as an impoundment [55].

Seepage lakes are landlocked and have no inlet nor outlet, and seldom overflow compared to drainage lakes. Their principal source of water is from precipitation and runoff that is supplemented by ground water. As a result, water depth commonly differs by seasons [55].

Spring lakes have no inlet, but have an outlet, with the primary source of ground water flow. Spring lakes form the headwaters of many streams.

Drained lakes have no inlet, but like spring lakes, have an outlet that flows continuously. Unlike spring lakes, drained lakes are not fed by ground water. Their primary source of water is from precipitation and direct drainage from adjacent land. Depending on the source, the depth in drained lakes fluctuates. During low precipitation, especially severe drought conditions, the outlet streams become intermittent.

As mentioned, an impoundments is human-constructed, and is hydrologically like a drainage lake given that each have an inlet and outlet that usually drains to a stream [55].

The water quality of a lake and species of fish present are significantly influenced by the type of hydrology [56]. Drainage lakes, for example, can support fish populations that may differ from their connected streams. Impoundments often have elevated nutrient levels compared to natural seepage or spring lakes. Since seepage lakes are not influenced by streams, they usually have lower diversity of aquatic species compared to most other lakes. Seepage lakes drain comparatively smaller land areas than most other lakes, so often have less nonpoint nutrient runoff and consequently lower concentrations of N, P, and other nutrients.

A particular problem for lakes is eutrophication when algal blooms that take on two- and three-dimensional changes, which affect light and other exchanges in ecosystems. When nutrient concentrations become sufficiently elevated, especially compounds and ions of nitrogen and phosphorous that lake is said to have become highly enriched. As such

---

c. In the late 1970s, I was the Clean Lakes Coordinator of the Kansas City Regional Office of the U.S. Environmental Protection Agency. It was interesting how the definition of a lake or pond varied by U.S. region. No matter the size, EPA was concerned about nutrient concentrations, especially nitrogen and phosphorous, in these water bodies. The States were also concerned, but the actual concentrations varied considerably. One state official, I hope jokingly, said his state was indeed concerned about these nutrient levels and solids content, but mainly because they were too low to support agriculture. Okay, maybe you had to be there.

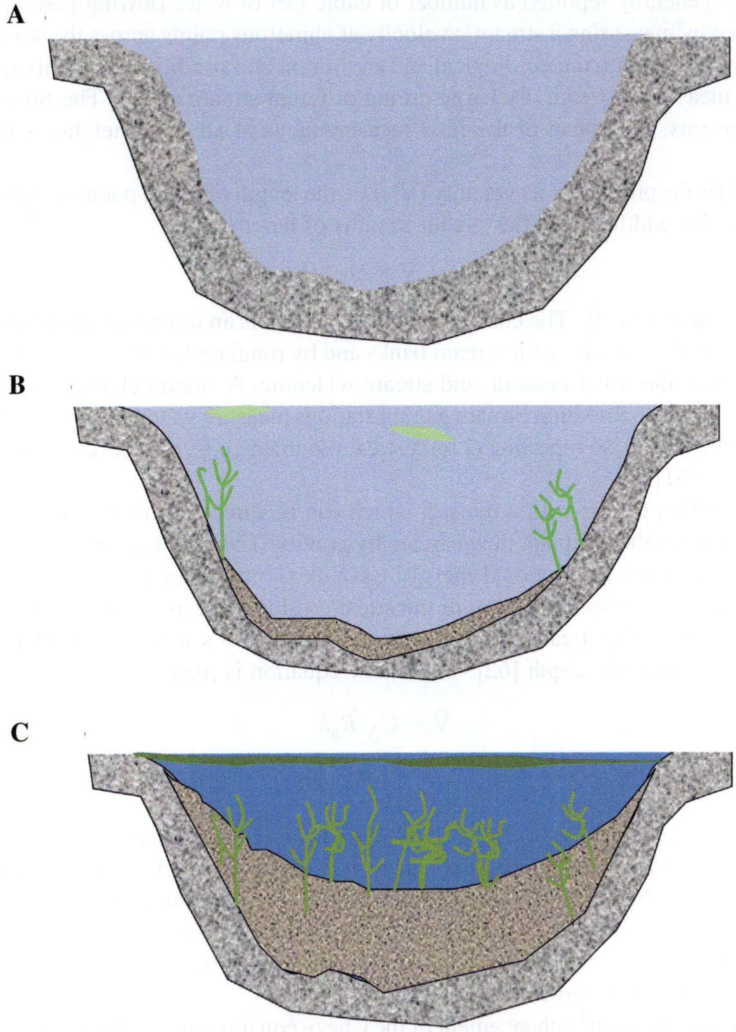

FIG. 10.4 Changes in trophic state brought on by eutrophication, leading to algal blooms. (A) Oligotrophic Lake which is low in nutrient concentrations (i.e., low enrichment), has low productivity, and has low turbidity (clear water), desirable habitat for sensitive species like game fish. (B) Mesotrophic Lake, which has higher nutrient concentrations (i.e., increasing enrichment), accumulates organic matter, sedimentation is increased, with occasional algal blooms, but still useful for fishery. (C) Eutrophic Lake: highly elevated nutrient concentrations (i.e., highly enriched); very productive (supports much plant life); dissolved oxygen becoming depleted; frequent algal blooms; poor fishery (rough fish common).

it is very productive, i.e., has a high net primary productivity (NPP). This is an example of the importance of context in environmental metrics. High NPP is often an indication of a healthy ecosystem, so long it is accompanied by commensurately acceptable biodiversity and is sufficiently resilient and sustainable. However, in this instance, the microbial and plant growth that depend on light energy from the sun, moisture and nutrient concentrations results in unhealthy NPP levels. The mass and energy that are distributed continuously among and within the lake lead to eutrophication [57,58]. Trophic interrelationships are altered by the introduction of nutrients that generates a cascade of changes in the lake's environmental conditions (see Fig. 10.4) [59].

4 Streams

Streams can range from small rivulets to tidal rivers with the source of flow strongly influenced by ocean tides to major freshwater rivers that cross continents. Surface water flow, also known as stream discharge (Q), has units of volume per time. Discharge is not limited to freshwater streams, but applies to other flowing systems, including engineered channels like storm sewers, and saltwater tidal rivers. The accepted discharge units are $m^3\ s^{-1}$ for most of the world, but stream discharge data in the U.S. is generally reported as number of cubic feet of water flowing past a point each second (cfs or $ft^3 s^{-1}$). Discharge is derived by measuring a stream's velocity at numerous points across the stream. Since water volumes and water levels of a stream change with meteorological and environmental conditions, stream-stage and stream-discharge relationships are found by measuring stream discharge during different stream stages. The flow of a stream is estimated based upon many measurements. The mean of the flow measurements at all stage heights is reported as the estimated discharge.

A stream's discharge (Q) is the product of its velocity (V), i.e., the length of travel per unit of time, times the depth of the water (D)—unit of length times width (W) of the stream in units of length [60]:

$$Q = \mathrm{V} \times D \times W \tag{10.4}$$

The cross-section of the streams is $D \times W$. The channel morphology also is an important factor since it affects the stream's velocity, which is decreased by friction along the stream banks and by roughness of the stream bed. Increasing Q results in higher stream velocity, deepens the water column, and stream widening. A stream channel that cannot accommodate an increasing Q will overflow and cause flooding. Stream gaging stations measure water level, i.e., stage, and Q. The values of V, D, and W are specific to the station, so reporting Q for a stream segment, e.g., a river reach, is often based on the sum of mean values for V, D, and W [61].

As mentioned, water pollution begins with a release, which can be directly from a point source or indirectly from a nonpoint source. The water carries the pollutant downstream by gravity. The gravitational forces gained by the water flow downslope must dissipate as heat from the frictional energy loss or into kinetic energy that moves the water volume. Stream velocity in a uniform channel is modeled according to the stream's slope and hydraulic radius (R_h), i.e., the ratio of the cross-sectional area through which the stream is flowing and to the stream's wetted perimeter (P). If the channel were rectilinear, R is the width plus twice the depth [62]. The Chézy equation is [63]:

$$\bar{\mathrm{V}} = C\sqrt{R_h S}$$

Where, $\bar{\mathrm{V}}$ is the mean stream velocity, C is the friction coefficient (nondimensional factor of flow resistance), and S is the team slope.

The flow in an open channel is governed by the ratio of inertial forces to gravitational forces, i.e., the Froude relation. The Froude number (Fr) is equal to V divided by the square root of the product of gravity (g) times D $[\mathrm{V} \times (gD)^{1/2}]$. If Fr equals 1, the flow is critical, whereas Fr > 1 is supercritical flow, i.e., very rapid, and Fr < 1 indicates a tranquil flow.

The value of C in the Chézy equation is a function of the flow Reynolds number (Re) and relative roughness of the stream. This must be determined experimentally and, for most rivers, ranges between $30\,\frac{\mathrm{m}^{0.5}}{\mathrm{s}}$ for a stream in a small, rough channel to $90\,\frac{\mathrm{m}^{0.5}}{\mathrm{s}}$ for a large smooth stream channel.

The Manning equation is an empirical[d] enhancement of the Chézy equation that is also used to calculate stream velocity as well as to derive C values:

$$Q = \bar{\mathrm{V}}A = \frac{k}{n}\left(\frac{A}{P}\right)^{\frac{2}{3}}\sqrt{S} \tag{10.5}$$

Where, k is a unit conversion factor, which is 1.49 for English units (feet and seconds) or 1.0 for SI units (meters and seconds) and n is the Manning roughness factor. For example, major rivers have n values around 0.035, a gravelly channel n is about 0.025, and heavy brush has n values around 0.075 [64].

Discharge can also be calculated as:

$$Q = AC\sqrt{R_h S} \tag{10.6}$$

Where, A is the stream's cross-sectional area.

Thus, the mean stream velocity in English units is simply:

$$\bar{\mathrm{V}} = \frac{1.49\,(R_h)^{\frac{2}{3}}\sqrt{S}}{n} \tag{10.7}$$

d. An empirical relationship. also known as phenomenological relationship, is an association or correlation that is supported by experiments or observations, but not necessarily supported by theory. Here, the Che′zy equation is theoretical and the Manning equation is empirical.

Stream flow example 1

The Tarhealian tributary to the Blue Devil River is 2 m wide, has a mean slope of 0.005, and has heavy brush growing in most of the bed. Last year, a thunderstorm the tributary water depth went from 0 to 15 cm. Assuming uniform, steady flow, estimate the mean velocity of the water immediately after the storm.

Begin with the hydraulic radius:

$$R_h = \frac{W \times D}{W + 2D} = \frac{2\text{m} \times 0.15\text{m}}{[2\text{m} + 2(15)\text{m}]} \cong 0.009\text{m} = 0.03 \text{ ft}$$

$$\overline{V} = \frac{1.49\,(R_h)^{\frac{2}{3}}\sqrt{S}}{n} = \frac{1.49\,(0.03)^{\frac{2}{3}}\sqrt{0.005}}{0.075} \cong \frac{1.49 \times 0.1 \times 0.07}{0.075} \cong 0.14 \text{ ft s}^{-1}.$$

Stream flow example 2

Recently, the town decided to clear the brush so that the tributary now has a gravelly bottom. Yesterday, another heavy rainfall raised the water depth again to 15 cm. How has this changed the flow, assuming again uniform, steady flow?

The only difference is the roughness, so the calculation is:

$$\overline{V} \cong \frac{1.49 \times 0.1 \times 0.07}{0.025} \cong 0.42 \text{ ft s}^{-1}.$$

Thus, by clearing the brush, the smoother bed caused the tributary to flow three times faster into the river.

When a pollutant moves with the water flow, this is known as advective transport. Fluid properties determine how the water will react to applied forces. Thus, fluid properties can be represented as a density (ρ) field:

$$\rho = \rho(x, y, z, t) \tag{10.8}$$

Where, ρ is the density of the fluid, x, y, z are the coordinates in space, and t is time.

A fluid can also be represented by a velocity field ($\vec{v}$):

$$\vec{v} = \vec{v}(x, y, z, t) \tag{10.9}$$

If the water's properties and the flow characteristics at each position do not vary with time, the fluid is said to be at steady flow:

$$\rho = \rho(x, y, z) \quad \text{or} \quad \frac{\partial \rho}{\partial t} = 0 \tag{10.10}$$

and

$$\vec{v} = \vec{v}(x, y, z) \quad \text{or} \quad \frac{\partial \vec{v}}{\partial t} = 0 \tag{10.11}$$

Time-dependent flow is an unsteady flow. A flow with unchanging magnitude and direction of the velocity vector $\vec{v}$ is a uniform flow. Water flow, then, can be classified according to the observable physical characteristics of flow fields. Laminar flow is much more common in ground water, whereas surface water flow is generally turbulent. Ground water flow is usually orders of magnitude less turbulent (Fr $\ll$1) than surface water flow. Given that water is nearly incompressible, its variations in density are assumed to be constant.

The time rate of change a fluid is the fluid velocity (V). This is a vector field quantity. Speed (V) is the magnitude of the vector velocity V at some given point in the fluid, and average speed ($\overline{V}$) is the mean fluid speed through a control volume's surface. Velocity, then, is a vector quantity since it has both magnitude and direction, whereas speed is a scalar quantity since it expresses only magnitude. However, velocity and speed are often used synonymously in environmental sciences and engineering, but the distinction is good to keep in mind when communicating with a mathematician or physicist.[e]

e. I apologize for the engineering humor.

Actually, there are times when engineers draw a distinction between velocity and speed. Since velocity is a vector quantity, it provides information about the direction of the fluid movement. For example, consider a suspended particle being carried in a straight tributary that connects south and north reaches of a river. If the particle is covering 20 m at every second, its speed is 20 meters per second (m s^{-1}). This speed provides no information where the particle is heading in the tributary. However, that the velocity of particle is 20 m s^{-1}. Thus, speed informs how fast the particle is moving, but velocity gives both the speed and the direction of the change, i.e., the particle's displacement. Note that the standard units of both velocity and speed are m s^{-1}.

Calculations of the concentration of a pollutant moving downstream after release into surface waters often assume that the pollutant is moving at the mean velocity of the stream. The mean amount of time for the pollutant's movement from the point of release to a specific point downstream, i.e., a defined stream segment (L_s), is the pollutant's travel time (τ). For a uniform velocity the travel time is [62]:

$$\tau = \frac{L_s}{V} \tag{10.12}$$

For a nonuniform velocity, the travel time is expressed as [62]:

$$\tau = \int_{x_2}^{x_1} \frac{dx}{V(x)} \tag{10.13}$$

Where, x_1 and x_2 are the endpoints of the stream segment or river reach and $V(x)$ is the magnitude of the mean velocity of the stream at any given point x.

A specific τ value is needed, for example, following a toxic spill or other discharge upstream from a water supply inlet, to estimate the amount of time before the pollutants make their way to the inlet [62]. It is also needed to estimate whether abiotic and biotic processes will degrade a biodegradable compound in sufficient time before reaching the inlet or the time needed to take remedial actions to prevent an oxygen sag in a stream.

A pollutant that is released into a stream forms a plume that moves with the flow. If the plume is less dense than the water, most will float atop the stream. However, most pollutants form a three-dimensional plume in the water. To estimate the plume's downstream travel, the mean velocity of the center of the plume is assumed to define the plume movement (see Fig. 10.5). At the point of entry to the stream (t_0) by a pulse of the pollutant begins the transport process, i.e., an instantaneous release of a substance uniformly throughout the stream's cross-section [62]. At successive times during the travel the plume continues advectively, shown as t_1, t_2, and t_3 in Fig. 10.5. The plume also widens, i.e., spreads in a perpendicular direction to the stream flow by mixing resulting from turbulent diffusion and dispersion from the nonuniform velocity across the stream's cross-section. Note that time (t) and distance (x) are proportional to each other. Travel time, τ, is the time between two plume mass center-points, so that the pollutant concentration [C] and τ can be calculated [62].

Velocity is a part of any pollutant transport, e.g., mixing rates after a pollutant is discharged to a stream, the rate at which an aquifer will become contaminated, and effectiveness of liners to retard the movement of landfill leachate. To measure surface water velocity, a velocity index can be related to mean velocity, and then multiplying that mean velocity by cross-sectional area. The index velocity may be measured at either a point or along a transect line, using hydrological monitoring equipment (see Figs. 10.6 and 10.7). The data are reported in digital and graphical formats (see Fig. 10.8). The measured velocity along a transect line usually is preferred to the point velocity because the former relates better to average stream velocity, so is a better index of mean velocity than is point velocity (see Fig. 10.9). The line velocity can be measured irrespective to flow direction [65]. Water velocity is commonly measured acoustically, such as using an acoustic Doppler current profiler (ADCP), one configuration of acoustic velocity meters (see Fig. 10.10).

With an understanding of velocity, the same concept can be described in three dimensions. The amount of water flowing in a stream (stream flow) is a common measurement around the world. The U.S. Geological Survey (USGS) maintains sites in hydrological regions, which are part of the standardized watershed classification system. Hydrological regions are comprised of hydrologic units, which are watershed boundaries organized in a nested hierarchy by sizes. The largest USGS units are regions. The finest units are local watersheds. Spatial datasets at scales for 1:2,000,000-scale hydrologic units (huc2m) and 1:250,000-scale hydrologic units are compiled by the National Water Quality Assessment (NAWQA) [66].

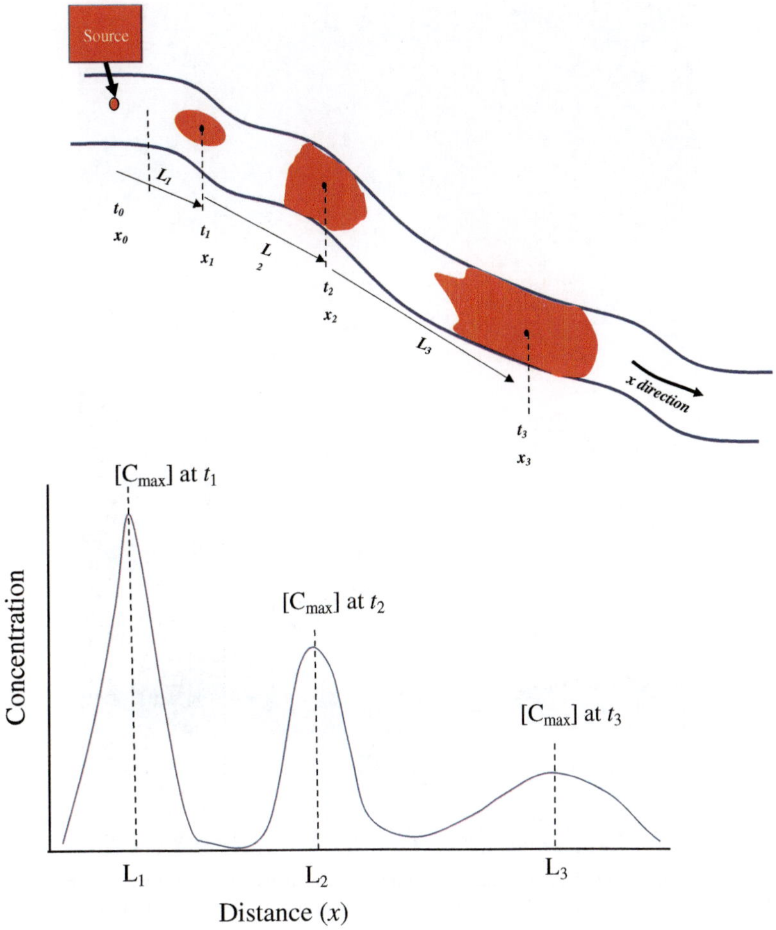

FIG. 10.5 Stream transport. Times (t) and distance (x) are at centers of mass. Lower panel shows the peak stream concentrations [C_{max}] at times t_1, t_2, and t_3. *(Data from H.F. Hemond, E.J. Fechner, Chemical Fate and Transport in the Environment, Elsevier, 2014.)*

Recall that stream flow or discharge (Q) in a channel section is velocity times area. The equations can be rearranged to find the total flux (J_{total}) of a pollutant across a stream cross-sectional area:

$$J_{total} = Q \times [C] \tag{10.14}$$

Where, $[C]$ is the mean pollutant concentration. Q is in units of volume per time (e.g., liters (L) s^{-1}) and concentration in this case is mass per volume (e.g., $\mu g\ L^{-1}$). Thus, J_{total} units are mass per area per time (e.g., $\mu g\ m^{-2}\ s^{-1}$). To be cost-effective, ratings are often used in flux measurements. These ratings are estimates of mean channel velocity and cross-sectional area by calibrating measurements with an *in situ* velocity, i.e., index velocity [67].

Most entities in the U.S. concerned with water pollution report Q in units of cubic feet of water flowing past a point each second (cfs). Discharge is calculated by measuring a stream's velocity at numerous points across the stream. Since water column heights and volume of water in a stream change constantly according to meteorological and other conditions, stream-stage/stream-discharge relationships are estimated by measuring stream discharge during different stream stages. The mean value of the flow measurements at all stage heights is reported as the estimated discharge [68].

The method most frequently used by the USGS for measuring discharge is the mechanical current-meter method. The stream channel cross-section is divided into vertical subsections (Fig. 10.11). Each subsection area is calculated by measuring its width and depth, and the stream velocity is measured using a current-meter. The discharge in each subsection is the product of the subsection area and the measured velocity. The total discharge is then calculated from the sum of each subsection's discharge [69].

FIG. 10.6 Streamgage station that provides hydrodynamic data, including mean velocity, of surface waters. *(From U.S. Geological Survey, USGS streamgaging network, USGS, Reston, Virginia. Available from: https://www.usgs.gov/mission-areas/water-resources/science/usgs-streamgaging-network, 2021. Photo Credit: Robert Swanson. Public domain.)*

FIG. 10.7 Streamgage station 01043500 on the Dead River, Maine after a heavy rainstorm saturated the soil, and a snowpack led to flooding in late December 2023. *(From U.S. Geological Survey, High flows on Dead River in Northwest Somerset, Maine, following flood event, USGS, Reston, Virginia. Available from: https://www.usgs.gov/media/images/high-flows-dead-river-northwest-somerset-maine-following-flood-event, 2023. Photo Credit: New England Water Science Center. Public domain.)*

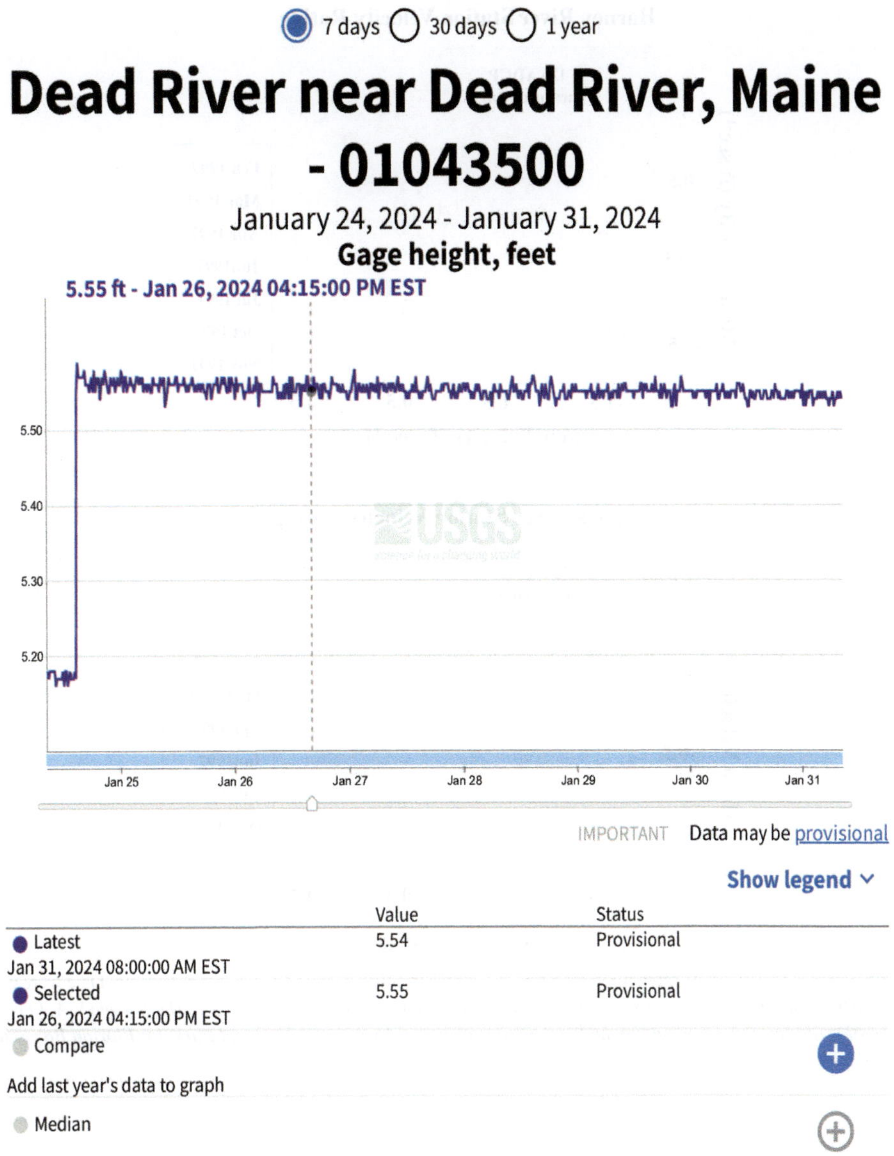

FIG. 10.8 Example of the data reported for streamgage station. Here, this shows that the Dead River gage height increased 0.35 ft on January 26, 2024 and remained at the elevated stage for at least a week. *(From U.S. Geological Survey, High flows on Dead River in Northwest Somerset, Maine, following flood event, USGS, Reston, Virginia. Available from: https://www.usgs.gov/media/images/high-flows-dead-river-northwest-somerset-maine-following-flood-event, 2023; Public domain.)*

Flow measurements taken from a subsection of a large stream can be used to estimate the discharge in the stream's cross-section. The calculation of discharge of the stream (Q) is the sum of the products of mean depth, mean width, and mean velocity [61,70]:

$$Q = \sum_{n=1}^{n} \frac{1}{2}(h_n + h_{n-1})(w_n + w_{n-1}) \times \frac{1}{2}(v_n + v_{n-1}) \tag{10.15}$$

Where, w_n is the nth distance from baseline or initial point of measurement (ft), h_n is the nth water depth (ft), and v_n is the nth velocity (ft s^{-1}) from velocity meter. Thus, Q is expressed in units of ft^3 s^{-1}. Or, if the values are meters instead of feet, Q would be expressed in m^3 s^{-1}.

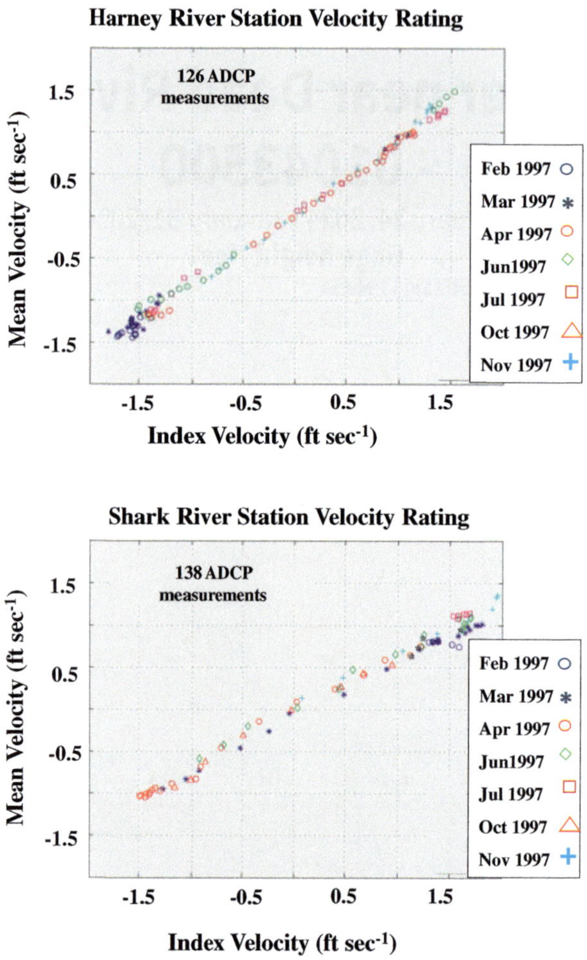

FIG. 10.9 Mean velocity versus index velocity from three monitoring stations in southwestern Florida estuary. *Note:* Data were collected using acoustic Doppler current profiler (ADCP) on fixed platforms to measure the detailed velocity profiles across the river. *(Data from V.A. Levesque, E. Patino (Eds.), Hydrodynamic characteristics of estuarine rivers along the South western coast of Everglades National Park, in: Florida Bay Science Conference, Key Largo, FL, April, 2001 (Abstract).)*

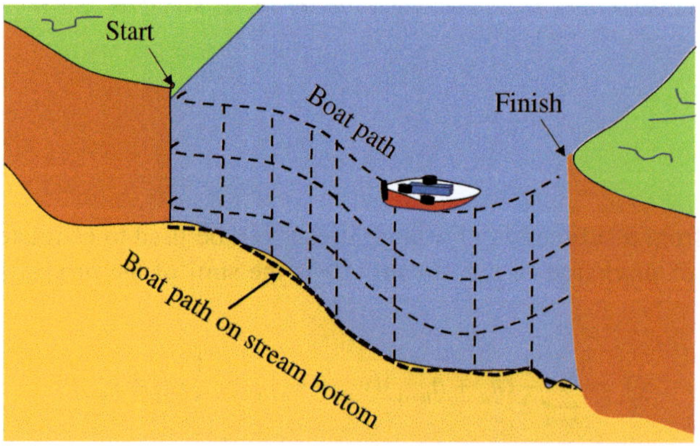

FIG. 10.10 Acoustic Doppler current profiler (ADCP) is mounted in a watercraft and used to measure stream discharge. The ADCP acoustic beams are pointed downward into the stream as the device is guided across the stream channel. *(Data from U.S. Geological Survey, How streamflow is measured—part 2: the discharge measurement. Available from: https://water.usgs.gov/edu/streamflow2.html, 2017.)*

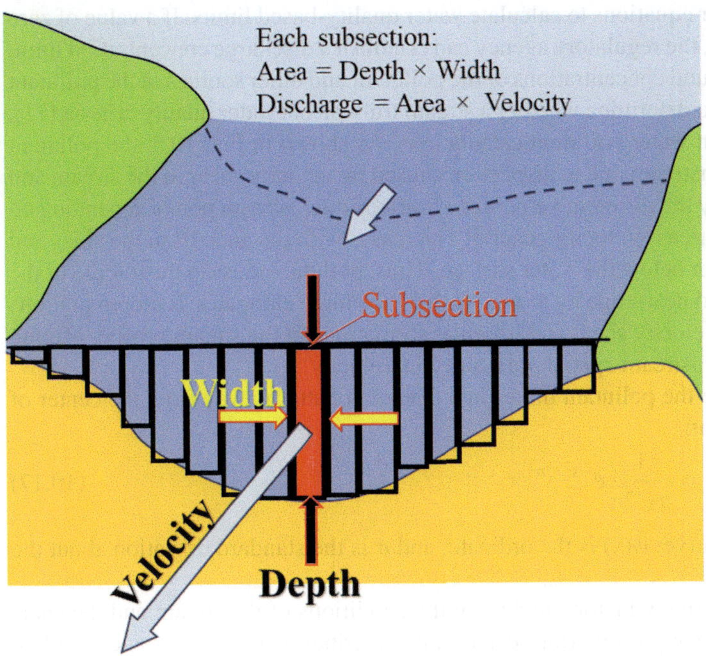

FIG. 10.11 Stream discharge measurement approach based upon stream flow measurements from subsections at various stream heights and the associated stream widths. Current-meters are used to measure discharge based on each subsection of the channel cross-section and summing these subsection discharges to arrive at total discharge. *(From U.S. Geological Survey, How streamflow is measured—part 2: the discharge measurement. Available from: https://water.usgs.gov/edu/streamflow2.html, 2017.)*

Stream discharge and velocity example

Wolfpack River's profile is being used to measure its flow. The profile is the river's width and depth. Measurements are taken from a boat traversing a 10-ft-wide stream at defined intervals and to measure the total depth and the velocity of the water at selected depths at each interval across the stream. Velocity measurements are taken at each point in the profile and each area of the measured interval is calculated. The water depth and velocity measurements obtained horizontally across the stream are 1, 3, 5, 7, and 9 ft. The water depth/velocity measurement obtained at a point 5 ft from the edge of the stream and at a total depth is 5 ft, the velocity readings are obtained at depths of a 1-ft and 4-ft box, representing an area that is midway between this measurement point and the measurement points on either side of the creek. The area is found to be 2 ft across and 5 ft high, or 10 ft^2.

The measured velocity at the top of the box is 2 ft s^{-1}, and 1 ft s^{-1} at the bottom of the box, so the box's average velocity is 1.5 ft s^{-1}. The volume of water flowing in the box each second is the product of box A and the velocity of the water:

$$2 \text{ ft wide} \times 5 \text{ ft high} = 10 \text{ ft}^2.$$

$$\therefore Q = 10 \text{ ft}^2 \times 1.5 \text{ ft s}^{-1} = 15 \text{ ft}^3 \text{ s}^{-1}.$$

The total stream discharge can be calculated using control cross-sectional areas between all the measurements and applying the average velocity of the water in every box. The total stream discharge is the sum of all the boxed areas.

This example is a simplification since a proper profile can only be drawn with many more measurement points averaged into the streamflow estimate. When an actual measurement is made, they should consist of about 20 points across the average-sized stream. Ideally, no single vertical cross-section will contain more than 5% of the total stream discharge. The USGS maintains flow monitoring sites across the nation. Data are recorded as frequently as daily and reported over varying lengths of time [69]. Therefore, stream discharge measurements can be compared to these generalized values.

The pollutant concentration in a stream (C_r) can be used to allocate waste loads discharged to surface waters and to calculate the concentration in the stream reach after complete mixing occurs [71]:

$$C_r = \frac{Q_d C_d + Q_s C_s}{Q_r} \tag{10.16}$$

Where: Q_d = waste discharge flow in units commonly used by U.S. environmental engineers, i.e., million gallons per day (MGD) or cubic feet per second (cfs); C_d = pollutant concentration in waste discharge in milligrams per liter (mg L^{-1}) [72]; Q_s = background stream flow in MGD or cfs above point of discharge; C_s = background in-stream pollutant concentration in mg L^{-1}; and, Q_r = resultant in-stream flow after discharge in MGD or cfs.

Government regulations use Eq. (10.16) and similar equations to calculate water quality-based limits. If a value of zero is used for the stream concentration for a pollutant (C_s), the regulatory agency can establish a discharge concentration limit. However, this would not account for upstream background concentrations of the pollutant and other sources of the pollutant like nonpoint runoff, so these would need to be added to determine whether a stream truly meets water quality criteria [71].

Mixing in a stream is an important consideration in water pollution calculations. As shown in Fig. 10.5, the pollutant load received by a stream spreads out as it travels downstream due to dispersion caused by velocity shear in the stream and turbulent diffusion [62]. The velocity shear is caused by the different velocities at various downstream points depending on the velocities at the stream's differing cross-sectional areas. Velocity generally increases with distance from the sides and bottom, with a maximum velocity near the stream center below the water surface. Thus, pollutant mass in the middle of the channel near the surface moves faster than in deeper and near-edge locations, so that the plume elongates. Random motions in the stream create the turbulent diffusion due to the velocity shear and irregular formations in the stream profile. Mixing intensifies as turbulent diffusion increases due to downstream spatial variability in velocity.

A method of estimating mixing is statistical, where the pollutant mixes in a Gaussian pattern away from the center of flow in a two-dimensional Cartesian coordinate system:

$$\phi(x) = \frac{1}{\sigma\sqrt{2\pi}} e^{-x^2/2\sigma^2} \tag{10.17}$$

Where, x is the abscissa of each point on the mixing curve, $\phi(x)$ is the ordinate, and σ is the standard deviation about the origin [62].

The extent and shape of plumes are determined by many factors, including the conditions of the stream and the characteristics of the transported substances. Assuming the simplest mixing scenario, the distribution of the substances will be random. Gaussian dispersion algorithms can be used to estimate current having low roughness and low relief. This means that the substances are assumed to be distributed randomly according to current vectors. In addition, standard deviations of the substances in three dimensions, i.e., along the x, y, and z axes, can be calculated to determine the location of the plume carrying these substances. For example, water column profiles and spatial characteristics affect mixing. As evidence, the mixing and dispersion of pollutants in a coastal zone are commonly dominated by turbulence induced by the fragmented wave-induced surfaces and beds that vary in space and time. For shallow waters in the coastal area, advective and diffusive processes govern pollutant transport in the two-dimensional Cartesian coordinate system [73].

The randomness of a pulse injection can be based on a Fickian mixing coefficient, D:

$$D = \frac{\sigma^2}{2t} \tag{10.18}$$

Where σ is the standard deviation of the pollutant distribution, σ^2 is the spatial variance, and t is the time since pulse of pollutants enters the stream. Ultimately, mixing of a substance C occurs within a defined, three-dimensional control volume and adheres to the first law of thermodynamics, expressed by the three major processes of stream flow in the advection-dispersion-reaction (ADR) equation:

$$\frac{dC}{dt} = -\mathrm{V}\cdot\frac{dC}{dx} + \frac{d}{dx}\left(D\cdot\frac{dC}{dx}\right) + r \tag{10.19}$$

Where, r represents all the chemical reactions in an infinitely small control volume and $\frac{dC}{dt}$ is the rate at which the pollutant changes at a fixed location in the flow.

Mixing models may be applied to more complex ecosystems characterized by vertical venting in forest areas, channeling down canyons, and both horizontal and vertical re-circulations that may occur at local sites. In fact, some of the recent computational models that have been developed for pollution in complex watersheds may be used in ecological risk assessments.

Another way of thinking about mixing is the overall dilution of a pollutant in the flow. Immediately following the discharge to a stream, the pollutant is progressively diluted until it reaches another source of the same pollutant during its travel. Beyond the immediate mixing zone, the dilution follows thermodynamic principles. This can be simplified to apply to a specific scenario, e.g., a worst-case of low flow. For example, the 7-day, 10-year low flow diluted pollutant concentration (C_d) can be estimated as [70]:

$$C_d = \frac{Q_s C_x + Q_w C_w}{Q_s + Q_w} \tag{10.20}$$

Where, Q_s (mg L^{-1}) is the stream flow upstream, of the effluent discharge entering the stream, C_s is the pollutant concentration of the upstream flow (ft^3 s^{-1}), Q_w is the effluent flow (ft^3 s^{-1}), and C_w is the pollutant concentration of the effluent (mg L^{-1}).

Pollutant dilution example

A hypothetical metal refinery pumps 25 $ft^3\ s^{-1}$ from Hermes Creek with a flow of 200 $ft^3\ s^{-1}$. The discharge from the refinery's ash pit is 22 $ft^3\ s^{-1}$. The total mercury (Hg) concentration in the upstream creek water is 0.05 mg L^{-1} and the total Hg in the effluent is 9.0 mg L^{-1}. Calculate the Hg concentration in Hermes Creek.

$$C_d = \frac{Q_s C_x + Q_w C_w}{Q_s + Q_w} = \frac{(200 - 25)(0.05) + (22)(9.0)}{(200 - 25) + 22} = 1.11 \text{ mg L}^{-1}.$$

The U.S. aquatic life criteria concentration for Hg is 1.4 μg L^{-1} for acute conditions and for 0.77 μg L^{-1} chronic conditions. These criteria are the highest concentration of water pollutants below which exposures are not expected to pose a significant risk to most species in a specific environment, in this case a freshwater stream [74]. So, Hermes Creek is more than three orders of magnitude above these water quality criteria, which means that the refinery needs to install technologies or take other actions to decrease these effluent concentrations.

5 Surface water quality

Water quality, as mentioned, is affected by various sources that are classified as either point sources or nonpoint sources. In a nonpoint source scenario, material in the runoff is collectively referred to as sediment, which when reaching high enough concentrations can substantially decrease the ability of light to penetrate the water column. The water's cloudiness or the amount of diminished light penetrations is known as turbidity. As a result, sediment loading diminishes the growth capacity of aquatic plants and disrupts normal processes, such as seasonal turnover and nutrient cycling in the surface waters. Like many other nations, the U.S. EPA sets drinking water standards for turbidity. For water supply systems that use conventional or direct filtration, turbidity must never be greater 1 nephelometric turbidity unit (NTU), and samples taken at the water treatment plant for turbidity must be $\leq$0.3 NTUs in at least 95% of the samples in any month. NTU correlates directly with the concentration of suspended particles in the water. Systems using filtration other than the conventional or direct filtration must follow state limits, i.e., turbidity at no time exceeding 5 NTUs.

Turbidity does not directly represent a health risk but indicates the potential presence of pathogens and pollutants in the water supply system. For example, pathogens can adhere to particles, which can reduce the efficiency of disinfection; elevated levels of turbidity in filtered water indicates inefficient pathogen removal in a water supply facility. Turbidity can also indicate the presence of breakage of supply mains if biofilms and oxide scales are sloughing or indicates the resulting entry of pollutants through these fissures [75]. In addition to water supply, turbidity is also one of the standards for surface water quality. Such standards are best tailored to local conditions. For example, studies conducted in Alaska and elsewhere indicated that increases 5 NPUs above ambient turbidity in clear cold-water habitats are relatively highly protective for salmonid fish resources in Alaska. However, stricter limits may be warranted to protect extremely clear waters, although may not be necessary to protect naturally turbid systems [76].

Household hazardous wastes, such as insecticides, pesticides, paint, solvents, used motor oil, and other auto fluids are also frequently part of the sediment load. These are often highly toxic to aquatic life and human populations.

Pathogenic microbes can be suspended or sorbed to suspended particles. Notably, *E. coli*, can cause major health problems and indicate the presence of enteric bacteria, i.e., those that grow in the intestines of animals, so they are also known as fecal coliform bacteria. This can be an indication of agricultural and human waste sources, e.g., septic tanks or even poorly treated effluent from municipal wastewater treatment systems.

Sediment loading can change the species composition and diversity of lakes and other surface waters. As mentioned, the increase input of nitrogen, phosphorus and other nutrients encourages the growth of algae, leading to algal blooms. When these algae die, they fall to the lake bottom. As they continue to decompose, more oxygen is demanded by the bacteria and abiotic processes, so the concentrations of dissolved oxygen (DO) decline. As a result, fish and other organisms at higher trophic levels in the food chain are stressed. This can result in fish kills and complete alterations to aquatic habitats.

5.1 Biochemical oxygen demand

Aquatic life depends on optimal ranges of dissolved molecular oxygen. Flowing water dissolves and releases oxygen constantly. Dissolved oxygen (DO) increases as air in the atmosphere mixes with the water by mechanical processes. Vascular plants and algae generate oxygen by photosynthesis. Fauna and bacteria use DO during respiration. Abiotic chemical reactions also deplete DO. Thus, the release of chemical pollutants, including nutrients, threatens to decrease DO concentrations below levels needed for various organisms. These thresholds vary by biological species. The DO sag curve represents a decline of DO with distance from a pollution source followed by an increase in DO concentrations. The DO sag is an indication of an impaired surface water ecosystem.

The biochemical oxygen demand (BOD) is the net amount of molecular oxygen that bacteria and other microbes will consume in the process of decomposing organic matter under aerobic conditions. The BOD is measured by incubating a sealed sample of water for 5 days and measuring the loss of oxygen by comparing the O_2 concentration of the sample at time = 0 (just before the sample is sealed) to the concentration at time = 5 days (known specifically as BOD_5). Samples are commonly diluted before incubation to prevent the bacteria from depleting all of the oxygen in the sample before the test is complete [77]. BOD_5 is merely the measured DO at the beginning time, i.e., the initial DO (D), which is measured immediately after it is taken from the source, minus the DO of the same water measured exactly 5 days after D_1, i.e., D_5:

$$BOD = \frac{D_1 - D_5}{P} \tag{10.21}$$

Where P = decimal volumetric fraction of water utilized. D units are in mg L^{-1}. If the dilution water is seeded, the calculation becomes:

$$BOD = \frac{(D_1 - D_5) - (B_1 - B_5)f}{P} \tag{10.22}$$

Where B_1 = initial DO of seed control; B_5 = final DO of seed control; and f = the ratio of seed in sample to seed in control $= \frac{\%\text{seed in } D_1}{\%\text{seed in } B_1}$. B units are in mg L^{-1}.

Biochemical demand example

Find the BOD_5 value for a 10 mL water sample added to 300 mL of dilution water with a measured DO of 7 mg L^{-1} and a measured DO of 4 mg L^{-1} 5 days later:

$$P = \frac{10}{300} = 0.03$$

$$BOD_5 = \frac{7 - 4}{0.03} = 100 \text{ mg L}^{-1}$$

Thus, the microbial population in this water is demanding 100 mg L^{-1} dissolved oxygen over the 5-day period. So, if a conventional municipal wastewater treatment system is achieving 95% treatment efficiency, the effluent discharged from this plant would be 5 mg L^{-1}.

Chemical oxygen demand (COD) does not differentiate between biologically available and inert organic matter, and it is a measure of the total quantity of oxygen required to oxidize all organic material completely to carbon dioxide and water. COD values always exceed BOD values for the same sample. COD (mg L^{-1}) is measured by oxidation using potassium dichromate ($K_2Cr_2O_7$) in the presence of sulfuric acid (H_2SO_4) and silver. By convention, 1 g of carbohydrate or 1 g of protein accounts for about 1 g of COD. On average, the ratio BOD:COD is 0.5. If the ratio is <0.3, the water sample likely contains elevated concentrations of recalcitrant organic compounds, i.e., compounds that resist biodegradation [78–80]. That is, there are numerous carbon-based compounds in the sample, but the microbial populations are not efficiently using them for carbon and energy sources. This is the advantage of having both BOD and COD measurements. Sometimes, however, COD measurements are conducted simply because they require only a few hours compared to the 5 days for BOD.

Since available carbon is a limiting factor, the carbonaceous BOD reaches a plateau, i.e., the ultimate carbonaceous BOD (see Fig. 10.12). However, carbonaceous compounds are not the only substances demanding oxygen. Microbial populations will continue to demand O_2 from the water to degrade other compounds, especially nitrogenous compounds, which account for the bump in the BOD curve. Thus, in addition to serving as an indication of the amount of molecular oxygen needed for biological treatment of the organic matter, BOD also provides a guide to sizing a treatment process, assign its efficiency and giving operators and regulators information about whether the facility is meeting its design criteria and is complying with pollution control permits.

If effluent with high BOD concentrations reaches surface waters, it may diminish DO to levels lethal to some fish and many aquatic insects. As the water body re-aerates because of mixing with air in the atmosphere and by algal photosynthesis, O_2 is added to the water causing the oxygen levels to increase gradually downstream. As mentioned, the drop and rise in DO concentrations downstream from a source of BOD is known as the DO sag curve, because the concentration of dissolved oxygen "sags" as the microbes deplete it. So, the falling O_2 concentrations fall with both time and distance from the point where the high BOD substances enter the water.

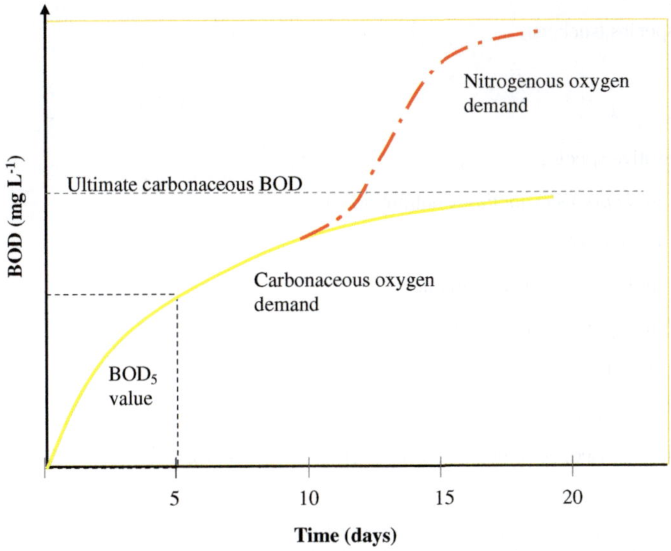

FIG. 10.12 Biochemical oxygen demand (BOD) curve showing ultimate carbonaceous BOD and nitrogenous BOD. *(Data from C. P. Gerba, I.L. Pepper, Wastewater treatment and biosolids reuse, in: Environmental Microbiology, second ed., Elsevier Inc, USA, 2009, pp. 503–530.)*

The stress from decreasing DO is usually indicated by the BOD. Like most environmental systems, the water bodies that receive sediment loads are complex in their response to increased input of materials. The DO will respond both positively and negatively to increased nutrient levels. Biota have unique optimal ranges of growth and metabolism that varies among species (e.g., algae will add some O_2 with photosynthesis, but use some O_2 for metabolism, whereas the bacteria will generally be net consumers of molecular oxygen).

The most widely applied environmental indices are those that follow the framework of an index of biological integrity. In biological systems, integrity is the capacity of a system to sustain a balanced and healthy community. This means the community of organisms in that system meets certain criteria for species composition, diversity and adaptability, often compared to a reference site that is used as a benchmark for integrity. Biological integrity indices integrate the relationships of chemical and physical parameters across levels of biological organization, from decomposers to top predator consumers. They assist scientists who evaluate the integrity of environmental systems using a range of metrics.

Environmental indices combine attributes to determine a system's condition (e.g., diversity and productivity) and to estimate stresses. The original index of biotic integrity developed by Karr [81] was based on fish fauna attributes and has provided predictions of how well a system will respond to a combination of stresses.[f] In fact, the index is completely biological, with no direct chemical measurements. However, the metrics (see Table 10.3) are indirect indicators of physicochemical factors (e.g., the abundance of game fish is directly related to dissolved oxygen concentrations). The metrics provide descriptions of a system's structure and function.

The information gleaned from these data is tailored to the physical, chemical, and biological conditions of an area, e.g., for large spatial regions. The information from a biologically based index can be used to evaluate a system, as shown in Fig. 10.13.

Systems involve scale and complexities in both biology and chemistry. For example, a fish's direct aqueous exposure (*AE* in μg day^{-1}) is the product of the organism's ventilation volume, i.e., the flow Q (in mL day^{-1}), and the compound's aqueous concentration, C_w (μg mL^{-1}). The fish's exposure by its diet (*DE*, in μg day^{-1}) is the product of its feeding rate, F_w (g wet weight day^{-1}), and the compound's concentration in the fish's prey, C_p (μg g^{-1} wet wt). If the fish's food consists of single type of prey that is at equilibrium with the water, the fish's aqueous and dietary exposures and the bioconcentration factor (*BC*) can be calculated when they are equal:

$$AE = DE; QC_w = F_wC_p; BCF = \frac{Q}{F_w} \tag{10.23}$$

The ventilation-to-feeding ratio for a 1 kg trout has been found [82,83] to be on the order of $10^{4.3}$ mL g^{-1}. Consider a study that finds that the quantitative structure activity relationship (QSAR) for the trout's prey indicates a bioconcentration factor

f. In more recent terminology, integrity indicates ecological resilience, i.e., the ability of a system to respond to a pollutant insult and return to its original condition.

TABLE 10.3 Biological metrics used in the original index of biological integrity (IBI).

Integrity aspect	Biological metric
Species richness and composition	Total number of fish species (total taxa)
	Number of *Catostomidae* species (suckers)
	Number of darter species
	Number of sunfish species
	Number of intolerant or sensitive species
Indicator species metrics	Percent of individuals that are *Lepomis cyanellus* (Centrarchidae)
	Percent of individuals that are omnivores
Trophic function metrics	Percent of individuals that are Insectivorous Cyprinidae
	Percent of individuals that are top carnivores or piscivores
	Percent of individuals that are hybrids
Reproductive function metrics	Abundance or catch per effort of fish
Abundance and condition metrics	Percent of individuals that are diseased, deformed, or that have eroded fins, lesions, or tumors (DELTs)

Data from J.R. Karr, Assessment of biotic integrity using fish communities, Fisheries 6 (6) (1981) 21–27. J.R. Karr, K.D. Fausch, P.L. Angermeier, P.R. Yant, I.J. Schlosser, Assessment of biological integrity in running waters: a method and its rationale, Illinois Natural History Survey, Champaign, Illinois, 1986.

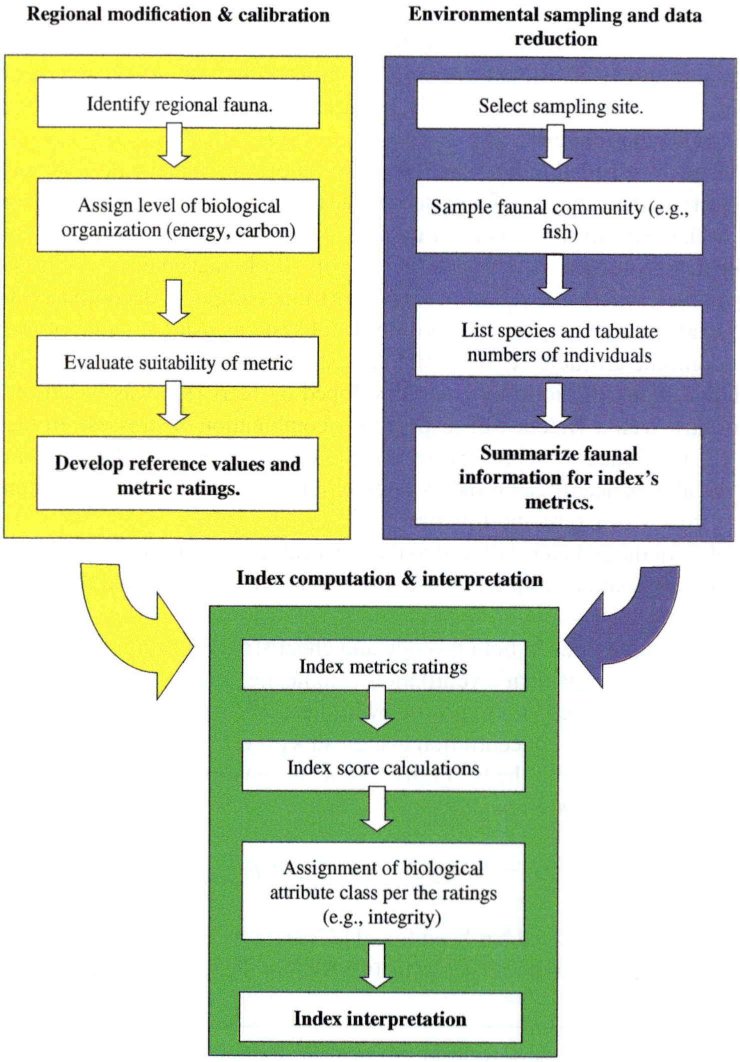

FIG. 10.13 Sequence of steps to calculate an Index of Biotic Integrity (IBI). *(Data from J.R. Karr, Assessment of biotic integrity using fish communities, Fisheries 6 (6) (1981) 21–27. J.R. Karr, K.D. Fausch, P.L. Angermeier, P.R. Yant, I.J. Schlosser, Assessment of biological integrity in running waters: a method and its rationale, Illinois Natural History Survey, Champaign, Illinois, 1986. M.C. Barber, Bioaccumulation and aquatic system simulator (BASS) user's manual beta test version 2.1, US Environmental Protection Agency, Athens, GA, 2001, p. 76. M.T. Barbour, Rapid bioassessment protocols for use in wadeable streams and rivers: periphyton, benthic macroinvertebrates and fish, US Environmental Protection Agency, Office of Water, 1999.)*

(*BCF*) = 0.048 times the octanol-water coefficient (K_{ow}). In this case, it appears that the trout's predominant route of exposure for any chemical with a $K_{ow} > 10^{5.6}$. Exposure must also account for the organism's assimilation of compounds in food, which for very lipophilic compounds, will probably account for most exposures compared to that from the water. Even though chemical exchange occurs from both food and water via passive diffusion (Fick's Law relationships, discussed in Chapter 7), the uptake from food, unlike direct uptake from water, does not necessarily relax the diffusion gradient into the fish. The difference between digestion and assimilation of food can result in higher contaminant concentrations in the fish's gut, i.e., an increase in the pollutant body burden in the fish.

Predicting expected uptake where the principal route of exchange is dietary can be further complicated by the fact that most fish species exhibit well-defined size-dependent, taxonomic, and temporal trends regarding their prey. Thus, a single *BCF* or bioaccumulation factor (*BAF*) may not universally be useful for risk assessments for all fish species.[g] Indeed, the *BAF* may not even apply to different sizes of the same species.

The systematic biological exchange of materials between the organism, in this case various species of fishes, is known as uptake, which can be expressed by the following three differential equations for each age class or cohort of fish [84]:

$$\frac{dB_f}{d_t} = J_g + J_i + J_{bt} \tag{10.24}$$

Where, B_f is the chemical compound body burden (μg fish^{-1}), J_g represents the net chemical exchange (μg day^{-1}) across the fish's gills from the water; J_i represents the net chemical exchange (μg day^{-1}) across the fish's intestine from food; and J_{bt} represents the compound's biotransformation rate (μg day^{-1}).

$$\frac{dW_d}{d_t} = F_d - E_d - R - EX - SDA \tag{10.25}$$

Where, a W_d = dry body weight (g dry wt fish^{-1}) of the average individual within the cohort; F_d = the fish's feeding; E_d = egestion (i.e., expulsion of undigested material); R = routine respiration; EX = excretion; and SDA = specific dynamic action (i.e., the respiratory expenditure exceeds the R required to assimilate food). All these parameters have units of g dry-weight day^{-1}. N is the cohort's population density (fish ha^{-1}).

$$\frac{dN}{d_t} = EM - NM - PM \tag{10.26}$$

Where, N is the cohort's population density (fish ha^{-1}); EM is the emigration/dispersal rate; NM is the nonpredatory mortality rate: and PM is the predatory mortality rate.

Physiologically based models for fish growth are often formulated in terms of energy content and flow (e.g., kcal fish^{-1} and kcal day^{-1}), Eq. (10.26) is basically the same as such bioenergetic models because energy densities of fish depend on their dry weight [85]. Obviously, feeding depends on the availability of suitable prey, so the mortality of the fish is a function of the individual feeding levels and population densities of its predators. Thus, the fish's dietary exposure is directly related to the organism's feeding rate and the concentrations chemicals in its prey.

5.2 Point and nonpoint sources of sediment

The amount of any substance (e.g., nutrient, pollutant, or microbes) that is discharged into a system is known as the load. Nutrient loading, pollutant loading, and microorganism loading can be quantified, not only by the gross mass of the substance entering a system, e.g., a water body, but also by the response of that system to the load. Thus, there is a direct relationship between a substance's entry and its impact on the receiving river, lake, wetland, estuary, ocean, or aquifer. However, these recipients of the pollutant can be further compartmentalized into the various media within them, e.g., soil, sediment, water, and biota.

The first major distinction of loads is between point sources and nonpoint sources. This is a distinction of convenience. What some may call a point source, others may consider to be a nonpoint source. For example, if a 1-ha field is near a small creek, this may be a relatively large and dispersed source of a pesticide being released to the creek. However, a 1-ha trickling filter system near a large river may be classified as a point source since most of the discharge to the river

g. For a water pollutant, the *BCF* is the concentration of a substance in an organism divided by the concentration of the substance in water. The *BAF* is the pollutant concentration in an organism from all routes of exposure especially from the water and diet. The biomagnification factor (*BMF*) is the concentration of a substance in a predator divided by the concentration in the predator's prey and other food. Thus, *BAF* results from both *BCF* and *BMF*.

comes through an outfall structure, i.e., a conduit that directly releases effluent from the trickling filter to the creek. However, it is highly likely that pollutants are being released from the treatment facility from sources other than the outfall structure.

Thus, nonpoint sources are generally associated with runoff, i.e., multiple sources in each two-dimensional space that leave these sources and are transported toward a receptor system. These receiving bodies can be above ground (e.g., streams and lakes) or below ground (aquifers). They are usually cascades of systems, such as a contaminant moving atop the soil, with some infiltration. The infiltrated fraction may find its way to the aquifer where the pollutant moves much more slowly than in the runoff above the ground. However, the contaminated ground water may recharge a stream flow. The same stream may have already been contaminated by the overland flow, so the stream is receiving a continuous flow (ground water source to the stream) and a plug or episodic flow (overland runoff). Thus, the net that reaches the water body is a function of numerous factors, including vegetative cover, soil porosity and permeability, ground water flow rates, soil texture, Henry's Law constant (See Chapter 5), and other equilibrium coefficients for the chemical compounds in the released material.

For biota, especially bacteria, the nonpoint and point contribution is even more complex than that of chemical compounds. Microbes may be directly loaded into a stream or other receiving body. In addition, they may accumulate as nonpoint source loads from a combination of sources. As they are transported overland and under the soil surface, the microbes meet hostile and accommodating conditions, which will decrease and increase their numbers, respectively. However, the hostile conditions may lead to the formation of cysts and spores, which are more likely than the bacteria themselves to be transported by advection in the air and water or carried by fauna. When the conditions change to become more accommodating, the spores will generate increased microbial populations. Thus, the actual loading to the water body consists of a complicated set of transport and fate processes.

Loading calculations must factor in background sources, sometimes referred to as "natural" sources. Atmospheric loading is similar, but usually the distinctions are made between mobile sources and stationary sources. Two systems, hydrologic and atmospheric, join at the interfaces between the earth's surface and water bodies and the atmospheric, i.e., terrestrial, and tropospheric fluxes. For example, atmospheric deposition is an important source of nutrient loading (e.g., N, P, S and K) to ecosystems. Ecosystems, e.g., wetlands, load nutrients back into the atmosphere, but often as different chemical species than were deposited. That is, the ecosystems are acting as control volumes in which reactions are taking place. Many of these reactions are biotic and are mediated by microbes and plants.

The linkages between biota and their environments are influenced by the availability and forms of nutrients. When rain containing nitrates falls to the wetland water and land surfaces, for instance, plant roots take it up and metabolize it into organic forms of nitrogen. Meanwhile, bacteria in sediments reduce it to ammonia. In the opposite direction, the reduced forms are oxidized by other bacteria (e.g., *Nitrosomonas* converts ammonium ions [NH_4^+] to nitrite [NO_2^-], *Nitrobacter* converts the NO_2^- to nitrate [NO_3^-]). Thus, two opposite reactions are at play constantly in loading scenarios. In this instance, these are ammonification (or deamination) and nitrification:

$$\text{Ammonification: } NH_4^+ + OH^- \Leftrightarrow NH_3 + H_2O \tag{10.27}$$

$$\text{Nitrification: } NH_4^+ + 2O_2 \rightarrow NO_3^- + 2H^+ + H_2O \tag{10.28}$$

Ammonification is written as an equilibrium reaction, and nitrification as an oxidation reaction. However, the loading depends on numerous environmental conditions, including pH. Thus, under acidic conditions, most of the ammonia nitrogen will ionize to ammonium and under basic conditions the nonionized ammonia concentrations will increase in proportion to the ammonium. This is important since nonionized ammonia is very toxic to aquatic biota, whereas ionized species are nutrients needed for plants and algae. Add to this, the decomposition of organic matter, and the loading becomes a complicated mix of reactions:

$$\text{Organic nitrogen} + O_2 \rightarrow NH_3 \text{ nitrogen} + O_2 \rightarrow NO_2^- \text{ nitrogen} + O_2 \rightarrow NO_3^- \text{ nitrogen} + O_2 \tag{10.29}$$

Even this is an oversimplification since the oxidation and reduction depend on the types of bacteria present. For example, the O_2 required for nitrification is theoretically 4.56 mg O_2 mg^{-1} NH_4^+, but this is an autotrophic reaction. Thus, O_2 is being produced by the nitrifying bacteria, decreasing the amount needed. However, the growth rate of nitrifying microbes is far less than that of heterotrophic microbes decomposing organic wastes. Therefore, when large amounts of organic matter are being degraded, the nitrifiers' growth rate will be sharply limited by the heterotrophs, which means the rate of nitrification will commensurately be decreased [86–89]. Furthermore, these same responses to conditions exist for all other nutrients in the system, e.g., organic phosphorous (P) will be oxidized and oxidized P species reduced, etc.. Thus, the kinetics of nutrient loading are quite complex.

5.3 Total maximum daily loading

In the United States, the Clean Water Act (CWA) restricts the total maximum daily load (TMDL) of pollutants released to water bodies that have been deemed to be impaired by the state in which technology-based and other engineering controls are not adequate to achieve water quality standards. Thus, the TMDL reflects the amount of a pollutant that can be discharged from point, nonpoint, and natural background sources, including a margin of safety (MOS) for any water quality-limited water body. The TMDL process consists of five steps:

1. Selection of the pollutant in need of consideration;
2. Estimation of the water body assimilative capacity (i.e., loading capacity);
3. Estimation of the pollutant loading from all sources to the water body;
4. Analysis of current pollutant load and determination of needed reductions to meet assimilative capacity; and,
5. Allocation, including the MOS, of the allowable pollutant load among the different pollutant sources in a manner that water quality standards are achieved.

A beneficial use that has been impaired is a function of a change in the chemical, physical or biological integrity of surface waters, such as:

- restrictions on fish and wildlife consumption
- tainting of fish and wildlife flavor
- degradation of fish wildlife populations
- fish tumors or other deformities
- bird or animal deformities or reproduction problems
- degradation of benthos
- restrictions on dredging activities
- eutrophication or undesirable algae
- restrictions on drinking water consumption, or taste and odor problems
- beach closings
- degradation of aesthetics
- added costs to agriculture or industry
- degradation of phytoplankton and zooplankton populations
- loss of fish and wildlife habitat.

Meeting water quality standards is a way to protect beneficial uses, including recreational activities, aquatic life, and water supply. For example, recreational beneficial use can be based on protection against fecal coliform, including *E. coli*, bacteriological criteria. For example, if a lake is designated for primary contact recreation, potential exposures to bacteria and a baseline level of disinfection will be required. This can be more straightforward for point sources, e.g., building wastewater treatment systems, than for nonpoint sources. The latter often requires management approaches, such as measures to prevent soil erosion since soil may carry coliform bacteria.

Let us consider two examples of impairments: the Cuyahoga River basin in northeast Ohio [90] and the Rock Creek basin in Maryland [91]. The Cuyahoga area of concern embodies the lower 45 miles of the river from the Ohio Edison Dam to the river's mouth, along with 10 miles of Lake Erie shoreline, and includes 22 miles of urbanized stream between Akron and Cleveland (see Fig. 10.14). The environmental degradation resulted from nutrient loading, toxic substances (including polychlorinated biphenyls [PCBs] and heavy metals), bacterial contamination, habitat change and loss, and sedimentation. Sources for these contaminants include municipal and industrial discharges, bank erosion, commercial/residential development, atmospheric deposition, hazardous waste disposal sites, urban stormwater runoff, combined sewer overflows (CSOs), and wastewater treatment plant bypasses.

In 1994, an advisory about eating fish was issued for Lake Erie and the Cuyahoga River due to elevated polychlorinated biphenyl (PCB) levels in fish tissue. The advisory restricted the consumption of white sucker, carp, brown bullhead, and yellow bullhead in the Cuyahoga River area of concern, and walleye, freshwater drum, carp, steelhead trout, white perch, Coho salmon, Chinook salmon, small mouth bass, white bass, channel catfish and lake trout in Lake Erie. Beginning at the Ohio Edison Gorge and extending downstream to Lake Erie, measures of fish population conditions ranged from fair to very poor and were below applicable Ohio warm water habitat aquatic life use criteria. Although fish communities have recovered significantly compared to the historically depleted segments of the Cuyahoga River, pollution-tolerant species continued to compose the dominant fish population. Wildlife and fish populations were impaired. Anecdotal information indicates some recovery of Great Blue Heron nesting in the Cuyahoga River watershed. Resident populations of black-crowned night herons

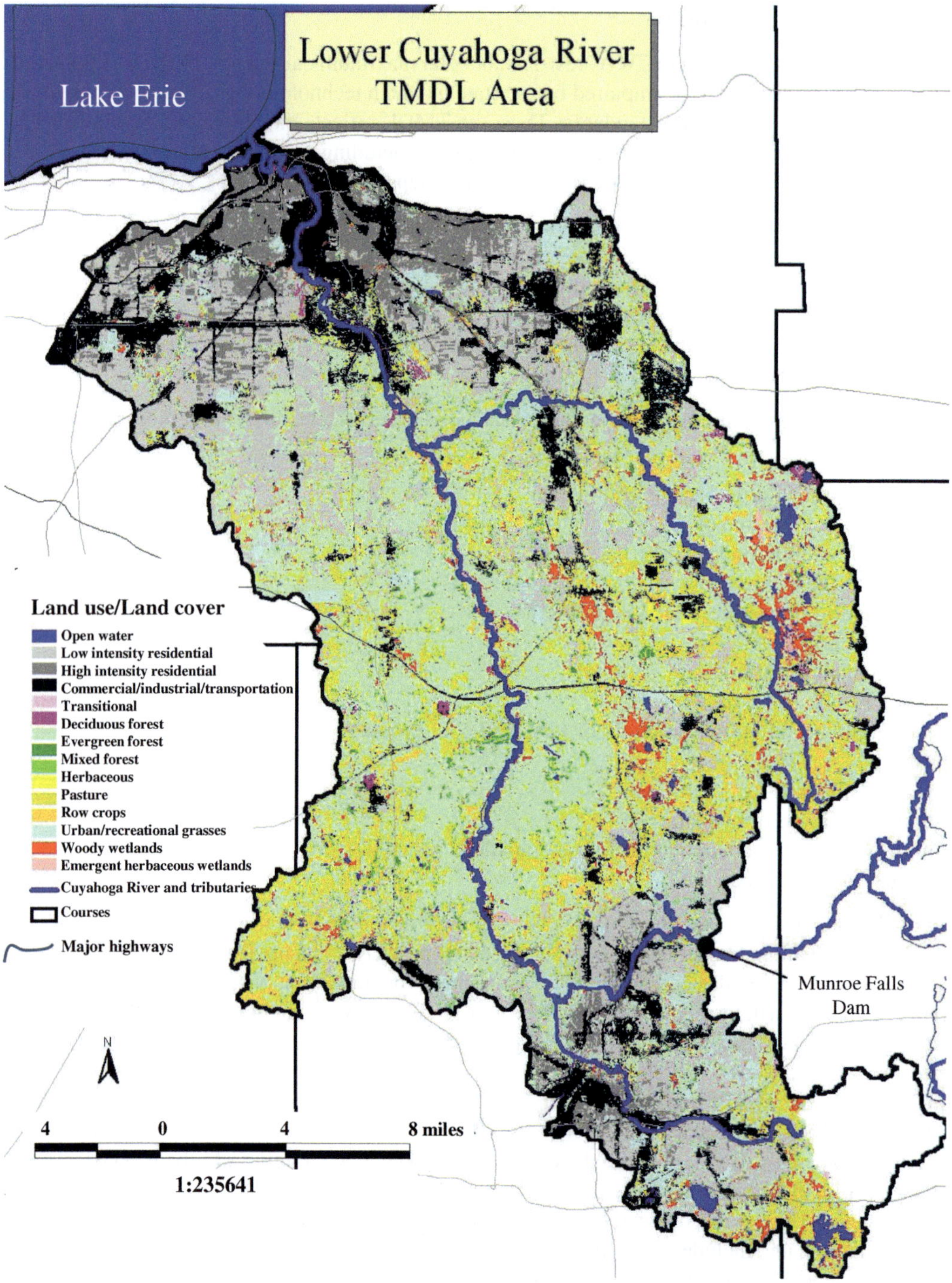

FIG. 10.14 Land use/land cover map of the Cuyahoga River basin in northeast Ohio. Drainage is northward to Lake Erie. Approximately 38% of the area is tree covered, 28% of is residential, 15% agricultural, 12% industrial and 3% wetlands. *(Data from Maryland Department of the Environment, Total maximum daily loads of fecal bacteria for the non-Tidal Rock Creek Basin in Montgomery County, Maryland, U.S. Environmental Protection Agency, Region III, Philadelphia, Pennsylvania. Available from: https://mde.maryland.gov/programs/Water/TMDL/DocLib_RockCreek_02140206/Rock_Creek_TMDL_060607_final.pdf, 2007.)* Map lines delineate study areas and do not necessarily depict accepted national boundaries.

had been noted in the navigation channel. Although the incidence of deformities, eroded fins, lesions, and external tumors (DELT anomalies) have declined throughout the watershed, significant impairments continued to be found in the headwaters to the nearshore areas of Lake Erie when the TMDL study was completed.

Macroinvertebrate populations living at or near the bottom (i.e., benthic organisms) of the Cuyahoga River remained impaired at certain locations, however, there are indications of substantial recovery, ranging from good to marginally good throughout most free-flowing sections of the river. Some fair and even poor designations were still seen, however.

The U.S. EPA restricts disposal of dredged sediment in most of the Cuyahoga basin due to high concentrations of heavy metals. Only a small amount of the dredged material contaminated sediments is transported and disposed of in a confined disposal facility in the Cleveland area.

The Cuyahoga navigation channel seemed to be impaired due to extreme oxygen depletion during summer months. The oxygen demand of sediment oxygen is a contributing factor. Polluted aquifers and surface waters may still be sources for individual supplies and wells. High bacterial counts following rain events periodically adversely affected the two beaches in the area of concern. Swimming advisories were issued after a storm, or if microbial counts exceed certain thresholds. According to some studies, phytoplankton populations in the river were impaired.

Channelization, nonexistent riparian cover, silt, bank reinforcement with concrete and sheet piling, alterations of littoral areas and shorelines, and dredging have been contributors to the impairment of fish and wildlife in the area of concern.

Complying with water quality standards in Ohio requires meeting aims at protecting various beneficial uses including recreational activities, aquatic life, and water supply. Ohio's recreational beneficial use attainment is based on fecal coliform or *E. coli* bacteriological criteria. Much of the lower Cuyahoga River is designated for primary contact recreation water where there is an intermediate potential exposure to bacteria and a baseline level of disinfection required.

The components needed to develop a TMDL plan are shown in Fig. 10.15. The microbial TMDL is expressed as most probable number (MPN) per day and is based on meeting the instream long-term geometric mean of enterococci bacteria. To determine the MPN, a set of tubes containing enriched broth are inoculated with different amounts of the water sample and incubated at a specific temperature for a predetermined time period [92–94]. If gas is produced in the tube, a sample of

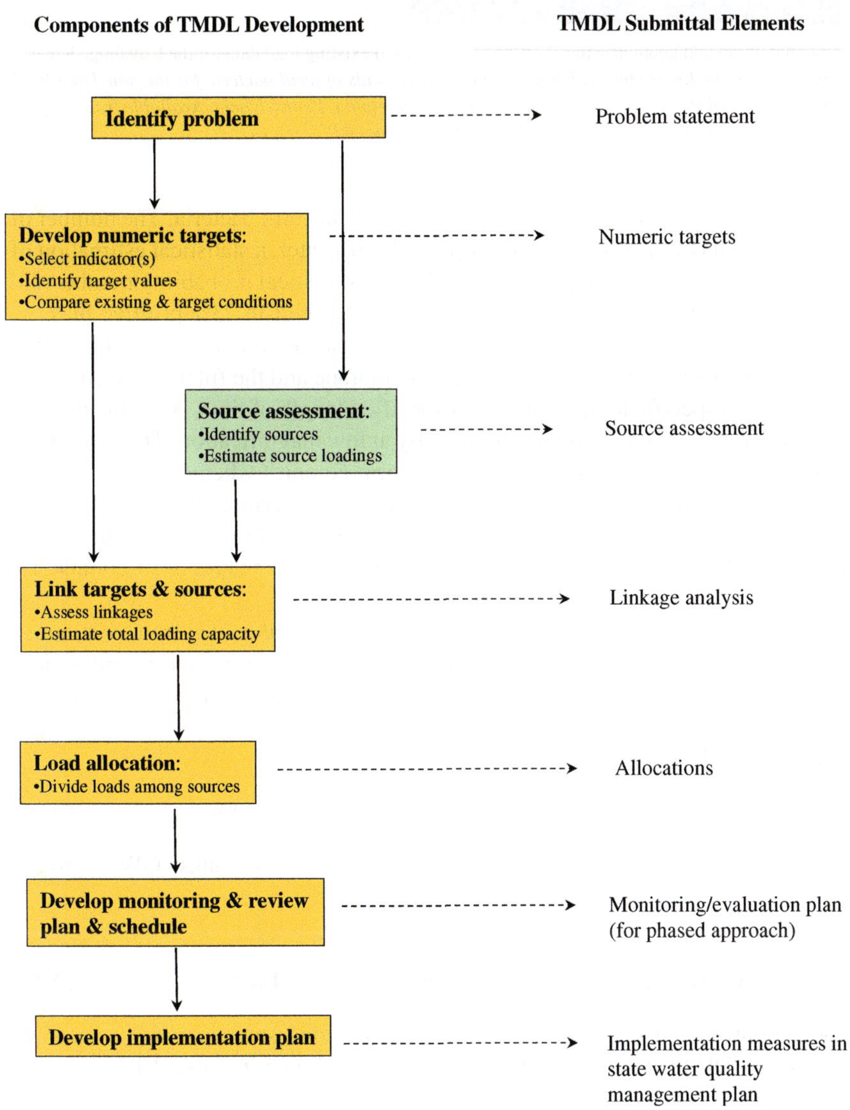

FIG. 10.15 Steps in developing a total maximum daily load (TMDL). *(Data from U.S. Environmental Protection Agency, Protocol for developing pathogen TMDLs, in: Report No.: EPA 841-R-00-002, 2001.)*

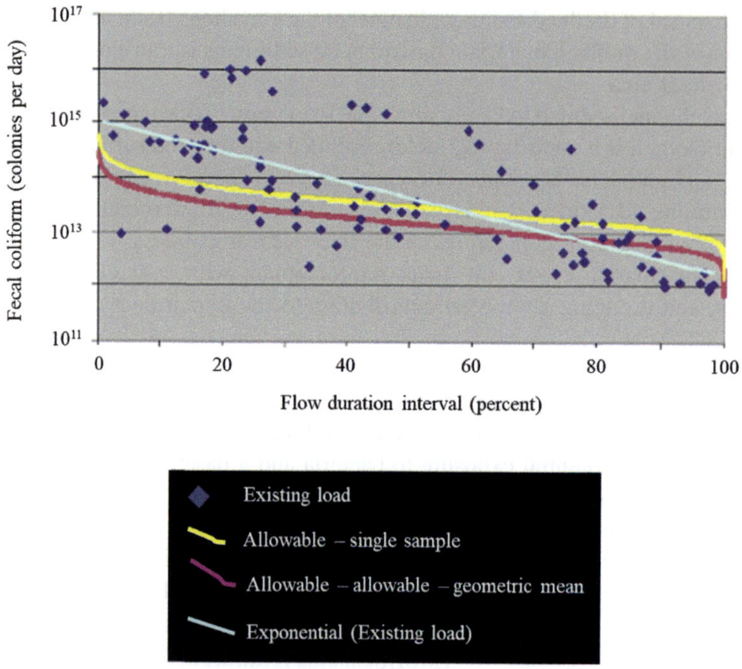

FIG. 10.16 Fecal coliform TMDLs for a single sample maximum and geometric mean criteria compared to existing load data for the Cuyahoga River at Independence gage station. *(Data from Maryland Department of the Environment, Total maximum daily loads of fecal bacteria for the non-Tidal Rock Creek Basin in Montgomery County, Maryland, U.S. Environmental Protection Agency, Region III, Philadelphia, Pennsylvania. Available from: https://mde.maryland.gov/programs/Water/TMDL/DocLib_RockCreek_02140206/Rock_Creek_TMDL_060607_final.pdf, 2007.)*

the bacteria in the broth is transferred to additional media to confirm the presence of fecal coliform bacteria. The number of tubes producing gas are converted to express the results of the test as the MPN per 100 mL water, a statistical estimation of the number of bacteria that would give the results shown by the laboratory test. This is a statistical probability number and not an actual empirical result. The test may give higher than actual results because of the built-in 23% positive bias.

The membrane filter technique is an alternate means of expressing MPN, using a measured amount of sample and filtering it through a membrane (pore size = 0.45 μm). Microbes are retained on the membrane and the filter is placed on the surface of a selective agar medium and incubated at a specific temperature for a specified length of time. After incubation, colonies that have formed by the growth of the microbes are counted microscopically at low magnification. The membrane filter technique thus provides an estimate of the number of coliform bacteria that form colonies when cultured (colony-forming units or CFU per 100 mL). Because colonies could be formed from more than one bacterium, the count is merely an estimate. The membrane filtration procedure is generally used in the U.S. for pathogens released to waterways because it is faster and more precise than the MPN technique; however, many local agencies may avoid it because of its complexity and the need to procure costly expertise to interpret the findings.

The TMDL for the Cuyahoga basin was determined using a load duration curve, which was calculated from continuous flow data from each gage site. It is a cumulative frequency distribution of the daily mean flow over a gage record period. Next, the load duration curve is generated from the product of flow duration curve and the applicable water quality criterion (see Fig. 10.16).

The quantification of the existing load per source is calculated as the in-stream load:

$$\text{Instream load} = \sum \text{WWTP} + \sum \text{Reservoirs} \sum \text{CSO} + \sum \text{ST} + \text{GW} + \text{RO} - \text{Loss} \tag{10.30}$$

Where, WWTP = wastewater treatment plants; CSO = combined sewer overflows; ST = septic tanks; GW = ground water; and RO = runoff.

Thus, the general description of the TMDL is:

$$\text{TMDL} = \sum \text{WLA} + \sum \text{LA} + \text{Background} + \text{Future Growth} + \text{MOS} - \text{Loss} \tag{10.31}$$

Where, WLA = waste load allocation of point sources and LA = load allocation for nonpoint sources.

The pollutant reductions required under the TMDL for the Cuyahoga River basin are shown in Table 10.4.

TABLE 10.4 Source category reductions at Independence, Ohio needed to meet the TMDL for fecal coliform and phosphorus.

	Fecal coliform			Total phosphorus		
Source	Existing Ave load (cfu/year)	Allocated Ave load (cfu/year)	% Reduction	Existing Ave load (lb/year)	Allocated Ave load (lb/year)	% Reduction
Runoff	3.47E+18	1.42E+18	59%	219,716	113,780	48%
Point sources	9.70E+13	9.70E+13	0%	170,580	120,101	30%
Akron CSOs & bypass	5.26E+16	1.05E+15	98%	34,629	4635	87%[a]
Septic	2.27E+15	1.45E+15	36%	28,831	20,181	30%
Lake Rockwell release	3.22E+13	3.22E+13	0%	22,043	15,847	28%
Ground water	7.27E+13	7.27E+13	0%	12,924	12,924	0%

[a]*The % reduction for TP associated with the Akron CSOs is expected only. The TP removal is incidental to the treatment methods proposed by Akron to treat for fecal coliform.*
Data from Ohio Environmental Protection Agency, Total maximum daily loads for the lower Cuyahoga River: final report, Ohio Environmental Protection Agency, Division of Surface Water, Columbus, Ohio, 2003.

The U.S. EPA requires that total maximum daily loads [91]:

1. Be designed to implement applicable water quality standards.
2. Include a total allowable load as well as individual waste load allocations (WLAs) and load allocations (LAs).
3. Consider the impacts of background pollutant contributions.
4. Consider critical environmental conditions.
5. Consider seasonal environmental variations.
6. Include a margin of safety to account of uncertainty about the relationship between the pollutant and the quality of the receiving waterbody.
7. Provide reasonable assurance that the TMDLs can be met.
8. Include public participation.

The total loads provide the sum of the individual waste load allocations (WLAs) for point sources and municipal separate stormwater systems (MS4) and the land-based load allocations for nonpoint sources. The TMDLs for fecal bacteria for Rock Creek are incorporated into the State's water quality management plan. Thus, the general description of the TMDL is:

$$\text{TMDL} = \sum \text{WLA} + \sum \text{LA} + \text{MOS} \tag{10.32}$$

Where, WLA is the waste load allocations for point sources in the basin; LA is the load allocation for nonpoint sources; and MOS is the added margin of safety.

The benchmarks of success are the water quality criteria. For example, in Maryland, *E. coli* must not exceed 125 MPN/100 mL steady state geometric mean indicator density; enterococci should not exceed 33 MPN/100 mL in fresh waters and 35 MPN/100 mL in marine water. The steady state geometric mean was calculated from samples collected during steady state flows with at least five representative sampling events. These are shown in Table 10.5. In addition to these sources, the county built a treatment facility that is allowed to release 60 billion MPN/day according to the provisions of a Clean Water Act (National Pollution Discharge Elimination System) permit. Much of the LA for nonpoint sources will be achieved through best management practices (BMPs).

Neither of these example TMDLs specifically identified animal feeding operations (AFOs) as point sources in the WLA fraction. This is revealing since both basins have substantial land use dedicated to agricultural operations. Perhaps, this fraction is included in the LA, but all allocations from AFOs must be more explicitly identified and calculated, especially when fecal coliform and other microbial populations are part of the contaminants of concern.

TABLE 10.5 Fecal bacteria summary for Rock Creek, Maryland TMDL allocations.

Subwatershed	Baseline	TMDL	WLA-WWTP	WLA-MS4	LA	MOS
	Billion MPN/100 mL enterococci/day					
NBR0002us	1786	37	0	13	24	5% Explicit
RCM0235us	497	32	0	12	20	5% Explicit
RC0111sub	1672	56	0	35	21	5% Explicit
Total	3955	125	0	60	65	

MS4 = municipal separate storm sewer system; MPN = most probable number; WLA-PS = wasteload allocation for non-MS4 systems (municipal or industrial); WLA-MS4 = wasteload allocation for MS4 systems.
Data from Ohio Environmental Protection Agency, Total maximum daily loads for the lower Cuyahoga River: final report, Ohio Environmental Protection Agency, Division of Surface Water, Columbus, Ohio, 2003.

After entering the water, microbial populations degrade and often detoxify water pollutants, but under certain conditions can worsen pollution. Bacteria, for example, can break down chemical pollutants to gain energy. Conversely, conditions in water bodies can lead to the excessive growth of pathogens. Since each water body is a thermodynamic control volume, it has fluids in gas and liquid phases carrying abiotic and biotic mass that enters the volume. Abiotic and biochemical reactions occur within the volume, and converted mass exits the volume.

For chemical pollutants, the exiting mass will be a combination of the parent chemical compound and degradation products. For microbes, the exiting mass will consist of net cellular biomass, i.e., the difference between the initial microbial population and net cell growth. Microbial populations undergo stages of growth and decrease [95].

Microbial population growth begins with a lag stage, followed by exponential growth until the microbial population levels off during the stationary stage, after which decay begins. Ultimately, the decay becomes exponential [8]. The duration of each stage varies for a given species and under different environmental conditions. For example, in the artificial environment created by an activated sludge wastewater treatment plant will prevent the decay stage to become exponential as new food and new bacteria are added to the microbial population [96–100]. The same is true for ground water *in situ* treatment, when biostimulation and bioaugmentation approaches are deployed. Biostimulation involves adding of nitrogen and phosphorus, oxygen, and other elements and compounds that serve as electron acceptors to stimulate the activity of microbial populations. [95]. Bioaugmentation is the process of adding microbes to the subsurface environment. The microbes are extracted from the site's contaminated soil, sediment, or water and mixed with the mentioned elements and compounds. Biostimulation is often favored over bioaugmentation because the subsurface usually abounds with indigenous species that have already developed enzyme production systems to break down the contaminants, which often have a better survival rates than those of nonindigenous microbes [101].

The microbial biomass exiting a control volume may increase or decrease, depending on the acceptability and hostility of the control volume. In a hostile control volume, the microbial population dies off comparatively rapidly so that exiting biomass is zero. The Monod equation can be used to calculate and express the rate of microbial biomass:

$$\mu = \frac{\mu_{max} S}{K_S + S} \tag{10.33}$$

where μ = the specific growth rate of the microbe; $\mu_{\max}$ = maximum specific growth rate; and K_s = the Monod growth rate coefficient representing the substrate concentration at which the growth rate is half the maximum rate. The $\mu_{\max}$ is reached at the higher ranges of substrate concentrations. K_s is an expression of the affinity of the microbe for a nutrient, i.e., as K_s decreases the more affinity that microbe has for that nutrient (as expressed by the concomitantly increasing μ). Thus, wastewater treatment must bring the waste into contact with the microbes and enhance the degradation environment, such as by activated sludge. Engineered systems may also add activated microbes when indigenous microbes are not already efficiently breaking down the chemical compound.

6 Conclusions

Surface waters can be polluted by many types of point sources and nonpoint sources. The pollution can be chemical or biological. This chapter has discussed the processes that lead to and exacerbate surface water pollution, including pollutant transport, which is determined by hydraulic and hydrological processes. Surfaces waters interact with other environmental compartments, including the atmosphere, sediment, soil, and ground water, the subject of the next chapter.

References

[1] F. Cordle, R. Locke, J. Springer, Risk assessment in a Federal Regulatory Agency: an assessment of risk associated with the human consumption of fish contaminated with polychlorinated biphenyls (PCBs), Environ. Health Perspect. 45 (1982) 171–182.
[2] C.R. Fremling, Ecological history of the upper Mississippi River, 1960.
[3] D.A. Vallero, Air Pollution Calculations: Quantifying Pollutant Formation, Transport, Transformation, Fate and Risks, Elsevier, 2019.
[4] J.D. Solomon, D.A. Vallero, From our partners—communicating risk and resiliency: special considerations for rare events, Center for Infrastructure Protection & Homeland Security, George Mason University, 2016. Available from: https://cip.gmu.edu/2016/06/01/partners-communicating-risk-resiliency-special-considerations-rare-events/.
[5] A. Androulakakis, N. Alygizakis, G. Gkotsis, M.-C. Nika, V. Nikolopoulou, E. Bizani, et al., Determination of 56 per-and polyfluoroalkyl substances in top predators and their prey from Northern Europe by LC-MS/MS, Chemosphere 287 (2022) 131775.
[6] G. Malarvannan, G. Poma, A. Covaci, Interspecies comparison of the residue levels and profiles of persistent organic pollutants in terrestrial top predators, Environ. Res. 183 (2020) 109187.
[7] J. Ramshaw, Investigating the effects of anthropogenic stressors on an aquatic top predator, 2018.
[8] I. Rodríguez-Jorquera, N. Vitale, L. Garner, D. Perez-Venegas, C. Galbán-Malagón, N. Duque-Wilckens, et al., Contamination of the upper class: occurrence and effects of chemical pollutants in terrestrial top predators, Curr. Pollut. Rep. 3 (3) (2017) 206–219.
[9] C.J. Martyniuk, N.J. Doperalski, M.S. Prucha, J.-L. Zhang, K.J. Kroll, R. Conrow, et al., High contaminant loads in Lake Apopka's riparian wetland disrupt gene networks involved in reproduction and immune function in largemouth bass, Comp. Biochem. Physiol. Part D Genomics Proteomics 19 (2016) 140–150.
[10] Arctic Monitoring Assessment Program, AMAP assessment 2002: persistent organic pollutants in the Arctic, 2004.
[11] F.A. Gobas, Assessing bioaccumulation factors of persistent organic pollutants in aquatic food-chains, in: Persistent Organic Pollutants: Environmental Behaviour and Pathways of Human Exposure, 2001, pp. 145–165.
[12] M. Long, M. Wielsøe, E.C. Bonefeld-Jørgensen, Time trend of persistent organic pollutants and metals in Greenlandic Inuit during 1994–2015, Int. J. Environ. Res. Public Health 18 (5) (2021) 2774.
[13] D. Mackay, A. Fraser, Bioaccumulation of persistent organic chemicals: mechanisms and models, Environ. Pollut. 110 (3) (2000) 375–391.
[14] O. Lindqvist, K. Johansson, L. Bringmark, B. Timm, M. Aastrup, A. Andersson, et al., Mercury in the Swedish environment—recent research on causes, consequences and corrective methods, Water Air Soil Pollut. 55 (1) (1991) xi–261.
[15] D. van de Meent, T. McKone, T. Parkerton, M. Matthies, M. Scheringer, F. Wania, et al., Persistence and transport potential of chemicals in a multimedia environment, in: SETAC Pellston Workshop on Criteria for Persistence and Long-Range Transport of Chemicals in the Environment, July 14–19, 1998, Society of Environmental Toxicology and Chemistry, Fairmont Hot Springs, British Columbia, Canada, 1999.
[16] M.A. Engle, M.T. Tate, D.P. Krabbenhoft, J.J. Schauer, A. Kolker, J.B. Shanley, et al., Comparison of atmospheric mercury speciation and deposition at nine sites across central and eastern North America, J. Geophys. Res. Atmos. (D18) (2010) 115.
[17] J.N. Galloway, J.D. Thornton, S.A. Norton, H.L. Volchok, R.A. McLean, Trace metals in atmospheric deposition: a review and assessment, Atmos. Environ. 16 (7) (1982) 1677–1700.
[18] J.A. Graydon, V.L. St Louis, S.E. Lindberg, K.A. Sandilands, J.W. Rudd, C.A. Kelly, et al., The role of terrestrial vegetation in atmospheric Hg deposition: pools and fluxes of spike and ambient Hg from the METAALICUS experiment, Global Biogeochem. Cycles 26 (1) (2012).
[19] J. Munthe, K. Kindbom, O. Kruger, G. Petersen, J. Pacyna, Å. Iverfeldt, Emission, deposition, and atmospheric pathways of mercury in Sweden, Water Air Soil Pollut. (2001).
[20] National Atmospheric Deposition Program, National Atmospheric Deposition Program: 2012 Annual Summary, National Atmospheric Deposition Program, 2013.
[21] E.M. Prestbo, D.A. Gay, Wet deposition of mercury in the US and Canada, 1996–2005: results and analysis of the NADP mercury deposition network (MDN), Atmos. Environ. 43 (27) (2009) 4223–4233.
[22] P.G. Meirmans, Hardy-Weinberg Equilibrium, 2019.
[23] O. Mayo, A century of Hardy–Weinberg equilibrium, Twin Res. Hum. Genet. 11 (3) (2008) 249–256.
[24] D.A. Vallero, Physics of ecosystems, in: Methods and Calculations in Environmental Physics, AIP Publishing LLC, Melville, New York, 2022, pp. 5-1–5-38.
[25] R. Elston, R. Forthofer, Testing for Hardy-Weinberg equilibrium in small samples, Biometrics (1977) 536–542.
[26] D.P. German, M.N. Weintraub, A.S. Grandy, C.L. Lauber, Z.L. Rinkes, S.D. Allison, Optimization of hydrolytic and oxidative enzyme methods for ecosystem studies, Soil Biol. Biochem. 43 (7) (2011) 1387–1397.
[27] J.H. Makoi, P.A. Ndakidemi, Selected soil enzymes: examples of their potential roles in the ecosystem, Afr. J. Biotechnol. 7 (3) (2008).
[28] J.J. Cole, Y.T. Prairie, The inorganic carbon complex, in: Wetzel's Limnology, Elsevier, 2024, pp. 301–323.
[29] S. Shao, C.T. Driscoll, T.J. Sullivan, D.A. Burns, B. Baldigo, G.B. Lawrence, et al., The response of stream ecosystems in the Adirondack region of New York to historical and future changes in atmospheric deposition of sulfur and nitrogen, Sci. Total Environ. 716 (2020) 137113.
[30] G.B. Lawrence, K.M. Roy, B.P. Baldigo, H.A. Simonin, S.B. Capone, J.W. Sutherland, et al., Chronic and episodic acidification of Adirondack streams from acid rain in 2003–2005, J. Environ. Qual. 37 (6) (2008) 2264–2274.
[31] A. Johnson, S. Andersen, T. Siccama, Acid rain and soils of the Adirondacks. I. Changes in pH and available calcium, 1930–1984, Can. J. For. Res. 24 (1) (1994) 39–45.
[32] C. Lajewski, H. Mullins, W. Patterson, C. Callinan, Historic calcite record from the Finger Lakes, New York: impact of acid rain on a buffered terrane, Geol. Soc. Am. Bull. 115 (3) (2003) 373–384.

[33] D.A. Vallero, Engineering aspects of climate change, in: Climate Change, second ed., Elsevier, 2015, pp. 547–568.
[34] National Oceanic and Atmospheric Administration, Carbon education tools, 2018, Available from: https://www.pmel.noaa.gov/co2/file/Carbon+Cycle+Graphics.
[35] Statista, Annual production of plastics worldwide from 1950 to 2021, National Academy Press, Washington, DC, 2023.
[36] H. Jones, Plastic pollution facts, 2023, Available from: https://www.clearcommunity.org/plastic-pollution-facts/?gclid=EAIaIQobChMI-8-FuoD-_QIVyu3Ch1ObgQDEAAYAiAAEgK1FvD_BwE.
[37] D.A. Vallero, T.M. Letcher, Unraveling Environmental Disasters, second ed., Elsevier Academic Press, Amsterdam, The Netherlands, 2023.
[38] L. Lebreton, The status and fate of oceanic garbage patches, Nat. Rev. Earth Environ. 3 (11) (2022) 730–732.
[39] X. Lim, Microplastics are everywhere—but are they harmful, Nature 593 (7857) (2021) 22–25.
[40] N. Welden (Ed.), Microplastics: emerging contaminants requiring multilevel management, in: Waste, Elsevier, 2019.
[41] Y. Sarkingobir, M. Dikko, S. Aliyu, U. Tambari, A. Adamu, I.A. Salau, et al., The dangers of plastics to public health: a review, NIPES J. Sci. Technol. Res. 2 (2) (2020) 195–200.
[42] S. Zazzini, Turbulent Dispersion in the Ocean Surface Boundary Layer, Sapienza Universita' di Roma, Rome, Italy, 2023.
[43] X. Qiu, S. Saovany, Y. Takai, A. Akasaka, Y. Inoue, N. Yakata, et al., Quantifying the vector effects of polyethylene microplastics on the accumulation of anthracene to Japanese medaka (Oryzias latipes), Aquat. Toxicol. 228 (2020) 105643.
[44] S. Xu, J. Ma, R. Ji, K. Pan, A.-J. Miao, Microplastics in aquatic environments: occurrence, accumulation, and biological effects, Sci. Total Environ. 703 (2020) 134699.
[45] M.A. Browne, P. Crump, S.J. Niven, E. Teuten, A. Tonkin, T. Galloway, et al., Accumulation of microplastic on shorelines worldwide: sources and sinks, Environ. Sci. Technol. 45 (21) (2011) 9175–9179.
[46] U.S. Environmental Protection Agency, Coastal waters, U.S. EPA, Washington, DC, 2023. Available from: https://www.epa.gov/report-environment/coastal-waters.
[47] J.-W. Oh, S.S.C. Pushparaj, M. Muthu, J. Gopal, Review of harmful algal blooms (HABs) causing marine fish kills: toxicity and mitigation, Plants 12 (23) (2023) 3936.
[48] U.S. Environmental Protection Agency, Climate change and harmful algal blooms, 2022, Available from: https://www.epa.gov/nutrientpollution/climate-change-and-harmful-algal-blooms. (Accessed 5 January 2022).
[49] M.B.S. Svendsen, N.R. Andersen, P.J. Hansen, J.F. Steffensen, Effects of harmful algal blooms on fish: insights from Prymnesium parvum, Fishes 3 (1) (2018) 11.
[50] G. Hallegraeff, Harmful algal blooms: a global overview, in: Manual on Harmful Marine Microalgae, vol. 33, 2003, pp. 1–22.
[51] J.M. Burkholder, Beyond algal blooms, oxygen deficits and fish kills: chronic, long-term impacts of nutrient pollution on aquatic ecosystems, in: Waters in Peril, Springer, 2001, pp. 103–125.
[52] P.K. Pandey, P.H. Kass, M.L. Soupir, S. Biswas, V.P. Singh, Contamination of water resources by pathogenic bacteria, AMB Express 4 (2014) 1–16.
[53] R. Fayer, J.M. Trout, Zoonotic protists in the marine environment, in: Oceans and Health: Pathogens in the Marine Environment, Springer, 2005, pp. 143–163.
[54] U.S. Environmental Protection Agency, Fresh surface waters, US EPA, Washington, DC, 2023. Available from: https://www.epa.gov/report-environment/fresh-surface-waters#:~:text=The%20state%20of%20fresh%20surface,of%20human%20and%20environmental%20stressors. (Accessed 14 July 2023).
[55] Dane County Conservation League, Lake variability and descriptions, Dane County Conservation League, Madison, Wisconsin, 2024. Available from: http://www.dccl.org/information/definitions-header.htm.
[56] T.C. Winter, The hydrology of lakes, in: The Lakes Handbook: Limnology and Limnetic Ecology, 2003, pp. 61–78.
[57] T.E. Graedel, On the concept of industrial ecology, Annu. Rev. Energy Environ. 21 (1) (1996) 69–98.
[58] D.A. Vallero, T.M. Letcher, Unraveling Environmental Disasters, Newnes, 2012.
[59] D.A. Vallero (Ed.), A systems approach to waste management, in: Waste, Elsevier, 2019.
[60] W.R. Dupre, J.C. Butler, I. Evans, Floods and flooding, University of Texas, Department of Geosciences, Houston, Texas, 2001. Available from: https://www.uh.edu/~jbutler/physical/flooding/streamdischarge.html.
[61] C. Lee, S.D. Lin, Handbook of Environmental Engineering Calculations, McGraw Hill, 2007.
[62] H.F. Hemond, E.J. Fechner, Chemical Fate and Transport in the Environment, Elsevier, 2014.
[[63]] A. Chézy, Formula to find the uniform velocity that the water will have in a ditch or in a canal of which the slope is known, Journal of the Association of Engineering Societies (1776) 18. In this issue.
[64] LMNO Engineering, Research, and Software, Ltd., Manning n values for open channel flow: Manning roughness coefficients, Available from: https://www.lmnoeng.com/manningn.htm.
[65] V.A. Levesque, E. Patino (Eds.), Hydrodynamic characteristics of estuarine rivers along the South western coast of Everglades National Park, in: Florida Bay Science Conference, Key Largo, FL, April, 2001 (Abstract).
[66] U.S. Geological Survey, Hydrologic unit maps, 2017, Available from: https://water.usgs.gov/GIS/huc.html.
[67] V.A. Levesque, K.A. Oberg, Computing discharge using the index velocity method, US Department of the Interior, US Geological Survey, 2012.
[68] D.A. Vallero, Translating Diverse Environmental Data Into Reliable Information: How to Coordinate Evidence From Different Sources, Academic Press, 2017.
[69] U.S. Geological Survey, How streamflow is measured—part 2: the discharge measurement, 2017, Available from: https://water.usgs.gov/edu/streamflow2.html.

[70] F.R. Spellman, N.E. Whiting, Handbook of Mathematics and Statistics for the Environment, CRC Press, Boca Raton, Florida, 2013.

[71] U.S. Environmental Protection Agency, Attachment B: water quality-based effluent limits—coal combustion waste impoundments, 2018. Washington, DC. Available from: U.S. Environmental Protection Agency, 2017 #978.

[72] U.S. Environmental Protection Agency, Coal ash (coal combustion residuals, or CCR), 2018, Washington, DC. Available from: https://www.epa.gov/coalash. (Accessed 1 March 2018).

[73] S. Abolfathi, S. Cook, A. Yeganeh-Bakhtiary, S. Borzooei, J. Pearson, Microplastics transport and mixing mechanisms in the nearshore region, Coast. Eng. Proc. (36) (2020) 63.

[74] U.S. Environmental Protection Agency, National recommended water quality criteria—aquatic life criteria table, U.S. EPA, Washington, DC, 2003. Available from: https://www.epa.gov/wqc/national-recommended-water-quality-criteria-aquatic-life-criteria-table. (Accessed 2 October 2023).

[75] World Health Organization, Water quality and health: review of turbidity, 2017. Report No.: WHO/FWC/WSH/17.01.

[76] D.S. Lloyd, Turbidity as a water quality standard for salmonid habitats in Alaska, N. Am. J. Fish Manag. 7 (1) (1987) 34–45.

[77] Georgia Department of Natural Resources, Development and implementation of Aucilla River watershed management plan, 2018.

[78] I.L. Pepper, C. Gerba, T. Gentry, EES642: environmental microbiology, in: Courses of Study, 2019, p. 141.

[79] I.L. Pepper, C.P. Gerba, T.J. Gentry, R.M. Maier, Environmental Microbiology, Academic Press, 2011.

[80] C.P. Gerba, I.L. Pepper, Wastewater treatment and biosolids reuse, in: Environmental Microbiology, second ed., Elsevier Inc, USA, 2009, pp. 503–530.

[81] J.R. Karr, Assessment of biotic integrity using fish communities, Fisheries 6 (6) (1981) 21–27.

[82] J.M. McKim, R.J. Erickson, Environmental impacts on the physiological mechanisms controlling xenobiotic transfer across fish gills, Physiol. Zool. 64 (1) (1991) 39–67.

[83] R.J. Erickson, J.M. McKim, A model for exchange of organic chemicals at fish gills: flow and diffusion limitations, Aquat. Toxicol. 18 (4) (1990) 175–197.

[84] M.C. Barber, Bioaccumulation and aquatic system simulator (BASS) user's manual beta test version 2.1, US Environmental Protection Agency, Athens, GA, 2001, p. 76.

[85] J.A. Kushlan, S.A. Voorhees, W.F. Loftus, P. Frohring, Length, mass, and calorific relationships of Everglades animals, Fla. Sci. (1986).

[86] V. Novotny, Integrated water quality management, Water Sci. Technol. 33 (4–5) (1996) 1–7.

[87] V. Novotny, G. Chesters, Delivery of sediment and pollutants from nonpoint sources: a water quality perspective, J. Soil Water Conserv. 44 (6) (1989) 568–576.

[88] K. Rahman, S. Barua, H. Imran, Assessment of water quality and apportionment of pollution sources of an urban lake using multivariate statistical analysis, Clean. Eng. Technol. 5 (2021) 100309.

[89] E. Dohner, A. Markowitz, M. Barbour, J. Simpson, J. Byrne, G. Dates, 5.2. Monitoring and assessing water quality, in: Volunteer Stream Monitoring: A Methods Manual, 1997.

[90] Ohio Environmental Protection Agency, Total maximum daily loads for the lower Cuyahoga River: final report, Ohio Environmental Protection Agency, Division of Surface Water, Columbus, Ohio, 2003.

[91] Maryland Department of the Environment, Total maximum daily loads of fecal bacteria for the non-Tidal Rock Creek Basin in Montgomery County, Maryland, U.S. Environmental Protection Agency, Region III, Philadelphia, Pennsylvania, 2007. Available from: https://mde.maryland.gov/programs/Water/TMDL/DocLib_RockCreek_02140206/Rock_Creek_TMDL_060607_final.pdf.

[92] M. Alexander, Most probable number method for microbial populations, in: Methods of Soil Analysis: Part 2. Chemical and Microbiological Properties, vol. 9, 1983, pp. 815–820.

[93] S. Sutton, The most probable number method and its uses in enumeration, qualification, and validation, J. Valid. Technol. 16 (3) (2010) 35–38.

[94] P.L. Woomer, Most probable number counts, in: Methods of Soil Analysis: Part 2. Microbiological and Biochemical Properties, vol. 5, 1994, pp. 59–79.

[95] S.S. Suthersan, J. Horst, M. Schnobrich, N. Welty, J. McDonough, Remediation Engineering: Design Concepts, CRC Press, 2016.

[96] D. Vallero, Environmental Biotechnology: A Biosystems Approach, Elsevier Science, 2015.

[97] B.E. Rittmann, P.L. McCarty, Environmental Biotechnology: Principles and Applications, Tata McGraw-Hill Education, 2012.

[98] B.E. Rittmann, Environmental biotechnology in water and wastewater treatment, J. Environ. Eng. 136 (4) (2009) 348–353.

[99] B.E. Rittmann, Microbial ecology to manage processes in environmental biotechnology, Trends Biotechnol. 24 (6) (2006) 261–266.

[100] M. Friedrich, I. Takács, A new interpretation of endogenous respiration profiles for the evaluation of the endogenous decay rate of heterotrophic biomass in activated sludge, Water Res. 47 (15) (2013) 5639–5646.

[101] D.A. Vallero, Hazardous wastes, in: Waste: A Handbook for Management, 2011, p. 393.

Chapter 11

Ground water

1 Introduction

Ground water provides about 25% of Canada's water supply [1] (see Fig. 11.1) and about 65% in the European Union Countries [2]. In the U.S., about 40% of the public water supplies are from ground water, with more than 40 million people, including most of the rural population, getting their drinking water from private wells [3]. As a result, ground water is an important source of drinking water in every State (Fig. 11.2). Ground water is also the major source of much agricultural irrigation. Ground water contributes substantially to the flow in many streams and is a crucial source of water needed by plants and animals in wetland habitats.

To provide context for the importance of protecting ground water, it is good to keep in mind that ground water quantity and quality depend on connections between the hydrosphere, atmosphere, and lithosphere. As mentioned in Chapter 10, a simple distinction between water bodies is whether they are on the earth's surface or under the surface. However, the environment is a highly interconnected system between surface and subsurface water systems, and among the air, soil, sediment, and biota within these systems. For example, when a pollutant is released into the environment, it can cause very rapid effects or effects that take a long time to occur. An immediate, acute response can range from the highly circumscribed with little impact, such as a compound that breaks down rapidly and in sufficiently low quantities and a sufficiently long distance between the point of release and the point where any vulnerable organism is exposed. By the time the water reaches this receptor, it is degraded into less hazardous compounds that cause no harm. The opposite can occur if the release is of sufficiently large quantities of a more recalcitrant pollutant that reaches a large population of human or ecological receptors. In the first example, the effect is isolated or mildly episodic, like a small number of spills of a rapidly degraded, nontoxic compound. In the latter example, large amounts of the released pollutants remain in the environment for an extended time or are continuous released into the environment, such as benzene, toluene, ethylbenzene, and xylenes (BTEX) leaking from an underground gasoline storage tank over decades into the ground water. This cross-media transport to and from water bodies is an example of biochemodynamics, wherein a contaminant moves and changes among discrete environmental media (Figs. 11.3 and 11.4).

The first two factors needed to understand ground water are porosity and permeability. The two terms are not synonymous. Indeed, a rock stratum may be very porous yet have very low permeability. Porosity (n) is the ratio of the voids to the total volume of a stratum:

$$n = \frac{V_{total} - V_{solids}}{V_{total}} = \frac{V_{voids}}{V_{total}} \tag{11.1}$$

Where V_{total} is the total volume of a material, V_{solids} = the volume of the solids in the material, and V_{voids} is the volume of the voids. Porosity is dimensionless and often expressed as a percentage by multiplying the ratio by 100. Porosity depends on sorting, i.e., grain size range, as well as the morphology of the subject material. Fine-grained materials are usually better sorted than coarse-grained materials, so have greater porosities [4]. For example, fine grained materials high in clay may have porosities around 50%, whereas as coarse sand about 25%, and gravel 20%.

As mentioned, porosity alone does not predict the permeability of material because physical properties like high amounts of adhesion, i.e., the molecular attraction of different substances, and cohesion, i.e., the molecular attraction of similar substances, decrease permeability. The water in an aquifer that is held as a film on the aquifer material or in very small openings is known as specific retention. Water in an aquifer that drains by gravity is specific yield or effective porosity. For example, a clayey material with 50% porosity may merely have a specific yield of 2% and specific retention of 48%, whereas gravel may have porosity of 20% but have a specific yield of 19% and specific retention of only 1%. Indeed, very fine-grained materials that contain large amounts of certain clays have a high porosity but low specific yield when compared to coarser-grained materials [4].

Fundamentals of Water Pollution. https://doi.org/10.1016/B978-0-443-28987-3.00008-4

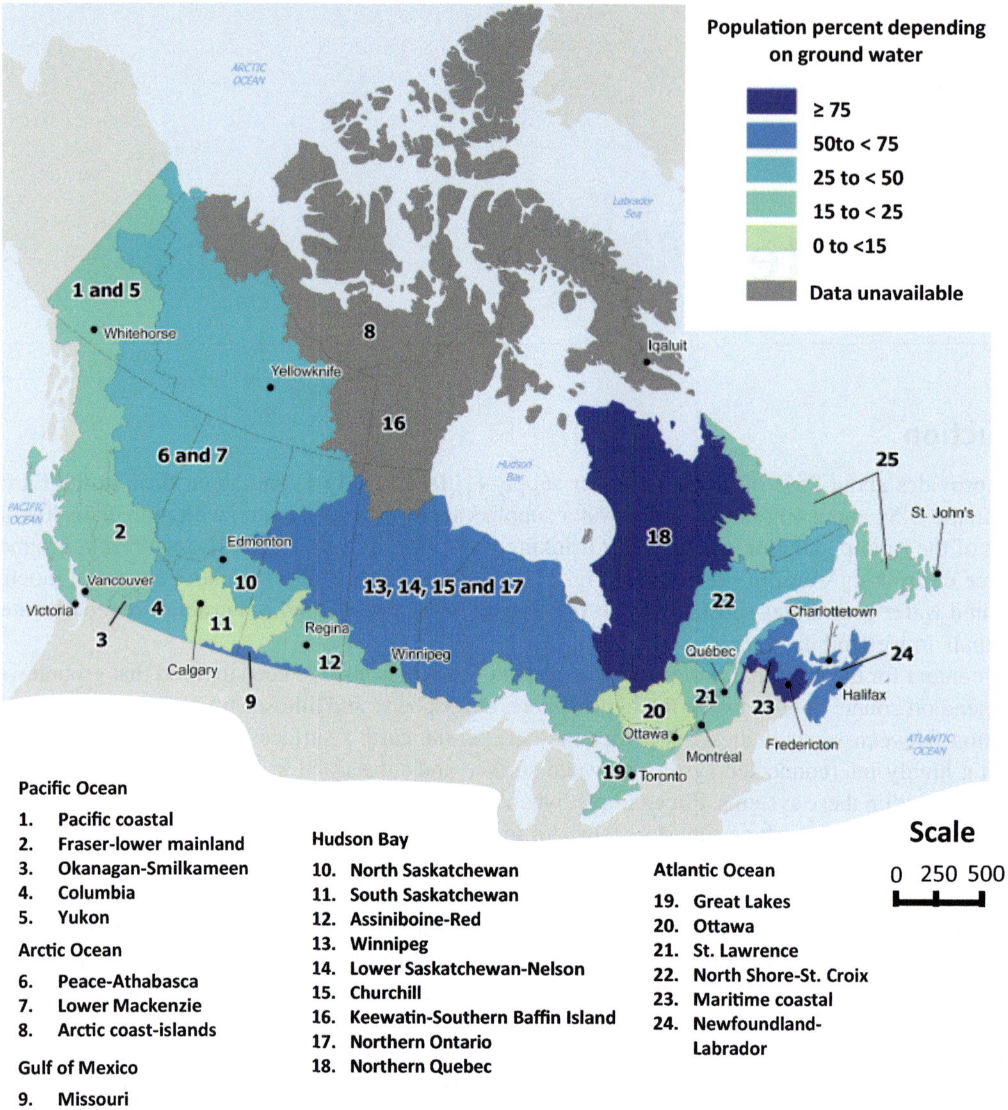

FIG. 11.1 Percentage of Canada's population depending on ground water in the year 2010. *(From Statistics Canada, Section 2: Canada's water supply—stocks and flows, in: Human Activities and the Environment. Statistics Canada, Ottawa, Ontario, 2010.)* Map lines delineate study areas and do not necessarily depict accepted national boundaries.

FIG. 11.2 Estimated percentage of each U.S. State using population using water as drinking water in the year 1995. *(From W.M. Alley, T.E. Reilly, O.L. Franke, Sustainability of ground-water resources, 0607930403. Available: https://pubs.usgs.gov/circ/circ1186/html/intro.html, 1999.)* Map lines delineate study areas and do not necessarily depict accepted national boundaries.

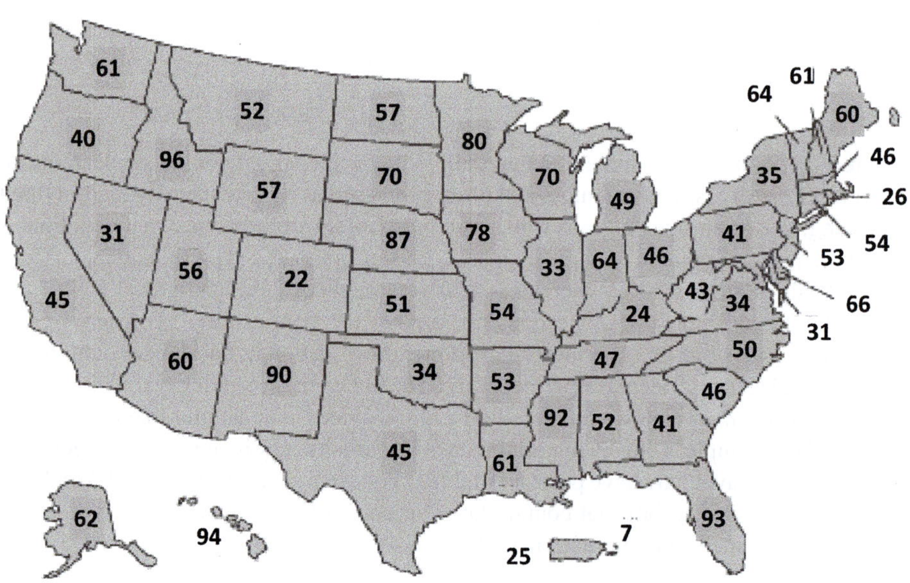

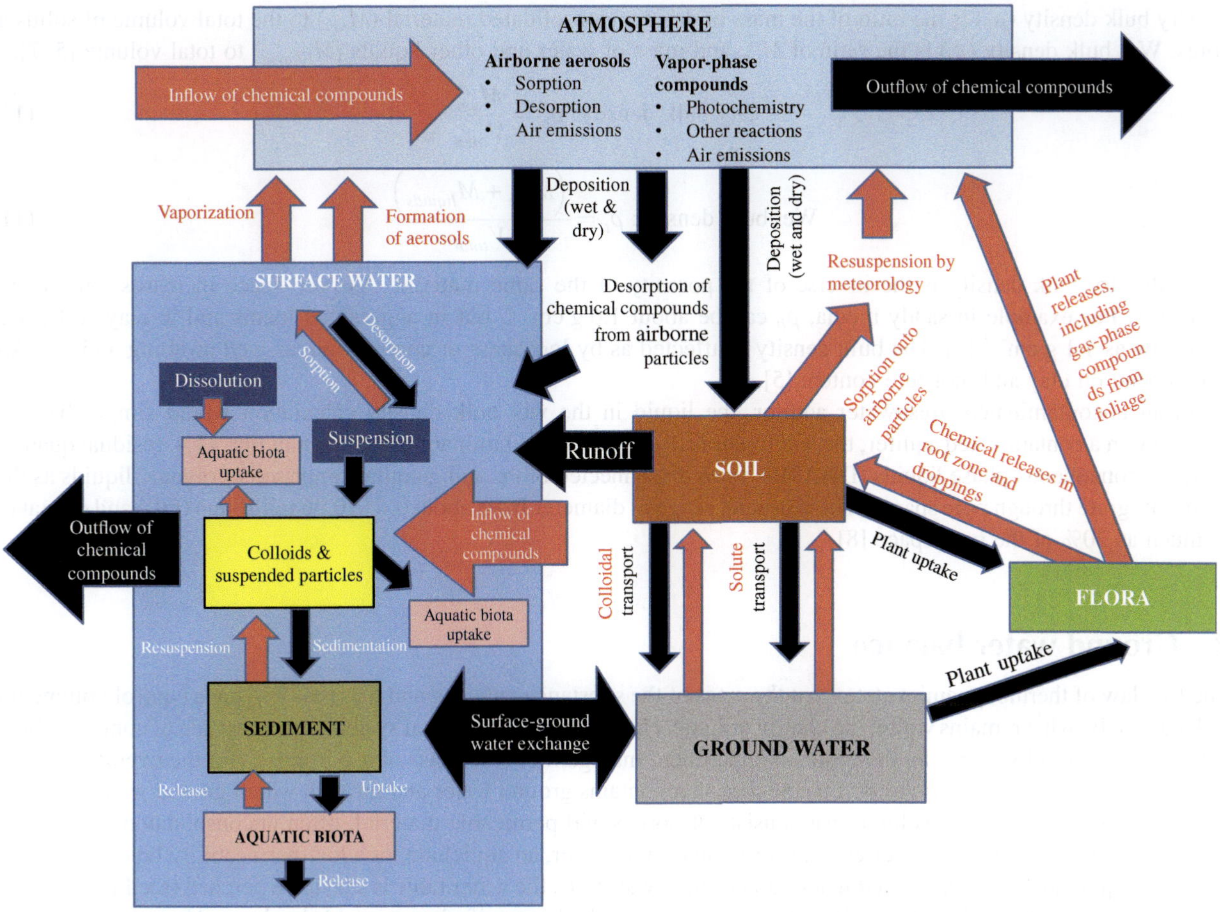

FIG. 11.3 Uptake, transport, transformation, and various fates of pollutants within and between environmental compartments. Partitioning coefficients must be developed for each arrow and steady-state conditions may not be assumed. Reaction rates and other chemical kinetics must be developed for each arrow and box. *(Data from D. Vallero, Environmental Biotechnology: A Biosystems Approach, Elsevier Science, 2015; used with permission.)*

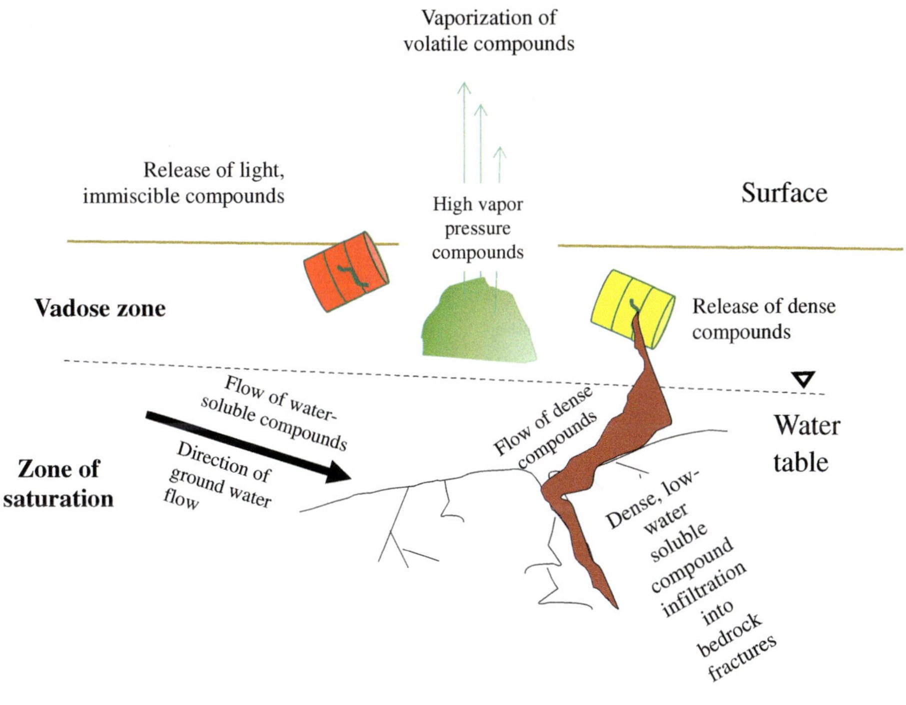

FIG. 11.4 Direct and indirect pollutant releases to the environment. Low-density substances with specific weight less than that of water may remain in the unsaturated (vadose) zone, with a portion eventually volatilizing to the atmosphere and others continuing to move within the ground water plume. Low-density compounds that readily mix with water will move further down-gradient, and denser compounds may continue to flow more perpendicularly to the surface until reaching an impermeable layer, whereupon they move gravitationally into pores and cracks in the bedrock or sit atop impervious strata.

Dry bulk density (ρ_b) is the ratio of the mass of dried unconsolidated material (M_{um}) to the total volume of solids and pores. Wet bulk density (ρ_t) is the ratio of M_{um} and mass of water and other liquids ($M_{liquids}$) to total volume [5–7]:

$$\text{Dry bulk density: } \rho_b = \frac{M_{um}}{V_{total}} \tag{11.2}$$

$$\text{Wet bulk density: } \rho_t = \frac{\left(M_{cm} + M_{liquids}\right)}{V_{total}} \tag{11.3}$$

Thus, the dry bulk density is the inverse of the porosity of the same material. As pore space increases, bulk density decreases. For example in sandy media, ρ_b can be about 1.6 g cm^{-3}, but in aggregated loams and in clay soils, it can be as low as 1.1 g cm^{-3} [7]. The bulk density is affected as by looseness or compaction, as well swelling and shrinkage that depend on clay and moisture content [5].

In an uncontaminated, freshwater aquifer, the liquid in the wet bulk density calculation would simply be water. However, in a contaminated aquifer, the wet fraction also contains contaminants. As shown in Fig. 11.4, residual quantities of dense nonaqueous-phase liquids (DNAPLs) form disconnected blobs and ganglia comprised of organic liquids as these liquids migrate through pore spaces. These usually range in diameter from about 0.6 to 0.06 milliliters (mL) and can take up as much as 20% of the pore space [8].

2 Ground water balance

The first law of thermodynamics states that the mass of the substance entering and the mass leaving a control volume must be balanced by what remains within the control volume. However, environmental systems are a cascade of control volumes. Within each control volume, an individual compartment may gain or lose mass of a substance, but the overall mass must balance. An aquifer is a system of permeable rock that contains ground water and through which ground water moves. As such, an aquifer is a control volume that consists of porous and permeable material, often unconsolidated like sand and gravel. The stratum below the aquifer is a nearly impervious layer, an aquiclude, or a low-permeability layer, the aquitard [9]. If the aquiclude or aquitard is a thin lens upon which water sits, the water table is said to be perched (see Fig. 11.5). This can be problematic for well-drillers who believe they have reached an aquifer's water table but instead have found a perched water table that will run dry must sooner than an aquifer.

To estimate a system's ground water balance, the change in ground water storage (ΔS) must be determined [10]. Irrigation canals provide a good example of this exchange. For an area where crops and other plants are irrigated with ground water, ΔS can be found as:

$$\Delta S = R_r + R_s + R_i + S_i + I_g - E_t - T_p - B_f - O_g \tag{11.4}$$

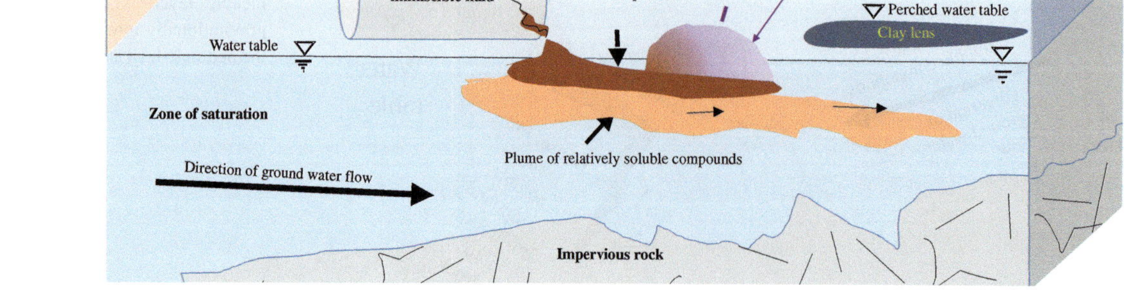

FIG. 11.5 Perched water table.

Where, R_r is the amount of recharge from precipitation, R_s is the amount of recharge from canal seepage, R_i is the amount of recharge from irrigation, S_i is the influent seepage from streams, I_g is the inflow from outside basins, E_t is the evapotranspiration from the aquifer, T_p is the draft from ground water, B_f is the baseflow (inflows to streams from the aquifer), and O_g is the outflow to outside basins.

As the water moves through irrigation canals, R_s represents the amount of water that escapes the system as canal seepage to join the ground water. Many factors influence R_s including the porosity of soils underlaying the canals and the head gradient between the canal and the connecting ground water. Most irrigation canals seep, with losses generally larger in developing countries [11].

During a rain event, some of the amount of precipitation (P) is evaporated and transpired from plants (E_t), some flows overland (O_f), and the remaining amount of precipitation recharges the aquifer (R_p). So, the inflow from rainfall seepage is calculated as:

$$R_p = P + R_i - E_t - O_f \tag{11.5}$$

Ground water balance example 1

To demonstrate this balance, calculate an aquifer's influent from rainfall seepage if the precipitation rate in the area is 85 cm year^{-1}. The evapotranspiration rate is 40 cm year^{-1}, the irrigation rate is 20 cm year^{-1}, the outflow to another basin is 25 cm year^{-1}, and the overland flow is 15 cm year^{-1}. The ground water inflow and outflow are unchanged.

$$R_p = 85 + 20 - 40 - 15 = 50 \text{ cm year}^{-1}.$$

Ground water balance example 2

Based on the above information, calculate the baseflow of that region if 21 cm of water are pumped per year (Q_p) and if 30 cm year^{-1} flows to streams.

$$I_g = O_g + R_f - R_r$$

$$I_g = 25 + 30 - 50 = 5 \text{ cm year}^{-1}$$

The inflow to ground water and the recharge from precipitation comprise the inflows. The baseflow and the discharge from pumping are the system's outflows:

$$B_f = I_g + R_r - O_g - Q_p$$

$$B_f = 5 + 50 - 25 - 21 = 9 \text{ cm year}^{-1}$$

Baseflow is also known as ground water recession flow, drought flow, low flow, low-water flow, fair weather runoff, and sustained weather runoff. Baseflow is derived from water that is stored in unconsolidated and porous material in bedrock near soils and in soil and unconsolidated and porous material at the interface between a stream and its adjacent land, i.e., the riparian zone. Baseflow water moves through the interstices, e.g., macropores and micropores, and through fractures in these layers and, ultimately, reaches the aquifer [12]. As such, this aquifer is providing the equivalent of 9 cm year^{-1} of rainfall the streams to which it is connected. This is the portion of streamflow that is sustained between precipitation events. During drought conditions, evapotranspiration is greatly increased, so that the water losses would decrease B_f below 9 cm year^{-1} until it would become negative, meaning that the aquifer would no longer provide such as baseflow.

3 Ground water interactions

As noted in the ground water balance discussion, surface waters and ground water systems continuously interact within the hydrological cycle. As such, an aquifer that is inseparably connected to a stream should be viewed as one system, with water constantly exchanged between the two components. The aquifer, which consists of soil and other particles and pore spaces between the particles, is recharged when precipitation infiltrates the unsaturated zone. Water in the unsaturated zone is a thin film around the particles and in the interstices between the particles but does not completely fill the interstices. Within the pore spaces are air, water, and other fluids. Water pollutants are among these other fluids. Gravity is the major force that causes the water to move through the pore spaces, along with the difference in pore water pressure, i.e., moisture potential, which varies depending on the pore size and moisture content of the unsaturated zone [10]. Thus, the total potential hydraulic head (h) is the sum of the gravity potential, Z, and the moisture potential $\Psi(\theta)$:

$$h = \sum Z + \Psi(\theta) \tag{11.6}$$

Since it is a measure of the energy per unit volume of water, h can also be calculated as the sum of elevation, pressure, and kinetic energy [9]:

$$h = z + \frac{p}{\rho g} + \frac{v^2}{2g} \tag{11.7}$$

Where, z is the height above an arbitrary reference elevation, p is the gauge pore water pressure, i.e., above the atmospheric pressure, ρ is the density of the water, g is the gravitational acceleration, and v is the water velocity.

When the ground water is not moving, i.e., under static conditions, h is constant in a continuous volume of water and equals the sum of z and $\frac{p}{\rho g}$. With decreasing (Δz), the water pressure will increase by the amount equal to $\rho g \Delta z$. Thus, the elevation term decrease is exactly offset by the increase in the pressure term, and the hydraulic head (see Fig. 11.6).

Incidentally, the soil above the water table often contains carbon dioxide (CO_2) and organic acids. This is due to the molecular oxygen (O_2) being taken up by soil bacteria and plant roots for metabolism. Thus, soil organic matter (SOM) and plant litter undergo decomposition by soil microbes and through plant roots and soil fauna. The CO_2 reacts with soil water to form carbonic acid (H_2CO_3). The organic acids are released from plant root exudates or produced by microbes in the soil or the rhizosphere, i.e., the area of soil near and around plant roots that influences their growth, respiration, and nutrient exchange [13,14]. Along with CO_2, organic acids may also be released from the decomposition of plant residues in soil [15,16]. Indeed, soil respiration reflects the condition of the physical and chemical environment of a soil. In the short term, a high rate of soil respiration may indicate an unstable soil system and loss of SOM. It is also an indication of nutrients being converted in organic matter to forms that have greater bioavailability for plant such as nutrient compounds and ions, e.g., nitrogen (N) as nitrate (NO_3^- (nitrate), phosphorous (P) as phosphate (PO_4^{3-}), and sulfur (S) as sulfate (SO_4^{2-}) [17–19].

The saturation zone consists of porous media in which the pores are filled with water and other fluids and compounds at comparatively low concentrations. The water table (▼) is the boundary between the saturated zone and the unsaturated zone. It is the depth at which the pressure of the pore water equals the atmospheric pressure [9]. As such, it is the top of the aquifer.

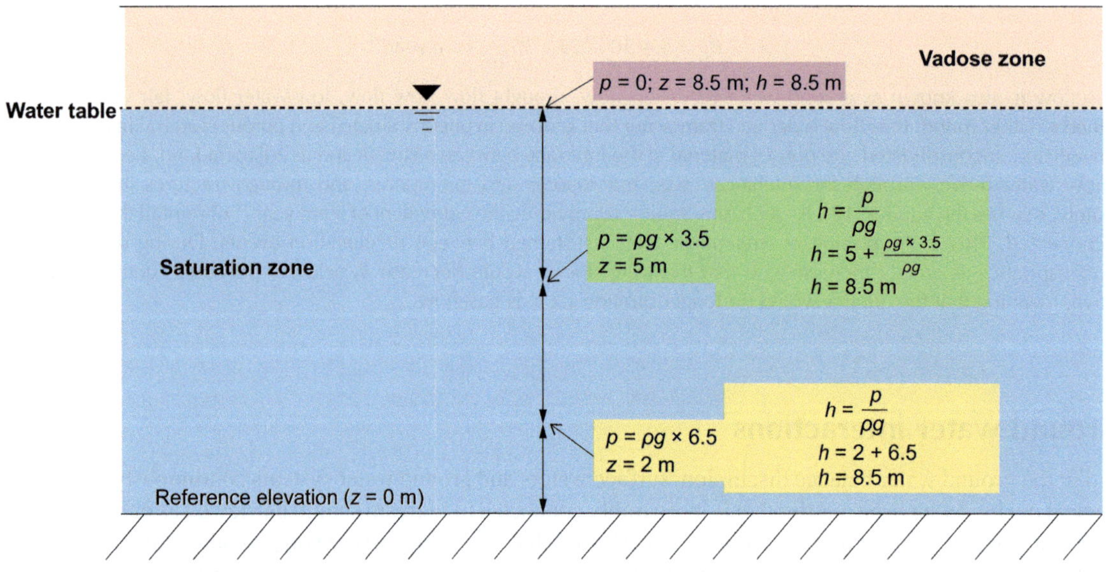

FIG. 11.6 Example of constant hydraulic head under static conditions at differ heights above reference elevation (z) and gauge pore pressure (p). Note: ρ is the density of the water, g is the gravitational acceleration. *(Data from H.F. Hemond, E.J. Fechner, Chemical Fate and Transport in the Environment, Elsevier, 2014.)*

A change in the infiltration rates, i.e., the amount of precipitation that enters the saturation zone, can be estimated by combining the flow equation with the principle of continuity [20], known as the Richards equation [21]. In one dimension, i.e., the vertical (z) dimension, this is:

$$\frac{\partial \theta}{\partial t} = \frac{\partial}{\partial t}\left[K(\Psi)\left(\frac{\partial \psi}{\partial z} + 1\right)\right] \tag{11.8}$$

This can be simplified by adding the specific water capacity of the porous medium, $C(\theta)$:

$$D(\theta) = \frac{K(\theta)}{C(\theta)} \tag{11.9}$$

Where $D(\theta)$ is the water diffusivity of the porous media, θ is the volumetric water content, and K is the hydraulic conductivity. Thus, the Richards equation can be restated as [22]:

$$\frac{\partial \theta}{\partial t} = \frac{\partial}{\partial t}\left[D(\theta)\left(\frac{\partial \theta}{\partial z} + K(\theta)\right)\right] \tag{11.10}$$

Incidentally, infiltration does not apply exclusively to water, but to any fluid that enters the subsurface environment. Infiltration rates are highest through media containing coarse particles that are all nearly the same size, i.e., well-sorted [22]. These have the highest hydraulic conductivity (K). Note that in previous chapters that K represented equilibrium constants and coefficients. In that case, it represents a chemical term, whereas here, K represents a geological transport term.

The interactions between ground water and other environmental media are well demonstrated by the compounds of sulfur (S), which exists as both reduced forms, for example, hydrogen sulfide (H_2S), and oxidized forms, for example, sulfur dioxide (SO_2). To demonstrate the movement and change of compounds above an aquifer, H_2S is oxidized in the air to SO_2 in three steps. A hydroxyl radical initiates the transformation:

$$H_2S + HO^\bullet \rightarrow HS^\bullet + H_2O \tag{11.11}$$

$$HS^\bullet + O_2 \rightarrow HO^\bullet + SO^\bullet \tag{11.12}$$

$$SO^\bullet + O_2 \rightarrow SO_2 + O^\bullet \tag{11.13}$$

The reactions of SO_2 are very complex and proceed through three different pathways to become the sulfate ion (SO_4^{2-}) in pore water and water droplets. Sulfur dioxide can react with the hydroxyl radical ($HO^\bullet$) to form the HSO_3 radical in the air, which then can react with another hydroxyl radical to form water and sulfur trioxide (SO_3) or H_2SO_4. Sulfur dioxide has sufficient aqueous solubility to dissolve in water droplets where it can react with oxygen gas to form SO_4^{2-}. The third pathway to sulfate occurs when SO_2 reacts with hydrogen peroxide (H_2O_2) to form sulfuric acid:

$$HO^\bullet + SO_2 \rightarrow HOSO_2^\bullet \tag{11.14}$$

$$HOSO_2^\bullet + HO^\bullet \rightarrow H_2SO_4 \tag{11.15}$$

$$SO_{2(\text{aqueous})} + [O]_{(\text{aqueous})} \rightarrow H_2SO_4 \tag{11.16}$$

$$SO_2 + H_2O_2 \rightarrow H_2SO_4 \tag{11.17}$$

With sufficient residence time in the atmosphere, S will be oxidized to the sulfate ion, usually as sulfuric acid (H_2SO_4). Ammonia (NH_3), the most common base in the atmosphere, reacts with H_2SO_4 to form ammonium bisulfate (NH_4HSO_4) and ammonium sulfate ($(NH_4)_2SO_4$). Sulfuric acid, NH_4HSO_4 and $(NH_4)_2SO_4$, all are hydroscopic substances, that is, readily dissolved in water. Thus, they wash out of the atmosphere during precipitation events [23], with a fraction reaching ground water.

The reduced form of sulfur H_2S is highly toxic Much of its concentrations in wetlands, sediment, and ground water are mediated biologically. Microbes, especially bacteria, reduce sulfur, using the S atom as energy sources through the acceptance of electrons. For example, sulfur-reducing bacteria can produce H_2S by chemically changing oxidized forms of sulfur, especially sulfate (SO_4) compounds. For this to happen, the bacteria must have access to the sulfur in water, whether in surface or ground water, soil water, or sediment. Anaerobes, bacteria that live in water where concentrations of molecular oxygen (O_2) are deficient, are particularly effective in chemical reduction. The bacteria remove the O_2 molecule from the sulfate, with only the S remaining, which then combines with hydrogen (H) to form gaseous H_2S [24–26]. The H_2S is formed from the anaerobic or nearly anaerobic decomposition of deposits of organic matter, for example, plant residues. In the other direction, redox principles can be used to treat H_2S contamination via oxidation using oxidants. Molecular oxygen and H_2O_2 are especially effective in oxidizing the reduced forms of S and other elements.

After a pollutant or a precursor of a pollutant is released into the environment, it reacts with other compounds and changes physically and chemically. It is transported advectively, during which it continues changing physically, chemically, and biologically. Compounds can be toxic or essential, depending on where they are in the environment and their chemical form. For example, ammonium nitrate (NH_4NO_3) is an essential nutrient for crops but can contaminate surface and ground water if the plants and microbes do not degrade it fast enough [27]. For example, the oxidized forms of sulfur like iron sulfate ($FeSO_4$) and nitrogen like NH_4NO_3 form acids when they react with water in any media, such as mining waste and ash that contaminate surface waters and ground water. Many ground water pollutants contain both N and S along with the organic elements carbon, hydrogen, and oxygen. Many originated as combustion byproducts, where the general form of the combustion reaction is

$$C_aH_bO_cN_dS_e + 4a + b - 2c \rightarrow aCO_2 + \left(\frac{b}{2}\right) H_2O + \left(\frac{d}{2}\right) N_2 + eS \tag{11.18}$$

This reaction shows that in the environment, pure reactions are rare. The environment is filled with mixtures and heterogeneous reactions. Reactions can occur in sequence, parallel, or both. For example, a feedstock to a landfill contains myriad types of wastes, from garbage to household chemicals, to commercial wastes, to even small (hopefully) amounts industrial wastes that may have been illegally dumped. These constituents contain a wide array of compounds, including those that contain N and S that may follow the stoichiometry of Reaction (11.18), such as during a landfill fire or respiration by microbes. In addition, the oxidation and reduction of nutrients including P, N and S occur in organism at every trophic state and energy level.

Within an aquifer, a contaminant may exist in various phases (e.g., dissolved in water and sorbed to a particle in the solid phase). Interphase reactions, or the physical interactions of the contaminant at the interface between each compartment, determine the amount of any substance in the environment. Thus, a pollutant may remain unchanged in an aquifer during its residence time, or it may move physically into another environmental compartment, and/or it may be transformed chemically into another substance. For example, a recalcitrant chemical compound may persist in the ground water given the very slow transport rate of most aquifers (some less than a meter per year), or may be transported downstream, evade into the soil and the atmosphere, may dissipate in concentration via dispersion and diffusion in the aquifer, or dramatically decline in concentration with the application of ground water treatment technologies.

Certainly, all ground water protection decisions are technical in content, but they vary in the mix of skills and content needed. For example, both structural and mechanical engineers address the laws of motion and thermodynamics but differ as to why. Mechanical engineers may be more interested in finding better ways to make something move, whereas structural engineers are more likely to make sure something never moves over its designed life, or at least moves in nanometers rather than meters. Even this distinction is variable, since a mechanical engineer working on a braking system may share more in common with the structural engineer than his mechanical engineering colleagues working on increasing an engine's torque and delivering more energy to turn the wheels. Often, however, the distinctions lie with the mix of physics, chemistry, and biology. Arguably, it is the hydrogeologists that have the basic skill set to address ground water.

The relationships between above ground and below ground hydrology requires a comprehensive skill set. Hydrologists, engineers, and hydrogeologists must especially understand the hyporheic zone, the porous areas of the stream bed and stream bank in which stream water mixes with upper layers of ground water (see Fig. 11.7). The chemical composition of surface and ground water can differ substantially from each other. Notably, surface waters usually have much higher dissolved O_2 concentrations than does ground water due to the reducing conditions of many ground water systems [28]. Conversely, CO_2 concentrations in ground water can be much higher than those in the upper layers of a surface water body and the atmosphere due less mixing with the air and because of the microbial oxidation of organic materials in ground water [28]. Even in the same system the stream and the ground water may have large differences in acidity and temperature. The hyporheic zone has gradients of the concentration of dissolved gases, redox, pH, and temperature. Thus, ground water, surface water and the atmosphere must be considered as a single system, with contaminants moving and transforming physically and chemically within and between each medium. For example, dust blowing off a coal mine can become dissolved or suspended in water vapor. Droplets return to the Earth as precipitation and reach surface and ground water. As metal compounds move among compartments the metal valence will change, which will change their mobility and could alter their toxicity [29].

Variation in hydraulic gradients can lead to downwelling flow, i.e., flow from surface water to ground water, or upwelling flow, i.e., flow from ground water to the surface water. The water flow over various streambed shapes and

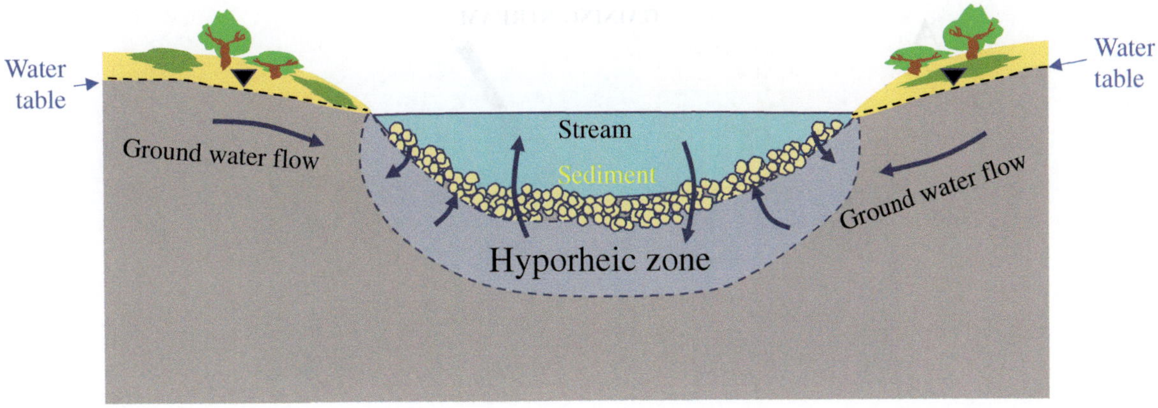

FIG. 11.7 Exchanges and reactions that can occur in ground water, sediment, and surface water. Some of the stream water moves into and out of the sediment and in shallow ground water (i.e., the hyporheic zone). The process can increase the mobility of dissolved metallic compounds.

changes in stream profile will alter hydraulic head gradients and affect hyporheic exchanges [30]. For example, drilling and mining operations can be a major source of ground water pollution when the oil, gas, hazardous wastes, or other substances enter aquifers from leaks in casings and cracks in confining strata that allow the oil, gas, and their impurities to enter other strata. For example, Fig. 11.8 shows the pressure within the unconfined aquifer is greater than the pressure at the top of the confining layer but lower than the pressure at the ground surface. Thus, to draw water to the surface, a pump is needed to increase the pressure differential between the surface and the confined aquifer, one that is isolated from the atmosphere at the point of discharge by impermeable geologic formations. If under normal conditions the pressure in the confined layer were higher than the atmospheric pressure at the ground surface, it would be an artesian system and would flow without the aid of a pump [31].

The connections between a stream and an aquifer vary, especially affected by meteorology. For example, in an area where evapotranspiration and other losses are small compared to precipitation rates, a stream is gaining water from the aquifer's inflow (see Fig. 11.9A). Conversely, if the stream is in a low moisture scenario, it will lose water to the aquifer (see Fig. 11.9B). The former is known as a gaining stream and the latter a losing stream. Indeed, there is a seasonal or episodic aspect to gaining and losing streams, for example, the same stream may be a gaining stream in the spring under high precipitation periods but become a losing stream later in the summer under low precipitation periods. The physiographic profiles of streams, combined with weather factors can change the direction of flow between the stream and aquifer, such as when the surface level of the stream rises and falls. Also, some streams are blocked from the aquifer by a sufficiently deep unsaturated zone (Fig. 11.9C). In this case, pumping water near the stream will not change the depth of the water table, as it would for a stream connected to the aquifer [3]. Incidentally, when the term "stream" is used it generally conveys a segment rather than the whole stream.

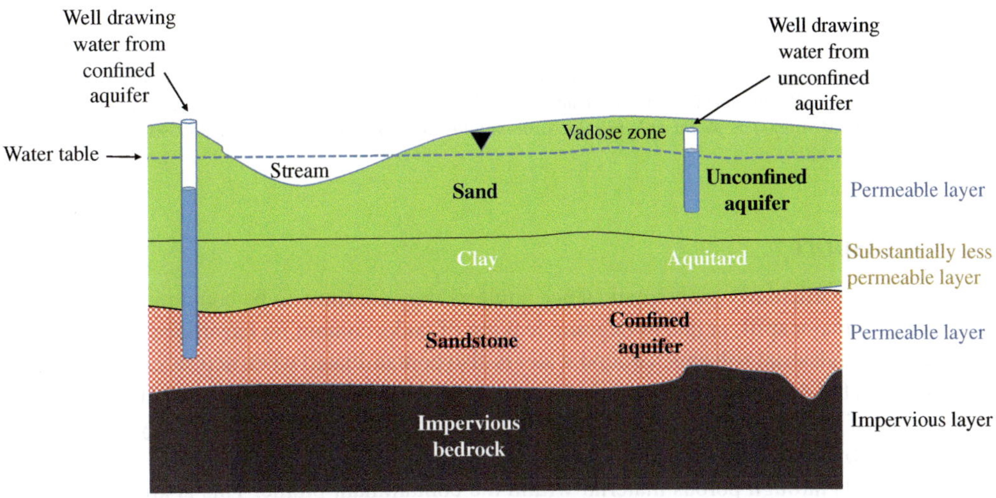

FIG. 11.8 Profile of aquifer between two impermeable strata, i.e., clay confining units.

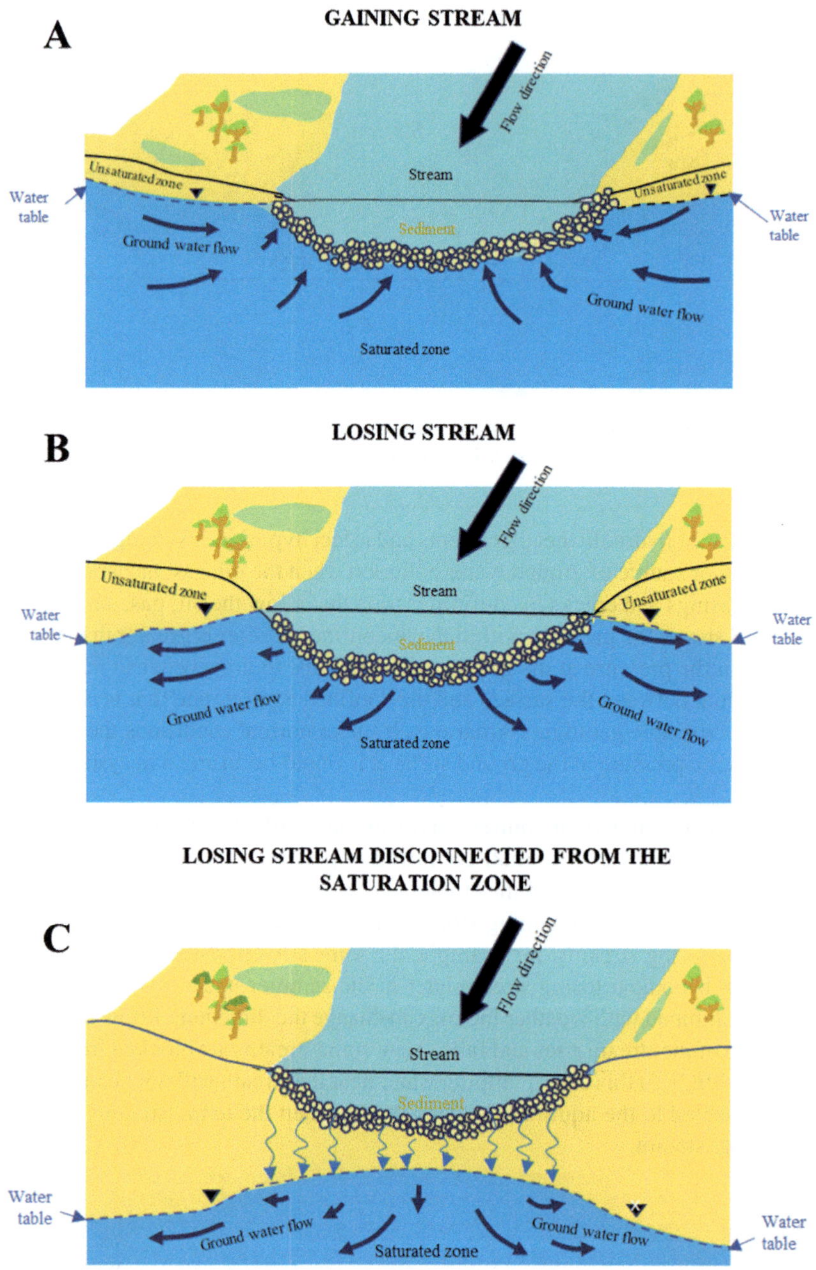

FIG. 11.9 Interactions between a stream and the underlying aquifer. (A) The gaining stream receives water from ground water; (B) The aquifer receives water from the losing stream; and (C) Water table is below the losing stream bottom, so that the aquifer only receives water passing through the vadose (unsaturated zone). *(Data from W.M. Alley, T.E. Reilly, O.L. Franke, Sustainability of ground-water resources, 0607930403. Available: https://pubs.usgs.gov/circ/circ1186/html/intro.html, 1999.)*

4 Pollutant transport

Ground water pollution is particularly problematic since aquifers may continue to be contaminated long after the original polluting event, such as a major spill, or due to ongoing leaking of tanks containing hazardous materials, such as gasoline, agricultural chemicals, or industrial solvents. Ground water is also contaminated when active, engineered systems fail, such as when pumping stations fail due to loss of electricity or compromised by allowing their contents to reach the ground water and threaten the quality of drinking water supplies or even find their way into residences. Many aquifers around the world have been contaminated by these and many other types of failure (see Fig. 11.10).

After release, pollutants through porous material within the contaminant plume. This is known as advective pollutant transport in a plume moving with the ground water flow. Pollutants are transported as either solutions or suspensions. In

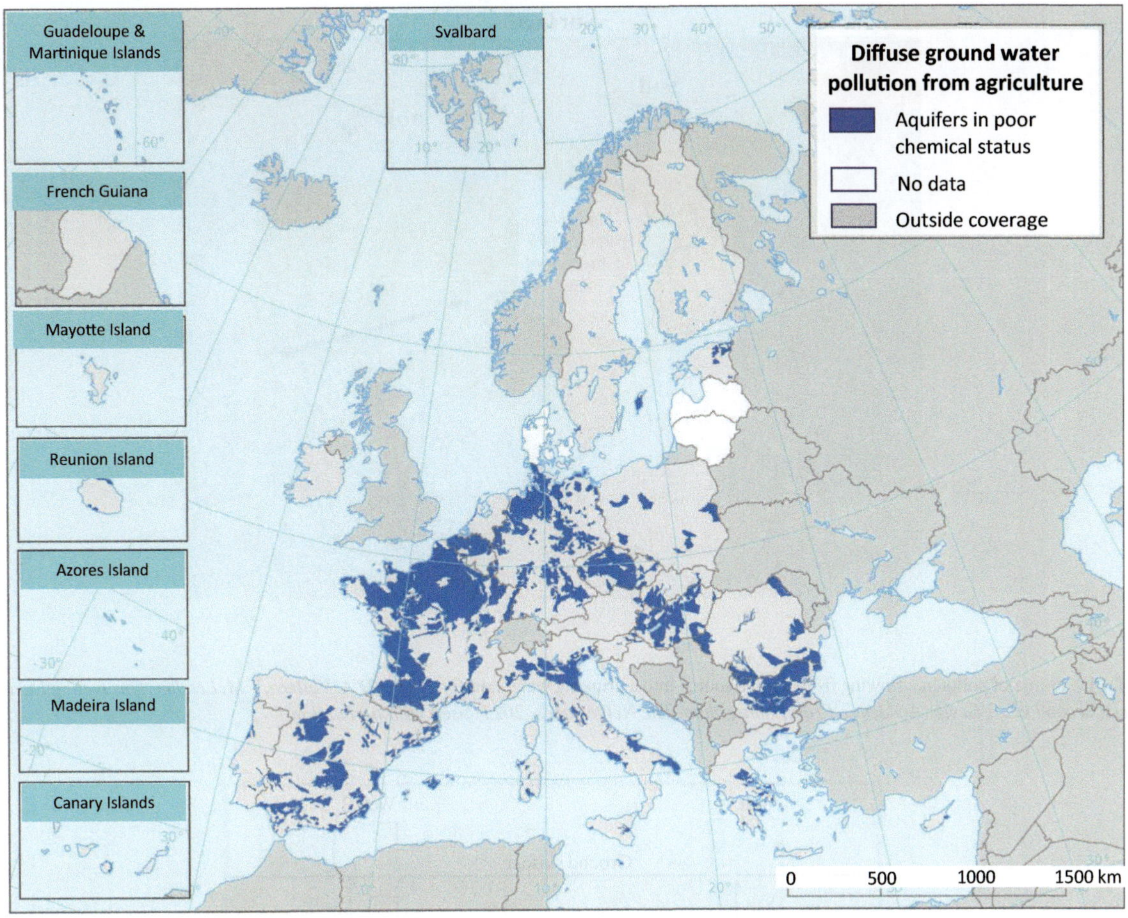

FIG. 11.10 Regions in Europe where ground water contains unacceptable concentrations of chemical pollutants. *(From European Environment Agency, Europe's groundwater—a key resource under pressure, EEA. https://www.eea.europa.eu/publications/europes-groundwater#:~:text=Groundwater%20stores%20almost%20one%20third,of%20water%20for%20agricultural%20irrigation (Accessed 11 January 2024); data from: A. Psomas, G. Bariamis, J. Rouillard, U. Stein, S. Roy, Study of the impacts of pressures on groundwater in Europe: analysis of groundwater associated aquatic ecosystems (GWAAEs) and groundwater dependent terrestrial ecosystems (GWDTEs), Service Contract No. 3415/B2020/EEA.58185, European Environment Agency, Copenhagen, Denmark, 2021.)* Map lines delineate study areas and do not necessarily depict accepted national boundaries.

solution transport, the pollutants become dissolved in surface or soil water and seep into the soil. As they reach layers of clay, silt, soil, sand, and other unconsolidated and porous materials, they may be sorbed to particles, with a portion suspended and carried with soil water. Pollutants with sufficiently high vapor pressures may also evaporate. Aquifers become increasingly contaminated as pollutants remain either dissolved or suspended and move with the ground water plume (see Fig. 11.11). The physical phase of these compounds ultimately depends upon both properties of the molecules themselves and the conditions in which they exist, including the hydrogeologic characteristics of the soil and aquifer.

Ground water usually moves by flowing through porous media, e.g., gravel, sand, and silt. The major exception is Karst topography, which has caves formed from eroded massive limestone and dolomite formations. The carbonate compounds in these sedimentary rocks dissolve over time along cracks that expand to caves in which water flows as underground streams. Since people have seen these impressive systems, they often incorrectly assume that this is how most ground water moves. To the contrary, nearly all subsurface water is stored in and moves through pores.

Water stands in a well at a particular level, known as hydraulic head (h). The units are length per length (e.g., m m^{-1}). If a pollutant source is upgradient from a well, the well water can be contaminated (see Fig. 11.12). Ground water flows in the direction of decreasing head. The change in head measurement is the hydraulic gradient ($\frac{dh}{dx}$), which is the change in elevation from one point (H_1 at x_1) to another (H_2 at x_2):

$$\frac{dh}{dx} = \frac{(H_1 - H_2)}{\Delta x} \tag{11.19}$$

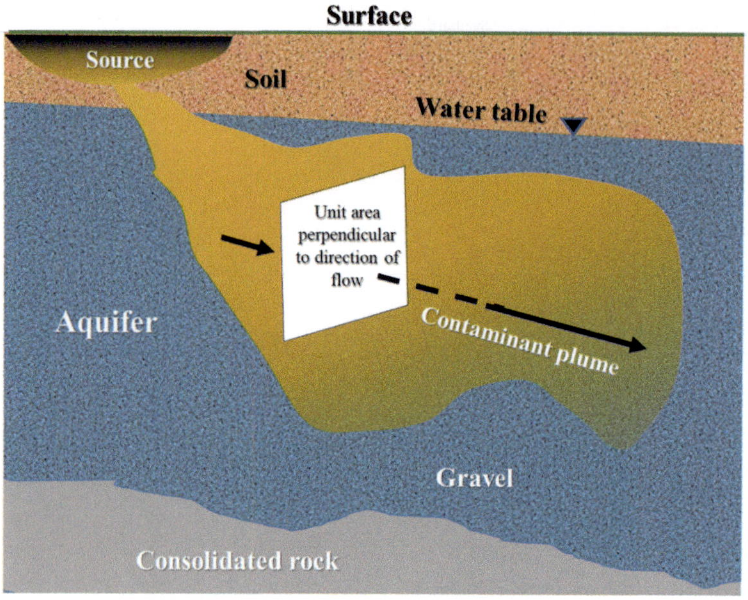

FIG. 11.11 Plume of pollutants flowing from a mine source into a drinking water aquifer. *(From D.A. Vallero, T.M. Letcher, Unraveling Environmental Disasters, second ed., Elsevier Academic Press, Amsterdam, The Netherlands, 2023; used with permission.)*

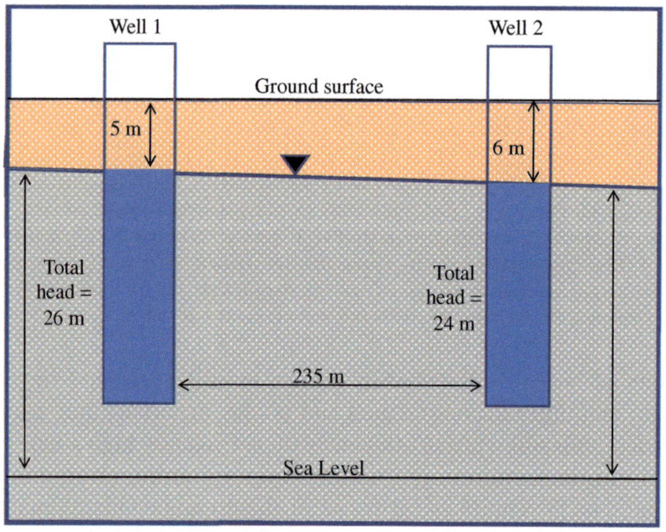

FIG. 11.12 Hydraulic gradient between two wells. The difference in head is 2 m over 235 m, so the hydraulic gradient $= \frac{26\text{m}-24\text{m}}{235\text{m}} = 0.0085\ \text{m}\ \text{m}^{-1}$. *(Data from North Carolina Department of Environment and Natural Resources, Groundwater, aquifers & confining beds. https://www.ncwater.org/?page=560 (Accessed 6 March 2018).)*

As mentioned, the capability of a medium to transmit a fluid, especially water, is known as hydraulic conductivity (K). The various strata under the earth's surface may be consolidated (not broken up) like solid limestone, dolomite, sandstone, or other rock layers. Strata may also consist of unconsolidated and porous materials, like soil, sand, or gravel. Darcy's Law is an empirically derived equation for the flow of fluids through a cross-section of porous media (A). It states that the specific discharge, Q, is directly proportional to the K and hydraulic gradient:

$$Q = -K\frac{dh}{dx} \tag{11.20}$$

In three dimensions, this includes a gradient operator (∇) in vector notation [9]:

$$\vec{q} = -K\nabla h \tag{11.21}$$

TABLE 11.1 Hydraulic conductivities for selected subsurface media.

Unconsolidated deposits	Hydraulic conductivity (m s^{-1})	Media	Hydraulic conductivity (m s^{-1})
Dense clay	10^{-13} to 10^{-8}	Dense sandstone	10^{-9} to 10^{-7}
Weathered clay	10^{-8} to 10^{-6}	Karstic sandstone	10^{-7} to 10^{-5}
Silt	10^{-7} to 10^{-5}	Dense limestone	10^{-9} to 10^{-7}
Alluvial deposits	10^{-5} to 10^{-3}	Karstic limestone	10^{-5} to 10^{-3}
Fine sand	10^{-5} to 10^{-4}	Dolomite	10^{-10} to 10^{-8}
Medium sand	5×10^{-4} to 5×10^{-3}	Dense crystalline rocks	10^{-13} to 10^{-12}
Coarse sand	10^{-4} to 10^{-3}	Fractured crystalline rocks	10^{-10} to 10^{-6}
Fine gravel	10^{-3} to 5×10^{-1}	Dense basalt	10^{-13} to 10^{-10}
Medium gravel	5×10^{-2} to 10^{-1}	Fractured basalt	10^{-7} to 10^{-4}
Coarse gravel	10^{-2} to 5×10^{-1}	Claystone	10^{-13} to 10^{-9}

Data from K. Spitz, J. Moreno, A Practical Guide to Groundwater and Solute Transport Modeling, John Wiley and Sons, 1996. Laboratory of Ecohydrology (ECHO), Darcy's law. https://echo2.epfl.ch/VICAIRE/mod_3/chapt_5/main.htm (Accessed 13 January 2024) (Chapter 5).

Where, $\vec{q}$ is the volumetric flux, i.e., the rate of volume flow across a unit area with dimensions of distance per time because volume (length^3 divided by area-time (length^2 × time)), e.g., m s^{-1}. Thus, it is the same as mean velocity.

Darcy's law assumes that the flow is laminar, and that inertia is neglected. For example, as shown in Fig. 11.12, if $K = 0.5$ m s^{-1}, $\frac{dh}{dx} = 0.0085$ m m^{-1}, and $A = 500$ m^2, then $Q = 2.125$ m^3 s^{-1}. So, in this case, if a leak has produced a plume, the aquifer is being contaminated at a rate of 2.125 m^3 s^{-1}.

Hydraulic conductivity (K) varies greatly among subsurface media (see Table 11.1).

Hydraulic conductivity is not a measure of water quality, but merely the amount of water that can flow within the aquifer. As such, a very small K can be a beneficial feature, such as when a low permeability clay is used as a liner of a hazardous waste pit or a cap on a landfill. In this case, little or no infiltration is ideal. Conversely, a high-K material is ideal in areas like a wetland or floodplain since it allows higher aquifer recharge and flood water infiltration.

The solubility of pollutants and other substances is vitally important to ground water pollution. The solubility of gases in water range over five orders of magnitude, from a low of one molecule of C_2F_6 in 1 million water molecules to 35 molecules of HCl in 100 molecules of water.

Thermodynamicists have a simple way[a] of understanding solubility by first defining an ideal solubility; a solute-solvent system that obey Raoult's and Henry's laws [23]. Raoult's law states that the partial vapor pressure of each component of an ideal mixture of liquids is equal to the vapor pressure of the pure component multiplied by its mole fraction in the mixture. Henry's law states that the solubility of a gas in a liquid is directly proportional to the partial pressure of the gas above the liquid. Thus, the enthalpy of an ideal solution or ideal mixture is zero, i.e., $\Delta H = 0$ (See Chapter 7). Ideal solubility implies that the interactions of the solute molecules, A, with the solute molecules, B, are the same as the A-A and B-B interactions. The ideal solubility for a solute, A, at a temperature T and at a partial pressure of p (1.0 atm) is given by:

$$\mathrm{B-B} \gg \mathrm{A-B} \leq \mathrm{A-A} \quad (11.22)$$

Where $p_A{}^*$ is the vapor pressure of the solute at temperature T.

The relationship between density and organic/aqueous phase partitioning is very important to pollutant transport through ground water, as shown in Fig. 7.6. The transport of the nonaqueous phase liquids (NAPLs) through the vadose zone assumes that the NAPLs have extremely high octanol-water coefficients (K_{ow}) values and extremely low water solubility. That is, they have a greater affinity for lipids than for water. As the aqueous solubility of a substance increases, its flow will increasingly follow the water flow lines.

When a dense, miscible fluid seeps into the zone of saturation, the contaminants move downward if their densities are greater than that of the water. When these contaminants reach the bottom of the aquifer, the shape dictates their continued movement and slope of the underlying bedrock or other relatively impervious layer, which will likely be in a direction other

a. My good friend and colleague, Trevor Letcher, shared this with me.

than the flow of the ground water in the aquifer. Solution and dispersion near the boundaries of the plume will have a secondary plume that will generally follow the direction of ground water flow. The physics of a system indicates that estimating the direction of the plume will entail more than the fluid densities and must include solubility and phase partitioning. So, monitoring wells will need to be installed upstream and downstream from the source to indicate plume movement.

Ground water, soil, and sediment are three-phase systems, consisting of solid particles, liquid water with varying amounts of dissolved or suspended chemical compounds, and gases (see Fig. 11.13).

Porous media can be completely dry, partially saturated, or saturated. Sediments consist predominantly of solids and liquids and are generally saturated unless they are dried near shores or exposed during drought or seasonal conditions. Several methods exist for measuring soil and other porous media water content [32]. Molecular water kinetics include both cohesion and adhesion, both which can be applied to ground water monitoring [33]. Water molecules are attracted to each other because of cohesion, forming droplets. Adhesion is the attraction of water to solid surfaces to solid surfaces, e.g., soil or sediment particles. In addition, water molecules also display surface tension.

The Derjaguin-Landau-Verwey-Overbeek (DLVO) theory explains colloidal stability and the aggregation of aqueous dispersions and the force between charged surfaces that interact through a liquid medium. It combines the effects of the van der Waals attraction and the electrostatic repulsion. In soils, sediment, and other unsaturated porous media, colloids can attach to stationary air-water interfaces in porous media. When a particle sorbs to an air-water interface, forces affect the particle as a result of surface tension and pressure differences [34]. These forces generate a strong attraction of the particle to stationary solid surfaces when the water saturation is low, and air–water interfaces pin the particles to solid surfaces [35–37]. With increasing water saturation, liquid films deepen, and particles are increasingly affected by repulsive forces that detach the particle from the solid surface [38].

Surface tension, σ, occurs at liquid surfaces, i.e., interfaces of liquid-liquid, liquid-gas, liquid-solid. Surface tension is the force in the liquid surface normal to a line of unit length drawn in the surface. Surface tension decreases with temperature and increases or decreases depending on the type of contact fluid. Water has a very high σ value (approximately $0.07\ \mathrm{N\ m^{-2}}$ at 20 °C). Of the environmentally important fluids, only mercury has a higher σ (see Table 11.2). The high

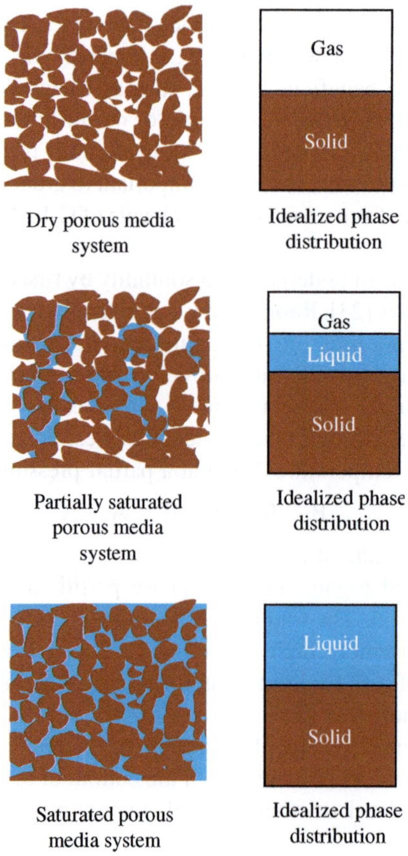

FIG. 11.13 Three-phase porous media system. *(Data from B.M. Das, N. Sivakugan, Introduction to Geotechnical Engineering, Cengage Learning, 2015. B.M. Das, Geotechnical Engineering Handbook. J. Ross Publishing, 2010.)*

TABLE 11.2 Surface tension of selected environmental fluids in the presence of air.

Fluid	Surface tension, σ (N m^{-1} at 20 °C)
Acetone	0.0236
Benzene	0.0289
Ethanol	0.0236
Glycerin	0.0631
Kerosene	0.0260
Mercury	0.519
n-Octane	0.0270
Tetrachloromethane	0.0236
Toluene	0.0285
Water	0.0728

Data from D. Vallero, Environmental Contaminants: Assessment and Control, Academic Press, 2010. C. Burr, Irrigation, in: Soil Water, University of Nebraska, Lincoln. https://passel2.unl.edu/view/lesson/bda727eb8a5a/3, 2021 (Accessed 2 December 2021) (Chapter 3).

surface tension creates a type of skin on a free surface, which is how an object that is denser than water (e.g., a steel needle) can "float" on a still water surface and the cause of insects sitting comfortably on water surfaces. Surface tension is somewhat dependent upon the gas that is contacting the free surface.

Capillarity results from a collective set of forces, i.e., cohesion, adhesion, and surface tension. Adhesive intermolecular forces between a water, or any liquid for that matter, and the solid surface are stronger than the cohesive intermolecular forces between water molecules. Capillarity is a particularly important fluid property of flow through porous media like soil and sediment, and into aquifers. The zone immediately above the zone of saturation in an aquifer is the capillary fringe (see Fig. 11.14). Irrespective of how tightly soil and sediment particles are packed, void or pore spaces exist between the particles. Pore spaces below the top of the zone of saturation, i.e., the water table, are filled completely with water. Above the water table, the spaces are filled with a mixture of air and water. As shown in Fig. 11.14 the spaces between particles are interconnected and behave like small conduits in their capacity to distribute water. Depending on the particle texture and

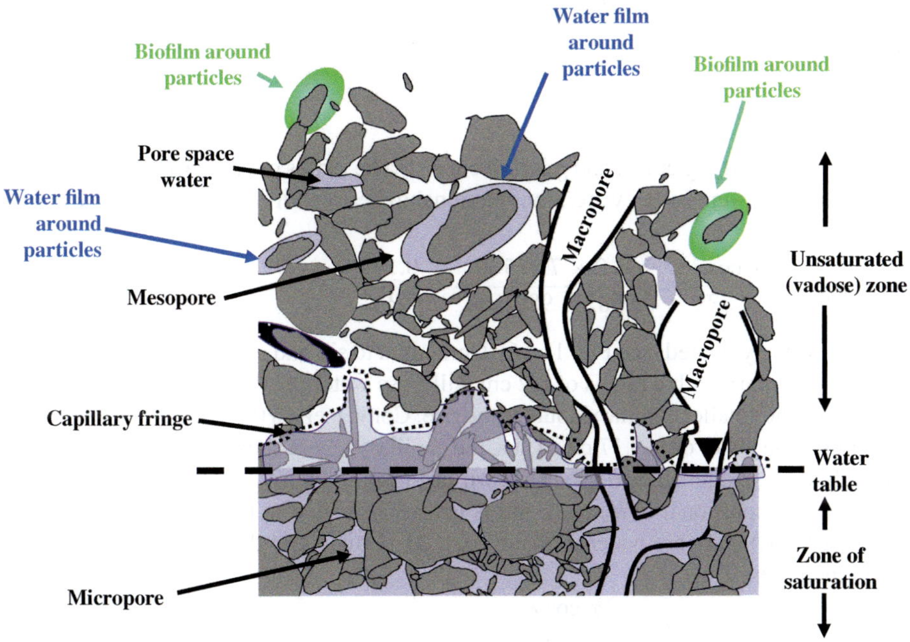

FIG. 11.14 Capillary fringe above a water table of an aquifer. *(Data from D. Vallero, Environmental Contaminants: Assessment and Control, Academic Press, 2010.)*

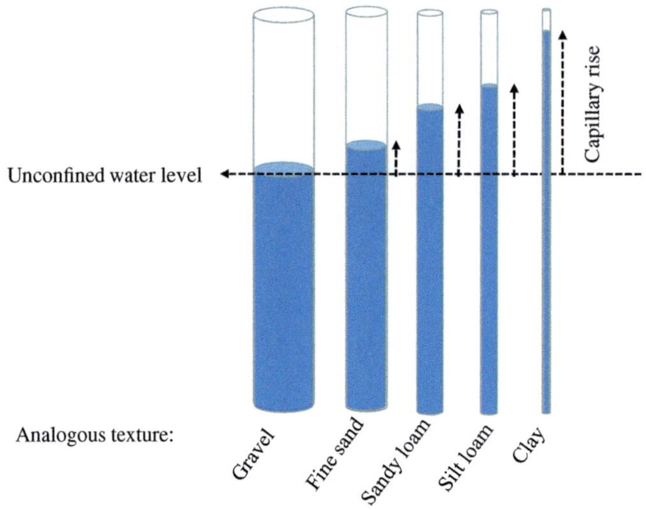

FIG. 11.15 Change in elevation of a fluid in a capillary and analogous soil and sediment texture types. *(Data from D. Vallero, Environmental Contaminants: Assessment and Control, Academic Press, 2010. C. Burr, Irrigation, in: Soil Water, University of Nebraska, Lincoln. https://passel2.unl.edu/view/lesson/bda727eb8a5a/3, 2021 (Accessed 2 December 2021) (Chapter 3).)*

density of packing, the conduits will vary in diameter, ranging from large pores (i.e., macropores), to medium pore sizes (i.e., mesopores), to extremely small pores (i.e., micropores) [6,39,40].

Fluid pressures above the water table are negative with respect to atmospheric pressure, creating tension. Water rises due to adhesion to a surface and due to the cohesion of water molecules to one another. Higher relative surface tension causes a fluid rise to in a tube (or a pore) is indirectly proportional to the diameter of the tube (see Fig. 11.15). Thus, capillarity increases with decreasing diameter of a tube (e.g., iced tea will rise higher in a thin straw in iced tea than a fatter straw). The rise is limited by the weight of the fluid in the tube. The rise ($h_{capillary}$) of the fluid in a capillary is expressed as:

$$h_{capillary} = \frac{2\sigma \cos \lambda}{\rho_w g R} = \frac{\varphi_O}{\rho_w g} \tag{11.23}$$

Where,

$\sigma =$ fluid surface tension (g s^{-2})
$\lambda =$ angle of meniscus (concavity of fluid) in capillary (degrees)
$\rho_w =$ fluid density (g cm^{-3})
$g =$ gravitational acceleration (cm s^{-1})
$R =$ radius of capillary (cm)
$\varphi_O =$ osmotic pressure

The density of water, ρ_w, at 1 atm and 20 °C is 0.9982 g cm^{-3}, and g is 980 cm s^{-1}. The radius of capillarity for water (R_w) is 72.75 dyne cm^{-1}. If a surface is completely covered with water, the relationship between R_w and capillary rise can be calculated as [32]:

$$h_{capillary} = \frac{2\sigma \cos \lambda}{\rho_w g R} = \frac{2 \times 72.75}{0.9982 \times 980 \times R_w} = \frac{0.149}{R_w} \tag{11.24}$$

Thus, height and radius of capillarity are inversely related (see Fig. 11.16). Compared to silt and sand, clay has the smallest pore radii and the highest capillary rise. A capillary with a radius of 0.2 cm will raise water 0.745 cm, whereas a capillary with a radius of 0.1 cm will rise 1.49 cm, i.e., the smaller capillary doubles the rise. Conversely, a 0.5-cm capillary raises the water only 0.30 cm, and 0.9-cm radius raises the water only 0.17 cm. Thus, the capillary fringe above the zone of saturation is extended upwardly mainly by the soils with smaller interstices, as shown in Fig. 11.14.

Note in Fig. 11.16 that matric suction is proportional to $h_{capillary}$ and is inversely related to pore size. In fact, matric suction is the difference in pore air pressure (u_a) and pore water pressure (u_w). As particle size decreases and hence pore size also decreases, shrink soil suction increases. Thus:

$$u_a - u_w = \frac{2\sigma \cos \lambda}{\rho_w g R} \tag{11.25}$$

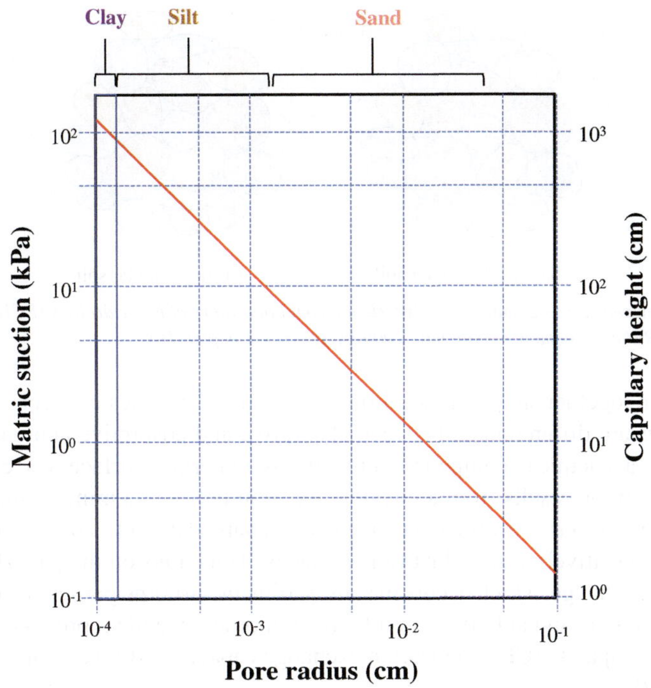

FIG. 11.16 Relationship among capillary height, pore radius, and matric suction for three soil textures. *(Data from Data from Laboratory of Ecohydrology (ECHO), Water potential in porous media, ECHO. https://echo2.epfl.ch/VICAIRE/mod_3/chapt_4/main.htm, 2006 (Accessed) (Chapter 4).)*

Soil compaction is also a factor in matric suction. Soil aggregates can retain more water than the same soil in a compact state. However, as compaction increases, a threshold is crossed where soil retention decreases in the aggregated soil compared to compacted soil. This is caused by the emptying of the aggregate spaces between the particles [41].

The contact angle indicates whether cohesive or adhesive forces dominate during capillary action. When λ values are greater than 90°, cohesive forces are dominant; when $\lambda < 90°$, adhesive forces dominate. Thus, λ is dependent upon both the type of fluid and the surface to which it comes into contact. For example, water-glass $\lambda = 0°$; ethanol-glass $\lambda = 0°$; glycerin-glass $\lambda = 19°$; kerosene-glass $\lambda = 26°$; water-paraffin $\lambda = 107°$; and mercury-glass $\lambda = 140°$. The base of the capillary fringe in the soil is saturated without regard to pore size. In the vadose zone, however, the capillary rise of water will be highest in the micropores, where relative surface tension and the effects of water cohesion are greatest.

The same action depicted in the tubes of Fig. 11.15 occurs in liquids held in porous media, including soil and sediment. The height to which the liquid rises depends on pore size and soil texture (grain size), i.e., the smaller the pores, the higher the rise. Thus, finely textured soils generally have smaller pores than coarsely textured soils. So, these soils have a greater holding capacity and ability to retain water in the soil's or sediment's interstitial spaces. As shown in Fig. 11.14, soils and sediment have both macropores and micropores. The macropores have much less potential for capillarity and water retention than micropores. The opposing forces between capillarity and gravity account for the extent of upward and horizontal movement of water within the soil or sediment profile. The difference in water content between upper and lower layers makes greater or lesser amounts of water available for plant roots.

The amount of resistance to flow when it is acted on by an external force, especially a pressure differential or gravity, is the fluid's viscosity, a crucial fluid property used in numerous environmental applications.

The physicochemical transfer of a chemical "A" from liquid to solid phase is expressed as [42]:

$$A_{(solution)} + \text{solid} = A - \text{solid} \tag{11.26}$$

The interaction of a substance in a fluid with the surface of a solid surface is complex and depends on the properties of the substance and the fluid. Engineers commonly discuss net sorption and net desorption between the fluid and the solid surface, but there are actually different mechanisms that determine the attraction [43]. Adsorption is the process wherein the chemical in solution attaches to a solid surface, which is a common sorption process in clay and organic constituents in soils. This simple adsorption mechanism can occur on clay particles where little carbon is available, such as in ground water. Absorption occurs in porous particles wherein the sorbed substance diffuses into the particle and becomes sorbed to interior surfaces of the particle. This commonly results from short-range electrostatic interactions between the surface and the contaminant. Since both adsorption and absorption occur together in many environmental scenarios, environmental

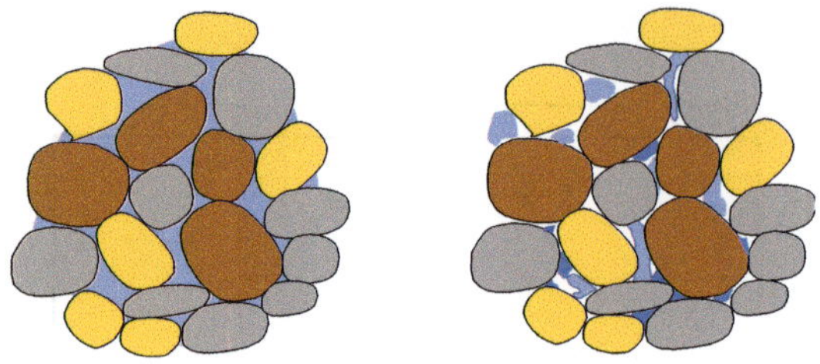

FIG. 11.17 Change in pore water movement as granular soils dry. *(Data from Laboratory of Ecohydrology (ECHO), Water potential in porous media, ECHO. https://echo2.epfl.ch/VICAIRE/mod_3/chapt_4/main.htm, 2006 (Accessed) (Chapter 4).)*

engineers consider them often together simply as sorption. However, the physics between them will differ, which is why pollution control technologies are differentiated between absorptive and adsorptive approaches [44–46].

Chemisorption integrates a chemical compound into porous materials surface via chemical reactions. The most common type of soil chemisorption results from a covalent reaction between a mineral surface and the contaminant.

Positively charged ions, i.e., cations, are attracted to negatively charged particle surfaces or negatively charged ions, i.e., anions, which are attracted to positively charged particle surfaces. Thus, ions on the particle surfaces become displaced. Such sorption is known is ion exchange, which occurs in soils, sediment, airborne particulate matter, or biota, e.g., cysts and pollen particles. Cation exchange is arguably the second most important chemical process on earth, after photosynthesis, given that the cation exchange capacity (CEC) and anion exchange capacity (AEC) are vital to the bioavailability of atmospheric nitrogen to plants [18].

In soil and sediment, a fraction of a chemical compound remains in soil solution and the rest is adsorbed in or adsorbed onto the surfaces of the particles. The strength of the attraction can vary widely. Sorption of chemicals on solid surfaces, therefore, is a key factor in fugacity to air or water, as discussed in Chapters 7 and 14. Sorption also decelerates the rate at which contaminants move downwardly through the soil profile.

Wetting and drying cycles undergo hysteresis, i.e., soil water desorption and sorption rates depend on the past desorption-sorption history. Water sorbed to soil and sediment particles is dried by increasing the amounts of tension and suction by the particles in the matrix. When a saturated soil starts drying, the larger pores drain first, causing the air-water interface to draw increasingly close to the solid soil particles. As the soil moisture decreases, the flow tubes become fragmented and the water travels exclusively through the laminar films around the particles (see Fig. 11.17). If the soils are more cohesive, decreasing soil moisture causes the adsorption complex to thin [41].

For cohesive soils the situation is different. For high degree of saturation, small flow tubes of water are available around the adsorption complex in the vicinity of particles. As the degree of saturation decreases, the adsorption complex becomes thinner, which causes the additional water to enter through a side of the adsorption complex of a soil particle and transferred at the other side to an adjacent particle (see Fig. 11.18). This is a reason for the increased capacity of cohesive soils to transfer water compared to granular soils when soil moisture is low [41].

Drier soil becomes gradually wetter by lowering tension and pressure [47].

The air from the atmosphere enters and exits soil at specific values determined by several factors, including temperature, relative humidity (RH), and atmospheric pressure. The principal means of exchange from the other sources to the atmosphere is by evaporation and transpiration, together known as evapotranspiration, with vapor pressure as the underlying inherent property of water driving these processes. Recall from Chapter 10 that evapotranspiration is a set of water fluxes (J), i.e., movements of energy or matter through a two-dimensional surface:

$$J = \frac{Q}{A} \tag{11.27}$$

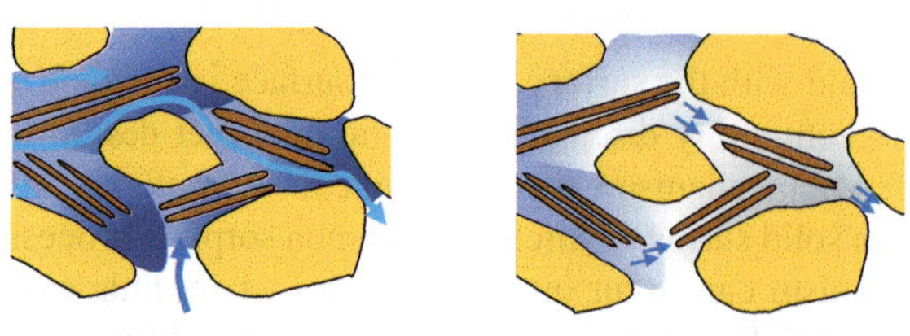

FIG. 11.18 Scheme of water movement in wet and dry cohesive soils. *(Data from Laboratory of Ecohydrology (ECHO), Water potential in porous media, ECHO. https://echo2.epfl.ch/VICAIRE/mod_3/chapt_4/main.htm, 2006 (Accessed) (Chapter 4).)*

Where, Q is the flow rate and A is the surface area (units of length squared) of the source. Q is expressed as volume or mass, e.g., mg s^{-1} or L h^{-1}. Thus, J is expressed in mass or volume per time per area, e.g., L h^{-1} m^2. Evaporative flux (E) is calculated as:

$$E = \frac{mR_n + \gamma \cdot 6.43(1 + 0536 \times U_2) \cdot \delta e}{\lambda_v \cdot (m + \gamma)} \tag{11.28}$$

Where, m is the slope of the saturation vapor pressure curve (kPa K^{-1}), R_n is the net irradiance (MJ m^{-1} day^{-1}), γ is the psychrometric constant (kPa K^{-1}), U_2 is wind speed (m s^{-1}), δe is vapor pressure deficit (kPa), and λ_v is the latent heat of vaporization (MJ kg^{-1}).

The psychrometric constant, γ, is calculated as:

$$\gamma = \frac{0.0016286 \cdot P}{\lambda_v} \tag{11.29}$$

Where P is atmospheric pressure in kPa.

As discussed in Chapter 5, the vapor pressure deficit (VPD) is the difference between the amount of water vapor in the air and the amount of water vapor that the air can hold:

$$\mathrm{VPD} = \overset{\circ}{p}_{saturated} - \overset{\circ}{p}_{actual} \tag{11.30}$$

And,

$$\overset{\circ}{p}_{actual} = \frac{RH \times \overset{\circ}{p}_{saturated}}{100} \tag{11.31}$$

Where, $\overset{\circ}{p}_{saturated}$ is the saturated vapor pressure and $\overset{\circ}{p}_{actual}$ is the actual vapor pressure. Eq. (11.29) is an expression of the air's drying power [48], because it is a major determinant of evapotranspiration, i.e., the combination of transpiration of water vapor from plant tissue, especially leaves, and abiotic evaporation. As such, VPD is an important factor in wetland's wet-dry cycles, as well as in crop management, landscaping, and environmental engineering [49].

Increasing VPD leads to the closure of plant stomata, moderates water loss, and avoids hydraulic breakdown like cavitation. In so doing, carbon uptake is diminished during photosynthesis. Simultaneously, the plant's transpiration rate will increase with increasing VPD until VPD either remains high or begins to increase, which exacerbates plant water stress This can cause widespread damage to forests and other ecosystems. Elevated levels of VPD also speeds up evaporation from moist soils, creating a cycle of drying and warming of soils, which also induces plant water stress. VPD is therefore an indicator of stress on plant life that involves water at the root zone and upper levels of the vadose zone.

The feedbacks between the abiotic and biotic components of an ecosystem can lead to subsequent, cumulative adverse effects [50]. When combined with other stressors, like elevated levels of tropospheric ozone, the dryness contributes to crop and ecosystem stresses. For example, ecological responses to climate change depends on myriad biotic and abiotic factors, e.g., temperature and precipitation patterns and the capacity of plants to acclimate and adapted to these changes. For example, short-term adjustment processes at the stomatal level may occur in response to long-term rise in evaporative demand, especially in mature forest ecosystems in semi-arid region. These adjustments have been shown to be independent of plant functional type [51]. Such stresses further damage the structure and function of the ecosystem, leading to irreversible crashes, such as the loss of diversity and net primary productivity [52].

Saturation vapor pressure is temperature-dependent, i.e., $\overset{\circ}{p}_{saturated}$ increasing with increasing air temperature. By extension, VPD is also temperature dependent.

Therefore, the total water potential (ϕ_T) is combination of several potentials [32]:

$$\phi_T = \phi_p + \phi_m + \phi_z + \phi_o + \phi_{ob} + \ldots \phi_n \tag{11.32}$$

Where, ϕ_p is the pressure potential, ϕ_m is the matric potential, ϕ_z is the gravitational potential, ϕ_o is the osmotic potential, ϕ_{ob} is the overburden potential, and ϕ_n represents the potentials induced by other factors.

Pressure potential is the height of water above a point in the soil, i.e., the water pressure exerted by the saturated column of water above that location in the soil. As mentioned, matric potential is due to capillary and sorptive forces. Gravitational potential is due to the force of gravity measured from a reference point. Osmotic pressure is the solute pressure caused by water's attraction to salts in the water film around the soil particles. Overburden potential is caused by the weight of soil on a given point in the soil column [32,41].

An important process in ground water plume migration is that of co-solvation, the process where a substance is first dissolved in one solvent and then the new solution is mixed with another solvent. As mentioned, with increasing aqueous solubility, a pollutant will travel along the flow lines of the ground- or surface water. However, even a substance with low aqueous solubility can follow the flow under certain conditions. Even a hydrophobic compound like a chlorinated benzene (a dense nonaqueous phase liquid, DNAPL), which has very low solubility in pure water, can migrate into and within water bodies if it is first dissolved in an alcohol or an organic solvent (e.g., toluene). So, a DNAPL will travel downward because its density is less than that of water and is transported in the plume. The DNAPL has undergone co-solvation with the water, so it is transported advectively. Likewise, the ordinarily lipophilic compound can be transported in the vadose zone or upper part of the zone of saturation where it undergoes co-solvation with water and a light nonaqueous phase liquid (LNAPL), e.g., toluene.

Even if a contaminant is not in solution, it can still be transported as a suspension. Fine particles can be held in suspension if the water flow has sufficient energy. When the energy falls below a certain level, the suspended particles will settle. Many suspended contaminants in ground water are small colloidal particles, i.e., so small that Brownian motion provides enough energy to keep them from settling.

When a contaminant enters soil, some of its mass remains in soil solution and some is adsorbed onto the surfaces of the soil particles. Sometimes this sorption is strong due to cations adsorbing to the negatively charged soil particles. In other cases, the attraction is weak. Sorption of chemicals onto solid surfaces needs to be understood because in this way solids can hold onto contaminants, not allowing them to move freely with the pore water or the soil solution. Therefore, sorption slows that rate at which substances move downwardly through the soil profile.

Compounds eventually establish a balance between the mass on the solid surfaces and the mass that is in solution. Molecules will migrate from one phase to another to maintain this balance. The properties of both the chemical compound and the soil (or other matrix) will determine how and at what rates the molecules partition into the solid and liquid phases. These physicochemical relationships, known as sorption isotherms, are found experimentally.

Other properties are important in ground water contamination. Notably, density, capillarity, and surface tension are among the fluid properties that can greatly affect the rate of contaminant movement into an aquifer, as well as which part of an aquifer will be most contaminated (e.g., strong surface tension could increase sorption onto soil particles). Hazardous waste sites are notorious for ground water contamination and pollutant transport. Among the first and most infamous occurred in upstate New York (see Discussion Box 11.1).

Bedrock is not the only layer in rock strata that underlay an aquifer. As mentioned the aquitard, is a geological formation of a semipervious nature that transmits water at slower rates than an aquifer [54]. An aquitard is the less-permeable bed in a stratigraphic sequence, which may be sufficiently permeable to transmit water in substantial quantities of regional ground water flow, but its permeability is not enough for productive water extraction. For example, in an interlayered sequence, the sand layers can be aquifers, but clays, shale, and silty clays and other less pervious layers would be aquitards acting as stratigraphic barriers [55].

As shown in Fig. 11.19, ground water can also exist in permeable rock like sandstone or unconsolidated materials like sand below an impervious layer and above another impervious rock layer. An aquifer between two impervious layers is known as a confined aquifer [54].

Flow into and through an aquifer can be a substantial transport mechanism. Hydraulic head (h) is the mechanical energy per unit weight of ground water. Ground water flows from positions of high hydraulic head to positions of lower head. The flow rate is directly proportional to the rate at which head decreases with distance along the path from high h to low h.

Thus, as mentioned, h at a particular point in an aquifer is expressed as the elevation of the point above sea level or other baseline added to the height to which the water will rise in a vertical tube whose lower end is open at the point and whose upper end is open to the atmosphere. At a point in the aquifer, h equals the elevation of the water surface in a well that is screened at that point.

Head and elevation can be equal or different, depending on the pressure in the aquifer. Fig. 11.19 indicates that h in the unconfined aquifer at point A equals the elevation or altitude of the water table. At point B, which is nearer to the bottom of the unconfined aquifer, h is the elevation of the water level in the well screened at point B. At point C in the confined aquifer h equals the elevation of the water level in the well which is screened at point C. The confined aquifer h at point C is above the upper boundary of the confined aquifer, which is characteristic of a confined aquifer. If the pressure in the confined aquifer were high enough that the head was higher than the ground surface, this is an artesian system, i.e., the well at point C would flow freely at the ground surface, obviating the need for a pump [56,57].

Water in an unconfined aquifer stands in a well at a level known as hydraulic head (h) and flows in the direction from higher to lower head. This change in head measurement is expressed as the hydraulic gradient in the in the x direction $\left(\frac{dh}{dx}\right)$. As mentioned, the ability to transmit water is known as hydraulic conductivity (K). The empirically-derived equation, i.e., Darcy's Law, expresses the flow of fluids through a cross-section of porous media (A). The law states that the specific

Discussion Box 11.1 The Love Canal disaster

The seminal and one of the most infamous cases of ground water contamination occurred at Love Canal, New York. The humanitarian beginnings of the case contradict its infamy. In the nineteenth century, William T. Love foresaw an opportunity for electricity generation from Niagara Falls and the potential for industrial development. To achieve this, Love planned to build a canal that would also allow ships to pass around the Niagara Falls and travel between the two great lakes: Erie and Ontario. The project started in the 1890s, but soon floundered due to inadequate financing and the development of alternating current which made it unnecessary for industries to locate near a source of power production. Hooker Chemical Company purchased the land adjacent to the canal in the early 1990s and constructed a production facility. In 1942, Hooker Chemical began disposal of its industrial waste in the Canal. This was war time in the United States, and there was little concern for possible environmental consequences. Hooker Chemical (which later became Occidental Chemical Corporation) disposed of more than 21,000 tons of chemical wastes including halogenated pesticides, chlorobenzenes, and other hazardous materials into the old Love Canal. The disposal continued until 1952 at which time the company covered the site with soil and deeded it to the City of Niagara Falls which wanted to use it for a public park. In the transfer of the deed, Hooker specifically stated that the site was used for the burial of hazardous materials and warned the city this should govern future decisions on the use of the land. Everything Hooker Chemical did during those years was legal and apparently aboveboard.

About this time, the Niagara Falls Board of Education was looking around for a place to construct a new elementary school and the old Love Canal seemed like a perfect spot. This area was a growing suburb with densely packed single-family residences on streets paralleling the old canal. A school on this site seemed to the board to be an ideal solution and so it was built.

In the 1960s, the first complaints began and intensified during the early 1970s. The ground water table rose during those years and brought to the surface some of the buried chemicals. Children in the school playground were seen playing with strange 55-gal drums that popped out of the ground. The contaminated liquids started to ooze into the basements of the nearby residents, causing odor and health problems. More importantly, perhaps, the contaminated liquid was found to have entered the storm sewers and was being discharged upstream of the water intake for the Niagara Falls water treatment plant.

In 1978, President Jimmy Carter declared the site to be an environmental emergency. This designation required that 950 families be evacuated from an area of 10 square blocks around the canal. In addition to the logistics of relocating and calming the people affected, the disaster response presented a difficult engineering problem. Excavating the waste would have been dangerous work and would probably have caused the death of some of the workers. Excavating the waste would also have exposed it to the atmosphere resulting in uncontrolled toxic air emissions. Finally, there was the question as to what would be done with the waste as it was removed. No single solution such as incineration would have been appropriate. The U.S. Environmental Protection Agency (EPA) finally decided that the only appropriate action was to isolate it and continue to monitor and treat the ground water. The contaminated soil on the school site was excavated, detoxified, and stabilized, and the school building itself was razed. All the sewers were cleaned, removing 62,000 tons of sediment that had to be treated and removed to a remote site. The EPA eventually removed the site from its National Priority Listing (NPL), i.e., the worst of the worst hazardous waste sites [53].

Costs were unprecedented. Occidental Chemical paid $129 million and continued to pay for oversight and monitoring. Total disaster response costs may have exceeded $250 million, with the difference being paid by the Federal Emergency Management Agency and by the U.S. Army, the latter which was found to have contributed waste to the canal.

The Love Canal story galvanized the American public into understanding the problems of hazardous waste and was the impetus for the passage of several significant pieces of legislation, such as the Resource Conservation and Recovery Act, the Comprehensive Environmental Response, Compensation, and Liability Act (so-called Superfund), and the Toxic Substances Control Act. In particular, a new approach to assessing and addressing ground water contamination problems had evolved.

discharge, Q, is directly proportional to aquifer's K and the hydraulic gradient. For example, assume that abandoned waste site is leaking into the ground. If $K = 0.02\ \text{m s}^{-1}$, $dh/dx = 0.05\ \text{m m}^{-1}$, and $A = 100\ \text{m}^2$, then $Q = 0.1\ \text{m}^3\ \text{s}^{-1}$. The rate at which the contaminants are flowing into the vadose zone $0.1\ \text{m}^3\ \text{s}^{-1}$.

In Fig. 11.19, h at point A is greater than h at B which is greater than h at C. Thus, given that flow is from higher h to lower h, the apparent ground water flow would be downward in the unconfined aquifer and confining bed to the confined aquifer. This apparent flow gradient condition could be the result of pumping water and consequently lowering the pressure in the confined aquifer. The gradient is apparent because the difference in pressure, i.e., piezometric, since the two aquifers are separated by the confining layer.

If water is pumped from the unconfined aquifer at a rate faster than the rate of that the aquifer is replenished with water, i.e., recharge, a cone of depression is formed. This is a radially symmetrical funnel of the water table (see Fig. 11.20).

In a confined aquifer, the cone of depression also a radially symmetrical funnel, but not an actual water table depression. Rather the cone is a potentiometric surface that reflects the loss of pressure near the point of discharge, e.g., the uncased part

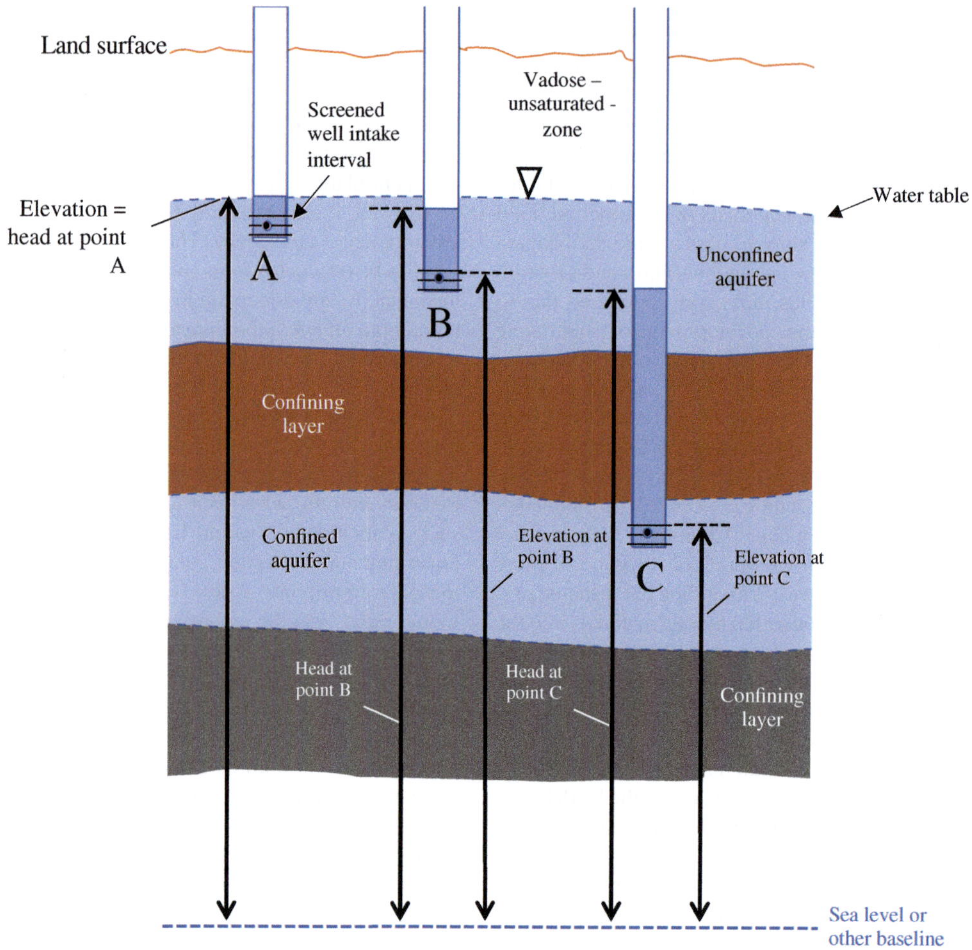

FIG. 11.19 Comparison of head and elevation for confined and unconfined aquifers. *(Data from D.A. Vallero, Soil and sediment physics, in: Methods and Calculations in Environmental Physics, AIP Publishing LLC, Melville, New York, 2022, pp. 8-1–8-34.)*

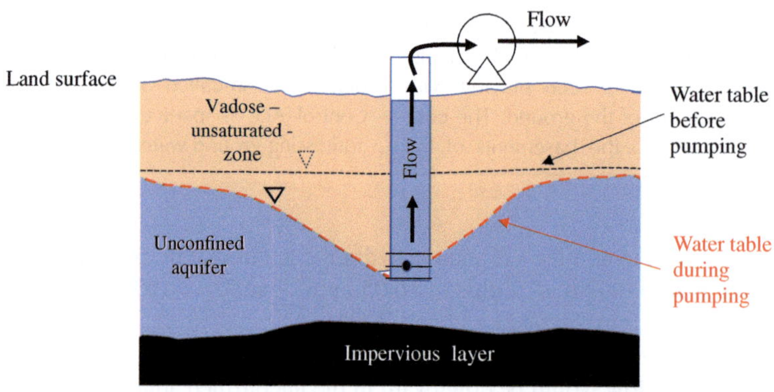

FIG. 11.20 Cone of depression in an unconfined aquifer. *(Data from D.A. Vallero, Soil and sediment physics, in: Methods and Calculations in Environmental Physics, AIP Publishing LLC, Melville, New York, 2022, pp. 8-1–8-34.)*

of the well tube (see Fig. 11.21). The pressure around the well is decreased by the drawdown, i.e., the change in h from before pumping compared to the h during pumping. Thus, drawdown will be at a maximum near the point of entry to the well. An aquifer can have several cones of depression at each pumping sites (Fig. 11.22), at which the cones may intersect, i.e., the drawdown is depressing the water table in the unconfined aquifer or expanding the cone of depression in the confined aquifer.

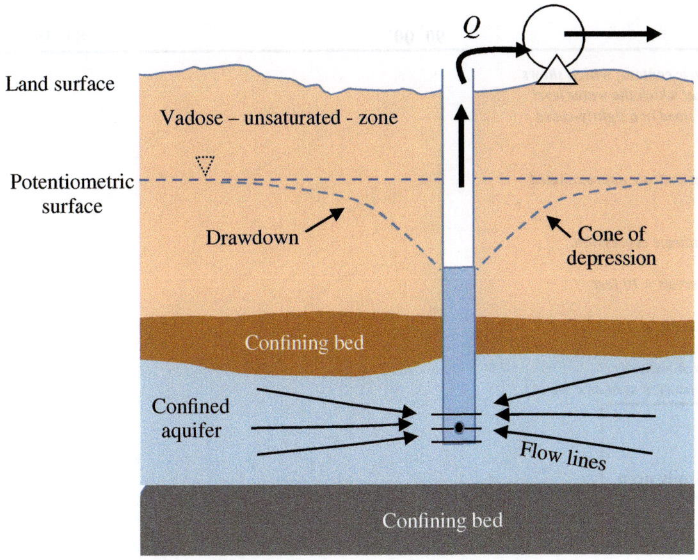

FIG. 11.21 Cone of depression in a confined aquifer. *(Data from D.A. Vallero, Soil and sediment physics, in: Methods and Calculations in Environmental Physics, AIP Publishing LLC, Melville, New York, 2022, pp. 8-1–8-34.)*

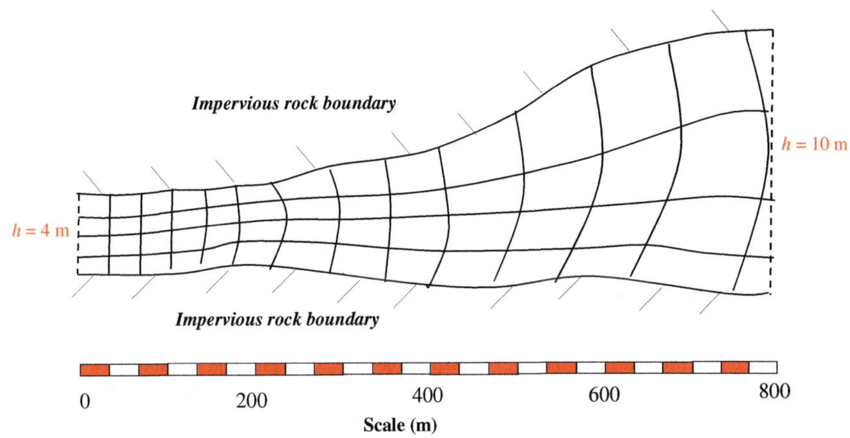

FIG. 11.22 Flow net of a hypothetical aquifer with a uniform thickness over its area between two confining rock layers. *(Data from H.F. Hemond, E.J. Fechner, Chemical Fate and Transport in the Environment, Elsevier, 2014.)*

Flow nets are visualization tools for ground water flow that include streamlines and isopotential lines. Streamlines follow the path of water parcels and isopotentials, which are perpendicular to streamlines, have constant h values [55,58–60]. Thus, flow nets represent the horizontal movement of ground water, and the plane of the flow net represents the horizontal aquifer surface where beneath each point on the surface flow all depths m of the aquifer are equal (see Fig. 11.23). Flow boundaries are considered to be streamlines since water is flowing parallel to them. When a boundary acts as a source or sink for water they are lines of known h. A line where h is constant, e.g., the shore of an estuary or lake, is an isopotential [9].

For most flow nets, the flow is predominately two-dimensional and isotropic, but anisotropic flow can be converted into isotropic flow. The media is assumed to be homogeneously porous and flow is assumed to be the same at all depths in the aquifer [9].

Transmissivity (T) is the unit width of an aquifer under a unit hydraulic gradient, expressed as the product of the average hydraulic conductivity and thickness (m) of the saturated portion of an aquifer [4]:

$$T = Km$$

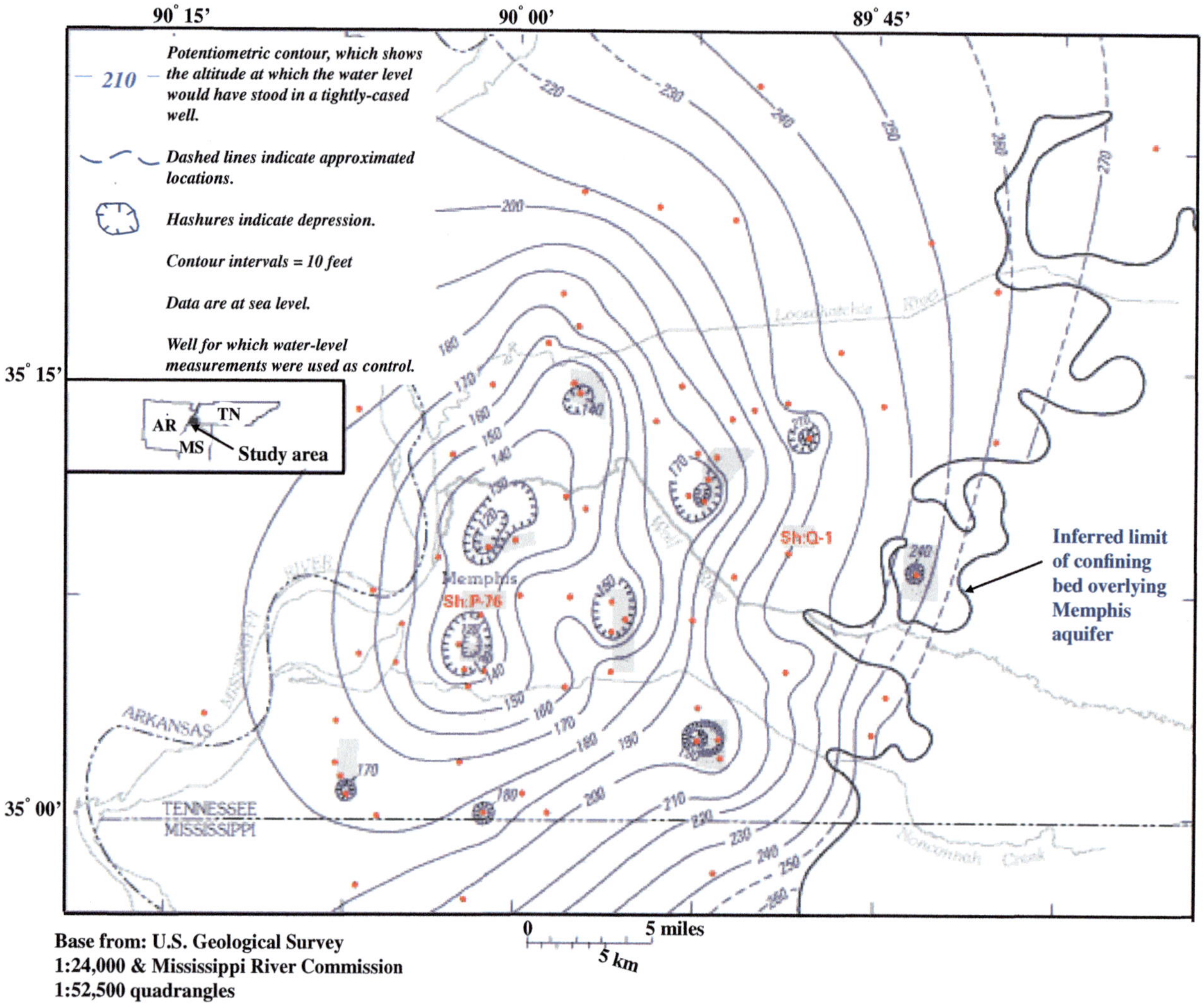

FIG. 11.23 Potentiometric surface of the Memphis aquifer in 1995 displaying cones of depression and locations of observation wells (Sh:P–76 and Sh:Q–1). *(Modified from J.A. Kingsbury, Altitude of the potentiometric surfaces, September 1995, and historical water-level changes in the Memphis and Fort Pillow aquifers in the Memphis area, Tennessee, 1996.)*

As shown in Fig. 11.24, for example, T is 1 m times the aquifer thickness (m), under a hydraulic gradient of 1 m per 1 m in a given time, often a day. Thus, T equals hydraulic conductivity (K) times aquifer thickness (m), and is expressed as length per time, e.g., m day^{-1}.

The ability of an aquifer to store water is also important for both water quantity and water quality. The storage coefficient (S) is the volume of water (Q) released from an aquifer per 1 unit of surface area (A) per 1 unit length change in head (h):

$$S = \frac{\text{Q in m}^3}{(A \text{ in m}^2) \times (h \text{ in m})} \tag{11.33}$$

Note that S is dimensionless. The storage coefficient differs for unconfined versus confined aquifers. In a confined aquifer, water derived from storage depends on the expansion of water as the aquifer is depressurized during pumping and the compression of the aquifer. As such, most confined aquifers have S values that range from 10^{-5} to 10^{-3} [4]. In an unconfined aquifer, the main source of water results from gravity with little expansion of water. Thus, the storage coefficient is much higher, ranging from about 0.1 to about 0.3. [4].

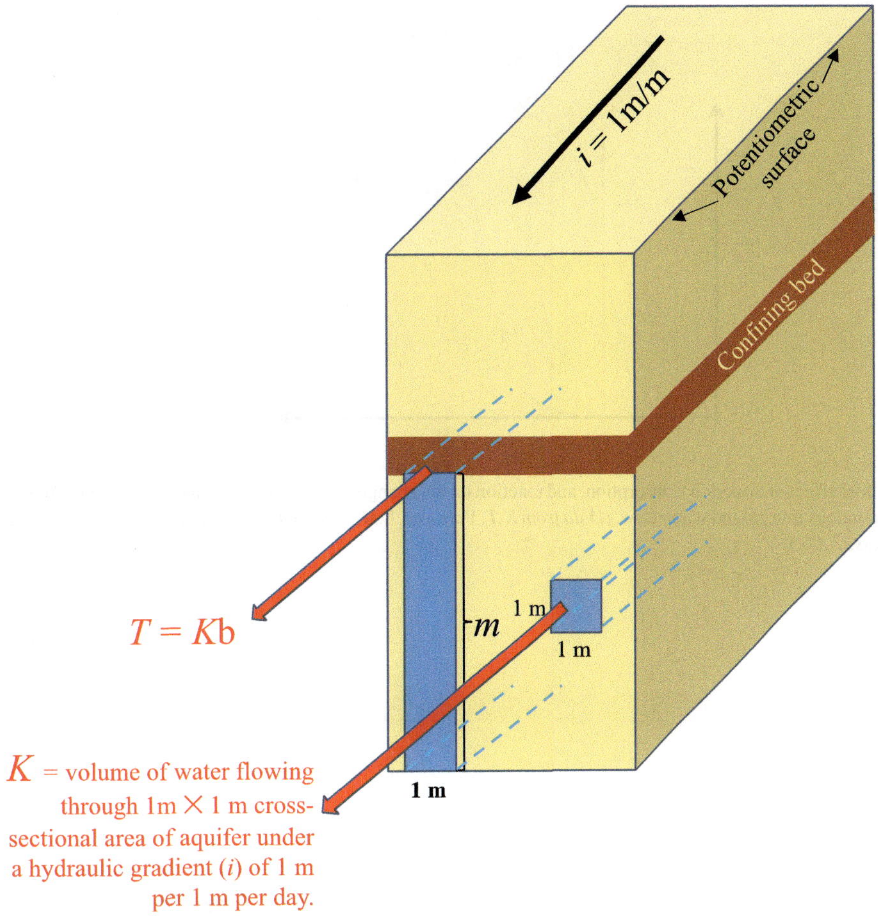

FIG. 11.24 Transmissivity of an aquifer. Note, K is hydraulic conductivity. *(Data from Utah Division of Water Rights, Basic hydrology, North Carolina Division of Water Resources. https://waterrights.utah.gov/wellinfo/theis/hydrogeology_discussion.asp#:~:text=If%20we%20think%20about%20our,)(ft.%2F1) (Accessed 16 January 2024). F.G. Driscoll, Groundwater and wells, Johnson Filtration Systems, U.S. Filter, Inc., St. Paul, Minnesota, 1986, p. 1986.)*

It is important to keep in mind that this, like many other hydrogeological calculations of flow, is somewhat steady state in the sense that the pollutant load is constant. The contaminant transport rate will be less than the water flow rate unless the contaminant always moves advectively with the flow. In fact, this is seldom the case, since other processes are occurring simultaneously, especially dispersion, adsorption, and reaction, that are also involved in pollutant transport. With steady state flow, the change in concentration is assumed to stay the same (dashed line in Fig. 11.25). Dispersion dilutes the concentration. Sorption to soil and other particles delays the transport. Reactions will decrease the concentrations of the parent compound but will produce other compounds that also have to be considered in the mass balance, and which may also reach the aquifer [61].

5 Ground water treatment

Ground water can be treated by drilling recovery wells to pump contaminated ground water to the surface. Commonly used ground water treatment approaches include air stripping, filtering with granulated activated carbon (GAC), and air sparging.

Air stripping moves volatile compounds from water to air within in a packed column or tower or in a low-profile sieve tray aeration system (see Fig. 11.26). In a tower filled with a permeable material, ground water moves by gravity downward while a stream of air flows upward. The configuration includes a spray nozzle, or a manifold connected to

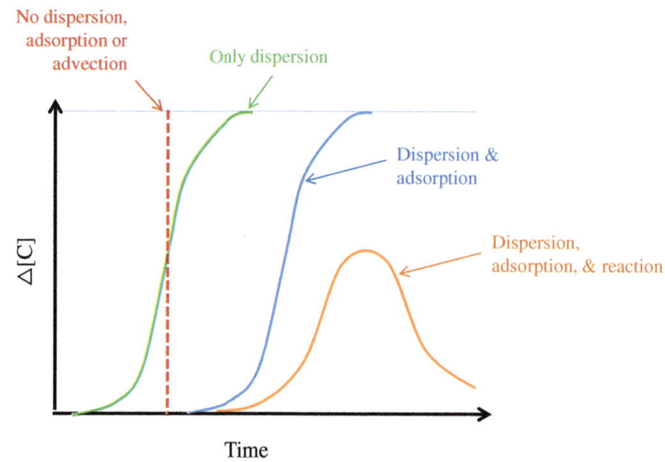

FIG. 11.25 Hypothetical effect of dispersion, absorption, and reaction on the change in the concentration of a pollutant, [Δ*C*] at a single point in the zone of saturation from a continuous into ground water flow. *(Data from K.T. Valsaraj, E.M. Melvin, Elements of Environmental Engineering: Thermodynamics and Kinetics, CRC Press, 2000.)*

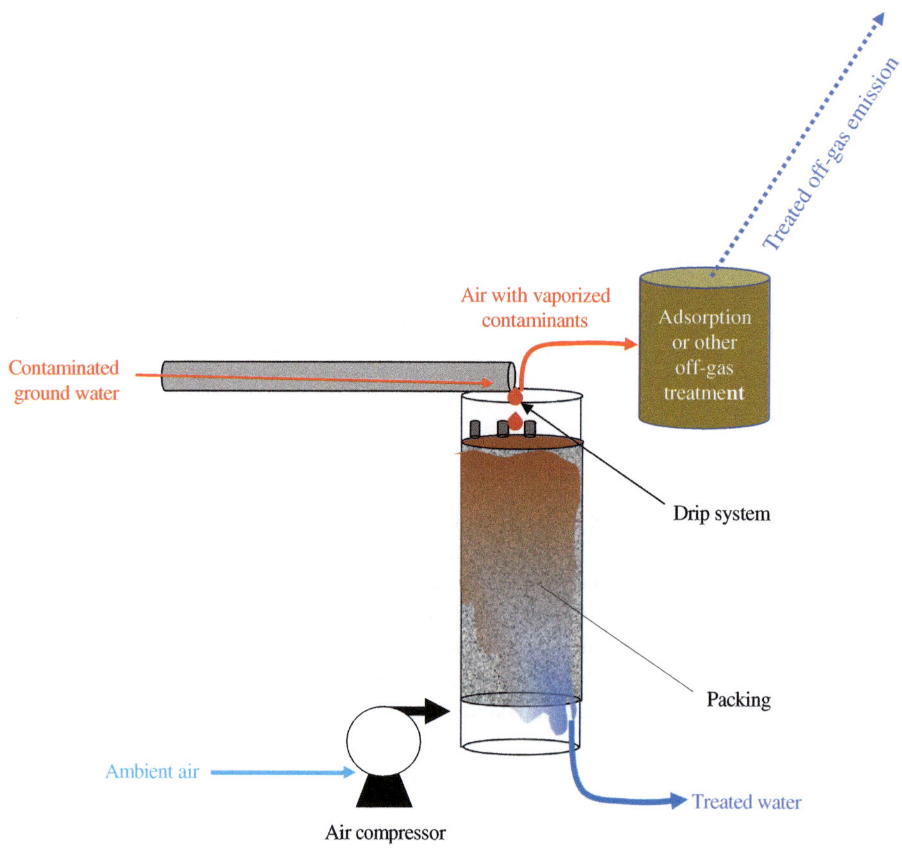

FIG. 11.26 Air stripping system to treat volatile compounds in water. *(Data from D. Vallero, Environmental Biotechnology: A Biosystems Approach, Elsevier Science, 2015.)*

several nozzles, at the top of the tower to spread contaminated water over the column packing in the column, a blower to move air in the countercurrent to the contaminated water flow, and a sump at the bottom to collect treated water. Low-profile sieve tray air strippers are configured as a horizontal tray or vertical box design with a series of stacked trays. Baffles and weirs inside the stripper trays move contaminated water twice or more along the length of the box tray [62].

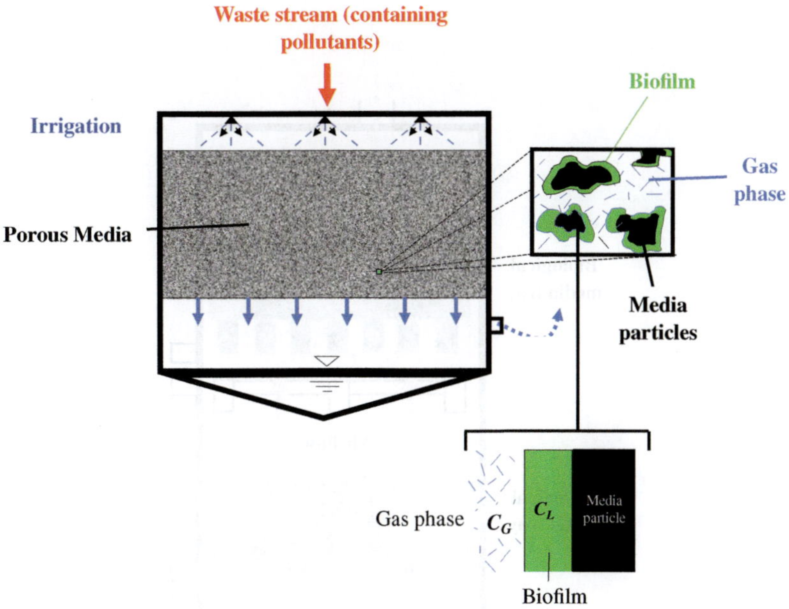

FIG. 11.27 Packed bed biological control system to treat volatile compounds. Air containing gas phase pollutants (C_G) moves through porous media. *(Data from D.A. Vallero, Fundamentals of Air Pollution, fifth ed., Elsevier Academic Press, Waltham, MA, 2014, p. 999 pages cm. S.J. Ergas, E.D. Schroeder, D.P. Chang, R.L. Morton, Control of volatile organic compound emissions using a compost biofilter, Water Environ. Res. 67 (5) (1995) 816–821. S.J. Ergas, K.A. Kinney, Biological control systems, in: Air pollution Engineering Manual, 2000, p. 55.)*

Air sparged through a vent pipe or through the bottom of the column passes in a countercurrent direction through the water to strip higher compounds with relatively high vapor pressures, e.g., volatile organic compounds (VOCs) like benzene or xylenes. These are considered single-pass systems although a portion of the effluent can be separated and recirculated into the column to improve removal efficiency and to keep the height of the packed tower reasonable, or to optimize the number of trays in a low-profile system. Diffused aeration strippers have aeration tanks in which volatile compounds are stripped by bubbling air into a tank through which the contaminated water is flowing. The forced air blower and an aeration-sparge manifold increase the contact of the water with air [63].

Contaminated ground water can also be filtered by pumping the water through the GAC to trap the contaminants. Air sparging consists of pumping air into the ground water *in situ*. A soil venting system is often combined with an air sparging system for vapor extraction [63].

Microbial populations can be added to the porous media to enhance removal of vapor phase compounds. The air carrying these volatile compounds is passed through a layer of biologically active media, where the microbes degrade the compounds to simpler compounds. The major difference between biofiltration and the trickling systems discussed in Chapter 10 owes to the liquid interfaces with the microbes. The liquid phase is stationary in a biofilter (see Fig. 11.27), where the soluble fraction of volatilized compounds in an air stream will partition into the biofilm (C_L) according to Henry's Law [64–66]:

$$C_L = \frac{C_G}{H} \tag{11.34}$$

Where H is the Henry's Law constant.

Wastewater treatment methods like trickling filter systems are also useful in treating ground water that has been pumped to the treatment facility. In this technology, the liquids move through the porous media (see Fig. 11.28). Bacteria and other microbes can reside in various porous media, such as compost, which usually has a very diverse microbial community of myriad species of bacteria and fungi [67–72].

As is the case for secondary wastewater treatment, ground water treatment efficiency depends on the flow rates and inherent properties of the pollutants being treated, e.g., sorption, vapor pressure, fugacity, and solubility. For biological treatment these and other factors influence the ability of microbes to produce biofilm that is needed for the target pollutant to enter the microbes' biofilm, and to achieve the desired pollutant loading rates [18]. For abiotic processes, such as the reduction of halogenated organic pollutant using zero-valent iron (Fe^0), the loading of reactants to produce less toxic end products must be optimized based on experimental studies specific to the contaminated site [73–76].

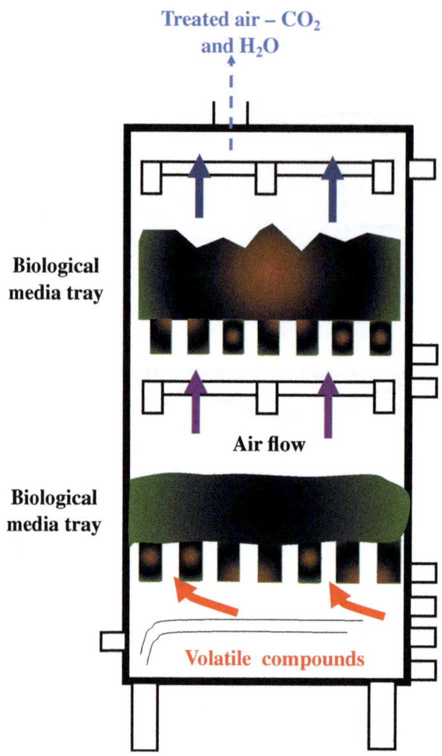

FIG. 11.28 Biofiltration without a liquid phase to treat vapor phase pollutants. Air carrying the volatilized contaminants upward through porous media (e.g., compost) that are hospitable to microbes that have acclimated to this microenvironment and decompose the contaminants. The wastes at the bottom of the system can be heated to an optimal temperature to increase the partitioning to the gas phase. Microbes in the biofilm surrounding each compost particle metabolize the contaminants into simpler compounds, ultimately converting them into carbon dioxide and water vapor. *(Data from D.A. Vallero, Fundamentals of Air Pollution, fifth ed., Elsevier Academic Press, Waltham, MA, 2014, p. 999 pages cm.)*

6 Conclusions

Ground water is a major source of drinking water and other uses throughout the world. Its quality and quantity are threatened by many human activities. This chapter discussed the principles underlying its movement, how it becomes contaminated, and how it can be remediated.

References

[1] Statistics Canada, Section 2: Canada's water supply—stocks and flows, in: Human Activities and the Environment, Statistics Canada, Ottawa, Ontario, 2010.

[2] European Environment Agency, Europe's groundwater—a key resource under pressure, EEA, 2022. https://www.eea.europa.eu/publications/europes-groundwater#:~:text=Groundwater%20stores%20almost%20one%20third,of%20water%20for%20agricultural%20irrigation. (Accessed 9 July 2024).

[3] W.M. Alley, T.E. Reilly, O.L. Franke, Sustainability of ground-water resources, 0607930403, 1999, Available: https://pubs.usgs.gov/circ/circ1186/html/intro.html.

[4] Utah Division of Water Rights, Basic hydrology, North Carolina Division of Water Resources, Raleigh, North Carolina, 2006. In this issue https://waterrights.utah.gov/wellinfo/theis/hydrogeology_discussion.asp#:~:text=If%20we%20think%20about%20our,)(ft.%2F1). (Accessed 9 July 2024).

[5] G. Blake, Bulk density, in: Methods of Soil Analysis: Part 1 Physical and Mineralogical Properties, Including Statistics of Measurement and Sampling, vol. 9, 1965, pp. 374–390.

[6] D. Hillel, J.L. Hatfield, Encyclopedia of Soils in the Environment, Elsevier, Amsterdam, The Netherlands, 2005.

[7] D. Hillel, Soil physics, in: R.A. Meyers (Ed.), Encyclopedia of Physical Science and Technology, third ed., Academic Press, New York, 2003, pp. 77–97.

[8] N.P. Cheremisinoff, Groundwater Remediation: A Practical Guide for Environmental Engineers and Scientists, John Wiley & Sons, 2017.

[9] H.F. Hemond, E.J. Fechner, Chemical Fate and Transport in the Environment, Elsevier, 2014.

[10] M. Karamouz, A. Ahmadi, M. Akhbari, Groundwater Hydrology: Engineering, Planning, and Management, CRC Press, Boca Raton, Florida, 2020.
[11] A.R. Lund, T.K. Gates, J. Scalia IV, Characterization and control of irrigation canal seepage losses: a review and perspective focused on field data, Agric. Water Manag. 289 (2023) 108516.
[12] A.D. Ward, S.W. Trimble, Environmental Hydrology, CRC Press, Boca Raton, Florida, 2003.
[13] E.A. Curl, B. Truelove, The Rhizosphere, Springer Science & Business Media, 2012.
[14] L.V. Rinaudi, W. Giordano, An integrated view of biofilm formation in rhizobia, FEMS Microbiol. Lett. 304 (1) (2010) 1–11.
[15] S.S. Sindhu, A. Sehrawat, B.R. Glick, The involvement of organic acids in soil fertility, plant health and environment sustainability, Arch. Microbiol. 204 (12) (2022) 720, https://doi.org/10.1007/s00203-022-03321-x.
[16] R.P. Schwarzenbach, P.M. Gschwend, D.M. Imboden, Organic acids and bases: acidity constant and partitioning behavior, Environ. Org. Chem. (1993) 245–274.
[17] V. Evangelou, Environmental Soil and Water Chemistry, John Wiley, 1998.
[18] D. Richter, Environ 721: Soil Resources, Mentioned in course lecture, Duke University, Durham, North Carolina, 1997.
[19] R.R. Schindelbeck, et al., Comprehensive assessment of soil quality for landscape and urban management, Landsc. Urban Plan. 88 (2–4) (2008) 73–80.
[20] S. Assouline, Infiltration into soils: conceptual approaches and solutions, Water Resour. Res. 49 (4) (2013) 1755–1772.
[21] L.A. Richards, Capillary conduction of liquids through porous mediums, Physics 1 (5) (1931) 318–333.
[22] A. Klute, Some theoretical aspects of the flow of water in unsaturated soils, Soil Sci. Soc. Am. J. 16 (2) (1952) 144–148.
[23] D.A. Vallero, T.M. Letcher, Unraveling Environmental Disasters, second ed., Elsevier Academic Press, Amsterdam, The Netherlands, 2023.
[24] D.A. Vallero, Soil and sediment physics, in: Methods and Calculations in Environmental Physics, AIP Publishing LLC, Melville, New York, 2022, pp. 8-1–8-34.
[25] D.A. Vallero, Physics of ecosystems, in: Methods and Calculations in Environmental Physics, AIP Publishing LLC, Melville, New York, 2022, pp. 5-1–5-38.
[26] D. Vallero, Air Pollution Calculations: Quantifying Pollutant Formation, Transport, Transformation, Fate and Risks, Elsevier, 2019.
[27] K. Lajtha, J. Jones, Trends in cation, nitrogen, sulfate and hydrogen ion concentrations in precipitation in the United States and Europe from 1978 to 2010: a new look at an old problem, Biogeochemistry 116 (2013) 303–334.
[28] R.L. Runkel, D.M. McKnight, H. Rajaram, Modeling hyporheic zone processes, Adv. Water Resour. 26 (9) (2003) 901–905, https://doi.org/10.1016/S0309-1708(03)00079-4.
[29] C. Gandy, J. Smith, A. Jarvis, Attenuation of mining-derived pollutants in the hyporheic zone: a review, Sci. Total Environ. 373 (2–3) (2007) 435–446.
[30] M.A.A. Mehedi, M.M.S. Yazdan, M.T. Ahad, W. Akatu, R. Kumar, A. Rahman, Quantifying small-scale hyporheic streamlines and resident time under gravel-sand streambed using a coupled HEC-RAS and MIN3P model, Eng 3 (2) (2022).
[31] D.A. Vallero, Methods and Calculations in Environmental Physics, AIP Publishing, Melville, New York, 2022.
[32] M.K. Shukla, Soil Physics: An Introduction, CRC Press, 2013.
[33] L. Tophoff, H. Schüttrumpf, F. Heimbecher, C. Vollmer, B. Philipp, Y. Abdullaeva, Field investigations and service life assessment of geosynthetic filters in tidally influenced revetments in cases of ochre clogging, Water 16 (13) (2024) 1840, https://doi.org/10.3390/w16131840. In this issue.
[34] N. Chatterjee, S. Lapin, M. Flury, Capillary forces between sediment particles and an air–water interface, Environ. Sci. Technol. 46 (8) (2012) 4411–4418.
[35] A. Bozon, et al., Effect of heterogeneous hydrophobic coating on floating of insoluble particles, Powder Technol. 395 (2022) 592–603.
[36] B. Gao, et al., Capillary retention of colloids in unsaturated porous media, Water Resour. Res. 44 (4) (2008).
[37] J. Wan, T.K. Tokunaga, Film straining of colloids in unsaturated porous media: conceptual model and experimental testing, Environ. Sci. Technol. 31 (8) (1997) 2413–2420.
[38] J. Shang, M. Flury, Y. Deng, Force measurements between particles and the air-water interface: implications for particle mobilization in unsaturated porous media, Water Resour. Res. 45 (6) (2009).
[39] D. Vallero, Environmental Contaminants: Assessment and Control, Academic Press, 2010.
[40] H. Peng, T. Lu, The mechanisms of water transport in the capillary fringe: sandbox experiments and numerical studies, Int. J. Environ. Sci. Technol. (2021) 1–12.
[41] Laboratory of Ecohydrology (ECHO), Water potential in porous media, in: Groundwater Hydrology, ECHO, 2006, https://echo2.epfl.ch/VICAIRE/mod_3/chapt_4/main.htm (Accessed 9 July 2024) (Chapter 4), In this issue.
[42] W. Lyman, Transport and transformation processes, in: Rand GM Fundamentals of Aquatic Toxicology. Effects, Environmental Fate, and Risk Assessment, second ed., Taylor & Francis, Washington, DC, 1995, pp. 449–492.
[43] D.A. Vallero, Fundamentals of Air Pollution, fifth ed., Elsevier Academic Press, Waltham, MA, 2014. p. 999 pages cm.
[44] K. Schifftner, Air Pollution Control Equipment Selection Guide, CRC Press, 2021.
[45] J. Sorrels, J. Coburn, K. Bradley, D. Randall, EPA Air Pollution Control Cost Manual, United States Environmental Protection Agency, Washington, DC, 2017.
[46] K.E. Noll, Adsorption Technology for Air and Water Pollution Control, CRC Press, 1991.
[47] H.D. Scott, Soil Physics: Agricultural and Environmental Applications, Iowa State University Press, Ames, Iowa, 2000.
[48] J. Monteith, M. Unsworth, Principles of Environmental Physics: Plants, Animals, and the Atmosphere, Academic Press, 2013.
[49] D.A. Vallero, Translating Diverse Environmental Data Into Reliable Information: How to Coordinate Evidence From Different Sources, Academic Press, 2017.

[50] C. Grossiord, VPD in the climate equation and its impact on vegetation, Swiss Federal Research Institute WSL, Birmensdorf, Switzerland, 2021. https://www.wsl.ch/en/about-wsl/locations/wsl-lausanne/vpd-in-the-climate-equation-and-its-impact-on-vegetation/. (Accessed 9 July 2024).
[51] C. Grossiord, et al., Tree water dynamics in a drying and warming world, Plant Cell Environ. 40 (9) (2017) 1861–1873.
[52] C. Grossiord, et al., Plant responses to rising vapor pressure deficit, New Phytol. 226 (6) (2020) 1550–1566.
[53] U.S. Environmental Protection Agency, Superfund Site Search Results: All Sites, 2024, https://cumulis.epa.gov/supercpad/CurSites/srchrslt.cfm?start=1 (Accessed 9 July 2024).
[54] R. Freeze, J. Cherry, Hydrogeology, Prentice-Hall, Englewood Cliffs, New Jersey, 1979.
[55] Z. Şen, Groundwater hydraulics and confined aquifers, in: Z. Şen (Ed.), Practical and Applied Hydrogeology, Elsevier, Oxford, 2015, pp. 99–208 (Chapter 3).
[56] R.C. Heath, Basic ground-water hydrology, US Geological Survey, 2004.
[57] R.C. Heath, Basic ground-water hydrology: US geological survey water-supply paper 2220, US Geological Survey, Washington, DC, 1983.
[58] E. Poeter, P. Hsieh, Graphical construction of groundwater flow nets. The groundwater project, Guelph, Ontario, Canada, in: Domain Editors, 2020, p. 5.
[59] M.P. Anderson, W.W. Woessner, R.J. Hunt, Applied Groundwater Modeling: Simulation of Flow and Advective Transport, Academic Press, 2015.
[60] S. Mahenthiran, M. Ganesan, L.V. Rajkumar, N. Sridhar, Quantification and classification of groundwater-lake water exchange using conventional method, Ecology, Environment and Conservation, vol. 27, EM International, Maharashtra, India, 2021, pp. S267–S273. ISSN 0971–765X; Feb Suppl. Issue, 2021. In this issue https://www.envirobiotechjournals.com/issues/article_abstract.php?aid=11157&iid=324&jid=3. (Accessed 9 July 2024).
[61] K.T. Valsaraj, E.M. Melvin, Elements of Environmental Engineering: Thermodynamics and Kinetics, CRC Press, 2000.
[62] U.S. Environmental Protection Agency, Air stripping (ex situ). https://frtr.gov/matrix/Air-Stripping-Ex-Situ/#Implementability-Considerations. (Accessed 22 January 2022).
[63] D. Vallero, Environmental Biotechnology: A Biosystems Approach, Elsevier Science, 2015.
[64] D.C. McKinney, M.-D. Lin, Pump-and-treat ground-water remediation system optimization, J. Water Resour. Plan. Manag. 122 (2) (1996) 128–136.
[65] J. Mercer, D. Skipp, D. Giffin, Basics of pump-and-treat ground-water remediation technology. Special report, GeoTrans, Inc., Herndon, VA (United States), 1990.
[66] J.F. Keely, Performance evaluations of pump-and-treat remediations, in: Ground Water Issue, US Environmental Protection Agency, Center for Environmental Research, Ada, Oklahoma, 1989. EPA/540/4-89/005. Available: https://www.epa.gov/sites/default/files/2015-06/documents/performance_eval_pump_treat.pdf. (Accessed 23 November 2021).
[67] C. Hort, S. Gracy, V. Platel, L. Moynault, Evaluation of sewage sludge and yard waste compost as a biofilter media for the removal of ammonia and volatile organic sulfur compounds (VOSCs), Chem. Eng. J. 152 (1) (2009) 44–53.
[68] C. Ima, D. Mann, Physical properties of woodchip: compost mixtures used as biofilter media, 2007.
[69] W. Namkoong, J.-S. Park, J.S. VanderGheynst, Biofiltration of gasoline vapor by compost media, Environ. Pollut. 121 (2) (2003) 181–187.
[70] R. Nicolai, K. Janni, Biofilter media mixture ratio of wood chips and compost treating swine odors, Water Sci. Technol. 44 (9) (2001) 261–267.
[71] B. Cárdenas-González, S.J. Ergas, M.S. Switzenbaum, Characterization of compost biofiltration media, J. Air Waste Manage. Assoc. 49 (7) (1999) 784–793.
[72] S.J. Ergas, E.D. Schroeder, D.P. Chang, R.L. Morton, Control of volatile organic compound emissions using a compost biofilter, Water Environ. Res. 67 (5) (1995) 816–821.
[73] W. Liang, et al., Recent advances of carbon-based nano zero valent iron for heavy metals remediation in soil and water: a critical review, J. Hazard. Mater. 426 (2022) 127993.
[74] A.N. Garcia, Y. Zhang, S. Ghoshal, F. He, D.M. O'Carroll, Recent advances in sulfidated zerovalent iron for contaminant transformation, Environ. Sci. Technol. 55 (13) (2021) 8464–8483.
[75] A.M. Zafar, M.A. Javed, A.A. Hassan, M.M. Mohamed, Groundwater remediation using zero-valent iron nanoparticles (nZVI), Groundw. Sustain. Dev. 15 (2021) 100694.
[76] H.-L. Lien, D.W. Elliott, Y.-P. Sun, W.-X. Zhang, Recent progress in zero-valent iron nanoparticles for groundwater remediation, J. Environ. Eng. Manag. 16 (6) (2006) 371.

Chapter 12

Water quality indicators

1 Introduction

The condition of an aquatic ecosystem depends on a complicated amalgam of myriad variables. Many physical, chemical, and biological factors are addressed throughout this book, but it is impossible to articulate every factor accurately using sound science. If that were possible, it would still not account for the systematic aspects of water quality, with many synergies and antagonisms among the factors. To address both, indicators can give a sense of the condition of water bodies. This chapter describes several composite or indirect indicators of environmental conditions of water bodies.

2 Oxygen demand

As discussed in Chapter 10, when molecular oxygen (O_2) concentrations decline in water it is an indication that the system is being stressed. Changes in dissolved O_2 levels are not only intrinsically important in supporting aquatic organisms, but they indirectly show that chemical, physical, and biological changes are occurring and give a qualitative picture of potential sources of the stress on surface water systems.

2.1 Biochemical oxygen demand

As discussed in Chapters 6, 10, and 14, biochemical oxygen demand (BOD) is a general indirect indicator of the quality of water at various locations in surface waters. It indicates the amount of O_2 that microbes consume as they decompose organic matter under aerobic conditions. The larger the BOD, the more stress that a water body is receiving from point and nonpoint sources of oxygen-demanding pollutants [1]. BOD_5 is merely the measured dissolved oxygen (DO) at time zero, i.e., the initial DO (D_1), measured immediately after it is taken from the effluent minus the DO of the same effluent measured exactly 5 days after D_1, i.e., D_5:

$$\mathrm{BOD} = \frac{D_1 - D_5}{P} \tag{12.1}$$

Where, P = decimal volumetric fraction of water in the sample container. D units are in $\mathrm{mg\,L^{-1}}$. If the dilution water is seeded, the calculation becomes:

$$\mathrm{BOD} = \frac{(D_1 - D_5) - (B_1 - B_5)f}{P} \tag{12.2}$$

Where, B_1 = initial DO of seed control; B_5 = final DO of seed control; and f = the ratio of seed in sample to seed in control $= \frac{\%\text{seed in } D_1}{\%\text{seed in } B_1}$. B units are in mg $\mathrm{L^{-1}}$.

Thus, the microbial population's demand for DO over the 5-day period indirectly indicates that there are upstream O_2-demading pollutants that have been discharged into a stream or other water body. Five days is the generally accepted time for BOD, in which usually O_2 demand is about 80% complete. Longer times will give more O_2 demand, but 100% completion is usually not observed even after 20 days [2].

When available carbon is a limiting factor, the carbonaceous BOD reaches a plateau, i.e., the water has reached the ultimate carbonaceous BOD (see Fig. 10.12 in Chapter 10). However, carbonaceous compounds are not the only substances demanding oxygen. Microbial populations will continue to demand O_2 from the water to degrade other compounds, especially nitrogenous compounds, which account for the upward bump toward the tail of the BOD curve. Thus, in addition to serving as an indication of the amount of O_2 needed for biological treatment of the organic matter, BOD gives a guide to

Fundamentals of Water Pollution. https://doi.org/10.1016/B978-0-443-28987-3.00002-3

sizing a wastewater treatment process, assign its efficiency, and give operators and regulators information about whether the facility is meeting its design criteria and is complying with pollution rules and regulations [3].

If effluent with high BOD concentrations reaches surface waters, it may diminish DO to levels lethal certain aquatic life. As the water body re-aerates due to mixing with the atmosphere and by algal photosynthesis, O_2 is added to the water, so the oxygen levels will slowly increase downstream. The decreases and increases in DO concentrations downstream from a source of BOD is known as the DO sag curve, i.e., the concentration of dissolved oxygen "sags" as the microbes deplete it. So, the O_2 concentrations fall with both time and distance from the point where the high BOD substances enter the water (see Fig. 12.1).

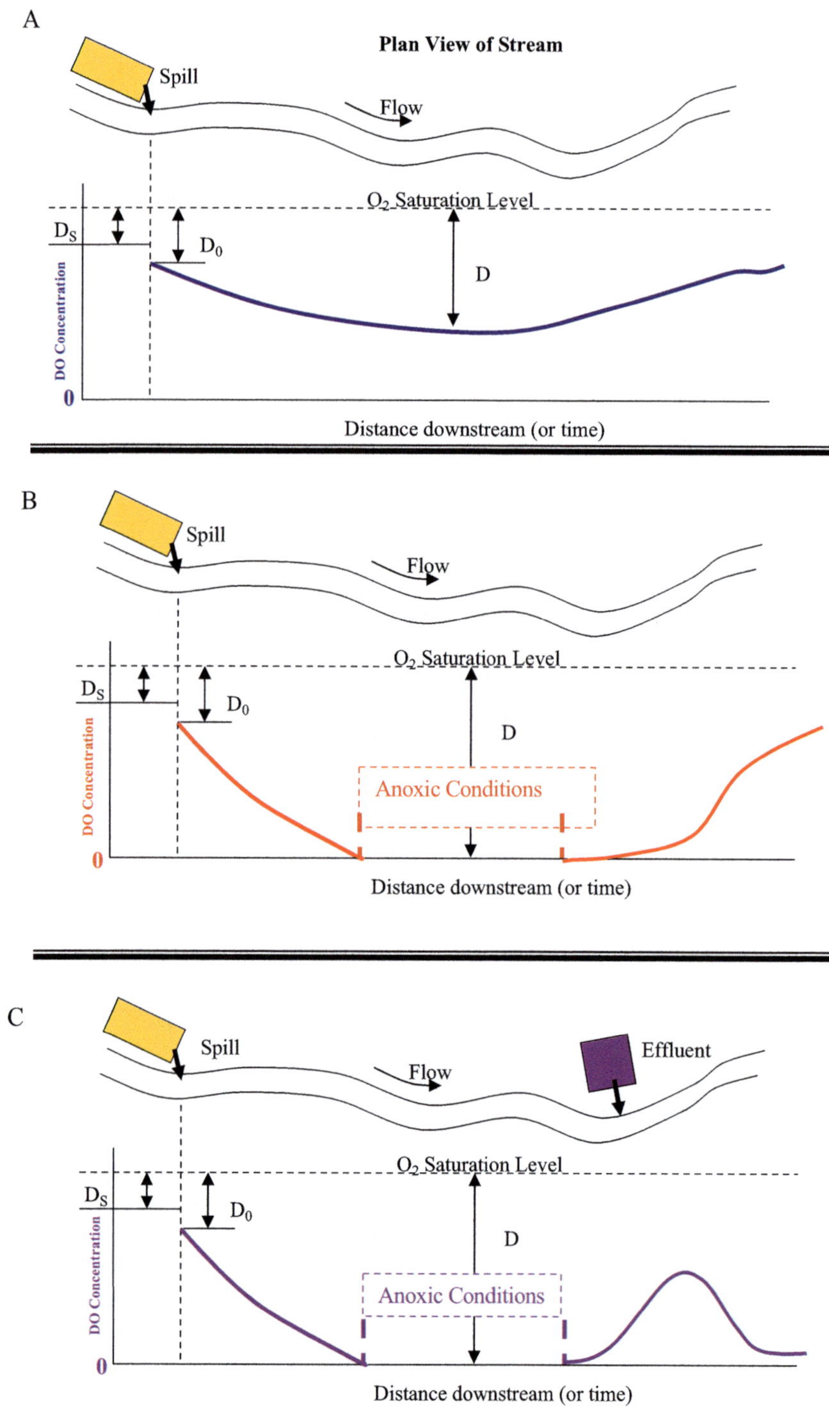

FIG. 12.1 Dissolved oxygen (DO) sag curve downstream from an O_2-depleting contaminant release to a stream. The concentration of dissolved oxygen in Curve A remains above 0. Although the available oxygen is reduced, the system remains aerobic. Curve B sags until dissolved oxygen falls to 0, and anaerobic conditions result and continue until the DO concentrations begin to increase. D_S is the background oxygen deficit before the pollutants enter the stream. D_0 is the oxygen deficit after the pollutant is mixed. D is the deficit for contaminant A which may be measured at any point downstream. This indicates both distance and time of microbial exposure to the DO demanding pollutant. For example, if the stream's average velocity is $5\,km\,h^{-1}$, D measured 10km downstream also represents 2h of microbial activity to degrade the pollutant. Curve C indicates the stream DO levels are recovering from anoxic conditions, but a downstream effluent source again increases the DO demand and begins to form a new sag.

The stress on aquatic life due to decreasing DO is indirectly indicated by the BOD. Water bodies that receive pollutant loads require that microbial populations adapt. Sometimes they deploy mechanisms to outlast the low DO conditions through dormancy, e.g., production of cysts. The DO concentrations are usually a net result of inputs and outputs of oxygen due to increased nutrient loads. Given that the various genera of biota vary in their tolerance ranges and metabolic processes to produce cells and to obtain energy, the net amount of oxygen in a waterbody varies in time and space. For example, both algae and bacteria demand O_2 for growth and respiration, drawing DO from the water, but algae also undergo photosynthesis, adding O_2.

2.2 Chemical oxygen demand

Chemical oxygen demand (COD) does not differentiate between biologically available and inert organic or abiotic matter, and it is a measure of the total quantity of oxygen required to oxidize all organic material completely to carbon dioxide (CO_2) and water. The driving principle of COD is that nearly all organic compounds can be completely oxidized to CO_2 with a strong oxidizing agent under acidic conditions [4,5].

COD values always exceed BOD values from the same sample. COD ($mg L^{-1}$) is measured by oxidation using potassium dichromate ($K_2Cr_2O_7$) in the presence of sulfuric acid (H_2SO_4) and silver. By convention, 1 g of carbohydrate or 1 g of protein accounts for about 1 g of COD. On average, the ratio BOD:COD is 0.5. If the ratio is <0.3, the water sample likely contains elevated concentrations of recalcitrant organic compounds, i.e., compounds that resist degradation [3]. That is, there are numerous carbon-based compounds in the sample, but the microbial populations are not efficiently using them for carbon and energy sources. This is the advantage of having both BOD and COD measurements. Sometimes, however, COD measurements are conducted simply because they require only a few hours compared to the 5 days or more needed for BOD. There are also real-time BOD simulations, e.g., by spectrophotometry, which identifies potential DO-demanding organic compounds by visible or ultraviolet light absorbance according to the Beer-Lambert Law, which states that the amount of light energy absorbed is proportional to chemical concentration.

2.3 Sediment oxygen demand

Many biogeochemical processes affect organic matter in surface water sediments. As mentioned, microbes decompose the organic matter and consume O_2. They also decompose and oxidize sediment organic matter and release the decomposition products into the surface water column. Benthic (bottom-dwelling) organisms alter the sediment structure by bioturbation, i.e., the physical movement of soil or sediment by animals forage and by plant roots [6]. Anoxic conditions can exist very near the top of the sediment, e.g., a few millimeters from the sediment interface with the water column. This produces strong concentration gradients that contribute to diffusion into the sediments as oxygen is used by microbes to fuel decomposition of organic matter, exerting a sediment oxygen demand (SOD) on the water column. Organic particle deposition rates and composition, electron acceptor availability, and benthic community composition are at work in sediment dynamics [7].

SOD measurements are conducted using 18–24 in.-diameter cylindrical opaque chambers. Smaller diameter opaque chambers or cores may also be utilized if SOD rates obtained are verified to be statistically similar to the larger chamber rates. Typically, two replicates for water column respiration and four replicates for sediment respiration rates are collected and analyzed. When conducting SOD measurements in small streams, one water column respiration chamber and two sediment contact chambers are usually deployed. Chambers should be placed on the bottom and sealed, with a DO probe placed inside the chamber. The DO concentration within the chamber is monitored and recorded every 5–15 minutes to measure the rate of sediment O_2 utilization. Generally, 12–24 readings over a period of 1.5–2 hours are sufficient as input to a regression plot to display the SOD rate [8].

Water samples may also be taken from the SOD chambers to estimate the rate of nutrient flux or nutrient exchange at the sediment-water column interface. Isolating a known volume of water over a known area of sediment gives nutrient flux based on periodic extraction of water samples from each chamber and analyzing and interpreting the change of concentration of nutrients over a sufficiently long incubation period. Nutrient flux estimates are typically made coincidentally with SOD measurement. The DO monitoring will likely need minor adjustments because of this extended time. Similar to SOD samples, the nutrient samples should be collected from blank chambers to account for water column changes in the calculation [8].

3 Conductivity

The ability of water to pass an electrical current is conductivity. Dissolved salts, i.e., metal-nonmetal ionic compounds, and other inorganic chemicals conduct electrical current, which means that salt concentrations are directly correlated with conductivity. In natural waters, the salts consist of various cations, but mostly sodium (Na^+), calcium (Ca^{2+}), magnesium (Mg^{2+}), and potassium (K^+). The anions are mainly chloride (Cl^-), sulfate (SO_2^{4-}), nitrate (NO_3^-), and bicarbonate (HCO_3^-). Only chemical compounds with highly electronegative atoms and polar covalent bonds ionize. If the local coordination polyhedra of the anions that surround the individual cations are symmetric and the overall charge distribution in the solid is approximately isotropic, the compound will develop polar covalent bonds [9]. So, since they are nonpolar, organic compounds generally are not good conductors of electrical current and, as a result, have a low conductivity in water. Conductivity is also affected by temperature. Increasing temperature increases conductivity [10].

Conductivity is useful as a general, screening level water quality indicator. Every water body has a relatively constant range of conductivity, which means that, once established, the conductivity for that water body can be used as a baseline for comparison with regular conductivity measurements. Significant changes in conductivity would indicate that a salt or inorganic pollution has increased due to an upstream source, whether point or nonpoint [10].

Conductivity can be measured in the field and in a laboratory. In most cases, it is preferable that the water samples be collected in the field and transported to a laboratory for more accurate and precise testing. For samples taken in the field for later measurement, the sample bottle should be made of inert material, e.g., glass or polyethylene, and washed in phosphate-free detergent and rinsed thoroughly with tap water, followed by distilled water. If factory-sealed, disposable sample containers, e.g., Whirl-pak® bags, are used for sampling, no preparation is needed [11].

Conductivity sampling equipment includes a conductivity meter and probe (for field testing), a conductivity standard solution appropriate for the range typical of the water body, and a data sheet or record-keeping device for recording results [11].

The conductivity meter used in the field or in the lab requires the following steps [11]:

1. Set up the conductivity meter per the manufacturer's directions.
2. Use a conductivity standard solution, e.g., potassium chloride (KCl) or sodium chloride (NaCl), to calibrate the meter for the range to be measured. The manufacturer's directions should describe the preparation procedures for the standard solution.
3. Rinse the probe with distilled or deionized water.
4. Select the appropriate range beginning with the highest range and working down. Read the conductivity of the water sample. If the reading is in the lower 10% of the range, use the next lower range. If the conductivity of the sample exceeds the range of the instrument, the sample may be diluted according to the manufacturer's directions. Incidentally, the dilution might not have a direct linear relationship to the conductivity.
5. Rinse the probe with distilled or deionized water and repeat step 4 until finished.
6. Send samples to the lab with a trusted delivery system and documented chain-of-custody protocol for conductivity analysis.
7. Refrigerate samples until testing.
8. Testing must be completed within 28days of collection.

4 Turbidity

Like conductivity, turbidity is a common standard for surface water quality. Such standards are best tailored to local conditions. Alaska, for example, has experienced increases of 5 nephelometric turbidity units (NTUs) above ambient turbidity in clear cold-water habitats, presenting a threat to salmonid fish species [12]. Turbidity is the measure of the relative clarity of water, i.e., the amount of light that can penetrate a water body. As such, turbidity is an optical characteristic of water's light scattering. As the intensity of scattered light increases, so does turbidity. Increased turbidity results from increasing suspended particles in the water as well as dissolved chemical compounds that diffract and otherwise change the distribution of sunlight. Particles can be abiotic, such silts and clays, and biotic, such as phytoplankton and other microscopic organisms [13].

Determining turbidity from water samples is entailed by shining a light through the water and reporting NTUs. During periods of base (low) flow, stream colors typically range from nearly clear to clear green. Since stream flow is directly related to turbidity, low flows generally have lower turbidity, e.g., <10 NTU. Conversely, during and immediately after a rain event, nonpoint sources of solids wash into the stream, increasing the total solids (TS) concentration and causing the water's color to become more brownish (see Fig. 12.2). Thus, TS and turbidity levels are directly correlated [13].

In the U.S., turbidity is one of the regulated parameters of drinking water quality. Since NTU correlates directly with the concentration of suspended particles, turbidity in water supply systems that use conventional or direct filtration must never be greater 1 NTU. Water distributed from a public water supply must have turbidity that is less than or equal to 0.3NTUs in

FIG. 12.2 High turbidity levels of tributaries and Paraná River, Argentina. *(Photo credit: National Aeronautics and Space Administration. Data from U.S. Geological Survey, Turbidity and water, Reston, Virginia, 2018. Available from: https://www.usgs.gov/special-topics/water-science-school/science/turbidity-and-water (Accessed 6 June 2018).)*

95% or more of the samples taken each month. Those supplies that use filtration other than the conventional or direct filtration must follow state limits, which typically require that turbidity at no time exceeding 5 NTUs [14].

Turbidity is also a rough screen for pathogens and pollutants in the water supply system. Pathogenic microorganisms can adhere to particles, decreasing the efficiency of disinfection. High turbidity in filtered water indicates inefficient removal of pathogens in a water supply facility. Turbidity may also indicate the presence of breakage of supply mains if biofilms and oxide scales are sloughing or if pollutants are entering the water distribution system through these fissures [14].

5 Biological indicators

The condition and health of organisms in a water body can be measured directly using organisms that integrate physical, chemical, and biological features of a habitat or ecosystem. Organisms can also be used as indicators of unsafe waterbodies. Measures of macroinvertebrate and fish diversity, benthic abundance, and habitat condition can be indicators of the condition of an ecosystem's functions and structures. Indeed, healthy ecosystems have a diverse mix of organisms.

One of the main advantages of bioindicators is that they are incorporators and integrators of synergistic and antagonistic processes, such as when an ecosystem is exposed to a pollutant. They also are early-warning diagnostic tools for potentially upcoming harm.

5.1 Biodiversity and species abundance

Biodiversity is defined as the:

> *"... composition, structure, and function (that) determine, and in fact constitute, the biodiversity of an area. Composition has to do with the identity and variety of elements in a collection, and includes species lists and measures of species diversity and genetic diversity. Structure is the physical organization or pattern of a system, from habitat complexity as measured within communities to the pattern of patches and other elements at a landscape scale. Function involves ecological and evolutionary processes, including gene flow, disturbances, and nutrient cycling" [15].*

It is a general rule that when an ecosystem or habitat loses species or when the abundance of species becomes unbalanced, that system is under stress. Generally, diversity indices are most valuable in comparing changes within the same ecosystem. For example, if a species of fish increases at the expense of another, e.g., carp for trout, the diversity indices would express that change. However, the indices are not as reliable comparing different systems. A biodiversity index is calculated as the number of species in area divided by the number of individuals in that area. It can also be described as species richness, which is the number of unique life forms, evenness, which expresses the equitability among the lifeforms, and heterogeneity, which is the dissimilarity among the lifeforms [16].

The diversity, abundance of biological species, and the efficiency of the survival and thriving of biota indicate ecosystem health [17,18]. The response of biota to pollutant exposures and changes in abiotic and biotic conditions can be estimated, in turn, to indicated how the pollution adversely affects the biotic conditions needed by higher-tier trophic levels of organisms.

Expressions of biodiversity are relative and merely indicators that describe the condition of habitat. As discussed in detail in Chapter 9, one of the commonly used diversity indices is the Shannon-Weiner index of community diversity [19]:

$$H' = \sum_{i=1}^{m} P_i \ln P_i \tag{12.3}$$

This can also be expressed as:

$$H' = -1.44 \sum_{i=1}^{m} \left(\frac{n_i}{N}\right) \ln\left(\frac{n_i}{N}\right) \tag{12.4}$$

Where, H' = index of community diversity; $P_i = n_i/N$; n_i = number (i.e., density) of the ith genera or species; N= total number (i.e., density) of all organisms in the sample; i=1, 2... m; and m=number of genera or species.

The index incorporates the number of species living in a habitat, i.e., species richness and their relative abundance, i.e., evenness. As the value of H' increases, the ecosystem's overall sustainability or stability also increases.

The Simpson index is a dominance index [20]. Again, P_i is the proportion (n/N) of individuals of one species counted in the community (n), which is divided by the total number of individuals in the community (N):

$$D = \frac{1}{\sum_{i=1}^{m} P_i^2} \tag{12.5}$$

The calculation of community diversity using diversity indices can screen for changes in the same habitat over time.

Therefore, an indicator of biodiversity is a genus, family, or order, or a selected group of species from a range of higher taxa whose overall species richness, number of rare species, levels of endemism[a], and other diversity factors reflects the condition of other higher taxa in an ecosystem [21].

5.1.1 Phytoplankton

Phytoplankton, also known as microalgae, contain chlorophyll and undergo photosynthesis to live and grow. Thus, most reside into the upper portion of surface, where light best infiltrates the water. Algae are sensitive to pollution, which is manifested in alteration to phytoplankton population levels and photosynthesis rates [22]. Their rapid reproduction rates and their continuous incorporation of nutrient molecules yield water quality changes faster than many other indicators of pollution's effects on aquatic ecosystems.

Phytoplankton are indicators of the condition of aquatic ecosystems given their sensitivity to changes in water quality. For example, low DO concentrations, excessive nutrient levels, toxic pollutants, and diminished predation adversely affect populations of phytoplankton. Plankton biomass, abundance, and species diversity are indicators of surface water condition [23–26].

Zooplankton are also indicators of water quality. They are a primary or supplemental food source for many marine and freshwater species, so if zooplankton biomass declines, this will affect several layers of tropic status (see Chapter 9). They are also indirect indicators of changes in salinity, competition, and predation [22].

5.1.2 Benthos

Benthic macroinvertebrates are small animals and insect larvae that spend at least some of their lives in the bottom regions of a surface waterbody. They are reliable indicators of the biological condition of waterbodies because they spend all lives in water, are easily collected, and vary in tolerances to pollution. They are readily identifiable in the laboratory and have relatively long lifespans (typically more than a year). Compared to fish and other swimming organisms, benthos are not mobile, so they are less likely to escape exposure to pollutants, meaning their exposure is less episodic and more continuous than that of moving creatures. As a result, macroinvertebrates can better integrate the effects over time. There is ample scientific basis for their use as bioindicators [27].

The abundance and diversity of benthic macroinvertebrates in a waterbody is an indicator of the biological condition of that waterbody. Healthy aquatic systems have biological conditions that support a wide variety and abundance of macroinvertebrate taxa, with a range of pollution tolerability [23,28]. Water samples that are comprised of most or all pollution-

a. Condition of organisms or species native to a geographic location such a desert, lake, river, island archipelago, or other defined zone.

tolerant species or that have little diversity or abundance indicates less than optimal conditions. Biological condition is the most comprehensive indicator of waterbody health [27].

5.2 Enterococci

Enterococci are bacteria found in the intestinal tracts of warm-blooded animals. Their presence in waterbodies indicates that the water is receiving fecal waste. Enterococci are released from wastewater treatment plant effluent, leaking septic systems, stormwater runoff, wildlife, improper land application and runoff of manure, feedlots, and other sources where animals are found.

The presence of enterococci is an indication of a threat to human health. Enterococci concentrations not only indicate the presence of fecal material in water, but also the potential presence of disease-causing bacteria, viruses, and protozoa. Exposure can result from consuming fish or shellfish harvested from waters with fecal contamination [29].

5.2.1 Disease-causing microbes

Coliforms and fecal streptococci indicate sewage contamination of surface and ground water because they are present in human and animal feces. Most species are typically not harmful, but their presence indicates the potential presence of pathogenic bacteria, viruses, and protozoa that also live in human and animal digestive tracts [30–32]. This means that when they are in streams, pathogenic microbes may threaten the safety of swimming and consuming shellfish and other aquatic fauna. Direct testing for the wide variety of pathogen is time-consuming and expensive so, instead, water is often tested for coliforms and fecal streptococci. In addition to health risks, the presence of elevated levels of fecal bacteria also is associated with elevated turbidity, noxious odors, and an increasing BOD.

The ratio of fecal coliforms to fecal streptococci (FC:FS) has been used to indicate the extent to which a water sample contains human wastes. An FC:FS value greater than 4 was thought to indicate human wastes. However, the use of this ratio has questionable diagnostic value. For example, warm shallow streams with elevated dissolved carbon would give misleading values. The value's representativeness would improve by increasing the number of samples taken from a water body and better knowledge of its characteristics [33].

Compared to the release of coliform bacteria into the wastewater of the entire population of a town or city, the number of people infected by and the number excreting pathogens is quite small [32]. The ratio of pathogenic to total coliforms is small. Thus, disease outbreaks are often localized. The survival rate of most pathogenic microbes other than viruses outside of the host is usually short unless they form cysts. Unfortunately, several pathogenic protozoa like *Giardia* and *Cryptosporidium* species do indeed form cysts [34].

Nonpoint and point contribute by complex and stepped processes to the pathogenic loading of waterbodies. Microbes may be directly loaded into a stream or other receiving body and accumulate as nonpoint source loads from a combination of sources. As they move overland and under the soil surface, microbes meet both hostile and accommodating conditions that can decrease or increase their numbers, respectively. The hostile conditions encourage the formation of cysts and spores that are more easily transported by advection and by wildlife than the bacteria themselves. When the conditions change to become more accommodating, the cysts and spores open and generate increased microbial populations. The load from point sources is directly into surface waters but nonpoint sources allow time and distance to increase the growth, metabolism and formation of spores and cysts until, in more accommodating environments, the microbes undergo metabolism and growth. During transport, diverse conditions vary in accommodation and hostility to the survival and growth of the microbial populations, e.g., insufficient or sufficient availability of nutrients and water, and environmental conditions like pH, and temperature [35].

5.3 Net primary productivity and eutrophication

Algal blooms affect light and other exchanges in ecosystems. When nutrient concentrations become sufficiently elevated, especially compounds and ions of nitrogen and phosphorous, the lake or other waterbody is said to become highly enriched. As such, the waterbody is very productive, i.e., has a high net primary productivity (NPP). Optimal NPP values indicate a healthy ecosystem, i.e., NPP must not be too high or too low for a given waterbody. NPP must be accompanied by commensurately acceptable biodiversity and be sufficiently resilient and sustainable. If the microbial and plant growth that depend on light energy from the sun become excessive, along with changes in moisture and nutrient concentrations, the concomitant in NPP levels indicates an unhealthy waterbody.

Mass and energy are distributed continuously among and within a lake or estuary. These distribution can lead to eutrophication [36,37]. Trophic status and interrelationships are altered when excessive concentrations of nutrients generate a cascade of changes in the lake's environmental conditions (see Fig. 10.4. in Chapter 10) [38].

An oligotrophic waterbody which is low in nutrient concentrations (i.e., low enrichment), has low productivity, and has low turbidity (clear water), desirable habitat for sensitive species like game fish. In a mesotrophic waterbody, nutrient loads lead to accumulation of organic matter and sedimentation increases, leading to an which occasional algal bloom. Fishing is generally not severely damaged. However, in a eutrophic waterbody, the water has highly elevated nutrient concentrations and very high NPP, so that DO concentrations are being depleted. This means that algal blooms are frequent, fish kills may occur, and the trophic status changes, e.g., the rough fish abundance increases at the expense of other, more economically value species.

Dissolved organic carbon (DOC) is a primary source of energy for aquatic microbes. Photosynthesis rates depend on the amount and quality of sunlight penetration. Photosynthesis rates can be correlated to CO_2 removal rates from the water or by the O_2 released into the water but balanced by O_2 removal by microbial respiration. NPP measurements should account for respiration and photosynthesis. Gross primary production (GPP) is the total carbon uptake by plankton. Net NPP takes the difference between the carbon uptake by photosynthesis and the carbon loss from respiration (R) [39]:

$$\mathrm{NPP} = \mathrm{GPP} - \mathrm{R} \tag{12.6}$$

Generally, an increase in DO concentrations when plankton are exposed to light is an indirect indicator of increasing NPP, so DO measurements of the water sample in the dark can be compared to DO measurements of the same sample in known light intensity and time of exposure. This would require incubation of the extracted water in a laboratory [39].

5.3.1 Chlorophyll-A

Chlorophyll's chemical formula is magnesium [methyl (3*S*,4*S*,21*R*)-14-ethyl-4,8,13,18-tetramethyl-20-oxo-3-(3-oxo-3-{[(2*E*,7*R*,11*R*)-3,7,11,15-tetramethyl-2-hexadecen-1-yl]oxy}propyl)-9-vinyl-21-phorbinecarboxylatato(2−)-$\kappa^2 N,N'$]. As mentioned, algae that reside in the water column and stream sediment contain chlorophyll within their biomass. The chlorophyll molecule allows algae to absorb energy from sunlight via photosynthesis. As such, chlorophyll is an indirect measure of algal content in surface waters (see Fig. 12.3). Chlorophyll-A is a common constituent of algal biomass, so it indicates patterns in phytoplankton growth [40]. Algal growth is often linked to nutrient concentrations in freshwaters. Nutrient concentrations, e.g., nitrogen (N) and phosphorous (P) in the water are directly linked to the production of phytoplankton biomass and the depletion of dissolved oxygen (DO) in nearly still waters like those in a lake or estuary [41]. However, chlorophyll-A is merely an indirect indicator of nutrient concentrations and the likelihood or presence of eutrophication [41]. Nutrient loading can affect several aspects of an aquatic system [42].

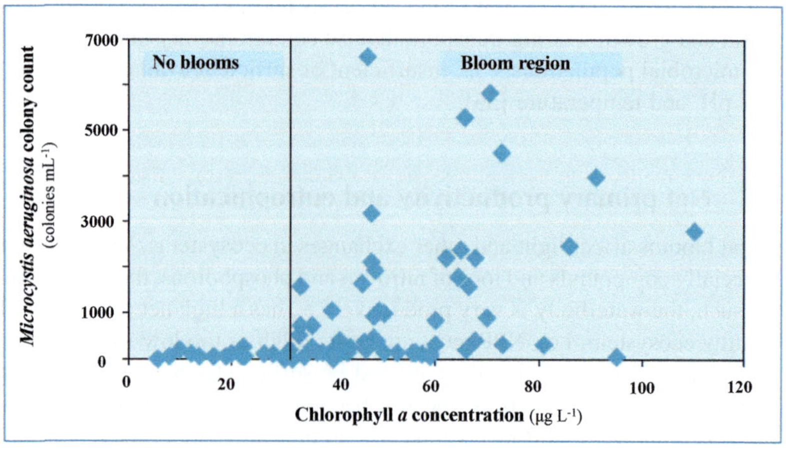

FIG. 12.3 Colony counts of the algal species *Microcystis aeruginosa* compared to the gradient of chlorophyll A, measured in Chesapeake Bay. The *vertical line* depicts the threshold between bloom and non-bloom conditions (approximately 500 colonies mL^{-1} and 30 $\mu g\ L^{-1}$ chlorophyll A. *(Data from: Maryland Department of National Resources (2003). Unpublished data.)*

6 Conclusions

Aquatic ecosystem condition is determined by myriad factors. It is impossible to articulate every factor. This chapter discusses some of the more widely used indicators, but there are many more potentially authoritative indicators of water quality. These include fish assemblage, fish tissue contamination, sediment diatoms, wetland vegetation, introduced species, habitat cover, sediment enzyme concentrations, sediment metal and metalloid concentrations, cyanobacteria abundance, and physical habitat complexity. Indicators give a sense of the synergies and antagonisms among the factors that affect or are affected by the condition of water bodies.

References

[1] D.M. Ashley, D.M. Word (Eds.), Developing watershed protection criteria for Georgia, Georgia Institute of Technology, 1991.
[2] C.H. Hach, R.L. Klein, C.R. Gibbs, Introduction to Biochemical Oxygen Demand, Technical Information Series - Booklet No. 7 (1997). https://cdn.bfldr.com/7FYZVWYB/at/wmhctr9b7rbszxn2w6cb5v/LIT7022.pdf. (Accessed 15 July 2024).
[3] C.P. Gerba, I.L. Pepper, Wastewater treatment and biosolids reuse, in: Environmental Microbiology, second ed., Elsevier Inc, USA, 2009, pp. 503–530.
[4] J.-L. Bertrand-Krajewski, S. Winkler, E. Saracevic, A. Torres, H. Schaar, Comparison of and uncertainties in raw sewage COD measurements by laboratory techniques and field UV-visible spectrometry, Water Sci. Technol. 56 (11) (2007) 17–25.
[5] D. Li, S. Liu, Water Quality Monitoring and Management: Basis, Technology and Case Studies, Academic Press, 2018.
[6] S. Ruiz, P.D. Hallett, D. Or, Bioturbation—physical processes, in: M.J. Goss, M. Oliver (Eds.), Encyclopedia of Soils in the Environment, second ed., Academic Press, Oxford, 2023, pp. 100–114.
[7] S. Arndt, B.B. Jørgensen, D.E. LaRowe, J. Middelburg, R. Pancost, P. Regnier, Quantifying the degradation of organic matter in marine sediments: a review and synthesis, Earth Sci. Rev. 123 (2013) 53–86.
[8] M. Parsons, Sediment oxygen demand, U.S. Environmental Protection Agency LSASD (Ed.), U.S. EPA, Region 4, Athens, Georgia, 2023.
[9] M. Jansen, J.C. Schön, 9.35—Structure prediction in solid-state chemistry as an approach to rational synthesis planning, in: J. Reedijk, K. Poeppelmeier (Eds.), Comprehensive Inorganic Chemistry II, second ed., Elsevier, Amsterdam, 2013, pp. 941–969.
[10] U.S. Environmental Protection Agency, Conductivity, U.S. EPA, Washington, DC, 2023. Available from: https://www.epa.gov/national-aquatic-resource-surveys/indicators-conductivity. (Accessed 20 June 2023).
[11] U.S. Environmental Protection Agency, 5.9. Conductivity, U.S. EPA, Washington, DC, 2012. Available from: https://archive.epa.gov/water/archive/web/html/vms59.html#:~:text=Conductivity%20is%20measured%20with%20a,calculate%20the%20conductivity%20per%20centimeter. (Accessed 6 March 2012).
[12] D.S. Lloyd, Turbidity as a water quality standard for salmonid habitats in Alaska, N. Am. J. Fish Manag. 7 (1) (1987) 34–45.
[13] U.S. Geological Survey, Turbidity and water, 2018, Reston, Virginia. Available from: https://www.usgs.gov/special-topics/water-science-school/science/turbidity-and-water. (Accessed 6 June 2018).
[14] World Health Organization, Water quality and health: review of turbidity, Report No.: WHO/FWC/WSH/17.01, 2017.
[15] R.F. Noss, Indicators for monitoring biodiversity: a hierarchical approach, Conserv. Biol. 4 (4) (1990) 355–364.
[16] Q. Yang, G. Liu, M. Casazza, F. Gonella, Z. Yang, Three dimensions of biodiversity: new perspectives and methods, Ecol. Indic. 130 (2021) 108099.
[17] S.T. Allen, C.L. Stagg, J. Brenner, K.L. Goodin, D. Faber-Langendoen, C.A. Gabler, et al., Ecological resilience indicators for salt marsh ecosystems, in: Ecological resilience indicators for five northern Gulf of Mexico ecosystems, NatureServe, Arlington, 2018, p. 53.
[18] B. Hermann, C. Kroeze, W. Jawjit, Assessing environmental performance by combining life cycle assessment, multi-criteria analysis and environmental performance indicators, J. Clean. Prod. 15 (18) (2007) 1787–1796.
[19] W. Strong, Biased richness and evenness relationships within Shannon–Wiener index values, Ecol. Indic. 67 (2016) 703–713.
[20] P.J. Somerfield, K.R. Clarke, R.M. Warwick, Simpson index, in: S.E. Jørgensen, B.D. Fath (Eds.), Encyclopedia of Ecology, Academic Press, Oxford, 2008, pp. 3252–3255.
[21] J.H. Lawton, K.J. Gaston, Indicator species, in: S.A. Levin (Ed.), Encyclopedia of Biodiversity, Elsevier, New York, 2001, pp. 437–450.
[22] T.K. Parmar, D. Rawtani, Y.K. Agrawal, Bioindicators: the natural indicator of environmental pollution, Front. Life Sci. 9 (2) (2016) 110–118.
[23] M. Burford, Phytoplankton dynamics in shrimp ponds, Aquac. Res. 28 (5) (1997) 351–360.
[24] M. Burford, P. Rothlisberg, Factors limiting phytoplankton production in a tropical continental shelf ecosystem, Estuar. Coast. Shelf Sci. 48 (5) (1999) 541–549.
[25] M.A. Burford, M.J. O'Donohue, A comparison of phytoplankton community assemblages in artificially and naturally mixed subtropical water reservoirs, Freshw. Biol. 51 (5) (2006) 973–982.
[26] E. Willén, Phytoplankton in water quality assessment—an indicator concept, in: Hydrological and Limnological Aspects of Lake Monitoring, vol. 2, 2000, pp. 58–80.
[27] U.S. Environmental Protection Agency, Indicators: Benthic Macroinvertebrates, U.S. EPA, Washington, DC, 2023. Available from: https://www.epa.gov/national-aquatic-resource-surveys/indicators-benthic-macroinvertebrates#:~:text=Benthic%20macroinvertebrates%20are%20commonly%20used,in%20their%20tolerance%20to%20pollution. (Accessed 27 October 2023).
[28] M. Casé, E.E. Leça, S.N. Leitão, E.E. Sant'Anna, R. Schwamborn, A.T. de Moraes Junior, Plankton community as an indicator of water quality in tropical shrimp culture ponds, Mar. Pollut. Bull. 56 (7) (2008) 1343–1352.

[29] U.S. Environmental Protection Agency, Indicators: Enterococci, U.S. EPA, Washington, DC, 2023. Available from: https://www.epa.gov/national-aquatic-resource-surveys/indicators-enterococci. (Accessed 9 June 2023).

[30] A. Dufour, Escherichia coli: the fecal coliform, in: Bacterial Indicators/Health Hazards Associated With Water, ASTM International, 1977.

[31] Y. Liu, Q. Dong, H. Shi, Distribution and population structure characteristics of microorganisms in urban sewage system, Appl. Microbiol. Biotechnol. 99 (18) (2015) 7723–7734.

[32] U.S. Environmental Protection Agency, 5.11 Fecal bacteria, U.S. EPA, Washington, DC, 2012. Available from: https://archive.epa.gov/water/archive/web/html/vms511.html. (Accessed 6 March 2012).

[33] M.S. Coyne, J.M. Howell, The fecal coliform/fecal streptococci ratio (FC/FS) and water quality in the bluegrass region of Kentucky, Soil Sci. News Views 15 (9) (1994) 1–4.

[34] G.M. Masters, W.P. Ela, Introduction to Environmental Science & Technology, third ed., Prentice Education, Inc., Upper Saddle River, New Jersey, 2008.

[35] D. Vallero, Environmental Biotechnology: A Biosystems Approach, Elsevier Science, 2015.

[36] T.E. Graedel, On the concept of industrial ecology, Annu. Rev. Energy Environ. 21 (1) (1996) 69–98.

[37] D.A. Vallero, T.M. Letcher, Unraveling Environmental Disasters, Newnes, 2012.

[38] D.A. Vallero (Ed.), A systems approach to waste management, in: Waste, Elsevier, 2019.

[39] D.C. Pierson, Light and primary production in lakes, in: L. Bengtsson, R.W. Herschy, R.W. Fairbridge (Eds.), Encyclopedia of Lakes and Reservoirs, Springer Netherlands, Dordrecht, 2012, pp. 485–492.

[40] P.H. Doering, R.H. Chamberlain, K.M. Haunert, Chlorophyll a and its use as an indicator of eutrophication in the Caloosahatchee Estuary, Florida, Fla Sci. (2006) 51–72.

[41] J.H. Ryther, W.M. Dunstan, Nitrogen, phosphorus, and eutrophication in the coastal marine environment, Science 171 (3975) (1971) 1008–1013.

[42] R. Harland, J.A. Howe, Dinoflagellate cysts and Holocene oceanography of the northeastern Atlantic Ocean, Holocene 5 (2) (1995) 220–228.

Chapter 13

Water quality measurements

1 Introduction

Water above and below the earth's surface is found in complex systems comprised of nonliving and living components. To determine the condition of the water in these systems requires the application of various means of measurement [1]. Many are direct physical and chemical measurements, e.g., temperature, density, viscosity, dissolved oxygen (DO) content, and pH. Others are indirect, such as light scattering as an indication of turbidity. This chapter discusses these and introduces some of the techniques available for sampling, analysis, and extrapolation and interpolation of measured results, using various types of models.

Environmental scientists and engineers who are searching for ways to reduce or eliminate risks posed by water pollution follow the scientific method to uncover the truth about the natural world [2]. A key part of this search for truth is reliable measurements in the real world. This means that addressing water pollution should be a collaborative effort that avoids alienation of points-of-view. Unfortunately, as the term "denier" has entered the scientific and journalism lexicons to label certain perspectives wrong without further scrutiny. Scientists and engineers often spend the workday with fellow truth-seekers, but even within the scientific community there is never complete consensus on anything, except arguably the first principles of physics until a new paradigm replaces these, such as Einstein's relativity over Newton's physics [3]. Indeed, what has advanced the sciences for millennia has been the ability to be objective in the face of existing paradigms of thought and power [4]. For example, there is little debate among all physicists regarding the existence of the greenhouse effect, but the consensus begins to dissipate when estimating the contribution of individual greenhouse gases (GHGs) to the overall warming of a very complex atmosphere. This lack of consensus is not with the scientific principles, but with the uncertainties about the meaning and importance of each input variable and model algorithm, the presence of synergistic and antagonistic factors, and, arguably most importantly, how matter and energy change in the earth systems [4]. Variability and error in measurements are key to environmental condition.

The lack of consensus grows as the stakeholder population becomes more diverse. Technologists must base their communications on fact-based information, but many stakeholders prefer to communicate based on perception [5]. Most stakeholders are less likely to understand fully the technical jargon of the scientists, engineers, and decision makers unless this is translated into more widely understood language. Indeed, social science recognizes that incoming data and information can be processed by carefully applying either deliberation or intuition [6,7]. Data and information can be processed differently by different sectors of the public whose knowledgebases are built from a wide variety of sources [7–10]. Having the same set of facts does not guarantee the same conclusion [11]. That said, having incorrect or incomplete facts will lead to wrong conclusions about the quality of one's drinking water or the swimmability of a stream.

The first step in assessing and managing human health and ecological risks is to determine the condition of the environment, which includes estimates of the amount of contaminants in each environmental medium [12]. Such estimates must be based on with reliable data and information, beginning by characterizing the release of a substance, e.g., a discharge from an outfall structure, the substance's transport and transformation in the environment, and its concentrations near or within an organism, i.e., the receptor.

Water quality measurement includes developing methods, applying those methods, deploying monitoring technologies, and interpreting the results from these technologies [13]. An environmental assessment addresses chemical, physical, and biological factors [14]. The key type of measurement is a water quality assessment discerning the concentrations of substances in water, whether above or below the ground surface. Such measurements are the basis for health, exposure, and risk assessments, which are augmented by companion information. As evidence, exposure and risk assessments combine measurements of pollutant concentrations with information about geospatial data [15–20].

A measurement of a pollutant at a central monitoring site is merely an indirect indicator of exposure measurement since it provides concentration data. For it to become exposure information requires companion data on activities and locational

Fundamentals of Water Pollution. https://doi.org/10.1016/B978-0-443-28987-3.00019-9

information. A personal monitor worn all day by a person or a monitor on an organism in a habitat would be a more precise and accurate measurement of exposure for that day.

Direct measurements are those in which the substance of concern is what is collected and analyzed. For example, a measurement of a water sample's pH would require a measurement with a pH meter onsite. To measure solids directly would require that the water sample be weighed, volume sized, and the water chemically analyzed. Direct measurements can be *in situ*, i.e., taken in the environment, or *ex situ*, collected and taken elsewhere for measurement. Similar terms are used for treatment and remediation, i.e., contaminated soil water or ground water may be treated *in situ*, e.g., by air stripping, or *ex situ*, e.g., removed and thermally treated in an incinerator. In these cases, measurements of the contaminated water would be taken and compared to measurements of concentrations of pollutants of concern following treatment [21,22].

An indirect measurement is one where the substance itself is not collected, but would be characterized by an indicator, e.g., light scattering in a nephelometer [23,24]. The amount and type of scattering would indicate the quantity and size of particles. For example, the standard turbidity measurement method in U.S. coastal waters is nephelometric analysis with formazin calibration. As such, field and laboratory measurements can be compared to water quality standards. Depending on the results, the measurements can support actions needed to decrease turbidity in a water body.

The representativeness of measurements can be estimated by using statistics. For example, regression analysis is a common way to compare various measurement methods to standard methods [25]. Testing the quality of water in streams, for instance, can use continuous in-stream water-quality monitors stream-gaging stations to provide real-time measurement of specific conductance, pH, water temperature, dissolved oxygen, turbidity, and total chlorophyll [26]. Linear regression equations using the least-squares method can compare sensor measurements of the water-quality monitors as explanatory variables and analytical results of manually collected water samples, i.e., the response variables, for the characteristics of each of the listed parameters. In the least-squares method, the equation can be the sum of the squared errors for all sample points [27]. The regression equations can be used to estimate chemical pollutants and biological pollutants, e.g., algal and bacterial concentrations [26]. The coefficient of determination (R^2) and the mean square error (MSE) can be applied to evaluate the regression equations. The R^2 is the fraction of the variance explained by regression, and the MSE is presented to evaluate the variance between measured and expected or estimated values. The relative percentage differences (RPDs) between measured and estimated concentrations are calculated based on the absolute value of the equation [27]:

$$\mathrm{RPD} = \frac{|\mathrm{B} - \mathrm{A}|}{\mathrm{A}} \times 100 \tag{13.1}$$

Where A is the measured concentration and B is the estimated concentration.

2 Sampling

The monitoring underpinning the assessment depends on the quality of sample collection, preparation, and analysis. Sampling is a geostatistical term. An environmental sample is a fraction of water or other media, e.g., sediment, soil, biota, or air that represents a larger population or body. For example, a sample of water may consist of bottle or bag that holds a defined quantity of water to be subsequently analyzed. The sample is representative of a portion of a water body. The number of samples must be collected, and their results aggregated to ascertain with defined certainty the quality of the surface or ground water, or water in sediment or soil. More samples will be needed for a large river than for a small creek. Intensive sampling is often needed for highly toxic contaminants and for sites that may be particularly critical, e.g., near a hazardous waste site or in an "at risk" neighborhood (such as one near a manufacturing facility that uses large quantities of potentially toxic materials). Like other statistical measures, environmental samples allow for statistical inference. Inferences can be made regarding the condition of one species, an ecological habitat, or the extent and severity of exposure, e.g., of a human population receiving drinking water from a surface waterbody or aquifer.

For example, to estimate the amount of a chemical compound in a lake near a chemical plant, an engineer gathers a 100 mL sample in the middle of the lake that contains 1 million liters of water. Thus, the sample represents only 1×10^{-7} of the lake's water. This is a grab sample, i.e., a single sample representing an entire water system. Such a sample is limited in location vertically and horizontally, so uncertainty is high. However, if 10 samples are taken at 10 spatially distributed sites, the inferences are improved. Moreover, if the samples are collected each season, there would be an enhanced understanding of seasonal, i.e., intraannual, variability. If the sampling is continued for several years, the interannual variability becomes better characterized. Grab samples are commonly used in surface and ground water, soil, sediment, and air [1].

2.1 Data quality objectives

A monitoring plan must be developed before samples are collected. The plan includes quality assurance provisions and the procedures to be employed to ensure collection methods and transportation to the laboratory are adequate, as well as good laboratory practice (GLP) during the analytical stages. The plan details the sampling equipment, e.g., real-time probes, sample bags, bottles, and sediment cores, the number of samples to be collected, and procedures for sample preparation, handling, and chain-of-custody, if the samples are to be used for regulatory or legal purposes. Sample quality and quantity requirements are articulated by clearly stated by data quality objectives (DQOs), which are defined by the objectives of the overall contaminant assessment plan. DQOs are qualitative, semi-quantitative, and quantitative statements that translate nontechnical project goals into scientific and engineering outputs needed to answer technical questions [28].

Quantitative DQOs specify a level of scientific and data certainty, whereas qualitative DQOs enumerate decisions goals narratively. DQOs must specify the decision that the data will ultimately support, but not the manner that the measurements will be taken. DQOs establish the data quality for both the sampling and analytical efforts. The U.S. Environmental Protection Agency has listed three examples of the range of detail of quantitative and qualitative DQOs [29,30]:

1. *Example of a less detailed, quantitative DQO*: Determine with greater than 95% confidence that contaminated water will not pose a human exposure hazard.
2. *Example of a more detailed, quantitative DQO*: Determine to a 90% degree of statistical certainty whether the concentration of a heavy metal in each sample is less than 96 parts per billion (ppb).
3. *Example of a detailed, qualitative DQO*: Determine the proper disposition of each sediment sample in real-time using a field method able to turnaround total heavy metal results on the sediment samples within 1 h of sample collection.

If the desired condition is tightly defined, e.g., the seasonal change in pH near a fish hatchery, a small number of samples using simple pH probes could be the DQO. Conversely, if the water quality assessment is more complex and larger in scale, e.g., chemical and biological characterization of year-round water quality for trout in the stream, the sampling plan's DQO may call for a specific number of samples at various points to be continuously sampled for inorganic and organic contaminants, turbidity, nutrients, ionic strength, and algal counts. As mentioned, DQOs can also be semi-quantitative such as rather than allocating an species abundance value for each microbial species, a ranking system is used, e.g., species counts falling into six categories ranging from very rare to extremely abundant [31].

Deciding on DQOs for biological monitoring is even more complicated than chemical monitoring, requiring microbiological and chlorophyll monitoring. For example, algae are among the phytoplankton that reside in the water column and stream sediment. They contain chlorophyll within their biomass. The chlorophyll molecule lets algae absorb energy from sunlight via photosynthesis, so chlorophyll is an indirect measure of algal content in rivers. Chlorophyll-A is a common constituent of algal biomass, so many marine and freshwater biologists and other scientists use it to indicate general patterns in phytoplankton growth [32]. Its chemical formula is magnesium [methyl (3*S*,4*S*,21*R*)-14-ethyl-4,8,13,18-tetramethyl-20-oxo-3-(3-oxo-3-{[(2*E*,7*R*,11*R*)-3,7,11,15-tetramethyl-2-hexadecen-1-yl]oxy}propyl)-9-vinyl-21-phorbinecarboxylatato (2−)-κ^2N,N']. Algal growth is often linked to nutrient enrichment in freshwaters. Indeed, there is a connection between nutrient input, e.g., nitrogen (N) and phosphorous (P) to a waterbody, which induces the production of phytoplankton biomass, leading to depletion of dissolved oxygen (DO) in nearly still waters like those in a lake or estuary [33]. However, estuarine systems respond unevenly to increased inputs of nutrients, so chlorophyll-A concentrations cannot absolutely and directly predict nutrient loading [33]. Nutrient loading can affect several aspects of an aquatic system [34]. The decrease in the growth of seagrass as a result of eutrophication, for example, can also increase shading of incoming solar radiation and the indirect disruption of submerged aquatic vegetation [35,36].

2.2 Sampling plan

The comprehensive sampling plan must include all environmental media, e.g., soil, air, water, and biota, however a water quality measurement plan can be more targeted, e.g., collection of only samples of the water column or a water column and sediment. The sampling and analysis plan must specify the methods to be used. For example, specific chemicals differ in the methods needed, with many having published methods that must be followed during sample preparation, staging, handling, and chemical analysis [37,38].

The geographic area where measurements are to be taken will affect monitoring since the type of measurement must account for distinctive physical features such as volume or area, e.g., metropolitan city limits, access within the property boundaries, coring needs by sediment depth, water column height, length along a shoreline, or the natural habitat range of a particular biotic species. For example, a sampling grid may show the location of where a sample is to be collected within each cell in the grid [39]. The target population may be divided into relatively homogeneous subpopulations within each area or subunit. This can reduce the number of samples needed to meet the tolerable limits for decision errors and to improve efficiency.

Time is an essential parameter that determines the type and extent of monitoring needed. Conditions vary over the course of a study due to changes in weather conditions, seasons, operation of equipment, and human activities. These include seasonal changes in ground water levels, seasonal differences in farming practices, daily or hourly changes in contaminant concentrations due to upstream discharges, and intermittent pollutant discharges from nonpoint sources. These and other variations must be considered during sample collection and in the interpretation of results. Some examples of environmental time-sensitivity are [40]:

- concentrations of reactive pesticides immediately after application versus a week later;
- cloud cover that decreases soil and upper-level water column temperatures;
- amount of agricultural chemical compounds on surfaces indicating greater variations in the summer given higher mean and daily temperatures and evapotranspiration rates;
- instruments that may not give accurate measurements when temperatures are colder;
- total solids (TS) measurements that lose accuracy when sampling is conducted in winter months rather than the drier summer months.
- contaminant concentrations that change with weather events, including loss of sample representativeness soon after a spill or disaster, e.g., dust samples needed to be taken from surfaces soon after the World Trade Center attacks [41–44];
- amount of pesticides on surfaces that show variations in the summer because of higher temperatures and volatilization [45–47];
- water quality measurements immediately after a large rainfall event that have higher concentrations of suspended solids and biochemical oxygen demand, but where the slug of pollutants may even move rapidly than the average stream concentration.

Financial and technical feasibility should also be considered. This includes gaining legal and physical access to the properties, equipment acquisition and operation, and environmental conditions, timing, and conditions when sampling is difficult or not representative of actual conditions, e.g., freezing temperatures [48,49].

2.3 Monitoring plan example

Consider a hypothetical monitoring plan to measure toxic substance concentrations and variations in concentrations as a function of distance from a roadway and relating these data to highway traffic flows including traffic count, vehicle types and speeds, and meteorological conditions such as wind speed and wind direction. Specifically, the monitoring plan has the following goals [50]:

1. Identify concentrations of toxic pollutants near roads and their potential flow to water systems.
2. Determine how vehicle operations and local meteorology influence near road water quality.
3. Collect data that will be useful in ground-truthing, evaluating, and refining models to determine the releases of pollutants near roadways.

The diverse skill set needed to implement the plan includes program management, contracts administration, field monitoring experience, laboratory expertise, and quality assurance oversight. The plan includes a site selection step to gather and analyze sufficient data that would lead to informed conclusions regarding the selection of the most appropriate site for the monitoring at a specific location.

Selecting a monitoring site must be based on scientific and feasibility factors, as shown in Table 13.1. Each step has varying degrees of complexity due to "real-world" issues. The first step is to determine site selection criteria (see Table 13.2). The subsequent steps include developing a list of candidate sites and supporting information, applying a site selection filter for target pollutants, e.g., direct versus indirect monitoring methods, visiting the candidate sites and recording pros and cons, selecting the candidate site, obtaining site access permission(s), and implementing the site logistics.

Next, a list of candidate sites based on these criteria is developed. Geographic information system (GIS) data, tools, and techniques and on-site visits will compare various sites that meet these criteria. Spatial tools are useful in making and explaining environmental decisions [18,51]. Until recently, the use of GIS and other spatial tools in decision processes have required the acquisition of large amounts of the data. In addition, the software has not been user-friendly. GIS data have recently become more readily available in both quantity and quality, and GIS exists in several common operating

TABLE 13.1 Example of steps in selecting an environmental monitoring site.

Step	Site selection steps	Method	Comment
1	Determine site selection criteria	Monitoring protocol	
2	Develop list of candidate sites	Geographic information system (GIS) data; on-site visit(s)	Additional sites added as information is developed
3	Apply coarse site selection filter	Team discussions, management input	Eliminate sites below acceptable minimums
4	Site visit	Field trip	Application of fine site selection filter
5	Select candidate site(s)	Team discussions, management input	
6	Obtain site access permissions	Contact property owners	If property owners do not grant permission, then the site is dropped from further consideration
7	Site logistics (i.e., physical access, utilities—electrical and communications)	Site visit(s), contact utility companies	

Data from S. Kimbrough, R.W. Baldauf, G.S. Hagler, R.C. Shores, W. Mitchell, D.A. Whitaker, et al., Long-term continuous measurement of near-road air pollution in Las Vegas: seasonal variability in traffic emissions impact on local air quality, Air Qual. Atmos. Health 6 (1) (2013) 295–305. S. Kimbrough, D. Vallero, R. Shores, A. Vette, K. Black, V. Martinez, Multi-criteria decision analysis for the selection of a near road ambient air monitoring site for the measurement of mobile source air toxics, Transp. Res. D Transp. Environ. 13 (8) (2008) 505–515.

TABLE 13.2 Example selection considerations and criteria.

Selection considerations	Monitoring protocol criteria
Essential criteria for this monitoring study:	
AADT (>150,000)	Only sites with more than 150,000 annual average daily traffic (AADT) are considered as candidates
Geometric design	The geometric design of the facility, including the layout of ramps, interchanges, and similar facilities, will be considered. Where geometric design impedes effective data collection on MSATs and $PM_{2.5}$, those sites will be excluded from further consideration. All sites have a "clean" geometric design
Topology (i.e., sound barriers, road elevation)	Sites located in terrain making measurement of pollutant concentrations difficult or that raise questions of interpretation of any results will not be considered. For example, sharply sloping terrain away from a roadway can affect results
Geographic location	Criteria applicable to representing geographic diversity
Availability of data (traffic volume data)	Any location where automated traffic monitoring data, meteorological, or pollutant concentration data are not readily available, or instrumentation cannot be properly sited to collect such data will not be considered for inclusion in the study
Meteorology	Sites will be selected based on their local climates to assess the impact of meteorology on dispersion of pollutant releases and environmental processes that affect chemical reactions and phase changes
Desirable, but not essential criteria:	
Downgradient sampling	Any location where proper siting of downstream sampling sites is restricted due to topology, existing structures, meteorology, etc., may exclude otherwise suitable sites for consideration and inclusion in this study
Potentially confounding pollutant sources	The presence of confounding pollution sources may exclude otherwise suitable sites for consideration and inclusion in this study, especially if these other pollutants interfere with the target pollutant measurements
Site access (admin/physical)	Any location where site access, is restricted or prohibited either due to administrative or physical issues, will not be considered for inclusion in the study

Data from S. Kimbrough, R.W. Baldauf, G.S. Hagler, R.C. Shores, W. Mitchell, D.A. Whitaker, et al., Long-term continuous measurement of near-road air pollution in Las Vegas: seasonal variability in traffic emissions impact on local air quality, Air Qual. Atmos. Health 6 (1) (2013) 295–305. S. Kimbrough, D. Vallero, R. Shores, A. Vette, K. Black, V. Martinez, Multi-criteria decision analysis for the selection of a near road ambient air monitoring site for the measurement of mobile source air toxics, Transp. Res. D Transp. Environ. 13 (8) (2008) 505–515.

TABLE 13.3 National scale map layers related to ecosystem services and biodiversity in the EnviroAtlas.

Select data by topic — clear all topics

National — EnviroAtlas Communities

Ecosystem Services and Biodiversity	
Carbon Storage	Pollutant Reduction: Air
Crop Productivity	Pollutant Reduction: Water
Ecosystem Markets	Protected Lands
Energy Potential	Species: At-Risk and Priority
Engagement with Outdoors	Species: Other
Health and Economic Outcomes	Water Supply, Runoff, and Flow
Land Cover: Near-Water	Water Use
Land Cover: Type	Weather and Climate
Landscape Pattern	Wetlands and Lowlands
Near-Road Environments	
Pollution Sources and Impacts	
EPA Regulated Facilities	Pollutants: Other
Impaired Waters	Pollutants: Nutrients
National Air Toxics Assessment	
People and Built Spaces	
Commuting and Walkability	Population Distribution
Employment	Quality of Life
Housing and Schools	Vacancy
Boundaries	
Ecological Boundaries	Political Boundaries
Hydrologic Features	

Screenshot from U.S. Environmental Protection Agency, EnviroAtlas data, U.S. EPA, Washington, DC. Available from: https://www.epa.gov/enviroatlas/enviroatlas-interactive-map#enviroatlas, 2023 (Accessed 30 November 2023).

system environments. Environmental regulatory agencies increasingly use data layers to assess and describe environmental conditions. For example, the U.S. EPA's EnviroAtlas is a system of interactive tools to support and to document the ecological benefits to humans from nature, including food supply, water supply, flood control, security, public health, and economy [52]. Table 13.3 describes the map layers that are foundational to the EnviroAtlas [53,54].

Even a well-designed monitoring plan needs adjustment during implementation. Investigators may discover barriers or differing conditions from what was observed in the planning phase. For example, different daily traffic counts or new road construction need to be documented since these can episodically change measurement results.

After applying site selection acceptance criteria, candidate sites are incrementally eliminated. For example, the first filter is minimum traffic counts; the next filter, the presence of extensive sound barriers, eliminates additional sites; and other filters, e.g., complex geometric design or lack of available traffic volume data. Next, feasibility considerations would eliminate additional candidate sites.

An important component of ground-truthing during a site visit is to obtain information from local sources. Local businesses and residents can provide important information needed in a decision process, such as abandoned or defunct businesses and industries, types of chemicals stored previously at a site, changes in vegetation or even ownership histories.

The GIS data layers that are commonly needed in environmental engineering include the location of suitable soils, wells, surface water sources, residential areas, schools, airports, roads, etc. From these data layers, queries are formulated to provide the most suitable sites, e.g., depth to water table may help identify sources of pollution. Usually, quantitative weighting criteria are associated with the siting criteria as well as elements of the data layers, e.g., certain types of soils would be more suitable to the study objectives than others and thus would have applicable quantitative values [55].

3 Environmental sampling approaches

As mentioned, the grab sample is simply a measurement at a site at a single point in time. Composite sampling physically aggregates multiple grab samples from different locations or times. This allows for physical, instead of mathematical, averaging, which is not the case for grab samples. The acceptable composite provides a single value of chemical compound concentration measurement that can be analyzed statistically.

Multiple composite samples improve sampling precision and decrease the total number of analyses needed compared to noncomposite sampling [56]. For a grab or integrated soil sample of x mass or y volume, the number of samples needed to determine statistical significance, and the minimum acceptable quality must be defined by the quality assurance (QA) plan, the sampling standard operating procedures (SOPs), and sample handling procedures after sample collection. As mentioned, the plan also describes how procedures will adhere to good laboratory practice (GLP), i.e., a quality system that describes how a study is to be planned, performed, monitored, recorded, procedures archived, and results are to be reported.

A weakness of composite sampling is the false negative effect. For example, the samples collected from an evenly distributed grid of homes to represent a neighborhood exposure to a contaminant, as shown in Fig. 13.1. In this hypothetical example, the measured pollutant concentrations are 3, 1, 2, 12, and $2\,mg\,L^{-1}$, so the mean contamination concentration is only $4\,mg\,L^{-1}$. If cleanup is needed above the threshold of $5\,mg\,L^{-1}$, the mean concentration would indicate the area does not need remediation and would be reported below the threshold level. However, the fourth home is well above the safety level. Mixing the five samples as a composite would cause a false positive because one of the sites does indeed need remediation of pollutant concentration from $12\,mg\,L^{-1}$ to below $5\,mg\,L^{-1}$. More precise measurements also assist in remediation decisions. For example, the whole neighborhood may not necessarily need cleanup if the source is isolated to a confined area in the yard of home 5. Additional samples around that measurement location would be needed to define a concentrations gradient.

Geographic composites may also not be representative if a grid is placed over a contaminated ground water plume (see Fig. 13.2) and does not account for horizontal and vertical impervious layers, unknown sources (e.g., tanks), and flow differences among strata, so that some of the plume is eliminated but pockets are left (as shown in Fig. 13.2B).

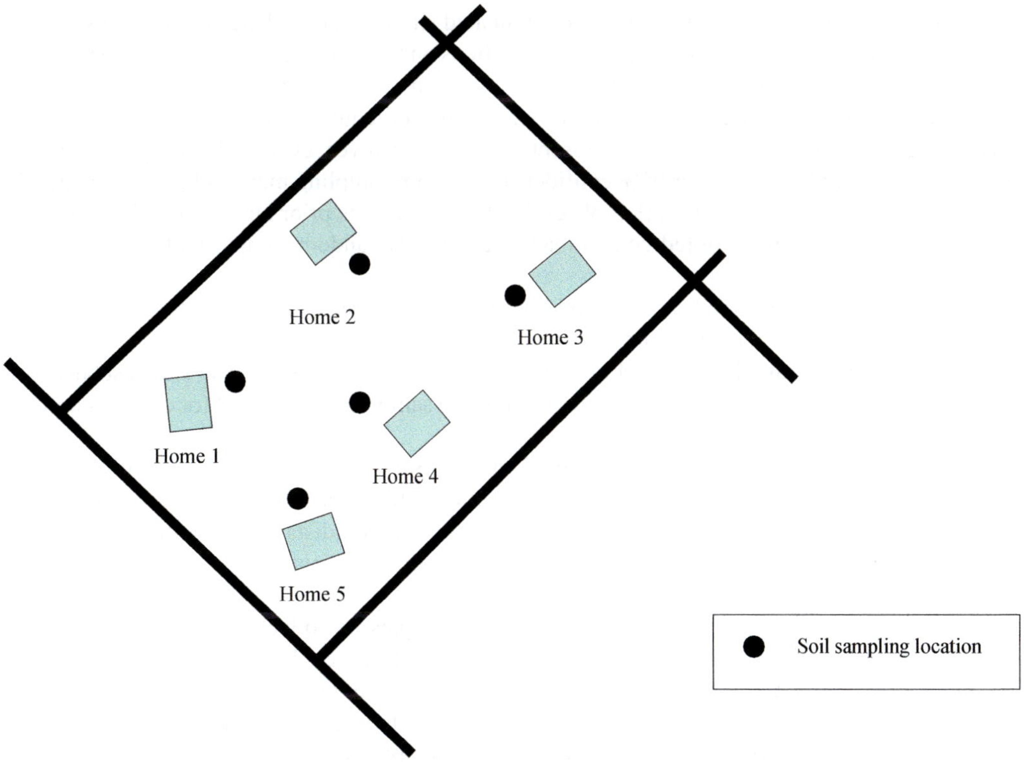

FIG. 13.1 Hypothetical composite sampling grid for a neighborhood. *(Data from D. Vallero, Environmental Biotechnology: A Biosystems Approach, Elsevier Science, 2015.)*

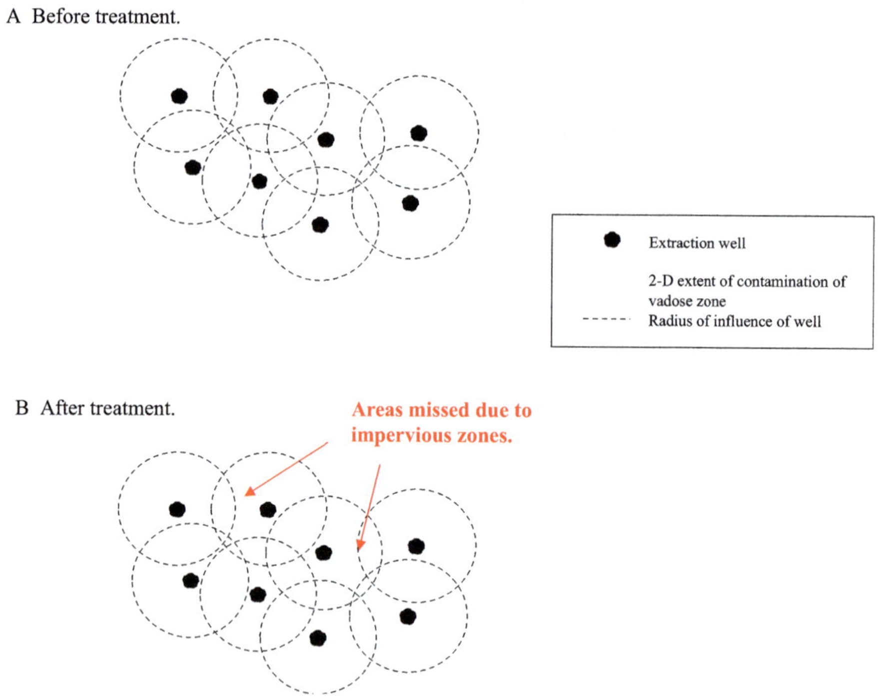

FIG. 13.2 Extraction well locations on a geometric grid, showing hypothetical remediation after 6 months. *(From D. Vallero, Environmental Biotechnology: A Biosystems Approach, Elsevier Science, 2015.)*

Soil is heterogeneous in its texture, chemical composition, moisture content, organic matter content, sorption coefficients, and other physical and biological characteristics, which may vary even a few meters away from a monitoring location [57–59]. It is good practice to assume that a contaminated site will have a heterogeneous distribution of contamination. Selecting appropriate sampling methods and considerations on their use are key parts of any study design and environmental assessment, since the results will be the basis for exposure models, risk assessments, feasibility studies, land use and zoning maps, and other information used by engineers, clients and regulators [60].

Random sampling has the value of statistical representativeness, However, even with enough samples for the defined confidence levels (e.g., x samples needed for 95% confidence), random sampling may lead to large areas of the site being missed for sampling due to the chance distribution of results. It also neglects prior knowledge of the site. For example, if maps show an old tank that may have stored contaminants, an entirely random sample will not give any preference to samples near the tank.

Stratified random sampling is accomplished by dividing the site into areas and randomly sampling within each area, avoiding the omission problems of random sampling alone.

In stratified sampling, chemical, physical, and biological parameters are targeted. The site is subdivided spatially, so that sampling patterns and densities vary in different areas. Stratified sampling can be used for complex and large sites, such as mining and construction sites.

A grid or systematic sampling covers the entire site of interest. Sampling locations are readily identifiable, which is valuable for follow-on sampling, if necessary. The grid does not have to be rectilinear. In fact, rectangles are not the best polygon to use in the value is to be representative of a cell. Circles provide equidistant representation but overlap. Hexagons are sometimes used as a close approximation to the circle. The U.S. Environmental Monitoring and Assessment Program (EMAP) has used a hexagonal grid pattern, for example [61,62].

Judgmental sampling is a qualitative type of sampling in which samples are collected based upon prior knowledge of the site conditions. This overcomes the problem of ignoring sources or sensitive areas but is vulnerable to bias of both inclusion and exclusion. Obviously, this would not be used for spatial representation, but for pollutant transport, plume characterization, or monitoring near a sensitive or potentially highly exposed site (e.g., a day care center).

At every stage of monitoring from sample collection through analysis and archiving, only qualified and authorized persons should be in possession of the samples. This is usually assured by requiring chain-of-custody manifests. Sample handling includes specifications of the temperature range needed to preserve the sample, the maximum amount of time the sample can be held before analysis, special storage provisions (e.g., some samples need to be stored in certain solvents and containers), and chain-of-custody provisions (only certain, authorized persons should be in possession of samples after collection).

Each person in possession of the samples must require that recipient sign and date the chain-of-custody form before transferring the samples. This is because samples have evidentiary and forensic content, so any compromising of the sample integrity must be avoided. Indeed, regulatory agencies and companies need such information for subsequent scrutiny by the courts and the public.

Often sample results must be compared to "clean" conditions or background concentrations of pollutants. Background sampling distinguishes site-related contamination from naturally occurring or other nonsite-related levels of chemical compounds and microbial populations. Two different types of background levels are important to water quality investigations. Natural background concentrations are those expected under ambient conditions that are present in the environment and that have not been influenced by human activities (e.g., naturally occurring metals like aluminum and manganese). Anthropogenic background concentrations are those that are present in the environment due to human-made, nonsite sources (e.g., industrial releases, mobile sources, agricultural activities, urban development, and landfills). Background scales can range from localized to general, ubiquitous concentrations. Most pesticides, for example, are not naturally occurring and may be ubiquitous in specific locations (e.g., intensive crops and agricultural areas). Roadway runoff in the winter in higher latitudes would be expected to increase sodium (Na) along transportation routes, i.e., high ubiquitous Na concentrations. Examples of ubiquitous anthropogenic chemicals in an industrialized nation would include polycyclic aromatic hydrocarbons (PAHs) and lead (Pb). However, both may be present at naturally occurring levels in the environment due to natural sources, e.g., wildfires may be a source of PAHs, and Pb is a natural component of some soils [63].

4 Measurement uncertainty and error

As indicated in Table 13.4, measurement errors and uncertainties will accompany the results and even be compounded as the data are translated into information [64], so it is important to include all necessary metadata to ensure others may deconstruct, quality assure, and ensure appropriate applications.

TABLE 13.4 Types of uncertainty and contributing errors in sampling and analysis.

Type of uncertainty	Type of error causing uncertainty	Description or example
Pollutant releases	Misclassification and miscalculation	Reliance on third-party and other sources of information with little or no metadata regarding quality. Confusing actual emission measurements with reported estimates
Transport and transformation	Incorrect model application	Applying a model that has incorrect chemical, biological, hydrological, hydrogeological, atmospheric, geomorphological, and other conditions, e.g., using a simple dispersion model in a complex terrain Applying a quantitative structural activity relationship (QSAR) to inappropriate compounds, e.g., for metals when the QSAR is only for organic compounds, or for semivolatile compounds when the QSAR is only for volatile organic species
Exposure scenario	Misclassification	Failure to adequately identify exposure routes, exposure media, and exposed population
Sampling or measurement (parameter uncertainty)	Measurement: random	Random errors in analytical devices (e.g., imprecision of continuous monitors that measure stack emissions)
	Measurement: systemic	Systemic bias (e.g., estimating inhalation from indoor ambient air without considering the effect of volatilization of contaminants from hot water during showers)
	Surrogate data	Use of alternate data for a parameter instead of direct analysis of exposure (e.g., use of population figures as a surrogate for population exposure)

Continued

TABLE 13.4 Types of uncertainty and contributing errors in sampling and analysis—cont'd

Type of uncertainty	Type of error causing uncertainty	Description or example
	Misclassification	Incorrect assignment of exposures of subjects in historical epidemiologic studies resulting from faulty or ambiguous information
	Random sampling error	Use of too small a sample of individuals to estimate risk to exposed workers
	Nonrepresentativeness	Developing exposure estimates for a population in a rural area based on exposure estimates for a population in a city
Observational or modeling	Relationship errors	Incorrectly inferring the basis of correlations between environmental concentrations and urinary output
	Oversimplification	Misrepresentations of reality (e.g., representing a three-dimensional aquifer with a two-dimensional mathematical model)
	Incompleteness	Exclusion of one or more relevant variables (e.g., relating a biomarker of exposure measured in a biological matrix without considering the presence of the metabolite in the environment)
	Surrogate variables	Use of alternate variables for ones that cannot be measured (e.g., using wind speed at the nearest airport as a proxy for wind speed at the facility site)
	Failure to account for correlations	Not accounting for correlations that cause seemingly unrelated events to occur more frequently than expected by chance (e.g., two separate components of a nuclear plant are missing a particular washer because the same newly hired assembler put them together)
	Model disaggregation	Extent of (dis)aggregation used in the model (e.g., separately considering subcutaneous and abdominal fat in the fat compartment of a physiologically based pharmacokinetic (PBPK) model)
	Biological plausibility	Applying a PBPK model to chemical compounds for which it has not been coded, e.g., a one-compartment model for a compound with known metabolites

Data from D.A. Vallero, Fundamentals of Air Pollution, fifth ed., Elsevier Academic Press, Waltham, MA, 2014, 999 pages cm p. U.S. Environmental Protection Agency, An examination of EPA risk assessment principles and practices, Washington, DC, 2004.

Contaminant assessments have numerous sources of uncertainty. There are two basic types of uncertainty: Type A and Type B. Type A uncertainties result from the inherent unpredictability of complex processes that occur in nature and are obtained as standard deviations of repeated measurement results. These uncertainties cannot be eliminated by increasing data collection or enhancing analysis. The scientist and engineer must simply recognize that Type A uncertainty exists, but must not confuse it with Type B uncertainties, which are estimated using methods other than statistical treatment of repeated measurement results. Type B uncertainties can be reduced by collecting and analyzing additional scientific data.

The first step in an uncertainty analysis is to identify and describe as many uncertainties as possible that may be encountered. Sources of Type B uncertainty take many forms [65]. There can be substantial uncertainty concerning the numerical values of the attributes being studied (e.g., contaminant concentrations, wind speed, discharge rates, ground water flow, and other variables). Modeling generates its own uncertainties, including errors in selecting the variables to be included in the model, such as surrogate contaminants that represent whole classes of compounds (e.g., does benzene represent the behavior or toxicity of other aromatic compounds?). The application of the findings, even if the results themselves have tolerable uncertainty, but lead to the propagation of uncertainties when ambiguity arises regarding their meaning. For example, a decision rule is a statement about which alternative will be selected, e.g., for cleanup, based on the characteristics of the decision situation. A "decision-rule uncertainty" occurs when there are disagreements or poor specification of objectives (i.e., is our study really addressing the client's needs?).

Variability and uncertainty must not be confused. Variability consists of measurable factors that differ across populations such as soil type, vegetative cover, or body mass of individuals in a population. Uncertainty consists of unknown or not fully known factors that are difficult to measure, such as the inability to access an ideal site that would be representative because it is on private property.

Modeling uncertainties, for example, may consist of extrapolations from a single value to represent a whole population, i.e., a point estimate (e.g., 70 kg as the weight of an adult male). Such estimates can be typical values for a population or an estimate of an upper end of the population's value, e.g., 70 years as the duration of exposure used as a worse-case scenario. Another approach is known as the Monte Carlo technique. The Monte Carlo-type assessments use probability distribution functions, which are statistical distributions of the possible values of each population characteristic according to the probability of the occurrence of each value [66]. These are derived using iterations of values for each population characteristic. While the Monte Carlo technique may help to deal with the point estimate limitation, it can suffer from confusing variability with uncertainty. This explains the uncertainty and variability in estimates by repeatedly sampling the probability distributions of the inputs and using these inputs to calculate a range of values [67] (see Fig. 13.3).

Other data interpretation uncertainties can result from the oversimplification of complex entities. For example, assessments consist of an aggregation of measurement data, modeling, and combinations of sampling and modeling results. However, these complicated models are providing only a snapshot of highly dynamic human and environmental systems. The use of more complex models does not necessarily increase precision, and extreme values can be improperly characterized. For example, a 50th percentile value can always be estimated with more certainty than a 99th percentile value.

The bottom line is that uncertainty is always present in sampling, analysis, and data interpretation, so the monitoring and data reduction plan should be transparent, systematic, and rigorous. The uncertainty analysis must be addressed for each step of the contaminant assessment process, including any propagation and enlargement of cumulative error (e.g., an incorrect pH value that goes into an index where pH is weighted heavily, and then used in another algorithm for sustainability). The characterization of the uncertainty of the assessment includes selecting and rejecting data and information ultimately used to make environmental decisions and includes both qualitative and quantitative methods.

Uncertainty factors are applied to the address both the inherent and study uncertainties upon which to establish safe levels of exposure to contaminants. Uncertainty can result from the variation in sensitivity among the members of the populations, including interhuman and intraspecies variability; the extrapolation of animal data to humans (i.e., interspecies variability), the extrapolation from data gathered in a study with less-than-lifetime exposure to lifetime exposure, i.e., extrapolating from acute or subchronic to chronic exposure; the extrapolation from different toxicity thresholds (see Chapter 17); and the extrapolation from an incomplete data base. Note that most of these sources of uncertainty have a component associated with measurement and analysis.

The uncertainties are directly related to the quality and representativeness of the sampling design and the analytical expressions described in Table 13.4. When these values are used in environmental models, they are known as parameter uncertainties that account for imprecision and inaccuracy associated with the measurement and analytical equipment, and systemic weaknesses in data gathering, i.e., bias.

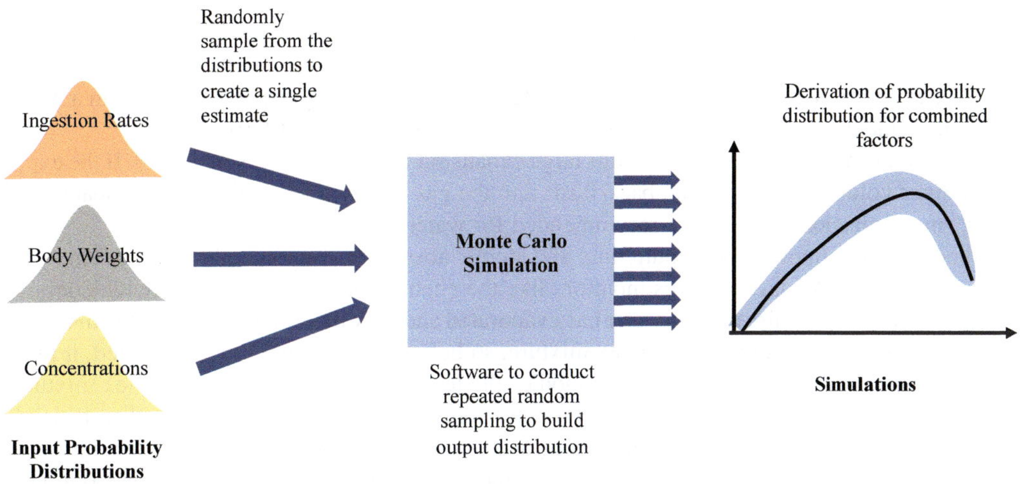

FIG. 13.3 Principles of the Monte Carlo method for aggregating data. *(Data from Australian Department of Health and Ageing, Environmental health risk assessment: guidelines for assessing human health risks from environmental hazards, 2002.)*

5 Laboratory analysis

Although real-time analysis of water is commonplace, many environmental samples must be brought to a laboratory to be analyzed after collection. This is especially true for emerging contaminants and any contaminant needing low levels of detection and quantitation. The steps to be taken to interpret the concentration of a chemical compound in the sample are known as "wet chemistry," which means the water or other environmental sample must be transported to a laboratory for chemical or biological analysis [1].

5.1 Extraction

When an environmental sample arrives at the laboratory, the next step may be extraction. Extraction is needed if the environmental sample is sediment or soil, where the chemicals of concern are sorbed to particles and must be freed for analysis to take place. Extraction is also called for when the actual collection may have been by trapping the chemical compounds onto sorbents which requires that the chemical compounds must first be freed from the sorbent matrix before being injected into a detection system. Many water pollutants have low vapor pressures and may not be readily dissolved in water. Thus, they may be found in various media, e.g., sorbed to particles, in the gas phase, or in the water column suspended to colloids or otherwise suspended in water. To collect such chemicals in the gas phase, a common method calls for trapping it on porous media like polyurethane foam (PUF). Thus, to analyze semivolatile organic compounds (SVOCs) like polychlorinated biphenyls (PCBs), the PUF must first undergo extraction and then chemically analyzed [68].

Extraction is both a physical and chemical process. For example, many compounds can be simply extracted with solvents, usually at elevated temperature. A common solvent extraction is the Soxhlet extractor, which removes sorbed chemicals by passing a boiling solvent through the media. Cooling water condenses the heated solvent so that the extract is collected over an extended period, usually several hours. Other automated techniques apply some of the same principals as solvent extraction, but allow for more precise and consistent extraction, especially when large volumes of samples are involved. For example, supercritical fluid extraction (SFE) brings a solvent, usually carbon dioxide, to the pressure and temperature near its critical point, where the solvent's properties are rapidly altered with very slight variations of pressure [69]. Solid phase extraction (SPE) uses a solid and a liquid phase to isolate a chemical from a solution and is often used to clean up a sample before analysis. Combinations of various extraction methods can enhance the extraction efficiencies, depending upon the chemical and the media in which it is found. Ultrasonic and microwave extractions may be used alone or in combination with solvent extraction. For example, the U.S. EPA Method 3546 provides a procedure for extracting hydrophobic or slightly water-soluble organic compounds from particles such as soils, sediments, sludges, and solid wastes [70,71]. In this method, microwave energy elevates the temperature and pressure conditions, i.e., 100–115 °C and 3.4–11.9 atm, in a closed extraction vessel containing the sample and solvent or mix of solvents. This combination can improve recoveries of chemical analytes and can decrease the time needed compared than the Soxhlet procedure alone.

5.2 Separation science

Not every sample needs to be extracted. For example, water samples may also be directly injected onto a gas chromatograph column, but care must be taken to ensure the material being injected is compatible with the detection system. For example, analyte expansion can occur. As evidence, 1.0 μL of methylene chloride will only expand to 361 μL in a 200 °C inlet with a 10 psi head pressure. Under the same conditions, 1.0 μL of water will expand to 1279 μL [72].

Considering that liners have specific volumes, this large expansion volume must be considered. If the expansion volume is larger than the liner volume, the vapor can back-flush into the gas supply lines, which are not heated, causing sample carryover. Thus, small injection volumes are recommended for water [72].

Surface methods, such as fluorescence, sputtering, and atomic absorption, require only that the sample be mounted on specific media (e.g., filters). Also, continuous monitors like the chemiluminescent system provide continuous measurements. These methods usually require that the water has evaporated and only the solids remain on the media to be analyzed.

An environmental or tissue sample is a complex mixture, so before a compound can be detected, it must first be separated from the mixture. Thus, separation science embodies the techniques for separating complex mixtures of analytes, which is the first stage of chromatography. The second is detection. Separation makes use of the chemicals' different affinities for certain surfaces under various temperature and pressure conditions. The first step, injection, introduces the extract to a column. Columns are of two major types, i.e., gas and liquid. The column in gas chromatography (GC) is a hollow tube coated inside with compounds with affinities for organic chemical compounds. The columns are in an oven, so after the extract is injected into the column, the temperature of the carrier gas (e.g., helium) increases, as does the pressure. Various organic compounds in the extract are released from the column surface differentially depending on their boiling points. As they exit the column, they are collected by the carrier gas and transported to the detector. Generally, the more volatile compounds are released first, i.e., they have the shortest retention times, followed by the SVOCs. So, boiling point is often a very useful indicator of the expected amount of time for the compound to exit the column. This is only a general expectation because many other characteristics such as polarity can influence a compound's resistance to be freed from the column surface. For this reason, GC columns vary in coatings, interior diameters, and lengths.

Instead of coated columns, liquid chromatography (LC) makes use of columns packed with various sorbing materials with differing affinities for compounds. Rather than a carrier gas, LC setup deploys solvent or blend of solvents to carry the compounds to the detector. In the high-performance LC (HPLC), pressures are also varied.

5.3 Detection

Detection is the final step for quantifying the chemicals in a sample. Energy absorption is arguably the most common type of detection. Chemical compounds absorb energy at various levels, depending upon their size, shape, bonds, and other structural characteristics. Some absorb very well in the ultraviolet (UV) range, but others do not. Diode arrays help to identify compounds by showing several absorption ranges in the same scan. Some molecules can be excited and will fluoresce. The Beer-Lambert Law states that energy absorption is proportional to chemical concentration:

$$A = eb[C] \tag{13.2}$$

Where, A is the absorbency of the molecule, e is the molar absorptivity (proportionality constant for the molecule), b is the light's path length, and $[C]$ is the chemical concentration of the molecule. Thus, the concentration of the chemical can be ascertained by measuring the light absorbed.

Using this relationship, in each separation-detection analytical system, solutions with several known concentrations of the chemical species of interest are injected, resulting in a calibration curve (see Fig. 13.4). The theoretical Beer-Lambert relationship is the idealized (blue) curve of the expected absorbance response, whereas there will be a deviation (red curve) response curve, due to inherent instrumentation error, imprecise known solutions, stray energy sources (e.g., light), unaccounted chemical reactions within the solvent, and other factors. As such, the most reliable part of the curve is its most linearly response range. Generally, at least five standards and one blank (i.e., same pure solvent absent the target chemical species) are needed to fit the standard curve. The owner's manual and websites for the instruments must be consulted, especially to see if a manual curvature correction function is available for the instrument.

The detector of choice depends upon the kinds of pollutants of interest. Detection gives the peaks that are used to identify compounds (see Fig. 13.5). For example, if hydrocarbons are of concern, GC with flame ionization detection (FID) may be used. GC-FID gives a count of the number of carbon atoms, so for example, long chains can be distinguished from short-chains. The short chains come off the column first and have peaks that appear before the long-chain peaks. However, if pesticides or other halogenated compounds are of concern, electron capture detection (ECD) is a better choice since they provide peaks for specific compounds.

The type of detector needed depends upon the kinds of pollutants of interest. Detection displays the peaks that are used to identify compounds (see Fig. 13.6).

The currently most popular detection methods for environmental pollutants are mass spectrometry (MS), which follows either GC or LC separation. The MS detection is highly sensitive using a stream of electrons to break organic compounds

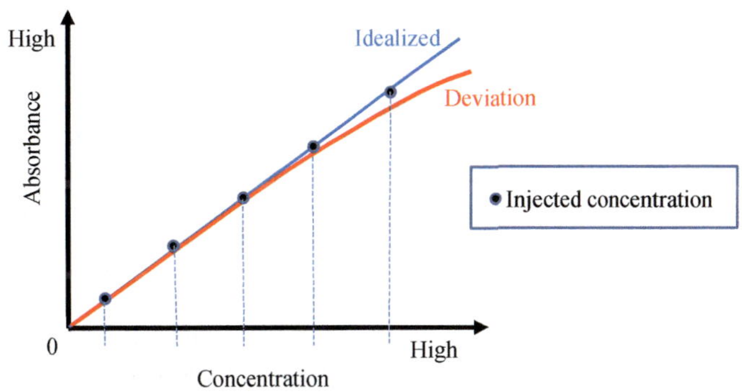

FIG. 13.4 Calibration curve for a hypothetical chemical compound.

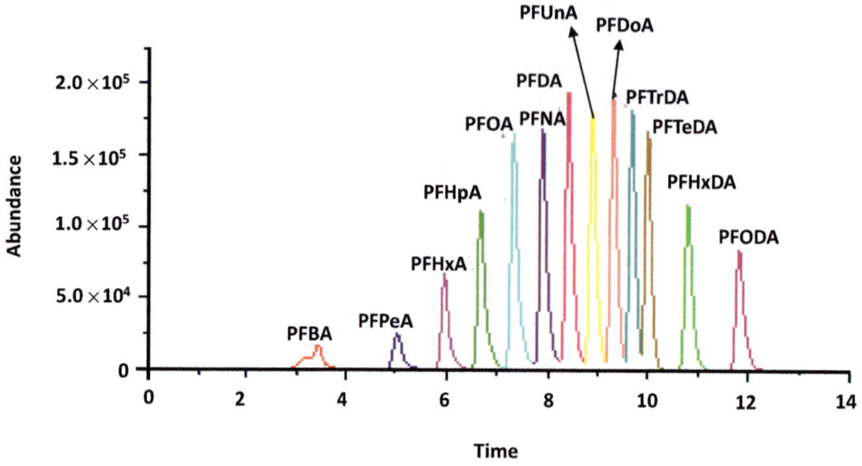

FIG. 13.5 Chromatograph of 14 per- and polyfluoroalkyl substances (PFAS) analyzed using liquid chromatography coupled to quadrupole time-of-flight mass spectrometry (LC-QToF).

into fragments. The positive ions resulting from the fragmentation separate according to their masses. This is referred to as the mass-to-charge ratio (*m/z*).

Residence time example (from Ref. [40])

An environmental analytical laboratory that has generated the chromatogram and table shown in Fig. 13.6, using an HPLC/UV at 254 nm using a 5 μm, C_{18}, 4.6 × 250 mm column from an extract. Find the retention times of compound A and B and decide which compound is present in a larger amount. Which compound has the higher boiling point? What would happen to the retention times of compounds A and B if the column temperature were raised? Can this information help to identify the source if compound B were suspected to be benzo(a)pyrene based on company records and the type of industries nearby?

Answers:

The retention time of compound A is 21.31 min, shown above of the peak and in the table's retention time column. The retention time of compound B is 24.53 min. One cannot tell from this table or chromatogram which compound is present in a larger amount, since the only way to do so is to have calibration curve from known concentrations of compound A and compound B (at least three, but preferably five). For example, the HPLC could be run successively with injections of pure solutions of 0.01, 0.1, 1, 10, and 100 μg L^{-1} concentrations of compound A, and again with pure solutions of the same concentrations of compound B. These concentrations would give peak areas associated with each known concentration. Then the HPLC software can construct the calibration curve.

Thus, for example, if an peak with an area of 200 is associated with 1 μg L^{-1} of compound A and a peak with an area of 2000 is associated with 10 μg L^{-1} of compound A, i.e., a linear calibration curve, at 21.31 min after the aliquot is injected into the HPLC, then when the unknown sample is run, showing a peak at 21.31 min with an area of 1000 would mean about 5 μg L^{-1} concentration of compound A is in the sample. The same procedure would be followed to draw a calibration curve for compound B at a retention time of 24.53 min.

The reason it is not sufficient to look at the percent area is that each compound is physically and chemically unique. According to the Beer-Lambert Law, the amount of energy absorbed (in this case, the UV light) is what gives the peak. If a molecule of compound A absorbs UV at this wavelength (i.e., 254 nm) at only 25% as that of compound B, compound A's concentration would be higher than that of compound B, because even though compound B has twice the percent area, its absorbance is four times that of compound B.

Compound A has the lower boiling point because it exits the column first. As mentioned, this is only true if other factors, especially polarity, are about the same. For example, if compound B has about the same polarity as the column being used, but compound A has a very different polarity, compound A will have a greater tendency to leave the column. Generally, however, retention time is a good indicator of boiling point; i.e., lower retention times mean lower boiling points.

If the column temperature were raised, both compounds A and B would come off the column in a shorter time. Thus, the retention times of both compounds A and B would be shorter than before the temperature was raised.

To determine whether the peak at 24.53 min is benzo(a)pyrene, one must first obtain a true sample of pure benzo(a)pyrene to place in a standard solution. This is the same process used to develop the calibration curve above. That is, this standard of known benzo(a)pyrene would be injected into the same HPLC and the same volume of injection. If the standard gives a peak at a retention time at about 25 min, there is a good chance it is benzo(a)pyrene. As it turns out, benzo(a)pyrene absorbs UV at 254 nm and does come off an HPLC column at about 25 min.

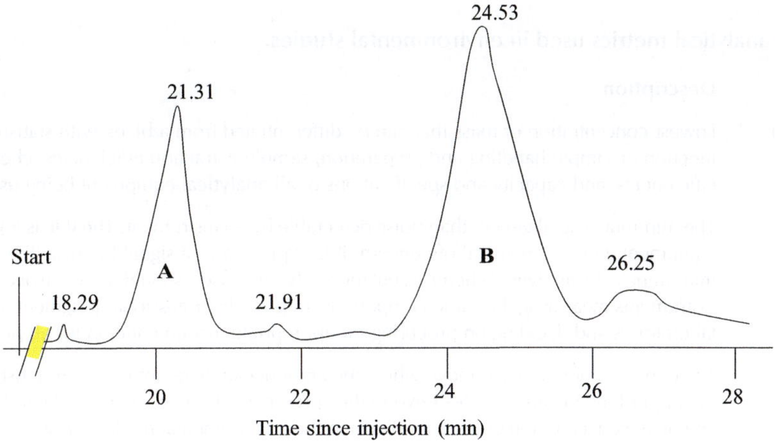

Retention Time	Area	Type	Area/Height	Area %
18.29	NA	NA	NA	0.1
21.31	NA	NA	NA	31.4
21.91	NA	NA	NA	0.2
24.53	NA	NA	NA	67.2
26.25	NA	NA	NA	1.1

FIG. 13.6 Hypothetical HPLC-UV curve and data table.

The column type also affects retention time and peak area. The one used by the laboratory is commonly used for polycyclic aromatic hydrocarbons, including benzo(*a*)pyrene. However, numerous columns can be used for semivolatile organic compounds, so both the retention time and peak area will vary somewhat. Another concern is coelution, i.e., two distinct compounds that have nearly the same retention times. One means of reducing the likelihood of coelution is to target the wavelength of the UV detector. For example, the recommended wavelength for benzo(*a*)pyrene is 254 nm, but 295 nm is preferred by environmental chromatographers because the interference peak in the benzo(*a*)pyrene window is decreased at 295 nm. Another way to improve detection is to use a diode array detection system with the UV detector. This gives several different chromatograms simultaneously at various wavelengths. Finally, there are times when certain detection systems are completely unable to detect a specific molecule. For example, if a molecule does not absorb UV light (i.e., it lacks a group of atoms in a molecule responsible for absorbing the UV radiation, known as chromophores), no UV detector is appropriate. In this case another detector, e.g., mass spectrometry, must be used.

For inorganic substances and metals, the additional extraction step may not be necessary. The water may first be filtered by surface techniques like atomic absorption (AA), X-ray fluorescence (XRF), inductively coupled plasma (ICP), or sputtering. As for organic compounds, the detection approaches can vary. For example, ICP may be used with absorption or MS. If elemental mass is all that is needed, for example to determine total lead or nickel in a sample, AA or XRF may be sufficient.

Standard solutions of metal ions are to calibrate the metal-detection instrument. The preparation of standards consists of several steps. For example, the steps for sample preparation using AA include [73]:

1. Selecting at least three concentrations of standard solution of the metal to be analyzed;
2. Aspiration of a blank solution and adjustment to zero;
3. Aspiration of each standard solution into flame;

TABLE 13.5 Types of analytical metrics used in environmental studies.

Type of limit	Description
Limit of detection (LOD)	Lowest concentration or mass that can be differentiated from a blank with statistical confidence. This is a function of sample handling and preparation, sample extraction efficiencies, chemical separation efficiencies, and capacity and specifications of all analytical equipment being used (see IDL below)
Instrument detection limit (IDL)	The minimum signal greater than noise detectable by an instrument. The IDL is an expression of the piece of equipment, not the chemical of concern. It is expressed as a signal to noise (S:N) ratio. This is mainly important to the analytical chemists, but the analyst and users should be aware of the different IDLs for various instruments measuring the same compounds, to provide professional judgment in contracting or selecting laboratories and deciding on procuring for appropriate instrumentation for all phases of remediation
Limit of quantitation (LOQ)	The concentration or mass above which the amount can be quantified with statistical confidence. This is an important limit because it goes beyond the "presence–absence" of the LOD and allows for calculating chemical concentration or mass gradients in the environmental media (air, water, soil, sediment, and biota)
Practical quantitation limit (PQL)	The combination of LOQ and the precision and accuracy limits of a specific laboratory, as expressed in the laboratory's quality assurance/quality control (QA/QC) plans and standard operating procedures (SOPs) for routine runs. The PQL is the concentration or mass that the analyst and users can consistently expect to have reported reliably

Data from D.A. Vallero, J.J. Peirce, Engineering the Risks of Hazardous Wastes, Butterworth-Heinemann, Burlington, MA, 2003, xxi, 306 pp.

4. Preparation of a calibration curve for absorbance versus concentration of standard solution;
5. Reading the prepared sample solution directly from the instrument; and,
6. Using the appropriate dilution factor for samples with elevated concentration of metal ions.

Unlike organic compounds, these metal analytical methods do not find individual chemical species, only total mass of the element. For speciation, e.g., finding the concentrations of a specific organometallic species of nickel (Ni) or tin (Sn), sample preparation is more involved, including a process known as derivatization. Derivatizing a sample is performed by adding a chemical agent that transforms the compound in question into one that can be recognized by the detector. This is done for both organic and inorganic compounds, for example, when the compound in question is too polar to be recognized by MS.

The physical and chemical characteristics of the compounds being analyzed must be considered before visiting the field and throughout all the steps in the laboratory. Although it is beyond the scope of this book to go into detail, it is worth mentioning that the quality of results generated about contamination depends upon the sensitivity and selectivity of the analytical equipment. Table 13.5 lists some of the most important analytical chemistry metrics.

6 Emerging trends in measurement

In addition to the traditional monitoring equipment discussed here, sensors have become important assets in environmental measurements. Remote sensing can provide real-time measurements, e.g., soil moisture, which is needed to cover large areas and distant regions, e.g., water and wastewater distribution systems [74].

Emerging technologies and software are improving the timeliness and spatial and temporal representation of environmental measurements. Local communities can use a variety of new tools tailored to neighborhoods, which can support their management of facilities, decisions on potential new businesses, and support of land use and other planning decisions [75].

Sensors are increasingly providing near real-time, portable measurements with resolutions near the resolution and quality those of the wet laboratory and are becoming more economical to deploy [76]. Examples in current use include infrared cameras that characterized emissions of toxic air pollutants rising from storage tanks and other equipment, as well as remote and fence line monitoring near pollution sources. Because many toxic substance releases are imperceptible by normal human senses, continuous monitoring can run constantly near potential pollutant sources [75]. They are also supporting green and sustainable operations, such as farmers who are increasingly using sensors to stay apprised of soil conditions, e.g., moisture and temperature, and to manage crops accordingly, e.g., to optimize irrigation water usage [77]. Users include government inspectors, but also the private sector who need to identify environmental conditions and pollution sources, i.e., "citizen science" [78–81].

7 Ethical conduct

Water quality measurements must be scientifically credible and conducted ethically. Any study that involves humans must ensure that the person is respected, that studies provide a tangible benefit, and that justice is ensured. This not only includes studies where a person knows that he or she is involved, but those where people may be indirectly affected. For example, permission is needed and those potentially involved must be informed, such as homeowners and other water users of a study that will be conducted to determine the quality of water in an aquifer that serves as a town's water supply. Often the city or town engineer is a good point of contact during the planning stages. This can prevent unnecessary effort or problems, such as finding that a similar study had already been recently conducted, or that the town has more than one source of drinking water. It is also a way to determine special monitoring needs, such as ways to sample in homes with vulnerable persons (e.g., the elderly) and historically underrepresented neighborhoods, i.e., environmental justice communities [82].

Measurement results must be trusted by the scientific community and the public [83–85]. Researchers and practitioners must apply physical, natural, and social scientific knowledge to address water pollution [86,87]. Water quality information must improve the public's awareness, and governmental and industrial decision making and behavior [88]. Scientists have implicit and engineers have explicit social contracts to solve water quality problems. This means that a consistent and fair process is needed to choose among actions to be taken from reasonable and available alternatives [89]. This selection process needs to consider the following [90]:

1. Proposed actions based on data and modeled water quality predictions must be accompanied by clear statements of uncertainty because the actual values of all key variables cannot be known completely given inherent gaps in knowledge and system variability.
2. Although continually improving, the physical, chemical, and biological sciences underpinning the processes that cause the water pollution are seldom fully understood, so models will propagate and compound errors due to physical, chemical, biological, and other interpretative uncertainties, calling for continuous enhancements to water quality measurements.
3. Water quality predictions are probabilistic (see Chapter 2), so are never perfect, so that the engineer and scientist must be sufficiently humble when presenting them to their clients and to the public.

A commonly used tool for water quality decisions and for selecting the best actions is the benefit-cost ratio (BCR), a comparison of gross benefits to gross costs. Certainly, the utility of a project is crucial to the decision to go forward, according to at least two criteria [12]:

- The usefulness of the project is the basis for its value to water quality.
- In pursuing this option, it must provide the most benefit for the greatest number (e.g., people, ecosystems, etc.).

The BCR is a useful metric due to its arithmetic simplicity and seeming transparency. To determine whether a project is worthwhile, all the benefits appear in the numerator and all the costs in the denominator. If the ratio is greater than 1, the option's benefits exceed its costs.

Accuracy in estimating the costs and risks is directly related to the track record of the actions, so that the BCR value is often limited to screening, without an absolute measure of potential or value. When projected benefits far outweigh by the costs (i.e., $BCR \gg 1$), this is evidence of support for an action, to the extent that the measurement and other data are representative and properly weighted. Conversely, a BCR much less than 1 may indicate potential problems in the future stemming from that action. Of course, this all depends on the availability of highly relevant and reliable data and models.

Any research involving humans must be meet rigorous ethical standards. For example, universities and research institutions must first receive approval of a committee whose sole purpose is to conduct ethical review of proposed research, known as an institutional review board [91]. Therefore, measurement studies must sufficiently protect people's privacy, properly obtain permissions, completely inform participants (including necessary consent forms), and does not allow personal information to be identified in any unauthorized way. A particularly useful resource for identifying potential ethical problems and improving protections of the confidentiality of individuals' information is the document: Scientific and Ethical Approaches for Observational Exposure Studies [92].

8 Conclusions

Accurate, precise, and representative measurements are needed as part of any effort to ensure that water is meeting quality standards for human health and the environment. This chapter introduced measurement concepts related to sampling and analysis, as well as ways to interpret results and address uncertainties.

References

[1] D.A. Vallero, Measurements in environmental engineering, in: M. Kutz (Ed.), Handbook of Environmental Engineering, John Wiley and Sons, Inc., Hoboken, New Jersey, 2018, pp. 13–43.
[2] I. Linkov, B.D. Trump, E. Anklam, D. Berube, P. Boisseasu, C. Cummings, et al., Comparative, collaborative, and integrative risk governance for emerging technologies, Environ. Syst. Decis. (2018).
[3] T.S. Kuhn, D. Hawkins, The structure of scientific revolutions, Am. J. Phys. 31 (7) (1963) 554–555.
[4] C.F. McSweeney, R.G. Jones, No consensus on consensus: the challenge of finding a universal approach to measuring and mapping ensemble consistency in GCM projections, Clim. Chang. 119 (2013) 617–629.
[5] D.A. Vallero, Waste governance, in: T.M. Letcher, A. Daniel (Eds.), Waste: A Handbook for Management, Elsevier, 2019, pp. 651–671 (Chapter 33).
[6] B. Bago, D.W. Rand, G. Pennycook, Reasoning about climate change, PNAS Nexus 2 (5) (2023) pgad100.
[7] C. Funk, M. Hefferon, US public views on climate and energy, Pew Research Center, 2019, p. 25.
[8] D. Ding, E.W. Maibach, X. Zhao, C. Roser-Renouf, A. Leiserowitz, Support for climate policy and societal action are linked to perceptions about scientific agreement, Nat. Clim. Chang. 1 (9) (2011) 462–466.
[9] Q. Ding, X. Shi, D. Zhuang, Y. Wang, Temporal and spatial distributions of ecological vulnerability under the influence of natural and anthropogenic factors in an eco-province under construction in China, Sustainability 10 (9) (2018) 3087.
[10] A.M. McCright, R.E. Dunlap, C. Xiao, Perceived scientific agreement and support for government action on climate change in the USA, Clim. Chang. 119 (2013) 511–518.
[11] C.B. Cooper, Media literacy as a key strategy toward improving public acceptance of climate change science, Bioscience 61 (3) (2011) 231–237.
[12] D. Vallero, Environmental Biotechnology: A Biosystems Approach, Elsevier Science, 2015.
[13] D.A. Vallero, Measurements in environmental engineering, in: M. Kutz (Ed.), Handbook of Measurement in Science and Engineering, vol. 3, John Wiley and Sons, Hoboken, New Jersey, 2016.
[14] U.S. Environmental Protection Agency, Environmental measurement, 2015, Available from: http://www.epa.gov/measurements.
[15] J. Dey, R. Vijay, A critical and intensive review on assessment of water quality parameters through geospatial techniques, Environ. Sci. Pollut. Res. 28 (31) (2021) 41612–41626.
[16] M. Hysenaj, Geospatial analysis of environment pollution, J. Ecol. Eng. 17 (1) (2016).
[17] B.R. Pickard, J. Daniel, M. Mehaffey, L.E. Jackson, A. Neale, EnviroAtlas: a new geospatial tool to foster ecosystem services science and resource management, Ecosyst. Serv. 14 (2015) 45–55.
[18] V. Sumathi, U. Natesan, C. Sarkar, GIS-based approach for optimized siting of municipal solid waste landfill, Waste Manag. 28 (11) (2008) 2146–2160.
[19] H. Sarptas, N. Alpaslan, D. Dolgen, GIS supported solid waste management in coastal areas, Water Sci. Technol. 51 (11) (2005) 213–220.
[20] F. Karimipour, M.R. Delavar, M. Kinaie, Water quality management using GIS data mining, J. Environ. Inf. 5 (2) (2005) 61–71.
[21] U.S. Environmental Protection Agency, Air stripping (ex situ), 2022, Washington, DC. Available from: https://frtr.gov/matrix/Air-Stripping-Ex-Situ/#Implementability-Considerations.
[22] S. Yan, M. Rakowska, X. Shen, T. Himmer, C. Irvine, R. Zajac-Fay, et al., Bioavailability assessment in activated carbon treated coastal sediment with in situ and ex situ porewater measurements, Water Res. 185 (2020) 116259.
[23] D.A. Vallero, Fundamentals of Air Pollution, fifth ed., Elsevier Academic Press, Waltham, MA, 2014. 999 pages cm p.
[24] K. Whitby, K. Willeke, Single particle optical counters: principles and field use, in: Aerosol Measurement, 1979, p. 145.
[25] G.J. Telesnicki, W.M. Goldberg, Comparison of turbidity measurement by nephelometry and transmissometry and its relevance to water quality standards, Bull. Mar. Sci. 57 (2) (1995) 540–547.
[26] V.G. Christensen, P.P. Rasmussen, A.C. Ziegler, Real-time water quality monitoring and regression analysis to estimate nutrient and bacteria concentrations in Kansas streams, Water Sci. Technol. 45 (9) (2002) 205–219.
[27] W.R. Ott, Environmental Statistics and Data Analysis, Lewis Publishers, CRC Press, Inc., Boca Raton, Florida, 1995.
[28] U.S. Environmental Protection Agency, Data quality objectives guidance, UEP Agency (Ed.), Washington, DC, 2006.
[29] D. Crumbling, Clarifying DQO terminology usage to support modernization of site cleanup practice, U.S. Environmental Protection Agency (Ed.), USEPA, Washington, DC, 2001.
[30] US Environmental Protection Agency, Method 1613: Tetra-through octa-chlorinated dioxins and furans by isotope dilution HRGC/HRMS (Rev. B), UEP Agency (Ed.), Washington, DC, 1994.
[31] N.V. Korde, The method of biological study of bottom sediments of lakes (field work and biological analysis), in: Zhizn'presnykh vod SSSR, Vol. 4, No. 1, Academy of Science of SSSR, Moscow, Leningrad, 1956, pp. 383–413. In this issue.
[32] P.H. Doering, R.H. Chamberlain, K.M. Haunert, Chlorophyll a and its use as an indicator of eutrophication in the Caloosahatchee Estuary, Florida, Fla. Sci. (2006) 51–72.
[33] J.H. Ryther, W.M. Dunstan, Nitrogen, phosphorus, and eutrophication in the coastal marine environment, Science 171 (3975) (1971) 1008–1013.
[34] R. Harland, J.A. Howe, Dinoflagellate cysts and Holocene oceanography of the northeastern Atlantic Ocean, Holocene 5 (2) (1995) 220–228.
[35] J.W. Day, R.R. Christian, D.M. Boesch, A. Yáñez-Arancibia, J. Morris, R.R. Twilley, et al., Consequences of climate change on the ecogeomorphology of coastal wetlands, Estuaries Coast. 31 (3) (2008) 477–491.
[36] R.R. Twilley, W.M. Kemp, K.W. Staver, J.C. Stevenson, W.R. Boynton, Nutrient enrichment of estuarine submersed vascular plant communities. 1. Algal growth and effects on production of plants and associated communities, Mar. Ecol. Prog. Ser. 23 (2) (1985) 179–191.

[37] U.S. Environmental Protection Agency, Method TO-9A in compendium of methods for the determination of toxic organic compounds in ambient air, second ed., 1999.

[38] U.S. Environmental Protection Agency, SW846 Method 8290A-1: Polychlorinated dibenzodioxins (PCDDs) and polychlorinated dibenzofurans (PCDFs) by high resolution gas chromatograph/high resolution mass spectrometry (HRGC/HRMS). Revision 1 ed, 2007.

[39] U.S. Environmental Protection Agency, Guidance for the data quality objectives process, EPA QA/G-4, Washington, DC, 2002.

[40] D.A. Vallero, Environmental Systems Science: Theory and Practical Applications, first ed., Elsevier Academic Press, 2021. Accessed 26 May 2021.

[41] J.D. Pleil, A.F. Vette, B.A. Johnson, S.M. Rappaport, Air levels of carcinogenic polycyclic aromatic hydrocarbons after the World Trade Center disaster, Proc. Natl. Acad. Sci. U. S. A. 101 (32) (2004) 11685–11688.

[42] P.J. Lioy, C.P. Weisel, J.R. Millette, S. Eisenreich, D. Vallero, J. Offenberg, et al., Characterization of the dust/smoke aerosol that settled east of the World Trade Center (WTC) in lower Manhattan after the collapse of the WTC 11 September 2001, Environ. Health Perspect. 110 (7) (2002) 703–714.

[43] D.A. Olson, G.A. Norris, Chemical characterization of ambient particulate matter near the World Trade Center: source apportionment using organic and inorganic source markers, Atmos. Environ. 42 (31) (2008) 7310–7315.

[44] L.-M. Yiin, J.R. Millette, A. Vette, V. Ilacqua, C. Quan, J. Gorczynski, et al., Comparisons of the dust/smoke particulate that settled inside the surrounding buildings and outside on the streets of southern New York City after the collapse of the World Trade Center, September 11, 2001, J. Air Waste Manage. Assoc. 54 (5) (2004) 515–528.

[45] A.E. Larsen, M. Patton, E.A. Martin, High highs and low lows: elucidating striking seasonal variability in pesticide use and its environmental implications, Sci. Total Environ. 651 (2019) 828–837.

[46] I. Delcour, P. Spanoghe, M. Uyttendaele, Literature review: impact of climate change on pesticide use, Food Res. Int. 68 (2015) 7–15.

[47] D.A. Vallero, S.S. Isukapalli, V.G. Zartarian, T.R. Mccurdy, T. Mckone, P. Georgopoulos, et al., Modeling and predicting pesticide exposures, in: R. Krieger (Ed.), Hayes Handbook of Pesticide Toxicology, Elsevier Science, New York, NY, 2010, pp. 995–1020.

[48] S. Kimbrough, R.W. Baldauf, G.S. Hagler, R.C. Shores, W. Mitchell, D.A. Whitaker, et al., Long-term continuous measurement of near-road air pollution in Las Vegas: seasonal variability in traffic emissions impact on local air quality, Air Qual. Atmos. Health 6 (1) (2013) 295–305.

[49] S. Kimbrough, R. Shores, D. Whitaker, B. Mitchell, R. Baldauf, A. Vette, et al., The national near-road mobile source air toxics study: Las Vegas, in: Proceedings, 103rd Annual Conference of the Air & Waste Management Association, Calgary, AB, Canada, June 22–25, 2010; Paper #553, Air & Waste Management Association, Pittsburgh, PA, 2010. In this issue.

[50] S. Kimbrough, D. Vallero, R. Shores, A. Vette, K. Black, V. Martinez, Multi-criteria decision analysis for the selection of a near road ambient air monitoring site for the measurement of mobile source air toxics, Transp. Res. D Transp. Environ. 13 (8) (2008) 505–515.

[51] J. Malczewski, GIS and Multicriteria Decision Analysis, John Wiley & Sons, 1999.

[52] U.S. Environmental Protection Agency, Ecosystem services in EnviroAtlas, 2015, Available from: http://www.epa.gov/enviroatlas/ecosystem-services-enviroatlas.

[53] U.S. Environmental Protection Agency, EnviroAtlas data layer matrix, 2015, Available from: http://www.epa.gov/enviroatlas/enviroatlas-data-layer-matrix.

[54] U.S. Environmental Protection Agency, EnviroAtlas data, U.S. EPA, Washington, DC, 2023. Available from: https://www.epa.gov/enviroatlas/enviroatlas-interactive-map#enviroatlas. (Accessed 30 November 2023).

[55] Environmental Systems Research Institute, Understanding GIS: The ARC/INFO Method: Self Study Workbook: Version 7 for UNIX and OpenVMS, Esri Press, 1995.

[56] U.S. Environmental Protection Agency, Test methods: frequent questions, 2015, Available from: http://www3.epa.gov/epawaste/hazard/testmethods/faq/.

[57] L. Wang, C. Liu, D.G. Alves, D.A. Frank, D. Wang, Plant diversity is associated with the amount and spatial structure of soil heterogeneity in meadow steppe of China, Landsc. Ecol. 30 (9) (2015) 1713–1721.

[58] D.A. Vallero, J.J. Peirce, Transformation and transport of vinclozolin from soil to air, J. Environ. Eng. 128 (3) (2002) 261–268.

[59] D.A. Vallero, Dicarboximide Fungicide Flux From an Aquic Hapludult Soil to the Lower Troposphere, Duke University, 2000.

[60] Environmental Health Australia, Environmental health risk assessment: guidelines for assessing human health risks from environmental hazards, Ageing DoHa (Ed.), Commonwealth of Australia, Canberra, ACT, 2012.

[61] J. Stoddard, D. Peck, S. Paulsen, J. Van Sickle, C. Hawkins, A. Herlihy, et al., An ecological assessment of western streams and rivers, US Environmental Protection Agency, Washington, DC, 2005. Available at http://www.epa.gov/nheerl/arm/documents/emap.w.assessmentfinal.pdf. (Accessed 10 July 2024. EPA 620/R-05/005, In this issue.

[62] D.B. Carr, A.R. Olsen, D. White, Hexagon mosaic maps for display of univariate and bivariate geographical data, Cartogr. Geogr. Inf. Syst. 19 (4) (1992) 228–236.

[63] U.S. Environmental Protection Agency, Risk assessment guidance for superfund: volume I—human health evaluation manual (part A): interim final, No. EPA/540/1-89/002, 1989, Washington, DC. Available from: https://www.epa.gov/sites/production/files/2015-09/documents/rags_a.pdf.

[64] U.S. Environmental Protection Agency, An examination of EPA risk assessment principles and practices, Washington, DC, 2004.

[65] A.M. Finkel, Confronting uncertainty in risk management; a guide for decision makers a report, Resources for the Future, Washington, DC, 1990.

[66] Environmental Health Australia, Australian exposure factor guidance, Ageing DoHa (Ed.), Commonwealth of Australia, Canberra, ACT, 2012.

[67] C.M. Thompson, B. Sonawane, H.A. Barton, R.S. DeWoskin, J.C. Lipscomb, P. Schlosser, et al., Approaches for applications of physiologically based pharmacokinetic models in risk assessment, J. Toxic. Environ. Health B Crit. Rev. 11 (7) (2008) 519–547.

[68] D. Kaya, F. Karakas, F.D. Sanin, I. Imamoglu, PCBs in sludge: development of a practical extraction procedure and its application in an urban water resource recovery facility, Water Environ. Res. 87 (2) (2015) 145–151.

[69] M.R. Ekhtera, G.A. Mansoori, M.C. Mensinger, A. Rehmat, B. Deville, Supercritical fluid extraction for remediation of contaminated soil, in: A. Abraham MaS (Ed.), Supercritical Fluids: Extraction and Pollution Prevention, vol. 670, American Chemical Society, Washington, DC, 1997, pp. 280–298.
[70] E. Swartz, L. Stockburger, D.A. Vallero, Polycyclic aromatic hydrocarbons and other semivolatile organic compounds collected in New York City in response to the events of 9/11, Environ. Sci. Technol. 37 (16) (2003) 3537–3546.
[71] J.J. Paré, J.M. Bélanger, B. Lesnik, R. Turpin, R. Singhvi, Final evaluation of US EPA method 3546: microwave extraction, a microwave-assisted process (MAPTM)* method for the extraction of contaminants under closed-vessel conditions, Soil Sediment Contam. 10 (4) (2001) 375–386.
[72] Phenomenex., Is water safe to inject in gas chromatography? 2024, Torrance, California. Available from: https://phenomenex.blog/2018/01/30/inject-water-gc/.
[73] B. Sharma, S. Tyagi, Simplification of metal ion analysis in fresh water samples by atomic absorption spectroscopy for laboratory students, J. Lab. Chem. Educ. 1 (3) (2013) 54–58.
[74] R.J. Gehl, C.W. Rice, Emerging technologies for in situ measurement of soil carbon, Clim. Chang. 80 (1–2) (2007) 43–54.
[75] T. Watkins, The US EPA roadmap for next generation air monitoring, in: EuNetAir Second Scientific Meeting, Queens' College, Cambridge, 2013.
[76] E.G. Snyder, T.H. Watkins, P.A. Solomon, E.D. Thoma, R.W. Williams, G.S. Hagler, et al., The changing paradigm of air pollution monitoring, Environ. Sci. Technol. 47 (20) (2013) 11369–11377.
[77] S. Khriji, D. El Houssaini, M.W. Jmal, C. Viehweger, M. Abid, O. Kanoun, Precision irrigation based on wireless sensor network, IET Sci. Meas. Technol. 8 (3) (2014) 98–106.
[78] D. Roman, N. Reeves, E. Gonzalez, I. Celino, S. Abd El Kader, P. Turk, et al., An analysis of pollution citizen science projects from the perspective of data science and open science, Data Technol. Appl. (2021).
[79] M. Ziolkowski, Citizen scientists documenting hazardous materials shipments on American railroads, Case Stud. Transp. Policy 8 (4) (2020) 1181–1190.
[80] Z. Tang, Y. Zhou, H. Yu, Y. Gu, T. Liu, Developing an interactive mobile volunteered geographic information platform to integrate environmental big data and citizen science in urban management, in: Seeing Cities Through Big Data, Springer, 2017, pp. 65–81.
[81] D.C. McKinley, A.J. Miller-Rushing, H.L. Ballard, R. Bonney, H. Brown, S.C. Cook-Patton, et al., Citizen science can improve conservation science, natural resource management, and environmental protection, Biol. Conserv. (2016).
[82] J. VanDerslice, Drinking water infrastructure and environmental disparities: evidence and methodological considerations, Am. J. Public Health 101 (S1) (2011) S109–S114.
[83] E.C. Hughes, Professions, Daedalus (1963) 655–668.
[84] D.E. Wueste, Linking academic integrity and ethics across the curriculum: groundwork for sustainability in practical and professional ethics, in: Ethics Across the Curriculum—Pedagogical Perspectives, Springer, 2018, pp. 303–326.
[85] D.A. Vallero, Biomedical Ethics for Engineers, The Biomedical Engineering Series, 2007.
[86] D. Vallero, New tools to address responsible conduct of nanobiotechnological research, J. Long Term Eff. Med. Implants 18 (1) (2008) 8.
[87] D.A. Vallero, Beyond responsible conduct in research: new pedagogies to address macroethics of nanobiotechnologies, J. Long Term Eff. Med. Implants 17 (1) (2007).
[88] National Science Foundation, Responsible conduct of research, Fed. Regist. 74 (160) (2009) 42126–42128.
[89] S.L. Derby, R.L. Keeney, Risk analysis: understanding "how safe is safe enough?", Risk Anal. 1 (3) (1981) 217–224.
[90] M.G. Morgan, Risk assessment: choosing and managing technology-induced risk: how much risk should we choose to live with? How should we assess and manage the risks we face? IEEE Spectr. 18 (12) (1981) 53–60.
[91] Department of Health, Education, and Welfare, The Belmont Report. Ethical principles and guidelines for the protection of human subjects of research, J. Am. Coll. Dent. 81 (3) (2014) 4.
[92] U.S. Environmental Protection Agency, Scientific and ethical approaches for observational exposure studies, National Exposure Research Laboratory, Office of Research and Development, U.S. Environmental Protection Agency, Research Triangle Park, NC, 2008.

Chapter 14

Environmental partitioning

1 Introduction

When pollutants are released into the environment, their effects are determined by their inherent properties and by the conditions of the environment into which they have been released. The interconnectedness of physical, chemical, and biological processes affects the fate and transport of substances as they move among environmental compartments (e.g., water, soil, sediment, or air). The compartment where they reside and the physicochemical properties of the pollutant determine the environmental fate of these substances. The properties of the pollutant are influenced and changed by the properties of the media. This dynamism is determined by numerous processes, including both abiotic and biotic media. Abiotic media include air, water, and the nonliving parts of soil and sediment. Biotic media include all levels of biological organization, from subcellular and cellular matter in organisms to communities within continental-scale biomes. Table 14.1 summaries a few of the volatilization processes that change the fate of released substances.

A substance's vapor pressure and aqueous solubility are key predictors of the likely movement of a substance throughout the environment. Upon reaching equilibrium, several partition coefficients influence the fate of a substance after it is released. Notably, the coefficients of sorption (distribution coefficient, K_D, or solid-water partition coefficient, K_p), dissolution or solubility, air-water partitioning (K_{AW}), Henry's law constant (K_H), and organic carbon-water (K_{oc}) are important. The octanol-water coefficient (K_{ow}) is particularly important to water pollution. Uptake and accumulation of the pollutant in water and other media are also expressed as the bioconcentration factor (*BCF*), biomagnification factor (*BMF*), and bioaccumulation factor (*BAF*).

A good place to start a discussion of water pollutant partitioning is with vapor pressure and solubility. Vapor pressure is an expression of the maximum amount of a chemical species that can be held in the gas phase and solubility is an expression of the maximum amount of a chemical species that can be held in the liquid phase. Given that both vapor pressure and solubility are described as the maximum amount that remains in a phase, i.e., gas phase for vapor pressure and liquid phase for solubility, both are "saturation properties" of a chemical species [1]. In chemistry, the equilibrium constant, *K*, is an expression of the relationship between the products and reactants of a reaction upon reaching equilibrium. Thus, a substance's vapor pressure and solubility combine with other properties and conditions as the basis for how a substance will partition and be distributed in various compartments in the environment.

2 Vapor pressure

Vapor pressure is the pressure exerted by a vapor in a confined space. Similarly, vaporization is the change of a liquid or solid to the vapor phase. The vapor pressure of a chemical species is the portion of the total atmospheric pressure due to that chemical species in its vapor phase. Thus, vapor pressure is one of the molecular characteristics of a compound that determines how a substance will behave physically, chemically, and biologically in the environment and within an organism. A compound's structure is a key predictor of its vapor pressure, including the geometry of the molecule, its electronic characteristics, and its arrangement in space, e.g., in a crystalline solid, gas, or liquid [2]. Some features are associated with vaporization potential. For example, a substance with low molecular weight will likely have a relatively high vapor pressure and a relatively low boiling point.

The vapor pressure of a chemical species is an inherent property, which means it is unique to that substance, notwithstanding the environmental conditions. For example, volatile organic compounds (VOCs) have an affinity to move from the liquid phase to the gas phase under normal environmental conditions. The VOCs have vapor pressures greater than 10^{-2} kilopascals (kP) or 9.9×10^{-4} atm [3]. Although the specific threshold that defines a VOC is debatable, all VOCs have relatively low boiling points and relatively high vapor pressures (see Table 14.1). For example, the European Union defines

Fundamentals of Water Pollution. https://doi.org/10.1016/B978-0-443-28987-3.00017-5

TABLE 14.1 Classification of organic compounds according to vapor pressure.

Category	Abbreviation	Boiling point range (°C)	Example compounds
Very volatile organic compounds	VVOC	<0 to 50–100	Propane, butane, methyl chloride
Volatile organic compounds	VOC	50–100 to 240–260	Formaldehyde, D-limonene, toluene, acetone, ethanol (ethyl alcohol) 2-propanol (isopropyl alcohol), hexanal
Semivolatile organic compounds	SVOC	240–260 to 380–400	Pesticides (DDT, chlordane, phthalates), fire retardants (including polybrominated biphenyls, PBBs), polychlorinated biphenyls, several polycyclic aromatic hydrocarbons (PAHs), chlorinated dioxins and furans

Data from U.S. Environmental Protection Agency, Technical overview of volatile organic compounds, general definition and classification. https://www.epa.gov/indoor-air-quality-iaq/technical-overview-volatile-organic-compounds, 2018 (Accessed 5 June 2018). World Health Organization, WHO guidelines for indoor air quality: selected pollutants, 2010. World Health Organization, Indoor air quality: organic pollutants, 0143-2060, 1989.

a VOC as any organic compound with an initial boiling point less than 260 °C at 1 atm [4]. Some compounds are so volatile that they are designated very volatile organic compounds (VVOCs). Substances will vary in the number of molecules that move from the liquid phase to the vapor phase, even under the same environmental conditions [5,6].

Vapor pressure example (from Ref. [6])

A study [7] measured the molecular makeup of conventional gasoline that contained 10% ethanol, E85 (85% ethanol and 15% gasoline), ultra-low sulfur diesel (ULSD), and B20 (20% soy-biodiesel and 80% ULSD) delivered by a fuel station in Michigan, USA (see Table 14.2). A customer purchases two fuel containers holding 4 L of gasoline and 4 L of E85 fuel and places them in his garage, forgetting to screw on the caps of both containers. The customer detects odors in the next morning and, after checking, replaces the caps. The temperature of the garage averaged 25 °C throughout the night. Which of the fuels will have lost the most toluene to the air?

TABLE 14.2 Chemical composition of volatile organic compounds in fuels collected from a Michigan filling station.

Fuel	Gasoline (mg L^{-1})	E85 (mg L^{-1})	E85/gasoline (%)	Diesel (mg L^{-1})	B20 (mg L^{-1})	(%)
Aromatics						
Benzene	6140	862	14	67	37	55
Toluene	15,400	4110	27	238	214	90
Ethylbenzene	3080	1990	65	124	186	150
p-Xylene, *m*-xylene	9120	6980	76	420	496	118
o-Xylene	4610	2790	60	185	212	115
Isopropyl benzene	351	156	45	44	70	160
n-Propyl benzene	2110	665	32	115	167	146
p-Isopropyl toluene	88	29	33	112	83	75
4-Ethyl toluene	8380	2650	32	400	464	116
2-Ethyl toluene	3460	928	27	194	264	136
1,3,5-Trimethylbenzene	4060	1030	25	202	150	74
1,2,4-Trimethylbenzene	10,600	3270	31	720	575	80
1,2,3-Trimethylbenzene	3950	975	25	2120	961	45
sec-Butylbenzene	159	65	41	90	117	130
n-Butylbenzene	822	198	24	375	111	30

TABLE 14.2 Chemical composition of volatile organic compounds in fuels collected from a Michigan filling station—cont'd

Fuel	Gasoline (mg L^{-1})	E85 (mg L^{-1})	E85/gasoline (%)	Diesel (mg L^{-1})	B20 (mg L^{-1})	(%)
Styrene	14	4	32	<0.02	<0.02	–
Naphthalene	2240	378	17	3000	1220	41
Alkanes						
n-Heptane	12,800	3330	26	174	138	79
n-Octane	2870	1550	54	481	612	127
n-Nonane	1790	1050	59	7020	4120	59
n-Decane	1390	262	19	7690	4310	56
n-Undecane	1120	121	11	7730	4560	59
n-Dodecane	822	85	10	8370	5000	60
n-Tridecane	644	80	12	13,400	7620	57
n-Tetradecane	213	19	9	10,100	7130	71
n-Pentadecane	62	<0.02	–	9030	7580	84
n-Hexadecane	18	<0.02	–	9300	5550	60
Cyclohexane	9830	880	9	191	69	36
Methyl cyclohexane	8280	778	9	426	270	63
Total measured VOCs	114,000	34,900	31	80,700	51,600	64

Data from J.-Y. Chin, S.A. Batterman, VOC composition of current motor vehicle fuels and vapors, and collinearity analyses for receptor modeling, Chemosphere 86 (9) (2012) 951–958.

From the table, we see that toluene's vapor pressure is about 4 kP and that of ethanol about 8 kP at 25 °C. The gasoline contains 1.5×10^4 mg L^{-1} toluene and <10% ethanol (<10^5 mg L^{-1}). The E85 contains 4.1×10^4 mg L^{-1} toluene and 15% (1.5×10^5 mg L^{-1}).

Therefore, the gasoline will have transported more toluene molecules into the air. The exact amount will be determined by the size of the spout, temperature range overnight, and the effect of mixtures (e.g., Raoult's law, which is discussed later in this chapter). The spout size is important because it determines the flux, i.e., the mass that moves across the spout's surface area per unit time. If the containers each have spouts the same size, we may disregard this variable since we are only interested in which fuel contributed the most to two VOCs to the air. However, if one spout has an opening twice as large as the other, the flux of toluene and ethanol would double. However, even if the spout were twice as large for the E85 fuel as for the gasoline, the gasoline would still have a much higher toluene flux since its toluene content is nearly four times greater than that of the E85.

The relatively nonvolatile organic compounds (NVOCs) have vapor pressures $<10^{-5}$ kP. These vapor pressures are predominantly found in substances that need significant energy added to increase their volatility. Thus, NVOCs have very few molecules that escape to the atmosphere. Between the extremes, the semivolatile organic compounds (SVOCs), have vapor pressures between 10^{-2} and 10^{-5} kP. As such SVOCs can exist in substantial concentrations in both the gas and aerosol phases in the ambient air, depending on the temperature, atmospheric pressure, the sorptive properties of the media where they reside, and other environmental factors. The SVOCs include some of the most important pollutants, including dioxins, polychlorinated biphenyls (PCBs), polycyclic aromatic hydrocarbons (PAHs), and numerous other persistent organic pollutants (POPs).

An inherent property like vapor pressure certainly is part of the prediction of where a compound will reside in the environment and its propensity to move, but certainly not the only factor in partitioning and transport. All inherent properties of a chemical species and those of the environmental compartment are continuously affecting each other under real-world conditions. Notably, vapor pressure interacts with the chemical species' solubility in water and its sorption to organic matter in soils. Thus, even a substance with a very high vapor pressure may remain in the water if it is extremely soluble in water, or in the soil if it has a strong attachment to a soil particle.

3 Aqueous solubility

The measure of the amount of substance that can dissolve in a liquid is called solubility. At the beginning of my half century of work in environmental science and engineering I heard the adage, "Dilution is not the solution to pollution." While the sentiment that we had to stop relying on natural processes to dilute the pollutants being released in the environment was and is true, the reality is that all standards of water quality reflect the need for dilution. Both the term dilution and solution are in this adage, although this definition of "solution" is not chemical or physical but implies a correction to the problem of environmental contamination. That is, pollutants concentrations must be kept below certain dilutions, such as 1 part per billion (1 ppb) or 10 nanograms per liter (10 ng L^{-1})). A water pollutant may either be in solution or in suspension. A pollutant is the solute and water is the solvent, or the pollutant may be in solid phase and suspended in the water column. Thus, the amount of a substance that is dissolved in water is indeed an indicator of quality.

Solubility is usually expressed in units of mass of solute (that which is dissolved) in the volume of solvent (that which dissolves). Solubility may also be expressed in mass per mass or volume per volume, represented as parts per million (ppm), parts per billion (ppb), or parts per trillion (ppt). Occasionally, solubility is expressed as a percent or in parts per thousand,[a] however, this is uncommon for pollutants, and is usually reserved for nutrients and essential gases (e.g., percent carbon dioxide in water or parts per thousand water vapor in the air).

The solubility of a chemical compound is important to environmental transport, including transport in surface waters and transport through porous media, as in a ground water aquifer. Water solubility, or aqueous solubility, is a generally reliable indication of the location of a pollutant in the environment. For example, the various solubilities of a few chemical compounds, including the most toxic form of dioxin, tetrachlorodibenzo-*para*-dioxin (TCDD), are provided in Table 14.3. Based on these solubility differences, any TCDD released into the environment would be expected to have a much greater affinity for sediment, organic particles, and the organic fraction of soils than its affinity to water. The low water solubilities indicate that dissolved TCDD in the water column should be at only extremely low concentrations. However, solubility of solvents other than water can increase the amount of a substance in the water column by a co-solvation, wherein the hydrophobic compound is first dissolved in an organic solvent and if that solvent has a sufficiently high aqueous solubility

TABLE 14.3 Solubility of tetrachlorodibenzo-*para*-dioxin in water and organic solvents.

Solvent	Solubility (mg L^{-1})
Water	1.93×10^{-5}
Water	6.90×10^{-4} (25 °C)
Methanol	10
Lard oil	40
n-Octanol	50
Acetone	110
Chloroform	370
Benzene	570
Chlorobenzene	720
Orthochlorobenzene	1400

Data from U.S. Environmental Protection Agency, 2,3,7,8-Tetrachlorodibenzo-p-dioxin—1746-01-6 | DTXSID2021315. https://www.epa.gov/chemical-research/chemistry-dashboardhttps://comptox.epa.gov/dashboard/chemical/details/DTXSID2021315, 2021 (Accessed 22 December 2021). Centers for Disease Control and Prevention, National report on human exposure to environmental chemicals. https://www.cdc.gov/biomonitoring/DioxinLikeChemicals_FactSheet.html, 2017 (Accessed). National Toxicology Program, Report on carcinogens: 2,3,7,8-tetrachlorodibenzo-p-dioxin, Research Triangle Park, NC. Available: https://ntp.niehs.nih.gov/ntp/roc/content/profiles/tetrachlorodibenzodioxin.pdf, 2016 (Accessed 20 December 2021). U.S. Environmental Protection Agency, SW846 method 8290A-1: polychlorinated dibenzodioxins (PCDDs) and polychlorinated dibenzofurans (PCDFs) by high resolution gas chromatograph/high resolution mass spectrometry (HRGC/HRMS). Available: http://www3.epa.gov/epawaste/hazard/testmethods/sw846/pdfs/8290a.pdf, 2007. California Office of Environmental Health Hazard Assessment, 2,3,7,8-Tetrachlorodibenzo-p-dioxin and related compounds. https://oehha.ca.gov/chemicals/2378-tetrachlorodibenzo-p-dioxin-and-related-compounds, 2010 (Accessed). National Toxicology Program, NTP report on carcinogens: background document for 2,3,7,8-tetrachlorodibenzo-p-dioxin (TCDD). Available: https://ntp.niehs.nih.gov/ntp/newhomeroc/other_background/tetrachl_tcdd_4apps_508.pdf, 1999. D.A. Vallero, Air Pollution Calculations: Quantifying Pollutant Formation, Transport, Transformation, Fate and Risks, Elsevier, 2019.

a. Up to a few years ago, parts per thousand was abbreviated as "ppt." However, since levels of detection and quantification have improved to parts per trillion or lower, ppt is now more commonly used to abbreviate parts per trillion.

the water, the concentration of the hydrophobic compound will increase. For example, from Table 14.3, if the TCDD is first dissolved in methanol, which is miscible (completely dissolved) in water, the methanol is effective in dissolving the TCDD (10 mg L^{-1}) and the amount of the very hydrophobic TCDD can reach much higher concentrations in water.

Polarity is an important physicochemical characteristic of a substance that determines its solubility. The polarity of a molecule is its unevenness in charge. Aqueous solubility owes to the fact that the water molecule's oxygen and two hydrogen atoms are aligned to give a slightly negative charge at the oxygen end and a slightly positive charge at the hydrogen ends. Since "like dissolves like," polar substances have an affinity to become dissolved in water, and nonpolar substances resist being dissolved in water.

Increasing a system's temperature, i.e., increased kinetic energy, increases the velocity of the system's molecules, which means that intermolecular forces are weakened. With increasing temperature, the molecular velocity increases sufficiently to overcome all intermolecular forces, so that the liquid vaporizes, i.e., it reaches its boiling point. Intermolecular forces may be relatively weak or strong. The weak forces in liquids and gases are often called van der Waals forces.

If a compound has high aqueous solubility, i.e., it is easily dissolved in water under normal environmental conditions of temperature and pressure, it is hydrophilic. If, conversely, a substance is not easily dissolved in water under these conditions, it is said to be hydrophobic. Since many pollutants are organic (i.e., consist of molecules containing carbon-to-carbon bonds and/or carbon-to-hydrogen bonds), the solubility can be further differentiated as to whether under normal environmental conditions of temperature and pressure the substance is easily dissolved in organic solvents. If so, the substance is said to be lipophilic (i.e., readily dissolved in lipids). If, conversely, a substance is not easily dissolved in organic solvents under these conditions, it is said to be lipophobic.

As mentioned, solubility is the second saturation property of a chemical substance. For water pollution, this usually refers to its solubility in water, i.e., aqueous solubility. However, other solvents are also involved in water pollution, e.g., solubility of an organic compound in octanol, half of the octanol-water partition coefficient (K_{ow}). Also, a water pollutant may be first dissolved in an organic solvent like benzene, then the benzene is dissolved in water, a process known as co-solvation. This is one means of substance with low aqueous solubility being found in surface and ground waters.

Hydrophobic compounds are often found in higher concentrations in sediment or animal tissues than in the water column of a lake, estuary, or other water body or the water is absorbed by an organism. Like vapor pressure, aqueous solubility does not fully predict the amount of a chemical species that will be in each compartment. That said, vapor pressure and aqueous solubility provide the first clues as to where a pollutant may reside and its likelihood to move to other environmental compartments. For example, the very hydrophobic compounds known as polychlorinated biphenyls (PCBs) and many polycyclic aromatic hydrocarbons (PAHs) would be expected to be found in soil and sediment, with less in water, given their low aqueous solubility and low vapor pressures.

Hydrophobic compounds are indeed found in water. Beyond this first screening, the fate of chemical substances is better predicted with partitioning. For example, a portion of the PCB and PAH molecules are sorbed to suspended particles or they are absorbed in tissues in organisms within the food chain. As a result of partitioning, substantial amounts of PCBs are found in the ocean. In addition, the abundance of water and biota in the ocean means that, even at low concentrations, many molecules of PCBs will be dissolved [8]. For example, at saturation, 0.45 mg of Aroclor 1232 (molecular weight: 232.2) is dissolved in water at 20 °C [9]. Thus, at saturation, if this PCB is dissolved in 1 million liters of water 450 g of this PCB would be found in the ocean water. This is equivalent to about 4.5×10^8 part per trillion (ppt). Incidentally, some guidelines set 1×10^4 ppt as the maximum to protect aquatic life [10], so this would be greater than four orders of magnitude above that level. Usually, ocean concentrations of PCBs are much lower than saturation, e.g., <1 pg/L (about 1 part per quadrillion) for several PCB congeners in the North Atlantic and Arctic Oceans [11].

Aqueous solubility and vapor pressure example (from Ref. [6])

Three compounds are found to be leaking into the ground water from an underground tank. Compound A has a vapor pressure of 12 kPa and density of 1.6 g mL^{-1}. Compound B has a vapor pressure of 59 kPa and density of 0.7 g mL^{-1}. Compound C has a vapor pressure of 0.003 kPa and density of 1.1 g mL^{-1}. These values are all at 20 °C. Based on these data alone, give the order of each compound likely to be emitted to the atmosphere through the soil.

These values are about the same as the values for three known solvents:

A = tetrachloromethane or carbon tetrachloride (CCl_4);

B = diethyl ether [$(C_2H_5)_2O$];

C = diethylene glycol [$(HOCH_2CH_2)_2O$]. Based entirely on the density and vapor pressure, the order would be that C is the least likely to be emitted given the relatively low vapor pressure and density less than that of water. The most likely to partition to the atmosphere is B, with a very high vapor pressure and low density. Thus, the order would be B, A, C.

However, A and C are each heavier than water, so they could settle toward the bottom of the aquifer if they have low aqueous solubilities so that few of their molecules would become dissolved in the water.

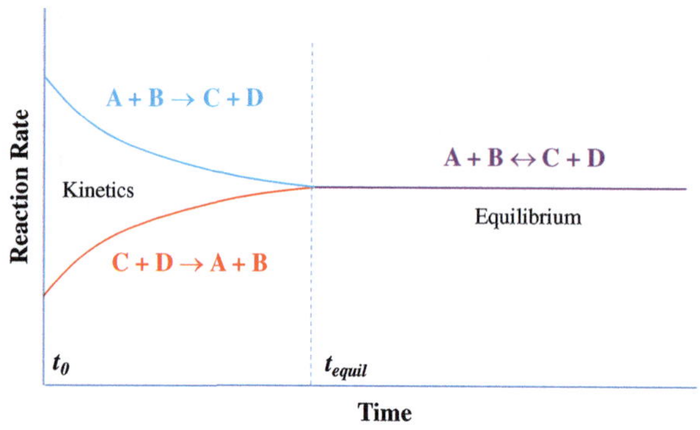

FIG. 14.1 The change in rate of forward and reverse chemical reactions leading to equilibrium in a closed thermodynamic system. *(Modified from D.A. Vallero, Fundamentals of Air Pollution, fifth ed., Elsevier Academic Press, Waltham, MA, 2014, p. 999 pages cm.)*

After the kinetics stage, the chemistry of a substance ultimately reaches equilibrium. Note that the arrow is in both directions in the reversible reaction:

$$aA + bB \leftrightarrow cC + dD \tag{14.1}$$

Fig. 14.1 shows that initially (t_0), the rate of the forward reaction is high, and the rate of the reverse reaction is low. The arrows in kinetic reactions on the left show a left-to-right direction, so the reaction is kinetic. As the reaction progresses, the rate of the forward reaction decreases, and the rate of the reverse reaction increases. This changes when the reaction rates approach equality, i.e., at t_{equil}, when the rate of the forward reaction is the same as the rate of the reverse reaction. At t_{equil}, the system has reached equilibrium.

4 The partition coefficient

When a substance dissolves in water after the kinetic stage, that substance has solubility product constant (K_{sp}), which is known as the equilibrium constant. The K_{sp} represents the point at which that the solute dissolves in solution. The K_{sp} increases in direct proportion to the aqueous solubility of a substance. Consider the aqueous reaction:

$$aA_{(s)} \leftrightarrow cC_{(aq)} + dD_{(aq)} \tag{14.2}$$

The K_{sp} of a reaction can be found by multiplying the molarities or concentrations of the products (cC and dD). The result is raised by the power of the coefficients:

$$K_{sp} = [C]^c[D]^d \tag{14.3}$$

Since the concentrations of solid-phase species $aA_{(s)}$ do not alter the expression, they are not part of the equilibrium constant calculations; any change in their concentrations is negligible. Thus, K_{sp} is the maximum extent that a solid-phase species can be dissolved in solution.

Equilibrium constant example

Find the K_{sp} for the hazardous compound cadmium fluoride.

Cadmium fluoride (CdF_2) is a salt that ionizes in water. The solubility of a salt is that amount of salt added to water to reach the solution's equilibrium. Therefore, the equilibrium reaction for cadmium fluoride is:

$$CdF_{2(s)} \leftrightarrow Cd^{2+}_{(aq)} + F^-_{(aq)}$$

Thus,

$$K_{sp} = \left[Cd^{2+}_{(aq)}\right]\left[F^-_{(aq)}\right]^2$$

Thus, K_{sp} is an example of an equilibrium or partitioning coefficient, i.e., from a solid-phase chemical species to aqueous product. We can now extend and generalize Eq. (14.3) to other equilibria.

The abruptness between the kinetic and equilibrium stages shown in Fig. 14.1 is a generalization. Rather, in the environment, or a small vial for that matter, the kinetics and equilibria are net phenomena. Indeed, the forward and reverse reactions continue at a microscopic scale, so it more properly stated that the system has reached dynamic equilibrium [6].

A reaction has a forward rate:

$$k_f\ [\mathrm{A}]^{\mathrm{a}}[\mathrm{B}]^{\mathrm{b}} \tag{14.4}$$

It also has a backward rate:

$$k_b\ [\mathrm{C}]^{\mathrm{c}}[\mathrm{D}]^{\mathrm{d}} \tag{14.5}$$

Thus, at equilibrium the forward rate equals the backward rate:

$$k_f\ [\mathrm{A}]^{\mathrm{a}}[\mathrm{B}]^{\mathrm{b}} = k_b\ [\mathrm{C}]^{\mathrm{c}}[\mathrm{D}]^{\mathrm{d}} \tag{14.6}$$

This means that an equilibrium constant, K_{eq}, can be found for this reaction:

$$K_{eq} = \frac{k_f}{k_b} = \frac{[\mathrm{C}]^{\mathrm{c}}[\mathrm{D}]^{\mathrm{d}}}{[\mathrm{A}]^{\mathrm{a}}[\mathrm{B}]^{\mathrm{b}}} \tag{14.7}$$

The ratio of the concentrations of the solute in two environmental media at equilibrium is known as a partition coefficient, P. However, the term "partition coefficient" has been used to mean something more specific, i.e., the ratio of the concentrations of a solute between two solvents. The term has an even more specific connotation when it is applied exclusively to the concentrations of a solute in a nonpolar solvent and in water. The water concentrations are in the denominator and the nonpolar solvent concentrations in the numerator. Thus, an increasing P indicates that a substance has decreasing water solubility and increasing lipophilicity [6].

In biomedical and pharmacological parlance, P is often the ratio of a solvent in unionized water and in unionized *n*-octanol [12], rendering it the same as the octanol-water coefficient (K_{ow}), discussed later in this chapter. The chemical authority, IUPAC, considers the term, partition coefficient, to be obsolete, preferring "distribution constant" instead [13]. The distribution coefficient differs from the partition coefficient by including ionized solvents. Indeed, this book will follow the approach of environmental venues, which considers the partition coefficient to an inclusive term for any type of partitioning at equilibrium, and K_{ow} to be specific to octanol-water partitioning [6].

5 Fugacity

The chemical potential of a compound is a form of energy that can be absorbed or released during a reaction because of a change in the particle number of that compound. When temperature and pressure are held constant, chemical potential is the partial molar Gibbs free energy (G) [6], discussed in detail in Chapter 7. At chemical equilibrium the sum of the product of chemical potentials and stoichiometric coefficients is zero, because the free energy is minimized. This is also true for equilibrium between physical phases, which implies that this also means the sum of the product of chemical potentials and stoichiometric coefficients is zero as equilibrium is reached between two environmental compartments, e.g., water and air.

The fugacity of a substance is its potential to escape from one compartment to another. Fugacity is like chemical potential but fugacity is proportional to concentration, whereas chemical potential is not [14]. In the gas phase, substances vary in their propensity to escape from one compartment to another, e.g., from the water to the air (see Fig. 14.2). This indicates that fugacity has units of pressure. When equilibrium is reached, the chemical compound has a common fugacity in all phases. For example, when the fugacity of toluene in water is equal to its fugacity in air, a toluene equilibrium is achieved between the air and water. This differs from having equal concentrations in both compartments. Equilibrium between air and water, for example, would have very different concentrations in these two compartments, given their densities, temperature, and other conditions.

If the fugacity in water is higher than the fugacity in air, toluene will evaporate, i.e., volatilize, until establishing a new equilibrium. The common units for fugacity calculations are mol m^{-3}, which are air concentration units. This necessitates unit conversions for the fleeing between other compartments. For example, water concentrations are expressed in mg L^{-1} whereas soil or sediment concentrations expressed in mg kg^{-1}, so these need to be converted to mol m^{-3}. The dry bulk density of soil is the dry weight of the soil divided by soil volume. The same soil that is loose will have a much smaller bulk density than a highly-compacted soil. That is why chemical concentration in solid matrices, like soil, are most often expressed as mass of chemical species per mass of matrix, as opposed to concentrations in air and water, which are usually expressed as mass per volume. Therefore, the densities of water and the solid media must be specified, e.g., bulk densities of different soils vary in any fugacity calculation.

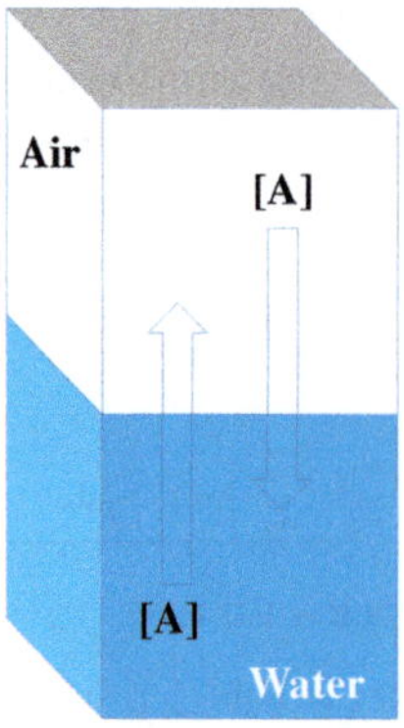

FIG. 14.2 Partitioning of a chemical compound between two phases at equilibrium. In this case, the fugacity (f) at equilibrium is: $f_{A,water} = f_{A,air}$. *(Data from D.A. Vallero, Air Pollution Calculations: Quantifying Pollutant Formation, Transport, Transformation, Fate and Risks, Elsevier, 2019.)*

Chemical concentration and fugacity directly relate to one another via the fugacity capacity constant, i.e., the *Z* value [15]:

$$C_i = Z_i \cdot f \tag{14.8}$$

where:

Ci = Concentration of substance in compartment i (mass per volume)
Z_i = Fugacity capacity (time^2 per length^2)
f = Fugacity (mass per length per time^2)

The *Z* value depends on pressure, temperature, physicochemical properties of the compound, and the characteristics of the environmental compartment where the compound resides. These factors are usually more sensitive than concentration, since concentration dependence is very weak in typical environmental conditions, i.e., highly diluted [14]. When Z is highest, the highest concentrations in an environmental compartment are reached.

The *Z* value expresses the capacity of a phase, or environmental medium, for a given chemical species. *Z* is large when the compound is readily soluble in a phase, i.e., the phase can hold a large quantity of the compound. Small *Z* values mean that the phase can accept only a small quantity of the compound. Deriving *Z* values for each chemical in each phase often begins in the air phase. In the air, the ideal gas law is applied:

$$PV = nRT \tag{14.9}$$

Where P is pressure, which can be substituted with fugacity since it has pressure units, V is the volume of the air, n is the number of moles of the compound, R is the gas constant $\left(8.314 \times \frac{(\mathrm{Pa \cdot m^3})}{\mathrm{mol \cdot K}}\right)$, and T is absolute temperature (K). Given that $\mathrm{C} = \frac{n}{V}$ and $\mathrm{C} = \mathrm{Z} \times f$, Eq. (14.9) can be rewritten in fugacity terms:

$$\mathrm{Z_A} = \frac{1}{RT} = \frac{C_A}{f_A} \tag{14.10}$$

Where the subscript A refers to the air phase.

Z_A under standard conditions, then, is approximately 4×10^{-4} mol m^{-3} Pa for all compounds in air. A partition coefficient is the ratio of the concentrations in two environmental media at equilibrium, so the partition is the ratio of the *Z* values of the two media. For example, the air-water partition coefficient (K_{AW}), which will be discussed in detail later, is:

$$K_{AW=} \frac{C_A}{C_W} = \frac{Z_A f_A}{Z_W f_W} \tag{14.11}$$

Because K_{AW} is measured when f_A is equal to f_W, then:

$$K_{AW=} \frac{Z_A}{Z_W} \tag{14.12}$$

And, the partition coefficient K can be generalized beyond air to be the fleeing between physical phase 1 and physical phase 2:

$$K_{1,2=}\frac{Z_1}{Z_2} \tag{14.13}$$

The logic of Eqs. (14.9) and (14.10) and Fig. 14.2 can be extended from the physics definitions of phases to environmental compartments. Fugacity is associated linearly with concentrations in the gaining and losing compartments:

$$C_{i,j}=Z_{i,j}\cdot f_{i,j} \tag{14.14}$$

where:

Cij = Concentration of ith substance in jth compartment (mol m^{-3})
$Z_{i,j}$ = Fugacity capacity of ith substance in jth compartment $\left(\frac{\text{mol}}{\text{m}^3\ \text{pascal}}\right)$
$f_{i,j}$ = Fugacity (pascal, Pa)

The Z constants are known as half-partition coefficients because, at equilibrium, half of the capacity is in each compartment. At equilibrium, the fugacity of the two environmental compartments (e.g., air and water) will be equal:

$$f_{\text{A,water}}=f_{\text{A,air}} \tag{14.15}$$

In addition, the ratio of concentration to fugacity capacity in the two compartments (i.e., 1 and 2) will be equal:

$$\frac{C_{A,1}}{Z_{A,1}}=\frac{C_{A,2}}{Z_{A,2}} \tag{14.16}$$

The rate constants also relate to the concentration and fugacity capacities:

$$k_{1,2}=\frac{C_{A,2}}{C_{A,1}}=\frac{Z_{A,2}}{Z_{A,1}} \tag{14.17}$$

To derive the Z constants, Raoult's Law can be applied. It states that the partial vapor pressure of each component of an ideal mixture of liquids is equal to the vapor pressure of the pure component multiplied by its mole fraction in the mixture. An ideal mixture is one where all intermolecular interactions are equal. The assumption of ideal conditions is often acceptable for environmental conditions because the concentrations of pollutants are very dilute, i.e., small number of pollutant molecules compared to the water, air, sediment, soil, or biota molecules. The relative lowering of vapor pressure of a dilute solution containing a nonvolatile solute is equal to the mole fraction of solute in the solution. Therefore, Raoult's law for a single component is:

$$p_i=p_i^*x_i \tag{14.18}$$

Where, p_i is the partial pressure of compound i in the gaseous mixture in the compartment above the liquid (known as. headspace), p_i^* is the inherent vapor pressure of pure compound, and x_i is the mole fraction of compound i in the mixture in the liquid.

Thus, Eq. (14.18) can be restated according to Raoult's Law to derive Z constants:

$$f_{i,j}=C_{i,j}\cdot v_j\cdot \gamma_{i,j}\cdot f_R \tag{14.19}$$

Where, v_j is the molar volume of jth compartment (mol m^{-3}), $\gamma_{i,j}$ is activity coefficient in the jth compartment, and f_R is the reference fugacity (Pa). The $\gamma_{i,j}$ activity coefficient is the ratio of the compound's fugacity to the ideal solution's f_R and can be derived from the phase density. Typical phase densities used on air models are shown in Table 14.4.

The affinities of a compound for each phase allows for predictions of the amount and rate of transformation, transport, and fate. The likely fate for a pollutant in the environment can be predicted using the partition coefficients. When equilibrium occurs among all phases and compartments, the chemical potential in each compartment has been reached.

6 Pollutant transport

The fugacity concepts can be used to estimate the affinity that a substance has for various environmental compartments. As such, fugacity is a first screen of the amount of the substance that will be transported. If a chemical species has a particularly

TABLE 14.4 Phase densities default values used in air pollution models.

Phase	Density (kg m^{-3})
Air	1.19
Water	1000
Sediment	1500
Colloids	1000
Biota	1000

Data from D. Mackay, S. Paterson, Evaluating the multimedia fate of organic chemicals: a level III fugacity model, Environ. Sci. Technol. 25 (3) (1991) 427–436.

strong affinity for an environmental compartment, it is less likely to flee that compartment. Less mass of the compound will partition to another compartment. At equilibrium, the fugacity of the system of all environmental compartments is:

$$f = \frac{M_{total}}{\sum_i (Z_i \cdot V_i)} \tag{14.20}$$

where: M_{total} = total number of moles of a substance in all the compartments in the environmental system; Z_i is the fugacity capacity of compound i, and V_i = volume of compartment i where the substance resides. M commonly represents molarity, i.e., the number of moles of solute in a volume, such as the number of liters of solution.

If a chemical substance obeys the ideal gas law, the fugacity capacity is the reciprocal of the gas constant (R) and absolute temperature (T). The ideal gas law can be arranged as:

$$\frac{n}{V} = \frac{P}{RT} \tag{14.21}$$

where:

n = Number of moles of a substance
$P°$ = Substance's vapor pressure

Rearranging and substituting terms:

$$P° = \frac{n}{V} \cdot RT = f \tag{14.22}$$

And,

$$C_i = \frac{n}{V} \tag{14.23}$$

Thus,

$$Z_{air} = \frac{1}{RT} \tag{14.24}$$

This allows for predicting a substance's behavior in the gas phase. The substance's affinity for other environmental media can then be predicted by relating the respective partition coefficients to the Henry's law constants (K_H). For water, the fugacity capacity (Z_{water}) can be found as the reciprocal of K_H:

$$Z_{water} = \frac{1}{K_H} \tag{14.25}$$

This is the dimensioned version of the Henry's law constant (length^2 per time^2).

Some chemical compounds have greater affinity for the air than for water or other compartments, for which Henry's Law can be fittingly applied. To estimate the amount of a substances that escapes and remains, fugacity must be calculated for every environmental compartment. For example, many persistent organic compounds spend part of their life cycles in sediment reservoirs beneath surface waters. As evidence, substances like polychlorinated aromatic compounds (e.g., PCBs) that have not been used for decades continue to be released into the environment. In sediment, the fugacity capacity is directly proportional to the substance's sorption potential, expressed as the solid-water partition coefficient (K_d), and the average sediment density ($\rho_{sediment}$). Sediment fugacity capacity is indirectly proportional to the chemical substance's Henry's law constant:

$$Z_{sediment} = \frac{\rho_{sediment} \cdot K_d}{K_H} \tag{14.26}$$

For biota, particularly fauna and especially fish and other aquatic vertebrates, the fugacity capacity is directly proportional to the density of the fauna tissue (ρ_{fauna}), and the pollutant's bioconcentration factor (*BCF*),[b] and inversely proportional to the pollutant's Henry's law constant:

$$Z_{fauna} = \frac{\rho_{fauna} \cdot BCF}{K_H} \tag{14.27}$$

As in the sediment fugacity capacity, a higher *BCF* value means that the fauna's fugacity capacity increases and the actual fugacity decreases. Again, this is logical, since the organism is sequestering the pollutant and blocking it from leaving if the organism has a large *BCF*. This is a function of both the species of organism, the characteristics of the pollutant, and the ecosystem in which the organism resides. Factors like temperature, pH, and ionic strength of the water and metabolic conditions of the organism will affect *BCF* and Z_{fauna}. This is one of the reasons for the wide ranges of published *BCF* values in the scientific literature [16]. Another reason is that *BCF* calculations vary by the species of organisms tested.

The total amount of partitioning of the environmental system is the aggregation of all the individual compartmental partitionings. The moles of the substance in each environmental compartment (M_i) are found as the product of the fugacity, volume, and fugacity capacity for each compartment:

$$M_i = Z_i \cdot V_i \cdot f \tag{14.28}$$

Environmental partitioning and fugacity example 1 (from Ref. [6])

A hypothetical lake has a three-compartment ecosystem of air, water, and fish. The lake has 7 billion liters of water and 3.5 cubic meters of fish swimming in it. 100 kg of dichloromethane (CH_2Cl_2) is spilled into this lake. The airshed above the lake contains 10 billion cubic meters of air (see Fig. 14.3). The air temperature above the lake is 77 °F. Calculate the partitioning of dichloromethane [CH_2Cl_2] among these three environmental media. Determine the relative affinity of this compound to the three compartments.

First, compatible units must be derived. To do so, the water volume is converted from liters to m^3:
7×10^9 L $= 7 \times 10^6$ m^3.

Next, find a bioconcentration factor. There are several reported *BCF* values for CH_2Cl_2. For example, the U.S. Environmental Protection Agency [17] reports one experimentally derived unitless *BCF* as 22.9, and predicted *BCF* values ranging from 2.63 to 15.1. The highest reported *BCF* by European Chemical Agency [18] is 40 L kg^{-1}. To be conservative, the European value will be used, but all reputable sources of these values must be reviewed for quality and relevance, including evaluating the metadata, i.e.,

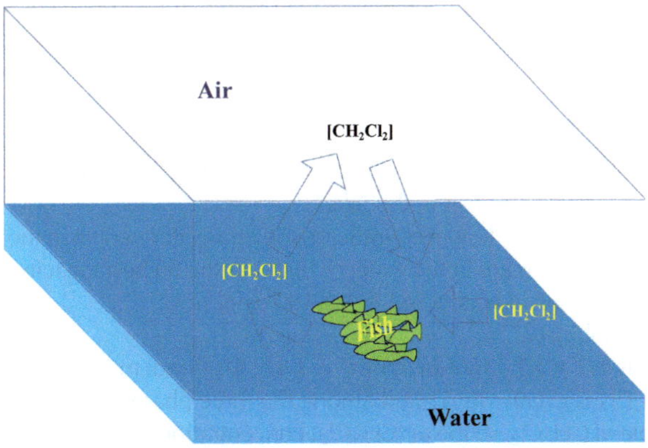

FIG. 14.3 Hypothetical 3-component lake ecosystem.

b. *BCF* is the quotient of the concentration of the substance in the organism and the concentration of the substance in the host compartment, usually water.

the information about the data such as why the study was done, where the experiments and fieldwork were conducted, and any limitations and uncertainties.

Partitioning is temperature dependent. We are given the air temperature of 77 °F (25 °C), so we will simplify the calculation by assuming all the compartments are that temperature. However, this is unlikely since the sediment below the lake will differ from that of the water column, and these will differ from the air temperature. So, these need to be more precise for real-world calculations.

Densities of biota vary considerably. For example, the size of fish affects the specific gravity of their tissue, e.g., a study of mesopelagic fish in the Pacific Ocean off the northeastern U.S. coastline [19] report mean specific gravities between 1.039 and 1.062 g cm^{-3}. Fish are generally somewhat denser than water, so we will assume the fish density to be 1.05 g cm^{-3}. Again, it is preferable to establish more precise mean or median body density of the species of interest for the calculations.

The vapor pressure of CH_2Cl_2 at 25 °C is 0.46 atm. Henry's Law constant is $3.25 \times 10^{-3}\ \frac{atm \cdot m^3}{mol}$. The molecular weight of CH_2Cl_2 is about 84.93 g. Converting the CH_2Cl_2 mass to moles:

$$100\text{kg} \times \frac{1000\text{g}}{\text{kg}} \times \frac{1\ \text{mol}}{84.93\ \text{g}} = 1177\ \text{mol}.$$

For air partitioning, we plug our data into Eq. (14.24):

$$Z_{air} = \frac{1}{RT} = \frac{1}{0.0821\ \text{L} \cdot \frac{\text{atm}}{\text{mol} \cdot \text{K}} \cdot 298\text{K}} \times \frac{1000\ \text{L}}{1\ \text{m}^3} = 40.9\ \frac{\text{mol}}{\text{atm} \cdot \text{m}^3}$$

For partitioning to water, we use Eq. (14.25):

$$Z_{water} = \frac{1}{K_H} = \frac{1}{3.25 \times 10^{-3}\ \frac{\text{atm} \cdot \text{m}^3}{\text{mol}}} = 308\ \frac{\text{mol}}{\text{atm} \cdot \text{m}^3}$$

And, for partitioning to the fish tissue, we use Eq. (14.27):

$$Z_{fauna} = \frac{\rho_{fauna} \cdot BCF}{K_H} = \frac{1000\frac{\text{g}}{\text{m}^3} \cdot 40\frac{\text{L}}{\text{kg}}}{3.25 \times 10^{-3}\ \frac{\text{atm} \cdot \text{m}^3}{\text{mol}}} \times \frac{1\ \text{kg}}{1000\ \text{g}} \times \frac{1\text{m}^3}{1000\ \text{L}} = 12.3\ \frac{\text{mol}}{\text{atm} \cdot \text{m}^3}$$

To calculate the system fugacity, we divide the mass of CH_2Cl_2 entering the system by the sum of the three compartment's partitions and volumes:

$$f = \frac{1177\ \text{mol}}{(40.9 \times 10^{10}) + [308 \times (7 \times 10^6)] + (12.3 \times 3.5)} = 2.9 \times 10^{-9}\ \text{atm}.$$

So, to find the moles of CH_2Cl_2 in each compartment, we use Eq. (14.28):

$$M_{air} = 2.9 \times 10^{-9} \times 10^{10} \times 40.9 = 1171\ \text{mol}$$

$$M_{water} = 2.9 \times 10^{-9} \times (7 \times 10^6) \times 308 = 6.2\ \text{mol}$$

$$M_{fish} = 2.9 \times 10^{-9} \times 3.5 \times 12.3 = 1.2 \times 10^{-7}\ \text{mol}$$

Note that this is an oversimplification of the total fugacity since we did not calculate the sediment-to-water fugacity and only one species of biota were calculated. Obviously, a lake system has large exchanges of substances from sediment to the water column, and myriad species of biota.

In terms of partitioning at equilibrium, dichloromethane would appear to have the greatest affinity for air. Indeed, the mass of CH_2Cl_2 fleeing to the air is over 2 orders of magnitude higher than water and 10 orders of magnitude higher than fish.

Environmental partitioning and fugacity example 2 (from Ref. [6])

What is the expected CH_2Cl_2 concentration in fish in the lake?

The Z values and chemical concentrations, however, indicate something different than the conclusions in example 1. Given the volumes, the highest CH_2Cl_2 concentrations are in the water, which has the highest fugacity capacity (i.e., $Z = 308$). The concentration in water is 6.2 mol in 7 million m^3 of water, i.e., 8.9×10^{-7} mol m^{-3}.

The air concentration is 1171 mol in 10^{10} m^3 = 1.2×10^{-7} mol m^{-3}.

The fish tissue concentration is 1.2×10^{-7} mol in 3.5 m^3 = 3.4×10^{-8} mol m^{-3}.

In general, however, the CH_2Cl_2 is evenly distributed among the three phases.

There are 84.93258 g per mole of CH_2Cl_2, so the air CH_2Cl_2, concentration is about 1×10^{-5} g m^{-3} = 10 μg m^{-3}; and the fish tissue CH_2Cl_2, concentration is about 3 μg m^{-3} or 3×10^{-5} μg L^{-1}. Given that the density of CH_2Cl_2 is about 1.23, this converts to 86.37 parts per billion by volume (ppbv).

Environmental partitioning and fugacity example 3 (from Ref. [6])

A later study of the lake's airshed showed that the air volume in which the CH_2Cl_2 is mixed is less than previous studies. Flight tests found the mixing layer to be half the altitude of the previous study, meaning that the volume of air should have been 5×10^9 m^3. How will this update affect the partitioning?

The *Z* values are not affected by volume. The air value is based on the gas laws, so it is not constrained by the air volume, nor by the inherent properties of the chemical compound. The water *Z* value is not affected by water volume, but is the reciprocal of the Henry's Law coefficient, which is unique for each chemical species. Likewise, the fauna *Z* value does not depend on the air nor the water volume but is inversely proportional to the Henry's Law coefficient and directly proportional to the *BCF* and the tissue density for each chemical compound.

The concentration calculations do depend on volume. Thus, the air concentration would double if the air volume were halved, so the 1171 mol fleeing to the atmosphere would be decreased by half in the smaller volume of air. In this case, the rounded air concentration is 2.3×10^{-7} mol m^{-3}.

Again, given there are 84.93258 g per mole of CH_2Cl_2, the air CH_2Cl_2, concentration is about 2×10^{-5} g m^{-3} = 20 μg m^{-3}.

Comparing the respective fugacity capacities for each physical phase or environmental compartment is useful for water pollution and environmental calculations. As evidence, if one compartment has a high fugacity and thus, a low fugacity capacity, for a pollutant, and the source of the pollutant no longer exists, the concentrations in that compartment would be expected to decrease precipitously with time given no other change in environmental conditions. Conversely, if a compartment has a very low fugacity, the pollutant will have a robust affinity for that compartment and only after sufficient energy is added would it move away from that compartment and decrease the chemical concentration of the pollutant in that compartment. Thus, a persistent compound like a PCB would remain in soil or sediment unless substantial amounts of energy force it to flee to the water column or air. Second, if a continuous source of the pollutant exists, and a compartment has a high fugacity capacity (and thus low fugacity), this compartment may serve as a conduit for delivering the pollutant to other compartments with relatively low fugacity capacities. Third, the higher relative fugacities of one set of compartments compared to another set in the same system allow for comparisons and estimates of sources and sinks (known as "hot spots") of the pollutant. These are important predictors of a pollutant's likely transport and fate which are useful in assessing potential human and modeling human and ecological exposures and risks [20,21].

Fugacity-based, multicompartmental environmental models take these relationships into account. The movement of a pollutant through the environment can be expressed by how equilibrium is achieved in each compartment. The processes that drive pollutant transport can be summarized into transfer coefficients or compartmental rate constants, known as *D* values [22]. By first calculating the *Z* values and then equating inputs and outputs of the pollutant to each compartment, *D* value rate constants can be derived. The actual transport process rate (*N*) is the product of fugacity and the *D* value:

$$N = D \cdot f \tag{14.29}$$

Because the chemical compound concentration is $Z{\cdot}f$, we can substitute and add a first-order rate constant k to give a first-order rate *D* value (D_R):

$$N = V[c]k = (V \cdot Z \cdot k) \cdot f = D_R \cdot f \tag{14.30}$$

Although the concentrations are shown as molar concentrations (i.e., in brackets), they may also be represented as mass per volume concentrations.[c]

Diffusive and nondiffusive transport processes follow Fick's laws, i.e., diffusive processes. They can also be expressed with their own *D* values (D_D), which are related to the mass transfer coefficient (*K*) applied to area A:

$$N = K\mathrm{A}[\mathrm{c}] = (K \cdot \mathrm{A} \cdot \mathrm{Z}.) \cdot f = D_D \cdot f \tag{14.31}$$

Nondiffusive transport, also known as advection or bulk flow, within a compartment with a flow rate (*G*) has a different *D* value (D_A) that is expressed as:

$$N = G[\mathrm{c}] = (G \cdot Z \cdot)f = D_A \cdot f \tag{14.32}$$

Thus, as a substance moves through the environment, having residence time in each compartment, it is affected by numerous physical transport processes. In addition, its fate is determined by rates of abiotic and biotic transformation processes. As it is degraded, the new compounds have their own transport characteristics. The processes are addressed by models with the respective *D* values, so that the total rate of transport and transformation is expressed as:

c. Bracketed values indicate molar concentrations, but they may be converted to mass per volume concentration values, as was done in the three previous examples. Mass per volume is the common expression of water pollutants, such as micrograms per liter (μg L^{-1}).

$$f \cdot (D_1 + D_2 + \ldots D_n) \tag{14.33}$$

Fast processes have large D values and slow processes have small D values.

Models, while imperfect, are very useful tools for estimating the movement of pollutants in the environment [14,21,23–26]. They are improved with more and better measurements of environmental conditions and outcomes. Indeed, measurements and models are mutually complementary. Compartmental model assumptions are verified with real-world measurements. Similarly, measurements usually only provide data and information about a confined geographical area based on a limited number of measurement locations. As such, measurement data need models to extrapolate conditions and results beyond these confined areas and to interpolate conditions and results between the measurement sites (See Chapter 13). Water pollutant transport models improve the understanding the mechanisms for the movement of pollutants into and within surface and ground water and throughout the environment.

Many possible outcomes can occur after a substance is released into the environment, but the possibilities generally fall into three basic categories:

1. the substance may remain where it is released and retain its physicochemical characteristics (at least within a specified time);
2. the substance may be transported to another location; or
3. the substance may be changed chemically, known as the transformation of the chemical.

Categories 2 and 3 often occur together, especially for compounds with low persistence. Every molecule of mass moving into and out of the control volume must be accounted for, as well as any chemical changes to the pollutant that take place within the control volume. A control volume may be a simple cube (Fig. 14.4A) through which pollutant fluxes and concentrations can be calculated. In the environment, a control volume can also be a cell within an organism. A control volume

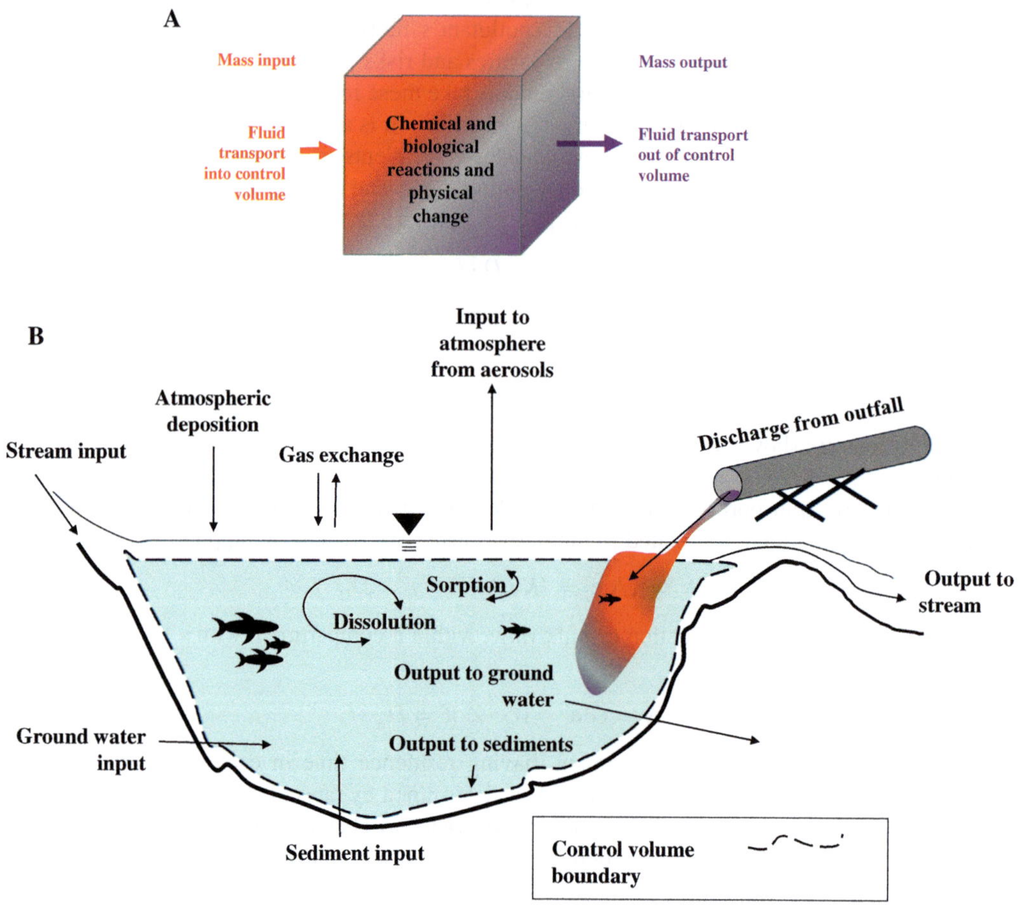

FIG. 14.4 (A) Control volume of an environmental matrix (e.g., soil, sediment, or other unconsolidated material) or fluid (e.g., water, air, or blood). (B) A pond. Both volumes have equal masses entering and exiting, with transformations and physical changes taking place within the control volume.

can also be the entire organism, including the human body, that takes up, absorbs, metabolizes, distributes, and eliminates the substance. A control volume can also be an entire ecosystem that receives the substance load, has abiotic and biotic reactions, and releases the substance or its transformation products out of the ecosystem (Fig. 14.4B). In fact, a control volume can be the earth's entire hydrosphere-atmosphere-lithosphere-biosphere complex.

Recall that first law of thermodynamics requires that any change in storage of a substance's mass in a control volume must equal the difference between the mass of the chemical transported into the system less the mass of the chemical transported out of the system. For a reactive compound, the actual chemical species transported into a control may be different from what was initially released and what has entered the control volume. Thus, the mass balance equation may be written as:

$$\text{Accumulation or loss of pollutant A} = \text{Mass of A transported in} - \text{Mass of A transported out} \pm \text{Reactions} \quad (14.34)$$

The reactions may be those that generate chemical compound A (i.e., sources), or those that destroy or remove chemical compound A (i.e., sinks). The entering mass transported equals the inflow to the system that includes pollutant discharges, transfer from other control volumes and other media. For example, the control volume's soil, water, and air may contribute mass of A and formation of A by abiotic chemical and biochemical transformation. The outflow is the mass transported out of the control volume, which includes uptake by biota, transfer to other compartments (e.g., volatilization to the atmosphere or deposition to surface waters and soil) and abiotic and biological degradation of chemical A.

The rate of change of mass in a control volume is equal to the rate of chemical compound A transported into the control volume minus the rate of chemical compound A transported out, plus the rate of production from sources, and minus the rate of elimination by sinks. Stated as a differential equation, the rate of change chemical compound A is:

$$\frac{d[A]}{dt} = -v \cdot \frac{d[A]}{dx} + \frac{d}{dx}\left(D \cdot \frac{d[A]}{dx}\right) + r \quad (14.35)$$

where:

v = fluid velocity
$\frac{d[A]}{dx}$ = concentration gradient of chemical A
r = internal sinks and sources within the control volume

Eq. (14.35) can be applied at any scale from cellular to planetary. At the human or other organism scale, it is the basis for pharmacokinetic and pharmacodynamic modeling. A basic concept of pollution is that substances tend to have affinity for certain compartments in abiotic and biotic systems. Thus, the specific partitioning relationships that control the "leaving" and "gaining" compartments must be known. Chemical compounds exit and enter water by precipitation from and evaporation to the atmosphere, and transfer from other surface and ground water systems, sediment, soil, and biota. Compounds move within unconsolidated materials, especially soil and sediment matrices, in and on particles, within the atmosphere, and within organic tissues. Furthermore, these partitioning relationships can be applied to estimate and to model where a substance will go within a designed system (e.g., a trickling filter or sedimentation basin) and after it is released (e.g., from surface waters to the atmosphere). Basic relationships between sorption, solubility, volatilization, and organic carbon-water partitioning are respectively expressed by coefficients of sorption (distribution coefficient, K_D, or solid-water partition coefficient, K_p), dissolution or solubility coefficients, air-water partitioning (often expressed as K_{AW} or the Henry's law (K_H) constant), octanol-water partitioning (K_{ow}) and organic carbon-water (K_{oc}) partitioning.

The common approach is to measure and model each compartment where a pollutant is found in discrete phases of water, air, soil, sediment, and biota. However, a complicating factor is that even within a single compartment, a pollutant may exist in various phases (e.g., dissolved in water and sorbed to a suspended solid-phase particle). Interphase reactions, i.e., physical interactions of the substance at the interface between each compartment, determine the concentration of that substance in the environment. Within a compartment, a pollutant may remain unchanged for a designated time-period, or it may move, or it may be transformed chemically into another substance.

Interactions of mechanisms occur within an environmental compartment. A mass fraction will remain unchanged in the compartment. Another fraction remains unchanged but is transported to a different compartment while another fraction is chemically transformed with all products remaining products staying in the compartment where the transformation products were generated. Finally, a fraction of the original pollutant mass is transformed and then moved to another compartment. Thus, upon release from a source, the pollutant moves because of thermodynamics. If a substance that is dissolved in water comes into contact with another substance, e.g., octanol, the substance will tend to move from the water to the octanol. Its octanol-water partitioning coefficient reflects just how much of the substance will move until the aqueous and

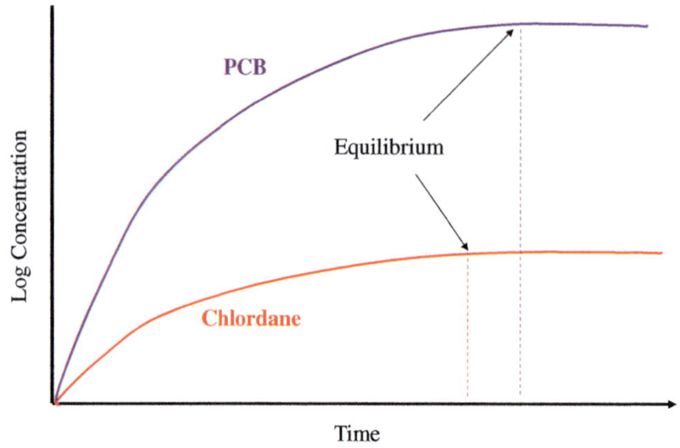

FIG. 14.5 Relative concentrations of a polychlorinated biphenyl (PCB) and chlordane in octanol in an environmental medium with time. *(From D. Vallero, Environmental Biotechnology: A Biosystems Approach, Elsevier Science, 2015 used with permission.)*

organic solvents (phases) will reach equilibrium. So, for example, in a spill of equal amounts of the polychlorinated biphenyl, decachlorobiphenyl (log K_{ow} of 8.23) and the pesticide chlordane (log K_{ow} of 2.78), the PCB has much greater affinity for the organic phases than does the chlordane (greater than five orders of magnitude).[d] This does not mean that a great amount of either of the compounds is likely to stay in the water column, since they are both hydrophobic, but it does mean that they will vary in the time and mass of each substance moving between phases. The rates of kinetics differ, which means that the time it takes for the PCB and chlordane to reach equilibrium will be different. This can be visualized by plotting the concentration of each compound with time (Fig. 14.5). At the time that the concentrations plateau, the compounds are at equilibrium with their phase.

Water pollution and other health studies increasingly rely on biomarkers as indicators of concentrations that occur in control volumes where humans and organisms live. Medical science has used markers for decades, e.g., cholesterol, fat, and blood sugar. More recently, environmental toxicologists have adopted numerous chemicals to indicate the presence of a pollutant and its effect. Some of the chemical compounds are concentrations solely of the parent pollutants to which the organism, including a person, has been exposed. An example of this is the metal lead (Pb). The amount of lead in a person's blood or hair indicates the extent to which that person has been exposed to Pb. In this case, total mass of Pb is a reliable indicator of the time and activities of the child within various control volumes, e.g., human microenvironments (Fig. 14.6). However, the r term in Eq. (14.35) means that the biomarkers can be chemical compounds other than the parent pollutant. Thus, biomarkers commonly are concentrations of the metabolites. For example, smokers inhale nicotine, much of which is metabolized to cotinine. Thus, cotinine and other biomarkers can be measures of exposure to nicotine and can represent the extent of tobacco use or environmental tobacco smoke [27–31].

7 Sorption

Sorption is an important transfer process that determines how bioavailable or toxic a compound will be in surface and ground waters and in other environmental media. Surface water-atmospheric exchanges are a means by which large quantities of pollutants reach water bodies. In the atmosphere, sorption occurs at the microscopic scale in water droplets. In addition, chemical transport can occur to and from a solid particle via any fluid. The physicochemical transfer [32] of chemical compound is an interaction of the solute (i.e., the chemical being sorbed) with the surface of a solid surface. These interactions can be complex and dependent upon the properties of the chemical compound and the water. Other fluids are often of such small concentrations that they do not determine the ultimate solid-liquid partitioning. There are four basic types or mechanisms of sorption:

1. Adsorption is the process wherein the chemical in solution attaches to a solid surface, which is a common sorption process in clay and organic constituents in soils. This simple adsorption mechanism can occur on clay particles where little carbon is available, such as in ground water.

d. Log values are often used for K_{ow} since the range of values for individual chemical compounds is several orders of magnitude.

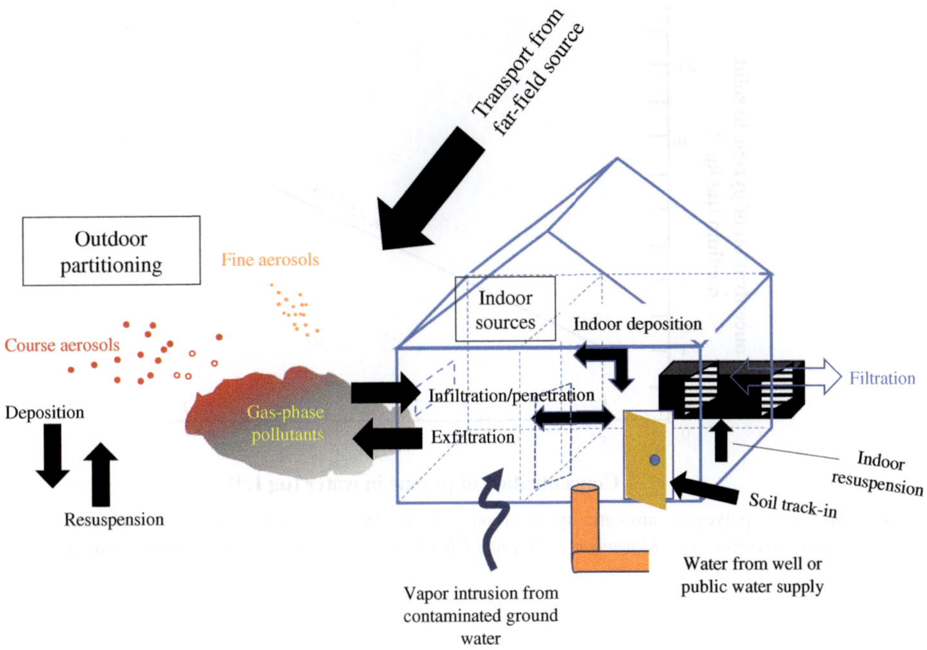

FIG. 14.6 Human residence as a microenvironmental control volume.

2. Absorption is the process that often occurs in porous materials so that the solute can diffuse into the particle and be sorbed onto the inside surfaces of the particle. This commonly results from short-range electrostatic interactions between the surface and the pollutant.
3. Chemisorption is the process of integrating a substance into a porous materials surface via a chemical reaction. In soil, this is usually the result of a covalent reaction between a mineral surface and the pollutant.
4. Ion exchange is the process by which positively charged ions (cations) are attracted to negatively charged particle surfaces or negatively charged ions (anions) are attracted to positively charged particle surfaces, causing ions on the particle surfaces to be displaced. Particles undergoing ion exchange can include soils, sediment, airborne particulate matter, or even biota, such as spores, cysts, and pollen particles [33]. Cation exchange is arguably the second most important chemical process on earth, after photosynthesis. This is because the cation exchange capacity (CEC), and to a lesser degree anion exchange capacity (AEC) in tropical soils, is how nutrients are made available to plant roots. Without this process, the atmospheric nutrients and the minerals in the soil would not come together to provide for the abundant plant life on planet earth [34].

These all consist of a combination of physical and chemical processes occurring at surfaces. The first two types are mainly controlled by physical factors, whereas the second two are combinations of chemical reactions and physical processes. Generally, sorption reactions affect water pollution partitioning in three ways, by [35]:

1. Modifying the chemical pollutant's transport potential due to distributions between the aqueous phase and particles;
2. Enhancing the aggregation and transport of the pollutant because of electrostatic properties of suspended solids; and,
3. Inducing and changing the rates of surface reactions, including dissociation, surface-catalysis, and precipitation of the chemical pollutant.

Sorption of chemical compounds on solid surfaces, like those in sediment, soil, and unconsolidated materials, is a means by which the chemical species is retained in that matrix. For example, when a chemical compound contacts a soil matrix, some of the compound remains in soil solution and the remainder is adsorbed onto the surfaces of the soil particles. This sorption can be weak or strong depending on the number of cations adsorbing to the negatively charged soil particles. Thus, sorption is a determinate of the residence time in the matrix. Strongly-sorbed compounds cannot move freely between a particle and the surrounding fluid, which can be liquid or gas. As such, sorption slows that rate at which substances move downwardly through the soil profile. Strongly sorbed chemical compounds can be a challenge for treatment. Indeed, one of the first steps in treating contaminated sediments and soils is often desorption [36–40].

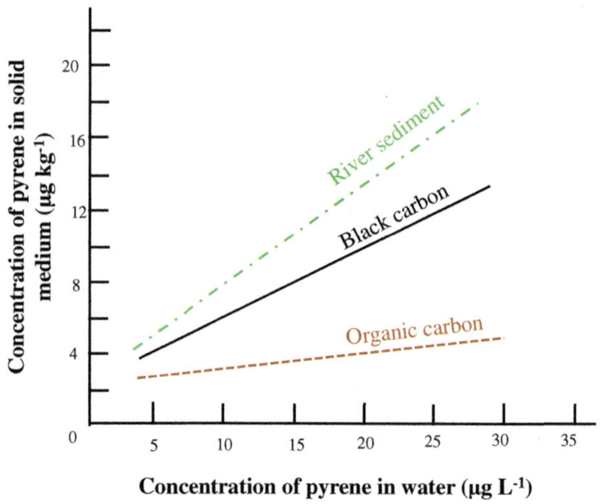

FIG. 14.7 Sorption isotherms for the polycyclic aromatic hydrocarbon, pyrene. *(Data from J.J. Hassett, W.L. Banwart, The sorption of nonpolar organics by soils and sediments, in: Reactions and Movement of Organic Chemicals in Soils, pp. 31–44, 1989, no. reactionsandmov.)*

Chemical compounds eventually establish a balance between the mass on the solid surfaces and the mass in solution. Molecules migrate from one phase to another to maintain this balance. The properties of both the chemical substance and the soil or other matrix determines how and at what rates the molecules partition into the solid and liquid phases. These physicochemical relationships, known as sorption isotherms, are found experimentally. Fig. 14.7 shows three isotherms for pyrene from experiments using different soils and sediments.

Fig. 14.7's x-axis shows the concentration of pyrene dissolved in water, and the y-axis shows the concentration of pyrene in the solid phase. In this case, pyrene appears to be strongly sorbed to river sediment compared to organic carbon. Each line represents the relationship between these concentrations for a single soil or sediment sample. As straight-line segment does not intersect zero, which indicates that low concentrations are below detection limits. The concentrations appear to be linear, however this is not always the case. If a linear relationship is assumed, the chemical partitioning can be expressed as:

$$S = K_D \cdot C_W \tag{14.36}$$

Where, S = concentration of pollutant in the solid phase (mass of solute per mass of soil or sediment); C_W = concentration of pollutant in the liquid phase (mass of solute per volume of pore water); and K_D = partition coefficient (volume of pore water per mass of soil or sediment) for the pollutant in this soil or sediment.

For many chemicals sorbed to soil and sediment, the partition coefficient can be estimated using:

$$K_D = K_{OC} \cdot C_{\text{org}} \tag{14.37}$$

where K_{OC} = organic carbon partition coefficient (volume of pore water per mass of organic carbon) and C_{org} = substrate organic matter (mass of organic carbon per mass of substrate, e.g., soil).

This relationship is applied to provide estimates of a chemical substance's unknown K_D from the known K_{OC} of the substance and the organic carbon content of the soil horizon or sediment layer of interest. K_{OC} values are readily available and published in handbooks and manuals. Users of these report values must keep in mind that these, like several other partitioning coefficients, are substrate dependent. Thus, the user must be certain that the substrate being used in these calculations are from experiments using the same substrate (e.g., carbon source = anthracite) or otherwise normalized. This is but another example of the importance of metadata. If the substrate is not reported, this may preclude the use of these values or, at a minimum, report that the substrate is unknown.

The actual derivation of K_D is:

$$K_D = C_S \,(C_W)^{-1} \tag{14.38}$$

where C_S is the equilibrium concentration of the solute in the solid phase and C_W is the equilibrium concentration of the solute in the water.

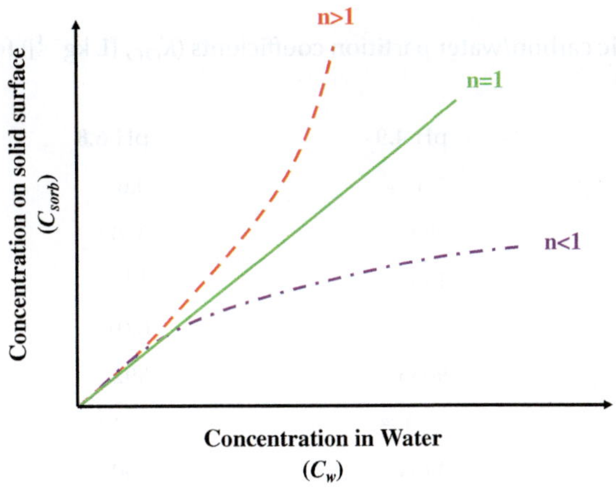

FIG. 14.8 Hypothetical Freundlich isotherms with exponents (n) less than, equal to, and greater than 1, as applied to the equation $C_{sorb} = K_F \cdot C^n$. *(Data from H.F. Hemond, E.J. Fechner, Chemical Fate and Transport in the Environment, Elsevier, 2014. P.M. Gschwend, Environmental Organic Chemistry, John Wiley & Sons, 2016.)*

Therefore, K_D is a direct expression of the partitioning between the aqueous and solid (soil or sediment) phases. A strongly sorbed chemical like a dioxin or the banned pesticide DDT can have a K_D value exceeding 10^6. Conversely, a highly hydrophilic, miscible substance like ethanol, acetone, or vinyl chloride, will have K_D values less than 1. This relationship between the two phases demonstrated by Eq. (14.38) is the Freundlich Sorption Isotherm:

$$C_{sorb} = K_F \cdot C^n \tag{14.39}$$

where C_{sorb} is the concentration of the sorbed pollutant, i.e., the mass sorbed at equilibrium per mass of sorbent, K_F is the Freundlich isotherm constant; and n is the exponent of the chemical concentration curve of the function of C_{sorb} *and* C_W in Fig. 14.8. The exponent determines the linearity or order of the reaction. Thus, if $n = 1$, then the isotherm is linear; meaning the more of the pollutant in solution, the more would be expected to be sorbed to surfaces. For values of $n < 1$, the amount of sorption is in smaller proportion to the amount of solution and, conversely, for values of $n > 1$, a greater proportion of sorption occurs with less pollutant in solution. Also note that if $n = 1$, then Eq. (14.38) and the Freundlich sorption isotherm are identical.

Increasing the organic matter content in substrates, i.e., soil and sediment, is accompanied by an increase in the amount of a pollutant that is sorbed, i.e., sorption is directly proportional to the soil/sediment organic matter content. This allows for a conversion of K_D values from those that depend on specific soil or sediment conditions to soil/sediment independent sorption constants, K_{OC}:

$$K_{OC} = K_D \cdot (f_{OC})^{-1} \tag{14.40}$$

where f_{OC} is the dimensionless weight fraction of organic carbon in the soil or sediment. The K_{OC} and K_D have units of mass per volume.

Incidentally, the straightforward relationship between soil organic carbon content and sorption observed for organic compounds differs for metals and inorganic constituents. Many geochemical processes affect the K_D for metals and other inorganic compounds, including pH; high clay content, soil organic matter content, presence of iron oxides, redox conditions; ionic strength, and the uniqueness of various compounds of the same metal. For organic compounds, variations in these and other factors can cause inorganic K_D values to differ by several orders of magnitude (see Table 14.5), which hinders attempts to derive generic K_D values for metals compared to those for organic compounds [41].

8 The octanol-water coefficient

The octanol-water partition coefficient (K_{ow}) is perhaps the most commonly applied partitioning coefficient for water systems, but it also has major applications in other environmental media. Most toxic compounds are organic, i.e., their molecular structure contains carbon-to-carbon bonds or carbon-to-hydrogen bonds. Their solubility can be differentiated as to whether under normal environmental conditions of temperature and pressure the substance is easily dissolved in

TABLE 14.5 Predicted soil organic carbon/water partition coefficients (K_{OC}, [L kg^{-1}]) for selected organic compounds as a function of pH.

Ionizing organics compound	pH 4.9	pH 6.8	pH 8.0
Benzoic acid	5.5	0.6	0.5
2-Chlorophenol	398	388	286
2,4-Dichlorophenol	159	147	72
2,4-Dinitrophenol	0.03	0.01	0.01
Pentachlorophenol	9055	592	410
2,3,4,5-tetrachlorophenol	17,304	4742	458
2,3,4,6-Tetrachlorophenol	4454	280	105
2,4,5-Trichlorophenol	2365	1597	298
2,4,6-Trichlorophenol	1040	381	131

Data from U.S. Environmental Protection Agency, Part 5: Technical background document, in: US EPA (Ed.), Soil Screening Guidance, vol. EPA/540/R-95/128, US EPA, Washington, DC, 1996.

organic solvents; then the compounds are described as lipophilic (i.e., readily dissolved in lipids). On the other hand, if a compound resists dissolution in organic solvents under these conditions, it is said to be lipophobic.

Table 14.6 provides the log K_{OC} values calculated from chemical structure and those measured empirically for several organic compounds, comparing them to the respective K_{ow} values.

This affinity for either water or lipids underpins an important indicator of environmental partitioning, i.e., the octanol-water partition coefficient (K_{ow}). The K_{ow} is the ratio of a substance's concentration in octanol ($C_7H_{13}CH_2OH$) to the substance's concentration in water at equilibrium (i.e., the reactions have all reached their final expected chemical composition in a control volume of the fluid). Octanol is a surrogate for lipophilic solvents in general because it has degrees of affinity for both water and organic compounds, that is, octanol is amphiphilic. Since the ratio forming the K_{ow} is [$C_7H_{13}CH_2OH$]:[H_2O], then the larger the K_{ow} value, the more lipophilic the substance. Table 14.7 shows the values for solubility in water and K_{ow} values of some important environmental compounds, along with their densities.

TABLE 14.6 Calculated and experimental organic carbon coefficients (K_{OC}) compared to octanol-water coefficients (K_{ow}) for selected pollutants found at hazardous waste sites.

		Calculated		Measured	
Chemical	log K_{ow}	log K_{oc}	K_{oc}	log K_{oc}	K_{oc} (geomean)
Benzene	2.13	1.77	59	1.79	61.7
Bromoform	2.35	1.94	87	2.10	126
Carbon tetrachloride	2.73	2.24	174	2.18	152
Chlorobenzene	2.86	2.34	219	2.35	224
Chloroform	1.92	1.60	40	1.72	52.5
Dichlorobenzene, 1,2- *(o)*	3.43	2.79	617	2.58	379
Dichlorobenzene, 1,4- *(p)*	3.42	2.79	617	2.79	616
Dichloroethane, 1,1-	1.79	1.50	32	1.73	53.4
Dichloroethane, 1,2-	1.47	1.24	17	1.58	38.0
Dichloroethylene, 1,1-	2.13	1.77	59	1.81	65
Dichloroethylene, *trans* -1,2-	2.07	1.72	52	1.58	38
Dichloropropane, 1,2-	1.97	1.64	44	1.67	47.0

TABLE 14.6 Calculated and experimental organic carbon coefficients (K_{OC}) compared to octanol-water coefficients (K_{ow}) for selected pollutants found at hazardous waste sites—cont'd

		Calculated		Measured	
Chemical	log K_{ow}	log K_{oc}	K_{oc}	log K_{oc}	K_{oc} (geomean)
Dieldrin	5.37	4.33	21,380	4.41	25,546
Endosulfan	4.10	3.33	2138	3.31	2040
Endrin	5.06	4.09	12,303	4.03	10,811
Ethylbenzene	3.14	2.56	363	2.31	204
Hexachlorobenzene	5.89	4.74	54,954	4.90	80,000
Methyl bromide	1.19	1.02	10	0.95	9.0
Methyl chloride	0.91	0.80	6	0.78	6.0
Methylene chloride	1.25	1.07	12	1.00	10
Pentachlorobenzene	5.26	4.24	17,378	4.51	32,148
Tetrachloroethane, 1,1,2,2-	2.39	1.97	93	1.90	79.0
Tetrachloroethylene	2.67	2.19	155	2.42	265
Toluene	2.75	2.26	182	2.15	140
Trichlorobenzene, 1,2,4-	4.01	3.25	1778	3.22	1659
Trichloroethane, 1,1,1-	2.48	2.04	110	2.13	135
Trichloroethane, 1,1,2-	2.05	1.70	50	1.88	75.0
Trichloroethylene	2.71	2.22	166	1.97	94.3
Xylene, *o*-	3.13	2.56	363	2.38	241
Xylene, *m*-	3.20	2.61	407	2.29	196
Xylene, *p*-	3.17	2.59	389	2.49	311

Data from U.S. Environmental Protection Agency, Soil screening guidance: user's guide, EPA/540/R-96/018, 1996. U.S. Environmental Protection Agency, Soil screening guidance: technical background document, EPA/540/R-95/128 PB96-963502, 1996.

TABLE 14.7 Solubility, octanol-water partitioning coefficient, and density values for selected water pollutants.

Chemical	Water solubility (mg L^{-1})	K_{ow}	Density (kg m^{-3})
Atrazine	33	724	
Benzene	1780	135	879
Chlorobenzene	472	832	1110
Cyclohexane	60	2754	780
1,1-Dichloroethane	4960	62	1180
1,2-Dichloroethane	8426	30	1240
Ethanol	Completely miscible	0.49	790
Toluene	515	490	870
Vinyl chloride	2790	4	910
Tetrachlorodibenzo-*para*-dioxin (TCDD)	1.9×10^{-4}	6.3×10^{6}	

Data from H.F. Hemond, E.J. Fechner, Chemical Fate and Transport in the Environment, Elsevier, 2014. TCDD data sources: National Toxicology Program, NTP report on carcinogens: background document for 2,3,7,8-tetrachlorodibenzo-p-dioxin (TCDD). Available: https://ntp.niehs.nih.gov/ntp/newhomeroc/other_background/tetrachl_tcdd_4apps_508.pdf, 1999. National Toxicology Program, NTP 11th report on carcinogens, in: Report on Carcinogens: Carcinogen Profiles, vol. 11, p. 1, 2004. U.S. Environmental Protection Agency, National Primary Drinking Water Regulations: dioxin (2,3,7,8-TCDD) EPA 811-F-95-0031-T, 1995.

The values in Table 14.7 illustrate aspects of solubility and organic/aqueous phase distribution. Water solubility is rather inversely related to K_{ow}. Various organic compounds are likely to have affinities for neither, either, or both the water and organic phases. Since most compounds are not completely associated with either phase, they have some amount of amphiphilicity.

This affinity for either water or lipids underpins an important indicator of environmental partitioning. The K_{ow} is the ratio of a substance's concentration in octanol ($C_7H_{13}CH_2OH$) to the substance's concentration in water at equilibrium (i.e., the reactions have all reached their final expected chemical composition in a control volume of the fluid). Octanol is a surrogate for lipophilic solvents in general because it has degrees of affinity for both water and organic compounds, that is, octanol is amphiphilic. Since the ratio forming the K_{ow} is $[C_7H_{13}CH_2OH]:[H_2O]$, then the larger the K_{ow} value, the more lipophilic the substance.

The relationship between density and organic/aqueous phase partitioning is very important to pollutant transport. For example, a nonaqueous-phase liquid (NAPL) will partition at different rates in ground water through the vadose (unsaturated) zone, depending on its density. A lighter compound may stay near the surface and be more likely to volatilize into the atmosphere than a denser compound, even if they have equal high K_{ow} values and water solubility.

Even minor structural changes to a molecule can make quite a difference in phase partitioning and in density. For example, the isomers (i.e., same chemical composition with a different arrangement) vary in their K_{ow} values and densities. As evidence, the "1,1" versus "1,2" arrangements of chlorine atoms on 1,1-dichloroethane, and 1,2-dichloroethane, causes the former to have a slightly decreased density but twice the K_{ow} value than the latter. The location of the chlorine atoms alone accounts for a large difference in water solubility of the two compounds.

Chemical transformation and movement within other environmental media often occur prior to a pollutant's release to water bodies. Water pollutants can be released as point sources from discharges directly into surface waters, but there are also sources that have their beginnings overland and below the ground. Leaking pipelines, underground tanks and contaminated soil, porous media and water can be surreptitious sources of pollutants. As mentioned, the relationship between density and organic/aqueous phase partitioning is very important to this type of pollutant transport. The transport of the lipophilic compound through the vadose zone assumes that the compound has extremely high K_{ow} values and extremely low water solubility. That is, they have a greater affinity for lipids than for water. As the aqueous solubility of a substance increases, its flow will increasingly follow the water flow lines [6,42].

When a dense, miscible fluid seeps into the zone of saturation, the dense pollutants move downward. When these pollutants reach the bottom of the aquifer, the shape of the bottom of the aquifer dictates their continued movement and slope of the underlying bedrock or other relatively impervious layer, which will likely be in a direction other than the flow of the ground water in the aquifer. Above the aquifer bottom, solution and dispersion mechanics near the boundaries of the plume will have a secondary plume that will follow the general direction of ground water flow. The physics of this system illustrates that deciding where the plume is heading will entail more than the fluid densities, but also aquifer dimensions, and the substance's solubility and phase partitioning. So, monitoring wells will need to be installed upstream and downstream from the pollutant source.

Low-density organic fluids often are highly volatile, i.e., their vapor pressures are sufficiently high to change phases from liquid to gas. So, this is another physicochemical property of environmental fluids, vapor pressure, which must be considered as a basis for partitioning [42].

Another complicating factor is that water may contain an amphiphilic compound (e.g., an alcohol) or a surfactant, which dissolves both hydrophilic and lipophilic compounds. As such, even a highly lipophilic compound may be dissolved by this solvent or surfactant which is carried with the water and ultimately is volatilized into the atmosphere. Such co-solvation occurs when a substance is first dissolved in one solvent and then the new solution is mixed with another solvent. In this instance, even a substance with low aqueous solubility will travel along the flow lines of the ground or surface water. For example, a dense nonaqueous phase liquid (DNAPL) like a chlorinated benzene which has very low solubility in pure water, can migrate into and within water bodies if it is first dissolved in an alcohol or an organic solvent (e.g., toluene). A DNAPL tends to migrate downward during it flow in ground water because its density is greater than that of water. Thus, the normally lipophilic compound can be transported in the vadose zone or upper part of the zone of saturation where it undergoes co-solvation with water and a light nonaqueous phase liquid (LNAPL), e.g., toluene.

Pollutants are often found in mixtures. The chemical compounds in these mixtures will partition at different rates. As mentioned, the dense substances move toward the bottom of the aquifer. With time, ground water plumes may partition between lighter and denser constituents. The lighter compounds are more likely than denser compounds to migrate upward and volatize to the atmosphere. Conversely, even semivolatile and nonvolatile compounds may reach to the atmosphere if they become sorbed to aerosols that migrate into the air. Meanwhile, compounds may differentially migrate because of co-solvation. Others may resist dissolution and continue to migrate downwardly in the aquifer. Thus, K_{ow} must be applied to each constituent of the pollutant plume. The rates of partitioning are specific to environmental conditions.

9 Henry's law constant

Henry's law constant (K_H) is the air-water partition coefficient, which is the ratio of vapor pressure to solubility in water. When K_H is a unitless value, it is reported as the air-water partitioning coefficient (K_{AW}). The conventional Henry's Law constant is calculated as:

$$K_H = \frac{P^o \text{ x MW}}{760 \times C_{sol}} \tag{14.41}$$

Where P^o = vapor pressure (torr); MW = molecular weight (g mol^{-1}); 760 = conversion factor (1 atmosphere [atm] = 760 torr), and C_{sol} = and water solubility (mg L^{-1}). The resulting units for K_H are atm for air to moles per cubic meter for water (atm m^{-3} mol^{-1}). The units of mg and liter (L) cancel given that there are 1000 mg in 1 g and 0.001 m^3 in 1 L.

The unitless Henry's Law Constant (K_{AW}) is calculated as:

$$K_{AW} = \frac{C_{air}}{C_{water}} = \frac{K_H}{R \cdot T} \tag{14.42}$$

Where R is the ideal gas constant of 8.205746×10^{-5} atm-m^3/K-mol; T is absolute temperature expressed in Kelvin (K); and K_H is expressed in atm-m^3 mol^{-1}. The equations require that all measured or calculated parameters have the same temperature (e.g., 25 °C).

10 Partitioning between air and tissue

The principal interest in partitioning between an organic substrate and water, i.e., the K_{ow}, is important when predicting the movement of water pollutants. For a complete treatment of partitioning of a compound, there is a need to estimate the movement into and from biota. K_{ow} is very useful for estimating movement of chemical compounds through the gills, but less robust for estimating movement through lungs. Thus, the octanol-air partitioning coefficient (K_{OA}) was devised for these estimates.

Models have been developed to combine K_{ow} and the Henry's Law Constant (K_H) to estimate the equilibrium between air and organic tissues. Henry's Law constant and the K_{OA} are physical-chemical properties that are often not measured but are needed to understand the environmental fate and transport of chemicals. When measured values of the movement are not available or when estimating potential long-range transport or terrestrial bioaccumulation is of pollutants, both K_H and K_{oa} can be applied.

The K_{OA} is calculated as the ratio of the concentration of a compound in an organic phase (represented by *n*-octanol) and air at equilibrium $\left(\frac{C_{air}}{C_{water}} = \frac{C_{octanol}}{C_{air}}\right)$, where $C_{octanol}$ and C_{air} represent equilibrium concentrations in air and *n*-octanol with the same units. The K_{OA} has been used to describe partitioning between air and aerosol particles, air and foliage, and air and soil [1,43,44]. Handbook values are available, but measured K_{OA} values may be substantially different than calculated values, so measured values are preferred. In the absence of sufficient data, K_{OA} values can be calculated based upon a quantitative structural activity relationship (QSAR)[e] [45,46]:

$$K_{OA} = \frac{K_{ow} \cdot R \cdot T}{K_H} \tag{14.43}$$

Where, K_{ow} is the *n*-octanol-water partition coefficient (unitless); R is the ideal gas constant = 8.205746×10^{-5} atm-m^3/K-mol; T is absolute temperature expressed in Kelvin (K), and K_H is the Henry's Law constant expressed in atm-m^{-3} mol^{-1}.

Some published values for some of these key partitioning coefficients are given in Table 14.8.

The ideal gas law assumptions can simplify the equation to a relationship between water and air partitioning, i.e., K_{ow} and K_{OA}:

e. QSAR is a model of the unknown properties of a chemical compounds based on the properties of a well-studied, known compound or compounds that are somewhat similar in structure to the target compound.

TABLE 14.8 Key air, water and organic tissue partitioning coefficients for hazardous air pollutants calculated from quantitative structural activity relationships (QSAR) and from published estimates.

Chemical name	Log K_{ow} from QSAR	Log K_{oa} from QSAR	Log K_{aw}	Log K_{oc} from QSAR	Log K_{oc} estimated from K_{ow}
Acetaldehyde	−0.17	2.387	−2.564	−0.2676	0.5078
Acetamide	−1.16	5.179	−6.339	0.475	0.2244
Acetonitrile	−0.15	2.753	−2.851	0.6693	1.1292
Acetophenone	1.67	5.067	−3.371	1.7147	1.9946
2-Acetylaminofluorene	3.12	11.225	−8.105	3.3436	2.6255
Acrolein	0.19	3.025	−2.302	−0.007	0.6903
Acrylamide	−0.81	5.808	−7.158	0.755	0.5507
Acrylic acid	0.44	5.368	−4.82	0.1583	0.3493
Acrylonitrile	0.21	2.459	−2.249	0.93	1.4556
Allyl chloride	1.93	1.786	−0.347	1.5977	1.6746
4-Aminobiphenyl	2.84	8.064	−5.224	3.3928	2.4854
Aniline	1.08	5.19	−4.083	1.8465	1.4013
o-Anisidine	1.16	6.495	−4.422	1.662	1.6121
Asbestos					
Benzene (including benzene from gasoline)	1.99	2.647	−0.644	2.1637	1.8482
Benzidine	1.92	10.595	−8.675	3.0757	1.6231
Benzotrichloride	3.9	5.873	−1.973	3.0005	3.3844
Benzyl chloride	2.79	3.858	−1.774	2.6495	1.9958
Biphenyl	3.76	5.531	−1.9	3.71	3.4799
Bis(2-ethylhexyl)phthalate (DEHP)	8.39	11.707	−4.957	5.0776	4.9977
Bis(chloromethyl)ether	0.57	2.929	−0.749	0.9867	1.1498
Bromoform	1.79	4.125	−1.66	1.5027	2.0826
1,3-Butadiene	2.03	1.527	0.478	1.5977	1.7267
Calcium cyanamide	−0.81	2.057	−7.975	0.6693	0.8637
Caprolactam (See Modification)	0.66	6.645	−5.985	1.3892	1.2646
Captan	2.74	9.467	−6.543	2.4018	2.4663
Carbaryl	2.35	9.242	−6.874	2.55	2.1344
Carbon disulfide	1.94	1.848	−0.23	1.337	1.6833
Carbon tetrachloride	2.44	2.424	0.052	1.6424	2.4558
Carbonyl sulfide	−1.33	−1.634	1.397	−0.2676	−0.0398
Catechol	1.03	9.653	−7.309	2.3899	1.7455
Chloramben	1.9	10.966	−8.801	1.3299	1.1851
Chlordane	6.26	8.802	−2.702	4.8296	5.3979
Chlorine	0.85	None	None	1.1211	0.7373
Chloroacetic acid	0.34	5.443	−6.422	0.1583	0.2774

TABLE 14.8 Key air, water and organic tissue partitioning coefficients for hazardous air pollutants calculated from quantitative structural activity relationships (QSAR) and from published estimates—cont'd

Chemical name	Log K_{ow} from QSAR	Log K_{oa} from QSAR	Log K_{aw}	Log K_{oc} from QSAR	Log K_{oc} estimated from K_{ow}
2-Chloroacetophenone	1.93	5.779	−3.849	1.9952	2.1882
Chlorobenzene	2.64	3.427	−0.896	2.369	2.4644
Chlorobenzilate	3.99	9.265	−5.529	3.1872	3.0699
Chloroform	1.52	2.401	−0.824	1.5027	1.7094
Chloromethyl methyl ether	0.32	2.226	−1.906	0.7261	1.0115
Chloroprene	2.53	2.169	0.361	1.7832	2.1954
Cresols/Cresylic acid (isomers and mixture)	2.06	6.657	−4.388	2.4777	2.165
o-Cresol	2.06	6.657	−4.309	2.4865	2.1706
m-Cresol	2.06	6.657	−4.456	2.4777	2.1761
p-Cresol	2.06	6.657	−4.388	2.4777	2.165
Cumene	3.45	3.817	−0.328	2.8438	3.1761
2,4-D, salts and esters	2.62	9.044	−5.839	1.4717	1.7659
DDE	6	8.842	−2.769	5.0701	5.6496
Diazomethane	2	1.861	0.139	1.1211	1.7354
Dibenzofurans	4.05	6.825	−2.775	3.9619	3.5146
1,2-Dibromo-3-chloropropane	2.68	4.285	−2.221	2.0636	2.5686
Dibutylphthalate	4.61	8.912	−4.131	3.0635	3.283
1,4-Dichlorobenzene(*p*)	3.28	4.197	−1.006	2.5743	2.9852
3,3-Dichlorobenzidene	3.21	12.145	−8.935	3.5038	2.8234
Dichloroethyl ether (Bis(2-chloroethyl)ether)	1.56	3.672	−3.158	1.508	1.548
1,3-Dichloropropene	2.29	2.289	−0.838	1.8583	1.7614
Dichlorvos	0.6	5.055	−4.63	1.7321	1.8194
Diethanolamine	−1.71	7.085	−8.801	−0.6817	−0.7323
N,N-Diethyl aniline (*N,N*-Dimethylaniline)	2.17	4.625	−2.634	1.8958	2.1377
Diethyl sulfate	1.14	4.867	−3.6	1.4502	1.7171
3,3-Dimethoxybenzidine	2.08	13.211	−11.131	2.7066	1.9949
Dimethyl aminoazobenzene	4.29	9.309	−5.019	3.3071	3.8243
3,3′-Dimethyl benzidine	3.02	11.61	−8.59	3.5038	2.1762
Dimethyl carbamoyl chloride	−0.72	3.809	−4.529	−0.0527	0.254
Dimethyl formamide	−0.93	4.59	−5.52	−0.247	0.0936
1,1-Dimethyl hydrazine	−1.19	4.356	−5.546	1.0773	0.2233
Dimethyl phthalate	1.66	6.698	−5.094	1.4996	1.6789
Dimethyl sulfate	0.16	4.133	−3.786	0.9289	1.175
4,6-Dinitro-*o*-cresol, and salts	2.27	8.174	−4.242	2.8776	2.7083

Continued

TABLE 14.8 Key air, water and organic tissue partitioning coefficients for hazardous air pollutants calculated from quantitative structural activity relationships (QSAR) and from published estimates—cont'd

Chemical name	Log K_{ow} from QSAR	Log K_{oa} from QSAR	Log K_{aw}	Log K_{oc} from QSAR	Log K_{oc} estimated from K_{ow}
2,4-Dinitrophenol	1.73	7.678	−5.454	2.6635	2.4538
2,4-Dinitrotoluene	2.18	7.602	−5.656	2.7601	2.4585
1,4-Dioxane (1,4-Diethyleneoxide)	−0.32	3.297	−3.707	0.4205	0.5946
1,2-Diphenylhydrazine	3.06	9.806	−4.709	3.1775	2.5081
Epichlorohydrin (L-Chloro-2,3-epoxypropane)	0.63	3.269	−2.906	0.9959	1.0834
1,2-Epoxybutane	0.86	2.922	−2.133	0.9959	1.3102
Ethyl acrylate	1.22	3.519	−1.858	1.0273	1.5896
Ethyl benzene	3.03	3.521	−0.492	2.6495	2.7335
Ethyl carbamate (Urethane)	−0.02	5.648	−5.58	1.0837	0.7596
Ethyl chloride (Chloroethane)	1.58	1.931	−0.343	1.337	1.2407
Ethylene dibromide (Dibromoethane)	2.01	3.285	−1.576	1.5977	1.7007
Ethylene dichloride (1,2-Dichloroethane)	1.83	2.136	−1.317	1.5977	1.2841
Ethylene glycol	−1.2	4.071	−5.61	−1.0381	−0.65
Ethylene imine (Aziridine)	−0.28	3.243	−3.306	0.9563	0.7266
Ethylene oxide	−0.05	2.259	−2.218	0.5102	0.6686
Ethylene thiourea	−0.49	4.372	−4.862	1.1128	0.8169
Ethylidene dichloride (1,1-Dichloroethane)	1.76	2.066	−0.639	1.5027	1.5531
Formaldehyde	0.35	2.77	−4.861	−0.4836	0.8894
Heptachlor	5.86	8.003	−1.92	4.6155	4.747
Hexachlorobenzene	5.86	7.298	−1.158	3.792	4.2389
Hexachlorobutadiene	4.72	5.075	−0.376	2.9269	4.1482
Hexachlorocyclopentadiene	4.63	5.694	0.043	3.1473	4.3738
Hexachloroethane	4.03	4.797	−0.799	2.294	3.5927
Hexamethylene-1,6-diisocyanate	3.2	5.907	−2.707	3.6829	2.7769
Hexamethylphosphoramide	−0.22	9.315	−6.087	0.3474	1.0526
Hexane	3.29	1.445	1.867	2.119	3.3844
Hydrazine	−1.47	4.76	−6.23	1.1211	−1.797
Hydrochloric acid	0.54	1.04	−0.5	1.1211	0.4683
Hydrogen fluoride (Hydrofluoric acid)	0.23	1.181	−0.951	1.1211	0.1992
Hydrogen sulfide (See Modification)	0.23	0.679	−0.449	1.1211	−1.1981
Hydroquinone	1.03	9.653	−8.714	2.3811	1.5851
Isophorone	2.62	5.188	−3.566	1.8139	2.061

TABLE 14.8 **Key air, water and organic tissue partitioning coefficients for hazardous air pollutants calculated from quantitative structural activity relationships (QSAR) and from published estimates—cont'd**

Chemical name	Log K_{ow} from QSAR	Log K_{oa} from QSAR	Log K_{aw}	Log K_{oc} from QSAR	Log K_{oc} estimated from K_{ow}
Lindane (all isomers)	4.26	None	None	3.4482	3.5927
Maleic anhydride	1.62	5.414	−3.794	−0.8957	1.3626
Methanol	−0.63	3.128	−3.73	−0.1968	0.0877
Methoxychlor	5.67	11.069	−5.081	4.4296	3.8468
Methyl bromide (Bromomethane)	1.18	1.64	−0.523	1.1211	1.0324
Methyl chloride (Chloromethane)	1.09	1.565	−0.443	1.1211	0.7894
Methyl chloroform (1,1,1-Trichloroethane)	2.68	3.438	−0.153	1.6424	2.1607
Methyl ethyl ketone (2-Butanone) (See Modification)	0.26	2.83	−2.633	0.6542	1.2811
Methyl hydrazine	−1	4.889	−5.889	1.1243	0.3225
Methyl iodide (Iodomethane)	1.59	2.238	−0.667	1.1211	1.3101
Methyl isobutyl ketone (Hexone)	1.16	3.484	−2.249	1.1003	1.8453
Methyl isocyanate	0.79	2.212	−1.422	1.5977	0.6852
Methyl methacrylate	1.28	3.507	−1.885	0.961	1.6228
Methyl *tert* butyl ether	1.43	2.513	−1.62	1.0631	1.3544
4,4-Methylene bis(2-chloroaniline)	3.47	12.341	−8.871	3.7557	3.0446
Methylene chloride (Dichloromethane)	1.34	1.768	−0.877	1.337	1.0845
Methylene diphenyl diisocyanate (MDI)	5.22	9.657	−4.437	5.4548	4.53
4,4′-Methylenedianiline	2.18	10.791	−8.611	3.3275	1.7614
Naphthalene	3.17	4.837	−1.745	3.1887	2.8637
Nitrobenzene	1.81	4.87	−3.008	2.3549	2.1675
4-Nitrobiphenyl	3.57	7.746	−4.176	3.9012	3.2571
4-Nitrophenol	1.91	8.954	−7.77	2.4636	2.3675
2-Nitropropane	0.87	3.388	−2.313	1.4886	1.6586
N-Nitroso-*N*-methylurea	−0.52	7.872	−8.392	1.0414	0.8765
N-Nitrosodimethylamine	−0.64	3.435	−4.128	1.3577	0.5662
N-Nitrosomorpholine	−0.43	5.809	−5.999	1.3524	0.5475
Parathion	3.73	8.647	−4.914	3.3842	3.1709
Pentachloronitrobenzene (Quintobenzene)	5.03	8.742	−2.743	3.7779	3.8551
Pentachlorophenol	4.74	10.032	−5.999	3.6954	4.0684
Phenol	1.51	6.149	−4.866	2.2724	1.8995
p-Phenylenediamine	−0.39	7.17	−7.56	1.5293	0.7159
Phosgene	−0.71	−0.272	−0.166	−0.102	0.3031
Phosphine	−0.27	−0.269	−0.001	1.1211	−0.2347

Continued

TABLE 14.8 Key air, water and organic tissue partitioning coefficients for hazardous air pollutants calculated from quantitative structural activity relationships (QSAR) and from published estimates—cont'd

Chemical name	Log K_{ow} from QSAR	Log K_{oa} from QSAR	Log K_{aw}	Log K_{oc} from QSAR	Log K_{oc} estimated from K_{ow}
Phosphorus	−0.27	−0.269	−0.001	1.1211	−0.2347
Phthalic anhydride	2.07	5.656	−6.176	0.1469	1.3515
Polychlorinated biphenyls (Aroclors)	7.62	10.173	−2.553	5.3208	5.1682
1,3-Propane sultone	−0.28	3.736	−4.016	0.9737	−0.9317
beta-Propiolactone	−0.8	1.7	−2.5	0.5508	0.417
Propionaldehyde	0.33	2.764	−2.523	−0.007	1.0222
Propoxur (Baygon)	1.9	8.763	−7.233	1.7778	1.7256
Propylene dichloride (1,2-Dichloropropane)	2.25	2.432	−0.938	1.7832	1.718
Propylene oxide	0.37	2.554	−2.546	0.7155	0.8511
1,2-Propylenimine (2-Methyl aziridine)	0.13	3.53	−3.388	1.1616	0.9534
Quinoline	2.14	6.691	−4.166	3.1887	1.7614
Quinone	0.25	7.552	−1.708	1.6148	1.9355
Styrene	2.89	3.838	−0.949	2.6495	2.5599
Styrene oxide	1.59	4.992	−3.19	2.0565	1.725
2,3,7,8-Tetrachlorodibenzo-*p*-dioxin	6.92	10.761	−2.689	5.3964	4.8308
1,1,2,2-Tetrachloroethane	2.19	3.399	−1.824	1.9775	2.0739
Tetrachloroethylene (Perchloroethylene)	2.97	3.141	−0.14	1.9775	2.9505
Titanium tetrachloride	1.47	None	None	1.6424	1.2754
Toluene	2.54	3.154	−0.566	2.369	2.369
2,4-Toluene diamine	0.16	7.677	−7.517	1.7434	0.9593
2,4-Toluene diisocyanate	3.74	7.083	−3.343	3.8706	3.2455
o-Toluidine	1.62	5.686	−4.092	2.0606	1.6336
Toxaphene (chlorinated camphene)	6.75	9.556	−2.806	4.8178	5.8579
1,2,4-Trichlorobenzene	3.93	4.978	−1.236	3.1322	3.2931
1,1,2-Trichloroethane	2.01	2.768	−1.473	1.7832	1.6399
Trichloroethylene	2.47	2.497	−0.395	1.7832	2.0999
2,4,5-Trichlorophenol	3.45	8.481	−4.179	3.2497	3.294
2,4,6-Trichlorophenol	3.45	8.481	−3.973	3.2497	3.2774
Triethylamine	1.51	3.967	−2.215	1.7059	1.6618
Trifluralin	5.31	7.372	−2.376	4.2147	4.2518
2,2,4-Trimethylpentane	4.09	2	2.094	2.3808	3.5493

TABLE 14.8 Key air, water and organic tissue partitioning coefficients for hazardous air pollutants calculated from quantitative structural activity relationships (QSAR) and from published estimates—cont'd

Chemical name	Log K_{ow} from QSAR	Log K_{oa} from QSAR	Log K_{aw}	Log K_{oc} from QSAR	Log K_{oc} estimated from K_{ow}
Vinyl acetate	0.73	2.05	−1.68	0.7469	1.2633
Vinyl bromide	1.52	1.819	−0.299	1.337	1.3622
Vinyl chloride	1.62	1.361	0.056	1.337	1.4056
Vinylidene chloride (1,1-Dichloroethylene)	2.12	2.005	0.028	1.5027	1.8482
Xylenes (isomers and mixture)	3.09	3.662	−0.532	2.5743	2.7769
o-Xylenes	3.09	3.662	−0.674	2.5831	3.7422
m-Xylenes	3.09	3.662	−0.532	2.5743	2.7769
p-Xylenes	3.09	3.662	−0.55	2.5743	2.7335

Data from U.S. Environmental Protection Agency, EPI Suite™-Estimation Program Interface, US EPA. https://www.epa.gov/tsca-screening-tools/epi-suitetm-estimation-program-interface, 2012 (Accessed 2 January 2024).

$$K_{OA} = \frac{K_{ow}}{K_{AW}} \tag{14.44}$$

Where K_{AW} is the unitless air-water partition coefficient.

Note that partitioning coefficients in Table 14.8 are reported as log values. This is because the ranges are so large that log-log relationships are commonly used, including for comparing chemicals' affinities. Thus, Eq. (14.44) can be restated as:

$$\log K_{OA} = \log K_{ow} - \log K_{AW} \tag{14.45}$$

Example K_{OA} calculation

Estimate the octanol-air partitioning of methyl isocyanate and formaldehyde.

The values in Table 14.8 indicate the variability of estimates depending on the method used. For example, the published log K_{OA} for methyl isocyanate (MIC) is 2.112, so taking its antilog gives a $K_{OA} = 129$. However, calculating K_{OA} from Eq. (14.45) gives:

$$\log K_{OA} = \log K_{ow} - \log K_{AW} = 0.79 - (-1.4222) \approx 1.5 \tag{14.46}$$

The calculated value is 0.7 less than the published value. The calculated K_{OA} is underreported by about one-third. If the calculated value is used, the affinity for tissue is understated. Thus, the MIC would appear to be less likely to exit from tissue into air than the calculated value indicates.

The published log K_{OA} for formaldehyde is 2.77. Calculating and estimate for K_{OA} from Eq. (14.45) gives:

$\log K_{OA} = \log K_{ow} - \log K_{AW} = 0.35 - (-4.861) \approx 4.81$.

This difference is a large discrepancy, i.e., two orders of magnitude between the two methods. As was the case for MIC, the calculated method underestimates formaldehyde's affinity for tissue and overestimates its likelihood to be transported to the atmosphere.

These differences may be explained in part because the calculation is a rule of thumb. Each partitioning coefficient is measured differently. For example, K_{ow} is not usually calculated as partitioning between pure octanol and pure water; rather the measurement reflects octanol saturated with water and water saturated with octanol [1]. When the inherent properties like solubility and vapor pressure are incorrect, derived properties like K_{ow}, K_{AW} and K_{OA} will propagate and amplify any error.

These differences also appear in the organic-carbon partitioning coefficients (K_{oc}) in Table 14.8, with large discrepancies between QSAR and K_{ow}-based K_{oc} values. Recall that octanol is merely a surrogate or indicator of the organic phase. A published K_{OA} based on credible methods is preferred whenever possible.

Physicochemical properties example

Three compounds are leaking into the ground water from an underground tank. At 20 °C, compound A has a vapor pressure of 12 kPa and density of 1.6 g mL^{-1}. Compound B has a vapor pressure of 59 kPa and density of 0.7 g mL^{-1}. Compound C has a vapor pressure of 0.003 kPa and density of 1.1 g mL^{-1}. Based on these data, B has the highest and C has the lowest expected mass emitted from soil to the air. However, since A and C are each heavier than water, they could settle toward the bottom of the aquifer. What other physicochemical properties could change this order?

Aqueous solubility and, sorption can certainly change the order of affinity for environmental compartments. If the compounds completely dissolve in water, then even denser compounds may not settle and may be transported with the water in a solution (e.g., a droplet containing the solvent). Searching physicochemical databases for compounds with similar vapor pressures and densities, compound A is similar to tetrachloromethane (CCl_4), also known as carbon tetrachloride. At 20 °C, the aqueous solubility of CCl_4 is 80 g kg^{-1}. Compound B has similar properties to diethyl ether [$(C_2H_5)_2O$]. The aqueous solubility of $(C_2H_5)_2O$ is 75,000 g kg^{-1}. Compound C has properties similar to diethylene glycol [$(HOCH_2CH_2)_2O$], which is completely miscible in water. So, even though the diethylene glycol is denser than water it will be dissolved in the water. To a lesser extent, a portion of the diethyl ether will also likely be dissolved, given its relatively high water solubility. The tetrachloromethane will not likely be dissolved, so that a fraction could settle and not be released, even given its high vapor pressure.

Sorption is also a complicating factor, since the clay and organic matter fractions of the soil and other substances situated between the troposphere and the aquifer could hold various compounds. Sorption varies depending on substrate and other conditions. The K_{oc} of CCl_4 has been found to be about 71 mL g^{-1}. Diethyl ether is about 14. Diethylene glycol is about 2. Thus, they all are within the same order of magnitude for this type of sorption, i.e., to organic matter, so this may not be a limiting step for these three compounds. However, for many semivolatile compounds, sorption can be very important in terms of half-life and transport. An increasing K_{oc} may mean a greater likelihood that some of the mass will aerosolize and be transported in the air. Higher K_{oc} may also slow the movement of pollutants in ground water through the vadose zone, thus decreasing the amount moving from soil to air.

The same physical and chemical processes in the open environment take place within an organism after uptake, albeit endogenously. Kinetics and equilibria will vary in scale and will differ in importance between the open environment and within organisms. For example, the degradation or activation of a pollutant may be much faster during an organism's absorption and metabolism stages in the presence of organic catalysts, i.e., enzymes in single-cellular [47–50] and multi-cellular organisms, including humans [51].

The reaction rates and equilibria differ between environmental and endogenous conditions, as do residence times and half-lives of chemical processes. However, the laws of thermodynamic and motion apply at every level of interest to water pollution, from cells to the planet.

11 Microbial partitioning

Microbes are present in soil, water, sediment, and air. These include fungi, bacteria and viruses that use a wide range of organic compounds as carbon and energy sources. Wastewater treatment processes are often based on acclimating microbes, especially bacteria, to degrade organic compounds as their food sources. For the metabolism that breaks down these compounds to occur, the organic compound and O_2 must first move from the vapor or solid phases to the aqueous phase to become bioavailable to the microorganisms. Thus, even though a pollutant may be a volatile organic compound, it is commonly treated exclusively in the aqueous phase. This change from gas phase to aqueous phase can be calculated as the volumetric mass transfer rate (mol m^{-3} s^{-1}) of the pollutants, molecular oxygen, and nutrients:

$$K_{1a_{G/A}}\left(\frac{S_G}{K_{G/A}} - S_A\right) \tag{14.47}$$

Where, $K_{1a_{G/A}}$ is the global volumetric mass transfer coefficient (h^{-1}), S_G and S_A are the substrate (e.g., benzene) concentrations (mol m^{-3}) in the bulk gas and aqueous phases, respectively, and $K_{G/A}$ is the substrate partition coefficient (dimensionless) between the gaseous and aqueous phases. $K_{G/A}$ is calculated as follows:

$$K_{\mathrm{G/A}} = \frac{S_{\mathrm{G}}}{S_{\mathrm{A}}^*} \tag{14.48}$$

Where, S_{A}^* is the substrate concentration at the gas/aqueous interface (mol m^{-3}).

In addition to the vapor and aqueous phases, there is also a nonaqueous (e.g., lipid) phase in the substrate (Fig. 14.9A). Reactions can only occur when reagents come into contact with the microbe, which often requires that a pollutant move

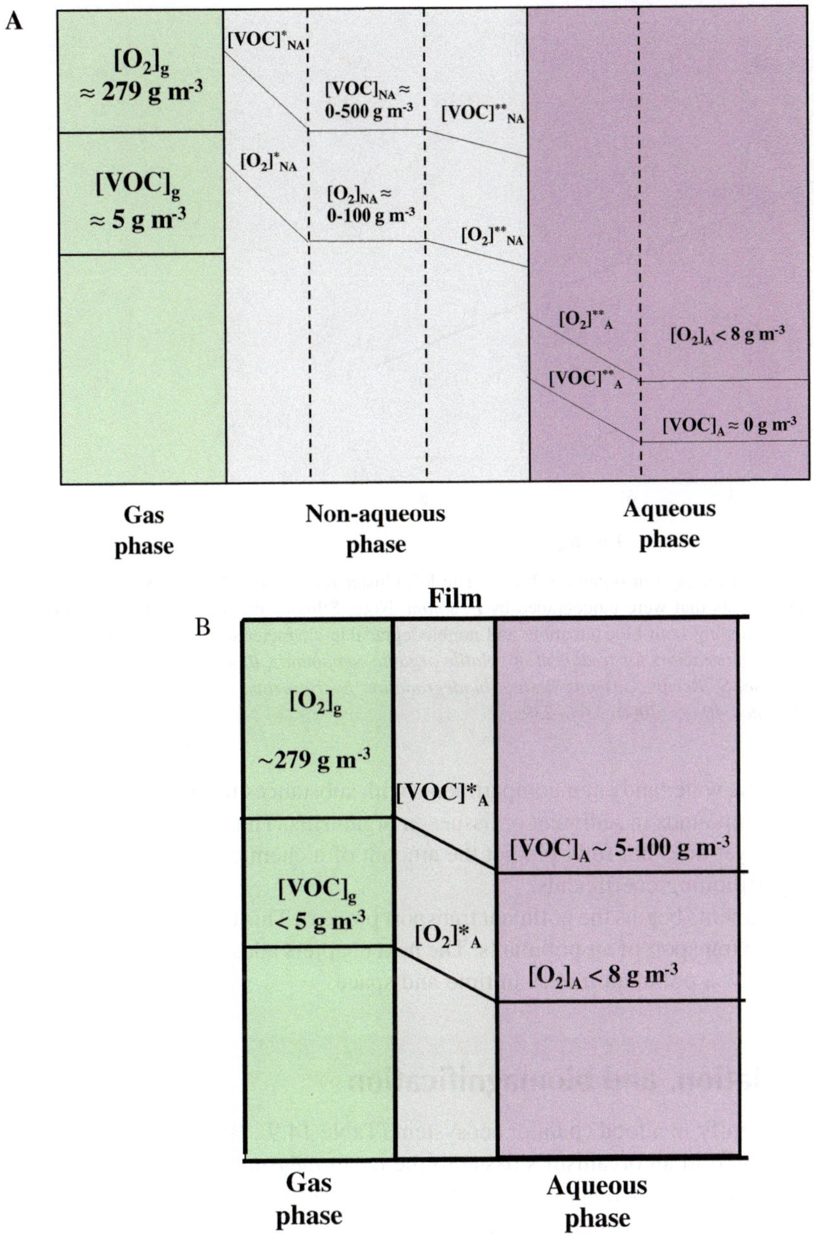

FIG. 14.9 Concentration profiles of lipophilic substrates (volatile organic compounds—VOCs—and O_2). (A) A single-phase system. (B) A two-phase partitioning bioreactor. [VOC] and $[O_2]$ = concentrations of volatile organic compounds and O_2, respectively, in the treated gas phase ([ ]g), aqueous phase ($[\ ]_A$), nonaqueous phase ($[\ ]_{NA}$), [ ]* and [ ]** represent the equilibrium concentrations at the gaseous/nonaqueous and nonaqueous/aqueous interfaces, respectively. Note: All concentrations based on air contaminated with 5 g VOC m^{-3}. *(Data from R. Muñoz, S. Villaverde, B. Guieysse, S. Revah, Two-phase partitioning bioreactors for treatment of volatile organic compounds, Biotechnol. Adv. 25 (4) (2007) 410–422.)*

between at least two phases (Fig. 14.9B). This can be seen as a film profile, with sectors based on the octanol–water coefficient (K_{ow}) and $K_{G/A}$ [52]. Some solvents are toxic whereas others are biocompatible, which allows the partitioning to occur, depending on the organism (Fig. 14.10) [52,53].

The inherent properties of a pollutant do not completely explain its affinity for an environmental compartment. Vapor pressure and molecular weight, as mentioned, are indicators of where a pollutant will reside, but are not definitive. Furthermore, the partitioning coefficients give even stronger clues as to where a pollutant will move. The physicochemical properties must be combined with environmental conditions to characterize and estimate the fate of the pollutant. After entering the air, water, soil, and biota, a chemical compound will either remain in a compartment or move to another. Indeed, a single compartment is seldom an exclusive residence, as some of the mass of a chemical compound remain and the remainder will move to one or more other compartments. The propensity for this distribution is the nature of partitioning and is why phase partitioning is also called "phase distribution."

Partitioning undoubtedly is influenced by inherency. A chemical compound with high aqueous solubility is easily dissolved in water under normal environmental conditions of temperature and pressure, i.e., it is hydrophilic. It is also likely to remain dissolved in water. If, conversely, a substance is not easily dissolved in water under these conditions, it is said to be

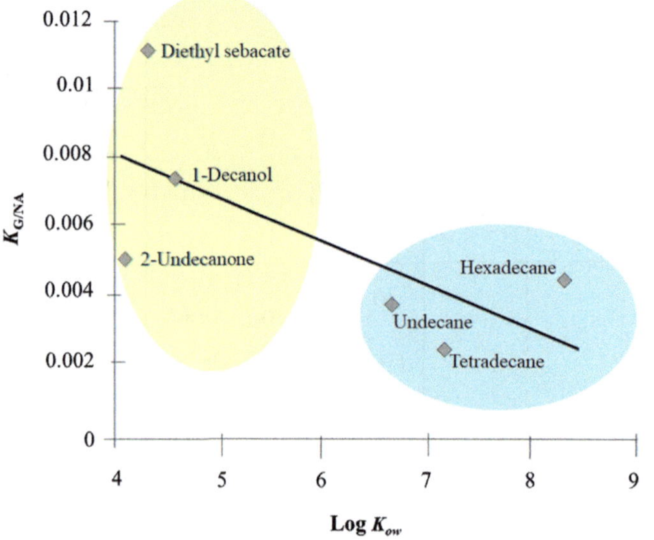

FIG. 14.10 Gaseous/nonaqueous hexane partition coefficient ($K_{G/NA}$) in organic solvents. The left cluster represents solvents toxic to the fungus, *Fusarium solani*. The right cluster shows biocompatible solvents that were biodegraded by *F. solani*. Note: Silicone oil ($K_{G/NA} = 0.0034$; unknown log K_{ow}) was the only nonaqueous phase substance tested showing both biocompatible and nonbiodegradable characteristics. *(Data from R. Muñoz, S. Villaverde, B. Guieysse, S. Revah, Two-phase partitioning bioreactors for treatment of volatile organic compounds, Biotechnol. Adv. 25 (4) (2007) 410–422. S. Arriaga, R. Muñoz, S. Hernández, B. Guieysse, S. Revah, Gaseous hexane biodegradation by Fusarium solani in two liquid phase packed-bed and stirred-tank bioreactors, Environ. Sci. Technol. 40 (7) (2006) 2390–2395.)*

hydrophobic, and many of its molecules will exit the water and enter compartments with substances for which the chemical compound has more affinity, e.g., hydrophobic compounds in sediment or tissues in organisms. Thus, an inherent property like aqueous solubility is a factor of partitioning, yet does not fully predict the amount of a chemical that will be in each compartment. Such prediction is afforded by partitioning coefficients.

Partitioning among the environmental compartments begins the pollutant transport process. This chapter has applied the laws of thermodynamics, including fugacity, to the transport of air pollutants. The next chapters adds the physical transport via the laws of motion. Together these explain how a pollutant moves in time and space.

12 Bioconcentration, bioaccumulation, and biomagnification

A substance can concentrate, accumulate, and magnify in a food chain or ecosystem (Table 14.9). Bioconcentration is the ratio of the mean concentration of a substance stored in an organism's tissue to the mean measured concentration in the organism's environment, usually water [57]. This includes chemical compound uptake via respiration and contact, but not via dietary sources. Bioconcentration factors (*BCF*s) are calculated by considering chemical compound tissue concentrations with respect to environmental chemical compound concentrations. *BCF* values >1 indicate that the concentration in the organism is greater than that of the medium (e.g., soil or water) from which the chemical compound was taken. *BCF*s can also be based on a total organism concentration, e.g., wet fish, that is normalized to the lipid content of the organism.

Bioaccumulation is a broader concept than bioconcentration in that the bioaccumulation factor (*BAF*) also includes the accumulation of a chemical compound via the consumption of food and other exposures. As such, *BAF* is the net uptake of a substance from the environment by all possible exposure routes, especially by food, respiration, and dermal contact. The source not only includes water, but also sediment and other biota [58,59]. Bioaccumulation factors (*BAF*) are calculated by considering chemical compound tissue concentrations with respect to environmental chemical compound concentrations. *BAF* values >1 indicate that the accumulation in the organism is greater than that of the water or other environmental medium from which the chemical was measured. These factors can be based on a total organism concentration or can be normalized to the lipid content of the organism [59].

Bioaccumulation can also include biomagnification, which is the increase of a chemical compound concentration in the tissue of an organism compared to the tissue concentrations of its prey [60]. As such, the biomagnification factor (*BMF*) represents the increase in concentrations in organisms moving from low trophic levels to highest trophic levels, e.g., top predators in an ecosystem The *BMF* is calculated by considering lipid normalized chemical compound tissue

TABLE 14.9 Definitions and commonly used testing methods to determine bioconcentration (*BCF*), bioaccumulation (*BAF*), and biomagnification factor (*BMF*) for chemical compounds.

Factor	Definition	Unit	Testing method
BCF	$\frac{\text{Concentration of substance in tissue}\left(\text{mg kg}^{-1}\right)}{\text{Concentration of substance in water}\left(\text{mg L}^{-1}\right)}$	L kg^{-1}	Bioaccumulation flow through fish test [54]
BAF	$\frac{\text{Concentration of substance in tissue}\left(\text{g kg}^{-1}\right)}{\text{Concentration of substance in surrounding medium}\left(\text{g kg}^{-1}\right)}$ Surrounding medium can be water, soil, or sediment	Unitless	Bioaccumulation in sediment-dwelling benthic oligochaetes [55] Bioaccumulation in terrestrial oligochaetes [56]
BMF	$\frac{\text{Concentration of substance in predator}\left(\text{mg kg}^{-1}\right)}{\text{Concentration of substance in prey or other food at steady state}\left(\text{g kg}^{-1}\right)}$	Unitless	Dietary exposure bioaccumulation fish test

Data from ChemSafetyPro, Differences between BCF, BAF and BMF. https://www.chemsafetypro.com/Topics/CRA/definition_differences_BCF_BAF_and_BMF.html, 2017 (Accessed 20 December 2023).

concentrations within an organism with respect to the lipid normalized concentration of that chemical compound in the prey of the organism. Factors >1 indicate the occurrence of biomagnification.

The *BCF*, *BAF* and *BMF* values are obtained from laboratory and field studies [61], and vary in the type and expressed in various units. For example, the European Union classifies chemical compounds as to those likely to accumulate versus those that are not likely to accumulate significantly.

The *BCF* value in fish represents a balance between the rate of chemical uptake from the water via the respiratory surface gills and skin of the organism and the excretion of the chemical from the organism. The *BCF* value can also be expressed as the ratio of the uptake rate constant to the depuration (cleansing) rate constant. It includes both the exposure from respiratory surface and the dietary exposure. For highly lipophilic and hydrophobic substances, having $\log K_{ow} > 5$ or high K_{oc} and aqueous solubility below ~0.01–0.1 mg L^{-1}, testing via aqueous exposure is challenging. For such compounds, a dietary study in fish is preferable. In this instance, the *BMF* value will be the result, rather than *BCF*, since *BAF* reflects both bioconcentration and biomagnification [62].

For example, a valid *BCF* or *BAF* > 2000 is considered to be bio-accumulative and a *BCF* or *BAF* > 5000 indicates a very bio-accumulative chemical compound [62].

Maximum concentration example (adapted from Ref. [63])

Frequently, an organism's likelihood of bioaccumulating a chemical substance is directly proportional to the substance's K_{ow}. With this in mind, consider a rainbow trout (*Oncorhynchus mykiss*) that is swimming in water the contains 1,2,4-trichlorobenzene (TCB). The TCB concentration is 10 nanomoles (nM). The water and air temperature are 25 °C. What is the maximum concentration in the fish?

Nonpolar organic compounds like TCB accumulate at rates that are indirectly proportional to their aqueous solubility. That is, a smaller K_{ow} means the compounds is likely to be soluble in water and a large K_{ow} means the compound will be hydrophobic and likely to have an affinity for lipids, i.e., it is highly lipophilic. Furthermore, given that they are equilibrium coefficients, calculating *BCF* and *BAF* requires an assumption that steady state has been reached and the chemical concentration in the tissue (C_T) and the concentration in the water (C_w) must be measured simultaneously.

The *BCF* estimates for *O. mykiss* and other fish species have been tested for K_{ow} based on their tissue concentrations compared to water concentrations, leading to a general relationship. Studies of the fathead minnow (*Pimephales promelas*) show this relationship between K_{ow} and *BCF* [64–67]:

$$\log BCF = 0.85(\log K_{ow}) - 0.70 \tag{14.49}$$

However, since most of the mass of an organic compound accumulates in fatty tissue and cell membranes [68] and this varies according the aquatic species, the total fish concentrations are best normalized to the lipid content of each species. For *O. mykiss*, this is:

$$\text{Lipid based } \log BCF = 0.85(\log K_{ow}) + 0.61 \quad (14.49)$$

These equations are derived empirically under different field and experimental conditions, so the published relationships vary somewhat. The equations indicate that lipids comprise about 5% of a fish's wet weight. The studies usually do not take into account the internal degradation rate of the chemical compound, so only apply to conditions when metabolism is not affecting the results [63,69]. Given these uncertainties, the estimates of *BCF* at 10 nM can be calculated from published literature [63], K_{ow} of 1,2,4-TCB is 10,000:

$$\text{Log}BCF = 0.85\ \log(10^4) - 0.7 \cong 2.7$$

Taking the antilog of 2.7 gives a *BCF* of 501 mL water g^{-1} wet fish.
On a lipid basis:

$$\text{Log}BCF = 0.85\ \log(10^4) + 0.61 \cong 4$$

Taking the antilog of 2.8 gives a *BCF* of 10,000 mL water g^{-1} lipid content. This demonstrates the drastic difference in basing *BCF* on whole fish versus lipid content.

Finally, to find the concentration in the whole fish (C_{trout}), we will multiply the *BCF* times the water concentration:

$$C_{trout} = BCF \times C_w$$

$$C_{trout} = BCF \times 500\left(\frac{\text{mL water}}{\text{g wet fish}}\right) \times \frac{10\ \text{nmol}}{10^{12}\text{mol}} \times \frac{\text{mol}}{\text{mL water}} = 5 \times 10^{-9}\ \text{mol g}^{-1}\ \text{wet fish}$$

Since the molecular weight of 1,2,4-TCB is 181.5 g mol^{-1}, and given that a mole is numerically equal to the molecular weight in grams, the maximum concentration of the fish is:

$\left(5 \times 10^{-9}\ \text{mol g}^{-1}\right) \times \left(181.5\ \text{g mol}^{-1}\right) \cong 9 \times 10^{-7}$ volume of TCB in water, or about 900 parts per billion (ppb) or about 6.7 μg L^{-1}.

A widely used model to estimate *BAF* is the U.S. Environmental Protection Agency's [59] K_{ow} (based) Aquatic BioAccumulation Model (KABAM), which is based on the food web bioaccumulation model for organic chemical compounds in aquatic ecosystems [70]. The K_{ow} in the model title represents the octanol-water coefficient, which is ratio of the equilibrium concentrations of a chemical compound in octanol to the concentration in water. Thus, as mentioned, K_{ow} values that are high represent compounds that are highly hydrophobic and, conversely, low K_{ow} values represent compounds more easily dissolved in water. The octanol represents the lipid phase in the two-phase system. K_{ow} is usually presented in logarithmic form, given the wide range of values for thousands of compounds. For example, the most toxic chlorinated dioxin, 2,3,7,8-tetrachlorodibenzo-*para*-dioxin, has a K_{ow} of about 5,000,000, but is usually presented as log K_{ow} of 6.7. Conversely, dimethyl sulfoxide (DMSO) has a K_{ow} is about 0.01 and a log K_{ow} of about −2.0.

High K_{ow} chemical compounds usually sorb more readily to organic in soil and sediment due to their low affinity for water. Compounds that exceed log K_{ow} values of 4.5 have an heightened potential to bioconcentrate and persist in the tissue of organisms; thus making K_{ow} values useful in first screening tool for bioaccumulation [66]. This is because the uptake of an organic compound often determined by its hydrophobicity. Below 4.5 K_{ow}, organic compounds tend to have less affinity for the lipids of an organism.

The KABAM model provides estimates of chemical concentrations and *BCF* and *BAF* values for aquatic ecosystems [59]. Each aquatic organism is assumed to reside in one compartment, e.g., surface water. Chemical compounds enter the organism via respiration and diet and leave the organism via respiration and fecal egestion. The chemical concentration in the organism can also be influenced by the growth of the organism as well as metabolism of the chemical within the organism. These processes that define uptake and loss of the chemical from aquatic organisms are described by rate constants and are incorporated into one equation that is used to define the concentration of the chemical in organism tissues:

$$C_B = \frac{k_1 \times (\text{m}_0 \times \Phi \times \text{C}_{\text{WTO}} + \text{m}_P * \text{C}_{\text{WDP}}) + \text{k}_D \times \sum(P_i \times C_{Di})}{k_2 + k_E + k_G + k_M} \tag{14.51}$$

Where, the C_B is the concentration of the chemical compound. The other symbols are described in Table 14.10.

As uptake constants (see Fig. 14.11) increase, so does the estimated chemical compound concentration in an organism.

With increasing values of elimination constants, i.e., k_2, k_E, k_G and k_M, the estimated chemical compound concentrations in an organism decrease. However, for respiration and diet, processes of uptake and elimination are linked. Therefore, factors that would influence uptake constants would also influence elimination constants, so these cannot be considered independently. In addition, as the freely dissolved fraction of the chemical compound in the water (φ) decreases, so will the estimated chemical compound concentrations in organisms. Rate constants defining the uptake of a chemical compound via

TABLE 14.10 Definition of symbols used in Eq. (14.51) and associated parameters to calculate chemical tissue residue for single trophic levels.

Parameter symbol	Definition	Units
C_B	Chemical compound concentration in the organism	g kg^{-1} (wet weight)
C_{BD}	Chemical compound concentration in the organism originating from uptake via diet $C_{BD} = C_B$ when $k_1 = 0$	g kg^{-1} (wet weight)
C_{BR}	Chemical compound concentration in the organism originating from uptake via respiration $C_{BR} = C_B$ when $k_D = 0$	g kg^{-1} (wet weight)
C_{Di}	Concentration of chemical compound in i (prey item)	g kg^{-1} (wet weight)
C_S	Concentration of the chemical in sediment (dry weight of sediment)	g kg^{-1} (dry sediment)
C_{WDP}	Freely dissolved chemical compound concentration in pore water of sediment	g L^{-1}
C_{WTO}	Total chemical compound concentration in water column above the sediment	g L^{-1}
k_1	Chemical compound uptake rate constant via respiratory area (i.e., gills, skin)	L kg^{-1} d^{-1}
k_2	Rate constant for elimination of the chemical compound via the respiratory area (i.e., gills, skin)	d^{-1}
k_D	Chemical compound uptake rate constant for uptake via ingestion of food	(kg of food) (kg of organism)$^{-1}$ d^{-1}
k_E	Rate constant for elimination of the chemical compound via excretion of contaminated feces	d^{-1}
k_G	Organism growth rate constant	d^{-1}
k_M	Rate constant for chemical compound metabolic transformation	d^{-1}
m_o	Fraction of respiratory ventilation involving overlying water	None
m_p	Fraction of respiratory ventilation that involves pore-water of sediment	None
P_i	Fraction of diet containing i (prey item)	None
Φ	Fraction of the overlying water concentration of the chemical compound that is freely dissolved and can be absorbed via membrane diffusion	None

Data from J.A. Arnot, F.A. Gobas, A food web bioaccumulation model for organic chemicals in aquatic ecosystems, Environ. Toxicol. Chem. 23 (10) (2004) 2343–2355.

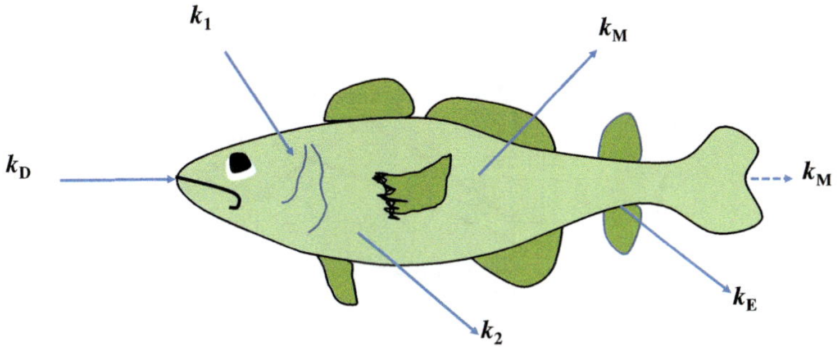

FIG. 14.11 Major routes of uptake and elimination of a chemical compound in an aquatic organism. k_D = dietary uptake rate constant; k_1 = gill uptake rate constant; k_2 = gill elimination rate constant; k_M = metabolic transformation rate constant; k_E = fecal egestion rate constant; k_G = growth dilution rate constant. *(Data from J.A. Arnot, F.A. Gobas, A food web bioaccumulation model for organic chemicals in aquatic ecosystems, Environ. Toxicol. Chem. 23 (10) (2004) 2343–2355. J.A. Arnot, D. MacKay, T.F. Parkerton, R.T. Zaleski, C.S. Warren, Multimedia modeling of human exposure to chemical substances: the roles of food web biomagnification and biotransformation, Environ. Toxicol. Chem. 29 (1) (2010) 45–55. Available: https://eurekamag.com/research/054/493/054493002.php.)*

respiration (k_1) and diet (k_D) and the elimination of a chemical via respiration (k_2) and fecal excretion (k_E) as well as growth dilution of a chemical (k_G) are estimated separately using several equations, with some described below. All equations are published by the U.S. EPA [59].

Use of Eq. (14.53) assumes that the organism is at steady state and that the chemical compound is homogenously distributed throughout organisms. A third assumption is that the effects of chemical partitioning into egg and sperm cells on chemical mass in parents is not a loss pathway. When data are lacking to define the metabolism rate constant for a chemical, metabolism is assumed not to occur and that the elimination rate constant for metabolism (k_M) is 0.

Uptake and elimination of a chemical compound from an organism is influenced by the body composition of the organism being modeled. Body composition includes lipid, nonlipid organic matter (NLOM) and water. NLOM can be carbohydrates, proteins, and other polar molecules. Chemical compounds are expected to partition differently to these components of an organism. Partitioning of a chemical into these components is related to the K_{ow}. Again, octanol is a surrogate for the lipid fraction of an organism. It is also assumed that the partitioning of a chemical to NLOM is less than to octanol, but that there is a relationship between the partitioning of the chemical to NLOM and to octanol and that octanol serves as a reasonable surrogate for estimating this parameter.

Eq. (14.53) is applied to an aquatic food web with several trophic levels (see Fig. 14.12). In increasing order of hierarchy, these trophic levels include [70]:

1. phytoplankton
2. zooplankton
3. benthic invertebrates
4. filter feeders
5. small (juvenile) fish
6. medium sized fish, and
7. large fish.

Concentrations in organisms are first calculated for phytoplankton, the lowest trophic level of the aquatic food chain [59]. Chemical compound concentrations are then calculated for zooplankton, including consideration that the diet of zooplankton includes phytoplankton, which contain chemical compound residues. Tissue residues are calculated for the next five trophic levels based on their diets of organisms from lower trophic levels.

The water column is a hypothetical section of a water body stretching from the surface to bottom sediments. It contains suspended organic matter in two phases, i.e., dissolved organic carbon (DOC) and particulate organic carbon (POC). Upon partitioning to organic carbon, chemical compounds cease to be bioavailable to aquatic organisms. The remaining fraction of mass of the chemical compound in in the water column that is freely dissolved (Φ), and thus bioavailable for uptake by aquatic fauna is estimated as [59]:

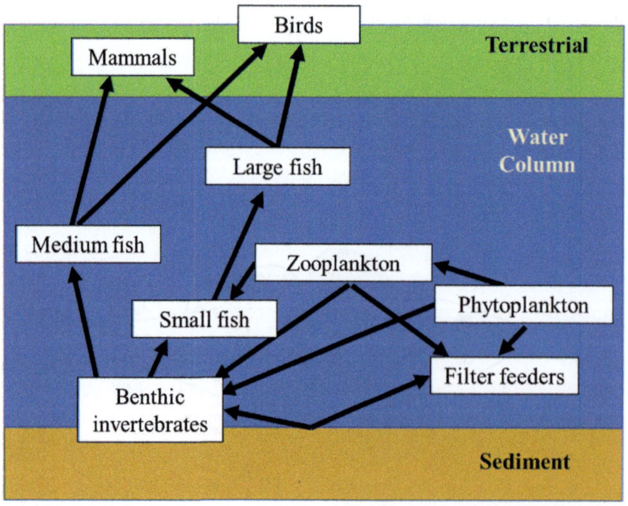

FIG. 14.12 Conceptual Model of Food Web modeled by the K_{ow} (based) Aquatic BioAccumulation Model (KABAM). Arrows show direction of trophic transfer of bioaccumulated chemical compounds from low levels to higher levels of the food web. Biomagnification generally increases from the bottom to the top of the web. *(Data from K. Gerber, User's guide and technical documentation KABAM version 1.0 (K OW (Based) Aquatic Bioaccumulation Model), in: Report of Environmental Fate and Effects Division Office of Pesticide Programs, US Environmental Protection Agency, Washington, DC, 2009.)*

$$\Phi = \frac{1}{1 + (X_{POC} \times \alpha_{POC} \times K_{OW}) + (X_{DOC} \times \alpha_{DOC} \times K_{OW})} \tag{14.52}$$

Where, X_{POC} is the concentration of particulate organic carbon in water, α_{POC} is the proportionality constant to describe the similarity of phase partitioning of POC in relation to octanol, X_{DOC} is the concentration of dissolved organic carbon in water, and α_{DOC} is the proportionality constant to describe the similarity of phase partitioning of DOC in relation to octanol. These and other factors important in this calculation are provided in Table 14.11.

This equation assumes that equilibrium exists between the chemical compound concentration in the water and in the organic carbon in the water column. It also assumes that partitioning between POC and water and DOC and water can be related to partitioning of the chemical between octanol and water. These relationships are defined using proportionality constants (α_{POC} and α_{DOC}) that are defined from the scientific literature.

The KABAM approach for calculating the fraction of bioavailable chemical compound in the water column differs from the Exposure Analysis Modeling System (EXAMS), which calculates the fraction of the overlying water concentration of the chemical compound that is freely dissolved and able to absorb via membrane diffusion [71]. The EXAMS equation for the chemical compound fraction in water (F) is:

$$F = 1/\left[1 + (X_{POC} * \alpha_{POC} * K_{OW}) + (X_{DOC} * \alpha_{DOC} * K_{OW}) + \left(X_{biota} * 0.436 * K_{OW}^{0.907}\right)\right]$$

$$F = \frac{1}{1 + (X_{POC} \times \alpha_{POC} \times K_{ow}) + (X_{DOC} \times \alpha_{DOC} \times K_{ow}) + \left(X_{biota} \times 0.436 \times K_{ow}^{0.907}\right)} \tag{14.53}$$

Where, X_{biota} is concentration of biota in water. Other parameters are described in Table 14.12.

Unlike KABAM, the EXAMS model accounts for decreases in bioavailable chemical compound in the water column due to sorption to biota. The values for α_{DOC} for the two approaches also differ somewhat. Despite these differences, for chemical compounds with Log K_{ow} values ranging between 4 and 8, the fractions of bioavailable chemical compound in the water column estimated by the two approaches differ by less than 0.02 in tested pesticides (Fig. 14.13). Therefore, utilizing estimated environmental concentrations (EECs) of the water column generated by EXAMS remains consistent with results that are generated using the KABAM model [59,70,71].

TABLE 14.11 Derivation of available chemical compound fraction in water (F) by EXAMS used in Eq. (14.52) and associated parameters.

Parameter symbol	Definition	From:	Units
X_{POC}	Concentration of particulate organic carbon in water	User defined	kg L^{-1}
X_{DOC}	Concentration of dissolved organic carbon in water	User defined	kg L^{-1}
K_{OW}	Octanol water partition coefficient	User defined	None
Φ	Fraction of the overlying water concentration of the chemical compound that is freely dissolved and can be absorbed via membrane diffusion	Calculated	None
α_{POC}	Proportionality constant to describe the similarity of phase partitioning of POC in relation to octanol	0.35	None
α_{DOC}	Proportionality constant to describe the similarity of phase partitioning of DOC in relation to octanol	0.08	None

Data from K. Gerber, User's guide and technical documentation KABAM version 1.0 (K OW (Based) Aquatic Bioaccumulation Model), in: Report of Environmental Fate and Effects Division Office of Pesticide Programs, US Environmental Protection Agency, Washington, DC, 2009. U.S. EPA, EXAMS version 2.98.04.06, Washington, DC. https://www.epa.gov/ceam/exams-version-2980406, 2005 (Accessed 22 December 2023).

TABLE 14.12 Parameters used in Eq. (14.53) and associated parameters used in the EXAMS model.

Parameter symbol	Definition	Derived by:	Units
F	Fraction of the overlying water concentration of the chemical compound that is freely dissolved and can be absorbed via membrane diffusion	Calculated	None
K_{ow}	Octanol water partition coefficient	User defined	None
OC_{SS}	Percent organic carbon in suspended sediment	User defined	Percent
X_{SS}	Concentration of suspended sediments in water column	User defined	kg/m^3
X_{POC}	Concentration of particulate organic carbon in water	X_{SS}*OC_{SS}	kg m^{-3}
X_{DOC}	Concentration of dissolved organic carbon in water	User defined	kg m^{-3}
X_{biota}	Concentration of biota in water	User defined	kg m^{-3}
α_{POC}	Proportionality constant to describe the similarity of phase partitioning of POC in relation to octanol	0.35	None
α_{DOC}	Proportionality constant to describe the similarity of phase partitioning of DOC in relation to octanol	0.074	None

Data from K. Gerber, User's guide and technical documentation KABAM version 1.0 (K OW (Based) Aquatic Bioaccumulation Model), in: Report of Environmental Fate and Effects Division Office of Pesticide Programs, US Environmental Protection Agency, Washington, DC, 2009. U.S. EPA, EXAMS version 2.98.04.06, Washington, DC. https://www.epa.gov/ceam/exams-version-2980406, 2005 (Accessed 22 December 2023).

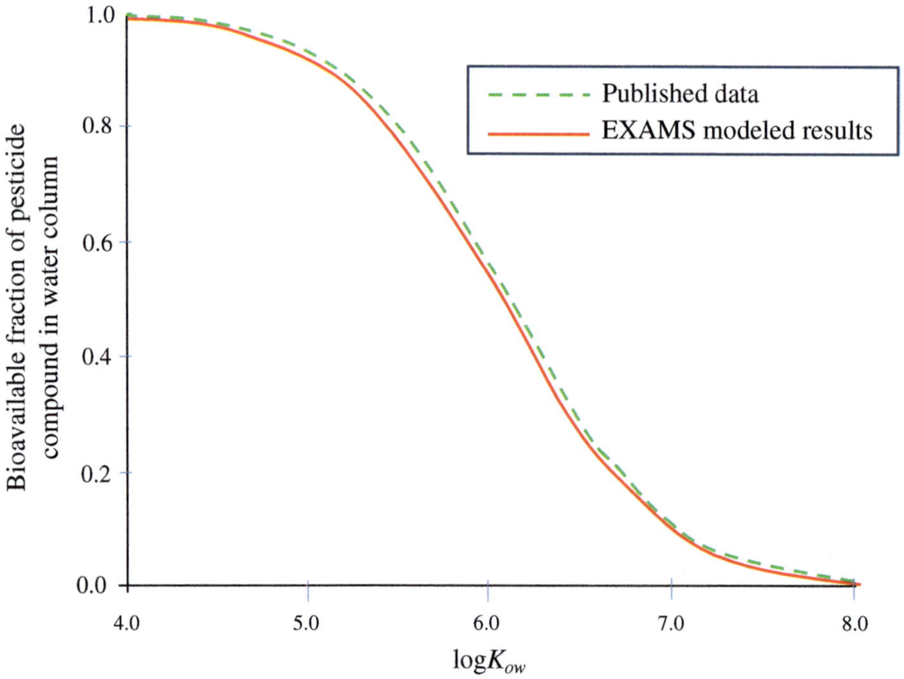

FIG. 14.13 Fraction of bioavailable pesticides in water column estimated using published data and EXAMS model results. *(Data from K. Gerber, User's guide and technical documentation KABAM version 1.0 (K OW (Based) Aquatic Bioaccumulation Model), in: Report of Environmental Fate and Effects Division Office of Pesticide Programs, US Environmental Protection Agency, Washington, DC, 2009.)*

13 Conclusions

Partitioning coefficients are ratios of concentrations between two different phases at equilibrium, such as octanol and water or air and water. They are part of calculations to estimate the fate, transport, exposure, and effects of relatively low concentrations of pollutants. The chapter discussed several of these coefficients, with an emphasis on how they affect the movement and fate of water pollutants.

References

[1] D. Mackay, W.-Y. Shiu, K.-C. Ma, S.C. Lee, Handbook of Physical-Chemical Properties and Environmental Fate for Organic Chemicals, CRC Press, 2006.

[2] S. Hilal, S. Karickhoff, L. Carreira, Prediction of Chemical Reactivity Parameters and Physical Properties of Organic Compounds From Molecular Structure Using SPARC, National Exposure Research Laboratory, Office of Research and Development, US Environmental Protection Agency, Research Triangle Park, NC, 2003.

[3] R. Lewis, S. Gordon, Sampling for Organic Chemicals in Air, Environmental Protection Agency, Research Triangle Park, NC, 1994.

[4] U.S. Environmental Protection Agency, Technical overview of volatile organic compounds, general definition and classification, 2018. https://www.epa.gov/indoor-air-quality-iaq/technical-overview-volatile-organic-compounds. (Accessed 5 June 2018).

[5] T. Boublík, V. Fried, E. Hála, The Vapour Pressures of Pure Substances, Elsevier, Amsterdam, NV, 1984.

[6] D.A. Vallero, Air Pollution Calculations: Quantifying Pollutant Formation, Transport, Transformation, Fate and Risks, Elsevier, 2019.

[7] J.-Y. Chin, S.A. Batterman, VOC composition of current motor vehicle fuels and vapors, and collinearity analyses for receptor modeling, Chemosphere 86 (9) (2012) 951–958.

[8] E. Jurado, R. Lohmann, S. Meijer, K.C. Jones, J. Dachs, Latitudinal and seasonal capacity of the surface oceans as a reservoir of polychlorinated biphenyls, Environ. Pollut. 128 (1) (2004) 149–162.

[9] O. Faroon, J.N. Olson, Toxicological Profile for Polychlorinated Biphenyls (PCBs), 2000.

[10] U.S. EPA, Ambient water quality criteria for polychlorinated biphenyls, 1980, EPA 440/5-80-068, vol. October 1980. Available: https://nepis.epa.gov/Exe/ZyPDF.cgi/00001M1G.PDF?Dockey=00001M1G.PDF. (Accessed 28 December 2023).

[11] R. Gioia, et al., Polychlorinated biphenyls in air and water of the North Atlantic and Arctic Ocean, J. Geophys. Res. Atmos. 113 (D19) (2008).

[12] Y. Kwon, Handbook of Essential Pharmacokinetics, Pharmacodynamics and Drug Metabolism for Industrial Scientists, Springer Science & Business Media, 2001.

[13] A.D. McNaught, A.D. McNaught, Compendium of Chemical Terminology, Blackwell Science Oxford, 1997.

[14] D. Mackay, S. Paterson, Evaluating the multimedia fate of organic chemicals: a level III fugacity model, Environ. Sci. Technol. 25 (3) (1991) 427–436.

[15] H.F. Hemond, E.J. Fechner, Chemical Fate and Transport in the Environment, Elsevier, 2014.

[16] L.P. Burkhard, Evaluation of published bioconcentration factor (BCF) and bioaccumulation factor (BAF) data for per-and polyfluoroalkyl substances across aquatic species, Environ. Toxicol. Chem. 40 (6) (2021) 1530–1543.

[17] U.S. Environmental Protection Agency, Chemistry dashboard, 2017. https://www.epa.gov/chemical-research/chemistry-dashboard. (Accessed 21 April 2017).

[18] European Chemicals Agency, Dichloromethane, 2011. https://echa.europa.eu/registration-dossier/-/registered-dossier/15182/2/3. (Accessed 11 July 2024).

[19] P. Davison, The specific gravity of mesopelagic fish from the northeastern Pacific Ocean and its implications for acoustic backscatter, ICES J. Mar. Sci. 68 (10) (2011) 2064–2074, https://doi.org/10.1093/icesjms/fsr140.

[20] D.A. Vallero, Fundamentals of Air Pollution, fifth ed., Elsevier Academic Press, Waltham, MA, 2014. p. 999 pages cm.

[21] X. Kong, F. Xu, W. He, W. Liu, B. Yang, Fugacity models, in: Ecological Model Types, 2016, pp. 181–210.

[22] D. Mackay, L. Burns, G. Rand, Fate modeling, in: Fundamentals of Aquatic Toxicology: Effects, Environmental Fate, and Risk Assessment, Taylor & Francis, Philadelphia, Pennsylvania, 1995, pp. 563–585.

[23] Y.-F. Li, et al., Treatment of particle/gas partitioning using level III fugacity models in a six-compartment system, Chemosphere 271 (2021) 129580.

[24] D. Mackay, J.A. Arnot, E. Webster, L. Reid, The evolution and future of environmental fugacity models, in: Ecotoxicology Modeling, 2009, pp. 355–375.

[25] D. Mackay, S. Patterson, Fugacity models, in: Practical Applications of Quantitative Structure-Activity Relations in Environmental Chemistry and Toxicology, 1990, pp. 433–460.

[26] S. Wang, Q. Wang, Z. Yuan, X. Wu, Application of the multimedia fugacity model in predicting the environmental behaviors of PCBs: based on field measurements and level III fugacity model simulation, Ecotoxicol. Environ. Saf. 263 (2023) 115286.

[27] C. Mesaros, J.S. Arora, A. Wholer, A. Vachani, I.A. Blair, 8-Oxo-2′-deoxyguanosine as a biomarker of tobacco-smoking-induced oxidative stress, Free Radic. Biol. Med. 53 (3) (2012) 610–617.

[28] O. Ghosheh, E.M. Hawes, N-glucuronidation of nicotine and cotinine in human: formation of cotinine glucuronide in liver microsomes and lack of catalysis by 10 examined UDP-glucuronosyltransferases, Drug Metab. Dispos. 30 (9) (2002) 991–996.

[29] N.L. Benowitz, Cotinine as a biomarker of environmental tobacco smoke exposure, Epidemiol. Rev. 18 (2) (1996) 188–204.

[30] M. Leventer-Roberts, et al., Environmental tobacco smoke exposure among children by urinary biomarkers and parent report, Acad. Pediatr. 21 (4) (2021) 663–669.

[31] N.L. Benowitz, Biomarkers of environmental tobacco smoke exposure, Environ. Health Perspect. 107 (Suppl. 2) (1999) 349–355.

[32] W. Lyman, Transport and transformation processes, in: Rand GM Fundamentals of Aquatic Toxicology. Effects, Environmental Fate, and Risk Assessment, second ed., Taylor & Francis, Washington, DC, 1995, pp. 449–492.

[33] D. Vallero, Environmental Biotechnology: A Biosystems Approach, Elsevier Science, 2015.

[34] D. Richter, in: D.A. Vallero (Ed.), Ion exchange, Duke University, Durham, North Carolina, 1996.

[35] J. Westfall, Adsorption mechanisms in aquatic surface chemistry, in: W. Stumm (Ed.), Aquatic Surface Chemistry: Chemical Processes at the Particle-Water Interface, vol. 87, John Wiley & Sons, 1987.

[36] C. Zhao, Y. Dong, Y. Feng, Y. Li, Y. Dong, Thermal desorption for remediation of contaminated soil: a review, Chemosphere 221 (2019) 841–855.

[37] EPA, Guide for conducting treatability studies under CERCLA: thermal desorption, 1992. EPA/540/R-92/074 B.

[38] M. Richter, E. Juritsch, O. Jann, Determination of recovery rates of adsorbents for sampling very volatile organic compounds (C1C6) in dry and humid air in the sub-ppb range by use of thermal desorption gas chromatography-mass spectrometry, J. Chromatogr. A 1626 (2020) 461389, https://doi.org/10.1016/j.chroma.2020.461389.

[39] F. Gharibzadeh, R.R. Kalantary, A. Esrafili, M. Ravanipour, A. Azari, Desorption kinetics and isotherms of phenanthrene from contaminated soil, J. Environ. Health Sci. Eng. 17 (2019) 171–181.

[40] P.P. Falciglia, et al., Remediation of petrol hydrocarbon-contaminated marine sediments by thermal desorption, Chemosphere 260 (2020) 127576.

[41] U.S. Environmental Protection Agency, Part 5: Technical background document, US EPA (Ed.), in: Soil Screening Guidance, vol. EPA/540/R-95/128, US EPA, Washington, DC, 1996.

[42] M.N. Sara, Ground-water monitoring system design, in: D.M. Nielsen (Ed.), Practical Handbook of Ground-Water Monitoring, Lewis Publishers, USA, 1991, pp. 17–68.

[43] D. Mackay, W. Shiu, K. Ma, Physical-Chemical Properties and Environmental Fate and Degradation Handbook. CRCnet-BASE 1999, Chapman & Hall CRCnetBASE, CRC Press LLC, Boca Raton, FL, 1999.

[44] T. Harner, M. Shoeib, Measurements of octanol–air partition coefficients (K OA) for polybrominated diphenyl ethers (PBDEs): predicting partitioning in the environment, J. Chem. Eng. Data 47 (2) (2002) 228–232.

[45] U.S. Environmental Protection Agency, Guidance for reporting on the environmental fate and transport of the stressors of concern in problem formulations, 2024. https://www.epa.gov/pesticide-science-and-assessing-pesticide-risks/guidance-reporting-environmental-fate-and-transport. (Accessed 11 July 2024. Last updated on February 13, 2024.

[46] U.S. Environmental Protection Agency, EPI Suite™-Estimation Program Interface, US EPA, Washington, DC, 2023. In this issue https://www.epa.gov/tsca-screening-tools/epi-suitetm-estimation-program-interface. (Accessed 11 July 2024. Last update on December 27, 2023.

[47] A. Berhanu, I. Mutanda, J. Taolin, M.A. Qaria, B. Yang, D. Zhu, A review of microbial degradation of per-and polyfluoroalkyl substances (PFAS): biotransformation routes and enzymes, Sci. Total Environ. (2022) 160010.

[48] C. Daccò, C. Girometta, M. Asemoloye, G. Carpani, A. Picco, S. Tosi, Key fungal degradation patterns, enzymes and their applications for the removal of aliphatic hydrocarbons in polluted soils: a review, Int. Biodeterior. Biodegradation 147 (2020) 104866.

[49] J.H. Makoi, P.A. Ndakidemi, Selected soil enzymes: examples of their potential roles in the ecosystem, Afr. J. Biotechnol. 7 (3) (2008).

[50] A.R. Othman, H.A. Hasan, M.H. Muhamad, N.I. Ismail, S.R.S. Abdullah, Microbial degradation of microplastics by enzymatic processes: a review, Environ. Chem. Lett. 19 (2021) 3057–3073.

[51] M. Bhateria, K. Karsauliya, A.K. Sonker, C. Yahavi, S.P. Singh, Cytochrome P450 isoforms contribution, plasma protein binding, toxicokinetics of enniatin A in rats and in vivo clearance prediction in humans, Food Chem. Toxicol. 164 (2022) 112988.

[52] R. Muñoz, S. Villaverde, B. Guieysse, S. Revah, Two-phase partitioning bioreactors for treatment of volatile organic compounds, Biotechnol. Adv. 25 (4) (2007) 410–422.

[53] S. Arriaga, R. Muñoz, S. Hernández, B. Guieysse, S. Revah, Gaseous hexane biodegradation by Fusarium solani in two liquid phase packed-bed and stirred-tank bioreactors, Environ. Sci. Technol. 40 (7) (2006) 2390–2395.

[54] OECD, Test No. 305: Bioconcentration: flow-through fish test, OECD, Paris, France, 1996.

[55] OECD, Test No. 315: Bioaccumulation in sediment-dwelling benthic oligochaetes, OECD, Paris, France, 2008.

[56] OECD, Test No. 317: Bioaccumulation in terrestrial oligochaetes, OECD, Paris, France, 2008.

[57] G. Rand, P. Wells, L. McCarty, Introduction to aquatic toxicology, in: Fundamentals of Aquatic Toxicology, CRC Press, 2020, pp. 3–67.

[58] A. Spacie, J. Hamelink, Appendix D, bioaccumulation, in: G.M. Rand (Ed.), Fundamentals of Aquatic Toxicology, Taylor & Francis, Washington, DC, 1995, pp. 1052–1082.

[59] K. Gerber, User's guide and technical documentation KABAM version 1.0 (K OW (Based) Aquatic Bioaccumulation Model), in: Report of Environmental Fate and Effects Division Office of Pesticide Programs, US Environmental Protection Agency, Washington, DC, 2009.

[60] B. Anderson, et al., White paper on methods for assessing ecological risk of pesticides with persistent, bioaccumulative and toxic characteristics, Office of Prevention, Pesticides and Toxic Substances, Office of Pesticide Programs, US EPA, Washington, DC, 2008, p. 221.

[61] Organisation for Economic Cooperation and Development, Guidance document on the validation of (Q)SAR models, in: Environmental Health and Safety Publications. Series on Testing and Assessment, vol. 69, 2007, p. 154. Available: https://www.oecd-ilibrary.org/docserver/9789264085442-en.pdf?expires=1703080750&id=id&accname=guest&checksum=6942D92F1AB6EA11D5BAAD9510977805.

[62] ChemSafetyPro, Differences between BCF, BAF and BMF, 2017. https://www.chemsafetypro.com/Topics/CRA/definition_differences_BCF_BAF_and_BMF.html. (Accessed 11 July 2024. Published April 25, 2017.

[63] R.P. Schwarzenbach, P.M. Gschwend, D.M. Imboden, Environmental Organic Chemistry: Illustrative Examples, Problems, and Case Studies, John Wiley & Sons, Inc., New York, New York, 1995.

[64] G.D. Veith, D.L. DeFoe, B.V. Bergstedt, Measuring and estimating the bioconcentration factor of chemicals in fish, J. Fish. Res. Board Can. 36 (9) (1979) 1040–1048.

[65] G.D. Veith, N. Austin, R. Morris, A rapid method for estimating log P for organic chemicals, Water Res. 13 (1) (1979) 43–47.

[66] ChemSafetyPro, n-Octanol/water partition coefficient (Kow/logKow), 2016. https://www.chemsafetypro.com/Topics/CRA/n_Octanol_Water_Partition_Coefficient_Kow.html. (Accessed 11 July 2024. Published January 13, 2016.

[67] A. Finizio, M. Vighi, D. Sandroni, Determination of n-octanol/water partition coefficient (Kow) of pesticide critical review and comparison of methods, Chemosphere 34 (1) (1997) 131–161.

[68] C.T. Chiou, Partition coefficients of organic compounds in lipid-water systems and correlations with fish bioconcentration factors, Environ. Sci. Technol. 19 (1) (1985) 57–62.

[69] R. Schwarzenbach, P. Gschwend, D. Imboden, Environmental Organic Chemistry, John Wiley & Sons, Hoboken, New Jersey, 2016.

[70] J.A. Arnot, F.A. Gobas, A food web bioaccumulation model for organic chemicals in aquatic ecosystems, Environ. Toxicol. Chem. 23 (10) (2004) 2343–2355.

[71] U.S. EPA, EXAMS version 2.98.04.06, Release Date: 2005, Washington, DC https://www.epa.gov/ceam/exams-version-2980406. (Accessed 11 July 2024. Last update on February 22, 2024.

Chapter 15

Water pollution engineering

1 Introduction

The best means of addressing pollution of any type is not to have it in the first place, i.e., prevent its existence through waste minimization and prevention approaches. These continue to improve but, unfortunately, pollution control technologies remain as the predominant methods for the removal and decontamination of water pollutants and their precursor substances. To do this, the wastewater or other polluted carrier fluid is expected to pass through a control device or system of technologies that collects or destroys the pollutant and releases the cleaned water to the land or water body, reduces the toxicity of sludge, and returns treated contaminated sediment and soil back to the land or renders the residues a valuable commodity. These are *ex situ* approaches, but remediation can also be *in situ*, e.g., removing contaminants from soil, sediment, or ground water at the pollution site.

The type of control is specific to the environmental media and the pollutant needing removal and treatment. As the water or other carrier fluid exits the treatment process, concentrated quantities have been collected in sludge, and slurries, on filters and traps, and in tanks. These residues must also be handled in environmentally acceptable ways. After the control technologies are constructed or installed, operation and maintenance (O&M) begins and continues through the system's operational life to meet effluent and release regulations, to prolong equipment life, to maintain efficiency, and to optimize pollution control variables against costs and feasibility factors [1].

The O&M program consist of: (1) an equipment and record-keeping and documentation system with equipment information, warranties, instruction manuals, and other operational documents and software[a]; (2) lubrication and cleaning schedules; (3) planning and scheduling of preventive maintenance; (4) storeroom and inventory system for spare parts and supplies; (5) listing of maintenance personnel; (6) costs and budgets for O&M; and (7) storage of special tools and equipment.

Residences, industries, and commercial operations use water, which adds waste that must be removed. Collection systems carry the wastewater from these locations by gravity and pumps to wastewater treatment plants.

2 Wastewater

Wastewater treatment, like all environmental treatment processes, is based on the principles of physics, chemistry and biology, i.e., biochemodynamics [2]. The type and sophistication of wastewater treatment needed depend on the content of the influent, i.e., the polluted water needing treatment, and the desired water quality after treatment. Treatment parameters include the removal of solids, biochemical oxygen demand (BOD), fecal coliform, and xenobiotic chemicals. Wastewater's hazards can be direct, owing to its biological or chemical components, e.g., pathogenic bacteria and heavy meals, respectively. Indirect hazards also occur, such as the amount and types of suspended and dissolved solids and organic substances. These may induce eutrophication of lakes and estuaries and/or lower receiving water's dissolved oxygen (DO) concentrations by enhancing the growth of microbes. The same contaminant may be both directly and indirectly hazardous, e.g., ammonia that kills fish directly or serves as a source of nitrogen that induces microbial growth, leading to eutrophication.

The solids content is expected to be measured to meet national, provincial, tribal, state, and local permit requirements, as well as operational performance standards, e.g., to prevent fouling and disruptions in the reliability of commercial and manufacturing systems [3–8]. Solids concentrations in effluent also indicate the efficiency of process controls. For example, the percent total solids (TS) indicates the amount of dewatering of sludge that is needed before efficient handling. A governmental effluent permit usually specifies the allowable percent of total suspended solids (TSS) that

a. This documentation system is also useful following system failures as data for root cause analyses. See Chapters 17 and 18.

Fundamentals of Water Pollution. https://doi.org/10.1016/B978-0-443-28987-3.00007-2

may be discharged to a stream. A plant engineer or operator is expected to measure the percent of total volatile solids (TVS) and total volatile suspended solids (TVSS) to indicate operation efficiency and effectiveness of a digester or trickling filter and to indicate proper food-to-mass (F:M) ratio loading rates for activated sludge treatment [9].

Total solids include, in descending size order, dissolved, suspended or settleable solids. Solids content in wastewater is ascertained by evaporating the water and weighing the residue. In the case of TS, a known volume of the unfiltered water is heated at 103–105 °C for 1 h [10]. The TS is then be filtered and evaporated to measure the total dissolved solids (TDS) and total suspended solids (TSS), i.e., the solids able to pass through the filter are TDS, and the solids held in and on the filter are TSS. The fixed solids remain after combusting the residue from TS. Volatile solids are those that are driven off during combustion. Oven and ignition temperatures, filter pore size, flow rates, and other treatment conditions are stipulated by governmental methods, e.g., U.S. Environmental Protection Agency Method 1684 [10], or standards handbooks, especially *Standard Methods for the Examination of Water and Wastewater* [11].

The removal of solids is mainly by physical processes, especially gravity and buoyancy. Chemical processes also are often used to enhance solids removal, such as adding chemical compounds that encourage finer, lighter particles and colloids to aggregate and precipitate. Finally, biological processes are involved in solids removal. As mentioned, secondary treatment involves microbial degradation of organic matter. Activated sludge, aerated lagoons, and biofilters are among the systems designed to achieve solids removal.

Finally, the solids become sludge, which is a mixture of abiotic and biological materials. In activated sludge systems, a defined mass of the sludge is returned to the treatment process and mixed with air. This causes the bacteria or other microbes to become activated, i.e., they grow and reproduce much faster as they consume the wastes at a higher rate. The remaining sludge must be dewatered, followed by thermal treatment, land application, and/or chemical alteration to manufacture commercial fertilizers. Biosolids and sludge are often considered to be synonymous, but they are distinguished in that biosolids are the nutrient-rich fraction of sludge that can be applied to land. Environmental agencies require removal and disinfection of pathogens and other pollution limits prior to allowing biosolids to be used [12–16].

Microbes can be beneficial or pathogenic. For example, nitrogen, phosphorus, and other nutrients encourage the growth of algae, leading to algal blooms. This is due to the increasing decomposition rates of organic matter so that more oxygen is demanded by the bacteria and abiotic processes, so the concentrations of DO decline. As a result, fish, and other organisms at higher trophic levels in the food chain are stressed. This can result in fish kills and complete alterations to aquatic habitats. In addition, pathogenic microbes may be suspended or sorbed to suspended particles in the wastewater. Coliform like *Escherichia coli* and other microbes can reach water supplies, where they have caused major human health problems.

As aerobic bacteria degrade organic compounds, nutrients, and other wastewater constituents, these aerobes demand oxygen. Wastewater microbes vary in their optimal ranges of growth and metabolism. Like solids content, the biochemical oxygen demand (BOD) is common indicator of treatment efficiency and water quality. BOD is the amount of oxygen is consumed with time. Most of the BOD is attributable to decomposing organic matter under aerobic conditions, but a fraction of the demand is abiotic, i.e., chemical reactions that do not involve microbes. Bacteria and other microbes will continue to demand O_2 from the water to degrade other compounds, especially nitrogenous compounds, which cause the second increase in the BOD curve (See Fig. 10.12 in Chapter 10) [17]. BOD, then, indicates the amount of organic matter to be removed by a treatment process, as well as an indication of the required removal efficiency.

Chemical oxygen demand (COD) is more inclusive than BOD, i.e., both biologically available and inert organic matter comprise COD. It represents the total concentration O_2 depleted to oxidize all organic material completely to carbon dioxide (CO2) and water. COD always exceeds BOD for the same sample [2]. The method for determining COD in water is by reacting all organic and oxidizable inorganic substances with a solution potassium dichromate ($K_2Cr_2O_7$) and sulfuric acid (H_2SO_4), 50% each by volume, followed by titration of the excess dichromate with ferrous ammonium sulfate [$(NH_4)Fe(SO_4)6H_2O$; also known as ferroin] as an indicator [18,19].

When elevated BOD effluents reach surface waters, they lessen the water's DO concentrations, potentially harming sensitive aquatic biota. By mixing with the air above a waterbody and by algal photosynthesis, O_2 concentrations increase in the water downstream. The decreasing and increasing concentrations in dissolved oxygen, i.e., initial demand just downstream from the high-BOD effluent point and the reaeration with distance creates a sag DO concentrations downstream from the effluent source, e.g., a storm or sanitary sewer (See Fig. 12.1 in Chapter 12).

The amount and chemical forms of nutrients in water affect aquatic biota in the food chain. For example, nutrients in rainwater reach a wetland. Plant roots take them up and metabolize nitrates, phosphates, and other oxidized forms of nutrients into organic forms. Simultaneously, microbes in sediments reduce the nitrogen compounds to ammonia. Conversely, other bacterial species are oxidizing these some of the reduced compounds, such as *Nitrosomonas* converting ammonium ions [NH_4^+] to nitrite [NO_2^-], and *Nitrobacter* converting NO_2^- to nitrate [NO_3^-]. Two opposite reactions occur constantly with nutrient loading [2]. During ammonification, amino groups ($-NH_2$) in organic forms of nitrogen are convert to NH_4^+ and then to ammonia (NH_3):

$$\text{Ammonification}: NH_4^+ + OH^- \Leftrightarrow NH_3 + H_2O \tag{15.1}$$

Conversely, during nitrification, NH_4^+ is oxidized to NO_2^- or NO_3^-:

$$\text{Nitrification}: NH_4^+ + 2O_2 \rightarrow NO_3^- + 2H^+ + H_2O \tag{15.2}$$

Ammonification is an equilibrium reaction, and nitrification is a kinetic oxidation reaction. Pollutant loading is always complicated by environmental conditions, including pH and temperature of the receiving waters. Under acidic conditions, for example, much of the ammonia nitrogen will ionize to ammonium but under basic conditions the nonionized ammonia concentrations increase proportionately with the ammonium concentration. Nonionized ammonia is toxic to aquatic biota, whereas ionized forms are needed for plants and algae.

The decomposition of organic matter, and the loading becomes a complicated mix of reactions [2]:

$$\text{Organic nitrogen} + O_2 \rightarrow NH_3 \text{ nitrogen} + O_2 \rightarrow NO_2 \text{ nitrogen} + O_2 \rightarrow NO_3 \text{ nitrogen} + O_2 \tag{15.3}$$

Oxidation and reduction rates depend on the genera of microbes and physicochemical properties of the water. Theoretically, the O_2 required for nitrification is 4.56 mg O_2 mg^{-2} NH_4^+ if by an autotrophic reaction. The O_2 is produced by the nitrifying bacteria, which decreases the amount of O_2 needed. However, the nitrifying microbial population growth rate is substantially less than that of heterotrophic microbial populations' rate of decomposing organic wastes. Thus, large amounts of organic matter being degraded means that the nitrifiers' growth rate will be severely limited by the heterotrophs. The rate of nitrification will commensurately be decreased [20]. Such conditions exist for all other nutrients in the system, e.g., organic phosphorus (P) or potassium (K) will be oxidized while oxidized P and K compounds will be reduced [2].

Wastewater is first treated by physical processes to remove solids. Total solids (TS) consist of dissolved solids and suspended solids. The smallest and lightest suspended solids are colloidal solids. Coarser, denser suspended solids are the settleable solids. Floatable solids are the least dense and most lipophilic of the solids, i.e., floatable solids do not settle during the wastewater's residence time in the primary treatment tank. The waste components in wastewater can be either organic, which are degraded by microbes that deplete the water's DO, or inorganic which consist of salts, metals, grit, and sand.

2.1 Treatment

As untreated wastewater first reaches a treatment plant, physical processes begin, including size separation, followed by settling, filtration, diffusion, osmosis, precipitation, coagulation, ion exchange, and sorption [21]. Physical processes continue throughout the treatment process but are subsequently joined by chemical and biological processes during the secondary and advanced treatment stages. Fig. 15.1 shows the stages of a wastewater treatment system. The earliest stages of wastewater treatment are mainly physical, i.e., removing solids, such as by settling. The next stages include physics, but also chemistry and biology, to transform the wastewater into better quality water [2].

Preliminary treatment begins with size separation using screens to remove the coarse fraction of suspended materials from the wastewater stream. Preliminary treatment often begins with a comminutor to shred the largest debris, e.g., rags, fibrous material like face masks and tissues.[b] Preliminary treatment often occurs in an aerated grit chamber through which wastewater flow at about 0.3 m s^{-1} to remove grit, i.e., dense and usually inorganic material. The aeration also removes a portion of the floatable solids. This is followed by hydrodynamic separation to remove particles from the wastewater and to separate wastewater constituents by density.

2.1.1 *Primary treatment*

Next, sedimentation removes the denser solids, i.e., primary clarification removes a large fraction of the screened materials in the carrier water entering a large tank, where fine particles lose their competence as water's velocity decreases in the tank. By gravitational acceleration (g), particles settle to the tank bottom, forming the beginning of sludge. Simultaneously, another hydrodynamic force, i.e., buoyance force (F_b), is competing with g. That is, F_b causes the lower density particles

b. One of the secondary problems posed by the COVID-19 epidemic was the dramatic increase in the improper disposal of protective gloves, masks, and disinfecting wipes into wastewater systems, rather than solid waste facilities. They have been entering sewer lines and reaching wastewater treatment plants. This increases the amount of material needing preliminary treatment and additional microplastics entering the treatment system, passing through, and exiting in the effluent [22].

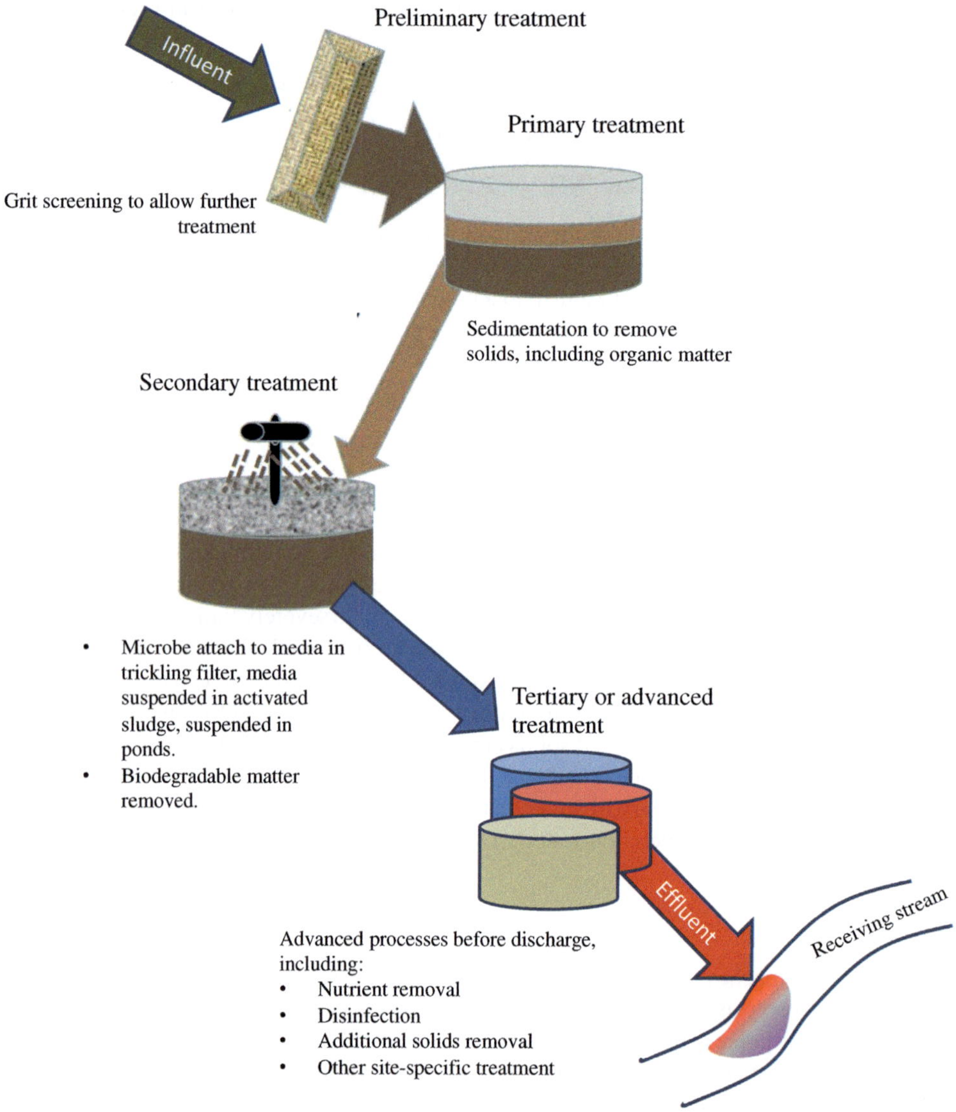

FIG. 15.1 Prototypical wastewater treatment flowchart. *(Data from D.A. Vallero, Wastewater, in: Waste, Elsevier, 2019, pp. 259–290.)*

and fluids like grease and oil float to the tank water surface to form scum. Thus, decreasing density (ρ) increases the buoyance force (F_b) which is also indirectly proportional to fluid volume (V_f) and gravitational acceleration (g):

$$F_b = \rho g \, V_f \tag{15.4}$$

Wastewater in the primary clarifier tank is predominantly water, but also contains other fluids, making it is a dilute fluid mixture. Thus, V_f differs slightly in composition from pure water. The specific gravity of wastewater (SG_{ww}), i.e., the ratio of wastewater density (ρ_{ww}) to water (ρ_w) generally is assumed to be nearly 1, but will increase with solids concentrations:

$$SG_{ww} = \frac{\rho_{ww}}{\rho_w} \approx 1 \tag{15.5}$$

Primary treatment with the typical, well-designed clarifier is expected to remove up to 99% of the influent's settleable solids, 60% of the suspended solids, and 15% of the total solids. Some of the solids have oxygen demand, so the clarifier can remove up to half of the biochemical oxygen demand (BOD) and up to 75% of the bacteria load [23]. The design of the clarifying tank is based on detention time (t_D), surface overflow rate (SOR), weir overflow rate (WOR), and solids loading rate (SLR) [24]. These design criteria are calculated as:

$$t_D = \frac{V_{tank}}{Q_{in}} \tag{15.6}$$

Where, t_D is detention time in hours (h), V_{tank} is the tank volume per time, which in the U.S., often expressed in units of million gallons per day (MGD), i.e., 10^6 gal × 24 h (or units of L × 24 h) and Q_{in} is the flow in units of volume the tank per time (e.g., MGD or L t^{-1}).

Hydrologic loading is represented by t_D. Clarifier tank designs are based on an ideal range where t_D values fall between 2 and 3 h. In other words, a well-designed clarifying allows a drop of water to move from the tank inlet to the tank outlet in 2–3 h, giving the optimal number of particles in its load enough time to either float to the surface or sink to the bottom [24].

For example, consider the city engineer who wants to base a new wastewater treatment plant design to handle a flow rate that has not exceeded 300,000 gal day^{-1} since the city has kept records. She decides to base the design on the t_D in a clarifying tank with a volume of 20,000 gal (90,922 L), receiving an inflow of wastewater of 350,000-gal day^{-1} (1,591,132 L day^{-1}). To decide whether is this an acceptable design specification, t_D is simply calculated as: $t_D = \frac{2\times10^4\,\text{gal}}{3.5\times10^5\,\text{gal day}^{-1}} \approx 0.06$ days or about 1 h and 22 min, which is well under the design criteria for detention time. Thus, if this were the only criteria, the flow rate (Q_{in}) could be doubled. The other consideration is whether the 50,000-gal day^{-1} factor of safety is sufficient.

This requires engineering judgment on how reliable the data are, how much development has occurred, and the extent to which trends in demographics and development are still represented by the data. For example, if the data show that the inflow is upward from 225,000 gal in the year 1990, to 250,000 in 2000, to 275,000 in 2010, to 300,000 in the latest data for the year 2020, a simple linear extrapolation would mean that the 350,000-gal day^{-1} tank capacity would be exceeded in 3 years. However, such a straight-line projection is an oversimplification without considering a more complete optimization that include per person usage rates, attention to upward and downward inflow rates, land use and zoning changes, and other demographic, socioeconomic, and population mobility variables. For example, mobility trends have changed substantially for many urban areas since the onset of the COVID-19 pandemic, with some smaller towns receiving large influxes of permanent and temporary residents moving from larger urban areas [25,26]. Future mobility is difficult to gauge given this new trend, but wastewater, water supply, and other infrastructural considerations must be factored into treatment capacities.

The detention time is expected to be sufficient to allow the solids to settle to the bottom of tank (see Fig. 15.2). This means that tank geometry is crucial. For example, the settling time shown in Fig. 15.2 requires a flow rate sufficiently slow and a tank length sufficiently long to allow for the solids vertical settling rate 8.5 $\times 10^{-4}$ m s^{-1}. In this case the length must be greater than 61 m, plus ample additional length to handle variations in the solids content of the influent. An adequate design optimizes sedimentation efficiency in terms of percent solids removed per hour. Insufficient time translates into less efficient solids removal, but too much time due to an overly long tank configuration means that there is little or no incremental increase in sedimentation efficiency, i.e., little or no increase in the absolute removal of solids [24].

The next design consideration is the surface overflow rate:

$$\text{SOR} = \frac{Q_{in}}{A_s} \tag{15.7}$$

Where, SOR units are volume per day per area (e.g., gal day^{-1} ft^{-2} or L day^{-1} m^{-2}), Q_{in} units are volume per day (e.g., gal day^{-1} or L day^{-2}), and surface area of the tank (A_s) units are length squared (e.g., ft^2 or m^2).

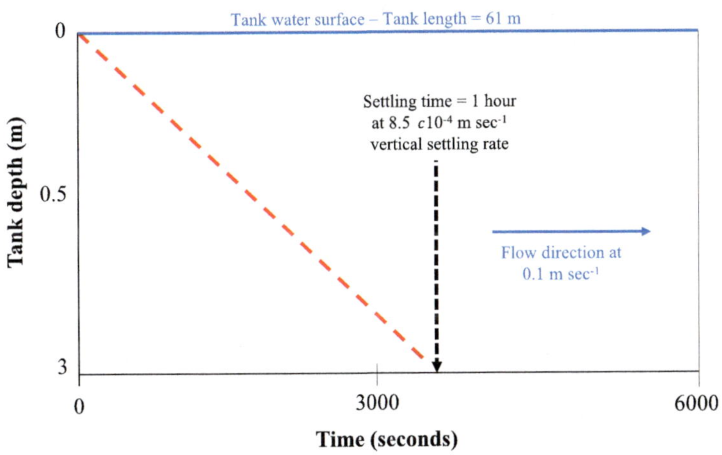

FIG. 15.2 Importance of tank geometry in primary wastewater treatment. The clarifying tank's flow rate, flow direction, tank length, and tank depth in this rectangular tank must be adequate to all for the solids to settle before reaching the tank outflow. *(Data from Michigan Department of Environmental Quality Operator Training and Certification Unit, Clarifier calculations. Available: https://www.michigan.gov/documents/deq/wrd-ot-clarifier-calculations_445211_7.ppt, 2022.)*

Wastewater treatment plants usually have SOR values between 400 and 700 gal day^{-1} ft^{-2} (16298–32,597 L day^{-1} - m^{-12}), based on mean flow conditions and 800–1600 gal day^{-1} ft^{-2}, based on peak flow [27]. Indeed, the size and geometry of the clarifying tank is largely based on the peak SOR.

Surface overflow rate example

Calculate the SOR for tank that has a surface area is 50 ft ×25 ft and the inflow to the tank is 350,000-gal day^{-1}. Decide if it is properly designed.

Using Eq. (15.7), the SOR is calculated as:

$$SOR = \frac{3.5 \times 10^5 \text{gal day}^{-1}}{(50 \times 25)\ \text{ft}^2} = 280\ \text{gal day}^{-1}\ \text{ft}^{-2}$$

Thus, the plant is possibly overdesigned, but may be appropriate depending on the trends in the amount of wastewater expected to be treated and any recent changes in population characteristics.

The weir overflow rate is another design criterion for primary treatment:

$$WOR = \frac{Q_{in}}{L_w} \tag{15.8}$$

Where, WOR units are volume per day per length of weir (e.g., gal day^{-1} ft^{-1}), Q_{in} units are volume per day (e.g., gal day^{-1}), and length of the weir (L_w) is in length units (e.g., ft or m).

The WOR specification is intended to decrease the potential for excessive velocities that cause short circuits within the tank The typical weir design ranges from about 10,000-gal day^{-1} ft^{-1} (124,193 L day^{-1} m^{-1}) to about twice that rate. Often, however, regulatory guidelines recommend that WOR not exceed 15,000 gal day^{-1} ft^{1} [28].

Another primary treatment design specification is the solids loading rate:

$$SLR = \frac{M_s \times t}{A_s} \tag{15.9}$$

Where, SLR units are mass per day per area (e.g., gal day^{-1} ft^{-2} or L day^{-1} m^{-2}), M_s is the mass of soils being loaded into the tank in units are mass per day (e.g., lbs or kg day^{-1}), and the surface area of the tank (A_s) units are length squared (e.g., ft^2 or m^2). SLR and maximum SOR are indicators of the solids content of the outflow from the primary clarifier to a secondary clarifier (for more solids removal) or to the secondary treatment processes. General design parameters for the final clarifier vary by the outflow expected in the next treatment component of the wastewater treatment plant on shown in Table 15.1.

TABLE 15.1 Final clarifier design parameters.

	Maximum surface overflow rate (gallons per day per square foot)		Solids loading rate (pound-mass per day per square foot)	
Process type	Mean design flow	Peak design flow	Mean design flow	Peak design flow
Trickling filter	600	1200	25	40
Activated sludge	800 (600 for plants receiving less than 1 MGD)	1200	30	50
Extended aeration	400	1000	25	35
Nitrification	400	800	25	35
Pure O_2	700	1200	25	40

Notes: MGD = million gallons per day. The surface overflow rate for final clarifiers in an activated sludge system must be based on influent wastewater flows and exclude all return activated sludge (RAS) flow. Solids loading rates assume that the sludge recycle rate is 100% of the mean design flow and the design mixed liquor suspended solids (MLSS).

Data from Tennessee Department of Environment and Conservation, Division of Water Resources, Design criteria for review of sewage works—construction plans and documents. DWR-NPDES-SOP-G-05-WW. Available: https://www.tn.gov/content/dam/tn/environment/water/policy-and-guidance/dwr-npdes-sop-g-05-ww-design-criteria-table-of-contents-072020.pdf, 2020.

2.1.2 Secondary treatment

After clarifying is secondary treatment, typically a type of bioreactor, e.g., activated sludge system or a trickling filter. Often, these steps are followed by another, i.e., tertiary treatment, which may include disinfection, nutrient removal, or additional solids removal beyond the sedimentation step [29].

Both primary and secondary treatment decrease the organic load of the wastewater, as expressed by BOD. However, primary treatment removes comparatively less organic matter. Indeed, microbial degradation in secondary treatment accounts for most of the BOD removal. This is accomplished by free energy, i.e., the measure of a system's ability to do work [17]. Reactants that have greater free energy than the products in a chemical reaction indicates that the energy is released from the reaction; i.e., reaction is exergonic. Conversely, if the products from the reaction have more energy than the reactants, energy is consumed; i.e., it is an endergonic reaction. Recall that Gibbs free energy (*G*) change is the thermodynamic basis for equilibrium constants in a complete reaction; expressed as:

$$G = H - TS \tag{15.10}$$

where *G* is the energy liberated or absorbed in the equilibrium by the reaction at constant *T*. *H* is the system's enthalpy and *S* is its entropy. Enthalpy is the thermodynamic property expressed as:

$$H = U + pV \tag{15.11}$$

where *U* is the system's internal energy, *p* is the pressure, and *V* is the volume of the system.[c]

The relationship between a change in free energy and equilibria can be expressed by:

$$\Delta G^{*} = \Delta G_{f}^{*0} + RT \ln K_{eq} \tag{15.12}$$

where ΔG_f^{*0} = free energy of formation at steady state (kJ g mol^{-1}).

Thus, enthalpy (*H*) describes the total energy in a system and the free energy (G) describes the usable energy. Cells need G for all biochemical reactions, especially growth, division, metabolism, and survival. Entropy (*S*) describes the unusable energy, i.e., an expression of disorder in the system. Since disorder tends to increase because of the many conversion steps outside and inside of the microbial cell, the cells have adapted ways of improving efficiencies. This is key to selecting among biological treatment technologies. Engineering improves these efficiencies beyond natural acclimation. Thus, microbial metabolic processes must be characterized to understand wastewater biotechnologies. All cells must synthesize new biomolecules to construct cellular components, i.e., biosynthesis,. Cells must also harvest energy. Metabolism needs the chemical reactions of both processes.

Secondary treatment systems have various configurations, including trickling filters, activated sludge systems, digesters, and lagoons. In addition, more passive processes include natural systems such as bioretention system (see Fig. 15.3), waste stabilization ponds, septic tanks, anaerobic sludge blanket (UASB) reactors, high-rate anaerobic ponds, Imhoff tanks, and constructed wetlands. These systems incorporate primary, secondary, and advanced treatment steps, but at slower rates using microbial and soil characteristics already existing in the natural environment. Passive systems are engineered, but require less operation, maintenance, and management compared to the active secondary and advanced systems.

Electromechanical systems have numerous components, depending on the type of pollutant being removed or detoxified. At present, the removal of microbial pathogens and highly toxic chemical compounds is typically achieved through electromechanical systems. This can be problematic for developing countries and low income regions, where electrical services are unavailable or less reliable and where personnel training and retention can be challenging [30].

Wastewater treatment systems are classified as either onsite, e.g., a septic tank, or centralized, e.g., a municipal wastewater treatment plant. In the U.S., a large fraction of residences treat wastewater onsite, i.e., 25% of U.S. population daily dispose of 1.5×10^7 m^3 wastewater by means of onsite disposal and treatment systems. Unfortunately, an estimated 10% of these systems do not function properly, according to the design specifications [31].

The efficiency and effectiveness of an onsite or centralized system is expressed as the rate and amount of destruction of pollutants. For onsite wastewater treatment systems (OWTSs) the principal contaminants needed to be treated are those produced by residential activities and product usage, e.g., pathogenic microbes, nutrients, and toxic substances (see Fig. 15.4). An efficient system keeps pollutants from reaching drinking water supplies, e.g., aquifers and surface waters. For example, efficiency decreases with declining retention time because the beneficial bacteria lack enough time to

c. *pV* is known as pressure energy (ε_p).

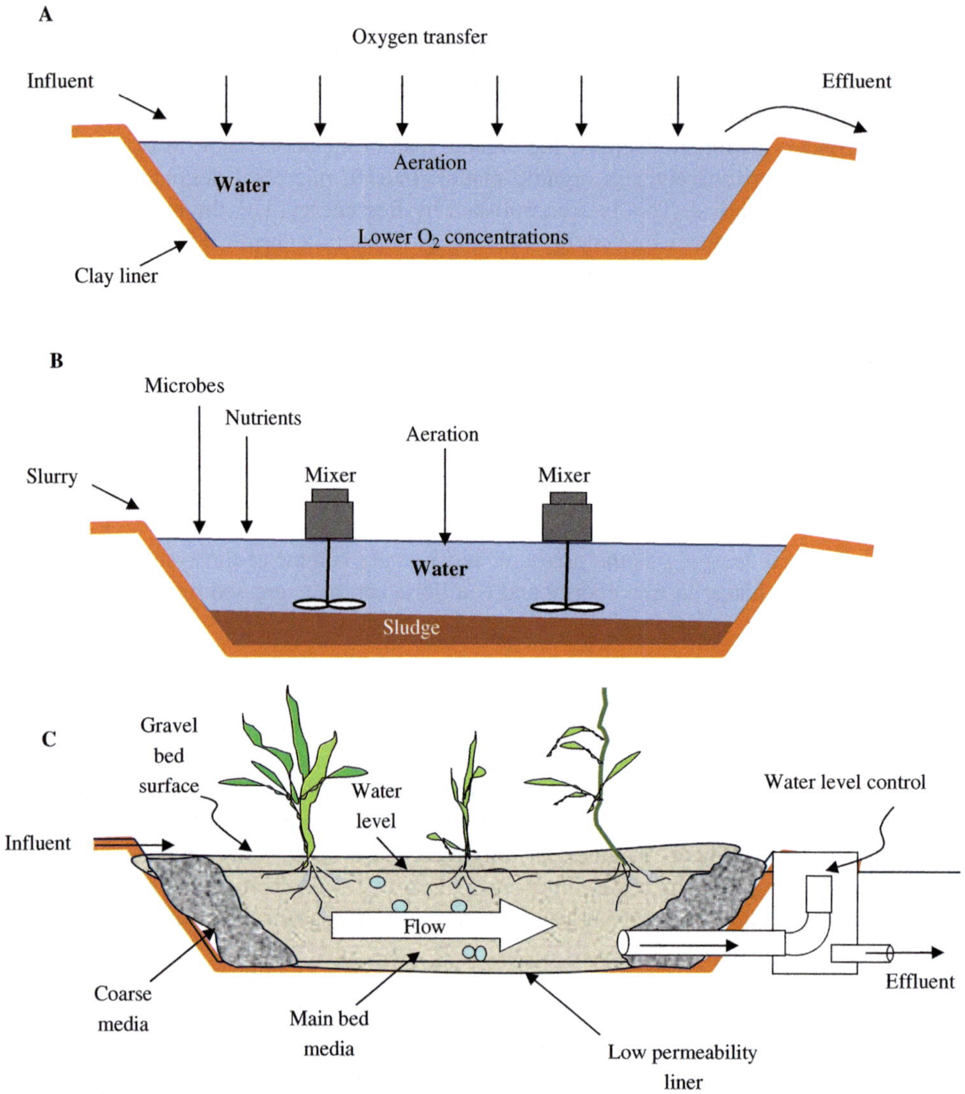

FIG. 15.3 Bioretention systems: (A) Passive system: aeration pond. (B) Active system: *in situ* slurry-phase lagoon. (C) Combined active and passive system: engineered and constructed wetland. All three systems make use of microbes, nutrients, and oxygen, but the active system increases the contact between the microbes and organic matter by mechanically mixing the sludge layer (which contains the organic contaminants to be degraded) with the water in the lagoon. The engineered wetland also incorporates plant processes in the degradation process. *(Data from Risk Reduction Engineering Laboratory, Office of Research and Development, U.S. EPA, Pilot-scale demonstration of slurry-phase biological reactor for creosote-contaminated soil: applications analysis report. EPA/540/A5-91/009, 1993. S. Wallace, Engineered wetlands lead the way, Land Water 48 (5) (2009). Available: http://www.landandwater.com/features/vol48no5/vol48no5_1.html.)*

metabolize the organic compounds in the wastewater. This can be because the soil is too permeable to allow an optimal biosorption rate.

At the municipal scale, the wastewater treatment plant employs the same physical and biological processes as those used in an onsite system, but for much larger volumes of wastewater. Following the previously discussed solids removal, secondary treatment mechanisms remove up to 90% of organic matter [17,32–35]. Trickling filters and other attached growth or fixed film processes are commonly used by towns and cities. In these systems, microbial populations adhere to various types of media which are coated with biofilm (see Fig. 15.5). Secondary treatment also includes suspended growth processes, especially activated sludge systems (see Fig. 15.6). Aerated lagoons, slow rate methods like land application, and constructed wetlands and other bioretention systems are used to remove organic matter.

In an activated sludge system, a tank of wastewater is injected with a mass of microorganisms. Pipes in the aeration basin supply O_2 as microorganisms meet, contact, are sorbed, and metabolize the organic content in the wastewater. The theoretical complete organic matter reactions approach is [36]:

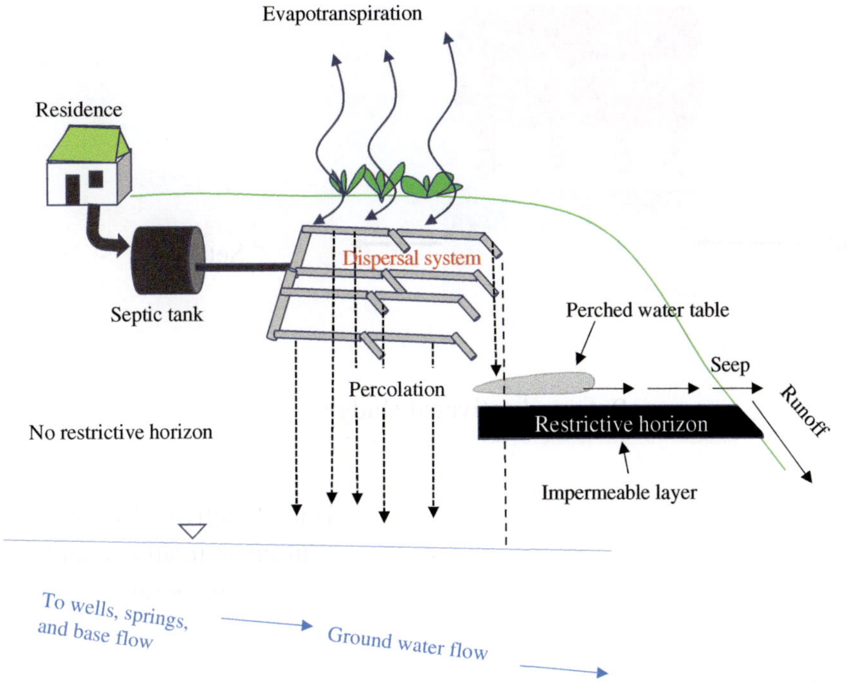

FIG. 15.4 Flow of wastewater and the constituents that it carries in an onsite wastewater treatment system (OWTS). *(Data from U.S. Environmental Protection Agency, Onsite Wastewater Treatment Design Manual. EPA/625/R-00/008, 2002.)*

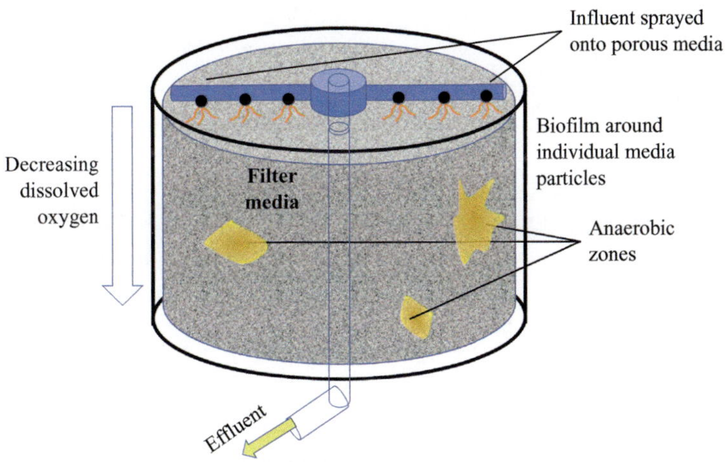

FIG. 15.5 Secondary treatment trickling filter system. O_2 concentrations available to the microbial populations generally decrease with depth in the porous media, although zones of varying O_2 concentrations exist at different depths, making a trickling filter to be a mixed reactor. *(Data from D.A. Vallero, J.J. Peirce, Engineering the Risks of Hazardous Wastes, Butterworth-Heinemann, Burlington, MA, 2003, p. xxi, 306 p.)*

$$CO_2 + H_2O + \text{microorganisms} + \text{energy} \tag{15.13}$$

The well-fed microbial communities increase in weight and density, flowing next into a quiescent tank where they settle by gravity, collect on the bottom of the tank, and are removed. Dictated by operating conditions, a specified volume of these biosolids consisting of the settled, hungry, and active microbes are returned to the aeration basin, i.e., return activated sludge (RAS). Contact with the RAS begins the microbial feeding anew. The liquid phase activated-sludge effluent can benefit from additional microbiological and chemical processing prior to release into a receiving stream to meet effluent limitations [28,37–39].

Other aerobic reactors include aeration lagoons or ponds, which treat liquid and dissolved contaminants for long time periods. Persistent organic molecules by definition do not readily degrade in a trickling filter or an activated sludge system,

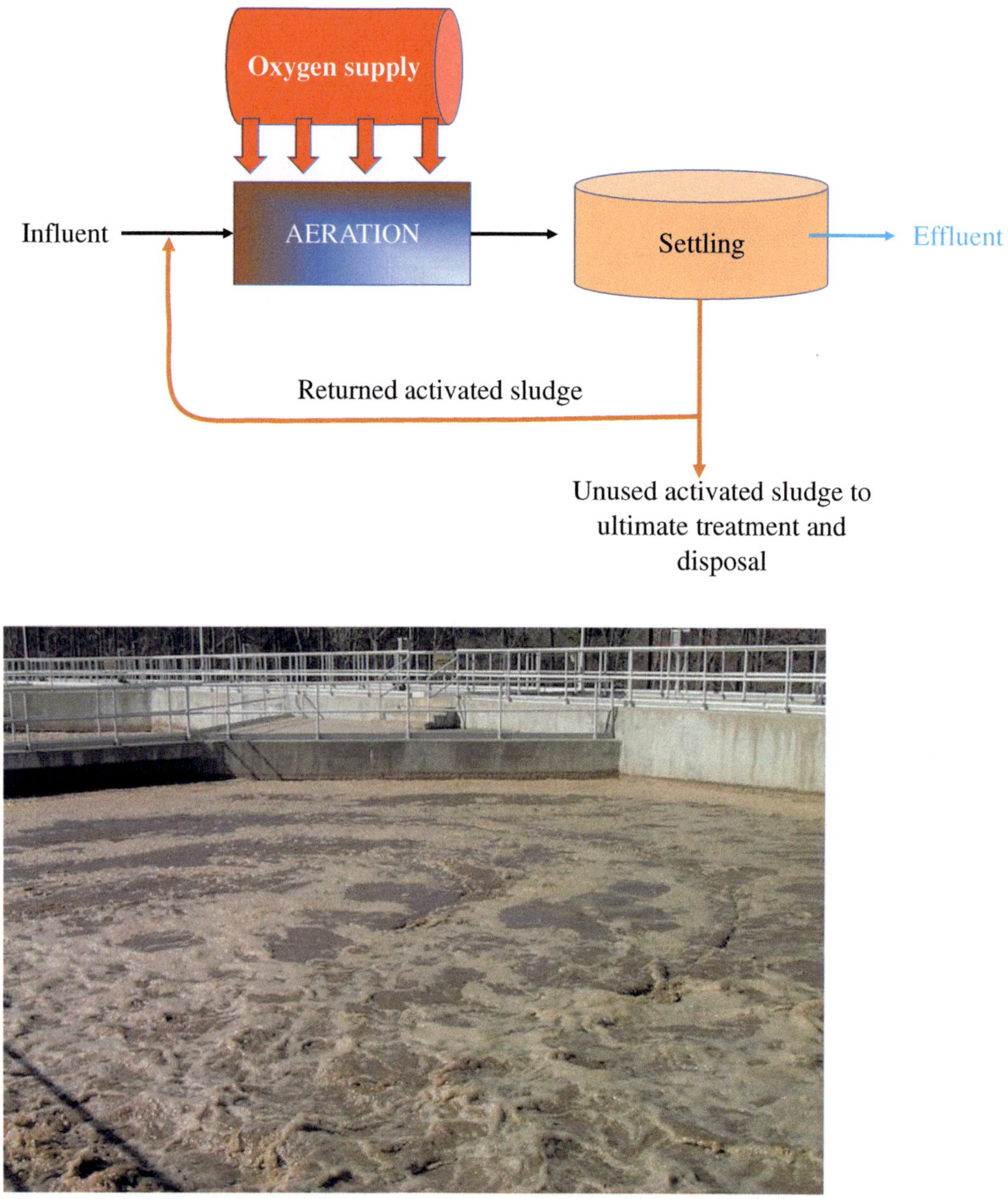

FIG. 15.6 Activated sludge treatment system. Influent mixes with recycled biomass and aerated to reach and maintain a specific dissolved oxygen (DO) content. Bacteria and other microbes consume the organic components, which decreases the organic dissolved and suspended solids in the wastewater. Optimal DO content can be maintained with tapered aeration, i.e., supplying the highest O_2 concentrations near the tank inlet to accommodate the oxygen demand from microbes as waste is introduced to the aeration tank. *(Data from D.A. Vallero, J.J. Peirce, Engineering the Risks of Hazardous Wastes, Butterworth-Heinemann, Burlington, MA, 2003, p. xxi, 306 p. Bottom photo, courtesy of D.J. Vallero.)*

but may be broken down by specific microbes into $CO_2 + H_2O$ + microorganisms + energy, if given enough time [40]. Usually, diffusion is provided naturally by air movement by wind and precipitation from the troposphere. Aerobic reactors vary in shape and size, but prototypical dimensions are [40]:

- Pond size: 1–8 ha (0.45–20 acres).
- Pond depth: 0.3–9 m (1–30 ft).
- Operation: In series with other treatment systems, other ponds, or stand-alone.
- Operation: The flow to the pond is either continuous or intermittent; and,
- Operation: Additional oxygen to the system may be needed, e.g., using active systems like blowers and diffusers.

An engineered wetland is an aerobic reactor with abiotic and biotic processes that render it a type of biofilter. The diverse microbial populations and plant life provide numerous processes to remove pollutants from the wetland's water. Microbial growth, nutrient extraction, photosynthesis, and ion exchange work together to slowly degrade a variety of pollutants. A radial-flow constructed wetland system, for example, has crushed concrete beds insulated with layers of mulch. Atop the mulch layer, native plant species are grown in separate sections of the wetland system. Depending on the size, substantial volumes of wastewater are passively treated. The insulated rock and mulch support continuous biological activity by modulating temperature and moisture content. As a passive treatment system, the engineered wetland can be highly cost-effective compared to active systems. Construction costs have been estimated to be about 20% of an active system that meets similar treatment criteria [41].

Often, effluent from secondary treatment alone does not clean the wastewater sufficiently to meet water quality standards and effluent limitation criteria, which calls for advanced treatment or tertiary treatment technologies. Following the removal of most of the solids and organic matter, for example, excessive nutrient concentrations, especially N and P, remain in the effluent. Elevated nutrient levels in surface waters increase the likelihood and severity of eutrophication due to the concomitant algal growth and blooms [40,42]. Dead algae fall to the bottom of the waterbody where they are decomposed by other microbes, which increases BOD. Up the food chain, other biota are stressed by these habitat changes [40,42,43]. In biotic nitrogen advanced treatment, nitrifying bacteria oxidize ammonia nitrogen to nitrate compounds, i.e., nitrification. The nitrate is then reduced to N_2 gas, i.e., denitrification.

A tell-tale sign that wastewater treatment processes are not operating efficiently is the emission of odorous sulfide compounds that have become chemically reduced because an aerobic system has become anaerobic in certain pockets of a system. With declining concentrations of O_2, *Desulfovibrio, Desulfotomaculum, Desulfobacter, Desulfuromonas* and *Desulfococcus* and other genera of anaerobic bacteria undergo fermentation to produce reduced compounds, especially hydrogen sulfide (H_2S), as well as methane (CH_4) [44–46]. The H_2S is particularly malodorous. However, many wastewater treatment systems are intentionally designed to operate anaerobically (see Fig. 15.7).

Anaerobic treatment processes combine waste and sludge. As in the aerobic activated-sludge systems, anaerobic biosolids are added to the wastewater and mixed in a reactor. This creates a once-through, completely mixed system, in which the hydraulic retention time (HRT) is equal to the solids retention time (SRT). Removal efficiency depends directly on the SRT. During operations, however, HRT and SRT can be decoupled. For example, an anaerobic upflow filter improves anaerobic degradation volumes and rates because the filter captures and sustains elevated biosolid concentrations. The resulting attenuated SRTs increase throughput, allowing improved degradation of low-strength organic wastes [38,47–52].

An anaerobic bed upflow reactor needs the biomass to sorb onto media surfaces. Wastewater moves upward through a bed of sand-texture media at a velocity high enough to fluidize and expand the sand bed somewhat [17,33]. The upflow anaerobic sludge blanket process (UASB) stimulates flocculation and sludge settling, so that loadings have a much higher HRT and gases H_2, CH_4, and other gases leave the sludge solids (see Fig. 15.7). The UASB separates solids and gases from the liquid, then degrades the organic matter, while biogas production enhances contact between substrate and anaerobes.

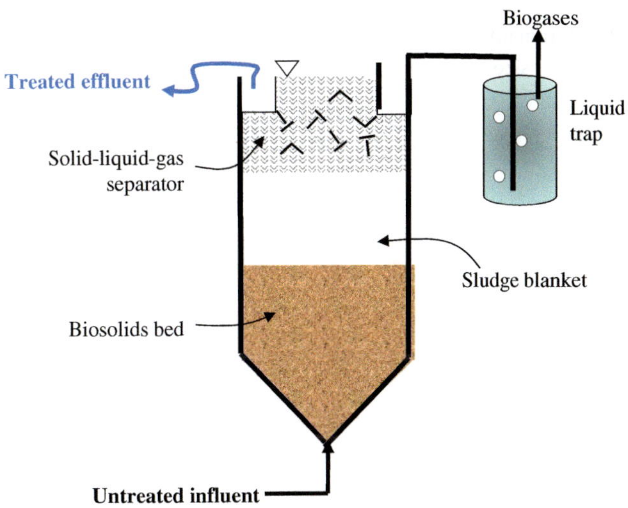

FIG. 15.7 Schematic of upflow anaerobic sludge blanket system. *(Data from D. Christensen, J. Gerick, J. Eblen, Design and operation of an upflow anaerobic sludge blanket reactor, J. Water Pollut. Control Fed. (1984) 1059–1062.)*

UASB reactors are designed to be completely mixed reactors. Agitation further enhances contact between biosolid particles and incoming organic liquids. This is achieved by the release of the gases being produced. Also, an inlet is expected to be designed to distribute the influent so that it reaches lower levels of media in the reactor.

For any biological treatment system, additional chemical or physical treatment may be required to extract inorganic materials from a waste stream prior to the biological treatment. If not, toxic substances my destroy beneficial microbial populations and/or may be present in the effluent or sludge, leading to the need for hazardous waste handling [36,48].

Advanced treatment processes combine biotic and abiotic processes [53]. O_2 is supplied by mechanical mixing or by diffused aeration, which keeps conditions aerobic. The O_2 maintains microbial contact with organic compounds dissolved in the secondarily treated effluent. Abiotic, physicochemical processes include deep-bed filtration, floating media filtration, and membrane filtration.

In addition, microbial populations can only live within an optimal pH range that must be maintained from the preliminary stage through advanced treatment to ensure that extreme acidic and basic conditions do not interfere with the biotic processes and to minimize damage to equipment [53].

Advanced wastewater technologies vary according to treatment objectives, but numerous approaches are employed, including:

1. coagulation-sedimentation;
2. granulated activated carbon (GAC) adsorption;
3. sand filtration;
4. ozone treatment;
5. ultraviolet (UV) treatment;
6. membrane bioreactors (MBR);
7. advanced oxidation processes (AOP);
8. nanotechnology; and,
9. automatic variable filtration

These and other advanced treatment processes can be used alone or in various combinations to remove myriad contaminants after secondary treatment [53].

After treatment, the resulting sludge must be removed and handled. This is accomplished by physical treatment that includes incineration, and solidification or stabilization, as well as various chemical treatment and biological methods. Deciding which approach depends upon the characteristics of the sludge [54]. For example, like other solid wastes, the efficiency and effectiveness of sludge treatment technologies is affected by particle size, solids content, and the concentrations of contaminants, with many being increased substantially in the sludge compared to their concentrations in the incoming wastewater (see Table 15.2).

TABLE 15.2 Effect of particle size, solids content and extent of contamination on decontamination efficiencies.

	Predominant particle size			Solids content		High contaminant concentration	
Treatment technology	Sand	Silt	Clay	High (slurry)	Low (*in situ*)	Organic compounds	Metals
Conventional incineration	N	X	X	F	X	F	X
Innovative incineration	N	X	X	F	X	F	F
Pyrolysis	N	N	N	F	X	F	F
Vitrification	F	X	X	F	X	F	F
Supercritical water oxidation	X	F	F	X	F	F	X
Wet air oxidation	X	F	F	X	F	F	X
Thermal desorption	F	X	X	F	X	F	N
Immobilization	F	X	X	F	X	X	N
Solvent extraction	F	F	X	F	X	X	N

TABLE 15.2 Effect of particle size, solids content and extent of contamination on decontamination efficiencies—cont'd

	Predominant particle size			Solids content		High contaminant concentration	
Treatment technology	Sand	Silt	Clay	High (slurry)	Low (*in situ*)	Organic compounds	Metals
Soil washing	F	F	X	N	F	N	N
Dechlorination	U	U	U	F	X	X	N
Oxidation	F	X	X	N	F	X	X
Bioslurry process	N	F	N	N	F	X	X
Composting	F	N	X	F	X	F	X
Contained treatment facility	F	N	X	F	X	X	X

Note: F—sediment characteristic favorable to the effectiveness of the process; N—sediment characteristic has no significant effect on process performance; U—effect of sediment characteristic on process is unknown; X—sediment characteristic may impede process performance or increase cost.
Source: D. Averett, Treatment technologies, in: Assessment and remediation of contaminated sediments (ARCS) program: remediation guidance document, Great Lakes National Program Office, U.S. Environmental Protection Agency, Chicago, Illinois, 1994 (Chapter 7).

As is the case for soil, sediment, and any porous matrix, sludge particle size is a key treatment parameter. Most treatment technologies work well on coarse textured sludges. Finer texture lessons the effectiveness of treatment by making dewatering more difficult. Also, when dried thermally, aerosol emissions increase with decreasing particle size. The larger specific surface area of fine textured sludges increases the rate of sorption of contaminants. Sludges heavy in clay particles are more difficult to handle than sludges with lower clay content [54].

Increased organic content means an associated total organic carbon (TOC) content and attendant caloric thermal value, favoring incineration and oxidation processes. However, elevated concentrations of heavy metal and halogenated organic compounds requires special handling to ensure safety of personnel and environmental protection. If the TOC contains halogenated compounds, there is a strong likelihood of increasing the toxicity of the incineration residue with the generation of dioxins, furans, and other persistent compounds [55,56].

3 On-site treatment technologies

As discussed in Chapter 11, ground water and other contaminated water can be treated by drilling recovery wells to pump contaminated ground water to the surface. Commonly used ground water treatment approaches include air stripping, filtering with granulated activated carbon (GAC), and air sparging.

Air stripping moves volatile compounds from water to air within in a packed column or tower or in a low-profile sieve tray aeration system (see Fig. 15.8). In a tower filled with a permeable material, ground water moves by gravity downward while a stream of air flows upward. The configuration includes a spray nozzle, or a manifold connected to several nozzles, at the top of the tower to spread contaminated water over the column packing in the column, a blower to move air in the countercurrent to the contaminated water flow, and a sump at the bottom to collect treated water. Low-profile sieve tray air strippers are configured as a horizontal tray or vertical box design with a series of stacked trays. Baffles and weirs inside the stripper trays move contaminated water twice or more along the length of the box tray [57].

Air sparged through a vent pipe or through the bottom of the column passes in a countercurrent direction through the water to strip higher compounds with relatively high vapor pressures, e.g., volatile organic compounds (VOCs) like benzene or xylenes. These are considered single-pass systems although a portion of the effluent can be separated and recirculated into the column to improve removal efficiency and to keep the height of the packed tower at a reasonable height or to optimize the number of trays in a low-profile system. Diffused aeration strippers have aeration tanks in which volatile compounds are stripped by bubbling air into a tank through which the contaminated water is flowing. A forced-air blower and an aeration-sparge manifold increase the contact of the water with air [17].

Contaminated water can also be filtered by pumping the water through the GAC to trap the contaminants. Air sparging consists of pumping air into the ground water *in situ*. A soil venting system is often combined with an air sparging system for vapor extraction [17].

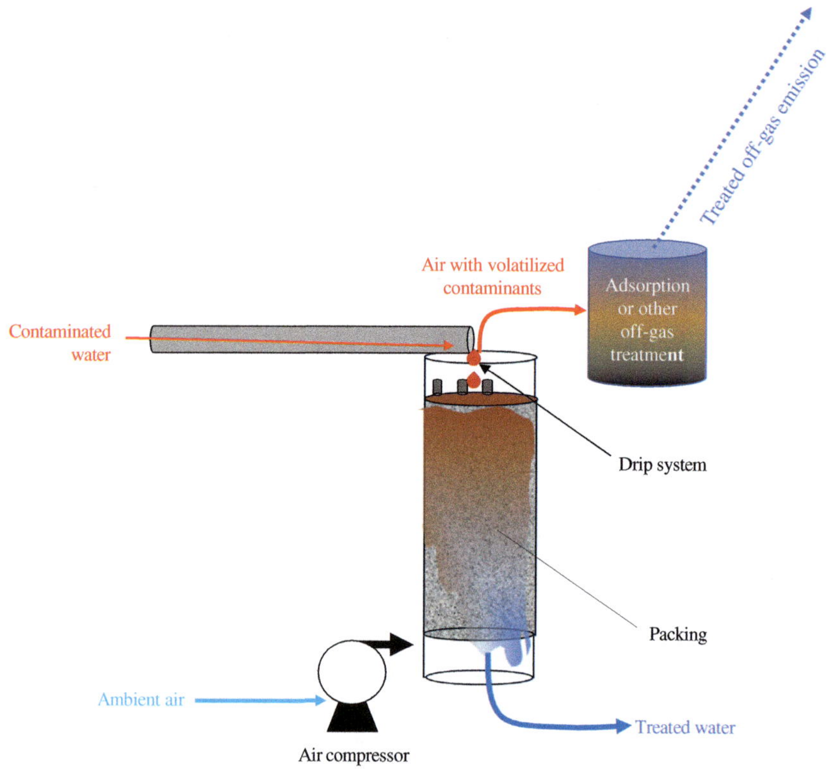

FIG. 15.8 Air stripping system to treat volatile compounds in water. *(Data from D. Vallero, Environmental Biotechnology: A Biosystems Approach, Elsevier Science, 2015.)*

Microbial populations can be added to the porous media to enhance removal of vapor phase compounds. The air carrying these volatile compounds is passed through a layer of biologically active media, where the microbes degrade the compounds to simpler compounds. The major difference between biofiltration and the previously discussed trickling systems owes to the liquid interfaces with the microbes. The liquid phase is stationary in a biofilter (see Chapter 11), but in a trickling system, the liquids move through the porous media [58]. Various porous media can house the microbes, such as compost, which usually has a very diverse microbial community of myriad species of bacteria and fungi [59–64]. In biotreatment systems where water passes through media, the microbial biofilm surrounding each media particle provides contact so that the microbe can metabolize the contaminants into simpler compounds, eventually converting them into carbon dioxide and water vapor [65].

As is the case for secondary wastewater treatment, ground water treatment efficiency depends on the flow rates and inherent properties of the pollutants being treated, e.g., sorption, vapor pressure, fugacity, and solubility. These and other factors influence the ability of microbes to produce biofilm that is needed for the target pollutant to enter the microbes' biofilm, and to achieve the desired pollutant loading rates [21].

4 Treating air pollutants

Although this book's focus is on water pollution, air pollution needs to be briefly discussed because water pollution treatment technologies often transfer substances to the air following treatment. Controlling air pollutants differ according to the type of pollutant [66]. The two phase-related classifications, i.e., particulate matter and gas-phase pollutants, are the predominant way that technologies to control air pollutants are differentiated [67–69].

4.1 Particle collection

Particles can be measured as either mass or count. Particle count is the number of particles in each band of mass, such as particles with aerodynamic diameters greater than 10 μm (coarse fraction), particles with diameters less than 10 μm, but greater than 2.5 μm (PM_{10} fraction), and particles with diameters less than 2.5 μm ($PM_{2.5}$ fraction, also known as the fine fraction). However, these bands can be further subdivided, e.g., ultrafines and engineered nanoparticles with aerodynamic diameters less than 100 nm. In the case of engineered nanoparticles, rather the aerodynamic diameter, any dimension of 100 nm or less defines an nanoparticle. Filtration is the most common method used to measure all these particle bands.

As result of these particle measurements, decisions on which type of control technology is expected to be deployed to meet air quality PM standards. There are numerous options, but all technologies that remove PM makes use of one or more physical mechanisms:

- Gravitational settling
- Inertial impaction
- Centrifugal inertial force
- Brownian motion
- Electrostatic attraction
- Thermophoresis
- Diffusiophoresis

Each mechanism combines forces to make a particle behave in an intentional manner. Air pollution equipment uses the particle behavior to transport the particle toward a surface. The physics to control this behavior is the sum of forces:

$$\sum F = m_p \cdot a_p = m_p \frac{dv_p}{dt} \tag{15.14}$$

Where, $\sum F$ is the sum of all forces acting on a particle (g cm s^{-2}); m_p is the particle mass (g); a_p is the particle acceleration (cm s^{-2}); v_p = particle velocity; and t is time (s). In English units, Eq. (15.14) is restated as:

$$\sum F = \frac{m_p \cdot a_p}{g_c} \tag{15.15}$$

Where, $\sum F$ is the sum of all forces acting on a particle (foot pounds, lb_f); m_p is the particle mass (international avoirdupois pound, lb_m) and a_p is the particle acceleration (ft s^{-2}). The foot-pound is the product of lb_m and g_0 (i.e., the standard gravitational field). 1 lb_f is equal to the force exert by lb_m in a gravitational field. The gravitational constant (g_c) is expressed as:

$$g_c = 32.2 \frac{\text{lb}_m \cdot \text{ft}}{\text{lb}_f \cdot \text{sec}^2} \tag{15.16}$$

These mechanisms together determine the particle movement. Air pollution engineers can account for efficient particle removal that involves three basic processes:

1. Initial capture of particles on surfaces
2. Gravity settling of solids into the hopper
3. Removal of solids from the hopper

Fabric filters, electrostatic precipitators, and other engineered systems apply these processes even though capture method and sequence vary considerably. Fabric filters, also known as baghouses, applies filtration for initial particle capture but precipitators use electrostatics, for example. Gravity settling also varies among technologies.

Cyclones and filter systems are widely used to remove PM. However, within the same type of pollution control equipment, mechanisms differ. For example, filtration does not rely exclusively on filter mechanisms. A pulse jet fabric filter uses inertial impaction, Brownian motion, and electrostatic attraction to capture particles in the size range of 100 μm to less than 0.01 μm onto the accumulated layers of dust on the exterior surfaces of the bags. Large aggregates of dust cake become dislodged from the filter surface and fall into the hopper.

Pressure differential is key to PM removal. Decreasing pressure well below that of the atmosphere causes air to move toward the pressure trough. Large pressure differences between air outside and inside will increase flow dramatically. The rapid air flow to the low-pressure region carries particles along with it. Increasing velocity is proportional to increasing mass and numbers of particles able to be carried. This is the same concept of competence that causes solids to settle in a wastewater treatment primary clarifier or at the delta of a stream. In all these cases, sedimentation of densest particles occurs first, whereas colloids remain suspended for much longer times.

In the 20th century, the earliest air quality standards were directed at total suspended particulates (TSP), which were measured using high-volume samplers that collect a wide range of particles with aerodynamic diameters up to 50 μm. Later in the decade, finer particles were identified as particular harmful to health. As mentioned, most recently, ultrafines [70–75] and nanoparticles [76–81] with aerodynamic diameters ≤100 nm have received much more attention. Surface area affects

TABLE 15.3 Relationship between spherical particle diameter, volume, and surface area.

Diameter (μm)	Volume (cm^3)	Area (cm^2)
0.1	5.23×10^{-16}	3.14×10^{-10}
1.0	5.23×10^{-13}	3.14×10^{-8}
10.0	5.23×10^{-10}	3.14×10^{-6}
100.0	5.23×10^{-7}	3.14×10^{-4}
1000.0	5.23×10^{-4}	3.14×10^{-2}

sorption, electrostatics, and other processes involved in PM removal, with each order of magnitude decrease in particle diameter associated with an increase by two orders of magnitude in surface area (see Table 15.3). This means that a 1-μm particle has six orders of magnitude greater surface area than a 1 mm particle. This difference in size-to-surface areas ratio affects sorption, which means that smaller particles will be more likely to sorb materials, including toxic compounds [82–84].

After initial capture, gravity is a key removal mechanism among the forces exerted on the particle determine the extent to which a particle or aggregates settle, i.e., gravitational force (F_G), buoyant force (F_B) and the drag force (F_D). The gravitational force causes the PM mass to drop:

$$F_G = m_p \cdot g = \rho_p \cdot V_p \cdot g \tag{15.17}$$

Where, F_G = gravitational force (g cm s^{-2}); m_p = particle mass (g); g = particle acceleration due to gravity (980 cm s^{-2}); ρ_p = particle density (g cm^{-3}) and V_p = particle volume (cm^3).

For a spherical particle, the volume term in Eq. (15.17) can be simplified as:

$$V_p = \frac{\pi d_p^3}{6} \tag{15.18}$$

Where, d_p is the physical diameter of the particle (cm). Thus, substituting this term yields:

$$F_G = \frac{\pi d_p^3 \rho_p g}{6} \tag{15.19}$$

As mentioned in the wastewater treatment discussion, buoyancy (B_F) is an opposite force to gravity. This force opposition means that by increasing B_F, the more likely a particle remains suspended in the air, expressed by:

$$F_B = m_g \cdot g = \rho_g \cdot V_p \cdot g \tag{15.20}$$

Where, F_B is the buoyancy force (g cm s^{-2}); m_g is the mass of displaced gas (g); g is the particle acceleration due to gravity (980 cm s^{-2}); ρ_g is the carrier gas density (g cm^{-3}) and V_p is the particle volume (cm^3). For a spherical particle, this is simplified as:

$$F_B = \frac{\pi d_p^3 \rho_g g}{6} \tag{15.21}$$

F_B depends on the density of the gas, which for air pollution control is usually air. Since pollution concentrations needing treatment rarely exceed a few percent of the volume of air, the physical properties are often used exclusively as the carrier gas. However, if the concentrations are sufficiently high, total gas densities can be calculated, e.g., using Raoult's law for vapor pressures. For air as the carrier gas, ρ_g is about 10^{-2} lb_m ft^{-3}. F_G depends on ρ_p, which is about four orders of magnitude larger than the gas density, i.e., about 10^2 lb_m ft^{-3}. Thus, since buoyant force is so much smaller than the gravitational force, it is often neglected in these calculations [65,85–89]. As the particle moves downward mainly by F_G, it meets a resistive force that increases with increasing downward velocity, i.e., drag:

$$F_D = \frac{A_g \cdot \rho_g \cdot V_p^2 \cdot C_D}{2} = \frac{\pi d_p^2 \rho_g v_p^2 \cdot C_D}{8} \tag{15.22}$$

TABLE 15.4 Morphology's effect on particle diameter (d) and density (ρ_p) for three particles with the same aerodynamic diameter (d_p).

	Shape	Density (ρ_p) and diameter (d)	Aerodynamic diameter (d_p)
	Solid sphere	$\rho_p = 2.0$ g cm^{-3} $d = 1.4$ µm	
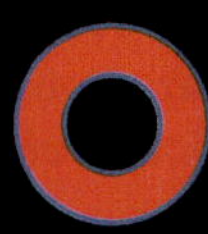	Hollow sphere	$\rho_p = 0.5$ g cm^{-3} $d = 2.8$ µm	$d_p = 2.0$ µm
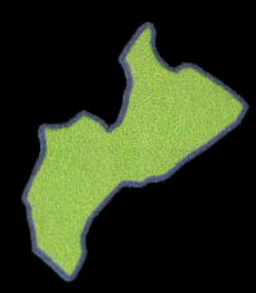	Irregular	$\rho_p = 2.3$ g cm^{-3} $d = 1.3$ µm	

Data from W.J. Franek, L. DeRose, Control of particulate matter emissions. Chapter 3: Particle sizing, in: U.S. Environmental Protection Agency, O. o. A. Q. P. a. S. Office of Air and Radiation, (Eds.), Student Manual: APTI 413, U.S. EPA, Research Triangle Park, North Carolina, 2021, sec. Washington, DC.

Where, F_D is the drag force (g cm s^{-2}); A_p is the cross-sectional area of the particle (cm), which is calculated from $\frac{\pi d_p^2}{4}$; ρ_g is the gas density (g cm^{-3}); and C_D is the drag coefficient (dimensionless).

Air or other carrier gas in front of the particle is displaced as the particle falls, impacting the momentum on the carrier gas. F_D equals the momentum imparted by the gas on the particle per unit time. The particle velocity (v_p) is also partially imparted to the gas velocity (v_g). The amount of energy that v_p adds to v_g relates to friction, which is the basis for the drag coefficient (C_D) [86,90].

Pollutant particles are seldom spheres, taking various dimensions and shapes. Depending on these, differently-shaped particles can have similar aerodynamic diameters, as shown in Table 15.4. On the other hand, certain particles that appear similar can have different aerodynamic diameters. Particle morphology includes spheres, hollow spheres, irregular shapes, flakes, fibers, aggregates and condensation flocs [91].

The principal PM removal mechanisms are inertial and gravitational separation, fabric filtration and electromagnetics. In addition to air pollutants, filtration is a key removal mechanism in treating wastewater and water treatment, water at point-of-use, hazardous wastes, and contaminated sediment. Filtration consists of four basic mechanisms: (1) diffusion, (2) interception, (3) inertial impaction, and (4) electrostatics. Molecular diffusion works for very small particles (<0.1 µm diameter) because the Brownian motion allows them to move in a random walk away from a nearly quiescent air stream. Interception is best for capturing particles with diameters between 0.1 and 1 µm. During interception, the particle stays in the air stream but encounters the filter medium, e.g., a fiberglass strand or fabric. Inertial impaction collects particles that are sufficiently coarse, usually diameters >1 µm, to exit the air stream by inertia [92,93].

Electrical interactions between the atoms in the filter and those in the particle at the point of contact, i.e., Van der Waal's force, are also used to collect particles. Electrostatic attraction is enabled by charge differences between particle and filter medium. Electrostatics is sometimes the main collection method, such as in electrostatic precipitators (ESPs) commonly employed to remove aerosols from coal-fired power plant emissions.

Other factors affecting filtration efficiency are the thickness and pore diameter or the filter, the uniformity of particle diameters and pore sizes, the solid volume fraction, the loading rate and velocity of particle encountering the filter, e.g., affecting the so-called particle bounce, the particle phase (liquid or solid), capillarity and surface tension (if either the particle or the filter media are coated with a liquid), and characteristics of air or other carrier gases, such as velocity, temperature, pressure, and viscosity.

Fabric filters remove particles as the air stream passing the air through a porous fabric at efficiencies up to 99%. Based solely on an extrapolation of air pollution control equipment, filtration is expected to be better using a cyclone, not considering operational efficiencies and reliability. Baghouses often require intricate operation and maintenance, diminishing system reliability with numerous bag changeouts and repairs. Efficiency is expressed as a percentage, e.g., a 99.99% HEPA filter's efficiency allows it to remove 99.99% particles from the air stream. Therefore, if 10,000 particles enter the filter, on average only one particle would pass all the way through the filter. U.S. federal standards, for example, require that the incineration of hazardous compounds be at least 99.99% efficient, and for extremely hazardous wastes the so-called "rule of six nines" applies, i.e., removal >99.9999%. The HEPA removal efficiency and the incineration destructive removal efficiency (DRE) calculations are the same [48,94,95]:

$$\text{DRE or HEPA efficiency} = \frac{M_{in} - M_{out}}{M_{in}} \times 100 \tag{15.23}$$

Thus, if 10 mg min^{-1} of a hazardous compound is fed into the incinerator, only 0.001 mg min^{-1} = 1 μg min^{-1} is allowed to exit the stack for a hazardous waste. If the waste is an extremely hazardous waste, only 0.00001 mg min^{-1} = 0.01 μg min^{-1} = 10 ng min^{-1} is allowed to exit the stack. This removal calculation is used throughout environmental engineering to calculate treatment and removal efficiencies.

Cyclones, baffles, louvers, and rotating impellers follow the principle that the particle mass in the carrying gas stream has a greater inertia than the gas. Recall that the drag forces on the particle are a function of the diameter squared and the inertial forces are a function of the diameter cubed. Thus, as the particle diameter increases, the inertial removal force increases correspondingly, which means that inertial collector efficiency is highest for coarse particles. Also, heavier particles are more efficiently removed by inertial collectors because inertia is also a function of the mass of the particle. Thus, an inertial collector will be highly efficient for removal of 10-μm rock dust and 75-μm wood particles, but inefficient in collecting 2.5-μm wood particles [86]. The most common inertial configuration is that of the cyclone, which is used in two basic forms: the tangential inlet and the axial inlet [93].

Within the cyclone, the radius of motion of the particle mass and the particle velocity determine the centrifugal force exerted on the particle:

$$F = MA \tag{15.24}$$

Where, F = force (centrifugal), M = mass of the particle and A:= acceleration (centrifugal), and

$$A = \frac{V^2}{R} \tag{15.25}$$

Where, V = velocity of particle and R = radius of curvature. Thus, considering the particle mass, particle velocity, and radius, centrifugal force is found as:

$$F = \frac{MV^2}{R} \tag{15.26}$$

Decreasing the radius of curvature and increasing the gas velocity both result in increased efficiency, but the changes also cause increased pressure-drop through the collector. Design of inertial collectors for maximum efficiency at minimum cost and minimum pressure drop is an optimization challenge.

Air in the cyclone is rapidly circulated, changing the direction of particles suspended in the air. The particle inertia causes them to exit the air stream. Coarser particles are efficiently collected by this direction change because they have relatively large mass, which means greater inertia. Conversely, very fine particles more likely will remain in the air stream and suspended. Particles in the air enter the cyclone from the top through the inlet is tangential to the cylindrical portion of the cyclone. The air turns downward to form a peripheral vortex, which creates centrifugal forces. Individual particles are carried toward the cyclone wall, impact the wall, and fall downward to be collected in a hopper. When the air reaches the end of the conical segment, it will change direction and move upward toward the outlet, forming an inner vortex. The upward airflow against gravitation allows for additional separation of particles [93]. The inertia of heavier particles (mainly coarse fraction, with some very dense fine particles) is high enough that they leave the air stream of the vortex created by the cyclone inside configuration. The particle next collides with the cyclone wall. The fine and ultrafine particles, having smaller mass, continue moving with the air stream, thus remaining suspended and not collected until the stream velocity decreases to a point where it can no longer carry the particle. If this velocity is never reached, the particle will exit the cyclone [86].

The geometry of air pollution control equipment often has circular dimensions to avoid pockets of zero to very low air flows. Such flows attract PM, increasing the frequency of cleaning. Also, inertial forces are often effective for PM-laden air flows in a circular route within a cylinder [86,96].

Filtration is important in both measuring particles and removing PM. Filter efficiency is a key metric of success in air pollution control. As mentioned in the HEPA and DRE, discussion, the general efficiency (*E*) equation can be applied to particle collection according to particle numbers and particle concentrations:

$$E = \frac{N_{in} + N_{out}}{N_{in}} \tag{15.27}$$

$$E = \frac{C_{in} + C_{out}}{C_{in}} \tag{15.28}$$

Where, N is the number of particles (count), and C is the mass concentration (subscripts: in = entering; out = exiting).

Air and other pollution control equipment often characterize efficiency using fraction entering versus that exiting the filter known as particle penetration (*P*):

$$P = \frac{N_{out}}{N_{in}} = 1 - E \tag{15.29}$$

$$P = \frac{C_{out}}{C_{in}} = 1 - E \tag{15.30}$$

P is the inverse of E, which means that inefficient filtration allows a too large a number or too much mass of particles to penetrate the filter. Thus, tolerances for filtration are key to any design of measurement or control technology. Inherent to penetration calculations is the velocity of air entering the system.

Scrubbers collect PM by applying two basic physical mechanisms already discussed, i.e., inertial impaction and Brownian motion. Scrubbing traps the particle as it travels from the supporting gaseous medium across the interface to the liquid scrubbing medium. Some relative motion is necessary for the particle and liquid-gas interface to contact each other. In the spray chamber, this motion is provided by spraying the droplets through the gas so that they impinge on and contact the particles [96]. In the bubbler, inertial forces and severe turbulence achieve this contact. In either case, the smaller the droplet or bubble, the greater the collection efficiency will be. In the scrubber, the smaller the droplet, the larger will be the surface area for a given weight of liquid, with a greater likelihood of wetting the particles. In a bubbler, smaller bubbles mean not only that more interface area is available but also that the particles have a shorter distance to travel before reaching an interface where they can be wetted.

The next wet collector process is to remove the wetted particles on a collecting surface, followed by their eventual removal from the device. The collecting surface is a bed or a wetted surface. One common combination configuration has the wetting section followed with an inertial collector which then separates the wetted particles from the carrying gas stream [96].

Thus, collection has two phases; first the particle is captured within a droplet; next the droplets aggregate into a slurry which is expected to be removed and treated as a solid waste and as a hazardous waste if it contains sufficient concentrations of toxic compounds. This is quite possible since the scrubber is removing particles from a polluted air stream. The particles may consist of hazardous chemicals, or of chemicals that have sorbed onto the particle surface. The chemicals are now concentrated in the scrubber slurry.

Increasing either the gas velocity or the liquid droplet velocity in a scrubber will increase the efficiency because of the greater number of collisions per unit time. The ultimate scrubber in this respect is the venturi scrubber, which operates at extremely high gas and liquid velocities with a very high pressure drop across the venturi throat [96].

The scrubbing liquid is recirculated to minimize the amount of liquid needing to be treated and discharged. The scrubbing liquid is collected in the sump of the scrubber and mist eliminator delivered by gravity to a recirculation tank having a liquid residence time of several minutes. This provides enough time to introduce alkali additives and to adjust the pH back to the proper range, if necessary. A centrifugal pump recirculates the liquid back to the scrubber vessel from the recirculation tank. Some scrubber systems use spray nozzles to atomize the scrubbing liquid. To project droplets across an entire circular area, a full-cone nozzle is used most frequently. This improves the effective gas-liquid contact. Full cone nozzles usually provide mean diameters of 100–1000 μm that are log-normally distributed [96].

The composition of particles and gases in the airstream often require pretreatment. For example, at sufficient concentrations of acid gases, e.g., H_2SO_4, HCl and HF, buffering is required. An alkali addition system is installed to collect acidic

PM or to treat gas streams that have acidic gases or vapors that could absorb in the liquid stream. In addition, CO_2 formed in most combustion processes can also be absorbed, with carbonic acid ($H_2CO_3^*$) lowering the pH. The most common alkalis used in scrubbers are liming agents, e.g., calcium carbonate ($CaCO_3$) or calcium hydroxide ($Ca(OH)_2$), soda ash (Na_2CO_3), and sodium hydroxide (NaOH) [96].

The amount of alkali needed depends on the quantities of acidic gases captured and the molar ratios necessary for the following reactions:

$$SO_3 + Ca(OH)_2 \rightarrow CaSO_4 + H_2O \tag{15.31}$$

$$2HCl + Ca(OH)_2 \rightarrow CaCl_2 + 2H_2O \tag{15.32}$$

$$2HF + Ca(OH)_2 \rightarrow CaF_2 + 2H_2O \tag{15.33}$$

Scrubber vessels generate water droplets that are carried in the gas stream as a mist, so mist eliminators from the gas stream are expected to operate to prevent these mists from being emitted to the atmosphere, to slow corrosion of air movers and other downstream equipment, and to minimize water losses [96]. Solids accumulate because of the impaction of solids-laden water droplets and the chemical precipitation of dissolved solids from the scrubbing liquid. Mist eliminators usually have at least one spray nozzle to remove these accumulated solids. Chevrons, mesh or woven pads, tube banks and cyclones are the most common types of mist eliminators. Static pressure drops across their surfaces range from 0.5 in. water column (WC) to more than 4 in. WC [96].

Mists usually contain liquid droplets from submicron size to about 10 μm. If the diameter exceeds 10 μm, the aerosol is usually referred to as a spray or simply as droplets [96]. Mists are nearly spherical because of their surface tension and are usually formed by nucleation and the condensation of vapors [97]. Larger droplets are formed from bursting bubbles, by entrainment from surfaces, by spray nozzles, or by splash-type liquid distributors. The large droplets tend to be elongated relative to their direction of motion because of the action of drag forces on the droplets [98].

Scrubber performance depends on many factors, including:

1. Particle size distribution
2. Mean and maximum gas flow rates
3. Mean and maximum inlet gas temperatures
4. Chemical composition of PM and gas stream, including concentrations of corrosive materials present in the inlet gas stream
5. Concentrations of potentially explosive materials present in the inlet gas stream
6. Make-up water availability
7. Purge liquid treatment and disposal requirements
8. Process type, raw materials, and fuels
9. Source operating schedule
10. Area available for scrubber and wastewater treatment
11. Alkali supply requirements
12. Emission test data [96].

The discussion has focused on wet scrubbers, but scrubbers can also operate in a dry mode, e.g., gravel bed filters that recirculate the gravel filter medium using an external cleaning or washing system. Some units combine other removal mechanisms such as sending an electrostatic field across the gravel bed to enhance removal of the particulate material and adding a fabric filter. Dry modes are generally much more efficient to operate when removing large quantities fly ash, for example, from hot gas streams.

Another novel configuration is the bioscrubber, which integrates suspended microbial population into the scrubbing process by adding a separate aerated vessel where suspended microbes degrade organic chemicals [99,100] and some inorganic compounds, notably H_2S [101,102]. Nutrients, acids, or bases are also added to the recirculated water to maintain conditions that support specific microbial populations [103].

4.2 Treating vapors and gases

If the air and the pollutants are both gas phase there are fewer theoretical complexities than in PM removal, Indeed, inertial mechanisms do not work for gases. Chemical reactions and physical properties like vapor pressure, solubility, and sorption are usually part of vapor phase pollutant collection and removal. The main physical processes are absorption into a liquid, adsorption onto a solid surface, and condensation from vapor phase to liquid phase. Chemical transformation to safer and less polluting substances complements these physical mechanisms.

All types of sorption, i.e., absorption, adsorption, chemisorption, and ion exchange, can be the physical basis for removing gas-phase pollutants. Often, it is best to use the inclusive term, sorption, because it is likely that two or more sorption types are working, but one may dominate in the removal.[d]

Absorption is accomplished by using a selective liquid in a matrix such as a wet scrubber, packed column, or bubble tower. Pollutant gases commonly controlled by absorption include sulfur dioxide (SO_2), H_2S, hydrogen chloride (HCl), Cl gas, ammonia (NH_3), oxides of nitrogen (NO_x), and volatile hydrocarbons [104–106]. The liquid solvent is expected to contain specified quantities of solvent. The solvent is expected to have a low vapor pressure to limit losses. Ideally, the gas is noncorrosive, nonflammable, chemically stable, and has a low freezing point [104–106]. The solvent must also not be designated as hazardous since some of its chemical constituents will be captured in the sorbent and will require special handling if hazardous. Thus, water is the most popular solvent used in absorption devices. The water may be treated with an acid or a base to enhance removal of a specific gas. If carbon dioxide is present in the gaseous effluent and water is used as the scrubbing liquid, the water solution will contain increasing concentrations carbonic acid ($H_2CO_3^*$).

Gases that readily dissolve in water, e.g., Cl_2, HCl, and hydrogen fluoride (HF), are examples of amenable to a water solvent, but its use is not universal and can have challenges. SO_2, for example, is only slightly soluble in water, so a scrubber using water would be quite large. However, SO_2 is readily soluble in an alkaline solution, so scrubbing solutions containing ammonia or amines are often preferable solvents [104–106].

During adsorption, pollutant gases are selectively retained on the surface or in the pores or interstices of prepared solids. This may be strictly a surface process with only molecular forces involved, or the process may also involve chemisorption, which is adsorption combined with a chemical reaction occurring at the surface once the gas and adsorber make contact. The solid materials used as adsorbents must be sufficiently porous and have extremely large surface-to-volume ratios. Activated carbon, for example, is excellent for removing low-density hydrocarbon molecules, many which are odorous. Silica gel is a polar, so is good at adsorbing hydrophilic gases [65,88,104–106]. Solid adsorbents must have enough structural integrity when packed into a tower to resist collapse and fracturing and, after saturation with gas molecules, be able to be regenerated to be reused. Most industrial adsorbers regenerate the adsorbent to recover the adsorbent.

The efficiency of most adsorbers approaches 100% when operation begins and remains high until a breakpoint when the adsorbent becomes saturated with adsorbate. At this breakpoint the slope of the percentage of mass of unadsorbed gaseous fluid increases with time. At the breakpoint, the adsorbent must be renewed or regenerated [88]. Industrial adsorption systems are designed to operate in an optimal range with enough time before the breakpoint. The optimal period is reset with each sorbent replacement.

Collection's next step in to condense the contaminant-laden vapor by either by decreasing the temperature or by increasing the pressure. However, this is most often accomplished by temperature decrease only. Contact condensers operate with the coolant, vapors, and condensate mixed. In surface condensers, the coolant contacts neither the vapors nor the condensate.

The control of organic gaseous emissions can occur by the oxidation of the combustible components to water and carbon dioxide. Thermal processes like incineration often work much better for homogeneous gas-phase reactions as opposed to heterogeneous reactions involving solids, liquids and gases [65]. Other systems such as the oxidation of H_2S to SO_2 and water are also used. However, the SO_2 produced is an air pollutant but is much less toxic and undesirable than the H_2S. The odor threshold for H_2S is about three orders of magnitude less than that for SO_2. If the SO_2 concentrations exceed emission or other operational limits, it will also have to be treated by scrubbing or other technology.

Afterburners are widely deployed to oxidize undesirable combustible gases. The two general types are direct-flame afterburners, in which the gases are oxidized in a combustion chamber at or above the temperature of autogenous ignition and catalytic combustion systems, in which the gases are oxidized at temperatures considerably below the autogenous ignition point [65]. Direct-flame afterburners are used to combust aerosols, vapors, gases, and odorous substances. Usual exit temperatures for the destruction of most organic materials range from 650 to 825 °C, with retention times at the elevated temperature of 0.3–0.5 °C [88].

A catalytic afterburner can destroy most pollutant gases within a temperature range of about 315–485 °C. Contamination and attrition of the catalyst requires replacement at regular intervals. Successful removal of vapor-phase

d. I recall a conversation with my Ph.D. advisor, Jeff Peirce, decades ago where he made this point. It is a common phenomenon in engineering that we look to microscopic and macroscopic explanations, knowing full well that many other mechanisms at Angstrom scales invisibly undergird what we are seeing empirically in the laboratory, field, and real-world. Jeff, now retired from Duke, is my mentor and among my favorite people. I always learn something every time we talk, just as I did when he advised by at Duke.

contaminants by catalytic systems depends on gas velocities through the system, amount of active catalyst surface, residence time, and preheat temperature necessary for complete oxidation of the emitted gases [88].

Biological treatment is a good candidate for VOC removal if an air stream has concentrations of VOCs that are sufficiently low so as not to diminish or kill the microbial communities in the reactor. Air pollution biological systems, including biofilters, biotrickling systems, and bioscrubbers, follow the same scientific principles that are the basis for treating wastewater. The biofilms of bacteria and fungi are grown on porous media. Air or other gas containing the VOCs is passed through the biologically active media, where the microbes break down the compounds to form simpler compounds, eventually degrading to water and CO_2 if aerobic, and water and CH_4, if anaerobic. The chief difference between biofiltration and trickling systems is the way the liquid interfaces with the microbes. The liquid phase is stationary in a biofilter, but liquids move through the porous media of a biotrickling system.

In these biosorption systems, air containing gas phase pollutants (C_G) traverses porous media. The soluble fraction of the volatilized compounds in the air stream moves into the biofilm (C_L) according to Henry's Law:

$$C_L = \frac{C_G}{H} \tag{15.34}$$

Where, H is the Henry's Law constant. Henry's Law defines the partitioning of the compounds from the gas phase in the carrier gas or air stream to the liquid phase, i.e., the biofilm.

Biofilters are also used to remove VOCs and odorous compounds from air streams. In addition to a wide array of volatile chain and aromatic organic compounds, biological systems have successfully removed vapor phase inorganics, such as ammonia, hydrogen sulfide and other sulfides including carbon disulfide, as well as mercaptans [65].

The bioscrubber is a two-unit setup. The first unit is an adsorption unit, which may be a spray tower, bubbling scrubber, or packed column. After this unit, the air stream enters a bioreactor with a design quite like an activated sludge system in a wastewater treatment facility. Bioscrubbers are much less common in the US than biofiltration systems.

Most biological systems have relatively low operating costs since they are operated near ambient temperature and pressure conditions. Power needs are generally for air movement and pressure drops are low (<10 cm H_2O m^{-1} or <1.2 in. WC ft^{-1} of packed bed). Other costs include amendments (e.g., nutrients) and humidification. Another advantage is the usual small number of toxic byproducts, as well as low rates of emissions of greenhouse gases (oxides of nitrogen and carbon dioxide), compared to thermal systems. Success is highly dependent on the degradability of the compounds present in the air stream, their fugacity and solubility needed to enter the biofilm (see Fig. 15.9), and pollutant loading rates. Care must to be taken to monitor the porous media for incomplete biodegradation, the presence of substances that may be toxic to the microbes, excessive concentrations of organic acids and alcohols, and pH. The system must also be checked for shock and the presence of dust, grease or other substances that may clog the pore spaces of the media [58,64].

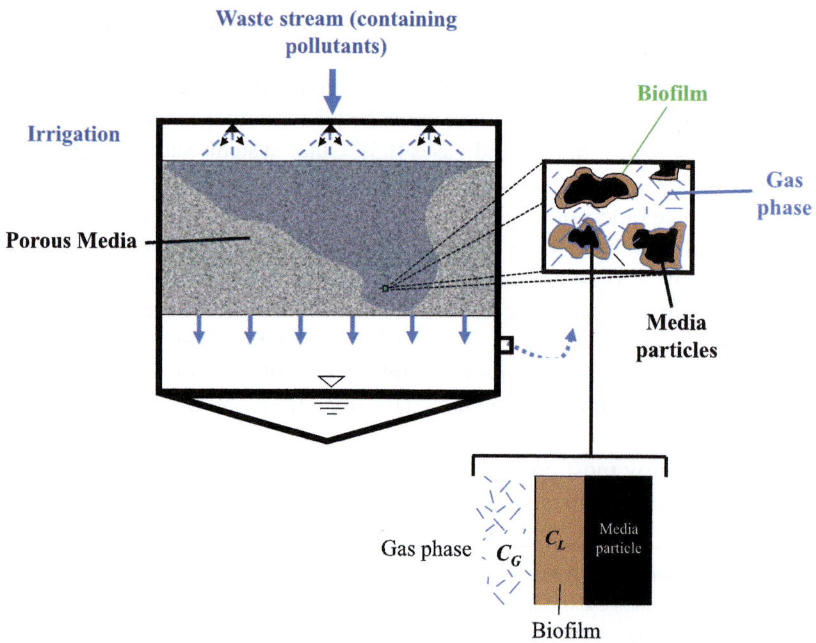

FIG. 15.9 Trickling filter system providing biotreatment of wastewater.

5 Water treatment

Public water must meet drinking water requirements. Communities have various strategies to treat and supply water. Water treatment plants clean and disinfect the water before it is distributed. The physical processes used to treat supplied water are often the same as those used in wastewater treatment plants. The main treatment steps are coagulation, flocculation, sedimentation, filtration, and disinfection.

Water that will be distributed from the water plant is first taken from an aquifer or surface waterbody. This water must be treated to ensure its safety when used by the population served by the water plant. Water plants are smaller than wastewater treatment plants for the same jurisdiction because wastewater requires much more intensive and extensive treatment. Commonly, water treatment employs coagulation and flocculation steps to remove suspended fine particles and colloids that are present in the aquifer or surface water source. Coagulation, typically the first step in water treatment, is accomplished, for example, by adding aluminum sulfate [$Al_2(SO_4)_3$], commonly called alum. The positive charge of the added chemical neutralizes the negative charge of soil particles and other particles in the water. The particles bind with the chemical compounds to form somewhat larger particles. In addition to alum, various salts, aluminum, and iron are common chemical compounds used to coagulate particles [107,108].

After coagulation, flocculation is deployed. Flocculation is a physical process of gently mixing the water to form coarser and denser particles called flocs [107,108]. Like coagulation, water treatment plants add chemical compounds during this step to enhance floc formation.

Sedimentation is a means of separating the floc and suspended solids from the water. The heavier aggregates settle to the bottom of the tank.

After the flocs have settled, the clearer water at the top of the tank is filtered to separate the finer-grained particles. The clearer water passes through filters with various pore sizes and different composition, e.g., sand, gravel, charcoal, and granulated activated carbon (GAC). These filters remove dissolved particles and germs, such as dust, chemicals, parasites, bacteria, and viruses. The GAC filters also remove organic compounds and odor-causing substances. Like wastewater, biological treatment technologies are also used to enhance water treatment [109–112].

Water treatment plants use ultrafiltration in addition to or instead of traditional filtration, depending on the pollutants in the water. Ultrafiltration consists of passing the water through a filter membrane with very small pores. This filter exclusively allows water and small molecules to pass through [113].

Reverse osmosis (RO) is another filtration method that removes additional particles and compounds from water. Indeed, this is used at water plants and at point-of-use in homes. RO and nanofiltration membrane separation technologies have proven to be effective methods for the removal of several recalcitrant contaminants, most recently poly- and perfluoroalkyl substances (PFASs), which are not effectively treated using the traditional water treatment approaches [114]. Water treatment plants also use RO to treat reused water and saltwater [115].

After the water has been filtered, water treatment plants often add one or more chemical disinfectants, e.g., chlorine, chloramines, and chlorine dioxide to kill parasites, bacteria, or viruses that may remain in the water. Disinfectants are also added at points in the distribution system to kill microorganisms that have infiltrated the system or otherwise grew in the system. Ultraviolet light and ozone (O_3) are also used to disinfect finished water at the plant but are not effective in the distribution system [116].

Water treatment plants commonly adjust the acidity and add fluoride after the disinfection step. Keeping pH at an optimal range, e.g., pH 6–8, improves taste, reduces corrosion of the distribution system, and extends the time for chemical disinfectants to continue working between the plant and the user's tap.

Communities vary on treatment approaches, depending on the source water quality. Among the other technologies to consider are activated alumina, which deploys porous aluminum-based material to absorb contaminants, anion exchange, which is a physicochemical process where ions are exchanged between solutions and solid resins that absorb anions and release chlorine to the water, and iron-based adsorptive media. Anion and cation exchanges can be combined in mixed bed exchange systems. Greensand filtration uses a manganese-coated filter to absorb and oxidize pollutants. These and other emerging techniques are available [117–120].

The other main water treatment plant process is disinfection to remove pathogens before distributing the water. Historically, this was accomplished by adding an agent that oxidizes microbes, e.g., one of the chemical species of chlorine (Cl), e.g., chloramines and chlorine dioxide (ClO_2). The problem with Cl is that during distribution, chlorinated carcinogenic compounds can form to which consumers may be exposed.

6 Conclusions

This chapter explored the physical principles involved in removing pollutants from wastewater. The treatment physics, chemistry, and biology are based on the laws of thermodynamics and motion, but their applications differ depending on the characteristics of the pollutant and the fluids where the pollutants reside. Preliminary, primary, secondary, and advanced wastewater treatment systems were discussed. Air pollutant treatments were described since wastewater treatment entails releases to air in its processes, including sludge incineration. Water supply technologies were also introduced.

References

[1] L. Theodore, A.J. Buonicore, Air Pollution Control Equipment: Selection, Design, Operation and Maintenance, 1982.

[2] D.A. Vallero, Wastewater, in: Waste, Elsevier, 2019, pp. 259–290.

[3] H. Zhong, Y. Tang, H. Yan, Y. Zhang, L. Dong, B. Wang, Corrosion of pipelines in urban water systems: current research status and future trends based on bibliometric analysis, J. Water Process Eng. 56 (2023) 104288.

[4] B.V. Salas, et al., H2S pollution and its effect on corrosion of electronic components, in: Air Quality—New Perspective, IntechOpen, London, 2012, pp. 263–285.

[5] R. Rathish, et al., Corrosion issues in electronic equipments—an overview, Int. J. Corros. Scale Inhib. 8 (4) (2019) 799–815.

[6] P. Rao, L. Mulky, Erosion-corrosion of materials in industrial equipment: a review, ChemElectroChem 10 (16) (2023) e202300152.

[7] A. Ghosh, H. Adams, S. Ash, M. Southard, Operators need to know how to maintain corrosion control conditions, Opflow 49 (7) (2023) 18–22.

[8] A. El-Meligi, Corrosion preventive strategies as a crucial need for decreasing environmental pollution and saving economics, Recent Pat. Corros. Sci. 2 (1) (2010).

[9] M. Johnson, Wastewater laboratory basics, in: Presented at the IWEA Laboratory Workshop, Springfield, Illinois, 2013. Available: https://www.iweasite.org/past_conference_slides.php.

[10] W. Telliard, Method 1684: total, fixed, and volatile solids in water, solids, and biosolids, US Environmental Protection Agency, Washington, 2001.

[11] Water Environment Federation, American Public Health Association, Standard Methods for the Examination of Water and Wastewater, American Public Health Association (APHA), Washington, DC, USA, 2005.

[12] M.C. Collivignarelli, M. Canato, A. Abba, M.C. Miino, Biosolids: what are the different types of reuse? J. Clean. Prod. 238 (2019) 117844.

[13] Q. Lu, Z.L. He, P.J. Stoffella, Land application of biosolids in the USA: a review, Appl. Environ. Soil Science 2012 (2012).

[14] C.P. Gerba, I.L. Pepper, Wastewater treatment and biosolids reuse, in: Environmental Microbiology, second ed., Elsevier Inc., USA, 2009, pp. 503–530.

[15] National Research Council, Biosolids Applied to Land: Advancing Standards and Practices, National Academies Press, 2002.

[16] J. Walker, L. Knight, L. Stein, Plain English Guide to the EPA Part 503 Biosolids Rule, 1994.

[17] D. Vallero, Environmental Biotechnology: A Biosystems Approach, Elsevier Science, 2015.

[18] U.S. Environmental Protection Agency, Method 410.3: chemical oxygen demand (titrametric, high level for saline waters) by titration, 1978. https://www.epa.gov/sites/default/files/2015-08/documents/method_410-3_1978.pdf;. (Accessed 13 July 2024).

[19] J.W. O'Dell, U.S. Environmental Protection Agency, The determination of chemical oxygen demand by semi-automated colorimetry, 1993. https://www.epa.gov/sites/default/files/2015-08/documents/method_410-4_1993.pdf, Revision 2.0. (Accessed 13 July 2024).

[20] V. Novotny, Water Quality: Diffuse Pollution and Watershed Management, John Wiley & Sons, 2003.

[21] H. Bridle, K. Jacobsson, A.C. Schultz, Sample processing, in: Waterborne Pathogens, Elsevier, 2021, pp. 63–109.

[22] S. Shukla, R. Khan, A. Saxena, S. Sekar, Microplastics from face masks: a potential hazard post Covid-19 pandemic, Chemosphere 302 (2022) 134805.

[23] K.D. Kerri, Operation of Wastewater Treatment Plants: A Field Study Training Program, California State University, 2004.

[24] Michigan Department of Environmental Quality Operator Training and Certification Unit, Clarifier calculations, 2022, Available: https://www.michigan.gov/documents/deq/wrd-ot-clarifier-calculations_445211_7.ppt.

[25] H.E. Åberg, S. Tondelli, Escape to the country: a reaction-driven rural renaissance on a Swedish Island post COVID-19, Sustainability 13 (22) (2021) 12895.

[26] R. Florida, A. Rodríguez-Pose, M. Storper, Critical commentary: cities in a post-COVID world, Urban Stud 60 (8) (2023) 1509–1531, https://doi.org/10.1177/00420980211018072. 00420980211018072.

[27] N. Ahuja, A.G. Griborio, Liquid stream fundamentals: sedimentation, 2017, Alexandria, Virginia, WSEC-2017-FS-022. Available: https://www.wef.org/globalassets/assets-wef/direct-download-library/public/03- - -resources/wsec-2017-fs-022-liquid-stream-fundamentals- -clarification-sedimentation_final.pdf. (Accessed 21 January 2022).

[28] Department of Environment and Conservation, Division of Water Resources, Design criteria for review of sewage works—construction plans and documents, 2020, DWR-NPDES-SOP-G-05-WW. Available: https://www.tn.gov/content/dam/tn/environment/water/policy-and-guidance/dwr-npdes-sop-g-05-ww-design-criteria-table-of-contents-072020.pdf.

[29] National Research Council, Use of Reclaimed Water and Sludge in Food Crop Production, National Academies Press, 1996.

[30] B. Jiménez, D. Mara, R. Carr, F. Brissaud, Wastewater treatment for pathogen removal and nutrient conservation: Suitable systems for use in developing countries, in: Wastewater Irrigation and Health, Routledge, 2009, pp. 175–196.

[31] EPA, Handbook for Managing Onsite and Clustered (Decentralized) Wastewater Treatment Systems: An Introduction to Management Tools and Information for Implementing EPA's Management Guidelines, 2005. EPA No. 832-B-05-001.

[32] D.A. Vallero, C.K. Gunsch, Applications and implications of emerging biotechnologies in environmental engineering, J. Environ. Eng. 146 (6) (2020) 03120005.

[33] B.E. Rittmann, P.L. McCarty, Environmental Biotechnology: Principles and Applications, Tata McGraw-Hill Education, 2012.

[34] B.E. Rittmann, Environmental biotechnology in water and wastewater treatment, J. Environ. Eng. 136 (4) (2009) 348–353.

[35] A.H. Scragg, Environmental Biotechnology, Oxford University Press, New York, 2005.

[36] D.A. Vallero, J.J. Peirce, Engineering the Risks of Hazardous Wastes, Butterworth-Heinemann, Burlington, MA, 2003. p. xxi, 306 p.

[37] A. Boudagga, H. Dimassi, S. Hadj-Said, F. M'Sahli, Robust state estimation for a nonlinear hybrid model of the alternating activated sludge process using filtered high-gain observers, Math. Probl. Eng. 2020 (2020).

[38] M. Dagnew, W.J. Parker, P. Seto, A pilot study of anaerobic membrane digesters for concurrent thickening and digestion of waste activated sludge (WAS), Water Sci. Technol. 61 (6) (2010) 1451–1458.

[39] A. Ford, S. Hawley, B. Rutherford, K. Uprety, C. Bott, Implementation of aeration control strategies and nitrate-based internal mixed liquor recycle control employing in-situ sensors and feedback PID controllers in an integrated fixed-film activated sludge wastewater treatment facility, Proc. Water Environ. Fed. 2018 (18) (2018) 251–262.

[40] F. Zhao, et al., New insights into eutrophication management: importance of temperature and water residence time, J. Environ. Sci. 111 (2022) 229–239.

[41] S. Wallace, Engineered wetlands lead the way, Land Water 48 (5) (2009). Available: http://www.landandwater.com/features/vol48no5/vol48no5_1.html.

[42] I. Chorus, E. Spijkerman, What Colin Reynolds could tell us about nutrient limitation, N: P ratios and eutrophication control, Hydrobiologia 848 (1) (2021) 95–111.

[43] D.A. Vallero, K.H. Reckhow, A.D. Gronewold, Application of multimedia models for human and ecological exposure analysis, in: International Conference on Environmental Epidemiology and Exposure, International Society of Exposure Analysis, Durham, NC, 2007.

[44] T. Hvitved-Jacobsen, Sewer Processes: Microbial and Chemical Process Engineering of Sewer Networks, CRC Press, 2001.

[45] M. Cyprowski, A. Stobnicka-Kupiec, A. Ławniczek-Wałczyk, A. Bakal-Kijek, M. Gołofit-Szymczak, R.L. Górny, Anaerobic bacteria in wastewater treatment plant, Int. Arch. Occup. Environ. Health 91 (5) (2018) 571–579, https://doi.org/10.1007/s00420-018-1307-6 (in Eng).

[46] Y. Liu, Q. Dong, H. Shi, Distribution and population structure characteristics of microorganisms in urban sewage system, Appl. Microbiol. Biotechnol. 99 (18) (2015) 7723–7734, https://doi.org/10.1007/s00253-015-6661-7 (in Eng).

[47] B. Li, et al., Recuperative thickening for sludge retention time and throughput management in anaerobic digestion with thermal hydrolysis pretreatment, Water Environ. Res. 92 (3) (2020) 465–477.

[48] D.A. Vallero, Hazardous wastes, in: T.M. Letcher, D.A. Vallero (Eds.), Waste, second ed., Academic Press, 2019, pp. 585–630.

[49] A. Bal, N. Dhagat, Upflow anaerobic sludge blanket reactor a review, Indian J. Environ. Health 43 (2) (2001) 1–82.

[50] D.A. Abramowicz, Aerobic and anaerobic PCB biodegradation in the environment, Environ. Health Perspect. 103 (Suppl. 5) (1995) 97.

[51] P. Elefsiniotis, The effect of operational and environmental parameters on the acid-phase anaerobic digestion of primary sludge, University of British Columbia, 1993.

[52] D. Christensen, J. Gerick, J. Eblen, Design and operation of an upflow anaerobic sludge blanket reactor, J. Water Pollut. Control Fed. (1984) 1059–1062.

[53] EPA, Wastewater technology fact sheet: package plants, 2000. EPA 832-F-00-016.

[54] U.S. Environmental Protection Agency, Assessment and remediation of contaminated sediments (ARCS) program: remediation guidance document, Chicago, Illinois, EPA 905-R94-003, 1994.

[55] U.S. Environmental Protection Agency, 2,3,7,8-Tetrachlorodibenzo-p-dioxin - 1746-01-6 | DTXSID2021315, 2021. https://www.epa.gov/chemical-research/chemistry-dashboardhttps://comptox.epa.gov/dashboard/chemical/details/DTXSID2021315. (Accessed 22 December 2021).

[56] National Toxicology Program, Report on carcinogens: 2,3,7,8-tetrachlorodibenzo-p-dioxin, 2016, Research Triangle Park, North Carolina. Available: https://ntp.niehs.nih.gov/ntp/roc/content/profiles/tetrachlorodibenzodioxin.pdf. (Accessed 20 December 2021).

[57] U.S. Environmental Protection Agency, Federal Remediation Technologies Roundtable, Air stripping (ex situ), 2020. https://frtr.gov/matrix/Air-Stripping-Ex-Situ/#Implementability-Considerations. (Accessed 12 July 2024. Technology Screening Matrix.

[58] S.J. Ergas, K.A. Kinney, Biological control systems, in: Air Pollution Engineering Manual, 2000, p. 55.

[59] C. Hort, S. Gracy, V. Platel, L. Moynault, Evaluation of sewage sludge and yard waste compost as a biofilter media for the removal of ammonia and volatile organic sulfur compounds (VOSCs), Chem. Eng. J. 152 (1) (2009) 44–53.

[60] C. Ima, D. Mann, Physical properties of woodchip: compost mixtures used as biofilter media, 2007.

[61] W. Namkoong, J.-S. Park, J.S. VanderGheynst, Biofiltration of gasoline vapor by compost media, Environ. Pollut. 121 (2) (2003) 181–187.

[62] R. Nicolai, K. Janni, Biofilter media mixture ratio of wood chips and compost treating swine odors, Water Sci. Technol. 44 (9) (2001) 261–267.

[63] B. Cárdenas-González, S.J. Ergas, M.S. Switzenbaum, Characterization of compost biofiltration media, J. Air Waste Manage. Assoc. 49 (7) (1999) 784–793.

[64] S.J. Ergas, E.D. Schroeder, D.P. Chang, R.L. Morton, Control of volatile organic compound emissions using a compost biofilter, Water Environ. Res. 67 (5) (1995) 816–821.

[65] D.A. Vallero, Fundamentals of Air Pollution, fifth ed., Elsevier Academic Press, Waltham, MA, 2014. p. 999 pages cm.

[66] H. Huo, Y. Wu, M. Wang, Total versus urban: well-to-wheels assessment of criteria pollutant emissions from various vehicle/fuel systems, Atmos. Environ. 43 (10) (2009) 1796–1804.

[67] K. Isaacs, J.A. Rosati, T.B. Martonen, Modeling deposition of inhaled particles, in: Aerosols Handbook: Measurement, Dosimetry, and Health Effects, CRC Press, 2012 (Chapter 5).
[68] C. Liu, S. Shi, C. Weschler, B. Zhao, Y. Zhang, Analysis of the dynamic interaction between SVOCs and airborne particles, Aerosol Sci. Technol. 47 (2) (2013) 125–136.
[69] D.A. Vallero, Air pollution, in: Kirk-Othmer Encyclopedia of Chemical Technology, John Wiley & Sons, Inc., 2000.
[70] R. Meier, et al., Differences in indoor versus outdoor concentrations of ultrafine particles, PM2. 5, PMabsorbance and NO2 in Swiss homes, J. Expo. Sci. Environ. Epidemiol. (2015).
[71] A.L. Moreno-Ríos, L.P. Tejeda-Benítez, C.F. Bustillo-Lecompte, Sources, characteristics, toxicity, and control of ultrafine particles: an overview, Geosci. Front. 13 (1) (2022) 101147.
[72] H.-S. Kwon, M.H. Ryu, C. Carlsten, Ultrafine particles: unique physicochemical properties relevant to health and disease, Exp. Mol. Med. 52 (3) (2020) 318–328.
[73] K. Donaldson, V. Stone, W. MacNee, The toxicology of ultrafine particles, in: Particulate Matter: Properties and Effects Upon Health, Garland Science, 2020, pp. 115–129.
[74] D.E. Schraufnagel, The health effects of ultrafine particles, Exp. Mol. Med. 52 (3) (2020) 311–317.
[75] S. Ohlwein, R. Kappeler, M.K. Joss, N. Künzli, B. Hoffmann, Health effects of ultrafine particles: a systematic literature review update of epidemiological evidence, Int. J. Public Health 64 (4) (2019) 547–559.
[76] X.D. Zhu, Y. Liu, J. Shen, Volatile organic compounds (VOCs) emissions of wood-based panels coated with nanoparticles modified water based varnish, Eur. J. Wood Wood Prod. 74 (4) (2016) 601–607, https://doi.org/10.1007/s00107-016-1012-7.
[77] D.H. Brouwer, et al., Occupational dermal exposure to nanoparticles and nano-enabled products: part 2, exploration of exposure processes and methods of assessment, Int. J. Hyg. Environ. Health 219 (6) (2016) 503–512.
[78] X. Zou, J. Shi, H. Zhang, Coexistence of silver and titanium dioxide nanoparticles: enhancing or reducing environmental risks? Aquat. Toxicol. 154 (2014) 168–175.
[79] J. García-Alonso, et al., Toxicity and accumulation of silver nanoparticles during development of the marine polychaete Platynereis dumerilii, Sci. Total Environ. 476 (2014) 688–695.
[80] A. Kumar, Making a case for human health risk-based ranking nanoparticles in water for monitoring purposes, Environ. Sci. Technol. 46 (10) (2012) 5267–5268.
[81] J.R. Lead, Manufactured nanoparticles in the environment, Environ. Chem. 7 (1) (2010) 1–2.
[82] C. Polakowski, M. Ryżak, A. Sochan, M. Beczek, R. Mazur, A. Bieganowski, Particle size distribution of various soil materials measured by laser diffraction—the problem of reproducibility, Minerals 11 (5) (2021) 465.
[83] T. Allen, Particle Size Measurement, Springer, 2013.
[84] L. Silverman, Particle Size Analysis in Industrial Hygiene, Elsevier, 2012.
[85] K. Schifftner, Air Pollution Control Equipment Selection Guide, CRC Press, 2021.
[86] D.A. Vallero, Air Pollution Calculations: Quantifying Pollutant Formation, Transport, Transformation, Fate and Risks, Elsevier, 2019.
[87] J. Sorrels, J. Coburn, K. Bradley, D. Randall, EPA Air Pollution Control Cost Manual, United States Environmental Protection Agency, Washington, DC, 2017.
[88] R.W. Boubel, D.L. Fox, B. Tuner, A.C. Stern, Fundamentals of Air Pollution, third ed., Academic Press, Amsterdam; Boston, 1995.
[89] D.R. Bullard, et al., APTI course SI:409: basic air pollution meteorology: student guidebook, U.S. Environmental Protection Agency, Air Pollution Training Institute, Research Triangle Park, NC, 1982 (in English).
[90] J. Richards, Control of Particulate Matter Emissions-Student Manual, United States Environmental Protection Agency, 2000.
[91] W.J. Franek, L. DeRose, Control of particulate matter emissions. Chapter 3: Particle sizing, U.S. Environmental Protection Agency, O. o. A. Q. P. a. S. Office of Air and Radiation (Eds.), in: Student Manual: APTI 413, U.S. EPA, Research Triangle Park, North Carolina, 2021. sec. Washington, DC.
[92] I.M. Hutten, Filtration mechanisms and theory, in: I.M. Hutten (Ed.), Handbook of Nonwoven Filter Media, Butterworth-Heinemann, Oxford, 2007, pp. 29–70 (Chapter 2).
[93] W.J. Franek, L. DeRose, Control of particulate matter emissions. Chapter 4: Particle collection mechanisms, U.S. Environmental Protection Agency, O. o. A. Q. P. a. S. Office of Air and Radiation (Eds.), in: Student Manual: APTI 413, U.S. EPA, Research Triangle Park, North Carolina, 2021. sec. Washington, DC.
[94] D.A. Vallero, Hazardous waste bioremediation, in: D. Yadav, P. Kumar, P. Singh, D.A. Vallero (Eds.), Hazardous Waste Management, Elsevier, 2022, pp. 347–379 (Chapter 14).
[95] D. Yadav, P. Kumar, P. Singh, D.A. Vallero, Hazardous Waste Management: An Overview of Advanced and Cost-Effective Solutions, 2021.
[96] W.J. Franek, L. DeRose, Control of particulate matter emissions. Chapter 8: Wet scrubbers, U.S. Environmental Protection Agency, O. o. A. Q. P. a. S. Office of Air and Radiation (Eds.), in: Student Manual: APTI 413, U.S. EPA, Research Triangle Park, North Carolina, 2021. sec. Washington, DC.
[97] C.G. Bell, W. Strauss, Effectiveness of vertical mist eliminators in a cross flow scrubber, J. Air Pollut. Control Assoc. 23 (11) (1973) 967–969.
[98] H. Perry, D. Green, Perry's Chemical Engineers Handbook, Mc-Graw Hill, Hoboken, New Jersey, 1997.
[99] P. Le Cloirec, P. Humeau, Bioscrubbers, in: Air Pollution Prevention and Control: Bioreactors and Bioenergy, 2013, pp. 139–153.
[100] J.W. Van Groenestijn, Bioscrubbers, in: Bioreactors for Waste Gas Treatment, Springer, 2001, pp. 133–162.

[101] P. San-Valero, J.M. Penya-Roja, F.J. Álvarez-Hornos, G. Buitrón, C. Gabaldón, G. Quijano, Fully aerobic bioscrubber for the desulfurization of H2S-rich biogas, Fuel 241 (2019) 884–891.

[102] S. Potivichayanon, P. Pokethitiyook, M. Kruatrachue, Hydrogen sulfide removal by a novel fixed-film bioscrubber system, Process Biochem. 41 (3) (2006) 708–715.

[103] I.L. Pepper, C. Gerba, T. Gentry, EES642: Environmental microbiology, in: Courses of Study, 2019, p. 141.

[104] P.H. Hunter, Control of Volatile Organic Compound (VOC) Air Pollutants, Virginia Tech, 2000.

[105] J. Joseph, D. Beachler, J. Jahnke, M. Leslie, J. Henry, APTI course 415: control of gaseous emissions. Instructor's guide, Northrop Services, Inc./ Environmental Sciences, Research Triangle Park, NC (USA), 1981.

[106] G.T. Joseph, D. Beechler, APTI course 415: control of gaseous emissions. Student manual, Northrop Services, Inc./Environmental Sciences, Research Triangle Park, NC (USA), 1981.

[107] U.S. Army Corps of Engineers, Engineering and Design: Precipitation/Coagulation/Flocculation, 2001, Department of Defense, EM 1110-1-4012. Available: https://www.publications.usace.army.mil/portals/76/publications/engineermanuals/em_1110-1-4012.pdf. (Accessed 9 November 2021) (Chapter 2).

[108] T.A. Malkoske, P.R. Bérubé, R.C. Andrews, Coagulation/flocculation prior to low pressure membranes in drinking water treatment: a review, Environ. Sci. Water Res. Technol. 6 (11) (2020) 2993–3023.

[109] W.A. Nugroho, J. Reungoat, J. Keller, The performance of biological activated carbon in removing pharmaceuticals in drinking water treatment, J. Appl. Sci. Environ. Sanit. 5 (2) (2010).

[110] B.E. Logan, J.R. Hunt, Bioflocculation as a microbial response to substrate limitations, Biotechnol. Bioeng. 31 (2) (1988) 91–101.

[111] M.J. Higgins, J.T. Novak, Characterization of exocellular protein and its role in bioflocculation, J. Environ. Eng. 123 (5) (1997) 479–485.

[112] J.P. Boltz, E.J. La Motta, J.A. Madrigal, The role of bioflocculation on suspended solids and particulate COD removal in the trickling filter process, J. Environ. Eng. 132 (5) (2006) 506–513.

[113] S. Al Aani, T.N. Mustafa, N. Hilal, Ultrafiltration membranes for wastewater and water process engineering: a comprehensive statistical review over the past decade, J. Water Process Eng. 35 (2020) 101241.

[114] T.F. Mastropietro, R. Bruno, E. Pardo, D. Armentano, Reverse osmosis and nanofiltration membranes for highly efficient PFASs removal: overview, challenges and future perspectives, Dalton Trans. 50 (16) (2021) 5398–5410.

[115] X. Wei, K.T. Sanders, A.E. Childress, Reclaiming wastewater with increasing salinity for potable water reuse: water recovery and energy consumption during reverse osmosis desalination, Desalination 520 (2021) 115316.

[116] G. Gelete, H. Gokcekus, D.U. Ozsahin, B. Uzun, T. Gichamo, Evaluating disinfection techniques of water treatment, Desalin. Water Treat. 177 (2020) 408–415.

[117] U.S. Environmental Protection Agency, National primary drinking water regulations, USEPA, 2024. https://www.epa.gov/ground-water-and-drinking-water/national-primary-drinking-water-regulations. (Accessed 4 January 2024).

[118] U.S. EPA, Per- and polyfluoroalkyl substances (PFAS): proposed PFAS national primary drinking water regulation, U.S. EPA, 2024. In this issue https://www.epa.gov/sdwa/and-polyfluoroalkyl-substances-pfas. (Accessed 12 July 2024. Last update: July 12, 2024.

[119] World Health Organization, Guidelines for drinking-water quality: incorporating the first and second addenda, World Health Organization, Geneva, Switzerland, 2022.

[120] G. Hu, et al., Human health risk-based life cycle assessment of drinking water treatment for heavy metal (loids) removal, J. Clean. Prod. 267 (2020) 121980.

Chapter 16

Ground water remediation

1 Introduction

As discussed in Chapter 11, ground water is the source of drinking water for much of the U.S. [1], Europe, [2], and most of the world [3]. Ground water can become contaminated in many ways and from myriad sources because the environment is a highly interconnected system between surface and subsurface water systems, and among the air, soil, sediment, and biota within these systems (see Fig. 16.1).

The first part of an aquifer to become polluted is the unsaturated zone, also known as the vadose zone. This zone is comprised of particles and pore spaces. Water in the unsaturated zone is a thin film around the particles and in the interstices between the particles but does not completely fill the particle interstices. In addition to the water in the pore spaces there are air and other fluids, including contaminants, but also nutrients needed by plants and soil microbes [4]. Chemical substances move in the fluids, mainly by gravity, through the pore spaces. The movement is also driven by the difference in pore water pressure, i.e., moisture potential. These potentials vary by factors like pore size and moisture content [5]. As mentioned in Chapter 11, the total potential head (h) is the sum of the gravity potential, Z, and the moisture potential $\Psi(\theta)$:

$$h = \sum Z + \Psi(\theta) \tag{16.1}$$

The vadose zone is a transition zone [4]. The water table (▼) bounds the saturated zone and the unsaturated zone. It is the depth at which the pressure of the pore water equals the atmospheric pressure [6]. Combining the flow equation with the principle of continuity [7], gives the Richards equation for flow in the vadose zone [8]. In one dimension, i.e., the vertical (z) dimension, the Richards equation is expressed as:

$$\frac{\partial \theta}{\partial t} = \frac{\partial}{\partial t}\left[K(\Psi)\left(\frac{\partial \psi}{\partial z} + 1\right)\right] \tag{16.2}$$

The equation can be simplified by adding the specific water capacity of the porous medium, $C(\theta)$:

$$D(\theta) = \frac{K(\theta)}{C(\theta)} \qquad ((16.3)$$

Where is the water diffusivity of the porous media. Thus, the Richards equation can be further restated as [9]:

$$\frac{\partial \theta}{\partial t} = \frac{\partial}{\partial t}\left[D(\theta)\left(\frac{\partial \theta}{\partial z} + K(\theta)\right)\right] \tag{16.4}$$

Natural processes are at work in the unsaturated zone to break down chemical compounds, such as in soil organic matter (SOM) and plant litter when these molecules undergo decomposition by soil microbes and move through plant roots and soil fauna where they are broken down by plant mechanisms like photosynthesis and respiration. As such, carbon dioxide (CO_2) reacts with soil water to form carbonic acid (H_2CO_3*).[a] The organic acids are released from plant root exudates or produced by microbes in the soil or the rhizosphere, i.e., the area of soil near and around plant roots that influences their growth, respiration, and nutrient exchange [10,11]. Organic acids, like CO_2, may also be released from the decomposition of plant residues in soil [12,13]. These and other processes are natural processes of biodegradation. Unfortunately, these processes are typically too slow or unable to break down persistent chemical compounds to protect the quality of ground

a. The asterisk (*) indicates that is an "apparent" compound, i.e., a composite of aqueous CO_2 and pure carbonic acid, H_2CO_3 without the *. The pure acid is much stronger than the apparent acid ($pK_a = 3.70$ versus $pK_a = 6.35$). Thus, the formula for apparent carbonic acid is $H_2CO_3^* = CO_2\,(aq) + H_2CO_3$ (See Chapter 8).

Fundamentals of Water Pollution. https://doi.org/10.1016/B978-0-443-28987-3.00022-9

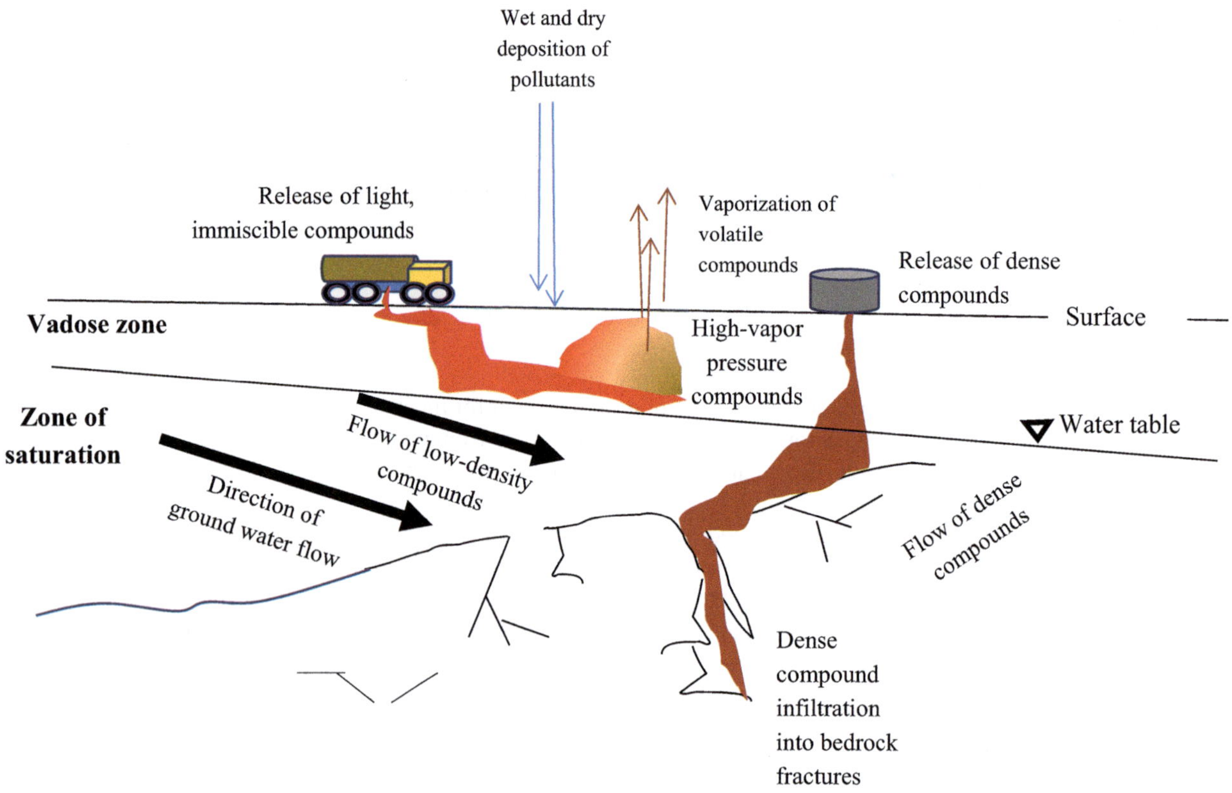

FIG. 16.1 Sources of ground water pollution.

water [14–16]. However, when an aquifer is contaminated with less persistent and more biodegradable compounds, natural abiotic and biotic degradation may be preferable treatment technologies. This is known as natural attenuation [17–23].

The environment is filled with mixtures and heterogeneous reactions. Reactions can occur in sequence, parallel, or both. For example, a feedstock to a landfill contains myriad types of wastes, from garbage to household chemicals, to commercial wastes, to industrial wastes. These constituents contain a wide array of compounds, many of which can reach an aquifer. Ground water pollutants exist in all phases, e.g., dissolved in water, sorbed to a particle, as vapors, and dissolved in other fluids. A pollutant may remain unchanged in an aquifer during its residence time, be chemically transformed in compounds that have different mobility and toxicity than the parent compound, or be transported to another environmental compartment. When a less toxic substance becomes transformed chemically into a toxic substance it is known as bioactivation.

A recalcitrant chemical compound can persist in the ground water for decades due to the slow transport rate of most aquifers (commonly $<\text{m year}^{-1}$). As discussed in Chapter 11, the capability of a medium to transmit a fluid is expressed as the medium's hydraulic conductivity (K). Aquifers mainly consist of strata of unconsolidated and porous materials like soil, sand, or gravel. In the saturated zone, assuming laminar flow and ignoring inertia, Darcy's Law describes the flow of fluids through a cross-section of porous media (A). It states that the specific discharge, Q, is directly proportional to K and hydraulic gradient [24]:

$$Q = -KA\frac{dh}{dt} \tag{16.5}$$

When measuring the rates of remediation of a contaminated aquifer, monitoring wells must be sited to account for variations in hydraulic gradients, which can cause downwelling flow, i.e., flow from surface water to ground water, or upwelling flow, i.e., flow from ground water to the surface water. The water flow over various streambed shapes and changes in stream profile will alter hydraulic head gradients and affect pollutant exchanges between surface and ground water [25]. Downwelling is particularly important for oil and gas operations as hazardous substances enter aquifers via leaks in casings and cracks in confining strata that allow the oil, gas, and their impurities to enter other strata. In these cases, the pressure within the unconfined aquifer can be greater than the pressure at the uppermost section of the confining layer with lower pressure at the ground surface [26]. Thus, to draw water to the surface, a pump is needed to increase the pressure differential between the confined aquifer and the surface, e.g., for pump-and-treat remediation [27]. If under normal conditions the pressure in the confined layer is higher than the atmospheric pressure at the ground surface, it is an artesian system and would flow without a pump [28].

Physicochemical factors are crucial to successful remediation. Notably, density and organic/aqueous phase partitioning are very important to pollutant transport through ground water, as shown in Fig. 16.1. Nonaqueous phase liquids (NAPLs) move through the vadose zone given their NAPLs given their high octanol-water coefficients (K_{ow}) and low water solubility. That is, they have a greater affinity for lipids than for water. As the aqueous solubility of a substance increases, its flow will increasingly follow the water flow lines. The first law of thermodynamics requires that the rate of change of mass in a control volume like a plume in an aquifer is equal to the rate of chemical compound A transported into the plume minus the rate of chemical compound A transported out, plus or minus chemical compound A from reactions in the plume:

$$\frac{d[A]}{dt} = -v \cdot \frac{d[A]}{dx} + \frac{d}{dx}\left(D \cdot \frac{d[A]}{dx}\right) + r \tag{16.6}$$

Where: v is the fluid velocity, $\frac{d[A]}{dx}$ is concentration gradient of chemical compound A, and r is the internal sinks and sources of A within the plume.

Eq. (16.6) is the basis for ground water contamination modeling. Chemical compounds exit and enter ground water by precipitation from the atmosphere, other surface and ground water systems, sediment, soil, and biota. As discussed in Chapter 14, basic relationships between sorption, solubility, volatilization, and organic carbon-water partitioning are respectively expressed by coefficients of sorption (distribution coefficient, K_D, or solid-water partition coefficient, K_p), dissolution or solubility coefficients, air-water partitioning (often expressed as K_{AW} or the Henry's law (K_H) constant), octanol-water partitioning (Kow) and organic carbon-water (K_{oc}) [29].

A substance's K_{ow} reflects just how much of the substance will move until the aqueous and organic solvents (phases) will reach equilibrium. This is a first clue to the type of remediation needed. For example, in a spill of equal amounts of the polychlorinated biphenyl, decachlorobiphenyl (log K_{ow} of 8.23) and the chemical 1,2-dichloropropane (log K_{ow} of 2.13), the PCB has much greater affinity for the organic phases than does the DCP (greater than five orders of magnitude). This does not mean that much of the mass of either of the compounds is likely to stay in the ground water, since they are both hydrophobic, and that it is more likely to be found in the organic fraction of soil. However, it does mean that the compounds will vary in transport time and mass of each substance in and out of the aquifer. The rates of kinetics differ, which means that the time it takes for the PCB and DCP to reach equilibrium will be different. Thus, in this case, technologies that remove semivolatile organic compounds (SVOCs) from soil are likely candidates for remediation.

When a dense, miscible fluid seeps into the zone of saturation, the contaminants move downward if their densities are greater than that of the water. When these contaminants reach the bottom of the aquifer, the shape dictates their continued movement and slope of the underlying bedrock or other relatively impervious layer, which will likely be in a direction other than the flow of the ground water in the aquifer. Solution and dispersion near the boundaries of the plume will have a secondary plume that will generally follow the general direction of ground water flow. The physics of system indicates that estimating the direction of the plume will entail more than the fluid densities and must include solubility and phase partitioning. So, monitoring wells will need to be installed upstream and downstream from the source [30,31].

An important process in ground water remediation is that of co-solvation, the process where a substance is first dissolved in one solvent and then the new solution is mixed with another solvent. As mentioned, with increasing aqueous solubility, a pollutant will travel along the flow lines of the ground water. However, even a substance with low aqueous solubility can follow the flow under certain conditions. Even a hydrophobic compound like a dense nonaqueous phase liquid, DNAPL with very low solubility in pure water, can migrate into and within ground water after it becomes dissolved in an alcohol or an organic solvent like toluene. A lipophilic compound can be transported in the vadose zone or upper part of the zone of saturation where it undergoes co-solvation with water and a light nonaqueous phase liquid (LNAPL), e.g., toluene. Thus, remediation in this case would need to consider treatment of both the soil and water that has been contaminated.

A lipophilic compound can also be transported as a suspension. Fine particles can be held in suspension if the water flow has sufficient energy. When the energy falls below a certain level, the suspended particles will settle. However, many suspended contaminants in ground water are colloidal particles, i.e., sufficiently small that Brownian motion provides enough energy to deter settling.

When a contaminant enters soil, some of the chemical remains in soil solution and some is adsorbed onto the surfaces of the soil particles. Sometimes this sorption is strong due to cations adsorbing to the negatively charged soil particles. In other cases, the attraction is weak. Sorption of chemicals onto solid surfaces needs to be understood because in this way solids can hold onto contaminants, not allowing them to move freely with the pore water or the soil solution. Therefore, sorption can retard the rate at which substances move downwardly through the soil profile. Thus, desorption is often needed as a first stage in remediation. Other properties are important in ground water contamination, including those discussed in Chapter 2. Notably, density, capillarity, and surface tension are among the fluid properties that can greatly affect the rate of contaminant movement into an aquifer, as well as which part of an aquifer will be most contaminated (e.g., strong surface tension could increase sorption onto soil particles).

2 Natural attenuation

As mentioned, natural attenuation is simply allowing the processes and mechanisms in the environment to degrade ground water pollutants. Pollutant concentrations are attenuated predominantly by microbial populations naturally present in the soil and aquifer. As such, it follows the same principles as those discussed in the bioremediation section below.

In addition, abiotic mechanisms can enhance natural attenuation [32]. Sorption retards contaminant plume spreading by particles holding on to pollutants. Ground water recharge can dilute concentrations of pollutants. Evapotranspiration can move pollutants from soil and upper regions of aquifers and move them to the upper soil horizons and to the air, after which biodegradation and photolysis, respectively, can continue the breakdown of pollutants. Abiotic chemical reactions, such as the redox can change the valence state of metals, rendering them less toxic and mobile. However, the abiotic processes that best support natural attenuation are those that improve the microbial populations' acclimation to use the contaminants as energy sources and those that encourage the growth and reproduction of beneficial microbes.

Natural attenuation is not the same as ignoring the pollution. Any acceptable plan will include ample monitoring of changes in pollutant concentrations with time, i.e., monitored natural attenuation (MNA) [32]. If monitoring indicates that natural attenuation is not progressing adequately, other remediation steps will be needed to enhance or replace it.

One enhancement is to pump air or molecular oxygen (O_2) into the aquifer. With time, declining O_2 concentrations can begin to disfavor aerobes and support the growth and reproduction anaerobic bacteria. The addition of nutrients, along with pumping air and adding other compounds that serve as electron acceptors stimulates the activity of microbial populations, i.e., biostimulation [33]. At low pollutant concentrations, reaction rates are often assumed to be first-order for microbial population biomass and for pollution concentration. Environmental engineers often describe first-order processes using a first-order rate as being constant because the rate of change relates directly to the constant. If the biodegradation rate is also operating as first-order, the rate constant for the combined effect can conveniently be calculated as the sum of the individual rate constants [20]. Fig. 16.2 gives a comparison of the first-order rate constants to the respective half-life units.

Natural attenuation is the forerunner of engineered bioremediation since they in many ways it mimicked natural systems, albeit in an accelerated manner. This was first accomplished via gradual exposure of soil bacteria and other microbes to the wastewater [34–37]. By the late 20th century, engineers began deploying increasingly aggressive biotechnologies by manipulating the genetic material of microbes to enhance their energy use and targeting the catabolic reactions toward specific organic compounds [38–43].

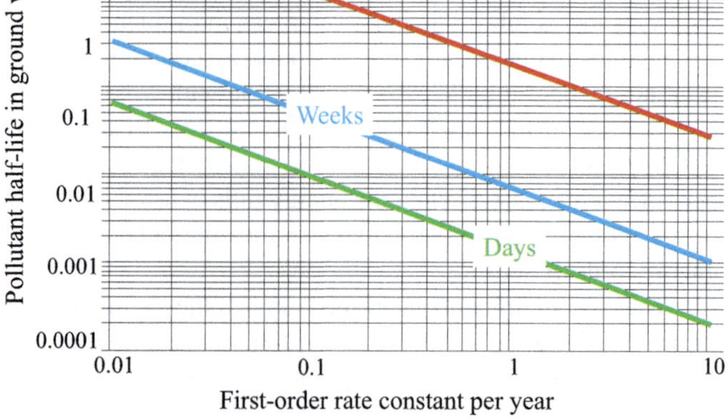

FIG. 16.2 Relationship between a first order rate constant in units of per year and half-lives in units of days, weeks, and years. *(Data from J. T. Wilson, K. Banks, R.C. Earle, Y. He, T. Kuder, C. Adair, Natural attenuation of the lead scavengers 1, 2-dibromoethane (EDB) and 1, 2-dichloroethane (1, 2-DCA) at motor fuel release sites and implications for risk management, US Environmental Protection Agency, Office of Research and Development, 2008.)*

TABLE 16.1 Particle size, solids content and extent of contamination in relation to decontamination efficiency.

	Predominant particle size			Solids content		High contaminant concentration	
Treatment technology	**Sand**	**Silt**	**Clay**	**High (slurry)**	**Low (in situ)**	**Organic compounds**	**Metals**
Conventional incineration	N	X	X	F	X	F	X
Innovative incineration	N	X	X	F	X	F	F
Pyrolysis	N	N	N	F	X	F	F
Vitrification	F	X	X	F	X	F	F
Supercritical water oxidation	X	F	F	X	F	F	X
Wet air oxidation	X	F	F	X	F	F	X
Thermal desorption	F	X	X	F	X	F	N
Immobilization	F	X	X	F	X	X	N
Solvent extraction	F	F	X	F	X	X	N
Soil washing	F	F	X	N	F	N	N
Dechlorination	U	U	U	F	X	X	N
Oxidation	F	X	X	N	F	X	X
Bioslurry process	N	F	N	N	F	X	X
Composting	F	N	X	F	X	F	X
Contained treatment facility	F	N	X	F	X	X	X

Note: F—sediment characteristic favorable to the effectiveness of the process. N—sediment characteristic has no significant effect on process performance. U—effect of sediment characteristic on process is unknown. X—sediment characteristic may impede process performance or increase cost.
Data from D. Averett, Treatment technologies, in: Assessment and Remediation of Contaminated Sediments (ARCS) Program: Remediation Guidance Document, Great Lakes National Program Office, U.S. Environmental Protection Agency, Chicago, Illinois, 1994 (Chapter 7).

3 Treatment basics

The most effective treatment technologies are customized around the characteristics of the contaminated site [44]. Contaminated unconsolidated materials in soils and aquifers have unique physical, chemical, and biological properties. These properties influence the choice of the type of treatment technology selected and will influence the efficacy of these technologies. Particle size, solids content, and contaminant concentration are examples of these properties (see Table 16.1). Technical guidance documents for such properties are available [44].

Particle size is a limiting characteristic for treatment technologies. Coarser media like sand and silt are responsive to several treatment technologies. The presence of fine-grained material worsens the effectiveness of treatment systems. Fine-grained clays have higher amounts of particulate generation when dried thermally, are more difficult to de-water, and have more sorption sites. These sites help the particles to hold onto chemical contaminants.

Solids content, by weight, ranges from high (30–60% solids) to low (10–30% solids). Treatment of slurries improves with decreasing solids contents. Adding water during processing improves the handling of high-solid media. Increasing the solids content is accomplished by evaporation and dewatering, using methods like those for municipal sludge handling. The efficiency of thermal and dehalogenation processes also decreases as the solids content is decreased. Adding water increases chemical costs and introduces the need to treat the newly contaminated water.

Elevated concentrations of organic compounds or heavy metals require safeguards. On the plus side, an increase in total organic carbon (TOC) content means more caloric value that could favor incineration and other oxidation processes. Both incineration and pyrolysis benefit from increasing TOC. Incineration destroys compounds in the presence of heat and oxygen, whereas pyrolysis destroys compounds in the presence of heat but with low or no oxygen.

Higher concentrations of metals or metalloids (especially arsenic [As] and antimony [Sb]) may detract from efficiencies of *ex situ* thermal treatment since metals are not "destroyed" by treatment but merely transformed to another chemical species. The transformation product can be more toxic and difficult to treat than the parent metal or metalloid [45].

Many physicochemical factors affect the efficiency of a treatment technology, with prominent ones shown in Table 16.2. Vitrification and supercritical water oxidation have been used mainly on small projects. Regulatory compliance and acceptance by the community are big factors when considering incineration. Disposal and handling of ash and other residues left after thermal treatment must be one of the factors in deciding on the type of treatment technology. Available land and other land use considerations are common concerns for solidification and solid-phase bioremediation projects, land application, and sludge farming. Cross-media considerations like the need to treat water contamination and air emissions, are also part of the ground water treatment and management processes [40,46–49].

Physical treatment separates the constituents of the water or media by altering their physical form. Screening, sedimentation, flotation, filtration, centrifugation, dialysis, membrane separations, ultrafiltration, distillation, solvent extraction, evaporation, and adsorption are examples of physical treatment technologies.

The segregation of valuable materials from contaminants within the treated water and media possibly allows for reuse. Another potential advantage is that chemicals that are separated may be more precisely targeted to allow for improved follow-on chemical or biological treatment, or destruction by incineration. This decreases the chance of producing toxic byproducts. Physical processes can separate chemical compounds from the soil and consolidated material matrix, so that a concentrated less-hazardous fraction of the water or media can be treated with simpler and less expensive technologies or even passive approaches like landfilling and land application is sufficiently safe. The remaining hazardous fraction can be concentrated and followed by destruction with an intensive biological or thermal treatment system.

3.1 Leachability testing prior to treatment

Leachability is a measure of the ease of which chemical compounds can move into the accessible environment. The increased leachability of metals is undesirable because their compounds may exit landfills or land-based storage systems [50]. The leachability of metals can be measured using the toxicity characteristic leaching procedure (TCLP) test. Ash from thermal treatment systems that fails the TCLP must be disposed of in a facility approved for hazardous wastes. Enhanced leachability is preferable only if designed to move rapidly toward an additional treatment step.

4 Thermal destruction

To initiate thermal destruction, contaminated media are mixed with O_2 and then heated to a high enough temperature so that within several seconds, produces the byproducts of gaseous CO_2 and water exit the top of the reaction vessel. The solid ash exits the bottom of the reaction vessel. Energy may also be produced during the reaction and the heat may be recovered.

Thermal treatment becomes increasing complicated for the media that contain halogens and/or heavy metals. Destroying the initial contaminant is less efficient, but more importantly, the process may generate hazardous off-gases containing halogenated hydrocarbons or may concentrate heavy metals in the ash and air emissions. Incineration of certain chlorinated hydrocarbons, for example, can form highly toxic compounds like chlorinated and brominated dioxins, furans, and hexachlorobenzene. Halogenated compounds can be thermally treated, but only with precise controls and operations, e.g., tight ranges of temperature, turbulence within the vessel, and reaction time [51,52].

The potential benefits of thermal systems include:

(1) potential for energy recovery;
(2) volume reduction to produce much smaller amount of residue compared to original waste;
(3) detoxification as molecules are transformed;
(4) well-understood thermodynamic principles, decades of engineering and design experience, and proven technologies;
(5) promising approach to treat large quantities of wastes that contain organic contaminants worldwide;
(6) scalability; and,
(7) less land surface needed compared to other hazardous-waste facilities like landfills.

Common drawbacks of thermal treatment include:

(1) capital intensity of equipment, e.g., refractory material lining of each combustion chamber wall, which must be replaced when cracked during and between heating and cooling periods;

TABLE 16.2 Selection factors for decontamination and treatment of hazardous wastes.

Treatment technology	Implementability at full scale	Regulatory compliance	Community acceptance	Land requirements	Residuals disposal	Wastewater treatment	Air emissions control
Conventional incineration		✓	✓				✓
Innovative incineration		✓	✓				✓
Pyrolysis		✓					✓
Vitrification	✓	✓					✓
Supercritical water oxidation	✓						
Wet air oxidation							
Thermal desorption					✓	✓	✓
Immobilization				✓			✓
Solvent extraction					✓	✓	
Soil washing					✓	✓	
Dechlorination							✓
Oxidation	✓						
Bioslurry process	✓						✓
Composting				✓			✓
Contained treatment facility				✓		✓	✓

Note: ✓—the factor is critical in the evaluation of the technology.
Data from D. Averett, Treatment technologies, in: Assessment and Remediation of Contaminated Sediments (ARCS) Program: Remediation Guidance Document, Great Lakes National Program Office, U.S. Environmental Protection Agency, Chicago, Illinois, 1994 (Chapter 7).

(2) need for skilled operators and increased costs to operate when fuel is needed to reach combustion and pyrolysis efficiencies;
(3) costly and difficult handling of ash, including disposal at certified hazardous waste, especially if the ash contains metals; and
(4) air pollutant emissions, especially products of incomplete combustion (PICs) and heavy metals in the flue gas, which must be monitored and controlled to meet emission standards.

Thermal processes include gasification, pyrolysis, hydrolysis, and combustion, which follow the general stoichiometry:

$$C_{20}H_{32}O_{10} + x_1O_2 + x_2H_2O \underset{\Delta}{\rightarrow} y_1C + y_2CO_2 + y_3CO + y_4H_2 + y_5CH_4 + y_6H_2O + y_7C_nH_m \tag{16.7}$$

The coefficients x and y must balance the compounds on either side of the equation. Commonly, C_nH_m includes the hydrocarbons, C_2H_2, C_2H_4, C_2H_6, C_3H_8, C_4H_{10}, C_5H_{12}, and benzene, C_6H_6. Incineration is the most common thermal process for destroying organic contaminants in industrial wastes. Flue gases exiting thermal systems are likely to include both organic and inorganic compounds that have been released as a result of temperature induced volatilization and/or newly transformed PICs with higher vapor pressures than the original contaminants.

Based on these underlying principles, seven general incineration guidelines can be highlighted:

1. Liquid phase, nearly pure organic contaminants are best candidates for combustion;
2. Halogen-containing organic materials deserve special consideration if in fact they are to be incinerated; i.e., special materials used in the construction of the incinerator, long (many seconds) combustion time, high temperatures (>1600 °C), with continuous mixing if the contaminant is in the solid or sludge form;
3. Feedstock containing heavy metals generally should not be incinerated;
4. Sulfur-containing organic material will emit sulfur oxides that must be controlled;
5. The formation of nitrogen oxides can be minimized if the combustion chamber is maintained above 1100 °C;
6. Destruction depends on the interaction of a combustion chamber's temperature, dwell time and turbulence; and,
7. Off-gases and ash must be monitored for chemical constituents; each residual must be treated appropriately so the entire combustion system operates within the requirements of the local, state and federal environmental regulators, and hazardous components of the off-gases, off-gas treatment processes; and the ash must reach ultimate disposal in a permitted facility [53].

Table 16.3 shows the principal reactions for commonly incinerated compounds. These generally follow the stoichiometry of Eq. (16.7), varying by the chemical composition of the fuel [54].

Wood is considered to have the composition of $C_{6.9}H_{10.6}O_{3.5}$. Therefore, the combustion consists simply of the oxidation of carbon and hydrogen:

$$C + O_2 \rightarrow CO_2, H_2 + 0.25O_2 \rightarrow 0.5H_2O$$

TABLE 16.3 Balanced combustion reactions for selected organic compounds.

Chlorobenzene	$C_6H_5Cl+7O_2 \rightarrow 6CO_2+HCl+2H_2O$
Tetrachloroethene (TCE)	$C_2Cl_4+O_2+2H_2O \rightarrow 2CO_2+HCl$
Hexachloroethane (HCE)	$C_2Cl_6+\frac{1}{2}O_2+3H_2O \rightarrow 2O_2+6HCl$
Postchlorinated polyvinyl chloride (CPVC)	$C_4H_5Cl_3+4.5O_2 \rightarrow 4CO_2+3HCl+H_2O$
Natural gas fuel (methane)	$CH_4+2O_2 \rightarrow CO_2+2H_2O$
PTEE Teflon	$C_2F_4+O_2 \rightarrow CO_2+4HF$
Butyl rubber	$C_9H_{16}+13O_2 \rightarrow 9CO_2+8H_2O$
Polyethylene	$C_2H_4+3O_2 \rightarrow 2CO_2+2H_2O$

Note: TCE is tetrachloroethene; HCE is hexachloroethane; CPVC is postchlorinated polyvinyl chloride and PFTE is polytetrafluoroethylene.
Data from D.A. Vallero, Fundamentals of Air Pollution, fifth ed., Elsevier Academic Press, Waltham, MA, 2014, p. 999 pages cm.

There are several stages in the incineration process where the generation of new compounds may need to be addressed. Ash and other residues may contain high levels of metals, at least higher than the original feed. As mentioned, the flue gases are likely to include both organic and inorganic compounds that need to be collected and treated.

Commonly referred to as "combustion" units, thermal treatment technologies should be called "thermal destruction" units, since combustion is the type that requires excess oxygen. Thermal destruction units range in size and characteristics, varying by the *3 T's*, i.e., time, temperature, and turbulence, to destroy hazardous waste constituents and to reduce the volume of wastes. If complete combustion is desired, excess oxygen is needed to allow the oxidative processes to predominate the reactions. The excess oxygen also helps deter pyrolytic processes that can produce toxic byproducts. The optimal turbulence, i.e., a well-mixed chamber, avoids pockets of oxygen deficit in the thermal vessel. Given waste constituent variabilities and thermal uncertainties, the O_2 concentrations should be 50–100% higher concentration than is theoretically required to decompose the feedstock. Excess air also promotes mixing and turbulence to ensure that air can reach all parts of the feedstock.

4.1 Thermal technologies

Commonly deployed thermal technologies prototypes include: (1) rotary kiln; (2) multiple hearth; (3) liquid injection; (4) fluidized bed; (5) moving grate, and (6) multiple chamber. For more specific information on these and other technologies, please see the book, *Fundamentals of Air Pollution* [54], the source of the following summaries.

The combustion chamber in a rotary kiln incinerator is a heated rotating cylinder mounted at an angle with baffles at times added to the inner face to provide the turbulence necessary for the target *3 T's*. Engineering design options include: (1) angle of the drum; (2) diameter and length of the drum; (3) presence and location of the baffles; (4) rotational speed of the drum; and (5) use of added fuel to increase the temperature of the combustion chamber as the specific contaminant requires. The liquid, sludge or solids are sent into the upper end of the rotating cylinder, rotates with the cylinder-baffle system, and then falls with gravity to the lower end of the cylinder. The heated upward moving off-gases are collected, monitored for chemical constituents, and subsequently treated as appropriate prior to release, while the ash falls with gravity to be collected, monitored for chemical constituents, and treated as needed before ultimate disposal. The newer rotary kiln systems consist of a primary combustion chamber, a transition volume, and a fired afterburner chamber. After exiting the afterburner, the flue gas is passed through a quench section followed by a primary air pollution control system (APCS). The primary APCS, for example, can be a venture scrubber followed by a packed-column scrubber. Downstream of the primary APCS, a backup secondary APCS, with a demister, an activated-carbon adsorber, and a high-efficiency particulate air (HEPA) filter can collect contaminants not destroyed by the incineration [56].

The rotary kiln can be used to incinerate most organic contaminants, it is well suited for solids and sludges, and in special cases liquids and gases can be injected through auxiliary nozzles in the side of the combustion chamber. Operating temperatures generally vary from 800 to 1650°C. Engineers use laboratory experiments to design residence times of seconds for gases and minutes or possibly hours for the incineration of solid material.

Multiple hearths usually treat contaminants in a solid or sludge form that is fed slowly through the top vertically stacked hearth; in special configurations hazardous gases and liquids can be injected through side nozzles. Early configurations were used to burn municipal wastewater treatment bio-solids, relying on gravity and scrapers working the upper edges of each hearth to transport the feedstock through holes from upper hotter hearths to lower cooler hearths. Heated upward-moving off-gases are collected, monitored for chemical constituents, and treated as appropriate prior to release; the falling ash is collected, monitored for chemical constituents, and subsequently treated prior to ultimate disposal.

Numerous waste types with varying organic content can be incinerated using a multiple hearth configuration. Operating temperatures generally vary from 300 to 980°C. These systems are designed with residence times of seconds if gases are fed into the chambers to several hours if solid materials are placed on the top hearth and allowed to eventually drop to the bottom hearth exiting as ash [57].

Liquid injection consists of nozzles mounted horizontally and/or vertical spraying of liquid wastes into liquid injection incinerators. That is, the wastes are atomized to match the waste being handled with the combustion chamber as determined in laboratory testing. The application obviously is limited to liquids that do not clog these nozzles, though some success has been experienced with hazardous waste slurries. Operating temperatures generally vary from 650°C to 1650°C. Liquid injection systems are designed with residence times of fractions of seconds as the upward moving off-gases are collected, monitored for chemical constituents, and treated as appropriate prior to release to the atmosphere [57].

In a fluidized bed system, contaminated feedstock is injected under pressure into a heated bed of agitated inert granular particles, which is usually sand. As the heat is transferred from the particles to the waste, the combustion process continues. External heat is applied to the bed prior to the injection of the waste. Heat is also continuously applied throughout the combustion operation as needed. Heated air is forced into the bottom of the particle bed and the particles become suspended among themselves during this continuous fluidizing process. The openings created within the bed permit the introduction and transport of the waste into and through the bed. The process enables the contaminant to come into contact with particles that maintain their heat better than, for example, the gases inside a rotary kiln. The heat maintained in the particles increases the time the contaminant is in contact with a heated element and thus the combustion process could become more complete with fewer harmful byproducts. Off-gases are collected, monitored for chemical constituents, and treated as appropriate prior to release, and the falling ash is collected, monitored for chemical constituents, and subsequently treated prior to ultimate disposal [58].

Most organic wastes can be incinerated in a fluidized bed, while the system is best suited for liquids. Operating temperatures generally vary from 750 to 900 °C.

Using moving grates to keep wastes and ash continuously moving through a thermal system is usually desirable. This offers more control compared to agitation, an advantage for biologicals, e.g., clinical and medical, wastes, that produce a high percentage of ash. It is usually less expensive to build and usually generate less PM emissions. There have been problems with some early designs that use moving grates [55].

In a multiple chamber configuration, the first step is to vaporize the waste on a grate in the ignition chamber. The gases created in this ignition chamber move through baffles to a secondary chamber where the combustion occurs. The secondary chamber is commonly located above the ignition chamber to promote natural advection of the hot gases through the system. Heat may be added to the system in either the ignition chamber or the secondary chamber as required for specific burns.

The application of multiple chamber incinerators is mainly restricted to solid wastes, wherein the waste enters the ignition chamber through an opened charging door in batch, not continuous, loading. Combustion temperatures are typically near 540 °C for most applications. Multiple chamber units normally are designed with residence times of minutes to hours for solid hazardous wastes as off-gases are collected, monitored for chemical constituents, and treated as appropriate prior to release to the lower troposphere. At the end of each burn period the system must be cooled so that the ash can be removed prior to monitoring for chemical constituents and subsequent treatment before ultimate disposal.

Incineration alone does not "destroy" some wastes. Elements, like the metals, are concentrated in the ash and other incineration residues. Thermal treatment changes the oxidation state (valence) of the metals. In fact, incineration can replace one problem with another, such as increasing the potential movement of metals via oxidation, allowing the metals to migrate into the environment. The mobility of metals can be substantially decreased by processes like slagging, i.e., operating at sufficiently high temperatures to melt and remove incombustible materials, and vitrification, i.e., producing nonleachable, basalt-like residues.

The chemical and physical characteristics of the waste determine how a waste is to be thermally treated. The key is obviously to make the waste less harmful. This often means making a waste less mobile and less toxic. For example, Australia [59] has identified the suitability of wastes to various thermal treatment process according to the waste's properties (see Table 16.4).

High-temperature incineration may not be needed to treat many remediation products, including most volatile organic compounds (VOCs). Also, in soils with heavy metals, the high temperatures will lead to an increase in the vapor pressure of metals, so they will find their way into the combustion flue gas. High concentrations of volatile trace metal compounds in the flue gas complicate pollution control strategies.

When successful in decontaminating soils to the necessary treatment levels, thermally desorbing contaminants from substrates has the benefits over incineration, including lower fuel consumption, no formation of slag, less volatilization of metal compounds, and less complicated air pollution control demands. Beyond monetary costs and ease of operation, a less energy (heat) intensive system can be more advantageous in terms of actual pollutant removal efficiency.

The technologies discussed thus far employ oxygen, but pyrolysis is also an option. Pyrolytic decomposition is induced in organic materials by heat in the absence of oxygen, i.e., it is the thermal conversion of organic matter in the total absence of oxygen at comparatively low temperatures of between 500 and 800 °C and short vapor residence times of 3–1500 s. Given the difficulty of achieving and maintaining a completely oxygen-free atmosphere; pyrolytic systems operate at very low O_2 concentrations, i.e., less than stoichiometric quantities of oxygen. This means oxidation will continuously occur, but the net results are pyrolytic. Desorption of compounds from the waste will also occur when volatile (i.e., vapor pressure $> 10^{-2}$ kPa) or semivolatile (i.e., vapor pressure between 10^{-2} and 10^{-5} kPa) compounds are present in the treated waste.

Organic compounds are converted to gaseous components during pyrolysis, as well as liquids and coke, i.e., the solid residue of fixed carbon and ash. Carbon monoxide (CO), hydrogen gas (H_2), and methane (CH_4) and other hydrocarbons are produced. Liquids form, leaving oily tar residues and water containing high concentrations of total organic carbon (TOC)

TABLE 16.4 Applicability of waste destruction by physical and chemical properties.

Waste Type	Cement Kiln	Rotary Kiln	Multiple Chamber	Multi-Hearth	Fluidized Bed	Liquid Injection
Solids						
Granular, homogeneous		X	X	X	X	
Irregular, bulky (e.g. pallets)		X	X			
Low melting point compounds (e.g., tars)	X	X	X	X	X	
Organics with fusible ash		X	X			
Gases						
Organic gases and vapors		X	X			X
Liquids						
Aqueous-phase, contaminated with organics	X	X	X			X
Organic liquids, including halogenated	X	X	X		X	X
Solids/liquids						
Waste with halogens	X	X	X		X	X
Aqueous/organic sludges	X	X	X	X	X	X

Data from Clear the Air, Incineration policy in Australia, Commonwealth of Australia. http://news.cleartheair.org.hk/?p=2040, 2018 (Accessed 8 May 2018).

when the gases cool and condense. The secondary vapors remaining in the gas phase require pollution controls and treatment, such as by a secondary combustion chamber, by flaring, and partial condensation. Particulates must be removed by additional air pollution controls, e.g., fabric filters or wet scrubbers.

Many of the same conventional thermal treatment methods used for incineration, such as rotary kiln, rotary hearth furnace, or fluidized bed furnace, can be used for waste pyrolysis. Kilns or furnaces used for pyrolysis may be of the same design as those used for combustion but operate at lower temperatures and with lower amounts of air than in combustion.

Pyrolysis allows for separating organic contaminants from various wastes and may be used to treat a variety of organic contaminants that chemically decompose when heated (i.e., "cracking"). Pyrolysis is not effective in either destroying or physically separating inorganic compounds that coexist with the organics in the contaminated medium.

Other promising thermal processes include high-pressure oxidation and vitrification. High-pressure oxidation combines two related technologies, i.e., wet air oxidation and supercritical water oxidation, which combine high temperature and pressure to destroy organics. Wet air oxidation can operate at pressures of about 10% of those used during supercritical water oxidation, an emerging technology that has shown some promise in the treatment of polychlorinated biphenyls (PCBs) and other stable compounds that resist chemical reaction. Wet air oxidation has generally been limited to conditioning of municipal wastewater sludges, but can degrade hydrocarbons (including polycyclic aromatic hydrocarbons; PAHs), certain pesticides, phenolic compounds, cyanides, and other organic compounds. Oxidation may benefit from catalysts.

Vitrification uses electricity to heat and destroy organic compounds and immobilize inert contaminants. A vitrification unit has a reaction chamber divided into two sections: the upper section introduces the feed material containing gases and pyrolysis products, and the lower section consists of a two-layer molten zone for the metal and siliceous components of the waste. Electrodes are inserted into the waste solids, and graphite is applied to the surface to enhance its electrical conductivity. A large current is applied, resulting in rapid heating of the solids, causing the siliceous components of the material to melt as temperatures reach about 1600 °C. The end product is a solid, glass-like material that is very resistant to leaching.

4.2 Destruction and removal efficiency

In the U.S., the Emergency Planning and Community Right-To-Know Act (EPCRA) requires certain emergency planning steps accompany the handling of extremely hazardous substances (EHSs) in quantities at or above the threshold planning quantity (TPQ). The same chemical compounds may also need to comply with requirements under the Comprehensive Environmental Response, Compensation and Liability Act (CERCLA) and the Clean Air Act. The U.S. EPA updates the so-called "list of lists" for these substances [60]. A permit is required whenever wastes containing hazardous substances are thermally treated. The owner or operator of an incinerator is allowed to burn only materials as specified in the permit and only under operating conditions specified for those wastes. When reliable data specific to new wastes is available, the permit's emission levels may be based on either trial burn results or on sufficiently relevant and sound alternative data that can establish appropriate conditions for each applicable requirement related to allowable waste feedstocks and operating conditions [48].

An incinerator burning a substance that contains principal organic hazardous constituents (POHCs) must be designed, constructed, and maintained to achieve a destruction and removal efficiency (DRE) of 99.99% for each POHC designated in its permit for each waste feed. For each such waste feed, the permit will specify acceptable operating limits including the following conditions [61]:

(1) Exhaust gas stack CO concentrations;
(2) Feed rate of waste;
(3) Combustion and pyrolytic temperature;
(4) Combustion gas velocity;
(5) Previously approved modifications of incinerator system design or changes to operating procedures; and
(6) Any other operating requirements needed to comply with the performance standards for hazardous waste incineration (e.g., in the U.S. this is within Section 264.343 of EPA regulations).
(7) Except those wastes exempted (e.g., in the U.S., this is Section 264.340 of EPA regulations) when an incinerator is started or shut down, the incinerator must not receive any hazardous waste unless the incinerator is running within the conditions of operation, e.g., temperature, air feed rate, etc., specified in the permit for that incinerator.
(8) Combustion zone fugitive emissions must be controlled by:
  (a) Complete sealing of the combustion zone against fugitive emissions; or
  (b) Keeping pressure in the combustion zone pressure below atmospheric pressure; or,
  (c) Delivering an alternate means of control of fugitive emissions at a rate equivalent to keeping the combustion zone pressure below atmospheric pressure. In the U.S., this is specified in Part B of the hazardous waste incinerator permit application.
(9) Automatic truncation of the waste feed to the incinerator whenever operating conditions deviate from prescribed limits.
(10) Ceasing incineration when changes in waste feed, incinerator design, or operating conditions exceed limits designated in its permit.

The DRE required for incinerators and other thermal control technologies is an expression of thermodynamic efficiency:

$$DRE = \frac{W_{in} - W_{out}}{W_{in}} \times 100 \tag{16.8}$$

Where, W_{in} = Rate of mass of waste flowing into the incinerator and W_{out} = rate of mass of waste flowing out of the incinerator.

Extremely hazardous waste constituents require at least 99.9999% DRE.

Destruction and removal efficiency example 1

The mass of waste from ground water remediation project of Chemical X (which is both POHC and EHS) is loaded into incinerator. The flue gas is measured at W_{in} = 10 mg min^{-1} and the mass flow rate of the compound measured downstream in the stack is W_{out} = 200 pg min^{-1} (in which pg refers to picograms or 10^{-12} g)?

$$DRE = \frac{10\ \text{mg min}^{-1} - 200 \times 10^{-9}\ \text{mg min}^{-1}}{10\ \text{mg min}^{-1}} \times 100 = 999999.98\%\text{removal}$$

Even though Chemical X is considered to be "extremely hazardous," this DRE is better than the "rule of six nines" so the incinerator is operating up to code.

Destruction and removal efficiency example 2

During the same project, the mass of Chemical Y loaded into incinerator is measured at the rate of 100 L min^{-1} and the mass flow rate of the compound measured downstream is 1 mL min^{-1}. Is the incinerator up to code for this POHC?

This is a lower removal rate since 100 L are in and 0.001 are leaving, so the DRE = 99.999. This is acceptable because its removal efficiency is greater than 99.99% by an order of magnitude. This is the case so long as Chemical Y is not designated to be an extremely hazardous compound. If it were, then the incinerator would not meet the rule of six nines, but it only has five.

If either of the compounds in the examples are halogenated, special precautions must be taken when dealing with such compounds, since even more toxic compounds that those being treated can end up being generated. Incomplete reactions are very important sources of environmental contaminants. For example, these reactions generate PICs, such as dioxins, furans, CO, polycyclic aromatic hydrocarbons (PAHs), and hexachlorobenzene.

Organic compounds have caloric, i.e., heat, value. Thermodynamically, the organic fraction of the waste can be completely destroyed, ultimately yielding CO_2 and water:

$$\text{hydrocarbons} + O_2 \rightarrow CO_2 + H_2O + \text{energy} \tag{16.9}$$

Molecular nitrogen (N_2) is also produced when combustion uses air as the oxygen source and/or when the waste contains nitrogen-containing compounds, e.g., combustion of methylamine. Incomplete combustion, on the other hand, can produce an assortment of PICs. Some are more toxic than the parent compounds being oxidized, e.g., benzo(A) pyrene and other PAHs, dioxins, furans, and CO.

5 Solidification and stabilization

Physical treatment systems can improve handling, change the physical characteristics of the waste, decrease the amount of surface area thereby reducing the flux rate of pollutants from the bulk material, and limit the solubility, mobility, and toxicity of hazardous constituents in the contaminate material. In solidification, these results are obtained primarily, but not exclusively, by producing a block of treated material with high structural integrity, thereby limiting the amount of flow through interstices. Stabilization techniques change the matrix of the material's contaminants. Thus, other physical and chemical characteristics of the material may not be changed. Stabilization usually involves the addition of substances to the matrix in an aim to render the hazardous constituents less soluble, less likely to be transferred, and less bioavailable.

6 Specific technologies

Ground water can be treated by drilling recovery wells to pump contaminated ground water to the surface. Commonly used ground water treatment approaches include air stripping, filtering with granulated activated carbon (GAC), and air sparging.

6.1 Air stripping

Air stripping is a common type of so-called pump-and-treat technologies. Air stripping apparatus moves volatile compounds from water to air within in an aboveground packed column or tower or in a low-profile sieve tray aeration system (see Chapter 11). The stripping happens when contaminated water is pumped into a tower filled with a permeable material in which the water moves by gravity downward while a stream of air flows upward. Air stripping is ideal for compounds with relatively high vapor pressures, i.e., volatile inorganic or organic compounds that readily evaporate, changing from a liquid to a vapor [62,63]. Air stripping is especially efficient treating VOCs like benzene or xylenes. The air passed through the packed column increases the evaporation rates. As the chemical-laden air exits the column it is collected and treated.

Air stripping configuration includes a spray nozzle, or a manifold connected to several nozzles, at the top of the tower that spray and spread the contaminated water over the column packing in the column, a blower to move air in the countercurrent to the contaminated water flow, and a sump at the bottom to collect treated water. Low-profile sieve tray air strippers are configured as a horizontal tray or vertical box design with a series of stacked trays. Baffles and weirs inside the stripper trays move contaminated water twice or more along the length of the box tray [62]. These are single-pass systems but some of the effluent can be separated and recirculated into the column to improve removal efficiency and to maintain the packed tower at a reasonable height or the to optimize the number of trays in a low-profile system.

Diffused aeration strippers have aeration tanks in which volatile compounds are stripped by bubbling air into a tank through which the contaminated water is flowing. A forced air blower and an aeration-sparge manifold increase the contact of the water with air [38].

During one air stripping configuration, ground water drips downward in a tower filled with a permeable material through which a stream of air flows upward. Another method bubbles pressurized air through contaminated water in a tank. The air leaving the tank (i.e., the off-gas) is treated by removing vapor phase pollutants. Filtering ground water with GAC entails pumping the water through the GAC to trap the contaminants.

6.2 Air sparging

Air sparging is a type of stripping also known as *in situ* volatilzation. In the process, air containing the volatilized pollutants is sparged through a vent pipe or through the bottom of the column passes in a countercurrent direction through the water to strip compounds with relatively high vapor pressures, e.g., volatile organic compounds (VOCs) like benzene or xylenes [38]. These are considered single-pass system although a portion of the effluent can be separated and recirculated into the column to improve removal efficiency and keeping the height of the packed tower at a reasonable height or to optimize the number of trays in a low-profile system. Contaminated ground water can also be filtered by pumping the water through the GAC to trap the contaminants. Air sparging consists of pumping air into the ground water *in situ*. A soil venting system is often combined with an air sparging system for vapor extraction[38]. Usually, a soil venting system is combined with an air sparging system to extract vapors, followed by treatment of the off-gas.

6.3 Fracturing remediation

Hydraulic fracturing or so-called "fracking" is best known as the process of pumping fluids under pressure into wells that reach strata that contain fossil fuels, especially natural gas. The principle of so-called “fracking” or hydraulic fracturing can also be used for ground water remediation. The fluid pressure causes unconsolidated and consolidated strata to fracture. Fracturing fluids are commonly liquid, especially water or water-based fluids, but they may also be gases, i.e., pneumatic fracturing. In the latter system, air is forced into pores in the strata to promote vaporization. The fracturing fluid often includes sand and other so-called propping agents to keep pores and fractures open during remediation [64].

This same process can be deployed to remediate ground water. Fracturing can be combined with pump-and-treat technologies. For example, blast-enhanced fracturing deploys dynamite or other explosives to fracture target rock strata. The fractures become reservoirs from which contaminated ground water is pumped and treated at the surface or off-site (*ex situ*).

During fracturing, pressure can be directed toward specific zones by using packers to seal the holes. However, this is not available for blast-enhanced fracturing [64].

In situ fracturing is usually employed when contaminants are not easily accessible due to depth and density of soil and rock layers, allowing the penetration of O_2 and other abiotic treatment chemical substances or microbes to reach contaminants in tight formations composed of fine-grained soils such as clay or silt, with low permeability. Fracturing can forego the need for rock excavation, and accelerate pollutant mass-transfer and degradation. Indeed, combinations of aerobic and anaerobic bacteria have been successfully activated in the open zones after injections of the microbial populations [65].

6.4 Bioremediation

Remediation can be abiotic or biotic [65]. Bioremediation takes advantage of microbes' natural ability to degrade organic chemical compounds to gain energy and to build biomass. The rate law discussed in Chapter 8 is the basis for bioremediation:

$$aA + bB \rightarrow gG + hH, \text{rate} = k[A]^x\,[B]^y \tag{16.10}$$

Where, k represents the rate constant for molar concentrations of the reactants and x and y represent the reaction rate order for the reactants. The overall order of the reaction is determined by summing the reactant orders. Experimental data are the basis for the rate law for a reaction. The first order reaction for compound A is:

$$\text{Rate} = k\,[A]^1 \text{ or simply } k[A]. \tag{16.11}$$

Therefore, any first order reaction rate for any chemical species is:

$$\text{Rate} = \frac{d[A]}{dt} = \frac{\Delta[A]}{\Delta t} + k[A] \tag{16.12}$$

Doubling [A] in a first order reaction double the reaction rate. Tripling the concentration will triple the reaction rate, and a hundred-fold increase in concentration of A multiples the reaction rate 100 times. Integrating the first order rate equation yields:

$$\ln\,[A] = \ln\,[A]_0 - k\Delta t \tag{16.13}$$

Correspondingly, integrating the rate equation for a second order reaction yields:

$$\frac{1}{[A]} = \frac{1}{[A]_0} + k\Delta t \tag{16.14}$$

Assuming first-order biodegradation rates is but a guide for expected rates of treating contaminated ground water. It is important to keep in mind that temperature or redox condition are not completely predictive of the treatment success. Many unknown factors influence the degradation rates at different sites [19,66,67]. Generally, however, if the microbes' food in the substrate is not limiting, i.e., there is enough pollutant and other organic mass in their biofilm, the pollutant degrades as a function of the logarithmic cellular density of the microbes. Thus, the biochemical reactions appear as zero-order kinetics, i.e., constant logarithmic growth and decay. When the rate of degradation of the chemical compound becomes directly proportional to the food mass, the decay will appear to follow first-order kinetics [47].

Microbial degradation models often assume nonlinear reactions, often due to the scarcity of reliable scientific information supporting stepwise degradation, particularly in natural systems. Thus, biodegradation models are often fit to experimental and field data and expressed as:

$$-d[A] = k[A]^n \frac{-d[A]}{dt} = k[A]_n \tag{16.15}$$

Where n is the rate curve fitting parameter.

Certainly, biological kinetics scenarios can also fluctuate between, zero, first and second orders, depending on environmental conditions, competition among species, and other factors. During bioremediation the degradation rate may cycle through various orders [33]. Physical factors like water and air dispersion, dilution from ground water recharge, volatilization, and sorption also affect the degradation rates. Thus, the rate orders will vary during remediation, depending on abiotic and biotic conditions [19,47,52,66–73].

Bioremediation can deploy aerobic, anaerobic, and facultative (features of both aerobic and anaerobic) microbes. Complete aerobic biodegradation results in all organic compounds ultimately oxidized to CO_2 and water. Complete anaerobic biodegradation results in all organic ultimately reduced to methane (CH_4) and water. However, in the environment, conditions almost always leave biomass untreated and containing carbonaceous and nitrogenous compounds. To increase the rates of biodegradation rates over those in nature, the surface area of biofilm must be greatly increased to improve electron transfer [74,75].

Anaerobic remediation is initiated when microbes hydrolyze the organic matter. The rate of hydrolysis depends on the time microbes are exposed to the compounds, as well as other factors like mixing rates, temperature, and pressure. The inherent properties of the pollutant, especially its persistence and bioaccumulation factors, also determine treatment success. Bioremediation usually involves step-wise reactions as microbial populations adjust to changes in the natural environment or in a bioreactor. As evidence, acidogenic microbes convert sugars and amino acids to CO_2, molecular hydrogen (H_2), ammonia (NH_3), and organic acids (R-COOH). Acetogenic microorganisms may then convert the organic acids into simpler organic acids, e.g., acetic acid and inorganic acids, leaving less of the biomass and generating H_2 and NH_3. Degradation can continue by methanogenesis, also known as biomethanation as methanogenic bacteria convert the remaining organic compounds into methane (CH_4) in a reduced environment and other microbes will convert remaining compound to CO_2 in an oxidized environment [17,52].

Therefore, successful bioremediation requires a defined range of temperature, available nutrients and food [17,76,77]. Proper conditions allow target microbes to grow and multiply, and as they continue to degrade the contaminants. Ideally, microbes will not grow too slowly or die prematurely. During *in situ* treatment, amendments can be pumped into the ground water, e.g., vegetable oil, air, and oxidizing chemical compounds, to improve treatment efficiency and shorten remediation time. This is known as bioaugmentation. For *ex situ* treatment, these amendments can be continuously added to the bioreactor. *Ex situ* bioremediation is needed when site conditions are unfavorable, e.g., soil is too dense to allow adequate transport or temperatures too low to support active microbial populations [76].

The basic *ex situ* bioremediation process begins when ground water is pumped into aboveground tanks. The water is mixed with amendments and returned to the aquifer. Alternatively, as mentioned, the treatment can be *in situ*, where the amendments are pumped directly into the aquifer. Bioremediation of ground water can require months to years to complete [76]. The time depends on the pollutant concentration, the accessibility of amended and microbe-laden water to parts of the pollutant plume that may be obstructed, the depth of the plume, temperature, available nutrients, difficulty in acclimating the microbial population to the pollutant, and cleanup of equipment (see Table 16.5).

Ideal conditions vary, but sufficient moisture always is needed for bioremediation success. This is best in the range of 25–85% water-holding capacity. The optimal ground water pH is usually between 6.5 and 8.0. The ideal nutrient ratio for carbon, nitrogen, and phosphorus (C:N:P) is 120:10:1. Microbes usually require temperatures between 15 and 45 °C, but the ideal range is commonly between 20 and 30 °C. Dissolved oxygen concentrations should be greater than $2\,mg\,L^{-1}$ for aerobes, with 10% air-filled pore space. Total heavy metals concentrations should not exceed 700 ppm (ppm), and hydrocarbons should be between 5% and 10% of the dry weight of the soil [77].

The treated water's redox potential (Eh) should be less than 50 mV [77]. Different metal ions have strong or weak oxidizing/reducing potential. Redox potential is a measure of ease at which an ion will lose electrons or retain electrons, not the likelihood that a specific oxidation or reduction reaction will happen. Redox is written as half-reactions with the most reducing reactions, i.e., donating electrons, having an Eh value that has a larger negative number and the most oxidizing reactions, i.e., receiving electrons, having an Eh value that is a larger positive number.

Redox kinetics frequently depend strongly on the number of electrons that must be transferred to form stable products, because many electron transfers are associated with complex changes in molecular configuration. Conversely, single electron transfers can involve less change in molecular structure. For example, when the ferrous aquo ion (Fe^{2+}) is oxidized

TABLE 16.5 Factors related to bioremediation success.

Factor	Condition
Organism	Aerobic, anaerobic and/or facultative
Biological mechanisms	Microbial metabolic rates (catabolism and anabolism)
Environmental	Temperature, oxygen (O_2) content, electron acceptor or donor, pH, redox, metal concentrations, contaminant concentrations
Nutrients	Carbon, nitrogen, phosphorous, potassium, micronutrients
Moisture	25–28% water-holding capacity
Matrix	Mix of soil textures (clay, silt), unconsolidated material (sand, gravel)

Data from D.A. Vallero, Hazardous waste bioremediation, in: D. Yadav, P. Kumar, P. Singh, D.A. Vallero (Eds.), Hazardous Waste Management, Elsevier, 2022, pp. 347–379 (Chapter 14). M. Alexander, Biodegradation and Bioremediation, Gulf Professional Publishing, 1999. S.M. Tripathi, S. Ram, Bioremediation of groundwater: an overview, Int. J. Appl. Eng. Res. 13 (24) (2018) 16825–16832. K. Sayed, L. Baloo, N.K. Sharma, Bioremediation of total petroleum hydrocarbons (TPH) by bioaugmentation and biostimulation in water with floating oil spill containment booms as bioreactor basin, Int. J. Environ. Res. Public Health 18 (5) (2021) 2226. M. Tyagi, M.M.R. da Fonseca, C.C. de Carvalho, Bioaugmentation and biostimulation strategies to improve the effectiveness of bioremediation processes, Biodegradation 22 (2011) 231–241.

to ferric aquo ion (Fe^{3+}) it can be relatively rapid and reversible [78]. The structure of both the oxidized and reduced forms of the iron complex is octahedral, so the central Fe atom binds to four water molecules at the corners of a square plane and two water molecules bind at either end of an axis oriented perpendicular to the square plane. The activation energy needed to transform Fe^{2+} to Fe^{3+} is not large, so the reaction can continue rapidly at typical ground water temperatures and pressure. Conversely, reductive dissolution of ferric oxyhydroxide minerals must have larger changes in the coordination environment for iron. The reduction of structural Fe^{3+} to form Fe^{2+} ions is a much slower reaction than the reaction between the aquo ions. Furthermore, the configurational deviations that are required in oxidizing aromatic organic compounds to CO_2 have even more complexity. The ring structure of benzene, e.g., is very different from the structure of the CO_2 gas molecule. The reaction mechanism occurs after several electron transfers with high activation energies needed for the substantial structural rearrangements that occur as electrons are removed from the C atoms in the benzene ring [19,78].

Many factors are involved in electron-transfer kinetics of redox processes in ground water [77]. Abiotic kinetics are slow given the high activation energy needed for the structural rearrangements in pollutant plumes for microbial catalysis to proceed. Enzymes stabilize intermediate compounds and aid the step-wise electron transfer to and from these intermediate compounds until the final chemical products are produced. Abiotic redox reactions involving dissolved iron, for example, often proceed well due to the presence of oxidants and reductants. Redox involving C, N, and sulfur (S) commonly need enzymatic catalysis because so many electron transfers have to occur before products are produced [78]. By decreasing the activation energy needed, a biochemical reaction can be initiated sooner and easier than if the enzymes were not present. Indeed, enzymes play a very large part in microbial metabolism. They facilitate each step along the metabolic pathway. Enzymes lower the reaction's activation energy, i.e., the minimum amount of free energy required for a molecule to undergo the reaction. While molecules meet to form, stretch, or break chemical bonds, the energy in the system is maximized, and then is decreased to the energy level of the products. The quantity of activation energy is the difference between the maximum energy and the energy of the products. This difference represents the energy barrier that must be overcome for a chemical reaction to take place. Catalysts speed up and increase the likelihood of a reaction by reducing the amount of energy, i.e., the activation energy, needed for the reaction.

An enzyme is a protein that has an active catalysis site, so it is limited to one or just a few compounds that it can catalyze. As evidence, RNA polymerase is specific to the formation of ribonucleic acid (RNA), but not deoxyribonucleic acid (DNA), which will be blocked from the reaction. Like other enzymes, RNA polymerase can bind to only a limited number of substrate molecules, in this case exclusively RNA. Other compounds do not fit the specific three-dimensional shape and structure of the active site. The resulting enzyme-substrate complex will yield a product and a free enzyme.

As mentioned, microbial cells need free energy (G) for cell growth, division, and metabolism. The unusable energy is entropy (S), i.e., the disorder in the system, which increases during the many conversion steps outside and inside of a system per the second law of thermodynamics.

Environmental engineers seek ways to improve treatment efficiencies beyond natural attenuation. Cells have two basic tasks to survive and grow. They must undergo biosynthesis, i.e., they must synthesize new biomolecules to construct cellular components. They must also harvest energy. Metabolism is the complement of the chemical reactions of both processes, so metabolism is the way that a cell derives energy from its surroundings and applies it to function and to construct more cellular material.

Energy that does chemical work is exemplified by cellular processes (see Fig. 16.3). Catabolism is the set of reactions with molecules as sources of energy, i.e., the microbes' food. Cells gain energy by degrading larger organic molecules. Anabolism is the set of reactions that synthesize cell biomass. Anabolic reactions use the energy gained from the catabolic reactions. Anabolism and catabolism occur simultaneously. During catabolism, cells use the food compounds as sources of energy. Anabolism is a synthesis process, whereas catabolism is a destructive process. Anabolism can only build cellular components by using the energy it receives from catabolism's destruction of organic compounds. So, as the cell grows, the food supply, i.e., toxic compounds in the case of remediation, must shrink. This is the goal of bioremediation, i.e., to acclimate the microbes to the pollutant as their food [34,36,79].

Free energy exchange occurs as a step-wise progression after the cell encounters a chemical compound. The reactant and product concentrations and environmental conditions, especially pH of the substrate, affect the observed ΔG^* values. In reactions with a negative ΔG^*, the free energy is released, the reaction will occur spontaneously, and the reaction is exergonic. In reactions with negative ΔG^* is positive, the reaction will not occur spontaneously. When the reverse reaction occurs, it is endergonic.

The most common coupling of exergonic and endergonic reactions occurs when high-energy molecules yield a net negative free energy of the nucleotide, adenosine triphosphate (ATP) with $\Delta G^* = -12$ to $-15\,\text{kcal}\,\text{mol}^{-1}$. Several other high-energy compounds also provide energy for reactions, including guanosine triphosphate (GTP), uridine triphosphate (UTP), cytosine triphosphate (CTP), and phosphoenolpyruvic acid (PEP). GTP, UTP, CTP, and PEP molecules store their energy

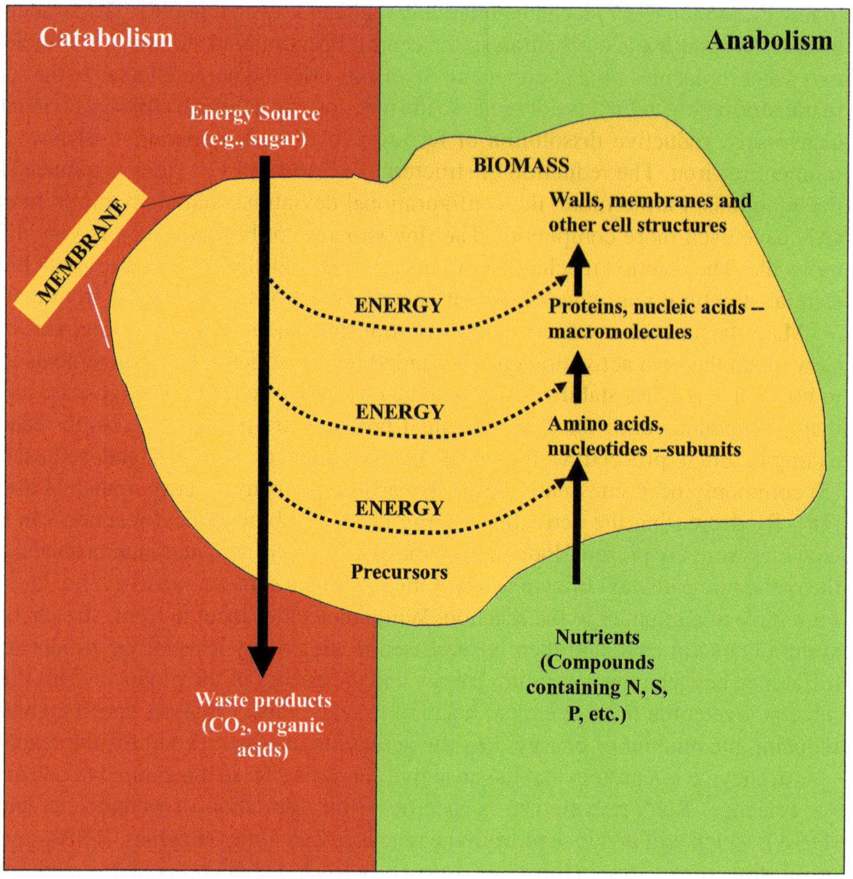

FIG. 16.3 Catabolic and anabolic reactions in a cell. *(Data from D. Vallero, Environmental Biotechnology: A Biosystems Approach, Elsevier Science, 2015. D.A. Vallero, Environmental Systems Science: Theory and Practical Applications, first ed., Elsevier Academic Press, 2021. D.A. Vallero, Estimating and predicting exposure to products from emerging technologies, in: Synthetic Biology 2020: Frontiers in Risk Analysis and Governance, Springer, 2020, pp. 107–142.)*

with high-energy bonds in the phosphate molecule (P_i). An example of free energy in microbial degradation is the possible first step in acetate metabolism by bacteria:

$$\text{Acetate} + \text{ATP} \rightarrow \text{acetyl} - \text{coenzyme A} + \text{ADP} + P_i \quad (16.16)$$

Here, the P_i represents a release of energy available to the cell. Conversely, to add the phosphate to the two P_i structure ADP to form the three P_i ATP requires energy (i.e., it is an endothermic process). Thus, the cell stores energy for later use when it adds the P_i to the ATP.

Environmental biotechnology is the use of microorganisms to decompose organic wastes into water, CO_2, and simple inorganic substances, or into simpler organic substances such as aldehydes and acids [80]. Contaminants, if completely organic in structure can theoretically be fully destroyed by microorganisms:

$$\text{Hydrocarbons} + O_2 + \text{microorganisms} + \text{energy} \rightarrow CO_2 + H_2O + \text{microorganisms} \quad (16.17)$$

Aerobic biological treatment involves mixing oxygen and aerobic microorganisms, sometimes in the presence of an external energy source in the form of added nutrition for the microorganisms. The biodegradation occurs in the biofilm around the microbes, as well as within the microbes. In time, the byproducts of gaseous CO_2 and water are produced which exit the top of the reaction vessel, while a solid mass of microorganisms is produced to exit the bottom of the reaction vessel.

If the water to be treated contains other chemical constituents, especially chlorine (Cl) and/or heavy metals, and if the microorganisms can withstand and flourish in such an environment, the simple input and output relationship is modified to become:

$$\begin{aligned}&\text{Hydrocarbons} + O_2 + \text{microorganisms} + \text{energy} + \text{Cl or heavy metal(s)} + H_2O + \text{inorganic salts} + \text{nitrogen compounds}\\&+ \text{sulfur compounds} + \text{phosphorus compounds} \rightarrow CO_2 + H_2O + \text{chlorinated hydrocarbons or heavy metal(s) inorganic salts}\\&+ \text{nitrogen compounds} + \text{sulfur compounds} + \text{phosphorus compounds}\end{aligned} \quad (16.18)$$

If the microbial population survives in this complex and potential hostile environment, new toxic products may be generated, e.g., containing newly-formed chlorinated hydrocarbons, higher heavy-metal concentrations, as well as more mobile or more toxic chemical forms of heavy metals.

All bioreactor systems rely on a population of microorganisms to metabolize organic contaminants, ideally into the harmless byproducts of CO_2+H_2O. Reactors also generate heat and other energy byproducts. The microorganisms must be either initially cultured in the laboratory to be able to metabolize a specific organic waste or must be given sufficient time to adapt to be able to digest the contaminant. The remediation manager must plan for and undertake extensive and continual monitoring and fine-tuning of each microbiological processing system during its operation [81].

Many microbially mediated treatment processes are redox reactions that transfer electrons from an organic compound to an electron acceptor. Oxygen is the electron acceptor if the process is aerobic, i.e., sufficient molecular oxygen is present. In aerobic microbial processes, trivalent iron (Fe^{3+}) as iron sulfate, and carbon dioxide are among the potential electron acceptors. Table 16.6 lists the redox reactions for three aromatic compounds commonly found in automobile junk yards and leak zones around underground gasoline storage tanks.

TABLE 16.6 Redox reactions for benzene, xylene, and ethylbenzene.

Type	Reaction	Electron acceptor
Benzene redox reactions:		
Oxidation	$C_6H_6+12H_2O \rightarrow 6CO_2+30H^++30e^-$	Oxygen
Reduction	$7.5O_2+30H^++30e^- \rightarrow 15H_2O$	
Reduction	$6NO_3^-+36H^++30e^- \rightarrow 3N_2+18H_2O$	Nitrate
Reduction	$15Mn^{4+}+30e^- \rightarrow 15Mn^{2+}$	Manganese
Reduction	$30Fe^{3+}+30e^- \rightarrow 30Fe^{2+}$	Iron
Reduction	$6SO_2^-+37.5H^++30e^- \rightarrow 3.75H_2S+15H_2O$	Sulfate
Reduction	$3.75CO_2+30H^++30e^- \rightarrow 3.75CH_4+7.5H_2O$	Methanogenic bacteria
Overall	$C_6H_6+7.5O_2 \rightarrow 6CO_2+3H_2O$	Oxygen
Overall	$C_6H_6+6H^++6NO_3^- \rightarrow 6CO_2+3N_2+6H_2O$	Nitrate
Overall	$C_6H_6+15Mn^{4+}+12H_2O^- \rightarrow 6CO_2+30H^++15Mn^{2+}$	Manganese
Overall	$C_6H_6+30Fe^{3+}+12H_2O^- \rightarrow 6CO_2+30H^++0Fe^{2+}$	Iron
Overall	$C_6H_6+3.75SO_4^{2-}+7.5H^+ \rightarrow 6CO_2+3.75H_2S+3H_2O$	Sulfate
Overall	$C_6H_6+4.5H_2O \rightarrow 2.25CO_2+3.75CH_4$	Methanogenic bacteria
Xylene and ethylbenzene redox reactions:		
Oxidation	$C_8H_{10}+16H_2O \rightarrow 8CO_2+42H^++ 42e^-$	Oxygen
Reduction	$10.5O_2+42H^++42e^- \rightarrow 21H_2O$	
Reduction	$8.4NO_3^-+50.4H^++42e^- \rightarrow 4.2N_2+25.2H_2O$	Nitrate
Reduction	$21Mn^{4+}+42e^- \rightarrow 21Mn^{2+}$	Manganese
Reduction	$42Fe^{3+}+42e^- \rightarrow 42Fe^{2+}$	Iron
Reduction	$5.25SO_2^-+52.5H^++42e^- \rightarrow 5.25H_2S+21H_2O$	Sulfate
Reduction	$5.25CO_2+42H^++42e^- \rightarrow 5.25CH_4+10.5H_2O$	Methanogenic bacteria
Overall	$C_8H_{10}+10.5O_2 \rightarrow 8CO_2+5H_2O$	Oxygen
Overall	$C_8H_{10}+8.4H^++8.4NO_3^- \rightarrow 8CO_2+4.2N_2+9.2H_2O$	Nitrate
Overall	$C_8H_{10}+21Mn^{4+}+16H_2O^- \rightarrow 8CO_2+42H^++21Mn^{2+}$	Manganese
Overall	$C_8H_{10}+42Fe^{3+}+16H_2O^- \rightarrow 8CO_2+42H^++ 42Fe^{2+}$	Iron

Continued

TABLE 16.6 Redox reactions for benzene, xylene, and ethylbenzene—cont'd

Type	Reaction	Electron acceptor
Overall	$C_8H_{10}+5.25SO_4^{2-}+10.5H^+ \rightarrow 8CO_2+5.25H_2S+5H_2O$	Sulfate
Overall	$C_8H_{10}+5.5H_2O \rightarrow 2.75CO_2+5.25CH_4$	Methanogenic bacteria

Although the three isomers of xylene and ethylbenzene all have the same formula, i.e., C_8H_{10}, the xylenes' methyl group (—CH_3) is located on at three different carbon sites on the benzene ring and ethylbenzene's methyl group is located on a carbon bonded with the ring (—C—CH_3). Thus, the reaction types are the same, but the rates differ.
Data from H. Rifai, C. Newell, J. Gonzales, S. Dendrou, L. Kennedy, J. Wilson, BIOPLUME III natural attenuation decision support system, version 1.0, user's manual, Air Force Center for Environmental Excellence (AFCEE), San Antonio, 1997.

The mathematics of the three orders of decay kinetics have been derived from experimental data. These are idealized curves that must be adapted to the heterogeneous and often poorly known conditions of the field. Thus, the expected growth and metabolism of microbes that catalyze contaminants will differ from handbook and textbook estimates [47].

Within a batch system employed by many biotreatment technologies, the rates will vary according to the microbial growth curve (see Chapter 5). For example, the lag stage is stable at the beginning, but growth increases rapidly before reaching the stationary stage.

6.5 Phyto-remediation

Phyto-remediation is bioremediation by way of plant life. It is usually *in situ* and is almost always dependent of available air, i.e., higher-order plants are exclusively aerobic.

Phyto-remediation takes advantage of plants' absorption of CO_2 for photosynthesis, the process whereby plants convert solar energy into biomass and release O_2 as a byproduct. Thus, the essential oxygen is the waste product of photosynthesis and is derived from carbon-based compounds. Respiration generates CO_2 as a waste product of oxidation that takes place in organisms, so there is a balance between green plants' uptake of CO_2 and release of O_2 in photosynthesis and the uptake of O_2 and release of CO_2 in respiration by animals, microbes, and other organisms.

Phyto-remediation [82] can be used for a wide range of contaminants and soil types. It is frequently used to remediate metal-contaminated sites (e.g., nickel and its compounds). Phyto-extraction (or phytoaccumulation) involves the uptake of contaminants from soil by roots into the transfer, i.e., translocation, to aboveground portions of the plants. Various species of plants, depending on the contaminant, soil type, climate, and local conditions, are planted and grown. Some are harvested like crops (e.g., grasses) and others maintained for removal and sequestration of the contaminants for longer time periods. The harvested materials can be treated and recycled by various methods, including composting and other bioremediation approaches and thermal processes, depending on the sequestered pollutants (e.g., content and form of halogenated and metallic compounds). This procedure may be repeated as necessary to decrease the concentrations of soil contaminants to target cleanup concentrations. For any harvest mass that is thermally treated, all emissions and ash residues must meet regulatory requirements, including handling and disposal as a possible hazardous waste.

Nickel, zinc, and copper are metals that have been particularly amenable to phytoextraction since hundreds of plant species have mechanisms for their uptake and absorption. This would seem to indicate numerous other metals would also be candidates since the micronutrient cycling and translocation mechanisms may be similar.

Rhizo-filtration (*rhizo*—root) is the adsorption or precipitation of dissolved compounds onto plant roots or absorption into the roots. Rhizo-filtration usually addresses contaminated ground water, while phytoextraction targets contaminated soil. Thus, the plants are usually grown hypoponically in greenhouses and exposed to water from the contaminated site once the plants grow mature root systems. This acclimation step allows the plant's endogenous processes to become acclimated to the chemicals in the water. After acclimation, the plants are transplanted in soil above the contaminated aquifer or perched water table, and when their uptake is considered to have reached a threshold they are harvested, and the biomass treated similarly to those from phytoextraction.

Phyto-stabilization immobilizes contaminants in the soil and ground water by sorption and accumulation by roots, by adsorption onto roots, or by precipitation within the root zone of plants (rhizosphere). The biochemodynamics hinder transport, e.g., to the ground water or air, as well as decrease bioavailability of the contaminants.

In addition to these physical translocation and filtering processes, plants can degrade pollutants, using mechanisms like microbial degradation. Phyto-degradation or phyto-transformation is the process of degrading compounds after uptake by the plant's metabolism. It may also involve the breakdown of compounds externally by exogenous secretions, such as enzymes. Complex organic molecules follow similar degradation pathways to those of microbes, discussed in this chapter. The metabolism breaks complex molecules into simpler compounds to gain energy and to build plant biomass, i.e., catabolism and anabolism, respectively.

Rhizo-degradation goes by a few names, including enhanced rhizosphere phyto-stimulation and planted-assisted bioremediation/degradation. It involves the same microbial process as those described earlier in this chapter, that is the degradation results from microbes using the contaminants as their energy and carbon sources. Microbes degrade soil-borne contaminants microbially, but these mechanisms are enhanced by the biochemodynamics of the root zone (the rhizosphere). This process is slower usually than phytodegradation. It can enhance co-metabolism because plant roots release natural substances, e.g., sugars, alcohols, and acid, that are food for soil microorganisms. The fixation and release of nutrients are also a natural type of bio-stimulation for the microbes. In addition, the roots provide conduits and physical loosening of soil, improving microbial contact with O_2 and nutrients—both a transport and transformation process. A plant takes up a compound from the soil or aquifer and transpires it to the troposphere. This can be solely the parent compound, if it resists degradation, i.e., metabolism and production of growth products, or it may be the metabolic degradation products, or both. For example, a poplar tree (genus *Populus*) may volatilize as much as 90% of the volatile organic compound taken up by the tree.

During *in situ* treatment, microbial and macrophytic processes occur simultaneously. Phyto-remediation has been used successfully on a variety of compounds in numerous locations. The local conditions and properties of the contaminants determine the degree of degradation [83,84].

The harvested plants contain elevated concentrations of compounds that have not been completely degraded if at all, especially metals. These plants must be treated onsite or transferred to a treatment facility. Indeed, one of the features sought in selecting plants is their ability to accumulate and concentration metals and other substances [85,86]. So-called hyperaccumulators can extract as much as 100 times as regular plants [87]. Certain plant species of the temperate herb pennycress (especially *Thlaspi caerulescens*) are excellent metal hyperaccumulators of metals from soil, with their shoots able to bioaccumulate 0.005 for lead (Pb) and 0.001 cadmium (Cd), and *Alyssum* able to accumulate 0.01 nickel (Ni). Unfortunately, the actual uptake rate is highly limited given both genera grow slowly and produce very little biomass, so the total amount of extracted metals is small [87]. That is, the rate-limiting step is not the bioaccumulation rate, but the storage capacity in the plant. Trees, on the other hand, have large amounts of biomass in which contaminants can be stored. And several species, e.g., willow trees (*Salix* spp.) are fast growers and tolerate wet conditions. Again, there is never an engineering panacea for hazardous substances, i.e., willows are subject to insect-borne and fungal diseases, so selection of the species must be matched to the waste site/tree stand conditions. Willows, as well as other promising genera, have also been considered to be invasive [88], so plans for containment should be part of the phytoremediation project.

Like all treatment approaches, phytoremediation should be seen from a systematic perspective. For example, synergies can be reached by combining abiotic and biotic processes [87]. Phytoremediation can be enhanced by bioaugmentation, i.e., adding lower-level technological interventions, which have been employed often in the past. Installing a barrier and adding nutrients and oxygen to enhance the growth of the soil microbial populations will have the added benefit of enhancing macrophytic growth needed for phytoremediation. Thus, both microbes and plants will be synergistically breaking down organic compounds and concentrating metals in plant tissue to be treated *ex situ*. Incidentally, this is known as biostimulation [33]. Often bioaugmentation is successful given that that soil bacteria that may have been eliminated in some of the contaminated soil may still be living in pockets and layers as spores and cysts, so that when these indigenous species are reintroduced to a less hostile environment, e.g., more nutrients, less toxins, more sorption sites, and more O_2, they will again increase in population and use their naturally developed enzymes to degrade the contaminants and provide a much better soil for plant growth [47]. Trees can benefit from such staging. Many species are sufficiently metal-tolerant to grow during the more hostile periods of high metal concentrations, e.g., by tree root avoidance of hostile pockets in the soil, known as "facultative tolerance" [87].

6.5.1 *Other biotreatment technologies*

There are several variations to bioremediation. Bioventing is an *in situ* bioremediation technology that stimulates aerobic degradation by enhancing the intrinsic ability of indigenous microbial populations to degrade organic chemical compounds that are adsorbed to soil and media particles. This is accomplished by introducing air or O_2 into the vadose zone, which is then circulated by direct air injection into the pollutant plume via vertical and horizontal wells [89].

Biosparging is also an *in situ* bioremediation technology. Biosparging consists of aeration under pressure in the zone of saturation below the water table, which increases O_2 concentrations in the ground water. This results in an increased pollutant degradation rate by indigenous microbes. This technique is like abiotic air sparging in its separation of volatile compounds by desorption and volatilization from the zone of saturation. The addition of pressurized air makes volatile compounds move upward toward the vadose zone, which also enhances the degradation process [89].

Bioaugmentation is the introduction of a naturally occurring or genetically engineered microbes that have better capabilities than the indigenous microbes in breaking down persistent pollutants in the ground water. It augments degradation rates during natural attenuation [90]. Bioaugmentation is commonly employed to remove chlorinated ethenes like tetrachloroethylene and trichloroethylene [89].

Bioslurping is an *in situ* bioremediation process, that combines the efficiency of vacuum-enhanced pumping and vapor extraction with bioventing to treat soil and ground water contaminated with volatile and semivolatile organic compounds [89].

Biofilters degrade contaminants in air emissions by immobilizing microorganisms on a solid support. When air pollutants are passed through biofilters, they become adsorbed onto microbial biofilms and are thus degraded sequentially. Biofilters are used for removing sulfur gases, hydrogen sulfide, ethyl benzene, dimethyl sulfides, nitrous oxide, etc. from the air [89].

During biostimulation, the biodegradation abilities and the growth of indigenous microbial populations are stimulated by adding aqueous solutions that contain nutrients or other amendments to the contaminated aquifer. It is designed to remediate soil and ground water containing hydrocarbons, VOCs, and many halogenated organic compounds [89].

Land farming involves bioremediating polluted soil at the earth's surface. Thus, contaminated soil and media must be excavated and mixed with microbes, nutrients, and other amendments. The mixture is then spread over a prepared bed to facilitate aerobic degradation. The receiving soil is periodically tilled until the waste is remediated completely [89].

Composting converts wastes in aerated piles, static piles, or continuous-fed reactors to a humus-like substance by microbial species at elevated temperature. Both aerobic and anaerobic conditions exist in different parts of the compost pile. Contaminated soil is added with nonhazardous organic waste, such as animal waste, agricultural waste, or manure, for the growth of microbe-rich population [89].

6.6 Combined technologies

In the U.S., Canada, and many western European countries, facilities can realize economies of scale by incorporating many treatment processes that might not be economical for individual generators. Comprehensive facilities are also able to exploit the synergistic opportunities made possible by having many different types of waste present at a single site. These include using waste acids and alkalines to neutralize one another, waste oxidants to treat cyanides, and organic contaminants in water, salts, and acids to remove organic compounds from wastewater, onsite incinerators to dispose of organic vapors generated by other onsite processes, ash and calcium and magnesium oxides to aid in stabilization processes, and combustible solids and liquids to produce blended liquid fuels.

The chemical structure of the pollutants can be altered to produce either an innocuous or a less hazardous material. This can be said of many remediation processes as well. Thus, it is possible to use abiotic and biotic reactions to treat organic and inorganic wastes [38].

Certainly, microbial activity in the tanks receiving the pumped ground water can even enhance the treatment of vapor phase compounds. In any biological system, including trickling filter systems, biofilms of microorganisms are grown on porous media. The air or other gas containing the VOCs is passed through the biologically active media, where the microbes break down the compounds to simpler compounds, eventually to CO_2 (if aerobic), CH_4 (if anaerobic), and water. The major difference between biofiltration and trickling systems is how the liquid interfaces with the microbes. The liquid phase is stationary in a biofilter, but liquids move through the porous media of the system. Compost can be used as the biofiltration system's porous media. Compost contains numerous species of beneficial microbes that are already acclimated to organic wastes. Industrial compost biofilters have achieved removal rates at the 99% level. Biofilters are also the most common method for removing VOCs and odorous compounds from air streams.

Success is highly dependent on the degradability of the compounds, their fugacity and solubility needed to enter the media and biofilm, and pollutant loading rates. Care must be taken in monitoring the porous media for incomplete biodegradation, the presence of substances that may be toxic to the microbes, excessive concentrations of organic acids and alcohols, and pH. The system must be inspected frequently for shock and the presence of dust, grease or other substances that may clog the pore spaces of the media [54].

Microbial populations can be added to the porous media to enhance removal of vapor phase compounds. The air carrying these volatile compounds is passed through a layer of biologically active media, where the microbes degrade the compounds to simpler compounds. The major difference between biofiltration and the trickling systems owes to the liquid interfaces with the microbes. The liquid phase is stationary in a biofilter, but in a trickling system, the liquids move through the porous media (see Chapter 11). Various porous media can house the microbes, such as compost, which usually has a very diverse microbial community of myriad species of bacteria and fungi [91–96]. Packed bed biological control systems are used for chemical compounds with relatively high vapor pressures. Air containing gas-phase pollutants (C_G) moves through porous media. The soluble fraction of volatilized compounds in an air stream partition into the biofilm (C_L) according to Henry's Law:

$$C_L = \frac{C_G}{H} \tag{16.19}$$

Where, H is the Henry's Law constant [97].

In biofiltration, air carries volatilized contaminants upward through compost or other media that contains microbes acclimated to degrade the pollutants in water pumped from the aquifer. The wastes at the bottom of the system can be heated to increase the partitioning to the gas phase. Microbes in the biofilm surrounding each individual compost particle metabolize the contaminants into simpler compounds, ideally converting them into CO_2 and water vapor [54]. Loading rates must be adjusted to ensure the ability of microbes to produce biofilm [19].

6.6.1 Ex situ and in situ treatment

Contaminated soil and sediment must often be removed and treated off-site (*ex situ*). Contaminated soil may be excavated and transported to kilns or other high-temperature operations, where the contaminated soil is mixed with combustible material. High-sand-content soils may be used as part of the mix. Contaminated soil may also be distributed onto an impermeable surface, allowing the more volatile compounds to evaporate. Microbial biodegradation can be accelerated and enhanced by adding nutrients and moisture to the soil. A faster process, thermal desorption, entails heating the soil to evaporate the contaminants and capture the compounds, and then burn them in a vapor-treatment device.

Treatment of media and water where they are found without first removing them is known as *in situ* remediation. Ground water is often treated by drilling recovery wells to pump contaminated water to the surface, with the water being treated on-site. *In situ* bioremediation begins with microbial acclimation. The acclimated microbes can then be taken from the laboratory and applied to the waste in the field. As mentioned, the most passive form of *in situ* bioremediation is natural attenuation, where no direct engineering intervention is used, and the contaminants are allowed to be degraded by resident microbes over time. The only role for the engineer is to monitor the soil and ground water to measure the rate at which the chemicals are degrading. Natural attenuation can work well for compounds that are found in the laboratory to break down under the conditions found at the site. For example, if a compound is degraded under reduced, low pH conditions in the laboratory, it may also degrade readily in soils with these same conditions (for example, in deeper soil layers where bacteria have adapted to these conditions naturally). Natural attenuation can be enhanced, for example, when soil moisture levels are increased.

In situ treatment can be a combination of several technologies. For example, the leachate collection systems of a hazardous waste landfill need to be treated provide a way to collect wastes that can then be treated. This is often via pump-and-treat systems that produce air pollutants that must be controlled. Leachate and ground water can be treated by drilling recovery wells to pump contaminated ground water to the surface.

7 Disposal

After treatment is completed, residue remains that must be disposed of safely. This is particularly true for *ex situ* and mobile incineration systems, which create ash and other substances. Most often, this residue has no economic value. There are options for disposing of hazardous residue:

1. Underground injection wells are steel- and concrete-encased shafts placed deep below the surface of the earth into which hazardous wastes are deposited by pressure. Some liquid waste streams are commonly disposed of in underground injection wells.
2. Surface impoundment involves natural or engineered depressions or diked areas that can be used to treat, store, or dispose of hazardous waste. Surface impoundments are often referred to as pits, ponds, lagoons, and basins.
3. Landfills are disposal facilities where hazardous waste is placed in or on land. Properly designed and operated landfills are lined to prevent leakage and contain systems to collect potentially contaminated surface water runoff. Most landfills isolate wastes in discrete cells or trenches, thereby preventing potential contact of incompatible wastes.

4. Land treatment is a disposal process in which hazardous waste is applied onto or incorporated into the soil surface. Natural microbes in the soil break down or immobilize the hazardous constituents. Land treatment facilities are also known as land-application or land-farming facilities.
5. Waste piles are noncontainerized accumulations of solid, nonflowing hazardous waste. While some are used for final disposal, many waste piles are used for temporary storage until the waste is transferred to its final disposal site.

Of the hazardous waste disposed of on land, nearly 60% is disposed of in underground injection wells, approximately 35% in surface impoundments, 5% in landfills, and less than 1% in waste piles or by land application. So, in addition to needing to dispose of wastes generated during ground water remediation projects, disposal is needed at thousands of other disposal sites.

Most countries have sites where hazardous waste has been disposed of improperly in the past and where cleanup operations are needed to restore the sites to their original state. Most of these contaminated sites require ground water remediation. In the U.S., about 20,000 of these sites have been identified and approximately 2000 sites require immediate action. In the U.S., these abandoned sites are treated differently from those regulated under the Resource Conservation and Recovery Act (RCRA). This is because there is often no clear means of removing wastes or remediating problems (e.g., the ownership is unknown, the former owner is bankrupt, or there are numerous potentially responsible parties). The Comprehensive Environmental Response, Compensation and Liability Act of 1980 (better known as Superfund) was enacted in response to the discovery of toxic waste dumps in the 1970s [98]. The law allows the EPA to clean up abandoned sites and to compel responsible parties to perform cleanups or reimburse the government for cleanups [99].

Canada has an estimated 1000 abandoned hazardous waste sites, including coal tar pits, leaky landfills, old plant sites and storage facilities. When a polluter cannot be charged with the task of remediation, Canada's federal government enters into agreements with individual provincial and territorial governments to clean up these high-risk properties. This happens when a responsible party cannot be found, or where the owner is unable or unwilling to finance a remediation project [100]. Many European sites have also been addressed. Denmark, the Netherlands, and Sweden have listed sites requiring immediate attention.

Environmental regulations lay out explicit steps for cleaning up contamination. In the U.S., the potential or confirmed releases of hazardous substances are evaluated and prioritized during the Superfund site assessment process. The federal government, states or tribal agency use the Hazard Ranking System (HRS) criteria to guide the process. The HRS is a numerically-based screen based on information from preliminary, limited investigations to assess the relative potential of sites to pose a threat to human health or the environment. After a site is identified, a series of assessments evaluate the potential need for and extent of remedial cleanup [101]. The assessment uses data to identify, evaluate and rank hazardous waste sites based on HRS criteria. It is the principal basis for placing uncontrolled waste sites on the National Priorities List (NPL). Sites with HRS scores of 28.5 or greater are eligible for placement on the NPL. Only sites on the NPL are eligible for Superfund Trust Fund-financed remedial actions [102].

The site assessment process [102] begins when a site is discovered and a site discovery date is logged. In the U.S., the site is also added to an active site inventory. Next, a preliminary assessment is conducted an, if needed, a site inspection or other more in-depth assessment is conducted to determine whether a site warrants short-term or long-term remediation. If the site does not require further attention, it is designated as "no further remedial action planned" (NFRAP). Sites requiring further attention are referred to the EPA's Emergency Response and Removal Program or other programs. Finally, the HRS model is applied to derive a preliminary site HRS score. Sites with preliminary HRS scores below 28.5 generally require no further Superfund remedial interest and are assigned a NFRAP decision. The NFRAP decision can also be made at sites with preliminary HRS scores of 28.5 or higher or based on other criteria to protect public health and ecosystems [102].

Cleaning up abandoned disposal sites involves isolating and containing contaminated material, removing, and redepositing contaminated soils, sediments, and other media, using situ and direct treatment technologies like those discussed in this chapter.

8 Conclusion

Various approaches are available to treat contaminated ground water. Ground water flow rates and other physical factors, along with the inherent properties of the pollutants being treated, determine the efficiency of the chosen treatment process. Sorption, vapor pressure, fugacity, and solubility are among the variables that must be considered when selecting a treatment technology. These and other factors influence abiotic degradation rates as well as biodegradation efficiency. Combinations of natural attenuation and engineered systems are available to address ground water contamination.

References

[1] Statistics Canada, Section 2: Canada's water supply—stocks and flows, in: Human Activities and the Environment, Statistics Canada, Ottawa, Ontario, 2010.

[2] European Environment Agency, Europe's groundwater—a key resource under pressure, EEA, 2022. https://www.eea.europa.eu/publications/europes-groundwater#:~:text=Groundwater%20stores%20almost%20one%20third,of%20water%20for%20agricultural%20irrigation. (Accessed 11 January 2024).

[3] W.M. Alley, T.E. Reilly, O.L. Franke, Sustainability of ground-water resources, 1999, Available: https://pubs.usgs.gov/circ/circ1186/html/intro.html.

[4] Laboratory of Ecohydrology (ECHO), n.d., Water Potential in Porous Media, ECHO, https://echo2.epfl.ch/VICAIRE/mod_3/chapt_4/main.htm (Chapter 4; Accessed on July 18, 2024).

[5] M. Karamouz, A. Ahmadi, M. Akhbari, Groundwater Hydrology: Engineering, Planning, and Management, CRC Press, Boca Raton, Florida, 2020.

[6] H.F. Hemond, E.J. Fechner, Chemical Fate and Transport in the Environment, Elsevier, 2014.

[7] S. Assouline, Infiltration into soils: conceptual approaches and solutions, Water Resour. Res. 49 (4) (2013) 1755–1772.

[8] L.A. Richards, Capillary conduction of liquids through porous mediums, Physics 1 (5) (1931) 318–333.

[9] A. Klute, Some theoretical aspects of the flow of water in unsaturated soils, Soil Sci. Soc. Am. J. 16 (2) (1952) 144–148.

[10] E.A. Curl, B. Truelove, The Rhizosphere, Springer Science & Business Media, 2012.

[11] L.V. Rinaudi, W. Giordano, An integrated view of biofilm formation in rhizobia, FEMS Microbiol. Lett. 304 (1) (2010) 1–11.

[12] S.S. Sindhu, A. Sehrawat, B.R. Glick, The involvement of organic acids in soil fertility, plant health and environment sustainability, Arch. Microbiol. 204 (12) (2022) 720, https://doi.org/10.1007/s00203-022-03321-x.

[13] R.P. Schwarzenbach, P.M. Gschwend, D.M. Imboden, Organic acids and bases: acidity constant and partitioning behavior, Environ. Org. Chem. (1993) 245–274.

[14] V. Evangelou, Environmental Soil and Water Chemistry, John Wiley, 1998.

[15] D. Richter, Environ 721: Soil Resources, Mentioned in course lecture, Duke University, Durham, North Carolina, 1997.

[16] R.R. Schindelbeck, et al., Comprehensive assessment of soil quality for landscape and urban management, Landsc. Urban Plan. 88 (2–4) (2008) 73–80.

[17] D.A. Vallero, Hazardous waste bioremediation, in: D. Yadav, P. Kumar, P. Singh, D.A. Vallero (Eds.), Hazardous Waste Management, Elsevier, 2022, pp. 347–379 (Chapter 14).

[18] P.R.M. Lopes, et al., Microbial bioremediation of pesticides in agricultural soils: an integrative review on natural attenuation, bioaugmentation and biostimulation, Rev. Environ. Sci. Biotechnol. 21 (4) (2022) 851–876.

[19] J. Regnery, A.D. Wing, M. Alidina, J.E. Drewes, Biotransformation of trace organic chemicals during groundwater recharge: how useful are first-order rate constants? J. Contam. Hydrol. 179 (2015) 65–75, https://doi.org/10.1016/j.jconhyd.2015.05.008 (in English).

[20] U.S. Environmental Protection Agency, Natural attenuation of the lead scavengers 1,2-dibromoethane (EDB) and 1,2-dichloroethane (1,2-DCA) at motor fuel release sites and implications for risk management, Ada, Oklahoma, September 2008, 2008, Available: https://nepis.epa.gov/Exe/ZyPDF.cgi/P1002UTI.PDF?Dockey=P1002UTI.PDF.

[21] J.T. Wilson, K. Banks, R.C. Earle, Y. He, T. Kuder, C. Adair, Natural attenuation of the lead scavengers 1, 2-dibromoethane (EDB) and 1, 2-dichloroethane (1, 2-DCA) at motor fuel release sites and implications for risk management, US Environmental Protection Agency, Office of Research and Development, 2008.

[22] C. Gandy, J. Smith, A. Jarvis, Attenuation of mining-derived pollutants in the hyporheic zone: a review, Sci. Total Environ. 373 (2–3) (2007) 435–446.

[23] H. Rifai, C. Newell, J. Gonzales, S. Dendrou, L. Kennedy, J. Wilson, BIOPLUME III natural attenuation decision support system, version 1.0, user's manual, Air Force Center for Environmental Excellence (AFCEE), San Antonio, 1997.

[24] D.A. Vallero, T.M. Letcher, Unraveling Environmental Disasters, second ed., Elsevier Academic Press, Amsterdam, The Netherlands, 2023.

[25] M.A.A. Mehedi, M.M.S. Yazdan, M.T. Ahad, W. Akatu, R. Kumar, A. Rahman, Quantifying small-scale hyporheic streamlines and resident time under gravel-sand streambed using a coupled HEC-RAS and MIN3P model, Eng 3 (2) (2022).

[26] K.E. Cognac, M.J. Ronayne, Changes to inter-aquifer exchange resulting from long-term pumping: implications for bedrock groundwater recharge, Hydrogeol. J. 28 (4) (2020) 1359–1370.

[27] S.G. Goddard, The chemical effects on a Cape Cod pond due to the upgradient reinjection of carbon-treated groundwater: case study, Snake Pond, Massachusetts Military Reservation, Falmouth, Massachusetts, Massachusetts Institute of Technology, 1997.

[28] D.A. Vallero, Methods and Calculations in Environmental Physics, AIP Publishing, Melville, New York, 2022.

[29] D.A. Vallero, Translating Diverse Environmental Data Into Reliable Information: How to Coordinate Evidence From Different Sources, Academic Press, 2017.

[30] B.M. Das, N. Sivakugan, Introduction to Geotechnical Engineering, Cengage Learning, 2015.

[31] B.M. Das, Geotechnical Engineering Handbook, J. Ross Publishing, 2010.

[32] U.S. Environmental Protection Agency, A citizen's guide to monitored natural attenuation, U.S. EPA, Office of Land and Emergency Management, Washington, DC, 2012. EPA 542-F-12-014. Available: https://www.epa.gov/sites/default/files/2015-04/documents/a_citizens_guide_to_monitored_natural_attenuation.pdf. (Accessed 13 February 2024).

[33] S.S. Suthersan, J. Horst, M. Schnobrich, N. Welty, J. McDonough, Remediation Engineering: Design Concepts, CRC Press, 2016.

[34] R. McKinney, Microbial relationships in biological wastewater treatment systems, in: The Structure and Function of Fresh-Water Microbial Communities, Virginia Polytechnic Institute and State Union, Blacksburg, Virginia, 1973, pp. 165–179.

[35] R.E. McKinney, in: D.A. Vallero (Ed.), Discussion of Population Dynamics in Biological Treatment Plants, 1966.
[36] R.E. Mckinney, Microbiology for Sanitary Engineers, McGraw Hill Book Co, New York, 1962.
[37] W. Eckenfelder, Biological treatment of waste water, Adv. Hydrosci. 3 (1966) 153–190.
[38] D. Vallero, Environmental Biotechnology: A Biosystems Approach, Elsevier Science, 2015.
[39] P.K. Jain, V.K. Gupta, V. Bajpai, Recent Advances in Environmental Biotechnology, 2011.
[40] B.E. Rittmann, Environmental biotechnology in water and wastewater treatment, J. Environ. Eng. 136 (4) (2009) 348–353.
[41] D.H. Pieper, W. Reineke, Engineering bacteria for bioremediation, Curr. Opin. Biotechnol. 11 (3) (2000) 262–270.
[42] N.P. Cheremisinoff, Biotechnology for Waste and Wastewater Treatment, Elsevier, 1997.
[43] S. Lindow, N. Panopoulos, B. McFarland, Genetic engineering of bacteria from managed and natural habitats, Science 244 (4910) (1989) 1300–1308.
[44] U.S. Environmental Protection Agency, Assessment and remediation of contaminated sediments (ARCS) program: remediation guidance document, Chicago, Illinois, EPA 905-R94-003, 1994.
[45] J. Beiyuan, Y.M. Awad, F. Beckers, D.C. Tsang, Y.S. Ok, J. Rinklebe, Mobility and phytoavailability of As and Pb in a contaminated soil using pine sawdust biochar under systematic change of redox conditions, Chemosphere 178 (2017) 110–118.
[46] T.M. Letcher, D.A. Vallero, Waste: A Handbook for Management, Academic Press, 2019.
[47] D.A. Vallero, Hazardous wastes, in: Waste: A Handbook for Management, 2011, p. 393.
[48] H.M. Freeman, Standard Handbook of Hazardous Waste Treatment and Disposal, 1989.
[49] P. Montague, Hazardous waste landfill: some lessons from New Jersey, Civ. Eng. 52 (9) (1982) 53–56.
[50] C.C. Lu, M.H. Hsu, Y.P. Lin, Evaluation of heavy metal leachability of incinerating recycled aggregate and solidification/stabilization products for construction reuse using TCLP, multi-final pH and EDTA-mediated TCLP leaching tests, J. Hazard. Mater. 368 (Apr 15) (2019) 336–344, https://doi.org/10.1016/j.jhazmat.2019.01.066.
[51] C. Zhao, Y. Dong, Y. Feng, Y. Li, Y. Dong, Thermal desorption for remediation of contaminated soil: a review, Chemosphere 221 (2019) 841–855.
[52] D.A. Vallero, Hazardous wastes, in: T.M. Letcher, D.A. Vallero (Eds.), Waste, second ed., Academic Press, 2019, pp. 585–630 (Chapter 31).
[53] US Environmental Protection Agency, Test methods for evaluating solid waste, physical/chemical methods, US EPA, Cincinnati, OH, USA, 1986.
[54] D.A. Vallero, Fundamentals of Air Pollution, fifth ed., Elsevier Academic Press, Waltham, MA, 2014. p. 999 pages cm.
[55] U.S. Environmental Protection Agency, Energy recovery from the combustion of municipal solid waste (MSW). https://www.epa.gov/smm/energy-recovery-combustion-municipal-solid-waste-msw. (Accessed 8 May 2018).
[56] J. Lee, D. Fournier, C. King, S. Venkatesh, C. Goldman, Evaluation of rotary kiln incinerator operation at low-to-moderate temperature conditions, in: Technical Results, vol. 1, 1996.
[57] U.S. Environmental Protection Agency, Locating and estimating air emissions from sources of benzene, Research Triangle Park, North Carolina, 1998.
[58] DOE/FE-0398, TIDD PFBC demonstration project, 1999.
[59] Clear the Air, Incineration policy in Australia, Commonwealth of Australia, 2018. http://news.cleartheair.org.hk/?p=2040. (Accessed 8 May 2018).
[60] U.S. Environmental Protection Agency, Facility pollution prevention guide, EPA/600/R-92/088, 1992.
[61] Federal Register, 40 CFR 264.342—Principal organic hazardous constituents (POHCs), in: 46 FR 7678, Jan. 23, 1981, as Amended at 48 FR 14295, Apr. 1, 1983, U.S. EPA, Washington, DC, 2010.
[62] U.S. Environmental Protection Agency, Air stripping (ex situ). https://frtr.gov/matrix/Air-Stripping-Ex-Situ/#Implementability-Considerations. (Accessed 22 January 2022).
[63] U.S. Environmental Protection Agency, Community guide to air stripping, U.S. EPA, Office of Land and Emergency Management, Washington, DC, 2021. EPA-542-F-21-001. Available: https://semspub.epa.gov/work/HQ/401577.pdf. (Accessed 13 February 2024).
[64] U.S. Environmental Protection Agency, Community guide to fracturing for site cleanup, U.S. EPA, Office of Land and Emergency Management, Washington, DC, 2021. EPA-542-F-21-009. Available: https://semspub.epa.gov/work/HQ/401593.pdf. (Accessed 13 February 2024).
[65] O. Lhotský, et al., The effects of hydraulic/pneumatic fracturing-enhanced remediation (FRAC-IN) at a site contaminated by chlorinated ethenes: a case study, J. Hazard. Mater. 417 (2021) 125883, https://doi.org/10.1016/j.jhazmat.2021.125883.
[66] I.S. Barkow, S.E. Oswald, H.J. Lensing, M. Munz, Seasonal dynamics modifies fate of oxygen, nitrate, and organic micropollutants during bank filtration-temperature-dependent reactive transport modeling of field data, Environ. Sci. Pollut. Res. Int. 28 (8) (2021) 9682–9700, https://doi.org/10.1007/s11356-020-11002-9 (in English).
[67] J. Greskowiak, E. Hamann, V. Burke, G. Massmann, The uncertainty of biodegradation rate constants of emerging organic compounds in soil and groundwater—a compilation of literature values for 82 substances, Water Res. 126 (2017) 122–133, https://doi.org/10.1016/j.watres.2017.09.017 (in English).
[68] T. Weidemeier, M. Swanson, J. Wilson, D. Kampbell, R. Miller, J. Hansen, Approximation of biodegradation rate constants for monoaromatic hydrocarbons (BTEX) in ground water, US Environmental Protection Agency Papers, 1996, p. 26.
[69] A. Mitra, Biology, genetic aspects, and oxidative stress response of Streptomyces and strategies for bioremediation of toxic metals, in: Microbial Biodegradation and Bioremediation, 2014, p. 287.
[70] Y. Ikada, H. Tsuji, Biodegradable polyesters for medical and ecological applications, Macromol. Rapid Commun. 21 (3) (2000) 117–132.
[71] S. Arriaga, R. Muñoz, S. Hernández, B. Guieysse, S. Revah, Gaseous hexane biodegradation by Fusarium solani in two liquid phase packed-bed and stirred-tank bioreactors, Environ. Sci. Technol. 40 (7) (2006) 2390–2395.
[72] M. Alexander, Biodegradation and Bioremediation, Gulf Professional Publishing, 1999.

[73] D.A. Abramowicz, Aerobic and anaerobic PCB biodegradation in the environment, Environ. Health Perspect. 103 (Suppl. 5) (1995) 97.
[74] EPA/600/R-07/060, Landfill bioreactor performance: second interim report: outer loop recycling & disposal facility—Louisville, Kentucky, 2007.
[75] A.L. Santos, D.B. Johnson, Design and application of a low pH upflow biofilm sulfidogenic bioreactor for recovering transition metals from synthetic waste water at a Brazilian copper mine, Front. Microbiol. 9 (2051) (2018), https://doi.org/10.3389/fmicb.2018.02051 (in English).
[76] U.S. Environmental Protection Agency, Community guide to bioremediation, U.S. EPA, Office of Land and Emergency Management, 2021. Washington, DC, EPA-542-F-21-004. Available: https://semspub.epa.gov/work/HQ/401583.pdf. (Accessed 13 February 2024).
[77] S.M. Tripathi, S. Ram, Bioremediation of groundwater: an overview, Int. J. Appl. Eng. Res. 13 (24) (2018) 16825–16832.
[78] T.H. Christensen, P.L. Bjerg, S.A. Banwart, R. Jakobsen, G. Heron, H.-J. Albrechtsen, Characterization of redox conditions in groundwater contaminant plumes, J. Contam. Hydrol. 45 (3) (2000) 165–241, https://doi.org/10.1016/S0169-7722(00)00109-1.
[79] R.E. McKinney, Civil engineering 876: wastewater treatment plant design, University of Kansas, Course Lecture, Spring Semester, 1984, 1984.
[80] D.A. Vallero, J.J. Peirce, Engineering the Risks of Hazardous Wastes, Butterworth-Heinemann, Burlington, MA, 2003. p. xxi, 306 pp.
[81] A. Dragt, J. van Ham, Biotechniques for Air Pollution Abatement and Odour Control Policies, Elsevier, 1992.
[82] D.A. Vallero, C. Brasier, Sustainable Design: The Science of Sustainability and Green Engineering, John Wiley & Sons, 2008.
[83] U.S. Environmental Protection Agency, Phytoremediation resource guide, in: Technologies for Cleaning up Contaminated Sites, 1999. Washington, USA, EPA 542-B-99-003.
[84] EPA, Introduction to phytoremediation, 2000.
[85] J.H. Lee, An overview of phytoremediation as a potentially promising technology for environmental pollution control, Biotechnol. Bioprocess Eng. 18 (3) (2013) 431–439.
[86] N. Rascio, F. Navari-Izzo, Heavy metal hyperaccumulating plants: how and why do they do it? And what makes them so interesting? Plant Sci. 180 (2) (2011) 169–181.
[87] I. Pulford, C. Watson, Phytoremediation of heavy metal-contaminated land by trees—a review, Environ. Int. 29 (4) (2003) 529–540.
[88] K.W. Cremer, Introduced willows can become invasive pests in Australia, Biodiversity 4 (4) (2003) 17–24.
[89] D. Kaur, A. Singh, K. Abhijit, S. Gupta, Genetic engineering approaches and applicability for the bioremediation of metalloids, in: D.K. Tripathi, et al. (Eds.), Plant Life Under Changing Environment, Academic Press, 2020, pp. 207–235 (Chapter 9).
[90] M. Andreolli, S. Lampis, P. Brignoli, G. Vallini, Bioaugmentation and biostimulation as strategies for the bioremediation of a burned woodland soil contaminated by toxic hydrocarbons: a comparative study, J. Environ. Manag. 153 (2015) 121–131.
[91] C. Hort, S. Gracy, V. Platel, L. Moynault, Evaluation of sewage sludge and yard waste compost as a biofilter media for the removal of ammonia and volatile organic sulfur compounds (VOSCs), Chem. Eng. J. 152 (1) (2009) 44–53.
[92] C. Ima, D. Mann, Physical Properties of Woodchip: Compost Mixtures Used as Biofilter Media, 2007.
[93] W. Namkoong, J.-S. Park, J.S. VanderGheynst, Biofiltration of gasoline vapor by compost media, Environ. Pollut. 121 (2) (2003) 181–187.
[94] R. Nicolai, K. Janni, Biofilter media mixture ratio of wood chips and compost treating swine odors, Water Sci. Technol. 44 (9) (2001) 261–267.
[95] B. Cárdenas-González, S.J. Ergas, M.S. Switzenbaum, Characterization of compost biofiltration media, J. Air Waste Manage. Assoc. 49 (7) (1999) 784–793.
[96] S.J. Ergas, E.D. Schroeder, D.P. Chang, R.L. Morton, Control of volatile organic compound emissions using a compost biofilter, Water Environ. Res. 67 (5) (1995) 816–821.
[97] S.J. Ergas, K.A. Kinney, Biological control systems, in: Air Pollution Engineering Manual, 2000, p. 55.
[98] M. Hammond, M. Goldin, R. Kumar, J. Croy, M. Sharma, S. Wilson, Comprehensive Environmental Response, Compensation, and Liability Act (Superfund); 42 U.S.C. §9601 et seq. (1980), United States, in: New Georgia Encyclopedia, University of Georgia Press, Athens, Georgia, 2013. ISSN 2765-8732. (Accessed 18 July 2024. Last modified August 27, 2013.
[99] U.S. Environmental Protection Agency, Superfund cleanup process, 2022. https://www.epa.gov/superfund/superfund-cleanup-process. (Accessed 24 January 2022).
[100] S. Meakin, Hazardous waste management: Canadian directions, Government of Canada: Depository Services Program, 1992.
[101] EPA/540/R-92/074 B, Guide for conducting treatability studies under CERCLA: thermal desorption, 1992.
[102] U.S. Environmental Protection Agency, Superfund site assessment process, 2018. https://www.epa.gov/superfund/superfund-site-assessment-process. (Accessed 8 May 2018).

Chapter 17

Risk assessment

1 Introduction

Risk depends on both the likelihood of a hazard and the likelihood of an exposure. Generally, the substances of greatest concern are those with high potential toxicity and high potential exposure profiles. Indeed, even when toxicity remains the same, those substances with higher exposure potential will lead to higher risk (see far-right of Fig. 17.1). When hazard concentrations and exposure potential intersect, populations are being exposed at concentrations most likely to lead to adverse outcomes. Similarly, exposure calculations are linked to risk estimates by metrics based on toxicity thresholds that are modified by uncertainty factors (UFs). The uncertainties are largely due to the differences between results found in animal testing and expected outcomes in human population.

To an assess the impact of a water pollution event on water quality, the event is characterized according to its potential to evoke an adverse outcome. Evidence-based risk assessments provide estimates of the amount of harm that could result from exposure to pollutant. A retrospective risk assessment as part of the root cause analysis following an adverse event includes every potential stressor and outcome that can be determined from data. A prospective risk assessment follows the root cause analysis to identify actual and potential causes of the adverse outcome. The risk (R) is commonly expressed as the product of a stressor's hazard (H) and exposure (E):

$$R = H \times E \tag{17.1}$$

Hazard is an inherent trait of a substance or microbe that expresses a potential adverse outcome. This inherency is based on peer-reviewed articles, government reports, and other scientific studies. The outcome can occur relatively rapidly, e.g., within 1 year of exposure is known as an acute hazard. Examples include the exposure to high concentrations of carbon monoxide (CO) leading to cardiovascular and neurological failures after a few minutes [1–3] or drinking water contaminated with the cryptosporidium protozoan that infects a humans and other vertebrates with acute gastroenteritis a few days after the exposure. Hazard may also occur long after exposure or after continuous exposure over time, which is known as a chronic hazard, such as neurotoxicity from exposure to the metal lead or carcinogenicity from the exposure to asbestos or benzene.

The same hazard agent may be both an acute and a chronic hazard depending on the concentration and exposure. For example, dimethyl mercury is acutely toxic at very low concentrations [4] and lethal at doses of approximately 5 mg per kg-body weight [5,6]. However, all forms of mercury (Hg) are also chronically toxic at even lower concentrations over longer durations of exposure [7–14].

Also, as mentioned CO at high concentrations is an acute hazard but it is also a chronic hazard at much lower concentrations for long-term exposure, such as tobacco smoking or side-stream (environmental) exposure to tobacco smoke for several years [15–17].

Although most environmental scientists would agree that Eq. (17.1) captures the general sense of risk, there is no unanimously accepted definition of risk within the greater scientific community. Most definitions include an adverse outcome and the probability of it occurring. Most definitions include both the probability of an event and the consequences that could result from that event, such as nuclear accidents or terrorist attacks. As evidence [18,19]:

> *In most formal discussions of risk assessment, risk is defined as the product of the probability that a certain event might occur … and the consequences that could result from such an event. The probability side of the equation is basically a combination of threats and vulnerabilities.*

Fundamentals of Water Pollution. https://doi.org/10.1016/B978-0-443-28987-3.00023-0

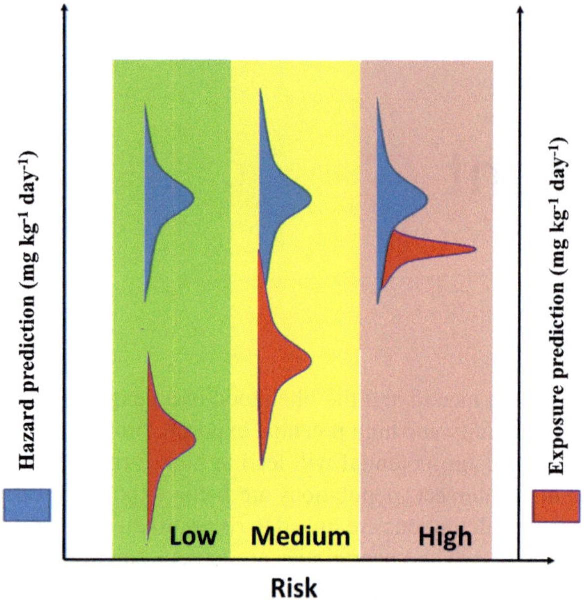

FIG. 17.1 Potential for human health risk extrapolated from constant toxicity and increasing exposure potential. *(Based on R.S. Judson, et al., Estimating toxicity-related biological pathway altering doses for high-throughput chemical risk assessment, Chem. Res. Toxicol. 24 (4) (2011) 451–462.)*

There is variation within engineering and technical professions regarding the definition of risk, with them recently placing greater emphasis on the magnitude of the adverse consequence has been proposed [20]:

> *Different events … can have adverse effects on human life, health, property, or the environment, and consequently constitute a risk. The concept of risk can be … considered as the product between the magnitude of such adverse effects, expressed numerically as the number of deaths, percentage increase in cancer cases, property value loss, and so on, and the likelihood that the event causing will occur or, alternatively, the subsequent likelihood that the consequences themselves will occur.… Risk* r *will simply be expressed as the product of magnitude* m *and likelihood* l*, so with the data coming from risk analysis* n *risks* $r_i = m_i l_i$ *can be calculated.*

This connotation of risk requires a choice as to whether to include the probability of the event or the consequence of the event. In many environmental risk assessments, it is the consequence, such as the added cancer cases or number of deaths (mortality rates) or falling below a threshold number of an endangered species that are included in the risk equation. Also, the shorthand in this text and numerous other environmental risk documents is to present risk as a unitless value, e.g., 1 in a million or 10^{-6} cases out of a population of 1 million. Even in these cases, the consequence is understood. Although the risk is stated as a unitless fraction, such as a cancer risk of 10^{-6}, that numerical expression implies that the units, in fact, are number of cancer cases, i.e., one added case per million people exposed. Similarly, the National Research Council's (NRC) Committee on Risk Perception and Communication has defined risk as [21]:

> *… the product of a measure of the size of the hazard and its probability of occurrence. Regardless of how numerical estimates are made the essence of the distinction between hazard and risk is that "risk" takes probability explicitly into account.*

This definition adds the hazard and its magnitude to the probability of harm [21]. Furthermore, the NRC's Committee on Risk Characterization defines risk as [22]:

> *A concept used to five meaning to things, forces, or circumstances that pose danger to people or to what they value. Descriptions of risk are typically stated in terms of the likelihood of harm or loss from a* hazard *and usually include: an identification of what is "at risk" and may be harmed or lost (e.g., health of human beings or an ecosystem, personal property, quality of life, ability to carry on an economic activity); the hazard that my occasion this loss; and a judgment about the likelihood that harm will occur.*

Both the type of adverse outcome and its probability of occurrence determine the amount of risk [23]:

> *The term risk usually expresses not only the potential for an undesired consequence but also how probable it is that such a consequence will occur….. A mathematical definition of risk commonly found in the literature is:*

$$\text{Risk (Consequence/unit time)} = \text{Frequency (event/unit time)} \times \text{Magnitude (Consequence/event)}. \tag{17.2}$$

Thus, there are several quantitative expressions of risk. All express a probability and either explicitly or implicitly express a consequence.

To ascertain possible risks posed during and after an event, the first step in a retrospective risk assessment is identify general a hazard and then determine the exposure scenarios that occurred. As mentioned, this can be followed by a prospective risk assessment wherein scenarios of every possible event that introduces a potential threat and that could lead to the effect in the future. Risk, then, is estimated based on the severity of the effect and its likelihood to occur in that scenario. This combination of each hazard and the exposure likelihood represents the harm to ecological condition and human health in the root cause analysis and the cumulative risk of a pending future adverse events in the prospective risk assessment [24,25].

The relationship between the severity and probability of a risk follows a general equation:

$$R = f(S, P) \tag{17.3}$$

where risk (R) is a function (f) of the severity (S) and the probability (P) of harm. The risk equation can be simplified to be a product of severity and probability:

$$R = S \times P \tag{17.4}$$

2 Hazard identification

Pollutant hazard primarily focuses on the toxicity of substance. For example, a compound that causes cancer or elicits respiratory problems in rats would be suspected of doing the same in humans, albeit at lower doses and longer time frames than what laboratory rats have been exposed. Also, the toxicity to other species is also important. For example, some species are sensitive to substances that may have relatively low human toxicity, e.g., copper compounds can be highly toxic to fish [26].

Ecosystem toxicity, i.e., ecotoxicity, is the hazard that a pollutant may cause to various species within an ecosystem. For example, a substance may have low toxicity to a particular fish species, but high toxicity to an organism on which the fish feeds. This would not kill the fish directly but would diminish the abundance and diversity of fish species in the ecosystem.

Besides toxicity, chemical hazards include other potential harm, including fire hazard, explosiveness, corrosivity, and chemical reactivity. Hazards may also be biological (biohazards) and physical (e.g., radioactive). Air pollution experts are primarily interested in chemical hazards, but must be aware of these other hazards, particularly in first response and emergency situations. Indeed, spills and accidents may at first be physical hazards, e.g., a highly reactive compound leaking from a ruptured tank. The reactions may form highly toxic vapors and gases that threaten human populations. In addition to stopping the leak, the immediate actions may be to evacuate nearby residences to prevent exposures to these toxic gases.

2.1 Thresholds

Health officials and occupational and environmental scientists and engineers often refer to *thresholds* of chemical compounds. The threshold level is the lowest amount needed to induce harmful effects in an organism. Establishing a toxicity threshold begins with a substance's inherent properties, e.g., cyanide is highly acutely toxic, whereas the metal iron is usually only acutely toxic in high doses. Next, the influence of factors such as a chemical's mobility in environmental media and tissues should be considered, along with its persistence in the organism and the environment. For example, a chlorinated compound is inherently harder to break down than is the unsubstituted compound. It is also much more likely to accumulate in living tissue.

The traditional risk assessment paradigm for chemical compounds in the environment (Fig. 17.1) is a step-wise process, starting with the identification of a hazard. Identification is comprised of a summary of an agent's physicochemical properties and routes and patterns of exposure and a review of toxic effects. The tools for hazard identification consider the chemical structures that are associated with toxicity, metabolic and pharmacokinetic properties, short-term animal and cell tests, long-term animal (*in vivo*) testing, and human studies (e.g., epidemiology, such as longitudinal and case-control studies). These comprise the core components of hazard identification, however additional hazard identification methods have been emerging that are increasingly provide improved reliability of characterization and prediction.

Biomarkers are useful in identifying hazards since many exposure biomarkers appear before tissue damage. Hazard identification is also being advanced by computational chemistry by enhancing the understanding molecular structure. Even similar substitutions on molecules (e.g., one versus two methyl groups, methyl versus ethyl substitutes, or halogenations (Cl versus F)) can lead to quite different levels of hazards. As mentioned, computational tools like QSAR models are increasingly used to identify and quantify hazards. They give clues to subtle and presymptomatic hazard information akin to the precision medicine's focus on a person's entire life, rather than the presence or absence of diseases [27]. New methods are constantly being developed to identify and characterize pollutant hazards.

Regulatory agencies differentiate cancer risk from noncancer risk. Carcinogens are classified according to the soundness of science that supports or rejects a substance's link to cancer (see Table 17.1). Noncancer hazard is generally based on studies that show thresholds above which a dose leads to an adverse outcome (e.g., neurological damage, reproductive and developmental disorder, and endocrine disruption). This threshold is known as the no observable adverse effect level (NOAEL). For dermal, ingestion and other exposure pathways, no observable adverse effect level (NOAEL) is commonly used as the threshold. The NOAEC is used for inhalation, so is most appropriate for air pollutants.

Cancer risk policies often assume that there is no threshold below which an exposure is acceptable. Another threshold, the "lowest observed adverse effect level" (LOAEL) is the area of the dose-response curve where studies have linked a dose to an effect. Cancer slope factors (CSFs, also known as q1*) are used to calculate the estimated probability of increased cancer incidence over a person's lifetime, known as the excess lifetime cancer risk (ELCR) [28]. Like the reference doses, slope factors relate to specific exposure pathways, i.e., water and food ingestion, inhalation and dermal [29–32].

As mentioned, for cancer risk assessments, the hazard is generally assumed to be the product of the CSF and the long-term exposure (E_{lt}):

$$\text{Cancer risk} = \text{CSF} \times E_{lt} \tag{17.5}$$

Since cancer is a chronic illness, with symptoms appearing after many years of exposure or at least after a latency period following episodic exposures, the corresponding E must also be chronic, e.g., LADD. In this case risk will be the product of LADD and potency represented by the CSF. To demonstrate, we can calculate the direct risk to the people living near a source of vinyl chloride a potent liver carcinogen. Assume that the mean LADD in the neighborhood is 100 ng kg^{-1} day^{-1} and, from EPA's Chemistry Dashboard [33], the vinyl chloride CSF of 7.2×10^{-1} (mg kg^{-1} day^{-1})$^{-1}$, then the mean lifetime cancer risk to the neighborhood exposed for an entire lifetime (exposure duration = 70 years) gives:

$$1 \times 10^2 \text{ ng kg}^{-1} \text{ day}^{-1} \times 10^{-6} \text{ mg ng}^{-1} \times 7.2 \times 10^{-1} \left(\text{mg kg}^{-1} \text{ day}^{-1}\right)^{-1} \approx 4.3 \times 10^{-4}.$$

TABLE 17.1 Cancer classes in the U.S., based on weight-of-evidence.

Cancer classification	Description
A	The chemical considered to be a human carcinogen
B	The chemical is a probable human carcinogen, with two subclasses
B1	Chemical compounds that have limited human data from epidemiological studies supporting their carcinogenicity
B2	Chemical compounds for which there is sufficient evidence from *in vivo* studies and for which but inadequate or no evidence from human epidemiological studies
C	The chemical is a possible human carcinogen
D	The chemical is not classifiable as to human carcinogenicity
E	There is evidence that the chemical does not induce cancer in humans

Data from U.S. Environmental Protection Agency, Quantitative Estimate of Carcinogenic Risk From Oral Exposure: Aniline, U.S. EPA. https://iris.epa.gov/ChemicalLanding/&substance_nmbr=350, 1988 (Accessed February 10, 2023). U.S. Environmental Protection Agency, EPA's Approach for Assessing the Risks Associated With Chronic Exposure to Carcinogens: Background Document 2—January 17, 1992, U.S. EPA. https://www.epa.gov/iris/epas-approach-assessing-risks-associated-chronic-exposure-carcinogens, 2020 (Accessed September 29, 2020). U.S. Environmental Protection Agency, Proposed Guidelines for Carcinogen Risk Assessment. Available: http://cfpub.epa.gov/ncea/cfm/recordisplay.cfm?deid=2833, 2015.

The neighborhood lifetime risk from exposures to vinyl chloride is about 4 per 10,000. Often, in cleanups, the threshold is that the risk should not exceed 1 in a million. For simple removals, the cleanup risk target is not to exceed 1 per 10,000. Both risk levels are exceeded in this hypothetical example. The combination of a steep slope factor and very high lifetime exposures leads to an elevated cancer risk.

The relatively high aqueous solubility and soil sorption rates of vinyl chloride should be considered in soil exposure dose and risk calculations. Plausible exposure scenarios would include the ingestion of drinking water for people on private wells drawing water from ground water that has been contaminated, and dermal exposures, e.g., children playing in the soil, are also plausible scenarios. Total risk from a single contaminant like vinyl chloride equals the sum of risks from all pathways, e.g., vinyl chloride in the air, water, and soil:

$$\text{Total Risk} = \sum \text{risks from all exposure pathways} \tag{17.6}$$

Cancer risk varies by the potency of the carcinogen and the exposure to that carcinogen. As for any air pollution risk, cancer risk is a function of hazard and exposure. As mentioned, since cancer is a chronic disease, the exposure must be long-term or at least must address the latency between exposure and cancer outcome. Thus, cancer risk assessments generally express hazard potency for the dose-response curves from animal research and human epidemiology. The quasi-linear segment of these curves provides what is known as the CSF and the long-term exposure is the lifetime average daily dose LADD.

An exposure concentration (*EC*) is a time-weighted average concentration derived from measured or modeled concentrations of a contaminant in the air but adjusted according to the characteristics of the exposure scenario. For an acute drinking water exposure, the *EC* and water concentration (C_W) values will be almost the same. However, estimating cancer risks from chronic exposure involves the both the C_W measured at a point of exposure point as well as scenario-specific parameters, such as the exposure duration and frequency:

$$EC = \frac{C_W \times ET \times EF \times ED}{AT} \tag{17.7}$$

where EF is the exposure frequency, ED is the exposure duration, and AT is the averaging time.

The value for AT for chronic exposures is ED in years × 365 days year^{-1} × 24 h day^{-1}. The AT units are hours, so AT here is multiplied by 24 to give an AT in days. Thus, *EC* usually is the same as C_W [34].

Total ingestion exposures occur when an individual comes intentional and inadvertently a substance which then enters the gastrointestinal (GI) tract. Depending on the properties of the substance, absorption can occur at various locations throughout the GI tract. A substance may directly reach the GI tissue or be subsequently enter the bloodstream. Ingestion exposure can be estimated as:

$$E_{ing} = C_{ing} \times IR \tag{17.8}$$

where: E_{ing} is ingestion exposure (mass per time), C_{ing} is concentration of the substance in food, water, or other exposure media (mass of substance per mass of medium or mass of substance per volume of medium), IR = ingestion rate (mass of medium ingested during the exposure per time). If more than one medium is being studied for ingestion exposure, the exposure from each medium is calculated separately and then summed. An exposure scenario is usually expressed as a frequency of ingestion event times the intake per event [35].

For hazard pollutants that are inhaled, the unit risk estimate (URE) is often used instead of equations based on the CSF. The URE is an upper-bound estimate of a person's chance of contracting cancer over a lifetime of exposure when exposed to 1 μg L^{-1} of drinking water or 1 μg m^{-3} air. An upper-bound risk is a likely limit to the true probability that a person will contract cancer over a 70-year lifetime due to a toxic substance [36–38].

Risks can be calculated by multiplying the actual concentration to which a person is exposed by the URE. For example, a URE of 3 ×10^{-6} per 1 μg L^{-1} of drinking water means that as many as three additional persons in an exposed population of 1 million would be expected to develop cancer during a lifetime to 1 μg of the carcinogen in 1 L of water drunk. Since the URE is an upper-bound estimate it is conservative, i.e., it avoids underestimating the cancer risk. The actual risk may be lower but is unlikely to be higher.

The exposure concentration is often combined with the inhalation-related URE to provide the inhalation unit risk (IUR). The IUR is an estimate of the increased cancer risk via inhalation exposure to a concentration of 1 per μg m^{-3} for a lifetime. Like the CSF, the IUR is a predictive cancer risk potency estimate. The IUR provides an upper-bound excess lifetime cancer risk estimated to result from continuous exposure to an agent at a concentration of 1 μg m^{-3} in air. For example, if the IUR for formaldehyde in a metropolitan area is 2×10^{-6} per μg m^{-3} predicts two excess cancer cases as upper bound estimate to develop per 1,000,000 people if exposed daily for a lifetime to 1 μg of the chemical in 1 m^3 of air.

2.2 Microbial hazard classification

In addition to chemical hazards, nonchemical pollutants, especially biological agents are of concern. These may include microbial organisms, e.g., viruses, bacteria, and fungi. They may also include parts of larger organisms, e.g., pet dander and insect antigens (as triggers for asthma), as well as pollen and spores. The concern has recently grown to include genetically modified organisms (GMOs). Risks and hazards from biological agents are often different that those posed by chemical compounds, so the risk assessment process is commensurately different. As evidence, the Safety in Biotechnology Working Party of the European Federation of Biotechnology has identified four risk classes for GMOs [39,40]:

Risk class 1. No adverse effect, or very unlikely to produce an adverse effect. Organisms in this class are considered to be safe.

Risk class 2. Adverse effects are possible but are unlikely to represent a serious hazard with respect to the value to be protected. Local adverse effects are possible, which can either revert spontaneously (e.g., owing to environmental elasticity and resilience) or be controlled by available treatment or preventive measures. Spread beyond the application area is highly unlikely.

Risk class 3. Serious adverse local effects are likely with respect to the value to be protected but spread beyond the area of application is unlikely. Treatment and/or preventive measures are available.

Risk class 4. Serious adverse effects are to be expected with respect to the value to be protected, both locally and outside the area of application. No treatment or preventive measures are available.

These classes indicate that even the safest genetic modifications carry some risk and that with more uncertainty about an organism, one cannot assume it to be safe. Thus, risk scenarios include not only the effects resulting from the intended purpose of the environmental application, but also downstream and side effects that are not part of the desired purpose.

The biological agent risk classes are analogous to the hazard classes of various human health risk paradigms, e.g., carcinogen listings or extremely hazardous designations. Thus, a microbe can be assigned to the appropriate risk class by based on its physical, chemical, and biological properties, notwithstanding the way it was selected or, in the case of genetically modification and synthetic biology, how a specific variant was generated. After which the potential risk is scored against a set of values to be protected [27]. As in the hazard identification process for chemical compounds, the specific microbe is classified according to its inherent properties. The next stage addresses the environmental conditions by characterizing aspects like the administered dose to various responses in different populations. Both the hazard identification and dose-response data are based on research that is the basis of the risk analysis. The virulence of strains of the same species or that of closely related species differ depending on the genetic makeup. For example, the bacteria *Salmonella enteritidis* and *S. pullorum* elicit various types of pathogenicity in different humans can vary by orders of magnitude [41,42]. Similarly, specific *E. coli* variants cause enteric diseases in humans but although most *E. coli* isolates are not considered to be pathogenic. Even two isolates with all the same virulence characteristics can lead to substantially dissimilar pathogenicity due to differences in the genetic expression of the toxicity or other characteristics [41].

Hazard classification of microbes is similar to that of chemical hazard identification processes in that available knowledge about other microbes with similar characteristics can be extrapolated, although this information is not sufficiently tested for the microbe of concern but has been observed in similar environmental conditions. A field study's results in one type of field may be extrapolated to an environmental remediation setting if the microbe in question had been observed in sufficient agricultural settings. Similarly, in chemical hazard identification, a chemical compound that has been sufficiently studied and has similar chemical structure to a les-studied compound a similarity model, like a quantitative structural activity relationship (QSAR) may be employed [43–47]. Such similarity models may be applied to microbial growth and pathogenicity [48].

2.3 Toxicokinetics

So-called "emerging" chemical contaminants have added uncertainties regarding adverse outcomes, exposure duration and timing, and predictions of the effectiveness of proposed actions. There is much greater uncertainty on the ability to meet an objective by regulatory agencies for substances like microplastics (see Discussion Box 17.1), nanoparticles [49–67] and perfluoroalkyl and polyfluoroalkyl substances (PFAS) [30,68,69], which at first have a scarcity of reliable data and peer-reviewed studies from which to make decisions.

Discussion Box 17.1 Applying ADMET to microplastics

Estimating the potential for ecotoxicity associated with microplastics is fraught with uncertainties. One approach is to predict endogenous activity within an organism, such as a sentinel species [70–73]. The biological activity within the organism is a first step toward characterizing potential adverse ecological outcomes. This is based on the specific rates and amounts of absorption, distribution, metabolism, elimination, and toxicity (ADMET) of various species exposed to different classes of microplastics [74,75]. Such models begin with the mass of a specific physicochemical form of a microplastic taken up by the organism. The inherent physicochemical properties and the potential for the microplastic to induce biological activity are key inputs to ADMET models [76,77].

The characteristics of nearly 1 million chemical compounds can be found on websites and dashboards, such as EPA's Chemistry Dashboard [78–80], so the building blocks of plastics, e.g., the base chemicals and monomers, are often well represented, but unfortunately there is a paucity of such data on microplastics themselves. Thus, such tools are often limited to a bottom-up approach to microplastics ADMET, i.e., estimating endogenous activities of simpler chemical structures that make up the plastic and using models to predict synergistic and antagonistic effects of the whole plastic. Such a screening approach may help to identify data gaps and steps toward closing knowledge gaps associated with data-poor microplastics by incrementally ranking hazard and exposure potentials [81,82]. Such screening tools can be also aid in developing of a hypothetical portfolio of products for an individual consumer of products, e.g., personal care, cleaning products, fuels, etc. A portfolio of products and an accompanying set of their chemical ingredients can allow decision makers to rank products according to potential risk, including the likelihood of the formation and transformation of pollutants [83]. These datasets are updated frequently as new studies and data become available.

In addition to the input on the chemical's characteristics, the model needs information about the organism. For example, the ADMET for a chemical needs information on the amount that is expected to be metabolized and the amount that will exit the human body. Toxicokinetic (TK) models often simulate based on estimates of the rate of clearance along with the unbound fraction of the absorbed chemical. For example, hepatic clearance is expressed as the rate at which the liver hepatocytes remove a substance. These modes have mainly targeted human subpopulations but can also apply to other species. In fact, data used as input to human models often begin with lower species, i.e., they are derived from *in vitro* to *in vivo* extrapolations (IVIVE). The fraction unbound in plasma is the soluble portion of the chemical.

The volume of distribution (Vd) is volume needed to remove the chemical compound if the organism's tissue concentration throughout the body were the same as the plasma concentration. Thus, Vd indicates the distribution rate of an organism's ADMET. The human steady-state plasma concentration at the time during which the plasma concentration of the chemical in the body stays consistent. A rule of thumb for a chemical to reach steady state is four to five half-lives with dosing at regular intervals, irrespective of the number of doses, the dose size, or the dosing interval. This dynamic equilibrium maintains chemical concentrations within defined limits for extended time periods. If these rates are similar to human ADMET, at a dose of 1 mg kg^{-1} day^{-1} and at the 95th percentile of distribution, which roughly would be expected for a healthy population of organisms, the Vd can be scaled linearly to estimate the blood plasma steady state concentration for any dose [84].

The days to steady state are assumed to be stable and relatively unchanged over time, i.e., the systemic concentration of the chemical. The pharmacokinetic (PK) half-life is the time needed for the body to eliminate half of the plasma concentration. Again, this is for the chemicals that make up the plastic substance, so it is very likely that the half-lives will differ for the whole plastic compared to the constituent chemicals. In addition, factors like sorption to aerosols in air, soil particles, and suspended solids in water will change the overall ADMET for an organism.

The TK model requires a variety of species-specific physiological parameters such as cardiac output and lung ventilation rate. These estimates vary by subpopulation characteristics, e.g., altered TK of older adults change the specific information needed to estimate metabolic clearances, chemical half-lives, body burden, and volume of distribution [85]. Many of these values are altered in the older adults compared to young adults for a variety of chemicals [86].

3 Dose-response

The hazard of a substance is expressed as the amount of adverse response that occurs after the amount of dose (see Fig. 17.2). Indeed, dose is a neutral pharmacological term that historically has been applied to medicine. An essential compound curve or substances like vitamins and certain metallic compounds is U-shaped (see the U-shaped curve for manganese in Fig. 6.4 in Chapter 6). The left-hand side represents the low dose region, i.e., deficiency. The right-hand side is the high dose region, i.e., toxicity or another hazard. The optimal region lies between these two adverse responses [87,88]. Between the low-dose and high-dose regions lies the optimal range, i.e., sufficiently high to avoid deficiency and sufficiently low to avoid toxicity. This applies to nutrients and vitamins. For metals, the curve must be drawn for specific valence states, for example, trivalent chromium (Cr^{3+}) has a U-shaped curve, but hexavalent chromium (Cr^{6+}) has a dose-response curve resembling the solid line curve in Fig. 17.3 because it is carcinogenic [88].

An acutely toxic substance is one that can cause damage with only a few doses in a short period of time. Conversely, chronic toxicity occurs when a person or population is exposed over a protracted period of time, with recurring

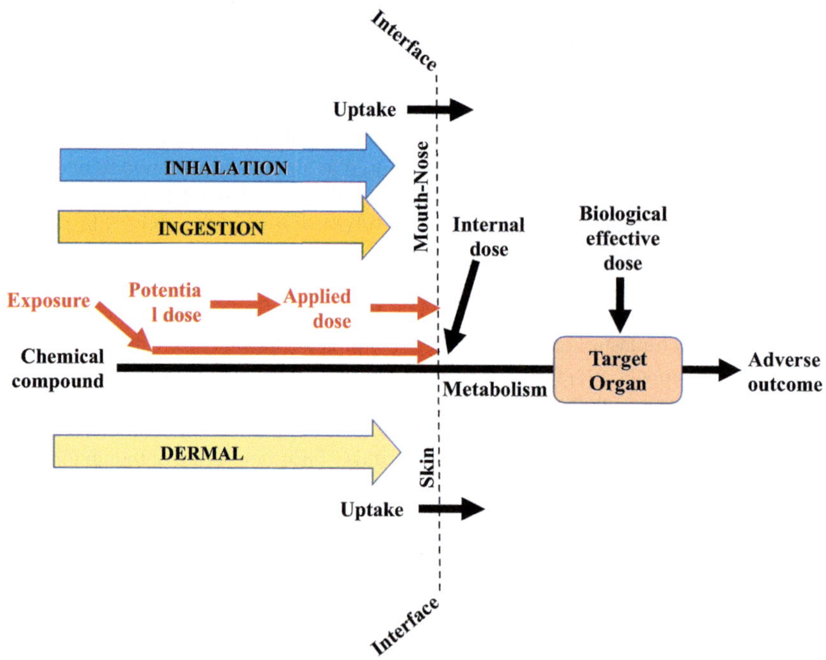

FIG. 17.2 Continuum of dose types from potential to biologically effective via inhalation, ingestion, and dermal exposure pathways. *(Based on information from U.S. Environmental Protection Agency, Rapid Chemical Exposure and Dose Research. https://www.epa.gov/chemical-research/rapid-chemical-exposure-and-dose-research, 2017 (Accessed March 28, 2017). U.S. EPA, Benchmark Dose Technical Guidance, US Environmental Protection Agency, 2012. J. Smith, Aflatoxins, in: Handbook of Plant and Fungal Toxicants, CRC Press, 2020, pp. 269–285. EPA, Exposure Factors Handbook—Soil and Dust Ingestion. Available: https://www.epa.gov/expobox/exposure-factors-handbook-chapter-5, 2017. D.A. Vallero, S.S. Isukapalli, V.G. Zartarian, T.R. Mccurdy, T.E. Mckone, P. Georgopoulos, C.C. Dary, Modeling and predicting pesticide exposures, in: R. Krieger (Ed.), Hayes Handbook of Pesticide Toxicology, Elsevier Science, New York, NY, 2010, pp. 995–1020 (Chapter 44). T. McCurdy, Modeling the dose profile in human exposure assessments: ozone as an example, Rev. Toxicol. 1 (1997) 3–23.)*

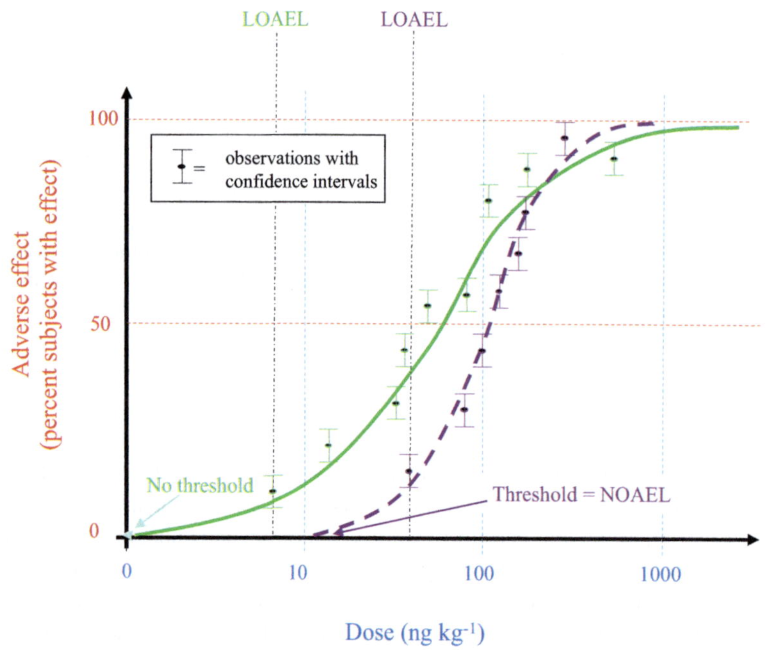

FIG. 17.3 Hypothetical dose-response curves. The solid line curve represents the "no-threshold" curve, which predicts an adverse response no matter how low the dose. Here, the low dose end of the curve, i.e., below which observational data are available, is assumed to be linear and crosses through the origin, i.e., zero effect-zero dose. Thus, the solid curve represents a linearized multistage model. The dashed curve represents an adverse effect only above a certain threshold, i.e., the no observable adverse effect level (NOAEL) or, no observable effect concentration (NOAEC) if inhaled. Below these thresholds, there is a dose that is not expected to elicit an adverse response; in this case the model predicts a NOAEL at about 10 ng kg^{-1}. The lowest observed adverse effect levels (LOAELs) are shown for both curves. Note that both curves have a part of the curve at doses below the LOAEL, so both curves require a model to extrapolate the unobserved curve below this dose. *(Data from: D. Vallero, Environmental Contaminants: Assessment and Control, Academic Press, 2010.)*

exposures [89]. The steepness of dose-response curves represents the contaminant's toxic potency or severity. The solid line curve is less steep than the dashed curve in Fig. 17.3 because the response elicited by the agent in the dashed is more potent than that of the agent in the solid curve. Potency is but one factor in articulating a hazard. A very potent agent that elicits a relatively innocuous effect, e.g., temporary loss of appetite headache, is less worrisome than agent with a gentle slope that causes cancer or birth defects. Thus, dose-response is best with comparing curves of different agents that elicit the same effect, comparing the dose-response curves for three different chemicals that cause a specific type of cancer.

The first of the three types of uncertainty are inherent biological differences between the species tested and that of humans (UF_{inter}) [90]. For example, how can the results of a study that shows toxicity in a bird's gizzard be translated to humans, who have no such organ? The UF_{intra} results from the differing sensitivities of human subpopulations when exposed to a substance than those of the general human population. The UF_{other} occurs because available data and science are lacking, including the uncertainty introduced when, the data may indicate a lowest observed adverse effect concentration (LOAEC), but not a NOAEC, i.e., the LOAEC-to-NOAEC (U_L). That is, data show a dose at which an effect is observed, but the "no effect" threshold cannot be delineated.

In toxicology, the first consideration of a chemical, biological, or physical pollutant is often the dose expected to elicit an adverse response, i.e., the mass or volume that encounters an organism. Toxicity is prominent hazard of environmental risk, whether to human populations or ecosystems. Other hazards include corrosiveness, explosiveness, and other environmental and health threats.

Dose is expressed in many ways, such as the applied dose, absorbed dose, or the biologically effective dose, which is the amount that reaches a particular organ (see Fig. 17.2). For example, a neurotoxin harms the central nervous system (CNS), which means only the mass or volume of that neurotoxin that reaches the brain and nerve cells will be the biological effective dose. The remaining absorbed dose is metabolized, stored or, in a best-case scenario, eliminated from the body before causing any harm.

Increasing the concentration of a chemical or pathogen that contacts an organism will increase the biologically effective dose and the probability that a response will occur, i.e., the expected adverse outcome. This gradient is known as "dose-response" curve (see Fig. 17.3). Increasing the amount of the substance encountering the receptor increases the absorbed and biologically effective dose and, consequently, the probability of an adverse outcome [43,82,91].

Assembling a dose-response curve follows a sequence of steps [34]. The dose-response data from *in vitro*, *in vivo* (in living organisms other than human), or human epidemiological studies fitted with a mathematical model. *In vitro* studies include bacterial mutagenicity studies [92,93], which can be a first screen for cancer and other adverse effects. The Ames *Salmonella*/microsome mutagenicity assay is a common *in vitro* method that displays the short-term reverse mutation in histidine-dependent *Salmonella* strains of bacteria [94,95]. The culture allows only those bacteria with genes that revert to histidine interdependence to form colonies. The mutagen population growth is a first dose-response curve, i.e., increasing the amount of chemical is associated with an increase in the number of mutagens and colony size on the plate. The test is arguably biologically plausible given that cancer is a genetic disease and a manifestation of mutations [92].

From these experimental and observational data, the upper confidence limit (e.g., 95%) line equation is expressed, and the confidence limit line is then extrapolated to a response point immediately below the lowest measured response point from the experimental data, i.e., the "point of departure" (POD). For example, for carcinogens the response is assumed to be a linear or nonlinear function of dose from the POD to zero response at zero dose. The last step is to calculate the dose on the line that is estimated to produce the response.

The sigmoidal-shaped curves in Fig. 17.3 are common for toxic substances, since increasing the dose at some point cannot increase the effect, i.e., mortality or morbidity rate cannot exceed the 100% asymptote [96]. There is often a distinction between physical and chemical agents that are suspected to cause cancer and those that are not expected to cause cancer. For suspected carcinogens, a linear no-threshold (LNT) model has been employed that resembles the solid line curve in Fig. 17.4. This applies to both ionizing radiation and chemicals. There is debate on what to do about the unobserved region, often on the shape of the curve below the LOAEL [89,97–101]. For example, this can be linear as suggested by the LNT, but may also be nonlinear. Thus, the arguments are not whether a single dose of radiation or a single molecule of a chemical carcinogen in fact can cause cancer, but how to arrive at a conservative, protective dose-response curve that may indeed have no threshold [34,102–106]. On the other hand, there are also arguments against assuming a threshold exists for every noncarcinogenic outcome, i.e., the assumption of a safe level at which the exposure to that substance will not elicit serious, but noncancer, outcomes like neurotoxicity, liver damage, and birth defects. This is the opposite argument against exclusive adherence to the no-threshold models for cancer. In the U.S., the National Academy of Science recommended a safe level of exposure for noncarcinogens only if strong evidence can support the threshold [107]. Otherwise, no safe level of exposure should be assumed in the risk model.

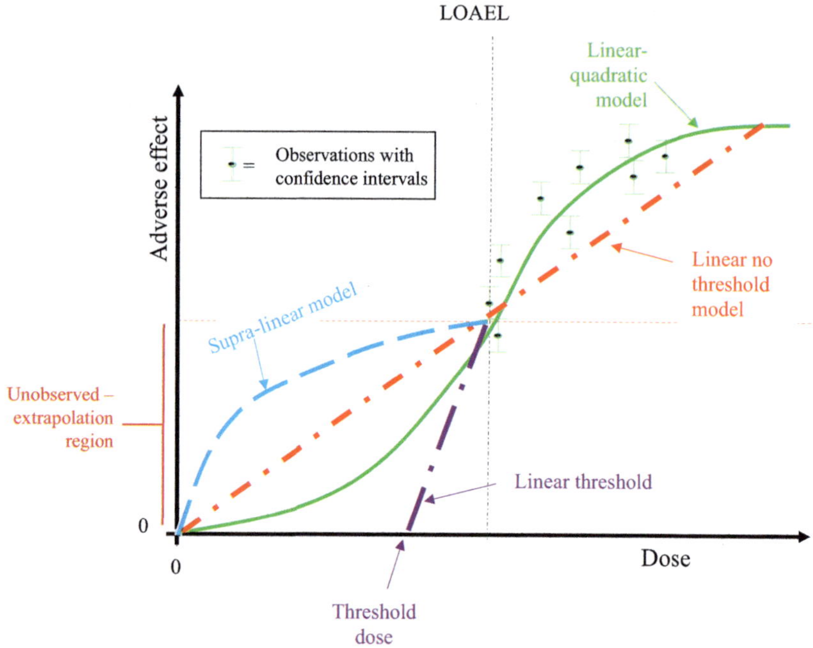

FIG. 17.4 Possible approaches to extrapolate dose-response from observed measurements. The adverse response to exposure to the pollutant is represented on the *y*-axis and the corresponding dose is represented on the *x*-axis. A "no threshold" model assumes that there is no safe level of exposure to that substance, so that the curve intercepts the *x*-axis at 0. *(Data from K.M. Seong, et al., Is the linear no-threshold dose-response paradigm still necessary for the assessment of health effects of low dose radiation? J. Korean Med. Sci. 31 (2016), S10–S23.)*

3.1 Reference dose

Safety thresholds for chemical hazards are known as reference doses (RfDs), usually based on oral exposure data, and applied to dermal exposures. For inhalation, the safe dose is known as the reference concentration (RfC), which is calculated in the same manner as the RfD, using units that apply to air (e.g., $\mu g\ m^{-3}$). Animal studies and epidemiology provide the knowledgebases for these values. Assays, i.e., *in vitro* studies, can begin the construction of RfDs and RfCs [78,108,109]. The main reasons for uncertainty include variability among the animals and human populations being tested, as well as differences in response to the compound by different species (e.g., one species may have decreased kidney function, while another may show neurological effects).

Lacking long-term human studies, the hazards and risks will be extrapolated from acute or subchronic studies of the same or similar agents, usually from secondary data sources that likely had different research objectives than those needed for an emerging contaminant[a] [30,110–117]. The routes and pathways of exposure used to administer the agent to subjects may differ from those applications in the real-world. For example, if the dose in a research study is administered orally, but the pollutant is more likely to be inhaled, this route-to-route extrapolation adds inconsistency, which translates into uncertainty [82].

The predicted no-effect concentration (PNEC) is a key toxicity metric used by public health and environmental scientists. The PNEC value is the concentration of a substance below which no effect is expected. In this sense, it is a forward-looking threshold analogous to the retrospective no adverse effect level (NOAEL) for water pollutants and no adverse effect concentration (NOAEC) for air pollutants. PNEC values are derived from acute toxicity data [118]. The PNEC and NOAEL values for a substance are directly proportional because the PNEC can be derived by dividing the NOAEL by a series of assessment factors to address safety and uncertainty. Thus, if uncertainties are high, very large safety factors are used, so that the PNEC is much smaller than the NOAEL for a substance (see Fig. 17.5). Unfortunately, assessment factors can be arbitrary and PNECs for many substances are not quantified, limiting their use exclusively to well-studied, data-rich substances [119]. Whereas the NOAEL is the highest dose at which no adverse effect has been identified, the lowest observed adverse effect level (LOAEL) is the lowest dose at which an adverse has been identified. As such, the LOAEL is always higher than the NOAEL for the same substance.

a. Emerging contaminants are those that have only recently appeared to present adverse effects or to which exposures have become known. They usually have limited data to support the amount of exposure or the extent of hazard, so they are commonly considered to be "data-poor" substances.

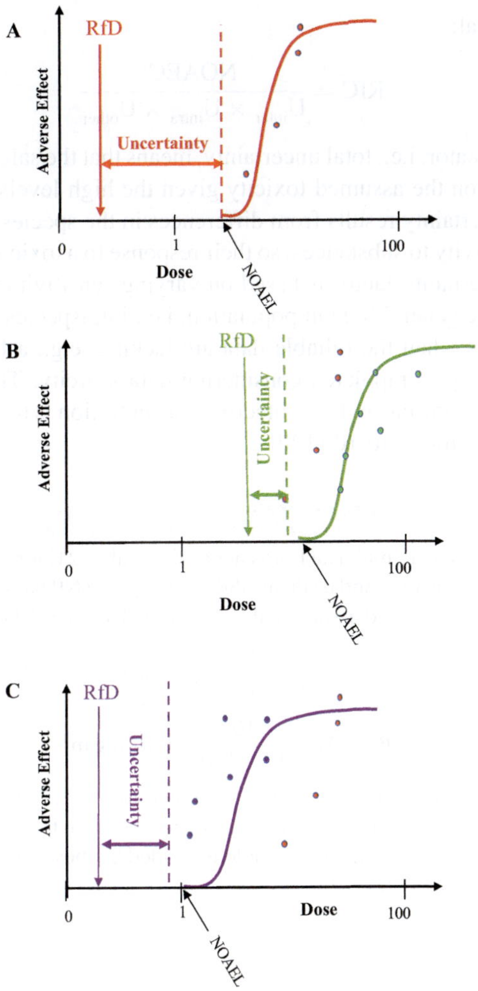

FIG. 17.5 Construction of a reference dose (RfD) with availability of new data. Curve A shows that based on fewer reliability the high level of uncertainty moves the RfD far from the dose-response curve's threshold, i.e., no observed adverse effect level (NOAEL) as a factor of safety. Curve B shows that new data improve the dose-response curve and show that the adverse effects are not as potent as assumed from the scarce measurements, so the curve shapes changes with the NOAEL moving to the right and the decreasing uncertainty moves the RfD closer to the NOAEL. Curve C shows what happens when studies provide new show that the adverse effects are more severe than assumed from the original dataset, so the curve changes shape, both the RfD and the NOAEL move to the left. However, since additional data decrease the uncertainty, the distance between the RfD and the NOAEL shortens. Note that RfD could move to the right while the NOAEL moves the left if the new data are so reliable that the uncertainty decreases more than the new data move the threshold.

Often the PNEC is compared to existing concentrations in one or more environmental media or to the predicted environmental concentration (PEC), i.e., the concentration of the chemical compound calculated to be in receiving waters that is estimated using a simple stream flow dilution model.[b] Ideally, the risk quotient $\frac{PNEC}{PEC}$ should be much less than 1.

The uncertainties in the RfD are often caused by the differences between results found in animal testing and expected outcomes in human population. As in other bioengineering operations, a factor of safety must be added to calculations to account for uncertainty factors (UFs). So, for environmental risk analyses and assessments, the safe level is expressed in the RfD, or in air the RfC. This is the dose or concentration below which regulatory agencies do not expect a specific unacceptable outcome. Thus, all UFs move the measured and modeled levels of threshold values, e.g., NOAELs and LOAELs, toward zero concentration. This is calculated as:

$$\text{RfD} = \frac{\text{NOAEL}}{\text{UF}_{\text{inter}} \times \text{UF}_{\text{intra}} \times \text{UF}_{\text{iother}}} \tag{17.9}$$

b. For water pollutants, the PEC is the same as the surface water concentration (SWC).

The calculation of the RfC is identical:

$$\mathrm{RfC} = \frac{\mathrm{NOAEC}}{\mathrm{U_{inter}} \times \mathrm{U_{intra}} \times \mathrm{U_{other}}} \tag{17.10}$$

The increase in the size of the denominator, i.e., total uncertainty, means that the safe dose continues to drop, so that the RfC and RfD continually decrease based on the assumed toxicity given the high levels of uncertainty.

The first of the three types of uncertainty results from differences in the species tested and compared to *Homo sapiens* (UF_{inter}). Species differ in their sensitivity to substances, so their response to a toxic substance differs from that of the tested species, i.e., UF_{inter}. The second uncertainty factor is based on varying sensitivities among human subpopulations to the effects of a substance compared to the general human population, i.e., intraspecies uncertainty factors (UF_{intra}). The third class of uncertainties (UF_{other}) results when the reliable data are lacking, e.g., a LOAEL is used rather than a NOAEL because a threshold cannot be properly extrapolated considering data scarcity. The UFs are in the denominator, which means that the greater the uncertainties, the closer the protective concentration is to zero. The UFs are commonly multiples of 10, although the UF_{other} can range from 2 to 10 [120].

Reference concentration example

Let us calculate the RfC by using the that the animal data from Landsiedel et al. [121] for a titanium oxide (TiO_2) nanoparticle (NP) show no NOAEC but have a LOAEC of 2 mg m^{-3} and a silicate (SiO_2) [122]. The NP has a NOAEC of 10 mg m^{-3}. Using the EPA's [120] most protective recommended U_L of 10 and assuming the remaining $UF_{inter} = 10$ for and $UF_{intra} = 100$ for both NPs, then:

$$\text{RfC-TiO}_2 = \frac{2}{10 \times 100 \times 10} = 0.0002 \text{ mg m}^3$$

$$\text{RfC-SiO}_2 = \frac{10}{10 \times 100} = 0.01 \text{ mg m}^3$$

Thus, although the thresholds may indicate that the toxicity is similar for the two NPs, given the lack of data to delineate a NOAEC, the TiO_2 NP requires a 50 times larger factor of safety. Since the UFs are in the denominator, the greater the uncertainties, the closer the safe level (i.e., the RfC) is to zero, i.e., the threshold is divided by these factors. The UFs are usually multiples of 10, although the UF_{other} can range from 2 to 10 [122].

Reference dose example

The U.S. Environmental Protection Agency was asked to re-register the use of the organophosphate pesticide, chlorpyrifos. The acute dietary scenario had a NOAEL of 0.5 mg kg^{-1} day^{-1} and the three UF values equaled 100. Thus, the acute RfD $= 5 \times 10^{-3}$ mg kg^{-1} day^{-1} but the more protective acute PAD 5×10^{-4} mg kg^{-1} day^{-1}. The chronic dietary scenario is even more protective since the exposure is long-term. The chronic NOAEL was found to be 0.03 mg kg^{-1} day^{-1}. Thus, the chronic RfD for chlorpyrifos $= 3 \times 10^{-4}$ mg kg^{-1} day^{-1} and the more protective acute PAD 5×10^{-5} mg kg^{-1} day^{-1}. This means that the allowable exposure would be three orders of magnitude higher if the NOAEL threshold been used alone without the RfD [16].

4 Exposure estimation

Hazard is one component of risk. The other is exposure. No matter the steepness of the hazard curve of a compound, it will only elicit risk when it encounters the receptor. Humans can contact a substance in various ways in numerous settings, so-called microenvironments. Thus, the exposure to a pollutant varies by activities. For example, the exposure potential is relatively high for workers whose occupations require them to work with a chemical compound, compared to the lower exposure potential for workers who may not work with it, but are nearby and breathe the vapors, and near zero for the rest of the population. Usually, worker exposure is based on a 5-workweek and 8–10 h of exposure per day. Often, the worker risk calculations include personal protection equipment PPE, so that the exposure estimate is decreased accordingly.

Conversely, environmental exposures are usually based on average lifetimes and continuous 24-h daily exposure that assumes no PPE is in use. Lifetime durations of exposure are especially needed for chronic disease risk estimates, like cancer. Thus, environmental regulations are often more stringent than occupational regulations when aimed at reducing exposure to a substance [24].

Chronic exposure may be less than the entire lifetime if the pollutant exists only for a specified time for the exposed population. For example, if it is known that the source of a pollutant only began manufacturing a chemical compound in

1990, went out of business in 2010, and cleanup of the source was documented in 2012, the risk calculation would include an exposure duration of about 22 years. However, if the exposure route were drinking water from contaminated ground water, even though remediation was successful in 2012, there may be residual contamination, especially if the aquifer had very low flow and the chemical compound is persistent (see Chapter 11). Ecosystem exposures assume the exposure duration is as long as the pollutant is deposited on and stored in the ecosystem's soil, water, sediment, and biota [123].

Estimates of the likelihood of exposure are part of the previously discussed uncertainty factors (UFs). The uncertainties are largely due to the differences between results found in animal testing and expected outcomes in human population. For environmental risk analyses and assessments, the safe level is expressed in the RfD for ingestion and dermal exposure and RfC for inhalation exposures. This is the dose or concentration below which regulatory agencies do not expect a specific unacceptable outcome. Thus, all the uncertainty factors adjust the actual measured levels of no effect (i.e., the threshold values, e.g., NOAELs and LOAELs) in the direction of a zero concentration.

After uptake, varying amounts of the parent substance are absorbed and distributed within the body [55]. In addition to chemical factors, physical properties affect toxicokinetics, e.g., for particles, the hydrophobic surface and large relative surface area of means that they are prone to absorb toxic substances like heavy metals and persistent organic compounds [124]. The low mass also means that the particle is likely to be colloidal and remain suspended for a long time. Such colloidal and nano-suspensions in surface waters mean that even lipophilic chemical compounds remain in the water column rather than settle onto the sediment. As a result, they are more likely to be oxidized than to be reduced, compared to the hypoxic and anoxic conditions of the sediment [125]. In the air, these same characteristics mean that the particles are more likely to remain airborne for longer time periods sufficient for some of the organic compounds to undergo chemical transformation [54,62,125,126], e.g., some of the metals may change their valence state which could make them more bioavailable and toxic [127,128].

Surface charge, surface chemistry, particle shape, particle mass, and particle density of the particles that make up a matrix like a consumer product, solid waste, or soil can mean that bulk materials will behave differently in an organism than their individual constituents in terms the potential to be absorbed, distributed, metabolized, and excreted, so that the bulk material toxicity will differ from the sum of the constituents' toxicities [129,130]. The composite fraction of a chemical compound or its transformation products that accumulates in lipids and other tissue substrates could be higher, and the amount excreted decreased, so that the difference results in bioaccumulation and increased body burden [131]. The same physical and chemical principles that are work outside of the organism occur within the organism but, for example, at different rates, using different catalysts, and under various endogenous redox conditions. [132].

From a high-throughput, high-tier screening perspective, the first order problem occurs when the likelihood of exposure *and* the likelihood of hazard coexist. As mentioned, no matter the steepness of the hazard curve of substance, it will only elicit risk when it encounters the receptor. For example, in the EPA's risk-based chemical prioritization efforts, the substances of greatest concern are those with high potential toxicity and high potential exposure profiles. Indeed, even when toxicity remains the same, those substances with higher exposure potential will lead to higher risk (see far-right of Fig. 17.1). The listing of contaminants of concern by this approach will differ from one based solely on toxicity, since even moderately toxic substances can have high risk potentials. Similarly, when hazard concentrations (e.g., greater than PNEC) and exposure potential intersect, there is greatest concern, since populations are being exposed at concentrations most likely to lead to adverse outcomes.

5 Human risk characterization

Fig. 17.6 shows the traditional chemical risk assessment paradigm [133]. It begins with the identification of a hazard, which is a summary of a chemical agent's physicochemical properties and routes and patterns of exposure and a review of toxic and other adverse effects. Hazard identification accounts for metabolic and pharmacokinetic[c] properties based on short-term animal and cell tests, long-term animal (*in vivo*) testing, and human studies (mainly epidemiological, such as longitudinal and case-control studies). Risk assessors can apply biomarkers of genetic damage (i.e., toxicogenomics), quantitative structure-activity relationship (QSAR) models based on stereochemistry, and other chemical descriptions of well-studied compounds to compare to lesser understood substances [45,134–140]. New methods such as cheminformatics [40,80,88,136,140–157] and computational chemistry are also increasingly applied [91,158–163].

c. The term pharmacokinetics has been supplanted by toxicokinetics in many environmental venues. The terms are synonymous, but toxicokinetics is exclusively applied in toxicology and environmental science to harmful agents, whereas pharmacokinetics can be applied to both harmful and beneficial agents.

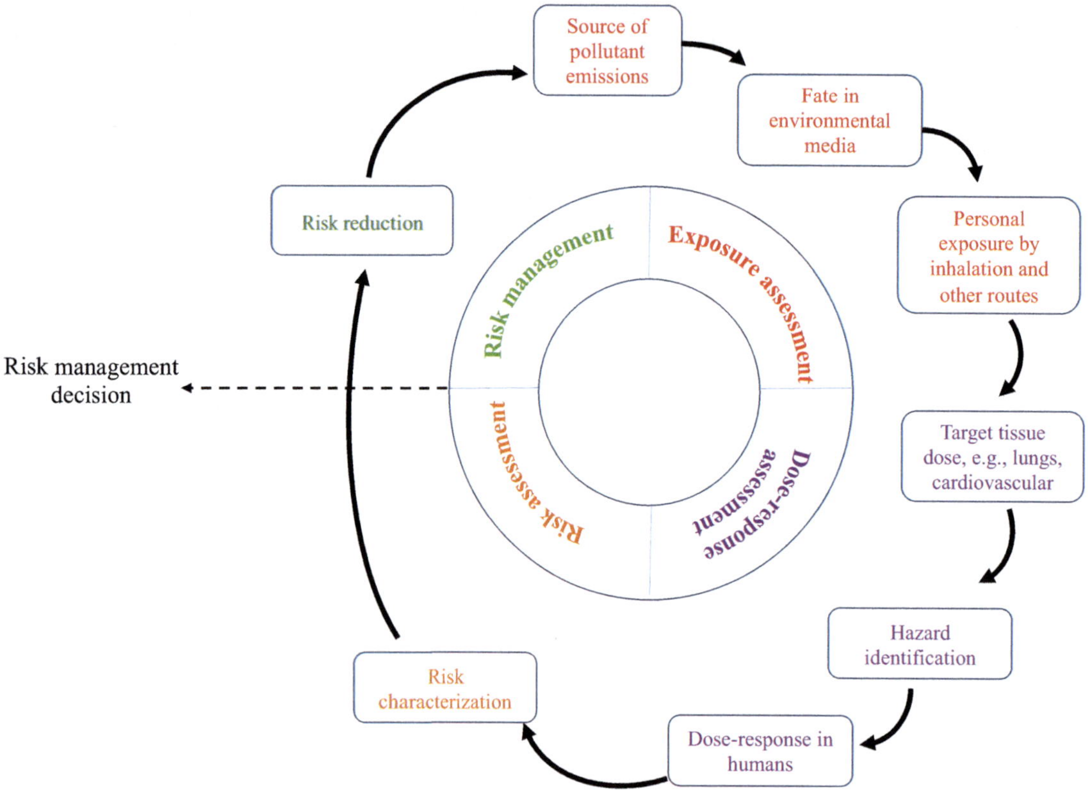

FIG. 17.6 Risk assessment and management paradigm as employed by environmental agencies in the United States. The inner circle includes the steps recommended by the National Research Council. The outer circle indicates the activities that are currently used by regulatory agencies to meet these required steps. *(Data from National Research Council, Risk Assessment in the Federal Government: Managing the Process. National Academy Press, Washington, DC, 1983. National Research Council (US) Committee on Improving Risk Analysis Approaches Used by the U.S. EPA, Science and Decisions: Advancing Risk Assessment, 2009.)*

Risk metrics are used to determine the success or failure in water pollution control and prevention efforts. The metrics relate to environmental and health hazards, exposures, and risk characterizations that are input into water pollution calculations to identify and select actions to address risks [164–169].

Like risk, the reliability of a system is a metric of accountability (see Chapter 18). A reliable system is one that will not fail during its designed operating period. Thus, reliability is the mathematical expression of success, whereas risk is the expression of the likelihood of failure. Reliability expresses the probability that the system will perform its intended function for a specified time and under specified conditions [170,171]. For example, as the reliability of a pollution control system increases, the better that system will be in decreasing effluent concentrations of water pollutants. This, in turn, means lower risks to humans and the environment. Like risk, reliability is a probability. Risk is the likelihood of an adverse and undesirable outcome, but reliability is the likelihood of a desired outcome. As probabilities, both have values between zero and less than or equal to one. They are often expressed as a percentage ranging from 0% to 100%. Complete reliability, i.e., 100%, and conversely, 0% risk are the ideal targets for water pollution control systems [171–175].

Systems are reliable when they perform as designed under the specified conditions for the intended. For example, each sampling and analytical device discussed in Chapter 13 requires a level of reliability to be useful to water quality measurement. Items within a system will indeed fail at some rate, e.g., x number of failures per hours of operation. For example, a pump that is expected to fail no more frequently than once every 1000 h of operation and a water testing group study using 10 pumps in the field for 11,000 h, they may not be sufficiently reliable even though the average pump lasted 1100 h, but if the range of endurance is 900–1200 h. With at least one pump failing more frequently than designed, at least one replacement extra pump is likely needed.

Reliability is the probability that something that is in operation at time *0* (t_0) will still be operating until the designed life (time t = (t_t)). As such, it is also a measure of the engineer's social accountability. People want to be assured that that water they drink and that they fish and swim in is healthful, which includes assurances that water and wastewater treatment systems are working as designed [171–175].

The probability of a failure per unit time is the hazard rate or failure density [$f(t)$]. This is a function of the likelihood that an adverse outcome will occur. It is not a function of the severity of the outcome [176]. The $f(t)$ is not affected by whether the outcome is very severe, e.g., lung cancer and loss of an entire species in an ecosystem, or relatively benign, e.g., slight odors or minor leaf damage. The likelihood that an item will fail at a given time interval is found by integrating $f(t)$ rate over a defined time interval:

$$P\left\{t_1 \leq T_f \leq t_1\right\} = \int_{t_1}^{t_2} f(t)dt \tag{17.11}$$

where T_f = time of failure.

Thus, the reliability function $R(t)$, of a system at time t is the cumulative probability that the system to perform without failure in the time interval from t_0 to t_t:

$$R(t) = P\left\{t_1 \leq T_f \leq t_1\right\} = 1 - \int_0^t f(x)dx \tag{17.12}$$

Environmental risk increases as the reliability of protective and control systems subsides. It is not a question of "if," but "when" an engineered system will fail. An engineer or manager may be able to extend the optimal time of a pollution control or prevention technology, i.e., increasing t_t to making the system more resilient under conditions that would otherwise lead to failure. For example, proper engineering design of a water treatment system should work well, not only during normal conditions, but resist failure during hostile conditions, e.g., a time-period of highly corrosive loading. A design flaw occurs if the wrong material is used, e.g., piping not matched to worst-case corrosivity conditions. However, even when the proper materials are used, the failure is not eliminated entirely. Selecting the right materials simply protracts the time before the failure occurs, i.e., increases T_f [57]. This means that the likelihood of the adverse outcome is reduced, which means that risk has also been decreased.

In addition to reliability, risk is also related to resilience. Reliable systems are also generally resilient systems. Like many other environmental systems, engineers and scientists can learn from nature. For example, ecosystems must be resilient. Ecosystem resilience and engineering resilience are mirror images of one another. The less resilient a system, the more that the engineer and manager must do to make it reliable. A very resilient natural system can withstand a wide array of insult. A system that lacks resilience is a sensitive and vulnerable system, needing controls to remain operational when conditions change. In this sense, aquatic ecosystems that are diverse have greater resilience than those with just a few dominant species, e.g., monoculture fish systems.

Resilient systems have wide tolerances, whereas sensitive system is easily harmed outside of very narrow tolerance ranges. Thus, an ecologically resilient system is in less need of engineering controls against insult than a nonresilient system. For example, atmospheric deposition of a pollutant onto or an effluent release into a resilient system may cause little impact and/or require a relatively short time to recover. Conversely, the same deposition onto or release into a nonresilient system could be devastating.

Unfortunately, human activities can transform resilient systems into nonresilient ones. For example, a wild, biodiverse area may be quite resilient in the face of drought and insect infestation. Human activities, e.g., building roads and infrastructure around and near this area, can reduce biodiversity and change biological processes, making it less able to withstand these insults. Thus, a natural system may have been able to withstand lower pH rainfall caused by sulfur dioxide (SO_2) emissions, but other human activities may have made the system more vulnerable to the acid rain.

Engineered systems, including pollution control technologies, must also be resilient. They must be able to return to the original performance after being after being stressed [174,177,178]. When such complete hysteresis exists, then risk levels remain stable [177,179,180]. Ecological risks can similarly be decreased when they are biodiverse and productive, and provide ecological resilience [181–186].

5.1 Unit risk estimates

Cancer models assume linearity in the extrapolation zone of the dose-response curve. The U.S. Environmental Protection Agency (EPA) often applies a linearized multistage model (LMS), which yields a slope factor known as q1*, to predict cancer risk at a particular dose. The LMS assumes a linear extrapolation with the zero dose threshold from the upper confidence level of the lowest dose that produces cancer in observed human populations using epidemiological methods or in animal studies [187].

Models vary in their estimates of cancer risk outcomes. For example, the concentration of organochlorine pesticide chlordane ($C_{10}H_6C_{18}$) in drinking water needed to cause a lifetime risk of 10^{-6} persons in a populations are estimated to be [188]:

- 0.03 μg L^{-1} using the one-hit model;
- 2 μg L^{-1} using the multi-hit model;
- 50 μg L^{-1} using the probit model; and,
- 0.07 μg L^{-1} using LMS.

Cancer risk can also be expressed in terms of risk per unit concentration of a substance in an exposure route, i.e., unit risks. The drinking water unit risk is based on a water consumption rate of 2 L day^{-1}. The air unit risk is calculated by dividing the slope factor by a mean body weight, usually 70 kg, and multiplying by the inhalation rate (20 m^3 day^{-1}). For a substance that is not completely absorbed, i.e., absorption fraction <1.0, when deriving the slope factor, an additional conversion factor is needed so that the unit risk is based on an administered dose. The exposure in a unit risk is assumed to be continuous over a lifetime. Thus, when no absorption conversion is required:

$$\text{Water unit risk} = \text{risk per } \mu\text{g L}^{-1} = \text{slope factor} \times \frac{1}{70}\text{ kg} \times \frac{2\text{ L}}{\text{day}} \times 10^{-3} \tag{17.13}$$

$$\text{Air unit risk} = \text{risk per } \mu\text{g m}^{-3} = \text{slope factor} \times \frac{1}{70}\text{ kg} \times \frac{20\text{ m}^3}{\text{day}} \times 10^{-3} \tag{17.14}$$

The multiplication by 10^{-3} is necessary to convert from mg to μg when the slope factor, or q1*, is given in the typical units of mg kg^{-1} day^{-1}.

5.2 Risk uncertainties

Hazard-related data for chemical substances must often be extrapolated from small-scale laboratory experiments and field studies to larger environmental exposure scenarios. As such, they rely on quantitative tools to extrapolate information from a data-rich substance to data-poor substances [44,135,136,141,189,190]. Uncertainly in this case is tied to any deviation from the behavior of the known to the unknown, since toxic agents always have some degree of uniqueness related to transport and transformation after they enter the environment or an organism's body [174].

Chemometrics is a systematic approach to using mathematical and statistical methods to evaluate chemical data, beyond univariant analysis, to describe chemical compounds quantitatively using existing data beginning with the physicochemical structure [88,169,191,192]. For example, if a data-poor compound has a similar chemical structure to a known compound, e.g., only one different functional group on a carbon ring, then statistical and mathematical structure-activity relationships (SARs) provide a first step in screening for potential hazards and exposure scenarios [44,134–136,138].

5.2.1 *Error*

Errors are a major source of potential risk assessment uncertainty [193]. Errors are classified as either Type 1 or Type 2. Experimental findings have a Type I error when they lead to the rejection of a null hypothesis that is in fact true. Conversely, a Type 2 error occurs if the findings lead to an acceptance of a null hypothesis that is in fact false. When an occurrence or event, such as a human disease outbreak or the loss of a species due to an environmental exposure, is not identified during an observation or measurement, but in fact the occurrence or event exists, this is a false negative. If sampling and analysis are not sufficiently sensitive and lack resolution it can lead to a false negative [194]. Other problems, e.g., level of detection (LoD) being too high, chain-of-custody with missing or erroneous documentation, improper interpretation of results, and poor documentation and data entry can also be sources of false negative findings.

The other major type of error is one that leads to a false positive, i.e., an occurrence or event is found but in fact does not exist [195]. This can be a major challenge for emerging contaminants. For example, consider the situation where previous evidence used by an agency to list a compound as a potential endocrine disruptor, but the agency has subsequently found that a wealth of new information is now showing that it has no such effect [196]. The decision makers must decide whether the new information is sufficient to remove protections that were only required based on the compound's apparently false positive endocrine disruption data. Sometimes, the false positive results from scientific shortfalls, e.g., based upon faulty models, or models that only work well for lower organisms (i.e., interspecies uncertainty). Often, this is rectified by scientific advancement, e.g., models that are subsequently developed consider newly available data that account for physical, chemical, and biological complexities of higher-level organisms, including humans.

False positives may also lead to overemphasis on less severe problems that cause public health officials to devote inordinate amounts of time and resources to address too many problems [197]. False positives also may discourage people to make use of potentially beneficial products. False positives, especially when they occur frequently, create credibility gaps between scientists and the public.

In addition to physical agents like ionizing radiation and chemical agents, pathogenic microbes introduce threats to human health. Wastewater, agricultural products, and other media contain populations of microorganisms that can reach food and drinking water that exposure people to pathogens [198–202]. Pathogens can also enter the food chain, where they can reside at various trophic levels and remain in reservoirs, e.g., fish and other aquatic biota that are consumed. Therefore, false positives about the presence of a pathogen can create undue concern, whereas false negatives about a pathogen can miss early warning signals that could have prevented diseases or allowed for interventions to avoid undue harm in an outbreak.

Biological hazards are part of the overall human health risks in populations. Cumulative risk is a function of all agents that cause an adverse outcome. Tools like life cycle analyses (LCAs) of chemical and physical hazards must be combined with biological stressors to gauge the cumulative of a population. For example, a quantitative microbial risk assessment (QMRA) can be conducted for waterborne pathogens and combined with risk assessments of chemical stressors. Pathogenic risk is often expressed as a burden of disease, e.g., disability-adjusted life years (DALYs) [199]. Thus, the DALYs can be combined with information about physical and chemical agents to depict cumulative risks [203–205].

5.3 Hazard index and hazard quotient

Recently, substances can be detected in the lower range of parts per quadrillion (ppq) range emerging chemical compounds like the so-called "forever chemicals," i.e., PFAS. For example, the U.S. recently proposed drinking water maximum contaminant levels (MCLs) for perfluorooctanoic acid (PFOA), perfluorooctane sulfonic acid (PFOS), perfluorononanoic acid (PFNA), hexafluoropropylene oxide dimer acid (HFPO-DA, known as GenX Chemicals), perfluorohexane sulfonic acid (PFHxS), and perfluorobutane sulfonic acid (PFBS) [206,207]. The recommended MCL for PFOA and PFOS is 4 ng L^{-1} [208] and a hazard index (HI) of 1 for the rest.

In risk assessments, the threshold concentrations of a pollutant are chosen to protect humans and ecosystems. For emerging contaminants, these concentrations aim to protect human health. The amount of protection is based on the potential of an unacceptable outcome [209]. Effluent limits apply to any source that releases the contaminant to a waterbody, e.g., not to exceed 100 mg L^{-1} of a chemical compound. Of course, the actual level of concern for emerging contaminants varies considerably given the large number of candidates of chemicals of concern. For example, the U.S. EPA publishes lists of contaminants that are not yet regulated but are known or anticipated to be found in the environment. If, for instance, a chemical or biological agent of concern is expected to reach public water systems and may need to be regulated under the Safe Drinking Water Act (SDWA), is placed on the Contaminant Candidate List (CCL). The most recent draft (CCL5) includes 66 chemicals, 3 chemical groups (PFAS, cyanotoxins, and disinfection byproducts) and 12 microbial contaminants [210].

The National Pollutant Discharge Elimination System promulgated under the Clean Water Act describes two types of effluent limits. The technology-based effluent limit (TBEL) allows the discharging source to use any available control technology to achieve the limit. The water quality-based effluent limitation applies to water bodies that need extra protection beyond the TBEL, based on total maximum daily loads (TMDLs) to determine wasteload allocations [209].

5.4 Estimating risks

Risk assumptions can seriously affect risk estimates [176]. Cancer risk, for instance, is the theoretical probability of a person or population getting a form of cancer when continually or episodically exposed to a carcinogen. Often, the probability used in this estimate is the upper confidence limit. The maximum risk may be the odds in a million of contracting cancer. There are two general ways that such risk characterizations are used to support environmental and public health decisions, i.e., direct risk assessments and risk-based treatment targets.

Population risk is the annual excess number of adverse events resulting from exposure to an agent, e.g., cancers in an exposed population after being exposed to a carcinogen. The agent may be physical, e.g., ionizing radiation, chemical, e.g., a compound or class of compounds like coke oven emissions, or biological, e.g., aflatoxins produced by fungi or a variant of the COVID-19 virus. For chemical compounds, the population risk is often the additive risk to a general population, e.g., a "residual" or increased probability above background or expected rates.

Risk can also be individual, i.e., the probability of one person having an adverse effect due to the exposure to a physical, chemical, or biological agent. As a worst-case scenario, individual risk might be calculated from exposure estimates based upon a maximum-exposed individual (MEI). The hypothetical MEI resides outdoors for an entire lifetime at the point where pollutant concentrations are highest. Such assumptions about exposure affect the risk estimates. For example, the cancer risk from power plants in the U.S. has been estimated to be that the risk to be 100- to 1000-fold lower for an average exposed individual than that calculated for the MEI [37].

As mentioned, common default approach for predicting cancer risk is a linear extrapolation from exposures observed in animal or human occupational studies. A nonlinear approach is recommended when there are sufficient supporting data to establish the carcinogen's mode of action (MOA) to conclude that it is not linear at low doses and the carcinogen does not demonstrate mutagenic or other activity consistent with linearity at low doses [211]. Thus, a straight line connects the beginning of point where reliable data are not available, known as the point-of-departure (POD), to the origin on the dose-response curve. The POD can be the lower-boundary of the dose for an estimated incidence or a change in response level from a dose-response model, such as the benchmark dose, BMD, or a NOAEL or LOAEL. The default linear extrapolation approach is generally considered to be conservatively protective of public health, including sensitive subpopulations [34]. The slope of this line is the inhalation unit risk (IUR), so the IUR can be presented as a linear extrapolation from a POD of 10% response:

$$\mathrm{IUR} = \frac{0.1}{\mathrm{LEC}_{10(\mathrm{HEC})}} \tag{17.15}$$

where, the $\mathrm{LEC}_{10(\mathrm{HEC})}$ is the lowest effective dose using a 10% response concentration, dosimetrically adjusted to a human equivalent concentration.

The air concentration must include all physical phases, i.e., both the pollutant in vapor phase and those in aerosol phase. The aerosol phase can simply be the same compound in liquid and solid phase, as well as any of the pollutant to this sorbed to particulate. At equilibrium a volatile or semivolatile compound will exist in gaseous and liquid phases. A fraction also can be adsorbed on particle surfaces or absorbed within the particle matrix. The concentrations from the three phases are often normalized and expressed in units of mass per volume (e.g., mg m^{-3}) or mass per mass (e.g., mg kg^{-1}).

Volatile organic compounds (VOCs) are almost completely in the gas phase in ambient air because their vapor pressures within most environmental conditions exceed 10^{-2} kPa [212]. For VOCs, then, the total air concentration is nearly equal to vapor-phase concentrations. Conversely, relatively nonvolatile organic compounds (NVOCs) with vapor pressures $<10^{-5}$ kPa are predominantly found in and on particles, unless significant energy is added to increase their volatility. For NVOCs, estimates do not expect a substance amount to reach the gas phase and the air concentration can will be entirely as the amount in and on particulate matter. Between VOCs and NVOCs are the semivolatile organic compounds (SVOCs) that under environmental conditions have vapor pressures between 10^{-2} and 10^{-5} kPa can exist as in both the gas and particle phases in the ambient air [212].

The hazard quotient (HQ) represents the risk of an adverse outcome other than cancer, i.e., noncancer risk [213]. The HQ is calculated by dividing the maximum daily dose (MDD) by the acceptable daily intake (ADI):

$$\text{Noncancer risk} = \mathrm{HQ} = \frac{\mathrm{MDD}}{\mathrm{ADI}} = \frac{\text{Exposure}}{\text{Reference dose or concentration}} \tag{17.16}$$

Recall that an MDD can be equal to or less than the LADD for a noncarcinogen. HQ is an index, not a probability, which means it indicates relative risk. If the noncancer risk is greater than 1, the potential risk may be significant, and if the noncancer risk is less than 1, the noncancer risk may be insignificant. Thus, the reference dose, RfD, and reference concentration, RfC, are types of ADI.

For example, assuming the chromic acid (Cr^{6+}) mist reference concentration (RfC) of 8×10^{-6} mg kg^{-1} day^{-1}, the actual respiratory exposure of people living near a metal processing plant was be calculated to be 4×10^{-5} mg kg^{-1} day^{-1}, the HQ for the noncancer risk of the chromic acid mist to the neighborhood near can be calculated from Eq. (17.16):

$$\mathrm{HQ} = \frac{\text{Exposure}}{\mathrm{RfC}} = \frac{4 \times 10^{-5}}{8 \times 10^{-6}} = 5.$$

Since the HQ is greater than 1, one would expect people chronically exposed at this level to show adverse effects from inhalation. However, at this same chronic exposure, the reference dose (RfD) for hexavalent chromic acid mists differs for dermal and oral routes, so additional adverse noncancer effects may even be larger than indicated by a single route of exposure.

The HQ value is ratio of a potential exposure to the applicable toxicity-based benchmark.

An HQ value <1 indicates that the specific contaminant is not likely to cause adverse ecological effects. If more than one potential ecological contaminant is present at the site, the HQs for receptors that could be simultaneously exposed to the contaminants should be summed and presented as the HQ [214].

A population that is exposed to more than one contaminant requires that the hazard index (HI) indicate the cumulative noncancer risk from pollutants 1 through n:

$$HI = \sum_{i}^{n} HQ \tag{17.17}$$

The HI is useful in comparing risks at various locations, e.g., benzene risks in Denver, St. Louis, Chicago, and Los Angeles. It can also provide an indication of the cumulative and additive risk in a single population exposed to more than one contaminant. For example, if the HQ for toluene is 0.2 (not significant), xylene is 0.5 (not significant), and tetrachloromethane is 0.4 (not significant), the cumulative risk of the three contaminants is 1.1 (potentially significant).

Estimations of both hazard and exposure are often worst-case scenarios, because the risk calculations can have large uncertainties. Effects can occur at very low doses. Human data gathered from even very high-quality epidemiological studies have some amount of error and do not completely address variability. Humans are highly diverse in their activities, introducing many confounding variables. Thus, uncertainty exists in every phase of risk assessment.

The hazard quotient (HQ) is a risk factor applied to noncarcinogenic chemicals that relates the delivered dose at the exposure point, e.g., the average daily dose (ADD) or median exposure (mg kg^{-1} day^{-1}) to an adverse health endpoint [215]:

$$Risk = \frac{ADD}{toxicity} \tag{17.18}$$

Therefore:

$$HQ = \frac{D}{MRL} \tag{17.19}$$

Or,

$$HQ = \frac{D}{RfD} \tag{17.20}$$

where, D is the exposure dose (mg kg^{-1} day^{-1}), and MRL is the minimum risk level (mg kg^{-1} day^{-1}). MRLs represent estimates of the daily human exposure to a contaminant that are not expected to cause noncancer health effects during a specified exposure duration. MRLs are set below concentrations that may cause adverse health effects in most humans, including sensitive populations. They are derived for acute (1–14 days), intermediate (15–364 days), and chronic (365 days and longer) exposure durations, and are available for oral and inhalation exposures [215].

The endpoint can be the RfD or RfC specific that that chemical compound. For example, if the exposure concentration (EC) is 5 μg L^{-1} and the RfD for a chemical is 10 μg L^{-1}, the HQ would be 0.5. Since $HQ < 1$, the hazard is not considered to be a major concern. If the EC were to double, the HQ would be 1, indicating a hazard concern.

Exposure factors can be based on a central tendency exposure (CTE), i.e., average or typical intake factors. CTE calculations use mean body weight. Reasonable maximum exposure (RME) refers to exposure factors in the upper end of the exposure distribution (approximately the 95th percentile). RME scenarios assess exposures higher than average but within a realistic exposure range [215].

If $HQ < 1.0$ the person is on average ingesting a safe dose. If $HQ > 1.0$ the person is ingesting a dose that is not safe. Neither of these are risk statements, nor have large amounts of uncertainty. Often the safe dose is not a clean bill of health and likely requires more thorough analysis. HQ alone is not a justification for a cleanup action or a reliable indication that remediation has been successful.

When exposures include more than one chemical, the hazard index (HI) is the sum of all HQs:

$$HI = HQ_A + HQ_B + HQ_C = \frac{EC_A}{RfD_A} + \frac{EC_B}{RfD_B} + \frac{EC_C}{RfD_C} \tag{17.21}$$

For example, if the EC of Chemical A is 5 $\mu g\ L^{-1}$ and the RfD for that chemical is 10 $\mu g\ L^{-1}$, the EC of Chemical B is 0.3 $\mu g\ L^{-1}$ and the RfD for that chemical is 1 $\mu g\ L^{-1}$, and the EC for Chemical C is 0.05 $\mu g\ L^{-1}$ and the RfD for that chemical is 0.5 $\mu g\ L^{-1}$, the HI would be:

$$\frac{5}{10}+\frac{0.3}{1}+\frac{0.05}{0.5}=0.9$$

So, the mixture does not reach a level of concern since HI < 1. However, if new data were to indicate that the RfD for Chemical C should be revised to be 0.2, the HI must be recalculated to 1.05. This mixture would then be a hazard of concern since HI > 1.

The HQ and HI method is also used at screening level for ecological risk assessments [216], especially to estimate risk to fauna in and around contaminated sites. This often requires uncertainty factors (UFs) associated with potential wildlife responses to contaminant exposure based laboratory-derived toxicity data [217].

6 Ecological risk assessment

In the United States, ecological risk assessment paradigms have differed from human health risk assessment paradigms. An ecological risk assessment is presently based mainly on characterizing exposure and ecological effects. Both exposure and effects are considered during problem formulation [218].

Compared to human risk assessments, the ecological risk assessment process focuses more intensely on the coherence and connections between risk assessment and risk management than on the technical analysis of risk [218]. Intellectual conflicts have arisen, especially in the early stages in the 1980s, regarding risk assessment versus risk management. For example, some of the criticism of federal responses to environmental disasters, such as those in Love Canal, New York [219] and Times Beach, Missouri [220], related to the mixing of scientifically sound studies (risk assessment) and decisions on whether to pursue certain actions (risk management). This led to actual and perceived political and financial motivations for buyouts and remediation when sound scientific evidence was incomplete or entirely missing. In fact, the final step of the human health risk assessment process is to characterize quantitatively and qualitatively the relevant risk analysis elements, including the scientific uncertainties that must considered by the risk manager [107]. In particular, the process allowed for an integration of research with risk assessment, which could underpin risk management decisions. The problem formulation step in the ecological framework has the advantage of providing an analytic-deliberative process early on. That is, it combines sound science with input from various stakeholders inside and outside of the scientific community [218].

Ecological risk assessment differs from chemical risk assessment as the former assessment requires that effects on ecosystems be characterized, whereas the latter assessment's first step is to identify the hazards of a specific chemical compound, especially its toxicity. This is because in chemical risk assessments hazard connotes either the intrinsic effect of a stressor, i.e., chemical inherency, or a margin of safety by comparing an effect with an estimate of exposure concentration. Margin of safety means that the hazard is at a concentration in the environment that could be expected to induce an adverse outcome.

Eco-risk example

Consider a hypothetical compound that is found at 10 ppm to reduce the ability of a species to reproduce, which in turn changes the predator-prey aspects of the ecosystem, leading to diminished biodiversity. This is an example of an ecological hazard. However, the threshold concentration denoting the hazard would not simply be 10 ppm, due to uncertainties in the data and science (e.g., measurement and modeling error and variability) underlying this hazard threshold. With an ample margin of safety, the hazard threshold is adjusted to a lower concentration, e.g., to 5 ppm, to account for these uncertainties.

Another aspect of ecological hazard is that of the food chain, which incorporates human health and ecosystem condition (see Fig. 17.7). A hazard to an ecosystem must be considered at multiple levels of biological organization [221]. The food chain or web is, therefore, a means by which an ecological hazard becomes a human health hazard. It is also the means by which a hazard can change physicochemical forms to become more hazardous, such as the methylation of mercury in sediment and aquatic faunal tissue.

After transport and transformation in the food web, the fate is mammalian tissue. Various modeling and decision tools are available to characterize the movement, transformation, uptake, and fate of the compound. From these, biochemodynamic paradigms can be constructed for individual chemical compounds and mixtures, as well as for microorganisms [43]. Although biological hazards are more complicated than those of purely chemical hazards, similar thresholds could exist for biota. For example, an introduced species may overwhelm another species in an ecosystem if introduced to the ecosystem or habitat. Seeds, spores, or cysts can be carried long distances by wind and water, as well attached to migrating organisms. In such cases,

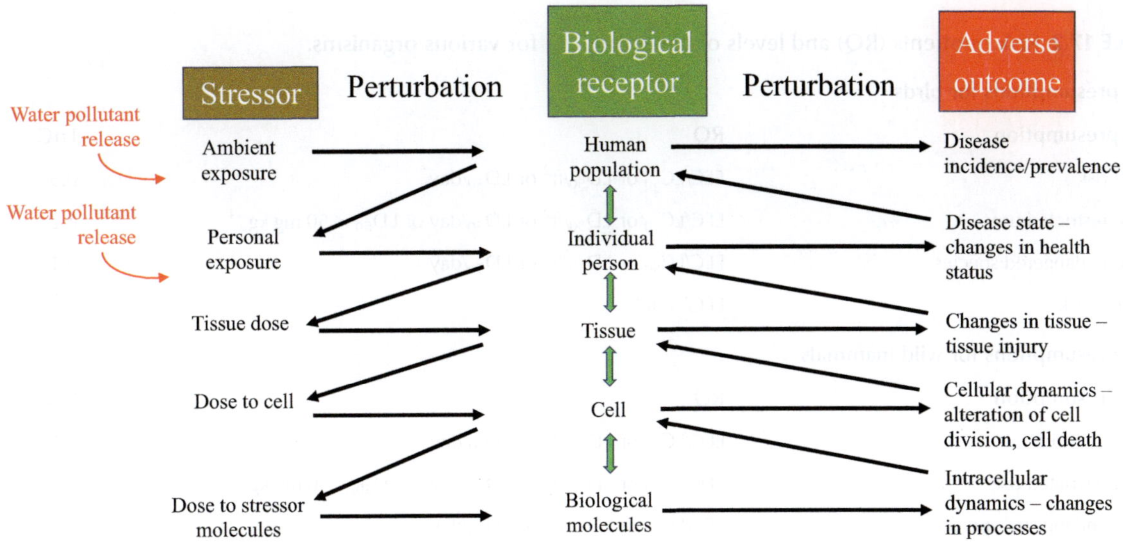

FIG. 17.7 Systems cascade of exposure-response processes. In this instance, scale and levels of biological organization are used to integrate exposure information with biological outcomes. The stressor (chemical or biological agent) moves both within and among levels of biological organization, reaching various receptors, thereby influencing and inducing outcomes. The outcome can be explained by physical, chemical, and biological processes (e.g., toxicogenomic mode-of-action information). *(Data from E.A.C. Hubal, et al., Exposure science and the US EPA National Center for Computational Toxicology, J. Expo. Sci. Environ. Epidemiol. 20 (3) (2010) 231–236.)*

the biological hazard would be identical to chemical hazard approach. Other biota, such as algae, would require a model that includes both direct and indirect risk factors, such as the increase in nutrient concentrations, changes in sunlight intensity and temperature, and decreases in flow rates that lead to algal blooms. These also result in decreasing dissolved oxygen concentrations and ultimately diminished health of the aquatic ecosystem and adverse outcomes like fish kills. In this case, the risk assessment had to include a succession of interim outcomes to reach the ultimate outcome of concern, i.e., the fish kill. Thus, there are layers of hazards and exposures that are step functions leading to the risk outcome [222].

Ecosystem hazard and exposure calculations differ from human health risk calculations. The risk quotient (RQ) is one tool used in eco-risk assessment. It is deterministic in that the RQ is found by dividing a point estimate of exposure by a point estimate of effects to compare risks [223]:

$$RQ = \frac{\text{Exposure}}{\text{Ecotoxicity}} \tag{17.22}$$

Ecotoxicity is an effect level or endpoint that is derived from toxicity tests that yield values such as lethal concentrations, e.g., the concentration at which 50% of the exposed fish die (LC_{50}). The bioassays may also indicate toxicity thresholds, e.g., no observed effect concentration (NOEC) or no observed effect level (NOEL). The ecotoxicity effect level is often the median effective concentration, i.e., EC_{50}, the concentration of test substance that decreases biological activity by 50%, e.g., algal growth or growth rate or Daphnia immobilization [224]. Since these organisms are waterborne, the EC_{50} units are mg L^{-1}. The estimated environmental concentration (EEC) relationship to RQ is:

$$RQ = \frac{EEC}{EC_{50}} \tag{17.23}$$

An ecological health criterion is based on the hazard quotient (HQ for an ecosystem. This differs from the human health version of HQ. For ecosystems, an EEC of the contaminant at the site, e.g., ng L^{-1} in ground water or ng kg^{-1} in soil, is divided by the threshold (e.g., no observed adverse effect level—NOAEL) in units that match the EEC:

$$HQ = \frac{EEC}{NOAEL} \tag{17.24}$$

The purpose of the RQ is to associate the exposure and toxicity ratio to a level of concern (LoC). After the risk quotient is calculated, it is compared to the LoC. An LoC is a relative concept used to interpret the RQ in terms of potential risk to a species. For example, the LoC is used to predict potential impacts on so-called nontarget organisms, i.e., those not intended to be killed by a pesticide. Decisions regarding environmental regulatory action are informed by the RQ and LoC [223]. Some representative LoCs are listed in Table 17.2.

TABLE 17.2 Risk quotients (RQ) and levels of concern (LoC) for various organisms.

Risk presumptions for birds		
Risk presumption	**RQ**	**LoC**
Acute risk	EEC/LC_{50} or LD_{50}/ft^2 or LD_{50}/day	0.5
Acute restricted use	EEC/LC_{50} or LD_{50}/ft^2 or LD_{50}/day or $LD_{50} < 50$ mg kg^{-1}	0.2
Acute endangered species	EEC/LC_{50} or LD_{50}/ft^2 or LD_{50}/day	0.1
Chronic risk	EEC/NOEC	1.0
Risk presumptions for wild mammals		
Risk presumption	**RQ**	**LoC**
Acute high risk	EEC/LC_{50} or LD_{50}/ft^2 or LD_{50}/day	0.5
Acute restricted use	EEC/LC_{50} or LD_{50}/ft^2 or LD_{50}/day or $LD_{50} < 50$ mg kg^{-1}	0.2
Acute endangered species	EEC/LC_{50} or LD_{50}/ft^2 or LD_{50}/day	0.1
Chronic risk	EEC/NOEC	1.0
Risk presumptions for aquatic animals		
Risk presumption	**RQ**	**LoC**
Acute high risk	EEC/LC_{50} or EC_{50}	0.5
Acute restricted use	EEC/LC_{50} or EC_{50}	0.1
Acute endangered species	EEC/LC_{50} or EC_{50}	0.05
Chronic risk	EEC/NOAEC	1.0
Acute endangered species	EEC/EC_{50} or NOEC	1.0
Risk presumptions for aquatic plants		
Risk presumption	**RQ**	**LoC**
Acute high risk	EEC/EC_{50}	1.0
Acute endangered species	EEC/EC_{05} or NOEC	1.0
Risk presumptions for terrestrial and semi-aquatic plants		
Risk presumption	**RQ**	**LoC**
Acute high risk	EEC/EC_{25}	1.0
Acute endangered species	EEC/EC_{05} or NOEC	1.0

Data from U.S. Environmental Protection Agency, Technical Overview of Ecological Risk Assessment: Risk Characterization. https://www.epa.gov/pesticide-science-and-assessing-pesticide-risks/technical-overview-ecological-risk-assessment-risk, 2017 (Accessed September 5, 2018).

Ecosystem risks may share the same or slightly adjusted parameters as those applied to human health risks. The predicted no-effect concentration (PNEC) used to assess harm to an ecosystem can be modified to be similar to the derived no-effect level (DNEL) for humans. Both express the concentration of a substance below which no effect is expected. Thus, both the DNEL and PNEC are forward-looking thresholds analogous to the retrospective human threshold concentrations (NOAEL or NOAEC). The DNEL is calculated by dividing a toxicological dose like the LD_{50} or NOAEL by an assessment factor (AF). For example, studies may show the incidence of adverse development effects in an animal species that would provide each an intraspecies AF $=10$, an interspecies AF $=5$, a duration AF $=2$, and route extrapolation and data quality AF $=2$. The product of the AF value (200) is the denominator of the DNEL equation. If the LD_{50} is 20 mg L^{-1}, then the DNEL $= 20/200 = 0.1$ mg L^{-1}.

PNEC values are derived from acute toxicity data from studies exposing bacteria, e.g., *Bacillus subtilis* and *Escherichia coli*, to a substance to obtain threshold concentrations for each species. For any species, the PNEC is directly proportional to the NOAEL because the PNEC can be derived by dividing the NOAEL by a series of assessment factors to address uncertainty and to be protective. Thus, if uncertainties are high, very large safety factors are applied, so that the PNEC becomes much smaller than the NOAEL for a substance. Assessment factors may be arbitrary and PNECs for many substances are not quantified, which limits their applications to species and substance that have been well-studied and for which reliable data are available [119].

The PNEC can be compared to existing concentrations in one or more environmental media or to the predicted environmental concentration (PEC). For acceptable risk, $\frac{\text{PNEC}}{\text{PEC}}$ should be much less than 1.

7 Conclusions

All the information discussed in previous chapters are part of environmental and public health risk assessments. This chapter has discussed the risk paradigm, beginning with identifying the hazard, followed by dose-response curves and interpretations, exposure assessment, and risk characterization. Several hazard, exposure, and risk metrics used in human and ecological risk assessments were explained. The chapter's main focus is chemical water pollutants, but also discussed biological and physical agents.

References

[1] B.S. Wankhade, et al., Neurological sequelae after acute carbon monoxide poisoning, Cureus 16 (1) (2024).
[2] A. Dorey, P. Scheerlinck, H. Nguyen, T. Albertson, Acute and chronic carbon monoxide toxicity from tobacco smoking, Mil. Med. 185 (1–2) (2020) e61–e67.
[3] M.T. Kleinman, Carbon monoxide, in: Environmental Toxicants: Human Exposures and Their Health Effects, 2020, pp. 455–486.
[4] D. HaMai, S.C. Bondy, Dimethylmercury, in: P. Wexler (Ed.), Encyclopedia of Toxicology, second ed., Elsevier, New York, 2005, pp. 54–57.
[5] R.E. Gosselin, R.P. Smith, H.C. Hodge, J.E. Braddock, Mercury, in: Clinical Toxicology of Commercial Products, fifth ed., vol. 1085, Williams & Wilkins, Baltimore, Maryland, 1984, pp. 262–275.
[6] D.W. Nierenberg, et al., Delayed cerebellar disease and death after accidental exposure to dimethylmercury, N. Engl. J. Med. 338 (23) (1998) 1672–1676.
[7] S. Saavedra, Á. Fernández-Recamales, A. Sayago, A. Cervera-Barajas, R. González-Domínguez, J.D. Gonzalez-Sanz, Impact of dietary mercury intake during pregnancy on the health of neonates and children: a systematic review, Nutr. Rev. 80 (2) (2022) 317–328.
[8] K. Dack, M. Fell, C.M. Taylor, A. Havdahl, S.J. Lewis, Mercury and prenatal growth: a systematic review, Int. J. Environ. Res. Public Health 18 (13) (2021) 7140.
[9] J. Wang, et al., Relation of prenatal low-level mercury exposure with early child neurobehavioral development and exploration of the effects of sex and DHA on it, Environ. Int. 126 (2019) 14–23.
[10] P. de Almeida Rodrigues, R.G. Ferrari, L.N. Dos Santos, C.A.C. Junior, Mercury in aquatic fauna contamination: a systematic review on its dynamics and potential health risks, J. Environ. Sci. 84 (2019) 205–218.
[11] M. Sakamoto, et al., Health impacts and biomarkers of prenatal exposure to methylmercury: lessons from Minamata, Japan, Toxics 6 (3) (2018) 45.
[12] C.-B. Jiang, et al., Mercury concentration in meconium and risk assessment of fish consumption among pregnant women in Taiwan, Sci. Total Environ. 408 (3) (2010) 518–523.
[13] Northeast Waste Management Officials' Association, Indoor Air Mercury. http://www.newmoa.org/prevention/mercury/MercuryIndoor.pdf. (Accessed July 26, 2018).
[14] B. Aberg, L. Ekman, R. Falk, U. Greitz, G. Persson, J.O. Snihs, Metabolism of methyl mercury (203Hg) compounds in man: excretion and distribution, Arch. Environ. Health 19 (4) (1969) 478–484.
[15] M. Leventer-Roberts, et al., Environmental tobacco smoke exposure among children by urinary biomarkers and parent report, Acad. Pediatr. 21 (4) (2021) 663–669.
[16] N.L. Benowitz, Biomarkers of environmental tobacco smoke exposure, Environ. Health Perspect. 107 (Suppl. 2) (1999) 349–355.
[17] N.L. Benowitz, Cotinine as a biomarker of environmental tobacco smoke exposure, Epidemiol. Rev. 18 (2) (1996) 188–204.
[18] C. Wormuth, Homeland Security Risk Assessments: Key Issues and Challenges. Testimony Before the Subcommittee on Intelligence, Information Sharing and Terrorism Risk Assessment Committee on Homeland Security United States House of Representatives, Center for Strategic and International Studies, 2005.
[19] U.S. House of Representatives. Serial No. 109-58, Homeland Security Risk Assessments: Key Issues and Challenges. Testimony Before the Subcommittee on Intelligence, Information Sharing and Terrorism Risk Assessment Committee on Homeland Security United States House of Representatives, 2005, Available: https://www.govinfo.gov/content/pkg/CHRG-109hhrg35939/html/CHRG-109hhrg35939.htm.
[20] E. Cameron, G.F. Peloso, Risk management and the precautionary principle: a fuzzy logic model, Risk Anal. 25 (4) (2005) 901–911.
[21] National Research Council, Improving Risk Communication, National Academy Press, Washington, DC, 1989.
[22] National Research Council, Understanding Risk: Informing Decisions in a Democratic Society, National Academies Press, Washington, DC, 1996.
[23] N.C. Rasmussen, The application of probabilistic risk assessment techniques to energy technologies, in: Readings in Risk, RFF Press, Washington, DC, 1990, pp. 195–206.
[24] D.A. Vallero, Hazardous wastes, in: Encyclopedia of Science and Technology, AccessScience, McGraw-Hill Education, New York, NY, 2017.
[25] D.A. Vallero, T.M. Letcher, Unraveling Environmental Disasters, second ed., Elsevier Academic Press, Amsterdam, The Netherlands, 2023.

[26] New Jersey Department of Health and Senior Services, Hazardous Substance Fact Sheet: Copper Acetoarsenate, 1999. https://nj.gov/health/eoh/rtkweb/documents/fs/0529.pdf. (Accessed 14 July 2024).
[27] National Research Council, Biosolids Applied to Land: Advancing Standards and Practices, National Academies Press, 2002.
[28] U.S. Environmental Protection Agency, Risk Assessment for Carcinogenic Effects, U.S. EPA, 2016. https://www.epa.gov/fera/risk-assessment-carcinogenic-effects. (Accessed January 30, 2024).
[29] D. Dawson, et al., Assessment of non-occupational 1, 4-dioxane exposure pathways from drinking water and product use, Environ. Sci. Technol. 56 (8) (2022) 5266–5275.
[30] N.M. DeLuca, M. Angrish, A. Wilkins, K. Thayer, E.A.C. Hubal, Human exposure pathways to poly- and perfluoroalkyl substances (PFAS) from indoor media: a systematic review protocol, Environ. Int. 146 (2021) 106308.
[31] C.L. Ring, et al., Consensus modeling of median chemical intake for the US population based on predictions of exposure pathways, Environ. Sci. Technol. 53 (2) (2018) 719–732.
[32] K.L. Dionisio, et al., Exploring consumer exposure pathways and patterns of use for chemicals in the environment, Toxicol. Rep. 2 (2015) 228–237.
[33] U.S. Environmental Protection Agency, Vinyl Chloride: 75-01-4 | DTXSID8021434, 2020. https://comptox.epa.gov/dashboard/chemical/executive-summary/DTXSID8021434?list=ATSDRLST. (Accessed January 18, 2022).
[34] U.S. Environmental Protection Agency, Proposed Guidelines for Carcinogen Risk Assessment, 2015, Available: http://cfpub.epa.gov/ncea/cfm/recordisplay.cfm?deid=2833.
[35] U.S. Environmental Protection Agency, Guidelines for Human Exposure Assessment Risk Assessment (No. EPA/100/B-19/001), Risk Assessment Forum, Washington, DC, 2019. Available: https://www.epa.gov/sites/default/files/2020-01/documents/guidelines_for_human_exposure_assessment_final2019.pdf.
[36] K. Huang, S.A. Sanchez, M.J. Eckelman, Using the US National Air Toxics Assessment to benchmark the USEtox inhalation-mediated carcinogenic impacts of air emissions, Int. J. Life Cycle Assess. 26 (7) (2021) 1417–1430.
[37] U.S. Environmental Protection Agency, National-Scale Air Toxics Assessment for 2005: Fact Sheet, 2013, Available: http://www.epa.gov/ttn/atw/nata2005/05pdf/nata2005_factsheet.pdf.
[38] M. Strum, et al., The 2011 National Air Toxics Assessment (2011 NATA): Emissions and Modeling Approach, 2015.
[39] OECD, The Adverse Outcome Pathway for Skin Sensitisation Initiated by Covalent Binding to Proteins: Part 1: Scientific Evidence. ENV/JM/MONO(2012)10/PART1, 2012.
[40] M. Hayashi, Y. Sakuratani, Report of the Workshop on Using Mechanistic Information in Forming Chemical Categories. ENV/JM/MONO(2011)8, 2011.
[41] R.L. Buchanan, J.L. Smith, W. Long, Microbial risk assessment: dose-response relations and risk characterization, Int. J. Food Microbiol. 58 (3) (2000) 159–172, https://doi.org/10.1016/S0168-1605(00)00270-1.
[42] M. Coleman, H. Marks, Topics in dose-response modeling, J. Food Prot. 61 (11) (1998) 1550–1559.
[43] D. Vallero, Environmental Biotechnology: A Biosystems Approach, Elsevier Science, 2015.
[44] U.S. Environmental Protection Agency, Quantitative Structure Activity Relationship, 2012. http://www.epa.gov/nrmrl/std/qsar/qsar.html. (Accessed March 19, 2015).
[45] U.S. Environmental Protection Agency, Quantitative Structure Activity Relationship, 2015, Available: http://www.epa.gov/nrmrl/std/qsar/qsar.html.
[46] D.E. Dawson, B.L. Ingle, K.A. Phillips, J.W. Nichols, J.F. Wambaugh, R. Tornero-Velez, Designing QSARs for parameters of high-throughput toxicokinetic models using open-source descriptors, Environ. Sci. Technol. 55 (9) (2021) 6505–6517.
[47] D. Wang, et al., A hybrid framework for improving uncertainty quantification in deep learning-based QSAR regression modeling, J. Cheminform. 13 (1) (2021) 69, https://doi.org/10.1186/s13321-021-00551-x.
[48] D.S. Esser, J.H.J. Leveau, K.M. Meyer, Modeling microbial growth and dynamics, Appl. Microbiol. Biotechnol. 99 (21) (2015) 8831–8846, https://doi.org/10.1007/s00253-015-6877-6.
[49] R. Maheshwari, G. Joshi, D.K. Mishra, R.K. Tekade, Bionanotechnology in pharmaceutical research, in: Basic Fundamentals of Drug Delivery, Elsevier, 2019, pp. 449–471.
[50] V. Stone, et al., Nanomaterials versus ambient ultrafine particles: an opportunity to exchange toxicology knowledge, Environ. Health Perspect. 125 (10) (2017) 106002.
[51] D.H. Brouwer, et al., Occupational dermal exposure to nanoparticles and nano-enabled products: part 2, exploration of exposure processes and methods of assessment, Int. J. Hyg. Environ. Health 219 (6) (2016) 503–512.
[52] D. Yang, et al., DNA materials: bridging nanotechnology and biotechnology, Acc. Chem. Res. 47 (6) (2014) 1902–1911.
[53] R. Falkner, N. Jaspers, Regulating nanotechnologies: risk, uncertainty and the global governance gap, Glob. Environ. Polit. 12 (1) (2012) 30–55.
[54] G.V. Lowry, K.B. Gregory, S.C. Apte, J.R. Lead, Transformations of Nanomaterials in the Environment, ACS Publications, 2012.
[55] D. Vallero, in: J. Weinstein (Ed.), Nano-Cerium Fuel Additives Phase Distribution, U.S. Environmental Protection Agency, National Exposure Research Laboratory, Research Triangle Park, North Carolina, 2012.
[56] L.S. Birnbaum, P. Jung, From endocrine disruptors to nanomaterials: advancing our understanding of environmental health to protect public health, Health Aff. 30 (5) (2011) 814–822.
[57] H. Becker, F. Herzberg, A. Schulte, M. Kolossa-Gehring, The carcinogenic potential of nanomaterials, their release from products and options for regulating them, Int. J. Hyg. Environ. Health 214 (3) (2011) 231–238.
[58] J.R. Lead, Manufactured nanoparticles in the environment, Environ. Chem. 7 (1) (2010) 1–2.

[59] K.D. Grieger, S.F. Hansen, A. Baun, The known unknowns of nanomaterials: describing and characterizing uncertainty within environmental, health and safety risks, Nanotoxicology 3 (3) (2009) 222–233.

[60] M. Baalousha, J.R. Lead, Overview of nanoscience in the environment, in: Environmental and Human Health Impacts of Nanotechnology, Wiley-Blackwell Publishing Ltd, Hoboken, NJ, 2009, pp. 1–25.

[61] F.F. Larese, et al., Human skin penetration of silver nanoparticles through intact and damaged skin, Toxicology 255 (1–2) (2009) 33–37.

[62] S.J. Klaine, et al., Nanomaterials in the environment: behavior, fate, bioavailability, and effects, Environ. Toxicol. Chem. 27 (9) (2008) 1825–1851.

[63] National Institute for Occupational Safety and Health, Approaches to Safe Nanotechnology: An Information Exchange With NIOSH, 2007. http://www.cdc.gov/niosh/topics/nanotech/safenano/. (Accessed August 2, 2007).

[64] U.S. Environmental Protection Agency, Nanotechnology White Paper, 2007.

[65] D.A. Vallero, Beyond responsible conduct in research: new pedagogies to address macroethics of nanobiotechnologies, J. Long Term Eff. Med. Implants 17 (1) (2007).

[66] H.-L. Lien, D.W. Elliott, Y.-P. Sun, W.-X. Zhang, Recent progress in zero-valent iron nanoparticles for groundwater remediation, J. Environ. Eng. Manag. 16 (6) (2006) 371.

[67] D.B. Warheit, Nanoparticles: health impacts? Mater. Today 7 (2) (2004) 32–35.

[68] US Environmental Protection Agency, PFAS Action Plan: Program Update February 2020, Office of Air Quality Planning & Standards, 2020. https://www.epa.gov/pfas/pfas-action-plan-program-update-february-2020. (Accessed July 31, 2020).

[69] US Environmental Protection Agency, Basic Information on PFAS, Office of Air Quality Planning & Standards, 2020. https://www.epa.gov/pfas/basic-information-pfas. (Accessed July 31, 2020).

[70] C. Biamis, K. O'Driscoll, G. Hardiman, Microplastic toxicity: a review of the role of marine sentinel species in assessing the environmental and public health impacts, Case Stud. Chem. Environ. Eng. (2020) 100073.

[71] M.J. Carvan, T.P. Dalton, G.W. Stuart, D.W. Nebert, Transgenic zebrafish as sentinels for aquatic pollution, Ann. N. Y. Acad. Sci. 919 (1) (2000) 133–147.

[72] M.C. Fossi, C. Panti, Sentinel species of marine ecosystems, in: Oxford Research Encyclopedia of Environmental Science, 2017.

[73] J. Zhu, et al., Cetaceans and microplastics: first report of microplastic ingestion by a coastal delphinid, Sousa chinensis, Sci. Total Environ. 659 (2019) 649–654.

[74] Y. Liu, et al., Uptake and depuration kinetics of microplastics with different polymer types and particle sizes in Japanese medaka (Oryzias latipes), Ecotoxicol. Environ. Saf. 212 (2021) 112007.

[75] X. Qiu, et al., Quantifying the vector effects of polyethylene microplastics on the accumulation of anthracene to Japanese medaka (Oryzias latipes), Aquat. Toxicol. 228 (2020) 105643.

[76] G.T. Ankley, et al., Adverse outcome pathways: a conceptual framework to support ecotoxicology research and risk assessment, Environ. Toxicol. Chem. 29 (3) (2010) 730–741.

[77] A. Lillicrap, et al., Alternative approaches to vertebrate ecotoxicity tests in the 21st century: a review of developments over the last 2 decades and current status, Environ. Toxicol. Chem. 35 (11) (2016) 2637–2646.

[78] A.L. Karmaus, D.L. Filer, M.T. Martin, K.A. Houck, Evaluation of food-relevant chemicals in the ToxCast high-throughput screening program, Food Chem. Toxicol. 92 (2016) 188–196.

[79] U.S. Environmental Protection Agency, Chemistry Dashboard, 2020. https://www.epa.gov/chemical-research/chemistry-dashboard. (Accessed October 12, 2020).

[80] A.J. Williams, et al., The CompTox Chemistry Dashboard: a community data resource for environmental chemistry, J. Cheminform. 9 (1) (2017) 61.

[81] M.D. Wood, K. Plourde, S. Larkin, P.P. Egeghy, A.J. Williams, V. Zemba, I. Linkov, D.A. Vallero, Advances on a decision analytic approach to exposure-based chemical prioritization, Risk Anal. 40 (1) (2020) 83–96.

[82] D.A. Vallero, Environmental Systems Science: Theory and Practical Applications, first ed., Elsevier Academic Press, 2021.

[83] P.T. Anastas, R.L. Lankey, Life cycle assessment and green chemistry: the yin and yang of industrial ecology, Green Chem. 2 (6) (2000) 289–295.

[84] U.S. Environmental Protection Agency, Benzo(a)pyrene: 50-32-8 | DTXSID2020139, 2020. https://comptox.epa.gov/dashboard/dsstoxdb/results?search=DTXSID2020139#adme. (Accessed October 12, 2020).

[85] D. Hattis, M. Lynch, Interspecies differences and human inter-individual variability in tissue-level pharmacokinetic parameters task 3 final report—description and analyses of the data bases, in: Report by Clark University to the US Environmental Protection Agency under Purchase Order no. EP05W002147, vol. 12 (19), 2006, p. 06. http://www2.clarku.edu/faculty/dhattis/. (Accessed).

[86] D. Hattis, A. Russ, Role of dosimetric scaling and species extrapolation in evaluating risks across life stages. II. Pharmacokinetic dosimetric considerations in old age, in: Report to US Environmental Protection Agency. RFQ no. DC-03-00009, National Center for Environmental Assessment, Washington, DC, 2003.

[87] E.J. Calabrese, L.A. Baldwin, The frequency of U-shaped dose responses in the toxicological literature, Toxicol. Sci. 62 (2) (2001) 330–338.

[88] D.A. Vallero, Translating Diverse Environmental Data Into Reliable Information: How to Coordinate Evidence From Different Sources, Academic Press, 2017.

[89] E.J. Calabrese, The linear no-threshold (LNT) dose response model: a comprehensive assessment of its historical and scientific foundations, Chem. Biol. Interact. 301 (2019) 6–25.

[90] U.S. Environmental Protection Agency, Guidance for Applying Quantitative Data to Develop Data-Derived Extrapolation Factors for Interspecies and Intraspecies Extrapolation. EPA/100/R-14/002F, 2014, Available: https://www.epa.gov/sites/default/files/2015-01/documents/ddef-final.pdf.

[91] National Research Council, Exposure Science in the 21st Century: A Vision and a Strategy, The National Academies Press (in English), Washington, DC, 2012, p. 196.
[92] M. Kamber, S. Flückiger-Isler, G. Engelhardt, R. Jaeckh, E. Zeiger, Comparison of the Ames II and traditional Ames test responses with respect to mutagenicity, strain specificities, need for metabolism and correlation with rodent carcinogenicity, Mutagenesis (2009) gep017.
[93] S. Flückiger-Isler, et al., Assessment of the performance of the Ames II™ assay: a collaborative study with 19 coded compounds, Mutat. Res. 558 (1) (2004) 181–197.
[94] B.N. Ames, J. McCann, E. Yamasaki, Methods for detecting carcinogens and mutagens with the Salmonella/mammalian-microsome mutagenicity test, Mutat. Res. 31 (6) (1975) 347–363.
[95] J. McCann, E. Choi, E. Yamasaki, B.N. Ames, Detection of carcinogens as mutagens in the Salmonella/microsome test: assay of 300 chemicals, Proc. Natl. Acad. Sci. U. S. A. 72 (12) (1975) 5135–5139.
[96] P.A. Schulte, et al., Occupational safety and health, green chemistry, and sustainability: a review of areas of convergence, Environ. Health 12 (1) (2013) 1.
[97] R. Golden, J. Bus, E. Calabrese, An examination of the linear no-threshold hypothesis of cancer risk assessment: introduction to a series of reviews documenting the lack of biological plausibility of LNT, Chem. Biol. Interact. 301 (2019) 2–5.
[98] D. Costantini, B. Borremans, The linear no-threshold model is less realistic than threshold or hormesis-based models: an evolutionary perspective, Chem. Biol. Interact. 301 (2019) 26–33.
[99] B.A. Ulsh, A critical evaluation of the NCRP COMMENTARY 27 endorsement of the linear no-threshold model of radiation effects, Environ. Res. 167 (2018) 472–487.
[100] M. Doss, The conclusion of the BEIR VII report endorsing the linear no-threshold model is no longer valid due to advancement of knowledge, J. Nucl. Med. 59 (11) (2018) 1777.
[101] W. Weber, P. Zanzonico, The controversial linear no-threshold model, J. Nucl. Med. 58 (1) (2017) 7–8.
[102] E.J. Calabrese, J.C. Hanekamp, D.Y. Shamoun, The EPA cancer risk assessment default model proposal: moving away from the LNT, Dose Response 16 (3) (2018) 1559325818789840.
[103] U.S. Environmental Protection Agency, Cancer Risk Calculations, 2014. http://www.epa.gov/oswer/riskassessment/sghandbook/riskcalcs.htm. (Accessed May 1, 2015).
[104] E.E. McConnell, H.A. Solleveld, J.A. Swenberg, G.A. Boorman, Guidelines for combining neoplasms for evaluation of rodent carcinogenesis studies, J. Natl. Cancer Inst. 76 (2) (1986) 283–289.
[105] International Agency for Research on Cancer, IARC monographs on the evaluation of carcinogenic risk of chemicals to man, in: IARC Monographs on the Evaluation of Carcinogenic Risk of Chemicals to Man, vol. 1, 1972.
[106] U.S. EPA, Guidelines for Carcinogen Risk Assessment. EPA/630/P-03/001F, 2005.
[107] National Research Council, Science and Decisions: Advancing Risk Assessment, The National Academies Press (in English), Washington, DC, 2009, p. 424.
[108] R. Judson, et al., In vitro and modelling approaches to risk assessment from the US Environmental Protection Agency ToxCast programme, Basic Clin. Pharmacol. Toxicol. 115 (1) (2014) 69–76.
[109] D.J. Dix, K.A. Houck, M.T. Martin, A.M. Richard, R.W. Setzer, R.J. Kavlock, The ToxCast program for prioritizing toxicity testing of environmental chemicals, Toxicol. Sci. 95 (1) (2007) 5–12.
[110] A.M. Evans, G.E. Rice, J.M. Wright, L.K. Teuschler, Exploratory cumulative risk assessment (CRA) approaches using secondary data, Hum. Ecol. Risk Assess. 20 (3) (2014) 704–723.
[111] P. Whaley, et al., Improving the quality of toxicology and environmental health systematic reviews: what journal editors can do, ALTEX 38 (3) (2021) 513–522.
[112] C.A. Weitekamp, L.J. Phillips, L.M. Carlson, N.M. DeLuca, E.A.C. Hubal, G.M. Lehmann, A state-of-the-science review of polychlorinated biphenyl exposures at background levels: relative contributions of exposure routes, Sci. Total Environ. (2021) 145912.
[113] E.A.C. Hubal, et al., Advancing systematic-review methodology in exposure science for environmental health decision making, J. Expo. Sci. Environ. Epidemiol. 30 (6) (2020) 906–916.
[114] E.A. Cohen Hubal, D.M. Reif, R. Slover, A. Mullikin, J.C. Little, Children's environmental health: a systems approach for anticipating impacts from chemicals, Int. J. Environ. Res. Public Health 17 (22) (2020) 8337.
[115] J. Mitchell, et al., A decision analytic approach to exposure-based chemical prioritization, PLoS One 8 (8) (2013) e70911, https://doi.org/10.1371/journal.pone.0070911.
[116] J. Mitchell, et al., Comparison of modeling approaches to prioritize chemicals based on estimates of exposure and exposure potential, Sci. Total Environ. 458 (2013) 555–567.
[117] J.F. Wambaugh, et al., High-throughput models for exposure-based chemical prioritization in the ExpoCast project, Environ. Sci. Technol. 47 (15) (2013) 8479–8488.
[118] G. Roman, P. Isnard, J.-M. Jouany, Critical analysis of methods for assessment of predicted no-effect concentration, Ecotoxicol. Environ. Saf. 43 (2) (1999) 117–125.
[119] X. Jin, J. Zha, Y. Xu, J.P. Giesy, K.L. Richardson, Z. Wang, Derivation of predicted no effect concentrations (PNEC) for 2, 4, 6-trichlorophenol based on Chinese resident species, Chemosphere 86 (1) (2012) 17–23.
[120] B. Benson, A. Mills, B. Wood, A review of the reference dose and reference concentration processes, in: Risk Assessment Forum EPA/630/P-02 F, vol. 2, 2002.
[121] R. Landsiedel, et al., Testing metal-oxide nanomaterials for human safety, Adv. Mater. 22 (24) (2010) 2601–2627.

[122] D.A. Vallero, Estimating and predicting exposure to products from emerging technologies, in: B.D. Trump, C.L. Cummings, J. Kuzma, I. Linkov (Eds.), Synthetic Biology 2020: Frontiers in Risk Analysis and Governance, Springer International Publishing, New York, NY, 2020, pp. 107–142, https://doi.org/10.1007/978-3-030-27264-7, ISBN: 978-3030272661, In this issue.
[123] I.M. Cozzarelli, J.R. McKelvie, A.L. Baehr, 11.12—Volatile hydrocarbons and fuel oxygenates, in: H.D. Holland, K.K. Turekian (Eds.), Treatise on Geochemistry, second ed., Elsevier, Oxford, 2014, pp. 439–480.
[124] N. Gao, L. Yang, X. Lu, Z. Duan, L. Zhu, J. Feng, A review of interactions of microplastics and typical pollutants from toxicokinetics and toxicodynamics perspective, J. Hazard. Mater. 432 (2022) 128736, https://doi.org/10.1016/j.jhazmat.2022.128736.
[125] D.A. Vallero, Mapping exposure onto nanoscale toxicity measures, in: J.R. Lead, S.H. Doak, M.J.D. Clift (Eds.), Nanotoxicology in Humans and the Environment, Springer International Publishing, Cham, 2021, pp. 141–191.
[126] M.R. Wiesner, et al., Decreasing uncertainties in assessing environmental exposure, risk, and ecological implications of nanomaterials, Environ. Sci. Technol. 43 (17) (2009) 6458–6462.
[127] O. Lindqvist, et al., Mercury in the Swedish environment—recent research on causes, consequences and corrective methods, Water Air Soil Pollut. 55 (1) (1991) xi–261, https://doi.org/10.1007/BF00542429.
[128] S. Yan, S.B. Subramanian, R. Tyagi, R.Y. Surampalli, T.C. Zhang, Emerging contaminants of environmental concern: source, transport, fate, and treatment, Pract. Period. Hazard. Toxic Radioact. Waste Manag. 14 (1) (2010) 2–20.
[129] S.-D. Li, L. Huang, Pharmacokinetics and biodistribution of nanoparticles, Mol. Pharm. 5 (4) (2008) 496–504.
[130] S.G. Elci, et al., Surface charge controls the suborgan biodistributions of gold nanoparticles, ACS Nano 10 (5) (2016) 5536–5542.
[131] U.S. Department of Health and Human Services, Agency for Toxic Substances and Disease Registry, Toxicological Profile for DDT, DDE, and DDD, 2002.
[132] D.A. Vallero, Fundamentals of Air Pollution, fifth ed., Elsevier Academic Press, Waltham, MA, 2014, p. 999.
[133] National Research Council, Risk Assessment in the Federal Government: Managing the Process, National Academy Press, Washington, DC, 1983.
[134] H. Liu, P. Gramatica, QSAR study of selective ligands for the thyroid hormone receptor β, Bioorg. Med. Chem. 15 (15) (2007) 5251–5261.
[135] S. Vilar, G. Cozza, S. Moro, Medicinal chemistry and the molecular operating environment (MOE): application of QSAR and molecular docking to drug discovery, Curr. Top. Med. Chem. 8 (18) (2008) 1555–1572.
[136] A. Lagunin, A. Zakharov, D. Filimonov, V. Poroikov, QSAR modelling of rat acute toxicity on the basis of PASS prediction, Mol. Inform. 30 (2–3) (2011) 241–250.
[137] T. Puzyn, et al., Using nano-QSAR to predict the cytotoxicity of metal oxide nanoparticles, Nat. Nanotechnol. 6 (3) (2011) 175.
[138] K. Roy, I. Mitra, Electrotopological state atom (E-state) index in drug design, QSAR, property prediction and toxicity assessment, Curr. Comput. Aided Drug Des. 8 (2) (2012) 135–158.
[139] U. Sahlin, Uncertainty in QSAR predictions, Altern. Lab. Anim. 41 (1) (2013) 111–125.
[140] U. Sahlin, N. Jeliazkova, T. Öberg, Applicability domain dependent predictive uncertainty in QSAR regressions, Mol. Inform. 33 (1) (2014) 26–35.
[141] T. Kimura, Y. Miyashita, K. Funatsu, S.-I. Sasaki, Quantitative structure–activity relationships of the synthetic substrates for elastase enzyme using nonlinear partial least squares regression, J. Chem. Inf. Comput. Sci. 36 (2) (1996) 185–189.
[142] Environmental Risk Management Authority, Preparing Information on Risks, Costs and Benefits for Applications Under the Hazardous Substances and New Organisms Act 1996. ER-TG-03-1 7/00, 2000.
[143] Chemical Computing Group, Molecular Operating Environment: Chemoinformatics and Structure Based Tools for High Throughput Screening, Montreal, Canada, 2013.
[144] J.-W. Lu, N.-B. Chang, L. Liao, Environmental informatics for solid and hazardous waste management: advances, challenges, and perspectives, Crit. Rev. Environ. Sci. Technol. 43 (15) (2013) 1557–1656.
[145] P. Ertl, Cheminformatics and Its Role in the Modern Drug Discovery Process, vol. 18, Novartis Institutes for BioMedical Research, Basel, Switzerland, 2014. https://infochim.u-strasbg.fr/CS3/program/material/Ertl.pdf. (Accessed 14 July 2024).
[146] U.S. Environmental Protection Agency, Integrated Risk Information System (IRIS), 2022. http://www.epa.gov/iris/. (Accessed 14 July 2024).
[147] D.A. Vallero, Exposure space: integrating exposure data and modeling with toxicity information, in: OpenTox USA 2015, 2015. Baltimore, MD.
[148] A.L. Washington, The interoperability of US federal government information: interoperability, Information Resources Management Association (Ed.), in: Big Data: Concepts, Methodologies, Tools, and Applications: Concepts, Methodologies, Tools, and Applications, IGI Global, Hershey, Pennsylvania, 2016, pp. 210–228.
[149] U.S. Environmental Protection Agency, Elements of a Quality Assurance Project Plan (QAPP) for Collecting, Identifying and Evaluating Existing Scientific Data/Information, 2017, Available: https://www.epa.gov/osa/elements-quality-assurance-project-plan-qapp-collecting-identifying-and-evaluating-existing.
[150] A. Takhom, S. Usanavasin, T. Supnithi, M. Ikeda, Collaborative ontology development approach for multidisciplinary knowledge: a scenario-based knowledge construction system in life cycle assessment, IEICE Trans. Inf. Syst. 101 (4) (2018) 892–900.
[151] T.I. Oprea, O. Taboureau, C.G. Bologa, Of possible cheminformatics futures, J. Comput. Aided Mol. Des. 26 (1) (2012) 107–112.
[152] E. Schymanski, D. Aurich, Cheminformatics and the Exposome in Health and Disease, 2021.
[153] E. Schymanski, Environmental Detective Work: Environmental Cheminformatics and the Exposome, 2022.
[154] J. Krier, et al., Discovering pesticides and their TPs in Luxembourg waters using open cheminformatics approaches, Environ. Int. 158 (2022) 106885.
[155] A. Mikolajczyk, N. Sizochenko, E. Mulkiewicz, A. Malankowska, B. Rasulev, T. Puzyn, A chemoinformatics approach for the characterization of hybrid nanomaterials: safer and efficient design perspective, Nanoscale 11 (24) (2019) 11808–11818.
[156] P. Willett, The Literature of Chemoinformatics: 1978–2018, vol. 21, MDPI, 2020, p. 5576.

[157] F. Lunghini, Comprehensive Data Analysis and Predictive Chemoinformatics Models for REACH Related Physicochemical and (Eco) Toxicity Properties, Université de Strasbourg, 2020.

[158] L. Hu, J.L. Au, M.G. Wientjes, Computational modeling to predict effect of treatment schedule on drug delivery to prostate in humans, Clin. Cancer Res. 13 (4) (2007) 1278–1287.

[159] C.A. Lipinski, F. Lombardo, B.W. Dominy, P.J. Feeney, Experimental and computational approaches to estimate solubility and permeability in drug discovery and development settings, Adv. Drug Deliv. Rev. 64 (2012) 4–17.

[160] P.P. Egeghy, et al., Computational exposure science: an emerging discipline to support 21st-century risk assessment, Environ. Health Perspect. 124 (6) (2016) 697.

[161] N. Brandon, K. Dionisio, K. Isaacs, D. Kapraun, W. Setzer, R. Tornero-Velez, A novel framework for characterizing exposure-related behaviors using agent-based models embedded with needs-based artificial intelligence (CSSSA2016), in: The Computational Social Science Society of the Americas, Santa Fe, NM, November 17–20, 2016, 2016. Available: https://computationalsocialscience.org/csssa-2016-program/.

[162] R.S. Thomas, et al., The next generation blueprint of computational toxicology at the US Environmental Protection Agency, Toxicol. Sci. (2019).

[163] B. Pecoraro, M. Tutone, E. Hoffman, V. Hutter, A.M. Almerico, M. Traynor, Predicting skin permeability by means of computational approaches: reliability and caveats in pharmaceutical studies, J. Chem. Inf. Model. 59 (5) (2019) 1759–1771.

[164] P. Babuji, S. Thirumalaisamy, K. Duraisamy, G. Periyasamy, Human health risks due to exposure to water pollution: a review, Water 15 (14) (2023) 2532.

[165] K. Rahman, S. Barua, H. Imran, Assessment of water quality and apportionment of pollution sources of an urban lake using multivariate statistical analysis, Clean. Eng. Technol. 5 (2021) 100309.

[166] Y. Liu, J. Zhang, Y. Zhao, The risk assessment of river water pollution based on a modified non-linear model, Water 10 (4) (2018) 362.

[167] D. Garrick, J.W. Hall, Water security and society: risks, metrics, and pathways, Annu. Rev. Environ. Resour. 39 (2014) 611–639.

[168] W.W. Piegorsch, H. Xiong, R.N. Bhattacharya, L. Lin, Nonparametric estimation of benchmark doses in environmental risk assessment, Environmetrics 23 (8) (2012) 717–728.

[169] G. Hanrahan, Environmental Chemometrics: Principles and Modern Applications, CRC Press, 2008.

[170] R. Vandoorne, P.J. Gräbe, Stochastic modelling for the maintenance of life cycle cost of rails using Monte Carlo simulation, Proc. Inst. Mech. Eng. F J. Rail Rapid Transit 232 (4) (2018) 1240–1251.

[171] P. O'Connor, A. Kleyner, Practical Reliability Engineering, John Wiley & Sons, 2012.

[172] D.A. Vallero, Economics and feasibility assessments, in: D.A. Vallero (Ed.), Environmental Systems Science, Elsevier, 2021, pp. 509–542 (Chapter 11).

[173] Institute of Water Pollution Control, Water Pollution Control, Institute of Water Pollution Control, 1967.

[174] J.D. Solomon, Communicating Reliability, Risk and Resiliency to Decision Makers, Duke University, 2017. October 12, 2017.

[175] U.M. Shamsi, Water Distribution Network Reliability, University of Pittsburgh, 1988.

[176] D.A. Vallero, Excerpted and Revised From: Fundamentals of Air Pollution, fifth ed., Elsevier Academic Press, Waltham, MA, 2014, p. 999.

[177] E. Hollnagel, D.D. Woods, N. Leveson, Resilience Engineering: Concepts and Precepts, Ashgate Publishing, Ltd., 2007.

[178] J.D. Solomon, D.A. Vallero, From Our Partners—Communicating Risk and Resiliency: Special Considerations for Rare Events, Center for Infrastructure Protection & Homeland Security, George Mason University, Fairfax, Virginia, 2016. https://cip.gmu.edu/2016/06/01/partners-communicating-risk-resiliency-special-considerations-rare-events/. (Accessed 14 July 2024).

[179] M. Lee, D. Basu, An integrated approach for resilience and sustainability in geotechnical engineering, Indian Geotech. J. 48 (2018) 207–234.

[180] H.R. Heinimann, A generic framework for resilience assessment, in: An Edited Collection of Authored Pieces Comparing, Contrasting, and Integrating Risk and Resilience With an Emphasis on Ways to Measure Resilience, 2016, p. 90.

[181] M. Dardonville, C. Bockstaller, O. Therond, Review of quantitative evaluations of the resilience, vulnerability, robustness and adaptive capacity of temperate agricultural systems, J. Clean. Prod. 286 (2021) 125456.

[182] D.A. Vallero, Effect of waste on ecosystems, in: T.M. Letcher, D.A. Vallero (Eds.), Waste, second ed., Academic Press, 2019, pp. 171–198 (Chapter 8).

[183] R.H. Day, S.T. Allen, J. Brenner, K. Goodin, D. Faber-Langendoen, K.W. Ames, Ecological resilience indicators for mangrove ecosystems, in: Ecological Resilience Indicators for Five Northern Gulf of Mexico Ecosystems, NatureServe, Arlington, Virginia, USA, 2018.

[184] S.T. Allen, et al., Ecological resilience indicators for salt marsh ecosystems, in: Ecological Resilience Indicators for Five Northern Gulf of Mexico Ecosystems, NatureServe, Arlington, 2018, p. 53.

[185] J. Fiksel, I. Goodman, A. Hecht, Resilience: navigating toward a sustainable future, Solutions 5 (5) (2014) 38–47.

[186] C. Reynolds, Resilience in aquatic ecosystems—hysteresis, homeostasis, and health, Aquat. Ecosyst. Health Manag. 5 (1) (2002) 3–17.

[187] K.S. Crump, The linearized multistage model and the future of quantitative risk assessment, Hum. Exp. Toxicol. 15 (10) (1996) 787–798.

[188] E. U. Health & Consumer Protection Directorate General, Training for the Health Sector, EU, 2004. https://ec.europa.eu/health/ph_projects/2003/action3/docs/2003_3_09_a23_en.pdf. (Accessed March 14, 2023).

[189] M.L. Card, et al., History of EPI Suite[trade mark sign] and future perspectives on chemical property estimation in US Toxic Substances Control Act new chemical risk assessments, Environ. Sci. Process. Impacts 19 (3) (2017) 203–212, https://doi.org/10.1039/C7EM00064B.

[190] OECD, Guidance Document on the Validation of (Quantitative) Structure-Activity Relationship [(Q)Sar] Models 69, 2007.

[191] M.F. Dupont, et al., Chemometrics for environmental monitoring: a review, Anal. Methods 12 (38) (2020) 4597–4620.

[192] J. Chapman, et al., Combining chemometrics and sensors: toward new applications in monitoring and environmental analysis, Chem. Rev. 120 (13) (2020) 6048–6069.

[193] A. Sutcliffe, G. Rugg, A taxonomy of error types for failure analysis and risk assessment, Int. J. Hum. Comput. Interact. 10 (4) (1998) 381–405.

[194] M.D. Adams, P.S. Kanaroglou, A criticality index for air pollution monitors, Atmos. Pollut. Res. 7 (3) (2016) 482–487, https://doi.org/10.1016/j.apr.2015.11.004.

[195] P. Boffetta, J.K. McLaughlin, C. La Vecchia, R.E. Tarone, L. Lipworth, W.J. Blot, False-positive results in cancer epidemiology: a plea for epistemological modesty, J. Natl. Cancer Inst. 100 (14) (2008) 988–995.

[196] J.L. Furman, K. Jensen, F. Murray, Governing knowledge in the scientific community: exploring the role of retractions in biomedicine, Res. Policy 41 (2) (2012) 276–290.

[197] D. Psaltopoulos, A.J. Wade, D. Skuras, M. Kernan, E. Tyllianakis, M. Erlandsson, False positive and false negative errors in the design and implementation of agri-environmental policies: a case study on water quality and agricultural nutrients, Sci. Total Environ. 575 (2017) 1087–1099.

[198] D.A. Vallero, Wastewater, in: Waste, Elsevier, 2019, pp. 259–290.

[199] R. Harder, S. Heimersson, M. Svanstrom, G.M. Peters, Including pathogen risk in life cycle assessment of wastewater management. 1. Estimating the burden of disease associated with pathogens, Environ. Sci. Technol. 48 (16) (2014) 9438–9445.

[200] C.P. Gerba, I.L. Pepper, Wastewater treatment and biosolids reuse, in: Environmental Microbiology, second ed., Elsevier Inc., USA, 2009, pp. 503–530.

[201] B. Jiménez, D. Mara, R. Carr, F. Brissaud, Wastewater treatment for pathogen removal and nutrient conservation: suitable systems for use in developing countries, in: Wastewater Irrigation and Health, Routledge, 2009, pp. 175–196.

[202] U.S. Environmental Protection Agency, Handbook for Managing Onsite and Clustered (Decentralized) Wastewater Treatment Systems: An Introduction to Management Tools and Information for Implementing EPA's Management Guidelines. EPA No. 832-B-05-001, 2005.

[203] O. Jolliet, A.S. Ernstoff, S.A. Csiszar, P. Fantke, Defining Product Intake Fraction to Quantify and Compare Exposure to Consumer Products, ACS Publications, 2015.

[204] O. Jolliet, A. Ernstoff, S. Csiszar, P. Fantke, Health impacts of consumer exposure during product use: near field exposure applied to risk assessment and LCA, in: International Society of Exposure Science 24th Annual Meeting, 2014, pp. 178–179.

[205] P. Fantke, A.S. Ernstoff, L. Huang, S.A. Csiszar, O. Jolliet, Coupled near-field and far-field exposure assessment framework for chemicals in consumer products, Environ. Int. 94 (2016) 508–518.

[206] M.G. Evich, et al., Per- and polyfluoroalkyl substances in the environment, Science 375 (6580) (2022) eabg9065.

[207] J.C. Quezada Davalos, M.A. Michaud, L.E. Lowe, E.N. Hanson, J.E. Owens, Per- and polyfluoroalkyl substances (PFASs) in the Fountain Creek Watershed, Colorado Springs, CO, USA: a yearlong investigation of PFAS levels in water, soils, and sediments, ACS ES&T Water 3 (1) (2023) 96–105, https://doi.org/10.1021/acsestwater.2c00440.

[208] J.A. Cotruvo, S.B. Goldhaber, A.J. Cohen, EPA's unprecedented interim drinking water health advisories for PFOA and PFOS, Groundwater (2023).

[209] M. Powers, M.F. Zamorano, M.J. Chimney, Assessment of a diel phosphorus pattern's potential to benefit phosphorus retention in the stormwater treatment areas, Ecol. Eng. 197 (2023) 107108, https://doi.org/10.1016/j.ecoleng.2023.107108.

[210] U.S. Environmental Protection Agency, Drinking water contaminant candidate list 5-draft: a proposed rule by the Environmental Protection Agency on 07/19/2021, in: Federal Register, 2021. Available: https://www.federalregister.gov/documents/2021/07/19/2021-15121/drinking-water-contaminant-candidate-list-5-draft.

[211] M. Bailey, et al., Risk Assessment Guidance for Superfund: Volume I: Human Health Evaluation Manual (Part F, Supplemental Guidance for Inhalation Risk Assessment), Environmental Protection Agency, 2009.

[212] R. Lewis, S. Gordon, Sampling for Organic Chemicals in Air, Environmental Protection Agency, Research Triangle Park, NC, USA, 1994.

[213] W.F. Bleam, Soil and Environmental Chemistry, Academic Press, 2016.

[214] C. Stephan, D. Mount, D. Hansen, J. Gentile, G. Chapman, W. Brungs, Guidelines for deriving numerical water quality criteria for the protection of aquatic organisms and their uses, in: USEPA PB85-227049, 1985.

[215] Agency for Toxic Substances and Disease Registry, Calculating Hazard Quotients and Cancer Risk Estimates, ATSDR, 2022. https://www.atsdr.cdc.gov/pha-guidance/conducting_scientific_evaluations/epcs_and_exposure_calculations/hazardquotients_cancerrisk.html#CommonNonCancerGuidelines. (Accessed 14 July 2024).

[216] J.M. Neff, S.A. Stout, D.G. Gunster, Ecological risk assessment of polycyclic aromatic hydrocarbons in sediments: identifying sources and ecological hazard, Integr. Environ. Assess. Manag. 1 (1) (2005) 22–33.

[217] L.D. Duke, M. Taggart, Uncertainty factors in screening ecological risk assessments, Environ. Toxicol. Chem. 19 (6) (2000) 1668–1680.

[218] U.S. Environmental Protection Agency, Guidelines for Ecological Risk Assessment, U.S. EPA, 2014. https://www.epa.gov/sites/default/files/2014-11/documents/eco_risk_assessment1998.pdf. (Accessed January 24, 2024).

[219] E. Sullivan, M. Sylvester, The evolution of environmental risk management, Risk Manag. 53 (1) (2006) 30–34.

[220] D. Wartenberg, C. Chess, The risk wars: assessing risk assessment, New Solut. 3 (2) (1993) 16–25.

[221] E.A. Cohen Hubal, et al., Exposure science and the US EPA national center for computational toxicology, J. Expo. Sci. Environ. Epidemiol. 20 (3) (2010) 231–236.

[222] Y. Li, et al., Ecological risk assessment of environmental stress and bioactive chemicals to riverine fish populations: an individual-based model of smallmouth bass Micropterus dolomieu, Ecol. Model. 438 (2020) 109322.

[223] U.S. Environmental Protection Agency, Technical Overview of Ecological Risk Assessment: Risk Characterization, 2024. https://www.epa.gov/pesticide-science-and-assessing-pesticide-risks/technical-overview-ecological-risk-assessment-risk. (Accessed 14 July 2024. Last update: January 2, 2024.

[224] ChemSafetyPRO, How to Calculate Hazard Quotient (HQ) and Risk Quotient (RQ). https://www.chemsafetypro.com/Topics/CRA/How_to_Calculate_Hazard_Quotients_(HQ)_and_Risk_Quotients_(RQ).html. (Accessed September 5, 2018).

Chapter 18

Water quality accountability

1 Introduction

Scientific, engineering, and technological soundness are the basis for water quality-related decisions. The acceptability of water pollution actions also depends on economics and feasibility. Every decision must be trustworthy, and every water pollution control and prevention project must be reliable. The design and operation of a water quality project must build in resilience to external stresses and be reasonable in the expectations that the project can be operationalized and properly managed. Furthermore, project planning must consider the appropriate means of communicating and collaborating with various stakeholders.

Water projects and management involve many processes. Basically, pollution control technologies replace natural systems with built systems that eliminate pollution faster and more efficiently than does nature. Flows within these systems are large and rapid. With this speed and size, unanticipated problems will certainly arise, calling for ample margins of safety in the design and operation [1,2].

A water quality project must also be cost effective and feasible. Privately-funded and managed projects share many feasibility aspects with public projects. Many public involve the private sector which is contracted for expertise and skills. Most private sector water projects have public interests and visibility. Public works and engineering departments usually do not and certainly should not approach water quality programs separate from other environmental, transportation, waste collection, hazardous materials handling, treatment, engineering, and remediation activities. Public agencies must search for synergies and economies of scale that solve and prevent multiple environmental, public health and other societal problems.

This chapter draws upon engineering economics to introduce approaches for evaluating the acceptability of proposed and new projects and upgrades and enhancements to existing projects.

Preserving water quality depends on knowing which sources are contributing to the deterioration. When the results of water pollution measurements are interpreted, one of the first questions asked by scientists, engineers, and policy makers is where did it come from? Sorting out the various sources of pollution is known as source apportionment. Various models and decision tools are used to try to locate the sources of pollutants. Dispersion models usually start from the source and estimate the direct where the plume and its contaminants may move based on scientific principles. Models can also be based upon measurements taken in a water body and from these observations, make use of algorithms and functions to determine pollution sources. Having these tools is important, but knowing when and where to apply them begins with identifying the causes of the pollution and its effects.

2 Cause and effect

Unlike an introductory statistics course or a purely academic discussion, water pollution decision making must infer causality. The risk of an adverse water quality outcome is increased when risk factors are increased in number and size. Risk factors must be present before the onset of the adverse event. Thus, the risk factor and the adverse event are linked, but causal links often cannot be shown directly. Instead, associations between risk factors and adverse events are employed. Such links must be supported by credible information to find causes.

The strength of association and the time between a cause and an effect are essential to risk and reliability assessments. Exposure to a pollutant must correlate with the onset of harm *and* the exposure must occur prior to the onset of the harm. The biomedical research and epidemiological communities use several criteria to determine the strength of an association with causality, first articulated by A. Bradford Hill as a means of linking risk factors to cancer [3]:

1. Strength of Association
2. Consistency

Fundamentals of Water Pollution. https://doi.org/10.1016/B978-0-443-28987-3.00010-2

3. Specificity
4. Temporality
5. Biologic Gradient
6. Plausibility
7. Coherence
8. Experimentation
9. Analogy

All have relevance to water pollution. Depending on the pollution scenario, some of Hill's criteria are more important than others. Strength of association and temporality are essential. Others, like coherence and plausibility, are desirable in most cases, but could stifle hypotheses since they could be reasons to resist necessary paradigm shifts, e.g., new hypotheses for diseases.

2.1 Strength of association

For an exposure to a physical, chemical, or biological agent to cause an effect, the exposure must be associated with that affect. Strong associations provide more certain evidence of causality than is provided by weak associations.. Common epidemiological metrics used in association include risk ratio, odds ratio, and standardized mortality ratio.

Descriptive epidemiology is often a first step in water pollution studies. For example, even before hypotheses are made regarding what may be causing or contributing to a health problem, epidemiologists must characterize differences. Sex, race, ethnicity, age and socioeconomic factors can show different outcomes for diseases in the same general population.

2.2 Temporality

This criterion requires that exposure to an agent must precede the effect. For example, in a retrospective study, the researcher must be certain that the manifestation of a disease was not already present before the exposure to the pollutant. If the disease was present prior to the exposure, it may not mean that the pollutant in question is not a contributing cause, but it does mean that it is not the sole cause of the adverse outcome (see "Specificity").

For water pollutants, the exposure begins the pollution progression process. Although the temporality is required for causality, it does not assure causality. That is, if the exposure to the water pollutant occurred after the onset of the adverse outcome, that study cannot be used as evidence linking the pollutant to the outcome. However, if the exposure occurred prior to the outcome, the study may be used, but only if other criteria support the causal link, since another agent or risk factor may be the actual cause of the outcome.

2.3 Biologic gradient

Biolgical gradient is known as the dose-response step in environmental risk assessment. AS discussed in Chapter 17, dose-response is an essential part of a pollutant's inherent hazard. The comparison of the dose of a pollutant to a biological response in humans, in other organisms or in ecosystems is an important aspect of hazard identification. Water pollution harms both human health and ecosystems, so dose-response curves can be drawn for either. If the intensity, duration, or magnitude of exposure to a pollutant is increased, a concomitant, progressive increase of the hazardous effect is expected.

2.4 Plausibility

Generally, an association needs to follow a well-defined explanation based on a known biological system. Nevertheless, so-called paradigm shifts in the understanding of key scientific concepts do change [4]. A noteworthy example is the change in the latter part of the 20th century of the understanding of how the endocrine, immune, neural systems function, from the view that these are exclusive systems to today's perspective that in many ways they constitute an integrated chemical and electrical set of signals in an organism [5].

2.5 Coherence

The criterion of coherence suggests that all available evidence concerning the natural history and biology of the disease should agree to form a cohesive whole. A proposed causal relationship should not conflict or contradict information from experimental, laboratory, epidemiologic, theory, validated models, or other knowledge sources. Sometimes this is merely a timing issue. For example, early epidemiolgical studies indicated adverse health effects in people exposed to arsenic, but this was not supported by animal studies until later. The coherence between human epidemiology and *in vivo* studies was not obvious for some time.

2.6 Experimentation

Laboratory evidence supporting a causal hypothesis may come in the form of clinical trials, *in vitro* laboratory experiments, animal models, and natural experiments.

2.7 Analogy

An analogue link implies a similarity in some respects among things that are otherwise different. It is considered weak evidence of causality. However, analogy is often an iterative process. For example, computation tools like quantitative structural relationship (QSAR) models are a type of analogical method in that they deploy similarity algorithms of known chemical compounds to less studied compounds. These enjoyed little trust in the scientific community when they were first introduced [6], but have gained much acceptance in recent decades [7]. Fuzzy logic, machine learning and artificial intelligence deploy anologies, so such tools range from anectodal information to semi-quantitiative data [8–12].

3 Reliability

Reliability is important for any water pollution enterprise, but is particularly crucial for water pollution engineering. Selecting a site, designing and installing control technologies and operating and maintaining a water or wastewater treatment system can be evaluated from the perspective of reliability engineering. The item is fundamental unit of reliability, i.e., a functional or structural unit of arbitrary complexity. For example, a wastewater treatment item can be a component of a device or a part in the device. It may also be an assembly, a piece of a supply or equipment, a subsystem, or an entire system [13]. The item's highest level of integration is the reliability system. Thus, for a system to be reliable, each item must be reliable, and every assemblage of items in a system must also be reliable.

A system is reliable according to four performance factors [14]:

1. Capability
2. Efficiency
3. Availability
4. Reliability

In addition to adequate functionality, the system must also be efficient, i.e., it must feasibly meet its design objectives.[a] In other words, an item must be "fit for purpose." Quality is the major influence on an item's capability and efficiency [15]. A reliable system must be capable and efficient and must also be available when needed, i.e., minimal downtime. Technologies that work well under laboratory or microcosm conditions may be unreliable at the operational scale, i.e., scale-up uncertainty. A new technology can be less reliable than existing technologies, for example, if when put into operation replacement parts are unavailable or less available than those for traditional technologies. Under actual treatment conditions, important variables may not be apparent at the outset. In addtion, proven technologies have track records, so to speak. New technologies do not. So, only after enough time in the stage of deployment do certain problems become apparent. As such, downtime predictions must be adjusted with each step in the item life cycle. Like risk, reliability is an expression of the probability. Reliability is the probability that a system or item will perform as designed under specified conditions and for a specified time period [16–18].

A water and wastewater treatment technology must be sufficiently reliable, available, maintainable, and safe (RAMS). Reliability engineering is an integral part of a comprehensive protocol to ensure that systems are cost and time effective. Thus, quality assurance (QA) must be integrated into performance tracking and concurrent engineering [15].

Reliability is a conditional quantity. For example, consider a surgical unit in a rural hospital that has a very low tolerance for electrical interruption. The hospital has an onsite water treatment system linked to a private well. So, both the surgery

---

a. Form follows function: The most efficient system has a design that allows for the optimal performance and the most efficient item within a system works optimally within the system.

unit and the water system will stop receiving well water and will begin receiving clean water only from a water storage system until power is restored. In this scenario, providing clean water may be delayed allowing surgeries to continue for as long as clean water can be stored. Thus, to decrease the probability of interrupted electricity, the hospital may maintain two energy systems, i.e., a primary and backup. The backup system temporarily takes nonessential electrical uses offline, including pumping water into the storage system designed to hold water at surgical quality for 1 month. Further, let us assume that the primary system fails after 10 years, i.e., it ceases to function as designed, and the backup system fails after 5 years. Can we conclude that the second system is less reliable than the first? As is almost always true for engineering, the correct answer is that "it depends" [19]. The longer-lasting system has a longer operational period but could be less reliable. Let us assume that the primary system was designed to last 15 years and the backup for 4 years. In this case, the first is unreliable and the second is reliable, based on the design time period. This could occur if the first was in a hospital that only operated at peak loads once a month and/or in a relatively noncorrosive environment; whereas the backup kicked on frequently for the 5-year period under corrosive conditions. However, the surgical water storage system is unreliable if the water it holds does not meet quality standards before 1 month. This is an example of a subsystem failure that is the genesis of a complete system failure. It is also a reminder of the importance of redundancy in engineering.

Reliability is compelled by and is a indicator of operational efficiency. For example, operational efficiency may call for replacement of parts well before the end of their expected operational life. Consider the hypothetical preventive maintenance decision to replace a pump used to extract a landfill's contaminated leachate every 3 months even though the part is expected to last 1 year. The site engineer has investigated the reliability of the pump and found that the 1-year expectation was based on drinking water scenarios, with less corrosivity than is possible in the pump-and-treat scenario at the landfill. The engineer also discovered that pump failures in the drinking water studies were in small wells where the pump is easily replaced and the time that the pump is off-line did not cause large problems so long as the pump was replaced within 2 days. The engineer concludes that this is not the case for this pump-and-treat application, so avoiding any breakdown is essential. The regular replacement would also allow for more flexibility in timing the replacement.

Emblematically, an automobile engine timing belt might last 100,000 miles, but a prudent mechanic may know that the realistic range is between 60,000 and 130,000 miles, so recommends replacing it at 60,000 miles to avoid almost completely the time period during which the likelihood of failure is rapidly increasing. It is better to build in maintenance than to be stranded on the side of the road or pay substantially more to repair an entire engine while driving a few miles with a torn belt.

Breakdown of water devices cause more than financial cost increases; they can threaten human health whenever the treatment is not efficient. The proper operation and maintenance (O&M) is one of the specified operational conditions required for efficiency. This can be likened to the maintenance of an automobile mentioned above. In additon to the timing belt example, there are many more parts than the belt, so the engine's reliability is highly uncertain. If proper O&M conditions are not met, the chance of breakdown is increased and time before failure shortened. Indeed, human factors almost always contribute to system failure, i.e., the O&M protocol is unreliable. Any engineered system must have a sufficient optimal operational timeframe, in which the failure rate is assumed to be minimal.

Unreliable systems can range from infrequent failures with little threat to health, safety or welfare, e.g., a simple nuisance of excessively long charging times, to catastrophic failures, e.g., loss of human life from an explosion or fire at a waste site or destruction of endangered species habitats from inadequate consideration of contaminant transport. People living near a wastewater treatment plant, landfill, or any engineered system must be ensured that it is reliable. If a failure were to occur during the operation or during disassembly, appropriate steps must be taken to prevent harm and that the failure would be properly managed according to a contingency plan approved and shared with all stakeholders before operations begin [20].

The hazard rate or failure density, $f(t)$, is the probability of a failure per unit time, also known as a function of the likelihood that an adverse outcome will occur. The $f(t)$ does not suggest the severity of the outcome, merely the chance of failure. The $f(t)$ is simply a probability that an event will occur, notwithstanding whether it will be an explosion where people are hurt or killed, or an odor. It predicts the probability that the hazard will occur during a designated time period. Thus $f(t)$ can be integrated over a defined time interval:

$$P\left\{t_1 \le T_f \le t_2\right\} = \int_{t_1}^{t_2} f(t)dt \tag{18.1}$$

where T_f = time of failure.

Eq. (18.1) does not state how T_f is reached. The system may have failed at the end or beyond expected life or early because of improper maintenance or defects. The reliability function, $R(t)$, of a system at time t is the cumulative probability that the system has not failed in the time interval from t_0 to t_t:

$$R(t) = P\left\{T_f \geq t\right\} = 1 - \int_0^t f(x)dx \tag{18.2}$$

Reliability improves when the optimal operating time is extended. A treatment technology's reliability improves by incorporating features that will increase its resistance to failure. For example, if cadmium (Cd) concentrations can remain in sealed components of a battery system after the system's operational life, the decommissioning process will not need to reclaim the metal [21]. Even if the panel parts inappropriately reach an engineered landfill they may remain at least somewhat contained and, hopefully, allow the landfill barrier to decrease the flow of Cd-contaminated water between the contents of the landfill and the surrounding aquifer, e.g., a velocity of a few microns per decade. However, the barrier does not completely eliminate failure, i.e., $R(t)$ would equal 0; it simply lengthens the time (increases T_f) before the failure occurs. Thus, other measures, e.g., a strong manifest system to show that the panels contain Cd, are needed to make the recycling and disposal systems more reliable.

Vulnerabilities inherent to a water pollution control or prevention system must be identified and addressed, such as unscientifically sound facility siting practices or the inclusion of inappropriate design criteria, e.g., cheapest land. Failure becomes more likely as mistakes or malpractice increase. Thus, efficiency is driven by reliability. Failure to account for and address inefficiencies accelerates and exacerbates failures, e.g., loss of life, loss of property, lawsuits, and violation of the engineer's code of ethics. The common graphical representation of engineering reliability is the "bathtub" curve (see Fig. 18.1). The U-shape indicates that failure has a higher probability at the introduction of a technology to the marketplace, known as "infant mortality." Early on, technology promoters and adopters seek information about the scale-up and make changes to improve the failure rate, i.e., "beta testing." After this stage, the technology remains on the market with increasing follow-up by a period of constant reliability. After this optimal period, the failure rate again begins to increase, with the indication that the technology is nearing the end of its useful life or that substantial repairs, retrofitting, and upgrades must be made to return to the ideal failure rate [22]. Indeed, failure can occur even before acquiring equipment or starting a new project. Many problems occur during the planning and idea stage. A promising energy technology may be terminated prematurely. The miscarriage stage has the highest rate of failure, $h(t)$. It occurs even before the system, process, or equipment has been made a reality. Exclusion of ideas that require a paradim shift in decision making or failure to get input about key scientific or social variables can create a high hazard [4,22–24].

Proper water and wastewater engineering provides ways to predict and manage events that lead to desirable outcomes, e.g., preventing a pollutant's concentration in the surface water, ground water, and soil from exceeding a standard concentration. This requires credible data and information from which to extrapolate and interpolate from what is known to what is

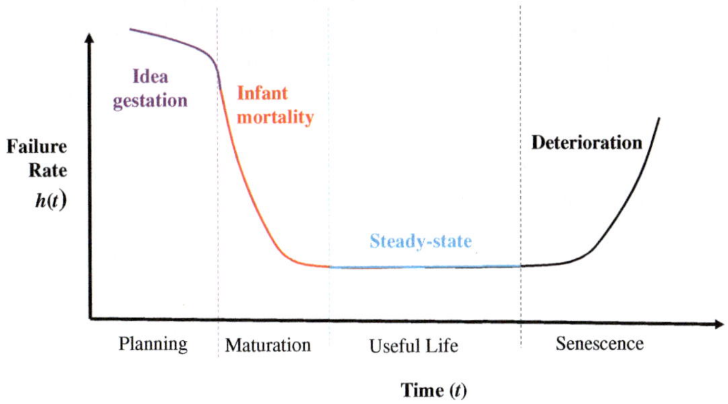

FIG. 18.1 Prototypical reliability curve of a new technology. The highest rates of failure, $h(t)$, occur during the earlies and latest stages. Well-designed systems extend the useful life. The environmental engineer must foresee potential problems and incorporate factors of safety accordingly, e.g., closer monitoring and more extensive safety measures in the early stages of adoption and toward the end of the steady-state period when the failure rates are likely higher. *(Data from D. Vallero, Paradigms Lost: Learning From Environmental Mistakes, Mishaps and Misdeeds, Butterworth-Heinemann, 2005.)*

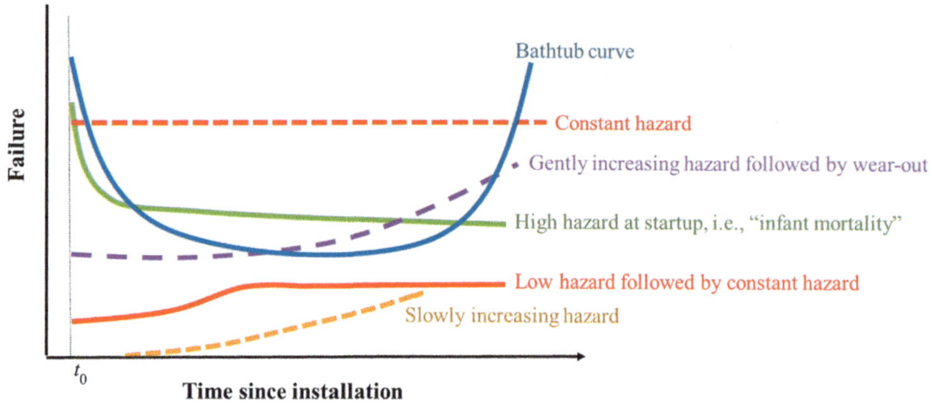

FIG. 18.2 Idealized patterns of reliability. *(Data from T.M. Letcher, D. Vallero, Excerpt From: Waste: A Handbook for Management, second ed., Academic Press, 2019.)*

less known, i.e., from the data-rich to the data-poor [25]. Risk management models, science-based decision making, and forward trajectories that begin from laboratory studies, can be enhanced with information from experiences of similar technologies elsewhere. Prevention of failure is best, so exhaustive literature reviews may uncover similar failures, especially if the failures have been document by root cause analyses, decision trees, deconstruction/reconstruction studies, and failure analyses [26]. An important and common technique is to assess the availability and reliability, $R(t)$. of a system is by calculating the mean time between failures (MTBF) [27]:

$$R(t) = \exp\left(-\frac{t}{MTBF}\right) = \exp(-\lambda t) \tag{18.3}$$

Where, t = operation time and λ = reciprocal of MTBF.

The survival analysis function is used to observe data up to the time of interest. Since our time of interest is the time until failure, the survival function [$S(t)$] is the same as $R(t)$. In reliability engineering, the hazard is the conditional probability of failure of a currently working system. Thus, failure can be expressed as a hazard function, i.e., the instantaneous rate at which events happen in the absence of previous events. The hazard function can be measured with time, e.g., the lifetime of a facility's components like the cells in a landfill or the tanks in a wastewater treatment plant. Fig. 18.1 represents the classic failure curve, however, the change of the hazard function with time produces various patterns (see Fig. 18.2), depending on the type of facility and equipment, as well as the operating conditions.

Reliability, availability, maintainability and safety (RAMS) software packages are available to calculate $R(t)$ [28]. Wastewater and water treatment fall under the domain of "industrial ecology" [29–31]. Industrial ecology systematically identifies potential waste streams, contaminants of concern, and approaches to address these concerns, e.g., conversion technologies. The process evaluates the economic and environmental effects of using wastes and transformations that may occur during transport, handling, and processing. Depending on the various pollutants to be addressed, the optimal methods will vary.

The idealized patterns in Fig. 18.2 more aptly applies to public health than to ecosystems. Certainly, ecosystem protection may also benefit from reliability assessments, e.g., carrying capacity, resilience and sustainability.

The acceptability of a treatment technology or other engineered system is determined and can be evaluated according to authoritative design and operation standards and criteria, i.e., functionality, safety, feasibility, sustainability, etc. The reliability can be compared in the forward direction using models, life cycle analyses, cost-effectiveness studies, sensitivity analysis, and uncertainty analyses. In the reverse direction, especially when something goes wrong, inverse methods are applied to investigate the contributing factors to the failure, e.g., root cause analysis [32].

4 Resilience

Engineers must optimize designs for risk reduction and system reliability, but with attention to resilience [33–35]. Indeed, resilience is a key part of risk and reliability. A system that is overly sensitive to perturbations is risky and unreliable, requiring that downstream users must constantly deploy controls to render the system sufficiently safe and reliable. Conversely, a more resilient system can withstand a wider range of stressor intensity that returns to optimal conditions relatively quickly. A system that lacks sufficient resilience is dangerously sensitive and vulnerable to changing conditions. Pollutants can vary considerably in their treatability. This calls for ongoing monitoring of all aspects of treatment processes. Over the operational life, environmental and operational conditions change, so that a resilient system is likely to be less expensive and in need of fewer repairs and replacements than a more inflexible system.

Resilience must be built into to risk and reliability specifications. The systems must be able to return to the original form or state after being bent, compressed, or stretched or the ability to recover return to the desired state readily following the application from some form of stress [34–36]. Of course, good practice and an eye to O&M has long been part of the engineer's tool kit, though they likely would not have readily thought of themselves as resilience engineers. The specifications of a technology or item includes tolerance ranges during the operational lifetime. For example, an item may perform its function for 10 years, so long as the environmental conditions, like temperature, pressure, corrosiveness and other operational factors, remain at specified tolerances. Thus, when the engineer finds ways not only to lengthen the lifetime, but also to broaden the tolerances, the design becomes more resilient. To achieve this, the item must become [37]:

1. more robust by widening the upper-to-lower performance limits;
2. more resistant to and recoverable from disruptions;
3. reconfigurable by incorporating adaptive properties; and,
4. more adaptable to changing conditions.

Thus, the engineer must be an eager and diligent learner, i.e., attentive to changes and disruptions to anticipate future problems early or even before they occur [37,38].

Resilient systems have wide tolerances, compared to an overly sensitive system that is readily harmed when condition fall outside of narrow tolerance ranges. For example, the proven treatment system in the marketplace must have much greater resilience than when one of its components was in nascent, highly controlled lab clean room stage. Real-world conditions vary during the operational life of the renewable.

Ecology has taught us that organisms in a resilient habitat are able to survive in broad ranges of moisture, temperature, light and nutrient concentrations. For example, exposure to a pollutant in a biologically diverse ecosystem may lead to little harm or a little time to recover, whereas the same pollutant concentration in a nonresilient, less diverse system, e.g., monoculture corn crop, could overwhelm the ecosystem [39]. The stressors on a habitat or ecosystem can be synergistic, e.g., although drought conditions could be withstood, the addition of the deposited pollutant cause an ecological disaster [40]. Metaphorically, a treatment technology that builds in resilience is in less need of intricate engineering controls against insult than a nonresilient system [33,41,42]. To be widely adopted, this is crucial. It is important to keep these in mind when selecting at treatment technology, such as those described in the following section.

5 Redundancy and robustness

At least two more "R-terms" can be added to this list of engineering metrics related to climate change, i.e., robustness and redundancy [43]. Though related, resilience and robustness differ in that resilience is reactive, whereas robustness is proactive. The time it takes for a system to recover from a failure is its resilience. If a system is inherently strong and resists insults well it is robust. Historically, surface waters, aquifers, soil systems, and the atmosphere have been robust and resilient, but in recent decades they have shown difficulty in recovering from the chemical loads of modern societies. Ways of improving its resilience by making the media more robust are currently limited. Geoengineering and other large-scale solutions being proposed all seem to do little to improve the robustness of the environmental medium itself, but there are ways to make adaptation and response to the threats more robust [44–50]. These include better land use planning and conservation to protect sensitive habitats, improving crop management to protect the food supply, decreased amounts of pollutant releases, and plans to address potential human population migrations due to environmental and other problems.

Engineering solutions must always expect failures to occur in time and space, so two more "R-terms" must be added to this discussion, i.e., robustness and redundancy [43]. Improved reliability lengthens the time to failure but does not eliminate it. An acceptably redundant system acknowledges this and builds in safeguards and redundant systems to meet the failures when they occur [46,48,51–53]. Though related, resilience and robustness differ in that resilience is reactive, whereas robustness is proactive. The time for a system to recover from a failure is its resilience. If a system is inherently strong and resists insults sufficiently, it is robust. Historically, the earth's water systems have been robust and resilient, but since the Industrial Revolution it has increasingly shown difficulty in recovering from the pollutant loads of modern societies. Building in robustness to systems to handle increasing pollutant is much less desirable from preventing pollutants from entering the environment. Assuming increasing amounts of pollutants and then applying large-scale solutions like geoengineering would do comparatively little to improve the robustness of the hydrosphere and atmosphere, but could very well potentially lead to irreversible adverse consequences [44–50]. Improved land use planning, cleaner fuels, and pollution prevention strategies are preferable for sustainable and resilient ecosystems.

Redundancy is also important to engineering resilience. Backup systems for pollution control equipment during low probability events, for example, prevent and shorten failures. Indeed, engineered systems must be prepared for failures to occur in time and space. Improved reliability lengthens the time to failure but does not eliminate it. An acceptably redundant system acknowledges this and builds in safeguards and redundant systems to meet the failures when they occur [46,48,51–53].

Engineers must continue to design and devise innovative and larger scale solutions to the direct and indirect problems brought on by environmental problems, many which we do not yet know are even problems. Engineers tend to be technological optimists and agents of change [54]. Binary perspectives can be obstacles to optimization. Engineers often can satisfy seemingly incongruent perspectives to protect the public *and* provide practical, sustainable, and cost-effective solutions to the problems now and those that will unfold in the future. After all, the first canon of the engineering profession is to hold paramount the safety, health, and welfare of the public. Engineers must be faithful agents to their clients [55]. Their principal client is the public and the public expects an environment that supports healthy human lives and sustainable ecosystems.

6 Engineering economics[b]

Once the decision is made to address a water pollution problem, engineering economics, also known as engineering economy, is used to determine whether a specific project is the most worthwhile compared to other candidates. Engineering economics is also used to prioritize various projects that may serve the same purpose or function, and to assign value or importance to design specifications. For example, engineering economics may help the city engineer determine whether to invest in new control technologies or keep repairing the old ones, which of the commercially equipment to purchase, which improvements to public assets and operations should be funded, and whether the added cost of materials and supplies needed to keep a technology operation are worth the price.

A principal economic concept is that time represents money, i.e., the so-called “time value of money.” Water quality management is an ongoing endeavor, so funding and other resources are viewed over several years. Expensive equipment may be purchased and immediately installed, yet paid for years, even if the benefit provided by the equipment ends before payoff. Therefore, the first step evaluating a project is the cash flow, which represents the time value of money. This requires that the cash receipts and disbursements (i.e., expenses) for each project be described over a specified time-period.

The costs include initial costs (e.g., purchase price, set-up, and installation), O&M, overhaul, and end-of-project life costs, e.g., decommissioning, disassembly, and recycling. The costs are somewhat fluid and interrelated. For example, a sludge pit will have decommissioning costs that would be high if hazardous substances accumulate. As an example of cross-media transfer, treating an air pollutant can introduce a water pollution problem, such as the cases of water contamination from power plant ash ponds in North Carolina (see Discussion Box 18.1). Therefore, O&M and surveillance costs upfront can prevent major costs over time. The disbursements include salvage value, i.e., what the equipment or property is worth after its operational life, and revenues (e.g., sales of fertilizer, compost or recycled materials).

6.1 Cash flow

The cost flow diagram (CFD) is used to evaluate costs and disbursements (see Fig. 18.3). The CFD shows the amount and timing of individual cash flows, and forms the basis for engineering economic analysis [61]. The CFD is developed by first drawing a segmented time-based horizontal line, divided into appropriate time unit. At each cash flow event, a vertical arrow points down to costs and up for revenues or benefits. For clarity, the CFD segments should be drawn to a relative scale.

The end of period t is the same as the beginning of period $(t+1)$. The beginning of period cash flows includes costs for rent, lease, and insurance payments, whereas the end-of-period cash flows are for O&M, salvages, revenues, and overhauls. Deciding where $t=0$ is arbitrary, e.g., when a project is analyzed, when funding is approved, or when construction begins. Note that one group’s cash outflow, represented as a negative value, is another group’s inflow, represented as a positive value. Incidentally, two or more cash flows in the same year should be diagrammed individually to indicate clearly the connection from the problem statement to each cash flow in the diagram.

Example: A city purchases a sludge combustion facility, including a water treatment system, for \$2 million, with O&M costs of \$30,000 in year 1 and increasing by \$5000 each year. Some of the sludge is converted to commercial-grade fertilizer with equipment provided by a local salvage company, providing an annual revenue of \$50,000. After year 7, the

b. This section includes excerpts from [15].

Discussion Box 18.1 Environmental and economic perils of decommissioning: ash pits

Coal combustion residuals (CCRs) represent the largest amount of industrial wastes generated in the U.S. The 10,000 tonnes[c] of CCRs generated per year from 470 electric generating plants consist of bottom ash, fly ash, boiler slag, and byproducts from FGD. Surface impoundments and landfills hold half of the CCRs [56,57]. If the effluent from ash is released into a surface waterbody in the U.S., the amount and type of effluent is limited by the National Pollutant Discharge Elimination System (NPDES) and requires an NPDES permit. Since this effluent very likely contains highly toxic metals and other contaminants, the permit process is complicated beyond a single effluent limitation threshold and requires that other factors, including background concentrations and characteristics of the receiving waterbody. For example, the U.S. Environmental Protection Agency (EPA) applies a formula to calculate waste load allocations and to find the in-stream pollutant concentration (C_r) in mg L^{-1} in the stream reach after complete mixing occurs [58]:

$$C_r = \frac{Q_d C_d + Q_s C_s}{Q_r} \qquad (18.4)$$

Where: Q_d = waste discharge flow in units commonly used by U.S. environmental engineers, i.e., million gallons per day (MGD) or cubic feet per second (cfs); C_d = pollutant concentration in waste discharge in milligrams per liter (mg L^{-1}) [57]; Q_s = background stream flow in mgd or cfs above point of discharge; C_s = background in-stream pollutant concentration in mg L^{-1}; and, Q_r = resultant in-stream flow after discharge in MGD or cfs.

This and similar equations are used by permitting authorities as part of the process in calculating water quality based limits. If a value of zero is used for the ambient concentration for a pollutant (C_s) in the equation, the permitting authority can establish an entire limit on of discharge. The resulting limit would not account for upstream background concentrations of the pollutant [58]. Depending on the pollutant, the background concentration is almost never zero, so the limit would not prevent an in-stream exceedence of the water quality criteria. In addition, some of the released pollutant is likely to come from nonpoint and other sources besides the direct discharge of the effluent, especially seepage from the ash storage system [58]. Thus, handling CCRs requires coordination among water quality, air quality and waste management authorities. Indeed, failures of ash storage systems can involve direct and indirect releases into surface and ground water.

On February 2, 2014, an estimated 39,000 tonnes of coal ash spilled from Duke Energy's Dan River Steam Station's ash pond into the Dan River in Eden, North Carolina. The day after the breach, the North Carolina Department of Environment and Natural Resources, the Virginia Department of Environmental Quality, the U.S. Fish and Wildlife Service and the EPA, along with the utility company, initatied a unified command address the spill, including technical assistance, oversight, data management and environmental sampling and analysis. Other agencies eventually became in involved in a massive effort to protect human health and the environment and guide the cleanup [57]. It was feared that toxic metals and other pollutants would reach drinking water supply intakes. The Dan River basin is also a recreation area for fishing, boating and wildlife-viewing, and provides vital habitat area for fish and wildlife. These include the endangered Roanoke logperch (*Percina rex*) and the James spinymussel (*Pleurobema collina*), along with a species currently under consideration for protection, i.e., the green floater (*Lasmigona subviridis*), a freshwater mussel [59].

On May 22, 2014, the EPA and utility company entered into an Administrative Order on Consent (AOC). Under the requirements of the AOC, the utility removed coal ash that had accumulated at the Schoolfield Dam, Town Creek Sand Bar, and at both the Danville and South Boston water treatment facilities. The removal action was completed in July 2014, and an estimated 4000 cubic yards of coal ash were removed.

Coal contains many metals. After combusting coal, the fly and bottom ash contain much higher concentrations of metals that do not volatilize during combution, so disposing of coal wastes in a pit or other storage system can allow toxic metals and metalloids to migrate to ground water, a major source of drinking water for much of the world. Drinking water quality is threatened if the ash pit or pond that is upgradient from a private and municipal well. As discussed in Chapter 11, water stands in a well at a particular level, known as hydraulic head (h). Ground water flows in the direction of decreasing head. The change in head measurement is the hydraulic gradient, $\frac{\partial h}{\partial x}$ [60]. The capability a fluid to move through underground materials is known as hydraulic conductivity (K). The various strata under the earth's surface may be consolidated rock or unconsolidated materials, like soil, sand or gravel. Most commonly, however, the movement of ground water is through the unconsolidated materials, i.e., porous media. Darcy's Law is an empirically derived equation for the flow of fluids through a cross-section of porous media (A). It states that the specific discharge, Q, is directly proportional to the K and hydraulic gradient:

$$\text{Darcy's Law}: Q = -KA\frac{\partial h}{\partial x} \qquad (18.5)$$

Thus, scientist can estimate the amount of pollutant that will flow through an aquifer and engineers can propose designs to abate the contamination. However, protecting water quality requires a combination of expertise and interests beyond science, engineering, and regulatory actions. It all needs a well-informed public.

[c] A tonne is a metric ton. 1 tonne = 1000 kg = 2205 pounds = 1.10231 U.S. tons.

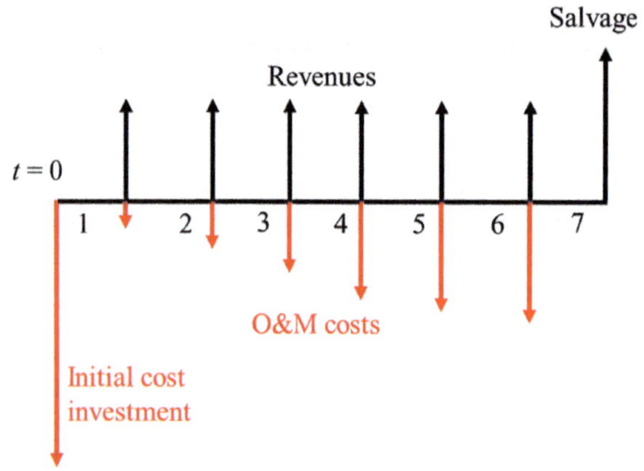

FIG. 18.3 Simple cash flow diagram. In this case, after the initial investment, revenues are constant, whereas operation and maintenance costs are increasing each year. After year 5, O&M costs exceed revenues. After year 7, the project is ended, and salvage is recovered.

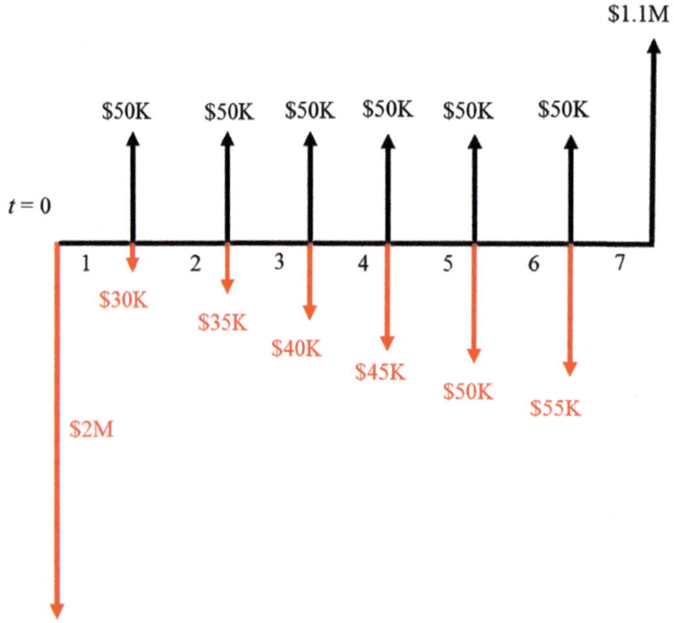

FIG. 18.4 Example cash flow diagram showing an initial $2 million expenditure. Following the initial investment, revenues are constant, whereas operation and maintenance costs are increasing each year. After year 5, O&M costs exceed revenues, so the project is ended, and salvage is recovered.

converter is salvaged, i.e., sold to another town for $110,000. To simplify, the year-end convention is applied, i.e., all expenses and revenue are assumed to occur at the end of the year. The CFD is shown in Fig. 18.4.

6.2 Discounting

Engineering economics dictates equivalence so that all values can be compared consistently. The present worth (P) is the present amount at $t = 0$. Future worth (F) is the equivalent value that will have accumulated after n periods (e.g., years) from P. The annual amount (A) is the uniform amount of money that repeats at the end of each period for n periods. If we do not assume a discrete payout, we also need to have a uniform gradient amount (G), which is the amount of money that repeats at the end of each period, beginning at the end of period 2 and ending at the end of period n.

Therefore, the CFDs can represent present value, future worth, uniform series, uniform gradient series, and exponential gradient series (see Fig. 18.5). Present value and future worth CFDs are used when determining whether a purchase is acceptable and feasible. Uniform series CFDs can be part of multiyear budgets. The two bottom CFDs in Fig. 18.5 may best apply to equipment and systems that demand greater maintenance and replacements. For example, if the reliability

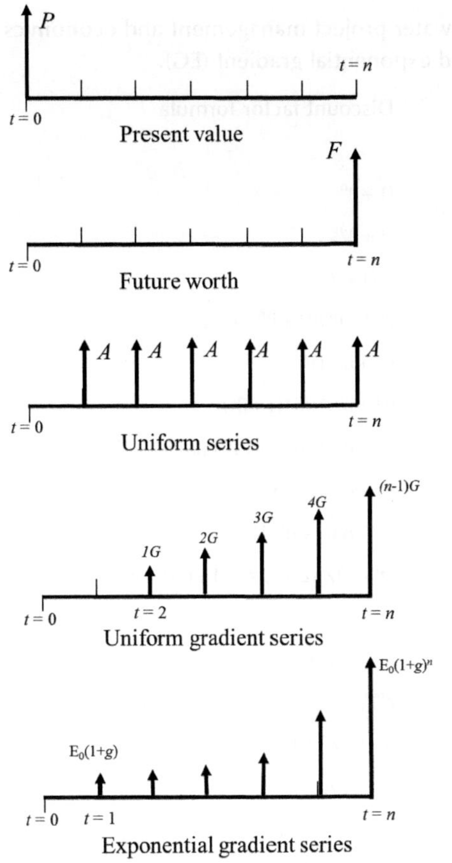

FIG. 18.5 Commonly used cash flow diagrams for public and privately-funded projects.

of a system declines with a linear slope, the uniform gradient series is used, but if the failure rates accelerate with time, the exponential (E) gradient CFD may be preferable.

Another important factor is the discount rate (i) that needs to be paid to borrow money, i.e., the effective rate per time-period. The annual i can be calculated as:

$$i = \left(1 + \frac{r}{k}\right)^{\frac{k}{p}} - 1 \tag{18.6}$$

Where, $r=$ the annual interest; $k=$ number of compounding periods per year, and $p=$ the number of periods per year corresponding to the basis for n. Thus, if payments are made monthly, $p=12$; if daily $p=365$.

The discounting principle states that investing P at $t=0$ will pay F at $t=n$. Thus, the discount formula is:

$$P = F(P/F, i\%, n) \tag{18.7}$$

Where $(P/F, i\%, n)$ is the discount factor.

Table 18.1 provides the discount factor conversion formulae for the most common conversions for P, F, A, G, and E used by waste managers and engineers. Numerous online calculators are available for these conversions, as well as spreadsheet programs, especially Microsoft Excel (see far-right column of Table 18.3). Microsoft software [63], for example provides support for calculating, P, F, and i at:

- *P=PV(i,N,A,F,Type)*: https://support.office.com/en-us/article/PV-function-23879d31-0e02-4321-be01-da16e8168cbd
- *F=FV(i,N,A,P,Type)*: https://support.office.com/en-us/article/FV-function-2eef9f44-a084-4c61-bdd8-4fe4bb1b71b3
- *i=RATE(N,A,P,F,Type,guess)*: https://support.office.com/en-us/article/RATE-function-9f665657-4a7e-4bb7-a030-83fc59e748ce

Where, "Type=0" means end-of-period cash payments; "Type=1" means beginning-of-period payments; and "guess" is the value of the interest rate.

TABLE 18.1 Conversions related to water project management and economics of present value (P), future worth (F), annual amount (A), gradient (G), and exponential gradient (EG).

Conversion	Symbol	Discount factor formula	Discount factor formula in excel
Discrete compounding			
P to F	(F/P,i%,n)	$(1+i)^n$	=FV(i,n,0,-1)
F to P	(P/F,i%,n)	$(1+i)^{-n}$	=PV(i,n,0,-1)
F to A	(A/F,i%,n)	$i/((1+i)^n\text{-}1)$	=PMT(i,n,0,-1)
P to A	(A/P,i%,n)	$i^*(1+i)^n/((1+i)^n\text{-}1)$	=PMT(i,n,-1)
A to F	(F/A,i%,n)	$((1+i)^n\text{-}1)/i$	=FV(i,n,-1)
A to P	(P/A,i%,n)	$((1+i)^n\text{-}1)/(i^*(1+i)^n)$	=PV(i,n,-1)
G to P	(P/G,i%,n)	$((1+i)^n\text{-}1)/(i^{2*}(1+i)^n)\text{-}n/(i^*(1+i)^n)$	{=NPV(i,(ROW(INDIRECT("1:"&n))-1))}
G to F	(F/G,i%,n)	$((1+i)^n\text{-}1)/i^2\text{-}(n/i)$	{=(P/G,i%,n) * (F/P,i%,n)}
G to A	(A/G,i%,n)	$(1/i)\text{-}n/((1+i)^n\text{-}1)$	{=(P/G,i%,n) * (A/P,i%,n)}
EG to P	(P/EG,z-1,n)	$(z^n\text{-}1)/(z^n(z\text{-}1))$, $z=(1+i)/(1+g)$	=PV(z-1,n,-1)
Discrete compounding			
F to P	(P/F,r%,n)	$e^{-r*n}=1/e^{r*n}$	=1/EXP(r*n)
P to F	(F/P,r%,n)	e^{r*n}	=EXP(r*n)
F to A	(A/F,r%,n)	$(e^r\text{-}1)/(e^{r*n}\text{-}1)$	=(EXP(r)-1)/(EXP(r*n)-1)
A to F	(F/A,r%,n)	$(e^{r*n}\text{-}1)/(e^r\text{-}1)$	=(EXP(r*n)-1)/(EXP(r)-1)
P to A	(A/P,r%,n)	$(e^r\text{-}1)/(1\text{-}e^{-r*n})$	=(EXP(r)-1)/(1−1/EXP(r*n))
A to P	(P/A,r%,n)	$(1\text{-}e^{-r*n})/(e^r\text{-}1)$	=(1−1/EXP(r*n))/(EXP(r)-1)

Note: Excel formulae for (F/G,i%,n) and (A/G,i%,n) are based on the algebraic equivalence of F/G=(P/G)*(F/P) and A/G=(P/G)*(A/P). The discount factor symbols (P/G,i%,n), (F/P,i%,n) and (A/P,i%,n) need to be replaced with the appropriate discount factor formula listed in the table.
Data from J. Wittwer, Discount factor table for excel: summary of discount factor formulas for TVM calculations in Excel®. https://www.vertex42.com/ExcelArticles/discount-factors.html, 2018 (Accessed 17 May 2018).

The symbols and nomenclature are not used consistently in engineering and waste management economics. For example, public agencies may also represent present value as *PV*:

$$PV = \frac{B_{y_i}}{(1+r)^{y_i - y_0}} \tag{18.8}$$

Where, B = the annual benefit, r = the discount rate, y_i = the year the benefit occurs, and y_0 = the year to which future dollars are discounted.

Present value is similarly determined for annual cost (C):

$$PV = \frac{C_{y_i}}{(1+r)^{y_i - y_0}} \tag{18.9}$$

Where, y_i = the year the cost occurs.

Present value example

Assume an electric power generating station has to change from an outdated cooling system to a modern cooling tower because the effluent water is heating an adjacent stream. The tower is estimated to cost $5,000,000. Construction and installation will occur in 2029. The year of analysis is 2024. If the discount rate is 3.6% for the year 2024, calcuate the present value of the construction cost.

Given: $AC = \$5{,}000{,}000$; $r = 0.036$, $n = 2029 - 2024 = 5$

Then, $PV = 5{,}000{,}000/(1+0.036)^5$

Therefore, $PV = \dfrac{5{,}000{,}000}{(1+0.036)^5} = \$4{,}189{,}587$ (see Fig. 18.6).

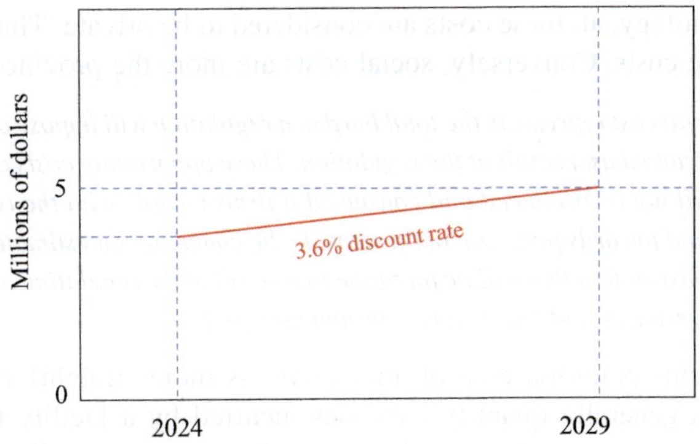

FIG. 18.6 Discount rate for hypothetical water cooling system. *(Data from Minnesota Department of Transportation, Benefit cost analysis for transportation projects. Available: https://www.dot.state.mn.us/planning/program/benefitcost.html, 2012.)*

The $5 million is the present value. If the project can wait until 2029, the money could be put into a bank where it earns interest, or put into other projects where it generates immediate benefits. The amount that needs to be deposited into the bank to have $5 million in 2029, using a 3.6% discount rate, is only $4.2 million—the present value of the cost of the project in 2024. The same logic applies to benefits.

Thus, if financial costs are the only criteria, the company would not change cooling systems in the example. However, large public projects or, in this case, highly regulated utility projects, fill societal needs, including increasing the capacity of existing facilities or systems, and/or improving public health, environmental quality, safety, and aesthetics of neighborhoods and enhancing the performance of other existing facilities or systems. In fact, the benefit-cost and other alternative selection techniques must include fines, penalties, and other monetized and non-monetized costs of discharging hot water into the stream.

The analysis of a large public project is conducted in stages:

1. Planning the analysis and defining its scope.
2. Performing engineering analyses of the alternatives.
3. Calculating the present value of project costs and benefits.
4. Evaluating the results—benefit-cost analysis.

Planning the benefit-cost analysis and performing the engineering analysis (the first two stages) require careful thought. The analysis should capture the appropriate benefits and cost differences between the base case and the identified alternatives [64]. These first two stages are the most complicated and require the most time and effort. The economic calculation stage is a relatively short and straightforward process. Evaluation and interpretation of the results requires judgment and experience. The process of conducting a benefit-cost analysis can be iterative: the process may require returning to a previous stage to verify results and explore subalternatives.

7 Comparing alternatives[d]

In most jurisdictions, the company, utility and other regulated entities have latitude in the strategies that they employ to meet water quality standards, especially for standards that are performance-oriented. Some standards are more prescriptive and technical, but even these are often open to more than one approach on how to comply [65]. This means that expenditures are seldom a one-time expense, usually requiring the comparison of alternatives with expenditures varying in amounts, with spending occuring at multiple times over a project's or equipment's useful life. Not only may the expenditures vary, but the timing of expenditures may also be different. This section addresses the selection of appropriate implementation and funding timeframes, addressing the time value of money, adjusting for prices over time, and selection of the appropriate measure of cost.

7.1 Social costs

It is important to keep in mind that the calculations in this chapter generally apply to monetizable costs and benefits, often the costs borne by a private or quasi-public entity. For instance, if the regulated entity pays for the installation and operation

d. This section includes excerpts from [15].

of a water pollution control technology, all these costs are considered to be private. Thus, selection of alternatives usually rests on minimizing these private costs. Conversely, social costs are more the province of macroeconomics:

> *From a regulatory standpoint, social cost represents the total burden a regulation will impose on the economy. It can be defined as the sum of all opportunity costs incurred as a result of the regulation. These opportunity costs consist of the value lost to society of all the goods and services that will not be produced and consumed if firms comply with the regulation and reallocate resources away from production activities and towards pollution abatement. To be complete, an estimate of social cost should include both the opportunity costs of current consumption that will be foregone as a result of the regulation, and also the losses that may result if the regulation reduces capital investment and thus future consumption [66].*

Comparing private costs of various pollution control alternatives is more straightforward than social cost assessment because private cost comparison generally quantify expenses incurred by a facility to purchase, finance, and operate the equipment. For example, if a governmental jurisdiciton wished to encourage pollution control for a power generation industrial sector, it could provide monetary support, e.g., grants, for 25% of the costs of a treatment system, e.g., scrubber. The private cost for the industry would be reduced to 75% of the cost of a scrubber. Other less direct monetary support strategies have been employed, e.g., eliminating or rebating sales and other taxes, which means that the private cost to the firm is lessened by the total that would otherwise need to include the tax.

The following sections stress private costs, but it is important to keep in mind that the basis for many of the costs are to provide a societal, macroeconomic value, i.e., cleaner water. Thus, the private costs lead to incremental improvements or, if the water is already clean, the maintenance of existing water quality [65].

7.2 Types of cost esimates

Uncertainty is highest during the conceptual phases of a project. Cost estimate accuracy depends on the decision being made. Simply knowing that releases have to be reduced may begin with order-of-magnitude cost estimates for water pollution control projects. For example, chemical engineers and water pollution control engineers [67] consider the order-of-magnitude to have an error between +50% and −30%.

7.3 Proposal review

Rather than simply deciding whether to accept a proposal, it is better to decide on a functional objective and then identify ways to provide that function in the most efficient and effective way. Both benefits and costs vary, so this analysis must find alternative means of maximizing net benefits and, similarly, minimizing costs to gain a desired amount of benefits, This is known as the benefit-cost analysis (BCA), which assumes:

1. Equivalence, i.e., benefits or costs are in the same units in all projects under review. Most often this monetized, e.g., expressed in dollars, pounds, yen, or euros.
2. The public is the "user" or "client" who will benefit from the selected project.
3. The "sponsor" is the level of government (e.g., town, city, county, state, province, nation, or even multinational organizations)

The analysis consists of:

1. Identifying all user benefits (positive outcomes) and "disbenefits" (negative outcomes);
2. Quantification and comparison of these benefits and disbenefits;
3. Identification and quantification of the sponsor's costs;
4. Finding the equivalent net benefits and net costs at the base period, with a discount rate specific to the project; and,
5. Determining that the project "acceptable" if the equivalent user's net benefits exceed the equivalent sponsor's net costs (however, a formal and defendable rule for benefit exceedance should be declared at the outset of the analysis and certainly before the project is accepted, e.g., some percentage above).

Basically, the BCA seeks to show whether the benefits are worth the economic investment. This can be viewed at various points in the design process. For example:

- Project planning: From an economic perspective, is better to install new pollution control equipment or repair and upgrade existing equipment?

- Design and environmental study: From an economic perspective, are the benefits of location "A" worth the project costs? How does location "A" compare to "B" or "C"?
- Construction planning: From an economic perspective, are the benefits of closing all industrial operations during installation of new equipment worth the delays and diversion costs, i.e., compared to keeping some open?

The BCA is best based on risk and reliability. It is more valuable as a tool for comparing and evaluating alternatives than for whether to provide a service. For example, if a risk assessment indicates that water is being contaminated, the BCA would not override the need to clean the water, but would seek the best and most feasible means of doing so. The cleanup would have to meet a risk-based or other environmetnal standard. Thus, the project would have to provide water that meets this standard, and the public works or other manager would decide which of the options for doing this are most cost-effective.

The BCA is a tool for maximizing benefits for a certain amount of costs, e.g., the basis for a water management or public works budget. So, the BCA needs to be more than a financial comparison. Indeed, various methods are used for comparing benefits and costs for a public project. After deciding which alternatives are feasible and viable, different techniques can be applied to compare them, including:

Net Present Value (NPV): NPV is the value resulting from subtracting the expected cost from the expected benefits. Projects having a NPV less than zero should be avoided.

Net present value is an expression of economic efficiency, which is useful for private cost comparisons and for estimating social costs. In the latter case, economic efficiency is the maximization of social welfare. An efficient market allows society to maximize the NPV of benefits, i.e., it is the difference between social benefits and social costs over time. The efficient level of production is "Pareto optimal," i.e., it is not possible to rearrange production or to reallocate goods so that someone is better off without making someone else worse off in the process. Economic efficiency requires only that net benefits be maximized, irrespective of to whom those net benefits accrue. It does not guarantee an "equitable" or "fair" distribution of benefits among consumers and producers, or between subgroups of consumers or producers [66]. This is an example of a societal zero-sum game.

Internal Rate of Return (IRR): The IRR is the discount rate at which the present value of costs equals the present value of benefits. Alternatively, IRR is the discount rate at which NPV = 0 and benefit-cost ratio (BCR) = 1.0. If IRR exceeds the opportunity cost of the capital, the project is economically sound and worth pursuing [2].

Benefit-Cost Ratio (BCR): The BCR is the ratio of the present value of benefits to the present value of costs. A BCR is a gross tool that may serve as a screen and is part of of a larger analysis. In its simplest form, it is ratio of benefits to cost of a project. For example, the city's public works department is asked to participate in an analysis of a proposed landfill. The project receives the scores shown in Table 18.2.

The scores are weighted according to importance, so the most important factors in this case are water quality and closing dumps. Thus, the benefits equal: 15 + 3 + 25 + 12 = 55. The costs equal: 15 + 4 + 6 = 25. Therefore, this benefit-to-cost ratio is

TABLE 18.2 Benefit and cost scores for a hypothetical landfill.

	Score	Weight	Total
Benefits			
Water quality	3	5	15
Gas sales	1	3	3
Removal of open dumps	5	5	25
Customer satisfaction	3	4	12
Costs			
Expenses	5	3	15
Fugitive dust	2	2	4
Odors	3	2	6

55/25 = 2.2. This may indicate that the landfill is acceptable. However, the scores and weightings are highly subjective. Indeed, the engineer from the air quality department was overruled on the scoring and weighting of fugitive dust (she thought they should be 5 and 5, respectively; incresing the cost side of the ratio by 21, i.e., B/C = 1.2). Such ratios are notorious for justifying previously ordained projects.

Ideally, the benefit-cost analysis (BCA) would project and evaluate all possibilities, but this is not possible or often not even desirable because of the large uncertainties. So, project alternatives need to be screen. Expert judgment can be a first screen to differentiate infeasible alternatives from realistic options and compare them to a base case. The base case is not necessarily the "no action" alternative, although, in most BCAs, one or more of the other options have substantial improvements over the base case. Linear programming and other optimization techniques can be used to quantify the economic feasibility of a project from a life cycle/systems perspective. Public projects must also differentiate economic costs and benefits from other societal costs and benefits. Thus, the BCA differs from an economic impact analysis (EIA). The U.S. Department of Transportation, for example, differentiates the two processes [68]:

> *A BCA measures the value of a project's benefits and costs to society, while an economic impact analysis measures the impact of increased economic activity within a region. Common metrics for measuring economic impacts include retail spending, business activity, tax revenues, jobs, and property values. Economic impact analyses often take a strictly positive view, (i.e., increased jobs, spending) and do not examine how the resources used for a project might have benefitted alternative societal uses of the resources (i.e., they do not assess the net effect on society).*

In this sense, an EIA would consider the initial investment in infrastructure as a stimulus to the local economy, as opposed to a cost to the local government. Also, economic impact analyses are smaller scale than the usual BCA, which usually consider broad, economy-wide and "societal" benefits and costs. Effects in one region may be offset by losses in a neighboring region, reflecting a transfer of spending or jobs that may be a net neutral summation. Thus, the two local jurisdictions would have different economic impact results, but the BCA would not change, since its focus is larger.

8 Replacement cost analysis

As any experienced engineer, plant operator, and manager can attest, things wear out. The need for replacement and repair will result; it is merely a matter of time. Often replacement applies to a "supply," i.e., something that is expected to be used and changed frequently. Repair is often seen as returning "capital equipment" to its designed efficiency.[e]

These are semantical distinctions. Of course, any piece of a system eventually needs to be replaced, including capital equipment. This is not always due to wear. Sometimes, the function has changed, e.g., labor-intensive sorting equipment may be fine until a town's population reaches a threshold where the equipment needs to be replaced by higher-capacity, automated equipment. Sometimes, there is a safety issue, e.g., the older equipment may have met previous worker safety regulations, but no longer does as safety rules change. Often, however, economics drives the decision, e.g., the mean time to repair (MTTR) may indicate the time to replace equipment. Any type of asset eventually loses value and must be replaced due to:

- Obsolesce: Fast for information technology, slower for mechanical equipment
- Depletion: Supplies
- Deterioration: Mechanical equipment

Replacement cost analysis (RCA) uses specific terminology. The "defender" is the existing piece. The "challenger" is the best available replacement equipment. Current market value is the selling price of the defender in the marketplace.

There are a few ways to compare useful life, but a popular one is the equivalent uniform annual cost (EUAC).

---

e. In government agencies, supplies are ordered routinely and are much lower cost that capital equipment. Capital equipment can be maintained periodically with subscription costs. Depending on the enterprise, capital equipment is not necessarily repaired when it fails. For example, if a computer fails after its warranty period, it is likely that there have been numerous upgrades to the failed device, so it would likely be replaced with the newer version. Conversely, a large pump is also capital equipment, but if it fails after the warranty period, it is likely to be repaired. Of course, the specific capital equipment dictates when repair or replacement is justified.

The RCA is done in steps:

1. Identify alternatives (reasonable and available), at least two: Defender and Best challenger.
2. Determine whether good defender marginal cost data are available.
3. If data are available, see if the marginal costs are increasing.
4. If increasing, then compare defender's next year marginal cost with challenger's EUAC. (Use Analysis Technique 1, described below)
5. If not increasing, then must find lowest EUAC for defender and compare it to the challenger's EUAC at its minimum cost life. (Use Analysis Technique 2, discussed below)
6. If defender marginal cost data not available, must find EUAC over given equipment life. (Use Analysis Technique 3; described below)
7. Compare defender's EUAC over its remaining life to the challenger's EUAC at its minimum cost life.
8. Calculate the EUAC for each value of the useful life (e.g., $n=1$, $n=2$, $n=3$, etc.).
9. Time period at which the EUAC is minimized is the minimum cost life (economic useful life)

Replacement cost analysis example

Assuming the conditions in Table 18.3, the engineer must decide how to address failing equipment.

Goal: Narrow replacement down to one challenger.

Approach: Evaluate each candidate challenger at its minimum cost life.

This can be done using the previously discussed alternative evaluation approaches, i.e., NPV, EUAC, IRR or BCR.

Next, find minimum cost life. An asset can have several possible lives, depending on frequency of asset replacement. The life that minimizes the asset's EUAC is known as the asset's minimum cost life.

Example:

Initial cost (P) is \$7500, arithmetic gradient maintenance cost (G) is \$900, uniform cost (A) is \$500, and arithmetic gradient operating cost (G) is \$400, at an 8% interest rate.

There is no salvage value.

The "n" is changing depending on the year, on an arithmetic gradient, i.e., it is changing each year ($n=1$; $n=2$, ...).

The 15-year spreadsheet is shown in Table 18.4.

The second column can be constructed in Excel as \$7500(A/P,8%,n). Third column: \$900(A/P,8%,n). Fourth column: \$500 + 400 (A/P,8%,n). The fifth column is the sum of the values in columns 2, 3 and 4.

Thus, the lowest total EUAC is year 4, so this is the economic useful life and minimum cost life is when $n=4$ years (see Fig. 18.7):

Next, find the year-by-year costs for keeping an asset (i.e., marginal costs). This includes loss in value of the asset for keeping it (usually per year) and the lost interest from tying up funds in this asset. The costs are directly related to the asset, e.g., expenses like insurance and O&M.

An example of marginal cost comparison is shown in Table 18.5.

In this example, the marginal costs of the defender constantly increasing, but the marginal costs of the challenger decrease for 3 years, then start to increase. By year 3, the challenger marginal costs become less than the defenders and stay that way until the end of the defender's useful life (year 5).

TABLE 18.3 Defender and challenger example from P.R. Rosenkrantz, Cal Poly Pomona [62].

	Where		Compare	
	Defender marginal cost	Defender marginal cost	Defender	Best challenger
1	Available	Increasing	Next-year Marginal Cost	EUAC at Minimum Cost Life
2	Available	Not increasing	EUAC at Minimum Cost Life	EUAC at Minimum Cost Life
3	Not available		EUAC over Remaining Useful Life	EUAC at Minimum Cost Life

TABLE 18.4 Spreadsheet for equivalent uniform annual cost (EUAC) example from P.R. Rosenkrantz, California Polytechnic—Pomona [62].

Year	EUAC of Capital Recovery Costs	EUAC of Maintenance and Repair Costs	EUAC of Operating Costs	EUAC Total	Interest rate
Initial year	-7500	0	-500		8%
Arithmetic gradient		-900	-400		
1	$8,100.00	$0.00	$500.00	$8,600.00	
2	$4,205.77	$432.69	$692.31	$5,330.77	
3	$2,910.25	$853.87	$879.50	$4,643.62	
4	$2,264.41	$1,263.56	$1,061.58	$4,589.55	<-----MIN
5	$1,878.42	$1,661.82	$1,238.59	$4,778.84	
6	$1,622.37	$2,048.71	$1,410.54	$5,081.62	
7	$1,440.54	$2,424.30	$1,577.47	$5,442.31	
8	$1,305.11	$2,788.67	$1,739.41	$5,833.19	
9	$1,200.60	$3,141.93	$1,896.41	$6,238.94	
10	$1,117.72	$3,484.18	$2,048.53	$6,650.43	
11	$1,050.57	$3,815.55	$2,195.80	$7,061.93	
12	$995.21	$4,136.17	$2,338.30	$7,469.68	
13	$948.91	$4,446.19	$2,476.08	$7,871.18	
14	$909.73	$4,745.75	$2,609.22	$8,264.69	
15	$876.22	$5,035.01	$2,737.78	$8,649.02	

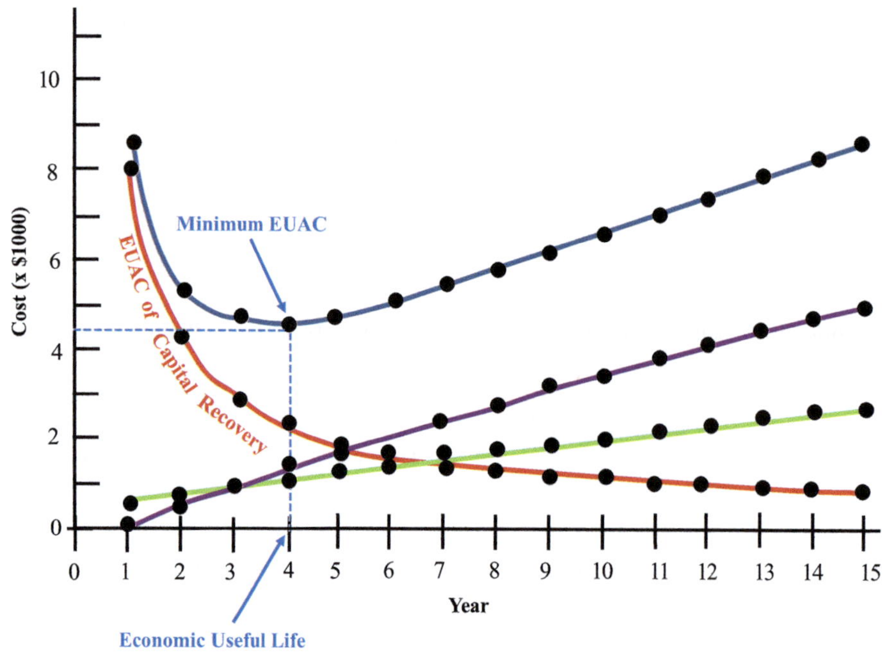

FIG. 18.7 Minimum equivalent uniform annual cost (EUAC) used to find the economic useful life in example from P.R. Data from Rosenkrantz, California Polytechnic—Pomona [62].

TABLE 18.5 Hypothetical cost estimates for new item versus defender item.

New item:

Expense	Cost	Years	Incremental cost thereafter
Investment Cost	$ 25,000.00	0	$ -
Annual O&M Cost	$ 2,000.00	1	$ 500.00
Annual Insurance Cost	$ 5,000.00	3	$ 1,500.00
Useful Life (years)	7		
MARR	15%		

Year	Market Value	Loss in Market Value	Foregone Interest	O&M Cost	Insurance Cost	Total Marginal Cost
0	$ 25,000.00					
1	$ 18,000.00	$ 7,000.00	$ 3,750.00	$ 2,000.00	$ 5,000.00	$ 17,750.00
2	$ 13,000.00	$ 5,000.00	$ 2,700.00	$ 2,500.00	$ 5,000.00	$ 15,200.00
3	$ 9,000.00	$ 4,000.00	$ 1,950.00	$ 3,000.00	$ 5,000.00	$ 13,950.00
4	$ 6,000.00	$ 3,000.00	$ 1,350.00	$ 3,500.00	$ 6,500.00	$ 14,350.00
5	$ 4,000.00	$ 2,000.00	$ 900.00	$ 4,000.00	$ 8,000.00	$ 14,900.00
6	$ 3,000.00	$ 1,000.00	$ 600.00	$ 4,500.00	$ 9,500.00	$ 15,600.00
7	$ 2,500.00	$ 500.00	$ 450.00	$ 5,000.00	$ 11,000.00	$ 16,950.00

Defender:

Expense	Cost	Years	Incremental cost thereafter
Investment Cost	$ 15,000.00	0	$ -
Annual O&M Cost	$ 10,000.00	1	$ 1,500.00
Annual Insurance Cost	$ -		$ -
Useful Life (years)	5		
MARR	15%		

Year	Market Value	Loss in Market Value	Foregone Interest	O&M Cost	Insurance Cost	Total Marginal Cost
0	$ 15,000.00					
1	$ 14,000.00	$ 1,000.00	$ 2,250.00	$ 10,000.00	$ -	$ 13,250.00
2	$ 13,000.00	$ 1,000.00	$ 2,100.00	$ 11,500.00	$ -	$ 14,600.00
3	$ 12,000.00	$ 1,000.00	$ 1,950.00	$ 13,000.00	$ -	$ 15,950.00
4	$ 11,000.00	$ 1,000.00	$ 1,800.00	$ 14,500.00	$ -	$ 17,300.00
5	$ 10,000.00	$ 1,000.00	$ 1,650.00	$ 16,000.00	$ -	$ 18,650.00
6						
7						

8.1 RCA techniques

Technique 1: If the defender marginal costs are increasing, replace when the marginal costs of ownership is more than the EUAC of the challenger. This is appropriate approach when replacement repeatability assumptions hold:

- Best challenger is available for all subsequent years and its costs will remain unchanged.
- The period of service is infinite.
- If these are too restrictive, a more realistic, known, and limited project life can be assumed.

Technique 2: If the replacement repeatability holds, compare the EUAC of the defender asset at its minimum cost life against the EUAC of the challenger at its minimum cost life.

Technique 3: Compare the defender's EUAC over the stated life against the minimum EUAC of the challenger. There may be a problem in defining first costs of the defender and challenger. Trade-in value is not suitable; market value is appropriate.

Any cost analysis is incomplete if it neglects tax. Thus, all relevant taxes will affect the remaining life of the defender, the economic life of the challenger, and defender versus challenger comparisons. After-tax replacement depends on the depreciation Modified Accelerated Cost Recovery System (MACRS) and the assets market over time [62].

9 Life cycle comparisons

The project evaluation techniques so far have helped the engineer or manager select from finite alternatives. Another aspect of evaluation is improving existing systems to provide the needed function or to eliminate problems that weaken that alternative. In life cycle assessment (LCA), alternatives assessment is a formal technique to characterize the hazards of substances and processes based on a full range of human health and environmental information [69]. Changes in the life cycle made based on these assessments ameliorate the potential for unintended consequences *a priori*.

The current approach for alternatives assessment of toxic chemicals is shown in Table 18.6, goes beyond hazard alternatives to address benefits and utility provided by a treatment option, i.e., steps in LCA stages may require helium (He), which is becoming less available. A thorough alternative assessment may identify steps where a different, more renewable chemical or even use a nonchemical technique to perform the same function. Thus, the steps in Table 18.6 can be applied to other substitutions and replacements, e.g., energy use, labor and skill mix, and social justice.

TABLE 18.6 Key steps in alternatives assessment for substituting toxic chemical compounds in the life cycle of a process.

Process	Action	Application to finite resources
Step 1: determine feasibility	Consider whether alternatives: 1. are commercially available and cost effective 2. have the potential for an improved health and environmental profile 3. are likely to result in lasting change and 4. to what extent stakeholders are interested	Consider whether the goods and services can be provided with lesser amounts of these resources
Step 2: collect information	Find data about and consider: 1. how well-characterized the possible alternatives are 2. how much is known about the chemical manufacturing process 3. the range of functional uses that the chemical serves 4. the feedstock of contaminants and residuals from the production process and 5. the work of others who are exploring alternatives for the chemical of concern, similar chemical compounds and functional uses Based on analysis of this information and preliminary stakeholder consultation, develop a proposed project scope and an approach for developing the alternatives assessment	Same, except collect market and production information

TABLE 18.6 Key steps in alternatives assessment for substituting toxic chemical compounds in the life cycle of a process—cont'd

Process	Action	Application to finite resources
Step 3: convene stakeholders	Stakeholders are drawn from entire supply-chain and all life-cycle stages of the chemical of concern. Involvement throughout the project helps to ensure that stakeholders contribute to, understand and support the outcome, enhancing credibility and promoting adoption of the safer alternatives. Typical stakeholders include: chemical manufacturers; product manufacturers; nongovernmental organizations; government agencies; academics; retailers; consumers; and waste and recycling companies. Chemical and technology innovators are critical members of the group	Same process, but include additional stakeholders, e.g., from similar industries
Step 4: identify viable alternatives	Collect information about viability on a range of potential alternatives. The focus is on finding alternatives that are functional with minimum disruption to the manufacturing process. To identify the most likely alternatives, may also include viability demonstrations by chemical and product manufacturers	Same
Step 5: conduct hazard assessment	Based on the best data that are available from the literature or can be modeled, assign and adjective descriptor of hazard concern level of high, moderate or low for each alternative across a range of endpoints, including acute and repeated dose toxicity, carcinogenicity and mutagenicity, reproductive and developmental toxicity, neurotoxicity, sensitization and irritation, acute and chronic aquatic toxicity, and persistence and bioaccumulation. In addition, we provide a qualitative description of potential endocrine activity	Rather than an assessment solely of potential hazards, include potential substitutions and utility of these substitutions in terms of providing the same goods and services as current processes

Data from D.A. Vallero, Translating Diverse Environmental Data Into Reliable Information: How to Coordinate Evidence From Different Sources, Academic Press, 2017; data for two left columns from U.S. Environmental Protection Agency, Design for the environment alternatives assessments. https://www.epa.gov/saferchoice/design-environment-alternatives-assessments, 2017 (Accessed 15 May 2017); right column D.A. Vallero, T.M. Letcher, Unraveling Environmental Disasters, Newnes, 2012. D.A. Vallero, T.M. Letcher, Engineering risks and failures: lessons learned from environmental disasters, Leadersh. Manag. Eng. 12 (4) (2012) 199–209.

Water quality managers and engineers increasingly apply systems thinking and seek greener, more sustainable approaches. They still rely heavily on proven pollution control, measurement, and models, but are increasing adding systems tools like design for the environment (DfE), design for disassembly (DfD), human exposure modeling, and multi-criteria decision analysis (MCDA). One of the means of achieving this is by expert elicitation, where professional judgment is used as a first step in screening for potential exposure and risk of large numbers of chemical compounds that are either pollutants or which may be transformed into pollutants [70].

Systems thinking begins with an appreciation that science and engineering are changing to address water quality needs. For example, personal exposure to most chemical compounds predominantly occurs within near-field exposure scenarios, e.g., exposure while using a product that contains the chemical or contact with an article treated with the chemical [71]. This is leading to new screening models and tools, including MCDA and expert elicitation to prioritize pollutants and find ways to address them [72].

Systems tools, like MCDA, allow for the consideration of many variables from various information sources. For near-field pollution, the two main categories of variables are: (1) those associated with product use; and (2) the physical and chemical properties of the pollutant.

In LCAs, human health characterization factors (CFs) have benefited from improved toxicity information [73]. The CFs are weighting factors used in the life cycle impact assessment (LCIA). A water pollution LCIA converts pollutant releases into impact scores for various impact categories [74]. An impact score is a weighted sum of the damage due to all water pollutant releases to one of these or other impact categories [75]. LCAs generally and LCIAs specifically are important tools for taking a systematic view of the costs and benefits of an entire process rather than a single stage of the life cycle. For water pollution, LCAs allow the engineer or process designer to compare among various alternatives. For example, a chemical compound may appear to be the best choice for manufacturing of product, but the LCIA may indicate that it will produce a toxic pollutant in later stages, requiring expensive treatment that adds costs and risks that can be prevented by changes in process design or selection of safer chemical compounds. The substitued compounds may obviate, or greatly reduce treatment costs. Likewise, the LCA may indicate that a choice of a substance may increase the potential exposure to other pollutants in earlier stages, such as those resulting from the extraction of ores, which may not be necessary if another substance with less extraction-related pollution is chosen.

Until recently, LCA's have underweighted exposure and relied more heavily on inherent toxicity estimates to indicate risks posed by substances in the human health CFs [76]. Knowing the toxicity of a pollutant is necessary, but not sufficient, for decision making. Reliable exposure predictions are also needed. Unlike the inherent physical and chemical properties of a pollutant, exposure predictions rely on numerous factors that differ among pollutants. These go beyond physical and biological sciences and incorporate the social sciences, e.g., the use of products or daily activities that increase exposures [70,77].

Alternative assessment and other life cycle approaches could supplant many of the single-approach, unsustainable approaches with more systematic, sustainable solutions. Note that Table 18.6 calls for stakeholder involvement and risk communication. For example, using expert elicitation techniques with other jurisdictions, universities, and governance to identify ways to eliminate the use of toxic chemical compounds before they reach the waste stream [72].

Although expert judgment is almost always applied at the outset of project evaluation, e.g., deciding which alternatives are not worthy of further scientific investigation, it is also useful as the evaluation continues. Project evaluation can be a protracted, lengthy, and iterative process. Indeed, there may come a point where decision makers ask the engineer or manager to "settle" on a less than optimal approach. In addition, after substantial time and psychic investment in narrowing down the alternative list to a few, well-understood options, it may be difficult to be completely objective. This may be a good time to bring in someone "cold" who can view the options more systematically and objectively and to advise whether the original functions and goals are being met.

Comparing water pollution control projects exists in two distinct domains: scientific credibility and economic feasibility. The chapters up to this one have addressed the former. This chapter has considered the ways to do the latter and to incorporate economic efficiencies with considerations of risk, reliability, and resilience. Economic acceptability is useless if a project fails to meet standards of performance based on public health, environmental quality, and societal acceptance.

9.1 Life cycle costing

Life cycle costing (LCC) focuses on financial flows for the production and consumption of goods and services. LCC sums the total costs of a product, process or activity discounted over its lifetime. It is associated with all costs, from purchase to disposal, and water quality and other environmental costs, i.e., externalities. As mentioned, the purchase price never comprises the full costs or even the highest costs of a product or process over its entire life cycle.

A robust LCC framework connects the life cycle assessment (LCA) to the monetary costing used by decision makers. Thus, sustainability and green purchasing and operations are built into alternative selections and other cost decisions. Combining the features of the LCA and LCC gives the proverbial triple bottom line, i.e., profit, people, and planet [78–81]. Regarding water quality LCCs, perhaps the order should be reversed, i.e., planet, people, and profit, especially for public works and environmental institutions. However, if prepared and followed correctly, all three should be mutually acceptable and achievable. After all, water pollution is the result of a financial, public health, and environmental wastes and inefficiencies.

LCC predates LCA, beginning in 1933 when an the Government Accounting (now Accountability) Office (GAO) purchased tractors. It was obvious 90 years ago to GAO that the purchase was merely the first cost that would be followed by continuous and new costs [82].

The LCC maps costs with each LCA stage, from extraction of ores to end-of-life actions like disposal and recycling. For water projects, it begins with planning to drilling wells and designing water supplies to constuction of facilities, access roads, sewers, distribution systems, treatment operations, sludge collection, disposal and reuse, and decommissioning and planning replacements of all these steps.

The LCC process follows the same costing steps discussed above, including cash flow analysis, discounting, and replacement cost analysis. However, these are conducted comprehensively and systematically for the entire life cycle of every LCA stage, rather than as stand-alone, project-specific cost analyses. When calculating the life cycle cost of a project to compare the proposed options and choose the optimal option among them, the discount rate used should be calculated for all options at a common base time [83,84].

10 Conclusions

Environmental scientists, engineers, managers, and decision makers have important roles in preventing and controlling the pollution of water below, on, and above the earth' s surface. They are accountable for ensuring clean water. For whatever water quality technologies and processes that are selected, purchased, operated, maintained, and decommisioned, each component and every process must be sufficiently available, maintainable, and safe [18,85–87]. Reliability and resilience engineering must ensure that the life cycle is successful in providing and protecting water quality and that even a seemingly insignificant item's failure does not threaten the entire system [15]. This approach includes selecting actions and projects not only based scientific soundness and relevance, but also financial and technical feasibility. This chapter provides tools for conducting the necessary analyses to address these objectives.

References

[1] L.D. James, Engineering economics revisited, Water Resources IMPACT 16 (3) (2014) 9–11. https://www.jstor.org/stable/wateresoimpa.16.3.0009. (Accessed 18 July 2024).

[2] D.A. Vallero, Economics and feasibility assessments, in: D.A. Vallero (Ed.), Environmental Systems Science, Elsevier, 2021, pp. 509–542 (Chapter 11).

[3] A.B. Hill, The environment and disease: association or causation? Proc. R. Soc. Med. 58 (5) (1965) 295.

[4] T.S. Kuhn, D. Hawkins, The structure of scientific revolutions, Am. J. Phys. 31 (7) (1963) 554–555.

[5] C.B. Pert, Molecules of Emotion: The Science Behind Mind-Body Medicine, Simon and Schuster, 2010.

[6] M.T. Cronin, T.W. Schultz, Pitfalls in QSAR, J. Mol. Struct.-THEOCHEM 622 (1–2) (2003) 39–51, https://doi.org/10.1016/S0166-1280(02)00616-4.

[7] F. Li, X. Wang, Y. Teng, Chapter 38—QSAR models in marine ecotoxicology and risk assessment, in: QSAR in Safety Evaluation and Risk Assessment, Academic Press, Amsterdam, NV, 2023, pp. 523–533, https://doi.org/10.1016/B978-0-443-15339-6.00027-8. ISBN: 978-0-443-15339-6, In this issue.

[8] A. Davison, Machine learning and theological traditions of analogy, Mod. Theol. 37 (2) (2021) 254–274.

[9] V. Claveau, M.-C. L'Homme, Structuring terminology using analogy-based machine learning, in: Proceedings of the 7th International Conference on Terminology and Knowledge Engineering, TKE, Citeseer, 2005, pp. 17–18.

[10] A. Salski, Fuzzy knowledge-based models in ecological research, Ecol. Model. 63 (1–4) (1992) 103–112.

[11] E. Cameron, G.F. Peloso, Risk management and the precautionary principle: a fuzzy logic model, Risk Anal. 25 (4) (2005) 901–911.

[12] F.-L. Fan, J. Xiong, M. Li, G. Wang, On interpretability of artificial neural networks: a survey, IEEE Trans. Radiat. Plasma Med. Sci. 5 (6) (2021) 741–760.

[13] A. Birolini, Reliability Engineering: Theory and Practice, Springer, 2017.

[14] M. Modarres, M.P. Kaminskiy, V. Krivtsov, Reliability Engineering and Risk Analysis: A Practical Guide, CRC Press, 2016.

[15] D.A. Vallero, Air Pollution Calculations: Quantifying Pollutant Formation, Transport, Transformation, Fate and Risks, Elsevier, 2019.

[16] P.R. Schulman, Reliability, uncertainty and the management of error: new perspectives in the COVID-19 era, J. Conting. Crisis Man. 30 (1) (2022) 92–101. https://doi.org/10.1111/1468-5973.12356.

[17] S. Jilke, Measuring technological uncertainty and technological complexity: scale development and an assessment of reliability and validity, Int. J. Innov. Sci. 13 (3) (2021) 381–400.

[18] E. Zio, Reliability engineering: old problems and new challenges, Reliab. Eng. Syst. Saf. 94 (2) (2009) 125–141.

[19] J.J. Peirce, Lecture: Hazardous waste engineering, in: D. Vallero (Ed.), Universally Correct Answer to Any Engineering Question: "It depends", Duke University, Durham, North Carolina, 1996.

[20] D.A. Vallero, T.M. Letcher, Unraveling Environmental Disasters, Newnes, 2012.

[21] Y. Xu, J. Li, Q. Tan, A.L. Peters, C. Yang, Global status of recycling waste solar panels: a review, Waste Manag. 75 (2018) 450–458.

[22] D. Vallero, Paradigms Lost: Learning From Environmental Mistakes, Mishaps and Misdeeds, Butterworth-Heinemann, 2005.

[23] A. Misuri, V. Cozzani, A paradigm shift in the assessment of Natech scenarios in chemical and process facilities, Process Saf. Environ. Prot. 152 (2021) 338–351.

[24] T. Kuhn, What Are Scientific Revolutions, MIT, Boston, 1981.

[25] D.A. Vallero, Translating Diverse Environmental Data Into Reliable Information: How to Coordinate Evidence From Different Sources, Academic Press, 2017.

[26] D.A. Vallero, T.M. Letcher, Engineering risks and failures: lessons learned from environmental disasters, Leadersh. Manag. Eng. 12 (4) (2012) 199–209.

[27] M. Eti, S. Ogaji, S. Probert, Integrating reliability, availability, maintainability and supportability with risk analysis for improved operation of the Afam thermal power-station, Appl. Energy 84 (2) (2007) 202–221.

[28] L. Sikos, J. Klemeš, RAMS contribution to efficient waste minimisation and management, J. Clean. Prod. 17 (10) (2009) 932–939.

[29] T.E. Graedel, On the concept of industrial ecology, Annu. Rev. Energy Environ. 21 (1) (1996) 69–98.

[30] R. Smith, D. Sengupta, S. Takkellapati, C. Lee, An industrial ecology approach to municipal solid waste management: I. Methodology, Resour. Conserv. Recycl. 104 (2015) 311–316.

[31] R. Smith, D. Sengupta, S. Takkellapati, C. Lee, An industrial ecology approach to municipal solid waste management: II. Case studies for recovering energy from the organic fraction of MSW, Resour. Conserv. Recycl. 104 (2015) 317–326.

[32] D.A. Vallero, Resilience and uncertainty in engineering decisions, Curr. Trends Civil Struct. Eng. 2643-6876, (2019), https://doi.org/10.33552/CTCSE.2019.03.000568.

[33] T. Elmqvist, et al., Response diversity, ecosystem change, and resilience, Front. Ecol. Environ. 1 (9) (2003) 488–494.

[34] E. Hollnagel, D.D. Woods, N. Leveson, Resilience Engineering: Concepts and Precepts, Ashgate Publishing, Ltd., 2007.

[35] J.D. Solomon, Communicating Reliability, Risk and Resiliency to Decision Makers, Duke University, 2017.

[36] J.D. Solomon, D.A. Vallero, From our partners—communicating risk and resiliency: special considerations for rare events, Center for Infrastructure Protection & Homeland Security, George Mason University, 2017. https://cip.gmu.edu/2016/06/01/partners-communicating-risk-resiliency-special-considerations-rare-events/. (Accessed 1 May 2017).

[37] H.R. Heinimann, A generic framework for resilience assessment. An edited collection of authored pieces comparing, contrasting, and integrating risk and resilience with an emphasis on ways to measure resilience, 2016, p. 90.

[38] P. Helm, Risk and resilience: strategies for security, Civ. Eng. Environ. Syst. 32 (1–2) (2015) 100–118.

[39] D. Vallero, Environmental Biotechnology: A Biosystems Approach, Elsevier Science, 2015.

[40] E. Paoletti, et al., Advances of air pollution science: from forest decline to multiple-stress effects on forest ecosystem services, Environ. Pollut. 158 (6) (2010) 1986–1989.

[41] J. Fiksel, I. Goodman, A. Hecht, Resilience: navigating toward a sustainable future, Solutions 5 (5) (2014) 38–47.

[42] M.Y. Reja, S.D. Brody, W.E. Highfield, G.D. Newman, Understanding the notion between resiliency and recovery through a spatial-temporal analysis of section 404 wetland alteration permits before and after hurricane Ike, World Acad. Sci. Eng. Technol. Int. J. Environ. Chem. Ecol. Geol. Geophys. Eng. 11 (4) (2017) 348–355.

[43] B. Minsker, et al., Progress and recommendations for advancing performance-based sustainable and resilient infrastructure design, J. Water Resour. Plan. Manag. 141 (12) (2015) A4015006.

[44] M. Dardonville, C. Bockstaller, O. Therond, Review of quantitative evaluations of the resilience, vulnerability, robustness and adaptive capacity of temperate agricultural systems, J. Clean. Prod. 286 (2021) 125456.

[45] J.K. Vanos, J.W. Baldwin, O. Jay, K.L. Ebi, Simplicity lacks robustness when projecting heat-health outcomes in a changing climate, Nat. Commun. 11 (1) (2020) 6079.

[46] J.M. Anderies, O. Barreteau, U. Brady, Refining the robustness of social-ecological systems framework for comparative analysis of coastal system adaptation to global change, Reg. Environ. Change 19 (2019) 1891–1908.

[47] J. Jäger, et al., Assessing policy robustness of climate change adaptation measures across sectors and scenarios, Clim. Change 128 (2015) 395–407.

[48] R. Wilby, S. Dessai, Robust adaptation to climate change, Weather 65 (7) (2010) 180–185, https://doi.org/10.1002/wea.543. In this issue.

[49] S. Dessai, M. Hulme, Assessing the robustness of adaptation decisions to climate change uncertainties: a case study on water resources management in the East of England, Glob. Environ. Change 17 (1) (2007) 59–72.

[50] J.A. Dunne, R.J. Williams, N.D. Martinez, Network structure and biodiversity loss in food webs: robustness increases with connectance, Ecol. Lett. 5 (4) (2002) 558–567.

[51] F.R. Siegel, Adaptations of Coastal Cities to Global Warming, Sea Level Rise, Climate Change and Endemic Hazards, Springer, 2019.

[52] C. Mellin, et al., Safeguarding nutrients from coral reefs under climate change, Nat. Ecol. Evol. 6 (12) (2022) 1808–1817.

[53] A. Sánchez-Arcilla, M. García-León, V. Gracia, R. Devoy, A. Stanica, J. Gault, Managing coastal environments under climate change: pathways to adaptation, Sci. Total Environ. 572 (2016) 1336–1352.

[54] C. Whitbeck, Ethics as design: doing justice to moral problems, Hastings Cent. Rep. 26 (3) (1996) 9–16.

[55] National Society of Professional Engineers, NSPE Code of Ethics for Engineers, 2016. https://www.nspe.org/resources/ethics/code-ethics. (Accessed 31 August 2018).

[56] J.S. Harkness, B. Sulkin, A. Vengosh, Evidence for coal ash ponds leaking in the southeastern United States, Environ. Sci. Technol. 50 (12) (2016) 6583–6592.

[57] U.S. Environmental Protection Agency, Coal ash (coal combustion residuals, or CCR), 2024. https://www.epa.gov/coalash. (Accessed 18 July 2024. Last update on June 25, 2024.

[58] U.S. Environmental Protection Agency, Attachment B: water quality-based effluent limits—coal combustion waste impoundments, U.S. Environmental Protection Agency, 2017 (Accessed 7 March 2018).

[59] U.S. Fish and Wildlife Service, Natural resource damage assessment and restoration for the Dan River coal ash spill, 2018. https://files.nc.gov/ncdeq/Coal%20Ash/documents/Coal%20Ash/20140609%20FINAL%20Dan%20River%20cooperative%20NRDAR%20factsheet.pdf. (Accessed 8 March 2018).

[60] North Carolina Department of Environment and Natural Resources, Groundwater, aquifers & confining beds, 2024. https://www.ncwater.org/?page=560. (Accessed 18 July 2024).

[61] T. Brown, Engineering Economics and Economic Design for Process Engineers, CRC Press, 2016.

[62] P.R. Rosenkrantz, Replacement, 2018. https://www.cpp.edu/~rosenkrantz/egr403site/egr403online/egr403_sv15_chapter12.ppt. (Accessed 17 May 2018; Chapter 12).

[63] Microsoft Support Office, Financial functions (reference), 2024. https://support.office.com/en-us/article/financial-functions-reference-5658d81e-6035-4f24-89c1-fbf124c2b1d8?ui=en-US&rs=en-US&ad=US. (Accessed 18 July 2024).
[64] Minnesota Department of Transportation, Benefit cost analysis for transportation projects, 2012, Available: https://www.dot.state.mn.us/planning/program/benefitcost.html.
[65] J.L. Sorrels, T.G. Walton, Cost estimation: concepts and methodology, in: EPA Air Pollution Control Cost Manual, vol. 27711, Office of Air Quality Planning and Standards, U.S. Environmental Protection Agency, Research Triangle Park, North Carolina, 2017 (Chapter 2).
[66] U.S. EPA, Guidelines for preparing economic analyses, EPA 240-R-00-003, US Environmental Protection Agency, Washington, DC, 2014. Available: https://www.epa.gov/sites/production/files/2017-08/documents/ee-0568-50.pdf. (Accessed 27 August 2018).
[67] K.B. Schnelle Jr., R.F. Dunn, M.E. Ternes, Air Pollution Control Technology Handbook, CRC Press, Boca Raton, Florida, 2015.
[68] U.S. Department of Transportation, Benefit-cost analysis guidance for TIGER and INFRA applications, 2017, Available: https://www.transportation.gov/sites/dot.gov/files/docs/mission/office-policy/transportation-policy/284031/benefit-cost-analysis-guidance-2017_1.pdf.
[69] C.P. Weis, The value of alternatives assessment, Environ. Health Perspect. 124 (2016) A40.
[70] D.A. Vallero, Air pollution monitoring changes to accompany the transition from a control to a systems focus, Sustainability 8 (12) (2016) 1216.
[71] J.F. Wambaugh, et al., High-throughput models for exposure-based chemical prioritization in the ExpoCast project, Environ. Sci. Technol. 47 (15) (2013) 8479–8488.
[72] M.D. Wood, K. Plourde, S. Larkin, P.P. Egeghy, A.J. Williams, V. Zemba, I. Linkov, D.A. Vallero, Advances on a decision analytic approach to exposure-based chemical prioritization, Risk Anal. 40 (2020) 83–96.
[73] P.A. Schulte, et al., Occupational safety and health, green chemistry, and sustainability: a review of areas of convergence, Environ. Health 12 (1) (2013) 1.
[74] J. Bare, TRACI 2.0: the tool for the reduction and assessment of chemical and other environmental impacts 2.0, Clean Techn. Environ. Policy 13 (5) (2011) 687–696.
[75] R. van Zelm, et al., European characterization factors for human health damage of PM 10 and ozone in life cycle impact assessment, Atmos. Environ. 42 (3) (2008) 441–453.
[76] J.C. Bare, D.A. Vallero, Incorporating exposure science into life-cycle assessment, in: AccessScience, McGraw-Hill Education; Yearbook of Science & Technology, 2014.
[77] D.A. Vallero, C. Brasier, Sustainable Design: The Science of Sustainability and Green Engineering, John Wiley & Sons, 2008.
[78] X. Pan, P. Sinha, X. Chen, Corporate social responsibility and eco-innovation: the triple bottom line perspective, Corp. Soc. Responsib. Environ. Manag. 28 (1) (2021) 214–228.
[79] W. Ahmed, B. Sarkar, Management of next-generation energy using a triple bottom line approach under a supply chain framework, Resour. Conserv. Recycl. 150 (2019) 104431.
[80] S. Hendiani, E. Sharifi, M. Bagherpour, S.F. Ghannadpour, A multi-criteria sustainability assessment approach for energy systems using sustainability triple bottom line attributes and linguistic preferences, Environ. Dev. Sustain. (2019) 1–35.
[81] E. Casey, S. Beaini, S. Pabi, K. Zammit, A. Amarnath, The triple bottom line for efficiency: integrating systems within water and energy networks, IEEE Power Energy Mag. 15 (1) (2017) 34–42.
[82] K. Lichtenvort, et al., Introduction—history of life cycle costing, its categorization, and its basic framework, in: Environmental Life Cycle Costing, 2008, pp. 1–6.
[83] S.S. Muthu, Life Cycle Costing, Springer Nature, 2023.
[84] M.H. Jahangir, R. Alimohamadi, Application of life cycle costing in building energy performance, in: Life Cycle Costing, Springer, 2023, pp. 1–26.
[85] H. Langdalen, E.B. Abrahamsen, J.T. Selvik, On the importance of systems thinking when using the ALARP principle for risk management, Reliab. Eng. Syst. Saf. (2020) 107222.
[86] U.D. Kumar, J. Crocker, J. Knezevic, M. El-Haram, Reliability, maintenance, and logistic support—introduction, in: Reliability, Maintenance and Logistic Support, Springer, 2000, pp. 1–12.
[87] U.M. Shamsi, Water Distribution Network Reliability, University of Pittsburgh, 1988.

Chapter 19

The future

Engineers are technological optimists for good reason. They have participated in solving big problems since the profession began. Water has been cleaned, waste sites remediated, open dumps eliminated, water supplies expanded, and technological advances occur daily. Environmental scientists have provided the required intellectual underpinning to support these advances. Technologists have improved instruments needed for measuring pollution and devices to improve water quality. Major legislation has been enacted around the world to improve wastewater treatment and to provide safe drinking water for the public. Regulatory agencies now set limits for many pollutants much lower than levels of detection of a decade ago. Measurement technologies, analytical approaches, and models continue to improve the ability to find and quantify pollutants.

There is still much to do. To name a few challenges, much of the world's population continues to drink unhealthy water. Water quality is a challenge for farmers who find it nearly impossible to avoid using dangerous pesticides and agriculture chemicals. Waterborne diseases affect millions. I am hopeful and optimistic that engineers and scientists will continue to help to confront and solve many of these challenges. New and emerging computational and analytical tools will underpin advanced technological and low-tech solutions to many of these problems.

Massive amounts of toxic compounds continue to be released into surface and ground water. Exposure to these pollutants present risks to human health, diminish the quality of habitats, and threaten the existence of sensitive species. Such problems are not readily resolved by the dependence on the old "command and control" approaches. Preventative strategies must continue to complement and, hopefully, eventually to replace the reliance on end-of-pipe pollution controls. It is preferable not to have polluted the water in the first place.

Research and systems thinking will continue to play crucial roles in clean water for all. This includes better understanding of the life cycles that include the processes and materials that lead to the formation of pollutants. We must find safer substitutes. Indeed, all economic sectors can improve their footprints on the environment. Agriculture can continue to become more sustainable and resilient by avoiding toxic chemical-dependence dictated by monoculture farming. Industries can consider every stage of the life cycle, from extraction of raw materials, to manufacturing, to product use, to end-of-product-life strategies.

Transportation can become more efficient and search for alternatives from a scientifically sound, systematic point of view. Too much conversation is polarized between those who become trapped in a metaphorical prison of two ideas. For example, comparisons of fossil fuel-burning vehicles and electric vehicles should be based on life cycle assessments, not the vested interests of advocates on either side of the argument. Neither is a perfect solution to the transportation-related environmental problems. Indeed, extraction and refining crude oil to provide gasoline and diesel fuel causes much pollution, but so does the extraction of rare earth elements, cobalt, and lithium for electric vehicles, not to mention that the electricity powering the electric vehicle still mainly comes from a fossil-fuel or nuclear-powered generating stations. Neither mode of transportation is ideal, so cleaner, less polluting alternatives must continue to be sought.

Innovations and market forces must also be part of the solution. New paradigms, such as pollution prevention, sustainable design, and green engineering are changing the perception and reality of more efficient, less polluting technologies and processes. The problems will best be addressed by systems thinking that is based on credible and sound science. The fundamentals of water pollution are based on the physical and biological sciences, but their application to addressing present and future water pollution problems must also rely on the innovations and effective communications.

The roles of the engineering profession as a whole differ from those of the individual engineering practitioner, i.e., the macro- versus microscales of engineering [1–3]. Environmental engineering success is based almost exclusively on the conduct and adherence to scientific principles that produce treatment and prevention outcomes based on the designs by an individual engineer [4,5]. Engineering designs must result in improvements in the quality of water and other

Fundamentals of Water Pollution. https://doi.org/10.1016/B978-0-443-28987-3.00014-X

environmental media. Recently, leaders in the engineering profession, especially the professional societies, have challenged engineers to address some "macro" issues that, even if all the microscale, individual projects are summed, would not be adequately addressed without institutional, top-down support. To support this line of thinking, the National Academy of Engineering articulated its "grand challenges" for the engineering profession this century [6]. These large, often global scale challenges include environmental problems. Addressing potential exposures to water pollutants must be among the strategies to combat these challenges [7].

Most environmental scientists and engineers are well trained in the physical and biological sciences, but much less so in the social sciences and humanities. Addressing water pollution calls for both, since most pollution results from human behavior and activities. Thus, the solutions must likewise go beyond solutions based solely on physical, chemical and biological purposes [8,9]. They must also be based on the feasibility to optimize the desired results of clean water and that these results be maintained and sustained, i.e., the risk is reduced and the solutions are sufficiently reliable and resilient to be ready to adapt to future uncertainties in both the natural and built environments [10–14].

The reader now has 18 previous chapters that, hopefully, will assist them in meeting these challenges. Indeed, I expect the present and future readers to add a few chapters of their own as new problems and opportunities for cleaner water emerge in the years ahead.

I was thirsty and you gave me something to drink.

—Matthew 25:35

References

[1] B. Allenby, Macroethical systems and sustainability science, Sustain. Sci. 1 (1) (2006) 7–13.
[2] D.A. Vallero, Macroethics and engineering leadership, Leadersh. Manag. Eng. 8 (4) (2008) 287–296.
[3] J.R. Herkert, Future directions in engineering ethics research: microethics, macroethics and the role of professional societies, Sci. Eng. Ethics 7 (3) (2001) 403–414.
[4] P.J. Lioy, K.R. Smith, A discussion of exposure science in the 21st century: a vision and a strategy, Environ. Health Perspect. 121 (4) (2013) 405.
[5] D.A. Vallero, P.J. Lioy, The 5 Rs: reliable postdisaster exposure assessment, Leadersh. Manag. Eng. 12 (4) (2012) 247–253.
[6] National Academy of Engineering, 14 Grand challenges for engineering in the 21st century, NAE, Washington, DC, 2018.
[7] D.A. Vallero (Ed.), Exposure space: integrating exposure data and modeling with toxicity information, in: OpenTox USA 2015, 2015. Baltimore, MD.
[8] T.L. Jacobs, P.A. Vesilind, Probabilistic environmental risk of hazardous materials, J. Environ. Eng. 118 (6) (1992) 878–889.
[9] P.A. Vesilind, A.S. Gunn, Hold Paramount: The Engineer's Responsibility to Society, Cengage Learning, 2010.
[10] I. Linkov, M.E. Bates, L.J. Canis, T.P. Seager, J.M. Keisler, A decision-directed approach for prioritizing research into the impact of nanomaterials on the environment and human health, Nat. Nanotechnol. 6 (12) (2011) 784–787.
[11] National Research Council, Exposure Science in the 21st Century: A Vision and a Strategy, The National Academies Press, Washington, DC, 2012. 196 pp.
[12] U.S. Environmental Protection Agency, Human health evaluation manual, supplemental guidance: update of standard default exposure factors OSWER directive 9200.1-120, U.S. EPA, Washington, DC, 2014. Available from: http://www.epa.gov/oswer/riskassessment/pdf/superfund-hh-exposure/OSWER-Directive-9200-1-120-ExposureFactors.pdf.
[13] World Health Organization, Environmental health criteria 214: human exposure assessment, Safety IPoC (Ed.), WHO, Geneva, Switzerland, 2000.
[14] P. Montague, Hazardous waste landfill: some lessons from New Jersey, Civil Eng. ASCE 52 (9) (1982) 53–56.

Glossary of water quality terms

Abiotic Description of chemical and physical processes occurring without the involvement of living organisms. In some cases, such processes do not involve microorganisms or plants at all, whereas in other cases, biological and abiotic processes occur simultaneously and/or serve to enhance each other.

Absorbed dose 1. Amount of a substance that enters the body of an organism, e.g., through the eyes, skin, stomach, intestines, or lungs. 2. Amount of active ingredient crossing exchange boundaries of a test organism or human (same as "internal dose").

Absorption 1. Process wherein a substance permeates another substance; a fluid is sorbed into a particle. 2. After uptake, the process by which a substance moves to tissues in an organism, e.g., absorption of a substance into the bloodstream. 3. Process by which incident radiated energy is retained in a medium, e.g., shortwave radiation from the sun is absorbed by soil and reradiated as longer wave radiation, e.g., infrared radiation (see *greenhouse effect*). Compare to *adsorption*.

Acceptable daily intake (ADI) The amount of a chemical a person can be exposed to on a daily basis over an extended period of time (usually a lifetime) without suffering deleterious effects.

Acclimation Adapting a microbe to food sources for carbon and energy to enhance biodegradation (e.g., in wastewater treatment).

Accuracy Degree of agreement between a measured value and the true value; usually expressed as $\pm\%$ of full scale. Compare to *precision*.

Acid rain Precipitation with depressed pH due to increases in concentrations of acid-forming compounds, such as oxides of sulfur and oxides of nitrogen, in the atmosphere. Term is usually limited to atmospheric deposition with pH <5.6, which is about the mean value for natural water (mainly due to carbonic acid content).

Action level 1. Concentration threshold above which actions must be taken, e.g., remediation, removal, treatment, or use restrictions and closures (e.g., beaches). 2. Regulatory level recommended by the U.S. Environmental Protection Agency for enforcement by Food and Drug Administration and U.S. Department of Agriculture when pesticide residues occur in food or feed commodities for reasons other than the direct application of the pesticide. Set for inadvertent residues resulting from previous legal use or accidental contamination. Compare to *tolerance*.

Activated carbon Carbon with a very high ratio of surface area to mass, rendering it a strong *adsorbent*. In granulated form, known as granulated activated carbon (GAC).

Activation energy Energy needed to bring all molecules in 1 mol of a substance to their reactive state at a given temperature.

Active ingredient Compound in a pesticide formulation that provides the biocidal mechanism of action. All other components are known as *inerts*.

Activity data Information on activities in which various individuals engage, length of time spent performing various activities, locations in which individuals spend time, and length of time spent by individuals within those various *microenvironments*.

Acute exposure A single exposure to a substance that results in severe biological harm or death. Acute exposures are usually characterized as lasting no longer than a day, as compared to longer, continuing *exposure* over a period of time.

Acute toxicity Any adverse effect that occurs within a short period of time following *exposure*, usually up to 24–96 h, resulting in biological harm and often death.

Acute Describing exposures, diseases, or responses with short durations.

Adaptive management (AM) Also known as "adaptive resource management" (ARM), a structured, iterative process of optimal decision making in the face of uncertainty, with an aim to reducing uncertainty over time via system monitoring. In this way, decision making simultaneously maximizes one or more resource objectives and, either passively or actively, accrues information needed to improve future management. AM is often characterized as "learning by doing."

Additive effect Response to exposure to multiple substances that equals the sum of responses of all the individual substances added together (compare with *antagonism* and *synergism*).

Adenosine 5′-triphosphate (ATP) The triphosphate of the nucleoside adenosine, which is a high energy molecule or has high phosphate group transfer potential and serves as the cell's major form of energy currency (see *photosynthesis*).

Adenosine diphosphate (ADP) The nucleoside diphosphate usually formed upon the breakdown of ATP when it provides energy for work (see *photosynthesis*).

Adsorption Process wherein a substance permeates another substance; a fluid is sorbed onto the surface of a particle. Compare to *absorption* (1).

Advanced waste treatment Physical, chemical, or biological treatment beyond that achieved by secondary treatment (e.g., additional removal of nutrients and solids). Synonymous with and preferred over *tertiary treatment*.

Advection Transport of a solute by the bulk motion of flowing ground water.

Adverse effect A biochemical change, functional impairment, or pathologic lesion that affects the performance of the whole organism or reduces an organism's ability to respond to an additional environmental challenge.

Adverse outcome pathway (AOP) A model that identifies the sequence of molecular and cellular events required to generate an adverse effect when an organism is exposed to a toxic substance.

Aerobe 1. Microorganism that can survive in the presence of molecular oxygen (O_2). 2. Microorganism that requires sufficient concentrations of O_2 to survive.

Aerobic respiration Metabolic process whereby microorganisms use oxygen as the final *electron acceptor* to generate energy.

Aerobic treatment Process by which microbes decompose complex organic compounds in the presence of oxygen and use the liberated energy for reproduction and growth.

Aerobic Conditions for growth or metabolism in which the organism is sufficiently supplied with molecular oxygen.

Aerodynamic diameter Diameter of a sphere with unit density that has aerodynamic behavior identical to that of the particle in question; an expression of aerodynamic behavior of an irregularly shaped particle in terms of the diameter of an idealized particle. Particles having the same aerodynamic diameter may have different dimensions and shapes (see also *Stokes diameter*).

Aerosol A suspension of liquid or solid particles in air.

Affinity Attraction, e.g., between an antigen and an antibody, or for a polar compound to the aqueous compartments of the environment.

Agar Polysaccharide complex gel derived from marine algae used to grow microbiological cultures. Gelling temperature usually ranges between 40 and 50 °C.

Agonist Substance that binds to a cell's receptor, often mimicking the actions of a normally occurring substance. Compare to *antagonist*.

Agonist Substance that blocks a cell's receptor, interfering with normal cellular response. Compare to *antagonist*.

Air purging Injection of air or oxygen through a contaminated aquifer to remove pollutants.

Air stripping Process of pushing air through contaminated water to remove volatile pollutants.

ALARP As low as reasonably possible.

Albedo Reflectivity of light; inverse of light absorption.

Algae Phototrophic *eukaryotic* microbes that can be either unicellular or multicellular. Plural of alga.

Algal bloom Masses of *algae*, plants, and other organisms that form scum at the top of surface waters; usually attributed to large inputs of nutrients to the waters.

Aliphatic compounds Acyclic or cyclic, saturated, or unsaturated carbon compounds, excluding aromatic compounds.

Alkane Single-bonded carbon chains or branched structures.

Alkene Carbon chains or branched structures that contain at least one double bond.

Alkyne Carbon chains or branched structures that contain at least one triple bond.

Allele One of several alternative forms of a gene that occupies a given locus on a chromosome.

Allergen Substance that causes an allergic reaction.

Allergenicity Capacity or potential that a substance will elicit and allergic reaction.

Alum 1. *Flocculant*, $K_2SO_4Al_2(SO_4)_3 \cdot 2H_2O$. 2. Aluminum sulfate alum, used to precipitate hydroxides for coagulation.

Ambient 1. Outdoor (e.g., ambient air). 2. Describing general environmental conditions, as contrasted with effluent or emission, e.g., ambient measurements versus effluent measurements. 3. Surrounding conditions.

Amendment Substrate introduced to stimulate the *in situ* microbial processes (vegetable oils, sugars, alcohols, etc.).

Amino acid Any of 20 basic building blocks of proteins with a free amino (NH_2) and a free carboxyl (COOH) group that have the basic formula NH_2C_RCOOH. The side group R is subdivided into polar or hydrophilic (serine, threonine, tyrosine, asparagine, and glutamine), nonpolar or hydrophobic (glycine, alanine, valine, leucine, isoleucine, proline, phenylalanine, tryptophan, and cysteine), acidic (aspartic acid and glutamic acid), and basics (lysine, arginine, and histidine). The sequence of amino acids determines the shape, properties, and the biological role of a protein.

Amino group NH_2 attached to carbon structures (e.g., in amines and amino acids).

Amphibole Inosilicate minerals, composed of double chain SiO_4 tetrahedra, linked at the vertices and generally containing iron and/or magnesium. Amphibole asbestos is a mixture of different fibers. For example, Libby, Montana, amphiboles comprise 85% winchite, 10% richterite, and 5% tremolite. Most toxic form of asbestos.

Amphiphilic Describing a chemical compound that has both *hydrophilic* and *lipophilic* properties.

Amphoteric Able to react as either a weak acid or a weak base.

Amplification 1. Treatment (e.g., use of chloramphenicol) designed to increase the proportion of plasmid *DNA* relative to that of bacterial (host) DNA. 2. Replicating a gene library in bulk. 3. Duplication of genes within a chromosomal segment. 4. Creation of numerous copies of a segment of DNA by the *polymerase chain reaction*.

Anabolism Synthesis of complex molecules from simpler molecules with the input of energy.

Anaerobe 1. Microorganism that cannot survive in the presence of molecular oxygen (O_2). 2. Microorganism that requires electron acceptors other than O_2 to survive.

Anaerobic respiration Process whereby microorganisms use a chemical other than oxygen as an electron acceptor. Common substitutes for oxygen are nitrate, sulfate, iron, carbon dioxide, and other organic compounds (fermentation).

Analogy Comparison of similarities between two things to a conclusion about an additional attribute common to both things. This is a type of inductive reasoning (see *inductive reasoning*). One of *Hill's criteria*.

Analyte 1. Substance measured in a scientific study. 2. Chemical for which a sample (e.g., water, air, or blood) is tested in a laboratory, e.g., to determine the amount of cadmium in soil, the specified amount of soil to be collected and analyzed in the laboratory.

Analytic epidemiology Evaluation of associations between exposure to physical, chemical, and biological agents and disease by testing scientific hypotheses.

Anion exchange capacity (AEC) Total of exchangeable anions that can be sorbed by a soil (units = centimoles of negative charge per kg soil).

Anion Negatively charged ion, e.g., chloride (Cl^{-1}) and fluoride (F^-), which are monovalent ions, and sulfate (SO_4^{2-}) and peroxide (O_2^{2-}), which are divalent anions.

Anisotropy Conditions under which one or more hydraulic properties of an aquifer vary with direction.

Anoxic An environment where there is no free oxygen and where microbial and chemical reactions use other chemicals in the environment to accept electrons.

Antagonism 1. Effect from a combination of two agents is less than the sum of the individual effects from each agent ($1+1<2$). Contrast with *synergism* and *additive effect*. 2. *Amensalism*.

Anthropocentrism Philosophy- or decision framework-based human beings. View that all and only humans have moral value. Nonhuman species and abiotic resources have value only in respect to that associated with human values. Contrast with *biocentrism* and *ecocentrism*.

Anthropogenic 1. Made, caused, or influenced by human activities. Contrast with *biogenic*. 2. Derived from human activities, as opposed to those occurring in natural environments without human influences.

Antibody Protein, e.g., immunoglobulin, manufactured by lymphocytes (a type of white blood cell) to neutralize an *antigen* or foreign protein. Microbes, pollens, dust mites, molds, foods, and other substances contain antigens which will trigger antibodies.

Antigen Foreign substance (e.g., protein, nucleoprotein, polysaccharide) to which lymphocytes respond. When immune system responds, it is known as an immunogen.

Antinutritional Factor, when present, stifles metabolism and growth of an organism (especially humans in the context of food biotechnologies). One of the three major risks associated with food biotechnology, along with toxicity and *allergenicity*.

Apoptosis Programmed cell death.

Applied mathematics Mathematical techniques typically used in the application of mathematical knowledge to domains beyond mathematics itself.

Aqueous phase Homogeneous fraction of a heterogeneous system consisting of the solution of a substance in water.

Aquifer A porous underground bed or layer of earth, sand, gravel, or porous stone that contains water. Geologic formation, group of formations, or part of a formation containing saturated permeable material that yields sufficient, economical quantities of ground water.

Artificial intelligence Computer systems that can perform tasks that normally require human intelligence, e.g., visual perception, speech recognition, decision making, and translation between languages.

Association Relationship, not necessarily causal, between two variables. The antecedent variable comes before and is associated with an outcome; however, it may or may not be the cause of the outcome. For example, mean birth weight of minority babies is less than that of babies of the general population. Ethnicity is an antecedent of low birth weight, but not the cause. Other factors, e.g., nutrition, smoking status, and alcohol consumption, may be the causal agents.

Atmospheric deposition Process by which a substance reaches the ground surface from the atmosphere.

Attenuation rate The rate at which a contaminant is removed. This is not a rate constant but a rate, with typical units of $\mu g\,L^{-1}\,year^{-1}$.

Attenuation The process by which a chemical compound's concentration decreases with time, through sorption, *degradation*, *dilution*, and/or *transformation*. The term applies to both destructive and nondestructive contaminant removal.

Attributable risk The rate of a disease in exposed individuals that can be attributed to the *exposure*. This measure is derived by difference between the rate (usually *incidence* or *mortality*) of the disease among persons not exposed to the suspected agent and the corresponding rate among exposed individuals.

Autotroph Organisms feeding on inorganic minerals, producing complex organic compounds from simple inorganic molecules using energy by *photosynthesis* or by inorganic chemical reactions.

Average daily dose (ADD) Dose rate averaged over a pathway-specific period of exposure expressed as a daily dose on a per-unit-body-weight basis. The ADD is usually expressed as $mg\,kg^{-1}\,day^{-1}$ or other mass-time units.

Average land cover condition Percentage of impervious cover needed to generate an equivalent amount of phosphorus (P) as the total land uses in a watershed.

B cell *Leucocyte* that makes antibodies; part of the immune system and develop from stem cells in the bone marrow.

***Bacillus thuringiensis* (Bt)** Bacterium that repels or kills insects; a major component of the microbial pesticide industry.

Bacillus Rod-shaped bacterium.

Bacteria Unicellular microorganisms that exist either as free-living organisms or as parasites, ranging from beneficial to harmful to humans.

Bacteriophage Virus (phage) that infects a bacterium.

Basin 1. System to impound stormwater. 2. Watershed of a stream.

Bayesian belief network Cause-and-effect tool represented by a probabilistic graphic model of a set of random variables and their conditional independencies.

Bayesian Statistical approach that addresses probability inference, named after Thomas Bayes (*An essay towards solving a problem in the doctrine of chances*, 1763), to decision making and inferential statistics that deals with probability inference (i.e., using knowledge of prior events to predict future events). In a Bayesian network, priors are updated by additional data that yield posterior probabilities that are often more robust than classical probabilities.

Benefit-cost analysis (or cost-benefit analysis) Method designed to determine the feasibility or utility of a proposed or existing project. Yields a benefit-cost ratio (see *benefit-cost ratio (BCR)*).

Benefit-cost ratio (BCR) Weighted benefits divided by weighted costs; used to compare and differentiate among project alternatives. Gross BCR <1 is undesirable. The greater the BCR, the more acceptable the alternative.

Benthic Pertaining to the bottom of a body of water. Often used to differentiate bottom organisms from those that swim and float.

Benthos Organisms living at or near the bottom of a surface water body.

Best management practice (BMP) Methods that have been determined to be the most effective, practical means of preventing or reducing pollution from *nonpoint sources*.

Best practice 1. Optimal service to the client. 2. Treatment is appropriate, accepted, and widely used according to expert consensus; embodies an integrated, comprehensive, and continuously improving approach to care (medicine). It is morally obligatory that healthcare practitioners provide patients with the best practice (also known as standard therapy or standard of care).

Bias 1. Systematic error in one direction, such as the positive bias of a scale that reads 1 mg too high (instrument error) or the negative bias in interpretations of lesions reported by a physician performing the procedure (operator bias) that consistently miss some lesions. Bias makes the reported values less accurate. 2. Any difference between the true value and that measured due to all causes other than sampling variability.

Binding site 1. Location on a cellular membrane where a *ligand* can bind. 2. Location on cellular *DNA* to which a chemical substance can bind. Typically, binding sites may be in the vicinity of genes and involved in activating *transcription* of that gene (i.e., promoter elements), in enhancing transcription of that gene (enhancer elements), or in reducing transcription of that gene (silencers).

Bio- Prefix indicating "life" (Greek).

Bioaccumulation factor (BAF) Ratio of a tissue concentration of substance in an organism to its concentration in the environment (usually water) where the organism lives. BAF indicates a compound's potential to accumulate in tissue through exposure to both food and water. Compare to *biomagnification* and *bioconcentration factor*.

Bioaccumulation The process whereby certain substances build up in living tissues.

Bioassay Test to assess the effects of certain substances on animals. Often used to estimate *acute toxicity*.

Bioaugmentation Addition of beneficial microorganisms into ground water to increase the rate and extent of *biodegradation*. Part of a *bioremediation* strategy.

Bioavailability The degree to which a substance becomes available to the target tissue after administration or exposure.

Biocatalysis Mediation of chemical reactions by biological systems, e.g., microbial communities, whole organisms or cells, cell-free extracts, or purified enzymes.

Biocentrism View that all life has moral value. Contrast with *anthropomorphism*.

Biochemical oxygen demand (BOD) A standard test to assess wastewater pollution due to organic substances. BOD_5 is based on the measurement of the oxygen used under controlled conditions of temperature (20 °C) and time (5 days). Compare to *chemical oxygen demand*.

Biochemodynamics The physical, chemical, and biological processes that transport and transform substances.

Bioconcentration factor (BCF) Ratio of the concentration of substance in an organism's tissue versus its concentration in the environment (usually water) in situations where the organism is exposed exclusively to water. BCF measures the compound's potential to accumulate in tissue through direct uptake from water (excludes uptake from food). Compare to *bioaccumulation factor*.

Biodegradation Breakdown of a contaminant, usually catalyzed by enzymes produced by organisms. Term usually applies only to microorganisms. Compare to *phytodegradation*.

Biodiversity Number and variety of organisms in an ecosystem (e.g., agricultural, wetland, or forest). Usually, lower biodiversity indicates that the system is stressed and in a poor condition.

Bioenergetics Energy flow and transformation through living systems. Can be within an organism, e.g., cellular energetics, or within and between levels of *biological organization*, e.g., *trophic states* (energy transfer from producers to first-order consumers, to second-order consumers, to and from decomposers, etc.).

Bioengineering See *biological engineering*.

Bioethics 1. Inquiry into ethical implications of biological research and applications. 2. Ethical inquiry into matters of life, especially biomedical and environmental ethics.

Biofilm Organized microbial system consisting of layers of microbial cells associated with surfaces, often with complex structural and functional characteristics. Influence microbial metabolic processes. Site where chemical degradation occurs both via extracellular enzymatic activity and by intracellular microbial processes. *Pseudomonas* and *Nitrosomonas* strains are notable for their ability to form a strong biofilm.

Biogenic Made, caused, or influenced by natural processes. Contrast with *anthropogenic*.

Biogeochemistry 1. Study of the fluxes, cycles, and other chemical and biological processes at various scales on earth. 2. Study of microbially mediated chemical transformations, especially with regard to nutrient cycling (N, P, S, and K, for example).

Bioinformatics Management and analysis of data using advanced computing techniques applied to biological research and inquiry.

Biolistic gun (particle gun) Method used to modify genes by directly shooting genetic information into a cell. *DNA* is bound to tiny particles of gold or tungsten and subsequently inserted into tissue or single cells under high pressure. The accelerated particles penetrate both the cell wall and membranes, slowing down upon impact. The DNA separates from the metal and can be integrated into the genetic material inside the nucleus.

Biolitic Formed by living organisms or their remains (e.g., sedimentary rocks).

Biological control Method of addressing problematic organisms by using a biochemical product or bioengineered or naturally occurring organism, e.g., introducing the European beetle (*Nanophyes marmoratus*) that feeds exclusively on the highly invasive purple loosestrife (*Lythrum salicaria*).

Biological criteria Measures of the condition of an environment, e.g., incidence of cancer in benthic fish species.

Biological engineering Combination of biomedical and biosystem engineering (see *biomedical engineering* and *biosystem engineering*) to develop useful biology-based technologies that can be applied across a wide spectrum of societal needs, including diagnosis, treatment, and prevention of disease; design and fabrication of materials, devices, and processes; and enhancement and sustainability of environmental quality.

Biological half-life The time required for a biological system (such as a human or animal) to eliminate, by natural processes, half the amount of a substance (such as a radioactive material) that has been absorbed into that system.

Biological magnification See *biomagnification*.

Biological organization Levels of living things, from *biomolecules* to planetary. The levels generally representing biological systems are molecule, cell, tissue, organ, organ system, organism, population, community, ecosystem, and biosphere.

Biological response Manner and type of effect in an organism (e.g., disease, change in metabolism, and homeostasis).

Biologically based dose-response (BBDR) model Predictive model that describes biological processes at the cellular and molecular level linking the target organ dose to the adverse effect.

Biomagnification (biological magnification) Process whereby certain substances, such as substances that move up the *food chain*, work their way into rivers or lakes and are eaten by aquatic organisms such as fish, which, in turn, are eaten by large birds, animals, or humans. The substances become increasingly concentrated in tissues or internal organs as they move up the chain.

Biomarker 1. Chemical, physical, or biological measurement that indicates biological condition. The biomarker may be a chemical to which an organism is exposed (e.g., lead in blood), a metabolite of the chemical (e.g., cotinine in blood as an indication of exposure to nicotine), or a biological response (e.g., an increase in body temperature as a result of exposure to a pathogen). 2. In geochemistry, organic compounds that are remnants of former living creatures (e.g., the suite of compounds that indicate the processes by which coal or petroleum has formed).

Biomass Material produced by the growth of microorganisms.

Biomedical engineering Application of engineering principles to medicine, including drug delivery systems, therapeutic systems, and medical devices.

Biomedical testing Investigations to determine whether a change in a body function might have occurred because of exposure to a hazardous substance.

Biomolecule Building block compounds of life that perform essential functions in living organisms, e.g., amino acids, carbohydrates, lipids, polysaccharides, proteins, and nucleic acids.

Biophile An element, arranged in myriad ways, that provides the structure for all living systems, e.g., oxygen, carbon, hydrogen, and nitrogen.

Bioprospecting Search for novel products from organisms in the natural habitats, usually plants and microbes. In its negative connotation, known as biopiracy.

Bioremediation Treatment processes that use microorganisms such as bacteria, yeast, or fungi to break down hazardous substances into less toxic or nontoxic substances. Bioremediation can be used to clean up contaminated soil and water. *In situ* bioremediation treats contaminated soil or ground water in the location in which it is found. For *ex situ* bioremediation processes, contaminated soil is excavated, or ground water is pumped to the surface before they can be treated.

Bioretention The act of holding water via a biological system such as a bioretention basin or bioretention filter.

Biosecurity Measures to control the transmission of microorganisms into or out of a specified area or population (see *physical containment*).

Biosensor A portable device that uses living organisms, such as microbes, or parts and products of living organisms, such as enzymes, tissues, and antibodies, to produce reactions to specific chemical contaminants.

Bioslurping Remediation that merges *bioventing* and *vacuum-enhanced free-product recovery*.

Biosolids 1. See *sludge*. 2. Organic product of wastewater treatment that can be beneficially used.

Biosparging *In situ* remediation technique that supplies oxygen and nutrients (if needed) to contaminated aquifers to promote aerobic biodegradation of the contaminants.

Biosphere Earth's zone that includes *biota*, extending from ocean sediment to mountaintops.

Biostatistics Application of statistical tools to interpret biological and medical data.

Biostimulation Adding chemical amendments, such as nutrients or electron donors, to soil or ground water to support bioremediation.

Biosystem or biosystems engineering (bioengineering) 1. Application of biological sciences to achieve practical ends. 2. Integration of physical, chemical, or mathematical sciences and engineering principles for the study of biology, medicine, behavior, or health to advance fundamental concepts; to create knowledge for the molecular to the organ systems levels; and to develop innovative biologics, materials, processes, implants, devices, and informatics approaches for the prevention, diagnosis, and treatment of disease; for patient rehabilitation; and for improving health.

Biosystem Living organism or a system of living organisms that can interact with other organisms directly or indirectly.

Biota 1. Any living creature, plant (flora), animal (fauna), or microbial. 2. Total of the living organisms of any designated area.

Biotechnology Use of living creatures to produce things of value to humans (e.g., hazardous waste cleanup, production of drugs, and improving agriculture and food supplies).

Bioterrorism Use of living agents to cause intentional harm to people (e.g., anthrax spores or pathogenic viruses) and society (e.g., agricultural pests).

Biotic Related to living systems.

Biotransformation Biologically catalyzed transformation of a chemical to some other product.

Bio-uptake 1. Process by which a compound enters an organism. 2. Amount of a substance that enters an organism.

Bioventing *In situ* remediation technology that stimulates the natural biodegradation of aerobically degradable compounds in soil by the injection of oxygen into the subsurface. Bioventing has been used to remediate releases of petroleum products, such as gasoline, jet fuels, kerosene, and diesel fuel. Bioventing stimulates the aerobic bioremediation of hydrocarbon-contaminated soils and vacuum-enhanced free-product recovery extracts *light nonaqueous-phase liquids* (LNAPL) from the capillary fringe and the water table.

Black swan event An outcome that would not have been predicted from past evidence due to the low probability of a possible confluence of events, but the event in fact occurred.

Blackbody radiator Idealized object that absorbs all electromagnetic radiation that reaches it. The earth behaves like a blackbody radiator when it absorbs incoming solar radiation (e.g., shortwave and visible light) and reemits it at longer wavelengths (e.g., infrared heat). See *greenhouse effect*.

Blastocyst Early stage embryo that consists of cells enclosing a fluid-filled cavity.

Body burden Total amount of a specific substance in an organism, including the amount stored, the amount that is mobile, and the amount *absorbed*.

Bottom-up View where fundamental components are first considered, working upward to larger perspectives. Contrast with *top-down*.

Boyle's law The pressure of a given mass of an ideal gas is inversely proportional to its volume at a constant temperature.

Brownfield Abandoned, idled, or underused industrial and commercial site where expansion or redevelopment is complicated by real or perceived environmental contamination. Generally applied to such sites that have been or are expected to be reused.

BTEX Term used for benzene, toluene, ethylbenzene, and xylene, which are volatile aromatic compounds typically found in petroleum products such as gasoline and diesel fuel.

Bulk density The mass of a soil per unit bulk volume of soil; mass is measured after all water has been extracted, and the volume includes the volume of the soil particles and pores.

Butterfly effect Sensitive dependence on initial conditions. Metaphor for the extreme sensitivity of chaotic systems (see Chaos), in which small changes or perturbations lead to drastically different outcomes. The phrase is derived from a butterfly flapping its wings in California, thereby initiating a change in weather patterns that results in the formation of a thunderstorm in Nebraska (from Edward Lorenz in his 1963 article, "Deterministic nonperiodic flow," *J. Atmos. Sci.* 20: 130–141; although in his presentation to the New York Academy, it was not a butterfly but a seagull's flapping of the wing that posited as the initial condition. Later in 1972, Lorenz used the butterfly in the example).

Calibration curve In analytic chemistry, a method to identify the concentration of an unknown substance by plotting data points of known substances at varying concentrations.

Calvin cycle Dominant pathway for the fixation (or reduction and incorporation) of CO_2 into organic material by *photoautotrophs* during photosynthesis. Also is found in *chemolithoautotrophs.*

Cancer slope factor (CSF) Linear part of cancer *dose-response curve* used to estimate the risk of cancer associated with exposure to a carcinogenic substance.

Cancer Disease of heritable, somatic mutations affecting cell growth and differentiation, characterized by an abnormal, uncontrolled growth of cells.

Capillary force Interfacial force between immiscible fluid phases, resulting in pressure differences between the two phases. Force due to capillary action that "pulls" water and/or waterborne contaminants toward a substance that attracts them, leading to the production of thin trails of contamination and the incorporation of contamination into the inner windings of a soil particle.

Capillary fringe Layer at the *water table* in which ground water seeps upwardly from an aquifer by capillary action to fill pores.

Carbonaceous biochemical oxygen demand (CBOD) Measure of the dissolved oxygen used for biological oxidation of C-containing compounds in a sample. See *biochemical oxygen demand.*

Carcinogen Physical, chemical, or biological agent that induces cancer.

Carcinogenesis Origin or production of a benign or malignant tumor. The carcinogenic event modifies the genome and/or other molecular control mechanisms of the target cells, giving rise to a population of altered cells.

CAS registration number An organization from Columbus, Ohio, which indexes information published in Chemical Abstracts by the American Chemical Society and provides index guides by which information about substances may be in the Abstracts when needed. CAS numbers identify specific chemicals.

Case study Evaluation of an actual occurrence of events to describe specific environmental and health conditions and past exposures.

Case-control study An epidemiologic study contrasting those with the disease of interest (cases) to those without the disease (controls). The groups are then compared with respect to exposure history, to ascertain whether they differ in the proportion exposed to the chemical(s) under investigation.

Catabolism Metabolism in which larger, more complex molecules are broken down into smaller, simpler molecules with the release of energy. Compare to *anabolism.*

Catch basin Inlet chamber at the curb line of a street or low area that collects runoff before it reaches the sewer or subdrain.

Cation exchange capacity (CEC) Ability of soil, sediment, or other solid matrix to exchange *cations* with a fluid. Very important measure of soil productivity and root behavior.

Cation Positively charged ion. The sodium ion (Na^+) and lithium ion (Li^+) are monovalent cations. The calcium ion (Ca^{2+}), lead ion (Pb^{2+}), and mercury ion (Hg^{2+}) are divalent cations. Metals can have more than one valence. For example, mercury can also have a monovalent form (Hg^+) and chromium has several, but the two most important for the environmental sciences are trivalent (Cr^{3+}) and hexavalent (Cr^{6+}).

Causation (causality) Relationship between causes and effects. Contrast with association.

Cell Basic unit of life; autonomous, self-replicating unit that either constitutes a unicellular organism or is a subunit of a multicellular organism; the lowest denomination of life.

Central nervous system Portion of the nervous system that consists of the brain and the spinal cord.

Channel stabilization Introduction of materials in a stream channel to minimize erosion of the channel bed or bank.

Chaos theory Exposition of the apparent lack of order in a system that nonetheless obeys specific rules. Condition discovered by the physicist Henri Poincaré around the year 1900 that refers to an inherent lack of predictability in some physical systems (i.e., Poincaré's concept of dynamical instability).

Charles' law Law stating that if the pressure on a sample of a dry gas is held constant, the Kelvin temperature and the volume will be in direct proportion.

Check dam System constructed in a channel to decrease flow velocity, minimize scour, and decrease stream competence to enhance sedimentation behind the dam.

Chemical oxygen demand (COD) Measure of the amount of oxygen required for the chemical oxidation of carbonaceous (organic) material in a waste, using inorganic dichromate or permanganate salts as oxidants in a 2-h test. Compare to *biochemical oxygen demand.*

Chemisorption Type of *adsorption* process wherein an adsorbate is held on the surface of an adsorbent by chemical bonds.

Chemotroph Organism that derives energy from inorganic reactions.

Chimera Organism, usually animal, that is a mixture of cells from two different embryonic sources.

Chloracne Skin eruption of blackheads, cysts, and nodules linked to exposure to chlorinated dioxins and furans.

Chlorinated ethene Chemical substances such as trichloroethene and tetrachloroethene that have been used in industry as solvents.

Chlorinated solvent Organic compounds with chlorine substituents that commonly are used for industrial degreasing and cleaning, dry cleaning, and other processes.

Chloromethanes Chemical substances such as carbon tetrachloride and chloroform that have been used in industry as solvents.

Chlorophyll-A Specific form of chlorophyll used as an indirect indicator of oxygenic photosynthesis and potential *eutrophication*.

Chloroplast Organelle containing chlorophyll, which carries out photosynthesis in plants and green algae.

Chromosome Structure within a cell's nucleus consisting of strands of *deoxyribonucleic acid* (DNA) coated with specialized cell proteins and duplicated at each mitotic cell division. Chromosomes transmit the genes of the organism from one generation to the next.

Chronic effect An adverse effect on a human or animal in which symptoms recur frequently or develop slowly over a long period of time.

Chronic exposure Multiple exposures occurring over an extended period of time, or a significant fraction of the animal's or the individual's lifetime.

Chronic lymphocytic leukemia Cancer that begins with white blood cells, i.e., lymphocytes, in the bone marrow.

Chronic toxicity Capacity of a substance to cause long-term adverse effects (usually applied to humans and human populations).

Chronic Having a persistent, recurring, or long-term nature. Contrast with *acute*.

Chrysotile Most common type of asbestos; a soft, fibrous silicate mineral in the serpentine group of phyllosilicates. Compare to *amphibole*.

Chute High velocity, open channel to convey water to a lower level that avoids erosion.

Ciliate Class of protozoans distinguished by short hairs on all or part of their bodies.

Cinnabar Bright red sulfide compound of mercury, HgS.

Cisgenesis Process by which genes are artificially transferred between organisms that could be conventionally bred (i.e., close related). Compare to *transgenesis*.

Citrus greening disease Huanglongbing; destructive plant disease in citrus trees caused by the phloem-limited bacterium *Candidatus* spp. and transmitted by *psyllids*, especially *Diaphorina citri* Kuwayama and *Trioza erytreae* (del Guercio).

Clarification Removal of suspended solids. Preferred terms are *sedimentation* or *settling*.

Clausius-Clapeyron equation Expression of the rate at which vapor pressure increases per unit increase in temperature for a substance's vapor pressure and temperature.

Clay Soil particle <0.002 mm in diameter. Compare to *silt* and *sand*.

Cleanup Actions taken to address a release or threat of release of a pollutant. Often used synonymously with *remediation* but also can consist of pollutant removals and other corrective actions that do not necessarily require *degradation* and *detoxification*.

Clone Line of cells genetically identical to the originating stem cell; group of genetically identical cells or organisms derived by asexual reproduction from a single parent. Act of generating these organisms is known as cloning.

Closure Procedure following a *remediation* project or the useful life of a landfill, e.g., installing a permanent cap.

Coccus Bacterial cell that is roughly spherical.

Code of ethics Established set of moral expectations of a group, especially of professional societies.

Coefficient of determination (r^2) Proportion of the variance of one variable predictable from another variable. The ratio of the explained variation to the total variation, which represents the percentage of the data nearest to the line of best fit. For example, if $r = 0.90$, then $r^2 = 0.81$, meaning that 81% of the total variation in one variable (y) can be explained by the linear relationship between the two variables (x and y) as described by the regression equation. Thus the remaining 19% of the total variation is unexplained.

Coenzyme Loosely bound nonprotein component of an enzyme required for catalytic activity that often dissociates from the enzyme active site after product has been formed.

Coherence Criterion for causality (i.e., *Hill's criteria*) based on the amount and degree of agreement among studies linking cause to effect, especially among various types of studies (e.g., animal testing, human epidemiological investigations, and *in vitro* studies).

Cohort study Epidemiologic study comparing those with an exposure of interest to those without the exposure. These two cohorts are then followed over time to determine the differences in the rates of disease between the exposure subjects. Also called a prospective study.

Coliform Gram-negative, nonsporing, facultative rod that ferments lactose with gas formation within 48 h at 35 °C.

Colloid Fine solid (<0.002 and >0.000001 mm) that does not readily settle; intermediate between true solutions and suspensions.

Combinational biology Introduction of genes from one microorganism into another microorganism to synthesize a new product or a modified product, especially in relation to antibiotic synthesis.

Combustion Sequence of rapid exothermic oxidation reactions that release heat and generate products.

Cometabolism A reaction in which microorganisms transform a contaminant even though the contaminant cannot serve as an energy source for growth, requiring the presence of other compounds (primary substrates) to support growth.

Commensalism Symbiosis where an organism lives on or within another organism with neither a positive nor a negative effect on the other organism.

Comparability Ability to describe likenesses and differences in the quality and relevance of two or more data sets.

Comparative risk An expression of the risks associated with two (or more) actions leading to the same goal; may be expressed quantitatively (e.g., ratio of 1.5) or qualitatively (one risk greater than another risk). Any comparison among the risks of two or more hazards with respect to a common scale.

Compartmental model Model that predicts or characterizes the transport and fate of a compound within an organism (e.g., moving from blood to tissues, transformed by metabolism, and detoxified or bioactivated during the path to elimination).

Compartmentalization Viewing a system by its individual components. This can be problematic when an engineer does not consider the system as a whole (e.g., when the structural engineer and soil engineer do not collaborate on selecting the best and safest combination of materials and structures suited to a soil type, or when a biomedical engineer does not work closely with various specialized healthcare professionals in a clinical setting to adapt a realistic device to the comprehensive needs of the patient). Compartmentalization can be good when it allows the engineer to focus adequate attention on the components (see *bottom-up*), so long as the design is properly built into a system.

Competence 1. Skill in practice. For professionals, competence is requisite to ethical practice. 2. Sufficient velocity for a fluid to carry a load (especially a stream's ability to carry solids).

Complementary DNA (cDNA) *DNA* copy of an *RNA* molecule.

Complexity Relative measure of uncertainty in achieving functional requirements or objectives. Designers are frequently expected to reduce the complexity of engineered systems.

Compliance monitoring Collection of data needed to evaluate the condition of the contaminated media against standards such as soil and/or water quality regulatory standards, risk-based standards of remedial action objectives.

Composite sample 1. Series of samples taken over a given period of time and weighted by flow rate or by other means to represent a concentration integrated with respect to time. 2. Soil sample that consists of soil taken from various depths or various locations.

Composite sample Combined individual environmental samples collected at preselected intervals to minimize the effect of the variability of the individual sample. Individual samples may be of equal volume or may be proportional to flow.

Compost Organic material produced from microbial degradation that is useful as soil conditioners and fertilizers. Process to produce such matter is known as composting.

Concentration 1. Quantity of substance per unit volume (fluid) or per unit weight (solid matrix, e.g., soil, sediment, tissue). 2. Method of increasing the dissolved solids per unit volume of solution, e.g., via evaporation of the liquid. 3. Increasing suspended solids per unit volume of *sludge* via sedimentation or dewatering.

Conceptual site model (CSM) A hypothesis about how releases occurred, the current state of the source zone, and current plume characteristics (plume stability).

Conductivity Ability to pass an electric current. An indirect method of measuring *total dissolved solids*.

Confidence level 1. Client's trust in a professional. 2. Amount of certainty that a statistical prediction is accurate. Physical sciences may differ from social sciences in what is considered acceptable confidence; e.g., the former may require 99%, while social scientific research may consider 95% to be acceptable. Depending on the application, engineering research ranges in acceptable confidence level (e.g., structural fatigue research may require higher confidence levels than environmental research).

Confined aquifer Underground strata beneath a confining layer of impervious rock that closes the aquifer to the surface.

Conflagration Destructive fire.

Confounder Factor that distorts or masks the true effect of risk factors in an epidemiologic study. A condition or variable that is a risk factor for disease and is associated with an exposure of interest. This association between the exposure of interest and the confounder (a true risk factor for disease) may make it falsely appear that the exposure of interest is associated with disease. For example, a study of low-birth-weight children in low-income families must first address the confounding effects of tobacco smoking before ascribing the actual risk associated with income.

Confounding factor Variable that may introduce differences between cases and controls, which do not reflect differences in the variables of primary interest. Factors that must be considered in *epidemiological* studies to ensure that the experimental variables are indeed the cause of an outcome (e.g., smoking can be a confounder in most cancer studies).

Conjugation In *genetic engineering*, transferring genetic material between bacteria through direct cell-to-cell contact, or through a bridge between the two cells.

Constitutive Quality of enzyme meaning that it is always synthesized and ready.

Constructed wetland Human-design and built system that emulates a natural wetland's structure and function.

Contact stabilization Enhanced *activated-sludge* process by adding a period of contact between wastewater and *sludge* for rapid removal of soluble *biochemical oxygen demand* by adsorption, followed by a longer period of aeration in a separate tank so that the sludge is oxidized and new biosolids are synthesized.

Contamination Contact with an admixture of an unnatural agent, with the implication that the amount is measurable. Increase in harmful or otherwise unwanted material in the environment. Often used synonymously with pollution.

Contingency plan Document setting out an organized, planned, and coordinated course of action to be followed in case of an emergency or episodic event that threatens public health or the environment (e.g., an oil spill, toxic release, or natural disaster).

Contingent probability Probability that an event will occur as a result of one or more previous events. Also known as conditional probability.

Continuous culture Microbial growth that is limited, and the effect of limiting the substrate or nutrients can be described by *Monod equation*.

Continuous sample A flow of water from a particular place in a plant to the location where samples are collected for testing; may be used to obtain grab or composite samples.

Contour Line of equal elevation.

Control group Group used as the baseline for comparison in epidemiologic studies or laboratory studies. This group is selected because it either lacks the disease of interest (case-control group) or lacks the exposure of concern (cohort study). Also known as a reference group.

Control volume Arbitrary volume in which the mass of the fluid remains constant at steady state so that as a fluid moves through, the mass entering the control volume is equal to the mass leaving the control volume.

Correlation coefficient (*r*) Statistical measurement of the strength and the direction of a linear relationship (association) between two variables.

Cost-benefit analysis A formal quantitative procedure comparing costs and benefits of a proposed project or act under a set of preestablished rules. To determine a rank ordering of projects to maximize the rate of return when available funds are unlimited, the quotient of benefits divided by costs is the appropriate form; to maximize absolute return given limited resources, benefits-costs is the appropriate form.

Crest Top of a structure, e.g., spillway, dike, dam, or weir.

Criteria Descriptive factors in setting standards for various pollutants. These factors are used to determine limits on allowable concentration levels and to limit the number of violations per year (plural of criterion).

Critical path Systems engineering of activities, decisions, and actions that must be completed on schedule and at a sufficient level of quality for the entire project to be successful.

Cross-resistance Mutation of a microbe that has mutated in such a way that it loses its susceptibility to more than one antibiotic simultaneously, not just the one to which it has been directly exposed.

Cross-sectional study Epidemiological study of observations representing a particular point in time. Contrast with *longitudinal study*.

Culture Intentional organic growth.

Cut Excavated area.

Cyanobacteria Large group of bacteria that carry out oxygenic photosynthesis using a system like that of photosynthetic *eukaryotes*.

Cyst Specialized microbial cell enclosed in a wall; formed by protozoa and a few bacteria. They may be dormant, resistant structures formed in response to adverse conditions or reproductive cysts that are a normal stage in the life cycle.

Cytochrome P450 (CYP) Enzymes that use iron to oxidize substances, often as part of the body's strategy to dispose of potentially harmful substances by making them more water soluble. Varying versions of CYP are used to identify different enzymatic activities so are important in *toxicodynamics* and *toxicokinetics modeling*.

Cytochrome Heme protein that carries electrons, usually as members of electron transport chains.

Cytokine Nonantibody protein released by a cell in response to inducing stimuli, which are mediators that influence other cells. Produced by lymphocytes, monocytes, macrophages, and other cells.

Dalton's law Total pressure exerted by a mixture of gases is equal to the sum of the pressures that would be exerted if each of the individual gases were to occupy the same volume by itself.

Damage Economic, social, and environmental loss or level of destruction caused by an event.

Darcy's law An empirically derived equation for the flow of fluids through porous media; it is based on assumptions that flow is laminar, and inertia can be neglected. It states that the specific discharge, q, is directly proportional to the hydraulic conductivity and the hydraulic gradient.

Dark field Microscope's optical system that makes small, clear, and colorless particles (e.g., many microbes) visible, by illuminating the object at an angle such that no light enters the microscope system except that which is diffracted by particles.

Data Plural of datum. Gathered facts from which conclusions can be drawn.

Decision tree Diagram indicating various steps to different outcomes. Supports the optimal course of action in situations where several possible alternatives have uncertain outcomes.

Declining growth phase Period of time in microbial population dynamics between the *log-growth phase* and the *endogenous phase*, where the amount of food is in short supply, leading to incrementally slowing growth rates.

Decomposer Organism that degrades complex materials into simpler ones.

Deductive reasoning A conclusion is necessitated by previously known facts. If the premises are true, the conclusion must be true. Starting from general knowledge and moving to specifics (e.g., from cause to effects). Contrast with inductive reasoning.

Deep ecology Environmental movement initiated in 1972 by Norwegian philosopher, Arne Naess, that advocates radical measures to protect the natural environment irrespective of their effect on the welfare of humans (opposite of *anthropocentrism*).

Dehydrohalogenation A process by which a halogenated alkane loses a halogen from one carbon atom and a hydrogen from the adjacent carbon atom, producing the alkene and an acid (e.g., 1,1,2,2-tetrachloroethane dehydrohalogenates to produce trichloroethene and HCl).

Demand Quantity of a goods or service that society chooses to buy at a given price.

Denitrification Reduction of nitrate to gas products, primarily nitrogen gas (N_2), during anaerobic respiration.

Denitrifier Diverse group of heterotrophic bacteria, mostly facultative and anaerobic, archaea, and fungi that oxidize substrates while reducing NO_3^- to either N_2O or N_2.

Dense, nonaqueous-phase liquid (DNAPL) An immiscible organic liquid that is denser than water (e.g., tetrachloroethene).

Deoxygenation constant Expression of the rate of the biochemical oxidation of organic matter under aerobic conditions. Value depends on the time unit involved (often 1 day) and varies with temperature and other environmental conditions.

Deoxyribonucleic acid (DNA) Double-stranded nucleic acid containing genetic information; polynucleotide composed of deoxyribonucleotides connected by phosphodiester bonds.

Deposition Removal of airborne substances to surfaces because of gravitational settling and diffusion, as well as electrophoresis and thermophoresis.

Depuration Cleansing of a previously dosed organism by ending the dosing completely.

Dermal exposure Contact between a chemical and the skin.

Dermal toxicity The ability of a pesticide or toxic chemical to poison people or animals by contact with the skin.

Descriptive epidemiology Study of the amount and distribution of a disease in a specified population by person, place, and time. Compare to *analytic epidemiology*.

Design storm A reference event of a specified amount, intensity, duration, and frequency from which to design and build stormwater or other facilities.

Desorption The converse of sorption, i.e., when a compound slowly releases from a surface(s) that it has previously accumulated upon or within.

Destruction and removal efficiency (DRE) Percentage of compound removed or destroyed by a process, usually thermal. Calculated as *efficiency*, where *I* equals the mass of compound to be treated and *E* equals the amount remaining after treatment; thus, DRE does not account for new compounds created during the treatment process. See *efficiency* and *treatment*.

Detection limit See *limit of detection.*

Detention basin System for the temporary impoundment of runoff.

Detritus Dead biota matter, usually at varying degrees of *decomposition* by microbes.

Developmental toxicity Adverse effects on developing organism that may result from exposure prior to conception (either parent), during prenatal development, or postnatally until the time of sexual maturation. The major manifestations of developmental toxicity include death of the developing organism, structural abnormality, altered growth, and functional deficiency.

Dewater 1. Remove or separate a portion of the water in a sludge or slurry to dry the sludge, so it can be handled and disposed. 2. Remove or drain the water from a tank or trench.

Diabetes Medical condition in which the body is unable to regulate and use glucose as a fuel source. Diabetes Mellitus, Type 2, also known as adult-onset diabetes, results in the body's cells resisting normal insulin effects, i.e., driving glucose in the blood into the cells. Type 1 Diabetes results because the pancreas does not make insulin when the body's immune system attacks the islet cells in the pancreas that make insulin.

Diatom Algal protist with siliceous cell wall (frustule); constitute a substantial group of phytoplankton.

Dichlorodiphenyltrichloroethane (DDT) Organochlorine pesticide, banned in many parts of the world due to associations with eggshell thinning, endocrine effects, and human effects. Still used to control mosquitoes and other disease *vectors*.

Diethylstilbestrol (DES) Synthetic nonsteroidal estrogenic *endocrine-disrupting compound* that caused cancer to children exposed *in utero*.

Diffused aeration Injection of air through submerged porous plates, perforated pipes, or other devices to form small air bubbles from which oxygen is transferred to the liquid as the bubbles rise to the water surface.

Diffusion 1. Process of net transport of solute molecules from a region of high concentration to a region of low concentration caused by their molecular motion and not by turbulent mixing. Graham's law of diffusion states that a gas diffuses at a rate inversely proportional to its density. Liquids also diffuse as a result of net spontaneous and random movement of molecules or particles from a region in which they are at a high concentration to a region of lower concentration. Diffusion will continue for a fluid until a uniform concentration is achieved throughout the region of the system. Diffusion is not a major mechanism of mass transport in rapidly flowing systems, such as air and surface waters, but is quite important in more quiescent systems, such as across cellular *membranes*, and in slow-moving regions of the environment, e.g., covered *sediment* and ground water. 2. Synonym for *dispersion*.

Digestion 1. In environmental biotechnology, the process of decomposing organic matter by microbial growth and metabolism. As such, organic matter is transformed and transferred to *sludge*, resulting in partial liquefaction, mineralization, and volume reduction. 2. Actions that occur in a *digester*.

Dilution A reduction in solute concentration caused by mixing with water at a lower solute concentration.

Dimer Molecule that consists of two identical simpler molecules: e.g., NO_2 can form the molecule $NO_2—O_2N$ or simply N_2O_4 that consists of two identical simpler NO_2 molecules.

Dinoflagellate Algal protist characterized by two flagella used in swimming in a spinning pattern; many are bioluminescent and an important group of marine phytoplankton. A few species are important marine pathogens.

Dioxin Highly *toxic*, *recalcitrant*, and *bioaccumulating* product of incomplete combustion and chlorination processes with a structure of two phenyl rings bonded by two oxygen atoms, with chlorine substitution. Most toxic form is 2,3,7,8-*tetrachlorodibenzo-para-dioxin*.

Diploid Cell with normal amount of *DNA* per cell, i.e., two sets of chromosomes or twice the *haploid* number.

Direct effects Effects that have a direct cause relationship with the occurrence of an event, most often represented as physical damage to people, goods, services, and/or the environment or by the immediate impact of social and economic activities.

Direct filtration A method of treating water that consists of the addition of coagulant chemicals, flash mixing, coagulation, minimal flocculation, and filtration. Sedimentation is not used.

Direct runoff Water that flows over the ground surface or through the ground directly into streams, rivers, and lakes.

Disaster A relative term meaning a catastrophic event that wreaks great destruction. However, the term is not exclusive to large-scale events, such as hurricanes or earthquakes, but can also include small-scale events with highly negative consequences, such as an engineering or medical failure where one or a few people are impacted but that has other implications (malpractice, bad publicity, blame, etc.).

Disaster Situation caused by a natural, technological, or human-induced phenomena, resulting in intense alterations for the people, goods, services, and/or the environment.

Discharge 1. Flow (*Q*) in a stream or canal or the outflow of a fluid from a source. Used in calculating liquid effluent from a facility or particulate or gaseous emissions into the air through designated venting mechanisms. 2. Any flow in an open or closed conveyance.

Discounting Process to determine the present value of a payment or a stream of payments that is to be received in the future.

Disease Abnormal and adverse condition in an organism.

Disparate effect Health outcome, usually negative, that is disproportionately high in certain members of a population, such as an increased incidence of certain cancers in minority groups.

Disparate exposure *Exposure* to a physical, chemical, or biological agent that is disproportionately high in certain members in a population, such as the higher-than-average exposure of minority children to lead.

Disparate susceptibility Elevated risk of certain members of a population (e.g., genetically predisposed) to the effects of a physical, chemical, or biological agent; can lead to disparate effects (see *disparate effect*).

Dispersion model Prediction tool of how a substance will behave after release.

Dispersion Spreading material in time and space.

Dispersion The spreading of a solute from the expected ground water flow path as a result of mixing of ground water.

Dissolution Act of going into solution; dissolving.

Dissolved oxygen (DO) Concentration of molecular O_2 in water.

Diversion System constructed to direct water to desired areas.

DNA ligase Enzyme that joins two *DNA* fragments together through the formation of a new phosphodiester bond.

DNA marker Cloned chromosomal locus with allelic variation that can be followed directly by a *DNA*-based assay such as Southern blotting or *PCR*.

Dose Amount of a substance available for interactions with metabolic processes or biologically significant receptors after crossing the outer boundary of an organism. Potential dose is the amount ingested, inhaled, or applied to the skin. Applied dose is the amount presented to an absorption barrier and available for absorption (although not necessarily having yet crossed the outer boundary of the organism). Absorbed dose is the amount crossing a specific absorption barrier (e.g., the exchange boundaries of the skin, lung, and digestive tract) through uptake processes. Internal dose is a more general term, denoting the amount absorbed without respect to specific absorption barriers or exchange boundaries. The amount of the chemical available for interaction by any organ or cell is termed the delivered or biologically effective dose for that organ or cell.

Dose-effect The relationship between dose (usually an estimate of dose) and the gradation of the effect in a population, that, is a biological change measured on a graded scale of severity, although at other times one may only be able to describe a qualitative effect that occurs within some range of exposure levels.

Dose-response assessment Process of characterizing the relationship between dose of an agent and the effect elicited by that dose.

Dose-response curve Curve with dose on the x-axis and response on the y-axis. The response is typically an adverse outcome for toxicology and a beneficial outcome for pharmacological efficacy and an adverse outcome for side effects.

Dose-response 1. Relationship between a quantified exposure (*dose*) and the proportion of subjects demonstrating specific biologically significant changes in incidence and/or in degree of change (response). 2. Correlation between a quantified exposure (*dose*) and the proportion of a population that demonstrates a specific *effect* (response).

Doubling time Time required for a population, e.g., bacterial, or cells to double in number or biomass.

Downgradient The direction that ground water flows; analogous to "downstream" for surface waters.

Drainage basin Land area that contributes stormwater and other runoff to a stream. Term usually denotes an area smaller than a watershed.

Drawdown Lowering of water table of an unconfined aquifer or the potentiometric surface of a confined aquifer caused by pumping of ground water from wells. Vertical distance between the original water level and the new water level. See *cone of depression*.

Drug Substance intended for use in the diagnosis, cure, mitigation, treatment, or prevention of disease, which is regulated by the U.S. Food and Drug Administration. Contrast with device and nutritional supplement.

Dry lab 1. *In silico* research (contrast with wet lab). 2. Walkthrough prior to actual laboratory work (step preceding wet lab). 3. Unethical practice of forging (making up) data.

Dual use 1. Science, engineering, and technology designed to provide both military and civilian benefits. 2. Research and technology that simultaneously benefit and place society at risk (e.g., biotechnological advances that improve vaccines but also increase the risks of bioterrorism).

Dynamical instability See *chaos theory*.

E. coli *Escherichia coli*, a Gram-negative, rod-shaped bacterium, often found in the intestine of warm-blooded animals.

Ecocentrism Perspective based on the whole ecosystem rather than a single species. Contrast with *anthropocentrism*.

Ecological impact Total effect of an environmental change, natural or of human origin, on the community of living things.

Ecological indicator A characteristic of the environment that, when measured, quantifies magnitude of stress, habitat characteristics, degree of exposure to a stressor, or ecological response to exposure. The term is a collective term for response, exposure, habitat, and stressor indicators.

Ecological risk assessment The application of a formal framework, analytical process, or model to estimate effects of human action(s) on a natural resource and to interpret the significance of those effects in light of the uncertainties identified in each component of the assessment process.

Ecology Science dealing with the relationship of all living things with each other and with their environment.

Ecosystem function Processes and interactions that operate within an ecosystem, including energy flow, nutrient cycling, filtering, and buffering of contaminants, and regulation of populations.

Ecosystem service Benefit to humans derived from ecosystems. *Anthropocentric* and *instrumental* value provided by natural ecosystems.

Ecosystem services Anthropocentric benefits of ecosystems, e.g., flood control, fisheries, and pharmaceutical sources.

Ecosystem structure Attributes related to instantaneous physical state of an ecosystem; examples include species population density, species richness or evenness, and standing crop biomass.

Ecosystem The interacting system of a biological community and its nonliving surroundings.

Ecotone A habitat created by the juxtaposition of distinctly different habitats, an edge habitat, or an ecological zone or boundary where two or more ecosystems meet.

Effect A biological change caused by an exposure.

Effectiveness Measure of the extent and degree to which a design achieves a goal.

Efficacy A measure of the probability and intensity of beneficial effects.

Efficiency Ratio of total energy or mass output to total energy or mass input, expressed as a percentage. Treatment or removal efficiency (η) is the product of the contaminant mass prior to treatment (I) times the contaminant mass after treatment (E) divided by I. To express efficiency as a percentage, these values are multiplied 100 times: $\eta = \frac{I \times E}{I} \times 100$.

Effluent Waste material discharged into the environment, treated or untreated. Generally refers to surface water pollution (analogous to *emission* in air pollution).

Effusion Escape of fluid into a body space or tissue. Effusion of a gas is inversely proportional to the square root of either the density or molecular weight of the gas. Compare to *diffusion*.

Electron acceptor 1. Oxidant. 2. Chemical substance, such as oxygen, nitrate, sulfate, and iron, which receives the electrons during microbial and chemical reactions. Microorganisms need these compounds to obtain energy. For MNA and EA, these electron acceptors often compete with chlorinated solvents and reduce the attenuation rates.

Electron donor 1. Reductant. 2. Chemical substance, such as molecular hydrogen or organic substrate, which yields an electron as they are oxidized, producing energy to sustain life and for the subsequent degradation of other chemicals, in this case, chlorinated solvents.

Electron A negatively charged subatomic particle that may be transferred between chemical species in chemical reactions.

Electron-transport chain Final steps of reactions occur in biological oxidation; composed of series of oxidizing agent (i.e., *electron acceptors*) arranged in sequence by increasing strength and terminating with oxygen (the strongest oxidizer).

Electrophoresis See *gel electrophoresis.*

Emission Release of a pollutant or other substance to the atmosphere (analogous to *effluent* in water pollution).

Endergonic Describing a reaction that does not spontaneously go to completion as written; the standard free-energy change is positive, and the equilibrium constant is less than 1.

Endocrine system Chemical messaging system in organisms used for regulation by secretion of hormones by glands that are sent through the circulatory system to cells where the hormones bind to receptors.

Endocrine-disrupting compound (EDC) or endocrine disruptor Chemical substance that interferes with endocrine system, as either an *agonist* or an *antagonist.*

Endogenous phase Microbial population growth period dominated by endogenous respiration.

Endoplasmic reticulum Organelle consisting of a network of membranes within the cytoplasm of cells, where proteins and lipids are synthesized.

Endospore Seed-like structure, formed by a microbe to survive during hostile conditions.

Endothermic Chemical reaction that absorbs energy from surroundings.

Endotoxin Heat-stable lipopolysaccharide in the outer membrane of the cell wall of Gram-negative bacteria that is released when the bacterium lyses, or during growth, and is toxic to the host.

Endpoint Observable or measurable biological event or chemical concentration (e.g., metabolite concentration in a target tissue) used as an index of an effect of a chemical.

Energy Capacity of a physical system to perform work.

Engineering 1. Application of scientific and mathematical principles to practical ends, especially design, manufacture, and operation of structures, machines, processes, and systems. 2. The profession that implements these applications.

Enhanced attenuation Any type of intervention that might be implemented in a source-plume system to increase the magnitude of attenuation by natural processes beyond that which occurs without intervention. Enhanced attenuation is the result of applying an enhancement that manipulates a *natural attenuation* process, leading to an increased reduction in mass of contaminants.

Enhanced bioremediation An engineered approach to increasing biodegradation rates in the subsurface.

Enteric bacteria 1. Members of the family Enterobacteriaceae, i.e., Gram negative, peritrichous, or nonmotile, facultatively anaerobic, straight rods with simple nutritional requirements. 2. Bacteria that live in the intestinal tract, e.g., enterococci.

Enthalpy of formation Standard enthalpy of formation of a substance is the enthalpy change for the formation of 1 mol of the substance from its elements, all in their standard states.

Enthalpy of reaction Heat of a reaction at constant pressure, usually 1 bar.

Entropy Entropy is a measure of disorder and can be determined by $\Delta S = k \ln W$, where k is the Boltzmann constant, and W is the number of possible arrangements of the molecule having the same energy.

Entropy, standard The entropy value of an ion or molecule for the standard state of the species, $S°$.

Environmental assessment (EA) Investigation whether a proposed action will adversely affect the environment. If so, in the United States, such action is usually followed by a formal *environmental impact statement.* If not, the agency will issue a "finding of no significant impacts" document.

Environmental engineering Subdiscipline of engineering (usually civil engineering) concerned with applications of scientific principles and mathematics to improve the condition of the environment.

Environmental impact statement (EIS) Document prepared by a government agency detailing the potential effects resulting from a major action being considered by that agency. In the United States, the EIS is required under the National Environmental Policy Act and is usually preceded by an *environmental assessment.*

Environmental science Systematic study of the environment and its components and processes (e.g., nutrient cycling, pollutant transport, and adverse effects).

Environmentalism 1. Advocacy in the protection of the environment. 2. Philosophy underpinning this advocacy. Such advocacy may or may not be scientifically based (i.e., differs from environmental science and environmental engineering).

Enzyme Protein catalyst with specificity for both the reaction catalyzed and its substrates.

Enzyme-linked immunosorbent assay (ELISA) A technique used for detecting and quantifying specific antibodies and antigens.

Epidemic Disease outbreak that occurs simultaneously or nearly simultaneously in a large area or in large percentage of a population.

Epidemiology 1. Study of the causes, distribution, and control of disease in populations. 2. The study of the distribution and determinants of health-related states or events in specified populations.

Epigenetics Concern with mechanisms that regulate gene activity.

Equilibrium constant Value representing relationship between compounds in a system that has reached equilibrium. Related to *partitioning coefficient*, e.g., *octanol-water coefficient*, *bioconcentration factor*, and *Henry's law constant*.

Equilibrium Condition in which a reaction is occurring at equal rates in its forward and reverse directions so that the concentrations of the reacting substances do not change with time.

Equivalent uniform annual cost (EUAC) Cost-per-year of owning, operating, and maintaining an asset over the course of its entire lifespan.

Erosion Wear and displacement of land surface materials by flowing water, wind, or other geological agents.

Error 1. Mistake. 2. In statistics, the difference between a reported value and the actual value (see *Bias*).

Erythrocyte Red blood cell made in bone marrow. Contains *hemoglobin*.

Estimated exposure dose (EED) The measured or calculated dose to which humans are likely to be exposed considering all sources and routes of exposure.

Ethics 1. Set of moral principles. 2. Study of morality and moral decision making.

Ethylenediaminetetraacetic acid (EDTA) Chelating agent that binds to and makes unavailable metal ions in a solution; because certain cations are essential for many enzymes to function, EDTA is applied to halt enzymatic and cellular activity (as such, is a common preservative).

Eukaryote Organism whose cell contains a distinct, membrane-bound nucleus.

Eutrophic Lake or estuary with low concentrations of dissolved oxygen due to high concentrations of nutrients and having high productivity.

Eutrophication Process by which water bodies receive excess nutrients, primarily nitrogen and phosphorus, which stimulate excessive algal and plant growth.

Evaporation Process by which a liquid becomes a vapor.

Evapotranspiration Process by which water moves from the earth's surface to the atmosphere by *evaporation* and plant *transpiration*.

Event tree Diagram of the flow of events following an initial event, showing subsequent possible events toward different outcomes. Each event has its own possible outcomes so that the critical path chosen will result in numerous potential outcomes.

Event Natural, technological, or human-induced phenomena, in terms of its characteristics, severity, location, and area of influence. It is the registration in time and space of a phenomenon that characterizes a threat.

Event Set of outcomes that are preceded and linked to an earlier set of outcomes (probability theory).

Ex situ Moved off-site (e.g., contaminated soil transported to an incinerator for treatment).

Ex vivo Outside the body, frequently the equivalent of *in vitro* (see *in vitro*).

Exergonic Describing a reaction that spontaneously goes to completion as written; the standard free-energy change is negative, and the equilibrium constant is greater than one.

Exfiltration Downward transport of runoff through the bottom of channel or other water conveyance into the soil.

Exothermic Chemical reaction that releases energy, usually heat.

Exotoxin Heat-labile, toxic protein produced by a bacterium as a result of its normal metabolism or because of the acquisition of a plasmid or prophage that redirects its metabolism, usually released into the bacterium's surroundings.

Experiment Investigation to support or reject a hypothesis or to increase knowledge about a phenomenon.

Exponential growth phase Microbial growth at a constant percentage per unit time, i.e., log phase growth.

Exposome Expression of all the exposures of an individual in a lifetime and how those exposures relate to health, beginning after conception as *in utero* exposure and including insults from environmental and occupational sources. An exposome describes one's biology, locations, and activities over lifetime.

Exposure assessment Identification and evaluation of the human population exposed to a toxic agent, describing its composition and size, as well as the type, magnitude, frequency, route, and duration of exposure.

Exposure concentration Concentration of a substance in its transport or carrier medium at the point of contact.

Exposure pathway Physical course a substance takes from the source to the organism exposed, e.g., through the air, drinking water, product use, or food consumption.

Exposure route The way a substance enters an organism after contact, e.g., by ingestion, inhalation, or dermal absorption.

Exposure scenario Set of facts, assumptions, and inferences regarding how exposure occurs to support risk assessors in evaluating, estimating, or quantifying exposures.

Exposure Contact made between a chemical, physical, or biological agent and the outer boundary of an organism. Exposure is quantified as the amount of an agent available at the exchange boundaries of the organism (e.g., skin, lungs, gut).

Expression (genetic expression) Effect on cell resulting from the gene's instructions in *transcription*.

Extended aeration Enhancement of activated-sludge process using long aeration periods to promote aerobic digestion of the biological mass by endogenous respiration; includes stabilization of organic matter under aerobic conditions and disposal of the gaseous end products into the air. Effluent contains both dissolved and fine, suspended matter.

Extrapolation 1. Estimate of the extent of conditions from measured data. 2. Estimate of the response at a point above or below the range of the experimental data, generally using a mathematical model. Compare to *interpolation*.

Fact That which can be shown to be true, to exist, or to have occurred.

Facultative anaerobe Microorganism, usually a bacterium, which grows equally well under *aerobic* and *anaerobic* conditions.

Facultative pond Most common type of treatment pond in current use. The upper portion (*supernatant*) is *aerobic*, and the bottom layer is *anaerobic*. *Algae* supply most of the oxygen to the supernatant.

Failure rate Anticipated number of times that an item will fail in a specified period of time. It is a calculated value that gives a measure of reliability for an item or product.

Failure Lack of success as indicated by design specifications and measures of success.

False negative Finding of the absence of a condition (e.g., disease) in a test when, in fact, the disease is present (e.g., a lung cancer screen shows that the patient has no cancer, but, at a level of detection below the screen, the cancer has cancer cells in the lung). See *type II error*.

False positive Positive finding of a test when, in fact, the true result was negative (e.g., a drug screen shows that a person has used opiates, even though the person has not). See *type I error*.

Fault tree analysis (FTA) Failure analysis in which an undesired state of a system is analyzed by combining a series of lower-level events. Mainly used in the field of safety engineering to find probability of a safety hazard.

Fecal coliform Coliform within the intestinal tract as its normal habitat and that can grow at 44.5 °C.

Feedstock Material entering a *reactor*.

Fermentation Energy-yielding process in microbes oxidizes an energy substrate without an exogenous electron acceptor. Usually, organic molecules serve as both electron donors and acceptors.

Finding of no significant impact (FONSI) Statement by an agency after completing an *environmental assessment* that a proposed action will not lead to significant impacts, so an *environmental impact statement* would not be required for this action.

Fire Exothermic process wherein matter is oxidized by rapid combustion, releasing heat, light, and reaction products.

Fishbone diagram See *Ishikawa diagram*.

Fixed phase reactor System in which solid-phase particles are fixed in position as the fluid phase passes through.

Flagellum Threadlike appendage on numerous prokaryotic and eukaryotic cells responsible for their motility.

Flare Device that combusts gaseous materials exiting a system; e.g., a landfill flare burns methane, and a chemical processing facility includes a flare backup system in the event of unplanned releases of otherwise toxic substances.

Floc Particles, including cells, which adhere to one another loosely to form clusters.

Flocculation Process by which *flocs* are formed; often enhanced in water treatment processes by addition of flocculant, e.g., alum—aluminum sulfate [$Al_2(SO_4)_3$].

Floodplain The area of land near a stream that would be covered with water after a given flood event. It is usually defined by the size of the event, e.g., a 100-year floodplain.

Flow cytometer Instrument with a laser detector and a very small orifice through which particles (including microbes) flow through one at a time. As they pass through a laser beam, biochemicals may be determined on a per-cell basis.

Flow splitter Engineered hydraulic structure that diverts a portion of flow, e.g., a portion of stormwater flow into a bypass system.

Fluidized bed reactor System that suspends small solid particles in an upwardly flowing stream of fluid. Fluid velocity must be sufficient to suspend the particles, but not so high as to transport them out of the vessel. Mixing occurs as the solid particles swirl around the bed rapidly. The material fluidized is nearly always solid, and the fluidizing medium can be either a liquid or a gas.

Flux Rate of flow of fluid, particles, or energy through a given two-dimensional surface.

Food chain A biological system in which individuals at a trophic level feed on organisms in the trophic level below theirs. As such, energy is transferred from level to level.

Food web Complex of interrelated *food chains* in an ecological community.

Force Capacity to do work. Interaction between two physical bodies, e.g., an object and its environment. Proportional to acceleration and the derivative of momentum with respect to time.

Free energy (ΔG) Intrinsic energy in a substance available to do work, especially to drive chemical transformations. Also known as Gibbs free energy.

Free energy, standard The standard free energy or standard Gibbs free energy is the standard free energy of reaction per mole of compound when it is formed from its elements, all in their standard states.

Freundlich sorption isotherm See *sorption isotherm.*

Fugacity Tendency of a substance to prefer one phase over another, and tendency to flee or escape one compartment (e.g., water) to join another (e.g., atmosphere).

Functional ecology Ecology, with an emphasis on the roles and behavior of species in ecosystems. Compare to *structural ecology.*

Fungus Achlorophyllous, heterotrophic, spore-bearing eukaryotes with absorptive nutrition, often with a walled thallus.

Fuzzy logic System dealing with the partial truths, assigning values ranging from completely true to completely false.

Game theory Decision making under conditions of uncertainty and interdependence; considering, the characteristics of players, strategies, actions, payoffs, outcomes at equilibrium.

Gamete Reproductive cell (i.e., an egg or a sperm).

Gas chromatography Separation and analysis of compounds in a mixture by injecting an analyte into a heated system and detecting compounds as they exit the heated column. Lower boiling point compounds exit first.

Gene flow Movement of genes from one population to another.

Gene Ordered sequence of nucleotides located in a particular position on a particular chromosome, representing the fundamental unit of heredity. *DNA* segment or sequence that codes for a polypeptide, *rRNA* or *tRNA*.

Generation time See *doubling time.*

Genetic engineering (GE) Modification of the structure of genetic material in a living organism, involving the production and use of *recombinant DNA.*

Genetic material *Deoxyribonucleic acid* and *ribonucleic acid.*

Genetically modified microbe (GMM) Subdivision of *genetically modified organisms* that includes bacteria, fungi, algae, and other microbes whose genetic material has been altered.

Genetically modified microbial pesticide Bacteria, fungi, viruses, protozoa, or algae, whose *DNA* has been modified to express pesticidal properties. The modified microorganism generally performs as a pesticide's active ingredient. For example, certain fungi can control the growth of specific types of weeds, while other types of fungi can kill certain insects.

Genetically modified organism (GMO) Organism whose genetic material has been changed in a way that does not occur under natural conditions through cross-breeding or natural recombination.

Genetics Scientific investigation of heredity.

Genome Entire genetic complement, i.e., all the hereditary material possessed by an organism.

Genomics 1. Study of genes, including their functions. 2. Study of the molecular organization of genomes, their information content, and the gene products they encode.

Genotype Combination of alleles, situated on corresponding chromosomes, that determine a specific trait of an individual.

Geographic information system (GIS) Mapping system that uses computers to collect, store, manipulate, analyze, and display data. For example, GIS can show the concentration of a contaminant in an ecosystem with respect to land cover, water depth, and potential sources of the contaminant.

Geometric mean The nth root of the product of n values.

Germ theory Paradigm that diseases are caused by singular, proximate, pathogenic microbes. Displaced miasma theory (see *Miasma*) in late 19th century.

***Gestalt* theory (German: Form)** View that perception and other psychological phenomena must be understood for their overall patterns and forms, as opposed to the individual components.

Gestation Time from fertilization of the ovum to birth. Also known as uterogestation.

Gibbs free energy See *free energy.*

Glycolysis Anaerobic conversion of glucose to lactic acid via the *Embden-Meyerhof-Parnas pathway.*

Glyoxylate cycle Modification of Kreb's cycle in certain bacteria wherein *acetyl coenzyme A* is generated directly by oxidization of light lipids (e.g., *fatty acids*).

Grab sample Single sample of environmental material collected without regard to flow or time.

Gram negative Describing a bacterial cell that retains crystal violet during staining processes and is then colored by a counterstain, e.g., *Thiobacillus* and *Pseudomonas.*

Gram positive Describing a bacterial cell that loses crystal violet during staining processes, e.g., *Bacillus.*

Gram stain Differential staining procedure that divides bacteria into Gram-positive and Gram-negative groups based on their ability to retain crystal violet when decolorized with an organic solvent such as ethanol.

Green engineering Design, commercialization, and use of processes and products, which are feasible and economical while minimizing the generation of pollution at the source and the risk to human health and the environment.

Greenhouse effect Physical process by which incoming solar radiation is reradiated as infrared wavelengths from the earth's surfaces. In turn, the heat is retained by radiant gases (i.e., *greenhouse gases*) so that the earth stays warm. Without these gases, virtually all the heat would be returned to space so that the diurnal heat variations would range from extremely hot during the day and extremely cold at night. Thus the greenhouse effect is essential to life on earth. However, the increase in greenhouse gas concentrations in the troposphere is causing concern within the scientific community, with fears of global warming and other changes in global climate.

Greenhouse gas (GHG) Gas released to the atmosphere that in turn retains heat that has been radiated from the earth.

Gross primary productivity Total amount of carbon compounds produced by photosynthesis of plants and microbes in an ecosystem in a given period of time. Compare to *net primary productivity.*

Ground water Water below the earth's surface, usually in an *aquifer.*

Guanine Purine derivative, 2-amino-6-oxypurine, found in nucleosides, nucleotides, and nucleic acids.

Half-life ($t_{1/2}$) Time needed for half the quantity of a substance taken up by a living organism to be metabolized and eliminated by normal biological processes. Also called biological half-life. Also applicable to any chemical reaction.

Haploid Cell with only one complete set of chromosomes.

Harm principle John Stuart Mill's recommendation that utilitarianism's (see *Utilitarianism*) premise of greatest good is restricted (e.g., by law) when others are harmed.

Harm Damage to another person or creature.

Hazard assessment 1. The process of determining whether exposure to an agent can cause an increase in the incidence of a particular adverse health effect (e.g., cancer, birth defect) and whether the adverse health effect is likely to occur in humans. 2. Process by which the probability that an event will occur is determined for a specific site and time.

Hazard assessment Process by which the probability that an event will occur is determined for a specific site and time.

Hazard characterization A description of the potential adverse health effects attributable to a specific environmental agent; the mechanisms by which agents exert their toxic effects; and the associated dose, route, duration, and timing of exposure.

Hazard rate Probability that a component will fail in a short interval of time given it has survived up till that moment per unit interval of time.

Hazard 1. Underlying danger associated with physical phenomena of natural, technological, or anthropogenic origin that can occur at a specific site and at a defined time, producing severe effects to the people, goods, services, and/or the environment; based on the probability that an event of a certain magnitude will occur, at a specific site and determined time. 2. Potential source of harm. Term is often formally defined and distinguished from nonhazards by regulatory agencies, e.g., hazardous and nonhazardous wastes.

Hemoglobin Protein in an erythrocyte that carries oxygen from the lungs to organs and tissues and carries carbon dioxide back to the lungs.

Henry's Law Relationship between the partial pressure of a compound and its equilibrium concentration in a dilute aqueous solution through a constant of proportionality, i.e., Henry's law constant. Expression of *fugacity.*

Herbicide-tolerant crop Crop that contains new genes that allow the plant to tolerate herbicides. The most common herbicide-tolerant crops (cotton, corn, soybeans, and canola) are those that are resistant to glyphosate, an effective herbicide used on many species of grasses, broadleaf weeds, and sedges.

Heterocyst Specialized cell produced by cyanobacteria; sites of nitrogen fixation.

Heterogeneous In chemistry, process involving more than one physical phase.

Heterologous encapsidation (transcapsidation) Generation of "new" viruses by surrounding one virus with the envelope protein of another virus; a natural process that can occur when plants are coinfected by different strains of viruses.

Heterotroph Organism needs carbon-energy compounds.

Heterozygous Having different alleles at a genetic locus.

Hill's criteria Minimal conditions necessary to establish causal relationship between two items; presented by British medical statistician Sir Austin Bradford Hill (1897–1991) as a means of finding causal links between a specific factor (e.g., exposure to air pollution) and specific adverse effects (e.g., asthma). These criteria, originally recommended for occupation setting but now applied to numerous health and environmental problems, are meant to be guidelines rather than inviolable rules of *epidemiology*.

Hodgkin's Disease Cancer of the immune system marked by the presence of the *Reed-Sternberg cell*. The two major types are classic Hodgkin lymphoma and nodular lymphocyte-predominant Hodgkin lymphoma. Signs and symptoms include painless, swollen lymph nodes, an enlarged spleen, fever, weight loss, fatigue, and night sweats. Also called Hodgkin lymphoma.

Homeostasis Ability of an organism to self-regulate functions; inherent trend toward stability.

Homogeneous In chemistry a reaction or other process involving substances all in the same physical phase.

Homozygous Having two identical alleles of a gene.

Horizontal gene transfer Process by which an organism incorporates genetic material from another organism without being the offspring of that organism.

Hormonally active agent Substances that possess hormone-like activity, regardless of mechanism. Synonymous with *endocrine disruptor*.

Hormone Chemical released by glands of the endocrine system (see *endocrine system*).

Host Organism that harbors another organism; microenvironment that shelters and supports the growth and multiplication of a parasitic organism.

Hours of retention (HRT) Common unit of time that a substance is held in a *bioreactor*.

Household hazardous waste (HHW) Discarded household products that contain corrosive, toxic, ignitable, or reactive ingredients.

Huanglongbing See *citrus greening disease*.

Humus Dark-colored organic material in soil and sediment that is a product of plant material decomposition.

Hybrid Offspring of genetically dissimilar parents or stock.

Hybridization 1. Act of mixing different species or varieties of organisms to produce *hybrids*. 2. Reaction by which pairing of complementary strands of nucleic acid occurs. Usually double stranded, when the strands of *DNA* are separated, they will rehybridize under appropriate conditions. Hybrids can form between DNA-DNA, DNA-RNA, and RNA-RNA.

Hydraulic conductivity A measure of the capability of a medium to transmit water.

Hydraulic fracturing Forcing open fissures in rock strata by injecting liquids under high pressure. Used to extract fossil fuels and for *biostimulation* and *bioaugmentation* in ground water remediation.

Hydraulic gradient Change in hydraulic head (per unit distance in a given direction) typically in the principal flow direction.

Hydraulic Related to moving liquids.

Hydric soil Soil that is sufficiently wet for enough time to have anaerobic conditions in the upper horizons.

Hydrograph Plot of measurements from a specific gaging station showing a stream's flow, depth, and/or discharge versus time.

Hydrolysis Decomposition or change of a chemical compound by reaction with water, such as the dissociation of a dissolved salt or the catalytic conversion of starch to glucose.

Hydrophilic Describing a polar substance with a strong affinity for water (i.e., high aqueous solubility).

Hydrophobicity Tendency of a substance to repel water or to be incapable of completely dissolving in water. Hydrophobic substances are readily soluble in many nonpolar solvents, such as octanol, but only sparingly soluble in water, a polar solvent. That is, most hydrophobic compounds are also *lipophilic*. The hydrophobicity of an organic contaminant influences the fate of the contaminant in the environment. In general, the more hydrophobic a contaminant is, the greater the likelihood it will be associated with nonpolar organic matter such as humic substances and lipids (fats). Can be predicted fairly well by the *octanol-water coefficient*.

Hypothetico-deductive method Method of logical deduction, attributed to Karl Popper (*The Logic of Scientific Discovery*, 1934), limiting scientific discovery to that which is testable; requiring an approach that formulates hypotheses, *a priori*, with the intent of rejecting these hypotheses. The method assumes that a hypothesis can never truly be proved but at best can be corroborated.

Hysteresis 1. Changes that occur depending on the direction taken in a pathway; e.g., a material may behave differently in the same temperature range when cooled than when heated in that same range. In mechanics, the changes of a body as it returns to its original shape after being stressed. 2. Failure to return to previous condition, such as due to an energy loss that always occurs under cyclical loading and unloading of a spring, proportional to the area between the loading and unloading load-deflection curves within the elastic range of a spring (engineering), or the failure of a variable to return to its initial equilibrium after a temporary shock (economics).

Ideal gas law The product of pressure and volume of a gas is equal to the product of amount of gas and temperature: $n/V = P^0/RT$, where V is the volume of the container, n is the number of moles of chemical, R is the molar gas constant, and n/V is the gas phase concentration ($\text{mol}\,\text{L}^{-1}$) of the chemical.

Imhoff tank Two-story vessel for both settling and *digestion* occurs in a waste treatment system, with one compartment below the other. This allows for a stepped and separate *aerobic* and *anaerobic* treatment.

Impoundment Artificial water storage, e.g., lake, pit, or reservoir.

In silico Based on information, usually using computational methods, rather than using actual materials being studied. To some extent, *in silico* research is an alternative to *in vivo* and *in vitro* research (see *in vivo* and *in vitro*), which is desirable in the case of limiting animal research and in reducing risks in humans who undergo *in vivo* procedures.

In situ Taking place where it is found (e.g., bioremediation of a hazardous waste site where it exists, rather than moving the materials off-site).

In utero In the womb (e.g., fetal alcohol syndrome results from the unborn child's exposure to alcohol and its metabolites during gestation (see *gestation*)).

In vitro Outside the organism (literally: "in glass," i.e., in a test tube).

In vivo Inside the organism (e.g., experiments within a rat to observe biochemical responses to a chemical dose).

Incidence or incidence rate Number of new cases in a defined population within a period of time (compare to *prevalence*).

Index of biological integrity (IBI) Method of indicating the quality of aquatic systems. Usually, the total number of organisms and the number of different species present are inventoried, followed by the application of an index, or scale, that lists organisms according to their sensitivity to pollution.

Indicator organism Organism whose abundance indicates the condition of a substance or environment. For example, the potential presence of pathogens in fecal pollution is indicated by *coliforms*.

Indirect effects Effects that indirectly influence the social and economic activities of the population as well as the environment.

Inductive reasoning Starting from a specific experience and drawing inferences from the specific set of facts or instances to a larger number of instances (generalization). Conclusions are drawn from the perspective that all individuals of a kind have a certain character on the basis that some individuals of the kind have that character. Contrast with *deductive reasoning*.

Industrial ecology Study of industrial system that focuses on material cycling, energy flow, and the ecological impacts of such systems.

Inert ingredient (inert) 1. Nonreactive ingredient. 2. Any ingredient in a pesticide formulation other than the ones that provide the mechanisms of biocidal action. An inert ingredient may or may not be reactive or may or may not be toxic. See *active ingredient*.

Inference Reasoning that one statement (the conclusion) is derived from one or more other statements (the premises). See *syllogism*.

Informatics Application of computational and other technologies to access and to enhance information; one means of turning data into information (see *data* and *information*).

Information Processed and organized *data*. Value-added data as a step toward knowledge.

Inhalation unit risk (IUR) Upper-bound excess lifetime cancer risk estimated to result from continuous exposure to an agent at a concentration of $1\,\mu\text{g}\,\text{m}^{-3}$ in air.

Initiation The first stage of carcinogenesis.

Inoculum Microbial culture used to initiate growth and introduced to a container.

Inorganic compound A compound that is not based on covalent carbon bonds, including most minerals, nitrate, phosphate, sulfate, and carbon dioxide.

Instrumental value Worth based on usefulness. In biomedical ethics, the perspective of whether a human life has value that depends on usefulness (Will the baby be loved? Will the elderly person continue to enjoy life?) is an instrumental viewpoint. In environmental ethics, the use of the term environmental resource implies that ecological value is based on the utility of the ecosystem (e.g., wetlands as breeding area for game fish, as retention areas to prevent floods, and as sinks for carbon to prevent global warming). Contrast with *intrinsic value*.

Intake rate (IR) Rate of inhalation, ingestion, and dermal contact depending on the route of exposure. For ingestion, the intake rate is simply the amount of food containing the contaminant of interest that an individual ingests during some specific time period (units of mass per time). For inhalation, the intake rate is the rate at which contaminated air is inhaled. Factors that affect dermal exposure are the amount of material that comes into contact with the skin, and the rate at which the contaminant is absorbed.

Intake Process by which a substance crosses the outer boundary of an organism without passing an absorption barrier, e.g., through ingestion or inhalation.

Integrated pest management (IPM) Combination of various strategies to reduce pests, rather than simply relying on the application of pesticides, including use of natural predators, physical removal of breeding areas (e.g., standing water for mosquitoes), and introduction of organisms that lead to sterile offspring, i.e., interruption of a pest or vector's life cycle.

Integrated Risk Information System (IRIS) Internet-based system that provides information related to the human health risk of hundreds of specific chemical compounds and classes of compounds.

Intensity Qualitative and quantitative measure of the severity of a phenomenon at a specific site.

Interspecies dose conversion The process of extrapolating from animal doses to human equivalent doses.

Intervention Direct involvement of corrective action to change existing condition for the better.

Intrinsic value Worth based on existence, not usefulness. All humans have intrinsic value in contemporary morality. In biomedical ethics, however, there is no unanimity of thought about the intrinsic value of an embryo or a fetus, or a person nearing end of life. Those subscribing to sanctity of life viewpoints see intrinsic value of any human being (beginning with the human zygote and ending in natural death). In environmental ethics, there is no unanimity of thought about nonhuman species. For example, the loss of a species is morally wrong based on the value of the existence of the species, not its actual or potential value (e.g., as a cure for cancer or as a food source for a food species). Contrast with instrumental value.

Intuition Direct perception of meaning without conscious reasoning. Compare to deductive and inductive reasoning.

Ion exchange Transfer of ions between two electrolytes or between an electrolyte solution and a complex. Physicochemical process in which ions held electrostatically on the surface of a solid phase are exchanged with ions of similar charge in a solution (e.g., drinking water).

Ionic strength Related to solution's total concentration and valences of ions.

Irreversible sorption A hysteresis effect in which a chemical species becomes more strongly bound over time. The term sometimes appears to be used to describe a situation where, once sorbed, the contaminant is removed from the plume and remains associated with the soil.

Ishikawa diagram Graphical technique for identifying cause-and-effect relationships among factors in a given situation or problem (also known as a fishbone diagram).

Isolation Separation of specific microorganisms from cultures.

Isotope One of the two or more atoms having the same atomic number but different mass numbers.

Key event One of the steps following the *molecular initiating event* that leads to an adverse outcome in an *adverse outcome pathway*.

Kinetics Rates of reactions and processes.

Kjeldahl nitrogen Ammonia and organic nitrogen in water.

Knock-in Targeted mutation in which an alteration in gene function other than a loss-of-function allele is produced.

Knock-out Targeted mutation in which a loss of function (often a null allele) is produced.

Knowledge Familiarity, awareness, or understanding gained through experience or study. A necessary step toward wisdom.

Kreb's cycle Oxidative pathway in respiration by which pyruvate, via *acetyl coenzyme A*, is decarboxylated to form CO_2.

Lag phase In microbial growth, period after inoculation and before exponential (log) growth.

Lag time Interval between the center of the mass of a precipitation event and the peak flow of the runoff caused by the event.

Lagoon Shallow surface water system used for wastewater treatment; often aerated mechanically to increase *aerobic* decomposition of waste material.

Laminar Describing flow of a viscous fluid in which particles of the fluid move in parallel layers, each of which has a constant velocity but is in motion relative to its neighboring layers. Also called streamline or viscous flow.

Land farming Addition of waste material, e.g., organic compound-laden waste, to the soil surface for *biodegradation*. The soil may be moistened or mixed to stimulate the desired degradation process.

Latency period The time between first exposure to an agent and manifestation or detection of a health effect of interest. Period of time between disease occurrence and detection, sometimes used interchangeably with induction.

Latency Period of time between disease occurrence and detection, sometimes used interchangeably with induction.

Law of diminishing returns Economic principle espoused by Thomas Malthus (*Essay on the Principle of Population*, 1798) stating that when a fixed input is combined in production with a variable input, using a given technology, increases in the quantity of the variable input will eventually depress the productivity of the variable input. Malthus proposed this as a law from his pessimistic idea that population growth would force incomes down to the subsistence level.

Law of supply and demand Economic principle stating that, in equilibrium, prices are determined so that demand equals supply; thus, changes in prices reflect shifts in the demand or supply curves.

Leachate collection system Assemblage of monitoring wells and above ground monitors to measure concentrations of substances migrating from a landfill.

Leachate 1. Percolated liquid through solid waste or other permeable material. 2. Extracted materials from this liquid.

Lethal concentration (LC_X) Concentration of a substance in air in which $X\%$ of test animals die, e.g., the median lethal concentration is LC_{50}: a common measure of *acute toxicity*.

Lethal dose (LD_X) Amount of a substance delivered to a test animal in a single dose that kills $X\%$, e.g., the median lethal dose is LC_{50}: a common measure of *acute toxicity*.

Lichen Organism composed of a fungus and either green algae or cyanobacteria in a symbiotic association.

Life cycle costing (LCC) Process of compiling all the costs the asset will incur over its lifespan.

Life Period from onset (i.e., conception) to end (i.e., death) of a unique organism. Antonym can be either death or nonliving.

Lifetime average daily dose (LADD) Estimated dose to an individual averaged over a lifetime of 70 years; used in assessments of *carcinogenic risk*.

Ligand Molecule that travels through the bloodstream as chemical messengers that will bind to a target cell's *receptor*.

Light nonaqueous-phase liquids (LNAPL) Immiscible organic liquid that is denser than water, e.g., gasoline and motor oils.

Lignin Organic polymer stored in plant cell walls of woody plants; an aromatic hydrocarbon compound forming a three-dimensional structural matrix.

Limit of detection (LOD) Lowest concentration of a chemical that can reliably be distinguished from a zero concentration (also known as detection limit).

Linearity Following the mathematical equation for a line ($y = mx + b$, where m is the slope and b is the y intercept). Also used to describe the degree to which data points approximate the line of best fit (linear regression).

Liner Clay or manufactured material that serves as a barrier against the movement of *leachate*. Liners have very low *hydraulic conductivity*.

Lipophilicity Tendency of a chemical compound to dissolve in fats, oils, lipids, and nonpolar solvents. See *hydrophilicity*.

Liquid chromatography Separation of molecules in a liquid mobile phase by using a solid stationary phase through which the mobile phase that contains the analyte passes.

Log-growth phase Also known as trophophase, portion of a population's life cycle when the number of organisms is increasing at logarithmic rate.

Logic Branch of philosophy addressing inference (e.g., using a syllogism (see *syllogism*) to determine the validity of an ethical argument).

Longitudinal study Epidemiological study using data gathered at more than one point in time, e.g., after an exposure or a medical intervention. Contrast with cross-sectional study.

Loss Any economic, social, or environmental value lost by a variable during a specific exposure time.

Lowest observed adverse effect level (LOAEL) Lowest exposure level at which there are biologically significant increases in frequency or severity of adverse effects between the exposed population and its appropriate control group. Compare to *NOAEL*.

Lymphocyte White blood cell. Immune cell made in the bone marrow and found in blood and in lymph tissue. Two main types are B lymphocytes and T lymphocytes. B lymphocytes make antibodies, and T lymphocytes help kill tumor cells and aid in controlling immune responses.

Machine learning Computer systems that can learn and adapt without following explicit instructions, using algorithms and statistical models to analyze and draw inferences from patterns in data.

Malaria Serious infectious disease caused by the parasitic protozoan *Plasmodium*. Characterized by bouts of high chills and fever that occur at regular intervals.

Manning's equation Equation that predicts water flow velocity in an open channel or in an enclosed conduit.

Mass balance Assessment includes a quantitative estimation of the mass loading to the dissolved plume from various sources, as well as the mass attenuation capacity for the dissolved plume.

Mass loading Contaminant released to the environment from the source material.

Mass spectroscopy Analytical method that measures the mass-to-charge ratio (*m/z*) of one or more molecules present in a sample and observing the peaks to determine the quantity of various organic compounds in a sample.

Mass transfer The irreversible transport of solute mass from the nonaqueous phase (i.e., *DNAPL*) into the aqueous phase, the rate of which is proportional to the difference in concentration.

Maximum contaminant level (MCL) Maximum concentration of a contaminant in water in a public water supply system permitted under the Safe Drinking Water Act (U.S.).

Maximum contaminant level goal (MCLG) The contaminant concentration in drinking water below which there is no known or expected risk to health. MCLGs allow for a margin of safety and are not enforceable public health goals.

Maximum Residual Disinfectant Level (MRDL) Highest concentration of a disinfectant allowed in drinking water.

Maximum Residual Disinfectant Level Goal (MRDLG) The drinking water disinfectant concentration below which there is no known or expected risk to health. Do not account for the benefits of the use of disinfectants to control microbial contaminants.

Mean Squared Error (MSE) Measure of nearness of a regression line to a set of data points.

Mechanism of action Specific biochemical interaction through which a substance produces its intended effect in the case of a pesticide or drug, or toxic reaction in the case of a toxic substance. The mechanism is usually characterized by specific molecular targets to which the substance binds, such as to an enzyme or to a receptor. For example, numerous pesticides' mechanism of action is by inhibition of neurotransmitters, e.g., acetylcholine. Organochlorine pesticides often alter the movement of ions across the nerve cell membranes, changing the ability of the nerve to fire. Organophosphate and carbamate pesticides act primarily at the synapses, altering the regulation of the transmission of signal between neurons.

Median value Value in a data set such that half the values are greater and half are less.

Medium 1. Substance in which organisms are grown. 2. Material in a bioreactor, e.g., *trickling filter*, on which microbes grow and produce *biofilm*. 3. Environmental compartment (e.g., air, water, soil, sediment, or biota).

Meiosis Process by which diploid germ cell precursors segregate their chromosomes into haploid nuclei within eggs and sperm.

Meltdown, nuclear Severe damage to a nuclear power plant's core, resulting in overheating to the point that it melts the materials (floor, soil) allowing the core to sink.

Membrane bioreactor (MBR) System that combines suspended growth with solids separation using ultrafine porous membranes; often follows aeration step in *activated-sludge* treatment.

Membrane 1. Film with pores. 2. Cellular amphiphilic layer that encloses the cell or separates parts within a cell.

Mendelism Heredity theory underlying classical genetics, proposed by Roman Catholic monk and scientist Gregor Mendel in 1866.

Mesotrophic Lake or estuary with an intermediate level of productivity.

Messenger RNA (mRNA) *RNA* containing sequences coding for a protein.

Metabolism Act of a living organism converting and degrading a substance from one form to another (known as a metabolite). Chemical reactions in living cells that convert food sources to energy and new cell mass. See *catabolism* and *anabolism*.

Metabonomics Study of total metabolite pool (especially using computational methods).

Methanogen Strictly anaerobic Archaebacteria able to use only a very limited substrate spectrum (e.g., molecular hydrogen, formate, methanol, carbon monoxide, or acetate) as substrates for the reduction of carbon dioxide to methane.

Michaelis constant Kinetic constant for an enzyme reaction equal to the substrate concentration required for the enzyme to operate at half-maximal velocity.

Microarray A multifaceted tray or array of *DNA* material. Microarrays are expected to revolutionize medicine by helping pinpoint a very specific disease or the susceptibility to it. Sometimes called "biochips," microarrays are commonly known as "gene chips."

Microbe Microorganism.

Microbial ecology Study of microorganisms in their natural environments, with a major emphasis on physical conditions, processes, and interactions that occur on the scale of individual microbial cells.

Microbiology Study of *microorganisms* (those too small to be seen with the naked eye). Special techniques are required to isolate and grow such organisms.

Microcosm A batch reactor used in a bench-scale experiment designed to resemble the conditions present in the ground water environment.

Microenvironment Location of a potential or actual exposure in which an activity occurs, e.g., a garage is an indoor microenvironment; the interior of an automobile is a mobile microenvironment.

Microinjection Process by which *DNA* or other materials are injected into fertilized eggs or blastocysts.

Microorganism An organism of microscopic or submicroscopic size, including bacteria.

Mineralization The complete degradation of an organic compound to carbon dioxide and other inorganic compounds, such as water and chloride ions.

Minimal risk level (MRL) Estimate of daily human exposure to a hazardous substance at or below which that substance is unlikely to pose a measurable risk of harmful, noncancerous effects. Calculated for a route of exposure (inhalation or oral) over a specified time period (*acute*, intermediate, or *chronic*). Not recommended as predictors of adverse health effects.

Minimax theorem Key convention of game theory holding that lowest maximum expected loss in a two-person zero-sum game equals the highest minimum expected gain. It is a useful technique to address uncertainties in decision making.

Miscibility Chemical property where two or more liquids or phases readily dissolve in one another, such as ethanol and water.

Mitigation Definition of intervention measures directed to reduce or decrease a risk. Mitigation is the result of a political decision of the level of acceptable risk, obtained by an extensive analysis and based on the criteria that the risk is not totally preventable.

Mitochondrion Eukaryotic organelle that is the site of electron transport, oxidative phosphorylation, and pathways such as the Krebs cycle; it provides most of a nonphotosynthetic cell's energy under aerobic conditions. Constructed of an outer membrane and an inner membrane, which contains the electron transport chain.

Mitosis Process in the nucleus of a eukaryotic cell that results in the formation of two new nuclei, each with the same number of chromosomes as the parent.

Mixed acid fermentation Process carried out by members of the family Enterobacteriaceae in which ethanol and a complex mixture of organic acids are produced.

Mixed liquor suspended solids (MLSS) Volume of *suspended solids* in *mixed liquor*.

Mixed liquor volatile suspended solids (MLVSS) Volume of organic solids from *mixed liquor* that will evaporate at relatively low temperatures (e.g., 55 °C); MLVSS is an indicator that microbial populations are active.

Mixed liquor Activated sludge and water containing organic matter undergoing *activated-sludge* treatment in aeration tank.

Mixotrophic Characteristic of microorganisms that combine autotrophic and heterotrophic metabolic processes (i.e., use both inorganic electron sources and organic carbon sources).

Mode of action Overall manner in which a substance acts, e.g., the way a pesticide kills an insect at the tissue or cellular level, or the way that a drug works at the cellular level.

Model A mathematical function with parameters that can be adjusted, so the function closely describes a set of empirical data. A mechanistic model usually reflects observed or hypothesized biological or physical mechanisms and has model parameters with real-world interpretation. In contrast, statistical or empirical models selected for particular numerical properties are fitted to data; model parameters may or may not have real-world interpretation. When data quality is otherwise equivalent, extrapolation from mechanistic models (e.g., biologically based dose-response models) often carries higher confidence than extrapolation using empirical models (e.g., logistic model).

Modifying factor (MF) A factor used in the derivation of a reference dose or reference concentration. The magnitude of the MF reflects the scientific uncertainties of the study and database not explicitly treated with standard uncertainty factors (e.g., the completeness of the overall database). An MF is greater than zero and less than or equal to 10, and the default value for the MF is 1. Use of a modifying factor was generally discontinued in 2004. Compare to *uncertainty factor (UF)*.

Molecular initiating event (MIE) Beginning step of an *adverse outcome pathway (AOP)*

Molecular pharming Use of *genetically modified organisms* to produce pharmaceuticals. Application of genetic engineering that introduces genes, primarily of human and animal origin, into plants or farm animals to produce medicinal substances. Premise is using plants as efficient chemical factories for producing antibodies, vaccines, blood proteins, and other therapeutically valuable proteins.

Monitor Device or system that provides information about environmental quality, such as a device placed in a well to continuously measure contaminants migrating from a landfill.

Monitored natural attenuation A system to measure progress of decontamination of an aquifer that is not receiving active treatment.

Monod equation Empirically derived expression of the rate of microbial biomass: $\mu = \frac{\mu_{\max} S}{K_s + S}$, where μ is the specific growth rate of the microbe, $\mu_{\max}$ is the maximum specific growth rate, and K_s is the Monod growth rate coefficient representing the substrate concentration at which the growth rate is half the maximum rate. The $\mu_{\max}$ is reached at the higher ranges of substrate concentrations. K_s is an expression of the affinity of the microbe for a nutrient, i.e., as K_s decreases the more affinity that microbe has for that particular nutrient (as expressed by the concomitantly increasing μ). Named in honor of French researcher, Jacques Monod.

Monte Carlo technique Repeated random sampling from the distribution of values for each of the parameters in a calculation (e.g., lifetime average daily exposure), to derive a distribution of estimates (e.g., of exposures) in the population.

Morbidity State of disease.

Mortality rate Proportion of a population that dies during a specified time period. Also called death rate.

Mosaic Individual consisting of cells of two or more *genotypes*.

Most probable number (MPN) Estimation of the probable population in a liquid by diluting and determining endpoints for microbial growth and conducting statistical tests.

Motile Capable of movement, e.g., used to characterize microorganisms.

Municipal solid waste (MSW) Garbage and trash.

Mutagen Substance that can induce an alteration in the structure of *DNA*.

Mutagenesis Generation of mutations; breeding whereby random mutations are induced in cell's *DNA* using chemicals or ionizing radiation.

Mutualism Symbiosis in which both partners gain from the association and are unable to survive without it. The mutualist and the host are metabolically dependent on each other.

Mycelium Branching hyphae found in fungi and some bacteria.

Mycoplasma Bacteria that are members of the class Mollicutes and order Mycoplasmatales, lacking cell walls and unable to synthesize peptidoglycan precursors; most require sterols for growth, smallest organisms capable of independent reproduction.

Myxobacteria Gram-negative, aerobic soil bacteria characterized by gliding motility, a complex life cycle with the production of fruiting bodies, and the formation of myxospores.

Nanoscale Having at least one dimension <100 nm.

Nanotechnology Emergent technologies at the *nanoscale*.

Nanotechnology Science and engineering addressing the design and production of extremely small (<100 nm diameter) devices and systems fabricated from individual atoms and molecules.

National Pollutant Discharge Elimination System (NPDES) Permit program addresses water pollution by regulating point sources that discharge pollutants to waters of the U.S. pursuant to the Clean Water Act.

Natural attenuation Naturally occurring processes in soil and ground water environments that act without human intervention to reduce the mass, toxicity, mobility, volume, or concentration of contaminants in those media. When analyzing data from a natural attenuation site, a key question often is whether the mechanisms that destroy or immobilize contaminants are sustainable for as long as the source area releases them to the ground water. More specifically, whether the rates of the protecting mechanisms will continue to equal the rate at which the contaminants enter the ground water may be a concern. Sustainability is affected by the rate at which the contaminants are transferred from the source area and whether the protecting mechanisms are renewable.

Natural disaster A disaster that is initiated by a natural event, e.g., volcano, earthquake, or hurricane. Although the trigger may have been natural, the damage is always determined by human activities.

Navier-Stokes equations Equations that explain motion of fluids, usually nonlinear partial differential equations.

Nephelometry Measurement of turbidity using light scattered at an angle to the incident beam; particularly sensitive at low turbidity.

Nerve Enclosed, cable-like bundle of nerve fibers or axons.

Net primary productivity Organisms' generation of organic compounds from carbon dioxide. Rate at which an ecosystem accumulates biomass and energy (*gross primary productivity*), but excluding the energy used for respiration.

Niche Function of an organism in a complex system, including place of the organism, the resources used in a given location, and the time of use.

Nicotinamide adenine dinucleotide (NAD) Coenzyme for dehydrogenases; reduced form is *nicotinamide adenine dinucleotide phosphate*. Formerly called DPN (diphosphopyridine nucleotide) and Coenzyme I.

Nicotinamide adenine dinucleotide phosphate (NADP) Coenzyme for dehydrogenases; reduced form is NADPH. Formerly called TPN (triphosphopyridine nucleotide) and Coenzyme II.

Nitrification Oxidation of reduced forms of nitrogen, e.g., ammonia to nitrate.

Nitrifying bacteria *Chemolithotrophic*, Gram-negative members of the family Nitrobacteraceae that convert ammonia to nitrate and nitrite to nitrate.

Nitrogen fixation Metabolic process by which atmospheric molecular nitrogen is reduced to ammonia; carried out by cyanobacteria, Rhizobium, and other nitrogen-fixing bacteria.

Nitrogen oxygen demand (NOD) Demand for oxygen in sewage treatment, caused by nitrifying microorganisms.

Nitrogenase Enzyme that catalyzes biological *nitrogen fixation*.

Nongovernmental organization (NGO) Entity that advocates or represents positions, including scientific, legal, and medical perspectives, without a governmental mandate. Examples include Doctors without Borders, Resources for the Future, and Engineers without Borders.

Non-Hodgkin lymphoma Large group of cancers of lymphocytes marked by enlarged lymph, fever, and weight loss. Categorized as either aggressive (fast-growing) or indolent (slow-growing) types. Can be formed from either *B cells* or *T cells*.

No-observed-adverse-effect level (NOAEL) Highest exposure level where there are no biologically significant increases in the frequency or severity of adverse effect between the exposed population and its appropriate control. Compare to *LOEAL*.

Nosocomial Describing infection acquired in a medical facility.

Nuclear transfer cloning Transfer of a nucleus into an enucleated egg cell.

Nuclease An enzyme which degrades nucleic acids.

Nucleotide Combination of ribose or deoxyribose with phosphate and a purine or pyrimidine base; a nucleoside plus one or more phosphates.

Nucleus Eukaryotic organelle enclosed by a double-membrane envelope that contains the cell's chromosomes.

Null mutation The complete elimination of the function of a gene.

Nutritional supplement A dietary supplement intended to be ingested in pill, capsule, tablet, or liquid form, not represented for use as a conventional food or as the sole item of a meal or diet and labeled accordingly. It is regulated by the U.S. Food and Drug Administration as a food and not as a drug.

Nymph Form of insects during metamorphosis after larvae and before adult stage.

Obligate Absolutely required, e.g., an obligate *aerobe* can live only in the presence of molecular oxygen and an obligate *anaerobe* can live only in the absence of molecular oxygen.

Ockham's razor Principle espoused by medieval nominalist William of Ockham that entities are not to be multiplied beyond necessity. The principle encourages asking whether any proposed kind of entity is necessary.

Octanol-water coefficient (K_{ow}) Ratio of the concentration of a chemical in octanol and in water at equilibrium and at a specified temperature. Octanol is used as a surrogate for natural organic matter. K_{ow} is used to help determine the fate of chemicals in the environment, e.g., to predict the extent a contaminant will *bioaccumulate* in aquatic biota.

Odds ratio Ratio of the odds of disease among the exposed compared with the odds of disease among the unexposed. For rare diseases, such as cancer, the odds ratio can provide an estimate of relative risk.

Offsetting behavior Inadvertent attenuation or reversal of a risk management action due to a reduction of care by those targeted for risk reduction. For example, if actions are taken to reduce bacterial growth in food supplies, and consumers consider the foods to be safer to the extent that they are less careful in food preparation, this behavior offsets at least some of the safety margin of the risk reduction efforts.

Oligonucleotide Short *DNA* sequence of up to 1000 nucleotides.

Oligotrophic Lake or estuary with low productivity so as to be clear and low in nutrient concentrations.

Omics Shorthand term for computational, biological subfields for describing very large-scale data collection and analysis, all with the suffix "omics" (e.g., genomics, proteomics, and metabonomics).

Onus Burden of responsibility.

Opportunity risk Likelihood that a better opportunity will present itself after an irreversible decision has been made (e.g., prohibiting research in an emerging technology may prevent exposure to a toxic substance to a few, but, in the process, the cure for a disease may be lost).

Optimal range Range of success, below and above which are unacceptable (e.g., trivalent chromium must be taken within the optimal range because intake at too low a dosage leads to a nutritional deficiency and too high a dosage leads to toxicity).

Optimization Selecting the best design for the conditions. The "best" is determined by the designer based on one or more variables (e.g., a heart valve may have three key variables: flow rate, reliability, and durability: the engineer would design the valve by optimizing these three variables to achieve the best performance).

Oral slope factor Upper bound, approximating a 95% confidence limit, on the increased cancer risk from a lifetime oral exposure to an agent. This estimate, usually expressed in units of proportion of a population affected per $mg\,kg^{-1}\,day^{-1}$, is generally reserved for use in the low-dose region of the dose-response relationship, that is, for exposures corresponding to risks less than 1 in 100.

Organ Completely differentiated unit of an organism that provides a certain, specialized function.

Organelle Structure within or on a cell that performs specific functions.

Organism Living entity; consisting of one or more cells (unicellular and multicellular organisms, respectively).

Organophosphate (OP) Phosphorous-containing synthetic pesticide active ingredient that acts on the nervous system by inhibiting acetylcholinesterase. Irreversible inhibition is characteristic of many OPs. Azamethiphos, chlorpyrifos, diazinon, dichlorvos, and malathion are OPs.

Organotroph Organism that uses reduced organic compounds as its electron source.

Orpiment As_2S_3, yellow, sulfur compound of arsenic.

Osmosis Movement of fluid through a partially (selectively) permeable membrane separating solutions of different concentrations. Movement of a fluid across a selectively permeable membrane from a dilute solution to a more concentrated solution. Cellular *membranes* and root systems, for example, take advantage of the separation by osmosis for nutrient transport.

Outlier Value that is markedly smaller or larger than other values in a data set. Can be problematic for researchers since it decreases the coefficient of determination (i.e., r^2).

Oxidation Loss of electrons from a compound.

Oxidation-reduction potential (ORP) Degree of completion of chemical reaction expressed as the ratio of reduced ions to oxidized ions.

Oxidation-reduction reaction Reaction involving electron transfer; the reductant donates electrons to an oxidant. Also called redox reaction.

Oxygen sag curve Decrease in dissolved oxygen plotted over a distance along a water body from a point at which pollutants have been discharged.

Ozonation Addition of ozone (O_3) to water and other *media* for disinfection and other oxidative processes.

PAH See *polycyclic aromatic hydrocarbon*.

Pandora's box Metaphor for a prolific source of problems (Greek: All gifted; from mythology of a box given to Pandora by Zeus who ordered her not to open it; Pandora succumbed to her curiosity and opened it; all the miseries and evils flew out to afflict humankind).

Paradox Argument appearing to justify a self-contradictory conclusion by using valid deductions from acceptable premises.

Parametrics Descriptors of an entire population, without the need of inference. Compare to *statistics*.

Pareto efficiency Resource allocation wherein there is no rearrangement that can make anyone better off without making someone else worse off.

Partial pressure Pressure exerted by a single gas in a mixture of gases.

Particulate matter (PM) Complex mixture of small solid particles and liquid droplets, consisting of numerous components, e.g., acids (such as nitrates and sulfates), organic chemicals, metals, and soil or dust particle. Compare to *aerosol*.

Partitioning Chemical equilibrium condition where a chemical's concentration is apportioned between two different phases according to the partition coefficient, which is the ratio of a chemical's concentration in one phase to its concentration in the other phase, e.g., *octanol-water coefficient*.

Pathogen Microbe capable of producing disease.

PBT Chemical that is persistent, bioaccumulates, and is toxic.

PCB See *polychlorinated biphenyl*.

PCR See *polymerase chain reaction*.

Pedagogy Instruction techniques used to promote learning.

Peptide bond Covalent bond between two *amino acids*, in which the carboxyl group of one amino acid (X_1-COOH) and the amino group of an adjacent amino acid (NH_2-X_2) react to form X_1-CO-NH-X_2 plus H_2O.

Peptide 1. Chain formed by two or more *amino acids* linked through *peptide bonds*: dipeptide, two *amino acids*; oligopeptide, small number of *amino acids*. 2. Molecule formed by peptide bonds covalently linking two or more *amino acids*. Larger peptides (i.e., *polypeptides*) are usually *expressed* from recombinant DNA.

Perception Information and knowledge gained through the senses.

Perched water table Aquifer above the regional water table, located atop an impermeable layer of rock or sediment (aquiclude) or relatively impermeable layer (aquitard) above the main water table/aquifer but below the land surface.

Performance monitoring Collection of information which, when analyzed, evaluates the performance of the system on the environmental contamination.

Permeability Ease at which a fluid moves through a substance.

Permeable reactive barriers Subsurface walls composed of reactive materials that will either degrade or alter the state of a contaminant when that contaminant in a ground water plume passes through the wall.

Permissible exposure level (PEL) Occupational limit for a contaminant.

Persistence Resistance to degradation in the environment.

Persistent organic compound (POP) Recalcitrant organic compounds, most of which are semivolatile and lipophilic.

Pesticide Substance used to control pesticide by using its toxic properties.

PFAS Per- and polyfluorinated substances (PFAS).

pH Negative logarithmic measure of the hydrogen ion concentration in water, ranging from 0 to 14 (acidic to basic).

Pharmacodynamics Manner in which a substance exerts its effects on living organisms (compare to *toxicodynamics*).

Pharmacokinetics Behavior of substances within an organism, especially by absorption, distribution, biotransformation, storage, and excretion (compare to *toxicokinetics*).

Phenotype Physical manifestation of the *genotype*.

Phosphatase, alkaline Enzyme that catalyzes the hydrolysis of phosphomonoesters of the 5′ nucleotides.

Phosphorylation Addition of phosphate monoester to macromolecule, catalyzed by a specific *kinase enzyme*.

Photolithotrophic autotroph Organism that uses light energy, an inorganic electron source (e.g., H_2O, H_2, H_2S), and CO_2 as a carbon source.

Photoorganotrophic heterotroph Microbe that uses light energy and organic electron donors, and simple organic molecules rather than CO_2 as their carbon source.

Photoreactivation Increase in survival rate or reduction in the frequency of mutation of a microbial population previously irradiated with ultraviolet light by exposure to light of 300–450 nm wavelength.

Photosensitization Increased sensitivity of microbes to oxygen and light by applying certain stains to cells (e.g., acridine orange, methylene blue).

Phototaxis Microbial movement (especially photosynthetic bacteria) toward light.

Phototroph Organism that can use light as an energy source. Green algae and higher plants produce oxygen. Photosynthetic bacteria produce sulfur (for instance). Blue-green algae are sometimes considered algae (Cyanophyta) and sometimes bacteria (Cyanobacteria).

Phycology Study of algae.

Physical containment Stopping the migration of an entity, e.g., a microbe being kept in a sealed containment system.

Physiologically based pharmacokinetic (PBPK) model Model used to characterize pharmacokinetic behavior of a chemical. Available data on blood flow rates and on metabolic and other processes that the chemical undergoes within each compartment are used to construct a mass-balance framework for the PBPK model.

Phytodegradation Process by which plants metabolically degrade a contaminant to a nontoxic form in roots, stems, or leaves.

Phytoextraction Removal of a substance from soils and ground water surrounding the roots of a plant through that plant's vascular system.

Phytoplankton Flora in the plankton community.

Phytoremediation Use of plants to clean up contamination.

Phytovolatilization Plants translocate contaminants into the atmosphere via normal transpiration.

Pica Deliberate ingestion of nonnutritive substances such as soil or paint chips.

Plankton Small, mainly microscopic, members of animal and plant communities in aquatic system. See *phytoplankton* and *zooplankton*.

Plant-incorporated protectants (PIPs) Proteins and other chemicals introduced to plants either through the conventional breeding of sexually compatible plants or through techniques of modern biotechnology, e.g., transferring specific genetic material from a bacterium to a plant to induce the plant to produce pesticidal proteins or other chemicals that the plant could not previously produce.

Plasma cell *Leucocyte* that makes a specific antibody. Develop from *B cells* that have been activated. Also called plasmacyte.

Plasmid Extrachromosomal *DNA* molecule separated from the chromosomal DNA capable of replicating independently of the chromosomal DNA; found in bacteria and protozoa. Since it can replicate autonomously, it is often used for the insertion of genetic material.

Plasmolysis Water loss with concomitant shrinkage of the cell contents and cytoplasmic membrane resulting from high osmotic pressure in a medium.

Pleiotropy Phenomenon in which one gene can influence two or more independent characteristics.

Ploidy Number of complete sets of chromosomes in a cell.

Plume A zone of dissolved contaminants, originating from a source and extending in the direction of flow. Volume of a substance moving from its source away from the source.

PM See *particulate matter*.

PM_{10} Particulate matter with aerodynamic diameters less than 10 μm. Also known as coarse particulates.

$PM_{2.5}$ Particulate matter with aerodynamic diameters less than 2.5 μm. Also known as fine particulates.

Point of departure The dose-response (see *dose-response*) point that marks the beginning of a low-dose extrapolation. This point is most often the upper bound on an observed incidence or on an estimated incidence from a dose-response model.

Pollen Powder from a plant's male anther that contains the male gamete.

Pollination Transfer of pollen from a plant's male anther to the female stigma to allow fertilization.

Pollinator Insect, bird, or other organism that conveys pollen.

Polychlorinated biphenyl (PCB) Highly toxic molecule of two benzene rings bonded to each other with chlorine substituents (209 structural variations, known as congeners); presently banned but manufactured by Monsanto for much of the 20th century.

Polycyclic aromatic hydrocarbon (PAH) Class of products of incomplete combustion consisting of fused aromatic rings. Several of them are suspected carcinogens (e.g., benzo(*a*)pyrene).

Polymerase chain reaction (PCR) Technique that enables the *in vitro* amplification of target *DNA* sequences.

Polymerase Enzyme that links individual nucleotides together into a long strand, using another strand as a template.

Pool An accumulation of DNAPL above a capillary barrier.

Porosity Percentage of void space in a solid matrix (e.g., soil, sediment, or gravel).

Porphyria cutanea tarda Rare skin disorder accompanied by painful blisters, change in color, increased hair growth, redness, swelling, and scarring.

Precautionary principle 1. Position that if the suspected health or environmental risk of an action is sufficiently widespread or irreversible, in the absence of scientific consensus that the action or policy is harmful, the burden of proof that it is *not* harmful falls on those taking the action. 2. Risk management approach taken when scientific knowledge is incomplete and the possible consequences could be substantial and irreversible (e.g., global climate change). The principle holds that scientific uncertainty must not be accepted as an excuse to postpone cost-effective measures to prevent a significant problem.

Precision Exactness and reproducibility. Usually represented by the number of significant figures.

Prevalence Proportion of cases that exist within a population at a specific point in time, relative to the number of individuals within that population at the same point in time.

Prevention Intentional modification of the characteristics of a phenomenon with the purpose of reducing risk and the intrinsic characteristics of a substance or process with the purpose of reducing vulnerability. The intervention has the purpose of modifying the risk factors. As such, prevention differs from controls and remediation in that preventive measures occur before a problem exists.

Primary treatment Clarification of wastewater influent (i.e., removal of suspended solids) by *sedimentation*.

Probabilistic uncertainty analysis Technique that assigns a *probability density function* to each input parameter, then randomly selects values from each of the distributions and inserts them into an equation. Repeated calculations produce a distribution of predicted values, reflecting the combined impact of variability in each input to the calculation. *Monte Carlo* is a common type.

Probability Measurement of the likelihood that an event will occur; ranging from 0 (no likelihood whatsoever) to 1 (absolute certainty).

Process monitoring The collection of information documenting the operation of a system's engineered components.

Profession Group (e.g., physicians or engineers) with a common mission, requiring substantial education and training, self-determination of professional requirements to enter, organized into an identifiable professional body, and requiring the adherence to standards of conduct.

Progenitor strain Original strain prior to *hybridization* or genetic modification. Unmodified strain.

Program evaluation review technique (PERT) chart Diagram depicting project tasks and their interrelationships.

Prokaryote Organism lacking a true nucleus and other membrane-bound cellular compartments and containing a single loop of stable chromosomal *DNA* in the nucleoid region and cytoplasmic structures.

Promoter An agent that is not carcinogenic itself, but when administered after an initiator of carcinogenesis, stimulates the clonal expansion of the initiated cell to produce a neoplasm.

Proteobacteria Bacteria, primarily Gram negative, that 16S *rRNA* sequence comparisons show to be phylogenetically related; proteobacteria contain the purple photosynthetic bacteria and their relatives and are composed of *a*, *b*, *g*, *d*, and *e* subgroups.

Proteome Complete collection of proteins that an organism produces.

Proteomics Study of proteins in the body, especially the protein complement of the genome *omics*.

Protist Eukaryote with unicellular organization, in the form of either solitary cells or colonies of cells lacking true tissues.

Protoplast Cell bounded by cytoplasmic membrane yet lacking rigid layer. Protozoa: Microscopic animals.

Protozoa *Motile* microbes that consume bacteria as carbon and energy sources.

Psychrophilic Describing a microbe with optimal temperatures between 0 and 20 °C.

Psyllid Jumping plant louse; usually host specific. See *citrus greening disease*.

Public Whole collection of people comprising a society.

Pump-and-treat Extraction of contaminated ground water to the surface and removing pollutants above ground, e.g., *air stripping*.

Pure culture Cell population with all members identical because they arise from a single cell.

Purine Basic, heterocyclic, nitrogen-containing molecule with two joined rings that occurs in nucleic acids and other cell constituents; most purines are oxy- or amino-derivatives of the purine skeleton. The most important purines are adenine and guanine.

Pyrethroid Natural (from the chrysanthemum family) or synthetic pesticide, of varying chemical structure, which acts on the nervous system by interfering with nerve conduction. Permethrin and D-phenothrin are synthetic pyrethroids.

Quantitative microbial risk assessment (QMRA) Systematic framework for estimating human health risks from exposure to pathogens via food, water, air, and other environmental routes.

Quantitative polymerase chain reaction (qPCR) Determination of a polynucleotide by including a known amount of readily distinguished template as an internal standard to compensate for variation in efficiency of amplification.

Quantitative structure-activity relationship (QSAR) Understanding and application of chemical structure to estimate the behavior of a compound, e.g., *toxicity* or *persistence*.

Quantum yield Number of photons required per molecule of carbon dioxide converted to sugar (or per oxygen molecule produced).

Radioactive half-life Time taken for half the initial number of nuclei to disintegrate, $t_{1/2}$.

Rainforest Forest with mean rainfall greater than 1750 mm per year.

Random sample Sample selected from a statistical population such that each sample has an equal probability of being selected.

Raoult's law A dissolved substance will lower the partial pressure of the solvent proportionally to the mole fraction of the dissolved substance.

Reactor System where physical, chemical, and biological reactions occur. Traditionally, these have been thought of vessels or other engineered systems, but reactors can be quite large and occur in natural systems (e.g., a wetland is a reactor in which *aerobic*, *anaerobic*, and mixed reactions occur). At the other end of the scale, any living cell is a reactor.

Realgar As_4S_4, red-orange, sulfur compound of arsenic.

Reasonable person standard Position (legal, engineering, etc.) expected to be held by a hypothetical person in society who exercises average care, skill, and judgment in conduct.

Reasoning Derivation of a conclusion from premises.

Rebound After contaminant concentrations in ground water have been reduced through *in situ* treatment and the treatment is terminated or reduced, concentrations return to elevated levels from the continued release of mass from a source zone beyond the natural attenuation capacity of the ground water system.

Recalcitrance 1. Resistance of a compound to degradation. 2. Inverse of degradability.

Receptor 1. Molecule in a cell or on its surface that binds to a specific substance and causes a specific physiologic effect in the cell. 2. The potentially or actually affected entity exposed to a physical, chemical, or biological agent.

Recharge Addition of water to an aquifer by the infiltration and transmission through soil and the vadose zone during and following a precipitation event.

Recombinant DNA (rDNA) Genetically engineered *DNA* prepared by transplanting or splicing genes from one individual to another, including from the cells of one species into the cells of a host organism of a different species. Thereafter, the rDNA becomes part of the host's genetic makeup and is replicated.

Recombinant Describing material produced by *genetic engineering*.

Redictable Ridiculously predictable (coined by Vallero DJ and Randall ACV). For example, if a microbe that has been genetically modified to degrade crude oil spilled from a tanker, the likelihood that the same microbe will degrade asphalt in adjacent roads is redictable.

Reductionism 1. Understanding complex phenomenon by reducing them to the interactions of their parts or to simpler or more fundamental processes and components. 2. Perspective that a complex system is merely the sum of its parts, and that an account of it can be reduced to accounts of individual constituents. Compare reductionist to *systematic*.

Reductive dechlorination The removal of chlorine from an organic compound and its replacement with hydrogen. Often, part of a two-step degradation process for *recalcitrant* halogenated compounds, i.e., the anaerobic step to remove halogens and the aerobic step to break aromatic rings and otherwise reach *ultimate degradation*.

Reed-Sternberg cell Large, abnormal *lymphocyte* that may contain more than one nucleus; found in people with Hodgkin lymphoma.

Reference concentration (RfC) An estimate (with uncertainty spanning perhaps an order of magnitude) of a continuous inhalation exposure to the human population (including sensitive subgroups) that is likely to be without an appreciable risk of deleterious effects during a lifetime. It can be derived from a NOAEL, LOAEL, or benchmark concentration, with uncertainty factors generally applied to reflect limitations of the data used. Generally used in noncancer health assessments. Durations include *acute*, *short-term*, *subchronic*, and *chronic*.

Reference dose (RfD) An estimate (with uncertainty spanning perhaps an order of magnitude) of a daily oral exposure to the human population (including sensitive subgroups) that is likely to be without an appreciable risk of deleterious effects during a lifetime. It can be derived from a NOAEL, LOAEL, or benchmark dose, with uncertainty factors generally applied to reflect limitations of the data used. Generally used in noncancer health assessments. Durations include *acute*, *short-term*, *subchronic*, and *chronic*.

Relative risk Ratio of the risk of disease or death among the exposed segment of the population to the risk among the unexposed. The relative measure of the difference in risk between the exposed and unexposed populations in a cohort study. The relative risk is calculated as the rate of an outcome in an exposed divided by the rate of the disease among the unexposed; e.g., a relative risk of two means that the exposed group has twice the disease risk as the unexposed group. Also known as *relative risk*.

Reliability Probability that a device or system will perform its specified function, without failure under stated environmental conditions, over a required lifetime.

Replacement cost analysis (RCA) Analysis method to assess whether it is more cost-effective to replace or maintain an existing asset.

Replicase DNA-duplication catalyzing enzyme.

Representativeness Degree to which a sample is characteristic of the whole medium, exposure, or dose for which the samples are being used to make inferences.

Resilience Attribute of ecosystem stability, expressing the system's ability to recover after disturbance.

Respiration Energy-yielding process in which the energy substrate is oxidized by means of an exogenous or externally derived electron acceptor.

Response boundary (control plane) A location within the source area, or immediately downgradient of the source area, where changes in the plume configuration are anticipated due to the implementation of the *in situ* bioremediation *DNAPL* source zone treatment.

Restriction enzyme Enzyme that cleaves *DNA* molecules at a precisely defined site; provides ubiquitous defense proteins in bacteria that cut DNA strands at specifically defined sequences.

Reverse osmosis High-pressure filtration to separate extremely fine particle and ions.

Reverse transcriptase Enzyme that synthesizes a strand of *DNA* complementary to the base sequence of an *RNA* template.

Reynolds number Dimensionless number associated with fluid flow that determines the transition point from laminar to turbulent flow. It represents the ratio of the momentum forces to the viscous forces in the fluid flow.

Rhizodegradation Plants promote a soil environment suitable for microbes that can degrade or sequester contaminants.

Rhizosphere Narrow region of soil directly influenced by soil microbes and plant root secretions.

Ribonucleic acid (RNA) Nucleic acid molecule similar to *deoxyribonucleic acid* but containing ribose rather than deoxyribose.

Ribosomal RNA (rRNA) Any of several *RNAs* that become part of the ribosome and thus are involved in translating *mRNA* and synthesizing proteins.

Richards equation Mathematical representation of the movement of water in unsaturated *unconsolidated media.*

Risk assessment The evaluation of scientific information on the hazardous properties of environmental agents (hazard characterization), the dose-response relationship (dose-response assessment), and the extent of human exposure to those agents (exposure assessment). The product of the risk assessment is a statement regarding the probability that populations or individuals so exposed will be harmed and to what degree (risk characterization).

Risk characterization The integration of information on hazard, exposure, and dose-response to provide an estimate of the likelihood that any of the identified adverse effects will occur in exposed people.

Risk homeostasis Defeat of built-in factors of safety by asserting new way to use the products.

Risk management Decision-making process that accounts for political, social, economic, and engineering implications together with risk-related information to develop, analyze, and compare management options and select the appropriate managerial response to a hazard.

Risk ratio (RR) See *relative risk.*

Risk shifting Taking an action that changes and reduces the risk of one population but increases the risk in a different population (e.g., banning DDT reduces the risk of cancer in developed nations but increases the risk of malaria in tropical and subtropical developing countries).

Risk tradeoff Eliminating or reducing one risk, but introducing or increasing a countervailing risk (e.g., reducing the pain of a headache by taking aspirin, but increasing the risk of Reye's syndrome; or removing mold in buildings may increase worker exposure to asbestos).

Risk 1. Likelihood of an adverse outcome. 2. Probability of adverse effects resulting from exposure to an environmental agent or mixture of agents. 3. Product of hazard times exposure.

Risk-benefit analysis Comparison of risks of various options to their benefits (e.g., health and wildlife risks of applying a pesticide compared to the benefits of crop protection).

Robustness Capacity of an engineered system to achieve the desired outcomes even when challenged by unlikely events.

Roughness coefficient Factor affecting the velocity and discharge in flow equations like the *Manning's equation.*

Runoff coefficient Fraction of total rainfall that becomes runoff.

Runoff Water running overland, often carrying solids and other pollutants to surface waters; important vehicle in *nonpoint* pollution.

Salinity Dissolved salt content of water. Contributes to *conductivity* and is a determinant of natural water chemistry and biology.

Sample Part of a population designed to show the nature or quality of the whole. For example, exposure-related measurements are usually samples of environmental or ambient media; exposures of a small subset of a population for a short time; or biological samples, from which the nature and quality of parameters can be inferred.

Sanctity of life View that human life is precious from conception to natural death, since it is created in the image of the Creator.

Sand Soil particle between 0.05 and 2.0 mm in diameter.

Sanitary sewer Engineered underground conveyance system for the transfer of sewage to a *wastewater treatment plant*.

Saprophyte Organism that derives its nutrients from decomposing organic matter.

Sarcoma Cancer that begins in bone or in the soft tissues of the body, including cartilage, fat, muscle, blood vessels, fibrous tissue, or other connective or supportive tissue.

Saturated zone See *zone of saturation*.

Saturation Ratio of the volume of a single fluid in the pores to pore volume expressed as a percentage or a fraction.

Scale Spatial extent from very small (molecular) to very large (planetary).

Science Systematic investigation, through experiment, observation, and deduction, to produce reliable explanations of the physical world and its processes.

Scientific method Progression of inquiry (1) to identify a problem you would like to solve, (2) to formulate a hypothesis, (3) to test the hypothesis, (4) to collect and to analyze the data, and (5) to draw valid conclusions.

Secondary treatment Clarification followed by a biological process with separate sludge collection and handling.

Sediment oxygen demand (SOD) Decrease in dissolved oxygen due to microbes decomposing organic matter and consuming O_2, releasing the decomposition products into the surface water column.

Sediment Matter that settles to the bottom of a surface water body.

Sedimentation Process in surface waters or in engineered systems wherein particles fall in direct proportion to a particle's mass and indirect proportion to flow. Most particle mass in a wastewater treatment facility is collected during *primary treatment*.

Senescence Organism's aging process.

Sensitivity 1. Ability of a test to detect a condition when it is truly present. 2. Smallest change in a physical quantity or parameter that can be detected by a measuring system. Determined by signal-to-noise ratio, system amplification, or quantizing limit. Compare to *specificity*.

Sewage treatment plant Wastewater treatment plant.

Sewage Untreated, predominantly liquid, waste in need of treatment; the fluid being transported to a wastewater treatment plant. Preferred term is wastewater.

Silage Fermented plant material with increased palatability and nutritional value for animals; often can be stored for extended periods.

Silt Soil particle with a diameter between 0.002 and 0.05 mm. Compare to *clay* and *sand*.

Sink Site where matter or energy is lost in a system; e.g., a wetland can be a sink for CO_2, but it may not be a net sink for carbon, for example, if it releases CH_4 from *anaerobic decomposition*, i.e., it would be a net carbon sink only if more is lost and stored than is released. If it releases more carbon than it takes up and stores it is a net *source*.

Slash and burn Cutting and using fire to clear land for agriculture.

Slime mold Organism that produces spores but moves with amoeba-like gliding motility; phenotypically like fungi and protozoa. Phylogenetically, slime molds are more closely related to amoeboid protozoa than to fungi. Cellular slime molds are composed of single amoeboid cells during their vegetative stage. Vegetative acellular slime molds are made up of plasmodia, amorphic masses of protoplasm.

Slope factor, cancer Dose-response curve for a substance, indicating cancer potency (units = mass per body mass per time, e.g., $mg\,kg^{-1}\,day^{-1}$). An upper bound, approximating a 95% confidence limit, on the increased cancer risk from a lifetime exposure to an agent. This estimate, usually expressed in units of proportion of a population affected per $mg\,kg^{-1}\,day^{-1}$, is generally reserved for use in the low-dose region of the dose-response relationship, that is, for exposures corresponding to risks less than 1 in 100.

Sludge Muddy aggregate at the bottom of tanks generated as particles settle. Also known as biosolids when a sludge contains large amounts of organic materials, e.g., microbes and their remnants.

Slurry Liquid mixture of aqueously insoluble matter, e.g., a lime (CaO, $CaCO_3$) slurry.

Social ecology View that environmental problems are firmly rooted in human social interactions. Social ecologists believe that an ecologically sustainable society can still be socially exploitative.

Society Collection of human beings that is distinguished from other groups by shared institutions and a common culture.

Solids retention time (SRT) Average time of retention of suspended solids in a treatment system, equal to the total weight of suspended solids exiting the system, per unit time.

Solubility The amount of substance that will dissolve in a given quantity of solvent, at a given temperature, to form a saturated solution.

Sorption isotherm Curves based on properties of both the chemical and the soil (or other matrix) that determine how and at what rates the molecules partition into the solid and liquid phases. Examples are the Freundlich sorption isotherms.

Sorption The uptake of a solute by a solid. General term that embodies *absorption*, *adsorption*, *chemisorption*, and *ion exchange*.

Source loading The flux of a substance leaving the original disposal location and entering the water migrating through the soil and aquifer.

Source zone The subsurface zone containing a contaminant reservoir sustaining a plume in ground water. The subsurface zone is or was in contact with *DNAPL*. Source zone mass can include sorbed and aqueous-phase contaminant mass as well as DNAPL.

Source Site from which matter or energy originates. Contrast with *sink*.

Spatial scale Geographic extent of a resource or problem. Global scale examples include pandemics, changes in climate, or nuclear threats. Continental scale examples are shifting biomes and border control between nations. Regional scale examples include the contamination of rivers or polluting the air. Local scale examples include crime, job loss, hazardous waste sites, and landfills.

Specialization Degree to which an individual professional concentrates his or her practice into a narrow range of expertise and activities.

Species abundance How common a species is in an ecological community. Compare to *biodiversity*.

Species Subdivision of a genus having members differing from other members of the same genus in minor details.

Specific gravity Ratio of the mass of a body to the mass of an equal volume of water at a specific temperature, typically 20 °C.

Specificity Ability of a test to exclude the presence of a condition when it is truly not present. Compare to *sensitivity*.

Spirochete Spiral-shaped bacterium with periplasmic flagella.

Spore Differentiated, specialized form that can be used for dissemination, for survival of hostile conditions because of its heat and desiccation resistance or for reproduction. Highly variable and usually unicellular. May develop into vegetative organisms or gametes and can be produced asexually or sexually.

Stakeholder A person other than regulators, owners, or technical personnel involved in the environmental activity of concern, who has a vested interest in decisions related to those activities.

Standard deviation Measure of the spread in a data set; wider spread means larger standard deviation.

Standard mortality ratio (SMR) Ratio of the number of deaths observed in the study group to the number of deaths expected based on rates in a comparison population, multiplied by 100.

Stationary phase Microbial growth period after rapid growth, in which cell multiplication is balanced by cell death.

Statistical significance The probability that a result is not likely to be due to chance alone. By convention, a difference between two groups is usually considered statistically significant if chance could explain it only 5% of the time or less. Study design considerations may influence the *a priori* choice of a different level of statistical significance.

Statistics Mathematics concerned with collecting and interpreting quantitative data and applying probability theory to estimate conditions of a universe or population from a sample.

Stoichiometric Related to chemical proportions exactly needed in a reaction.

Stokes diameter Diameter of a spherical particle with the same density and settling velocity as the particle of interest.

Stokes law At low velocities, frictional force on a spherical body moving through a fluid at constant velocity equals 6π times the product of the velocity, the fluid viscosity, and the radius of the sphere.

Stomata Small pores in the epidermis of the leaf or stem of a plant, forming a gap that allows movement of gases and vapors into and out of intercellular spaces. See *transpiration*.

Storage In hydrogeology, volume of water stored in the aquifer that potentially will be released or stored; determined by the saturated thickness, the drainage base of the aquifer, and ground water storage characteristics.

Storm sewer Engineered conveyances in which stormwater moves from a drainage area. Compare to *sanitary sewer*.

Strain 1. Geometrical expression of deformation caused by the action of stress (see *stress*) on a physical body. 2. In biological taxonomy, group of organisms bred within a closed colony to maintain certain defining characteristics.

Stratified random sample Separation of a sample into several groups and randomly assigning subjects to those groups.

Stratospheric ozone Layer just above the troposphere in the earth's atmosphere with high concentrations of ozone (O_3) that absorbs much of the incoming ultraviolet light.

Stress 1. Applied force or system of forces that are apt to strain (see *Strain*) or deform a physical body. 2. The internal resistance of a physical body to such applied force or system of forces.

Structural ecology Ecology, with an emphasis on the components (e.g., niches, associations, canopies) of ecosystems. Compare to *functional ecology*.

Subchronic exposure Repeated exposure by the oral, dermal, or inhalation route for more than 30 days, up to approximately 10% of the lifespan in humans (more than 30 days up to approximately 90 days in typically used laboratory animal species) (see also *longer-term exposure*).

Substrate Molecule that can transfer an electron to another molecule. Substance upon which an enzyme or microorganism acts. For example, organic compounds, such as lactate, ethanol, or glucose, are commonly used as substrates for bioremediation of chlorinated ethenes.

Sulfate reducer A microorganism that exists in anaerobic environments and reduces sulfate to sulfide.

Supernatant 1. Liquid stratum in a bioreactor, e.g., a sludge digester. 2. Layer of liquid above a precipitate (e.g., sediment) after settling.

Supply curve Relationship between a good's quantity supplied and the good's price.

Supply Quantity of a goods or service that a seller would like to sell at a particular price.

Surface water Any water body at the earth's surface, from small rivulets to the ocean.

Surfactant Surface-active agent that concentrates at interfaces, forms micelles, increases solution, lowers surface tension, increases adsorption, and disperses otherwise sorbed substances into the aqueous phase. All have common chemical structure, i.e., a *hydrophobic* and a *hydrophilic* moiety. The hydrophobic part usually consists of an alkyl chain, which is then linked to a hydrophilic group. Each part of the surfactant molecule interacts differently with water. The hydrophilic group is surrounded by water molecules, leading to enhanced aqueous solubility. Simultaneously, the hydrophobic moieties are repulsed by strong interactions between the water molecules. This combination allows for otherwise insoluble substances to be desorbed from soil and sediment.

Susceptibility 1. Extent to which an individual is prone to the effects of an agent. 2. Increased likelihood of an adverse effect, often discussed in terms of relationship to a factor that can be used to describe a human subpopulation (e.g., life stage, demographic feature, or genetic characteristic).

Susceptible subgroups May refer to life stages, for example, children or the elderly, or to other segments of the population, for example, asthmatics or the immune-compromised, but are likely to be somewhat chemical specific and may not be consistently defined in all cases.

Suspended solids (SS) 1. Floating solids with low aqueous solubility. Common measure of water pollution and treatment efficiency, e.g., 20 mg SS L^{-2} prior to treatment and 1 mg SS L^{-1} after treatment.

Sustainability 1. Processes and activities that are currently useful and that do not diminish these same functions for future generations. 2. The ability of a system to maintain the important attenuation mechanisms through time. In the case of reductive dechlorination, sustainability might be limited by the amount of electron donor, which might be used up before remedial goals are achieved.

Sustainable design Application of principles of sustainability (see *sustainability*) to structures, products, and system; an aspect of green engineering.

Sustainable enhancement An intervention action that continues until such time that the enhancement is no longer required to reduce contaminant concentrations or fluxes.

Symbiosis Two dissimilar organisms living together or in close association.

Synergism Effect from a combination of two agents is greater than the sum of the additive effects from both agents ($1+1>2$). Contrast with *antagonism*.

System 1. Combination of organized elements comprising a unified whole. 2. In thermodynamics, a defined physical entity containing boundaries in space, which can be open (i.e., energy and matter can be exchanged with the environment) or closed (no energy or matter exchange).

Systematic Perspective that includes relationships of numerous factors simultaneously. Compare to *reductionist*.

Systemic effect Toxic effect because of absorption and distribution of a toxicant to a site distant from its entry point. Also known as systemic toxicity.

Systems biology 1. Discipline that seeks to study the relationships and interactions between various parts of a biological system (e.g., metabolic pathways, organelles, cells, and organisms) and to integrate this information to understand how biological systems function. 2. Treatment of biological entities as systems composed of defined elements interacting in distinct ways to enable the observed function and behavior of that system. Properties of such systems are embedded in a quantitative model that guides further tests of systems behavior.

T cell *Leucocyte* that makes antibodies; part of the immune system and develop from stem cells in the bone marrow and help protect the body from infection.

Target organ The biological organ(s) most adversely affected by exposure to a chemical, physical, or biological agent.

Taxonomy Classification system.

Technology 1. Application of scientific knowledge. 2. The apparatus that results from such applications.

Temporal scale Range of complexity associated with time. Extremely short temporal scale may be measured in nanoseconds, e.g., nuclear reactions, whereas long-temporal scales may be measured in millions of years, e.g., fossilization of plants to coal.

Temporality Criterion for causality requiring that the cause (e.g., exposure to an infectious agent) precedes the effect (e.g., disease). One of *Hill's criteria.*

Teratogen Substance that causes defects in development between conception and birth. A teratogen is a substance that causes a structural or functional birth defect.

Teratogenic Structural developmental defects due to exposure to a chemical agent during formation of individual organs.

Terrorism Unlawful use of, or threatened use of, force or violence against individuals or property to coerce or intimidate governments or societies, often to achieve political, religious, or ideological objectives.

Tertiary treatment Removal of nutrients, especially phosphorus and nitrogen, along with most suspended solids; generally synonymous with *advanced waste treatment*, which is becoming the preferred term.

Tetrachlorodibenzo-*para*-dioxin (TCDD) Most toxic *dioxin* form, especially when 2,3,7,8-TCDD.

Texture Particle size classification of soil in terms of the U. S. Department of Agriculture system, which uses the term loam for a soil having equal properties of *sand*, *silt*, and *clay*. The basic textural classes, in order of their increasing proportions of fine particles, are sand, loamy sand, sandy loam, loam, silt loam, silt, sandy clay loam, clay loam, silty clay loam, sand clay, silty clay, and clay. The sand, loamy sand, and sandy loam classes may be further divided by decreasing size, i.e., coarse, fine, or very fine.

Thallus Body devoid of root, stem, or leaf; characteristic of certain algae, many fungi, and lichens.

Thermodynamics Principles addressing the physical relationships between energy and matter, especially those concerned with the conversion of different forms of energy.

Thermophilic Describing microbes that thrive in temperatures >45 °C, e.g., *Bacillus licheniformis.*

Threshold limit value (TLV) Occupational standard of the concentration of airborne substances to which a healthy person may be exposed during a 40-h workweek with adverse effects.

Threshold The dose or exposure below which no deleterious effect is expected to occur. An example is the *no observable adverse effect level.*

Time of concentration Time needed for water to flow from the hydrologically most distant point of a drainage area to a designated point.

Tissue Collection of interconnected cells that carry out a similar function in an organism.

Tolerance Established concentration of a substance (e.g., in pesticide residues) occurring as a direct result of proper usage. Compare to *action level.*

Top-down Starting at the upper levels of organization and working downward to the details. Contrast with *bottom-up*.

Total dissolved solids (TDS) Common measure of the aqueously soluble content in water or wastewater, estimated from electrical conductivity since pure water is a poor conductor and the solids are often electrolytes and other conducting substances.

Total maximum daily load (TMDL) Maximum amount of a pollutant that a water body can receive and still meet water quality standards, and an allocation of that amount to the pollutant's sources; varies according to specific watersheds. Sum of the allowable loads of a single pollutant from all contributing point and nonpoint sources. Must include a margin of safety to ensure that the water body can be used for the purposes the State has designated. Calculation must also account for seasonal variation in water quality.

Total solids (TS) Total dissolved and suspended solids in water.

Total suspended particulate (TSP) Entire quantity of particulate matter in the air; i.e., all PM. Compare to $PM_{2.5}$ and PM_{10}.

Total suspended solids (TSS) Fraction of *suspended solids* collected by a filter in water.

Toxic Release Inventory (TRI) Database of annual releases from specified manufacturers in the United States, which includes almost 400 chemicals, as part of the Community Right to Know regulations. Data are self-reported by the manufacturers annually.

Toxic substance A chemical, physical, or biological agent that may cause an adverse effect or effects to biological systems.

Toxicity 1. Deleterious or adverse biological effects elicited by a chemical, physical, or biological agent. 2. Extent and degree of biological harm of a chemical, physical, or biological agent, ranging from *acute* to *chronic*.

Toxicodynamics The determination and quantification of the sequence of events at the cellular and molecular levels, leading to a toxic response to an environmental agent. Often used synonymously with the broader term *pharmacodynamics*, but toxicodynamics is exclusively applied to adverse agents rather than the efficacy of substance.

Toxicokinetics The determination and quantification of the time course of absorption, distribution, biotransformation, and excretion of chemicals. Often used synonymously with the broader term *pharmacokinetics*, but toxicokinetics is exclusively applied to adverse agents rather than the efficacy of substance.

Toxicology Study of harmful interactions between chemical, physical, or biological agents and biological systems.

Tragedy of the Commons Term coined by Garrett Hardin [*Science* (1968), volume 162] characterizing the degradation of commonly held resources as a result to selfish self-interest in maximizing utility of each individual using the resource.

Transcapsidation See *heterologous encapsidation.*

Transcription Transfer of information in *DNA* sequences to produce complementary *messenger RNA* (*mRNA*) sequences. It is the beginning of the process by which the genetic information is translated to functional peptides and proteins.

Transduction Gene transfer between bacteria by bacteriophages.

Transect 1. Cross-section through which ground water flows. 2. Straight line placed on a surface along which measurements are taken.

Transfection Method by which experimental *DNA* is inserted into a cultured mammalian cell.

Transfer RNA (tRNA) Small *RNA* that binds an amino acid and delivers it to the ribosome for incorporation into a polypeptide chain during protein synthesis, using an *mRNA* as a guide.

Transformation 1. Chemical change to a substance. 2. Process by which bacterium acquires a plasmid and gains antibiotic resistance. Commonly refers to bench procedure that introduces experimental plasmids into bacteria. 3. Change in cell morphology and behavior, especially related to carcinogenesis. Transformed cell sometimes referred to as transformed phenotype.

Transgenic Referring to cells or organisms containing integrated sequences of cloned *DNA* transferred using techniques of *genetic engineering*.

Translation Decoding of *messenger RNA* (*mRNA*) occurs after transcription to produce a specific polypeptide according to the rules specified by the genetic code.

Translocation 1. Transfer, usually in plants, of compounds from the *rhizosphere* by *capillarity* to roots, and ultimately to other tissue, especially leaves. 2. Chromosome rearrangement in which two nonhomologous chromosomes are each broken and then repaired in such a way that the resulting chromosomes each contain material from the other chromosome (a reciprocal translocation).

Transmissivity Rate at which a fluid of the prevailing kinematic *viscosity* is transmitted through a unit width of the aquifer under a unit hydraulic gradient.

Transpiration Process by which water is released to the atmosphere by plants, usually through the *stomata* in leaves.

Travel time Time needed for water to flow from the outlet of a drainage subbasin to the larger drainage basin of interest.

Treatment In environmental engineering, the general term for any process that removes or detoxifies a substance.

Treatment Technique (TT) Process needed to decrease the concentration of a contaminant in drinking water.

Tricarboxylic acid cycle (TAC) Cycle of the oxidation of *acetyl coenzyme A* to CO_2, and the generation of NADH and FADH2 for oxidation in the electron transport chain; the cycle also supplies carbon skeletons for biosynthesis.

Trickling filter Beds of rock and other media covered *biofilm* that aerobically degrades organic waste during secondary sewage treatment.

Trophic state Level of biological organization.

Truth Conformity to fact.

Turbidity Scattering and adsorption of light in a fluid, usually caused by suspended matter.

Turbulence (turbulent flow) 1. Fluid property characterized by irregular variation in the speed and direction of movement of individual particles or elements of the flow. 2. State of flow of which the fluid is agitated by cross-currents and eddies, as contrasted with *laminar* flow.

Two-film model Resistance model in which dissolved compound moves from the bulk fluid through a liquid film to the evaporating surface and then diffuses through a stagnant air film to well-mixed atmosphere above; assumes that chemical is well mixed in the bulk solution below the liquid film and that mass transfer across each film is proportional to the concentration difference.

Type I error Error of rejecting a true null hypothesis. Contrast with *type II error*.

Type II error Error of accepting (not rejecting) a false null hypothesis. Contrast with *type I error*.

U-shaped dose-response curve Curve that shows deficiency at low dose and toxicity at high dose. Often seen in nutrients, vitamins, and some metal compounds.

Ultimate biochemical oxygen demand (BOD_u) 1. Total quantity of oxygen needed for complete degradation of organic material in the first stage of biochemical oxygen demand. 2. Quantity of oxygen required to completely degrade all biodegradable material in a wastewater.

Ultimate degradation The final breakdown products of reactions of contaminants; usually carbon dioxide and water but also methane and water for anaerobic systems.

Ultrafines or ultrafine particulate Very small particles that can infiltrate deeply into lung tissues. Smaller than $PM_{2.5}$.

Uncertainty factor (UF) One of several, generally 10-fold, default factors used in operationally deriving the *RfD* and *RfC* from experimental data. The factors are intended to account for (1) variation in *susceptibility* among the members of the human population (i.e., inter-individual or intraspecies *variability*); (2) uncertainty in extrapolating animal data to humans (i.e., interspecies uncertainty); (3) uncertainty in extrapolating from data obtained in a study with less-than-lifetime exposure (i.e., extrapolating from subchronic to chronic exposure); (4) uncertainty in extrapolating from a *lowest observed adverse effect level* rather than from a *no-observed-adverse-effect level*; and (5) uncertainty associated with extrapolation when the database is incomplete.

Uncertainty Difference between what is known and what is actually the truth. Scientific uncertainty includes error and unknowns, such as those resulting for selecting variables, undocumented variability, and limitations in measurements and models. In science, there is almost always uncertainty. The goal is to decrease uncertainty and to document known uncertainties.

Unconfined aquifer Aquifer directly open at the surface of the ground.

Unconsolidated material or media Soil, sediment, sand, gravel, and other material with interconnected pore spaces.

Unicellular Describing an organism that consists of a single cell.

Unit risk estimate (URE) Upper-bound excess lifetime cancer risk estimated to result from continuous exposure to an agent at a concentration of $1\,\mu g\,m^{-3}$ in air or drinking water concentration of $1\,\mu g\,L^{-1}$ for an adult weighing 70 kg and consuming 2 L per day.

Unit risk Upper-bound excess lifetime cancer risk estimated to result from continuous exposure to an agent at a concentration of $1\,\mu g\,L^{-1}$ in water or $1\,\mu g\,m^{-3}$ in air. The interpretation of unit risk would be as follows: if unit risk $= 2 \times 10^{-6}$ per $\mu g\,L^{-1}$, two excess cancer cases (upper bound estimate) are expected to develop per 1,000,000 people if exposed daily for a lifetime to $1\,\mu g$ of the chemical per liter of drinking water.

Upper bound A plausible upper limit to the true value of a quantity. This is usually not a true statistical confidence limit.

Utilitarianism Theory proposed by Jeremy Bentham (*Principles of Morals and Legislation*, 1789) and James Mill (*Utilitarianism*, 1863) that action should be directed toward achieving the greatest happiness for the greatest number of people (if applied to nonhuman species, this is referred to merely as "greatest number").

Utility 1. Useful outcome. 2. Level of enjoyment an individual attains from choosing a certain combination of goods.

Vacuum-enhanced free-product recovery Technology that extracts *LNAPLs* from the *capillary fringe* and *water table*.

Vadose zone Unsaturated zone of soil or other unconsolidated material above the water table (i.e., above the *zone of saturation*). Includes root zone, intermediate zone, and capillary fringe. Pore spaces contain water, as well as air and other gases at less than atmospheric pressure. Saturated bodies, such as perched ground water, may exist in the unsaturated zone, and water pressure within these may be greater than atmospheric. Also known as unsaturated zone.

Valuation Quantifying or otherwise placing value on goods and services. Monetized valuation uses monetary currency (e.g., gross domestic product), whereas many environmental and quality of life resources are not readily conducive to monetized valuation, e.g., old growth forests have nonmonetized value (e.g., habitat, ecological diversity) but little monetized value since they are not used for timber.

Value engineering (VE) Systematic application of recognized techniques by a multidisciplinary team to identify the function of a product or service, establish a worth for that function, generate alternatives using creative thinking, and provide the needed functions to accomplish the original purpose of the project, reliably, and at the lowest life cycle cost without sacrificing safety, necessary quality, and environmental attributes of the project.

Value 1. Principle, standard, or quality that is good for a person to hold. 2. Worth.

Valve Device in a pipe that controls the magnitude and direction of flow.

Vapor pressure Pressure exerted by a vapor in a confined space.

Vapor 1. Gas. 2. Gas phase of compound that under standard conditions would not be a gas.

Vaporization Change of a liquid or solid to the vapor phase.

Variability 1. True heterogeneity or diversity. For example, among a population that is exposed to airborne pollution from the same source and with the same contaminant concentration, the risks to each person as a result of breathing the polluted air will vary; among a population that drinks water from the same source and with the same contaminant concentration, the risks from consuming the water may vary, i.e., different people drinking different amounts of water and having different body weights, different exposure frequencies, and different exposure durations. Variability exists in every aspect of environmental data, e.g., in transport, fate, exposures, and effects. Overall variability may be due to differences in exposure (i.e., different people drinking different amounts of water and having different body weights, different exposure frequencies, and different exposure durations) as well as differences in response (e.g., genetic differences in resistance to a chemical dose). Differences among individuals in a population are referred to as interindividual variability, and differences for one individual over time are referred to as intraindividual variability. 2. In modeling, the differences in responses and other factors among species, i.e., *interspecies variability*, and within the same species, i.e., *intraspecies variability*, are part of the *uncertainty factors*, important to calculating *reference dose and concentrations.*

Variety Distinct population within a species with distinguishing, heritable traits.

Vector 1. Vehicle by which genetic material is inserted into a cell (usually a *plasmid* or *virus*). 2. Agent that carries *pathogens* among hosts (e.g., mosquito). 3. In physics, straight line segment with length representing magnitude and orientation in space representing direction.

Vegetative In a growth mode, e.g., a vegetative bacterial cell is one that is growing and feeding, in contrast with its *spore*.

Ventilation rate (VR) Alternative term for inhalation rate or breathing rate. Usually measured as minute volume, i.e., volume of air exhaled per minute.

Vertical gene transfer Crossing of organisms sexually and passing their genes on to following generations (usually called "crossing"). For example, gene transfer via pollen between plants of the same or related species takes place in the wild, so the transfer of disease and pest resistance from cultivated plants to related wild species and vice versa takes place irrespectively to the resistance genes that were initially acquired.

Vesicle Arbuscular mycorrhizal fungi's intracellular structures, usually spherical in shape.

Vibrio 1. *Anaerobes* of the genus *Vibrio*. 2. Rod-shaped, curved bacterial cell.

Virology Branch of microbiology that is concerned with viruses and viral diseases.

Virulence Degree of pathogenicity, i.e., the greater the virulence, the more pathogenic the microbe is.

Virus Submicroscopic organism typically containing a protein coat surrounding a nucleic acid core, only able to grow in a living cell.

Viscosity Molecular attractions with a fluid that evokes a tendency to deform under applied force. Internal friction within a fluid that causes it to resist flow. Absolute or dynamic viscosity is a measure of a fluid's resistance to tangential or shear stress (typical units are centipoises). Kinematic viscosity is the ratio of dynamic viscosity to mass density, obtained by dividing dynamic viscosity by the fluid density (typical units are centistokes).

Volatile acid Fatty acid containing $\leq$6 carbon atoms, with relatively high *aqueous solubility*. Often produced by *anaerobes*.

Volatile organic compound (VOC) Organic compound that readily evaporates, e.g., benzene, methylene chloride.

Volatile solids (VS) Materials, generally organic, that can be driven off from a sample by heating, usually to 550°C (1022°F); nonvolatile inorganic solids (ash) remain.

Volatile Capacity to change to the vapor phase, often expressed as *vapor pressure*.

Volatilization The transfer of a chemical from its liquid phase to the gas phase.

Vulnerability assessment The process by which the level of exposure is determined.

Vulnerability elements Social, material, or environmental context represented by the people and the resources and services that can be affected during an event. This corresponds to human activities, the systems created by humans, such as houses, roads, infrastructure, production centers, services, the people who use these systems, and the environment.

Vulnerability 1. Condition of an individual or population determined by physical, social, economic, and environmental variables, wherein susceptibility to a hazard increases (e.g., asthmatics are more vulnerable to the effects of some pollutants than is the average person). 2. Recently, the term has been a combination of *susceptibility* and high-end exposure; i.e., vulnerable subpopulations are those that are both susceptible and receive sufficient levels of exposure for an adverse outcome.

Waste-activated sludge Solids taken from the activated-sludge systems to prevent accumulation and an in sludge return to add microbes to increase waste degradation.

Wastewater treatment plant Facility design to remove pollutants from sewage.

Water table Upper surface of the free ground water in the *zone of saturation*.

Watershed Topographic area drained by surface water, wherein all outflows are discharged through a single outlet.

Weight-of-evidence Strength of data and information supporting a conclusion. When little reliable data are available, the weight of evidence is lacking, whereas when numerous studies provide reliable information to support a particular position (e.g., exposure to a chemical associated with a health effect), the weight of evidence is strong. Used by regulatory agencies to characterize the extent to which the available data support a hypothesis, e.g., whether a specific agent causes cancer in humans. Under the U.S. EPA's 1986 risk assessment guidelines, the WOE was described by categories "A through E," i.e., Group A for known human carcinogens through Group E for agents with evidence of noncarcinogenicity. The approach outlined in EPA's guidelines for carcinogen risk assessment (2005) considers all scientific information in determining whether and under what conditions an agent may cause cancer in humans and provides a narrative approach to characterize carcinogenicity rather than categories. Five standard weight-of-evidence descriptors are used as part of the narrative.

Wetland Lands inundated by water at least part of the year, indicated by plants that require or at least tolerate saturated soil conditions for substantial time periods.

Wilderness A large, natural (or nearly natural) region, not controlled by humans.

Willingness to pay Economic concept meaning the most money that people will give for a goods or service; depicted by the total area under a demand curve.

Wisdom Insight, erudition, and enlightenment resulting from the accumulation of knowledge and the ability to discern what is meaningful from what is not.

Work Product of a force times the distance through which the force acts.

Xenobiotic 1. Compound that is not normally found in natural systems. 2. A general term for an *anthropogenic* substance that is not easily degraded (see *recalcitrant*) by native microbial populations.

Xenotransplantation Transplantation of cells, tissue, and organs between nonrelated species.

Xerophile Organism adapted to dry conditions.

Yeast Unicellular fungus with a single nucleus that reproduces either asexually by budding or fission, or sexually through spore formation.

Zone of saturation Underground layers below the water table where the void space is filled with water. See *aquifer*. Compare to *vadose zone*.

Zooplankton Fauna in the plankton community.

Zygote *Diploid* cell that results from the fertilization of an egg cell by a sperm cell.

Appendix

Key equations, formulae, and nomenclature

Water and wastewater science, engineering, and management employ many equations. Several are included in this Appendix. The complete descriptions of variables and coefficients can be found in the relevant chapters of this book.

Chemistry

Conversion of air concentration from ppm to μg m^{-3} (25°C and 760 mmHg):	$1\ \text{ppm (volume)} = \dfrac{\left(\dfrac{1\ \text{L}}{22.4}\right) \times \text{MW} \times 10^6 \times \mu\text{g} \times \text{g} \times \text{m}^{-1}}{10^6\ \text{L} \times 298\ \text{K}^\circ \times 10^{-3}\ \text{m}^3 \times \text{L}^{-1}}$
Metric prefixes:	T tera—10^{12} (trillion 1,000,000,000,000)
	G giga—10^9 (billion 1,000,000,000)
	M mega—10^6 (million 1,000,000)
	k kilo—10^3 (1000)
	h hecto—10^2 (100)
	da deka—10^1 (10)
	d deci—10^{-1} (0.1)
	c centi—10^{-2} (0.01)
	m milli—10^{-3} (0.001)
	μ micro—10^{-6} (one-millionth 0.000001)
	n nano—10^{-9} (one-billionth 0.000000001)
Molarity:	M = moles solute per liter of solution
Normality:	N = equivalents of solute per liter of solution
Weight percentage:	Wt% = (mass of solute per mass of solution) × 100%
Parts per million:	ppm = (mass of solute per mass of solution) × 10^6
Mass per volume:	mg L^{-1} = mass of solute per liter of solution
Density:	unit mass per unit volume
Density of seawater:	1.03 g mL^{-1}
Mass per volume in seawater:	$\text{mg L}^{-1}{}_{\text{seawater}} = 1.00\ \text{mg L}^{-1}{}_{\text{water}} \times (1\ \text{mL}\ 1.03\ \text{g}^{-1}) \times 1\ \text{L}\ 1000\ \text{mL}^{-1} \times 1000\ \text{mg g}^{-1} = 0.971\ \text{mg kg}^{-1} = 0.971\ \text{ppm}$
Molality:	m = moles of solute per mass of solvent
Mole fraction:	χ = moles of solute per total moles

Continued

cont'd

Synthesis reaction:	$A+B \rightarrow AB$
Decomposition reaction:	$AB \rightarrow A+B$
Single replacement reaction:	$A+BC \rightarrow AC+B$
Double replacement reaction:	$AB+CD \rightarrow AD+CB$
Reversal reaction (equilibrium):	$aA+bB \leftrightarrow cC+dD$
Concentration and time gradient proportionality:	$\frac{dC}{dx} \approx \frac{dC}{dt}$
Complete combustion:	$a(CH)_x + bO_2 \xrightarrow{\Delta} cCO_2 + dH_2O$
Combustion with nitrogen:	$4CH_3NH_2\,(l) + 9O_2(\,g) \xrightarrow{\Delta} + 4CO_2\,(\,g) + 10H_2O\,(g) + 2N_2\,(\,g)$
Combustion with nitrogen and sulfur:	$C_aH_bO_cN_dS_e + 4a + b - 2c) \rightarrow aCO_2 + \left(\frac{b}{2}\right)H_2O + \left(\frac{d}{2}\right)N_2 + eS$
Kinetic reaction:	$r = -\frac{\Delta(A)}{\Delta t}$
Rate law:	$r = k[A]^m[B]^n$
First-order reaction rate:	$\frac{d[\,A]}{dt} = \frac{\Delta[A]}{\Delta t} + k[\,A]$
First-order rate equation:	$\ln[A] = \ln[A]_0 - k\Delta t$
Second-order rate:	$\frac{1}{[A]} = \frac{1}{[A]_0} + k\Delta t$
Differential rate law:	$\text{Rate} = -\frac{d[A]}{dt} = k[A]^n$
Integrated rate law:	$\ln[A] = -kt + \ln[A]_0$
Solubility product constant:	$K_{sp} = [C]^c[D]^d$
Generalized equilibrium constant:	$K_{eq} = \frac{k_f}{k_b} = \frac{[C]^c[D]^d}{[A]^a[B]^b}$
Indoor air concentration:	$C_{indoor} = \frac{\sum s_1 \ldots s_n}{V_{ME} \times ACH}$
Water dissociation constant:	$K_w = \frac{[H_3O^+][OH^-]}{[HOH]^2}$
Water dissociation constant at 25°C:	$K_w = [H_3O^+][OH^-] = 1.0 \times 10^{-14}\ \text{mol L}^{-1}$
"p" value:	$pX = -\log X$
pH:	$pH = -\log\,[H^+]$ or $pH = -\log[H_3O^+]$
Acid equilibrium constant for acetate:	$K_a = \frac{[H_3O^+][CH_3COO^-]}{[CH_3COOH]}$
Dissociation of carbon dioxide:	$CO_2 + H_2O \leftrightarrow H_2CO_3^*$
Henderson-Hasselbalch equation:	$pK_a = pH + \log\frac{[HA]}{[A^-]}$
Arrhenius equation:	$k = A \cdot e^{\frac{E_a}{RT}}$
NO_x:	$NO_y = NO + NO_2$
NO_y:	$NO_y = NO_2 + NO + HNO_3 + PAN + 2N_2O_5 + HONO + NO_3 + NO_3^-$ compounds $+ NO_3^-$ aerosols
Time-weighted concentration:	$\overline{C} = \frac{M}{Q \times t}$
Beer-Lambert law:	$A = eb[C]$

Physical laws and properties

Planck's law:	$E = hv = \frac{hc}{\lambda}$
First law of thermodynamics:	$\Delta U = q + w$
Mass balance:	$\frac{dM}{dt} = M_{in} - M_{out} \pm R$
Change of mass in control volume:	$\frac{d[A]}{dt} = -v \cdot \frac{d[A]}{dx} + \frac{d}{dx}\left(D \cdot \frac{d[A]}{dx}\right) + r$
Entropy:	$q = m.\ SH.\ \Delta T$
Boyle's law:	$V \propto P^{-1}$
Charles law:	$V \propto T$
Gay-Lussac's law:	$P \propto T$
Ideal gas law:	$PV = nRT$
Universal gas constant:	$R = \frac{0.08212\ \mathrm{L \cdot atm}}{\mathrm{mol \cdot K}}$ or $\frac{8.314\ \mathrm{joules}}{\mathrm{mol \cdot K}}$
Specific gas constant:	$R_{gas} = \frac{R}{m_{gas}}$
Advection-dispersion-reaction equation:	$\frac{dC}{dt} = -V \cdot \frac{dC}{dx} + \frac{d}{dx}\left(D \cdot \frac{dC}{dx}\right) + r$
Flux:	$j = \frac{Q}{A}; q = \frac{m}{A\Delta t} = \frac{m}{V} \times \frac{V}{A \cdot t} = C \cdot u$
Flux density:	$F = vC_A$
Flux density for dry deposition:	$F_d = v_d \cdot C_A$
Diffusive flux:	$j = C_A \cdot u' - C_B \cdot u' = u'\Delta C; j = -D\frac{\Delta C}{\Delta x}$
Fick's law:	$j = -D_{0}\,.\, i_c; j = -AD\frac{\Delta C}{\Delta x}; j = \frac{i_c}{R}$
Advective flux:	$J_{Advection} = \overline{v}\eta_e[c]$
Two-dimensional flux:	$J = \frac{Q}{A}$
Evaporative flux:	$E = \frac{mR_n + \gamma \cdot 6.43(1 + 0536 \times U_2) \cdot \delta e}{\lambda_v \cdot (m + \gamma)}$
Psychrometric constant:	$\gamma = \frac{0.0016286 \cdot P}{\lambda_v}$
Vapor pressure deficit:	$VPD = \mathring{p}_{saturated} - \mathring{p}_{actual}$
Saturated vapor pressure:	$\mathring{p}_{actual} = \frac{RH \times \mathring{p}_{saturated}}{100}$
Clausius-Clapeyron equation:	$\frac{dp}{dT} = \frac{\lambda}{T(V_2 - V_1)}$
Total water potential:	$\phi_T = \phi_p + \phi_m + \phi_z + \phi_o + \phi_{ob} + \cdots \phi_n$
Advective transport:	$N = QC$
Advective and diffusive flux:	$q = Cu + j$
Mean residence time:	$\tau = \frac{M}{J_{out}}$
Kinetic energy of gas molecule:	$E_g = \frac{mu^2}{2} = \frac{3RT}{2}$
Virtual temperature:	$T_v = \frac{T}{1 - \left(\frac{e}{P}\right)(1 - \varepsilon)}$
Newton's second law of motion:	$\mathbf{F} = \frac{d\mathbf{p}}{dt}$

Continued

cont'd

Newton's third law of motion:	$\mathbf{F}_{reacting} = -\mathbf{F}_{acting}$
Velocity of flow:	$u = f(x, y, z)$
Mean mass flow:	$\frac{dm}{dt} = \dot{m} = \rho \times A \times u$
Mass continuity steady state:	$\frac{dm}{dt} = \rho_1 \times A_1 \times u_1 = \rho_2 \times A_2 \times u_2 = \text{constant}$
Reference evapotranspiration (ET_0);	$ET_L = ET_0 \times K_L$
Plant adjustment factors (K_L):	$K_L = K_S \times K_D \times K_{MC}$

Partitioning and flow

Discharge:	$Q = V \times D \times W$
Stream discharge:	$Q = \sum_{n=1}^{n} ½\ (hn + hn - 1)\ (wn + wn - 1) \times ½\ (vn + vn - 1)½(h_n + h_{n-1})$
Stream pollutant concentration after complete mixing:	$C_r = \frac{Q_d C_d + Q_s C_s}{Q_r}$
Total maximum daily load:	$TMDL = \sum WLA + \sum LA + \text{Background} + \text{Future Growth} + MOS - \text{Loss}$
Mean stream velocity:	$\overline{V} = \frac{1.49(R_h)^{\frac{2}{3}}\sqrt{S}}{n}$
Travel time (uniform velocity):	$\tau = \frac{L_s}{V}$
Travel time (nonuniform velocity):	$\tau = \int_{x_2}^{x_1} \frac{dx}{V(x)}$
Chézy equation:	$V = C\sqrt{R_h S_0}$
Manning equation:	$Q = VA = \frac{k}{n}\left(\frac{A}{P}\right)^{\frac{2}{3}}\sqrt{S}$
Richards equation (simplified):	$D(\theta) = \frac{K(\theta)}{C(\theta)}$
Richards equation (restated):	$\frac{\partial\theta}{\partial t} = \frac{\partial}{\partial t}\left[D(\theta)\left(\frac{\partial\theta}{\partial z} + K(\theta)\right)\right]$
Richards equation (vadose zone, vertical dimension):	$\frac{\partial\theta}{\partial t} = \frac{\partial}{\partial t}\left[K(\Psi)\left(\frac{\partial\psi}{\partial z} + 1\right)\right]$
Storage coefficient:	$S = \frac{Q \text{ in } m^3}{(A \text{ in } m^2) \times (h \text{ in } m)}$
Transmissivity:	$T = Km$
Porosity:	$n = \frac{V_{total} - V_{solids}}{V_{total}} = \frac{V_{voids}}{V_{total}}$
Dry bulk density:	$\rho_b = \frac{M_{um}}{V_{total}}$
Wet bulk density:	$\rho_t = \frac{\left(M_{cm} + M_{liquids}\right)}{V_{total}}$
Bulk modulus, *E*:	$E = \frac{\text{stress}}{\text{strain}} = -\frac{dp}{dV/V_1}$
Ground water balance:	$\Delta S = R_r + R_s + R_i + S_i + I_g - E_t - T_p - B_f - O_g$
Inflow from rainfall seepage:	$R_p = P + R_i - E_t - O_f$
Fugacity:	$Z_A = \frac{1}{RT} = \frac{C_A}{f_A}$
Fugacity capacity constant:	$C_i = Z_i \cdot f$

cont'd

Total compartmental fugacity:	$f = \frac{M_{total}}{\sum_i (Z_i \cdot V_i)}$
Air-water partitioning coefficient:	$K_{AW} = \frac{C_A}{C_W} = \frac{Z_A f_A}{Z_W f_W}$
Raoult's law:	$p_i = p_i^* x_i$
Sorption:	$S = K_D \cdot C_W$
Soil or sediment partitioning:	$K_D = K_{OC} \cdot C_{org} = C_S (C_W)^{-1}$
Freundlich sorption isotherm:	$C_{sorb} = K_F \cdot C^n$
Partial pressure:	$p_a = K_H [c]$
Henry's law constant:	$K_H = \frac{P^{\circ} \text{ x MW}}{760 \times C_{sol}}$
Henry's law constant (unitless):	$K_{AW} = \frac{C_{air}}{C_{water}} = \frac{K_H}{R \cdot T}$
Octanol-air partitioning:	$K_{OA} = \frac{K_{ow} \cdot R \cdot T}{K_H} = \frac{K_{ow}}{K_{AW}}$; $\log K_{OA} = \log K_{ow} - \log K_{AW}$
Volumetric mass transfer rate:	$K_{1a_{G/A}} \left(\frac{S_G}{K_{G/A}} - S_A \right)$
Substrate partition coefficient:	$K_{G/A} = \frac{S_G}{S_A^*}$
In-stream pollutant concentration:	$C_r = \frac{Q_d C_d + Q_s C_s}{Q_r}$
Hydraulic head:	$h = z + \frac{p}{\rho g} + \frac{v^2}{2g}$
Hydraulic gradient:	$i = \frac{h_2 = h_1}{L} = \frac{\Delta h}{L}$
Hydraulic gradient:	$\frac{dh}{dx} = \frac{(h_1 - h_2)}{\Delta x}$
Darcy's law:	$Q = -KA \frac{\partial h}{\partial x}$
Capillary rise:	$h_{capillary} = \frac{2\sigma \cos \lambda}{\rho_w g R} = \frac{\varphi_O}{\rho_w g}$
Matric suction:	$u_a - u_w = \frac{2\sigma \cos \lambda}{\rho_W g R}$
Actual vapor pressure:	$p^{\circ}_{actual} = \frac{RH \times p^{\circ}_{saturated}}{100}$
Scavenging ratio:	$S = C_p \cdot \rho \cdot C_{aerosol} - 1$
Hydrologic continuity equation:	$\frac{dS}{dx} = I - O$ and $\frac{dS}{dx} = P - R - E - T - I$
Humidity ratio:	$\omega = 0.62198 \cdot \frac{p_w}{p_{at} - p_w}$
Degree of saturation:	$\mu = \frac{\omega}{\omega_{sat}}$
Mobilization factor:	$\text{MF} = \frac{E_h}{E_n}$
Enrichment factor (cadmium):	$\text{EF} = \frac{\left(\frac{[Cd]}{[Al]}\right)_{air}}{\left(\frac{[Cd]}{[Al]}\right)_{crust}}$
Half-life:	$t_{1/2}$ = (original amount)(0.5)x, where x = number of half-lives

Pollution treatment

Pollutant transport:	$N=D\cdot f=V[c]k=(V\cdot Z\cdot k)\cdot f=D_R.f$
Biochemical oxygen demand (BOD):	$y_t=L(1-e^{-k_1 t})$ where y_t=BOD in mg L^{-1}, L=ultimate BOD in mg L^{-1}, t=time in days, k_1=deoxygenation rate constant in $days^{-1}$
BOD:	$\mathrm{BOD}=\frac{D_1-D_5}{P}$ where P=decimal volumetric fraction of water, D units in mg L^{-1}
BOD:	$\mathrm{BOD}=\frac{(D_1-D_5)-(B_1-B_5)f}{P}$
Monod equation:	$\mu=\frac{\mu_{max}S}{K_s+S}$
Michaelis–Menten equation:	$\frac{dC}{dt}=-\frac{V_{max}C}{K_m+C}$
Biodegradation rate (first-order):	$R=\ln\frac{C_d}{C_u}(Dv)$
Target remediation concentration:	$C=\frac{\mathrm{Risk}\cdot\mathrm{BW}\cdot\mathrm{AT}}{\mathrm{CR}\cdot\mathrm{EF}\cdot\mathrm{ED}\cdot\mathrm{AF}\cdot\mathrm{SF}}$
Fuel stoichiometric combustion reaction:	$\underbrace{C_mH_n}_{fuel}+\left(m+\frac{n}{4}\right)\underbrace{(O_2+3.76N_2)}_{air}\rightarrow mCO_2+\frac{n}{2}H_2O+3.76\left(m+\frac{n}{4}\right)N_2$
Solid/liquid fuel stoichiometric combustion reaction:	$\underbrace{C_aH_bO_cN_dS_e}_{fuel}+x\underbrace{(O_2+3.76N_2)}_{air}\rightarrow aCO_2+\frac{b}{2}H_2O+\left(\frac{d}{2}+3.76x\right)N_2+eSO_2$
Combustion efficiency:	$CE=\left[\frac{C_{CO2}}{C_{CO2}+C_{CO}}\right]\times 100$
Destruction and removal efficiency:	$DRE=\left[\frac{W_{in}-W_{out}}{W_{in}}\right]\times 100$
Molecular momentum:	$\Delta mu=(\Delta mu_x)_f-(\Delta mu_x)_i=-(mu_x)-(mu_x)=-2mu_x$
Transfer rate for gas molecules:	$F_{total}=\frac{m}{l}(u_{x1}+u_{x2}+\cdots u_{xn})$
Pressure:	$p=\frac{F}{A}$
Pressure:	$P=\frac{F_{total}}{A}=\frac{m}{l^3}(u_{x1}+u_{x2}+\cdots u_{xn})=\frac{mN}{V}\left(\frac{u_{x1}+u_{x2}+\cdots u_{xn}}{N}\right)$
Stress:	$\sigma(P)=\lim_{\delta A\to 0}\frac{\delta\mathbf{F}}{\delta\mathbf{A}}$
Density field:	$\rho=\rho(x,y,z,t)$
Velocity field:	$\vec{v}=\vec{v}(x,y,z,t)$
Static pressure of flow:	$p+\frac{1}{2}p\,V^2+pgh=\mathrm{constant}$
Bernoulli equation:	$\left(\frac{p}{\rho g}+\frac{V^2}{2g}+z\right)_1=\left(\frac{p}{\rho g}+\frac{V^2}{2g}+z\right)_2$
Bernoulli equation for pump flow:	$\left(\frac{p}{\rho g}+\frac{V^2}{2g}+z\right)_1+h_p=\left(\frac{p}{\rho g}+\frac{V^2}{2g}+z\right)_2+h_i$
Bernoulli equation considering conservation of mass:	$p_1+\frac{1}{2}p\,V_1^2+pgh_1=p_2+\frac{1}{2}p\,V_2^2+pgh_2$
Lagrangian velocity field:	$\vec{v}=[x(t),y(t),z(t)]$
Volumetric flow rate:	$Q=Vt^{-1}$

cont'd

Aerodynamic diameter (aerosol):	$D_{pa} = D_{ps}\sqrt{\rho_p}$
Incompressible fluid continuity:	$\nabla \cdot \boldsymbol{u} = 0$
Navier-Stokes equation:	$\rho\left[\frac{\partial u}{\partial t} + (u \cdot \nabla)u\right] = -\nabla p + \eta\nabla^2 u + \mathrm{f}$
Continuity equation (cylinder):	$\frac{1}{r}\frac{\partial}{\partial r}(ru_r) + \frac{1}{r}\frac{\partial}{\partial \theta}u_\theta + \frac{1}{r}\frac{\partial}{\partial z}u_z = 0$
Airway volume (lung physiology):	$V_{aw} = 1.018 \times \text{Height (cm)} - 76.2$
Stokes number:	$\mathrm{Stk} = \frac{t_0 u_0}{l_0}; \mathrm{Stk} = \frac{\rho_p d_p^2 \dot{u}}{18\mu_g d}$
Impaction factor (dimensionless):	$\Psi = \frac{C_C d_p^2 3 y_p \rho_p}{18\mu_g D_C}$
Pollutant gradient:	$\frac{d[A]}{dt} = -v \cdot \frac{d[A]}{dx} + \frac{d}{dx}\left(D\frac{d[A]}{dx}\right) + r$
Maximum use concentration:	$\mathrm{MUC} = \mathrm{PEL} \times \mathrm{APF}$
Sum of forces on a particle:	$\sum F = m_p \cdot a_p = m_p \frac{dv_p}{dt}$ and $\sum F = \frac{m_p \cdot a_p}{g_c}$
Gravitational constant:	$g_c = 32.2\frac{\mathrm{lb_m} \cdot \mathrm{ft}}{\mathrm{lb_f} \cdot \mathrm{sec}^2}$
Gravitational force:	$F_G = m_p \cdot g = \rho_p \cdot V_p \cdot g$
Volume of a spherical particle:	$V_p = \frac{\pi d_p^3}{6}$
Gravitational force of a spherical particle:	$F_G = \frac{\pi d_p^3 \rho_p g}{6}$
Buoyancy force:	$F_B = \frac{\pi d_p^3 \rho_g g}{6}$
Drag force:	$F_D = \frac{A_g \cdot \rho_g \cdot V_p^2 \cdot C_D}{2} = \frac{\pi d_p^2 \rho_g v_p^2 \cdot C_D}{8}$
Reynolds number:	$\mathrm{Re} = \frac{\text{Inertial Forces}}{\text{Viscous Forces}}$
Reynolds number:	$\mathrm{Re} = \frac{d_e v \rho}{\mu}$
Reynolds number:	$\mathrm{Re} = \frac{d_e v}{\upsilon}$
Particle Reynolds number (dimensionless):	$\mathrm{Re}_p = \frac{d_p v_p \rho_g}{\mu_g}$
Drag coefficient (laminar):	$C_D = \frac{24}{\mathrm{Re}_p}$
Drag coefficient (transition):	$C_D = \frac{18.5}{\mathrm{Re}_p^{0.6}}$
Drag coefficient (turbulent):	$C_D = 0.44$
Cunningham slip factor:	$C_C = 1 + \frac{6.21 \times 10^{-4} T}{d_p}$
Drag coefficient with Cunningham slip factor:	$C_D = \frac{24}{\mathrm{Re}_p \cdot C_C}$

Continued

cont'd

Drag force (laminar):	$F_D = \frac{3\pi\mu_g v_p d_p}{C_C}$
Drag force (transition):	$F_D = 2.30(d_p \cdot v_p)^{1.4} \cdot \mu_g^{0.6} \cdot \rho_g^{0.4}$
Drag force (turbulent):	$F_D = 0.05\pi(d_p \cdot v_p)^2 \cdot \rho_g$
Terminal velocity (laminar flow):	$v_t = \frac{g \cdot C_C \cdot \rho_p \cdot d_p^2}{18\mu_g}$
Terminal velocity (transition flow):	$v_t = \frac{0.153 g^{0.71} \cdot \rho_p^{0.71} \cdot d_p^{1.14}}{\mu_g^{0.43} \cdot \rho_g^{0.29}}$
Terminal velocity (turbulent flow):	$v_t = 1.74\left(\frac{g \cdot \rho_p \cdot d_p}{\cdot \rho_g}\right)^{0.5}$
Settling characteristic value:	$K = d_p\left(\frac{g \cdot \rho_p \cdot \rho_g}{\mu_g^2}\right)^{0.33}$
Collection efficiency:	$\eta_i = 1 - e^{-X}$ and $\eta_i = 1 - e^{-\frac{v_t LW}{Q}}$
Particle residence time:	$t_r = \frac{L}{v_g}$
Worst-case (longest path) particle residence time:	$t_r = \frac{H}{v_t}$
Volumetric gas flow:	$Q = \frac{v_g}{W \times H}$
Centrifugal force:	$F = \frac{MV^2}{R}$
Particle diffusivity:	$X_p = \frac{C_C kT}{3\pi\mu_g d_p}$
Force on a particle in electric field:	$F_E = n \cdot e \cdot E$
Face velocity at filter:	$U_0 = \frac{Q}{A}$
Pressure drop	$\Delta p = \frac{8\mu LQ}{\pi r^4} = \frac{8\pi\mu LQ}{A^2}$
Pressure drop across air filter:	$\Delta P_{total} = \Delta P_{media} + \Delta P_{dust\text{-}cake} = (k_1 \cdot v_f) + (k_2 \cdot c_i \cdot v_f^2 \cdot t)$
Filter drag:	$S_f = \frac{\Delta P_{total}}{v_f} = k_1 + (k_2 \cdot c_i \cdot v_f \cdot t)$
Ammonification:	$NH_4^+ + OH^- \Leftrightarrow NH_3 + H_2O$
Nitrification:	$NH_4^+ + 2O_2 \rightarrow NO_3^- + 2H^+ + H_2O$

Geometry

Area of a circle:	$A = \pi r^2 = \pi\left(\frac{D}{2}\right)^2 = \frac{\pi D^2}{4}$
Area of a cylinder:	$A = 2\pi rh + 2\pi(r)^2$
Volume of a cylinder:	$V = h\pi(r)^2$
Volume of a sphere:	$V = \frac{4}{3}\pi r^3$
Aspect ratio:	$AR = \frac{width}{height}$

Probability and statistics

Sample space:	$\Omega=(s_1, s_2, \ldots, s_n)$
Root mean square speed:	$u_{rms}=(\overline{u}^2)^1/2=\left(\frac{3P}{\rho}\right)^1/2$
Wet-dry gas concentration conversion:	$C_{dry}=\frac{C_{wet}}{1-w}$
Light interaction with atmosphere:	$-dI=b_{ext}I\cdot dx$
Extinction coefficient:	$b_{ext}=b_{rg}+b_{ag}+b_{scat}+b_{ap}$
Arithmetic mean:	$\overline{x}=\frac{\sum_{i=1}^{n}x_i}{n}$
Standard deviation:	$s=\sqrt{\left[\sum_{n=1}^{n}\left[\frac{(Y_1-\overline{Y})^2}{n-1}\right]\right]}$
Geometric mean:	$GM=\mu/\left(1+0.5*\left(\frac{\sigma}{\mu}\right)^2\right)$
Geometric standard deviation:	$GSD=\exp\left(\frac{\ln\left(\frac{xc}{GM}\right)}{z-score}\right)$
	$GSD=\exp\left(\sqrt{2*\ln\left(\frac{\mu}{GM}\right)}\right)$
Relative percent difference:	$RPD=\frac{(C_1-C_2)}{C_1+C_2)/2}\times 100\%$
Relative standard deviation:	$RSD=\frac{s}{\overline{Y}}\times 100\%$
Percent recovery:	$\%R=\frac{S-U}{C}\times 100\%$ or $\%R=\frac{C_M}{C_{SRM}}\times 100\%$
Standard mortality rate:	$SMR=\frac{D_{observed}}{D_{expected}}=\frac{P_{exposed}}{P_{unexposed}}$
Standard morbidity rate:	$SMbR=\frac{D_{observed}}{D_{expected}}=\frac{P_{exposed}}{P_{unexposed}}$
Bayes theorem:	$P(A\|B)=\frac{P(B\|A)P(A)}{P(B)}$
Bayesian inference:	$P(\theta\|D)=\frac{P(D\|\theta)P(\theta)}{P(D)}$

Risk assessment

Risk function:	$R=f(S, P)$
Reference dose:	$RfD=\frac{NOAEL}{UF_{inter}\times UF_{intra}\times UF_{iother}}$
Reference concentration:	$RfC=\frac{NOAEC}{U_{inter}\times U_{intra}\times U_{other}}$
Average daily dose:	$ADD=\frac{C_{medium}\times IR\times EF\times ED}{BW\times AT}$
Exposure:	$E=D/t$
Exposure:	$E=\int_{t=t_1}^{t=t_2}C(t)dt$

Continued

cont'd

Exposure to particles in air:	$E=\frac{(C)\cdot(PC)\cdot(IR)\cdot(RF)\cdot(EL)\cdot(AF)\cdot(ED)\cdot(10^{-6})}{(BW)\cdot(TL)}$
Exposure to vapors:	$E=\frac{(C)\cdot(IR)\cdot(EL)\cdot(AF)\cdot(ED)}{(BW)\cdot(TL)}$
Drinking water exposure:	$E=\frac{(C)\cdot(CR)\cdot(ED)\cdot(AF)}{(BW)\cdot(TL)}$
Soil exposure:	$E=\frac{(C)\cdot(SA)\cdot(BF)\cdot(FC)\cdot(SDF)\cdot(ED)\cdot(10^{-6})}{(BW)\cdot(TL)}$
Daily intake:	$\mathrm{DI}\left(\mathrm{mg\ kg^{-1}\ day^{-1}}\right)=\frac{\mathrm{Concentration}\left(\mathrm{mg\ volume^{-1}}\right)\times\mathrm{Intake}\left(\mathrm{volume\ day^{-1}}\right)}{\mathrm{body\ mass\ (kg)}}$
Chemical intake:	$I=\frac{C\cdot CR\cdot EF\cdot ED\cdot AF}{BW\cdot AT}$
Daily intake:	$\mathrm{DI}\left(\mathrm{mg\ kg^{-1}day^{-1}}\right)=\frac{\mathrm{Concentration}\left(\mathrm{mg^2\ volume^{-1}}\right)\times\mathrm{Intake}\left(\mathrm{volume\ day^{-1}}\right)}{\mathrm{body\ mass\ (kg)}}$
Serum concentration:	$CP=\frac{DP}{kP*Vd}$
Chemical intake from air pathway:	$I=C_A\times\frac{IR}{BW}\times\frac{ET\times EF\times ED}{AT}$
Exposure concentration:	$EC=\frac{C_A\times ET\times EF\times ED}{AT}$
Cancer risk:	$\mathrm{R}=\mathrm{CSF}\times E_I$
Inhalation unit risk:	$\mathrm{IUR}=\frac{\mathrm{POD}}{\mathrm{LEC}_{10(\mathrm{HEC})}}$, where POD is point of departure
Noncancer risk hazard quotient:	$\mathrm{HQ}=\frac{\mathrm{MDD}}{\mathrm{ADI}}=\frac{\mathrm{Exposure}}{\mathrm{RfD}}$
Noncancer risk hazard index:	$\mathrm{HI}=\sum_1^n\mathrm{HQ}$
Eco-risk quotient:	$\mathrm{RQ}=\frac{\mathrm{Exposure}}{\mathrm{Ecotoxicity}}=\frac{\mathrm{EEC}}{\mathrm{EC}_{50}}$
Risk-based target concentration:	$\mathrm{C}=\frac{\mathrm{Risk}\cdot\mathrm{BW}\cdot\mathrm{AT}}{\mathrm{CR}\cdot\mathrm{EF}\cdot\mathrm{ED}\cdot\mathrm{AF}\cdot\mathrm{SF}}$
Aquatic life exposure:	$AE=DE;QC_w=F_wC_p;BCF=\frac{Q}{F_w}$
Fish pollutant uptake:	$\frac{dB_f}{d_t}=J_g+J_i+J_{bt}$
Pollutant biotransformation rate:	$\frac{dW_d}{d_t}=F_d-E_d-R-EX-SDA$
Fish body burden:	$\frac{dN}{d_t}=EM-NM-PM$

Contaminant dispersion modeling

Box model mass balance:	$\frac{dyCV}{dt}=QA+(\overline{u}\cdot C_{in}\cdot\mathrm{W}\cdot H_{mix})-(\overline{u}\cdot C_{in}\cdot\mathrm{W}\cdot H_{mix})$
Gaussian concentration:	$C(x,y,z)=\frac{Q}{2\pi u\sigma_y\sigma_z}e^{-\frac{y^2}{2\sigma_y^2}}\left(e^{-\frac{(z+H)^2}{2\sigma_z^2}}+e^{-\frac{(z-H)^2}{2\sigma_z^2}}\right)$
Gaussian concentration (stable conditions):	$C=Q(1/u)\{g_1/[(2\pi)^{0.5}\sigma_y]\}\{g_2/[(2\pi)^{0.5}\sigma_z]\}$
Gaussian concentration (unstable or neutral conditions):	$C=Q(1/u)\{g_1/[(2\pi)^{0.5}\sigma_y]\}(1/L)$

cont'd

Maximum concentration along centerline (Gaussian):	$C_{max} = \frac{2Q}{\pi \cdot u \cdot e \cdot H^2} \cdot \frac{\sigma_z}{\sigma_y}$
Mean concentration based on mean emission rate:	$\overline{C}(x,y,z) = \frac{\overline{Q}}{\pi\sigma_y\sigma_z u} \exp\left[-\frac{1}{2}\left(\frac{y^2}{\sigma_y^2} + \frac{z^2}{\sigma_z^2}\right)\right]$
Advection rate:	$G_i = Vt^{-1}$
Model uncertainty:	Y=h(U,V)

Feasibility assessment

Failure likelihood:	$P\{t_1 \le T_f \le t_1\} = \int_{t_1}^{t_2} f(t)dt$
Reliability function:	$R(t) = P\{t_1 \le T_f \le t_1\} = 1 - \int_0^t f(x)dx$
Rollback model:	$R = \frac{g(P) - D}{g(P) - B} \times 100$
Annual discount rate:	$i = \left(1 + \frac{r}{k}\right)^{\frac{k}{p}} - 1$
Discount rate (spreadsheet):	$i = RATE(N, A, P, F, Type, guess)$
Discount factor:	$P/F,\ i\%,\ n$
Discount formula:	$P = F(P/F, i\%, n)$
Present value:	$PV = \frac{B_{y_i}}{(1+r)^{y_i - y_0}}$ and $PV = \frac{C_{y_i}}{(1+r)^{y_i - y_0}}$
Present worth (spreadsheet):	$P = PV(i, N, A, F, Type)$
Future worth (spreadsheet):	$F = FV(i, N, A, P, Type)$
Equivalent uniform annual worth:	EUAW=EUAB−EUAC
Availability time interval:	$T = u + d$
Average availability:	$\overline{a} = \frac{1}{T}\int_0^T a(t)dt$
Limiting availability:	$a = \lim_{t \to \infty} a(t)$
Mean availability:	$A = \frac{u}{u + d}$
Nines of availability:	$A = \left[1 - \frac{\text{MTTR}}{\text{MTBF}}\right] \times 100\%.$
Reliability function:	$R(t) = \exp\left(-\frac{t}{\text{MTBF}}\right) = \exp(-\lambda t)$
Net goodness	NG=Σ(goodness of each consequence)×(importance)×(likelihood)

Life-cycle analysis and sustainability

Impact of human activity on the environment	I=P×A×T, where P=population, A=affluence, T=technology
Kaya identity	$F = P \times \frac{G}{P} \times \frac{E}{G} \times \frac{F}{E}$, where F=global anthropogenic CO_2 emissions, P=global population, G=global gross national product, E=global energy consumption
Characterization factor	CF=EF×FF×XF, where EF=effect factor, FF=fate factor, XF=exposure factor
Intake fraction	iF =FF×XF

Abbreviations and symbols

The following abbreviations and symbols are used in this text and/or cited sources. The list is not exhaustive since there is much variability in the use of symbols. Water pollution draws from numerous scientific and professional disciplines.

A	magnetic vector potential (tesla meter—T m)
A	area, surface area (length^2)
A	availability (engineering economics)
a	acceleration
A	activity rate (in emission calculation)
A	air (subscript)
A	amplitude
A	aqueous phase
A	atomic mass number
A	uniform cost
A/C	air-to-cloth ratio (filtration; length time^{-1})
AA	atomic absorption detection
A_{aw}	airway surface area (length^2)
abs	absorbed
AC	activated carbon
AC	alternating current (electricity)
ACH	air changes per hour
ADD	average daily dose
ADE	average daily exposure
ADME	absorption, distribution, metabolism, and excretion (or elimination)
ADMET	absorption, distribution, metabolism, excretion, and toxicity
ADI	acceptable daily intake
ADP	adenosine diphosphate
A-D-R	advection-dispersion-reaction equation
AER	air exchange rate
AF	absorption or absorption factor
AF	assessment factor
AHERA	Asbestos Hazard and Emergency Response Act
AIRS	Aerometric Information Retrieval System (U.S.)
$Al_2(SO_4)_3$	aluminum sulfate (alum)
ALARP	as low as reasonably practicable (risk)
amu	atomic mass unit
A_p	cross-sectional area of the particle (length^2) $=\frac{\pi d_p^2}{4}$
AP	averaging period
a_p	particle acceleration
aq	aqueous phase
As	arsenic
A_s	exposed body surface area
A_s	clarifier tank surface area
ASCE	American Society of Civil Engineers
ASME	American Society of Mechanical Engineers
ASPECT	airborne spectral photometric environmental collection technology
AT or $t_{averaging}$	averaging time for exposure
ATC	allowable threshold concentration
atm	atmosphere (unit of pressure)

ATP	adenosine triphosphate
ATSDR	Agency for Toxic Substances and Disease Registry (U.S.)
AWQC	ambient water quality criteria
AWWA	American Water Works Association
b	saturated thickness of an aquifer (also *m*)
B	bioavailability factor (exposure)
BAC	biological activated carbon
BACT	best available control technology
BAF	bioaccumulation factor
BAF	biological activate filter
B(a)P	benzo(a)pyrene
BAT	best available technology
BCA	benefit-cost analysis
B/C or BCR	benefit-to-cost ratio
BCF	bioconcentration factor
BDL	below detection limit
BF	bioavailability (percent of substance absorbed per day)
BMD	benchmark dose
BMP	best management practices
BNA	base-neutral and acid extractable organics
BOD	biochemical oxygen demand
BTEX	benzene, toluene, ethylbenzene, and xylenes (indication of gasoline contamination)
Btu or BTU	British Thermal Unit (power) = 1060 J
BW	body weight (mass)
C	coulomb (ampere · second—A · s)
C	chemical concentration ([mass mass^{-1}], [volume volume^{-1}] or [mass volume^{-1}])
C	heat capacity (J · K^{-1})
C_a	airborne chemical concentration (mass volume^{-1} of air in breathing zone)
$\overline{C}$	time-weighted concentration
c	speed of light (299,792,458 m s^{-1})
c	specific heat capacity (J kg^{-1} K^{-1})
cal	calorie
C_c	Cunningham slip correction factor
CCR	coal combustion residuals
Cd	cadmium
C_D	drag coefficient (dimensionless)
CDC	Centers for Disease Control and Prevention (U.S.)
CDD	chlorinated dibenzodioxin
CDE	carbon dioxide equivalents
CDF	chlorinated dibenzofuran
CERCLA	Comprehensive Environmental Response, Compensation and Liability Act (also known as Superfund; U.S.)
CF	characterization factor (LCA)
CFC	chlorofluorocarbon
CFD	computational fluid dynamics
CFM	cubic feet per minute
CFR	Code of Federal Regulations (U.S.)
C_6F_{14}	perfluorohexane
cfu	colony forming unit
C_2H_3Cl	vinyl chloride
CH_2F_2	difluoromethane (hydrofluorocarbon-32)
CI	confidence interval
Ci	curie (radioactivity unit)
C_{ij}	concentration of exposure by person *i* in microenvironment *j*
CNS	central nervous system
CO	carbon monoxide
CO_2	carbon dioxide
CoC	contaminant of concern
COD	chemical oxygen demand

COHb	carboxyhemoglobinemia
COPD	chronic obstructive pulmonary disease
$\mathbf{C_{org}}$	substrate organic matter (mass of organic carbon per mass of substrate)
cos	cosine
c_p	specific heat at constant pressure
CP	serum concentration
CPA	critical path analysis
CPM	critical path method
cps	cycles per second (1 cps = 1 Hz)
Cr	chromium
C_S	equilibrium concentration of the solute in the solid phase
CSF	cancer slope factor [(mass mass^{-1} $\text{time}^{-1})^{-1}$] (also SF)
c_v	specific heat at constant volume
CVD	cardiovascular disease
$\mathbf{C_xH_yO_x}$	organic compound (hypothetical or representative)
d	day
d	distance
d	diameter (length)
D	electric flux density (coulomb per square meter—C m^{-2})
D	dose
D	index of community diversity
D	intermedia or intercompartmental transport term (fugacity models)
D	mass of pollutant per body mass
D	transfer coefficient or compartmental rate constant (for fugacity)
$\underline{D}$	dispersion tensor (length time^{-1})
d_0	proportionality constant (diffusion)
D_A	nondiffusive transfer coefficient for fugacity
DAF	dissolved air floatation
dB	decibel
DBCP	1,2-dibromo-3-chloropropane
DBP	disinfection by-product
D_C	diameter of the collection object (length)
DC	direct current (electricity)
DCA	dichloroethane
DCE	dichloroethylene
D_D	diffusive transfer coefficient for fugacity
DDD	2,2-bis(*p*-chlorophenyl)-1,1-dichloroethane
DDE	dichlorodiphenyldichloroethylene
DDT	1,1,1-trichloro-2,2-bis(*p*-chlorophenyl)ethane
D_f	Feret's particle diameter
d_i	percentage difference (precision)
DI	daily intake
DIAL	differential absorption lidar
dL	deciliter (volume)
D_m	Martin's particle diameter
DNA	deoxyribonucleic acid
DNAPL	dense (high density) nonaqueous phase liquid
DNEL	derived no-effect level
DO	dissolved oxygen
DOAS	differential optical absorption spectroscopy
DOC	dissolved organic carbon
DP	degradation product
d_p	physical particle diameter (length)
$\mathbf{D_{pa}}$ **or** d_{pa}	aerodynamic diameter of a particle (length)
$\mathbf{D_{ps}}$ **or** d_{ps}	Stokes particle diameter (length)
DQO	data quality objective
DRE	destruction and removal efficiency
DRL	differential rate law

DTW	depth to water
DU	Dobson unit (ozone)
E	electric field
e	eccentricity
e	charge of electron (4.8×10^{-10} statcoulombs)
E	electrical field strength (potential; voltage length^{-1})
E	bulk modulus (fluid stress divided by strain)
E	emission
E	energy required in a reaction
e	Euler's number = 2.71828183...
E	evaporative flux
E	event (probability and statistics)
E	exposure
E	particle collection efficiency; inverse of penetration (filtration)
$\boldsymbol{E_0}$	strength of electrical field (volts length^{-1})
E/P	evaporation/percolation
$\boldsymbol{E_a}$	Arrhenius activation energy
EC	average exposure concentration
EC	elemental carbon
$\mathbf{EC_{50}}$	median effective concentration
ECD	electron capture detection
$\boldsymbol{E_D}$	exposure dose
ED	exposure duration (time)
EDB	ethylene dibromide (1,2-dibromoethane)
EDC	ethylene dichloride (1,2-dichloroethane)
EDS	energy dispersive spectroscopy
EEC	estimated environmental concentration
$\mathbf{E_F}$	emission factor (in emission calculation)
EF	exposure frequency (time)
EF	enrichment factor (biogeochemistry)
EF	effect factor (LCA)
EHS	extremely hazardous substance
EHW	extremely hazardous waste
$\boldsymbol{E_{ing}}$	ingestion exposure (mass per time)
$\boldsymbol{E_{inh}}$	inhalation exposure (mass per time)
EIA	environmental impact assessment
EIS	environmental impact statement
EJ	environmental justice
EL	exposure length (time per time)
ELCR	excess lifetime cancer risk
ELF	extremely low frequency
EMP	elongate mineral particulate
EMR	electromagnetic radiation
EPA or U.S. EPA	Environmental Protection Agency (U.S.)
EPH	extractable petroleum hydrocarbons
EPS	extracellular polymeric substance
$\mathbf{E_R}$	percent overall emission reduction efficiency (in emission calculation)
erg	energy and work unit (10^{-7} J)
EROEI	energy return on energy investment
ESA	Endangered Species Act (U.S.)
ESP	electrostatic precipitator
EU	European Union
EUAB	equivalent uniform annual benefit
EUAC	equivalent uniform annual cost
EUAW	equivalent uniform annual worth
exp(x)	exponential (e^x)
$\boldsymbol{E_\upsilon}$	energy at frequency υ
F	force

f	Coriolis parameter
F	energy flux
F	favorable
F	force (mass length time^{-2})
F	Faraday constant
f	frequency (probability and statistics)
f	friction
f	fugacity
f	fugacity (subscript)
$f(x)$	function of x
F_B	buoyancy force (mass length time^{-2})
F_C	centrifugal force (mass length time^{-2})
F_D	drag force (mass length time^{-2})
FDA	Food and Drug Administration (U.S.)
F_E	electrostatic charge (force)
Fe	iron
FF	fate factor (LCA)
FGD	flue gas desulfurization
FID	flame ionization detection
FIFRA	Federal Insecticide, Fungicide and Rodenticide Act (U.S.)
f_{OC}	weight fraction of organic carbon in compartmental matrix (dimensionless)
FP	flashpoint
FR	Federal Register (U.S.)
FS/PA	feasibility study and preliminary assessment (U.S. Superfund)
ft	foot (length)
FTIR	Fourier-transform infrared
g	gas or vapor phase (subscript or parenthetic)
G	gas-phase emissions (mass time^{-1})
G	Gibbs free energy
G	shear modulus
g	gram
g	gravitational force
G	nondiffusive flow rate (for fugacity)
G	universal gravitational constant
g_0	standard gravitational field
GAC	granulated activated carbon
gal	gallon
GC	gas chromatography
g_c	gravitational constant $= 32.2 \frac{\text{lb}_\text{m} \cdot \text{ft}}{\text{lb}_\text{f} \cdot \text{sec}^2} = 980\,\text{cm}\,\text{s}^{-2}$
GHG	global greenhouse gas
GHz	gigahertz (10^9 Hz)
G_i	advection rate (fugacity models)
GI	gastrointestinal tract
GIS	geographic information system
GLP	good laboratory practice
GMM	genetically modified microbe
GMO	genetically modified organism
gpd	gallons per day
gpm	gallons per minute
GPS	global positioning system
grad	gradient; grad $f = \nabla f(x, y)$ in a plane and $\nabla f(x, y, z)$ in space
GTP	global temperature potential
GWP	global warming potential
ΔG_f^{*0}	free energy formation at steady state
H	hazard
h or hr	hour
h	elevation
h	hydraulic head

$\frac{dh}{dx}$	hydraulic gradient
H	enthalpy
H	heat (subscript)
H	Henry's Law (subscript) (also K_H)
H	horizontal (subscript)
h	Planck's constant (6.62×10^{-27} erg s)
HC	hydrocarbon
$h_{capillary}$	rise of the fluid in a capillary
HEPA	high efficiency particulate air (filter)
Hg	mercury
HHM	hierarchical holographic modeling
HI	hazard index
HLRW	high-level radioactive waste
HMIWI	hospital/medical/infectious waste incinerators
HON	hazardous organic NESHAPs
hPa	hectopascal (unit of pressure)
HPLC	high performance liquid chromatography
HPV	high production volume (chemical class)
HQ	hazard quotient
HR	hazard ratio
HRS	hazard ranking system
HW	hazardous waste
HVAC	heating, ventilation, and air conditioning
HYSPLIT	Hybrid Single Particle Lagrangian Integrated Trajectory (model)
Hz	hertz (cycles s^{-1})
hv	photon
i	group, species, subgroup of total (subscript)
i	flow of electrons
I	immobilized
I	*in situ*
I	intake dose
I	moment of inertia
i	intensity
i	imaginary unit
i_c	concentration gradient
IBI	Index of Biotic Integrity
ICP	inductively coupled plasma
ICP	inductively coupled plasma detection
ICR	information collection request
ICS	Incident Command System
ID	interior diameter
IDL	instrument detection limit
IDLH	immediately dangerous to life and health
IAEA	International Atomic Energy Agency
IH	imminent hazard
in Hg	inches of mercury (unit of pressure)
IN	mass or energy inflow to control volume; influent (subscript)
IPCC	Intergovernmental Panel on Climate Change
IPM	integrated pest management
IR	intake rate ([mass time^{-1}] or [volume per time])
IR	inhalation rate
IR	infrared
IR	ingestion rate
IR	infrared
IR-DIAL	infrared-differential absorption lidar
IRIS	Integrated Risk Information System
IRL	integrated rate law
IRR	internal rate of return

ISCLT	Industrial Source Complex Long Term (model)
IUR	inhalation unit risk
J	joule (energy unit)
J	flux
k	Boltzmann's constant ($1.37 \times 10^{-16}\,\mathrm{erg\,K^{-1}}$)
K	kinetic energy
K	partitioning coefficient
K	hydraulic conductivity
K	Kelvin (temperature)
k	reaction rate constant
$K_{1a_{G/A}}$	global volumetric mass transfer coefficient (time^{-1})
k_1	filter media resistance constant
k_2	filter dust cake resistance constant
K_a	acid equilibrium constant
K_{AW}	air-water partitioning coefficient
K_b	base equilibrium constant
K_D	soil sorption partitioning coefficient
K_x, K_y, K_z	hydraulic conductivity in x, y, and z direction
k_E	unitless exposure factor
K_F	Freundlich isotherm constant
kg	kilogram
$K_{G/A}$	partition coefficient between the gaseous and aqueous phases
K_H	Henry's Law partitioning coefficient (also H)
k_{inh}	inhalation rate (volume of air per unit time)
km	kilometer (length)
K_m	Michaelis constant (Michaelis-Menten equation)
K_{oa}	octanol-air partitioning coefficient
K_{oc}	organic-carbon partitioning coefficient
K_{ow}	octanol-water partitioning coefficient
k_p	rate of chemical energy storage by primary producers
kPa	kilopascal (unit of pressure)
K_s	Monod growth rate coefficient
K_{sp}	solubility product constant
K_w	water dissociation equilibrium constant
kWh	kilowatt-hour
L or l	liter
L	angular momentum
l	liquid phase (subscript or parenthetic)
LADD	lifetime average daily dose
LADSCR	lifetime average daily soil contact rate (normalized to bodyweight)
LADSIR	lifetime average daily soil intake rate (normalized to bodyweight)
lb	pound (weight)
$\mathbf{lb}_f$	foot-pound (force)
$\mathbf{lb}_m$	international avoirdupois pound (weight)
LC	liquid chromatography
$\mathbf{LC_{50}}$	lethal concentration-50 (lethal to 50% of tested organisms)
LCA	life cycle analysis or life cycle assessment
LCI	life cycle inventory
LCRA	life cycle risk assessment
$\mathbf{LD_{50}}$	lethal dose-50 (lethal to 50% of tested organisms)
LEL	lower explosive limit
LEV	low emissions vehicle
lidar	remote sensing of distance using a laser (combination of light and radar)
LLWR	low-level radioactive waste
ln(x)	natural log of x
LNAPL	light (low density) nonaqueous phase liquid
LOAEC	lowest observed adverse effect concentration
LOAEL	lowest observed adverse effect level

LoC	level of concern
LoD	limit of detection (also known as DL, detection limit)
LOEL	lowest-observed-effects-level
log	logarithm
LoQ	limit of quantitation
LOSU	level of scientific understanding
LPG	liquefied propane gas
LRTAP	long-range transboundary air pollutants
LTP	local temperature and pressure
L_v	distance at which a black object is just barely visible
m	saturated thickness of an aquifer (also b)
m	mass
m	meter
M	mole (gram molecular weight)
M	relatively nonreactive molecule in a reaction
$\mathbf{M_A}$	total soil mass adhered to skin
MACRS	Modified Accelerated Cost Recovery System
MACT	maximum achievable control technology
max	maximum (subscript)
MB	mixed bed
MCE	mixed cellulose ester
MCL	maximum contaminant level (drinking water)
MCLG	maximum contaminant level goal
MDL	method detection limit
Me	metal
MEI	maximum exposed individual
MEK	methyl ethyl ketone
MF	mobilization factor
MF	modifying factor
MFAA	Millipore cellulose ester membrane
Mg	magnesium
m_g	mass of displaced gas
MGD	million gallons per day
MHz	megahertz (10^6 Hz)
MIC	methylisocyanate
min	minimum (subscript)
min	minute (time)
MIR	maximum individual risk
mm	millimeter
mM	millimole
mmBTU	10^6 BTU
mm Hg	millimeters of mercury (unit of pressure)
MMT	methylcyclopentadienyl manganese tricarbonyl
mmt	million metric tons
MMTCDE	million metric tons carbon dioxide equivalents
Mn	manganese
MOA	mode of action
MOE	margin of exposure
mol	mole (gram molecular weight)
m_p	particle mass
MPN	most probable number (microbes)
MQO	model quality objective
MRF	materials recovery facility
MRL	minimum risk level
MRT	mean residence time
MS	mass spectroscopy
MSAT	mobile source air toxic
MSDS	material safety data sheet

MSL	mean sea level
MSW	municipal solid waste
MSWLF	municipal solid waste landfill
MTBE	methyl tertiary-butyl ether
MTTR	mean time to repair
MUC	maximum use concentration
mV	millivolts (potential)
MW	megawatt
MW	molecular weight
n	number of charges (electrostatics)
N	concentration (number)
N	Newton (unit of force)
N	not significant
n	number (e.g., number of moles)
N	transport process rate (fugacity × compartmental rate constant)
N_2	molecular nitrogen
N-100	respirator protection efficiency against particulate aerosols (99.97%), but not resistant to oil
NA	nonaqueous phase
NA	nonattainment
NAA	neutron activation analysis
NAAQS	National Ambient Air Quality Standard (U.S.)
NADP	National Atmospheric Deposition Program (U.S.)
NAPL	nonaqueous phase liquid
NATA	National Air Toxics Assessment (U.S.)
ND, n.d.	not detected
NDIR	nondispersive infrared
NF	nanofiltration
ng	nanogram
NHANES	National Health and Nutrition Examination Survey (U.S.)
N_{in}	number of particles (count) entering a system (e.g., filter)
Ni	nickel
NIOSH	National Institute for Occupational Safety and Health
NIST	National Institute of Standards and Technology (U.S.)
nM	nanomole
NMHC	nonmethane hydrocarbon
NMOG	nonmethane organic gases
NMVOC	nonmethane volatile organic compound
NOAA	National Oceanic and Atmospheric Administration (U.S.)
NOAEC	no observed adverse effect concentration
NOAEL	no observed adverse effect level
NOEL	no-observed-effect-level
NOHSCP/NCP	National Oil and Hazardous Substances Contingency Plan
N_{out}	number of particles (count) exiting a system (e.g., filter)
NO_x	oxides of nitrogen; specifically NO and NO_2
NO_y	oxides of nitrogen; specifically NO_x and NO_z
NO_z	$HNO_3 + HONO + N_2O_5 + HO_2NO_3 + PAN + $organic nitrates
NPDES	National Pollutant Discharge Elimination System (U.S., Clean Water Act)
NPL	National Priorities List (Superfund)
NPL	National Physics Laboratory (U.K.)
NPP	net primary product (or productivity)
NPS	nonpoint source
N_R	Reynolds number (also Re)
N_{Rp}	dimensionless particle Reynolds number
NRC	Nuclear Regulatory Commission (U.S.)
NSR	no significant risk
NTP	National Toxicology Program (U.S.)
NTP	Normal temperature and pressure
NVOC	nonvolatile organic compound

o	octanol (subscript)
O_2	molecular oxygen
O_3	ozone
O&M	operation and maintenance
OA	organic acid
OC	organic carbon (subscript)
OC/EC	organic carbon-to-elemental carbon ratio
OECD	Organisation for Economic Co-operation and Development
OMI	ozone monitoring instrument
OP	organophosphorus (pesticide class)
OP-FTIR	open-path Fourier transform infrared
OSC	on-scene coordinator (Superfund)
OSHA	Occupational Safety and Health Administration (U.S.)
OTC	once-through cooling
OUT	mass or energy outflow from control volume; effluent (subscript)
OWTS	on-site wastewater treatment system (septic system)
P	skin permeability coefficient
P* or *p	pressure
p	momentum
p	partial
p	particle (subscript)
P	particle penetration; inverse of efficiency (filtration)
P	particulate matter emissions (mass $time^{-1}$)
P	probability
P	power
P	initial cost
P^0	vapor pressure
P_I	net primary productivity (or production)
P-100	oil-proof respirator protection efficiency against particulate aerosols (99.97%)
p_a	partial pressure of chemical substance gas in inhaled air
p_{a}	partial pressure of gas a
Pa	pascal (unit of pressure)
PA	preliminary assessment
PAC	powdered activated carbon
PAD	population adjusted dose
PAH	polycyclic aromatic hydrocarbon (sometimes polyaromatic hydrocarbon)
PAL	point, area, and line (model)
PAL-DS	point, area, and line model with deposition and settling
PAMS	photochemical assessment monitoring station
Pb	lead
p_b	partial pressure of chemical substance gas in blood
PBB	polybrominated biphenyl
PBPD	physiologically based pharmacodynamic
PBPK	physiologically based pharmacokinetic
PBT	persistent, bioaccumulating, and toxic
PC	particle concentration
PC	polycarbonate
PCA	principal component analysis
PCB	polychlorinated biphenyl
PCDD/F	pentachlorodibenzodioxins and furans
PCE	tetrachloroethene (perchloroethylene)
pCi	picocurie (radioactivity unit)
PCM	phase contrast microscopy
PCP	pentachlorophenol
PCR	polymerase chain reaction
PDF	probability density function
PEL	permissible exposure limit (OSHA)
PFAS	per- and polyfluoroalkyl substances

PFOA perfluorooctanoic acid
PFOS perfluorooctane sulfonate
pH negative log of H^+ or H_3O^+ concentration
P_i phosphate molecule with high energy bonds
PIC product of incomplete combustion
PID photoionization detector
pK_a negative log of the acid dissociation constant
p_L supercooled liquid vapor pressure
PLM polarized light microscopy
PM particulate matter
PM_{10} Particulate matter, <10-μm diameter
$PM_{2.5}$ Particulate matter, <2.5-μm diameter
PMR proportionate mortality rate
PNEC predicted no-effect concentration
PNS peripheral nervous system
POD point of departure
pOH negative log of OH^- concentration
POM polynuclear organic material
POP persistent organic pollutant
POTW publicly owned treatment works
POU point of use
PPA potentially productive aquifer
ppb parts per billion
ppbV parts per billion by volume
PPE personal protection equipment
ppm parts per million
ppmV parts per million by volume
ppt parts per trillion
PQAO primary quality assurance organization
PQL practical quantitation limit
PRP potentially responsible party (Superfund, U.S.)
PRTR Pollutant Release and Transfer Register (Mexico)
PSA particle size analysis
PSD particle size distribution
PSF particle size fraction
psi pounds per square inch (pressure)
PUF polyurethane foam
P_v velocity pressure
PV present value
PVC polyvinyl chloride
PWS public water system
Q flow; volumetric flow rate
Q stream discharge
Q heat passing through boundary
q electric charge on particle
q flux
q*1* cancer slope value (U.S. Cancer Assessment Group)
Q ionic product
QA quality assurance
QAPP quality assurance project plan
QC quality control
Q_g gas flow rate ($length^2\ time^{-1}$)
QM-A Whatman QM-A, quartz microfiber filter
QSAR quantitative structure-activity relationships
R or r correlation coefficient
r chemical production or degradation rate
R gas constant
R radial position of particle (centrifugal force calculation)

r	radius
R	remote
R	removed
R	resistance of the compartment to flux
R or *R*	risk
r	thermodynamic loss term
R	universal gas constant ($8.315\,JK^{-1}mol^{-1}$)
R-100	oil-resistant respirator protection efficiency against particulate aerosols (99.97%)
R^2 or r^2	coefficient of determination
RACT	reasonably available control technology
RAGS	risk assessment guidelines (Superfund)
RAM	release abatement measure
RC	reportable concentration
RCA	replacement cost analysis
RCA	root cause analysis
RCRA	Resource Conservation and Recovery Act (U.S.)
rDNA	ribosomal deoxyribonucleic acid
Re	Reynolds number (also R_e and N_R)
REACH	Registration, Evaluation, Authorisation and Restriction of Chemicals (Europe)
REL	recommended exposure limit
REL-TWA	recommended exposure limit (time-weighted average)
Re_p	dimensionless particle Reynolds number
RFA	RCRA facility assessment (U.S.)
RfC	reference concentration
RfD	reference dose
RFI	RCRA facility investigation (U.S.)
RH	relative humidity
r_i	mole ratio
RI	remedial investigation
RI/FS	remedial investigation and feasibility study (U.S. Superfund)
RIDS	response information data sheet (OSHA)
RIOPA	Relationship Between Indoor, Outdoor and Personal Air Study (U.S.)
R_m	intake rate for a medium
RMCL	recommended maximum contaminant level
rms	root mean square
R_n	net irradiance
RNA	ribonucleic acid
RO	reverse osmosis
RO_2	radical made of a chain of organic compounds with O_2 substituted for H
ROD	record of decision
ROS	reactive oxygen species
RP	respirable particulates
RPGen	residential population generator
RQ	risk quotient
RQ	reportable quantity
RR	relative risk
rRNA	ribosomal ribonucleic acid
RRT	Regional Response Team (Superfund)
R_s	intake rate of contaminated soil
R(t)	reliability function
RT	residence time
RT	retention time
RTECS	Registry of Toxic Effects of Chemical Substances
RTP	room temperature and pressure
RX	organic compound
s or sec	second (time)
S	entropy
S	severity of risk

s	solid or sorbed phase (subscript or parenthetic)
S	solid-phase concentration
S	solubility
S	sorption
S	concentration of limiting substrate (Monod equation)
S	source or sink
S	storage (aquifer)
S^2	variance
s^2	variance of the source
S_A	substrate concentrations in the aqueous phase
SA	skin surface area
SAR	structure-activity relationships
SATB	surface area of the tracheobronchial region
SCBA	self-contained breathing apparatus
SCR	selective catalytic reduction
SD or σ	standard deviation
SE	standard error
SEM	scanning electron microscopy
SES	socioeconomic status
S_f	filter drag force
SF	slope factor
SF_6	sulfur hexafluoride
S_G	substrate concentrations in the bulk gas phase
SG_{ww}	gravity of wastewater
S_h	horizontal surface
SHEDS	Stochastic Human Exposure and Dose Simulation (model)
SIC	standard industrial classification
sin	sine
SLR	solids loading rate
SMR	standard morbidity ratio
SMR	standard mortality ratio
S/N	signal-to-noise ratio
SNCR	selective noncatalytic reduction
SOC	synthetic organic chemical
SOF	soluble organic fraction
SOM	soil organic matter
SOP	standard operating procedure
SOR	surface overflow rate
SO_x	oxides of sulfur
SO_2	sulfur dioxide
SPMS	special purpose monitoring station
SQG	small quantity generator
S_s	specific storage of an aquifer
SS	suspended solids
SS	steady state (subscript)
Stk	Stokes number
STP	standard temperature and pressure
SVE	soil vapor extraction
SVOC	semivolatile organic compound
SW	surface water
SW	solid waste
SWO	storm water overflow
S_y	specific yield of an aquifer
T	temperature
T	transmissivity (aquifer)
t	ton
T	total (subscript)
$\bar{t}$	averaging time (exposures)

t_0	initial time
$t_{1/2}$	half-life
tan	tangent
TC	total carbon
TCDD	tetrachlorodibenzo-*para*-dioxin
TCDF	tetrachlorodibenzofuran
TCE	trichloroethene
TCLP	toxicity testing leaching procedure
TCM	tetrachloromercurate
TCM	transportation control measure
t_D	detention time
TDL	tunable diode laser
TDS	total dissolved solids
t_E	exposure duration
TEA	terminal electron acceptor
TEF	toxicity equivalency factor
TEL	tetraethyl lead
TEL	threshold effects level (air)
TEM	transmission electron microscopy
TEQ	toxicity equivalent
T_f	time of failure
THM	trihalomethane
T_{ij}	time spent in microenvironment *j* by person *i*
TK	toxicokinetic
TL	typical lifetime
TLV	threshold limit value
TLV-C	threshold limit value (ceiling)
TLV-STEL	threshold limit value-short-term exposure limit
TMDL	total maximum daily load
TNT	trinitrotoluene
TOC	total organic carbon
TOMS	total ozone mapping spectrometer
TOV	total organic vapors
TPTH	triphenyltin hydroxide
tpy	tons per year
T-R	transformer-rectifier
TRS	total reduced sulfur
TS	total solids
TS	total sulfur
TSCA	Toxic Substances Control Act (U.S.)
TSDF	treatment, storage, and disposal facility
TSP	total suspended particulates
TSS	total suspended solids
TWA	T wave alternans
TWA	time-weighted average
u	velocity
u	wind speed
U	epistemic uncertainty
U	internal energy
U	potential energy
U	unknown
u_a	pore air pressure
UAM	urban airshed model
UASB	upflow anaerobic sludge blanket
UCC	ultra-clean coal
UCL	upper concentration limit
UCL	upper confidence limit
UDL	upper detection limit

UEL	upper explosive limit
UF_{inter}	interspecies uncertainty factor
UF_{intra}	intraspecies uncertainty factor
UF_{other}	RfC or RfC uncertainty other than UF_{inter} or UF_{intra}
UFP	ultrafine particles
UNEP	United Nations Environmental Programme
UNFCCC	United Nations Framework Convention on Climate Change
UR	unit risk value
u_{rms}	root-mean-square (rms) speed (gas molecules)
USDA	U.S. Department of Agriculture
USEPA	U.S. Environmental Protection Agency
USFS	U.S. Forestry Service
USGS	U.S. Geological Service
UST	underground storage tank
UTM	Universal Transverse Mercator
UTS	universal treatment standards
UV	ultraviolet
UV-DOAS	differential optical absorption spectroscopy in UV range
u_w	pore water pressure
V	fluid velocity
V	voltage
V	shear force
V	speed
V	stochastic or aleatory uncertainty
v	vapor phase (subscript or parenthetic)
V	volume
$\bar{v}$	average linear velocity (length $time^{-1}$)
$\bar{V}$	average speed (length $time^{-1}$)
V_{aw}	airway volume ($length^3$)
VC	vinyl chloride (C_2H_3Cl)
$\mathbf{V_d}$	volume of distribution
$\mathbf{V_e}$	minute volume (volume $time^{-1}$)
v_f	velocity through the filter (length $time^{-1}$)
VFA	volatile fatty acid
V_{ff}	filter fiber volume
v_g	gas velocity
v_g	geostrophic velocity
VI	vapor intrusion
VI	volume ingested
vis	visible light range
V_{max}	maximum reaction rate
VMT	vehicle miles traveled
VOC	volatile organic compound
v_p	particle velocity
$\mathbf{V_p}$	particle volume
VPD	vapor pressure deficit
VPH	volatile petroleum hydrocarbons
VR	volume respired
VS	volatile solids
V_{total}	total filter volume
w	work
W	body weight
W	width (also w or w)
W	water (subscript)
W	watt (unit of power; $J\,s^{-1}$)
w_B	body weight (mass)
WC	water column
WET	whole effluent toxicity test

WHO	World Health Organization
WLA	waste load allocation
WOE	weight of evidence
WOR	weir overflow rate
WoSE	weight of scientific evidence
WTP	water treatment plant
WUE	water use efficiency
WW	wastewater
WWTP	wastewater treatment plant
x	at distance x from $x=0$ (subscript)
x	distance (horizontal)
XAD	sorbing material; ion exchange resin
XF	exposure factor (LCA)
XRD	X-ray diffraction spectrometry
XRF	X-ray fluorescence
y	cross-sectional distance, i.e., at distance y from $y=0$ (subscript)
y	width
Y	modeling result/outcome
y	year
$\bar{y}$	sample mean
y_p	difference in velocity between the particle and collection object (length time^{-1})
z	at distance z from $z=0$ (subscript)
Z	atomic number
z	height
Z	fugacity capacity
Z	zenith angle (degrees)
Z	electrical impedance (ohm—Ω)
α	compressibility of aquifer matrix
α	packing density (solidity)
α	alpha particle
β	compressibility of water
β	beta particle
γ	gamma ray
γ	specific weight
γ	psychrometric constant
$\sum F$	sum of forces
Δ	change
Δ	displacement
Δ	heat in a chemical reaction
ΔG^*	standard change in Gibbs free energy
$\Delta P_{dust\text{-}cake}$	pressure drop across dust cake on filter
ΔP_{media}	pressure drop across filter media
ΔP_{total}	total pressure drop
ε_0	permittivity of free space (8.85×10^{-12} F m^{-1})
μ	specific growth rate of microbe
μ	dynamic viscosity
μ_{max}	maximum specific growth of microbe
η	hour angle (degrees)
θ	volumetric water content of soil
θ_0	initial soil water content
λ	log slope of soil water retention curve (pore-size distribution coefficient)
λ	average or expected counts or events per unit time (Poisson equation)
λ	degradation term in chemical decay
λ	wavelength
λ_s	capillary length scale
ν_p	relative velocity between the particle and the target object
ν_τ	tangential velocity of gas
π	3.14159...

ρ	density (volume length^{-3})
ρ_0	standard density (1 g m^{-3})
ρ_b	dry bulk density
ρ_g	gas density (mass length^{-3})
ρ_p	particle density
ρ_w	fluid density (water)
ρ_{ww}	fluid density (wastewater)
σ	Stefan-Boltzmann constant
τ	tortuosity
τ	mean residence time at steady state
τ_s	shear stress (mass length^{-1} time^{-2})
υ or υ_v	kinematic viscosity (length^2 time^{-1}) = dynamic viscosity divided by density ($\mu\,\rho^{-1}$)
υ	specific volume (reciprocal of density; length^3 mass^{-1})
X	leaving group from organic compound
$\mathbf{X_p}$	particle diffusivity (length^2 time^{-1})
Ω	migration velocity (electrostatics)
Ω	angular speed of the earth's rotation
Ω	electrical resistance
η_e	effective porosity (dimensionless)
σ or σ°	surface tension (forces length^{-1})
σ_y	cross-sectional dispersion coefficient
σ_z	dispersion coefficient in the vertical direction
σ or SD	standard deviation
σ^2	variance
ζi	mass ratio
μ	degree of saturation
µg	microgram
µm	micron or micrometer
σ_n	normal shear stress (mass length^{-1} time^{-2})
Ω	humidity ratio
ψ	surface deposition
Ψ	impaction factor (dimensionless)
Ψ_D	diffusional collection parameter (dimensionless)
Ψ_p	particle diffusivity
φ_O	osmotic pressure
φ	angle (degrees or radians)
φ	filter porosity
φ	fugacity coefficient
φ	latitude (degrees)
φ°	fraction, ratio, or coefficient
ω	solar azimuth (degrees)
ω	spherical particle mass
ϕ	fluid potential
ϕ_p	pressure potential
ϕ_m	matric potential
ϕ_zI	gravitational potential
ϕ_o	osmotic potential
ϕ_{ob}	overburden potential
$\xrightarrow{\Delta}$	energy (heat) required for reaction
$\frac{dy}{dx}$	ordinary derivative of x with respect to y
S_A^*	substrate concentration at the gas/aqueous interface
$\bar{\bar{Y}}$	source mean
τ	mean residence time
$\frac{\partial x}{\partial y}$	partial derivative of x with respect to y
$[C]$	molar concentration
–	mean or average (capped symbol or letter)
∇ (nabla)	gradient of a scalar field, especially in Navier-Stokes equations
∇ (nabla)	water table

$\cong$	approximately equal to
$\approx$	almost equal to
—	single chemical bond
$=$	double chemical bond
$\equiv$	triple chemical bond
$\rightarrow$	kinetic reaction
$\leftrightarrow$	equilibrium reaction
°C	degrees Celsius or Centigrade (temperature)
°F	degrees Fahrenheit (temperature)
°lat	latitude (degrees)
°long	longitude (degrees)
°R	degrees Rankine (temperature)
μ **or** μ_v	dynamic viscosity (mass length^{-1} time^{-1})
μ	chemical potential (mass length^{2} time^{-2})
μE	microenvironment (exposure science)
μ_g	gas viscosity (mass length^{-1} time^{-1})
μ_m	arithmetic mean
μM	micromole
−	anion (superscript)
+	cation (superscript)
0	initial (subscript)
1	first (subscript)
1,4-D	1,4-dioxane

Index

Note: Page numbers followed by *f* indicate figures, *t* indicate tables, and *b* indicate boxes.

D

E

N

O

P

Q

R

S

X

Z

9780443289873